Plant and Process Engineering 360°

Note from the Publisher

This book has been compiled using extracts from the following books within the range of Plant and Process Engineering books in the Elsevier collection:

Crowder, (2005) Electric Drives and Electromechanical Systems, 9780750667401
Laughton and Warne, (2002) Electrical Engineer's Reference Book, 9780750646376
Parr, (2006) Hydraulics and Pneumatics, 9780750644198
Zhang, (2008) Industrial Control Technology, 9780815515715
Warne, (2005) Newnes Electrical Power Engineer's Handbook, 9780750662680
Mobley, (2001) Plant Engineer's Handbook, 9780750673280
Girdhar, Moniz and Mackay, (2004) Practical Centrifugal Pumps: Design, Operation and Maintenance, 9780750662734
Barnes, (2003) Practical Variable Speed Drives and Power Electronics, 9780750658089

The extracts have been taken directly from the above source books, with some small editorial changes. These changes have entailed the re-numbering of Sections and Figures. In view of the breadth of content and style of the source books, there is some overlap and repetition of material between chapters and significant differences in style, but these features have been left in order to retain the flavour and readability of the individual chapters.

End of chapter questions
Within the book, several chapters end with a set of questions; please note that these questions are for reference only. Solutions are not always provided for these questions.

Units of measure
Units are provided in either SI or IP units. A conversion table for these units is provided at the end of the book.

Plant and Process Engineering 360°

Amsterdam · Boston · Heidelberg · London · New York · Oxford
Paris · San Diego · San Francisco · Sydney · Tokyo
Butterworth-Heinemann is an imprint of Elsevier

Butterworth-Heinemann is an imprint of Elsevier
The Boulevard, Langford Lane, Kidlington, Oxford OX5 1GB, UK
30 Corporate Drive, Suite 400, Burlington, MA 01803, USA

First edition 2010

British Library Cataloguing in Publication Data
A catalogue record for this book is available from the British Library

Library of Congress Cataloging-in-Publication Data
A catalog record for this book is availabe from the Library of Congress

ISBN–13: 978-1-85617-840-2

For information on all Butterworth-Heinemann publications
visit our web site at books.elsevier.com

Printed and bound by CPI Group (UK) Ltd, Croydon, CR0 4YY

Transferred to Digital Print 2012

Contents

Section One

Introduction

Chapter 1.1

1.1

Instrumentation and transducers

Parr

1.1.1 Introduction

1.1.1.1 Definition of terms

Accurate measurement of process variables such as flow, pressure and temperature is an essential part of any industrial process. This chapter describes methods of measuring common process variables.

Like most technologies, instrumentation has a range of common terms with precise meanings.

Measured variable and *process variable* are both terms for the physical quantity that is to be measured on the plant (e.g. the level in tank 15). The *measured value* is the actual value in engineering units (e.g. the level is 1252 mm).

A *primary element* or *sensor* is the device which converts the measured value into a form suitable for further conversion into an instrumentation signal, i.e. a sensor connects directly to the plant. An orifice plate is a typical sensor. A *transducer* is a device which converts a signal from one quantity to another (e.g. a PT 100 temperature transducer converts a temperature to a resistance). A *transmitter* is a transducer which gives a standard instrumentation signal (e.g. 4–20 mA) as an output signal, i.e. it converts from the process measured value to a signal which can be used elsewhere.

1.1.1.2 Range, accuracy and error

The *measuring span*, *measuring interval* and *range* are terms which describe the difference between the lower and upper limits that can be measured (e.g. a pressure transducer which can measure from 30 to 120 bar has a range of 90 bar). The *rangeability* or *turndown* is the ratio between the upper limit and lower limits where the specified accuracy can be obtained. Assuming the accuracy is maintained across the range, the pressure transmitter above has a turndown of 4:1. Orifice plates and other differential flowmeters lose accuracy at low flows, and their turndown is less than the theoretical measuring range would imply.

The *error* is a measurement of the difference between the measured value and the true value. The *accuracy* is the maximum error which can occur between the process variable and the measured value when the transducer is operating under specified conditions. Error can be expressed in many ways. The commonest are absolute value (e.g. ±2 °C for a temperature measurement), as a percentage of the actual value, or as a percentage of full scale. Errors can occur for several reasons; calibration error, manufacturing tolerances and environmental effects are common.

Many devices have an inherent coarseness in their measuring capabilities. A wire wound potentiometer, for example, can only change its resistance in small steps and digital devices such as encoders inherently measure in discrete steps. The term *resolution* is used to define the smallest steps in which a reading can be made.

In many applications the accuracy of a measurement is less important than its consistency. The consistency of a measurement is defined by the terms *repeatability* and *hysteresis*.

Repeatability is defined as the difference in readings obtained when the same measuring point is approached several times from the same direction.

Hysteresis occurs when the measured value depends on the direction of approach as Figure 1.1.1. Mechanical backlash or stiction are common causes of hysteresis.

Electrical Engineer's Reference Book; ISBN: 9780750646376

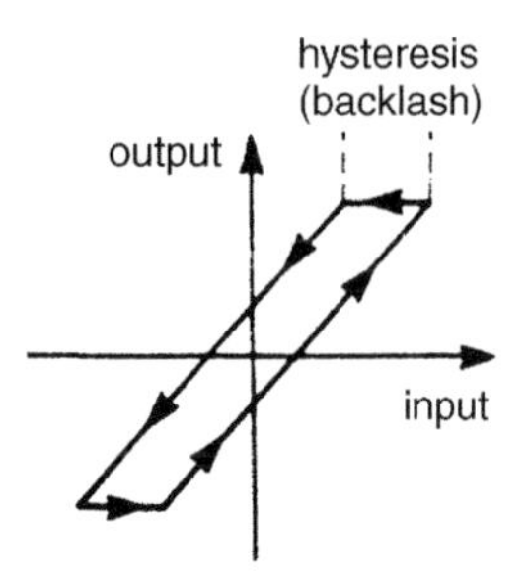

Fig. 1.1.1 Hysteresis, also known as backlash. The output signal is different for an increasing or decreasing input signal.

The accuracy of a transducer will be adversely affected by environmental changes, particularly temperature cycling, and will degrade with time. Both of these effects will be seen as a *zero shift* or a change of sensitivity (known as a *span error*).

1.1.1.3 Dynamic effects

A sensor cannot respond instantly to changes in the measured process variable. Commonly the sensor will change as a first order lag as Figure 1.1.2 and represented by the equation:

$$T\frac{dx}{dt} + x = a$$

Here T is called the time constant. For a step change in input, the output reaches 63% of the final value in time T. Table 1.1.1 shows the change at later times.

It follows that a significant delay may occur for a dynamically changing input signal.

A second order response occurs when the transducer is analogous to a mechanical spring/viscous damper. The response of such a system to a step input of height a is given by the second order equation:

$$\frac{d^2x}{dt^2} + 2b\omega_n\frac{dx}{dt} + \omega_n^2 x = a$$

where b is the damping factor and ω_n the natural frequency. The final steady state value of x is given by

$$x = \frac{a}{\omega_n^2}$$

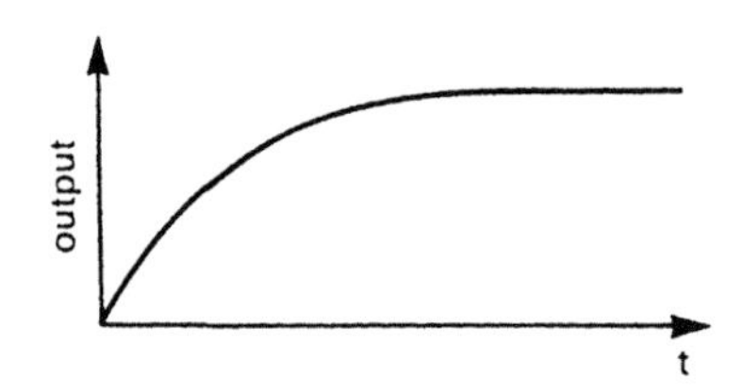

Fig. 1.1.2 Response of a first order lag to a step input signal.

Table 1.1.1

Time	% Final value
T	63
2T	86
3T	95
4T	98
5T	99

The step response depends on both b and ω_n, the former determining the overshoot and the latter the speed of response as shown in Figure 1.1.3. For values of $b < 1$ damped oscillations occur. The case where $b = 1$ is called *critical damping*. For $b > 1$, the system behaves as two first order lags in series.

Intuitively one would assume that $b = 1$ is the ideal value. This may not always be true. If an overshoot to a step input signal can be tolerated a lower value of b will give a faster response and settling time within a specified error band. The signal enters the error band then overshoots to a peak which is just within the error band as shown in Figure 1.1.4. Many instruments have a damping factor of 0.7 which gives the fastest response time to enter, and stay within, a 5% settling band. Table 1.1.2 shows optimum damping factors for various settling bands. The settling time is in units of $1/\omega_n$.

1.1.1.4 Signals and standards

The signals from most primary sensors are very small and in an inconvenient form for direct processing. Commercial transmitters are designed to give a standard output signal for transmission to control and display devices.

The commonest electrical standard is the 4–20 mA loop. As its name implies this uses a variable current with

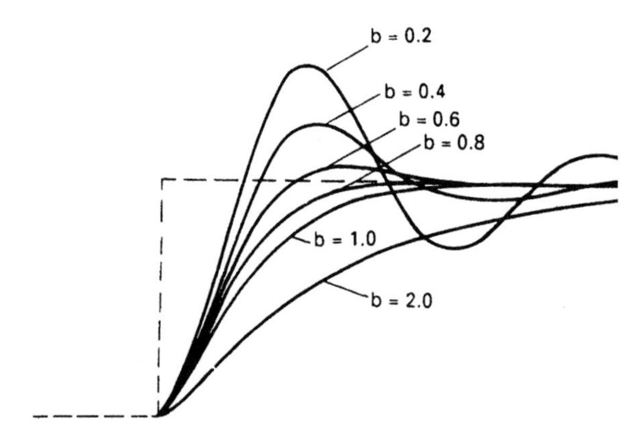

Fig. 1.1.3 Response of a second order lag to a step input signal for various values of damping factor.

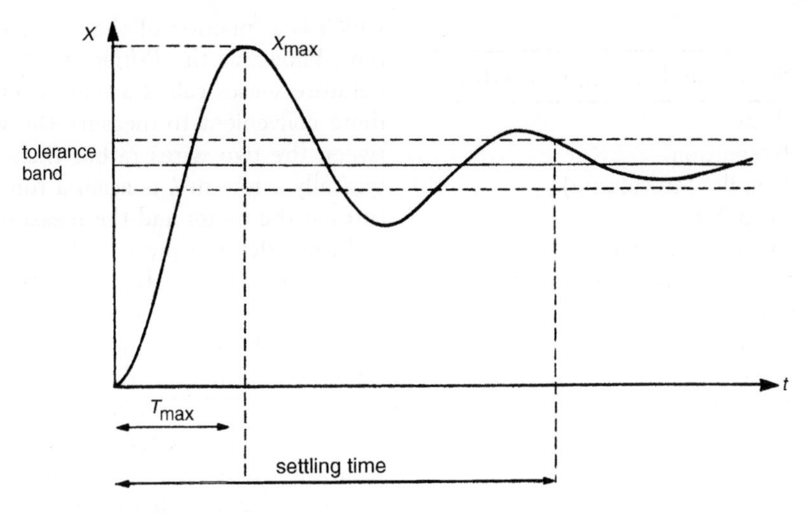

Fig. 1.1.4 Overshoot on a second order system and definition of settling time.

4 mA representing one end of the signal range and 20 mA the other. This use of a current (rather than a voltage) gives excellent noise immunity as common mode noise has no effect and errors from different earth potentials around the plant are avoided. Because the signal is transmitted as a current, rather than a voltage, line resistance has no effect.

Several display or control devices can be connected in series provided the total loop resistance does not rise above some value specified for the transducer (typically 250 to 1 kΩ).

Transducers using 4–20 mA can be current sourcing (with their own power supply) as Figure 1.1.5(a) or designed for two wire operation with the signal line carrying both the supply and the signal as Figure 1.1.5(b). Many commercial controllers incorporate a 24 V power supply designed specifically for powering two wire transducers.

The use of the offset zero of 4 mA allows many sensor or cable faults to be detected. A cable break or sensor fault will cause the loop current to fall to zero. A cable short circuit will probably cause the loop current to rise above 20 mA. Both of these fault conditions can be easily detected by the display or control device.

Table 1.1.2

Settling band (%)	Optimum '*b*'	Settling time
20	0.45	1.8
15	0.55	2
10	0.6	2.3
5	0.7	2.8
2	0.8	3.5

1.1.1.5 P&ID symbols

Instruments and controllers are usually part of a large control system. The system is generally represented by a *Piping & Instrumentation Drawing* (or P&ID) which shows the devices, their locations and the method of interconnection. The description *Process & Instrumentation Diagram* is also used by some sources. The letters P&ID should not be confused with a PID controller described in Chapter 1.2. The basic symbols and a typical example are shown in Figure 1.1.6.

The devices are represented by circles called balloons. These contain a unique tag which has two parts. The first is two or more letters describing the function. The second part is a number which uniquely identifies the device, for example FE127 is flow sensor number 127.

The meanings attached to the letters are:

First letter	Second and subsequent letters
A Analysis	Alarm (often followed by H for high or L for low)
B Burner	
C Conductivity	Control function
D Density	
E Voltage or misc. electrical	Primary element (sensor)
F Flow	
G Gauging	Glass sight tube (e.g. level)
H Hand	High (with A = alarm)
I Current	Indicator
J Power	
K Time	Control station
L Level	Light. Also low (with A = alarm)
M Moisture	
O	Orifice

(*Continued*)

(*contd*)

First letter	Second and subsequent letters
P Pressure	Point
Q Concentration or quantity	Integration (e.g. flow to volume)
R Radioactivity	Recorder
S Speed	Switch or contact
T Temperature	Parameter to signal conversion
U Multivariable	Multifunction
V Viscosity	Valve
W Weight, force	Well
X	Signal to signal conversion
Y	Relay
Z Position	Drive

It is good engineering practice for plant devices to have their P&ID tag physically attached to them to aid maintenance.

1.1.2 Temperature

1.1.2.1 General

Accurate knowledge and control of temperature is required in the majority of industrial processes.

1.1.2.2 Thermocouples

If two dissimilar metals are joined together as shown in Figure 1.1.7(a) and one junction is maintained at a high temperature with respect to the other, a current will flow which is a function of the two temperatures. This current, known as the *Peltier effect*, is the basis of a temperature sensor called a thermocouple. In practice, it is more convenient to measure the voltage difference between the two wires rather than current. The voltage, typically a few mV, is again a function of the temperatures at the meter and the measuring junction.

In practice the meter will be remote from the measuring point. If normal copper cables were used to link the meter and the thermocouple, the temperature of these joints would not be known and further voltages, and hence errors, would be introduced. The thermocouple cables must therefore be run back to the meter. Two forms of cable are used: *extension cables*, which are essentially identical to the thermocouple cable, or *compensating cables*, which match the thermocouple characteristics over a limited temperature range. Compensating cables are much cheaper than extension cables.

Because the indication is a function of the temperature at both ends of the cable, correction must be made for the local meter temperature. A common method, called *Cold Junction Compensation*, measures the local temperature by some other method (such as a resistance thermometer) and adds in a correction as Figure 1.1.7(b).

Thermocouples are non-linear devices and the voltage can be represented by an equation of the form

$$V = a + bT + cT^2 + dT^3 + eT^4 + \cdots$$

where a, b, c, d, etc. are constants (not necessarily positive) and T is the temperature. Linearising circuits must be provided in the meter if readings are to be taken over an extended range.

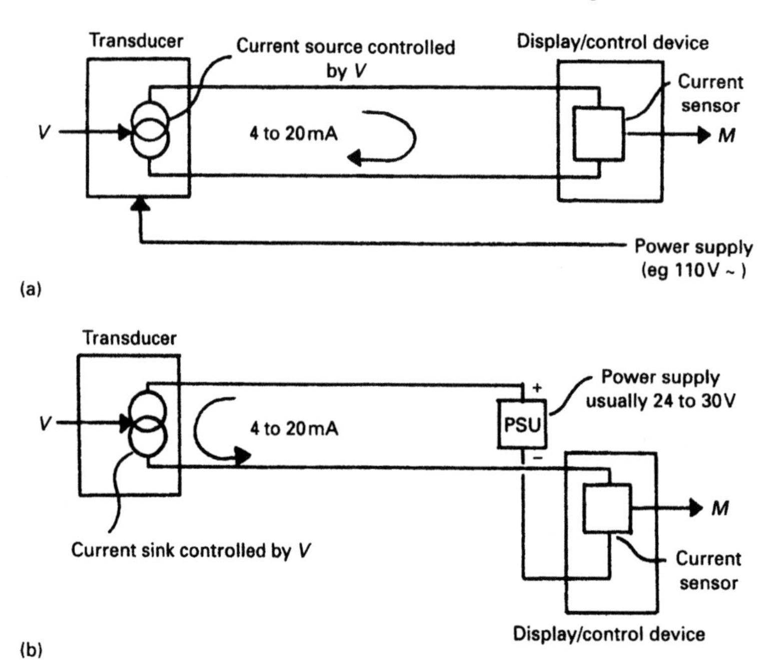

Fig. 1.1.5 Current loop circuits. (a) Current sourcing with a self-powered transducer; (b) current sinking with a loop powered (two wire) transducer.

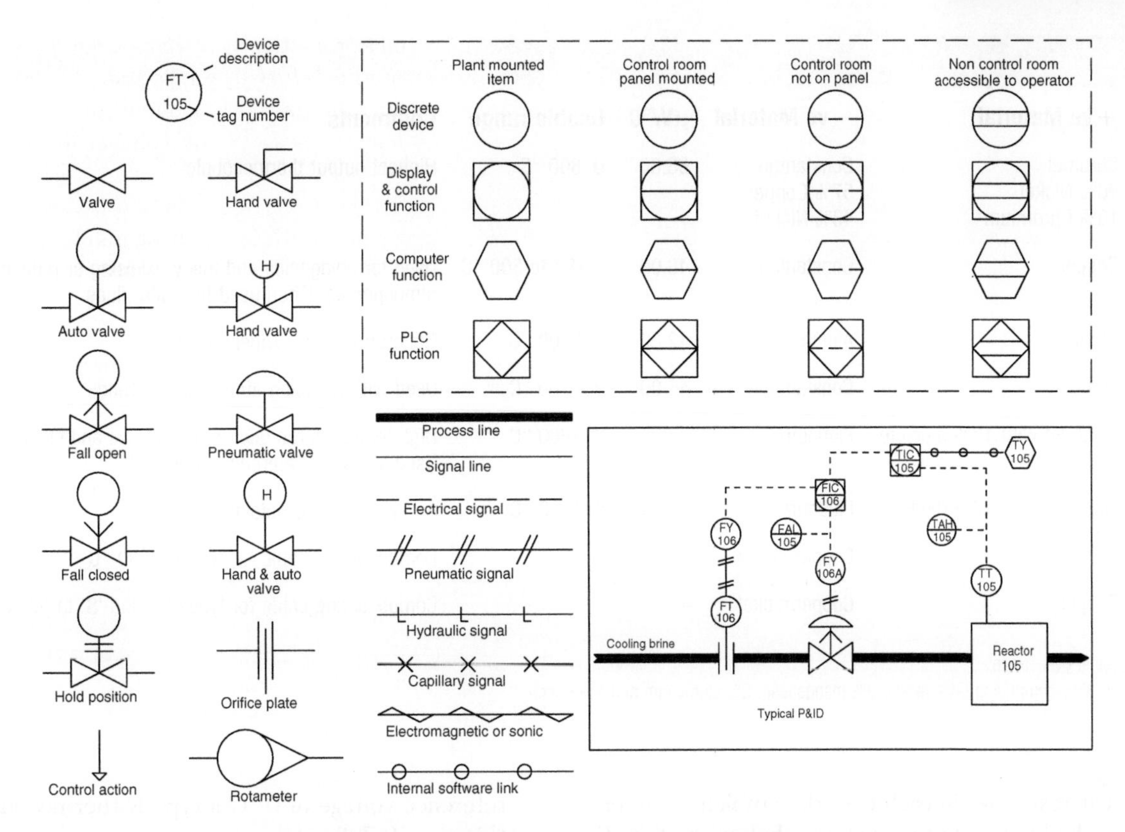

Fig. 1.1.6 Common piping & instrumentation diagram (P&ID) symbols and simple schematic.

Although a thermocouple can be made from any dissimilar metals, common combinations have evolved with well-documented characteristics. These are identified by single letter codes. Table 1.1.3 gives details of common thermocouple types.

Thermocouple tables give the thermocouple voltage at various temperatures when the cold junction is kept at a defined reference temperature, usually 0 °C. A typical table for a type K thermocouple is shown on Table 1.1.4. This has entries in 10 °C steps, practical tables are in 1 °C steps.

Thermocouple tables are used in two circumstances: for checking the voltage from a thermocouple with a millivolt-metre or injecting a test voltage from a millivolt source to test an indicator or controller. Because a thermocouple is essentially a differential device with

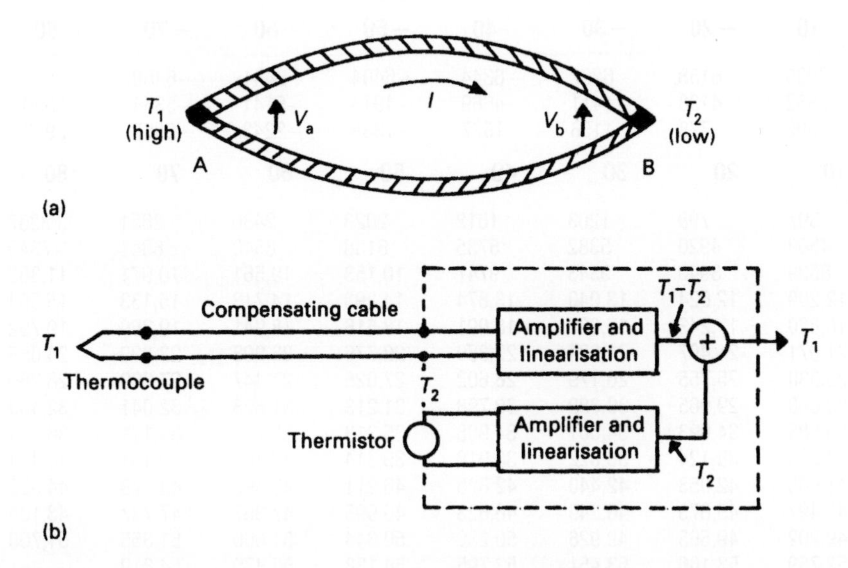

Fig. 1.1.7 Thermocouple circuits. (a) Basic thermocouple; (b) cold junction compensation.

Table 1.1.3 Common thermocouple types

Type	+ve Material	−ve Material	µV/°C	Usable range	Comments
E	Chromel 90% Nickel 10% Chromium	Constantan 57% Copper 43% Nickel	68.00	0–800 °C	Highest output thermocouple
T	Copper	Constantan	46.00	−187 to 300 °C	Used for cryogenics and mildy oxidising or reducing atmospheres. Often used for boiler flues
K	Chromel	Alumel	42.00	0–1100 °C	General purpose. Widely used
J	Iron	Constantan	46.00	20–700 °C	Used with reducing atmospheres. Tends to rust
R	Platinum with 13% rhodium	Platinum	8.00	0–1600 °C	High temperatures (e.g. iron foundries and steel making). Used in UK in preference to type 'S'
S	Platinum with 10% rhodium	Platinum	8.00	0–1600 °C	As Type R. Used outside the UK
V	Copper	Copper/nickel	–		Compensating cable for type K to 80 °C
U	Copper	Copper/nickel	–		Compensating cable for types 'R' and 'S' to 50 °C

µV/°C is typical over range.
Alumel is an alloy comprising 94% nickel, 3% manganese, 2% aluminium and 1% silicon.

a non-linear response, in each case the ambient temperature must be known. In the examples below the type K thermocouple from Table 1.1.4 is used.

To interpret the voltage from a thermocouple there are four steps:

(1) Measure the ambient temperature and read the corresponding voltage from the thermocouple table. For an ambient temperature of 20 °C and reference voltage of 0 °C a type K thermocouple this gives 0.798 mV.

(2) Measure the thermocouple voltage. Let us assume it is 12.521 mV.

(3) Add the two voltages; 12.521 + 0.798 = 13.319 mV.

(4) Read the temperature corresponding to the sum. The thermocouple table shows that the voltage corresponding to 330 °C is 13.040 mV and

Table 1.1.4 Voltages for a type K thermocouple, voltages in microvolts with reference junction at 0 °C

°C	0	−10	−20	−30	−40	−50	−60	−70	−80	−90	µV/°C
−200	−5891	−6035	−6158	−6262	−6344	−6404	−6441	−6458			
−100	−3554	−3852	−4138	−4411	−4669	−4913	−5141	−5354	−5550	−5730	23.4
0	0	−392	−778	−1156	−1527	−1889	−2243	−2587	−2920	−3243	35.5
°C	**0**	**10**	**20**	**30**	**40**	**50**	**60**	**70**	**80**	**90**	
0	0	397	798	1203	1612	2023	2436	2851	3267	3682	41.0
100	4096	4509	4920	5382	5735	6138	6540	6941	7340	7739	40.4
200	8138	8539	8940	9343	9747	10,153	10,561	10,971	11,382	11,795	40.7
300	12,209	12,209	12,624	13,040	13,874	14,293	14,713	15,133	15,554	15,975	41.9
400	16,397	16,820	17,243	17,667	18,091	18,516	18,941	19,366	19,792	20,218	42.5
500	20,644	21,071	21,497	21,924	22,350	22,776	23,203	23,629	24,055	24,480	42.6
600	24,905	25,330	25,755	26,179	26,602	27,025	27,447	27,869	28,289	28,710	42.2
700	29,129	29,548	29,965	30,382	30,798	31,213	31,628	32,041	32,453	32,865	41.5
800	33,275	33,685	34,093	34,501	34,908	35,313	35,718	36,121	36,524	36,925	40.5
900	37,326	37,725	38,124	38,522	38,918	39,314	39,708	40,101	40,494	40,885	39.5
1000	41,276	41,665	42,053	42,440	42,826	46,211	43,595	43,978	44,359	44,740	38.4
1100	45,119	45,497	45,873	46,249	46,623	46,995	47,367	47,737	48,105	48,473	37.2
1200	48,838	49,202	49,565	49,926	50,286	50,644	51,000	51,355	51,708	52,060	35.7
1300	52,410	52,759	53,106	53,451	53,795	54,138	55,479	54,819			

interpolation with 41.9 μV/°C gives a temperature of just over 336 °C.

To determine the correct injection voltage there are three steps:

(1) As before measure the ambient temperature at the instrument or controller terminals and read the corresponding voltage from the tables. As before we will assume an ambient of 20 °C which gives a voltage of 0.798 mV.

(2) Find the table voltage corresponding to the required test temperature, say 750 °C which gives 31.213 mV.

(3) Subtract the ambient voltage from the test temperature voltage. The result of 30.415 mV is the required injection voltage. As before the μV/°C slope can be used to work out voltages for temperatures between the 10 °C steps of Table 1.1.4.

Note that if the local input terminals at the meter are shorted together, the local ambient temperature (from the cold junction compensation) should be displayed. This is a useful quick check.

The tables show that the voltage from a thermocouple is small, typically less than 10 mV. High gain, high stability amplifiers with good common mode rejection are required and care taken in the installation to avoid noise.

In critical applications a high resistance voltage source is connected across the thermocouple so that in the event of a cable break the meter will indicate a high temperature.

1.1.2.3 Resistance thermometers

If a wire has resistance R_0 at 0 °C its resistance will increase with increasing temperature giving a resistance R_t at temperature T given by

$$R_t = R_0\left(1 + aT + bT^2 + cT^3 + \cdots\right)$$

where a, b, c, etc. are constants. These are not necessarily positive.

This change in resistance can be used to measure temperature. Platinum is widely used because the relationship between resistance and temperature is fairly linear. A standard device is made from a coil of wire with a resistance of 100 Ω at 0 °C, giving rise to the common name of a *Pt100 sensor.* These can be used over the temperature range −200 to 800 °C. At 100 °C a Pt100 sensor has a resistance of 138.5 Ω, and the 38.5 Ω change from its resistance at 0 °C is called the *fundamental interval.*

The current through the sensor must be kept low to avoid heating effects. Further errors can be introduced by the resistance of the cabling to the sensor as shown in Figure 1.1.8(a). Errors from the cabling resistance can be overcome by the use of the three and four wire connections of Figures 1.1.8(b)–(d). Three wire connections are usually used with a bridge circuit and four wire connections with a constant current source.

A *thermistor* is a more sensitive device. This is a semiconductor crystal whose resistance changes dramatically with temperature. Devices are obtainable

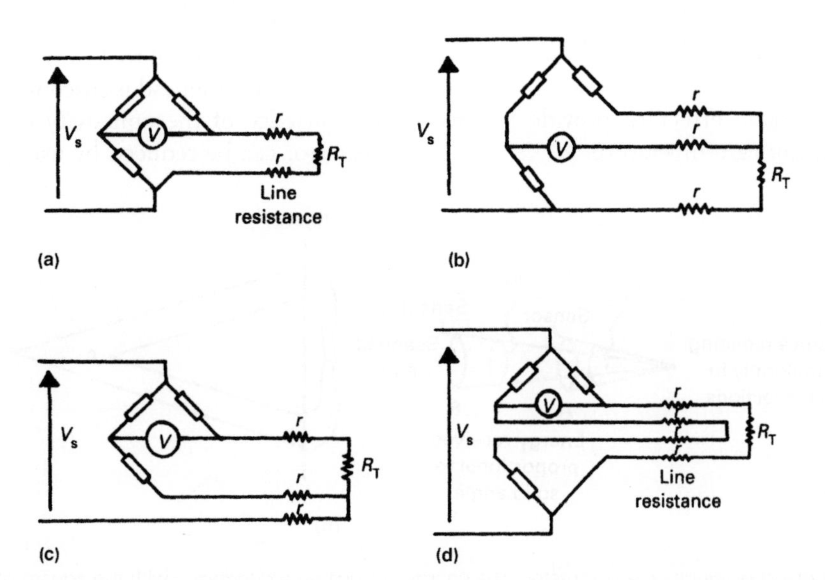

Fig. 1.1.8 The effect of line resistance on RTDs. (a) Simple two wire circuit introduces an error of 2r; (b) three wire circuit places line resistance into both legs of the measuring bridge; (c) alternative three wire circuit; (d) four wire circuit.

which decrease or increase resistance for increasing temperature. The former (decreasing) is more common.

The relationship is very non-linear, and is given by

$$R = R_0 \exp\left(B\left(\frac{1}{T} - \frac{1}{T_0}\right)\right)$$

where R_0 is the defined resistance at temperature T_0, R the resistance at temperature T, and B is a constant called the *characteristic temperature*. A typical device will go from 300 kΩ at 0 °C to 5 kΩ at 100 °C.

Although very non-linear, they can be used for measurement over a limited range. Their high sensitivity and low cost makes them very useful for temperature switching circuits where a signal is required if a temperature goes above or below some preset value. Electric motors often have thermistors embedded in the windings to give early warning of motor overload.

1.1.2.4 Pyrometers

A heated object emits electromagnetic radiation. At temperatures below about 400 °C this radiation can be felt as heat. As the temperature rises the object starts to emit visible radiation passing from red through yellow to white as the temperature rises. Intuitively we can use this radiation to qualitatively measure temperature as below:

Temperature (°C)	Colour
500	Barely visible dull red glow
800	Bright red glow
950	Orange
1000	Yellow
1200	White
1500	Dazzling white, eyes naturally avert

Pyrometers use the same effect to provide a non-contact method of measuring temperature.

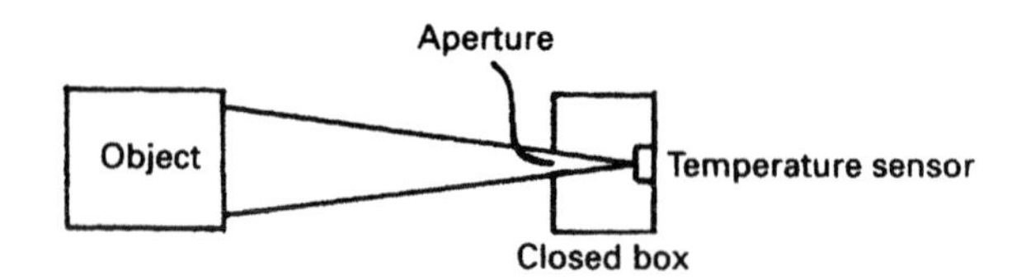

Fig. 1.1.9 The principle of an optical pyrometer.

A pyrometer is, in theory, a very simple device as shown in Figure 1.1.9. The object whose temperature is to be measured is viewed through a fixed aperture by a temperature measuring device. Part of the radiation emitted by the object falls on the temperature sensor causing its temperature to rise. The object's temperature is then inferred from the rise in temperature seen by the sensor. The sensor size must be very small, typically of the order of 1 mm diameter. Often a circular ring of thermocouples connected in series (called a *thermopile*) is used. Alternatively a small resistance thermometer (called a *bolometer*) may be used. Some pyrometers measure the radiation directly using photo-electric detectors.

A major (and surprising) advantage of pyrometers is the temperature measurement is independent of distance from the object provided the field of view is full. As shown in Figure 1.1.10 the source is radiating energy uniformly in all directions, so the energy received from a point is proportional to the solid angle subtended by the sensor. This will decrease as the square of the distance. There is, however, another effect. As the sensor moves away the scanned area also increases as the square of the distance. These two effects cancel giving a reading which is independent of distance.

Although all pyrometers operate on radiated energy there are various ways in which the temperature can be deduced from the received radiation. The simplest measures the total energy received from the object (which is proportional to T^4 where T is the temperature in kelvin). This method is susceptible to errors from lack of knowledge of the emissivity of the object's surface. This error can be reduced by using filters to restrict the

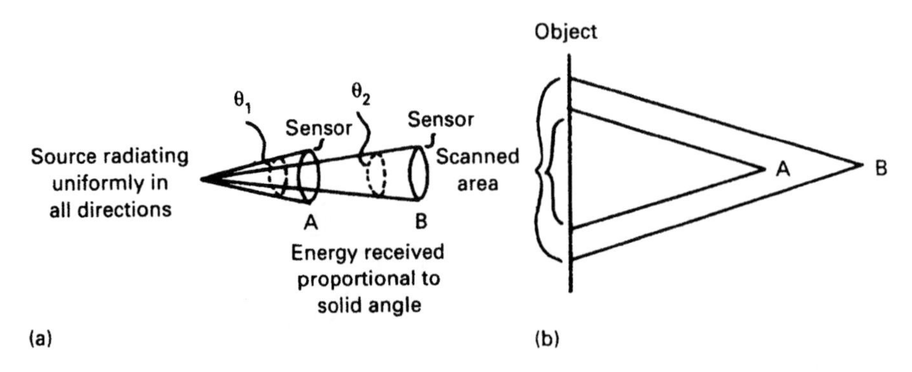

Fig. 1.1.10 The effect of object distance on a pyrometer. The energy per unit area *decreases* with the square of the distance; however, the scanned area *increases* as the square of the distance. Neglecting other influences (such as atmospheric absorption) these effects cancel as the pyrometer reading is independent of distance.

measuring range to frequencies where the object's emissivity approaches unity.

An alternative method takes two measurements at two different frequencies (i.e. two different colours) and compares the relative intensities to give an indication of temperature. This method significantly reduces emissivity errors.

Pyrometers have a few major restrictions which should be appreciated. The main problem is they measure surface temperature and only surface temperature. Lack of knowledge of the emissivity is also a major source of error.

1.1.3 Flow

1.1.3.1 General

The term 'flow' can generally be applied in three distinct circumstances:

Volumetric flow is the commonest and is used to measure the volume of material passing a point in unit time (e.g. m^3/s). It may be indicated at the local temperature and pressure or normalised to some standard conditions using the standard gas law relationship:

$$V_n = \frac{P_m V_m T_n}{P_n T_m}$$

where suffix 'm' denotes the measured conditions and suffix 'n' the normalised condition.

Mass flow is the mass of fluid passing a point in unit time (e.g. kg/s).

Velocity of flow is the velocity with which a fluid passes a given point. Care must be taken as the flow velocity may not be the same across a pipe, being lower at the walls. The effect is more marked at low flows.

1.1.3.2 Differential pressure flowmeters

If a constriction is placed in a pipe as Figure 1.1.11 the flow must be higher through the restriction to maintain equal mass flow at all points. The energy in a unit mass of fluid has three components:

(1) Kinetic energy given by $mv^2/2$.

(2) Potential energy from the height of the fluid.

(3) Energy caused by the fluid pressure, called, rather confusingly, *flow energy*. This is given by P/ρ where P is the pressure and ρ the density.

In Figure 1.1.11 the pipe is horizontal, so the potential energy is the same at all points. As the flow velocity increases through the restriction, the kinetic energy will increase and, for conservation of energy, the flow energy (i.e. the pressure) must fall:

$$\frac{mv_1^2}{2} + \frac{P_1}{\rho_1} = \frac{mv_2^2}{2} + \frac{P_2}{\rho_2}$$

This equation is the basis of all differential flowmeters.

Flow in a pipe can be smooth (called *streamline* or *laminar*) or *turbulent*. In the former case the flow velocity is not equal across the pipe being lower at the walls. With turbulent flow the flow velocity is equal at all points across a pipe.

For accurate differential measurement the flow must be turbulent.

The flow characteristic is determined by the *Reynolds number* defined as

$$R_e = \frac{vD\rho}{\eta}$$

where v is the fluid velocity, D the pipe diameter, ρ the fluid density and η the fluid viscosity. Sometimes the *kinematic viscosity* ρ/η is used in the formula. The Reynolds number is a ratio and has no dimensions. If $R_e < 2000$ the flow is laminar. If $R_e > 10^5$ the flow is fully turbulent.

Calculation of the actual pressure drop is complex, especially for compressible gases, but is generally of the form:

$$Q = K\sqrt{\Delta P} \quad (1.1.1)$$

where K is a constant for the restriction and ΔP the differential pressure. Methods of calculating K are given

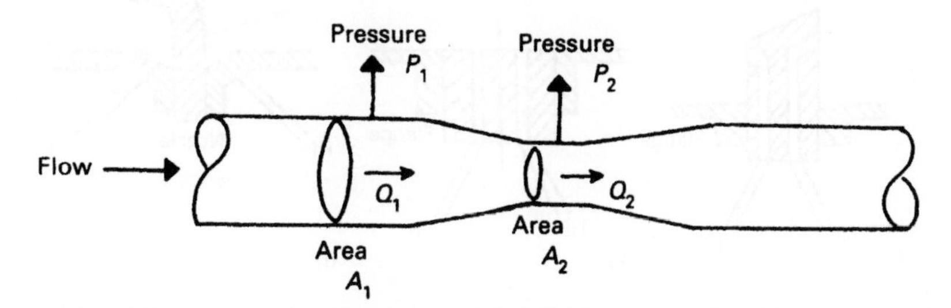

Fig. 1.1.11 The basis of a differential flowmeter. Because the mass flow must be equal at all points (i.e. $Q_1 = Q_2$) the flow velocity must increase in the region of A_2. As there is no net gain or loss of energy, the pressure must therefore decrease at A_2.

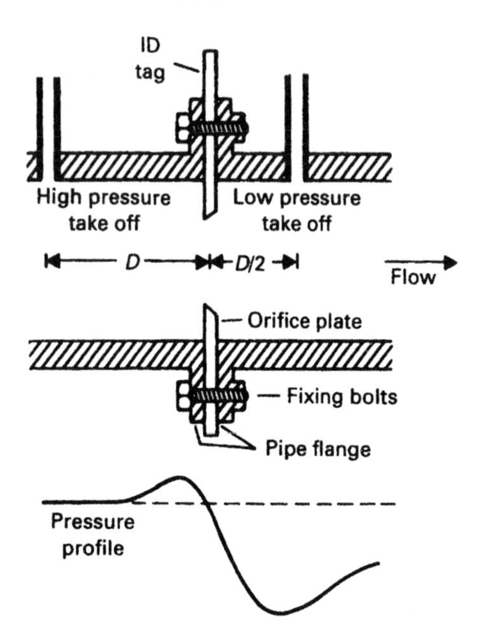

Fig. 1.1.12 Mounting of an orifice pipe between flanges with $D - D/2$ tappings.

in British Standard BS 1042 and ISO 5167:1980. Computer programs can also be purchased.

The commonest differential pressure flowmeter is the orifice plate shown in Figure 1.1.12. This is a plate inserted into the pipe with upstream tapping point at D and downstream tapping point at $D/2$ where D is the pipe diameter. The plate should be drilled with a small hole to release bubbles (liquid) or drain condensate (gases). An identity tag should be fitted showing the scaling and plant identification.

The $D - D/2$ tapping is the commonest, but other tappings shown in Figure 1.1.13 may be used where it is not feasible to drill the pipe.

Orifice plates suffer from a loss of pressure on the downstream side (called the *head loss*). This can be as high as 50%. The venturi tube and Dall tube of Figure 1.1.14(b) have lower losses of around 5% but are bulky and more expensive. Another low loss device is the pitot tube as shown in Figure 1.1.15. Equation (1.1.1) applies to all these devices, the only difference being the value of the constant K.

Conversion of the pressure to an electrical signal requires a differential pressure transmitter and a linearising square root unit. This square root extraction is a major limit on the turndown as zeroing errors are magnified. A typical turndown is 4:1.

The transmitter should be mounted with a manifold block as shown in Figure 1.1.16 to allow maintenance. Valves B and C are isolation valves. Valve A is an equalising valve and is used, along with B and C, to zero the

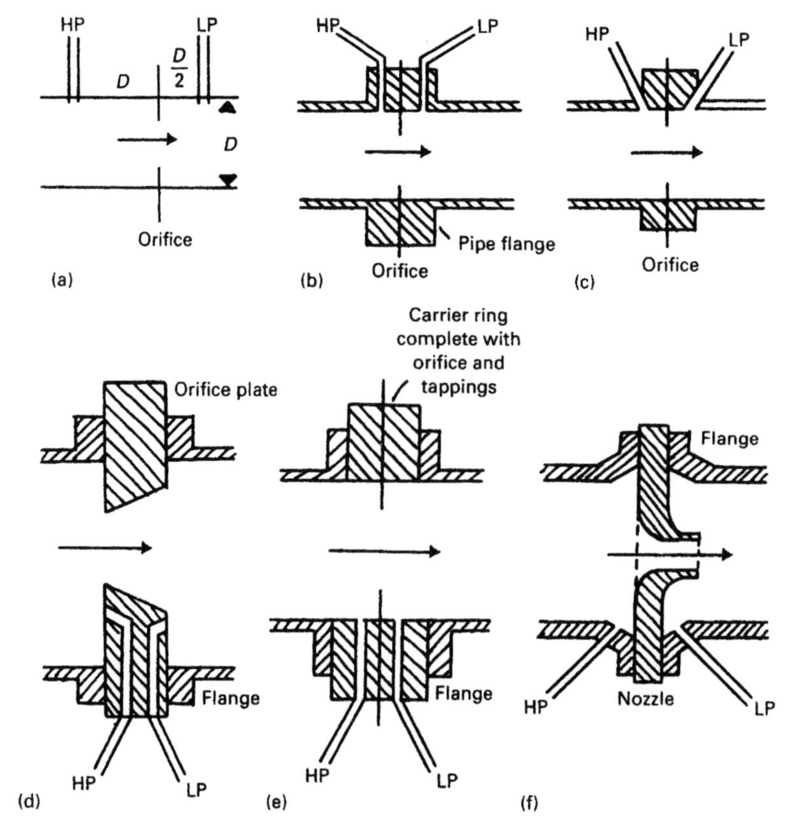

Fig. 1.1.13 Common methods of mounting orifice plates. (a) $D - D/2$, probably the commonest; (b) flange taps used on large pipes with substantial flanges; (c) corner taps drilled through flange; (d) plate taps, tappings built into the orifice plate; (e) orifice carrier, can be factory made and needs no drilling on site; (f) nozzle, gives smaller head loss.

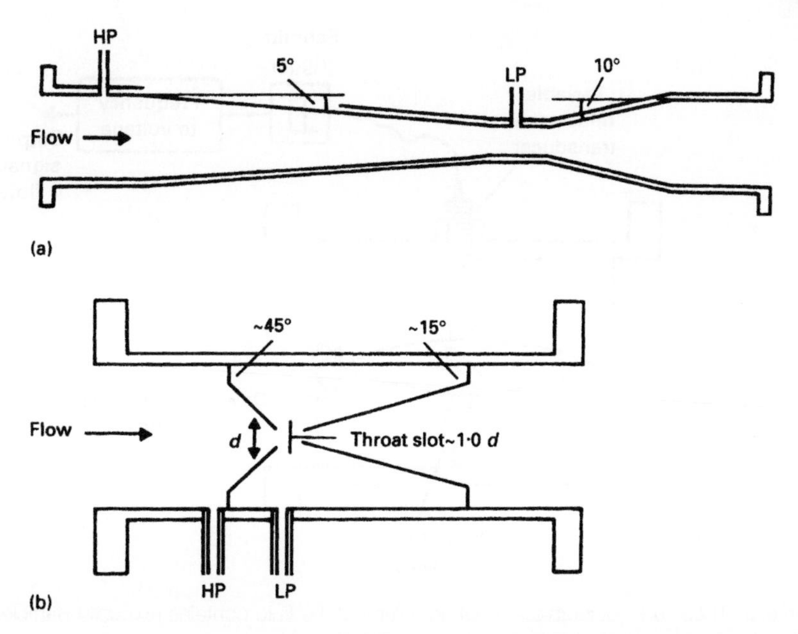

Fig. 1.1.14 Low loss differential pressure primary sensors. Both give a much lower head loss than an orifice plate but at the expense of a great increase on pipe length. It is often impossible to provide the space for these devices. (a) Venturi tube; (b) Dall tube.

transmitter. In normal operation A is closed and B and C are open. Valve A should always be opened before valves B and C are closed prior to removal of the transducer to avoid high pressure being locked into one leg. Similarly on replacement, valves B and C should both be opened before valve A is closed to prevent damage from the static pressure.

Gas measurements are prone to condensate in pipes, and liquid measurements are prone to gas bubbles. To avoid these effects, a gas differential transducer should be mounted above the pipe and a liquid transducer below the pipe with tap off points in the quadrants as shown later in Section 1.1.4.6.

Although the accuracy and turndown of differential flowmeters is poor (typically 4% and 4:1) their robustness, low cost and ease of installation still makes them the commonest type of flowmeter.

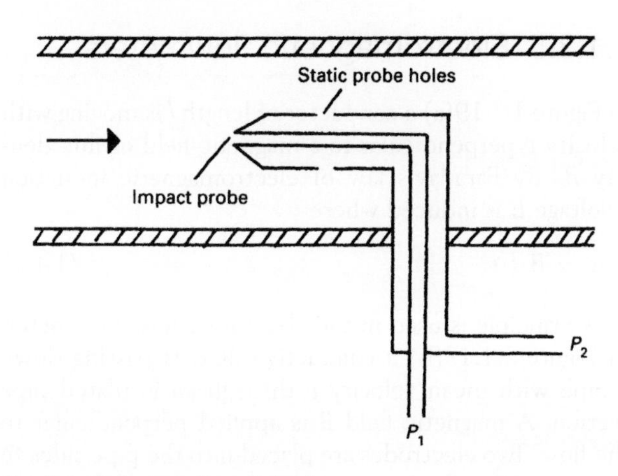

Fig. 1.1.15 An insertion pitot tube.

1.1.3.3 Turbine flowmeters

As its name suggests a turbine flowmeter consists of a small turbine placed in the flow as Figure 1.1.17. Within a specified flow range (usually with about a 10:1 turndown for liquids, 20:1 for gases), the rotational speed is directly proportional to flow velocity.

The turbine blades are constructed of ferromagnetic material and pass below a variable reluctance transducer

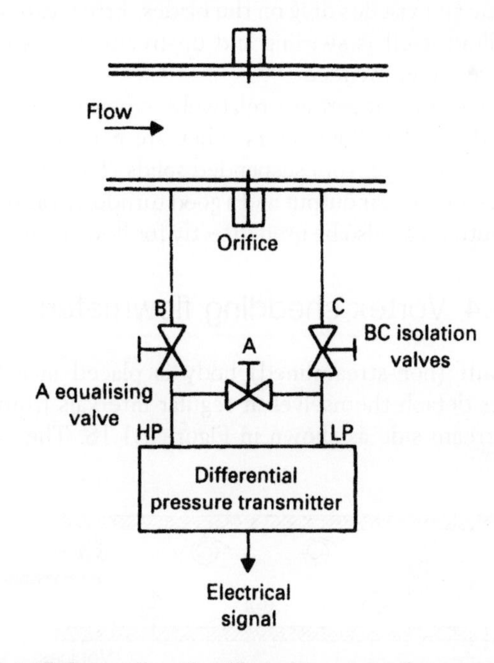

Fig. 1.1.16 Connection of a differential pressure flow sensor such as an orifice plate to a differential pressure transmitter. Valves B and C are used for isolation and valve A for equalisation.

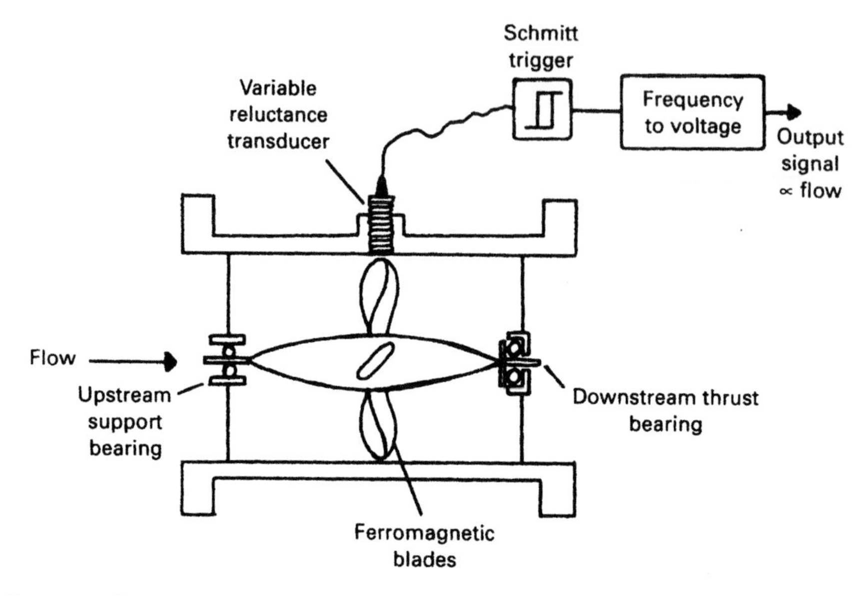

Fig. 1.1.17 A turbine flowmeter. These are vulnerable to bearing failures if the fluid contains any solid particles.

producing an output approximating to a sine wave of the form:

$$E = A\omega \sin(N\omega t)$$

where A is a constant, ω is the angular velocity (itself proportional to flow) and N is the number of blades. Both the output amplitude and the frequency are proportional to flow, although the frequency is normally used.

The turndown is determined by frictional effects and the flow at which the output signal becomes unacceptably low. Other non-linearities occur from the magnetic and viscous drag on the blades. Errors can occur if the fluid itself is swirling and upstream straightening vanes are recommended.

Turbine flowmeters are relatively expensive and less robust than other flowmeters. They are particularly vulnerable to damage from suspended solids. Their main advantages are a linear output and a good turndown ratio. The pulse output can also be used directly for flow totalisation.

1.1.3.4 Vortex shedding flowmeters

If a bluff (non-streamlined) body is placed in a flow, vortices detach themselves at regular intervals from the downstream side as shown in Figure 1.1.18. The effect can be observed by moving a hand through water. In flow measurement the vortex shedding frequency is usually a few hundred Hz. Surprisingly at Reynolds numbers in excess of 10^3 the volumetric flow rate, Q, is directly proportional to the observed frequency of vortex shedding f, i.e.

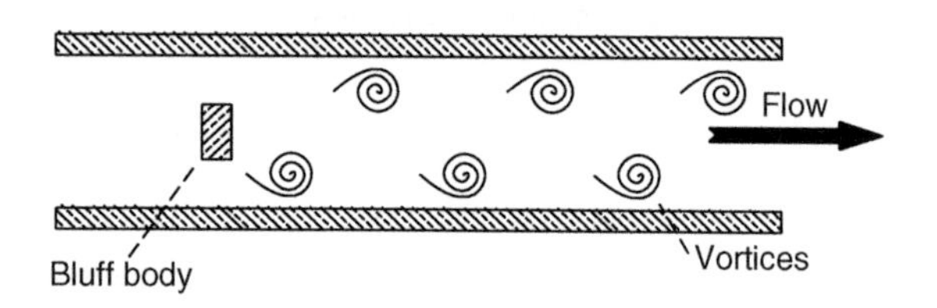

Fig. 1.1.18 Vortex shedding flowmeter.

$$Q = Kf$$

where K is a constant determined by the pipe and obstruction dimensions.

The vortices manifest themselves as sinusoidal pressure changes which can be detected by a sensitive diaphragm on the bluff body or by a downstream modulated ultrasonic beam.

The vortex shedding flowmeter is an attractive device. It can work at low Reynolds numbers, has excellent turndown (typically 15:1), no moving parts and minimal head loss.

1.1.3.5 Electromagnetic flowmeters

In Figure 1.1.19(a) a conductor of length l is moving with velocity v perpendicular to a magnetic field of flux density B. By Faraday's law of electromagnetic induction a voltage E is induced where

$$E = B \cdot l \cdot v \qquad (1.1.2)$$

This principle is used in the electromagnetic flowmeter. In Figure 1.1.19(b) a conductive fluid is passing down a pipe with mean velocity v through an insulated pipe section. A magnetic field B is applied perpendicular to the flow. Two electrodes are placed into the pipe sides to form, via the fluid, a moving conductor of length D

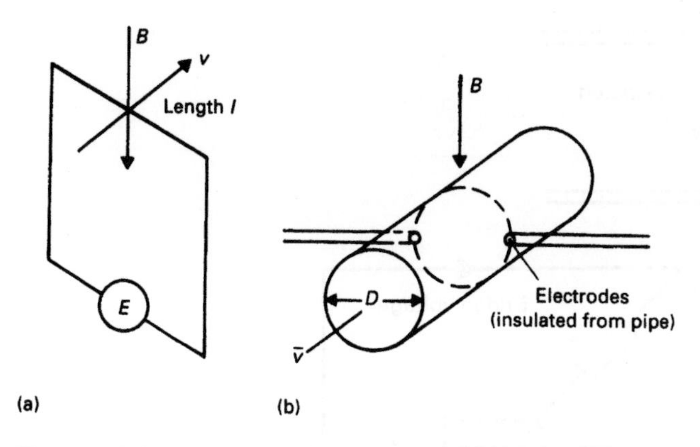

Fig. 1.1.19 Electromagnetic flowmeter. (a) Electromagnetic induction in a wire moving in a magnetic field; (b) The principle applied with a moving conductive fluid.

relative to the field where D is the pipe diameter. From Equation (1.1.2) a voltage will occur across the electrodes which is proportional to the mean flow velocity across the pipe.

Equation (1.1.2) and Figure 1.1.19(b) imply a steady d.c. field. In practice an a.c. field is used to minimise electrolysis and reduce errors from d.c. thermoelectric and electrochemical voltages which are of the same order of magnitude as the induced voltage.

Electromagnetic flowmeters are linear and have an excellent turndown of about 15:1. There is no practical size limit and no head loss. They do, though provide a few installation problems as an insulated pipe section is required, with earth bonding either side of the meter to avoid damage from any welding which may occur in normal service. They can only be used on fluids with a conductivity in excess of 1 mS/m which permits use with many (but not all), common liquids but prohibits their use with gases. They are useful with slurries with a high solids content.

1.1.3.6 Ultrasonic flowmeters

The Doppler effect occurs when there is relative motion between a sound transmitter and receiver as shown in Figure 1.1.20(a). If the transmitted frequency is f_t Hz, V_s is the velocity of sound and V the relative velocity, the observed received frequency, f_r, will be

$$f_t\frac{(V+V_s)}{V}$$

A Doppler flowmeter injects an ultrasonic sound wave (typically a few hundred kHz) at an angle θ into a fluid moving in a pipe as shown in Figure 1.1.20(b). A small part of this beam will be reflected back off small bubbles, solid matter, vortices, etc. and is picked up by a receiver mounted alongside the transmitter. The frequency is subject to two changes, one as it moves upstream against

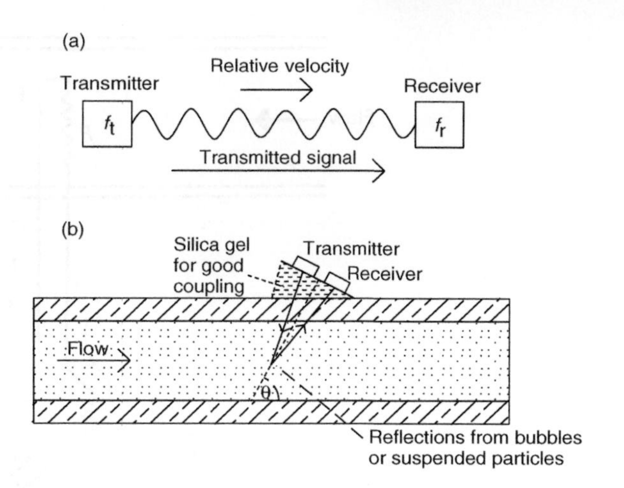

Fig. 1.1.20 An Ultrasonic Flowmeter. (a) Principle of operation; (b) Schematic of a clip on ultrasonic flowmeter.

the flow, and one as it moves back with the flow. The received frequency is thus

$$f_r = f_t\frac{(V_s + V\cos(\theta))}{(V_s - V\cos(\theta))}$$

which can be simplified to

$$\Delta f = \frac{2f_t}{V_s}V\cos(\theta)$$

The Doppler flowmeter measures mean flow velocity, is linear, and can be installed (or removed) without the need to break into the pipe. The turndown of about 100:1 is the best of all flowmeters. Assuming the measurement of mean flow velocity is acceptable it can be used at all Reynolds numbers. It is compatible with all fluids and is well suited for difficult applications with corrosive liquids or heavy slurries.

1.1.3.7 Hot wire anemometer

If fluid passes over a hot object, heat is removed. It can be shown that the power loss is

$$P = A + B\sqrt{v} \tag{1.1.3}$$

where v is the flow velocity and A and B are constants. A is related to radiation and B to conduction.

Figure 1.1.21 shows a flowmeter based on Equation (1.1.3). A hot wire is inserted in the flow and maintained at a constant temperature by a self-balancing bridge. Changes in the wire temperature result in a resistance change which unbalances the bridge. The bridge voltage is automatically adjusted to restore balance.

The current, I, though the resistor is monitored. With a constant wire temperature the heat dissipated is equal to the power loss from which

$$v = (I^2R - A)/B^2$$

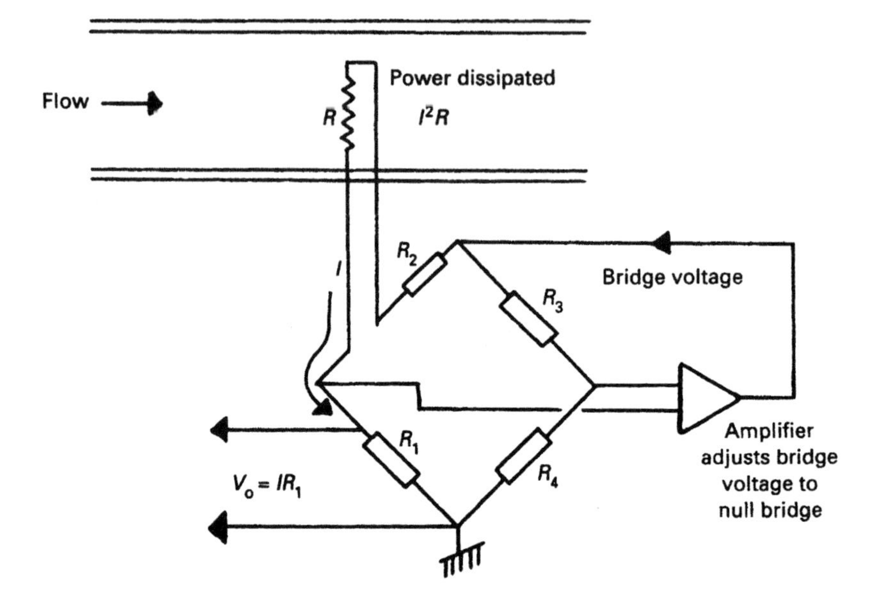

Fig. 1.1.21 The hot wire anemometer.

Obviously the relationship is non-linear, and correction will need to be made for the fluid temperature which will affect constants A and B.

1.1.3.8 Mass flowmeters

The volume and density of all materials are temperature dependent. Some applications will require true volumetric measurements, some, such as combustion fuels, will really require mass measurement. Previous sections have measured volumetric flow. This section discusses methods of measuring mass flow.

The relationship between volume and mass depends on both pressure and absolute temperature (measured in kelvin). For a gas,

$$\frac{P_1V_1}{T_1} = \frac{P_2V_2}{T_2}$$

The relationship for a liquid is more complex, but if the relationship is known and the pressure and temperature are measured along with the delivered volume or volumetric flow the delivered mass or mass flow rate can be easily calculated. Such methods are known as *Inferential* flowmeters.

In Figure 1.1.22 a fluid is passed over an in-line heater with the resultant temperature rise being measured by two temperature sensors. If the specific heat of the material is constant, the mass flow, F_m, is given by

$$F_m = \frac{E}{C_p\theta}$$

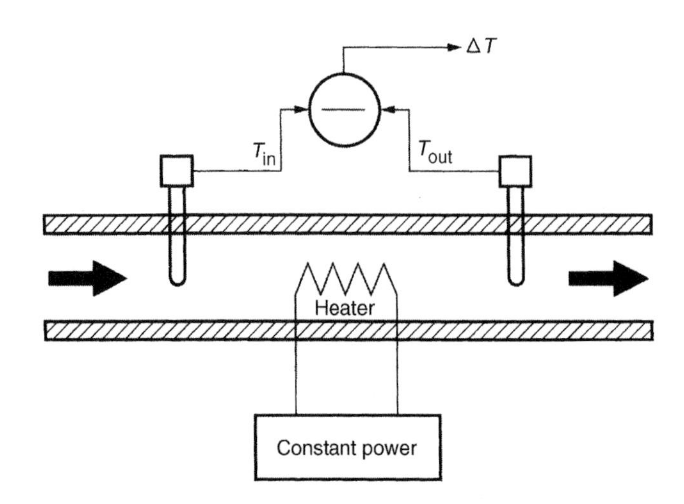

Fig. 1.1.22 Mass flow measurement by noting the temperature rise caused by a constant input power.

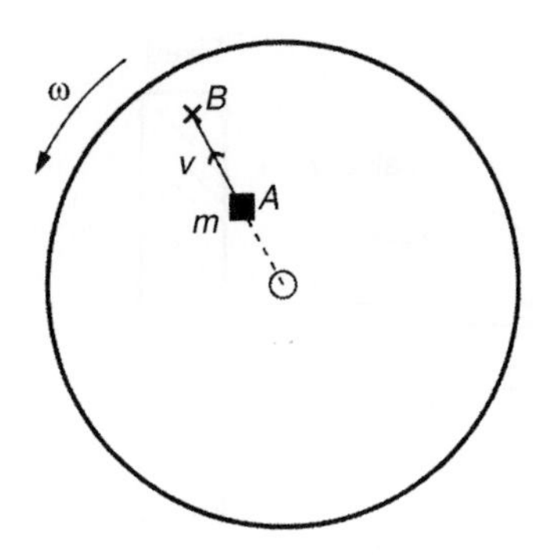

Fig. 1.1.23 Definition of Coriolis force.

where E is the heat input from the heater, C_p is the specific heat and θ the temperature rise. The method is only suitable for relatively small flow rates.

Many modern mass flowmeters are based on the Coriolis effect. In Figure 1.1.23 an object of mass m is required to move with linear velocity v from point A to point B on a surface which is rotating with angular velocity. If the object moves in a straight line as viewed by a static observer, it will appear to veer to the right when viewed by an observer on the disc.

If the object is to move in a straight line as seen by an observer on the disc a force must be applied to the object as it moves out. This force is known as the *Coriolis force* and is given by

$$F = 2m\omega v$$

where m is the mass, ω is the *l*angular velocity and v the linear velocity. The existence of this force can be easily experienced by trying to move along a radius of a rotating children's roundabout in a playground.

Coriolis force is not limited to pure angular rotation, but also occurs for sinusoidal motion. This effect is used as the basis of a Coriolis flowmeter shown in Figure 1.1.24. The flow passes through a 'C' shaped tube which is attached to a leaf spring and vibrated sinusoidally by a magnetic forcer. The Coriolis force arises not, as might be first thought, because of the semi-circle pipe section at the right-hand side, but from the angular motion induced into the two horizontal pipe sections with respect to the fixed base. If there is no flow, the two pipe sections will oscillate together. If there is flow, the flow in the top pipe is in the opposite direction to the flow in the bottom pipe and the Coriolis force causes a rolling motion to be induced as shown. The resultant angular deflection is proportional to the mass flow rate.

The original meters used optical sensors to measure the angular deflection. More modern meters use a coil rather than a 'C' section and sweep the frequency to determine the resonant frequency. The resonant frequency is then related to the fluid density by

$$f_c = \sqrt{\frac{K}{\text{density}}}$$

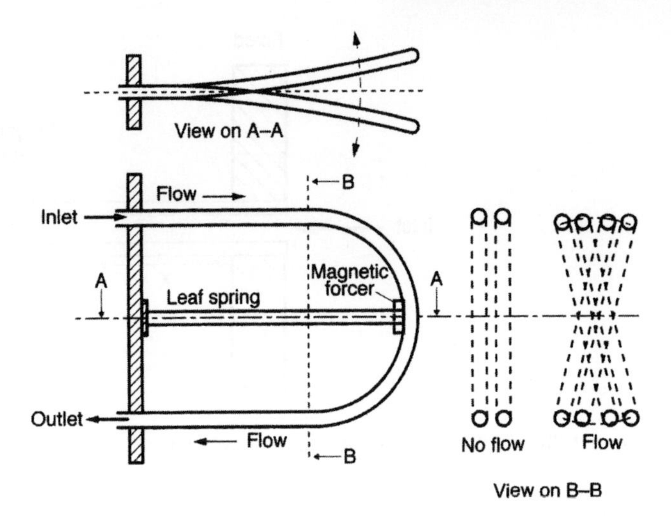

Fig. 1.1.24 Simple Coriolis mass flowmeter. A multi-turn coil is often used in place of the 'C' segment.

where K is a constant. The mass flow is determined either by the angular measurement or the phase shift in velocity. Using the resonant frequency maximises the displacement and improves measurement accuracy.

Coriolis measurement is also possible with a straight pipe. In Figure 1.1.25 the centre of the pipe is being deflected with a sinusoidal displacement, and the velocity of the inlet and outlet pipe sections monitored. The Coriolis effect will cause a phase shift between inlet and outlet velocities. This phase shift is proportional to the mass flow.

1.1.4 Pressure

1.1.4.1 General

There are four types of pressure measurement.

Differential pressure is the difference between two pressures applied to the transducer. These are commonly used for flow measurement as described in Section 1.1.3.2.

Gauge pressure is made with respect to atmospheric pressure. It can be considered as a differential pressure measurement with the low pressure leg left open. Gauge pressure is usually denoted by the suffix 'g' (e.g. 37 psig). Most pressure transducers in hydraulic and pneumatic systems indicate gauge pressure.

Absolute pressure is made with respect to a vacuum.

Absolute pressure = Gauge pressure + Atmospheric pressure

Atmospheric pressure is approximately 1 bar, 100 kPa or 14.7 psi.

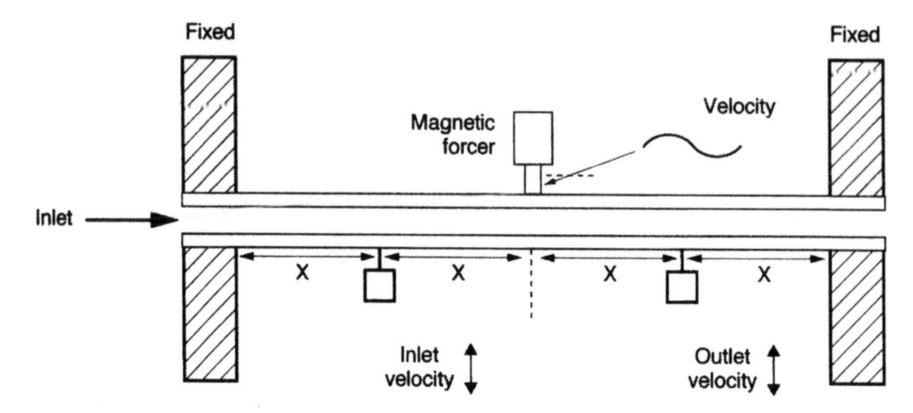

Fig. 1.1.25 Vibrating straight pipe mass flowmeter.

Head pressure is used in liquid level measurement and refers to pressure in terms of the height of a liquid (e.g. inches water gauge). It is effectively a gauge pressure measurement, but if the liquid is held in a vented vessel any changes in atmospheric pressure will affect both legs of the transducer equally giving a reading which is directly related to liquid height. The head pressure is given by

$$P = \rho g h \quad (1.1.4)$$

where P is the pressure in pascals, ρ is the density (kg/ m^2), g is acceleration due to gravity (9.8 m/s^2) and h the column height in metres.

In imperial units, pounds are a term of weight (i.e. force) not mass so Equation (1.1.4) becomes

$$P = \rho h$$

where P is the pressure in pounds per square unit (inch or foot), ρ the density in pounds per cubic unit and h is the height in units.

1.1.4.2 Manometers

Although manometers are not widely used in industry they give a useful insight into the principle of pressure measurement. If a U-tube is part filled with liquid, and differing pressures applied to both legs as Figure 1.1.26, the liquid will fall on the high pressure side and rise on the low pressure side until the head pressure of liquid matches the pressure difference. If the two levels are separated by a height h then

$$h = (P_1 - P_2)/\rho g$$

where P_1 and P_2 are the pressures (in pascals), ρ is the density of the liquid and g is the acceleration due to gravity.

1.1.4.3 Elastic sensing elements

The Bourdon tube, dating from the mid-nineteenth century, is still the commonest pressure indicating device. The tube is manufactured by flattening a circular cross-section tube to the section shown in Figure 1.1.27 and bending it into a C shape. One end is fixed and connected to the pressure to be measured. The other end is closed and left free.

If pressure is applied to the tube it will try to straighten causing the free end to move up and to the right. This motion is converted to a circular motion for a pointer with a quadrant and pinion linkage. A Bourdon tube inherently measures gauge pressure.

Bourdon tubes are usable up to about 50 MPa, (about 10,000 psi). Where an electrical output is required the tube can be coupled to a potentiometer or linear variable displacement transformer (LVDT).

Diaphragms can also be used to convert a pressure differential to a mechanical displacement. Various arrangements are shown in Figure 1.1.28. The displacement

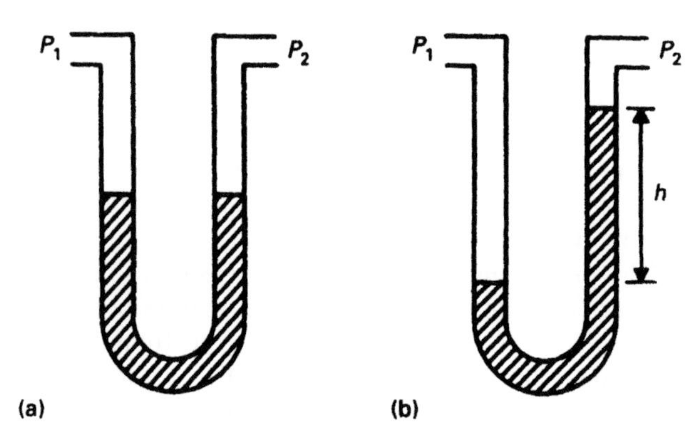

Fig. 1.1.26 The U-tube manometer. (a) Pressures P_1 and P_2 are equal; (b) Pressure P_1 is greater than P_2 and the distance h is proportional to $(P_1 - P_2)$.

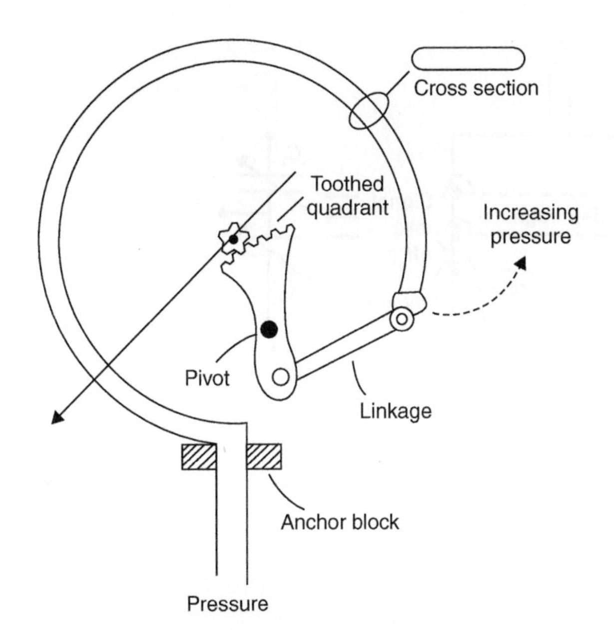

Fig. 1.1.27 The Bourdon tube pressure gauge.

can be measured by LVDTs (see Section 1.1.6.4), strain gauges (see Section 1.1.6.4). The diaphragm can also be placed between two perforated plates as Figure 1.1.29. The diaphragm and the plates form two capacitors whose value is changed by the diaphragm deflection.

1.1.4.4 Piezo elements

The piezo-electric effect occurs in quartz crystals. An electrical charge appears on the faces when a force is applied.

This charge is directly proportional to the applied force. The force can be related to pressure with suitable diaphragms. The resulting charge is converted to a voltage by the circuit of Figure 1.1.30, which is called a *charge amplifier.* The output voltage is given by

$$V = -q/c$$

Because q is proportional to the force, V is proportional to the force applied to the crystal. In practice the charge leaks away, even with FET amplifiers and low leakage capacitors. Piezo-electric transducers are thus unsuitable for measuring static pressures, but they have a fast response and are ideal for measuring fast dynamic pressure changes.

A related effect is the piezo-resistive effect which results in a change in resistance with applied force. Piezo-resistive devices are connected into a Wheatstone bridge.

1.1.4.5 Force balance systems

Friction and non-linear spring constants can cause errors in elastic displacement transducers. The force balance

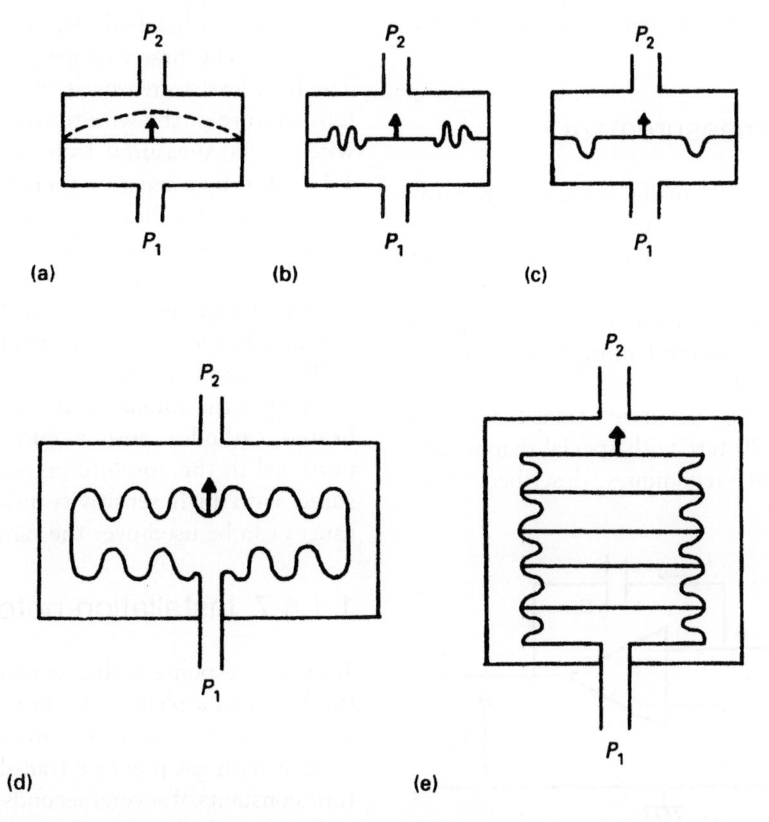

Fig. 1.1.28 Various arrangements of elastic pressure sensing elements. (a) diaphragm; (b) corrugated diaphragm; (c) catenary diaphragm; (d) capsule; (e) bellows.

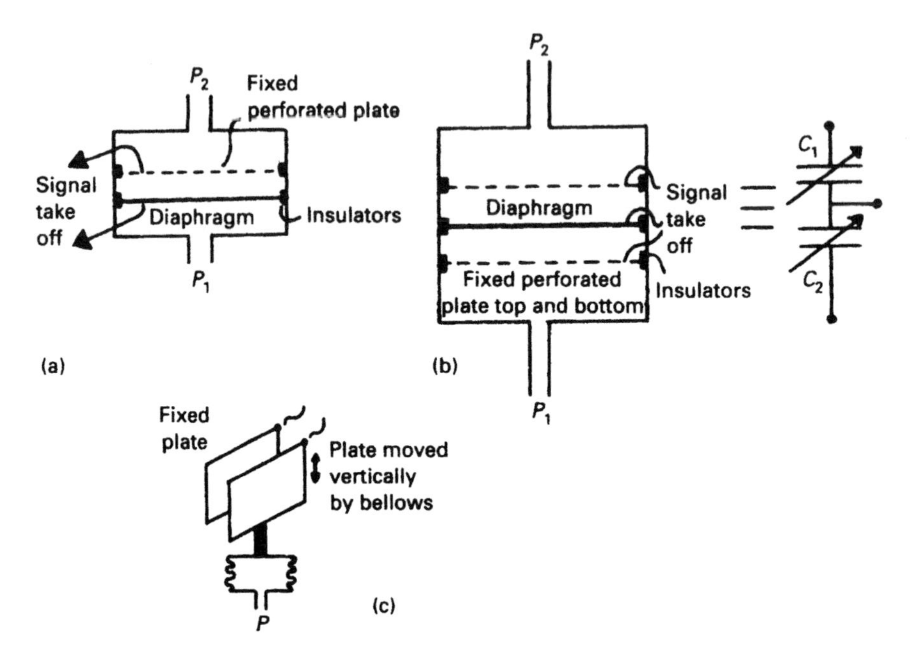

Fig. 1.1.29 Variable capacitance pressure sensing elements. (a) Single capacitor variable spacing; (b) double capacitor variable spacing; (c) variable area.

system uses closed loop control to balance the force from the pressure with an opposing force which keeps the diaphragm in a fixed position as shown in Figure 1.1.31. As the force from the solenoid is proportional to the current, the current in the coil is directly proportional to the applied pressure. LVDTs are commonly used for the position feedback.

1.1.4.6 Vacuum measurement

Vacuum measurement is normally expressed in terms of the height of a column of mercury which is supported by the vacuum (e.g. mmHg). Atmospheric pressure (approximately 1 bar) thus corresponds to about 760 mmHg, and an absolute vacuum is 0 mmHg. The term '*torr*' is generally used for 1 mmHg. Atmospheric pressure is thus about 760 torr.

Conventional absolute pressure transducers are usable down to about 20 torr with special diaphragms. At lower pressures other techniques, described below, must be used. At low pressures the range should be considered as logarithmic, i.e. 1–0.1 torr is the same 'range' as 0.1–0.01 torr.

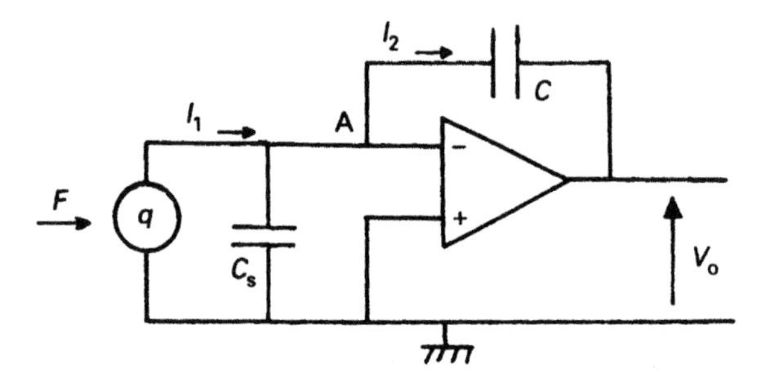

Fig. 1.1.30 The charge amplifier.

The first method, called the *Pirani gauge,* applies constant energy to heat a thin wire in the vacuum. Heat is lost from a hot body by conduction, convection and radiation. The first two losses are pressure dependent. The heat loss decreases as the pressure falls causing the temperature of the wire to rise. The temperature of the wire can be measured from its resistance (see Section 1.1.2.3) or by a separate thermocouple.

The range of a Pirani gauge is easily changed by altering the energy supplied to the wire. A typical instrument would cover the range 5 to 10^{-5} torr in three ranges. Care must be taken not to overheat the wire which will cause a change in the heat loss characteristics.

The second technique, called an *ionisation gauge,* is similar to a thermionic diode. Electrons emitted from the heated cathode cause a current flow which is proportional to the absolute pressure. The current is measured with high sensitivity microammeters. Ionisation gauges can be used over the range 10^{-3}–10^{-12} torr.

1.1.4.7 Installation notes

To avoid a response with a very long time constant the pipe run between a plant and a pressure transducer must be kept as short as possible. This effect is particularly important with gas pressure transducers which can exhibit time constants of several seconds if the installation is poor.

Entrapped air can cause significant errors in liquid pressure measurement, and similarly condensed liquid

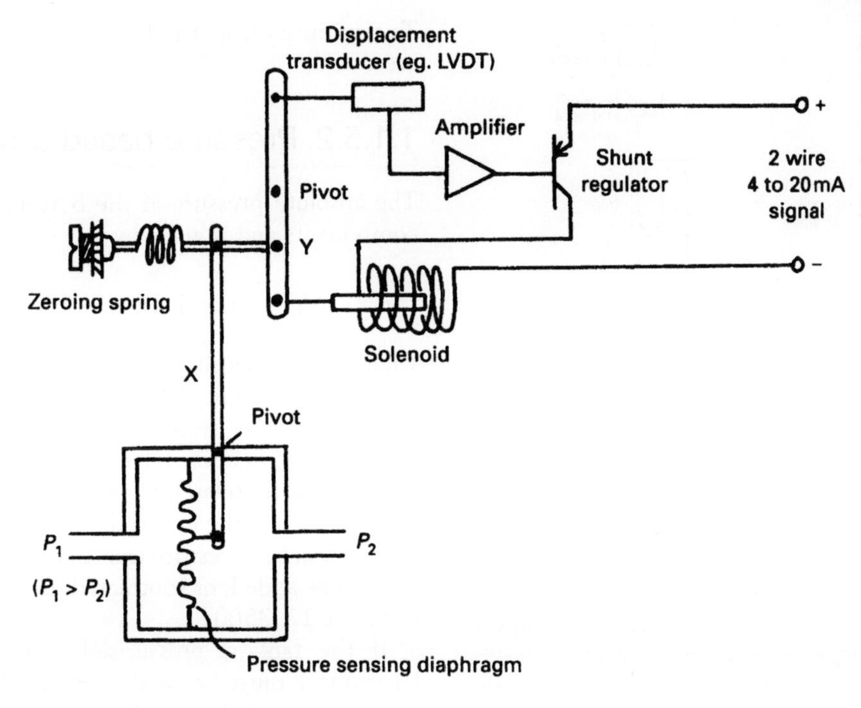

Fig. 1.1.31 A force balance pressure transducer.

will cause errors in gas pressure measurement. Ideally the piping should be taken from the top of the pipe and rise to the transducer for gas systems, and be taken from the lower half of the pipe and fall to the transducer for liquid pressure measurement. If this is not possible, vent/drain cocks must be fitted. Care must be taken to avoid sludge build up around the tapping with liquid pressure measurement. Typical installations are shown in Figure 1.1.32. Under no circumstances should the piping form traps where gas bubbles or liquid sumps can form.

Steam pressure transducers are a special case. Steam can damage the transducer diaphragm, so the piping leg to the transducer is normally arranged to naturally fill with water by taking the tapping of the top of the pipe and mounting the transducer (with unlagged pipes) below the pipe as Figure 1.1.33(a). If this is not possible, a steam trap is placed in the piping as Figure 1.1.33(b).

In both cases there will be an offset from the liquid head. A similar effect occurs when a liquid pressure transmitter is situated below the pipe. In both cases the

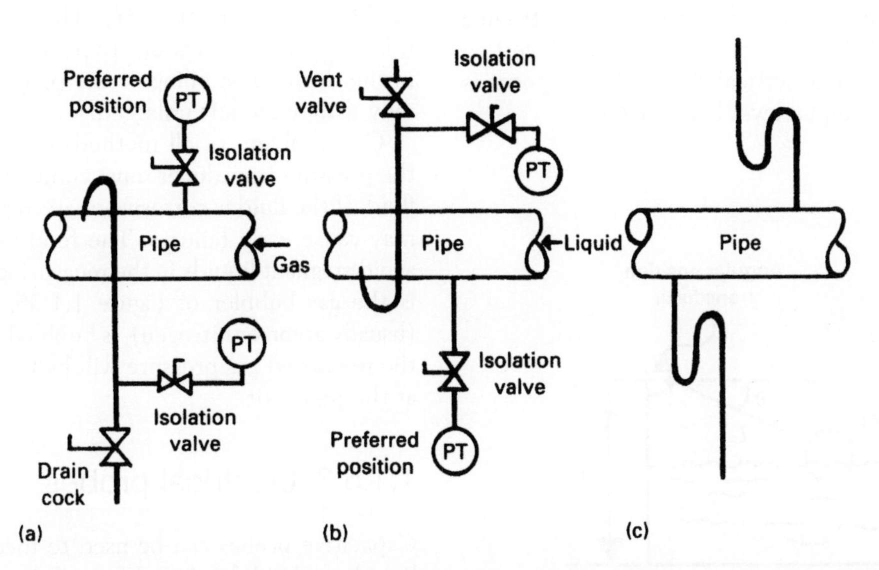

Fig. 1.1.32 Installation of pressure transducers. Similar considerations apply to differential pressure transducers used with orifice plates and similar primary flow sensors. (a) Gas pressure transducer; (b) liquid pressure transducer; (c) faulty installations with potential traps.

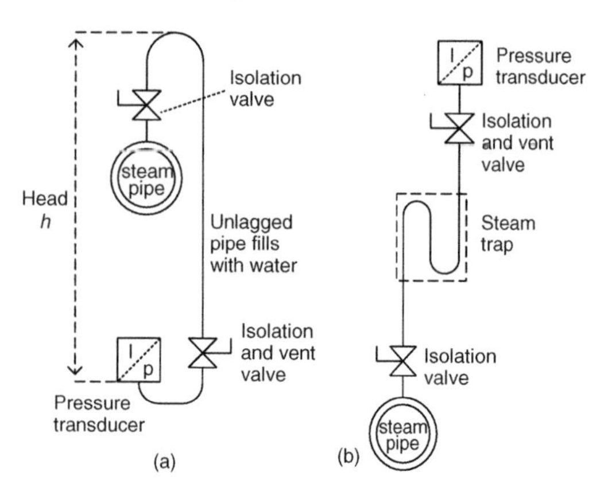

Fig. 1.1.33 Piping arrangements for steam pressure transducers. (a) Preferred arrangement; (b) alternative arrangement.

offset pressure can usually be removed in the transducer setup.

Pressure transducers will inevitably need to be changed or maintained on line, so isolation valves should always be fitted. These should include an automatic vent on closure so pressure cannot be locked into the pipe between the transducer and the isolation valve.

Where a low pressure range differential transducer is used with a high static pressure, care must be taken to avoid damage by the static pressure getting 'locked in' to one side. Figure 1.1.16 (Section 1.1.3.2) shows a typical manifold for a differential transducer. The bypass valve should always be opened first on removal and closed last on replacement.

1.1.5 Level transducers

1.1.5.1 Float based systems

A float is the simplest method of measuring or controlling level. Figure 1.1.34 shows a typical method of converting the float position to an electrical signal. The response is non-linear, with the liquid level being given by

$$h = H - l\sin(\theta)$$

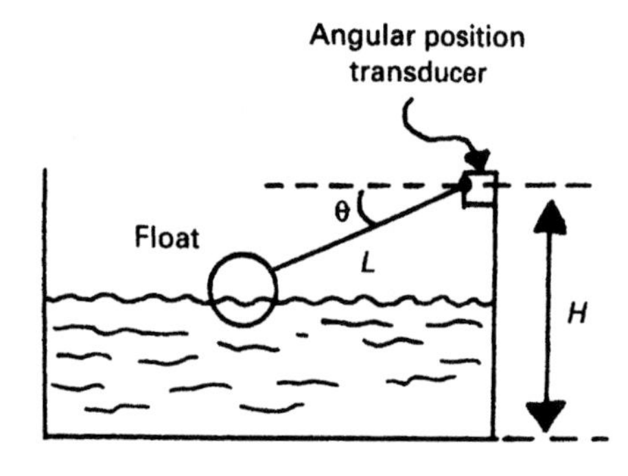

Fig. 1.1.34 Lever arm float based level measurement system.

To minimise errors the float should have a large surface area.

1.1.5.2 Pressure based systems

The absolute pressure at the bottom of a tank has two components and is given by

$$P = \rho gh + \text{atmospheric pressure}$$

where ρ is the density, g is gravitational acceleration and h the depth of the liquid. The term ρgh is called the head pressure. Most pressure transducers measure gauge pressure (see Section 1.1.4.1), so the indicated pressure will be directly related to the liquid depth. It should be noted that 'h' is the height difference between the liquid surface and the pressure transducer itself. An offset may need to be added, or subtracted, to the reading as shown in Figure 1.1.35(a).

If the tank is pressurised, a differential pressure transmitter must be used as Figure 1.1.35(b). This arrangement may also require a correction if the pressure transducer height is not the same as the tank bottom. Problems can also occur if condensate can form in the low pressure leg as the condensate will have its own, unknown, head pressure.

In these circumstances the problems can be overcome by deliberately filling both legs with fluid as Figure 1.1.35(c). Note that the high pressure leg is now connected to the tank top. The differential pressure is now

$$\Delta P = (D - h)g(\rho_1 - \rho_2)$$

where ρ_1 is the liquid density and ρ_2 the vapour density (often negligible). Note that the equation is independent of the transducer offset H_1. This arrangement is often used in boiler applications to measure the level of water in the drum. The filling of the pipes often occurs naturally if they are left unlagged.

One problem of all methods described so far is that the pressure transducer must come into contact with the fluid. If the fluid is corrosive or at a high temperature this may cause early failures. The food industries must also avoid stagnant liquids in the measuring legs. One solution is the gas bubbler of Figure 1.1.36. Here an inert gas (usually argon or nitrogen), is bubbled into the liquid, and the measured gas pressure will be the head pressure ρgh at the pipe exit.

1.1.5.3 Electrical probes

Capacitive probes can be used to measure the depth of liquids and solids. A rod, usually coated with PVC or PTFE, is inserted into the tank (Figure 1.1.37(a)) and the

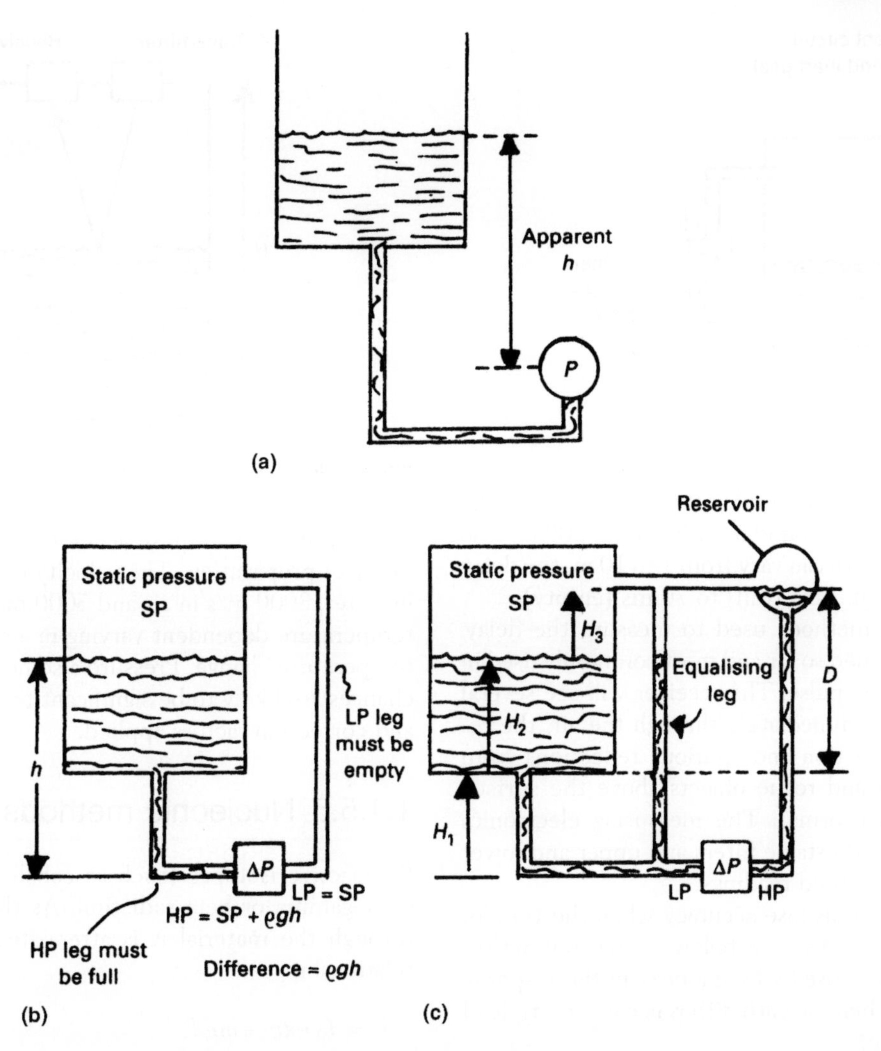

Fig. 1.1.35 Level measurement from hydrostatic pressure. (a) Head error arising from transducer position. This error can be removed by a simple offset; (b) differential pressure measurement in a pressurised tank; (c) level measurement with condensable liquid. This method is commonly used for drum level in steam boilers.

capacitance measured to the tank wall. This capacitance has two components, C_1 above the surface and C_2 below the surface. As the level rises C_1 will decrease and C_2 will increase. These two capacitors are in parallel, but as liquids and solids have a higher di-electric constant than vapour, the net result is that the capacitance rises for increasing level.

The effect is small, however, and the change is best measured with an a.c. bridge circuit driven at about 100 kHz (Figure 1.1.37(b)). The small capacitance also means that the electronics must be close to the tank to prevent errors from cable capacitance.

The response is directly dependent on the dielectric constant of the material. If this changes, caused, say, by bubbling or frothing, errors will be introduced.

If the resistance of a liquid is reasonably constant, the level can be inferred by reading the resistance between two submerged metal probes. Stainless steel is often used to reduce corrosion. A bridge measuring circuit is again used with an a.c. supply to avoid electrolysis and plating effects. Corrosion can be a problem with many liquids, but the technique works well with water. Cheap probes using this principle are available for stop/start level control applications.

1.1.5.4 Ultrasonic transducers

Ultrasonic methods use high frequency sound produced by the application of a suitable a.c. voltage to a piezo-electric crystal. Frequencies in the range 50 kHz to 1 MHz can be used, although the lower end of the range is more common in industry. The principle is shown in Figure 1.1.38. An ultrasonic pulse is emitted by a transmitter. It reflects off the surface and is detected by a receiver. The time of flight is given by

$$t = \frac{2d}{v}$$

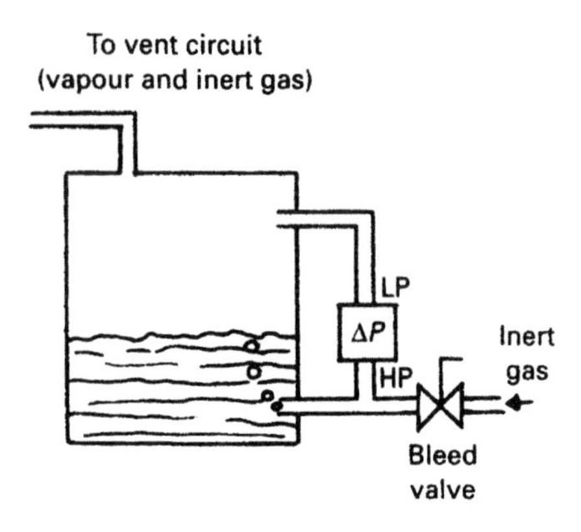

Fig. 1.1.36 Level measurement using a gas bubbler.

where v is the velocity of sound in the medium above the surface. The velocity of sound in air is about 3000 m/s, so for a tank whose depth can vary from 1 to 10 m, the delay will vary from about 7 ms (full) to 70 ms (empty).

There are two methods used to measure the delay. The simplest, assumed so far and most commonly used in industry, is a narrow pulse. The receiver will see several pulses, one almost immediately through the air, the required surface reflection and spurious reflections from sides, the bottom and rogue objects above the surface (e.g. steps and platforms). The measuring electronics normally provides adjustable filters and upper and lower limits to reject unwanted readings.

Pulse driven systems lose accuracy when the time of flight is small. For a distance below a few millimetres a swept frequency is used where a peak in the response will be observed when the path difference is a multiple of the wavelength, i.e.

$$d = \frac{v}{2f}$$

where v is the velocity of propagation and f the frequency at which the peak occurs. Note that this is ambiguous as peaks will also be observed at integer multiples of the wavelength.

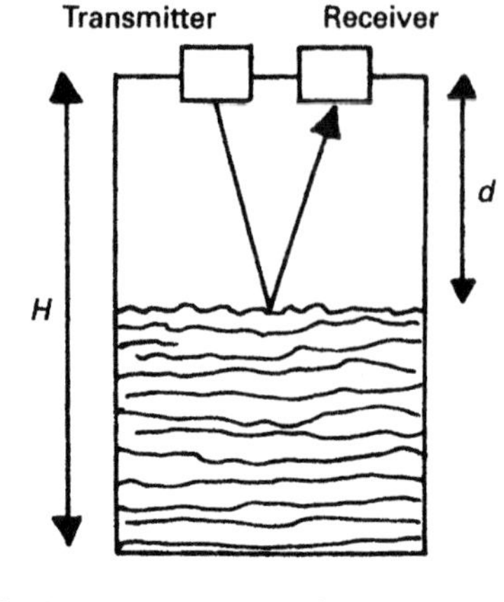

Fig. 1.1.38 Basic arrangement for ultrasonic level measurement.

Both methods require accurate knowledge of the velocity of propagation. The velocity of sound is 1440 m/s in water, 3000 m/s in air and 5000 m/s in steel. It is also temperature dependent varying in air by 1% for a 30 °C temperature change. Pressure also has an effect. If these changes are likely to be significant they can be measured and correction factors applied.

1.1.5.5 Nucleonic methods

Radioactive isotopes (such as cobalt 60) spontaneously emit gamma or beta radiation. As this radiation passes through the material it is attenuated according to the relationship

$$I = I_0 \exp(-m\rho d)$$

where I_0 is the initial intensity, m is a constant for the material, ρ is the density and d the thickness of the material. This equation allows a level measurement system to be constructed in one of the ways shown in Figure 1.1.39. In each case the intensity of the received radiation is dependent on the level, being a maximum when the level is low.

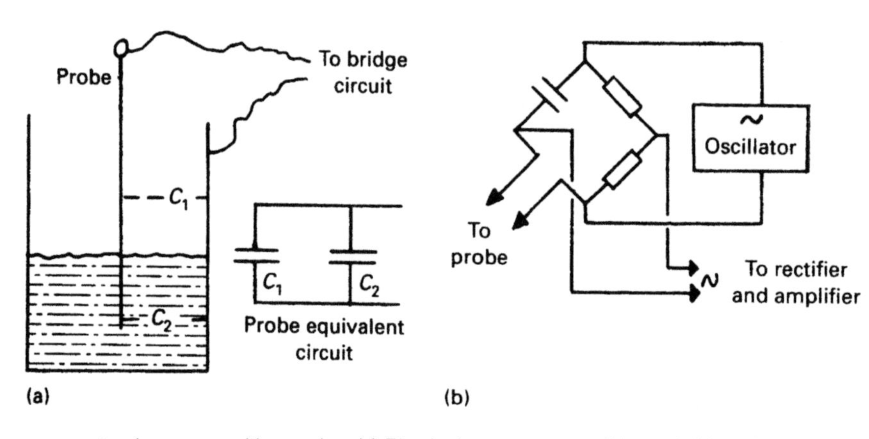

Fig. 1.1.37 Level measurement using a capacitive probe. (a) Physical arrangement; (b) a.c. bridge circuit.

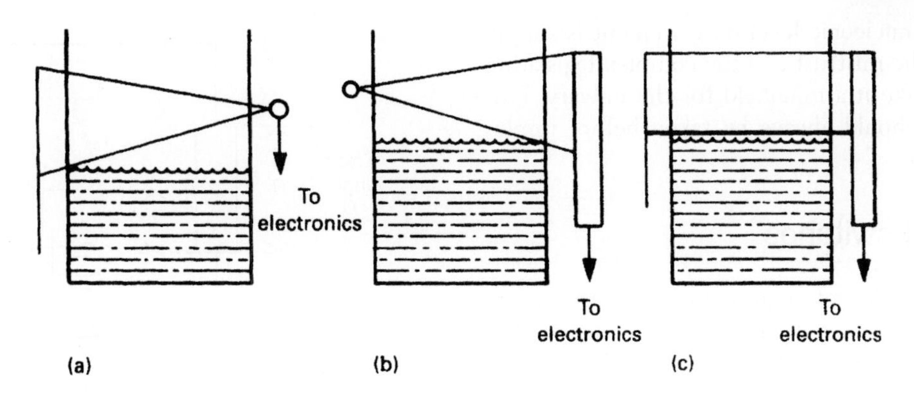

Fig. 1.1.39 Various methods of level measurement using a radioactive source. (a) Line source, point detector; (b) point source, line detector; (c) collimated line source, line detector.

The methods are particularly attractive for aggressive materials or extreme temperatures and pressures as all the measuring equipment can be placed outside the vessel.

Sources used in industry are invariably sealed, i.e. contained in a lead or similar container with a shuttered window allowing a narrow beam (typically $40^\circ \times 7^\circ$) to be emitted. The window can usually be closed when the system is not in use and to allow safe transportation. Source strength is determined by the number of disintegrations per second (dps) at the source. There are two units in common use: the curie (Ci 3.7×10^{10} dps) and the SI derived unit the bequerel (Bq 1 dps). The typical industrial source would be 500 mCi (18.5 GBq) caesium 137 although there are variations by a factor of at least ten dependent on the application.

The strength of a source decays exponentially with time and this decay is described by the *half life* which is the time taken for the strength to decrease by 50%. Cobalt 60, a typical industrial source, for example has a half life of 3.5 years and will hence change by about 1% per month. Sources thus have a built in drift and a system will require regular calibration.

The biological effects of radiation are complex and there is no real safe level; all exposure must be considered harmful and legislation is based around the concept of ALARP; '*as low as reasonably practical*'. Exposures should be kept as low as possible, and figures quoted later should not be considered as design targets.

The *adsorbed dose* (AD) is a measure of the energy density adsorbed. Two units are again used; the rad and the SI derived unit the gray. These are more commonly encountered as the millirad (mrad) and the milligray (mGy):

$$1 \text{ mrad} = 6.25 \times 10^4 \text{ MeV/g}$$

$$1 \text{ mGy} = 6.25 \times 10^6 \text{ MeV/g}$$

The AD is not, however, directly related to biological damage as it ignores the differing effect of α, β and γ radiations. A quality factor Q is defined which allows a *dose equivalent* (DE) to be calculated from

$$\text{DE} = \text{Q} \times \text{AD}$$

There are, yet again, two units in common use; the rem and SI derived unit the sievert, again these are more commonly encountered as the millirem and millisievert (mSv);

$$\text{mrem} = \text{Q} \times \text{mrad}$$

$$\text{mSv} = \text{Q} \times \text{mGy}$$

The so called *dose rate* is expressed as DE/time (e.g. mrem/h). Dose rates fall off as the inverse square of the distance, i.e. doubling the distance from the source gives one-quarter of the dose rate.

British legislation recognises three groups of people:

Classified workers are trained individuals wearing radiation monitoring devices (usually film badges). They must not be exposed to dose rates of more than 2.5 mrem/h and their maximum annual exposure (DE) is 5 rem. Medical records and dose history must be kept.

Supervised workers operate under the supervision of classified workers. Their maximum dose rate is 0.75 mrem/h and maximum annual exposure is 1.5 rem.

Unclassified workers refers to others and the general public. Here the dose rate must not exceed 0.25 mrem/h and the annual exposure must be below 0.5 rem.

It must be emphasised that the principle of ALARP applies and these are not design criteria.

Nucleonic level detection also requires detectors. Two types are commonly used, the *Geiger–Muller* (GM) tube and the *scintillation counter.* Both produce a semi-random pulse stream, the number of pulses received in a given time being dependent on the strength of the radiation. This pulse chain must be converted to a d.c. voltage by suitable filtering to give a signal which is dependent on the level of material between the source and the detector.

Conceptually, nucleonic level measurement is simple and reliable. Public mistrust and the complex legislation can, however, make it a minefield for the unwary. Professional advice should always be taken before implementing a system.

1.1.5.6 Level switches

Many applications do not require analog measurement but just a simple material present/absent signal. For liquids the simplest is a float which hangs down when out of a liquid and inverts when floating. The float orientation is sensed by two internal probes and a small mercury pool. Solids can be detected by rotating paddles or vibrating reeds which seize solid when submerged.

Two non-moving detectors are shown in Figure 1.1.40. The first is a heated tube whose heat loss will be higher (and hence the sensed temperature lower) when submerged. The second is a light reflective probe which will experience total internal reflection when submerged.

1.1.6 Position transducers

1.1.6.1 Introduction

The measurement of position or distance is of fundamental importance in many applications. It also appears as an intermediate variable in other measurement transducers where the measured variable (such as pressure or force) is converted to a displacement.

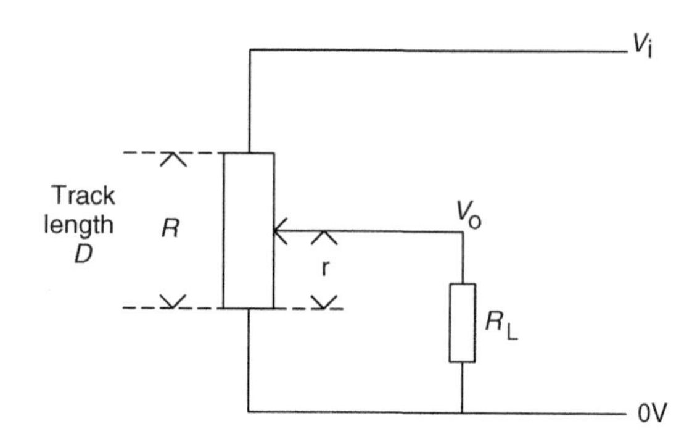

Fig. 1.1.41 Potentiometer used for position measurement.

1.1.6.2 The potentiometer

The simplest position transducer is the humble potentiometer. They can directly measure angular or linear displacements. Figure 1.1.41 shows a potentiometer with track resistance R, connected to a supply V_i and a load R_L. If the span is D and the slider at position d, we can define a fractional displacement $x = d/D$. The output voltage is then

$$V_0 = \frac{xV_i}{1 + x(1-x)(R/R_L)}$$

The error is zero at both ends of the travel and maximum at $x = 2/3$. If R_L is significantly larger than R, the error is approximately $25R/R_L$% of value or $15R/R_L$% of full scale. A 10 K potentiometer connected to a 100 K load will thus have an error of about 1.5% FSD. If $R_L \gg R$, then $V_0 = xV_i$.

Loading errors are further augmented by linearity and resolution errors. Potentiometers have finite resolution

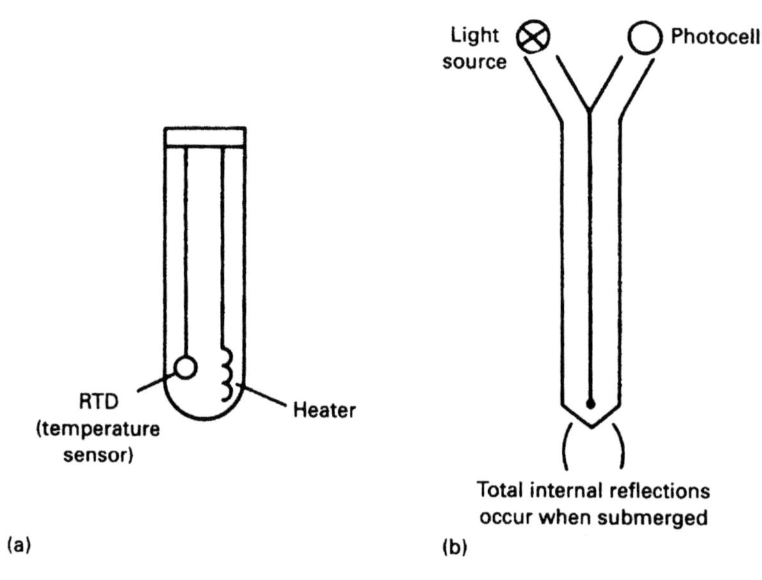

Fig. 1.1.40 Non-contact level switches. (a) Using heat loss from submerged sensor; (b) optical level switch based on total internal reflection.

from the wire size or grain size of the track. They are also mechanical devices which require a force to move and hence can suffer from stiction, backlash and hysteresis.

The failure mode of a potentiometer also needs consideration. A track break can cause the output signal to be fully high above the failure point and fully low below it. In closed loop position control systems this manifests itself as a high speed dither around the break point.

Potentiometers can be manufactured with track resistance to give a specified non-linear response (e.g. trigonometric, logarithmic, square root, etc.).

1.1.6.3 Synchros and resolvers

Figure 1.1.42(a) shows a transformer whose secondary can be rotated with respect to the primary. At angle θ the output voltage will be given by

$$V_0 = KV_i \cos\theta$$

where K is a constant. This relationship is shown in Figure 1.1.42(b). The output amplitude is dependent on the angle, and the signal can be in phase (from $\theta = 270$–$90°$) or antiphase (from $\theta = 90$–$270°$). This principle is the basis of synchros and resolvers.

A synchro link in its simplest form consists of a transmitter and receiver connected as Figure 1.1.43. Although this looks superficially like a three-phase circuit it is fed from a single-phase supply, often at 400 Hz. The voltage applied to the transmitter induces in-phase/antiphase voltages in the windings as described above and causes currents to flow through the stator windings at the receiver. These currents produce a magnetic field at the receiver which aligns with the angle of the transmitter rotor.

The receiver rotor also produces a magnetic field. If the two receiver fields do not align, torque will be produced on the receiver shaft which causes the rotor to rotate until the rotor and stator fields align, i.e. the receiver rotor is at the same position as the transmitter rotor.

Such a link is called a *torque link* and can be used to remotely position an indicator. A more useful circuit, shown in Figure 1.1.44 can be used to give an electrical signal which represents the difference in angle of the two shafts. The receiving device is called a *control transformer*.

The transmitter operates as described above and induces a magnetic field in the control transformer. An a.c. voltage will be induced in the control transformer's rotor if it is not perpendicular to the field. The magnitude of this voltage is related to the angle error and the phase to the direction.

The a.c. output signal must be converted to d.c. by a phase sensitive rectifier. The *Cowan rectifier* (Figure 1.1.45) is commonly used. Another common circuit is the positive/negative amplifier (Figure 1.1.46) with the electronic polarity switch driven by the reference supply.

Resolvers have two stator coils at right angles and a rotor coil as shown in Figure 1.1.47. The voltages induced in the two stator coils are simply

$$V_1 = KV_i \cos\theta$$
$$V_2 = KV_i \sin\theta$$

Resolvers are used for co-ordinate conversion and conversion from rectangular to polar co-ordinates. They are also widely used with solid state digital converters which

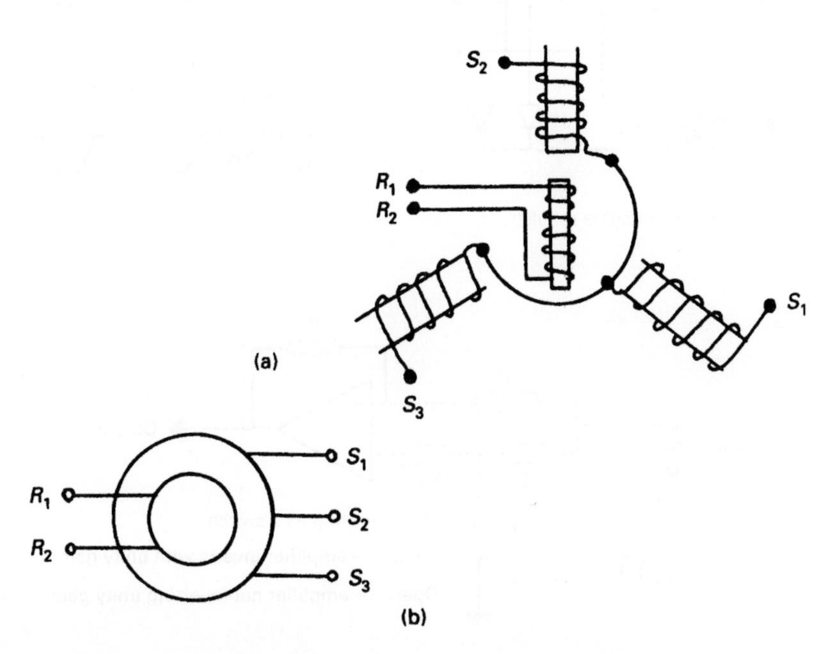

Fig. 1.1.42 The synchro torque transmitter, (a) construction; (b) schematic representation.

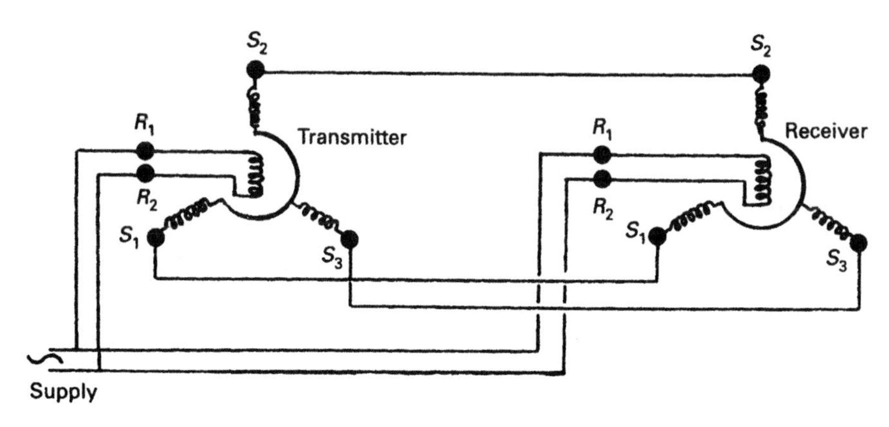

Fig. 1.1.43 A synchro transmitter receiver link.

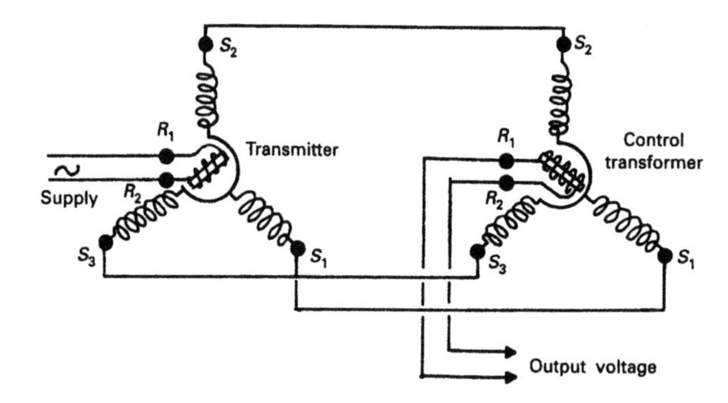

Fig. 1.1.44 Connection of a position control system using a control transformer. The output voltage is zero when the rotor on the control transformer is at 90° to the rotor on the transmitter. Note there are two zero positions 180° apart.

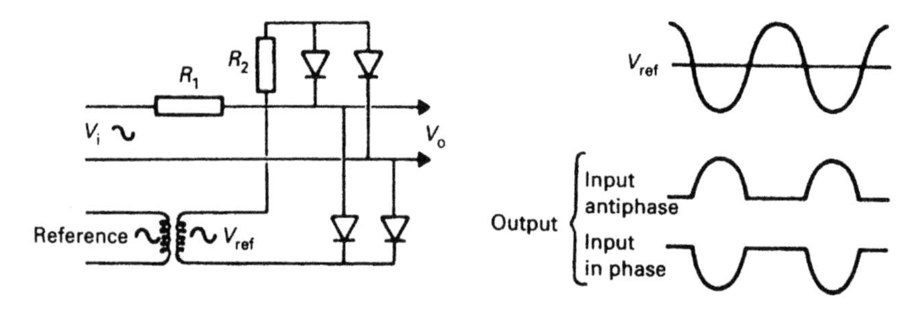

Fig. 1.1.45 The Cowan half wave phase sensitive rectifier.

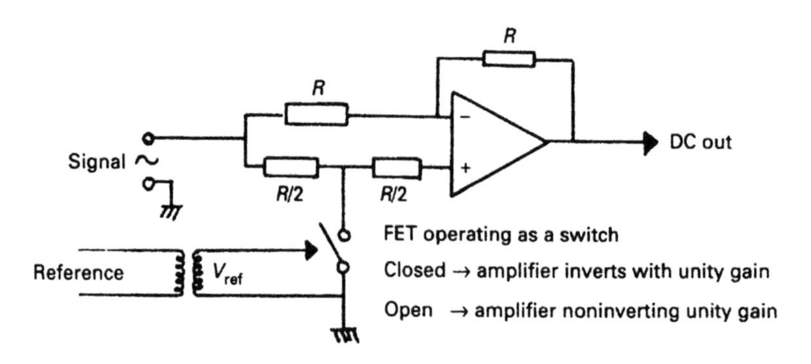

Fig. 1.1.46 Full wave phase sensitive rectifier based on an operational amplifier. The switch represents a FET driven by a reference signal.

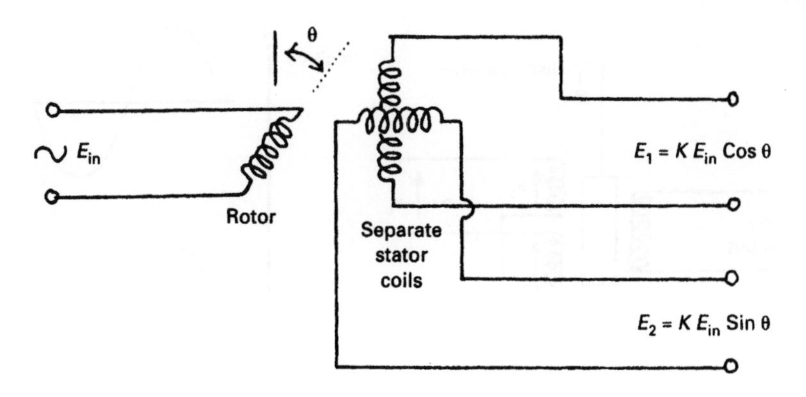

Fig. 1.1.47 The resolver; the reference voltage is applied to the rotor and the output signals read from the two stator windings.

give a binary output signal (typically 12 bit). One useful feature is that $(V_1^2 + V_2^2)$ is a constant which makes an open circuit winding easy to detect.

Resolvers can also be used with a.c. applied to the stator windings and the output signal taken from the rotor (Figure 1.1.48). One stator signal is shifted by 90°. Both signals are usually obtained from a pre-manufactured quadrature oscillator. The output signal is then

$$V_0 = KV_i(\sin \omega t \cos \theta + \cos \omega t \sin \theta)$$

which simplifies to

$$V_0 = KV_i(\sin \omega t + \theta)$$

This a.c. signal can be converted to d.c. by a phase sensitive rectifier.

Complete rotary transducers comprising quadrature oscillator, resolver and phase sensitive rectifier can be obtained which give a d.c. output signal proportional to shaft angle. Because there is little friction these are low torque devices.

1.1.6.4 Linear variable differential transformer

An LVDT is used to measure small displacements, typically less than a few mm. It consists of a transformer with two secondary windings and a movable core as shown in Figure 1.1.49(a). At the centre position the voltages induced in the secondary windings are equal but of opposite phase giving zero output signal. As the core moves away from the centre one induced voltage will be larger, giving a signal whose amplitude is proportional to the displacement and whose phase shows the direction as Figure 1.1.49(b). Again a d.c. output signal can be obtained with a phase sensitive rectifier. Figure 1.1.49(c) shows the same idea used to produce a small displacement angular transducer.

1.1.6.5 Shaft encoders

Shaft encoders give a digital representation of an angular position. They exist in two forms.

An *absolute* encoder gives a parallel output signal, typically 12 bits (one part in 4095) per revolution. Binary coded decimal (BCD) outputs are also available. A simplified four bit encoder would therefore operate as in Figure 1.1.50(a). Most absolute encoders use a coded wheel similar to Figure 1.1.50(b) moving in front of a set of photocells.

A simple binary coded shaft encoder can give anomalous readings as the outputs change state. Suppose a four bit encoder is going from 0111 to 1000. It is unlikely that all photocells will change together, so the output states could go 0111 > 0000 > 1000 or 0111 > 1111 > 1000 or any other sequence of bits.

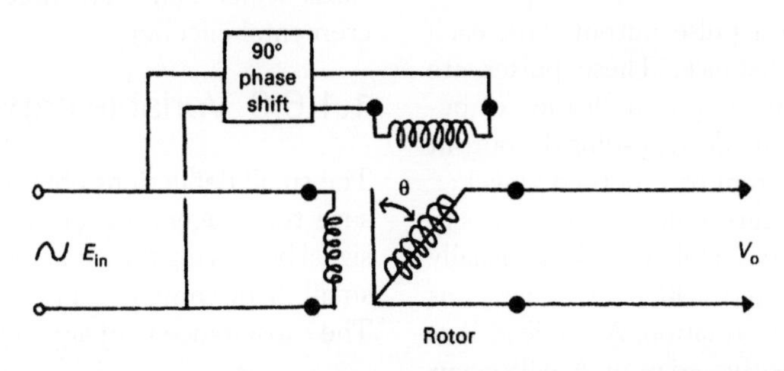

Fig. 1.1.48 Alternative connection for a resolver. Two reference voltages shifted by 90° are applied to the stator windings and the output read from the rotor. The output signal is a constant amplitude sine wave with phase shift determined by the rotor angle.

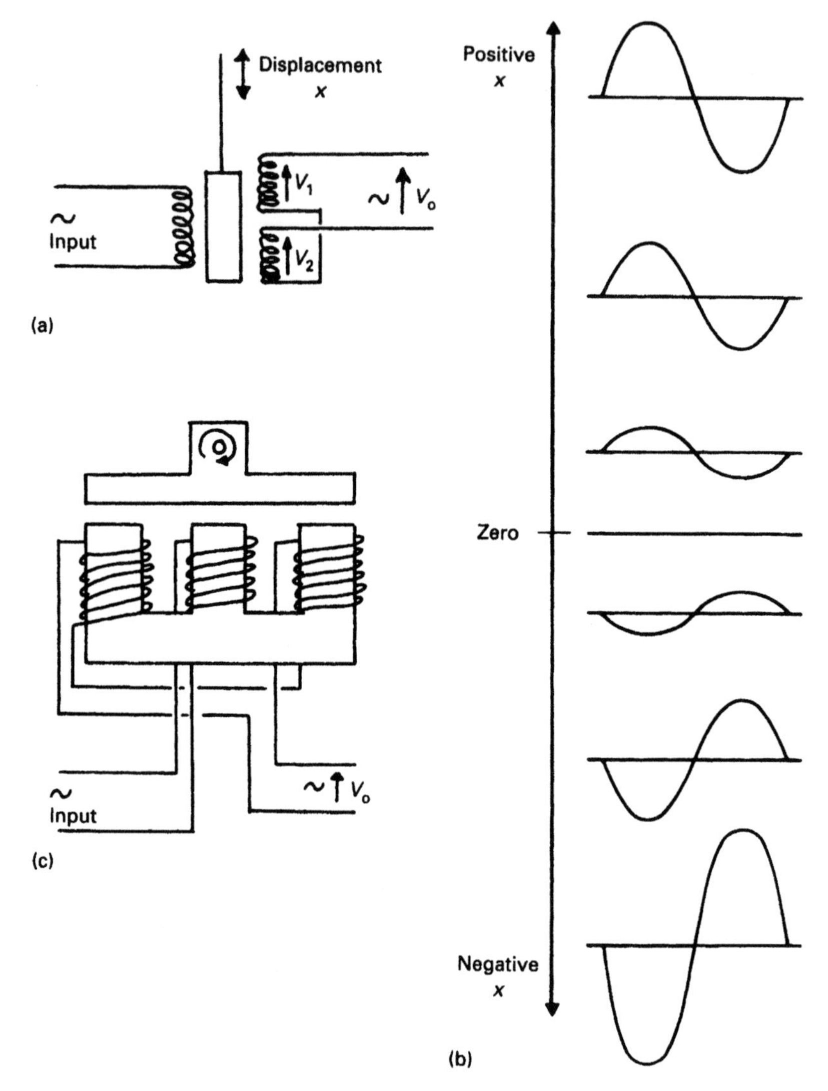

Fig. 1.1.49 The LVDT. (a) Construction; (b) output signal for various positions; (c) small displacement angular device.

There are two solutions to this problem. The first uses an additional track, called an *anti-ambiguity track*, which is used by the encoder's internal logic to inhibit changes around transition points.

The second solution uses a *unit distance code*, such as the *Gray code* which has no ambiguity. The conversion logic from Gray to binary is usually contained within the encoder itself.

Incremental encoders give a pulse output with each pulse representing a fixed distance. These pulses are counted by an external counter to give indication of position. Simple encoders can be made by reading the output from a proximity detector in front of a toothed wheel.

A single pulse output train carries no information as to direction. Commercial incremental encoders usually provide two outputs shifted 90° in phase as Figure 1.1.51(a). For clockwise rotation, A will lead B as Figure 1.1.51(b) and the positive edge of A will occur when B is low. For anti-clockwise rotation B will lead A as Figure 1.1.51(c) and the positive edge of A will occur when B is high. The logic (A and not B) can be used to count up, and (A and B) to count down. The two output incremental encoder can thus be used to follow reversals without cumulative error, but can still lose its datum position after a power failure.

Most PLC systems have dedicated high speed counter cards which can read directly from a two channel incremental encoder.

1.1.6.6 Variable capacitance transducers

The small deflections obtained in weigh systems or pressure transducers are often converted into an electrical signal by varying the capacitance between two plates. Very small displacements can be detected by these methods. The capacitance C of a parallel plate capacitor is given by

$$C = \frac{\varepsilon A}{d}$$

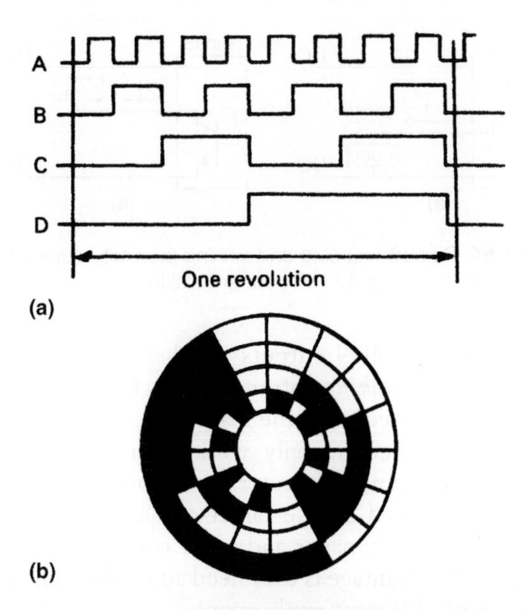

Fig. 1.1.50 Absolute position shaft encoder. (a) Output from a four bit encoder. Real devices use 12 bits and employ unit distance coding such as the Gray code; (b) the wheel on a four bit encoder.

where ε is the permitivity of the material between the plates, A the area and d the separation.

Variation in capacitance can be obtained from varying e (sliding in a dielectric, or moving the plates horizontally (change A) or apart (change d)). Although the change is linear for e and A, the effect is small. A common arrangement uses a movable plate between two fixed plates as Figure 1.1.52(a). Here two capacitors are formed which can be directly connected into an a.c. bridge as Figure 1.1.52(b). If the fixed plates are distance D apart, d is the null position separation and the centre plate is displaced by x, the output voltage change can be shown to be

$$V_0 = V_i x / 2d$$

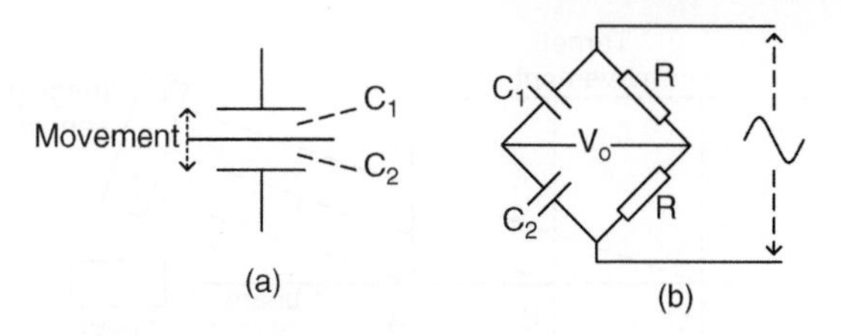

Fig. 1.1.52 Capacitive based displacement transducer. (a) Physical arrangement; (b) a.c. bridge.

1.1.6.7 Laser distance measurement

Lasers can give very accurate measurement of distance and are often used in surveying. There are two broad classes of laser distance measurement.

In the first, shown in Figure 1.1.53 and called a *triangulation laser*, a laser beam is used to produce a very bright spot on the target object. This is viewed by an imaging device which can be considered as a linear array of tiny photocells. The position of the spot on the image will vary according to the distance as shown, and the distance found by simple triangulation trigonometry.

The second type of distance measurement is used at longer distances and times how long it takes a laser pulse to reach the target and be reflected back. For very long distances (in surveying for example), a pulse is simply timed. At shorter distances a continuous amplitude modulated laser beam is sent and the phase shift between transmitted and received signals used to calculate the distance. This latter method can give ambiguous results, so often the two techniques are combined to give a coarse measurement from time of flight which is fine tuned by the phase shift.

Lasers can blind, and are classified according to their strength. Class 1 lasers are inherently safe. Class 2 lasers

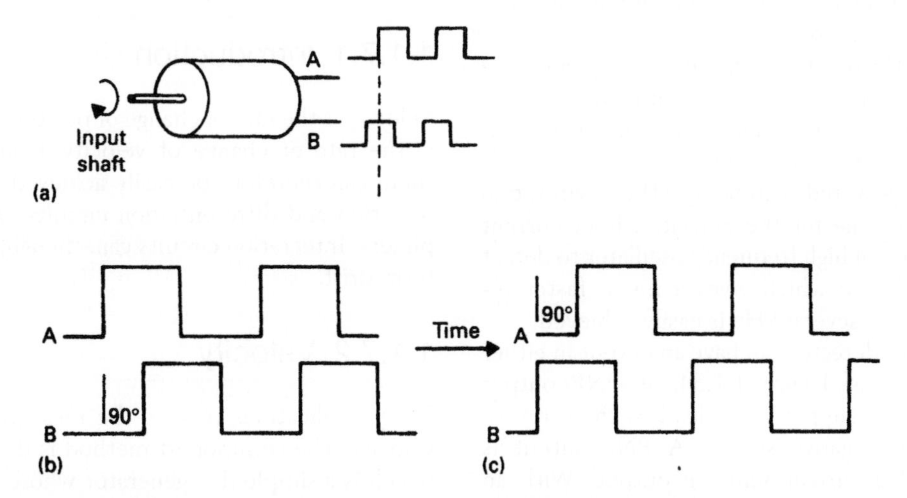

Fig. 1.1.51 The incremental encoder. (a) physical arrangements—the two phase shifted outputs give directional information; (b) clockwise rotation; (c) anti-clockwise rotation.

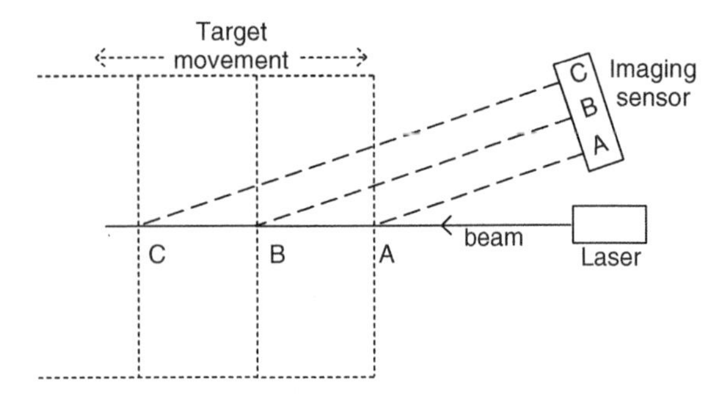

Fig. 1.1.53 Triangulation laser position measurement.

can cause retinal damage with continuous viewing but the natural reaction to blink or look away gives some protection.

Class 3 lasers can cause damage before the natural reaction occurs. Risk assessments, safe working practices and suitable guarding must be provided for Class 2 and 3 lasers. Optical devices such as binoculars must never be used whilst working around lasers.

1.1.6.8 Proximity switches and photocells

Mechanical limit switches are often used to indicate the positional state of mechanical systems. These devices are bulky, expensive and prone to failure where the environment is hostile (e.g. dust, moisture, vibration, heat).

Proximity switches can be considered to be solid state limit switches. The commonest type is constructed around a coil whose inductance changes in the presence of a metal surface. Simple two wire a.c. devices act just like a switch, having low impedance and capable of passing several hundred mA when covered by metal, and a high impedance (typical leakage 1 mA) when uncovered. Sensing distances up to 20 mm are feasible, but 5–10 mm is more common.

The leakage in the off state (typically 1 mA) can cause problems with circuits such as high impedance PLC input cards. In these circumstances a dummy load must be provided in parallel with the input.

Direct current powered switches use three wires, two for the supply and one for the output. Direct current sensors use an internal high frequency oscillator to detect the inductance change which gives a much faster response. Operation at several kHz is easily achieved.

A d.c. proximity detector can have an output in either of the two forms of Figure 1.1.54. A PNP output switches a positive supply to a load with a return connected to the negative supply. A PNP output is sometimes called a *current sourcing* output. With an NPN output the load is connected to the positive supply and the proximity detector connects the load to the negative supply. NPN outputs are therefore sometimes called a *current sinking* output. It is obviously important to match the detector to the load connection.

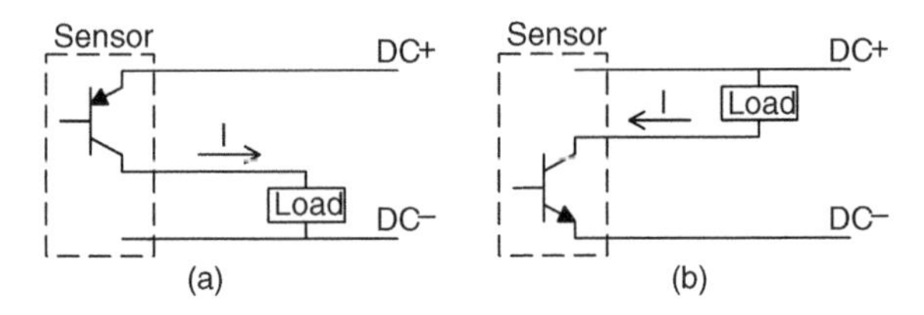

Fig. 1.1.54 Current sourcing and sinking proximity detectors. (a) PNP current sourcing; (b) NPN current sinking.

Inductive detectors only work with metal targets. Capacitive proximity detectors work on the change in capacitance caused by the target and accordingly work with wood, plastic, paper and other common materials. Their one disadvantage is they need adjustable sensitivity to cope with different applications.

Ultrasonic sensors can also be used, operating on the same principle as the level sensor described in Section 1.1.5.4. Ultrasonic proximity detectors often have adjustable near and far limits so a sensing window can be set.

Photocells (or PECs) are another possible solution. The presence of an object can be detected either by the breaking of a light beam between a light source and a PEC or by bouncing a light beam off the object which is detected by the PEC. In the first approach the PEC sees no light for an object present. In the second approach (called a *retro-reflective* PEC) light is seen for an object present. It is usual to use a modulated light source to avoid erratic operation with changes in ambient light and prevent interaction between adjacent PECs.

1.1.7 Velocity and acceleration

1.1.7.1 Introduction

Velocity is the rate of change of position, and acceleration is the rate of change of velocity. Conversion between them can therefore be easily achieved by the use of integration and differentiation circuits based on d.c. amplifiers. Integration circuits can, though, be prone to long term drift.

1.1.7.2 Velocity

Many applications require the measurement of angular velocity. The commonest method is the tacho-generator which is a simple d.c. generator whose output voltage is proportional to speed. A common standard is 10 V per 1000 rpm. Speeds up to 10,000 rpm can be measured,

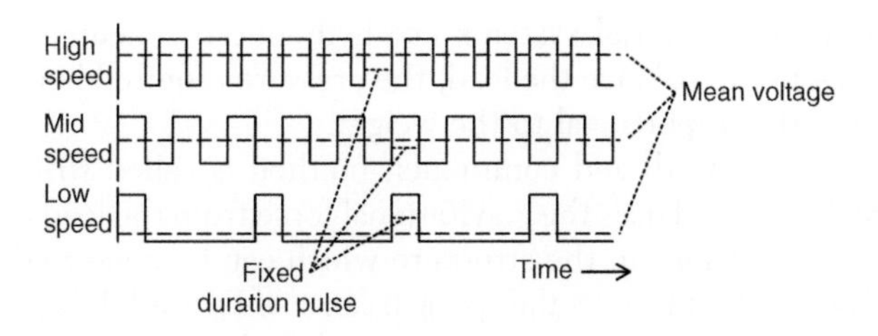

Fig. 1.1.55 Speed measurement using incremental pulse encoder. The pulses from the encoder fire a fixed duration monostable. The resulting pulse train is filtered to give a mean voltage proportional to speed.

the upper limit being centrifugal force on the tacho commutator.

Pulse tachometers are also becoming common. These are essentially identical to incremental encoders described in Section 1.1.6.5 and produce a constant amplitude pulse train whose frequency is directly related to rotational speed. This pulse train can be converted into a voltage in three ways.

The first, basically analog, method used fires a fixed width monostable as Figure 1.1.55 which gives a mark/space ratio which is speed dependent. The monostable output is then passed through a low pass filter to give an output proportional to speed. The maximum achievable speed is determined by the monostable pulse width.

The second, digital, method counts the number of pulses in a given time. This effectively averages the speed over the count period which is chosen to give a reasonable balance between resolution and speed of response.

Counting over a fixed time is not suitable for slow speeds as adequate resolution can only be obtained with a long duration sample time. At slow speeds, therefore, the period of the pulses is often directly timed giving an average speed per pulse. The time is obviously inversely proportional to speed.

Digital speed control systems often use both of the last two methods and switch between them according to the speed.

Linear velocity can be measured using *Doppler shift*. This occurs when there is relative motion between a source of sound (or electromagnetic radiation) and an object. It is commonly experienced as a change of pitch of a car horn as the vehicle passes.

Suppose an observer is moving with velocity v towards a source emitting sound with a frequency f. The observer will see each wavefront arrive early and the perceived frequency will be

$$f_v = (c + v)/\lambda$$

where c is the velocity of propagation. As the original frequency is c/l, the frequency shift is

$$\delta f = (f_v - f) = v/\lambda = fv/c \qquad (1.1.5)$$

i.e. proportional to speed.

The Doppler shift thus allows remote measurement of velocity. The principle is shown in Figure 1.1.56. A transmitter (ultrasonic, radar or light) emits a constant frequency signal. This is bounced off the target. Two Doppler shifts occur as the object is acting as both moving observer and moving (reflected) source. The frequency shift is thus twice that for the simple case of Equation (1.1.5).

The transmitted and received frequencies are mixed to give a beat frequency equal to the frequency shift and hence proportional to the target object's speed. The beat frequency can be measured by counting the number of cycles in a given time.

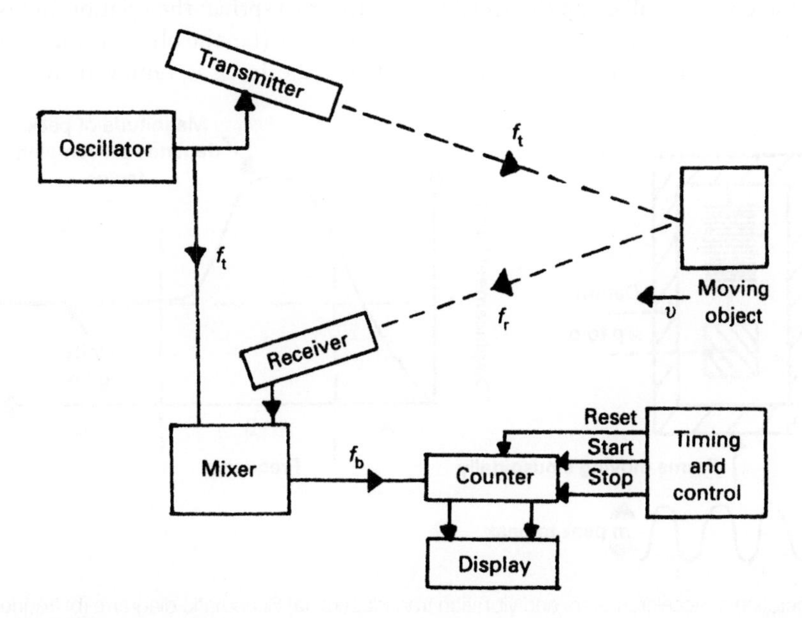

Fig. 1.1.56 Doppler shift velocity measurement.

1.1.7.3 Accelerometers and vibration transducers

Both accelerometers and vibration transducers consist of a seismic mass linked to a spring as Figure 1.1.57(a). The movement will be damped by the dashpot giving a second order response defined by the natural frequency and the damping factor, typically chosen to be 0.7.

If the frame of the transducer is moved sinsusoidally, the relationship between the mass displacement and the frequency will be as Figure 1.1.57(b).

In region A, the device acts as an accelerometer and the mass displacement is proportional to the acceleration. In region C, the mass does not move, and the displacement with respect to the frame is solely dependent on the frame movement (but is shifted by 180°). In this region the device acts as a vibration transducer. Typically region A extends to about one third of the natural frequency and region C starts at about three times the natural frequency.

In both cases the displacement is converted to an electrical signal by LVDTs or strain gauges.

1.1.8 Strain gauges, load cells and weighing

1.1.8.1 Introduction

Accurate measurement of weight is required in many industrial processes. There are two basic techniques in use. In the first, shown in Figure 1.1.58(a) and called a *force balance system*, the weight of the object is opposed by some known force which will equal the weight. Kitchen scales where weights are added to one pan to balance the object in the other pan use the force balance system. Industrial systems use hydraulic or pneumatic pressure to balance the load, the pressure required being directly proportional to the weight.

The second, and commoner, method is called *strain weighing*, and uses the gravitational force from the load to cause a change in the structure which can be measured. The simplest form is the spring balance *of* Figure 1.1.58(b) where the deflection is proportional to the load.

1.1.8.2 Stress and strain

The application of a force to an object will result in deformation of the object. In Figure 1.1.59 a tensile force F has been applied to a rod of cross-sectional area A and length L. This results in an increase in length ΔL. The effect of the force will depend on the force and the area over which it is applied. This is called the *stress*, and is the force per unit area:

$$\text{strees} = F/A$$

Stress has the units of N/m^2 (i.e. the same as pressure, so pascals are sometimes used).

The resulting deformation is called the *strain*, and is defined as the fractional change in length:

$$\text{strain} = \Delta L/L$$

Strain is dimensionless. Because the change in length is small, microstrain (μstrain defined as strain* 10^6) is often used. A 10 m rod exhibiting a change in length of 0.0045 mm because of an applied force is exhibiting 45 μstrain.

The strain will increase as the stress increases as shown in Figure 1.1.60. Over region AB the object behaves as a spring; the relationship is linear and there is no hysteresis (i.e. the object returns to its original dimension when the force is removed. Beyond point B the object

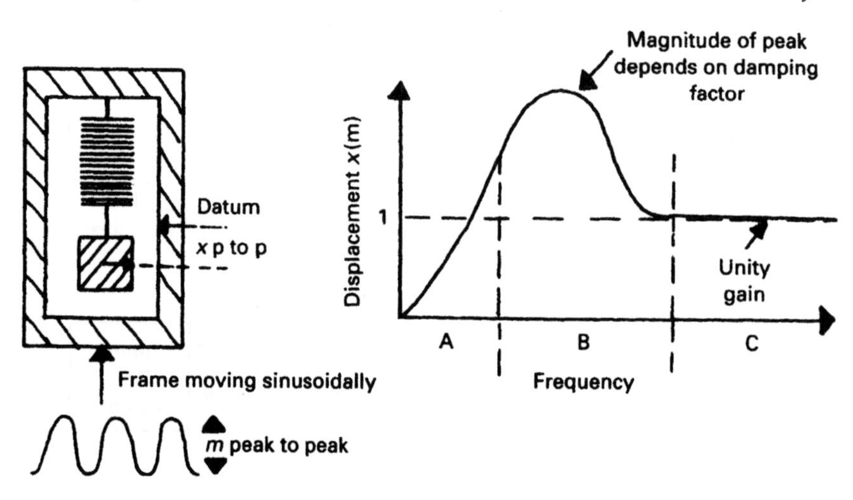

Fig. 1.1.57 Principle of operation of accelerometers and vibration transducers. (a) Schematic diagram; (b) frequency response. In region A the device acts as an accelerometer. In region C the device acts as a vibration transducer.

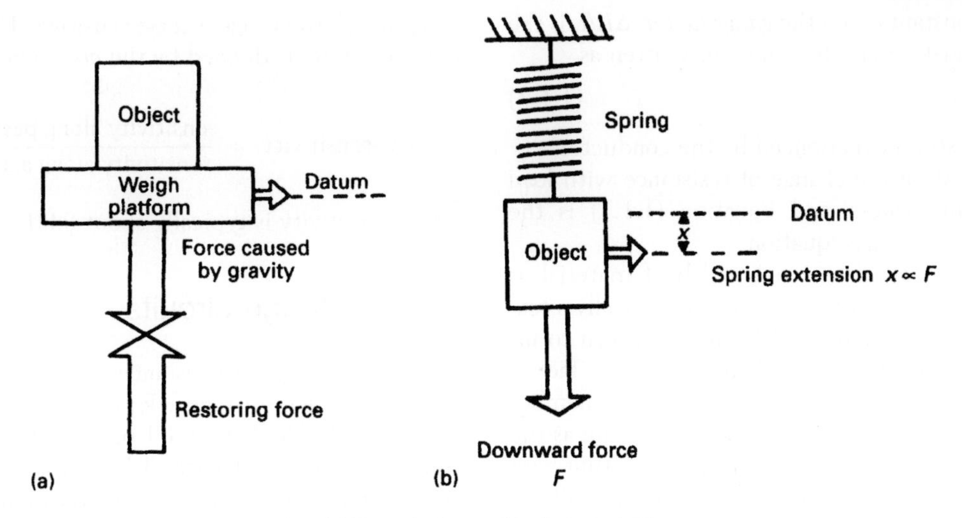

Fig. 1.1.58 The two different weighing principles. (a) Force balance; (b) strain weighing.

suffers deformation and the change is not reversible. The relationship is now non-linear, and with increasing stress the object fractures at point C. The region AB, called the *elastic region*, is used for strain measurement. Point B is called the *elastic limit*. Typically AB will cover a range of 10,000 μstrain.

The inverse slope of the line AB is sometimes called the *elastic modulus* (or *modulus of elasticity*). It is more commonly known as *Young's modulus* defined as

$$\text{Young's modulus} = \text{stress/strain} = (F/A)/(\Delta L/L)$$

Young's modulus has dimensions of N/m^2, i.e. the same as pressure. It is commonly given in pascals. Typical values are:

Steel	210 GPa
Copper	120 GPa
Aluminium	70 GPa
Plastics	30 GPa

When an object experiences strain, it displays not only a change in length but also a change in cross-sectional area. This is defined by *Poisson's ratio*, denoted by the Greek letter ν. If an object has a length L and width W in its unstrained state and experiences changes ΔL and ΔW when strained, Poisson's ratio is defined as

$$\nu = (\Delta W/W)/(\Delta L/L)$$

Typically ν is between 0.2 and 0.4. Poisson's ratio can be used to calculate the change in cross-sectional area.

1.1.8.3 Strain gauges

The electrical resistance of a conductor is proportional to the length and inversely proportional to the cross-section, i.e.

$$R = \rho L/A$$

where ρ is a constant called the *resistivity* of the material.

When a conductor suffers stress, its length and area will both change resulting in a change in resistance. For tensile stress, the length will increase and the cross-sectional area decrease both resulting in increased resistance. Similarly the resistance will decrease for compressive stress.

Ignoring second order effects the change in resistance is given by

$$\Delta R/R = G \cdot \Delta L/L \qquad (1.1.6)$$

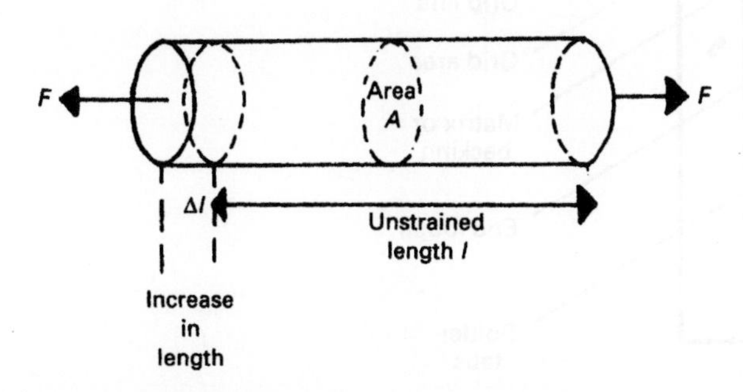

Fig. 1.1.59 Tensile strain.

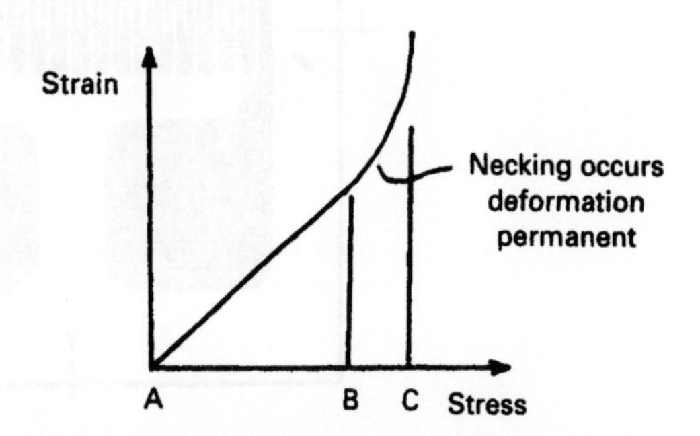

Fig. 1.1.60 The relationship between stress and strain.

where G is a constant called the *gauge factor.* $\Delta L/L$ is the strain, E, so Equation (1.1.6) can be re-written as

$$\Delta R = G \cdot R \cdot E \tag{1.1.7}$$

where E is the strain experienced by the conductor.

Devices based on the change of resistance with load are called *strain gauges*, and Equation (1.1.7) is the fundamental strain gauge equation.

Practical strain gauges are not a slab of material as implied so far but consist of a thin small (typically a few mm) foil similar to Figure 1.1.61 with a pattern to increase the conductor length and hence the gauge factor. The gauge is attached to some sturdy stressed member with epoxy resin and experiences the same strain as the member. Early gauges were constructed from thin wire but modern gauges are photo-etched from metallised film deposited onto a polyester or plastic backing. Normal gauges can experience up to 10,000 μstrain without damage. Typically the design will aim for 2000 μstrain under maximum load.

A typical strain gauge will have a gauge factor of 2, a resistance of 120 Ω and experience 1000 μstrain. From Equation (1.1.7) this will result in a resistance change of 0.24 Ω.

Strain gauges must ignore strains in unwanted directions. A gauge has two axis, an active axis along which the strain is applied and a passive axis (usually at 90°) along which the gauge is least sensitive. The relationship between these is defined by the *cross-sensitivity*

$$\text{cross-sensitivity} = \frac{\text{sensitivity along passive axis}}{\text{sensitivity along active axis}}$$

Cross-sensitivity is typically about 0.002.

1.1.8.4 Bridge circuits

The small change in resistance is superimposed on the large unstrained resistance. Typically the change will be 1 part in 5000. In Figure 1.1.62 the strain gauge R_g is connected into a classical Wheatstone bridge. In the normal laboratory method, R_a and R_b are made equal and calibrated resistance box R_c adjusted until V_0 (measured by a sensitive millivolt-metre) is zero. Resistance R_c then equals R_g.

With a strain gauge, however, we are not interested in the actual resistance, but the change caused by the applied load. Suppose R_b and R_c are made equal, and R_a is made equal to the unloaded resistance of the strain gauge. Voltages V_1 and V_2 will both be half the supply voltage and V_0 will be zero. If a load is applied to the strain gauge such that its resistance changes by fractional change x (i.e. $x = \Delta R/R$) it can be shown that

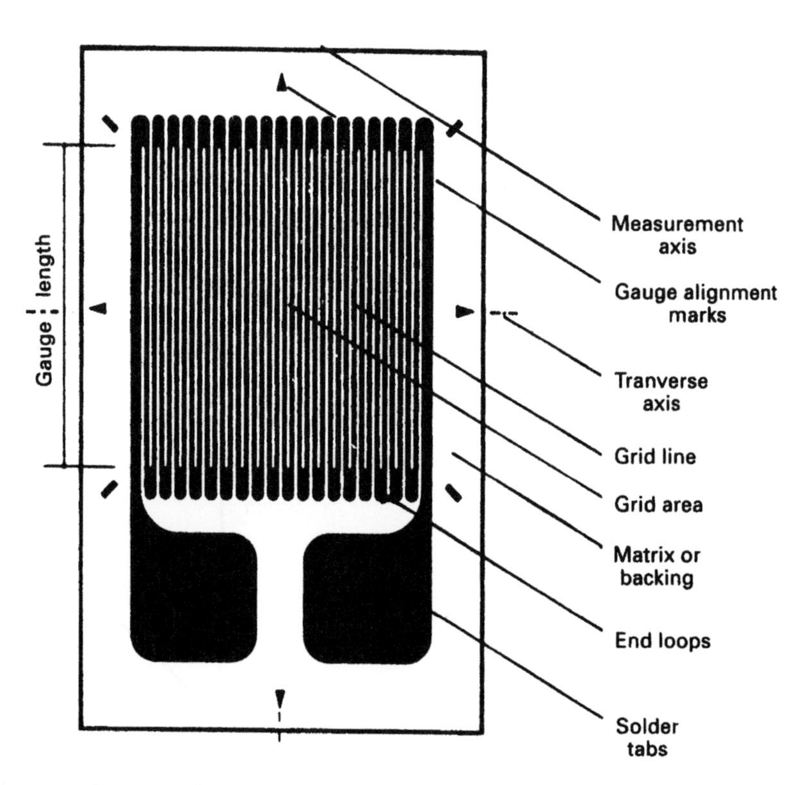

Fig. 1.1.61 A typical strain gauge (courtesy of Welwyn strain measurements).

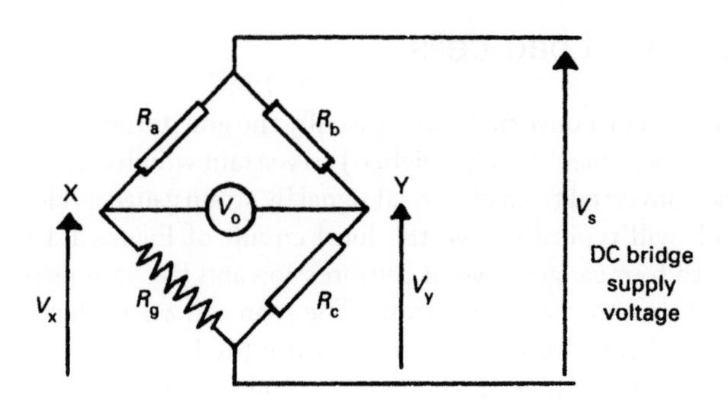

Fig. 1.1.62 Simple measurement of strain with a Wheatstone bridge.

$$V_0 = V_s \cdot x/2(2 + x) \quad (1.1.8)$$

In a normal circuit x will be very small compared with 2. For the earlier example x has the value $0.24/120 = 0.002$, so Equation (1.1.8) can be simplified to

$$V_0 = V_s \cdot \Delta R \cdot x/4 \cdot R$$

But $\Delta R = E \cdot G \cdot R$ where E is the strain and G the gauge factor giving

$$V_0 = E \cdot G \cdot V_s/4 \quad (1.1.9)$$

For small values of x, the output voltage is thus linearly related to the strain.

It is instructive to put typical values into Equation (1.1.9). For a 24 V supply, 1000 μstrain and gauge factor 2 we get 12 mV. This output voltage must be amplified before it can be used. Care must be taken to avoid common mode noise so the differential amplifier circuit of Figure 1.1.63 is commonly used.

The effect of temperature on resistance was described in Section 1.1.2.3 and resistance changes from temperature variation are of a similar magnitude to resistance changes from strain. The simplest way of preventing this is to use two gauges arranged as Figure 1.1.64(a). One gauge has its active axis and the other gauge the passive axis aligned with the load. If these are connected into a bridge circuit as shown in Figure 1.1.64(b) both gauges will exhibit the same resistance change from temperature and these will cancel leaving the output voltage purely dependent on the strain.

Temperature errors can also occur from dimensional changes in the member to which the strain gauges are attached. Gauges are often temperature compensated by having coefficients of linear expansion identical to the material to which they are attached.

Many applications use gauges in all four arms of the bridge as shown in Figure 1.1.64(c) with two active gauges and two passive gauges. This provides temperature compensation and doubles the output voltage giving

$$V_0 = E \cdot G \cdot V_s/2$$

In the arrangement of Figure 1.1.64(d) four gauges are used with two gauges experiencing compressive strain and two experiencing tensile strain. Here all four gauges are active, giving

$$V_0 = E \cdot G \cdot V_s \quad (1.1.10)$$

Again temperature compensation occurs because all gauges are at the same temperature.

Gauges are manufactured to a tolerance of about 0.5%, i.e. about ±0.6 Ω for a typical 120 Ω gauge. As this is larger than the resistance change from strain some form of zeroing will be required. Three methods of achieving this are shown in Figure 1.1.65. Arrangement b and c are preferred if the zeroing is remote from the bridge.

Equation (1.1.10) shows that the output voltage is directly related to the supply voltage. This implies that a large voltage should be used to increase the sensitivity. Unfortunately a high voltage supply cannot be used or I^2R heating in the gauges will cause errors and ultimately failure. Bridge voltages of 15–30 V and currents of 20–100 mA are typically used.

Equation (1.1.10) also implies that the supply voltage must be stable. If the bridge is remote from the supply and electronics, as is usually the case, voltage drops down the cabling can introduce significant errors. Figure 1.1.66

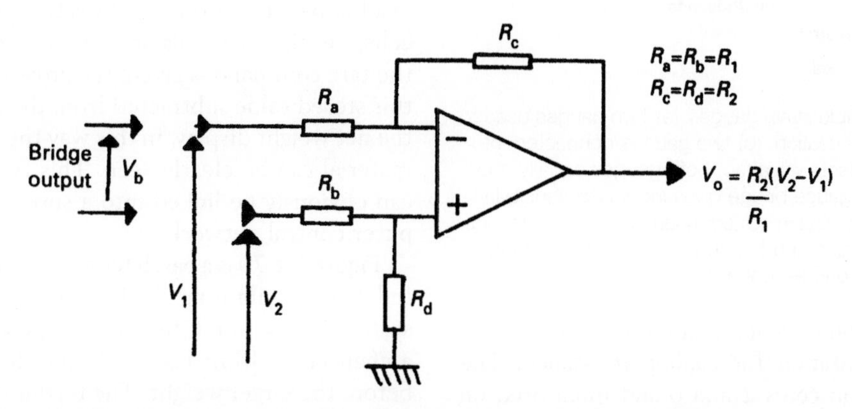

Fig. 1.1.63 Differential amplifier. The output voltage is determined by the difference between the two input voltages.

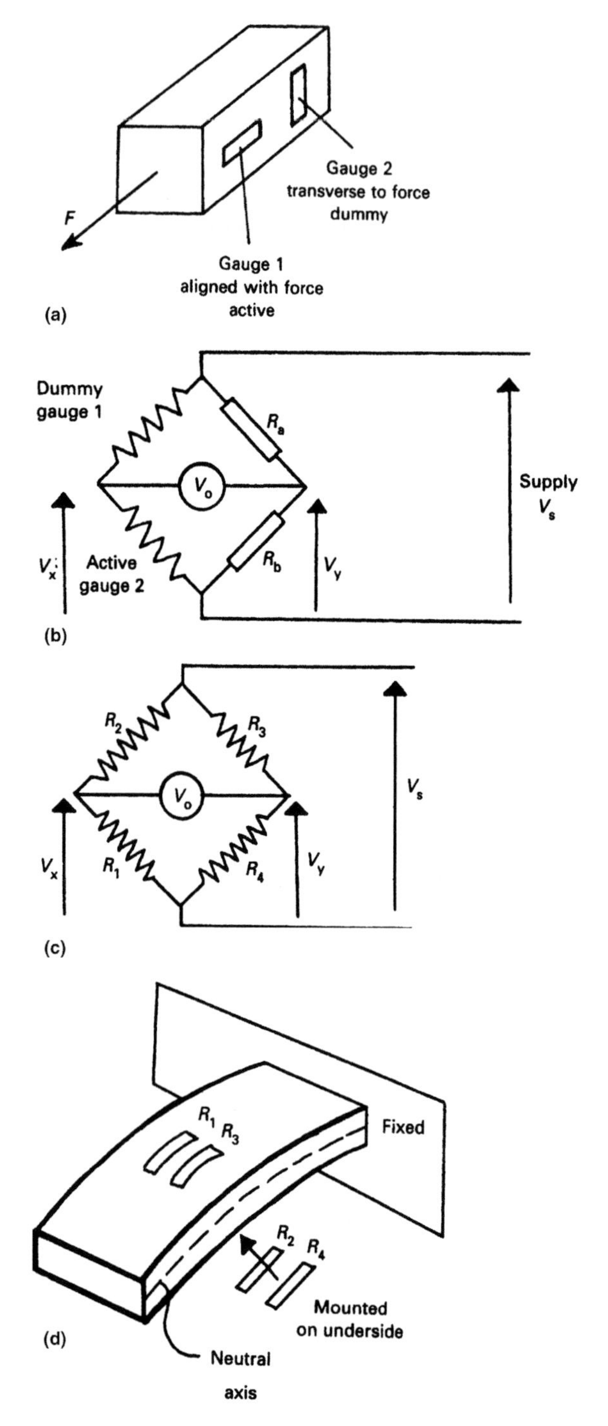

Fig. 1.1.64 Use of multiple strain gauges. (a) Two gauges used to give temperature compensation; (b) two gauges connected into bridge. Temperature effects will affect both gauges equally and have no effect; (c) four gauge bridge gives increased sensitivity and temperature compensation; (d) arrangement for four gauges. Two will be in compression and two in tension for load changes. Temperature affects all gauges equally.

shows a typical cabling scheme used to give remote zeroing and compensation for cabling resistance. The supply is provided on cores 2 and 6 and monitored on cores 1 and 7 to keep constant voltage at the bridge itself.

1.1.8.5 Load cells

A load cell converts a force (usually the gravitational force from an object being weighed) to a strain which can then be converted to an electrical signal by strain gauges. A load cell will typically have the local circuit of Figure 1.1.67 with four gauges, two in compression and two in tension, and six external connections. The span and zero adjust on test (AOT) are factory set to ensure the bridge is within limits that can be further adjusted on site. The temperature compensation resistors compensate for changes in Young's modulus with temperature, not changes of the gauges themselves which the bridge inherently ignores.

Coupling of the load requires care, a typical arrangement being shown in Figure 1.1.68. A pressure plate applies the load to the proof-ring via a knuckle and avoids error from slight misalignment. A flexible diaphragm seals against dust and weather. A small gap ensures that shock overloads will make the load cell bottom out without damage to the proof-ring or gauges. Maintenance must ensure that dust does not close this gap and cause the proof-ring to carry only part of the load. Bridging and binding are common causes of load cells reading below the load weight.

Multi-cell weighing systems can be used, with the readings from each cell being summed electronically. Three cell systems inherently spread the load across all cells. With four cell systems the support structure must ensure that all cells are in contact with the load at all times.

Usually the load cells are the only route to ground from the weigh platform. It is advisable to provide a flexible earth strap, not only for electrical safety but also to provide a route for any welding current which might arise from repairs or later modification.

1.1.8.6 Weighing systems

A weigh system is usually more than a collection of load cells and a display. Figure 1.1.69 is a taring weigh system used when a 'recipe' of several materials is to be collected in a hopper. The gross weight is the weight from the load cells, i.e. the materials and the hopper itself. Each time the tare command is given, the gross weight is stored and this stored value subtracted from the gross weight to give the net weight display. In this way the weight of each new material can be clearly seen. This type of weigh system can obviously be linked into a supervisory PLC or computer control network.

Figure 1.1.70 is a batch feeder system where material is fed from a vibrating feeder into a weigh hopper. The system is first tared, then a two speed system used with a changeover from fast to dribble feed a preset weight before the target weight. The feeder turns off just before the correct weight to allow for the material in flight from

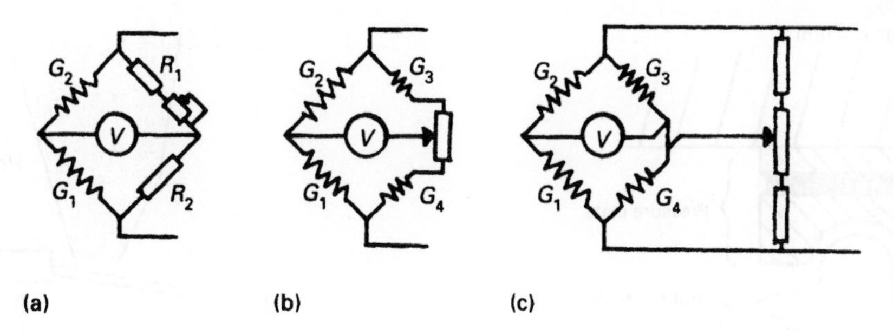

Fig. 1.1.65 Common methods of bridge balancing. (a) Single leg balancing. R_1, is less than R_2; (b) apex balancing; (c) parallel balancing.

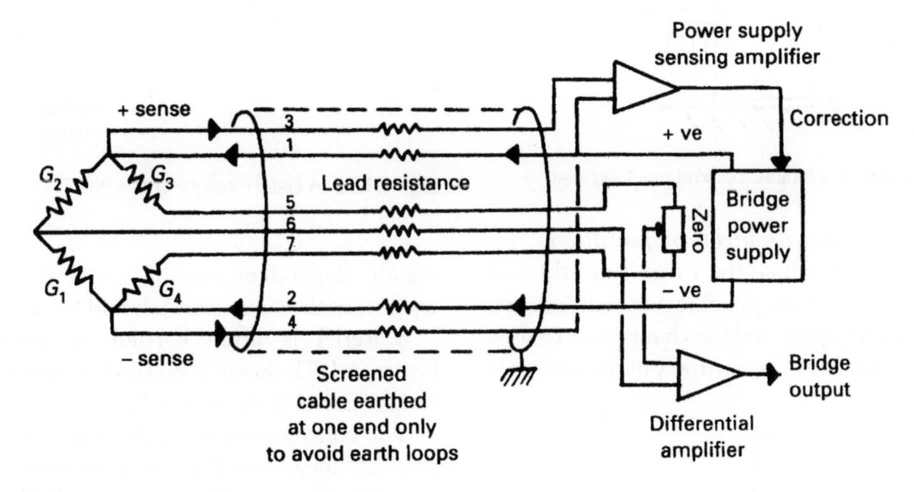

Fig. 1.1.66 Typical cabling scheme for a remote bridge.

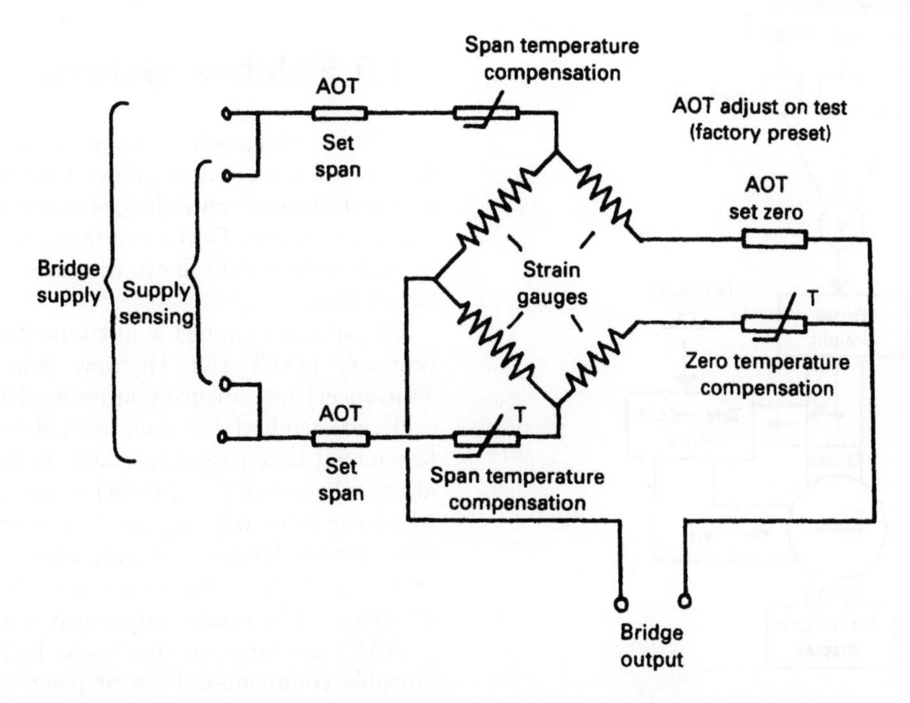

Fig. 1.1.67 Connection diagram for a typical commercial load cell. Note that a six core cable is required.

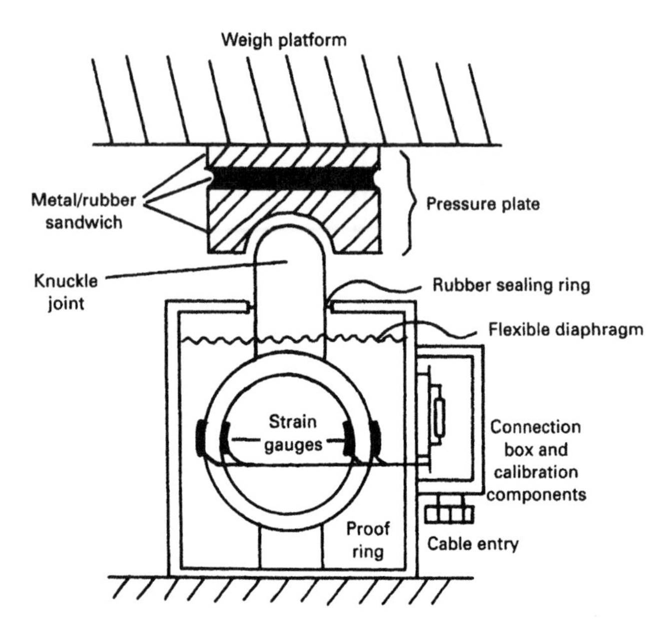

Fig. 1.1.68 Construction of a typical commercial load cell.

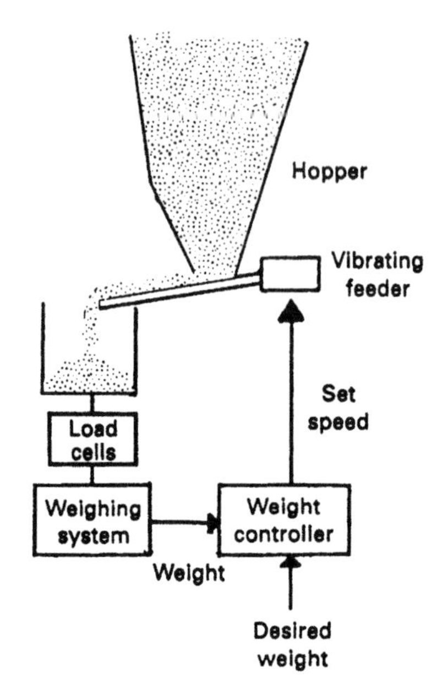

Fig. 1.1.70 A batch weighing system.

the feeder. This is known as *'in flight compensation'*, *'anticipation'* or *'preact'*. To reduce the weigh time the fast feed should be set as fast as possible without causing avalanching in the feed hopper and the changeover to slow feed made as late as possible. The accuracy of the system is mainly dependent on the repeatability of the preact weight, so the slow speed should be set as low as possible.

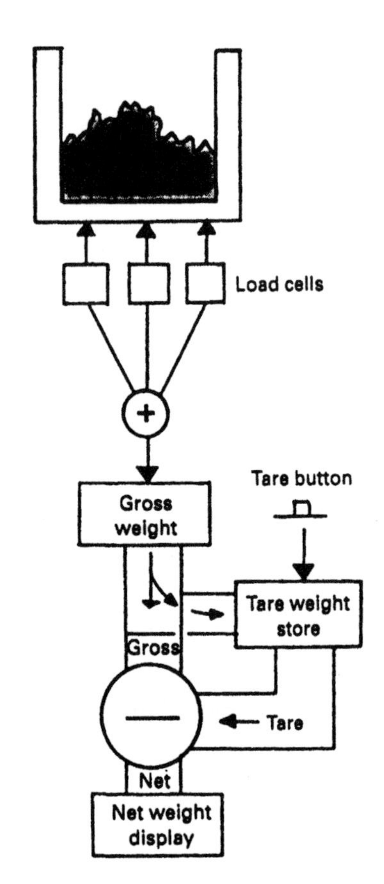

Fig. 1.1.69 Taring weighing system.

Material is often carried on conveyor belts and Figure 1.1.71 shows a method of providing feed rate in weight per unit time (e.g. kg/min).

The conveyor passes over a load cell of known length and the linear speed is obtained from the drive motor, probably from a tacho-generator. If the weigh platform has length L m, and is indicating weight W kg at speed V m/s, the feed rate is simply $W \cdot V/L$ kg/s.

1.1.9 Fieldbus systems

A fieldbus, discussed in more detail in Chapter 1.3, Section 1.3.5 is a way of interconnecting several devices (e.g. transducers, controllers, actuators) via a simple and cheap serial cable. Devices on the network are identified by addresses which allows messages to be passed between them.

Of particular interest in instrumentation is the HART protocol. HART (for Highway Addressable Remote Transducer) was originally introduced by Rosemount as a closed standard for their own devices. Generously, Rosemount have passed the rights to the Hart Communication Foundation and HART is now an open protocol which can be used by anyone. It is a very simple master/slave system; devices only speak when requested and the operation is always master request, slave replies. Up to 15 slaves can be connected to each master.

HART can work in two ways. In its simplest, and probably commonest, form of point to point it superimposes the serial communication data on to a standard

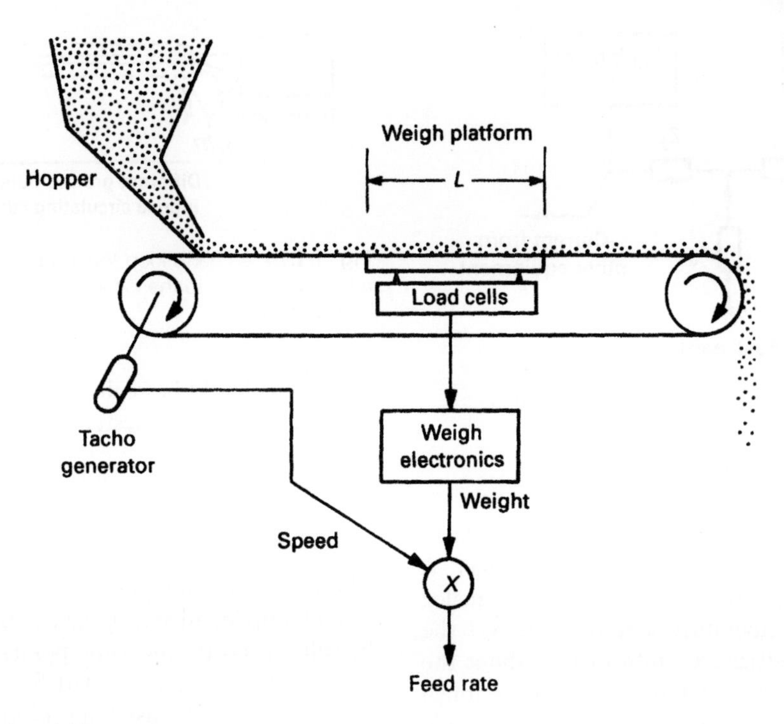

Fig. 1.1.71 Continuous belt weighing system.

4–20 mA loop signal as shown in Figure 1.1.72(a). Frequency shift keying (FSK) is used with frequencies of 1200 Hz for a '1' and 2200 Hz for a '0'. These frequencies are far too high to affect the analog instrumentation so the analog signal is still used in the normal manner. The system operates with one master (usually a computer, PLC or hand held programming terminal) and one slave (a transducer or actuator). The attraction of this approach is that it allows HART to be retro-fitted onto existing cabling and instrumentation schemes. The disadvantage is that the full cabling benefits of a fieldbus system are not realised.

The serial data, though, allows much more information to be conveyed in addition to the basic analog signal. HART devices can all be remotely configured and monitored allowing very simple diagnostics

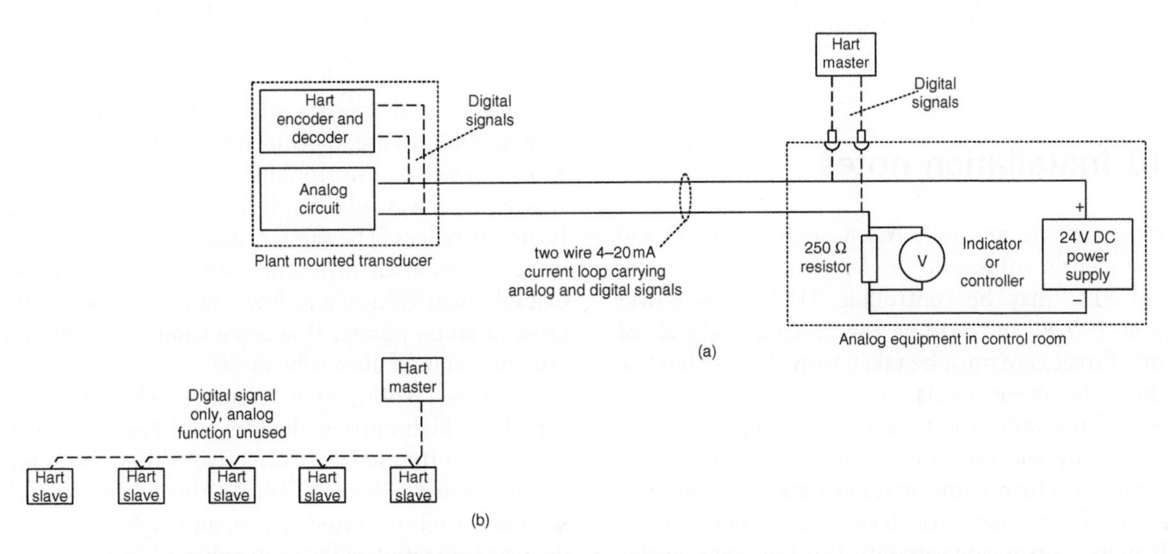

Fig. 1.1.72 The Hart communication system. (a) Hart digital data superimposed onto a two wire 4–20 mA analog loop to give additional data. A Hart master (e.g. a PLC or hand held terminal) is used to read the digital data. The transducer can also be configured by the master; (b) a Hart based fieldbus system. All data transfer is done via digital communications. Hart is not as efficient in this mode of operation as normal Fieldbus systems.

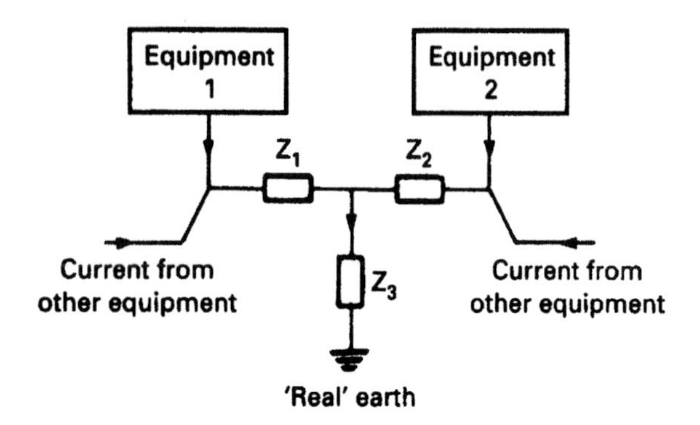

Fig. 1.1.73 A poorly installed earthing system which leads to interaction between signals.

and quick replacement after a failure. In addition much more plant data can be passed from the transducer. A flow transducer giving flow as the 4–20 mA analog signal can also give temperature and pressure via the HART link along with diagnostic information about the transducer status. With a HART programming terminal connected to the line, the transducer can also be made to send fixed currents to aid loop checks and fault diagnosis.

The second method of using HART is in a normal fieldbus multidrop system as Figure 1.1.72(b). Here each device has an identifying address and sends data to the master on request. Usually devices (and their relevant parameters) are polled on a regular cyclic scan. The multidrop brings all the cost savings from simple cabling and makes the system easy to modify and expand. By comparison with other fieldbus systems, however, HART is rather slow, and this slow response time can cause difficulties in some applications. The speed is generally adequate for simple monitoring systems where the process variables cannot change quickly.

1.1.10 Installation notes

Analog systems are generally based on low voltages and are consequently vulnerable to electrical noise. In most plants, a PLC may be controlling 415 V high power motors at 100 A, and reading thermocouple signals of a few mV. Great care must be taken to avoid interference from the high voltage signals.

The first precaution is to adopt a sensible earthing layout. A badly laid out system, as Figure 1.1.73, will have common return paths, and currents from the high powered load returning through the common impedance Z_e will induce error voltages into the low level analog

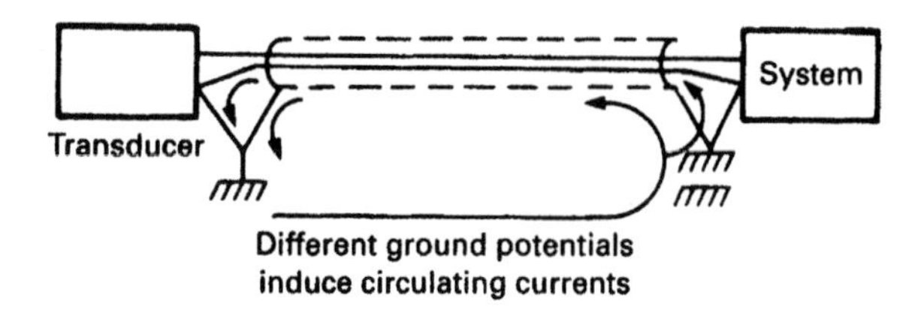

Fig. 1.1.74 An earth loop formed by connecting both ends of a screened cable to the local earth. Screens should be earthed at one point only.

circuit. It should be realised that there are at least three distinct 'earths' in a system:

- A safety earth (used for doors, frames etc.).
- A dirty earth (used for high voltage/high current signals).
- A clean earth (for low voltage analog signals).

These should meet at one point and one point only (which implies all analog signals should return, and hence be referenced to, the same point).

Screened cable is needed for all analog signals, with foil screening to be used in preference to braided screen. The screen should NOT be earthed at both ends as any difference in earth potential between the two points will cause current to flow in the screen as Figure 1.1.74, and induce noise onto the signal lines. A screen must be earthed at one point only, ideally the receiving end. When a screened cable goes through intermediate junction boxes, screen continuity must be maintained, AND the screen must be sleeved to prevent it touching the frame of the junction boxes. Earthing faults in screened cables can cause very elusive faults.

High voltage and low voltage cables should be well separated, most manufacturers suggest at least 1 m between 415 V and low voltage cables but this can be difficult to achieve in practice. In any case, separation can only be achieved until some other person, not knowing the system well, straps a 415 V cable to the same cable tray as a multicore thermocouple cable. It is therefore good practice to use trunking or conduit for low voltage signals as a way of identifying low voltage cables for future installers. The same result can also be achieved by using cables with different coloured PVC sleeves. Inevitably high voltage and low voltage cables will have to cross at some points. If spacing cannot be achieved the crossings should always be at 90°.

In an ideal world, separate cubicles should be provided for 110 V/high current devices, and low voltage signals, but this is often not cost effective. Where both types of signals have to share a cubicle the cables should take separate, well separated routes, and high and low voltage devices separated as far as possible.

Chapter 1.2

1.2

Control systems

Flower and Parr

1.2.1 Introduction

Examples of the conscious application of feedback control ideas have appeared in technology since very early times: certainly the float-regulator schemes of ancient Greece were notable examples of such ideas. Much later came the automatic direction-setting of windmills, the Watt governor, its derivatives, and so forth. The first third of the 1900s witnessed applications in areas such as automatic ship steering and process control in the chemical industry. Some of these later applications attracted considerable analytical effort aimed at attempting to account for the seemingly capricous dynamic behaviour that was sometimes found in practice.

However, it was not until during, and immediately after, World War II that the fundamentals of the above somewhat disjointed control studies were subsumed into a coherent body of knowledge which became recognised as a new engineering discipline. The great thrust in achieving this had its main antecedents in work done in the engineering electronics industry in the 1930s. Great theoretical strides were made and the concept of feedback was, for the first time, recognised as being all pervasive. The practical and theoretical developments emanating from this activity, constitute the classical approach to control which are explored in some detail in this chapter.

Since the late 1940s, tremendous efforts have been made to expand the boundaries of control engineering theory. For example, ideas from classical mechanics and the calculus of variations have been adapted and extended from a control-theoretic viewpoint. This work is based largely on the state-space description of systems (this description is briefly described in Section 1.2.11). However, it must be admitted that the practical uses and advantages of many of these developments have yet to be demonstrated. Most control system design work is still based on the classical work mentioned previously. Moreover, nowadays these applications rely, very heavily, on the use of computer techniques; indeed, computers are commonly used as elements in control loops.

Techniques from the 'classical' period of control engineering development is easily understood, wide-ranging in application and, perhaps most importantly, capable of coping with deficiencies in detailed knowledge about the system to be controlled.

These techniques are easily adapted for use in the computer-aided design of control systems, and have proved themselves capable of extension into the difficult area of multi-variable system control; however, this latter topic is beyond the scope of this chapter. So with the above comments in mind, a conventional basic approach to control theory is presented, with a short discussion of the state-space approach and a more extensive forage into sampled-data systems. These latter systems have become important owing to the incorporation of digital computers, particularly microcomputers, into the control loop. Fortunately, an elementary theory for sampled data can be established which nicely parallels the development of basic continuous control theory.

The topics covered in this introduction, and extensions of them, have stood practitioners in good stead for several decades now, and can be confidently expected to go on delivering good service for some decades to come.

Electrical Engineer's Reference Book; ISBN: 9780750646376

1.2.2 Laplace transforms and the transfer function

In most engineering analysis it is usual to produce mathematical models (of varying precision) to predict the behaviour of physical systems. Often such models are manifested by a differential equation description. This appears to fit in with the causal behaviour of idealised components, e.g. Newton's law relating the second derivative of displacement to the applied force. It is possible to model such behaviour in other ways (for example, using integral equations), although these are much less familiar to most engineers. All real systems are non-linear; however, it is fortuitous that most systems behave approximately like linear ones, with the implication that superposition holds true to some extent. We further restrict the coverage here in that we shall be concerned particularly with systems whose component values are not functions of time—at least over the time-scale of interest to us.

In mathematical terms this latter point implies that the resulting differential equations are not only linear, but also have constant coefficients, e.g. many systems behave approximately according to the equation

$$\frac{d^2x}{dt^2} + 2\zeta\omega_n\frac{dx}{dt} + \omega_n^2 x = \omega_n^2 f(t) \tag{1.2.1}$$

where x is the dependent variable (displacement, voltage, etc.), $f(t)$ is a forcing function (force, voltage source, etc.), and ω_n^2 and ζ are constants the values of which depend on the size and interconnections of the individual physical components making up the system (spring-stiffness constant, inductance values, etc.).

Equations having the form of Equation (1.2.1) are called 'linear constant coefficient ordinary differential equations' (LCCDE) and may, of course, be of any order. There are several techniques available for solving such equations but the one of particular interest here is the method based on the Laplace transformation. This is treated in detail elsewhere, but it is useful to outline the specific properties of particular interest here.

1.2.2.1 Laplace transformation

Given a function $f(t)$, then its Laplace transformation $F(s)$ is defined as

$$L[f(t)] = F(s) = \int_0^\infty f(t)\exp(-st)dt$$

where, in general, s is a complex variable and of such a magnitude that the above integral converges to a definite functional value.

A list of Laplace transformation pairs is given in Table 1.2.1.

The essential usefulness of the Laplace transformation technique in control engineering studies is that it transforms LCCDE and integral equations into algebraic ones and, hence, makes for easier and standard manipulation.

1.2.2.2 The transfer function

This is a central notion in control work and is, by definition, the Laplace transformation of the output of a system divided by the Laplace transformation of the input, with the tacit assumption that all initial conditions are at zero.

Thus, in Figure 1.2.1, where $y(t)$ is the output of the system and $u(t)$ is the input, then the transfer function $G(s)$ is

$$L[y(t)]/L[u(t)] = Y(s)/U(s) = G(s)$$

Supposing that $y(t)$ and $u(t)$ are related by the general LCCDE

$$a_n\frac{d^ny}{dt^n} + a_{n-1}\frac{d^{n-1}y}{dt^{n-1}} + \ldots + a_0y = b_m\frac{d^mu}{dt^m} + b_{m-1}\frac{d^{m-1}u}{dt^{m-1}} + \ldots + b_0u \tag{1.2.2}$$

then, on Laplace transforming and ignoring initial conditions, we have (see later for properties of Laplace transformation)

$$\left(a_ns^n + a_{n-1}s^{n-1} + \ldots + a_0\right)Y(s) = \left(b_ms^m + b_{m-1}s^{m-1} + \ldots + b_0\right)U(s)$$

whence

$$\frac{Y(s)}{U(s)} = G(s) = \sum_{i=0}^{m} b_is^i \Big/ \sum_{i=0}^{n} a_is^i$$

There are a number of features to note about $G(s)$.

(1) Invariably $n > m$ for physical systems.

(2) It is a ratio of two polynomials which may be written

$$G(s) = \frac{b_m(s - z_1)\ldots(s - z_m)}{a_n(s - p_1)\ldots(s - p_n)}$$

$z_1,\ldots, z_m$ are called the *zeros* and $p_1,\ldots, p_n$ are called the *poles* of the transfer function.

Table 1.2.1 Laplace transforms and *z* transforms

$f(t)$	$F(s)$	$F(z)$
0	0	0
$f(t-nT)$	$\exp(-nsT)F(s)$	$z^{-n}F(z)$
$\delta(t)$	1	1
$\delta(t-nT)$	$\exp(-nsT)$	z^{-n}
$\sum_{n=0}^{\infty}\delta(t-nT)$	$[1-\exp(-st)]^{-1}$	$z(z-1)^{-1}$
$h(t)$	s^{-1}	$z(z-1)^{-1}$
$u_T(t)$	$[1-\exp(-st)]s^{-1}$	—
A	As^{-1}	$Az(z-1)^{-1}$
t	s^{-2}	$Tz(z-1)^{-2}$
$f(t)t$	$-\mathrm{d}F(s)/\mathrm{d}s$	—
$(t-nT)h(t-nT)$	$\exp(-nsT)s^{-2}$	$Tz^{-(n-1)}(z-1)^{-2}$
t^2	$2s^{-3}$	$T^2z(z+1)(z-1)^{-3}$
t_n	$n!s^{-(n+1)}$	—
$\exp(\alpha t)$	$(s-\alpha)^{-1}$	$z(z-\exp(\alpha T))^{-1}$
$f(t)\exp(\alpha t)$	$F(s-\alpha)$	$F[z\exp(-\alpha T)]$
$\delta(t)+\alpha\exp(\alpha t)$	$s(s-\alpha)^{-1}$	—
$t\exp(\alpha t)$	$(s-\alpha)^{-2}$	$TZ\exp(\alpha T)[z-\exp(\alpha T)]^{-2}$
$t^n\exp(\alpha t)$	$n!(s-\alpha)^{-(n+1)}$	—
$\sin\omega t$	$\frac{\omega}{s^2+\omega^2}$	$\frac{z\sin\omega T}{z^2-2z\cos\omega T+1}$
$\cos\omega t$	$\frac{s}{s^2+\omega^2}$	$\frac{z(z-\cos\omega T)}{z^2-2z\cos\omega T+1}$
$\frac{t}{2\omega}\sin\omega t$	$\frac{s}{(s^2+\omega^2)}$	—
$\frac{1}{2\omega}(\sin\omega t-\omega t\cos\omega t)$	$\frac{\omega^2}{(s^2+\omega^2)^2}$	—
$\frac{1}{\cos\delta}\sin(\omega t+\delta)$	$\frac{A}{s^2+\omega^2}(s+\frac{\omega}{A})$ where $\tan\delta=A$	—
$\frac{1}{\cos\delta}\cos(\omega t+\delta)$	$\frac{1}{s^2+\omega^2}(s-A\omega)$	—
$\exp(\alpha t)\sin\omega t$	$\frac{\omega}{(s-\alpha)^2+\omega^2}=\frac{\omega}{(s-\alpha+j\omega)(s-\alpha-j\omega)}$	$\frac{z\exp(\alpha T)\sin\omega T}{z^2-2z\exp(\alpha T)\cos\omega T+\exp(2\alpha T)}$
$\exp(\alpha t)\cos\omega t$	$\frac{s-\alpha}{(s-\alpha)^2+\omega^2}$	$\frac{z[z-\exp(\alpha T)\cos\omega T]}{z^2-2z\exp(\alpha T)\cos\omega T+\exp(2\alpha T)}$
$\frac{t}{2\omega}\exp(\alpha t)\sin\omega t$	$\frac{s-\alpha}{[(s-\alpha)^2+\omega^2]^2}$	—
$\frac{1}{2\omega}\exp(\alpha t)(\sin\omega t-\cos\omega t)$	$\frac{\omega^2}{[(s-\alpha)^2+\omega^2]^2}$	—
$\frac{1}{\cos\delta}\exp(\alpha t)\sin(\omega t+\delta)$	$\frac{A}{(s-\alpha)^2+\omega^2}(s-\alpha+\frac{\omega}{A})$ where $\tan\delta=A$	—
$\frac{1}{\cos\delta}\exp(\alpha t)\cos(\omega t+\delta)$	$\frac{1}{(s-\alpha)^2+\omega^2}(s-\alpha-A\omega)$	—
$\sinh\omega t$	$\omega(s^2-\omega^2)^{-1}$	—
$\cosh\omega t$	$s(s^2-\omega^2)^{-1}$	—
$f'(t)$	$sF(s)-f(0-)$	—

(Continued)

Table 1.2.1 Laplace transforms and z transforms (continued)

$f(t)$	$F(s)$	$F(z)$
$f''(t)$	$s^2F(s) - sf(0-) f'(0-)$	—
$f^n(t)$	$s^nF(s) - s^{n-1}f(0-) - s^{n-2}f'(0-) \ldots - f^{n-1}(0-)$	—
$f^{-1}(t)$	$\frac{F(s)}{s} + \frac{f^{-1}(0-)}{s}$	—
$f(t)$ $t \to 0$	$sF(s)$ $s \to \infty$	$F(z)$ $z \to \infty$
$f(t)$ $t \to \infty$	$sF(s)$ $s \to 0$	$(z-1)z^{-1}F(z)$ $z \to 1$

$\delta(t)$, the unit impulse function.
$h(t)$, the unit step function.
$u_T(t)$, the unit step function followed by a unit negative step at $t = T$, where T is the sampling period.

(3) It is not an explicit function of input or output, but depends entirely upon the nature of the system.

(4) The block diagram representation shown in Figure 1.2.1 may be extended so that the interaction of composite systems can be studied (provided that they do not load each other); see below.

(5) If $u(t)$ is a delta function $\delta(t)$, then $U(s) = 1$, whence $Y(s) = G(s)$ and $y(t) = g(t)$, where $g(t)$ is the *impulse response* (or weighting function) of the system.

(6) Although a particular system produces a particular transfer function, a particular transfer function does not imply a particular system, i.e. the transfer function specifies merely the input–output relationship between two variables and, in general, this relationship may be realised in an infinite number of ways.

(7) Although we might expect that all transfer functions will be ratios of finite polynomials, an important and common element which is an exception to this is the pure-delay element. An example of this is a loss-free transmission line in which any disturbance to the input of the line will appear at the output of the line without distortion, a finite time (say τ) later. Thus, if $u(t)$ is the input, then the output $y(t) = u(t - \tau)$ and the transfer function $Y(s)/U(s) = \exp(-s\tau)$. Hence, the occurrence of this term within a transfer function expression implies the presence of a pure delay; such terms are common in chemical plant and other fluid-flow processes.

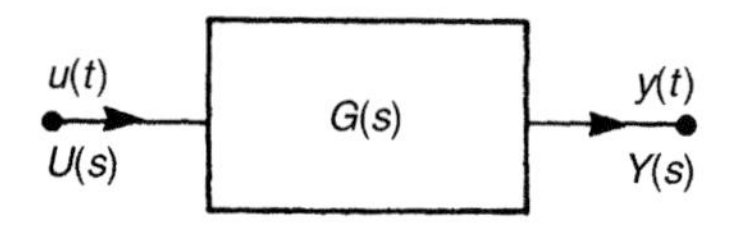

Fig. 1.2.1 Input–output representation.

Having performed any manipulations in the Laplace transformation domain, it is necessary for us to transform back to the time domain if the time behaviour is required. Since we are dealing normally with the ratio of polynomials, then by partial fraction techniques we can arrange $Y(s)$ to be written in the following sequences:

$$Y(s) = \frac{K(s - z_1)(s - z_2)\ldots(s - z_m)}{(s - p_1)(s - p_2)\ldots(s - p_n)}$$

$$Y(s) = K\left[\frac{A_1}{s - p_1} + \frac{A_2}{s - p_2} + \cdots + \frac{A_n}{s - p_n}\right]$$

and by so arranging $Y(s)$ in this form the conversion to $y(t)$ can be made by looking up these elemental forms in Table 1.2.1.

Example Suppose that

$$Y(s) = \frac{5(s^2 + 4s + 3)}{s^3 + 6s^2 + 8s} = \frac{5(s^2 + 4s + 3)}{s(s + 2)(s + 4)}$$

$$= 5\left[\frac{3}{8s} + \frac{1}{4(s + 2)} + \frac{3}{8(s + 4)}\right]$$

Then

$$y(t) = \frac{5}{4}\left[\frac{3}{2}\{1 + \exp(-4t)\} + \exp(-2t)\right]$$

1.2.2.3 Certain theorems

A number of useful transform theorems are quoted below, without proof.

(1) *Differentiation*

If $F(s)$ is the Laplace transformation of $f(t)$, then

$$L[d^n f(t)/dt^n] = s^n F(s) - s^{n-1} f(0)$$
$$= s^{n-2} f'(0) - \cdots - f^{n-1}(0)$$

For example, if $f(t) = \exp(-bt)$, then

$$L\left[\frac{d^3}{dt^3}\exp(-bt)\right] = \frac{s^3}{s+b} - s^2 + bs - b^2$$

(2) *Integration*

If $L[f(t)] = F(s)$, then

$$L\left[\int_0^1 f(t)dt\right] = \frac{F(s)}{s} + f(0)$$

Repeated integration follows in a similar fashion.

(3) *Final-value theorem*

If $f(t)$ and $f'(t)$ are Laplace transformable and if $L[f(t)] = F(s)$, then if the limit of $f(t)$ exists as t goes towards infinity, then

$$\lim_{s \to 0} sF(s) = \lim_{t \to \infty} f(t)$$

For example

$$F(s) = \frac{b-a}{s(s+a)(s+b)}$$

then

$$\lim_{s \to 0} \frac{s(b-a)}{s(s+a)(s+b)} = \frac{b-a}{ab} = \lim_{t \to \infty} f(t)$$

(4) *Initial-value theorem*

If $f(t)$ and $f'(t)$ are Laplace transformable and if $L[f(t)] = F(s)$, then

$$\lim_{s \to \infty} sF(s) = \lim_{t \to 0} f(t)$$

(5) *Convolution*

If $L[f_1(t)] = F_1(s)$ and $L[f_2(t)] = F_2(s)$, then

$$F_1(s) \cdot F_2(s) = L\left[\int_0^\infty f_1(t-\tau) \cdot f_2(\tau)d\tau\right]$$

1.2.3 Block diagrams

It is conventional to represent individual transfer functions by boxes with an input and output (see note (4) in Section 1.2.2.2). Provided that the components represented by the transfer function do not load those represented by the transfer function in a connecting box, then simple manipulation of the transfer functions can be carried out. For example, suppose that there are two transfer functions in cascade (see Figure 1.2.2): then we may write $X(s)/U(s) = G_1(s)$ and $Y(s)/X(s) = G_2(s)$. Eliminating $X(s)$ by multiplication, we have

$$Y(s)/U(s) = G_1(s)G_2(s)$$

which may be represented by a single block. This can obviously be generalised to any number of blocks in cascade.

Another important example of block representation is the prototype feedback arrangement shown in Figure 1.2.3. We see that $Y(s) = G(s)E(s)$ and $E(s) = U(s) - H(s)Y(s)$. Eliminating $E(s)$ from these two equations results in

$$\frac{Y(s)}{U(s)} = \frac{G(s)}{1 + H(s)G(s)} = W(s)$$

In block diagram form we have Figure 1.2.4. If we eliminate $Y(s)$ from the above equations, we obtain

$$\frac{E(s)}{U(s)} = \frac{1}{1 + H(s)G(s)}$$

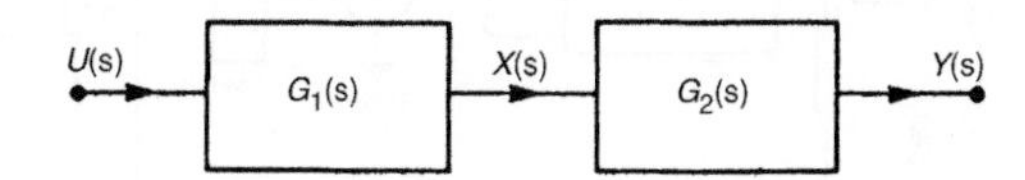

Fig. 1.2.2 Systems in cascade.

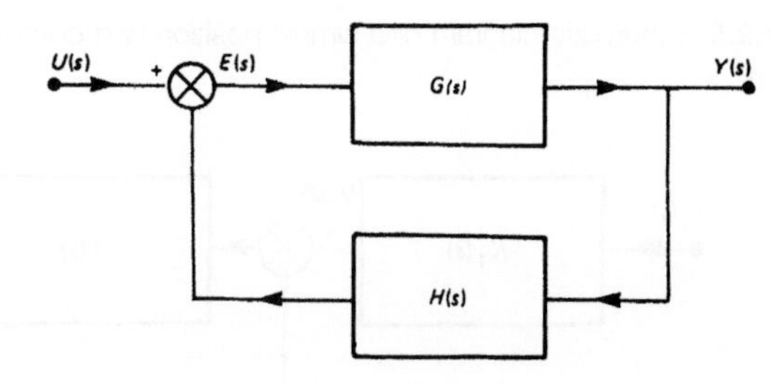

Fig. 1.2.3 Block diagram of a prototype feedback system.

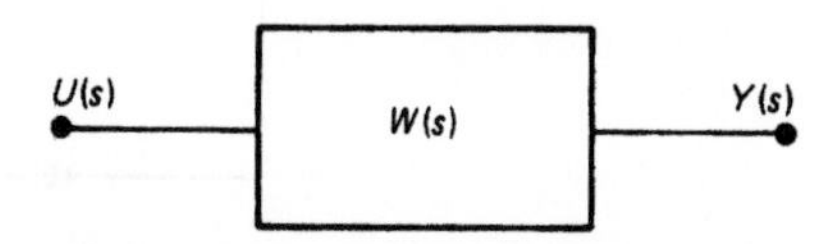

Fig. 1.2.4 Reduction of the diagram shown in Figure 1.2.3 to a single block.

1.2.4 Feedback

The last example is the basic feedback conceptual arrangement, and it is pertinent to investigate it further, as much effort in dealing with control systems is devoted to designing such feedback loops. The term 'feedback' is used to describe situations in which a portion of the output (and/or processed parts of it) are fed back to the input of the system. The appropriate application may be used, for example, to improve bandwidth, improve stability, improve accuracy, reduce effects of unwanted disturbances, compensate for uncertainty and reduce the sensitivity of the system to component value variation.

As a concrete example consider the system shown in Figure 1.2.5, which displays the arrangements for an angular position control system in which a desired position θ_r is indicated by tapping a voltage on a potentiometer. The actual position of the load being driven by the motor (usually via a gearbox) is monitored by θ_o, indicated, again electrically, by a potentiometer tapping. If we assume identical potentiometers energised from the same voltage supply, then the misalignment between the desired output and the actual output is indicated by the difference between the respective potentiometer voltages. This difference (proportional to error) is fed to an amplifier whose output, in turn, drives the motor. Thus, the arrangement seeks to drive the system until the output θ_o and input θ_r are coincident (i.e. the error is zero).

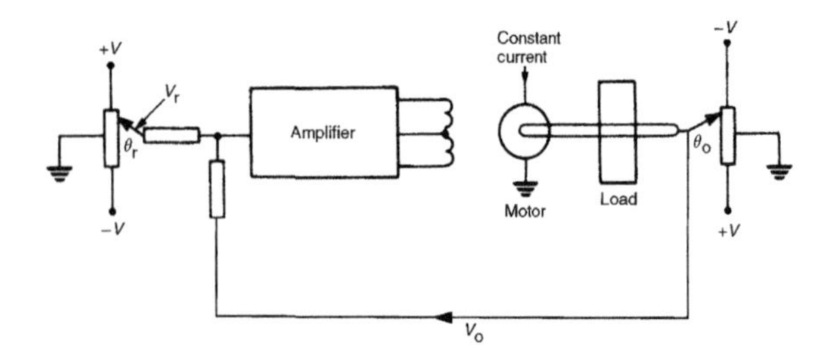

Fig. 1.2.5 Schematic diagram of a simple position and control system.

In the more general block diagram form, the above schematic will be transformed to that shown in Figure 1.2.6, where $\theta_r(s)$ and $\theta_o(s)$ are the Laplace transforms of the input and output positions, respectively; $K_1(s)$ and $K_2(s)$ are the potentiometer transfer functions (normally taken as straight gains); $V_r(s)$ is the Laplace transform of the reference voltage; $V_o(s)$ is the Laplace transform of the output voltage; $G_m(s)$ is the motor transfer function; $G_l(s)$ is the load transfer function; and $A(s)$ is the amplifier transfer function.

Let us refer now to Figure 1.2.3 in which $U(s)$ is identified as the transformed input (reference or demand) signal, $Y(s)$ is the output signal and $E(s)$ is the error (or actuating) signal. $G(s)$ represents the *forward transfer function* and is the product of all the transfer functions in the forward loop, i.e. $G(s) = A(s)G_m(s)G_l(s)$ in the above example.

$H(s)$ represents the *feedback transfer function* and is the product of all transfer functions in the feedback part of the loop.

We saw in Section 1.2.3 that we may write

$$\frac{Y(s)}{U(s)} = \frac{G(s)}{1 + H(s)G(s)}$$

and

$$\frac{E(s)}{U(s)} = \frac{1}{1 + H(s)G(s)}$$

i.e. we have related output to input and the error to the input.

The product $H(s)G(s)$ is called the *open-loop transfer function* and $G(s)/[1 + H(s)G(s)]$ the *closed-loop transfer function*. The open-loop transfer function is most useful in studying the behaviour of the system, since it relates the error to the demand. Obviously it would seem desirable for this error to be zero at all times, but since we are normally considering systems containing energy storage components, total elimination of error at all times is impossible.

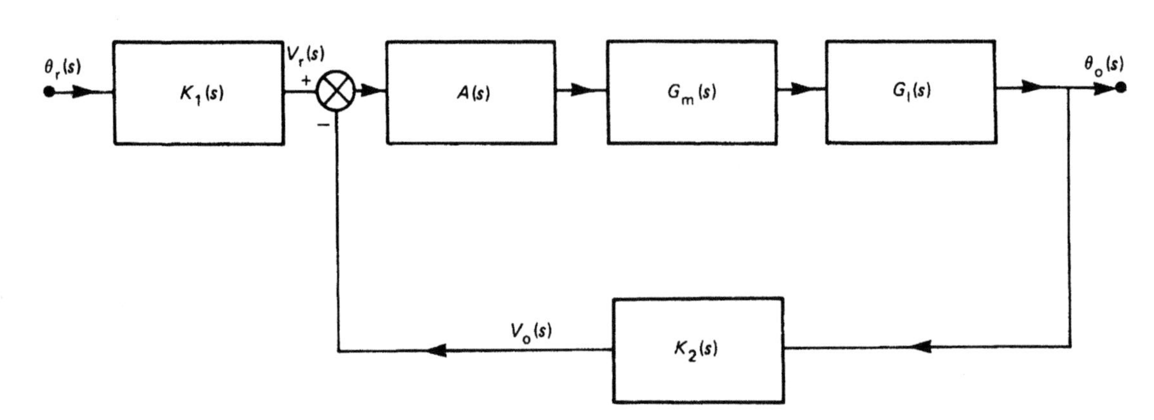

Fig. 1.2.6 Block diagram of the system shown in Figure 1.2.5

1.2.5 Generally desirable and acceptable behaviour

Although specific requirements will normally be drawn up for a particular control system, there are important general requirements applicable to the majority of systems. Usually an engineering system will be assembled from readily available components to perform some function, and the choice of these components will be restricted. An example of this would be a diesel engine–alternator set for delivering electrical power, in which normally the most convenient diesel engine–alternator combination will be chosen from those already manufactured.

Even if such a system were assembled from customer-designed components, it would be fortuitous if it performed in a satisfactory self-regulatory way without further consideration of its control dynamics. Hence, it is the control engineer's task to take such a system and devise economical ways of making the overall system behave in a satisfactory manner under the expected operational conditions.

For example, a system may oscillate, i.e. it is unstable; or, although stable, it might tend to settle after a change in input demand to a value unacceptably far from this new demand, i.e. it lacks static accuracy. Again, it might settle to a satisfactory new steady state, but only after an unsatisfactory transient response. Alternatively, normal operational load disturbances on the system may cause unacceptably wide variation of the output variable, e.g. voltage and frequency of the engine–alternator system.

All these factors will normally be quantified in an actual design specification, and fortunately a range of techniques is available for improving the behaviour. But the application of a particular technique to improve the performance of one aspect of behaviour often has a deleterious effect on another, e.g. improved stability with improved static accuracy tends to be incompatible. Thus, a compromise is sought which gives the 'best' acceptable all-round performance. We now discuss some of these concepts and introduce certain techniques useful in examining and designing systems.

1.2.6 Stability

This is a fairly easy concept to appreciate for the types of system under consideration here. Equation (1.2.2) with the right-hand side made equal to zero governs the free (or unforced, or characteristic) behaviour of the system, and because of the nature of the governing LCCDE it is well known that the solution will be a linear combination of exponential terms, *viz.*

$$y(t) = \sum_{i=1}^{n} A_i \exp(\alpha_i t)$$

where the α_i values are the roots of the so-called 'characteristic equation'.

It will be noted that should any α_i have a positive real part (in general, the roots will be complex), then any disturbance will grow in time. Thus, for stability, no roots must lie in the right-hand half of the complex plane or s plane. In a transfer function context this obviously translates to 'the roots of the denominator must not lie in the right-hand half of the complex plane'.

For example, if $W(s) = G(s)/[1 + H(s)G(s)]$, then the roots referred to are those of the equation

$$1 + H(s)G(s) = 0$$

In general, the determination of these roots is a non-trivial task and, as at this stage we are interested only in whether the system is stable or not, we can use certain results from the theory of polynomials to achieve this without the necessity for evaluating the roots.

A preliminary examination of the location of the roots may be made using the *Descartes rule of signs*, which states: if $f(x)$ is a polynomial, the number of positive roots of the equation $f(x) = 0$ cannot exceed the number of changes of sign of the numerical coefficients of $f(x)$, and the number of negative roots cannot exceed the number of changes of sign of the numerical coefficients of $f(-x)$. 'A change of sign' occurs when a term with a positive coefficient is immediately followed by one with a negative coefficient, and vice versa.

Example Suppose that $f(x) = x^3 + 3x - 2 = 0$; then there can be at most one positive root. Since $f(-x) = -x^3 - 3x - 2$, the equation has no negative roots. Further, the equation is cubic and must have at least one real root (complex roots occur in conjugate pairs); therefore, the equation has one positive-real root.

Although Descartes' result is easily applied, it is often indefinite in establishing whether or not there is stability, and a more discriminating test is that due to Routh, which we give without proof.

Suppose that we have the polynomial

$$a_0 s^n + a_1 s^{n-1} + \ldots + a_{n-1}s + a_n = 0$$

where all coefficients are positive, which is a necessary (but not sufficient) condition for the system to be stable, and we construct the following so-called 'Routh array':

$$\begin{array}{rllllll}
s^n: & a_0 & a_2 & a_4 & a_6 & \ldots \\
s^{n-1}: & a_1 & a_3 & a_5 & a_7 & \ldots \\
s^{n-2}: & b_1 & b_2 & b_3 & \ldots & \\
s^{n-3}: & c_1 & c_2 & c_3 & \ldots & \\
s^{n-4}: & d_1 & d_2 & \ldots & &
\end{array}$$

where

$$b_1=\frac{a_1a_2-a_0a_3}{a_1},\quad b_2=\frac{a_1a_4-a_0a_5}{a_1},\quad b_3=\frac{a_1a_6-a_0a_7}{a_1},\ldots$$
$$c_1=\frac{b_1a_3-a_1b_2}{b_1},\quad c_2=\frac{b_1a_5-a_1b_3}{b_1},\ldots$$
$$d_1=\frac{c_1b_2-b_1c_2}{c_1},\ldots$$

This array will have $n+1$ rows.

If the array is complete and *none* of the elements in the first column vanishes, then a sufficient condition for the system to be stable (i.e. the characteristic equation has all its roots with negative-real parts) is for all these elements to be positive. Further, if these elements are not all positive, then the number of changes of sign in this first column indicates the number of roots with positive-real parts.

Example Determine whether the polynomial $s^4+2s^3+6s^2+7s+4=0$ has any roots with positive-real parts. Construct the Routh array:

$$\begin{array}{llll}
s^4: & 1 & 6 & 4\\
s^3: & 2 & 7 & \\
s^2: & \dfrac{(2)(6)-(1)(7)}{2}=2.5 & \dfrac{(2)(4)-(1)(0)}{2}=4 & \\
s: & \dfrac{(2.5)(7)-(2)(4)}{2.5}=3.8 & & \\
s^0: & 4 & &
\end{array}$$

There are five rows with the first-column elements all positive, and so a system with this polynomial as its characteristic would be stable.

There are cases that arise which need a more delicate treatment.

(1) Zeros occur in the first column, while other elements in the row containing a zero in the first column are non-zero. In this case the zero is replaced by a small positive number, ϵ, which is allowed to approach zero once the array is complete.

For example, consider the polynomial equation

$s^5+2s^4+2s^3+4s^2+11s+8=0$:

$$\begin{array}{llll}
s^5: & 1 & 2 & 11\\
s^4: & 2 & 4 & 8\\
s^3: & \epsilon & 5 & 0\\
s^2: & \alpha_1 & 8 & \\
s^1: & \alpha_2 & 0 & \\
s^0: & 8 & &
\end{array}$$

where

$$\alpha_1=\frac{4\epsilon-10}{\epsilon}\simeq-\frac{10}{\epsilon}\quad\text{and}\quad\alpha_2=\frac{5\alpha_1-8\epsilon}{\alpha_1}\simeq 5$$

Thus, α_1 is a large negative number and we see that there are effectively two changes of sign and, hence, the equation has two roots which lie in the right-hand half of this plane.

(2) Zeros occur in the first column and other elements of the row containing the zero are also zero. This situation occurs when the polynomial has roots that are symmetrically located about the origin of the s plane, i.e. it contains terms such as $(s+j\omega)(s-j\omega)$ or $(s+v)(s-v)$.

This difficulty is overcome by making use of the auxiliary equation which occurs in the row immediately before the zero entry in the array. Instead of the all-zero row the equation formed from the preceding row is differentiated and the resulting coefficients are used in place of the all-zero row.

For example, consider the polynomial $s^3+3s^2+2s+6=0$:

$$\begin{array}{lll}
s^3: & 1 & 2\\
s^2: & 3 & 6\quad\left(\text{auxiliary equation } 3s^2+6=0\right)\\
s^1: & 0 & 0
\end{array}$$

Differentiate the auxiliary equation giving $6s=0$, and compile a new array using the coefficients from this last equation, *viz.*

$$\begin{array}{lll}
s^3: & 1 & 2\\
s^2: & 3 & 6\\
s^1: & 6 & 0\\
s^0: & 1 &
\end{array}$$

Since there are no changes of sign, the system will not have roots in the right-hand half of the s plane.

Although the Routh method allows a straightforward algorithmic approach to determining the stability, it gives very little clue as to what might be done if stability conditions are unsatisfactory. This consideration is taken up later.

1.2.7 Classification of system and static accuracy

1.2.7.1 Classification

The discussion in this section is restricted to unity-feedback systems (i.e. $H(s)=1$) without seriously affecting generalities. We know that the open-loop system has a transfer function $KG(s)$, where K is a constant and we may write

$$KG(s)=\frac{K(s-z_1)(s-z_2)\ldots(s-z_m)}{s^l(s+p_1)(s+p_2)\ldots(s-p_3)}=\frac{K\sum_{k=0}^{m}b_ks^k}{s^l\sum_{k=0}^{n-1}a_ks^k}$$

and for physical systems $n\geq m+1$.

The *order* of the system is defined as the degree of the polynomial in s appearing in the denominator, i.e. n.

The *rank* of the system is defined as the difference in the degree of the denominator polynomial and the degree of the numerator polynomial, i.e. $n - m \geq 1$.

The *class* (or *type*) is the degree of the s term appearing in the denominator (i.e. l), and is equal to the number of integrators in the system.

Example

(1) $$G(s) = \frac{s+1}{s^4 + 6s^3 + 9s^2 + 3s}$$

implies order 4, rank 3 and type 1.

(2) $$G(s) = \frac{s^2 + 4s + 1}{(s+1)(s^2 + 2s + 4)}$$

implies order 3, rank 1 and type 0.

1.2.7.2 Static accuracy

When a demand has been made on the system, then it is generally desirable that after the transient conditions have decayed the output should be equal to the input. Whether or not this is so will depend both on the characteristics of the system and on the input demand. Any difference between the input and output will be indicated by the error term $e(t)$ and we know that for the system under consideration

$$E(s) = \frac{U(s)}{1 + KG(s)}$$

Let $e_{ss} = \lim_{t\to\infty} e(t)$ (if it exists), and so e_{ss} will be the steady-state error. Now from the final-value theorem we have

$$e_{ss} = \lim_{t\to\infty} e(t) = \lim_{s\to 0} [sE(s)]$$

Thus,

$$e_{ss} = \lim_{s\to 0}\left[\frac{sU(s)}{1 + KG(s)}\right]$$

1.2.7.2.1 Position-error coefficient K_p

Suppose that the input is a unit step, i.e. $R(s) = 1/s$; then

$$e_{ss} = \lim_{s\to 0}\left[\frac{1}{1 + KG(s)}\right] = \frac{1}{1 + \lim_{s\to 0}[KG(s)]} = \frac{1}{1 + K_p}$$

where $K_p = \lim_{s\to 0}[KG(s)]$ and this is called the *position-error coefficient.*

Example For a type-0 system

$$KG(s) = \left[K \sum_{k=0}^{m} b_k s^k\right] \Big/ \left[\sum_{k=0}^{n} a_k s^k\right]$$

Therefore $K_p = K(b_0/a_0)$ and $e_{ss} = 1/(1 + K_p)$.

It will be noted that, after the application of a step, there will always be a finite steady-state error between the input and the output, but this will decrease as the gain K of the system is increased.

Example For a type-1 system

$$KG(s) = \left[K \sum_{k=0}^{m} b_k s^k\right] \Big/ \left[s \sum_{k=0}^{n-1} a_k s^k\right]$$

and

$$K_p = \lim_{s\to 0}\left[K \sum_{k=0}^{m} b_k s^k\right] \Big/ \left[s \sum_{k=0}^{n-1} a_k s^k\right] \to \infty$$

Thus,

$$e_{ss} = \frac{1}{1+\infty} \to 0$$

i.e. there is no steady-state error in this case and we see that this is due to the presence of the integrator term $1/s$. This is an important practical result, since it implies that steady-state errors can be eliminated by use of integral terms.

1.2.7.2.2 Velocity-error coefficient, K_v

Let us suppose that the input demand is a unit ramp, i.e. $u(t) = t$, so $U(s) = 1/s^2$. Then

$$e_{ss} = \lim_{s\to 0}[sE(s)] = \lim_{s\to 0}\left[\frac{1}{s + sKG(s)}\right] = \frac{1}{\lim_{s\to 0}[sKG(s)]} = \frac{1}{K_v}$$

where $K_v = \lim_{s\to 0} [sKG(s)]$ is called the *velocity-error coefficient.*

Examples 1 For a type-0 system $K_v = 0$, whence $e_{ss} \to \infty$.

For a type-1 system $K_v = K(b_0/a_0)$ and so this system can follow but with a finite error.

For a type-2 system

$$K_v = \lim_{s\to 0}\left[\frac{K}{s}\frac{b_0}{a_0}\right] \to \infty$$

whence $e_{ss} \to 0$ and so the system can follow in the steady state without error.

1.2.7.2.3 Acceleration-error coefficient K_a

In this case we assume that $u(t) = t^2/2$, so $U(s) = 1/s^3$ and so

$$e_{ss} = \lim_{s\to 0}[sE(s)] = \lim_{s\to 0}\left[\frac{1}{s^2 + s^2KG(s)}\right] = \frac{1}{\lim_{s\to 0}[s^2KG(s)]} = \frac{1}{K_a}$$

where $K_a = \lim_{s\to 0}[s^2\ KG(s)]$ is called the *acceleration-error coefficient* and similar analyses to the above may be performed.

These error-coefficient terms are often used in design specifications of equipment and indicate the minimum order of the system that one must aim to design.

1.2.7.3 Steady-state errors due to disturbances

The prototype unity-feedback closed-loop system is shown in Figure 1.2.7 modified by the intrusion of a disturbance $D(s)$ being allowed to affect the loop. For example, the loop might represent a speed-control system and $D(s)$ might represent the effect of changing the load. Now, since linear systems are under discussion, in order to evaluate the effects of this disturbance on $Y(s)$ (denoted by $Y_D(s)$), we may tacitly assume $U(s) = 0$ (i.e. invoke the superposition principle)

$$Y_D(s) = D(s) - KG(s)Y_D(s)$$
$$Y_D(s) = D(s)/[1 + KG(s)]$$

Now $E_D(s) = -Y_D(s) = -D(s)/[1 + KG(s)]$, and so the steady-state error, e_{ssD} due to the application of the disturbance, may be evaluated by use of the final-value theorem as

$$e_{ssD} = -\lim_{s\to 0}\left[\frac{sD(s)}{1 + KG(s)}\right]$$

Obviously the disturbance may enter the loop at other places but its effect may be established by similar analysis.

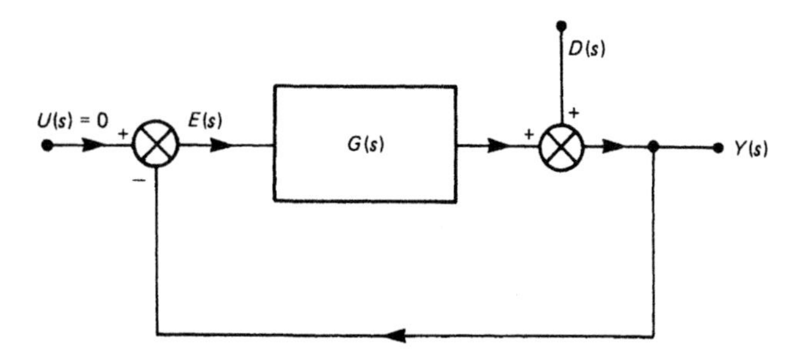

Fig. 1.2.7 Schematic diagram of a disturbance entering the loop.

1.2.8 Transient behaviour

Having developed a means of assessing stability and steady-state behaviour, we turn our attention to the transient behaviour of the system.

1.2.8.1 First-order system

It is instructive to examine first the behaviour of a first-order system (a first-order lag with a time constant T) to a unit-step input (Figure 1.2.8).

Now

$$\frac{Y(s)}{U(s)} = G(s) = \frac{1}{1 + sT}$$

where $U(s) = 1/s$

$$Y(s) = \frac{1}{s(1 + sT)} = \frac{1}{Ts[s + (1/T)]} = \frac{1}{s} - \frac{1}{[s + (1/T)]}$$

or $y(t) = 1 - \exp(-t/T)$; note also that $dy/dt = (1/T)\exp(-t/T)$.

Figure 1.2.8 shows this time response for different values of T where it will be noted that the corresponding trajectories have slopes of $1/T$ at time $t = 0$ and reach approximately 63% of their final values after T.

Suppose now that such a system is included in a unity-feedback arrangement together with an amplifier of gain K (Figure 1.2.9); therefore

$$\frac{Y(s)}{U(s)} = \frac{K/(1 + sT)}{1 + K/(1 + sT)} = \frac{K}{(1 + K)(1 + s(T/(1 + K))}$$

For a unit-step input the time response will be

$$y(t) = \frac{K}{1 + K}[1 - \exp\{-(1 + K)(t/T)\}]$$

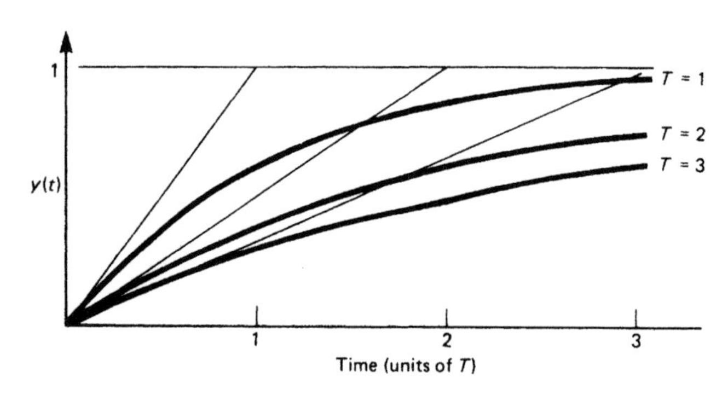

Fig. 1.2.8 First-order lag response to a unit step (time constant = 1, 2, 3 units).

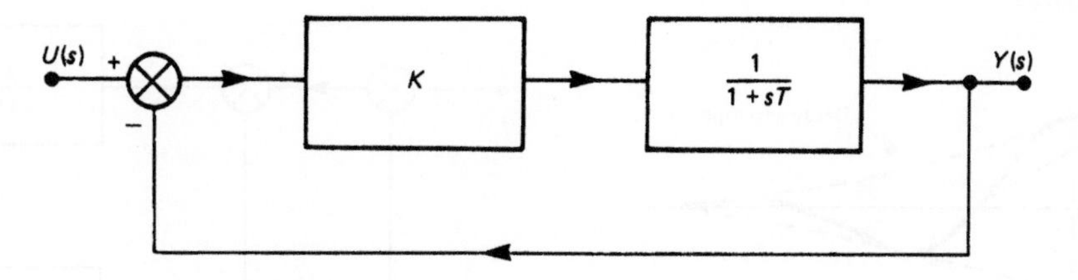

Fig. 1.2.9 First-order lag incorporated in a feedback loop.

This expression has the same form as that obtained for the open loop but the effective time constant is modified by the gain and so is the steady-state condition (Figure 1.2.10). Such an arrangement provides the ability to control the effective time constant by altering the gain of an amplifier, the original physical system being left unchanged.

1.2.8.2 Second-order system

The behaviour characteristics of second-order systems are probably the most important of all, since many systems of seemingly greater complexity may often be approximated by a second-order system because certain poles of their transfer function dominate the observed behaviour. This has led to system specifications often being expressed in terms of second-order system behavioural characteristics.

In Section 1.2.2 the importance of the second-order behaviour of a generator was mentioned, and this subject is now taken further by considering the system shown in Figure 1.2.11.

The closed-loop transfer function for this system is given by

$$W(s) = \frac{KG(s)}{1 + KG(s)} = \frac{K}{s^2 + as + K}$$

and this may be rewritten in general second-order terms in the form

$$W(s) = \frac{\omega_n^2}{s^2 + 2\zeta\omega_n s + \omega_n^2}$$

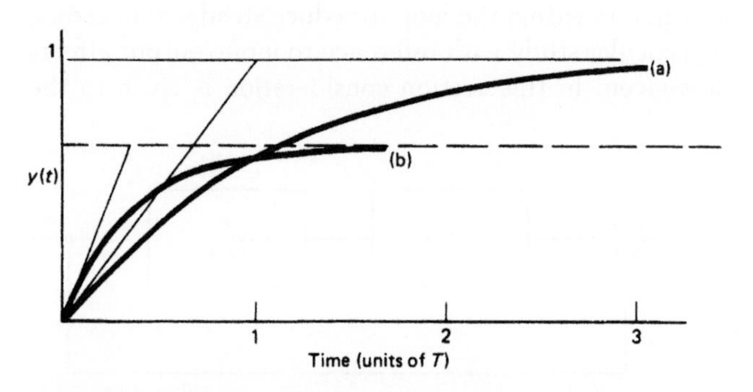

Fig. 1.2.10 Response of first-order lag: (a) open-loop condition ($T = 1$); (b) closed-loop condition ($T = 1$, $K = 2$).

where $K = \omega_n^2$ and $\zeta = a/(2\sqrt{K})$. The unit-step response is given by

$$y(t) = 1 - \exp(-\zeta\omega_n t)[\cos(\gamma\omega_n t) - (\zeta/\gamma)\sin(\gamma\omega_n t)]$$

where $\gamma = \sqrt{(1 - \zeta^2)}$. This assumes, of course, that $\zeta < 1$, so giving an oscillating response decaying with time.

The *rise time* t_r will be defined as the time to reach the first overshoot (note that other definitions are used and it is important to establish which particular definition is being used in a particular specification):

$$t_r = \pi/(\gamma\omega_n) = \pi/\sqrt{\left[K - (a/2)^2\right]}$$

i.e. the rise time decreases as the gain K is increased.

The *percentage overshoot* is defined as:

$$\begin{aligned}&\text{Percentage overshoot}\\&= \frac{100(\text{Max. value of } y(t) - \text{Steady-state value})}{\text{Steady-state value}}\\&= 100\exp(-\zeta\pi/\gamma)\\&= 100\exp\left[-a\pi/\sqrt{\left(4K - a^2\right)}\right]\end{aligned}$$

i.e. the percentage overshoot increases as the gain K increases.

The *frequency of oscillation* ω_r is immediately seen to be

$$\omega_r = \omega_n\gamma = \sqrt{\left[K - (a/2)^2\right]}$$

i.e. the frequency of oscillation increases as the gain K increases.

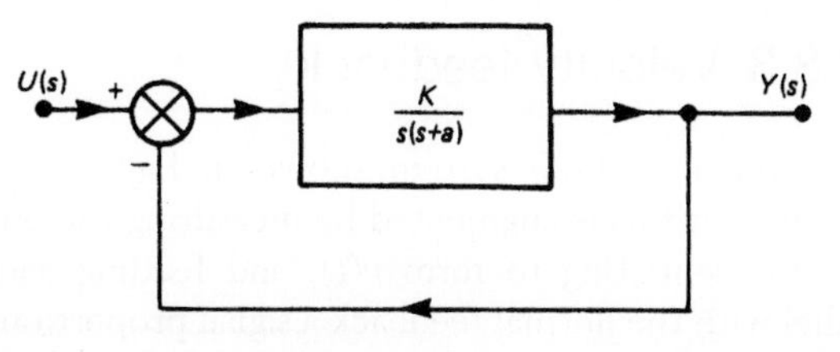

Fig. 1.2.11 Second-order system.

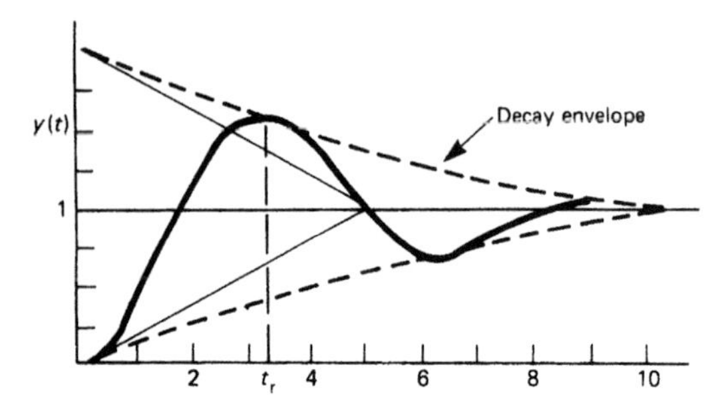

Fig. 1.2.12 Step response of the system shown in Figure 1.2.11. The rise time t_r, is the time taken to reach maximum overshoot. The predominant time constant is indicated by the tangents to the envelope curve.

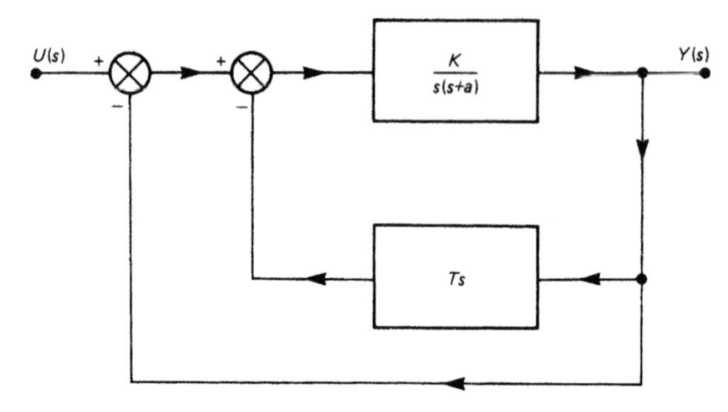

Fig. 1.2.13 Schematic diagram showing the incorporation of velocity feedback.

The *predominant time constant* is the time constant associated with the envelope of the response (Figure 1.2.12) which is given by $\exp(-\zeta\omega_n t)$ and thus the predominant time constant is $1/\zeta\omega_n$:

$$\frac{1}{\zeta\omega_n} = \frac{1}{(a/2\sqrt{K})\sqrt{K}} = \frac{2}{a}$$

Note that this time constant is unaffected by the gain K and is associated with the 'plant parameter a', which will normally be unalterable, and so other means must be found to alter the predominant time constant should this prove necessary.

The *settling time* t_s is variously defined as the time taken for the system to reach 2–5% (depending on specification) of its final steady state and is approximately equal to four times the predominant time constant.

It should be obvious from the above that characteristics desired in plant dynamical behaviour may be conflicting (e.g. fast rise time with small overshoot) and it is up to the skill of the designer to achieve the best compromise. Overspecification can be expensive.

A number of the above items can be directly affected by the gain K and it may be that a suitable gain setting can be found to satisfy the design with no further attention. Unfortunately, the design is unlikely to be as simple as this, in view of the fact that the predominant time constant cannot be influenced by K. A particularly important method for influencing this term is the incorporation of so-called *velocity feedback*.

1.2.8.3 Velocity feedback

Given the prototype system shown in Figure 1.2.11, suppose that this is augmented by measuring the output $y(t)$, differentiating to form $\dot{v}(t)$, and feeding back in parallel with the normal feedback a signal proportional to $\dot{y}(t)$: say $T\dot{y}(t)$. The schematic of this arrangement is shown in Figure 1.2.13. Then, by simple manipulation, the modified transfer function becomes

$$W'(s) = \frac{K}{s^2 + (a + KT)s + K}$$

whence the modified predominant time constant is given by $2/(a + TK)$. The designer effectively has another string to his bow in that manipulation of K and T is normally very much in his command.

A similar effect may be obtained by the incorporation of a *derivative* term to act on the error signal (Figure 1.2.14) and in this case the transfer function becomes

$$W'(s) = \frac{K(1 + Ts)}{s^2 + (a + KT)s + K}$$

It may be demonstrated that this derivative term when correctly adjusted can both stabilise the system and increase the speed of response. The control shown in Figure 1.2.14 is referred to as *proportional-plus-derivative* control and is very important.

1.2.8.4 Incorporation of integral control

Mention has previously been made of the effect of using integrators within the loop to reduce steady-state errors; a particular study with reference to input/output effects was given. In this section consideration is given to the

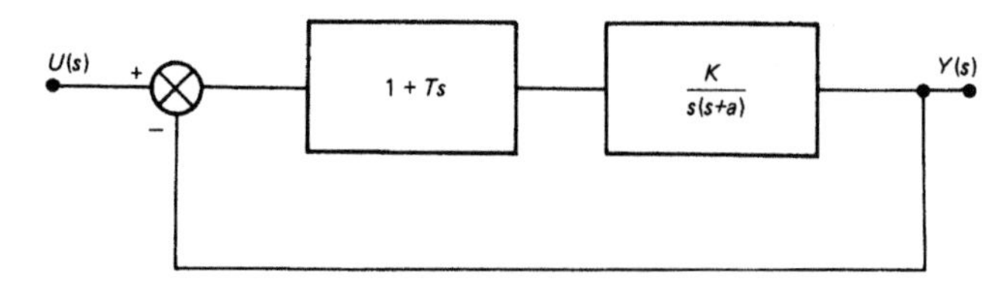

Fig. 1.2.14 Schematic diagram of the proportional-plus-derivative control system.

effects of disturbances injected into the loop, and we consider again the simple second-order system shown in Figure 1.2.11 but with a disturbance occurring between the amplifier and the plant dynamics. Appealing to superposition we can, without loss of generality, put $U(s) = 0$ and the transfer function between the output and the disturbance is then given by

$$\frac{Y(s)}{D(s)} = \frac{1}{s(s+a)+K}$$

Assuming that $d(t)$ is a unit step, $D(s) = 1/s$, and using the final-value theorem, $\lim_{t\to\infty} y(t)$ is obtained from

$$\lim_{t\to\infty} y(t) = \lim_{s\to\infty}\left[\frac{s}{[s(s+a)+K]s}\right] = \frac{1}{K}$$

and so the effect of this disturbance will always be present. By incorporating an integral control as shown in Figure 1.2.15, the output will, in the steady state, be unaffected by the disturbance, *viz*.

$$y(s) = \frac{Ts}{Ts^2(s+a)+K(1+Ts)}D(s)$$

and so

$$y_{ss} \to 0$$

This controller is called a *proportional-plus-integral* controller.

An unfortunate side-effect of incorporating integral control is that it tends to destabilise the system, but this can be minimised by careful choice of T. In a particular case it might be that *proportional-plus-integral-plus-derivative* (*PID*) *control* may be called for, the amount of each particular control type being carefully proportioned.

In the foregoing discussions we have seen, albeit by using specific simple examples, how the behaviour of a plant might be modified by use of certain techniques. It is hoped that this will leave the reader with some sort of feeling for what might be done before embarking on more general tools, which tend to appear rather rarefied and isolated unless a basic physical feeling for system behaviour is present.

1.2.9 Root-locus method

The root locus is merely a graphical display of the *variation of the poles of the closed-loop system* when some parameter, often the gain, is varied. The method is useful since the loci may be obtained, at least approximately, by straightforward application of simple rules, and possible modification to reshape the locus can be assessed.

Considering once again the unity-feedback system with the open-loop transfer function $KG(s) = Kb(s)/a(s)$, where $b(s)$ and $a(s)$ represent mth- and nth-order polynomials, respectively, and $n > m$, then the closed-loop transfer function may be written as

$$W(s) = \frac{KG(s)}{1+KG(s)} = \frac{Kb(s)}{a(s)+Kb(s)}$$

Note that the system is nth order and the zeros of the closed loop and the open loop are identical for unity feedback. The characteristic behaviour is determined by the roots of $1 + KG(s) = 0$ or $a(s) + Kb(s) = 0$. Thus, $G(s) = -(1/K)$ or $b(s)/a(s) = -(1/K)$.

Let s_r be a root of this equation; then

$$\text{mod}\left[\frac{b(s_r)}{a(s_r)}\right] = \frac{1}{K}$$

and

$$\text{phase}\left[\frac{b(s_r)}{a(s_r)}\right] = 180° + n360°$$

where n may take any integer value, including $n = 0$. Let $z_1, \ldots, z_m$ be the roots of the polynomial $b(s) = 0$, and $p_1, \ldots, p_n$ be the roots of the polynomial $a(s) = 0$. Then

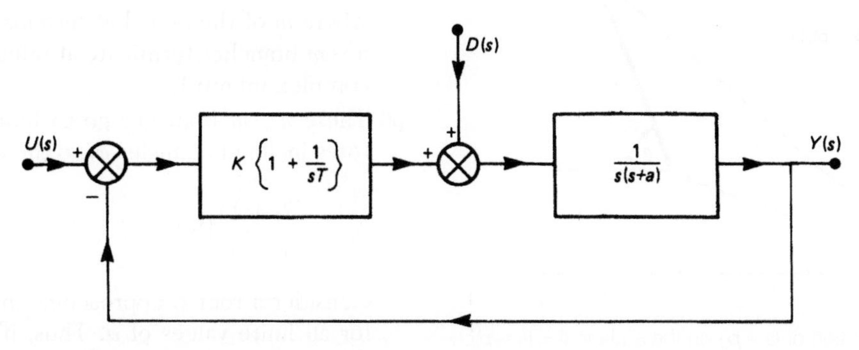

Fig. 1.2.15 Schematic diagram of the proportional-plus-integral control system.

$$b(s) = \prod_{i=1}^{m}(s - z_i)$$

and

$$a(s) = \prod_{i=1}^{n}(s - p_i)$$

Therefore

$$\frac{\prod_{i=1}^{m}|s_r - z_i|}{\prod_{i=1}^{n}|s_r - p_i|} = \frac{1}{K}, \quad \textit{the magnitude condition}$$

and

$$\sum_{i=1}^{m}\text{phase}(s_r - z_i) - \sum_{i=1}^{n}\text{phase}(s_r - p_i)$$
$$= 180° + n360°, \quad \textit{the angle or phase condition}$$

Now, given a complex number p_j, the determination of the complex number $(s - p_j)$, where s is some point in the complex plane, is illustrated in Figure 1.2.16, where the mod$(s - p_j)$ and phase$(s - p_j)$ are also illustrated. The determination of the magnitudes and phase angles for all the factors in the transfer function, for any s, can therefore be done graphically.

The complete set of all values of s, constituting the root locus may be constructed using the angle condition alone; once found, the gain K giving particular values of s_r may be easily determined from the magnitude condition.

Example Suppose that $G(s) = K/[(s + a)(s + b)]$, then it is fairly quickly established that the only sets of points satisfying the angle condition

$$-\text{phase}(s_r + a) - \text{phase}(s_r + b) = 180 + n360°$$

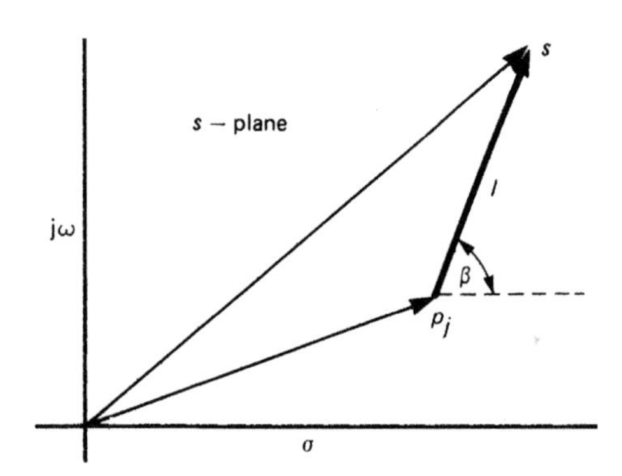

Fig. 1.2.16 Representation of $(s - p_j)$ on the s plane ($l = |s - p_j|$; $\beta = \angle(s - p_j)$).

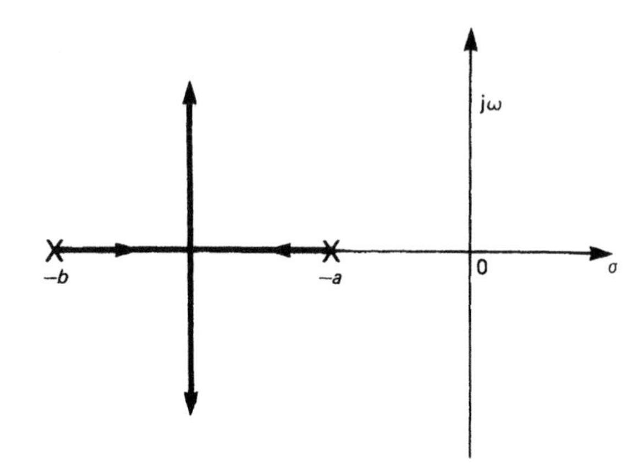

Fig. 1.2.17 Root-locus diagram for $KG(s) = K/[(s + a)(s + b)]$.

are on the line joining $-a$ to $-b$ and the perpendicular bisector of this line (Figure 1.2.17).

1.2.9.1 Rules for construction of the root locus

(1) The angle condition must be obeyed.

(2) The magnitude condition enables calibration of the locus to be carried out.

(3) The root locus on the real axis must be in sections to the left of an odd number of poles and zeros. This follows immediately from the angle condition.

(4) The root locus must be symmetrical with respect to the horizontal real axis. This follows because complex roots must appear as complex conjugate pairs.

(5) Root loci always emanate from the poles of the open-loop transfer function where $K = 0$. Consider $a(s) + Kb(s) = 0$; then $a(s) = 0$ when $K = 0$ and the roots of this polynomial are the poles of the open-loop transfer function. Note that this implies that there will be n branches of the root locus.

(6) m of the branches will terminate at the zeros for $K \to \infty$. Consider $a(s) + Kb(s) = 0$, or $(1/K)a(s) + b(s) = 0$, whence as $K \to \infty$, $b(s) \to 0$ and, since this polynomial has m roots, these are where m of the branches terminate. The remaining $n - m$ branches terminate at infinity (in general, complex infinity).

(7) These $n - m$ branches go to infinity along asymptotes inclined at angles ϕ_i to the real axis, where

$$\phi_i = \frac{(2_i + 1)}{n - m}180°, \quad i = 0, 1, ..., (n - m - 1)$$

Consider a root s_r approaching infinity, $(s_r - a) \to s_r$ for all finite values of a. Thus, if ϕ_i is the phase s_r, then each pole and each zero term of the transfer function term will contribute approximately ϕ_i and $-\phi_i$, respectively. Thus,

$$\phi_i(n-m) = 180° + i360°$$

$$\phi_i = \frac{(2i+1)}{n-m}180°,\ i = 0, 1, \ldots, (n-m-1)$$

(8) The centre of these asymptotes is called the '*asymptote centre*' and is (with good accuracy) given by

$$\sigma_A = \left(\sum_{i=1}^{n} p_i - \sum_{j=1}^{m} z_i\right) \Big/ (n-m)$$

This can be shown by the following argument. For very large values of s we can consider that all the poles and zeros are situated at the point σ_A on the real axis. Then the characteristic equation (for large values of s) may be written as

$$1 + \frac{K}{(s+\sigma_A)^{n-m}} = 0$$

or approximately, by using the binomial theorem,

$$1 + \frac{K}{s^{n-m} + (n-m)s^{n-m-1}\sigma_A} = 0$$

Also, the characteristic equation may be written as

$$1 + \frac{K \prod_{i=1}^{m}(s+z_i)}{\prod_{i=1}^{m}(s+p_i)} = 0$$

Expanding this for the first two terms results in

$$1 + \frac{K}{s^{n-m} + (a_{n-1} - b_{m-1})s^{n-m-1}} = 0$$

where

$$b_{m-1} = \sum_{i=1}^{m} z_i \quad \text{and} \quad a_{n-1} = \sum_{i=1}^{n} p_j$$

whence

$$(a_{n-1} - b_{m-1}) = (n-m)\sigma_A$$

$$\sigma_A = \frac{a_{n-1} - b_{m-1}}{n-m}$$

as required.

(9) When a locus breaks away from the real axis, it does so at the point where K is a local maximum. Consider the characteristic equation $1 + K[b(s)/a(s)] = 0$; then we can write $K = p(s)$, where $p(s) = -[a(s)/b(s)]$. Now, where two poles approach each other along the real axis they will both be real and become equal when K has the maximum value that will enable them both to be real and, of course, coincident. Thus, an evaluation of K around the breakaway point will rapidly reveal the breakaway point itself.

Example Draw the root locus for

$$KG(s) = \frac{K(s+1)}{s(s+2)(s+3)}$$

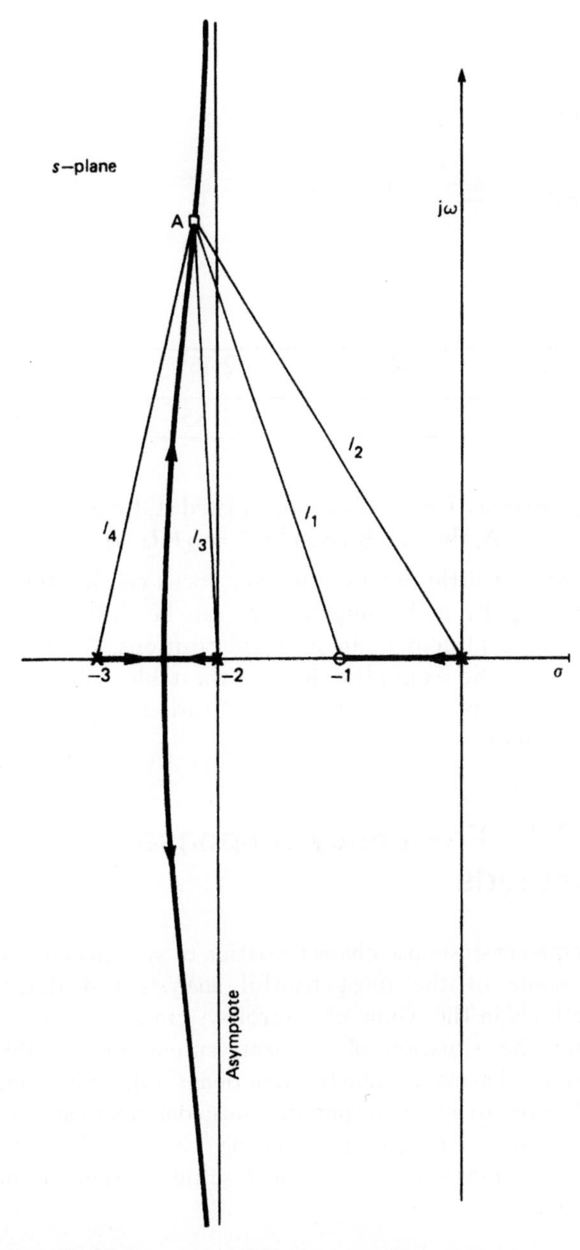

Fig. 1.2.18 Root-locus construction for $KG(s) = [K(s+1)]/[s(s+2)(s+3)]$.

Procedure (Figure 1.2.18):

(1) Plot the poles of the open-loop system (i.e. at $s = 0$, $s = -2$, $s = -3$).

(2) Plot the zeros of the system (i.e. at $z = -1$).

(3) Determine the sections on the real axis at which closed-loop poles can exist. Obviously these are between 0 and -1 (this root travels along the real axis between these values as K goes from $0 \rightarrow \infty$), and between -2 and -3 (two roots are moving towards each other as K increases and, of course, will break away).

(4) Angle of asymptotes

$$\phi_1 = \tfrac{1}{2} + 180° = 90°$$
$$\phi_2 = \tfrac{3}{2} \times 180° = 270°$$

(5) Centroid σ_A is located at

$$\sigma_A = \frac{-2 - 3 + 1}{2} = -2$$

(6) Breakaway point, σ_B

σ_B	-2.45	-2.465	-2.48
K	0.418	0.4185	0.418

(7) Modulus. For a typical root situated at, for example, point A, the gain is given by $K = l_2 l_3 l_4 / l_1$.

After a little practice the root locus can be drawn very rapidly and compensators can be designed by pole-zero placement in strategic positions. A careful study of the examples given in the table will reveal the trends obtainable for various pole-zero placements.

1.2.10 Frequency-response methods

Frequency-response characterisation of systems has led to some of the most fruitful analysis and design methods in the whole of control system studies. Consider the situation of a linear, autonomous, stable system, having a transfer function $G(s)$, and being subjected to a unit-magnitude sinusoidal input signal of the form $\exp(j\omega t)$, starting at $t = 0$. The Laplace transformation of the resulting output of the system is

$$C(s) = G(s)/(s - j\omega)$$

and the time domain solution will be

$$c(t) = G(j\omega)\exp(j\omega t) + \begin{pmatrix}\text{Terms whose exponential terms} \\ \text{correspond to the roots} \\ \text{of the denominator of } G(s)\end{pmatrix}$$

Since a stable system has been assumed, then the effects of the terms in the parentheses will decay away with time and so, after a sufficient lapse of time, the steady-state solution will be given by

$$c_{ss}(t) = G(j\omega)\exp(j\omega t)$$

The term $G(j\omega)$, obtained by merely substituting $j\omega$ for s in the transfer function form, is termed the *frequency-response function*, and may be written

$$G(j\omega) = |G(j\omega)| \angle G(j\omega)$$

where $|G(j\omega)| = \mathrm{mod}\, G(j\omega)$ and $\angle G(j\omega) = \text{phase } G(j\omega)$. This implies that the output of the system is also sinusoidal in magnitude $|G(j\omega)|$ with a phase-shift of $\angle G(j\omega)$ with reference to the input signal.

Example Consider the equation of motion

$$m\ddot{y} + bz + ky = f(t)$$

$$\frac{Y(s)}{F(s)} = G(s) = \frac{1}{ms^2 + bs + k}$$

If $f(t) = F_0 \exp(j\omega t)$, then

$$y_{ss}(t) = \frac{F_0 \exp(j\omega t)}{(k - \omega^2 m)^2 + j\omega b}$$

whence

$$y_{ss}(t) = \frac{F_0 \exp[j(\omega t - \phi)]}{\left(k - \omega^2 m\right)^2 + (b\omega)^2}$$

where $\phi = \arctan b\omega/(k - m\omega^2)$.

Within the area of frequency-response characterisation of systems three graphical techniques have been found to be particularly useful for examining systems and are easily seen to be related to each other. These techniques are based upon:

(1) The *Nyquist plot*, which is the locus of the frequency-response function plotted in the complex plane using ω as a parameter. It enables stability, in the closed-loop condition, to be assessed and also gives an indication of how the locus might be altered to improve the behaviour of the system.

(2) The *Bode diagram*, which comprises two plots, one showing the amplitude of the output frequency response (plotted in decibels) against the frequency ω (plotted logarithmically) and the other of phase angle θ of the output frequency response plotted against the same abscissa.

(3) The *Nichols chart*, a direct plot of amplitude of the frequency response (again in decibels) against the phase angle, with frequency ω as a parameter, but further enables the closed-loop frequency response to be read directly from the chart.

In each of these cases it is the *open-loop* steady-state frequency response, i.e. $G(j\omega)$, which is plotted on the diagrams.

1.2.10.1 Nyquist plot

The closed-loop transfer function is given by

$$\frac{C(s)}{R(s)} = \frac{G(s)}{1 + H(s)G(s)}$$

and the stability is determined by the location of the roots of $1 + H(s)G(s) = 0$, i.e. for stability no roots must have positive-real parts and so must not lie on the positive-real half of the complex plane. Assume that the open-loop transfer function $H(s)G(s)$ is stable and consider the contour C, the so-called 'Nyquist contour' shown in Figure 1.2.19, which consists of the imaginary axis plus a semicircle of large enough radius in the right half of the s plane such that any zeros of $1 + H(s)G(s)$ will be contained within this contour. This contour C_n is mapped via $1 + H(s)G(s)$ into another curve γ into the complex plane s'. It follows immediately from complex variable theory that the closed loop will be stable if the curve γ does not encircle the origin in the s' plane and unstable if it encircles the origin or passes through the origin. This result is the basis of the celebrated Nyquist stability criterion. It is rather more usual to map not $1 + H(s)G(s)$ but $H(s)G(s)$; in effect this is merely a change of origin from $(0, 0)$ to $(-1, 0)$, i.e. we consider curve γ'_n.

The statement of the stability criterion is that the closed-loop system will be stable if the mapping of the contour C_n by the open-loop frequency-response function $H(j\omega)G(j\omega)$ does not enclose the so-called critical point $(-1, 0)$. Actually further simplification is normally possible, for:

(1) $|H(s)G(s)| \to 0$ as $|s| \to \infty$, so that the very large semicircular boundary maps to the origin in the s' plane.

(2) $H(-j\omega)G(-j\omega)$ is the complex conjugate of $H(j\omega)G(j\omega)$ and so the mapping of $H(-j\omega)G(-j\omega)$ is merely the mirror image of $H(j\omega)G(j\omega)$ in the real axis.

(3) Note: $H(j\omega)G(j\omega)$ is merely the frequency-response function of the open loop and may even be directly measurable from experiments. Normally we are mostly interested in how this behaves in the vicinity of the $(-1, 0)$ point and, therefore, only a limited frequency range is required for assessment of stability.

The mathematical mapping ideas stated above are perhaps better appreciated practically by the so-called *left-hand rule* for an open-loop stable system, which reads as follows: if the open-loop sinusoidal response is traced out going from low frequencies towards higher frequencies, the closed loop will be stable if the critical

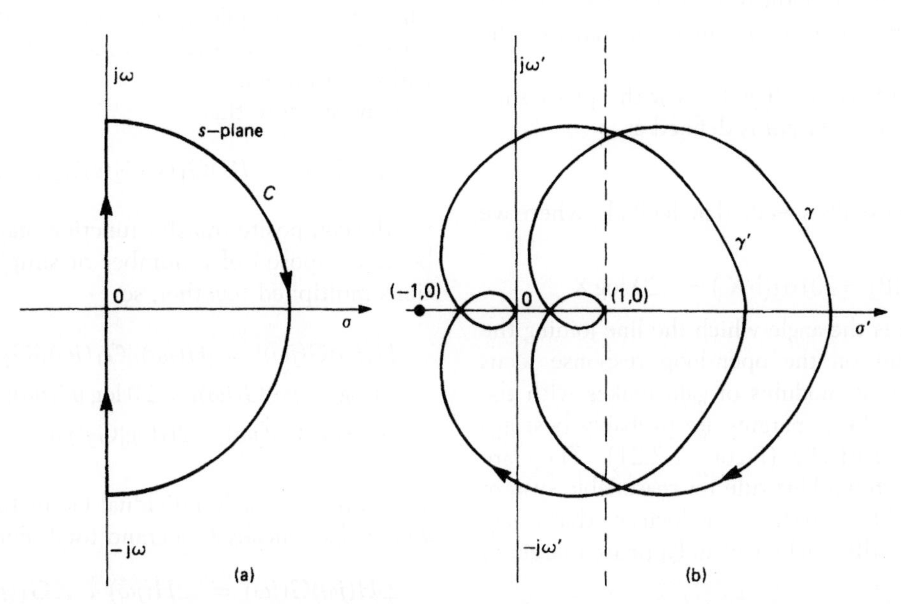

Fig. 1.2.19 Illustration of Nyquist mapping: (a) mapping contour on the s plane; (b) resulting mapping of $1 + H(s)G(s) = 0$ and the shift of the origin.

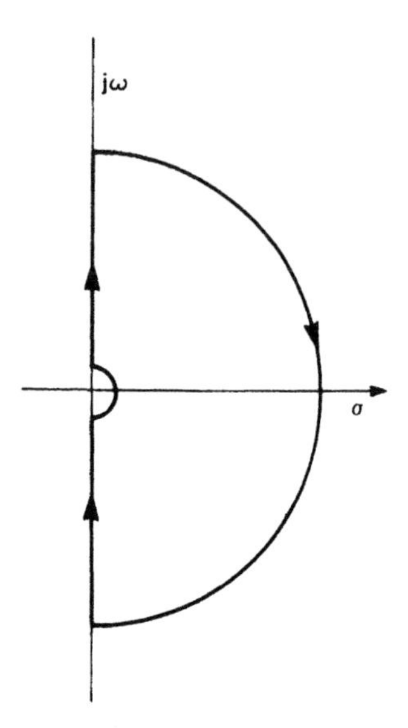

Fig. 1.2.20 Modification of the mapping contour to account for poles appearing at the origin.

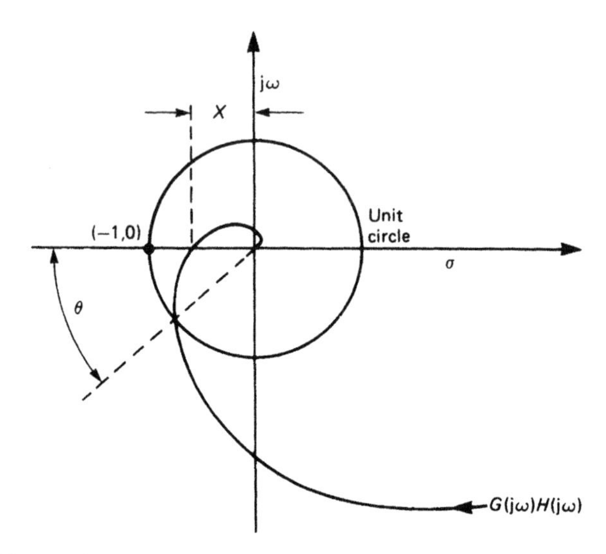

Fig. 1.2.21 Illustration of the gain and phase margins. Gain margin = 1/X; phase margin = 0.

point (−1, 0) lies on the left of all points on $H(j\omega)G(j\omega)$. If this plot passes through the critical point, or if the critical point lies on the right-hand side of $H(j\omega)G(j\omega)$, the closed loop will be unstable.

If the open loop has poles that actually lie on the imaginary axis, e.g. integrator $1/s$, then the contour is indented as shown in Figure 1.2.20 and the above rule still applies to this modification.

1.2.10.1.1 Relative stability criteria

Obviously the closer the $H(j\omega)G(j\omega)$ locus approaches the critical point, the more critical is the consideration of stability, i.e. we have an indication of relative stability, given a measure by the gain and phase margins of the system.

If the modulus of $H(j\omega)G(j\omega) = X$ with a phase shift of 180°, then the *gain margin* is defined as

$$\text{Gain margin} = 1/X$$

The gain margin is usually specified in decibels, where we have

$$\text{Gain margin (dB)} = 20\log(1/X) = -20\log X$$

The *phase margin* is the angle which the line joining the origin to the point on the open-loop response locus corresponding to unit modulus of gain makes with the negative-real axis. These margins are probably best appreciated diagrammatically (Figure 1.2.21). They are useful, since a rough working rule for reasonable system damping and stability is to shape the locus so that a gain margin of at least 6 dB is achieved and a phase margin of about 40°.

Examples of the Nyquist plot are shown in Figure 1.2.22. Although from such plots the modifications necessary to achieve more satisfactory performance can be easily appreciated, precise compensation arrangements are not easily determined, since complex multiplication is involved and an appeal to the Bode diagram can be more valuable.

1.2.10.2 Bode diagram

As mentioned above, the Bode diagram is a logarithmic presentation of the frequency response and has the advantage over the Nyquist diagram that individual factor terms may be added rather than multiplied, the diagram can usually be quickly sketched using asymptotic approximations and several decades of frequency may be easily considered.

Now suppose that

$$H(s)G(s) = H(s)G_1(s)G_2(s)G_3(s)\ldots$$

i.e. the composite transfer function may be thought of as being composed of a number of simpler transfer functions multiplied together, so

$$|H(j\omega)G(j\omega)| = |H(j\omega)||G_1(j\omega)||G_2(j\omega)||G_3(j\omega)|\ldots$$
$$20\log|H(j\omega)G(j\omega)| = 20\log|H(j\omega)| + 20\log|G_1(j\omega)| + 20\log|G_2(j\omega)| + 20\log|G_3(j\omega)| + \cdots$$

This is merely each individual factor (in decibels) being *added* algebraically to a grand total. Further,

$$\angle H(j\omega)G(j\omega) = \angle H(j\omega) + \angle G_1(j\omega) + \angle G_2(j\omega) + \angle G_3(j\omega) + \ldots$$

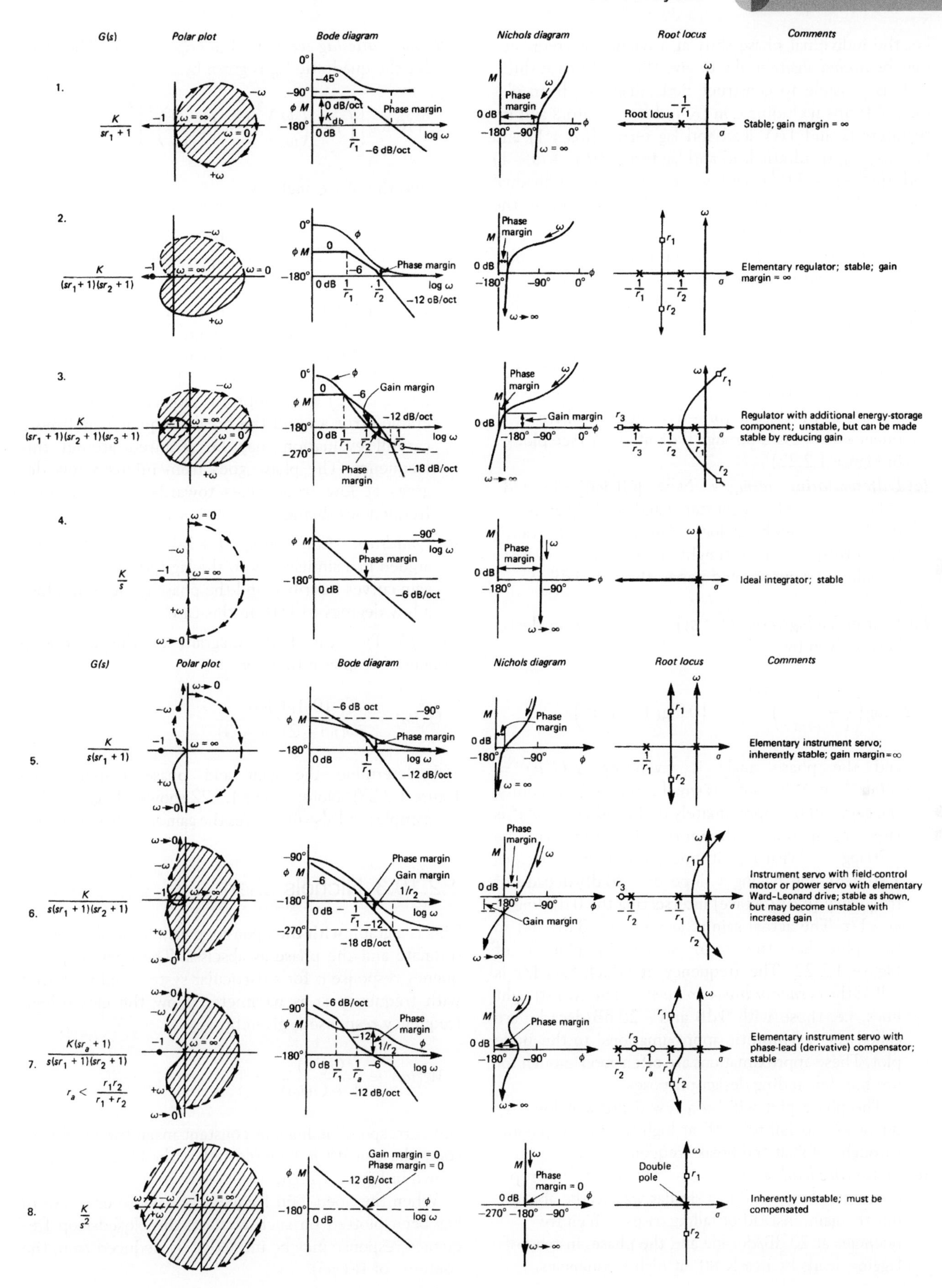

Fig. 1.2.22 Transfer function plots for typical transfer functions.

i.e. the individual phase shift at a particular frequency may be *added* algebraically to give the total phase shift.

It is possible to construct Bode diagrams from elemental terms including gain (K), differentiators and integrators (s and $1/s$), lead and lag terms ($(as+1)$ and $(1+as)^{-1}$), quadratic lead and lag terms ((bs^2+cs+1) and $(bs^2+cs+1)^{-1}$), and we consider the individual effects of their presence in a transfer function on the shape of the Bode diagram.

(a) *Gain term, K* The gain in decibels is simply $20 \log K$ and is *frequency independent;* it merely raises (or lowers) the combined curve $20 \log K$ dB.

(b) *Integrating term, 1/s* Now $|G(j\omega)| = 1/\omega$ and $\angle G(j\omega) = -90°$ (a constant) and so the gain in decibels is given by $20 \log(1/\omega) = -20 \log \omega$. On the Bode diagram this corresponds to a straight line with slope -20 dB/decade (or -6 dB/octave) of frequency and passes through 0 dB at $\omega = 1$ (see plot 4 in Figure 1.2.22).

(c) *Differentiating term, s* Now $|G(j\omega)| = \omega$ and $\angle G(j\omega) = -90°$ (a constant) and so the gain in decibels is given by $20 \log \omega$. On the Bode diagram this corresponds to a straight line with slope 20 dB/decade of frequency and passes through 0 dB at $\omega = 1$.

(d) *First-order lag term,* $(1+s\tau)^{-1}$ The gain in decibels is given by

$$20 \log\left(\frac{1}{1+\omega^2\tau^2}\right)^{1/2} = -10 \log\left(1+\omega^2\tau^2\right)$$

and the phase angle is given by $\angle G(j\omega) = -\tan^{-1} \omega\tau$. When $\omega^2\tau^2$ is small compared with unity, the gain will be approximately 0 dB, and when $\omega^2\tau^2$ is large compared with unity, the gain will be $-20 \log \omega\tau$. With logarithmic plotting this specifies a straight line having a slope of -20 dB/decade of frequency (6 dB/octave) intersecting the 0 dB line at $\omega = 1/\tau$. The actual gain at $\omega = 1/\tau$ is -3 dB and so the plot has the form shown in plot 1 of Figure 1.2.22. The frequency at which $\omega = 1/\tau$ is called the *corner or break frequency.* The two straight lines, i.e. those with 0 dB and -20 dB/decade, are called the 'asymptotic approximations' to the Bode plot. These approximations are often good enough for not too demanding design purposes.

The phase plot will lag a few degrees at low frequencies and fall to $-90°$ at high frequency, passing through $-45°$ at the break frequency.

(e) *First-order lead term,* $1+\omega\tau$ The lead term properties may be argued in a similar way to the above, but the gain, instead of falling, rises at high frequencies at 20 dB/decade and the phase, instead of lagging, leads by nearly 90° at high frequencies.

(f) *Quadratic-lag term,* $1/(1+2\tau\zeta s+\tau^2 s^2)$ The gain for the quadratic lag is given by

$$-10 \log\left[\left(1-\left(\frac{\omega}{\omega_n}\right)^2\right)^2+\left(2\zeta\frac{\omega}{\omega_n}\right)^2\right]$$

and the phase angle by

$$\angle G(j\omega) = \arctan\left[-\frac{2\zeta(\omega/\omega_n)}{1-(\omega/\omega_n)^2}\right]$$

where $\tau = 1/\omega_n$. At low frequencies the gain is approximately 0 dB and at high frequencies falls at -40 dB/decade. At the break frequency $\omega = 1/\tau$ the actual gain is $20 \log(1/2\zeta)$. For low damping (say $\zeta < 0.5$) an asymptotic plot can be in considerable error around the break frequency and more careful evaluation may be required around this frequency. The phase goes from minus a few degrees at low frequencies towards $-180°$ at high frequencies, being $-90°$ at $\omega = 1/\tau$.

(g) *Quadratic lead term,* $1+2\tau\zeta s+\tau^2 s^2$ This is argued in a similar way to the lag term with the gain curves inscribed and the phase going from plus a few degrees to 180° in this case.

Example Plot the Bode diagram of the open-loop frequency-response function

$$G(j\omega) = \frac{10(1+j\omega)}{j\omega(j\omega+2)(j\omega+3)}$$

and determine the gain and phase margins (see Figure 1.2.23). Note: Figure 1.2.22 shows a large number of examples and also illustrates the gain and phase margins.

1.2.10.3 Nichols chart

This is a graph with the open-loop gain in decibels as coordinate and the phase as abscissa. The open-loop frequency response is for a particular system and is plotted with frequency ω as parameter. Now the closed-loop frequency response is given by

$$W(j\omega) = \frac{G(j\omega)}{1+G(j\omega)}$$

and corresponding lines of constant magnitude and constant phase of $W(j\omega)$ are plotted on the Nichols chart as shown in Figure 1.2.24.

When the open-loop frequency response of a system has been plotted on such a chart, the closed-loop frequency response may be immediately deduced from the contours of $W(j\omega)$.

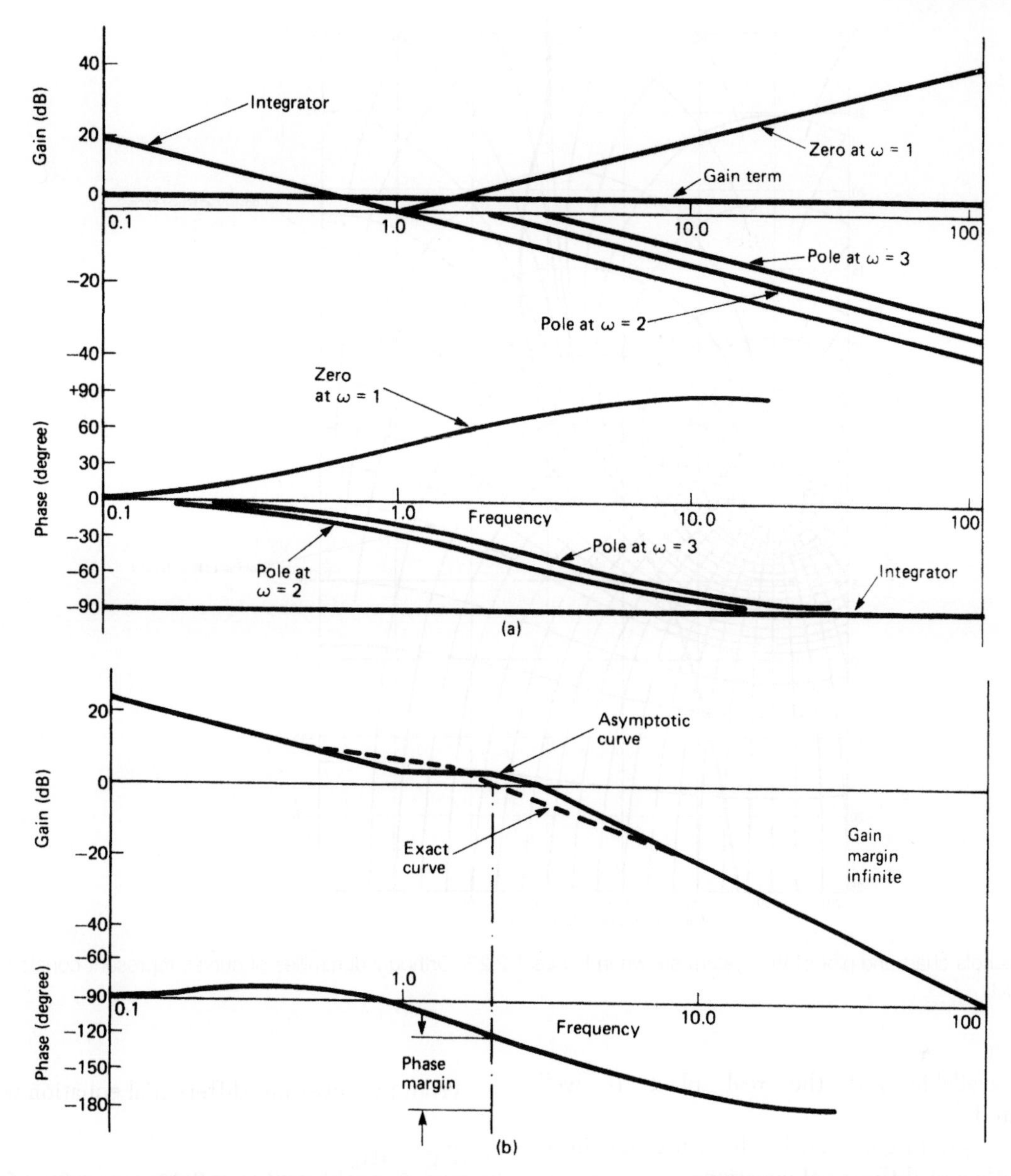

Fig. 1.2.23 (a) Gain and phase curves for individual factors (see Figure 1.2.18). (b) Composite gain and phase curves. Note that the phase margin is about 60°, and the gain margin is infinite because the phase tends asymptotically to −180°.

1.2.11 State-space description

Usually in engineering, when analysing time-varying physical systems, the resulting mathematical models are in differential equation form. Indeed, the introduction of the Laplace transformation, and similar techniques, leading to the whole edifice of transfer-function-type analysis and design methods are, essentially, techniques for solving, or manipulating to advantage, differential equation models. In the state-space description of systems, which is the concern of this section, the models are left in the differential equation form, but rearranged into the form of a set of first-order simultaneous differential equations. There is nothing unique to systems analysis in doing this, since this is precisely the required form that differential equations are placed in if they are to be integrated by means of many common numerical techniques, e.g. the Runge–Kutta methods. Most of the interest in the state-space form of studying control systems stems from the 1950s, and intensive research work in this area has continued since then; however, much of it is of a highly theoretical nature. It is arguable that these methods have yet to fulfill the hopes and aspirations of the research workers who developed them. The early expectation was that they would quickly supersede classical techniques. This has been very far from true, but they do have a part to play, particularly if there are good mathematical models of

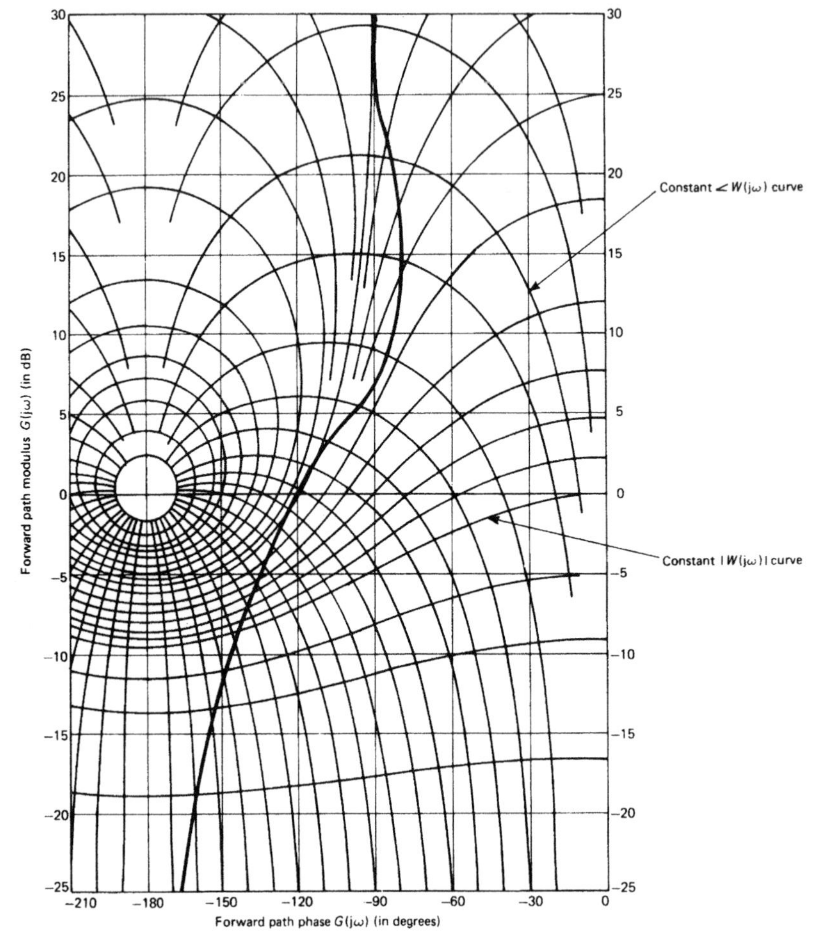

Fig. 1.2.24 Nichols chart and plot of the system shown in Figure 1.2.23. Orthogonal families of curves represent constant *WfJJ* and constant *LW*(*jω*).

the plant available and the real plant is well instrumentated.

Consider a system governed by the nth order linear constant-coefficient differential equation

$$\frac{d^n y}{dt^n} + \cdots + a_1 \frac{dy}{dt} + a_0 y = ku(t)$$

where y is the dependent variable and $u(t)$ is a time-variable forcing function.

Let $y = x_1$, then

$$\frac{dy}{dt} = \frac{dx_1}{dt} = x^2$$

say, and

$$\frac{d^2 y}{dt^2} = \frac{dx_2}{dt} = x_3$$

$$\frac{d^{n-1} y}{dt^{n-1}} = \frac{dx_{n-1}}{dt} = x_n$$

From the governing differential equation we can write

$$\frac{d^n y}{dt^n} = \frac{dx_n}{dt} = -a_0 x_1 - a_1 x_2 - \cdots - a_{n-1} x_n + ku(t)$$

i.e. the nth order differential equation has been transformed into n first-order equations. These can be arranged into matrix form:

$$\begin{bmatrix} \dot{x}_1 \\ \dot{x}_2 \\ \vdots \\ \dot{x}_{n-1} \\ \dot{x}_n \end{bmatrix} = \begin{bmatrix} 0 & 1 & 0 & \cdots & & \\ 0 & 0 & 1 & 0 & \cdots & \\ & & & 0 & 1 & \\ -a_0 & -a_1 & \cdots & & -a_{n-1} \end{bmatrix} \begin{bmatrix} x_1 \\ x_2 \\ \vdots \\ x_n \end{bmatrix} + \begin{bmatrix} 0 \\ 0 \\ 0 \\ k \end{bmatrix} u(t) \qquad (1.2.3)$$

which may be written in matrix notation as

$$\dot{\mathbf{x}} = \mathbf{A}\mathbf{x} + \mathbf{b}u(t)$$

where $\mathbf{x} = [x_1, \ldots, x_n]^T$ and is called the 'state vector', $\mathbf{b} = [0, 0, \ldots, k]^T$ and $\mathbf{A}$ is the $n \times n$ matrix pre-multiplying $\mathbf{x}$ on the right-hand side of Equation (1.2.3).

It can be shown that the eigenvalues of **A** are equal to the characteristic roots of the governing differential equation which are also equal to the poles of the transfer function $Y(s)/U(s)$. Thus the time behaviour of the matrix model is essentially governed by the position of the eigenvalues of the **A** matrix (in the complex plane) in precisely the same manner as the poles govern the transfer function behaviour. Hence, if these eigenvalues do not lie in acceptable positions in this plane, the design process is to somehow modify the **A** matrix so that the corresponding eigenvalues do have acceptable positions (cf. the placement of closed-loop poles in the s plane).

Example Consider a system governed by the general second-order linear differential equation

$$\frac{d^2y}{dt^2} + 2\zeta\omega_n\frac{dy}{dt} + \omega_n^2 y = \omega_n^2 u$$

Let $y = x_1$, then

$$\frac{dy}{dt} = \frac{dx_1}{dt} = x_2$$

and so

$$\frac{dx_2}{dt} = -\omega_n^2 x_1 - 2\zeta\omega_n x_2 + \omega_n^2 u$$

or

$$\begin{bmatrix} \dot{x}_1 \\ \vdots \\ \dot{x}_2 \end{bmatrix} = \begin{bmatrix} 0 & 1 \\ -\omega_n^2 & -2\zeta\omega_n \end{bmatrix} \begin{bmatrix} x_1 \\ x_2 \end{bmatrix} + \begin{bmatrix} 0 \\ \omega_n^2 \end{bmatrix} u \qquad (1.2.4)$$

The eigenvalues of the **A** matrix are given by the solution to the equation $\lambda^2 + 2\zeta\omega_n\lambda + \omega_n^2 = 0$, i.e.

$$\lambda_{1,2} = -\zeta\omega_n \pm \omega_n\sqrt{\zeta^2 - 1}$$

Now let $u = r - k_1x_1 - k_2x_2$ where r is an arbitrary, or reference, value or input, and k_1 and k_2 are constants. Note this is a feedback arrangement, since u has become a linear function of the state variables which, in a dynamic system, might be position and velocity. Substituting for u in Equation (1.2.3), gives

$$\begin{bmatrix} \dot{x}_1 \\ \vdots \\ \dot{x}_2 \end{bmatrix} = \begin{bmatrix} 0 & 1 \\ -\omega_n^2(1+k_1) & -\omega_n(2\zeta + \omega_n k_2) \end{bmatrix} \begin{bmatrix} x_1 \\ x_2 \end{bmatrix} + \begin{bmatrix} 0 \\ \omega_n^2 \end{bmatrix} r$$

The eigenvalues of the **A** matrix are given by the roots of $\lambda^2 + (2\zeta\omega_n + \omega_n^2 k_2)\lambda + \omega_n^2(1+k_1) = 0$ and, by choosing suitable values for k_1 and k_2 (the feedback factors), the eigenvalues can be made to lie in acceptable positions in the complex plane. Note that, in this case, k_1 alters the effective undamped natural frequency, and k_2 alters the effective damping of the second-order system.

If the governing differential equation has derivatives on the right-hand side, then the derivation of the first-order set involves a complication. Overcoming this is easily illustrated by an example. Suppose that

$$\frac{d^2y}{dt^2} + a_1\frac{dy}{dt} + a_2 y = b_0 u + b_1\frac{du}{dt}$$

Let $y = x_1$, and

$$\frac{dy}{dt} = \frac{dx_1}{dt} = x_2 + b_1 u$$

then

$$\frac{d^2y}{dt^2} = \frac{dx_2}{dt} + b_1\frac{du}{dt}$$

$$= -a_1(x_2 + b_1u) - a_0x + b_0u + b_1\frac{du}{dt}$$

$$\begin{bmatrix} \dot{x}_1 \\ \dot{x}_2 \end{bmatrix} = \begin{bmatrix} 0 & 1 \\ -a_0 & -a_1 \end{bmatrix} \begin{bmatrix} x_1 \\ x_2 \end{bmatrix} + \begin{bmatrix} b_1 \\ b_0 - a_1b_1 \end{bmatrix} u$$

Note that care may be necessary in interpreting the x derivatives in a physical sense.

The state-space description is also a convenient way of dealing with multi-input/multi-output systems. A simple example is shown in Figure 1.2.25, where $U_1(s)$ and $U_2(s)$ are the inputs and $Y_1(s)$ and $Y_2(s)$ are the corresponding outputs, and so

$$Y_1(s) = \frac{k_1}{s+a_1}U_1(s) + \frac{k_3}{s+a_3}U_2(s)$$

and

$$Y_2(s) = \frac{k_2}{s+a_1}U_1(s) + \frac{k_4}{s+a_4}U_2(s)$$

The first of these two equations may be written as

$$\left[s^2 + s(a_1 + a_2) + a_1a_3\right]Y_1(s) = k_1U_1(s) + k_3U_2(s)$$

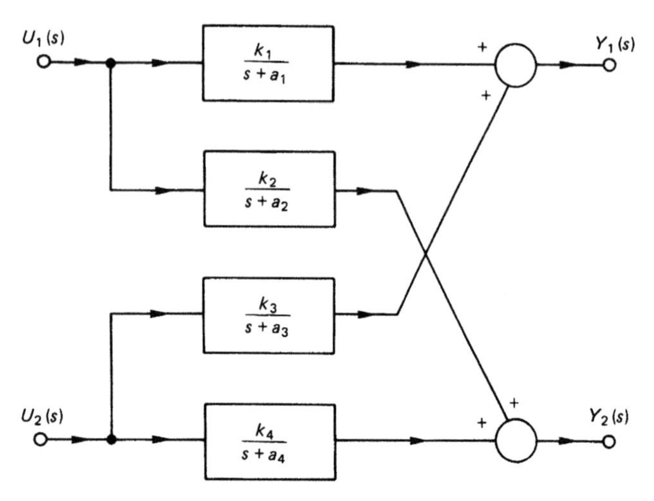

Fig. 1.2.25 Block diagram of a two-input/two-output multivariable system.

or

$$\frac{d^2y_1}{dt^2} + (a_1 + a_2)\frac{dy_1}{dt} + a_1a_3y_1 = k_1u_1 + k_2u_2$$

let $y_1 = x_1$, then

$$\frac{dy_1}{dt} = \frac{dx_1}{dt} = x_2$$

and

$$\frac{d^2y_1}{dt^2} = \frac{dx_2}{dt}$$
$$= -(a_1 + a_2)x_2 - a_1a_3x_1 + k_1u_1 + k_3u_2$$

Similarly, for the second of the two equations, writing

$$y_2 = x_3 \quad \text{and} \quad \frac{dy_2}{dt} = \frac{dx_3}{dt} = x_4$$

leads to

$$\frac{d^2y_2}{dt^2} = \frac{dx_4}{dt} = -(a_2 + a_4)x_4 - a_2a_4x_3 + k_2u_1 + k_4u_2$$

Whence the entire set may be written as

$$\begin{bmatrix} \dot{x}_1 \\ \dot{x}_2 \\ \dot{x}_3 \\ \dot{x}_4 \end{bmatrix} = \begin{bmatrix} 0 & 1 & 0 & 0 \\ -a_1a_3 & -(a_1 - a_3) & 0 & 0 \\ 0 & 0 & 0 & 1 \\ 0 & 0 & -a_2a_4 & -(a_2 + a_4) \end{bmatrix} \times \begin{bmatrix} \dot{x}_1 \\ \dot{x}_2 \\ \dot{x}_3 \\ \dot{x}_4 \end{bmatrix} + \begin{bmatrix} 0 & 0 \\ k_1 & k_3 \\ 0 & 0 \\ k_2 & k_4 \end{bmatrix} \begin{bmatrix} u_1 \\ u_2 \end{bmatrix}$$

The problem is now how to specify u_1 and u_2 (e.g. a linear combination of state variables similar to the simple second-order system above), so as to make the plant behave in an acceptable manner. It must be pointed out that the theory of linear matrix-differential equations is an extremely well developed mathematical topic and has been extensively plundered in the development of state-space methods. Thus a vast literature exists, and this is not confined to linear systems. Such work has, among other things, discovered a number of fundamental properties of systems (for example, controllability and observability); these are well beyond the scope of the present treatment. The treatment given here is a very short introduction to the fundamental ideas of the state-space description.

1.2.12 Sampled-data systems

Sampled-data systems are ones in which signals within the control-loop are sampled at one or more places. Some sort of sampling action may be inherent in the very mode of operation of some of the very components comprising the plant, e.g. thyristor systems, pulsed-radar systems and reciprocating internal combustion engines. Moreover, sampling is inevitable if a digital computer is used to implement the control laws, and/or used in condition monitoring operations. Nowadays, digital computers are routinely used in control-system operation for reasons of cheapness and versatility, e.g. they may be used not only to implement the control laws, which can be changed by software alterations alone, but also for sequencing control and interlocking in, say, the start up and safe operation of complex plant. Whatever the cause, sampling complicates the mathematical analysis and design of systems.

Normally most of the components, comprising the system to be controlled, will act in a continuous (analog) manner, and hence their associated signals will be continuous. With the introduction of a digital computer it is necessary to digitise the signal, by an analog-to-digital converter before the signal enters the computer. The computer processes this digital sequence, and then outputs another digital sequence which, in turn, passes to a digital-to-analog converter. This process is shown schematically in Figure 1.2.26.

In this diagram the sampling process is represented by the periodic switch (period T), which at each sampling instant is closed for what is regarded as an infinitesimal time. The digital-to-analog process is represented by the hold block. Thus the complete system is a hybrid one, made up of an interconnection of continuous and discrete devices. The most obvious way of representing the system mathematically is by a mixed difference-differential equation set. However, this makes a detailed analysis of the complete system difficult.

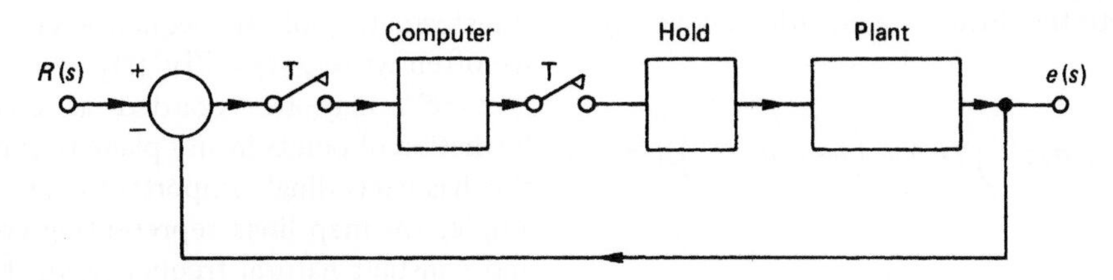

Fig. 1.2.26 General arrangement of a sampled-data system.

Fortunately, provided the investigator or system designer is prepared to accept knowledge of the system's behaviour at the instants of sampling only, a comparatively simple approach having great similarity to that employed for wholly continuous systems is available. At least for early stages of the analysis or design proposal, the added complications involved in this process are fairly minor. Further, the seemingly severe restriction of knowing the system's behaviour at the instants of sampling only is normally quite acceptable; for example, the time constants associated with the plant will generally be much longer than the periodic sampling time, so the plant effectively does not change its state significantly in the periodic time. The sampling time period is a parameter which often can be chosen by the designer, who will want sampling to be fast enough to avoid aliasing problems; however, the shorter the sampling period the less time the computer has available for other loops. Suffice it to say that the selection of the sampling period is normally an important matter.

If we take a continuous signal $y(t)$, say, and by the periodic sampling process convert it into a sequence of values $y(n)$, where n represents the nth sampling period, then the sequence $y(n)$ becomes the mathematical entity we manipulate, and the values of $y(t)$ between these samples will not be known. However, if at an early stage it is essential to know the inter-sample behaviour of the system with some accuracy, then advance techniques are available for this purpose.[1] In addition, it is now fairly routine to simulate control system behaviour before implementation, and a good simulation package should be capable of illustrating the inter-sample behaviour.

We need techniques for mathematically manipulating sequences, and these are discussed in the following section.

1.2.13 Some necessary mathematical preliminaries

1.2.13.1 The *z* transformation

This transformation plays the equivalent role in sampled-data system studies as the Laplace transformation does in the case of continuous systems; indeed, these two transformations are mathematically related to each other. It is demonstrated below that the behaviour of sampled-data systems at the sampling instant is governed mathematically by difference equations, e.g. a linear system might be governed by the equation

$$\begin{aligned} y(n) + a_1 y(n-1) + a_2 y(n-2) \\ = b_1 x(n) + b_2 x(n-1) \end{aligned}$$

where, in the case of $y(n)$, the value of a variable at instant n is in fact dependent on a linear combination of its previous two values and the current and previous values of an independent (forcing variable) $x(n)$. In a similar way to using the Laplace transformation to convert linear differential equations to transfer-function form, the z transformation is used to convert linear difference equations into the so-called 'pulse transfer-function form'. The definition of the z transformation of a sequence $y(n)$, $n = 0, 1, 2, \ldots$, is

$$Z[y(n)] = Y(z) = \sum_{n=0}^{\infty} y(n) z^{-n}$$

The z transformations of commonly occurring sequences are listed in Table 1.2.1, and a simple example will illustrate how such transformations may be found.

Suppose $y(n) = nT$ $(n = 0, 1, 2, \ldots)$ such a sequence would be obtained by sampling the continuous ramp function $y(t) = t$, at intervals of time T. Then, by definition,

$$\begin{aligned} Z[y(n)] &= y(z) = 0 + Tz^{-1} + 2Tz^{-2} + \cdots \\ &= T(z^{-1} + 2z^{-2} + \cdots) \\ &= \frac{Tz}{(z-1)^2} \end{aligned}$$

It can also be shown that

$$Z[y(n-1)] = z^{-1} Z[y(n)] = z^{-1} Y(z)$$

and

$$Z[y(n-2)] = z^{-2} Z[y(n)] = z^{-2} Y(z)$$

Then, applying this to the difference equation above, we have

$$Y(z) = -\left(a_1z^{-1} + a_2z^{-2}\right)Y(z) + \left(b_0 + b_1z^{-1}\right)X(z)$$

or

$$Y(z) = \frac{(b_0 + b_1z^{-1})X(z)}{(1 + a_1z^{-1} + a_2z^{-2})}$$

So that, if $x(n)$ or $X(z)$ is given, $Y(z)$ can be rearranged into partial fraction form, and $y(n)$ determined from the table. For example, suppose that

$$Y(z) = \frac{z(z - 0.25)}{(z - 1)(z - 0.5)}$$

then

$$\frac{Y(z)}{z} = \frac{1.5}{z - 1} - \frac{0.5}{z - 0.5}$$

or

$$Y(z) = \frac{1.5z}{z - 1} - \frac{0.5z}{z - 0.5}$$

Whence, from the tables we see that

$$y(n) = 1.5 - 0.5\exp(-0.60n)$$

The process of dividing $Y(z)$ by z before taking partial fractions is important, as most tabulated values of the transformation have z as a factor in the numerator, and the partial function expansion process needs the order of the denominator to exceed that of the numerator.

An alternative method of approaching the z transform is to assume that the sequence to be transformed is a direct consequence of sampling a continuous signal using an impulse modulator. Thus a given signal $y(t)$ is sampled with periodic time T, to give the assumed signal $y^*(t)$, where

$$y^*(t) = y(o)\delta(t) + y(T)\delta(t - T) + y(2T)\delta(t - 2T) + \cdots$$

where $\delta(t)$ is the delta function.

Taking the Laplace transformation of $y^*(t)$ gives the series

$$\mathscr{L}[y^*(t)] = y(o)y(T)e^{-sT} + y(2T)e^{-2sT} + \cdots$$

On making the substitution $e^{sT} = z$, then the resulting series is identical to that obtained by taking the z transformation of the sequence $y(n)$. For convenience, we often write $Y(z) = Z[y^*(t)]$.

$z = e^{sT}$ may be regarded as constituting a transformation of points in an s plane to those in a z plane, and this has exceedingly important consequences. If, for example, we map lines representing constant damping ζ, and constant natural frequency ω_n, for a system represented in an s plane onto a z plane, we obtain Figure 1.2.27.

There are important results to be noted from this diagram.

(1) The stability boundary in the s plane (i.e. the imaginary axis) transforms into the unit circle $|z| = 1$ in the z plane.

(2) Points in the z plane indicate responses relative to the periodic sampling time T.

(3) The negative real axis of the z plane always represents half the sampling frequency ω_s, where $\omega_s = 2\pi/T$.

(4) Vertical lines (i.e. those with constant real parts) in the left-half plane of the s plane map into circles *within* the unit circle in the z plane.

(5) Horizontal lines (i.e. lines of constant frequency) in the s plane map into radial lines in the z planes.

(6) The mapping is not one-to-one; and frequencies greater than $\omega_s/2$ will coincide on the z plane with corresponding points below this frequency. Effectively this is a consequence of the Nyquist sampling theorem which states, essentially, that faithful reconstruction of a sampled signal cannot be achieved if the original continuous signal contained frequencies greater than one-half the sampling frequency.

A vitally important point to note is that *all the roots* of the denominator of a pulse transfer function of a system must fall *within* the unit circle, on the z plane, if the system is to be stable; this follows from (1) above.

1.2.14 Sampler and zero-order hold

The sampler produces a series of discrete values at the sampling instant. Although in theory these samples exist for zero time, in practice they can be taken into the digital computer and processed. The output from the digital computer will be a sequence of samples with, again in theory, each sample existing for zero time. However, it is necessary to have a continuous signal constructed from this output, and this is normally done using a *zero-order hold.* This device has the property that, as each sample (which may be regarded as a delta function) is presented to its input, it presents the strength of the delta function at its output until the next sample

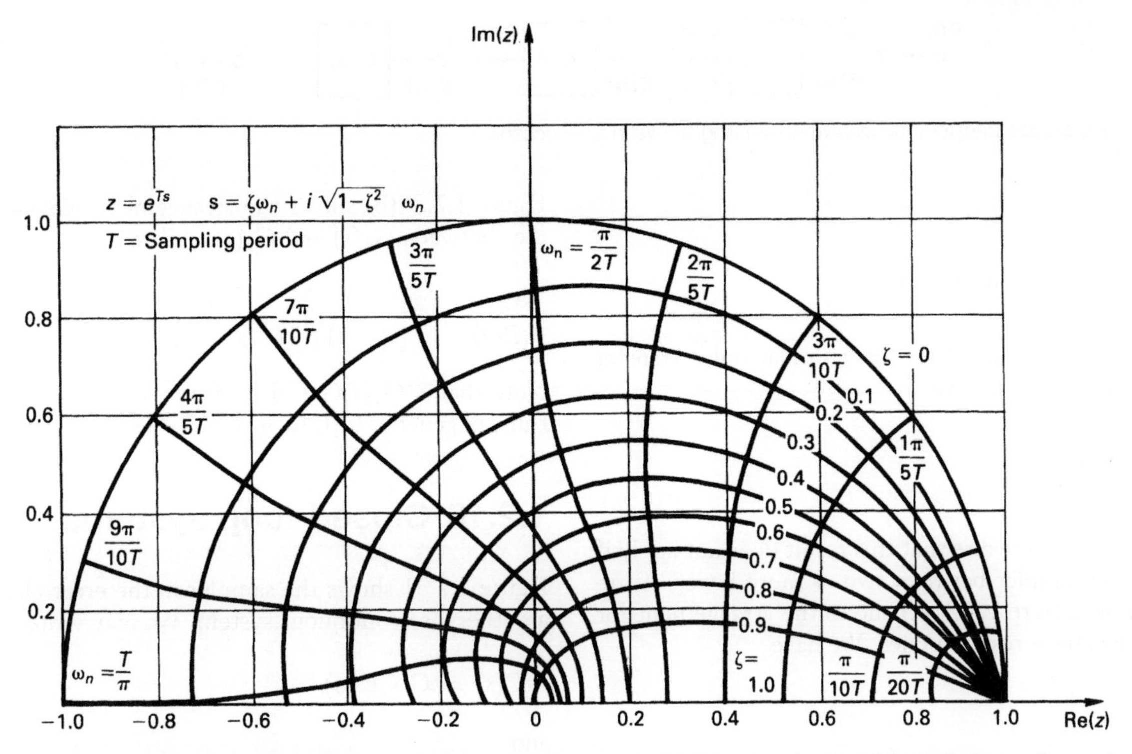

Fig. 1.2.27 Natural frequency and damping loci in the z plane. The lower half is the mirror image of the half shown. (Reproduced from Franklin et al.,[2] courtesy of Addison-Wesley.)

arrives, and then changes its output to correspond to this latest value, and so on.

This is illustrated diagrammatically in Figure 1.2.28. Thus a unit delta function $\delta(t)$ arriving produces a positive unit-value step at the output at time t. At time $t = T$, we may regard a negative unity-value step being superimposed on the output. Since the transfer function of a system may be regarded as the Laplace transformation of the response of that system to a delta function, the zero-order hold has the transfer function

$$\frac{1}{s}\left[1 - e^{-sT}\right]$$

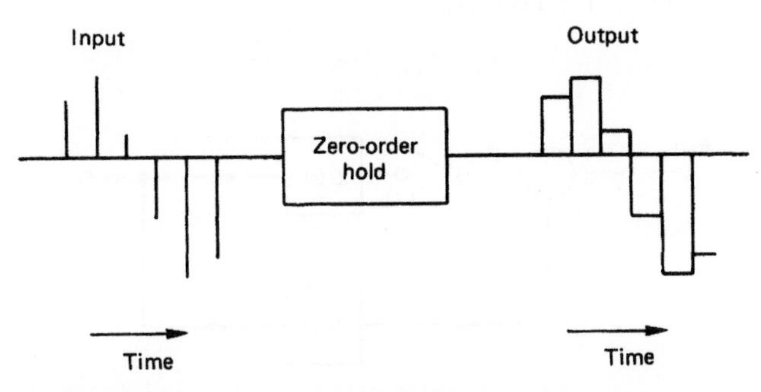

Fig. 1.2.28 Diagrammatic representation of input/output for zero-order hold.

1.2.15 Block diagrams

In a similar way to their use in continuous-control-system studies, block diagrams are used in sampled-data-system studies. It is convenient to represent individual pulse transfer functions in individual boxes. The boxes are joined together by lines representing their z transformed input/output sequences to form the complete block diagrams. The manipulation of the block diagrams may be conducted in a similar fashion to that adopted for continuous systems. Again, it must be stressed that such manipulation breaks down if the boxes load one another.

Consider the arrangement shown in Figure 1.2.29. Here we have a number of continuous systems, represented by their transfer functions, in cascade. However, a sampler has been placed in each signal line, and so for each box we may write

$$C_1(s) = G_1(s)R^*(s) \rightarrow C_1^*(s) = G_1^*(s)R^*(s)$$

$$C_2(s) = G_2(s)C_1^*(s) \rightarrow C_2^*(s) = G_2^*(s)C_1^*(s)$$

$$C(s) = G_3(s)C_2^*(s) \rightarrow C^*(s) = G_1(s)C_2^*(s)$$

Thus

$$C^*(s) = G_1^*(s)G_2^*(s)G_3^*(s)R^*(s)$$

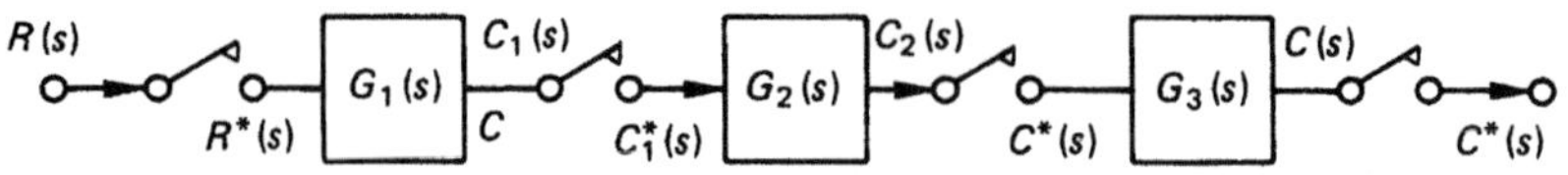

Fig. **1.2.29** Cascade transfer functions with sampling between connections.

i.e.

$$\frac{C(z)}{R(z)} = G_1(z)G_2(z)G_3(z)$$

This, of course, generalises for n similar pulse transfer functions in series to give

$$\frac{C(z)}{R(z)} = \prod_{i=1}^{n} G_i(z)$$

It is necessary to realise that this result does not apply if there is no sampler between two or more boxes. As an illustration, Figure 1.2.30(a) shows the arrangement for which the above result applies. We have

$$G_1(s) = \frac{1}{s}$$

whence (see Table 1.2.1)

$$G_1(z) = \frac{z}{z-1}$$

and

$$G_2(s) = \frac{1}{s+1}$$

whence (see Table 1.2.1)

$$G_2(z) = \frac{z}{z-e^{-T}}$$

Therefore,

$$\frac{C(z)}{R(z)} = G_1(z)G_2(z) = \frac{z^2}{(z-1)(z-e^{-T})}$$

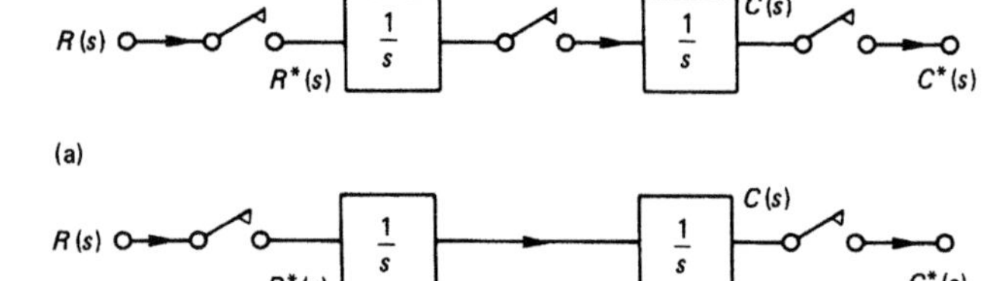

Fig. 1.2.30 Two transfer functions with (a) sampler interconnection and (b) with continuous signal connecting transfer functions.

Figure 1.2.30(b) shows the arrangement *without* a sampler between $G_1(s)$ and $G_2(s)$, and so

$$\frac{C(z)}{R(z)} = Z\left[\frac{1}{s(s+1)}\right] = \frac{z(1-e^{-T})}{(z-1)(z-e^{-T})}$$

Note that $Z[G_1(s)G_2(s)]$ is often written $G_1G_2(z)$, and thus, in general, $G_1(z)G_2(z) \neq G_1G_2(z)$.

1.2.16 Closed-loop systems

Figure 1.2.31 shows the sampler in the error channel of an otherwise continuous system. We may write

$$C(s) = G(s)E^*(s)$$

and

$$E(s) = R(s) - H(s)C(s)$$

or

$$E(s) = R(s) - H(s)G(s)E^*(s)$$

and

$$E^*(s) = R^*(s) - HG^*(s)E^*(s)$$

and so

$$E^*(s) = \frac{R^*(s)}{1+HG^*(s)}$$

Thus

$$\frac{C^*(s)}{R^*(s)} = \frac{G^*(s)}{1+HG^*(s)}$$

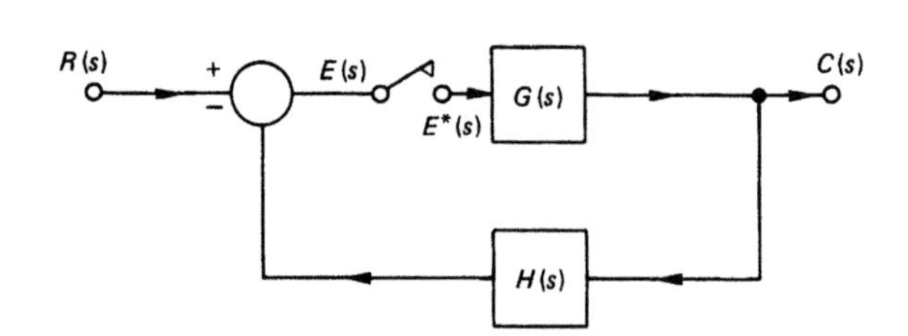

Fig. 1.2.31 Prototype sampled system with a sampler in the error channel.

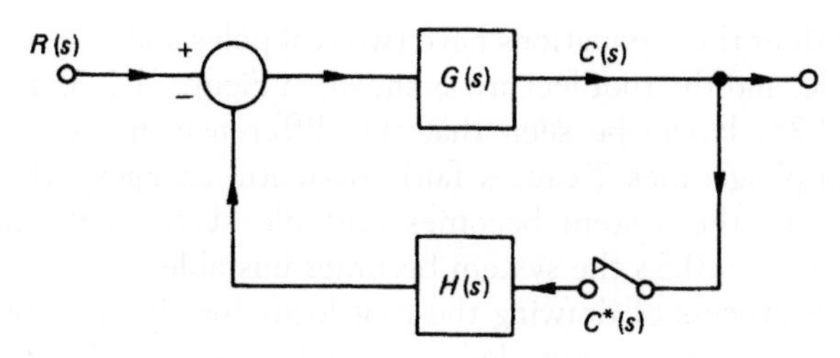

Fig. 1.2.32 Prototype sampled system with a sampler in the feedback channel.

or

$$\frac{C(z)}{R(z)} = \frac{G(z)}{1+HG(z)}$$

If the sampler is in the feedback loop, as shown in Figure 1.2.32, a similar analysis would show that

$$C(z) = \frac{GR(z)}{1+HG(z)}$$

Note that, in this case, it is not possible to take the ratio $C(z)/R(z)$. We may conclude that the position of the sampler(s) within the loop has a vitally important effect on the behaviour of the system.

Example Consider the arrangement shown in Figure 1.2.33. To calculate the pulse transfer function it is necessary to determine

$$L\left[\frac{1}{s}\frac{1}{s}(1-e^{-sT})\right]$$

Consider (from Table 1.2.1)

$$Z\left[\frac{1}{s^2}\right] = \frac{Tz}{(z-1)^2}$$

and, therefore,

$$Z\left[e^{-sT}\frac{1}{s^2}\right] = z^{-1}Z\left[\frac{1}{s^2}\right] = \frac{T}{(z-1)^2}$$

Thus

$$Z\left[\frac{1}{s^2}\left(1-e^{-Ts}\right)\right] = \frac{T}{(z-1)} = G(z)$$

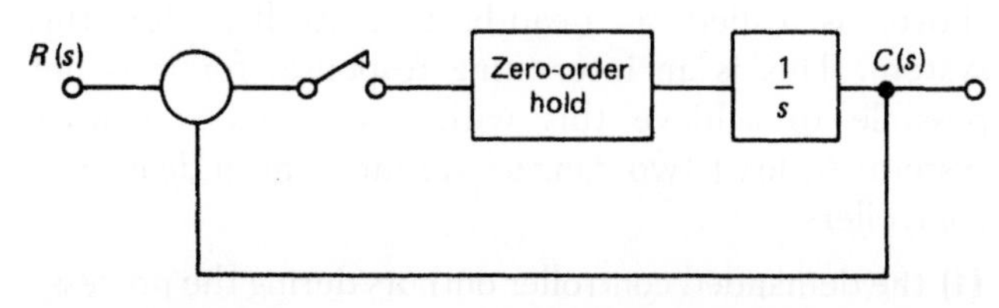

Fig. 1.2.33 Arrangement used in the example in Section 1.2.16.

and

$$\frac{C(z)}{R(z)} = \frac{G(z)}{1+G(z)} = \frac{T}{z+(T-1)}$$

1.2.17 Stability

It should be appreciated from the above that, in general, $C(z)/R(z)$ results in a ratio of polynomials in z in a similar way as, for continuous systems, $C(s)/R(s)$ results in a ratio of polynomials in s. Thus, just as the equation $1+G(s)H(s)=0$ is called the 'characteristic equation' for the continuous system, $1+GH(z)=0$ is the characteristic equation for the sampled-data system. Both of these characteristic equations are polynomials in their respective variables, and the positions of the roots of these equations determine the characteristic behaviour of the corresponding closed-loop systems. Mathematically, the process of determining the roots is identical in the two cases. The difference between the two characteristic equations arises because of the need to interpret the effects of the location of the roots, when they are plotted in their respective s and z planes, on the two plants. For continuous systems, if any of these poles are located in the right-half s plane, then the system is unstable. Similarly, since the whole of the left-hand s plane maps into the unit-circle of the z plane under the transformation $z=e^{sT}$, then in the simple-data case, for stability *all* of the roots of $1+GH(z)=0$ must lie within the unit circle.

Much of the design process of control systems is to arrange for the roots of the characteristic equation to locate at desired positions in either the s or z plane. It will be recalled, from continuous theory, that the locus of these roots, as a particular parameter is varied, may be determined by using the root-locus technique. Thus, since the characteristic equation of the sample-data system has a similar form (i.e. a polynomial), the root-locus technique may be applied to $1+GH(z)=0$ in exactly the same way. Only once the root-locus has been determined is there a difference in interpreting the effects of pole positions between the two cases.

1.2.18 Example

Consider the system shown in Figure 1.2.34, and suppose that the requirement is to draw the root-locus diagrams for, say, sampling periods of 1 and 0.5 s.

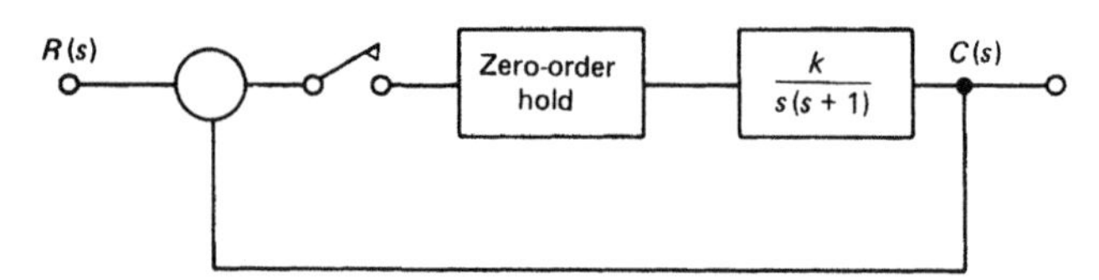

Fig. 1.2.34 Arrangement used in the example in Section 1.2.18.

The first requirement is to determine the pulse-transfer function for the open loop, i.e. $G(z)$:

$$G(z) = Z\left[\frac{K\left(1-e^{-Ts}\right)}{s^2(s+1)}\right] = K\left(1-z^{-1}\right)Z\left[\frac{1}{s^2(s+1)}\right]$$

Consider

$$Z\left[\frac{1}{s^2(s+1)}\right] = Z\left[\frac{1}{s^2}-\frac{1}{s}+\frac{1}{s+1}\right]$$

where, from Table 1.2.2, we have

$$Z\left[\frac{1}{s^2(s+1)}\right] = \frac{Tz}{(z-1)^2}-\frac{z}{z-1}-\frac{z}{(z-e^{-T})}$$
$$= z\left[\frac{z\left(T+e^{-T}-1\right)+1-e^{-T}(1+T)}{(z-1)^2\left(z-e^{-T}\right)}\right]$$

and so

$$G(z) = \left[\frac{K\left[z\left(T+e^{-T}-1\right)+1-e^{-T}(1+T)\right]}{(z-1)\left(z-e^{-T}\right)}\right]$$

Thus, when $T=1$ s,

$$G_1(z) = \frac{0.368K(z+0.718)}{(z-1)(z-0.368)}$$

and when $T=0.5$ s

$$G_2(s) = \frac{0.107K(z+0.841)}{(z-1)(z-0.606)}$$

Table 1.2.2

Settling band(%)	Optimum 'b'	Settling time
20	0.45	1.80
15	0.55	2.00
10	0.60	2.30
5	0.70	2.80
2	0.80	3.50

Both of these equations have two real poles and one zero pole, and the root loci are as shown in Figures 1.2.35 and 1.2.36. It can be seen that the difference in the two sampling times T causes fairly dramatic changes; when $T=1$ s the system becomes unstable at $K=1.9$, and when $T=0.5$ s the system becomes unstable at $K=3.9$. The process of drawing the root locus for either a continuous plant or a sampled-data plant is identical. It is the interpretation of the positions of the roots that is different, although in both cases the design is to place the roots in acceptable locations in the two planes. It is possible to use Bode diagrams in sampled-data design work.

1.2.19 Dead-beat response

Consider the system shown above where $T=1$ s and $K=1$; suppose that compensation of the form

$$D(z) = \frac{1.582(z-0.368)}{(z+0.418)}$$

is inserted immediately after the sampler. Then it is easy to show that

$$\frac{C(z)}{R(z)} = \frac{0.582(z+0.71)}{z^2}$$

If

$$R(z) = \frac{z}{z-1}$$

i.e. $r(t)$ is a unit step function, then

$$C(z) = \frac{0.582(z+0.718)}{z(z-1)} = \frac{1}{z}\left[\frac{0.582z+0.418}{z-1}\right]$$
$$= \frac{1}{z}\left[0.582+\frac{1}{z}+\frac{1}{z^2}+\cdots\right]$$

i.e. $c(0)=0$, $c(1)=0.582$ and $c(n)=1$, for $n=2, 3, \ldots$.

The implication is that $c(t)$ has reached its target position after two sample periods. If an nth order system reaches its target position in, at most, nth sampling instants, then this is called a 'dead-beat response'; a controller that achieves this, such as $D(z)$ above, is called a 'dead-beat controller' for this system. This is an interesting response, for it is not possible to achieve this with a continuous control system. At least two dangers are inherent in dead-beat controllers:

(1) the demanded controller outputs during the process may be excessive; and

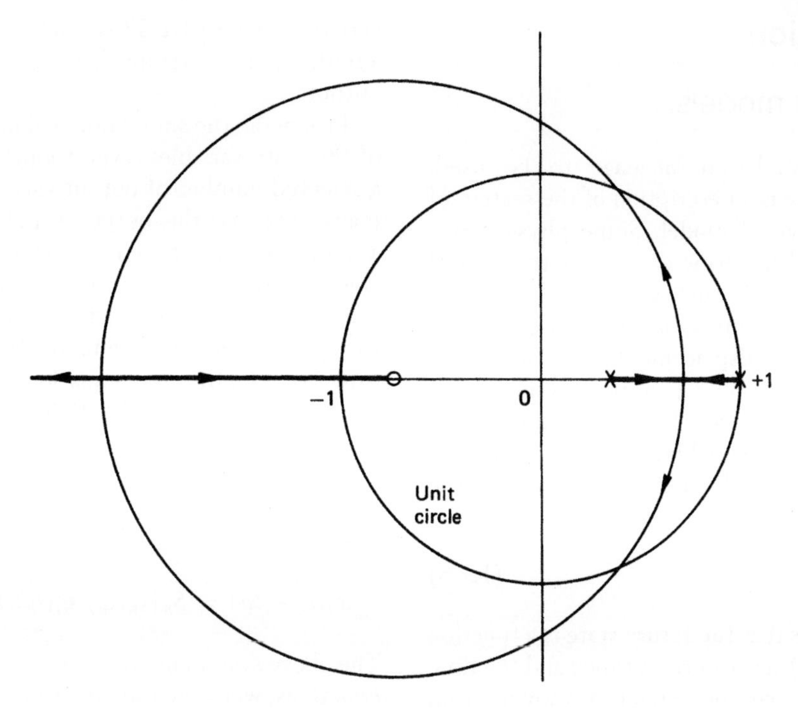

Fig. 1.2.35 Root locus plot: $G(z) = [0.368K(z + 0.718)]/[(z - 1)(z - 0.368)]$

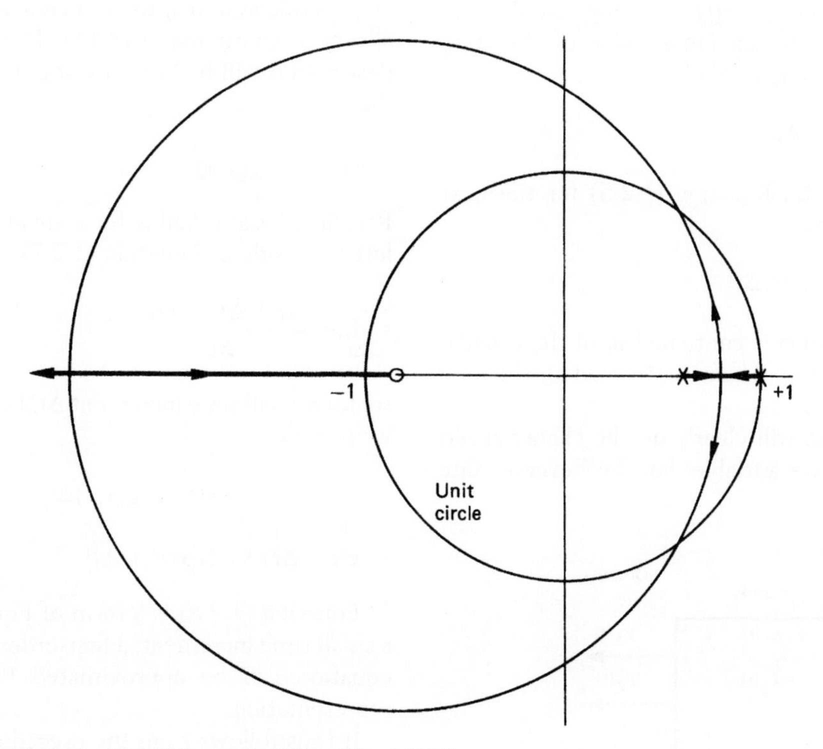

Fig. 1.2.36 Root-locus plot: $G(z) = [0.107K(z + 0.841)]/[(z - 1)(z - 0.606)]$.

(2) there may be an oscillation set up which is not detected without further analysis.

In fact, the system is only 'dead beat' at the sampling instants. Indeed, in the above example, there is an oscillation between sampling instants of about 10% of the step value. However, theoretically it is possible for a sampled-data system to complete a transient of the above nature in finite time.

1.2.20 Simulation

1.2.20.1 System models

Regardless of the simulation language to be used, a necessary prerequisite is a description of the system of interest by a mathematical model. Some physical systems can be described in terms of models that are of the state transition type. If such a model exists, then given a value of the system variable of interest, e.g. voltage, charge position, displacement, etc., at time t, then the value of the variable (state) at some future time $t+\Delta t$ can be predicted. The prediction of the variable of interest (state variable) $x(t)$ at time $t+\Delta t$, given a state transition model S, can be expressed by the state equation:

$$x(t+\Delta t) = S[x(t), t, \Delta t] \tag{1.2.5}$$

Equation (1.2.5) shows that the future state is a function of the current state $x(t)$ at the current time t and the time increment Δt Thus, once the model is known, from either empirical or theoretical considerations, Equation (1.2.5), given an initial condition (value), allows for the recursive computation of $x(t)$ for any number of future steps in time. For an initial value of the state variable $\bar{x} = x(t_1)$ at time t_1, then

$$x(t_1+\Delta t) = S[\bar{x}, t_1, \Delta t]$$

then letting $t_2 = t_1 + \Delta t$, Equation (1.2.5) for the next time step Δt, becomes

$$x(t_2+\Delta t) = S[x(t_2), t_2, \Delta t]$$

Obviously, this operation is continued until the calculation of the state variable has been performed for the total time period of interest.

Systems of interest will clearly not be characterised only by a single state variable but by several state variables. Figure 1.2.37 is a schematic representation of a multi-variable system that has r inputs, n states and m outputs.

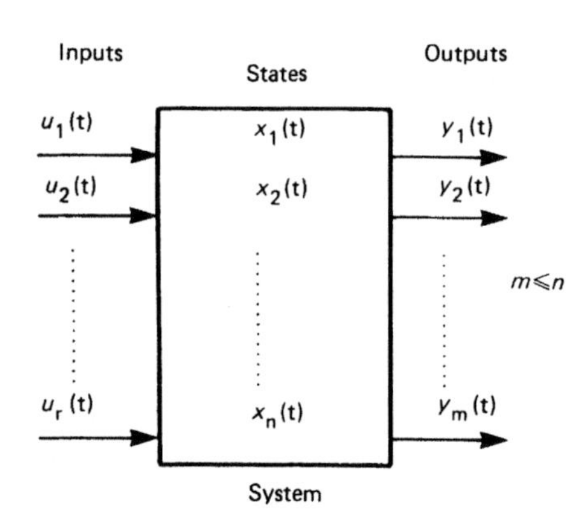

Fig. 1.2.37 Schematic representation of a multi-variable system.

In general, the simulation will involve calculation of all of the state variables, even though the response of only a selected number of output variables is of interest. For many systems, the output variables may well exhibit a simple one-to-one correspondence to the state variables. As shown by the representation in Figure 1.2.37, the values of the state variables depend on the inputs to the system. For a single interval, between the k and $k+1$ time instants, the state equations for the n state variable system for a change in the jth input $(j \leq r)u_j(t)$ is written as

$$\begin{aligned} x_1(t_k+\Delta t) &= S_1[x_1(t_k), u_j(t_k), t_k, \Delta t] \\ x_2(t_k+\Delta t) &= S_2[x_2(t_k), u_j(t_k), t_k, \Delta t] \\ &\vdots \\ x_n(t_k+\Delta t) &= S_n[x_n(t_k), u_j(t_k), t_k, \Delta t] \end{aligned} \tag{1.2.6}$$

The above system of equations, a collection of difference equations, would be used to predict the state variables x_1, $x_2, \ldots, x_n$ at time intervals of Δt from the initial time t_0 until the total time duration of interest $T = t_0 + K\Delta t$. For engineering systems, the dependent variable will generally be a continuous variable. In this case the system description will be in terms of a differential equation of the form

$$dx/dt = g(x, t) \tag{1.2.7}$$

Recalling basic calculus for a small time increment, the left-hand side of Equation (1.2.7) can be expressed as

$$\lim_{\Delta t \to 0} \frac{x(t+\Delta t) - x(t)}{\Delta t}$$

so, for a small time increment Δt, Equation (1.2.7) can be written as

$$\begin{aligned} x(t+\Delta t) &= x(t) + [g(x,t)\Delta t] \\ \text{or} \\ x(t+\Delta t) &= G[x(t), t, \Delta t] \end{aligned} \tag{1.2.8}$$

Equation (1.2.8) is a form of Equation (1.2.5), so for a small time increment, a first-order ordinary differential equation can be approximated by a state transition representation.

It thus follows from the preceding discussion that, in digital continuous system simulation, the principal numerical task is the approximate integration of Equation (1.2.7). For a small time increment DT, the integration step size, the computation involves the evaluation of the difference equation

$$x(t+DT) = x(t) + [g(xt)]DT) \tag{1.2.9a}$$

which can be written explicitly as

$$x(t_{k+1}) = x(t_k) + \int_{t_k}^{t_{k+1}} g[x(t_k), t_k]DT) \tag{1.2.9b}$$

where $DT = t_{k+1} - t_k$. The calculation starts with a known value of the initial state $x(0)$ at time t_0 and proceeds successively to evaluate $x(t_1)$, $x(t_2)$, etc. The computation involves successive computation of $x(t_{k+1})$ by alternating calculation of the derivative $g[x(t_k),t_k]$ followed by integration to compute $x(t_{k+1})$ at time $t_{k+1} = t_k + DT$.

Obviously, most physical systems will be described by second- or higher-order ordinary differential equations so the higher-order equation must be re-expressed in terms of a group of first-order ordinary differential equations by introducing state variables. For an nth-order equation,

$$\frac{d^n z}{dt^n} = f\left[z, \frac{dz}{dt}, \frac{d^2 z}{dt2} \cdots \frac{d^{n-1} z}{dt^{n-1}}; t\right] \tag{1.2.10}$$

the approach involves the introduction of new variables as state variables to yield the following first-order differential equations

$$\begin{aligned} \frac{dx_1}{dt} &= x_2 \\ \frac{dx_2}{dt} &= x_3 \\ \frac{dx_3}{dt} &= x_4 \\ &\vdots \\ \frac{dx_{n-1}}{dt} &= x_n \\ \frac{dx_n}{dt} &= f(x_1, x_2, x_3, \ldots, x_n; t) \end{aligned} \tag{1.2.11}$$

It should be noted that this equation can be expressed in shorthand notation as a vector-matrix differential equation. In an analogous manner, Equation (1.2.6) can be expressed as a vector different equation. There is no unique approach to the selection of state variables for system representation, but for many systems the choice of state variables will be obvious. In electric circuit problems, capacitor voltages and inductor currents would be logical choices, as would position, velocity and acceleration for mechanical systems.

1.2.20.2 Integration schemes

The simple integration step, embodied by the first-order Euler form in Equation (1.2.9a) only provides a satisfactory approximation of the solution of the differential equation, within specified error limits, for a very small integration step size DT. Since the small integration interval leads to substantial computing effort and to round-off error accumulation, all digital simulation languages use improved integration schemes. Despite the wide variety of different integration schemes that are available in the many different simulation languages, the calculational approach can be categorised into two groups. The types of algorithm are:

(1) *Multi-step formulae* In such algorithms, the value of $x(t + DT)$ is not calculated by the simple linear extrapolation of Equation (1.2.9a). Rather than use only $x(t)$ and one derivative value, the algorithms use a polynomial approximation based on past values of $x(t)$ and $g[x(t),t]$, that is at times $t - DT$, $t - 2DT$, etc.

(2) *Runge–Kutta formulae* In Runge–Kutta type algorithms, the derivative value used for the calculation of $x(t + DT)$ is not the point value at time t. Instead, two or more approximate derivative values in the interval t, $t + DT$ are calculated and then a weighted average of these derivative values is used instead of a single value of the derivative to compute $x(t + DT)$.

1.2.20.3 Organisation of problem input

Most simulation language input is structured into three separate sections, although in some programs the statement can be used with limited sectioning of the program. A typical structure and the type of statements, functions or parts of the simulation program that appear are as follows.

(1) *Initialisation*
Problem documentation (e.g. name, date).
Initial conditions for state variables.
Parameter values (problem variables that may not be constant, problem time, integration order, integration step size, etc.).
Problem constants.

(2) *Dynamic*
Derivative statements.
Integration statements (including any control parameters not given in the initialisation section).

(3) *Terminal*
Conditional statements (e.g. total time, variable(s), value(s)).
Multiple run parameters.
Output (print/plot/display) option(s).
Output format (e.g. designation of independent variable; increment for independent variable; dependent variable(s) to be output; maximum and minimum values of variable(s); or automatic scaling; total number of points for the independent variable or total length of time).

It should be understood that the specific form of the statements within each section is not exactly the same for all digital simulation languages. However, from the continuous system modelling package (CSMP) simulation programs presented in the next section, with the aid of the appropriate language manual, there should be no difficulty in formulating a simulation program using any continuous system simulation language (CSSL)-type digital simulation program.

1.2.20.4 Illustrative example

Simulation programs are presented, using the CSMP language, that would be suitable for investigating system dynamic behaviour. The system model, although relatively simple in nature, is typical of those used for system representation.

1.2.20.4.1 Example

Frequently, it will be found that system dynamic behaviour can be described by a differential equation of the form

$$y^n + a_1 y^{n-1} + a_2 y^{n-2} + a_{n-1} y^1 + a_n y = b_0 r^m + b_1 r^{m-1} + b_{m-1} r^1 + b_m r \quad (1.2.12)$$

where

$$y^n = \frac{d^n y}{dt^n} \quad \text{and} \quad r^m = \frac{d^m r}{dt^m}$$

Use of CSMP for studying the dynamic behaviour of a system described by a high-order differential equation is illustrated here using a simulation program for the differential equation

$$y^3 + 2.5y^2 + 3.4y^1 + 0.8y = 7.3r \quad (1.2.13)$$

with the initial conditions

$$y^2(0) = 0; \quad y^1(0) = -4.2; \quad y^0 = 2.5$$

Development of the simulation program follows logically by rewriting Equation (1.2.13) as

$$\frac{d^3y}{dt^3} = -2.5\frac{d^2y}{dt^2} - 3.4\frac{dy}{dt} - 0.8y + 7.3r$$
$$\left.\frac{d^2y}{dt^2}\right|_{t=0} = 0; \quad \left.\frac{dy}{dt}\right|_{t=0} = -4.2; \quad y|_{t=0} = 2.5 \quad (1.2.14)$$

A block diagram showing the successive integrations to be solved for the dependent variable y is given in Figure 1.2.38. As can be seen from the labelling on the diagram, the output of the integration blocks is successive derivative values and the dependent variable. In fact, the output of each integration block is a state variable. This becomes obvious by introducing new variables, x_1, x_2, x_3 defined as

$$x_1 = y$$
$$\frac{dx_1}{dt} = x_2$$
$$\frac{dx_2}{dt} = x_3$$

which allows Equation (1.2.14) to be expressed as

$$\frac{dx_1}{dt} = x_2$$
$$\frac{dx_2}{dt} = x_3 \quad (1.2.15)$$
$$\frac{dx_3}{dt} = -2.5x_3 - 3.4x_2 - 0.8x_1 + 7.3r$$

with the initial conditions

$$x_3(0) = 0; \quad x_2(0) = -4.2; \quad x_1(0) = 2.5$$

A program for solving Equation (1.2.15) is given in Figure 1.2.39. Examination of the program shows that the value of the forcing function r is not constant but varies with time. The variation is provided using the quadratic interpolation function, NLFGEN. Total simulation time is set for 6 min with the interval for tabular output specified as 0.2 min. The time unit is determined by the problem parameters. It is to be noted that the program does not include any specification for the method of integration. The CSMP language does not require that a method of integration be given, but a particular method may be specified. If a method is not given,

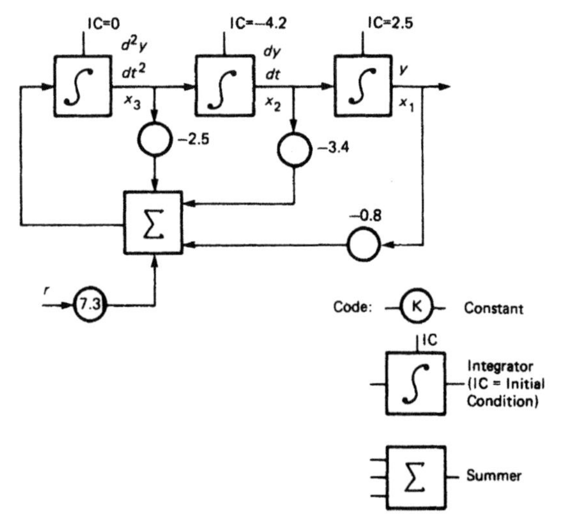

Fig. 1.2.38 CSMP block diagram for a third-order differential equation.

```
LABEL THIRD ORDER DIFFERENTIAL EQUATION
INITIAL
CONSTANT A1=-2.5,A2=-3.4,A3--0.8,B0=7.3,   ...
         X1INIT=2.5,X2INIT=-4.2,X3INIT-0.0
FUNCTION FCHG-(0.5,4.8),(1.0,6.3),(1.5,2.8),(2.0,3.9),   ...
              (2.5,4.8),(3.0,3.2),(3.5,2.1),(4.0,5.6),   ...
              (4.5,6.8),(5.0,3.7),(5.5,4.6),(6.0,3.4)
DYNAMIC
      R=NLFGEN(FCHG,TIME)
      X1=INTGRL(X1INIT,X2)
      X2=INTGRL(X2INIT,X3)
      X3=INTGRL(X3INIT,DHX3)
      DHX3=A1*X3+A2*X2+A3*X1+B0*R
TERMINAL
TIMER FINTIM=6.0,PRDEL=0.2
PRINT R,X1,X2,X3
END
STOP
ENDJOB
```

Fig. 1.2.39 Simulation program for studying the dynamic behaviour of a system described by a third-order differential equation.

then by default the variable step size fourth-order Runge–Kutta method is used for calculation. The initial step size, by default, is taken as 1/16 of the PRDEL (or OUTDEL) value. Minimum step size can be limited by giving a value for DELMIN as part of the TIMER statement. If a DELMIN value is not given then, by default, the minimum step size is FINTIM $\times$ 10^{-7}.

1.2.21 Multivariable control

Classical process control analysis is concerned with single loops having a single setpoint, single actuator and a single controlled variable. Unfortunately, in practice, plant variables often interact, leading to interaction between control loops. A typical interaction is shown in Figure 1.2.40, where a single combustion air fan feeds several burners in a multi-zone furnace. An increase in air flow, via V_1 say to raise the temperature in zone 1, will lead to a reduction in the duct air pressure P_d, and a fall in air flow to the other zones. This will lead to a small fall in temperature in the other zones which will cause their temperature controllers to call for increased air flow which affects the duct air pressure again. The temperature control loops interact via the air valves and the duct air pressure.

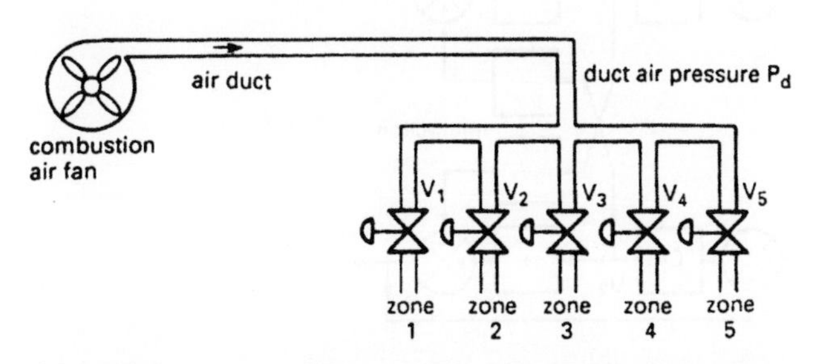

Fig. 1.2.40 A typical example of interaction between variables in multi-variable control. The air flows interact via changes in the duct air pressure.

Where interaction between variables is encountered an attempt should always be made to remove the source of the interaction, as this leads to a simpler, more robust, system. In Figure 1.2.40, for example, the interaction could be reduced significantly by adding a pressure control loop which maintains duct pressure by using a VF to set the speed of the combustion airfan. Often, however, the interaction is inherent and cannot be removed.

Figure 1.2.41 is a general representation of two interacting control loops. The blocks C_1 and C_2 represent the controllers comparing setpoint R with process variable V to give a controller output U. The blocks K_{ab} represent the transfer function relating variable a to controller output b. Blocks K_{11} and K_{12} are the normal forward control path, with blocks K_{21} and K_{12} representing the interaction between the loops.

The process gain of process 1 can be defined as $\Delta V_1/\Delta U_2$ where Δ denotes small change. This process gain can be measured with loop 2 in open-loop (i.e. U_2 fixed) or loop 2 in closed-loop control (i.e. V_2 fixed) we can thus observe two gains

$$K_{2OL} = \frac{\Delta V_1}{\Delta U_1} \quad \text{for loop 2 open loop}$$

and

$$K_{2CL} = \frac{\Delta V_1}{\Delta U_1} \quad \text{for loop 2 closed loop}$$

The gains will, of course vary with frequency and have magnitude and phase shift components. We can now define a relative gain λ for loop 1

$$\lambda = \frac{K_{2OL}}{K_{2CL}}$$

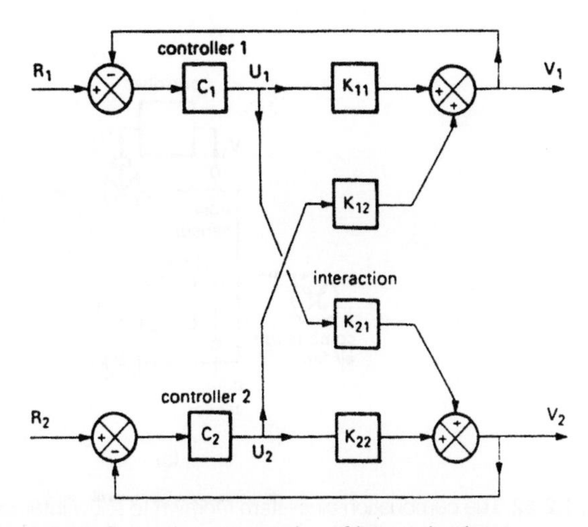

Fig. 1.2.41 General representation of interacting loops.

If λ is unity, changing from manual to auto in loop 2 does not affect loop 1, and there is no interaction between the loops.

If $\lambda < 1$, the interaction will apparently increase process 1 gain when loop 2 is switched to automatic. If $\lambda > 1$, process 1 gain will apparently be decreased when loop 2 is in automatic.

This apparent change in gain can be seen with loop 2 in manual, U_2 is fixed, so K_{2OL} is simply K_{11}. To find K_{2CL} we must consider what happens when loop 2 effectively shunts K_{11}. We have

$$V_1 = K_{11}U_1 + K_{12}U_2 \qquad (1.2.16)$$

and

$$V_2 = K_{22}U_2 + K_{21}U_1 \qquad (1.2.17)$$

Re-arranging Equation (1.2.17) gives

$$U_2 = \frac{V_2 - K_{21}U_1}{K_{22}}$$

which can be substituted in Equation (1.2.16) giving

$$V_1 = K_{11}U_1 + \frac{K_{12}}{K_{22}}(V_2 - K_{21}U_1)$$

The process 1 gain with loop 2 in auto is

$$K_{2CL} = \frac{dV_1}{dU_1} = \frac{K_{11}K_{22} - K_{12}K_{21}}{K_{22}}$$

The relative gain, λ, is

$$\lambda = \frac{K_{2OL}}{K_{2CL}} = \frac{1}{1 - K_{12}K_{21}/K_{11}K_{22}}$$

It should be remembered that the gains K_{ab} are dynamic functions, so λ will vary with frequency.

The term $(K_{12}K_{21}/K_{11}K_{22})$ is the ratio between the interaction and forward gains. This should be in the range 0–1. If the term is greater than unity, the interactions have more effect than the supposed process, and the process variables are being manipulated by the wrong actuators!

It is possible to determine the range of λ from the relationship $(K_{12}K_{21}/K_{11}K_{22})$. If this is positive, λ will be greater than unity, and loop 1 process gain will decrease when loop 2 is switched to auto. This will occur if there is an even number of K_{ab} blocks with negative sign (0, 2 or 4). If the relationship is negative, λ will be less than unity and loop 1 process gain will increase when loop 2 is closed. This occurs if there is an odd number of blocks with negative sign (1 or 3).

The combustion air flow system of Figure 1.2.40 is redrawn in Figure 1.2.42(a). Increasing U_1 obviously decreases V_2, and increasing U_2 similarly decreases V_1. The interaction block diagram thus has the signs of Figure 1.2.42(b). There are two negative blocks, so λ is greater than unity.

If λ is greater than unity, the interaction can be considered benign as the reduced process gain will tend to increase the loop stability (albeit at the expense of response time). The loops can be tuned individually in the knowledge that they will remain stable with all loops in automatic control.

If λ is in the range $0 < \lambda < 1$, care must be taken as it is possible for loops to be individually stable but collectively unstable requiring a reduction in controller gains to maintain stability. The closer λ gets to zero, the greater the interaction and the more de-gaining will be required.

The calculation of dynamic interaction is difficult, even for the two variable case. With more interacting variables, the analysis becomes exceedingly complex and computer solutions are best used. Ideally, though, interactions once identified, should be removed wherever possible.

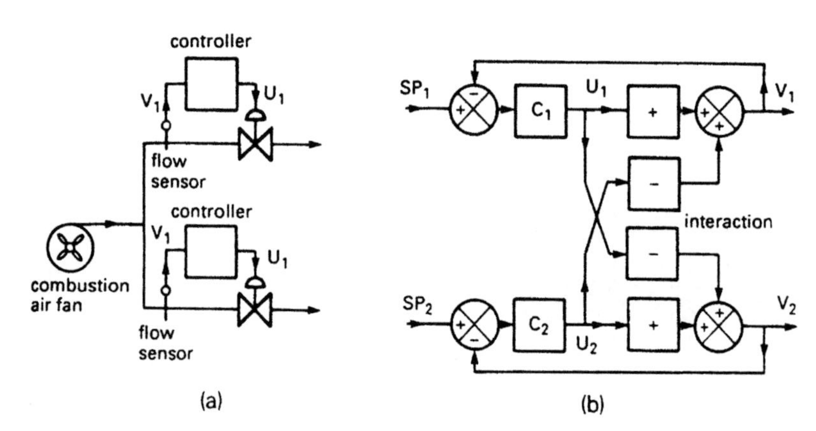

Fig. 1.2.42 The combustion air system redrawn to show interactions: (a) block diagram; (b) interaction diagram, with two negative blocks the interaction decreases the apparent process gain and the interaction is benign.

1.2.22 Dealing with non-linear elements

1.2.22.1 Introduction

All systems are non-linear to some degree. Valves have non-linear transfer functions, actuators often have a limited velocity of travel and saturation is possible in every component. A controller output is limited to the range 4–20 mA, say, and a transducer has only a restricted measurement range.

One of the beneficial effects of closed-loop control is the reduced effect of non-linearities. The majority of non-linearities are therefore simply lived with, and their effect on system performance is negligible. Occasionally, however, a non-linear element can dominate a system and in these cases its effect must be studied.

Some non-linear elements can be linearised with a suitable compensation circuit. Differential pressure flow meters have an output which is proportional to the square of flow. Following a non-linear differential pressure flow transducer with a non-linear square root extractor gives a linear flow measurement system.

Cascade control can also be used around a non-linear element to linearise its performance as seen by the outer loop. Butterfly valves are notoriously non-linear. They have an S shaped flow/position characteristic, suffer from backlash in the linkages and are often severely velocity limited. Enclosing a butterfly valve within a cascade flow loop, for example, will make the severely non-linear flow control valve appear as a simple linear first-order lag to the rest of the system.

There are two basic methods of analysing the behaviour of systems with non-linear elements. It is also possible, of course, to write computer simulation programs and often this is the only practical way of analysing complex non-linearities.

1.2.22.2 The describing function

If a non-linear element is driven by a sine wave, its output will probably not be sinusoidal, but it will be periodic with the same frequency as the input, but of differing shape and possibly shifted in phase as shown in Figure 1.2.43. Often the shape and phase shift are related to the amplitude of the driving signal.

Fourier analysis is a technique that allows the frequency spectrum of any periodic waveform to be calculated. A simple pulse can be considered to be composed of an infinite number of sine waves.

The non-linear output signals of Figure 1.2.43 could therefore be represented as a frequency spectrum, obtained from Fourier analysis. This is, however, unnecessarily complicated. Process control is generally concerned with only dominant effects, and as such it is only necessary to consider the fundamental of the spectrum. We can therefore represent a non-linear function by its gain and phase shift at the fundamental frequency. This is known as the *describing function*, and will probably be frequency and amplitude dependent.

Figure 1.2.44 shows a very crude bang/bang servo system used to control level in a header tank. The level is sensed by a capacitive probe which energises a relay when a nominal depth of probe is submerged. The relay energises a solenoid which applies pneumatic pressure to open a flow valve. This system is represented by Figure 1.2.45.

The level sensor can be considered to be a level transducer giving a 0–10 V signal over a 0.3 m range. The signal is filtered with a 2 s time constant to overcome noise from splashing, ripples, etc. The level transducer output is compared with the voltage from a setpoint control and the error signal energises or de-energises the relay. We shall assume no hysteresis for simplicity although this obviously would be desirable in a real system.

The relay drives a solenoid assumed to have a small delay in operation which applies 15 psi to an instrument air pipe to open the valve. The pneumatic signal takes a finite time to travel down the pipe, so the solenoid valve and piping are considered as a 0.5 s transit delay. The valve actuator turns on a flow of 150 m^3/min for an applied pressure of 15 psi. We shall assume it is linear for other applied pressures. The actuator/valve along with the inertia of the water in the pipe appear as a first-order lag of 4 s time constant. The tank itself appears as an integrator from flow to level.

This system is dominated by the non-linear nature of the level probe and the solenoid. The rest of the system can be considered linear if we combine the level comparator, relay and solenoid into a single element which switches 0–15 psi according to the sign of the error signal (15 psi for negative error, i.e. low level).

This non-linear element will therefore have the response of Figure 1.2.46. when driven with a sinusoidal error signal. The output will have a peak to peak amplitude of 15 psi regardless of the error magnitude.

From Fourier analysis, the fundamental component of the output signal is a sine wave with amplitude $4 \times 7.5/\pi$ psi as shown. The phase shift is zero at all frequencies. The non-linear element of the comparator/relay/solenoid can thus be considered as an amplifier whose gain varies with the amplitude of the input signal.

For a 1 V amplitude error signal the gain is

$(4 \times 7.5)/(\pi \times 1) = 9.55$

For a 2 V amplitude error signal the gain is

$(4 \times 7.5)/(\pi \times 2) = 4.78$

In general, for an E volt error signal the gain is

$$(4 \times 7.5)/(\pi \times E) = 9.55/E \qquad (1.2.18)$$

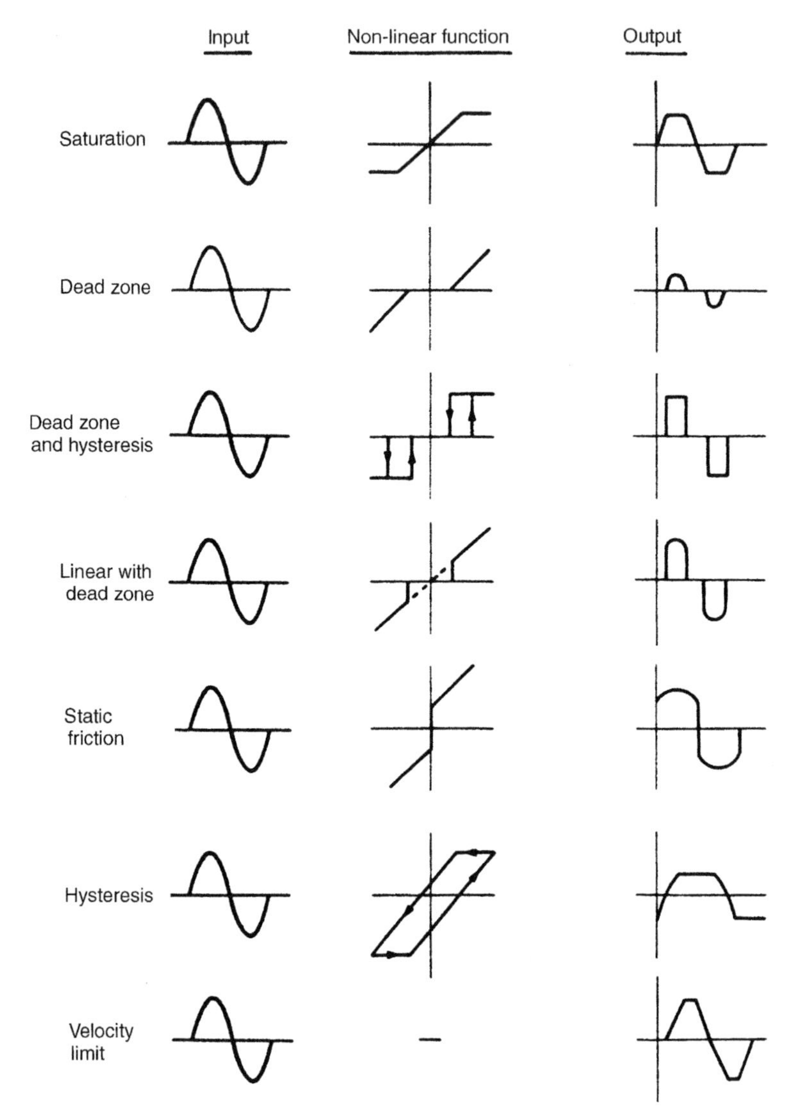

Fig. 1.2.43 Common non-linearities.

Figure 1.2.47 is a Nichols chart for the linear parts of the system. This has 180° phase shift for $\omega = 0.3$ rad/s, so if it was controlled by a proportional controller, it would oscillate at 0.3 rad/s if the controller gain was sufficiently high. The linear system gain at this frequency is −7 dB, so a proportional controller gain of 7 dB would just sustain continuous oscillation.

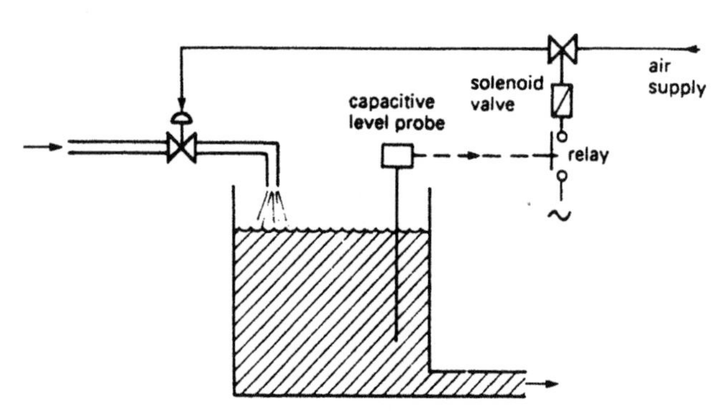

Fig. 1.2.44 Bang–bang level control system.

Let us now return to our non-linear level switch. This has a gain which varies inversely with error amplitude. If we are, for some reason, experiencing a large sinusoidal error signal the gain will be low. If we have a small sinusoidal error signal the gain will be high.

Intuitively we know the system of Figure 1.2.44 will oscillate. The non-linear element will add just sufficient gain to make the Nichols chart of Figure 1.2.47 pass through the 0dB/−180° origin. Self-sustaining oscillations will result at 0.3 rad/s. If these increase in amplitude for some reason, the gain will decrease causing them to decay again. If they cease, the gain will increase until oscillations recommence.

The system stabilises with continuous constant amplitude oscillation.

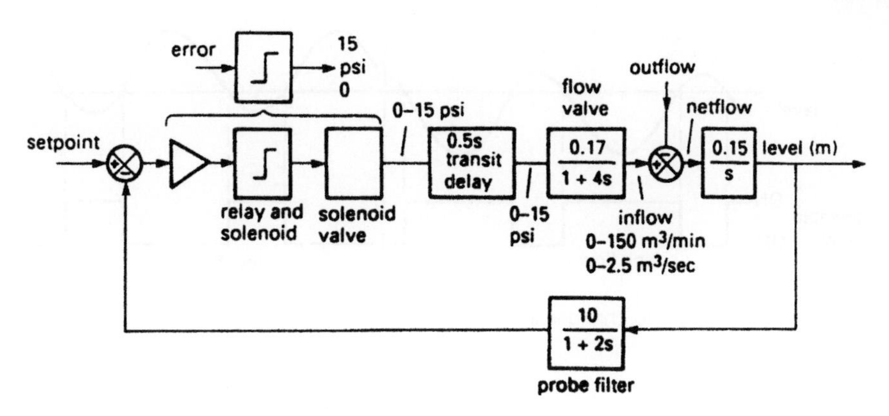

Fig. 1.2.45 Block diagram of bang–bang level control system.

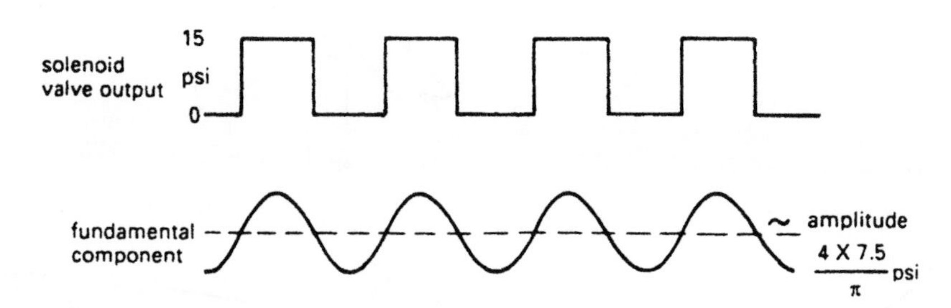

Fig. 1.2.46 Action of solenoid valve in level control system.

To achieve this the non-linear element must contribute 7 dB gain, or a linear gain of 2.24. From Equation (1.2.18) above, the gain is $9.55/E$ where E is the error amplitude. The required gain is thus given by an error amplitude of $9.55/2.24 = 4.26$ V. This corresponds to an oscillation in level of 0.426 m.

The system will thus oscillate about the set level with an amplitude of 0.4 m (the assumptions and approximations give more significant figures a relevance they do not merit) and an angular frequency of 0.4 rad/s (period fractionally over 20 s).

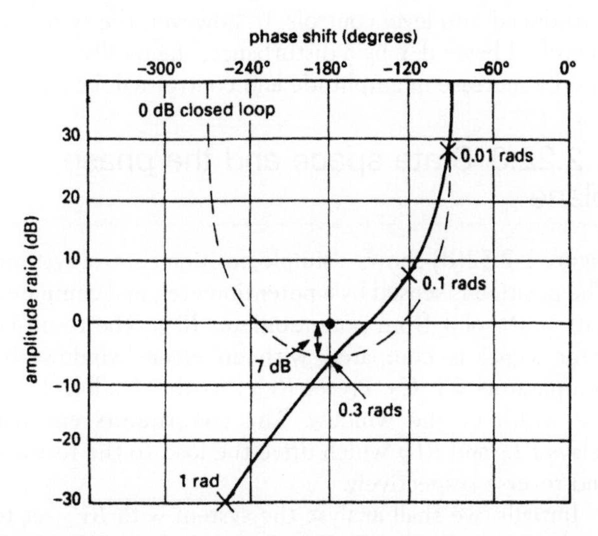

Fig. 1.2.47 Nichols chart for linear portion of level control system.

There is a hidden assumption in the above analysis that the outgoing flow is exactly half the available ingoing flow to give equal mark/space ratio at the valve. Other flow rates will give responses similar to Figure 1.2.48, exhibiting a form of pulse width modulation. The relatively simple analysis however has told us that our level control system will sustain constant oscillation with an amplitude of around half a metre and a period of about 20 s at nominal flow.

Similar techniques can be applied to other non-linearities; a limiter, shown in Figures 1.2.49(a) and (b), for example, will have unity gain for input amplitudes less than the limiting level. For increasing amplitude the apparent gain will decrease. The describing function when limiting occurs has a gain dependent on the ratio between the input signal amplitude and the limiting value as plotted in Figure 1.2.49(c). There is no phase shift between input and output.

Hysteresis, shown in Figure 1.2.50, introduces a phase shift, and a flat top to the output waveform. This is not the same waveform as the limiter; the top is simply levelled off at $2a$ below the peak where a is half the dead zone width. If the input amplitude is large compared to the dead zone, the gain is unity and the phase shift can be approximated by

$$\phi = \sin^{-1}(a/V_i)$$

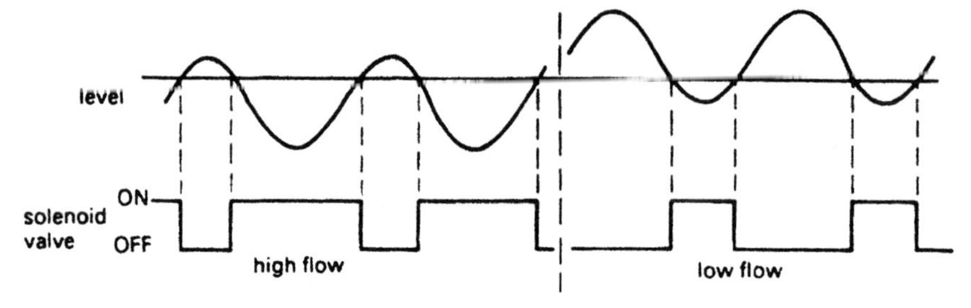

Fig. 1.2.48 Response of level control system to changes in flow.

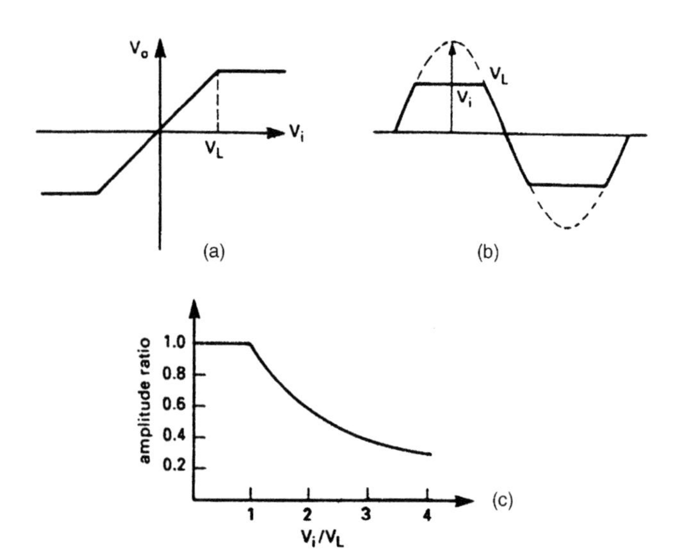

Fig. 1.2.49 The limiter circuit: (a) relationship between input and output; (b) effect of limiting on a sine wave input signal; (c) 'gain' of a limiter related to the input signal amplitude.

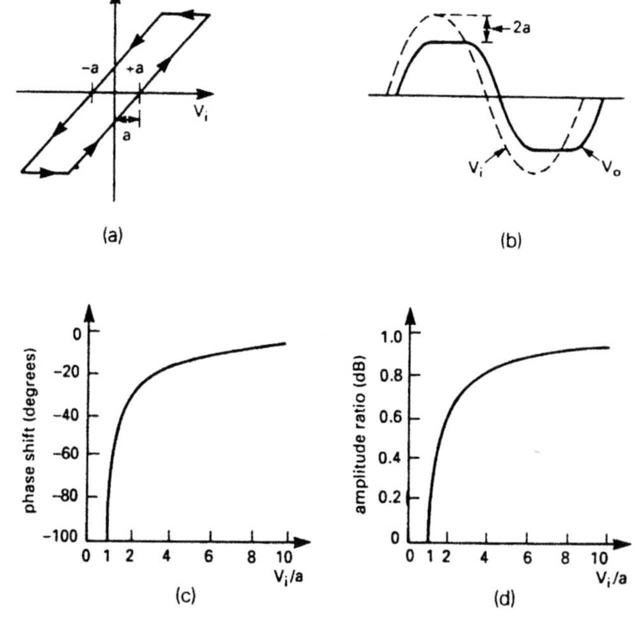

Fig. 1.2.50 The effect of hysteresis: (a) relationship between input and output signals; (b) the effect of hysteresis on a sine wave input signal; (c) the relationship between phase shift and signal amplitude; (d) the relationship between 'gain' and signal amplitude.

As the input amplitude decreases, the gain increases and becomes zero when the input peak to peak amplitude is less than the dead zone width. The exact relationship is complex, but is shown in Figures 1.2.50(c) and (d).

Non-linear elements generally have gains and phase shift which increase or decrease with input amplitude (usually a representation of the error signal). Figure 1.2.51 illustrates the two gain cases. For a loop gain of unity, constant oscillations will result. For loop gains greater than unity, oscillations will increase in amplitude, for loop gain less than unity oscillations will decay.

In Figure 1.2.51(a), the gain falls off with increasing amplitude. The system thus tends to approach point X as large oscillations will decay and small oscillations increase. The system will oscillate at whatever gain gives unity loop gain. This is called *limit cycling*. Most non-linearities (bang-bang servo, saturation, etc.) are of this form.

Where loop gain increases with amplitude as Figure 1.2.51(b), decreasing gain gives increasing damping as the amplitude decreases, so oscillations will quickly die away. This response is sometimes deliberately introduced into level controls. If, however, the system is provoked beyond Y by a disturbance, the oscillation will rapidly increase in amplitude and control will be lost.

1.2.22.3 State space and the phase plane

Figure 1.2.52(d) shows a simple position control system. The position is sensed by a potentiometer, and compared with a setpoint from potentiometer RV_1. The resulting error signal is compared with an error 'window' by comparators C_1, C_2. Preset RV_2 sets the deadband, i.e. the width of the window. The comparators energise relays RL_F and RL_R which drive the load to the forward and reverse respectively.

Initially, we shall analyse the system with RV_2 set to zero, i.e. no deadband. This has the block diagram of

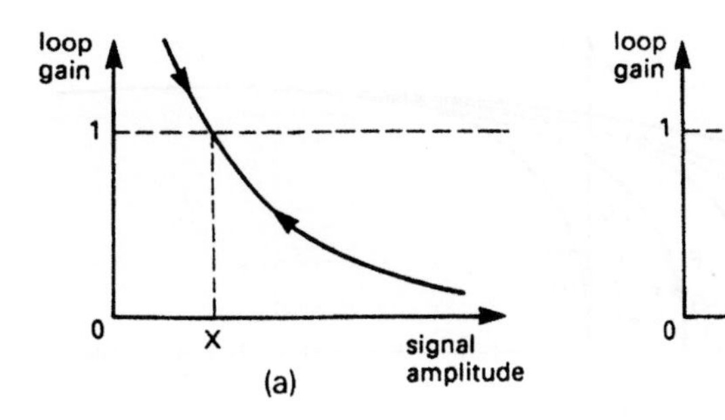

Fig. 1.2.51 Possible relationships between gain and signal amplitude: (a) gain decreases with increasing amplitude; (b) gain increases with increasing amplitude.

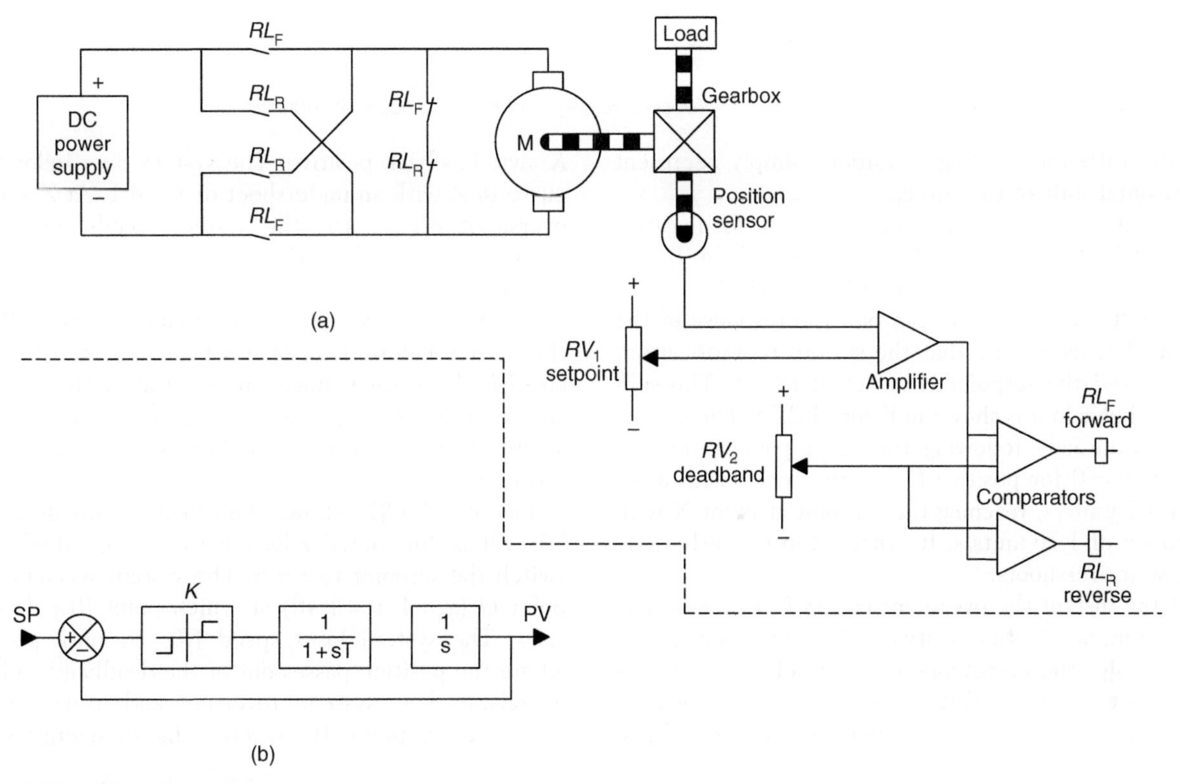

Fig. 1.2.52 A non-linear position control system: (a) system diagram; (b) block representation.

Figure 1.2.52(b), with a first-order lag of time constant T arising from the inertia of the system, and the integral action converting motor velocity to load position.

The system is thus represented by

$$x = \frac{\pm K}{s(1 + sT)} \tag{1.2.19}$$

where K represents the acceleration resulting from the motor torque and inertia with the sign of K indicating the sign of the error. This has the solution

$$x = x_0 - TK + TV_0 + Kt + T(K - V_0)e^{-t/T} \tag{1.2.20}$$

where x_0 and V_0, respectively, represent the initial position and velocity.

Differentiating gives the velocity, V

$$V = K - (K - V_0)e^{-t/T} \tag{1.2.21}$$

Equations (1.2.20) and (1.2.21) fully describe the behaviour of the system. These can be plotted graphically as Figure 1.2.53 with velocity plotted against position for positive K for various times from $t = 0$. Each curve represents a different starting condition; curve D, for example, starts at $x_0 = -5$ and $v_0 = -2$.

In each case, the curve ends towards $v = 2$ units/s as t gets large. The family of curves have an identical shape,

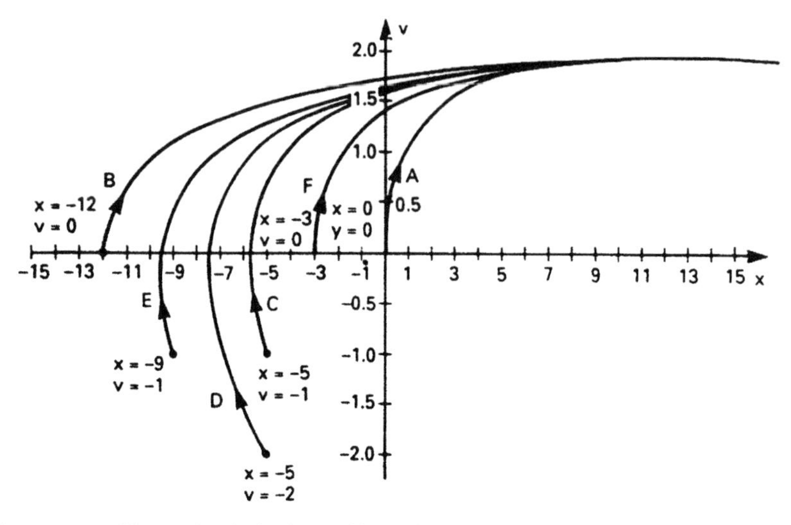

Fig. 1.2.53 Relationship between position and velocity for positive values of K for various initial conditions

and the different starting conditions simply represent a horizontal shift of the curve.

A similar family of curves can be drawn for negative values of K. These are sketched in Figure 1.2.54. In this case, the velocity tends towards $V = -2$ units/s.

Given these curves, we can plot the response of the system. Let us assume that the system is stationary at $x = -5$, and the setpoint is switched to $+5$. The subsequent behaviour is shown in Figure 1.2.55. The system starts by initially following the curve passing through $x = -5$, $V = 0$ for positive K, crossing $x = 0$ with a velocity 1.5 units/s, reaching the setpoint at point X with a velocity of 1.76 units/s. It cannot stop instantly however, so it overshoots.

At the instant the overshoot occurs K switches sign. The system now has a velocity of $+1.76$, with K negative, so it follows the corresponding curve of Figure 1.2.54 from point X to point Y. It can be seen that an overshoot to $x = 7$ occurs. At point Y, another overshoot occurs and K switches back positive. The system now follows the curve to Z with an undershoot of $x = 4.1$. At Z another overshoot occurs and the system spirals inwards as shown. The predicted step response is shown in Figure 1.2.56.

In Figure 1.2.57(a), the deadband control (RV_2 in the earlier Figure 1.2.52) has been adjusted to energise RL_F for error voltages more negative than -1 unit and energise RL_R for error voltages above $+1$ unit. There is thus a dead-band 2 units wide around the setpoint.

Figure 1.2.57(b) shows the effect of this deadband. We will assume initial values of $x_0 = 0$, $v_0 = 0$ when we switch the setpoint to $x = 5$. The system accelerates to point U ($x = 4$, $v = 1.40$) at which point RL_1 de-energises. The system loses speed ($K = 0$) until point V, where the position passes out of the deadband and RL_R energises. The system reverses, and re-enters the deadband at point W, where RL_F de-energises. An

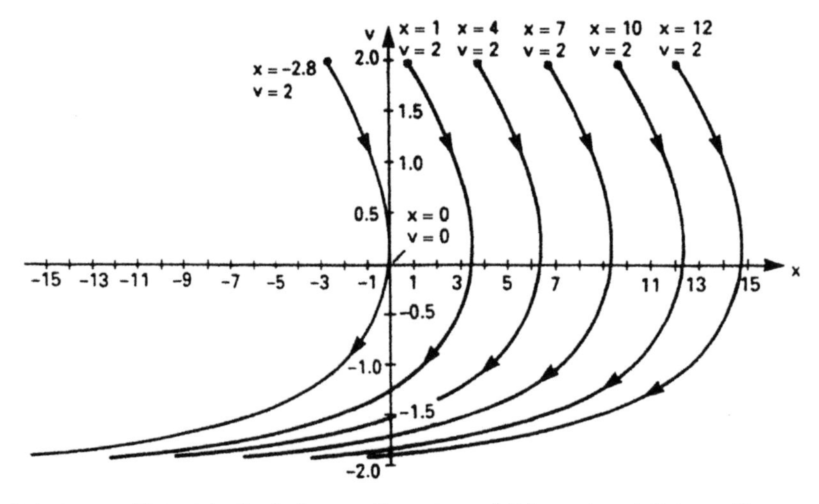

Fig. 1.2.54 Relationship between position and velocity for negative values of K *for* various initial conditions.

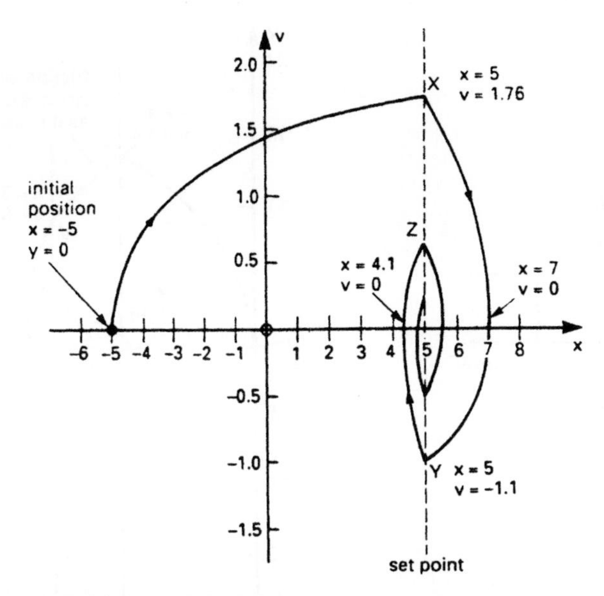

Fig. 1.2.55 System behaviour following change of setpoint from $x = -5$ to $x = +5$.

undershoot then occurs (X to Y) where the deadband is entered for the last time, coming to rest at point Z ($x = 4.75$, $v = 0$).

The position x and velocity dx/dt completely describe the system and are known as *state variables*. A linear-system can be represented as a set of first-order differential equations relating the various state variables. For a second-order system there are two state variables, for higher-order systems there will be more.

For the system described by Equation (1.2.19), we can denote the state variables by x (position) and v (velocity). For a driving function K, we can represent the system by Figure 1.2.58 which is called a *state space model*. This

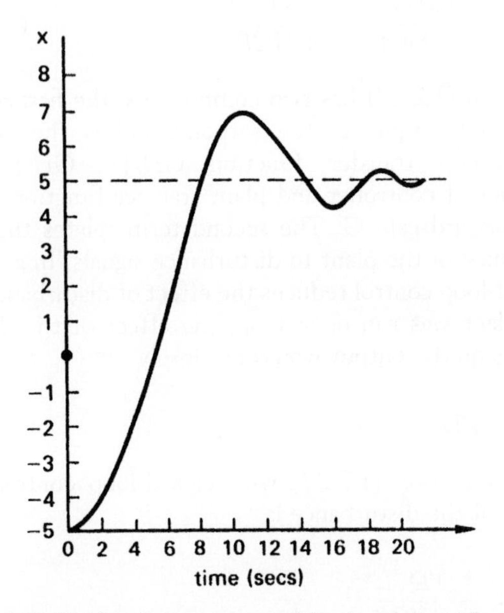

Fig. 1.2.56 Predicted step response following change of setpoint.

describes the position control system by the two first-order differential equations:

$$T\frac{dv}{dt} = K - x$$

and

$$v = \frac{dx}{dt}$$

Figures 1.2.56 and 1.2.57 plot velocity against position, and as such are plots relating state variables. For two state variables (from a second-order system) the plot is known as a *phase plane*. For higher-order systems, a multi-dimensional plot, called *state space*, is required. Plots such as Figures 1.2.53 and 1.2.54 which show a family of possible curves are called *phase plane portraits*.

Similar phase planes can be drawn for other non-linearities such as saturation, hysteresis, etc. Various patterns emerge, which are summarised in Figure 1.2.59. The system behaviour can be deduced from the shape of the phase trajectory.

In a linear closed-loop system stability is generally increased by adding derivative action. In a position control system this is equivalent to adding velocity (dx/dt) feedback. The behaviour of a non-linear system can also be improved by velocity feedback. In Figure 1.2.60(a) velocity feedback has been added to our simple Bang/Bang position servo.

The switching point now occurs where

$$S_P - x - Lv = 0$$

or

$$v = \frac{2}{L}(S_P - x)$$

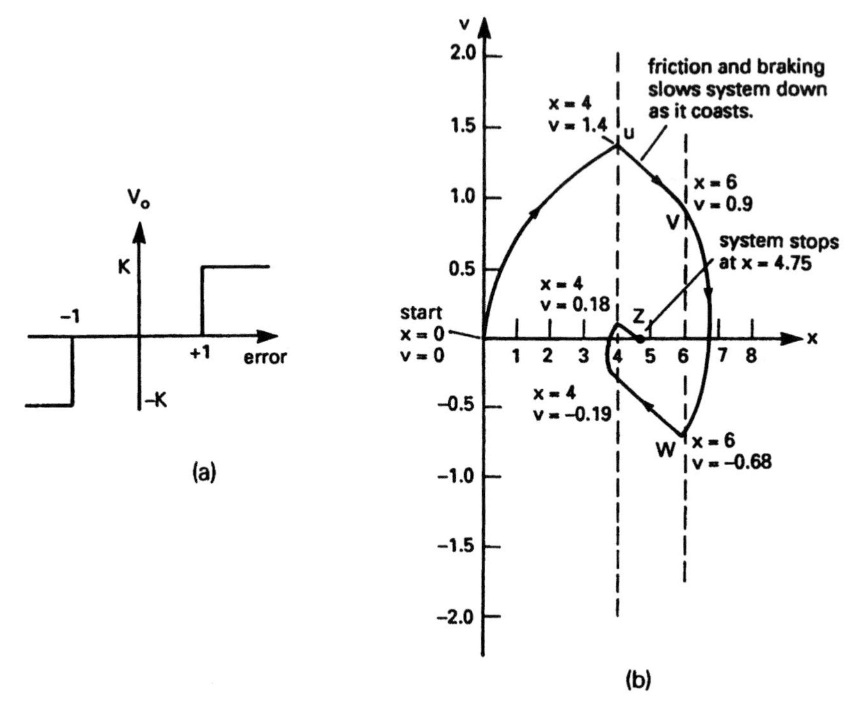

Fig. 1.2.57 System with deadband and friction: (a) deadband response; (b) position/velocity curve for setpoint change from $x = 0$ to $x = 5$. Note system does not attain the setpoint.

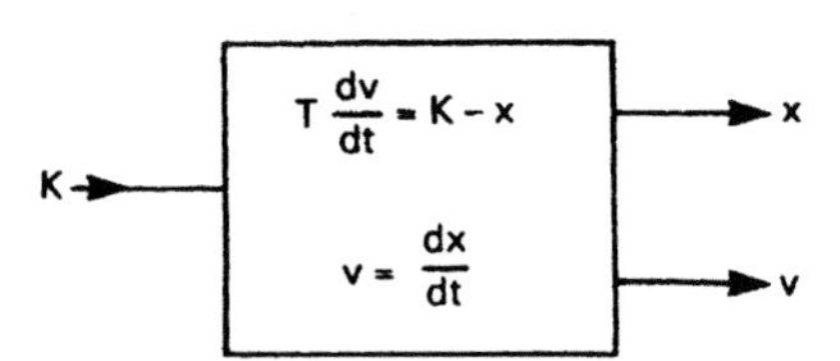

Fig. 1.2.58 State variables for position control system.

This is a straight line of slope $-1/L$, passing through $x = S_P$, $v = 0$ on the phase plane. Note that L has the units of time. The line is called the switching line, and advances the changeover as shown in Figure 1.2.60(b), thereby reducing the overshoot. Too much velocity feedback as Figure 1.2.60(c) simulates an overdamped system as the trajectory runs to the setpoint down the switching line.

1.2.23 Disturbances

1.2.23.1 Introduction

A closed-loop control system has to deal with the malign effects of outside disturbances. A level control system, for example, has to handle varying throughput, or a gas fired furnace may have to cope with changes in gas supply pressure. Although disturbances can enter a plant at any point, it is usual to consider disturbances at two points; supply disturbances at the input to the plant and load/demand disturbances at the point of measurement as shown in Figure 1.2.61(a).

The closed-loop block diagram can be modified to include disturbances as shown in Figure 1.2.61(b). A similar block diagram could be drawn for load disturbances or disturbances entering at any point by subdividing the plant block. By normal analysis we have

$$V = \frac{CPS_P}{1 + HCP} + \frac{PD}{1 + HCP} \tag{1.2.22}$$

Equation (1.2.22) has two components; the first relates the plant output to the setpoint and is the normal closed-loop transfer function $GH/(1 + GH)$. The product of controller and plant transfer function CP is the forward gain G. The second term relates the performance of the plant to disturbance signals. In general, closed-loop control reduces the effect of disturbances. If the plant was run open loop, the effect of the disturbances on the output would be simply

$$V = PD$$

From Equation (1.2.22) with closed-loop control, the effect of the disturbance is

$$V = \frac{PD}{1 + HCP}$$

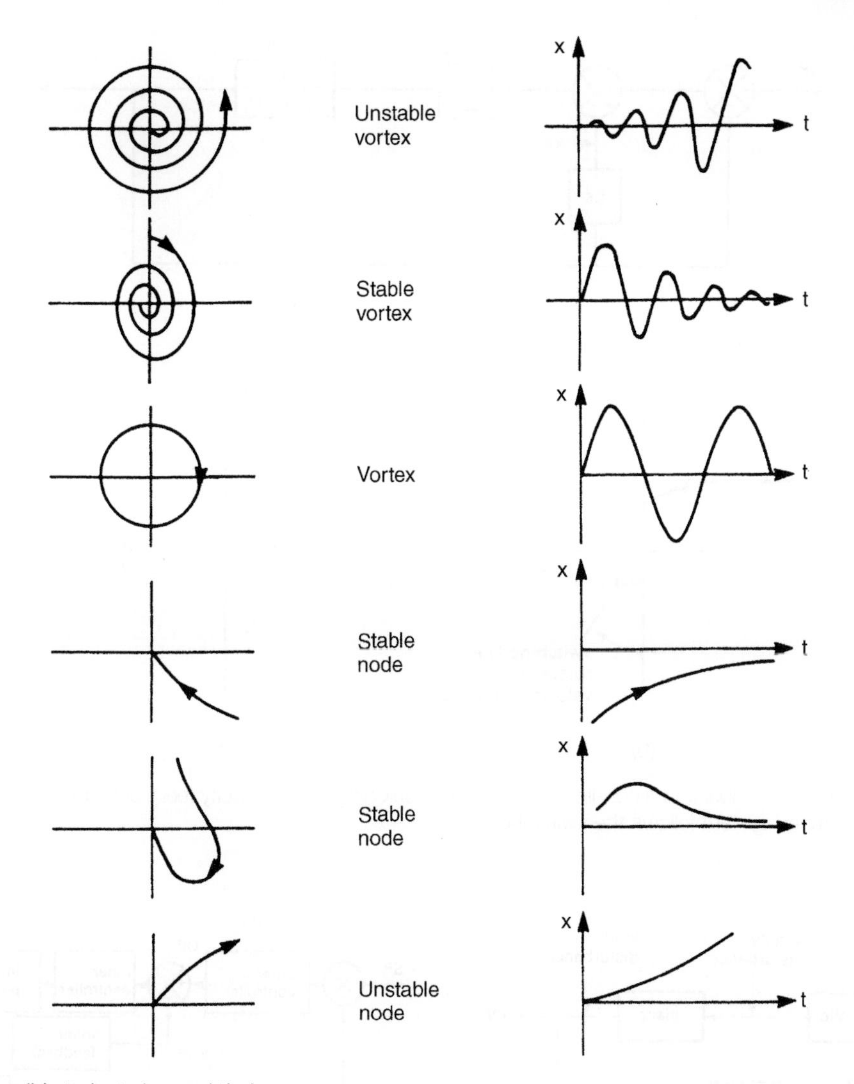

Fig. 1.2.59 Various possible trajectories and their response.

i.e. it is reduced providing the magnitude of $(1 + HCP)$ is greater than unity. If the magnitude of $(1 + HCP)$ becomes less than unity over some range of frequencies, closed-loop control will magnify the effect of disturbances in that frequency range. It is important, therefore, to have some knowledge of the frequency spectra of expected disturbances.

1.2.23.2 Cascade control

Closed-loop control gives increased performance over open-loop control, so it would seem logical to expect benefits from adding an inner control loop around plant items that are degrading overall performance. Figure 1.2.62 shows a typical example, here the output of the outer loop controller becomes the setpoint for the inner controller. Any problems in the inner loop (disturbances, non-linearities, phase lag, etc.) will be handled by the inner controller, thereby improving the overall performance of the outer loop. This arrangement is known as *cascade control*.

To apply cascade control, there must obviously be some intermediate variable that can be measured (P_{Vi} in Figure 1.2.62) and some actuation point that can be used to control it.

Cascade control brings several benefits. The secondary controller will deal with disturbances before they can affect the outer loop. Phase shift within the inner loop is reduced, leading to increased stability and speed of response in the outer loop. Devices with inherent integral action (such as a motorised valve) introduce an inherent $-90°$ integrator phase lag. This can be removed by adding a valve positioner in cascade. Cascade control will also reduce the effect of non-linearities (e.g. non-linear gain, backlash) in the inner loop.

There are a few precautions that need to be taken, however. The analysis so far ignores the fact that

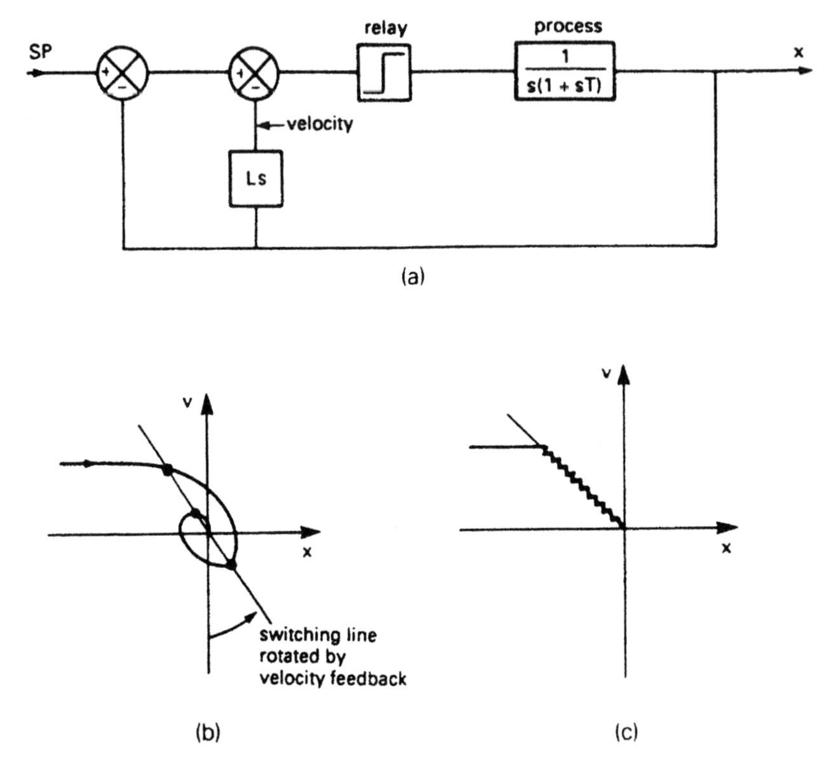

Fig. 1.2.60 Addition of velocity feedback to a non-linear system: (a) block diagram of velocity feedback; (b) system behaviour on velocity/position curve; (c) overdamped system follows the switching line.

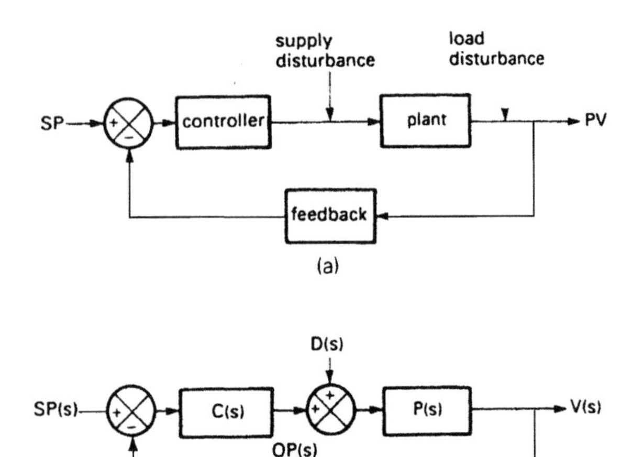

Fig. 1.2.61 The effect of disturbances: (a) points of entry for disturbances; (b) block diagram of disturbances.

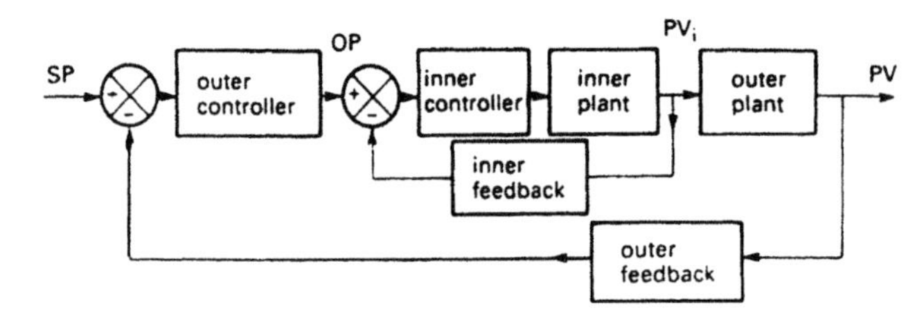

Fig. 1.2.62 A system with cascade control.

components saturate and stability problems can arise when the inner loop saturates. This can be overcome by limiting the demands that the outer loop controller can place on the inner loop (i.e. ensuring the outer loop controller saturates first) or by providing a signal from the inner to the outer controller which inhibits the outer integral term when the inner loop is saturated.

The application of cascade control requires an intermediate variable and control action point, and should include, if possible, the plant item with the shortest time constant. In general, high gain proportional only control will often suffice for the inner loop, any offset is of little concern as it will be removed by the outer controller. For stability, the inner loop must always be faster than the outer loop.

Tuning a system with cascade control requires a methodical approach. The inner loop must be tuned first with the outer loop steady in manual control. Once the inner loop is tuned satisfactorily, the outer loop can be tuned as normal. A cascade system, once tuned, should be observed to ensure that the inner loop does not saturate, which can lead to instability or excessive overshoot on the outer loop. If saturation is observed, limits must be placed on the output of the outer loop controller, or a signal provided to prevent integral windup as described later in Section 1.2.27.6

1.2.23.3 Feedforward

Cascade control can reduce the effect of disturbances occurring early in the forward loop, but generally cannot deal with load/demand disturbances which occur close to, or affect directly the process variable as there is no intermediate variable or accessible control point.

Disturbances directly affecting the process variable must produce an error before the controller can react. Inevitably, therefore, the output signal will suffer, with the speed of recovery being determined by the loop response. Plants which are difficult to control tend to have low gains and long integral times for stability, and hence have a slow response. Such plants are prone to error from disturbances.

In general a closed-loop system can be considered to behave as a second-order system, with a natural frequency ω_n and a damping factor. At frequencies above ω_n the closed-loop gain falls off rapidly (at 12 dB/octave). Disturbances occurring at a frequency much above $2\omega_n$ will be uncorrected. If the closed-loop damping factor is less the unity (representing an underdamped system), the effect of disturbances with frequency components around ω_n can be magnified.

Figure 1.2.63(a) shows a system being affected by a disturbance. Cascade control cannot be applied because there is no intermediate variable between the point of

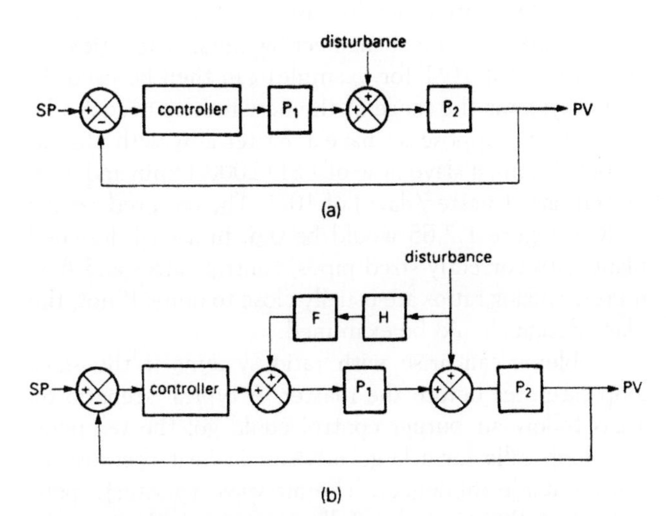

Fig. 1.2.63 Effect of a disturbance reduced by feedforward: (a) a system to which cascade control cannot be applied being subject to a disturbance; (b) correcting signal derived by measuring the disturbance.

entry and the process variable. If the disturbance can be measured, and its effect known (even approximately), a correcting signal can be added to the controller output signal to compensate for the disturbance as shown in Figure 1.2.63(b). This is known as *feedforward* control.

This correcting signal, arriving by blocks H, F, and P_1 should ideally exactly cancel the original disturbance, both in the steady state and dynamically under changing conditions. The transfer functions of the transducer H and plant P_1 are fixed, with F a compensator block designed to match H and P_1.

In general, the compensator block transfer function will be

$$F = -\frac{1}{HP_1}$$

If the plant acts as a simple lag with time constant T (i.e. $1/(1+sT)$), the compensator will be a simple lead $(1+sT)$. In many cases a general purpose compensator $(1+sT_a)/(1+sT_b)$is used.

The feedforward compensation does not have to match exactly the plant characteristics; even a rough model will give a significant improvement (although a perfect model will give perfect control). In most cases a simple compensator will suffice.

Cascade control can usually deal with supply disturbances and feedforward with load or demand disturbances. These neatly complement each other so it is very common to find a system where feedforward modifies the setpoint for the inner cascade loop.

1.2.24 Ratio control

1.2.24.1 Introduction

It is a common requirement for two flows to be kept in precise ratio to each other; gas/oil and air in combustion control, or reagents being fed to a chemical reactor are typical examples.

1.2.24.2 Slave follow master

In simple ratio control, one flow is declared to be the master. This flow is set to meet higher level requirements such as plant throughput or furnace temperature. The second flow is a slave and is manipulated to maintain the set flow ratio.

The controlled variable here is ratio, not flow, so an intuitive solution might look similar to Figure 1.2.64 where the actual ratio A/B is calculated by a divider module and used as the process variable for a controller which manipulated the slave control valve.

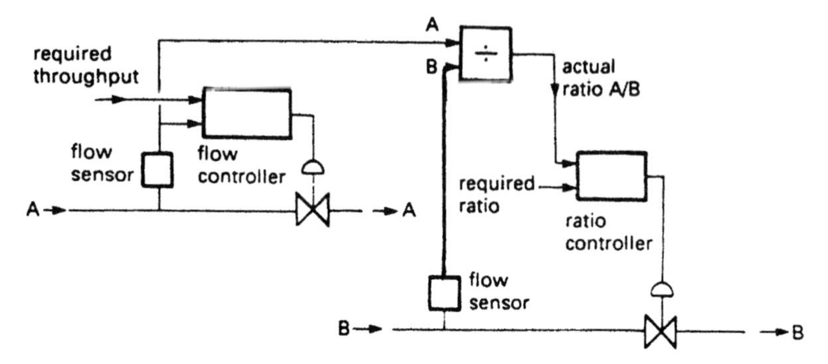

Fig. 1.2.64 An intuitive, but incorrect, method of ratio control. The loop gain varies with throughput.

This scheme has a hidden problem. The slave loop includes the divider module and hence the term A. The loop gain varies directly with the flow A, leading to a sluggish response at low flows and possible instability at high flow. If the inverse ratio B/A is used as the controller variable the saturation becomes worse as the term $1/A$ now appears in the slave loop giving a loop gain which varies inversely with A, becoming very high at low flows. Any system based on Figure 1.2.64 would be impossible to tune for anything other than constant flow rates.

Ratio control systems are often based on Figure 1.2.65. The master flow is multiplied by the ratio to produce the setpoint for the slave flow controller. The slave flow thus follows the master flow. Note that in the event of failure in the master loop (a jammed valve for example) the slave controller will still maintain the correct ratio.

The slave flow will tend to lag behind the master flow. On a gas/air burner, the air flow could be master and the gas loop the slave. Such a system would run lean on increasing heat and run rich on decreasing heat. To some extent this can be overcome by making the master loop slower acting than the slave loop, possibly by tuning.

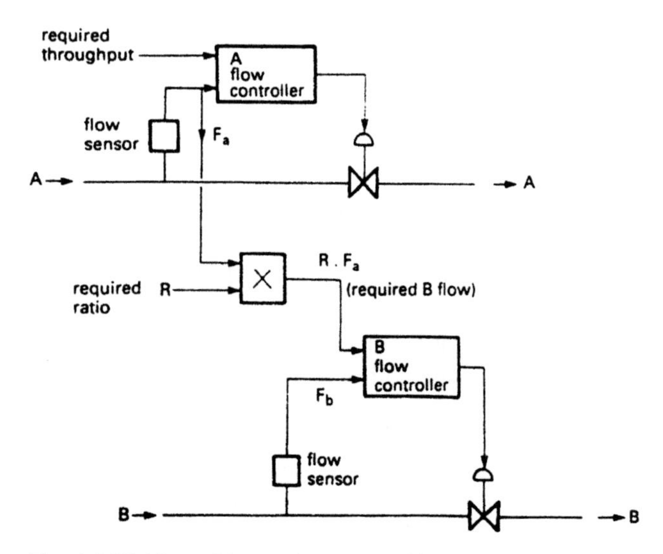

Fig. 1.2.65 Master/slave ratio system with stable loop gains.

In a ratio system, a choice has to be made for master and slave loops. The first consideration is usually safety. In a gas/air burner, for example, air master/gas slave (called *gas follow air*) is usually chosen as most failures in the air loop cause the gas to shut down. If there are no safety considerations, the slowest loop should be the master and the fastest loop the slave to overcome the lag described above. Since 'fuel' (in both combustion and chemical terms) is usually the smallest flow in a ratio system and consequently has smaller valves/actuators, the safety and speed requirement are often the same.

The ratio block is a simple multiplier. If the ratio is simply set by an operator this can be a simple potentiometer acting as a voltage divider (for ratios less than unity) or an amplifier with variable gain (for ratios greater than unity). In digital control systems, of course, it is a simple multiply instruction. If the ratio is to be changed remotely (a trim control from an automatic sampler on a chemical blending system for example) a single quadrant analog multiplier is required.

Ratio blocks are generally easier to deal with in digital systems working in real engineering units. True ratios (an air/gas ratio of 10/1 for example) can then be used. In analog systems the range of the flow meters needs to be considered. Suppose we have a master flow with FSD of 12,000 l/min, a slave flow of FSD 2000 l/min and a required ratio (master/slave) of 10/1. The required setting of R in Figure 1.2.65 would be 0.6. In a well-designed plant with correctly sized pipes, control valves and flow meters, analog ratios are usually close to unity. If not, the plant design should be examined.

Problems can arise with ratio systems if the slave loop saturates before the master. A typical scenario on a gas follow air burner control could go; the temperature loop calls for a large increase in heat (because of some outside influence). The air valve (master) opens fully, and the gas valve follows correctly but cannot match the requested flow. The resulting flame is lean and cold (flame temperature falls off rapidly with too lean a ratio) and the temperature does not rise. The system is now locked with the temperature loop demanding more heat and the air/gas loops saturated, delivering full flow but no temperature rise. The moral is; the master loop must saturate before the slave. If this is not achieved by pipe sizing the output of the master controller should be limited.

1.2.24.3 Lead–lag control

Slave follow Master is simple, but one side effect is that the mixture runs lean for increasing throughput and rich for decreasing throughput because the master flow must always change first before the slave can follow. There is also a possible safety implication because a failure of the slave valve or controller could lead to a gross error in the actual ratio such as the fuel valve wide open and the air valve closed.

Better performance can be obtained with a system called *Lead–lag control* shown for an air/fuel burner in Figure 1.2.66. This uses cross linking and selectors to provide an air setpoint which is the highest of the external power demand signal or ratio'd fuel flow. The fuel setpoint is the lowest of the external power demand or ratio'd air flow.

This cross linking provides better ratio during changes, both air and fuel will change together. There is also higher security; a jammed open fuel valve will cause the air valve to open to maintain the correct ratio and prevent an explosive atmosphere of unburned fuel forming.

1.2.25 Transit delays

1.2.25.1 Introduction

Transit delays are a function of speed, time and distance. A typical example from the steel industry is the tempering process of Figure 1.2.67 where red hot rolled steel travelling at 15 m/s is quenched by passing beneath high pressure water sprays. The recovery temperature, some 50 m downstream, is the controlled variable which is measured by a pyrometer and used to adjust the water flow control valve. There is an obvious transit delay of $50/15 = 3.3$ s in the loop. A transit delay is a simple time shift which is independent of frequency.

Transit delays give an increasing phase shift with rising frequency which is de-stabilising. If conventional controllers are used significant detuning (low gain, large T_i) is necessary to maintain stability. The effect is shown on the Nichols charts of Figure 1.2.68 for a simple system of two first-order lags controlled by a PI controller. The destabilising effect of the increasing phase shift can clearly be seen. Derivative action, normally a stabilising influence, can also adversely affect a loop in which a transit delay is the dominant feature.

1.2.25.2 The Smith predictor

The effects of a transit delay can be reduced by the arrangement of Figure 1.2.69 called a *Smith predictor*. The plant is considered to be an ideal plant followed by a transit

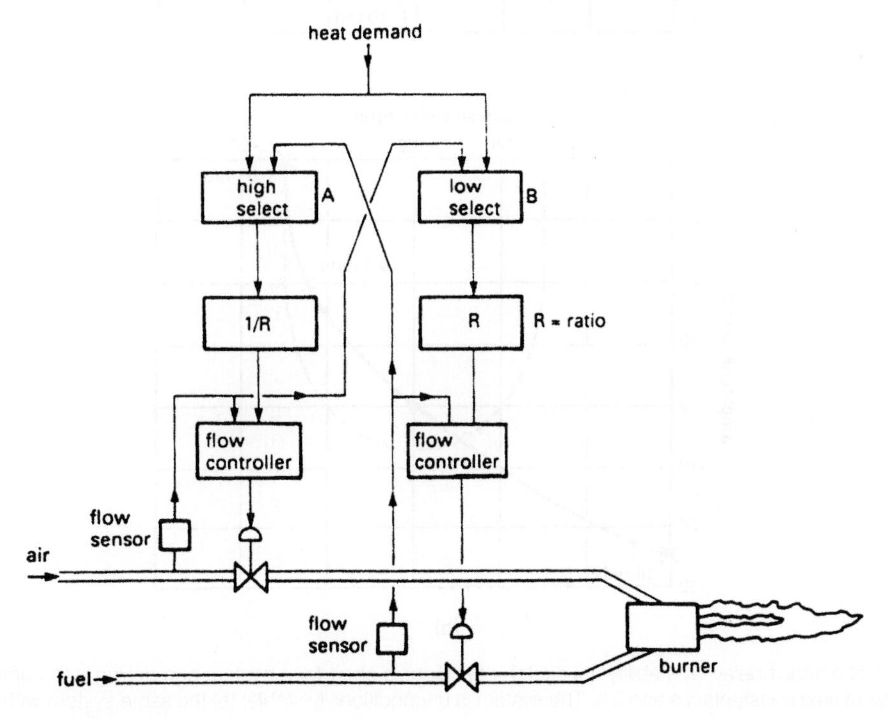

Fig. 1.2.66 Lead/lag combustion control.

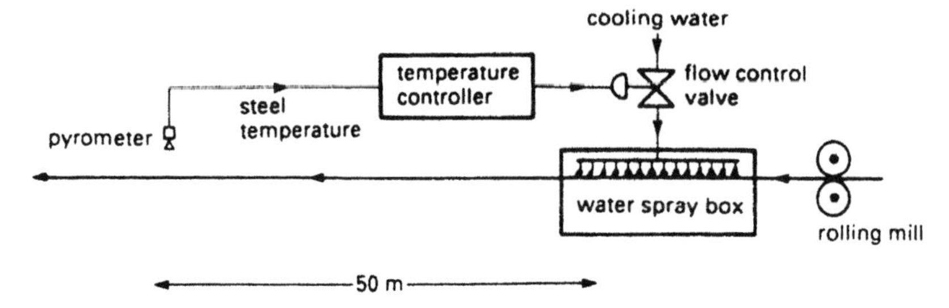

Fig. 1.2.67 A tempering system dominated by a transit delay.

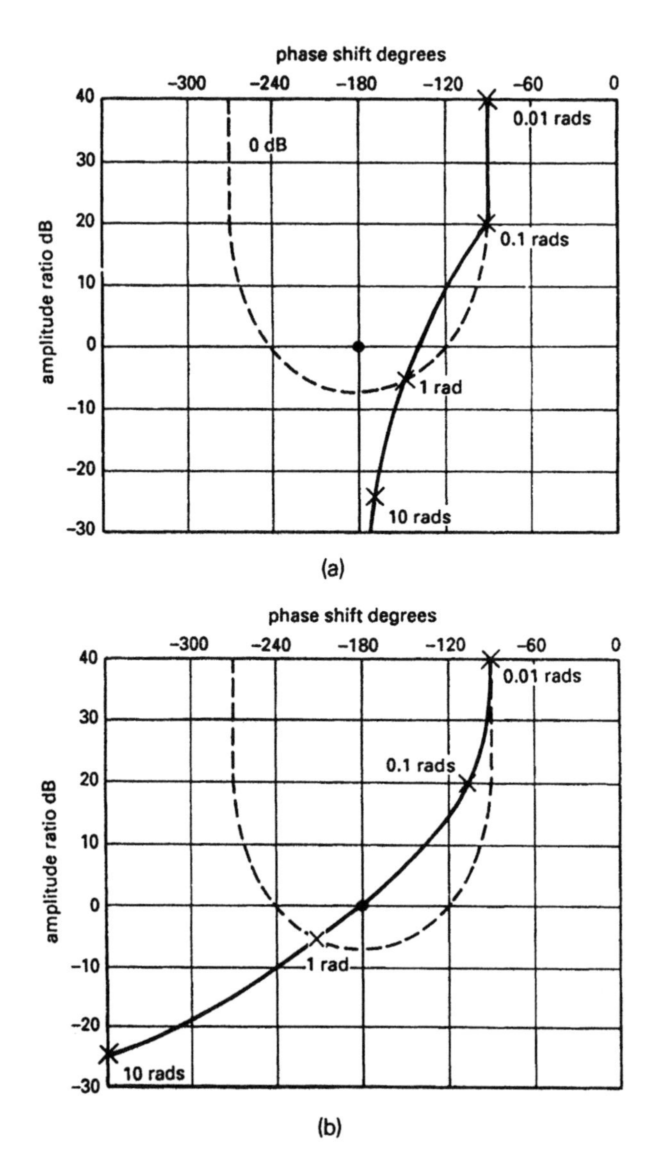

Fig. 1.2.68 The effect of a transit delay on stability: (a) sketch of a Nichols chart for a system comprising a PI controller ($K = 5$, $T_i = 5$ s) and two first-order lags of time constants 5 s and 2 s. The system is unconditionally stable; (b) the same system with a one second transit delay. The transit delay introduces a phase shift which increases with rising frequency and makes the system unstable.

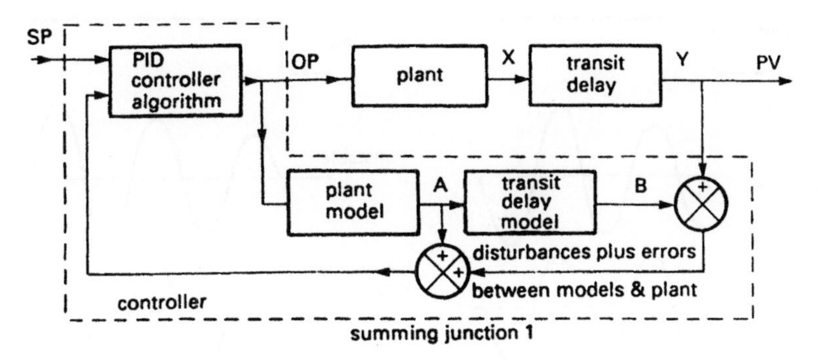

Fig. 1.2.69 The Smith predictor used to reduce the effect of transit delays.

delay. (This may not be true, but the position of the transit delay, before or after the plant, makes no difference to the plant behaviour.) The plant and its associated delay are modelled as accurately as possible in the controller.

The controller output, O_P is applied to the plant and to the internal controller model. Signal *A* should thus be the same as the notional (and unmeasurable) plant signal X, and the signal *B* should be the same as the measurable controlled variable signal *Y*.

The PID controller however, is primarily controlling the model, not the plant, via summing junction 1. There are no delays in this loop, so the controller can be tuned for tight operation. With the model being the only loop, however, the plant is being operated in open-loop control, and compensation will not be applied for model inaccuracies or outside disturbances.

Signal *Y* and *B* are therefore compared by a subtractor to give an error signal which encompasses errors from both disturbances and the model. These are added to the signal *A* from the plant model to give the feedback signal to the PID control block.

Discrepancies between the plant model and the real plant will be compensated for in the outer loop, so exact modelling is not necessary. The poorer the model, however, the less tight the control that can be applied in the PID block as the errors have to be compensated via the plant transit delay.

Smith predictors are usually implemented digitally, analog transit delays being difficult to construct. A digital delay line is simply a shift register in which values are shifted one place at each sample.

The Smith predictor is not a panacea for transit delays; it still takes the delay time from a setpoint change to a change in the process variable, and it still takes the delay time for a disturbance to be noted and corrected. The response to change, however, is considerably improved.

Systems with transit delays can benefit greatly from feedforward described previously in Section 1.2.23.3. Feedforward used in conjunction with a Smith predictor can be a very effective way of handling control systems with significant transit delays.

1.2.26 Stability

1.2.26.1 Introduction

At first sight it would appear that perfect control can be obtained by utilising a large proportional gain, short integral time and long derivative time. The system will then respond quickly to disturbances, alterations in load and setpoint changes.

Unfortunately life is not that simple, and in any real life system there are limits to the settings of gain T_i and T_d beyond which uncontrolled oscillations will occur. Like many engineering systems, the setting of the controller is a compromise between conflicting requirements.

1.2.26.2 Definitions and performance criteria

It is often convenient, (and not too inaccurate), to consider that a closed-loop system behaves as a second-order system, with a natural frequency ω_n and a damping factor β:

$$\frac{d^2x}{dt^2} + 2\beta\omega_n\frac{dx}{dt} + \omega_n^2x = f(t)$$

It is then possible to identify five possible performance conditions, shown for a setpoint change and a disturbance in Figures 1.2.70(a) and (b).

An unstable system exhibits oscillations of increasing amplitude. A marginally stable system will exhibit constant amplitude oscillations. An underdamped system will be somewhat oscillatory, but the amplitude of the oscillations decreases with time and the system is stable. (It is important to appreciate that oscillatory does not necessarily imply instability.) The rate of decay is determined by the damping factor. An often used performance criteria is the '*quarter amplitude damping*' of Figure 1.2.70(c) which is an under-damped response with each cycle peak one-quarter of the amplitude of the previous. For many

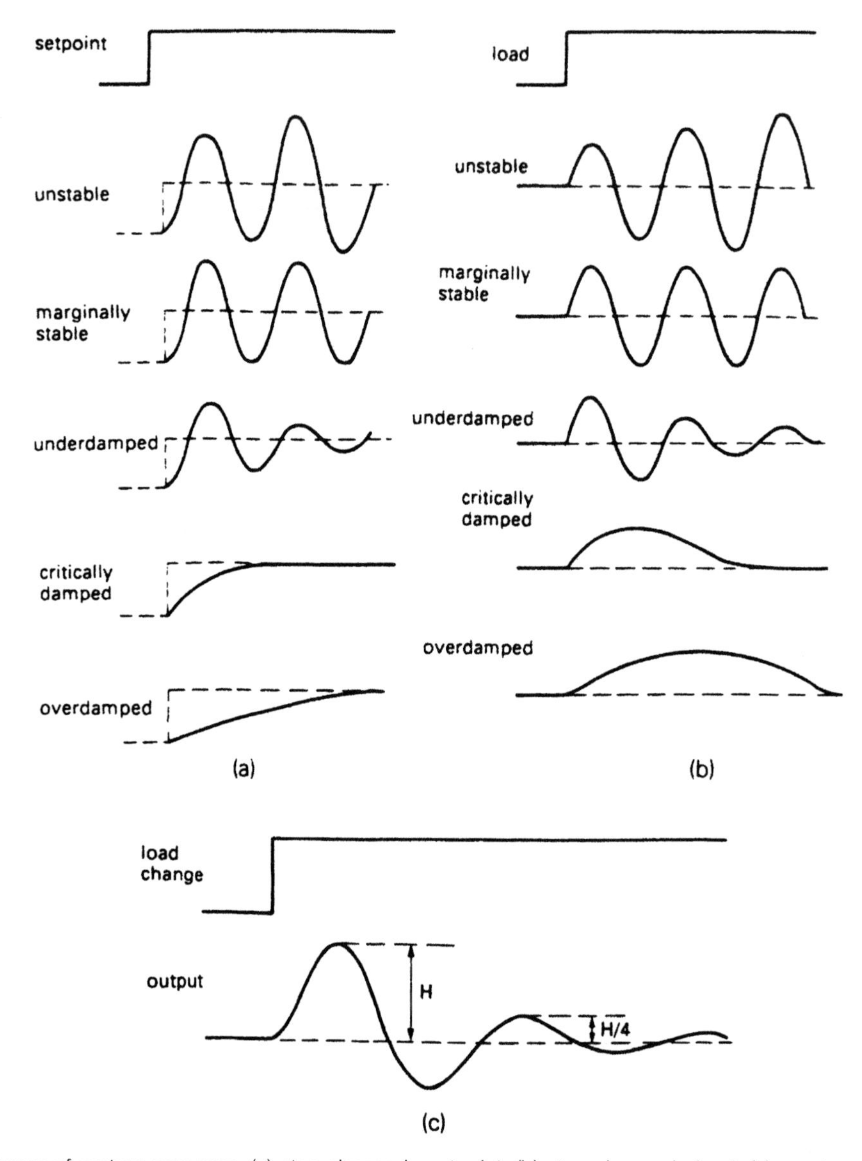

Fig. 1.2.70 Various forms of system response: (a) step change in setpoint; (b) step change in load; (c) quarter amplitude damping.

applications this is an adequate, and easily achievable response.

An overdamped system exhibits no overshoot and a sluggish response. A critical system marks the boundary between underdamping and overdamping and defines the fastest response achievable without overshoot.

For a simple system the responses of Figures 1.2.70(a) and (b) can be related to the gain setting of a *P* only controller, overdamped corresponding to low gain with increasing gain causing the response to become underdamped and eventually unstable.

It is impossible for any system to respond instantly to disturbances and changes in set point. Before the adequacy of a control system can be assessed, a set of performance criteria is usually laid down by production staff. Those defined in Figure 1.2.71 are commonly used.

The *'rise time'* is the time taken for the output to go from 10% to 90% of its final value, and is a measure of the speed of response of the system. The time to achieve 50% of the final value is called the *'delay time'*. This is a function of, but not the same as, any transit delays in the system. The first overshoot is usually defined as a percentage of the corresponding setpoint change, and is indicative of the damping factor achieved by the controller.

As the time taken for the system to settle completely after a change in set point is theoretically infinite, a *'settling band, 'tolerance limit'* or *'maximum error'* is usually defined. The settling time is the time taken for the system to enter, and remain within, the tolerance limit. Surprisingly an under-damped system may have a better settling time than a critically damped system if the first

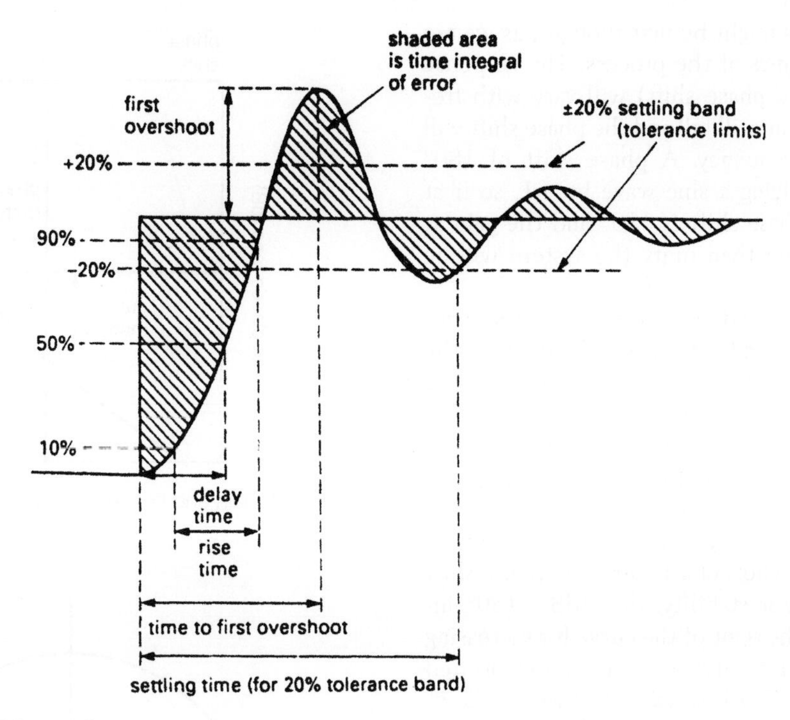

Fig. 1.2.71 Common definitions of system response.

overshoot is just within the settling band. Table 1.2.2 shows optimum damping factors for various settling bands. The settling time is defined in units of $1/\omega_n$.

The shaded area is the integral of the error and this can also be used as an index of performance. Note that for a system with a standing offset (as occurs with a P only controller) the area under the curve will increase with time and not converge to a final value. Stable systems with integral action control have error areas that converge to a finite value. The area between the curve and the setpoint is called the *integrated absolute error* (IAE) and is an accepted performance criterion.

An alternative criterion is the integral of the square of the instantaneous error. This weights large errors more than small errors, and is called *integrated squared error* (ISE). It is used for systems where large errors are detrimental, but small errors can be tolerated.

The performance criteria above were developed for a setpoint change. Similar criteria can be developed for disturbances and load changes.

1.2.26.3 Methods of stability analysis

The critical points for stability are open-loop unity gain and a phase shift of $-180°$. It is therefore reasonable to give two figures of 'merit':

(a) The *gain margin* is the amount by which the open loop gain can be increased at the frequency at which the phase shift is $-180°$. It is simply the inverse of the gain at this critical frequency, for example if the gain at the critical frequency is 0.5, the gain margin is two.

(b) The *phase margin* is the additional phase shift that can be tolerated when the open-loop gain is unity. With $-140°$ phase shift at unity gain, there is a phase margin of 40°.

For a reasonable, slightly underdamped, closed-loop response the gain margin should be of the order of 6–12 dB and the phase margin of the order of 40–65°.

Any closed-loop control system can be represented by Figure 1.2.72 where G is the combined block transfer function of the controller and plant and H the transfer function of the transducer and feed back components. The output will be given by

$$P_V = \frac{G}{1+GH} S_P$$

The system will be unstable if the denominator goes to zero or reverses in sign, i.e. $GH \leq -1$. This is not as

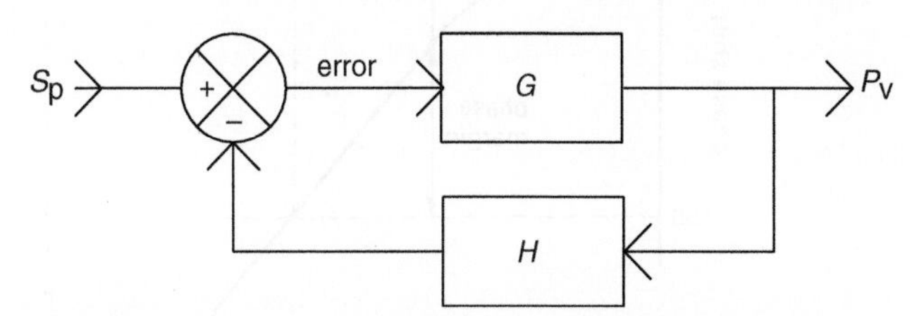

Fig. 1.2.72 General block diagram of a closed-loop control system.

simple a relationship as might be first thought, as we are dealing with the dynamics of the process. The response of the system (gain and phase shift) will vary with frequency; generally the gain will fall and the phase shift will rise with increasing frequency. A phase shift of 180° corresponds to multiplying a sine wave by −1, so if at some frequency the phase shift is 180° and the gain at that frequency is greater than unity the system will be unstable.

There are several methods of representing the gain/phase shift relationship, and inferring stability from the plot. Figure 1.2.73 is called a Bode diagram and plots the gain (in dB) and phase shift on separate graphs. Log-Lin graph paper (e.g. Chartwell 5542) is required. For stability, the gain curve must cross the 0 dB axis before the phase shift curve crosses the 180° line. From these two values, the gain margin and the phase margin can be read as shown.

Figure 1.2.74 is a Nichols chart and plots phase shift against gain (in dB). For stability, the 0 dB/−180° intersection must be to the right of the curve for increasing frequency. Nichols charts are plotted on pre-printed graph paper (Chartwell 7514 for example) which allows the closed-loop response to be read directly. If for example the curve is inside the closed-loop 0 dB line damped oscillations will result. The gain and phase margins can again be read from the graph.

The final method is the Nyquist diagram of Figure 1.2.75. This plots gain again phase shift as a polar diagram (gain represented by distance from the origin). Chartwell graph paper 4001 is suitable. For stability the −180° point must be to the left of the graph for increasing frequency. Gain and phase margin can again be read from the graph.

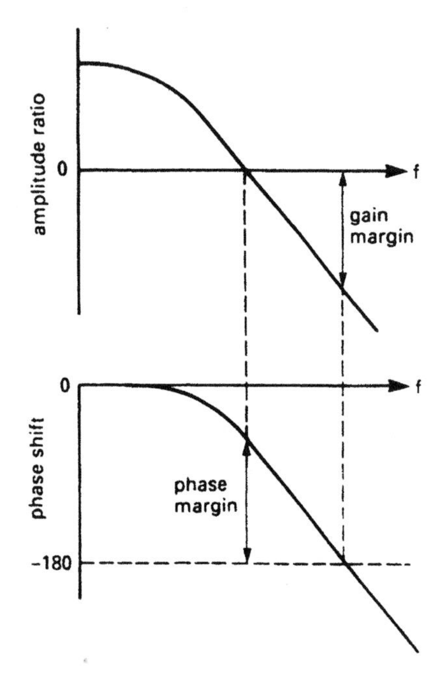

Fig. 1.2.73 Gain and phase margins on the Bode diagram.

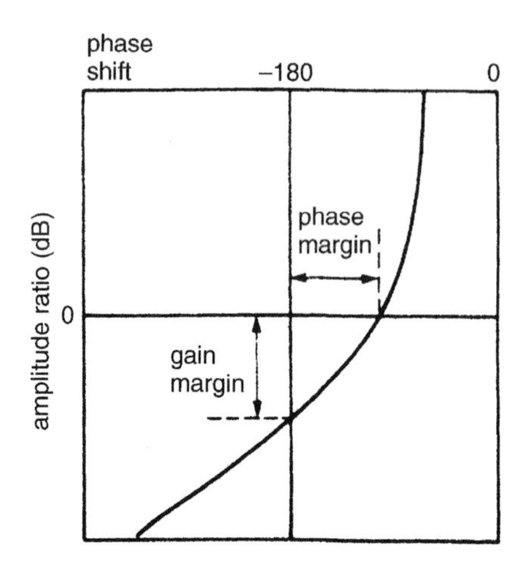

Fig. 1.2.74 Gain and phase margins on a Nichols chart.

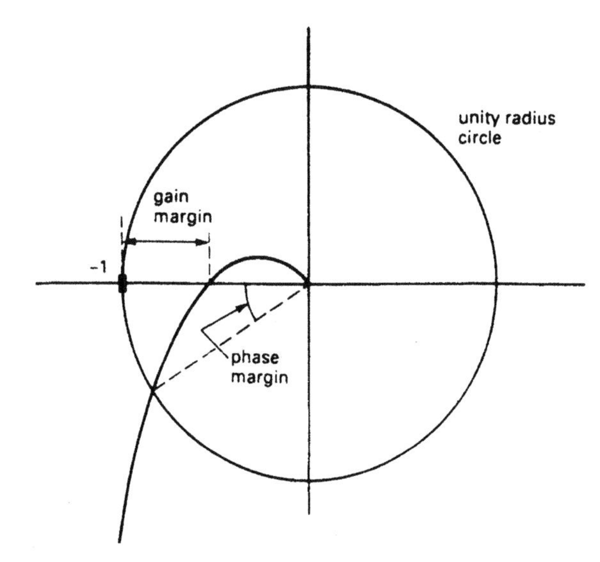

Fig. 1.2.75 Gain and phase margins on a Nyquist diagram.

1.2.27 Industrial controllers

1.2.27.1 Introduction

The commercial three term controller is the workhorse of process control and has evolved to an instrument of great versatility. This section describes some of the features of practical modern microprocessor based controllers.

1.2.27.2 A commercial controller

The description in this section is based on the 6360 controller manufactured by Eurotherm Process Automation Ltd of Worthing, Sussex.

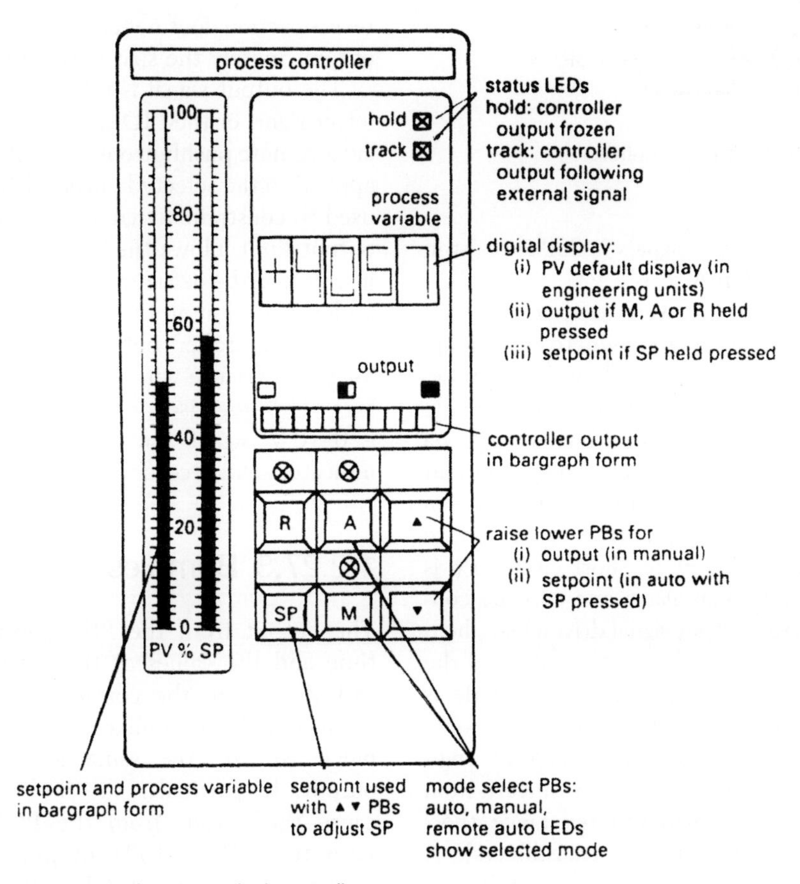

Fig. 1.2.76 Front panel operator controls on a typical controller.

The controller front panel is the 'interface' with the operator who may have little or no knowledge of process control. The front panel controls should therefore be simple to comprehend. Figure 1.2.76 shows a typical layout.

The operator can select one of three operating modes—manual, automatic or remote—via the three push buttons labelled M, A, R. Indicators in each push button show the current operating mode.

In manual mode, the operator has full control over the driven plant actuator. The actuator drive signal can be ramped up or down by holding in the M button and pressing the ▲ or ▼ buttons. The actuator position is shown digitally on the digital display, whilst the M button is depressed and continuously in analog form on the horizontal bar graph.

In automatic mode the unit behaves as a three term controller with a setpoint loaded by the operator. The unit is scaled into engineering units (i.e. real units such as °C, psi, litres/min) as part of the setup procedure so that the operator is working with real plant variables. The digital display shows the setpoint value when the SP button is depressed and the value can be changed with the ▲ and ▼ buttons.

The set point is also displayed in bar-graph form on the right-hand side of the dual vertical bar graph.

Remote mode is similar to automatic mode except the set point is derived from an external signal. This mode is used for ratio or cascade loops (see Sections 1.2.23 and 1.2.24) and batch systems where the setpoint has to follow a predetermined pattern. As before the setpoint is displayed in bar-graph form and the operator can view, but not change, the digital value by depressing the SP button.

The process variable itself is displayed digitally when no push button is depressed, and continually on the left-hand bar graph. In automatic or remote modes the height of the two left-hand bar graphs should be equal, a very useful quick visual check that all is under control.

Alarm limits (defined during the controller set up), can be applied to the process variable or the error signal. If either move outside acceptable limits, the process variable bar graph flashes, and a digital output from the controller is given for use by an external annunciator audible alarm of data logger.

Figure 1.2.77 shows a simple block diagram representation of a controller.

Input analog signals enter at the left-hand side. Common industrial signal standards are 0–10 V, 1–5 V, 0–20 mA and 4–20 mA. These can be accommodated by two switchable ranges 0–10 V and 1–5 V plus suitable

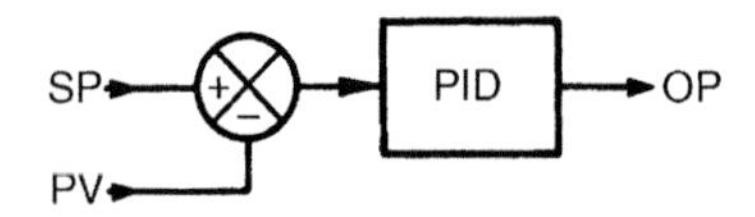

Fig. 1.2.77 Block diagram of a typical controller.

burden resistors for the current signals (a 250 Ω resistor, for example, converts 4–20 mA to 1–5 V).

Signals of 4–20 and 0–20 mA were used on two wire loops require a DC power supply somewhere in the loop. A floating 30 V power supply is provided for this purpose.

Open circuit detection is provided on the main P_V input. This is essentially a pull up to a high voltage via a high value resistor. A comparator signals an open circuit input when the voltage rises. Short circuit detection can also be applied on the 1–5 V input (the input voltage falling below 1 V). Open circuit or short circuit P_V is usually required to bring up an alarm and trip the controller to manual, with the output signal driven high, held at last value, or driven low according to the nature of the plant being controlled. The open circuit trip mode is determined by switches as part of the setup procedure.

The P_V and remote S_P inputs are scaled to engineering units and linearised. Common linearisation routines are thermocouples, platinum resistance thermometers and square root (for flow transducers). A simple adjustable first-order filter can also be applied to remove process or signal noise. The set point for the PID algorithm is selected from the internal set point or the remote set point by the from panel auto and remote push button contacts A, R.

The error signal is obtained by a subtractor (P_V and S_P both being to the same scale as a result of the scaling and engineering unit blocks). At this stage two alarm functions are applied. An absolute input alarm provides adjustable high and low alarm limits on the scaled and linearised P_V signals, and a deviation alarm (with adjustable limits) applied to the error signal. These alarm signals are brought out of the controller as digital outputs.

The basic PID algorithm is implemented digitally and includes a few variations to deal with some special circumstances. These modifications utilise the additional signals to the PID block (P_V, hold, track, output balance) and are described later.

The PID algorithm output is the actuator drive signal scaled 0–100%. The PID algorithm assumes that an increasing drive signal causes an increase in P_V. Some actuators, however, are reverse acting, with an increasing drive signal reducing P_V. A typical example is cooling water valves which are designed to fail open delivering full flow on loss of signal. Before the PID algorithm can be used with reverse acting actuators (or reverse acting transducers) its output signal must be reversed. A setup switch selects normal or inverted PID output. Note that reverse action does not alter the polarity of the controller output, merely the sign of the gain.

The output signal is selected from the manual raise/lower signal or the PID signal by the front panel manual/auto/remote pushbuttons M, A, R. At this stage limits are applied to the selected output drive. This limiting can be used to constrain actuators to a safe working range. The output limit allows the controller output to be limited just before the actuator's ends of travel, keeping the P_V under control at all times.

Two controller outputs are provided, 0–10 V and 4–20 mA for use with voltage and current driven actuators. The linearised P_V signal is also retransmitted as a 0–10 V signal for use with the separate external indicators and recorders.

1.2.27.3 Bumpless transfer

The output from the PID algorithm is a function of time and the values of the set point and the process variable. When the controller is operating in manual mode it is highly unlikely that the output of the PID block will be the same as the demanded manual output. In particular the integral term will probably cause the output from the PID block to eventually saturate at 0% or 100% output.

If no precautions are taken, therefore, switching from auto to manual, then back to auto again some time later will result in a large step change in controller output at the transition from manual to automatic operation.

To avoid this 'bump' in the plant operation, the controller output is fed back to the PID block, and used to maintain a PID output equal to the actual manual output. This balance is generally achieved by adjusting the contribution from the integral term.

Mode switching can now take place between automatic and manual modes without a step change in controller output. This is known as *manual/auto balancing, preload* or (more aptly) *bumpless transfer.*

A similar effect can occur on setpoint changes. With a straightforward PID algorithm, a setpoint change of ΔS_P will produce an immediate change in controller output of $K \cdot \Delta S_P$ where K is the controller gain. In some applications this step change in output is unacceptable. In Figure 1.2.78 a term $K \cdot S_P$ is subtracted from the PID block output. The controller now responds to errors caused by changes in P_V in the normal way, but only reacts to changes in S_P via the integral and derivative terms. Changes in S_P thus result in a slow change in controller output. This is known as *setpoint change balance*, and is a switch selectable setup option.

This balance signal fed back from the output to the PID block is also used when the controller output is forced to follow an external signal. This is called *track mode.*

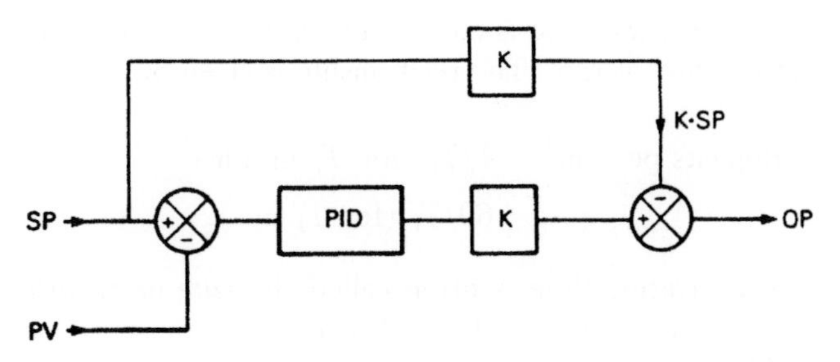

Fig. 1.2.78 Setpoint change balance, the controller only follows setpoint changes on the integral term giving a ramped response to a change of set point.

As before, the PID algorithm needs to be balanced to avoid a bump when transferring between track mode and automatic mode. The feedback output signal achieves this balance as described previously.

1.2.27.4 Integral windup and desaturation

Large changes in S_P or large disturbances to P_V can lead to saturation of the controller output or a plant actuator. Under these conditions the integral term in the PID algorithm can cause problems.

Figure 1.2.79 shows the probable response of a system with unrestricted integral action. At time A a step change in set point occurs. The output O_P rises first in a step (K × setpoint change) then rises at a rate determined by the integral time. At time B the controller saturates at 100% output, but the integral term keeps on rising.

At the time C P_V reaches, and passes, the required value, and as the error changes sign the integral term starts to decrease, but it takes until time D before the controller desaturates. Between times B and D the plant is uncontrolled, leading to an unnecessary overshoot and possibly even instability.

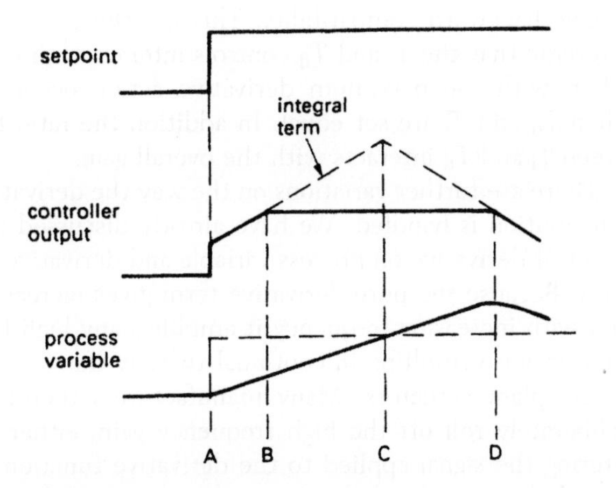

Fig. 1.2.79 The effect of integral windup.

This effect is called '*integral windup*' and is easily avoided by disabling the integral term once the controller saturates either positive or negative. This is naturally a feature of all commercial controllers, but process control engineers should always by suspicious of 'home brew' control algorithms constructed (or written in software) by persons without control experience.

In any commercial controller, the integral term would be disabled at point B in Figure 1.2.80 to prevent integral windup. The obvious question now is at what point it is re-enabled again. Point C is obviously far too late (although much better than point D in the unprotected controller).

A common solution is to desaturate the integral term at the point where the rate of increase of the integral action equals the rate of decrease of the proportional and derivative terms. This occurs when the slope of the PID output is zero, i.e. when

$$e = -T_i\left(\frac{de}{dt} + T_d\frac{d^2e}{dt^2}\right) \tag{1.2.23}$$

with e being the error and T_i and T_d the controller constants.

Equation (1.2.23) brings the controller out of saturation at the earliest possible moment, but this can, in some cases, be too soon leading to an unnecessarily damped response. Some controllers allow adjustment

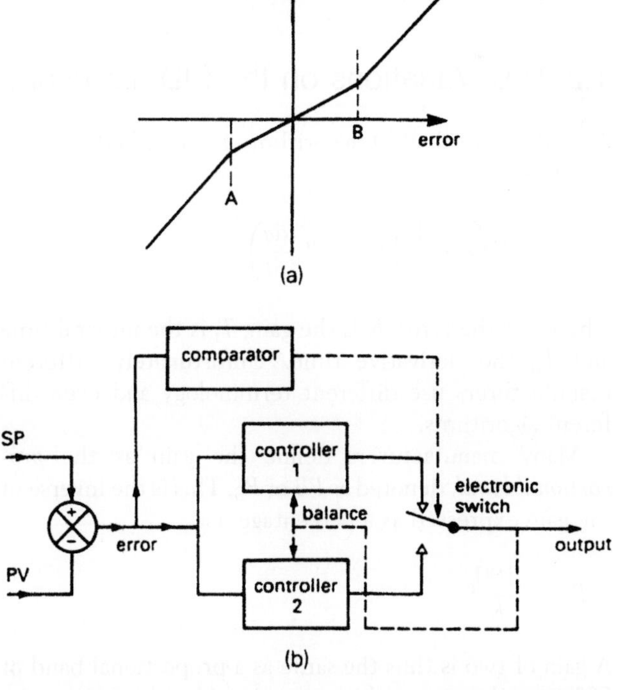

Fig. 1.2.80 Variable gain controller: (a) system response; (b) block diagram.

of the desaturation point by adding an error limit circuit to delay the balance point to Equation (1.2.23) forcing the controller to remain in saturation for a longer time. The speed of desaturation and the degree of overshoot can thus be adjusted by the commissioning engineer.

1.2.27.5 Selectable derivative action

The term T_d (de/dt) in the three term controller algorithm can be rearranged as

$$T_d\left(\frac{dS_P}{dt}-\frac{dP_V}{dt}\right)$$

where S_P is the set point and P_V the process variable. The derivative term thus responds to changes in both the set point and the plant feedback signal.

This is not always desirable; in particular, a step change in set point leads to an infinite spike controller output and a vicious 'kick' to the actuator. Commercial controller therefore include a selectable option for the derivative term to be based on true error (S_P–P_V) or purely on the value of P_V alone.

There is generally no noticeable difference in plant performance between these options; stability or the ability to deal with disturbances or load changes are unaffected, and derivative on P_V is normally the preferred choice. The only occasion when true derivative on error is advantageous is where the P_V is required to track a continually changing S_P.

1.2.27.6 Variations on the PID algorithm

The theoretical PID algorithm is described by the equation

$$O_P = K\left(e+\frac{1}{T_i}\int e\,dt+T_d\frac{de}{dt}\right)$$

where e is the error, K is the gain, T_i is the integral time and T_d the derivative time. Unfortunately different manufacturers use different terminology and even different algorithms.

Many manufacturers define the gain as the *proportional band*, denoted as *PB* or P_B. This is the inverse of the gain expressed as a percentage, i.e

$$P_B = \frac{100}{K}\%$$

A gain of two is thus the same as a proportional band of 50%, and decreasing the proportional band increases the gain.

The integral time is commonly expressed as '*Repeats per Minute*' or rpm. The relationship is given by

$$\text{Repeats per min} = 1/T_i \ \text{(for } T_i \text{ in min)}$$
$$= 60/T_i \ \text{(for } T_i \text{ in s)}$$

The derivative time is often called the *rate* or *pre-act* term but these are all identical to T_d.

More surprisingly there are variations on the basic algorithm. Some manufacturers use a so called '*non-interacting*' or '*parallel*' equation which can be expressed as

$$O_P = K_e+\frac{1}{T_i}\int e\,dt+T_d\frac{de}{dt}$$

or

$$O_P = K_e+K_i\int e\,dt+K_d\frac{de}{dt}$$

In these the three terms are totally independent. In the second version K_i is called the integral gain and K_d the derivative gain. Note that increasing K_i has the same effect as decreasing T_i. It is tempting to think that the non-interacting equations are simpler to use, but in practice the theoretical model is more intuitive. In particular, as the gain K is reduced in the non-interacting equation, any integral action has more effect and contributes more phase shift. Increasing or decreasing the gain with a non-interacting controller can thus cause instability.

There is yet a third form of PID algorithm known as the '*series*' equation. This can be expressed as

$$O_P = K\left(e+\frac{1}{T_i}\int e\,dt\right)\left(1+T_d\frac{de}{dt}\right)$$

This algorithm is based on pneumatic and early electronic controllers, and some manufacturers have maintained it to give backward compatibility. This has the odd characteristic that the T_i and T_d controls interact with each other, with the maximum derivative action occurring when T_d and T_i are set equal. In addition the ratio between T_i and T_d interacts with the overall gain.

There are further variations on the way the derivative contribution is handled. We have already discussed the effect of derivative on process variable and derivative on error. Because the pure derivative term gives increasing gain with increasing frequency it amplifies any high frequency noise resulting in continual twitchy movements of the plant actuators. Many manufacturers therefore deliberately roll off the high frequency gain, either by filtering the signal applied to the derivative function or directly limiting the derivative action.

1.2.27.7 Incremental controllers

Diaphragm operated actuators can be arranged to fail open or shut by reversing the relative positions of the drive pressure and return spring. In some applications a valve will be required to hold its last position in the event of failure. One way to achieve this is with a motorised actuator, where a motor drives the valve via a screw thread.

Such an actuator inherently holds its last position but the position is now the integral of the controller output. An integrator introduces 90° phase lag and gain which falls off with increasing frequency. A motorised valve is therefore a destabilising influence when used with conventional controllers.

Incremental controllers are designed for use with motorised valves and similar integrating devices. They have the control algorithm

$$\mathrm{Op} = K\left(\frac{1}{T_\mathrm{i}}e + \frac{\mathrm{d}e}{\mathrm{d}t} + \frac{1}{T_\mathrm{d}}\frac{\mathrm{d}^2e}{\mathrm{d}t^2}\right)$$

which is the time derivative of the normal control algorithm.

Incremental controllers are sometimes called *boundless controllers* or *velocity controllers* because the controller output specifies the actuator rate of change (i.e. velocity) rather than actual position.

Incremental controllers cannot suffer from integral windup per se, but it is often undesirable to keep driving a motorised valve once the end of travel is reached. End of travel limits are often incorporated in motorised valves to prevent jamming. The controller also has no real 'idea' of the valve true position, and hence cannot give valve position indication. If end of travel signals are available, a valve model can be incorporated into the controller to integrate the controller output to give a notional valve position. This model would be corrected whenever an end of travel limit is reached. Alternatively, a position measuring device can be fitted to the valve for remote indication.

Pulse width modulated controllers are a variation on the incremental theme. Split phase motor drive valves require logic raise/lower signals, and normal proportional control can be simulated by using time proportional raise/lower outputs.

1.2.27.8 Scheduling controllers

Many loops have properties which change under the influence of some measurable outside variable. The gain of a flow control valve (i.e. the change in flow for change in valve position) varies considerably over the stroke of a valve. The levitation effect of steam bubbles in a boiler drum causes the drum level control to have different characteristics under start-up, low load and high load conditions.

A scheduling controller has a built-in look up table of control parameters (gain, filtering, integral time, etc.) and the appropriate values selected for the measured plant conditions.

1.2.27.9 Variable gain controllers

Process variable noise occurs in many loops; level and flow being possibly the worst offenders. This noise causes unnecessary actuator movement, leading to premature wear and inducing real changes in the plant state. Noise can, of course, be removed by first- or second-order filters, but these reduce the speed of the loop and the additional phase shift from the filters can often act to destabilise a loop.

A controller with gain K will pass a noise signal $K \cdot n(t)$ to the actuator where $n(t)$ is the noise signal. One obvious way to reduce the effect of the noise is to reduce the controller gain, but this degrades the loop performance. Usually the noise signal has a small amplitude compared with the signal range, if it has not the process will be practically uncontrollable. What is intuitively required is a low gain when the error is low, but a high gain when the error is high.

Figure 1.2.80(a) shows how such a scheme operates. The noise amplitude lies in the range AB, so this is made a low gain region. Outside this band the gain is much higher. The gain in the region AB should be low, but not zero, to keep the process variable at the set point. With a pure deadband (i.e. zero gain in region AB) the process variable would cycle between one side of the centre band and the other.

Figure 1.2.80(b) shows a possible implementation. A comparator switches between a low gain and high gain controller according to the magnitude of the error. Note that integral balancing is required between the two controllers to stop integral windup in the unselected controller.

Figure 1.2.80 has two gain regions. It is possible to construct a controller whose gain varies continuously with error. Such a controller has a response

$$\mathrm{Op} = Kf(e)\left(e + \frac{1}{T_\mathrm{i}}\int e\,\mathrm{d}t + T_\mathrm{d}\frac{\mathrm{d}e}{\mathrm{d}t}\right)$$

where $f(e)$ is a function of error.

A common function is

$$f(e) = abs\left(\frac{m + (1-m)e}{100}\right) \quad (1.2.24)$$

where e is expressed as a percentage (0–100%) and m is a user set linearity adjustment ($0 < m < 1$). The *abs* operation (which always returns a positive sign) is necessary to prevent the controller action changing sign on negative error.

With $m = 1$, $f(e) = 1$ and Equation (1.2.24) behaves as a normal three term controller. With $m = 0$. the proportional part of Equation (1.2.24) follows a square law. Like Figure 1.2.80(a), this has low gain or small error (zero gain at zero error) but progressively increasing gain as the error increases.

Position control systems often need a fast response but cannot tolerate an overshoot. These often use Equation (1.2.24) with m at a low value approximating to the quadratic curve. This gives a high take off speed, but a low speed of approach.

1.2.27.10 Inverse plant model

The ideal control strategy, in theory, is one which mimics the plan behaviour. Given a totally accurate model of the plant, it should be possible to calculate what controller output is required to follow setpoint change, or compensate for a disturbance. The problem here is, of course, having an accurate plant model, but even a rough approximation should suffice as the controller output will converge to the correct value eventually.

One possible solution is shown in Figure 1.2.81. The process is represented by a block with transfer function $K{\cdot}f(s)$ where K is the DC (low frequency) gain. Following a change in set point, the signal A should mimic exactly the process variable B, leading to a constant output from the controller exactly correct to bring the plant to the set point without overshoot. With a perfect model, the change at A should match the change at B as the set point is approached.

The inverse plant model is usually implemented with a sampled digital system. The problem with this simple, and apparently ideal, controller is that it will probably demand actuation signals which will drive the controller output, the actuator or parts of the plant, into saturation. It also requires an accurate plant model. A more gentle version of this technique aims to get a fraction, say 0.1 of the way from the current value to the desired value of the process variables at each sample time. This approximates to an exponential response.

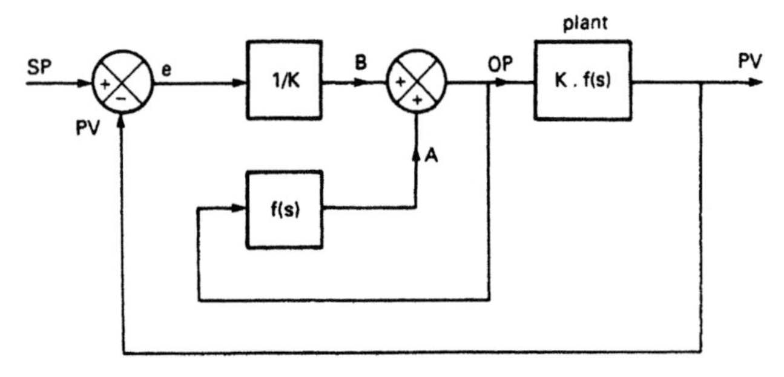

Fig. 1.2.81 The inverse plant model.

1.2.28 Digital control algorithms

1.2.28.1 Introduction

So far we have assumed that controllers deal with purely analog signals. Increasingly controllers are digital, with the analog signal from the transducer being sampled by an ADC, the control algorithm being performed by software and the analog output being obtained from a DAC. The system does not therefore continually control but takes 'snapshots' of the system state. Such an approach is called a *sampled system*.

1.2.28.2 Shannon's sampling theorem

A sampled system only knows about the values of its samples. It cannot infer any other information about the signals it is dealing with. An obvious question, therefore, is what sample rate we should choose if our samples are to accurately represent the original analog signals.

In Figure 1.2.82(a) a sine wave is being sampled at a relatively fast rate. Intuitively one would assume this sampling rate is adequate. In Figure 1.2.82(b) the sample rate and the frequency are the same. This is obviously too slow as the samples imply a constant unchanging output.

In Figure 1.2.82(c) the sample rate is lower than the frequency and the sample values are implying a sine wave of much lower frequency than the signal. This is called '*aliasing*'. A visual effect of aliasing can be seen on cinema screens where moving wheels often appear to go backwards. This effect occurs because the camera samples the world at about 50 times per second.

Any continuous signal will have a bandwidth of interest. The sampling frequency should be at least twice the bandwidth of interest. This is known as *Shannon's sampling theorem*. Any real life system will not, however, have a well-defined bandwidth and sharp cut-off point. Noise and similar effects will cause any real signal to have a significant component at higher frequencies. Aliasing may occur with these high frequency components and cause apparent variations in the frequency band of interest. Before sampling, therefore, any signal should be passed through a low pass *anti-aliasing filter* to ensure only the bandwidth of interest is sampled.

Most industrial control signals have a bandwidth of a few Hz, so sampling within Shannon's limit is usually not a problem. Normally the critical bandwidth is not known precisely so a sample rate of about 5–10 times the envisaged bandwidth is used.

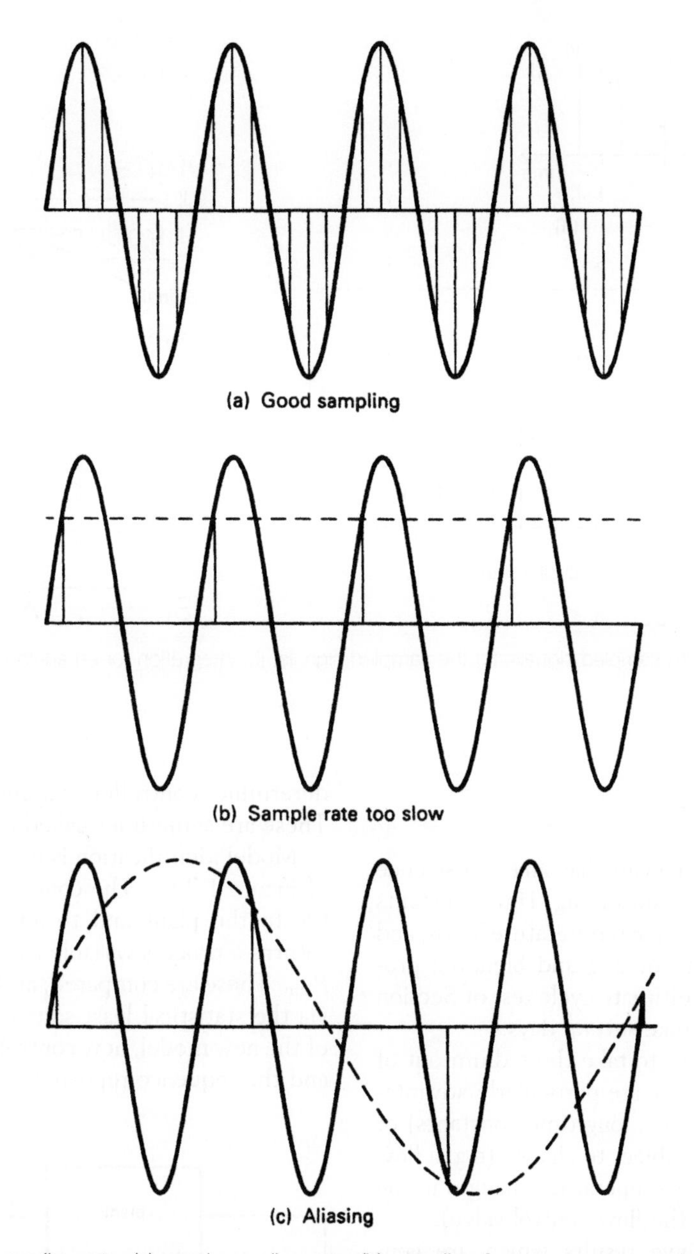

Fig. 1.2.82 The effect of the sampling rate: (a) good sampling rate; (b) sampling frequency same as signal frequency, too slow; (c) sampling rate much too slow, aliasing is occurring.

1.2.28.3 Control algorithms

To achieve three term control with sampled signals we must find the derivative and the integral of the error. As shown in Figure 1.2.83 we are dealing with a set of sampled signals, y_n, y_{n-1}, y_{n-2}, etc. where y_n is the most recent. If the sample time is Δt, the slope is then given by

$$\text{slope} = \frac{y_n - y_{n-1}}{\Delta t}$$

Integration is equivalent to finding the area under a curve as shown for an analog and digital signal in Figure 1.2.83(b). The trapezoid integration of Figure 1.2.83(c) is commonly used where the area is given by

$$\text{area} = \frac{\Delta t(y_n + y_{n-1})}{2}$$

Combining these gives a digital sampled PID algorithm

$$\text{Op} = K\left(e_n + \frac{1}{T_i}\sum\frac{\Delta t(e_n + e_{n-1})}{2} + \frac{T_d}{\Delta t}(e_n - e_{n-1})\right)$$

where e_n is the error for sample n and Δt the sample time as before.

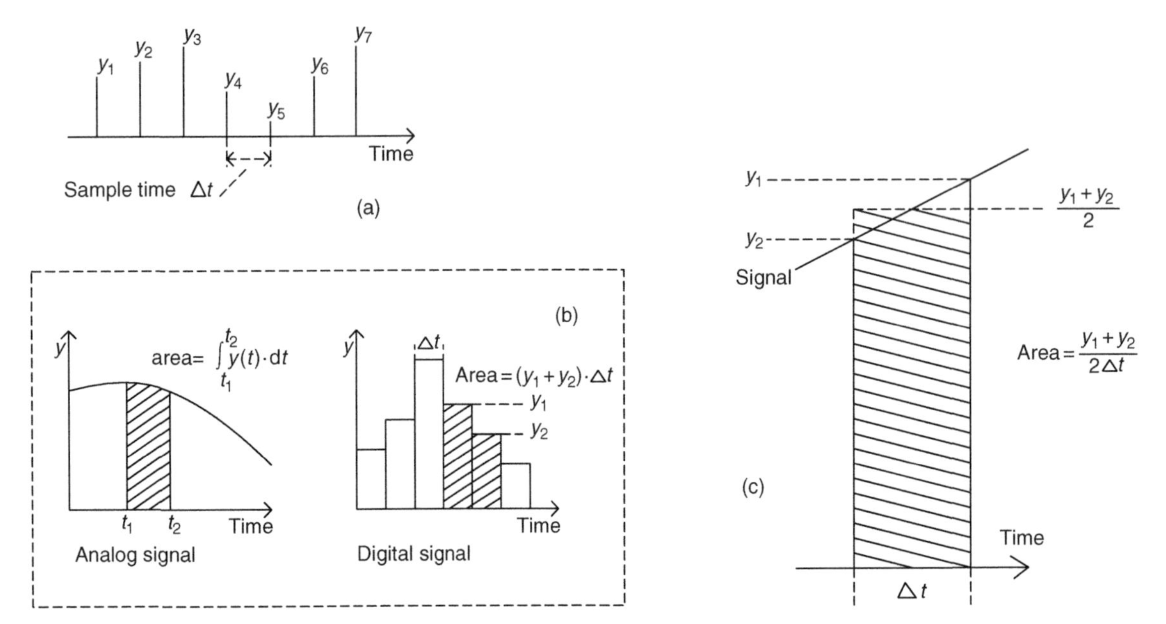

Fig. 1.2.83 Control algorithm with sampled signals: (a) the sampled signals; (b) integration for an analog and digital signal; (c) trapezoid integration.

1.2.29 Auto-tuners

Tuning a controller is more of an art than an exact science and can be unbelievably time consuming. Time constants of tens of minutes are common in temperature loops, and lags of hours occur in some mixing and blending processes. Performing, say, the ultimate cycle test of Section 1.2.30.2 on such loops can take several days.

Self-tuning controllers aim to take the tedium out of setting up a control loop. They are particularly advantageous if the process is slow (i.e. long time constants) or the loop characteristics are subject to change (e.g. a flow control loop where pressure/temperature changes in the fluid alter the behaviour of the flow control valve).

Self-tuning controllers give results which are generally as good, if not slightly better, than the manual methods of Section 1.2.30 (possibly because self-tuning controllers have more patience than humans!). In the author's experience, however, the results from a self-tuner should be viewed as recommendations or initial settings in the same way as the results from the manual methods described in the following sections. One early decision to be made when self-tuners are used is whether they should be allowed to alter control parameters without human intervention. Many engineers (of whom the author is one) view self-tuners as commissioning aids to be removed before a plant goes into production.

There are essentially two groups of self-tuners. *Modelling self-tuners* try to build a mathematical model of the plant (usually second order plus transit delay) then determine controller parameters to suit the model. These are sometimes called *explicit self-tuners*.

Model identification is usually based on the principles of Figure 1.2.84. The controller applies a control action O_P to the plant and to an internal model. The plant returns a process variable P_V and the model a prediction P_{Vm}. These are compared, and the model updated (often via the statistical least squares technique). On the basis of the new model, new control parameters are calculated, and the sequence repeated.

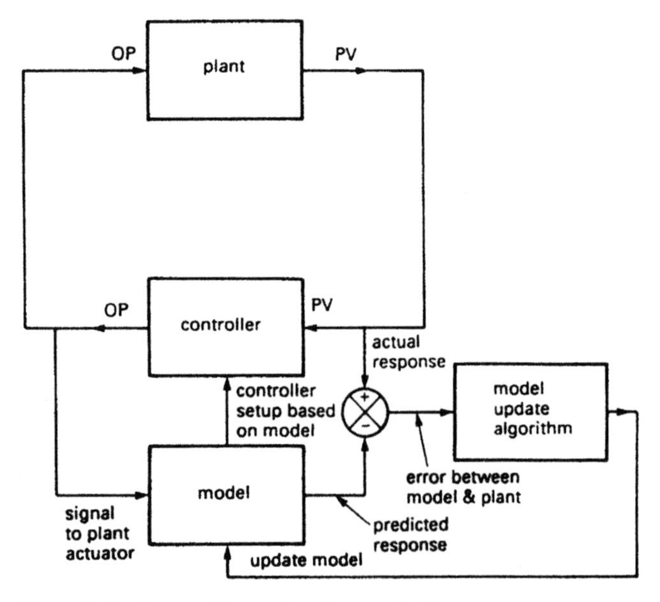

Fig. 1.2.84 A modelling self-tuning controller.

A model building self-tuner requires actuation changes to update its model, so it follows that self-tuners do not perform well in totally static conditions. In a totally stable unchanging loop, the model, and hence the control parameters, can easily drift off to ridiculous values. To prevent this, most-self tuners are designed to 'kick' the plant from time to time, with the size and repetition rate of the kick being set by the control engineer. Less obviously, model building self-tuners can be confused by outside disturbances which can cause changes in P_V that are not the result of the controller output.

The second group of self-tuners (sometimes called *implicit tuners*) use automated versions of the manual tests described in Section 1.2.30, and as such do not attempt to model the plant. A typical technique will vary the controller gain until a damped oscillatory response is observed. The control parameters can then be inferred from the controller gain, the oscillation period and the oscillation decay rate.

The useful bang/bang test of Section 1.2.30.3 can be performed automatically by a controller which forces limit cycling in the steady state via a comparator. A limit block after the comparator restricts the effect on the plant.

Implicit self-tuners, like their modelling brothers, do not perform well on a stable unchanging loop, and can be equally confused by outside disturbances.

1.2.30 Practical tuning methods

1.2.30.1 Introduction

Values must be set for the gain and integral/derivative times before a controller can be used. In theory, if a plant model is available, these values can be determined from Nichols charts or Nyquist diagrams. Usually, however, the plant characteristics are not known (except in the most general terms) and the controller has to be tuned by experimental methods.

It should be noted that all these methods require pushing the plant to the limit of stability. The safety implications of these tests must be clearly understood. Tuning can also be very time consuming. With large chemical plants tuning of one loop can take days.

Most of the tests aim to give a quarter cycle decay and assume the plant consists of a transit delay in series with a second-order block (or two first-order lags) plus possible integral action.

In conducting the tests, it is useful to have a two pen recorder connected to the P_V (process variable) and O_P (controller output) as shown in Figure 1.2.85. The range of the pens (e.g. 0–10 V or 1–5 V) should be the same.

In the tests below, the gain is expressed as proportional band (P_B) per cent. Time is used for integral and derivative action. Conversion to gain or repeats per minute is straightforward.

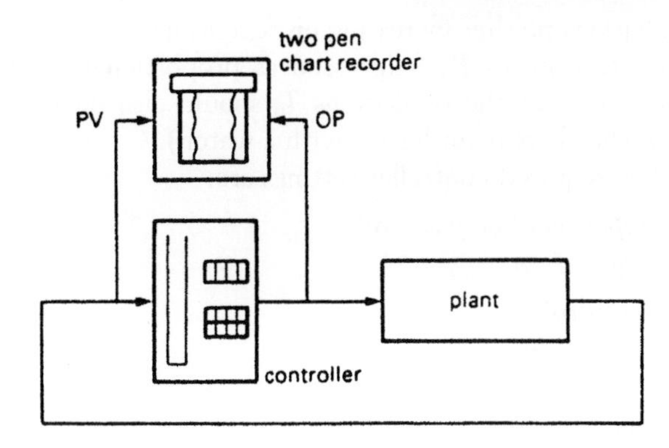

Fig. 1.2.85 Suggested equipment setup for controller tuning.

1.2.30.2 Ultimate cycle methods

The basis of these methods is determining the controller gain which just supports continuous oscillation, i.e. point *A* and gain *K* on the Nichols chart and Nyquist diagram of Figure 1.2.86. The method is based on work by J. G. Ziegler and N. B. Nichols and is often called the Ziegler Nichols method.

The integral and derivative actions are disabled to give proportional only control, and the control output manually adjusted to bring P_V near the required value. Auto control is selected with a low gain.

Step disturbances are now introduced and the effect observed. One way of doing this is to go back into manual, shift Op by, say 5%, then reselect automatic control. At each trial the gain is increased. The increasing gain will give a progressively underdamped response and eventually continuous oscillation will result. Care must be taken in these tests to allow all transients to die away before each new value of gain is tried.

If the value of gain is too high, the oscillations will increase. The value of gain which gives constant

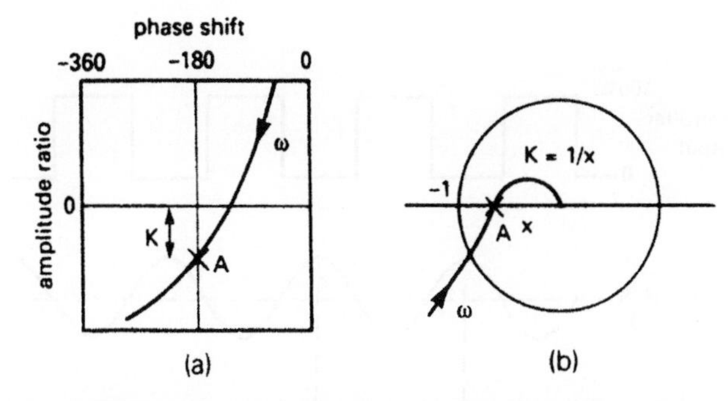

Fig. 1.2.86 Basis of the ultimate cycle test. Point A determines the frequency at which continuous oscillations will occur when gain K is applied: (a) Nichols chart; (b) Nyquist diagram.

oscillations neither increasing or decreasing is called the ultimate gain, or P_u (expressed as proportional band). The period of the oscillations T_u should also be noted from the chart recorder (or with a watch).

The required controller settings are:

Proportional only control

P_B $2P_u\%$

PI control

P_B $2.2P_u\%$
T_i $0.8.\ T_u$

PID control

P_B $1.67.\ P_u\%$
T_i $T_{u/2}$
T_d $T_{u/8}$

$T_i = 4\ T_d$ is a useful rule of thumb.

Other recommended settings for a PID controller are:

P_B $2P_u\%$
T_i T_u
T_d $T_u/5$

and

P_B $2P_u\%$
T_i $0.34T_u$
T_d $0.08\ T_u$

All of these values should be considered as starting points for further tests.

1.2.30.3 Bang/bang oscillation test

This is the fastest, but most vicious, test. It can, though, be misleading if the plant is non-linear. Integral and derivative actions are disabled and the controller gain set as high as possible (ideally infinite) to turn the controller into a bang-bang controller. The controller output is set manually to bring the process value near the set point then the controller switch into automatic mode.

Violent oscillations will occur as shown in Figure 1.2.87. The period of the oscillations T_o is noted along with the peak to peak height of the process variable oscillations as a percentage $H_o\%$ of full scale.

The required controller settings are:

Proportional control

P_B $2.H_o\%$

PI control

P_B $3.H_o\%$
T_i $2.T_o$

PID control

P_B $2.H_o\%$
T_i T_o
T_d $T_o/4$

1.2.30.4 Reaction curve test

This is an open-loop test originally proposed by American engineers Cohen and Coon. It assumes the plant consists of a measurable transit delay and a dominant time constant. It cannot be applied to plants with integral action (e.g. level control systems).

A chart recorder must be connected to the plant as shown earlier in Figure 1.2.85 to perform the test. The controller output is first adjusted manually to bring the plant near to the desired operating point. After the transients have died away a small manual step ΔO_P is applied which results in a small change ΔP_V as shown in Figure 1.2.88.

The process gain K_p is then simply $\Delta P_V/\Delta O_P$.

A tangent is drawn to the process variable curve at the steepest point from which an apparent transit delay T_t and time constant T_c can be read. The settings for the controller are then given by

Proportional

P_B $100.K_p.T_t/T_c\%$

PI

P_B $110.K_p.T_t/T_c\%$
T_i $3.3\ T_t$

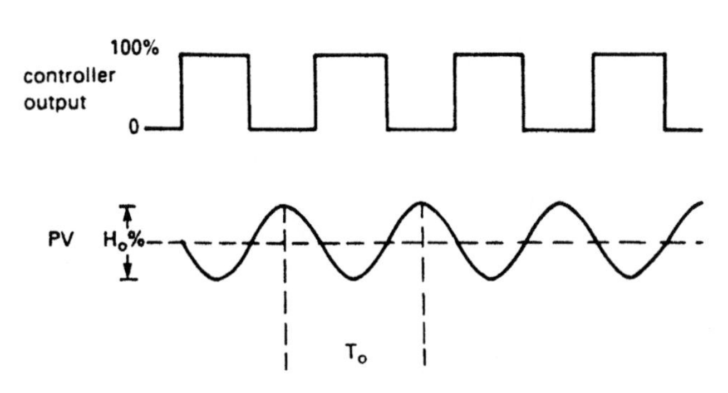

Fig. 1.2.87 The bang–bang oscillation test.

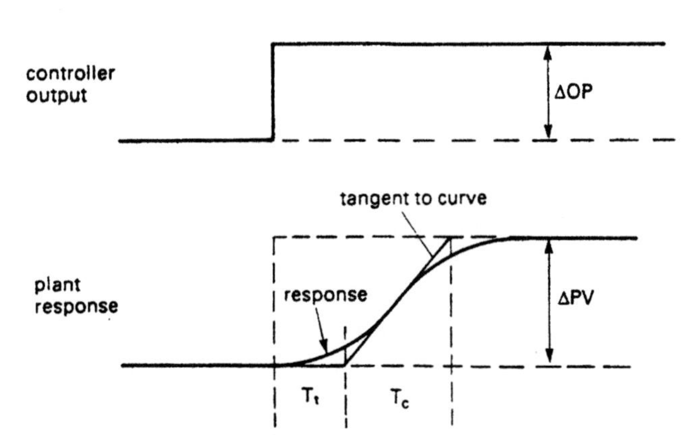

Fig. 1.2.88 The reaction curve test.

PID

P_B $80.K_p.T_t/T_c\%$
T_i $2.5\ T_t$
T_d $0.4\ T_t$

Because the test uses an open-loop trial with a small change in the controller output it is the gentlest and least hazardous tuning method.

1.2.30.5 A model building tuning method

The closed-loop tuning methods described so far require the plant to be pushed to (and probably beyond) the edge of instability in order to set the controller. An interesting gentle tuning method was described by Yuwana and Seborg in the journal *AIChE* Vol 28 no 3 in 1982.

The method assumes the plant has gain K_M and behaves as a dominant first-order lag T_M in series with a transit delay D_M. This assumption can give gross anomalies with plants with integral action such as level or position controls, but is commonly used for many manual and automatic/adaptive controller tuning methods. With the above warning noted, the suggested controller settings can be found from

$$K = A(D_M/T_M)^{-B}/K_M$$
$$T_I = CT_M(D_M/T_M)^D$$
$$T_D = T_M E(D_M/T_M)^F$$

where *A, B, C, D, E, F* are constants defined:

Mode	*A*	*B*	*C*	*D*	*E*	*F*
P	0.490	1.084				
PI	0.859	0.997	1.484	0.680		
PID	1.357	0.947	1.176	0.738	0.381	0.99

These apparently random equations and constants come from experimental work described by Miller *et al* in *Control Engineering* Vol 14 no 12.

The method of finding the plant gain, time constant and transit delay is based on a single quick test with the plant operating under closed-loop control. The test is performed on the plant operating under proportional only control, with a gain sufficient to produce a damped oscillation as Figure 1.2.89 when a step change in set point from R_0 to R_1 is applied.

The subsequent process maximum C_{P1}, minimum C_{M1} and next maximum C_{P2} are noted along with the time D_{T2} between C_{P1} and C_{P2}. The controller proportional band used for the test, P_B, is also recorded, from which the controller gain $K_{PB} = 100/P_B$ is found.

Given the values from the test the method estimates the value of the plant steady state gain K_M, lag time constant T_M and the transit delay time D_M. The background mathematics is given at length in the original paper.

The equations above are not very practical for manual use on site, so the original paper was developed into a program for the Hewlett Packard HP-67 calculator by Jutan and Rodriguez and published in the magazine *Chemical Engineering September* 1984. The nomenclature used in the above equations is based on this article.

1.2.30.6 General comments

The above test procedures do not give guaranteed results and should be viewed as a method of putting the engineer in the right area. They should be viewed as the starting point for further trials. The important thing in these trials is only change one thing at once.

With values set as above the effect of changing the gain should be tried first. It is always useful to have the proportional gain as high as possible to give the largest initial control action to changes and disturbances. However, a large gain can give undesirable changes in the controller output if the process variable is noisy. The gain should be adjusted to give the desired overshoot and damping.

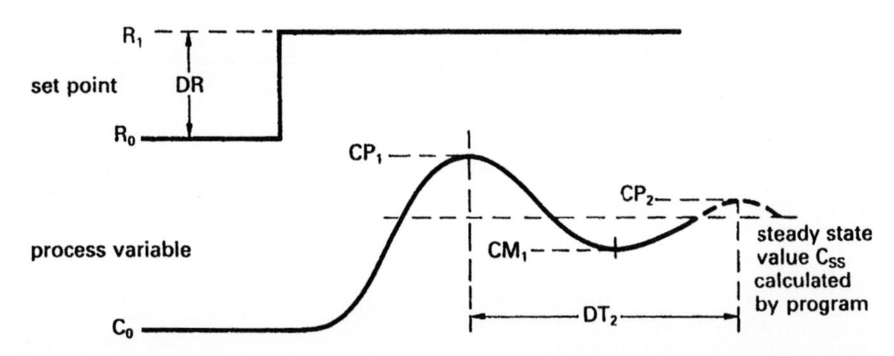

Fig. 1.2.89 Test performed for model building tuning test. Note that R_0 and C_0 need not be the same and DR can be positive or negative.

Integral action should be adjusted next to give best removal of offset error. During these trials it is best to disable any derivative action. Decreasing the integral time reduces the time taken to remove the offset error. It may be necessary to reduce the gain again as integral time is decreased. A useful rule of thumb is that the ratio of T_i/Gain is an 'index' of stability for a given system, i.e. a T_i of 12 s and a gain of 2 will give a similar damping to a T_i of 24 s and a gain of 4.

The derivative action should be adjusted last. Many systems do not benefit from derivative action, particularly those with a noisy process variable signal which causes large controller output swings. Where derivative action is required, $T_d = T_i/4$ is a good starting point. Many controllers allow the user to select derivative action on error or derivative action on process variable. The former is best for tracking systems, but gives large controller output swings for step changes in the set point. Derivative action on process variable is usually the best choice.

One final observation, based on experience rather than theory, is that a P_B of 200% (gain of 0.5), T_i, of 20 s and no derivative action is a good starting point for a majority of plants. Adjust the gain to give the required overshoot then adjust T_i to be as small as possible. Finally set T_d, if needed, to $T_i/4$.

References

1 Saucedo, R. and Schiring, E. E., *Introduction to Continuous and Digital Control Systems*, Macmillan, New York (1968)

2 Franklin, G. F., Powell, J. D. and Emami-Naeini, A., *Feedback Control of Dynamic Systems*, Addison-Wesley, New York (1986)

Bibliography

The authors have found the following books useful for basic control engineering studies. This list is by no means exhaustive.

Anand, D. K., *Introduction to Control Systems*, Pergamon Press, Oxford (1974)

Chen, C. F. and Haas, I. J., *Elements of Control Systems Analysis*, Prentice Hall, Englewood Cliff, NJ (1968)

Distefano, J.J., Stubberud, A.R. and Williams, I.J., *Theory and Problems of Feedback and Control Systems*, Schaum's Outline Series, McGraw-Hill, New York (1990)

Dorf, R. C., *Modern Control Systems*, Addison-Wesley, New York (1980)

Douce, J. L., *The Mathematics of Servomechanisms*, English Universities Press, London (1963)

Elgard, O. I., *Control Systems Theory*, McGraw-Hill, New York (1967)

Golten, J. and Verwer, A., *Control System Design and Simulation*, McGraw-Hill (1991)

Healey, M., *Principles of Automatic Control*, Hodder and Stoughton, New York (1975)

Jacobs, O. L. R., *Introduction to Control Theory*, Oxford University Press, Oxford (1974)

Langill, A. W., *Automatic Control Systems Engineering*, Vols I and II, Prentice Hall, Englewood Cliffs, NJ (1965)

Marshall, S. A., *Introduction to Control Theory*, Macmillan, New York (1978)

Power, H. M. and Simpson, R. J., *Introduction to Dynamics and Control*, McGraw-Hill, New York (1978)

Raven, F. H., *Automatic Control Engineering*, McGraw-Hill, New York (1961)

Shinskey, F. G., *Process Control Systems*, McGraw-Hill (1988)

Chapter 1.3

1.3

Programmable controllers

Parr

1.3.1 Introduction

1.3.1.1 The computer in control

A computer can be considered as a device that follows predetermined instructions to manipulate input data in order to produce new output data as summarised on Figure 1.3.1(a). Early computer systems tended to be based on commercial functions: payroll, accountancy, banking and similar activities. The operations tended to be batch processes, a daily update of stores stock for example.

A computer can also be used as part of a control system as Figure 1.3.1(b). The input data will be the operator's commands and signals from the plant (limit switches, flows, temperatures). The output data are control actions to the plant and status displays to the operator. The instructions will define what action is to be taken as the input data (from both the plant and the operator) changes.

The first industrial computer application was probably a system installed in an oil refinery in Port Arthur USA in 1959. The reliability and mean time between failure of computers at this time meant that little actual control was performed by the computer, and its role approximated to a simple monitoring subsystem.

1.3.1.2 Requirements for industrial control

Industrial control has rather different requirements than other computer applications. It is worth examining these in some detail.

A conventional computer takes data, usually from a keyboard, and outputs data to a screen or printer. The data being manipulated will generally be characters or

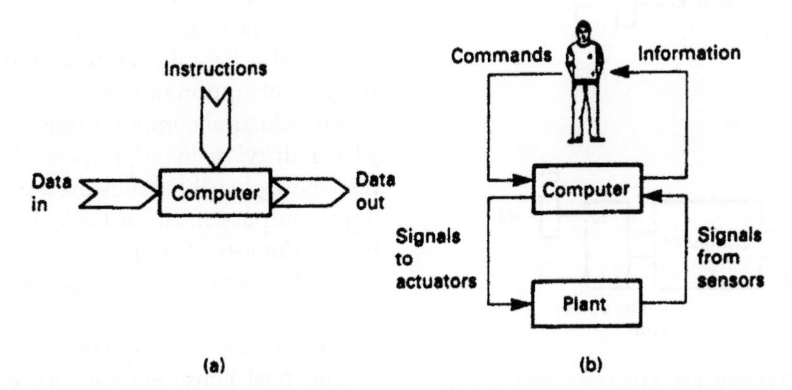

Fig. 1.3.1 The computer as part of an industrial control system: (a) a simple overview of a computer; (b) the computer as part of a control system.

Electrical Engineer's Reference Book; ISBN: 9780750646376

numbers (e.g., item names and quantities held in a stores stock list).

An industrial control computer is very different. Its inputs come from a vast number of devices. Although some of these will be numeric (flows, temperature, pressures and similar analog signals) the majority will be single bit, on/off, digital signals representing valves, limit switches, motor contactors, etc.

There will also be a similar large amount of digital and analog output signals. A very small control system may have connections to about 20 input and output signals; figures of over 200 connections are quite common on medium sized systems.

Although it is possible to connect this quantity of signals into a conventional machine, it requires non-standard connections and external boxes. Similarly, although programming for a large amount of input and output signals can be done in Pascal, BASIC or C, the languages are being used for a purpose for which they were not really designed, and the result can be very ungainly.

In Figure 1.3.2(a), for example, we have a simple motor starter. This could be connected as a computer driven circuit as Figure 1.3.2(b). The two inputs are identified by addresses 1 and 2, with the output (the relay starter) being given the address 10.

If we assume a program function bitread (N) exists which gives the state (on/off) of address N, and a function bitwrite (M, var) which sends the state of program variable var to address M, we could give the actions of Figure 1.3.2 by

```
Repeat
  start:= bitread(1);
  stop:= bitread(2);
  run:= ((start) or (run)) & stop;
  bitwrite (10,run);
until hellfreezesover
```

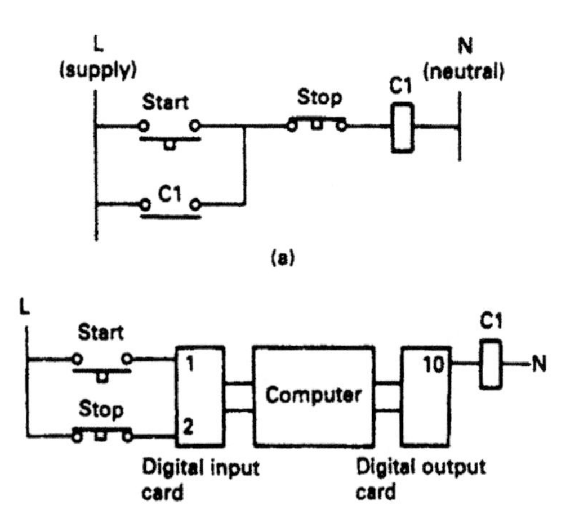

Fig. 1.3.2 Comparison of hardwire and computer-based systems: (a) hardwire motor starter; (b) computer-based motor starter.

where start, stop and run are one bit variables. The program is not very clear, however, and we have just three connections.

An industrial control program rarely stays the same for the whole of its life. There are always modifications to cover changes in the operations of the plant. These changes will be made by plant maintenance staff, and must be made with minimal (preferably none) interruptions to the plant production. Adding a second stop button and a second start button into Figure 1.3.2 would not be a simple task.

In general, computer control is done in real time, i.e., the computer has to respond to random events as they occur. An operator expects a motor to start (and more important to stop!) within a fraction of a second of a button being pressed. Although commercial computing needs fast computers, it is unlikely that the difference between a one second and two second computation time for a spreadsheet would be noticed by the user. Such a difference would be unacceptable for industrial control.

Time itself is often part of the control strategy (e.g., start air fan, wait 10 s for air purge, open pilot gas valve, wait 0.5 s, start ignition spark, wait 2.5 s, if flame present open main gas valve). Such sequences are difficult to write with conventional languages.

Most control faults are caused by external items (limit switches, solenoids and similar devices) and not by failures within the central control itself. The permission to start a plant, for example, could rely on signals involving cooling water flows, lubrication pressure and temperatures all being within allowable ranges. For quick fault finding the maintenance staff must be able to monitor the action of the computer program whilst it is running. If, as is quite common, there are 10 interlock signals which allow a motor to start, the maintenance staff will need to be able to check these quickly in the event of a fault. With a conventional computer, this could only be achieved with yet more complex programming.

The power supply in an industrial site is shared with many antisocial loads; large motors stopping and starting, thyristor drives which put spikes and harmonic frequencies onto the mains supply. To a human these are perceived as light flicker; to a computer they can result in storage corruption or even machine failure.

An industrial computer must therefore be able to live with a 'dirty' mains supply, and should also be capable of responding sensibly following a total supply interruption. Some outputs must go back to the state they were in before the loss of supply, others will need to turn off or on until an operator takes corrective action. The designer must have the facility to define what happens when the system powers up from cold.

The final considerations are environmental. A large mainframe computer generally sits in an air-conditioned room at a steady 20 °C with carefully controlled humidity. A desk top PC will normally live in a fairly constant office environment because human beings do not work well at extremes. An industrial computer, however, will probably have to operate away from people in a normal electrical substation with temperatures as low as −10 °C after a winter shutdown, and possibly over 40 °C in the height of summer. Even worse, these temperature variations lead to a constant expansion and contraction of components which can lead to early failure if the design has not taken this factor into account.

To these temperature changes must be added dust and dirt. Very few industrial processes are clean, and the dust gets everywhere. The dust will work itself into connectors, and if these are not of a highest quality, intermittent faults will occur which can be very difficult to find.

In most computer applications, a programming error or a machine fault can often be humorous (bills and reminders for 0*p*) or at worse expensive and embarrassing. When a computer controlling a plant fails, or a programmer misunderstands the plants operation, the result could be injuries or fatalities. It behooves everyone to take extreme care with the design.

Our requirements for industrial control computers are very demanding, and it is worth summarising them:

- They should be designed to survive in an industrial environment with all that this implies for temperature, dirt and poor-quality mains supply.
- They should be capable of dealing with bit form digital input/output signals at the usual voltages encountered in industry (24 V d.c. to 240 V a.c.) plus analog input/output signals. The expansion of the I/O should be simple and straightforward.
- The programming language should be understandable by maintenance staff (such as electricians) who have no computer training. Programming changes should be easy to perform in a constantly changing plant.
- It must be possible to monitor the plant operation whilst it is running to assist fault finding. It should be appreciated that most faults will be in external equipment such as plant mounted limit switches, actuators and sensors, and it should be possible to observe the action of these from the control computer.
- The system should operate sufficiently fast for real-time control. In practice, 'sufficiently fast' means a response time of around 0.1 s, but this can vary depending on the application and the controller used.
- The user should be protected from computer jargon.
- Safety must be a prime consideration.

1.3.1.3 Enter the PLC

In the late 1960s the American motor car manufacturer General Motors was interested in the application of computers to replace the relay sequencing used in the control of its automated car plants. In 1969 it produced a specification for an industrial computer similar to that outlined at the end of the previous section.

Two independent companies, Bedford Associates (later called Modicon) and Allen Bradley (now owned by Rockwell), responded to General Motors specifications. Each produced a computer system similar to Figure 1.3.3 which bore little resemblance to the commercial mini-computers of the day.

The computer itself, called the central processor, was designed to live in an industrial environment, and was connected to the outside world via racks into which input, or output cards could be plugged.

Each input or output card could connect to 16 signals. A typical rack would contain eight cards and the processor could connect to eight racks, allowing connection to 1024 devices. It is very important to appreciate that the card allocations were the user's choice, allowing great flexibility.

The most radical idea, however, was a programming language based on a relay schematic diagram, with inputs (from limit switches, pushbuttons, etc.) represented by relay contacts, and outputs (to solenoids, motor starters, lamps, etc.) represented by relay coils. Figure 1.3.4(a) shows a simple hydraulic cylinder which can be extended or retracted by pushbuttons. Its stroke is set by limit switches which open at the end of travel, and the solenoids can only be operated if the hydraulic pump is running. This would be controlled by the computer program of Figure 1.3.4(b) which is identical to the relay circuit needed to control the cylinder. These programs look like the rungs on a ladder, and were consequently called 'Ladder Diagrams'.

The program was entered via a programming terminal with keys showing relay symbols (normally open/normally closed contacts, coils, timers, counters, parallel branches, etc.), with which a maintenance electrician would be familiar. Figure 1.3.5 shows the programmer's keyboard for an early PLC. The meaning of the majority of the keys should be obvious to any maintenance electrician. The program, shown exactly on the screen as Figure 1.3.4(b), would highlight energised contacts and coils allowing the programming terminal to be used for simple fault finding.

The name given to these machines was *Programmable Controllers* or PCs. The name *Programmable Logic Controller* or PLC was also used, but this is, strictly, a registered trade mark of the Allen Bradley Company, now part of Rockwell. Unfortunately in more recent

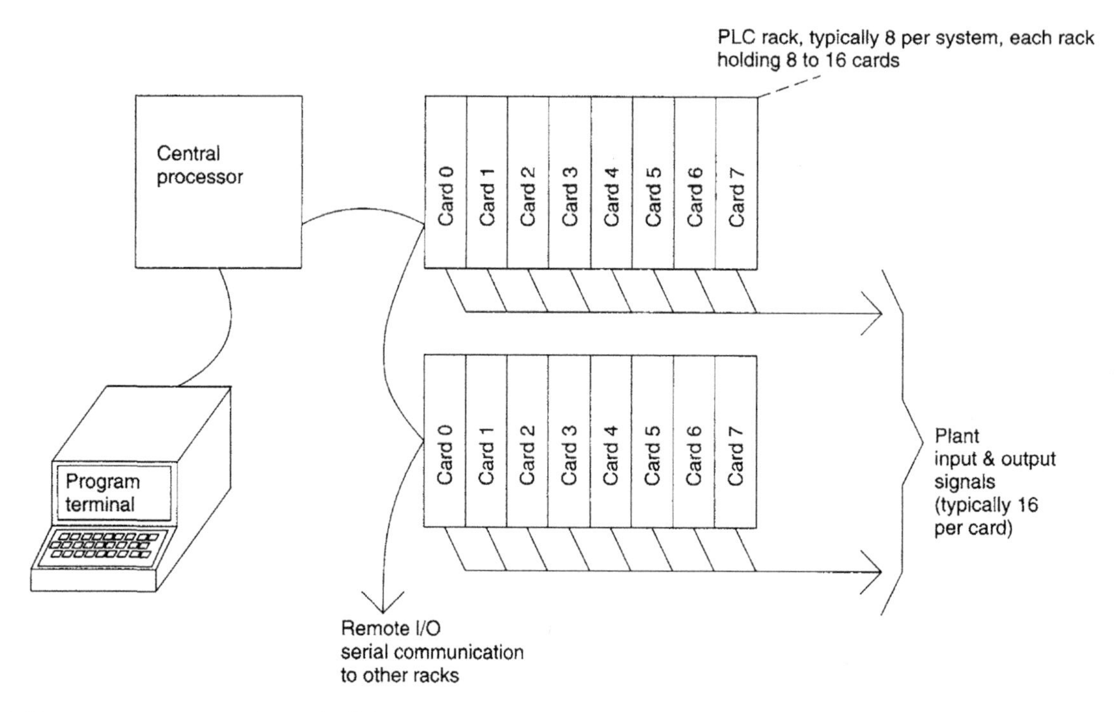

Fig. 1.3.3 The component parts of an early PLC system.

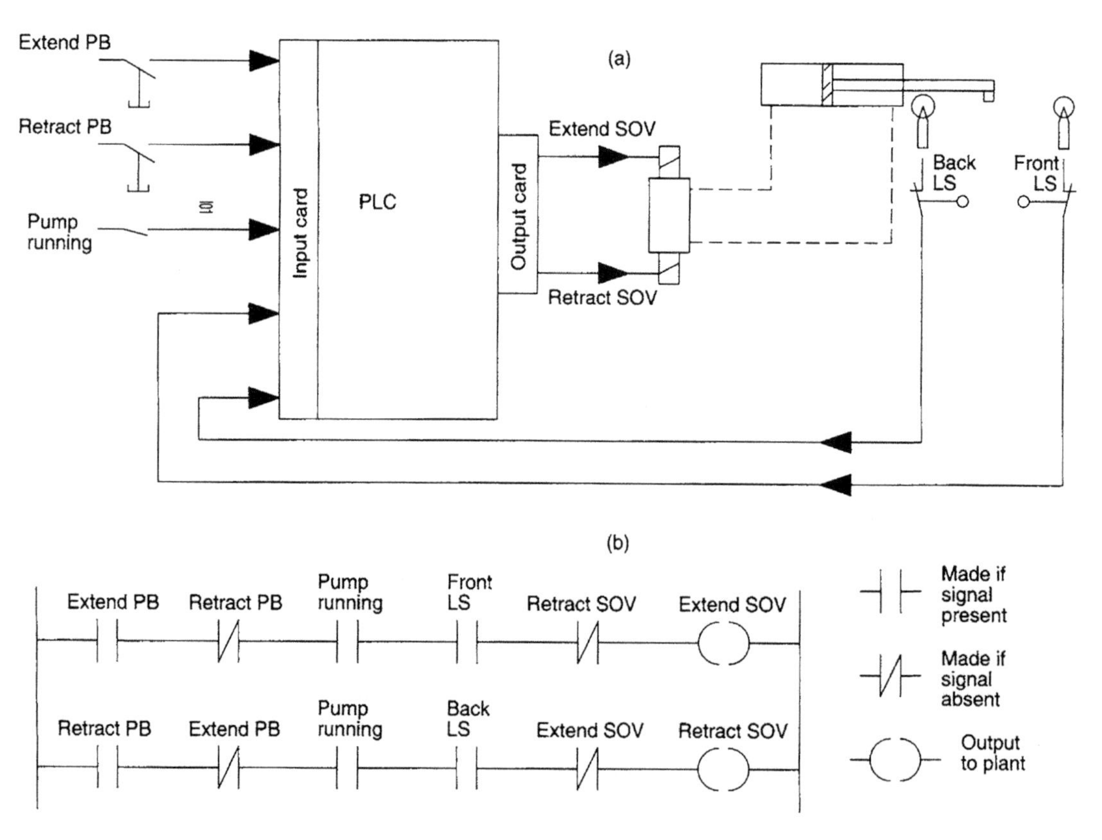

Fig. 1.3.4 A simple PLC application: (a) a hydraulic cylinder controlled by a PLC; (b) the 'Ladder Diagram' program used to control the cylinder.

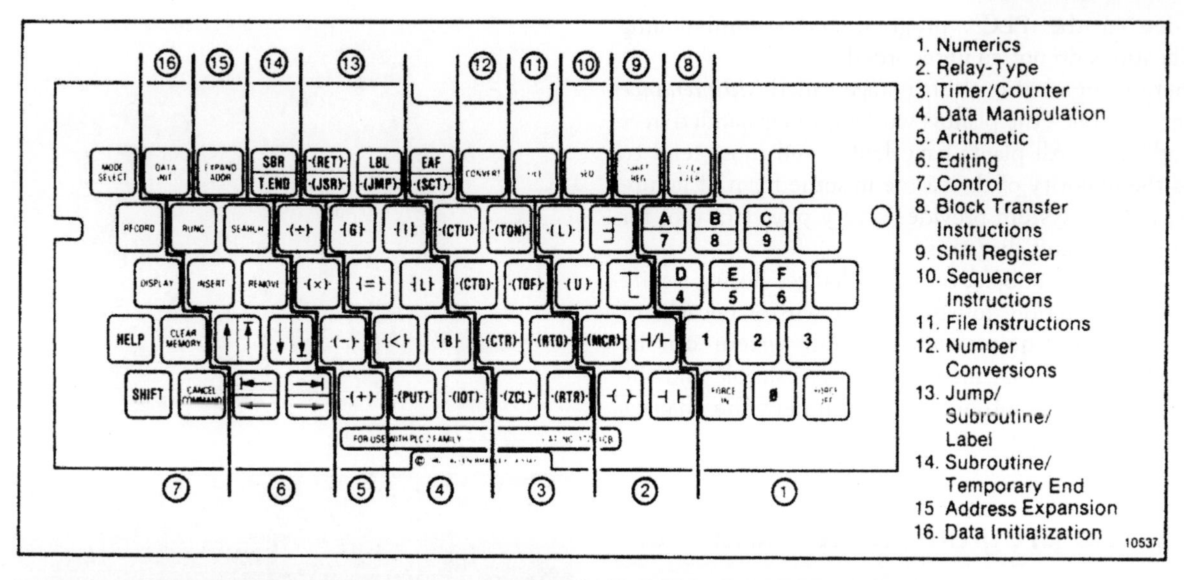

1. Numerics—provides addresses and decimal or hexadecimal values for instructions. It also provides force instructions.
2. Relay-Type—examines and controls the status of individual bits in specified memory areas.
3. Timer/Counter—allows the user to select various time-incremented and count-incremented and decremented functions.
4. Data Manipulation—used to transfer and compare BCD or octal values in the user program.
5. Arithmetic—performs the four indicated math functions.
6. Editing—used to locate, display and change instructions in the user program.
7. Control—directs the operation of the industrial terminal and its communication with the PLC-2 family processors and peripherals. Also provides HELP information.
8. Block Transfer Instructions—used to program block transfer Instructions in block format.
9. Shift Register—used to shift a word (all 16 bits) up or down one word in the shift register file.
   - used to shift a bit in the shift register file to the left or right one position.
   - used to create FIFO stacks.
10. Sequencer Instructions—used to establish and maintain user sequencer tables.
11. File Instructions—used to establish and manipulate user files.

Fig. 1.3.5 A programming keyboard from an early PLC programming terminal. The link between the keys and relay symbols can be clearly seen. Figure courtesy of Allen Bradley.

times the letters PC have come to be used for Personal Computer, and confusingly the worlds of programmable controllers and personal computers overlap where portable and laptop computers are now used as programming terminals. To avoid confusion, we shall use PLC for a programmable controller and PC for a personal computer.

1.3.1.4 The advantages of PLC control

Any control system goes through several stages from conception to a working plant.

The first stage is *Design* when the required plant is studied and the control strategies decided. With conventional systems every '*i*' must be dotted before construction can start. With a PLC system all that is needed is a possibly (usually!) vague idea of the size of the machine and the I/O requirements (so many inputs and outputs). The cost of the input and output cards is cheap at this stage, so a healthy spare capacity can be built in to allow for the inevitable omissions and future developments.

Next comes *Construction*. With conventional schemes, every job is a 'one-off' with inevitable delays and costs. A PLC system is simply bolted together from standard parts.

The next stage is *Installation*; a tedious and expensive business as sensors, actuators, limit switches and operator controls are cabled. A distributed PLC system (discussed in Section 1.3.5) using serial links and pre-built and tested desks can simplify installation and bring huge cost benefits. The majority of the PLC program is usually written at this stage.

Finally comes *Commissioning*, and this is where the real advantages are found. No plant ever works first time. Human nature being what it is, there will be some oversights. (We need a limit switch to only allow feeding when the discharge valve is 'shut' or 'Whoops, didn't we say the loading valve is energised to UNLOAD on this system' and so on.) Changes to conventional systems are time consuming and expensive. Provided the designer of the PLC systems has built in spare memory capacity, spare I/O and a few spare cores in multi-core cables, most changes can be made quickly and relatively cheaply. An added bonus is that all changes are inherently recorded in the PLC's program and commissioning modifications do not go unrecorded.

There is an additional fifth stage called *Maintenance* which starts once the plant is working and is handed over to production. All plants have faults, and most tend to spend the majority of their time in some form of failure mode. A PLC system provides a very powerful tool for assisting with fault diagnosis.

A plant is also subject to many changes during its life to speed production, to ease breakdowns or because of changes in its requirements. A PLC system can be changed so easily that modifications are simple and the PLC program will automatically document the changes that have been made.

1.3.2 The programmable controller

1.3.2.1 Modern PLC systems

This chapter is written around five manufacturers' ranges:

- The Allen Bradley PLC-5 series. Allen Bradley, now owned by Rockwell, were one of the original PLC originators (and actually has the US copyright on the name PLC). They have been responsible for much of the development of the ideas used in PLCs and have succeeded in maintaining a fair degree of upward compatibility from their earliest machine without restricting the features of the latest.
- The Siemens Simatic 55 range which is probably the most common PLC in mainland Europe.
- The British GEM-80, originally designed by GEC from a long association with industrial computers dating back to English Electric. This part of GEC is now known as CEGELEC and is part of a French group in which Alsthom are a major shareholder.
- The ASEA Master System, now manufactured by the ABB company formed by the merger of ASEA and Brown Boveri. The Master system has features more akin to a conventional computer system and its

Fig. 1.3.6 Four medium sized PLCs: (a) the Allen Bradley PLC-5.

Fig. 1.3.6 (b) the Siemens 115U; (c) the CEGELEC GEM-80; (d) the ABB Master.

programming language has some interesting and powerful features.

The above four PLCs are shown in Figure 1.3.6. Many PLC systems are now very small, and as an example of this bottom end of the market we shall also consider the Japanese Mitsubishi F2-40.

1.3.2.2 I/O connections

Internally, a computer usually operates at 5 V d.c. The external devices (solenoids, motor starters, limit switches, etc.) operate at voltages up to 110 V a.c. The mixing of these two voltages will cause irreparable damage to the PLC electronics. A less obvious problem can occur from electrical 'noise' introduced into the PLC from voltage spikes, caused by interference on signals lines, or from load currents flowing in a.c. neutral or d.c. return lines. Differences in earth potential between the PLC cubicle and outside plant can also cause problems.

There are obviously very good reasons for separating the plant supplies from the PLC supplies with some form of barrier to ensure that the PLC cannot be adversely affected by anything happening on the plant. Even a cable fault putting 415 V a.c. onto a d.c. input would only damage the input card; the PLC itself (and the other cards in the system) would not suffer.

1. Numerics—provides addresses and decimal or hexadecimal values for instructions. It also provides force instructions.
2. Relay-Type—-examines and controls the status of individual bits in specified memory areas.
3. Timer/Counter—allows the user to select various time-incremented and count-incremented and decremented functions.
4. Data Manipulation—used to transfer and compare BCD or octal values in the user program.
5. Arithmetic—performs the four indicated math functions.
6. Editing—used to locate, display and change instructions in the user program.
7. Control—directs the operation of the industrial terminal and its communication with the PLC-2 family processors and peripherals. Also provides HELP information.
8. Block Transfer Instructions—used to program block transfer Instructions in block format.
9. Shift Register—used to shift a word (all 16 bits) up or down one word in the shift register file.
   - used to shift a bit in the shift register file to the left or right one position.
   - used to create FIFO stacks.
10. Sequencer Instructions—used to establish and maintain user sequencer tables.
11. File Instructions—used to establish and manipulate user files.

This isolation is achieved by optical isolators consisting of a linked light emitting diode and photoelectric transistor. When current is passed through the diode, it emits light causing the transistor to switch on. Because there is no electrical connections between the diode and the transistor, very good electrical isolation (typically 1–4 kV) is achieved.

A d.c. input can be provided as Figure 1.3.7(a). When the pushbutton is pressed, current will flow through Dl causing TR1 to turn on passing the signal to the PLC internal logic. Diode D2 is a light emitting diode used as a fault finding aid to show when the input signal is present. Such indicators are present on almost all PLC input and output cards. The resistor R sets the voltage range of the input. The d.c. input cards are usually available for three voltage ranges: 5 V (TTL), 12–24 V, 24–50 V.

A possible a.c. input circuit is shown in Figure 1.3.7(b). The bridge rectifier is used to convert the a.c. to full wave-rectified d.c. Resistor R2 and capacitor Cl act as a filter (typically 50 ms time constant) to give a clean signal to the PLC logic. As before a neon LP1 acts as an input signal indicator for fault finding, and resistor Rl sets the voltage range.

Output connections also require some form of isolation barrier to limit damage from the inevitable plant faults and to stop electrical 'noise' corrupting the processor's operations. Interference can be more of a problem on outputs because higher currents are being controlled by the cards and the loads (solenoids and relay coils) are often inductive.

In Figure 1.3.8, eight outputs are fed from a common supply, which originates local to the PLC cubicle (but separate from the supply to the PLC itself). This arrangement is the simplest and the cheapest, to install. Each output has its own individual fuse protection on the card and a common circuit breaker. It is important to design the system so that a fault, say, on load 3 blows the fuse FS3 but does not trip the supply to the whole card shutting down every output. This is known as *'discrimination'*.

Contacts have been shown on the outputs in Figure 1.3.8. Relay outputs can be used (and do give the required isolation) but are not particularly common. A relay is an electromagnetic device with moving parts and hence a finite limited life. A purely electronic device will have greater reliability. Less obviously, though, a relay-driven inductive load can generate troublesome interference and lead to early contact failure.

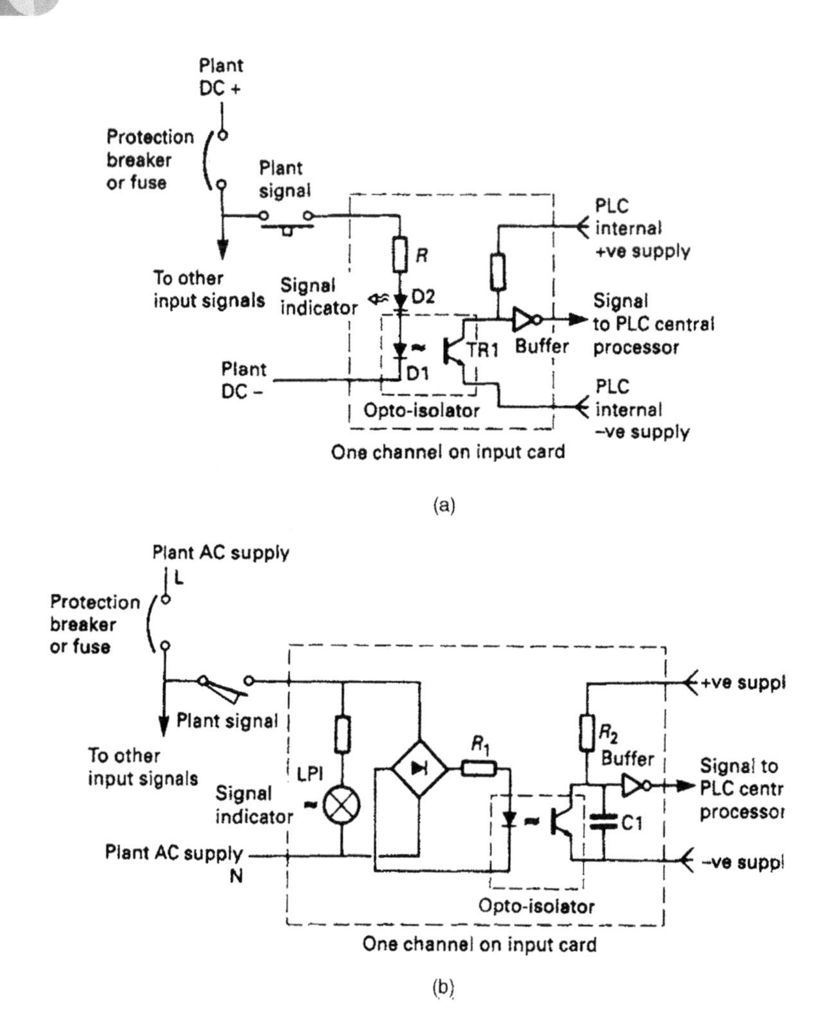

Fig. 1.3.7 Optical isolation of input signals: (a) d.c. input; (b) a.c. input.

A transistor output circuit is shown in Figure 1.3.9(a). Opto-isolation is again used to give the necessary separation between the plant and the PLC system. Diode Dl acts as a spike suppression diode to reduce the voltage spike encountered with inductive loads as shown in Figure 1.3.9(b). The output state can be observed on LED1. Figure 1.3.9(a) is a current-sourcing output. If NPN transistors are used, a current sinking card can be made as Figure 1.3.9(c).

The a.c. output cards invariably use triacs, a typical circuit being shown in Figure 1.3.10. Triacs have the advantage that they can be made to turn on at zero voltage and inherently turn off at zero current in the load. The zero current turn off eliminates the spike interference caused by breaking the current through an inductive load. If possible, all a.c. loads should be driven from triacs rather than relays.

An output card will have a limit to the current it can supply, usually set by the printed circuit board tracks rather than the output devices. An individual output current will be set for each output (typically 2 A) and a total overall output (typically 6 A). Usually, the total allowed for the card current is lower than the sum of the allowed individual outputs.

1.3.2.3 Remote I/O

So far we have assumed that a PLC consists of a processor unit and a collection of I/O cards mounted in local racks. Early PLCs were arranged like this, but in a large and scattered plant, all signals had to be brought back to some central point in expensive multi-core cables. This also makes commissioning and fault finding rather difficult, as signals can only be monitored effectively at a point distant from the plant device being tested.

In all bar the smallest and cheapest systems, PLC manufacturers therefore provide the ability to mount I/O racks remote from the processor, and linked with simple

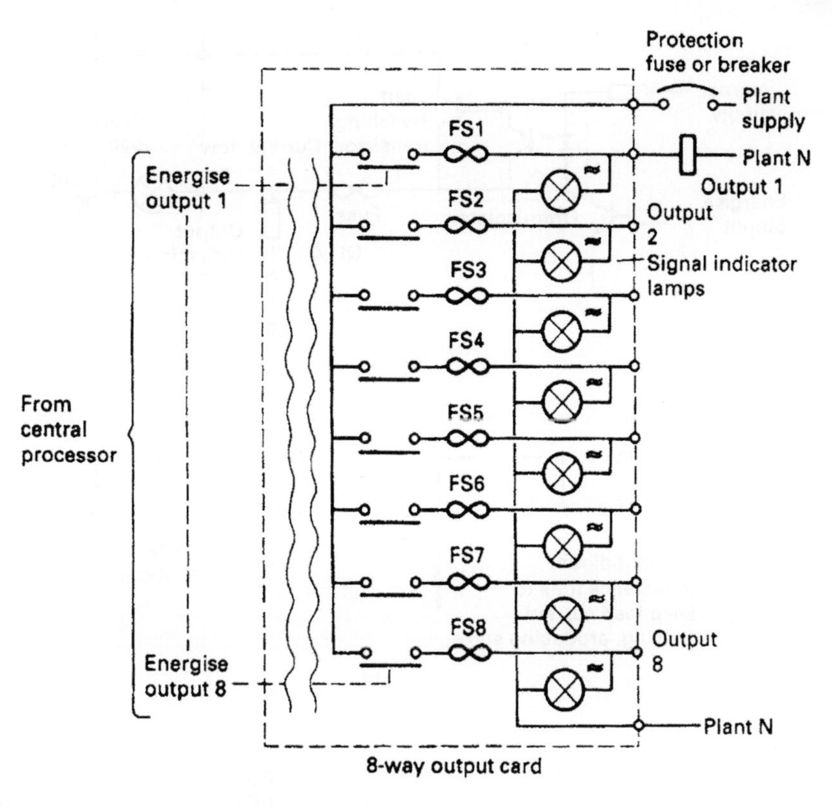

Fig. 1.3.8 Schematic of an 8-way output card with common supply.

(and cheap) screened single pair or fibre optic cable. Racks can then be mounted up to several kilometres away from the processor.

There are many benefits from this. It obviously reduces cable costs as racks can be laid out local to the plant devices and only short multi-core cable runs are needed. The long runs will only be the communication cables (which are cheap, easy to install and only have a few cores to terminate at each end) and hardwire safety signals.

Less obviously, remote I/O allows complete plant units to be constructed, wired to a built in PLC rack, and tested off site prior to delivery and installation. Typical examples are hydraulic skids, desks and even complete control pulpits. The use of remote I/O in this way can greatly reduce installation and commissioning time and cost.

The use of serial communication for remote I/O means some form of sequential scan must be used to read input and update outputs. This scan, typically 30–50 ms, introduces a small delay in the response to signals discussed further in the following section.

If remote I/O is used, provision should be made for a program terminal to be connected local to each rack. It negates most of the benefits if the designer can only monitor the operation from a central control room several hundred metres from the plant. Fortunately, manufacturers have recognised this and most PLCs have programming terminals which can be remotely connected to the processor.

1.3.2.4 The program scan

A PLC program can be considered to behave as a permanent running loop similar to Figure 1.3.11(a). The user's instructions are obeyed sequentially, and when the last instruction has been obeyed the operation starts again at the first instruction. A PLC does not, therefore, communicate continuously with the outside world, but acts, rather, by taking 'snapshots'.

The action of Figure 1.3.11(a) is called a *program scan*, and the period of the loop is called the *program scan time*. This depends on the size of the PLC program and the speed of the processor, but is typically 2–5 ms per K of program. Average scan times are usually around 10–50 ms.

Figure 1.3.11(a) can be expanded to Figure 1.3.11(b). The PLC does not read inputs as needed (as implied by Figure 1.3.11(a)) as this would be wasteful of time. At the start of the scan it reads the state of all the connected inputs and stores their state in the PLC memory. When the PLC program accesses an input, it reads the input state as it was at the start of the current program scan.

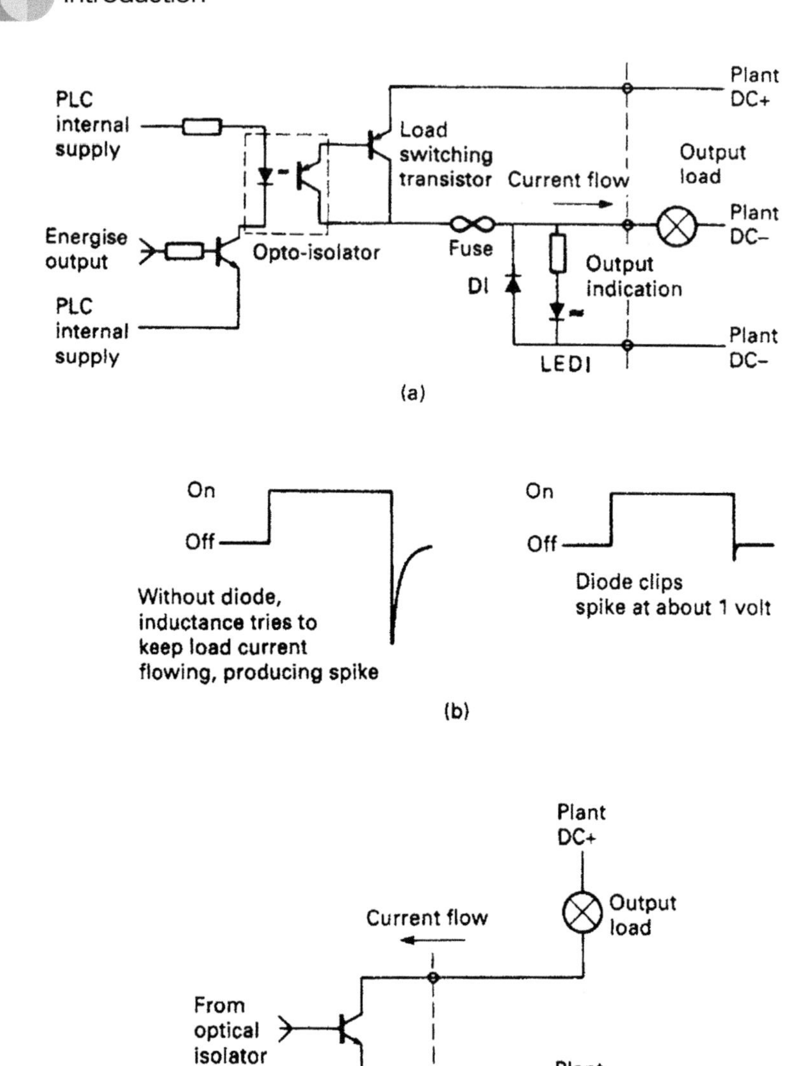

Fig. 1.3.9 The d.c. output circuits: (a) isolated output circuit, current sourcing; (b) the effect of an inductive load and the reason for including diode D1; (c) current sinking output.

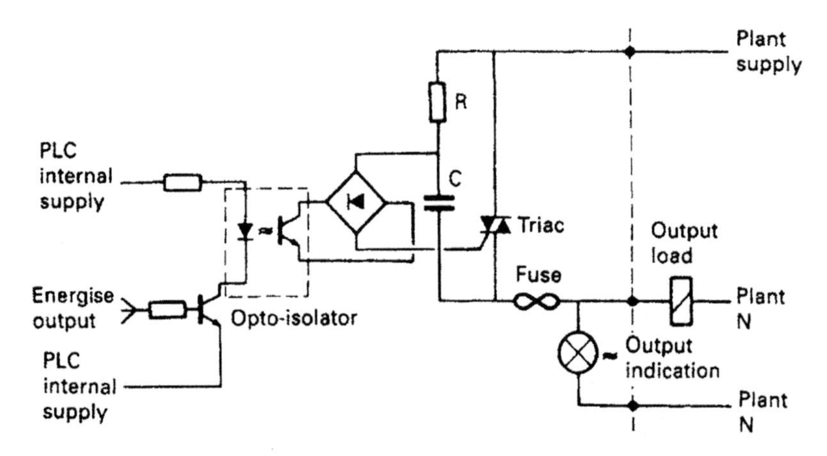

Fig. 1.3.10 The a.c. isolated output. The triac switches on at zero voltage and off at zero current which minimises interference.

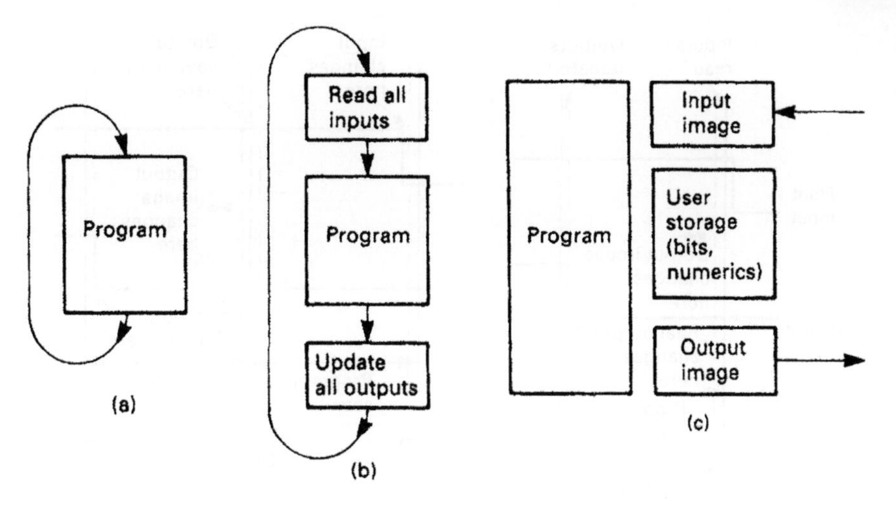

Fig. 1.3.11 The program scan and memory organisation: (a) simple view of PLC operation; (b) more detailed view of PLC operation; (c) memory organisation.

As the PLC program is obeyed through the scan, it again does not change outputs instantly. An area of the PLC's memory corresponding to the outputs is changed by the program, then ALL the outputs are updated simultaneously at the end of the scan. The action is thus:

```
Read Inputs,
Scan Program,
Update Outputs.
```

The PLC memory can therefore be considered to consist of four areas as shown on Figure 1.3.11(c). The inputs are read into an input mimic area at the start of the scan, and the outputs updated from the output mimic area at the end of the scan. There will be an area of memory reserved for internal signals which are used by the program but are not connected directly to the outside world (timers, counters, storage bits (e.g. fault signals) and so on). These three areas are often referred to as the *data table* (Allen Bradley) or the *database* (ASEA/ABB).

This data area is smaller than may be at first thought. A medium size PLC system will have around 1000 inputs and outputs. Stored as individual bits in a PLC with a 16-bit word this corresponds to just over 60 storage locations. An analog value read from the plant or written to the plant will take one word. Timers and counters take two words (one for the value, and one for the preset) and 16 internal storage bits take just one word. The majority of the store, therefore, is taken up by the fourth area, the program itself.

The program scan limits the speed of signals to which a PLC can respond. In Figure 1.3.12(a) a PLC is being used to count a series of fast pulses, with the pulse rate slower than the scan rate. The PLC counts correctly. In Figure 1.3.12(b) the pulse rate is faster than the scan rate and the PLC starts to miscount and miss pulses. In the extreme case of Figure 1.3.12(c) whole blocks of pulses are totally ignored.

In general, any input signal a PLC reads must be present for longer than the scan time; shorter pulses

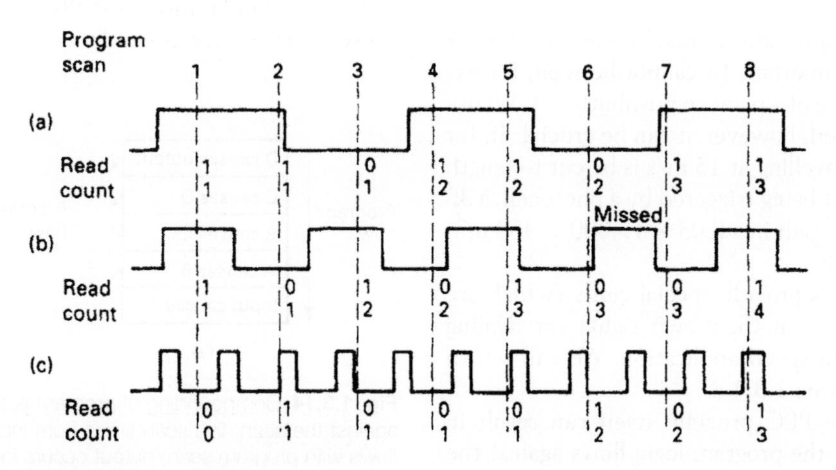

Fig. 1.3.12 The effect of program scan on a fast pulse train.

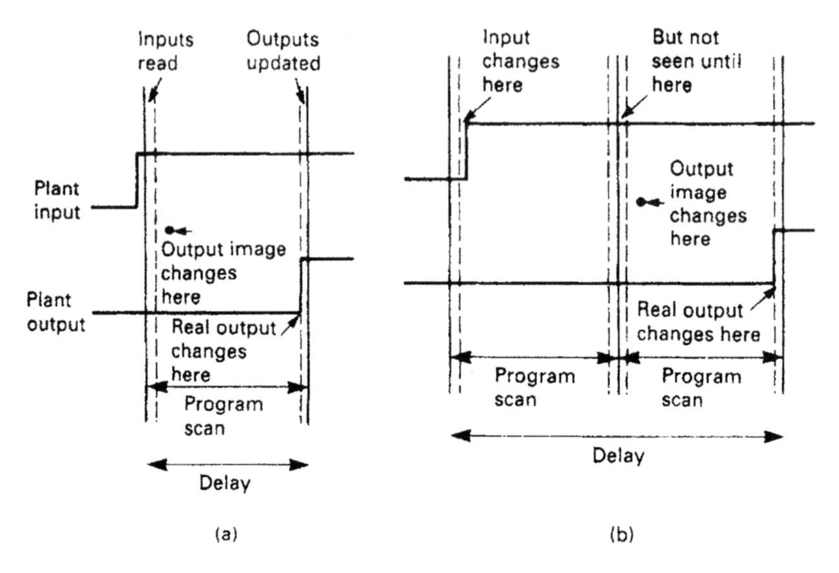

Fig. 1.3.13 The effect of program scan on response time: (a) best case; (b) worst case.

may be read if they happen to be present at the right time but this cannot be guaranteed. If pulse trains are being observed, the pulse frequency must be slower than $1/(2 \times \text{scan period})$. A PLC with a scan period of 40 ms can, in theory, just about follow a pulse train of $1/(2 \times 0.04) = 12.5$ Hz. In practice, other factors such as filters on the input cards have a significant effect and it always advisable to be conservative in speed estimates.

Less obviously, the PLC scan can cause a random 'skew' between inputs and outputs. In Figure 1.3.13 an input I is to cause an 'immediate' output O. In the best case of Figure 1.3.13(a), the input occurs just at the start of the scan, resulting in the energisation of the output one scan period later. In Figure 1.3.13(b) the input has arrived just after the inputs are read, and one whole scan is lost before the PLC 'sees' the input, and the rest of the second scan passes before the output is energised. The response can thus vary between one and two scan periods.

In the majority of applications this skew of a few tens of milliseconds is not important (it cannot be seen, for example, in the response of a plant to pushbuttons). Where fast actions are needed, however, it can be crucial. If, for example, material travelling at 15 m/s is be cut to length by a PLC with the cut being triggered by a photocell, a 30 ms scan time would result in a $0.03 \times 15{,}000 = 450$ mm variation in cut length.

PLC manufacturers provide special cards (which are really small processors in their own right) for dealing with this type of high-speed application. We will return to these later in Section 1.3.4.8.

The layout of the PLC program itself can result in undesirable delays if the program logic flows against the PLC program scan. The PLC starts at the first instruction for each scan, and works its way through the instructions in a sequential manner to the end of the program when it does its output update, then goes to read its inputs and run through the program again.

In Figure 1.3.14(a), an input I again causes an output O, but it goes through five steps first (it could be stepping a counter or seeing if some other required conditions are present). The program logic, however, is flowing against the scan. On the first scan the input I causes event A. On the next scan event A causes event B and so on until after five scans event D causes the output to energise. If the program had been arranged as Figure 1.3.14(b) the whole sequence would have occurred in one single scan.

The failings of Figure 1.3.14(a) are self-evident, but the effect can often occur when the layout of the program is not carefully planned. The effect can also be used deliberately to ensure sequences operate correctly.

The effect of scan times can become even more complex when remote serially scanned I/O racks are present. These are generally read by an I/O scanner as

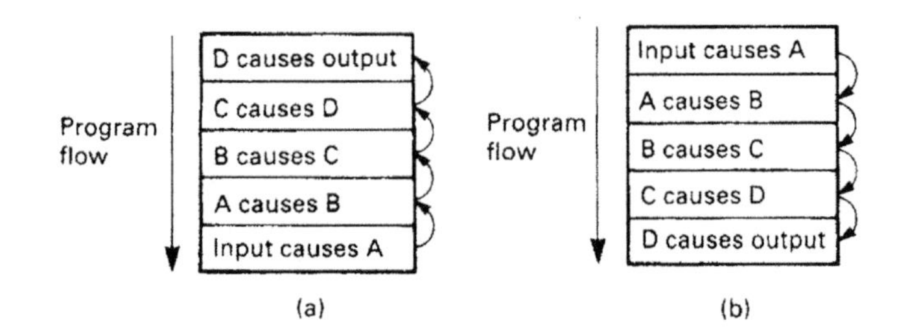

Fig. 1.3.14 Compounding of program scan delays: (a) logic flows against the scan, five scan times from input to output; (b) logic flows with program scan, output occurs in same program scan as input.

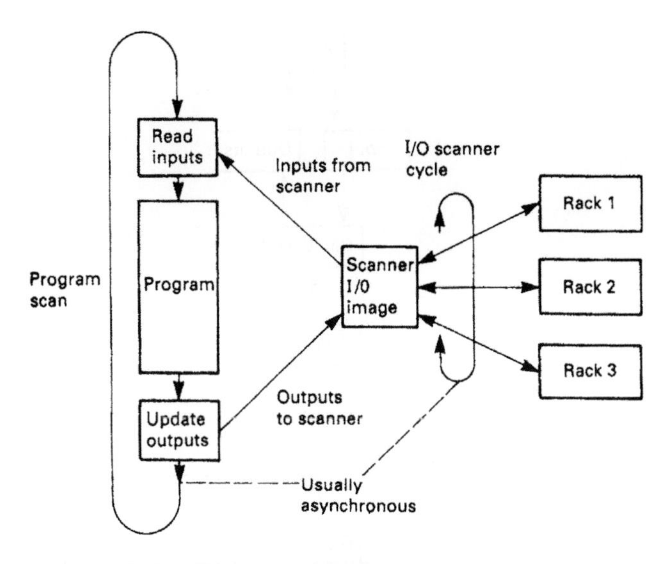

Fig. 1.3.15 The effect of remote input/output scan times. The remote I/O scan usually free-runs and is not synchronised with the program scan.

Figure 1.3.15 but the remote I/O scan is not usually synchronised to the program scan. In this case with, say, a program scan of 30 ms and a remote I/O scan of 50 ms the fastest response to an input could be 30 ms, but the slowest response (with an input just missing the I/O scan and the I/O scan just missing the program scan and the programming scan just missing the I/O scan to update the output) could be 180 ms.

PLC manufacturers offer many facilities to reduce the effect of scan times. Typical are intelligent high-speed-independent I/O cards and the ability to sectionalise the program into areas with different scan rates.

1.3.3 Programming methods

1.3.3.1 Introduction

The programming language of a PLC will be used by engineers, technicians and maintenance electricians. It should therefore be based on techniques used in industry rather than techniques used in computer programming. In this section we shall look at the various ways of programming PLCs from different manufacturers.

1.3.3.2 I/O identification

The PLC program is concerned with connections to the outside plant, and these input and output devices need to be identified inside the program. Before we can examine how the program is written we will first discuss how various manufacturers treat the I/O.

It is shown in Figure 1.3.3 that a medium-sized PLC system consists of several racks each containing cards, with each card interfacing generally with 8, 16 or 32 devices. I/O addressing is usually based on this rack/card/bit idea.

The Allen Bradley PLC-5 family has a range of processors which can address up to 64 racks. Its medium size 5/25 can have up to eight racks. The rack containing the processor is automatically defined as rack 0, but the designer can allocate addresses of the other racks (in the range 1–7) by set-up switches. The racks other than rack 0 connect to the processor via a remote I/O serial communications cable.

Each rack contains 16 card positions which are grouped in pairs called a 'slot'. A 16-card rack thus contains eight slots, numbered 0–7. A slot can contain one 16-way input card and one 16-way output card OR two 8-way cards usually (but not necessarily) of the same type.

The addressing for inputs is

```
I: Rack Slot/Bit
```

with bit being two digits. Allen Bradley use octal addressing for bits, so allowable numbers are 00–07 and 10–17. The address I: 27/14 is input 14 (octal remember) on slot 7 in rack 2.

Outputs are addressed in a similar manner:

```
O:rack slot/bit
```

so O : 3 5/06 is output 6 in slot 5 of rack 2. Note that if 16-way cards are used an input and an output can have the same rack/slot/bit address, being distinguished only by the I: or the O:. With 8-way cards there can be no sharing or rack/slot/bit addressing.

The digital I/O in Siemens 115 PLCs is arranged into groups of 8 bits, called a Byte. A signal is identified by its bit number (0–7) and its byte number (0–127).

Inputs are denoted

```
I<byte>.<bit >
```

and outputs by

```
Q<byte>.<bit>
```

I9.4 is thus an input with bit address 4 in byte 9, and Q63 . 6 is an output with bit address 6 in byte 63.

Like Allen Bradley, Siemens use card slots in one or more racks. The cards are available in 16 bit (2 byte) or 32 bit (4 byte) form. A system can be built with local racks connected via a parallel bus cable or as remote racks with a serial link.

The simplest form of addressing is fixed slot where four bytes are assigned sequentially to each slot; 0–3 to the first slot, 4–7 to the next slot and so on. Input I12.4 is thus input bit 4 on the first byte of the card in slot 3 of the first rack. If 16 bit (2 byte) cards are used with fixed (4 byte) addressing the upper 2 bytes in each slot are lost.

In all, barring the simplest system, the user has the ability to assign byte addresses. This is known as variable slot addressing. The first byte address and the range (2 byte for 16 bit cards or 4 byte for 32 bit cards) can be set independently for each slot by switches in the adaptor module in each rack. Although any legitimate combination can be set up, it is recommended that a logical order is used.

Siemens use different notations in different countries with multi-lingual programming terminals. A common European standard is German, where E (for Eingang or input) is used for inputs (e.g. E4 . 7) and A (for Ausgang) used for outputs (e.g. A3 . 5) .

The GEM-80 again configures its I/O in terms of bits and slots within racks. The processor rack can contain 8 card positions, and additional I/O can be connected into 12 position racks local to the processor connected via ribbon cable (called Basic I/O) or remotely via a serial link.

The I/O is addressed in terms of 16 bit words, one word corresponding to one or two card positions, and the prefix A being used for inputs and B for outputs. The bit addressing runs in decimal from 0 to 15.

`A3 . 12` is thus input bit 12 in word 3 and
`B5 . 04` is output bit 4 in word 5

A word can only be an input or an output; duplication of word addresses is not allowed. I/O cards are available in 8 bit, 16 bit and 32 bit form, so one slot can be half a word, one word or two words according to the cards being used. Individual slot addresses are set by rotary switches on the back plane of each rack. The user has a more or less free choice in this allocation, but as usual it is best to use a logical sequential progression.

The ABB (originally ASEA) Master system is a more complex system than any we have discussed so far. Its organisation brings the user closer to the computer, and its language is more akin to the ideas used by programmers. If the PLCs discussed so far are taken to be represented by the home computer language BASIC, the ABB Master is analogous to PASCAL or C. This comparison is actually closer than might, at first, be thought. BASIC is quick and easy to use, but can de-generate into a web of spaghetti programming if care is not taken. PASCAL and C are more powerful but everything has to be declared and the language forces organisation and structure on the user.

The I/O cards are NOT identified by position in the rack, but by an address set on the card by a small plug with solder links. The I/O addressing does not, therefore, relate to card position, and a card can, in theory, be moved about without changing its operation.

The processor memory is arranged as Figure 1.3.16(a). The I/O is connected to a processor database, but unlike PLCs described earlier, the designer can specify different scan rates for different cards.

The designer also has considerable power over how the PLC program is organised. This is heavily modularised as we shall see later, and the user can also specify different scan rates for different modules of the program.

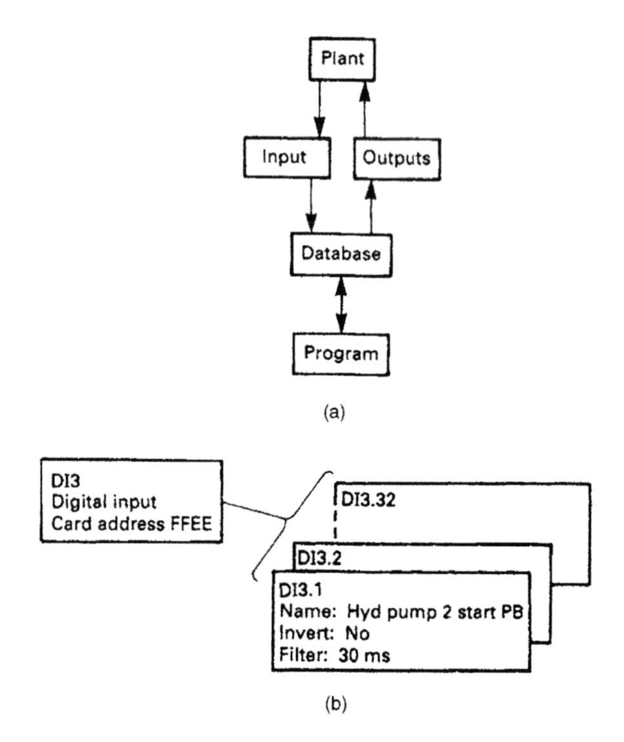

Fig. 1.3.16 The ABB Master system: (a) organisation of the memory; (b) definition of a digital input in the database.

Figure 1.3.16(b) indicates the database for one input card. There are two levels of the definition, the top level relating to details of the board itself such as address and scan rate, then lower levels relating to details of each channel on the board such as its name and whether the signal is to be inverted. The database holds details for all the I/O which can then be referenced by the program either by its database identification (e.g., DI3.1) or by its unique name (e.g., HydPump2StartPB).

The Mitsubishi F2 range is typical of small PLCs with input/output connection, power supply and processor all contained in one unit. The smallest unit, the F2-40 M, has 24 inputs and 16 outputs. (It is a characteristic of process control systems that the ratio input:outputs is generally 3:2.)

The 24 inputs are designated X400–X427 in octal notation and the 16 outputs Y430–Y447. The apparently arbitrary numbers are directly related to the storage locations used to hold the image of the inputs and output. Further addresses are used in larger PLCs in the series.

1.3.3.3 Ladder logic

Early PLCs, designed for the car industry, replaced relay control schemes. The symbols used in American relay drawings, –] [– for a normally open (NO) contact, –] / [–for a normally closed (NC) contact, and – ( ) – for a plant output, were the basis of the language. Figure 1.3.5 shows the keyboard for a programmer for this type of PLC; the relationship to relay symbolism is obvious.

Suppose we have a hydraulic unit, and we wish to give a healthy lamp indication when

The Pump is running (sensed by an auxiliary contact on the pump starter).
There is oil in the tank (sensed by a level switch which makes for good level).
There is oil pressure (sensed by a pressure switch which makes for adequate pressure).

With conventional relays, we would wire up a circuit as Figure 1.3.17(a).

To use a PLC, we connect the input signals to an input card, and the lamp to an output card as Figure 1.3.17(b). The I/O notation used is Allen Bradley.

The program to provide the function is shown in Figure 1.3.17(c). The line on the left can be considered to be a supply, and the line on the right a neutral. The output is represented by a coil – ( ) – and is energised when there is a route from the left-hand rail. Output 0 : 22/01 will come on when signals I:21/00, T:21/01 and I:21/02 are all present.

The program is entered from a terminal with keys representing the various relay symbols. The terminal can also be used to monitor the state of the inputs and outputs, with 'energised' inputs and outputs being shown highlighted on the screen.

In Figure 1.3.18(a), a hydraulic cylinder can be extended or retracted by operation of two pushbuttons. The notation this time is for a GEM-80. It is undesirable to allow both solenoids to be operated together; this will almost certainly result in blown fuses in the supply to the output card, so some protection is needed. The program to achieve this is shown in Figure 1.3.18(b).

Normally closed contacts –] / [– have been used here. Output B2 . 9, the extend solenoid, will be energised when the extend pushbutton is pressed, provided the retract solenoid is not energised or the retract button pressed, and the extend limit switch has not been struck.

There are two points to note in Figure 1.3.18. Contacts can be used from outputs as well as inputs, and contacts can be used as many times as needed in the program. Figure 1.3.18 also shows the origin of the name *'Ladder Program'*. A program in this form looks like a ladder, with each instruction statement forming a *'rung'* and the power rail and neutral the supports. The term 'rung' is invariably applied to the contacts leading to one output.

Let us return to the hydraulics healthy light of Figure 1.3.17 and add a lamp test pushbutton (a useful feature that should be present on all panels. It not only allows lamps to be tested, but can also be used to check

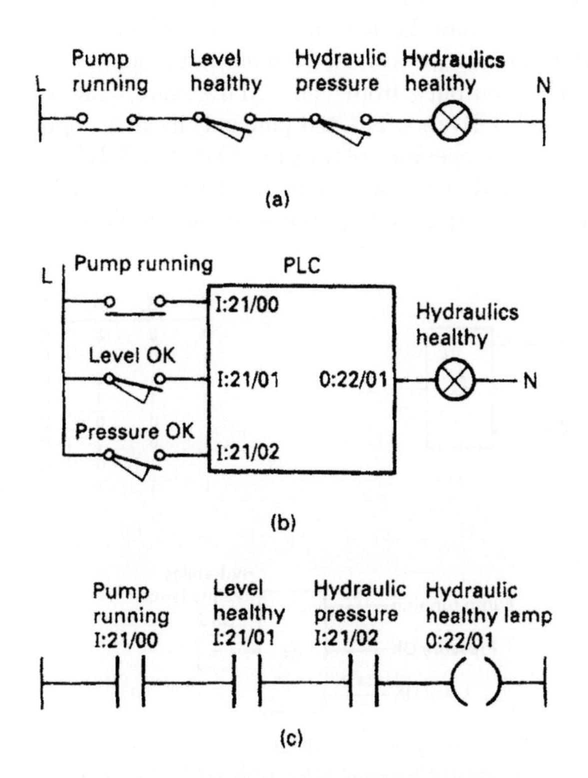

Fig. 1.3.17 From a relay circuit to a PLC program: (a) basic non-PLC circuit; (b) wiring of I/O to a PLC; (c) the corresponding PLC program.

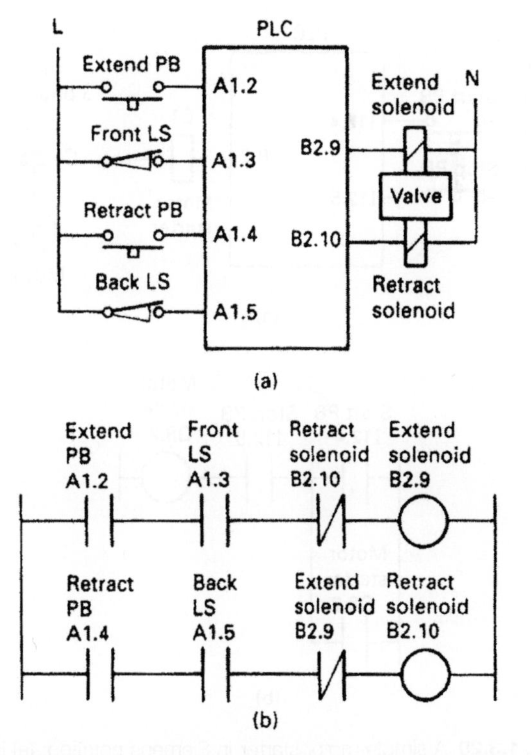

Fig. 1.3.18 Ladder diagram in GEM-80 notation: (a) input/output connections; (b) GEM-80 ladder diagram.

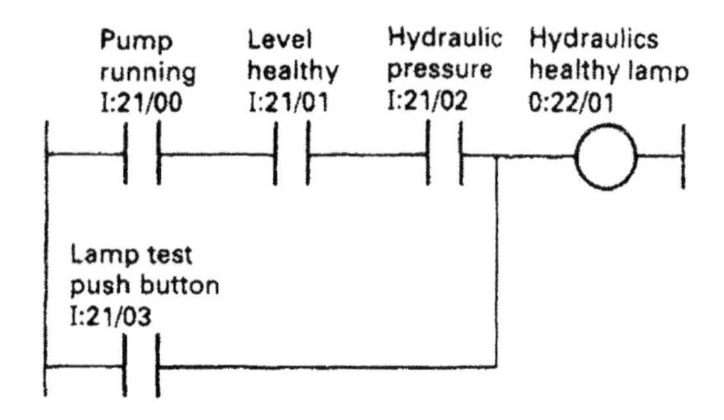

Fig. 1.3.19 Adding a lamp test pushbutton with a branch.

the PLC and the local rack are healthy). To do this we add the lamp test pushbutton to the PLC and modify the program to Figure 1.3.19.

Here we have added a branch, and the output will energise if our three plant signals are all present OR the lamp test button is pressed. The way in which the branch is programmed need not concern us here as it varies between manufacturers. Some use start branch and end branch keys (the keypad shown earlier in Figure 1.3.5 uses this method; the corresponding keys can readily be identified). Others use a branch from/to approach. All are simple to use.

A further use of a branch is shown in Figure 1.3.20. This is probably the most common control circuit, a motor starter, shown using Siemens notation. The operation is simple, pressing the start pushbutton causes

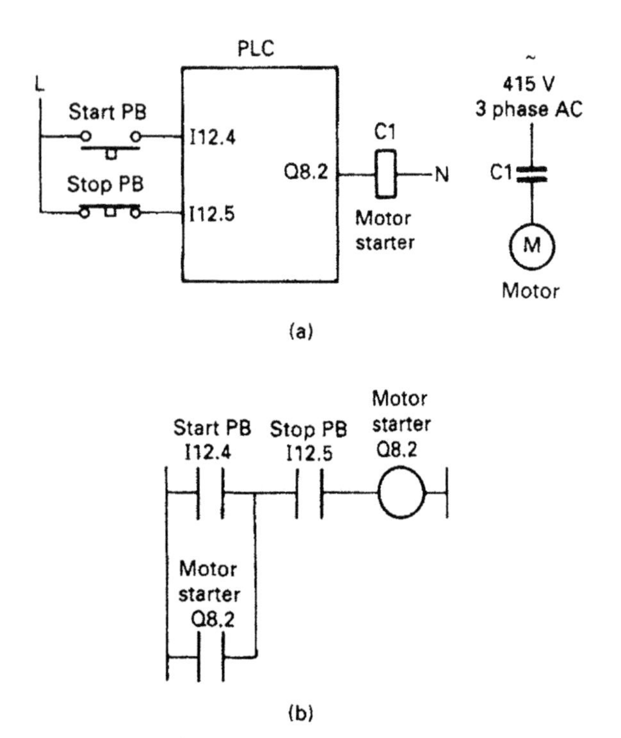

Fig. 1.3.20 A simple motor starter in Siemens notation: (a) input/output connections; (b) the ladder diagram. Note how the stop button appears in the program.

the output QS . 2 to energise, and the contact of the output in the branch keeps the output energised until the stop button is pressed. The program, like its relay equivalent, remembers which button was last pressed.

There is, however, a very important point to note about the pushbutton wiring and the program. For safety, a normally closed stop button has been used giving an input signal on 112 . 5 when the stop button is not pressed. A loss of supply to the button, or a cable fault, or dirt under the contacts will cause the signal to be lost making the program think the stop PB has been pressed causing the motor to stop. If a normally open stop PB has been used, the PLC program could easily be made to work, but a fault with the stop button or its circuit could leave the motor running with the only way of stopping it being to turn off the PLC or the motor supply.

This topic is discussed further in Section 1.3.7.4, but note the effect on the program in Figure 1.3.20. The sense of the stop button input (I12.5) inside the program is the opposite of what would be expected in a relay circuit. The input is really acting as *'Permit to Run'* rather than *'Stop'*.

1.3.3.4 Logic symbols

Logic gates are widely used in digital systems (including the boards used inside PLCs). The circuits on these boards are represented by logic symbols, and these symbols can also be used to represent the operations of a PLC program. Logic symbols are used by Siemens and ABB; initially we will use Siemens notation.

The output from an AND gate, shown in Figure 1.3.21(a), is TRUE if (and only if) all its inputs are TRUE. The operation of the gate of Figure 1.3.21(a) can be represented by the table of Figure 1.3.21(b). In Figure 1.3.21(c) we have the hydraulics healthy lamp of

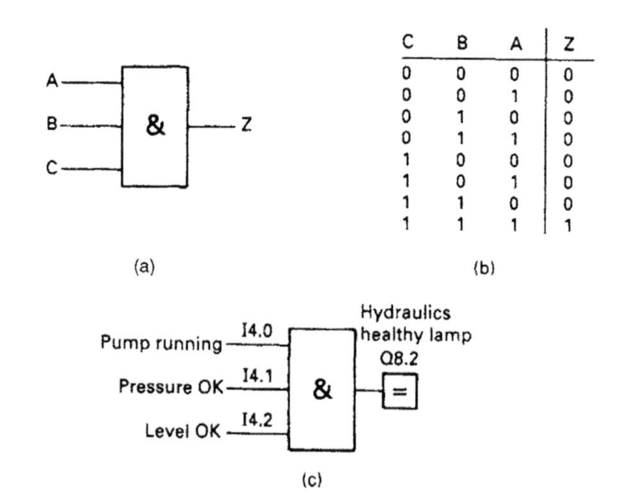

C	B	A	Z
0	0	0	0
0	0	1	0
0	1	0	0
0	1	1	0
1	0	0	0
1	0	1	0
1	1	0	0
1	1	1	1

(b)

Fig. 1.3.21 PLC programming using logic symbols: (a) an AND gate; (b) truth table for a three-input AND gate; (c) the healthy lamp of Figure 1.3.17 using a logic symbol in Siemens notation.

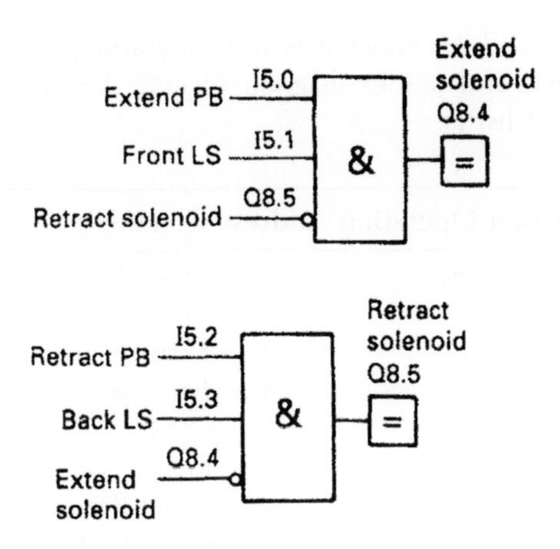

Fig. 1.3.22 The hydraulic cylinder of Figure 1.3.18 in logic notation and Siemens addressing. Note the use of inverted inputs (denoted by small circles).

Figure 1.3.19 programmed using logic symbols for a Siemens PLC. The output block, denoted by equals =, is energised when its input is true, so the lamp Q8 . 2 is energised (lit) when all the inputs to the AND gate are true.

Often a test has to be made to say a signal is NOT true. This is denoted by a small circle 'o'. Figure 1.3.18 illustrats the control of a hydraulic cylinder with a program which prevented the extend and retract solenoids from being energised simultaneously. This is shown programmed with logic symbols for a Siemens PLC in Figure 1.3.22. Note the NOT inputs on each AND gate.

The output of an OR gate, Z in Figure 1.3.23(a), is TRUE if any of its inputs are TRUE. The inverse of a signal can be tested, as before, with a small circle 'o'. The output Z of the gate in Figure 1.3.23(b) is TRUE if A is TRUE or B is FALSE or C is TRUE. In Figure 1.3.23(c) we have used an OR gate to add a lamp test to our hydraulic healthy lamp.

The circuit of Figure 1.3.23(c) is an AND/OR combination. The ABB Master has logic combination blocks as well as the basic gates. Figure 1.3.24(a) is the Master block corresponding to Figure 1.3.23(c) (with a Master program referring to the names in its database). Similarly, for an OR/AND combination the OR/AND block of Figure 1.3.24(b) can be used in a Master program.

1.3.3.5 Statement list

A statement list is a set of instructions which superficially resemble assembly language instructions for a computer. Statement lists, available on the Siemens and Mitsubishi range, are the most flexible form of programming for the experienced user but are by no means as easy to follow as ladder diagrams or logic symbols.

Figure 1.3.25 shows a simple operation in both ladder and logic formats for a Siemens PLC. The equivalent statement list would be:

Instruction	Operation	Address number	Comment
00	:A	I 3.7	Forward Pushbutton
01	:A	I 3.2	Front Limit OK
02	:AN	Q 4.2	Reverse Solenoid
03	:=	Q 4. 11	Output to Forward Solenoid

Here :A denotes AND, :AN denotes AND-NOT, and := sends the result to the output address Q4.11.

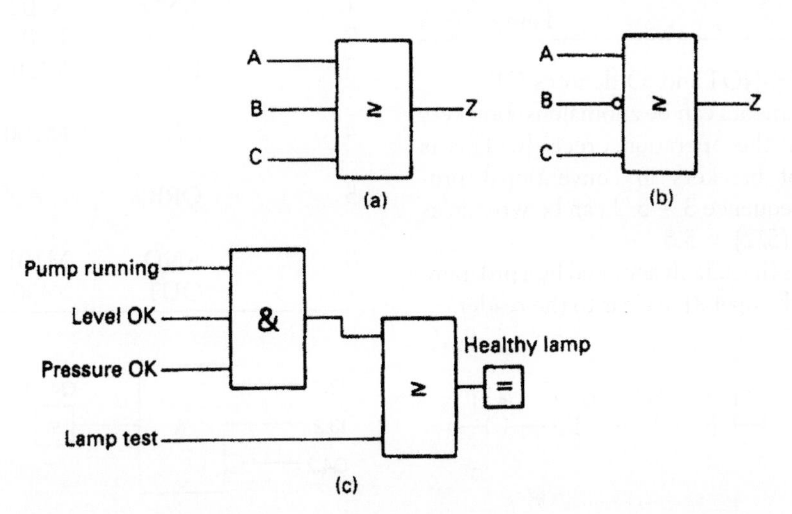

Fig. 1.3.23 The OR gate: (a) logic symbol; (b) OR gate with inverted input; (c) lamp test added to Figure 1.3.21(c).

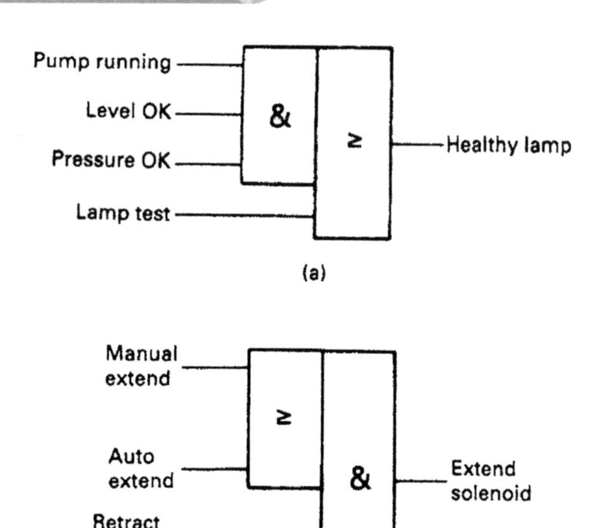

Fig. 1.3.24 ABB Master composite gates: (a) AND/OR gate (equivalent to Figure 1.3.23(c)); (b) OR/AND gate.

An OR operation is shown in Figure 1.3.26. The equivi lent statement list is:

Instruction	Operation	Address number	Comment
00	:ON	I 2.7	Local Pump Running
01	:O	F 3.6	Remote Pump Running
02	:ON	Q 4.2	Local Pump Starter
03	: =	Q 4.4	Pump Healthy Lamp

where :ON denotes OR-NOT and :O denotes OR.

Where a set of statements can be anomalous, brackets can be used to define the operation precisely. This is similar to the use of brackets in conventional programming where the sequence 3 + 5/2 can be written as (3 + 5)/2 = 4 or 3 + (5/2) = 5.5.

Although the latter is the default assumed by a program, the brackets do make the operation clear to the reader.

Figure 1.3.27 shows a typical operation, as usual in both logic and ladder diagram format. The equivalent statement list is:

Instruction	Operation	Address	Comments
00	:A(		Open First Set of Brackets
01	:O	F 3.3	Forward from desk 1
02	:O	F 3.4	Forward from desk 2
03	:)		Result of first set of brackets
04	:A(		AND Result with second set of brackets
05	:A	I 2.0	Motor 1 Selected
06	:A	I 2.1	Motor 2 Selected
07	:)		Now at point X
08	:A	I 4.1	Front Limit Switch Healthy
09	:AN	Q 5.5	Reverse Starter
10	: =	Q 5.6	Output to Forward Starter

Computer programmers will recognise this as being similar to the operation of a stack with the brackets pushing data down, or lifting data up, the stack.

The Mitsubishi PLC also uses statement lists, although the manual recommends the designer to construct a ladder diagram first then translate it into a statement list. The PLC system shown in Figure 1.3.28 with Mitsubishi notation becomes the statement list:

Instruction	Operation	Address	Comments
0	LD	X401	LD starts rung or branch
1	AND	X402	Xnnn are inputs
2	AN I	X403	ANI is And-Not
3	LD	Y430	LD starts a new branch leg
4	AN	M100	Mnnn are internal storage
5	ORB		OR the two branch legs
6	AND	M101	
7	OUT	Y430	End of Rung

```
I3.7   I3.2    Q4.2    Q4.11
|--] [----] [----]/[----( )--|
```

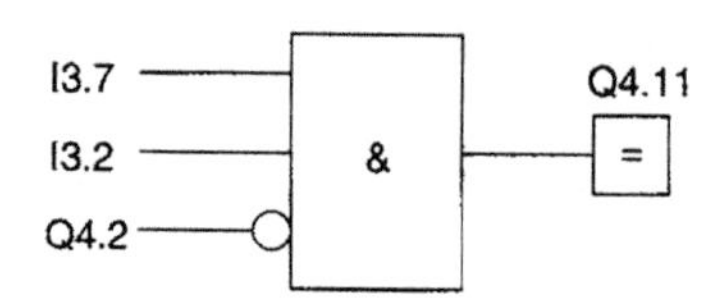

Fig. 1.3.25 Equivalent ladder and logic statements in Siemens notation.

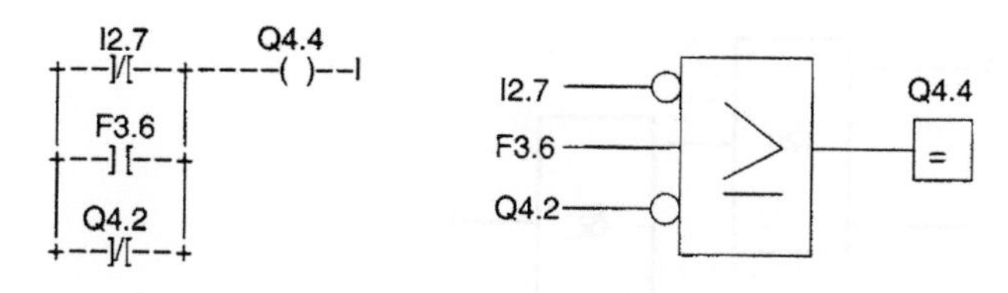

Fig. 1.3.26 OR gate equivalence in Siemens notation.

1.3.3.6 Bit storage

Besides inputs and outputs, the PLC will need to hold internal signals for data such as 'Standby Pump Running', 'System Healthy', 'Lubrication Fault' and so on. It would be very wasteful to allocate real outputs to these signals, so all PLCs provide some form of internal bit storage. These are known variously as Auxiliary Relays (Mitsubishi), Flags (Siemens), General Work space (GEM-80) and Bit Storage (Allen Bradley). The notation used within the programs vary, of course, from manufacturer to manufacturer.

Mitsubishi use Mnnn with nnn representing numbers within the predefined area M100 to M377 octal. Like most small PLCs the memory layout is fixed and cannot be defined by the user. In the other, larger, PLCs we discuss, the user can define how many storage bits are needed.

The Siemens notation is F <Byte>.<Bit> (e.g. F27 . 06).

The GEM-80 has a variety of general work space. The most common is called the G<Word>.<Bit> table, and appears in programs as G<Word>.<Bit> (e.g., G52.14). The G table is cleared when the PLC goes from a stopped state to a run state. Storage in the R table (e.g., R12 . O3) retains its state with the processor halted or with power removed.

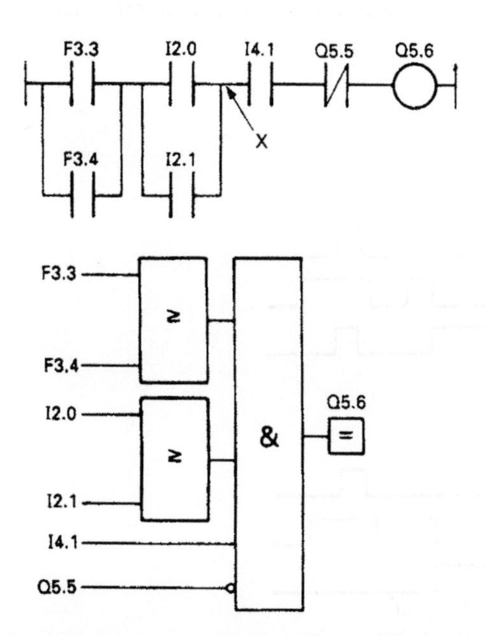

Fig. 1.3.27 More complex statements in ladder and logic notations.

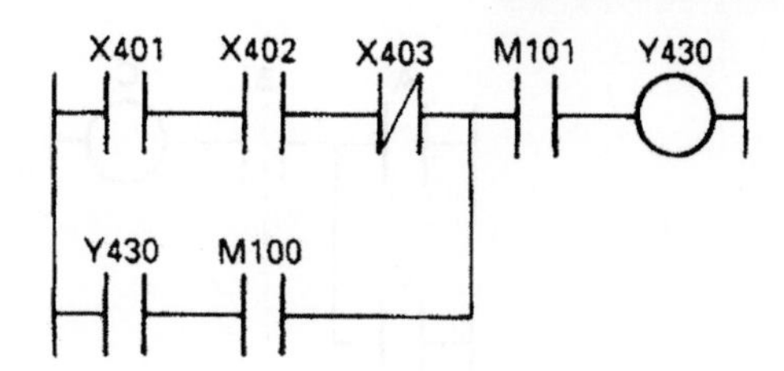

Fig. 1.3.28 A rung in a Mitsubishi ladder program.

Bit storage in the PLC-5 is denoted by B3/N where *n* denotes the signal (e.g., B3/192). The B denotes bit storage and the 3 is mandatory and arises out of the way the PLC-5 holds data in files. Bit storage is file 3; timers are file 4 (T4) and counters file 5 (C5) as we shall see later.

The ABB Master programming language does not really require internal storage bits, the function being provided by elements and connections within its database and the programming language.

Some form of memory circuit is needed in practically every PLC program. Typical examples are catching a fleeting alarm and the motor starter of the earlier Figure 1.3.20 where the rung remembers which button (start or stop) has been last pressed. These are known, for obvious reasons, as storage circuits.

The most common form is shown in ladder and logic form in Figure 1.3.29(a). Here output C is energised when input A is energised, and stays energised until input B is de-energised.

The operation is summarised in Figure 1.3.29(b). As can be seen input B overrides input A, the action required of a start/stop circuit. In some circuits, however, the start is required to override the stop. We all have a typical example in our motor cars; the windscreen wipers run when we switch them on, but continue to run to the park position when we turn them off. The PLC equivalent is Figure 1.3.29(c), where A would be the run switch, B the park limit switch and C the wiper motor. B has again been shown energised to allow running. The operation is summarised in Figure 1.3.29(d).

Storage is provided in digital systems by a device called a flip flop shown in Figure 1.3.30(a). This has two inputs, S (for Set) and R (for Reset). The device remembers which input was last energised. If both inputs occur together, the top (S) input wins. Such a circuit is called an SR flip flop. If the device is drawn with the R input at the top, as Figure 1.3.30(b), the reset input will override the set input if both are present together.

The flip flop is used in logic symbol PLC programming. A motor starter using a Siemens PLC is shown in Figure 1.3.31. Note that the RS version has been used

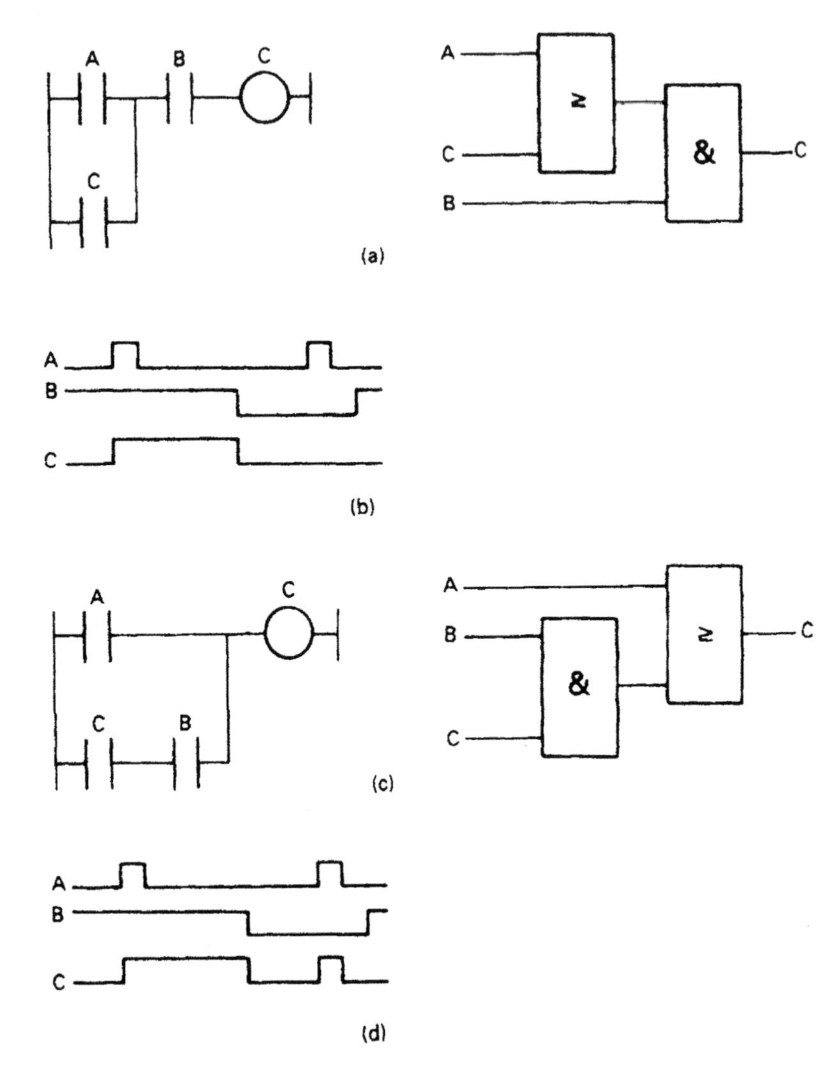

Fig. 1.3.29 Bit storage programs: (a) most common storage program, stop B overrides start A; (b) operation of (a), (c) program where start A overrides stop B; (d) operation of (c).

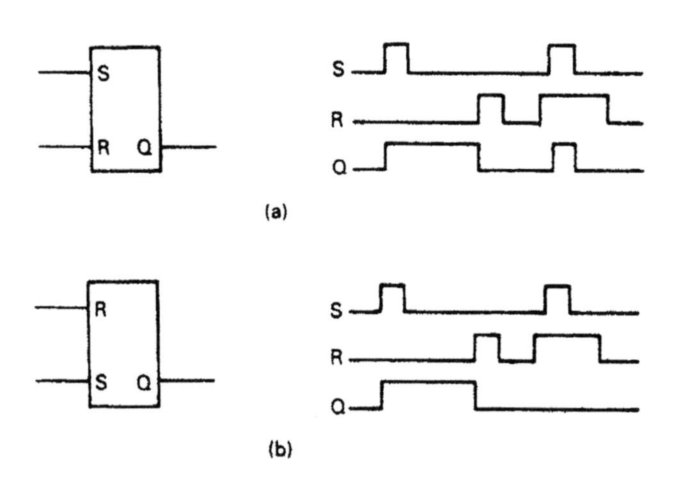

Fig. 1.3.30 The two types of flip flop storage: (a) the SR flip flop, set overrides R; (b) the RS flip flop, reset overrides set.

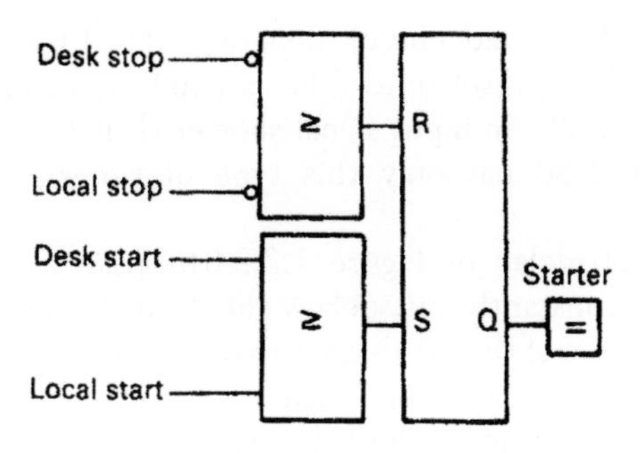

Fig. 1.3.31 Flip flop storage is commonly preceded by logic gates. Here either stop button will reset the flip flop. Note the circles on the stop button inputs denoting inverted inputs. These are necessary because the stop buttons give a signal in the not pressed state.

to ensure the stop logic overrides the run logic, and the stop signal acts as a permit to run.

The ABB Master uses an almost identical symbol for the flip flop, with the addition that there are five versions. The first of these is the simple SR type shown in Figure 1.3.30. The other versions are based on the fact that flip flops are invariably preceded by AND/OR combination of which Figure 1.3.31 is typical. The additional flip flops are one unit blocks consisting of a flip flop with built-in AND/OR gates of user-defined size. Figure 1.3.32, for example, is an ABB SRAO with an AND gate on the set input and an OR gate on the reset inputs. Other units are SRAA (AND/AND), SROA and SROO.

In Allen Bradley ladder diagrams, program clarity can be improved by the use of latch and unlatch outputs shown in Figure 1.3.33(a). These work on the same bit, setting the bit when the latch – (L) – is energised and resetting the bit when the – (U) – is energised. When both latch and unlatch are de-energised the bit holds its last state.

The Mitsubishi F2 uses a similar idea, but calls them S and R outputs as shown in Figure 1.3.33(b). This would be coded into a statement list:

```
0    LD     X400
1    OR     X401
```

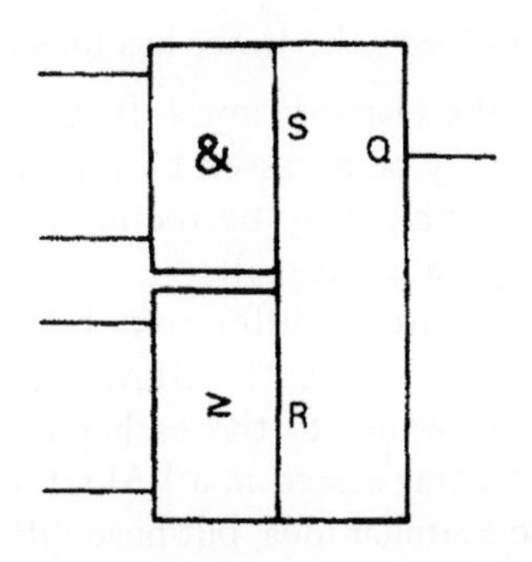

Fig. 1.3.32 An ABB master SRAO composite flip flop.

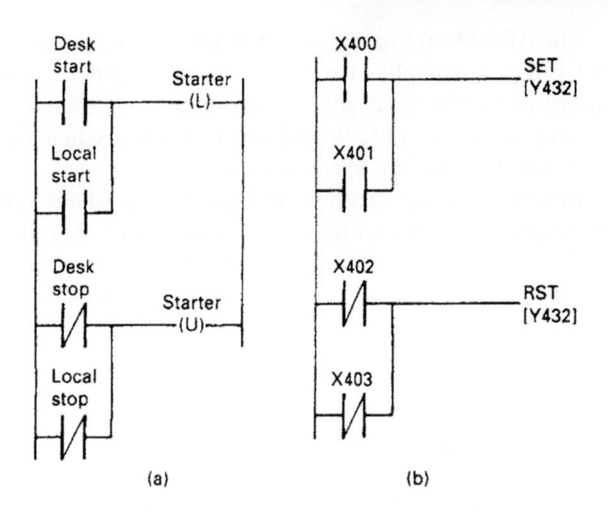

Fig. 1.3.33 Other forms of storage: (a) the Allen Bradley latch/unlatch; (b) the Mitsibushi set/reset.

```
2    S      Y432    Set Output
3    LDI    X402
4    ORI    X403
5    R      Y432    Reset Output
```

With both the Allen Bradley latch/unlatch, and the Mitsibushi set/reset, the priority goes to whichever is last in the program because of the program scan. Both the examples of Figure 1.3.33 correctly give priority to the stop signals.

Power failure or halting of the PLC can cause a problem with storage. When the PLC restarts should a memory bit hold the state it was in before the PLC halted, or should the memory be cleared? This is always a question of safety and convenience. A water pump in a pump house by a river 5 km from the main site should probably be allowed to restart itself if it was running before the power fail, an automatic stamping machine should almost certainly not restart.

The PLC manufacturers therefore allow the designer to choose whether a storage bit holds its state after a power fail (called *retentive memory)* or is cleared when the PLC is first run (called *non-retentive memory)*.

In the Allen Bradley PLC-5, this is determined by the circuit; the simple coil of Figure 1.3.29 is nonretentive and, the latch/unlatch of Figure 1.3.33(a) is retentive.

Other PLCs use the bit address. On a Siemens 115, flag addresses F0.O–F127.7 can be made retentive. On the Mitsubishi PLC, auxiliary relays N100–277 are nonretentive, and M300–M377 are retentive. In the GEM-80, the general bit storage G Table is nonretentive, a similar R Table is retentive, so a circuit similar to Figure 1.3.29 constructed with R3.4 as the coil and retaining contact would hold its state after a power failure.

The ABB Master uses a very structured PLC language, and forces a disciplined style on the programmer. The nature of subelements such as memories and their behaviour when the PLC is first run is defined when the program elements are first declared.

Retentive storage can be very hazardous as plants can unexpectedly leap into life after a power fail. The designer should take care that the design does not accidentally introduce retentive features by an inadvertent selection of bit addresses.

1.3.3.7 Timers

Time is nearly always a part of a control system. A typical example is: *'Lift Parking Brake, wait 0.5 s for brake to lift, drive to forward limit and stop drive, wait 1 s and apply parking brake'*. A PLC system must therefore include timers as part of its programming language. There are many types of timer, some of which are shown in Figure 1.3.34.

By far the most common is the on-delay of Figure 1.3.34(a). All the other timer blocks can be built with this block and a bit of thought. A 0 to 1 transition is delayed for a preset time T, but a 1 to 0 transition is not delayed at all. An input signal shorter than T is ignored. The GEM-80 has only this type of timer, calling it a *delay*.

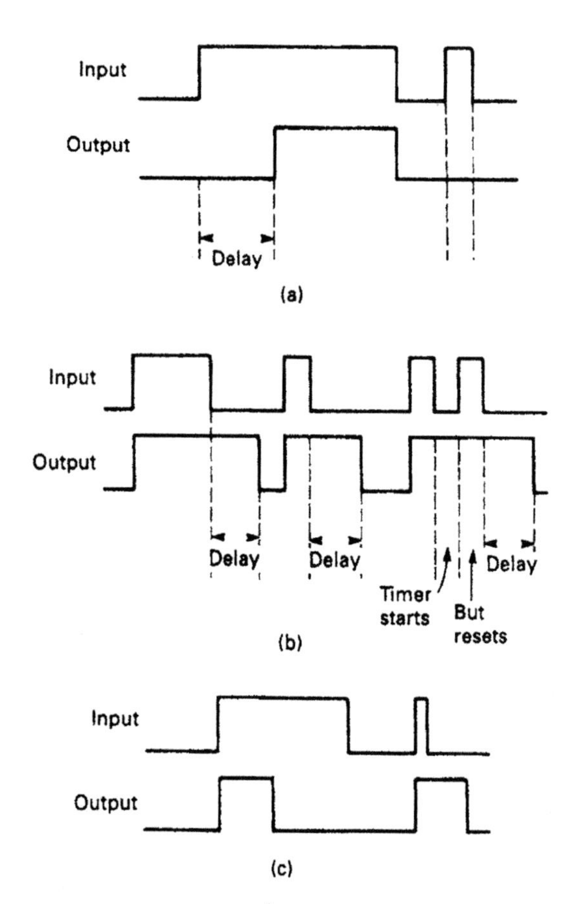

Fig. 1.3.34 Different forms of timer: (a) the on-delay. This is the most common timer and is often the only type available in many smaller PLCs; (b) the off-delay; (c) the fixed width pulse, often called a mono-stable.

The off-delay of Figure 1.3.34(b) passes a 0 to 1 transition instantly but delays the 1 to 0 transition. A common use of the off-delay is to remove contact bounce or noise from an input signal. An off-delay can be obtained from an on-delay by using the inverse of the input signal and taking the inverse of the timer output signal (although the resulting program lacks some clarity).

Figure 1.3.34(c) is an edge-triggered pulse timer; this gives a fixed width pulse for every 0–1 transition at the timer input. The PLC-5 has a one-scan pulse timer which produces a pulse lasting one (and only one) program scan. Pulses are useful for resetting counters or gating some information from one location to another.

A timer of whatever type has some values that need to be set by the user. The first of these is the basic unit of time (i.e., what units the time is measured in). Common units are 10 ms, 100 ms, 1 s, 10 s and 100 s. The base unit does not affect the accuracy of the timer; normally, the accuracy is similar to the program scan.

Next the timer duration (often called the *preset*) is defined. This is normally set in terms of the time base; a timer with a preset of 15 and a time base of 100 ms will last 1.5 s for example. In small PLCs this preset can only be set by the programmer, in the larger PLCs the duration can be changed from within the program itself. A off-delay timer used to apply a parking brake, for example, could have different preset times depending on whether the drive concerned is travelling at low speed or high speed.

When a timer is used, there are several signals that may be available. Figure 1.3.35 shows the signals given for a PLC-5 on-delay timer (called a TON) and a off-delay timer (called a TOF).

EN (for enable) is a mimic of the timer input.

TT (for timer timing) is energised whilst the time is running.

DN (for done) says the timer has finished.

In larger PLCs the elapsed time (often called the *Accumulated Time*) may be accessed by the program for use elsewhere (a program may be required to record how long a certain operation takes).

PLC manufacturers differ on how a timer is programmed. Some, such as the GEM-80, treat the timer as a delay block similar to the earlier Figure 1.3.34(a) with the preset being stored in a VALUE block.

Siemens use a similar idea, but have different types of timer. The PLC-5, however, uses the timer as a terminator

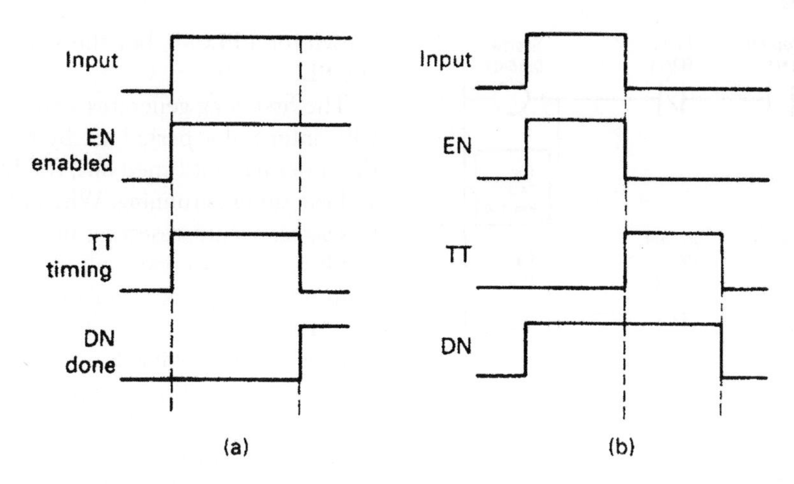

Fig. 1.3.35 Allen Bradley timer notations: (a) EN, TT and DN for an on-delay (TON) timer; (b) EN, TT and DN for an off-delay (TOF) timer.

for a rung, with the timer signals being available as contacts for use elsewhere.

Figure 1.3.36 shows a typical application programmed for a PLC-5 and a GEM-80 in ladder logic and a Siemens 115-U using logic symbols. The program controls a motor starter which is started and stopped via pushbuttons. The motor starter has an auxiliary contact which makes when the starter is energised, effectively saying the motor is running. If the drive trips because of an overload, or because an emergency stop is pressed, or there is a supply fault, the auxiliary contact signal will be lost. The contact cannot, however, be checked until 0.5 s after the starter has been energised to allow time for the contact to pull in. The program in each case checks the auxiliary contact and signals a drive fault if there is a problem. Note the difference in the way the timer is used and the fault signal is stored.

The accumulated time in the timers discussed so far goes back to zero each time the input goes to a zero. This is known as a *non-retentive* timer. Most PLC timers are of this form. Occasionally it is useful to have a timer which holds its current value even though the input signal has gone. When the input occurs again, the timer continues from where it stopped. This, not surprisingly, is known as a *retentive* timer. A separate signal must be used to reset the timer to zero. If a retentive timer is not available on a particular PLC, the same function can be provided with a counter, a topic discussed in the next section.

A typical timer can count up to 32,767 base time units (corresponding to 15 binary bits). Some older PLCs working in BCD can only count to 999. With a 1 s time base the maximum time will be just over 546 min or about 9 h. Where longer times are needed (or times with a resolution better than one second), timers and counters can be used together as described in the next section.

1.3.3.8 Counters

Counting is a fundamental part of many PLC programs. The PLC may be required to count the number of items in a batch, or record the number of times some event occurs. With large motors, for example, the number of starts have to be logged. Not surprisingly, all PLCs include some form of counting element.

A counter can be represented by Figure 1.3.37(a), although not all PLCs will have all the facilities we will describe. There will be two numbers associated with the counter. The first is the count itself (often called the *accumulated value)* which will be incremented when a $0 \rightarrow 1$ transition is applied to the count-up input, or decremented when a $0 \rightarrow 1$ transition is applied to the count-down input. The accumulated value (i.e., the count) can be reset to 0 by applying a 1 to the reset input. Like the elapsed time in a timer, the value of the count can be read and used by other parts of the program.

The second number is the *preset* which can be considered as the target for the counter. If the count value reaches the preset value, a *count complete* or *count done* signal is given. The preset can be changed by the program; a batching sequence, for example, may require the operator to change the number of items in a batch by a keypad or VDU entry.

Similarly a signal *zero count* is sometimes available. The operation can be summarised in Figure 1.3.37(b).

PLC manufacturers handle counters, like timers in slightly different ways. The PLC-5 and the Mitsubishi use count up (CTU) count down (CTD) and reset (RES) as rung terminators with the count-done signal (e.g., C 5 : 4 . DN) available for use as a contact.

The Siemens S5, ABB Master and the GEM-80 treat a counter as an intermediate block in a logic diagram or rung from which the required output signals can be used.

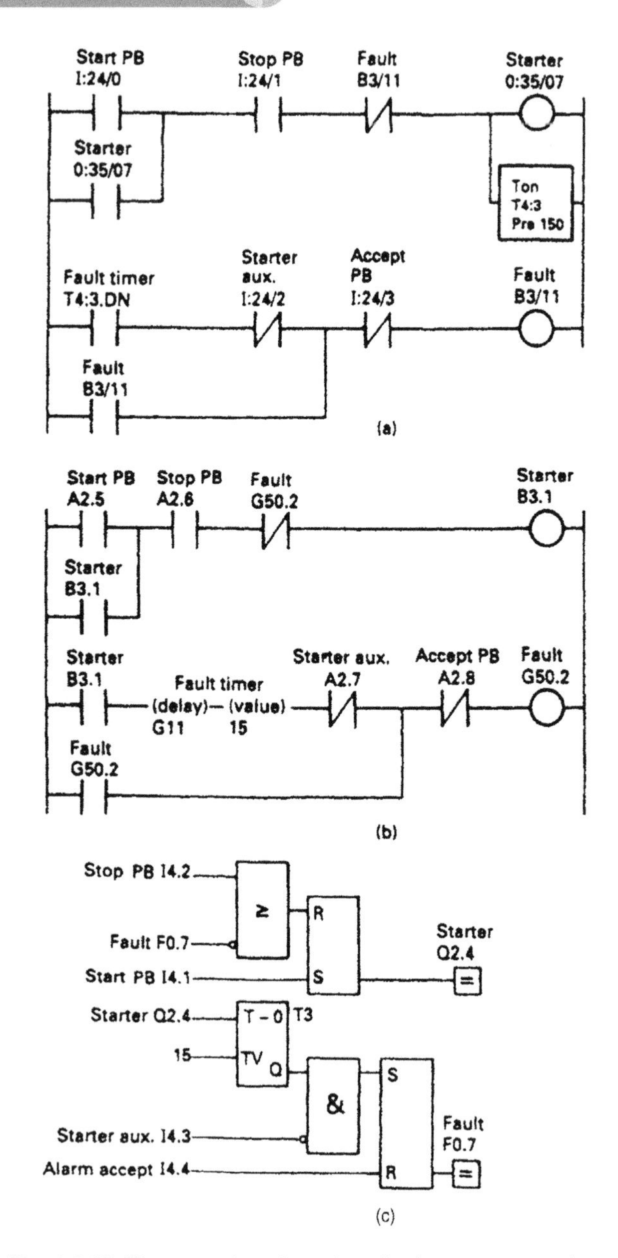

Fig. 1.3.36 The same timer-based application programmed on three different machines: (a) Allen Bradley PLC-5 TON Timer; (b) GEM-80 delay block; (c) Siemens S5 in logic notation.

Figure 1.3.38 shows a simple count application performed by a PLC-5, a Siemens S5 and a GEM-80. Items passing along a conveyor are detected by a photocell and counted. When a batch is complete, the conveyor is stopped and a batch complete light is lit for the operator to remove the batch. When he does this, a restart button sets the sequence running again.

As we saw with timers, most PLCs allow a counter to count up to 32,767. Where larger counts are needed, counters can be cascaded with the complete (or *done*) signal from the first counter being used to step the second counter and reset the first. Figure 1.3.39 is a variation on the same idea used to give a very long timer. It is shown for a PLC-5, but the same idea could be used on any PLC.

The first rung generates a free running one scan pulse with inter-pulse period set by the timer period. (When the timer has not timed out, the DN signal is not present and the timer is running. When it reaches the preset, the DN signal occurs, resetting and restarting the timer.) The resulting one second pulse is counted by successive counters to give accumulated seconds/minutes/hours/days. As each counter reaches its preset, it steps the next counter and resets itself.

Long duration timers built from counters are normally retentive (i.e., they hold their value when the controlling event is not present). They can be made nonretentive by resetting the counters when the controlling event is not present, but this is rarely required.

1.3.3.9 Combinational logic

Any control system based on digital signals can be represented by Figure 1.3.40(a), where a system has a set of outputs Z, Y, X, W, etc. whose state is determined by inputs A, B, C, D, etc. The control scheme can operate in a combination of two basic manners.

The simplest of these is *combinational logic* where the scheme can be broken down into smaller blocks as Figure 1.3.40(b) with one output per block, with each output state being determined solely by the corresponding input states. The loading valve for a hydraulic pump, for example, is to be energised when

```
The pump is running
AND(Raise is selected AND top limit SW is
not struck)
OR (Lower is selected AND bottom limit SW is
not struck)
```

The operation of this loading valve can be implemented with the simple ladder or logic program of Figure 1.3.41, but it is worth developing a standard way of producing a combinational logic program.

The first stage is to break the control system down into a series of small blocks, each with one output and several inputs. For each output we now draw up a so-called *truth table* in which we record all the possible input states and the required output state. In Figure 1.3.42(a) we have an output Z controlled by four inputs A, B, C, D. There are 16 possible input states, and Z is energised for four of these. This can be translated directly into the ladder diagram of Figure 1.3.42(b) or the logic circuit of Figure 1.3.42(c), with each rung branch and or gate corresponding to one row in the truth table. The use of a truth table method for the design of combinational logic circuits leads directly to an AND/OR arrangement called, technically, a *Sum of Products* (S of P) circuit.

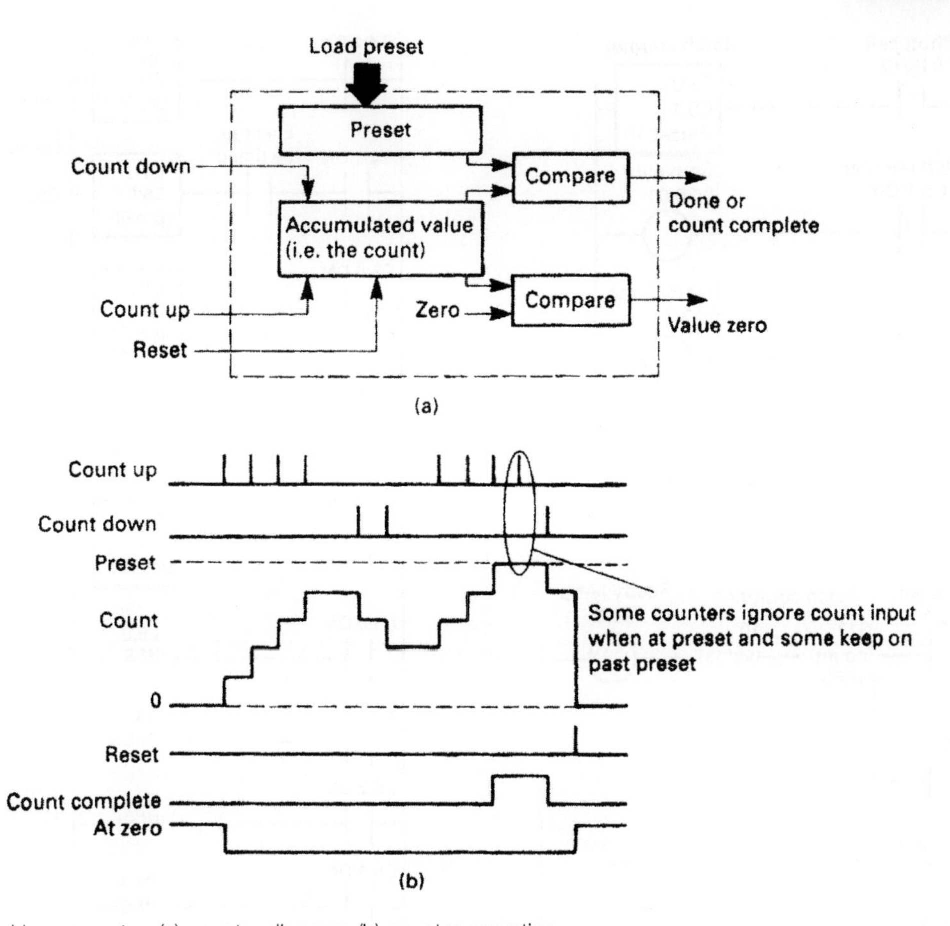

Fig. 1.3.37 The up/down counter: (a) counter diagram; (b) counter operation.

1.3.3.10 Event driven logic and SFCs

The states of outputs in combinational logic are determined solely by the input signals. In event-driven logic (also known as a sequencer) the state of an output depends not only on the state of the inputs, but also on what was occurring previously. It is not therefore possible to draw a truth table from which the required logic can be deduced.

Consider, for example, the simple motor starter circuit of Figure 1.3.43(a). With neither button pressed, the motor could be running or stopped depending on what occurred last. The operation can be described by Figure 1.3.43(b) which is known as a *state transistion diagram* (often shortened to *state diagram)*.

The square boxes are the states the system can be in; the motor can be running or stopped, and the arrows are the transitions that cause the system to change states. If the motor is running, pressing the stop button will cause the motor to stop. A bar above a signal (e.g., above stop PB OK) means signal not present; note the wiring of the stop PB and the signal sense. It is a useful convention to label states with numbers and transitions with letters.

State transition diagrams can be constructed from storage elements, with one less storage element than there are states, and the one default state being inferred by the absence of others. It therefore requires just one storage element (latch, SR flip flop or whatever) to implement the motor starter of Figure 1.3.43.

Figure 1.3.44 is a more complex example (based on a real silo). A preset weight of material is fed into a weigh hopper ready for the next discharge, which is initiated by a Discharge pushbutton. A hood then lowers (to reduce dust emissions) and the material discharges. After the discharge, the hood retracts and the weigh hopper refills. An abort pushbutton stops a discharge, and a feed permit switch stops the feed.

There are two fault conditions: failure to get the batch weight in a given time (probably caused by material jamming in the feeder) and failure to get zero weight from the discharge (again in a given time and again probably caused by a material jam). Both of these trip the system from automatic to manual operation to allow the cause of the fault to be determined.

We can now draw the state diagram of Figure 1.3.44(b). The default state is the state that the system will enter from manual, and care needs to be taken in its selection.

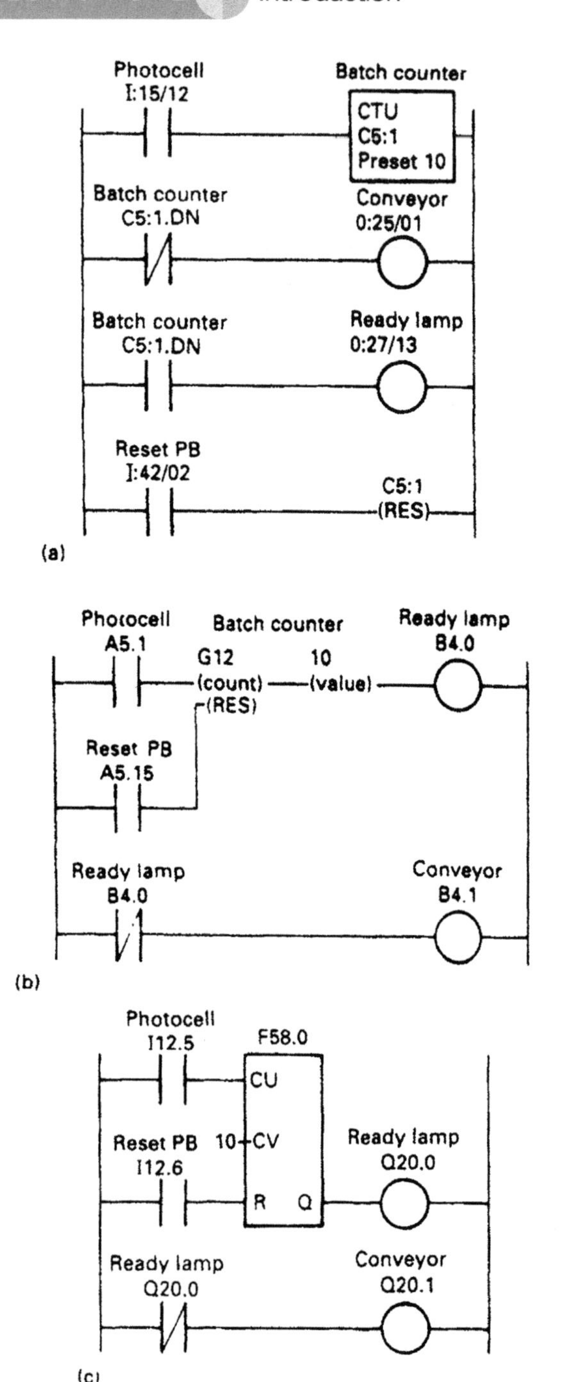

Fig. 1.3.38 A simple batch counter programmed on three different machines: (a) Allen Bradley PLC-5; (b) GEM-80; (c) Siemens S5 in logic notation.

Here feed is the sensible choice; if the hopper if already full the system will immediately pass to state 1 (ready), if not, the hopper will be filled. The choice of any other state as default could lead to a wasted cycle through all the states with no material in the weigh hopper.

We can now construct a table linking the outputs to the states. This is straightforward and is given in Figure 1.3.44(c).

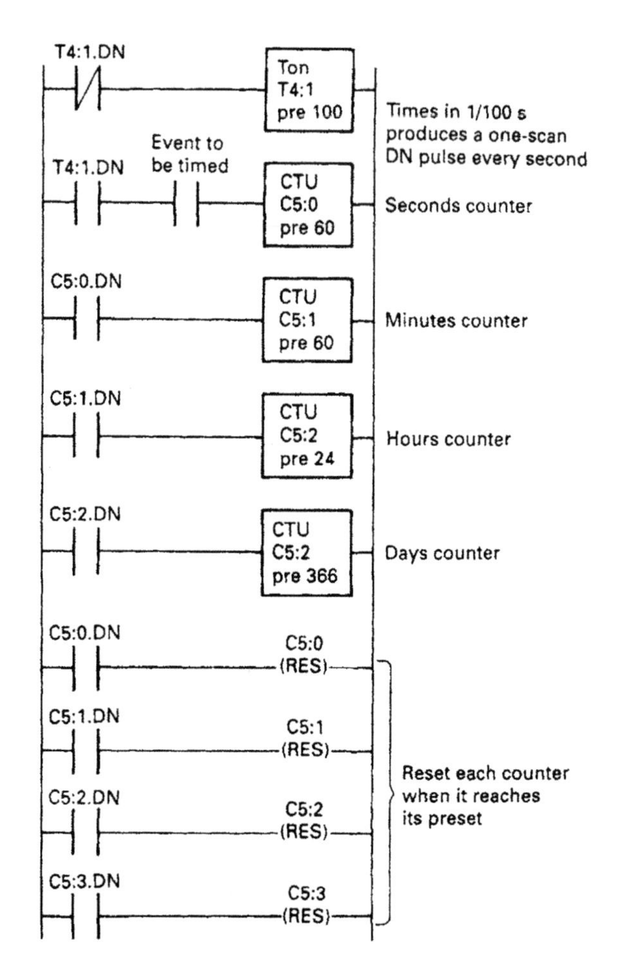

Fig. 1.3.39 Cascading counters to give a long delay. Allen Bradley PLC-5 notation has been used.

The next stage is to translate this state diagram into a PLC program. The programming method relies very much on the idea of the program scan, described earlier in Section 1.3.2.4. By breaking down the program for our state diagram into four areas as Figure 1.3.45, we can control the order in which each stage operates. The actual layout is not critical, but it is essential for transitions and states to be kept separate and not mixed.

Automatic/manual selection comes first; this is achieved with the simple rung of Figure 1.3.46. Automatic mode is only allowed if there are no faults and the hood is raised.

Next come the transitions, some of which are shown in Figure 1.3.47. These are straightforward and need little comment. Note that the first contact in each rung is a state, so inputs are only examined at the correct point in a sequence.

Some of the states themselves are given in Figure 1.3.48. With the exception of state 0, simple latches have been used throughout for the states and the auto/man selection so that after a power failure the system will resume in manual mode. Note that these are set and reset by the transitions.

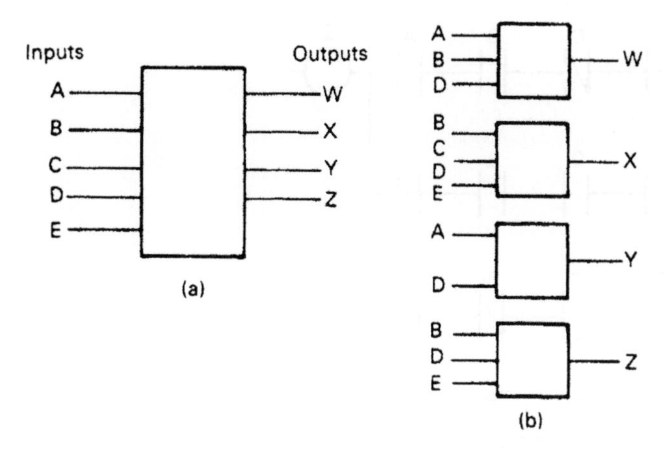

Fig. 1.3.40 Combinational logic: (a) top level view; (b) broken down into smaller blocks, each with one output, for programming.

Finally we have the outputs themselves in Figure 1.3.49. An output is energised during the corresponding state(s) in automatic or from the manual maintenance pushbutton in manual.

The state diagram technique is very powerful, but it can lead to confusion if the basic philosophy is not understood. The often-quoted argument is it takes more rungs or logic elements than a direct approach programmed around the outputs.

This is true, but programming around the outputs can lead to very twisted and difficult-to-understand programs. Figure 1.3.50 is one rung roughly corresponding to state 2 of our state diagram. It mixes manual and automatic operation and its action is by no means clear (known colloquially as *Spaghetti programming)*. Problems can arise where transitions go against the program scan as transition E on the earlier Figure 1.3.44(b). If care is not taken, a sequence based purely on outputs can easily end up doing two things at once, or nothing at all because of the way the program scan operates. Modifications are also tricky with a direct approach, but simple with a state diagram.

State diagrams are being formalised by the International Eiectrotechnical Commission and the British Standards Institute, and already exist with the French Standard Grafset. These are basically identical to the approach outlined above, but introduce the idea of parallel routes which can be operated at the same time. Figure 1.3.5l(a) is called a *divergence,* state 0 can lead to state 1 for condition s OR to state 2 for condition – with transitions s and – mutually exclusive. This is the form of the state diagrams described so far.

Figure 1.3.5l(b) is a *simultaneous divergence,* where state 0 will lead to state 1 and state 2 simultaneously for transition *u*. States 1 and 2 can now run further sequences in parallel.

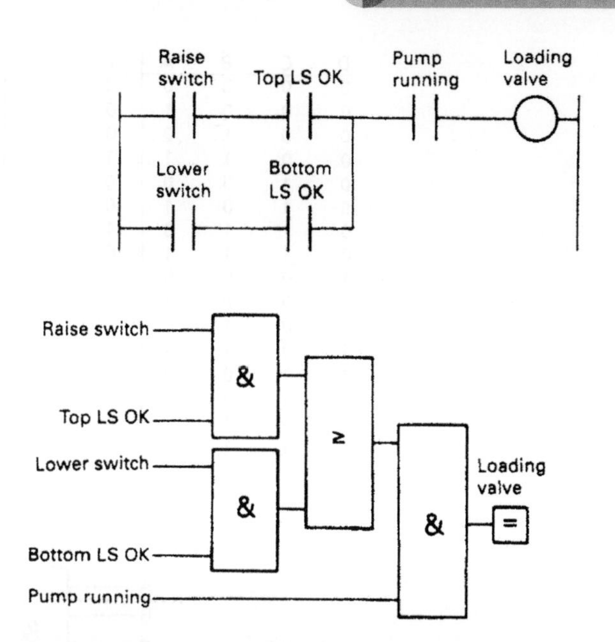

Fig. 1.3.41 Combinational logic in ladder and logic notations. Both perform the same function.

Figure 1.3.51(c) again corresponds to the state diagrams described earlier, and is known as a *convergence.* The sequence can go from state 5 to state 7 if transition v is true or from state 6 to state 7 if transition w is true.

Figure 1.3.51(d) is called a *simultaneous convergence* (note again the double horizontal line) state 7 will be entered if the left-hand branch is in state 5 AND the right-hand branch is in state 6 and transition x is true.

The state diagram is so powerful that most medium size PLCs include it in their programming language in one form or another. Telemecanique give it the name Grafcet (with a 'c'), others use the name Sequential Function Chart (SFC) (Allen Bradley) or Function Block (Siemens).

Even the simple Mitsubishi F2 supports state diagrams with its STL (Stepladder) instruction. These have the prefix S and can range from S600 to S647. They have the characteristic that when one or more are set, any others energised are automatically reset. A RET instruction ends the sequence. The state diagram of Figure 1.3.52(a) thus becomes the ladder diagram of Figure 1.3.52(b).

Where there are no branches and the sequence is a simple ring (operating rather like a uniselector), a sequence can be driven by a counter which selects the required step. The counter is stepped when the transitions for the current step are met. The GEM-80 has a SEQR (sequence) instruction which acts as a 16-step uniselector.

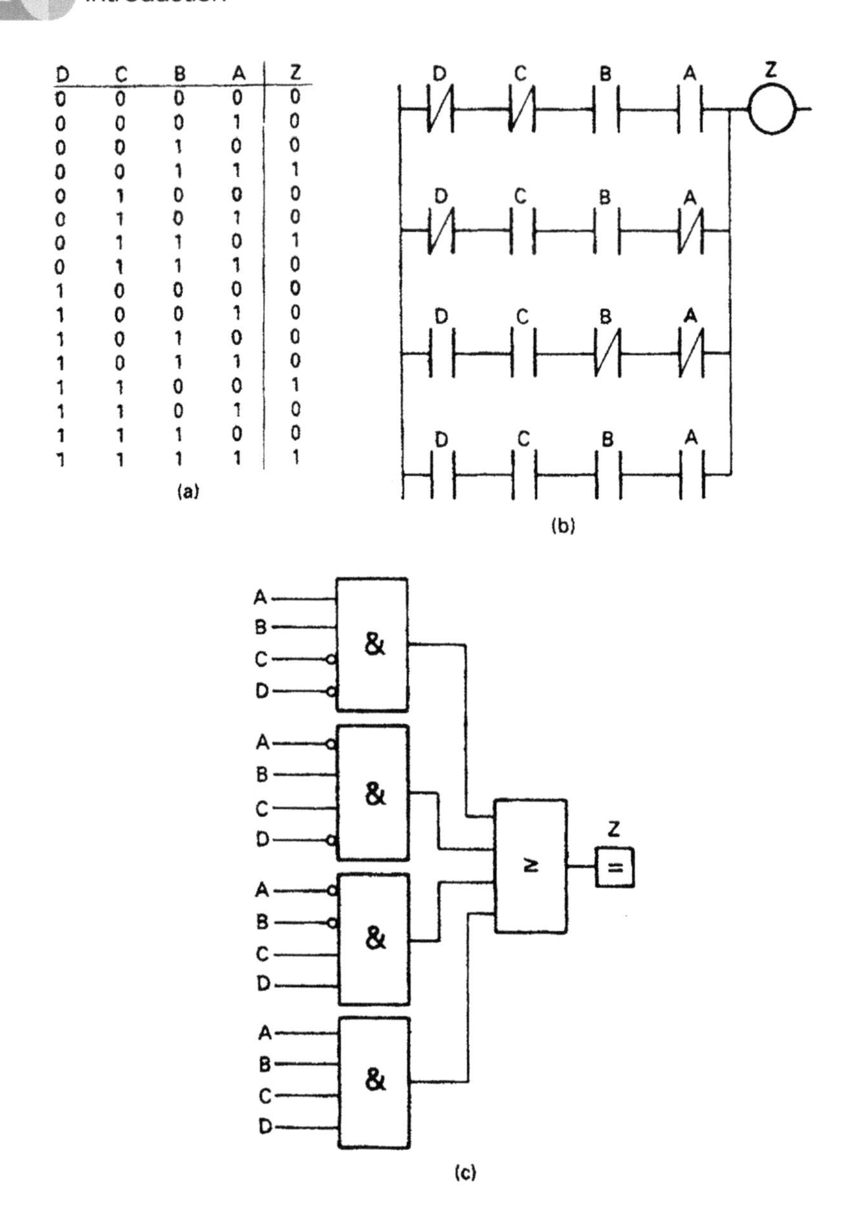

D	C	B	A	Z
0	0	0	0	0
0	0	0	1	0
0	0	1	0	0
0	0	1	1	1
0	1	0	0	0
0	1	0	1	0
0	1	1	0	1
0	1	1	1	0
1	0	0	0	0
1	0	0	1	0
1	0	1	0	0
1	0	1	1	0
1	1	0	0	1
1	1	0	1	0
1	1	1	0	0
1	1	1	1	1

Fig. 1.3.42 Building combinational logic from a truth table: (a) truth table; (b) direct conversion to a ladder program. Each row in the truth table which makes $Z = 1$ is represented by one level on the branch; (c) direct conversion to a logic diagram. Each row in the truth table which makes $Z = 1$ is represented by an AND gate. The AND gate outputs are then OR'd together.

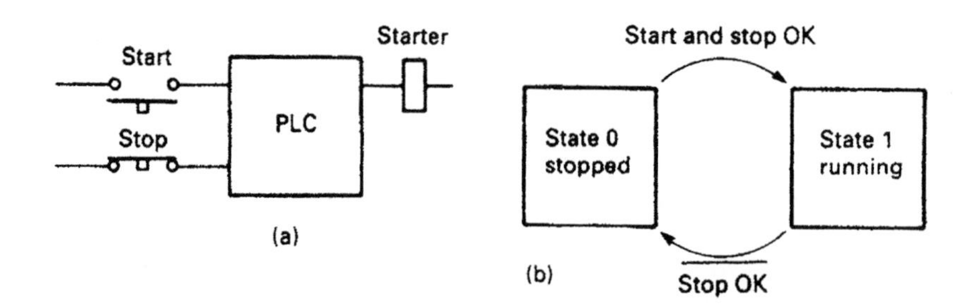

Fig. 1.3.43 A simple state transition diagram: (a) a motor starter; (b) state transition diagram. Note that with no buttons pressed the system can be in either state.

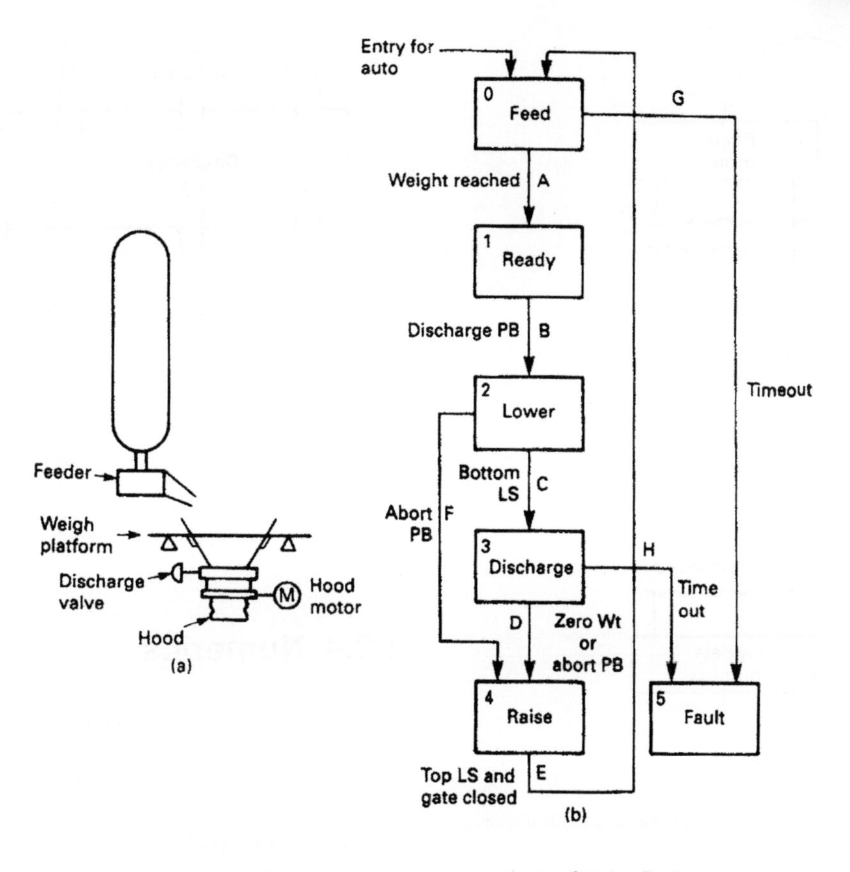

	State	Feed	Lower	Raise	Disch	Auto Permit	Ready Lamp	Fault Lamp
Feed	0	1	0	0	0	1	0	0
Ready	1	0	0	0	0	1	1	0
Lower	2	0	1	0	0	1	0	0
Discharge	3	0	0	0	1	1	0	0
Raise	4	0	0	1	0	1	0	0
Fault	5	0	0	0	0	0	0	1

(c)

Fig. 1.3.44 A more complex state transition diagram for a real plant: (a) physical layout; (b) state transition diagram; (c) output table.

1.3.3.11 IEC 1131

We have seen that PLCs can be programmed in several different ways. In recent years the International Eiectrotechnical Commission (IEC) have been working towards defining standard architectures and programming methods for PLCs. The result is IEC 1131, a standardised approach which will help at the specification stage and assist the final user who will not have to undergo a mind-shift when moving between different machines.

The earliest, and probably still the most common, programming method described is the *Ladder Diagram* (or LD in IEC 1131).

Function Block Diagrams (FBDs) use logic gates (AND/OR, etc.) for digital signals and numeric function blocks (arithmetic, filters, controllers, etc.) for numeric signals. FBDs are similar to PLC programs for the ABB Master and Siemens SIMATIC families. There is a slight tendency for digital programming to be done in LD, and analog programming in FBD.

Many control systems are built around state transition diagrams, and IEC 1131-3 calls these *Sequential Function Charts* (SFCs). The standard is based on the French Grafcet standard shown earlier in Figure 1.3.51.

Finally are text-based languages. *Structured Text* (ST) is a structured high-level language with similarities to Pascal and C. *Instruction List* (IL) contains simple mnemonics such as LD, AND, ADD, etc. IL is very close to the programming method used on small PLCs where the user draws a program up in ladder

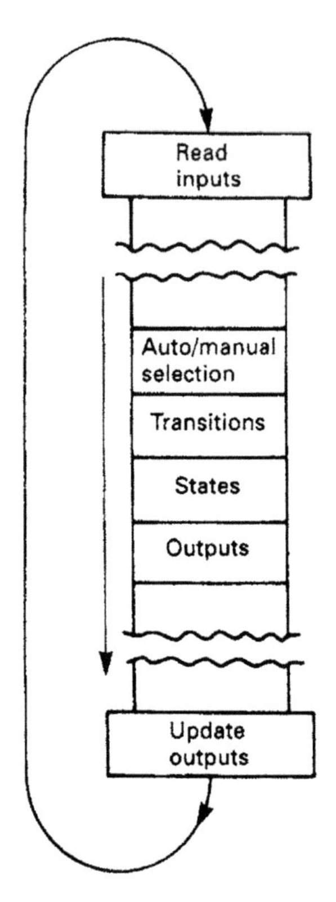

Fig. 1.3.45 The program scan and the layout of a state transition diagram program.

form on paper, then enters it as a series of simple instructions.

Figure 1.3.53 illustrates all of these programming methods.

A given project does not have to stick with one method; they can be intermixed. A top level, for example, could be an SFC, with the states and transitions written in ladder rungs or function blocks as appropriate.

It will be interesting to see the effect of IEC 1131-3. Most attempts at standardisation fail for reasons of national and commercial pride. MAP, and latterly fieldbus, have all had problems in gaining wide acceptance. A standard will be useful at the design stage, and could be accepted by the end user if programming terminals presented a common face regardless of the connected machine.

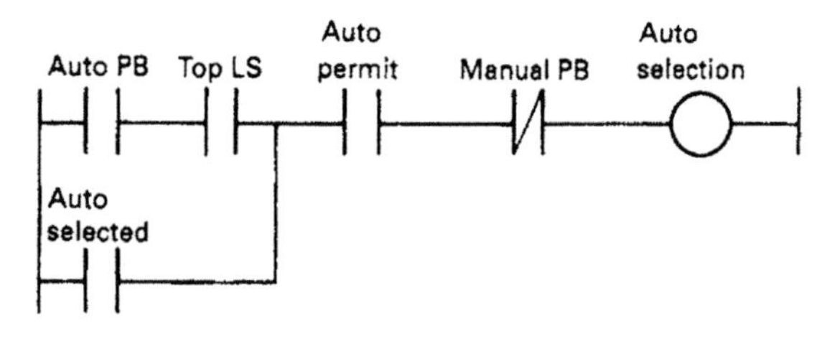

Fig. 1.3.46 Auto/manual selection.

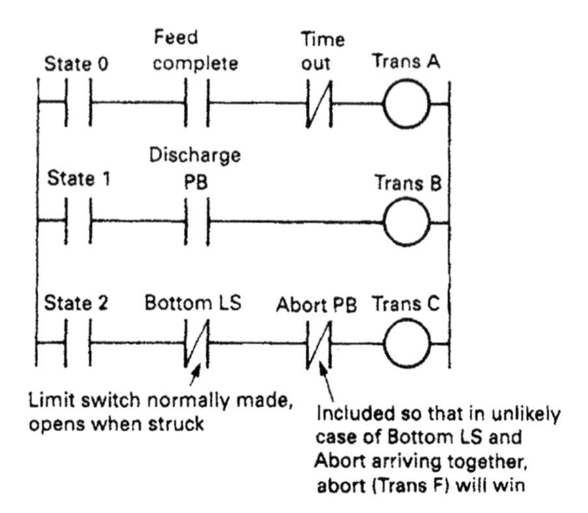

Fig. 1.3.47 The first three transitions.

1.3.4 Numerics

1.3.4.1 Numerical applications

So far we have been primarily discussing single bit operations. Numbers are also often part of a control scheme; a PLC might need to calculate a production rate in units per hour averaged over a day, or give the amount of liquid in a storage tank. Such operations require the ability to handle numeric data.

1.3.4.2 Numeric representations

Most PLCs work with a 16-bit word, allowing a positive number in the range 0 to +65,535 to be represented, or a signed (positive or negative) number in the range −32,768 to +32,767. In the latter case, known as *2's complement*, the most significant bit represents the sign, being 1 for negative numbers and 0 for positive numbers.

Numbers such as these are known as integers, and can only represent whole numbers in the above range. Where larger whole numbers are required, two 16 words can be used allowing a range −2147,483,648 to +2147,483,647. This type of integer is available in the ABB Master (where it is known as a *long integer)* and the 135-U and 155-U in the Siemens family (where the term *double word integer* is used).

Where decimal fractions are needed (to deal with a temperature of 45.6 °C for example), a number form similar to that found on a calculator may be used. These are known as *real or floating point* numbers, and generally consist of two 16 bit words which contain the *mantissa* (the numerical portion) and the *exponent*. In base ten, for example, the number 74,057 would have a mantissa of 7.4057 and an exponent of 4 representing 10^4. PLCs, of

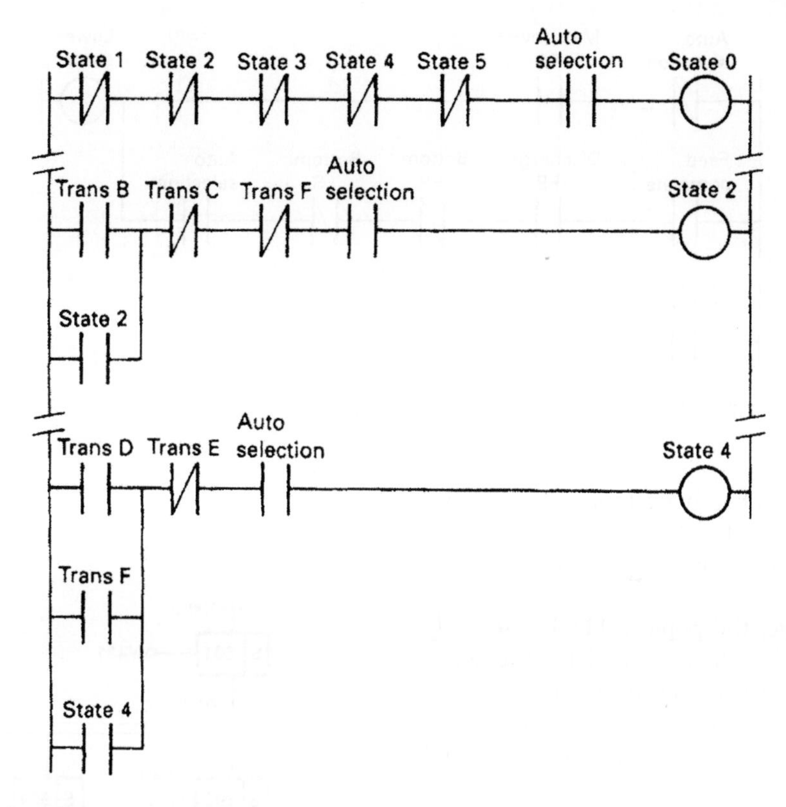

Fig. 1.3.48 Three of the six states.

course, work in binary and represent mantissa and exponent in 2's complement form.

Real numbers are very useful but their limitations should be clearly understood. There are two common problems. The first occurs when large numbers and small numbers are used together. Suppose we had a system operating to base ten with four significant figures, and we wish to add 857,800 (stored as 8.578E5) and 96 (stored as 9.600E1). Because the smaller number is outside the range (four significant figures) of the larger, it will be ignored giving the result 857,800 + 96 = 857,800.

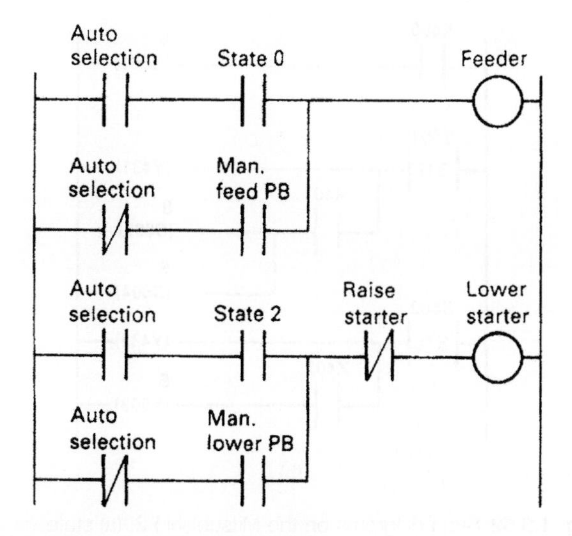

Fig. 1.3.49 Two of the plant outputs.

The second problem occurs when tests for equality are made on real numbers. The conversion of decimal numbers to binary numbers can only be made to the resolution of the floating point format. If real numbers must be used for comparison, a simple equates (=) is very risky. The composites >=, (greater than or equals), and <=, (less than or equals), are safer, but it is generally better practise to use integers for tests if at all possible.

The final representation, BCD for *Binary Coded Decimal*, is used for connection to outside world devices such as digital displays or thumbwheel switches. Such devices are arranged in a decimal format, with 4 binary bits per decade, for example

1001 0101

can be interpreted in BCD as 95.

This representation is wasteful, as six 'numbers' are not used per four bits (10–15 inclusive). It is, however, a convenient form to use with external wiring. Most PLCs therefore have instructions which convert BCD to the internal binary format of the PLC, and binary back to BCD.

The types of numbers available in each PLC range vary considerably according to the model (and obviously

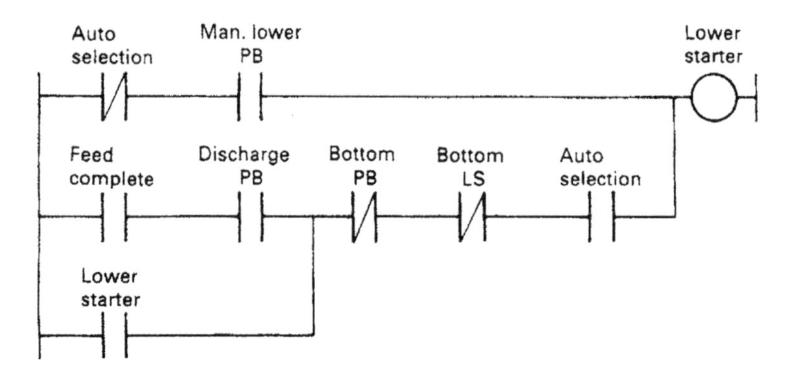

Fig. 1.3.50 An example of spaghetti programming approximating to state 2.

the price). The Mitsubishi F2, for example, purely allows movement, comparison and output of numerical data from counters or timers, making it essentially a bit operation machine.

In the Siemens range, the popular 115-U uses only 16 bit integer numbers but the next model in the range, the 135-U, can handle 16 bit and 32 bit integers and floating point numbers. A similar spread of capabilities will be found amongst the Allen Bradley, GEM-80 and ABB families.

1.3.4.3 Data movement

Numbers are often moved from one location to another; a timer preset may be required to be changed according to plant conditions, a counter value may need to be sent to an output card for indication on a digital display or the result of some calculations may be used in another part of a program.

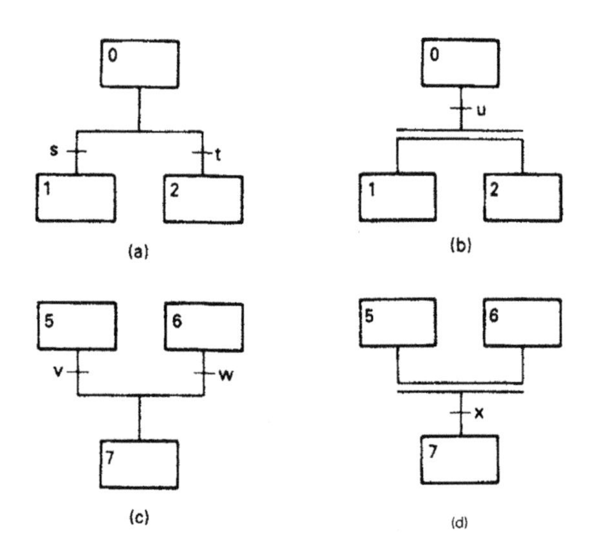

Fig. 1.3.51 Grafset symbols: (a) divergence; (b) simultaneous divergence; (c) convergence (d) simultaneous convergence.

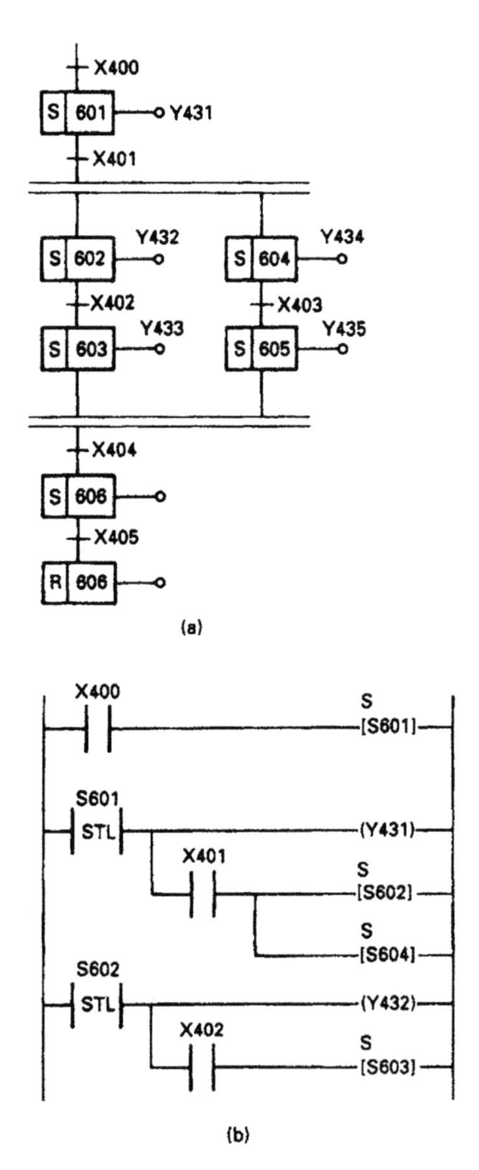

Fig. 1.3.52 State diagrams on the Mitsubishi F2: (a) state diagram; (b) part of the ladder diagram corresponding to (a).

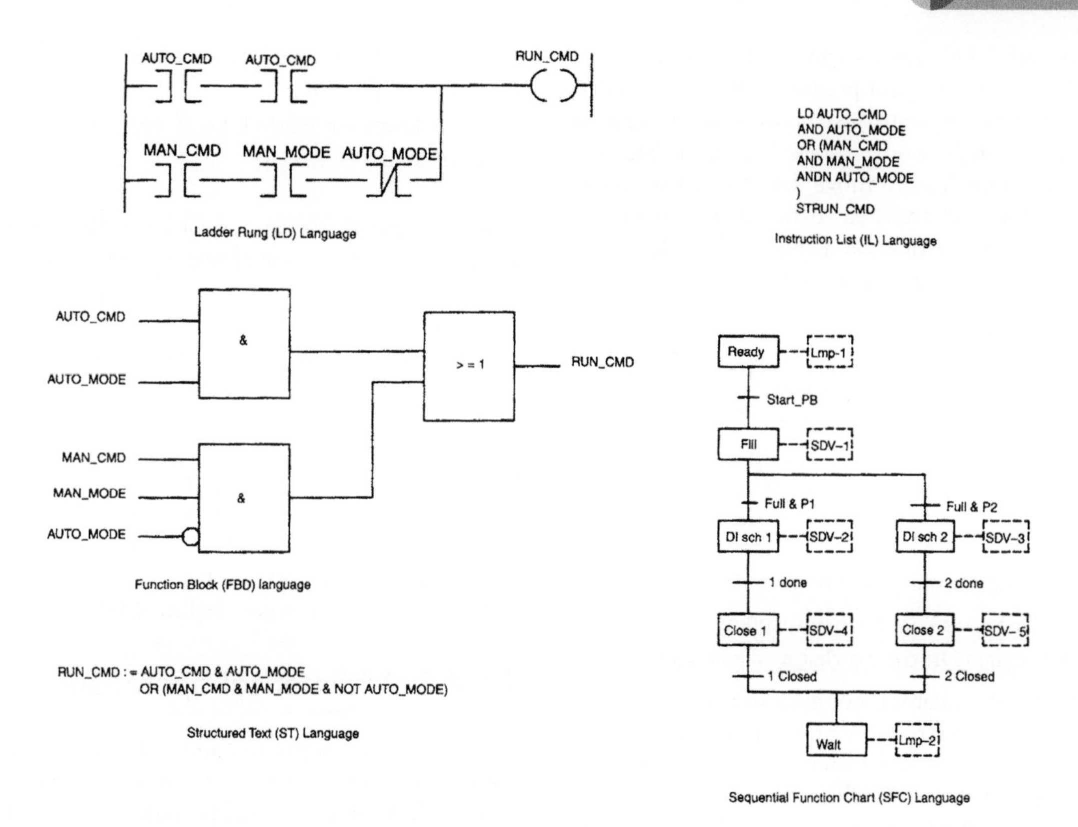

Fig. 1.3.53 The five programming methods defined in IEC 1131-3.

The Allen Bradley PLC-5 uses one rung per move operation, and is possibly the simplest to explain first. Its simplicity of one rung per operation is continued in all the arithmetic functions we shall consider, but it can lead to more rungs being used for a given operation than in other machines.

Figure 1.3.54(a) shows the form of the rung. It starts with some binary conditions; if these are all made the output MOV (for MOVE) is obeyed, transferring data from the source to the destination. The source and destination can be any location where numerical data can occur, for example

Integer number (e.g. `N7 : 26`)
Floating point number (e.g. `F8 : 33`)
Counter or timer preset (e.g. `C5:17.PRE` or `T4:52.PRE`)
Counter or timer accumulated value (eg `C5 : 22 . ACC` or `T4:6.ACC`)
I/O word data (e.g. `I:23` which is all 16 bits from inputs on card 3 in rack 2)

If data are transferred between integer and floating point forms, the conversion is performed automatically; however, care must be taken transferring floating point numbers to integers as an error can occur if the floating point number is outside the integer range. Finally, as a source only, a constant (such as 3, 17 or 4057) can be used.

The example of Figure 1.3.54(a) thus moves the number held in N7 : 34 to the preset of timer T4 : 6 when the rung conditions are met.

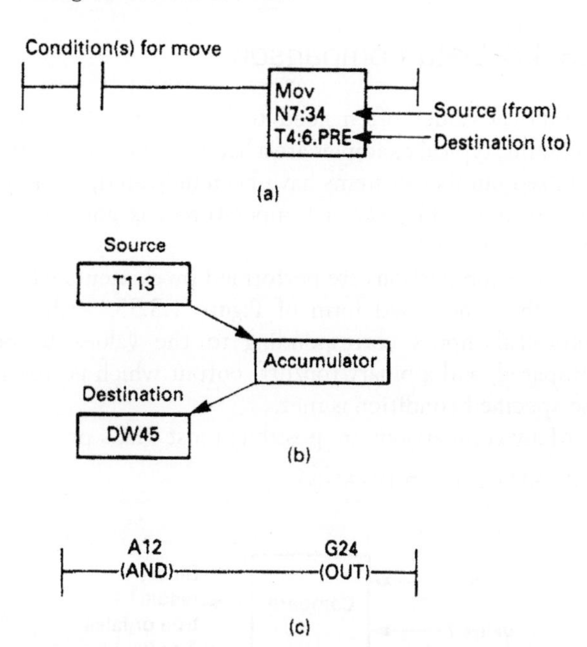

Fig. 1.3.54 Data Movement: (a) Allen Bradley PLC-5; (b) Siemens S5; (c) GEM-80.

Siemens and GEC use a slightly different approach which leads to more compact programs. Both treat a data movement as two separate instructions via a separate accumulator (a single word storage location). Siemens use the instructions load to move data from a source to the accumulator, and transfer to move data from the accumulator to the destination as Figure 1.3.54(b). The data can come from (or go to) any data storage area, some of which are

IW	a 16 bit input word
QW	a 16 bit output word
T	a timer word
C	a counter word
DW	a 16 bit data storage word

Figure 1.3.54(b) would thus be programmed as

```
:L T113 (timer value to accumulator)
:T DW45 (accumulator to data word 45)
```

The use of the accumulator is not obvious in the GEM-80. The – <AND> – instruction puts the binary number from the specified location (again internal storage or I/O) into the rung, and the – <OUT> – instruction puts the value from the rung to the specified address. In Figure 1.3.54(c) the (binary) value from 16 bit input word A12 is placed into 16 bit storage word G24.

BCD/binary conversion is available with – <BCDIN> and– <BCDOUT> – instructions, the direction of the conversion being obvious.

In the **ABB** Master, the points between which data are to be transferred are simply linked on the logic diagram.

1.3.4.4 Data comparison

Numerical values often need to be compared in PLC programs; typical examples are a batch counter saying the required number of items have been delivered, or alarm circuits indicating, say, a temperature has gone above some safety level.

These comparisons are performed by elements which have the generalised form of Figure 1.3.55, with two numerical inputs corresponding to the values to be compared, and a binary (on/off) output which is true if the specified condition is met.

Many comparisons are possible; most PLCs provide

```
A Greater Than B (A>B)
A Greater Than or Equal to B (A>=B)
AEqualsB (A = B)
A Less Than or Equal to B (A<=B)
A Less Than B (A<B)
```

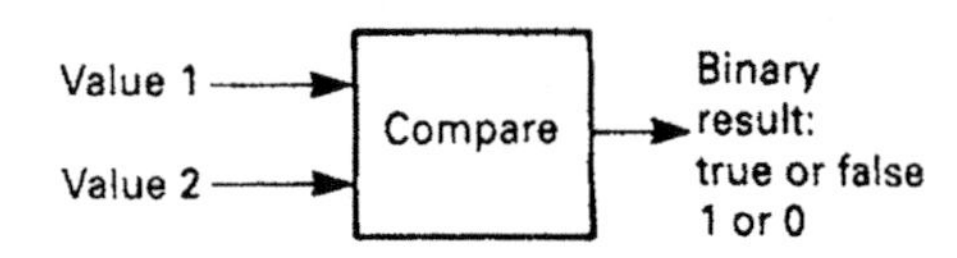

Fig. 1.3.55 Basic idea of data comparison.

where A and B are numerical data. With real (floating point) numbers, the equal test should be avoided for the reasons given in the previous section. There are many other possible comparisons; a PLC-5, for example has a limit instruction which tests for A lying between B and C and the GEM and Siemens have a not equal test.

Figure 1.3.56 shows the setting and resetting of an alarm flag B3/21 (for a PLC-5 ladder diagram) and F21/02 (for Siemens logic symbols). The alarm bit is set if temperature (read from an analog input card in format NN.N° and held in N7:15 in the PLC-5 or DW42 in the Siemens 115-U) goes above 50.0° Once set, the alarm is stored until the temperature goes below 40.0°

1.3.4.5 Arithmetical operations

Numerical data imply the ability to do arithmetical operations, and all PLCs we are considering (apart from the simple F2) provide the ability to do at least four function maths (add, subtract, multiply and divide).

In Section 1.3.4.2 we discussed integer and floating point numbers. Care needs to be taken with integer operations. The range of a 16 bit two's complement number

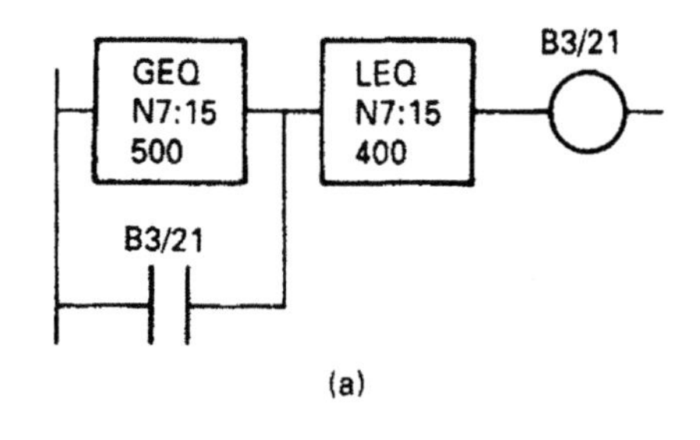

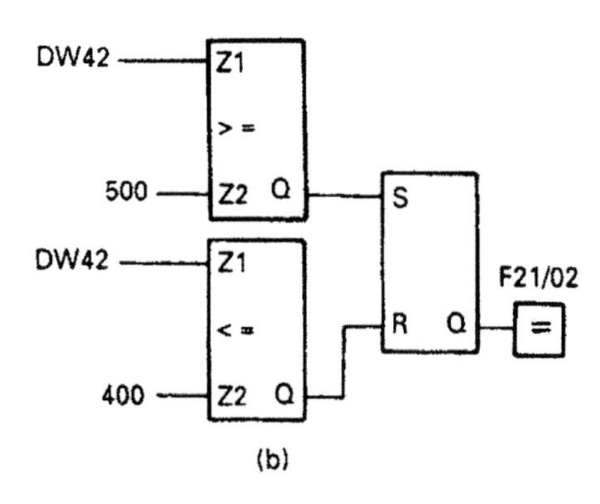

Fig. 1.3.56 Use of data comparison for a high-temperature alarm: (a) Allen Bradley PLC-5; (b) Siemens S5 in logic notation.

is −32,768 to +32,767. f an arithmetical operation goes outside this range, the number will overspill, for example

```
26732
+8647
-----
-30157 in 16 bit 2's complement
```

which is not quite the expected result. The PLCs have an overspill flag which can be examined and used to flag an alarm, or set the result to, say, zero with a move instruction. Similar precautions need to be taken with subtraction and multiplication (the latter being particularly vulnerable to giving an overspill; for example 200 × 200 = 40,000, well over-range).

Even greater care needs to be taken with division. A fault condition on external plant or a PLC input card or a programming error can lead to a divide by zero error. This will stop many PLCs dead in their tracks with a 'Program Fault'. It is therefore good practice to proceed any vulnerable divide instruction with a limit check to ensure it will only be obeyed when a sensible result is obtained.

Each PLC manufacturer handles arithmetic in a slightly different way with varying degrees of ease and readability. None are as simple as a high-level language such as BASIC or Pascal, and the facilities are generally limited to four function Maths plus square root in all bar the most expensive machines.

A PLC-5 uses maths blocks such as ADD, SUB, MULT, DIV, giving a simple, if somewhat lengthy, program. Figure 1.3.57 shows how a simple calculation could be performed for a self-correcting length cutting program. More powerful PLC-5s (such as the 5–40) have a block compute instruction (CPT) which allows a mathematical expression to be evaluated in a single instruction.

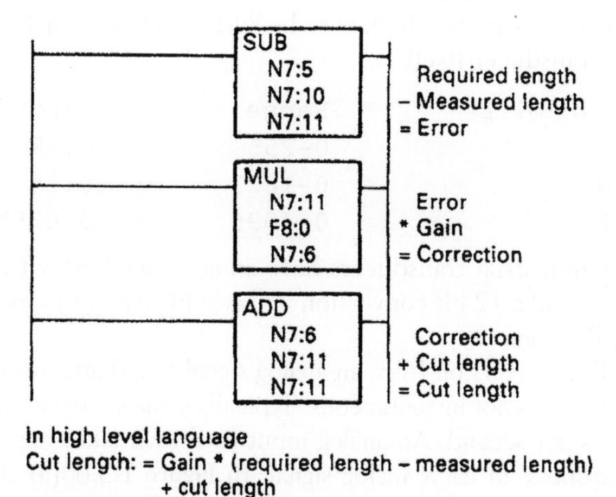

Fig. 1.3.57 Arithmetic in the PLC-5.

The 115-U only evaluates arithmetic instruction in STL (statement list) format. It will be remembered from our discussion of the accumulator that the load, (L) and transfer (T) instructions use an internal accumulator. There are, in fact, two accumulators, and a load instruction moves the contents of accumulator 1 to accumulator 2 then moves the contents of the source to accumulator 1, shown in Figure 1.3.58(a). An arithmetic instruction (add, subtract, etc.) works on the contents of both accumulators. Figure 1.3.58(b) thus adds two numbers and transfers the result to storage.

The Siemens equivalent of Figure 1.3.57 would be

```
LDW30 (required length)
LDW31 (measured length)
SUB (leaving error in Acc 1)
LDW32 (gain)
MULT (leaving correction)
LDW40 (the old cut length)
ADD (add change to give new length)
TDW40 (put back to store)
```

The most understandable form of representation is possibly the GEM-80 ladder and the ABB Master formats shown in Figure 1.3.59(a) and (b) respectively.

All maths operations, particularly those involving floating point numbers, are time consuming, and it is good programming practice to only obey instructions when they are needed, and not waste time repetitively obeying them on every PLC scan.

1.3.4.6 Analog signals

So far we have considered signals that are essentially digital (on/off) in nature plus simple numerical data from timers and counters. Often, though, a PLC will be required to measure, or control, plant signals which can assume any value in some predetermined range. Typical signals of this type are temperatures, flows, pressure,

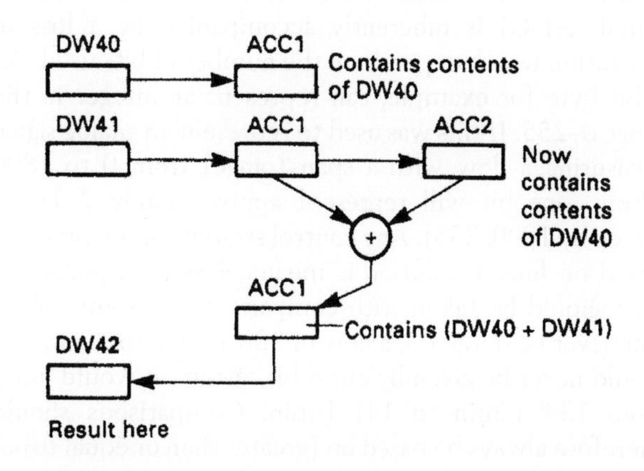

Fig. 1.3.58 Arithmetic in a Siemens S5.

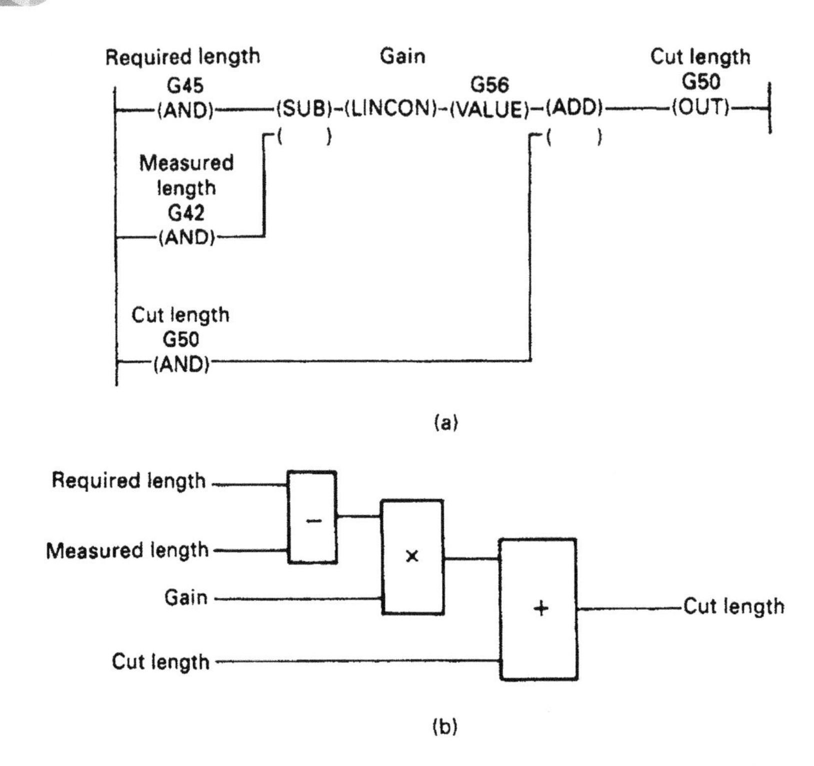

Fig. 1.3.59 The same mathematical function in a GEM-80 and ABB Master: (a) GEM-80 program. LINCON is an arithmetic function used to avoid truncation errors with integer arithmetic; (b) ABB master program using function blocks. Variables are accessed by database names.

speeds, etc. These are known as analog signals. In a similar way a PLC may have to produce analog output signal to drive meters and proportional valves or provide a speed reference for a motor drive controller.

To meet these requirements a PLC needs analog input and output cards. These have somewhat different characteristics to the simple digital cards we have discussed so far. This section considers analog signals and the way they are handled.

An analog input card converts a continuously varying analog signal to a digital form that can be used inside a PLC program. The analog signal is generally represented initially, at least, as an integer number.

This analog to digital conversion (usually known by the initials ADC) is inherently accompanied by a loss of resolution which depends on the number of bits used. An 8 bit byte for example, can represent an integer in the range 0–255. If this was used to represent an analog signal measuring a flow with a span (range) from 0 to 1800 l/min, one bit will represent approximately 7 l/min (given by 1800/255). Any control strategy in the program based on finer resolution is meaningless (and particular care should be taken with comparisons, as some values can never be obtained; a flow of 138 l/min, for example, would never be given by our 8 bit system, it would jump from 134 l/min to 141 l/min. Comparisons should therefore always be based on (greater than or equal to) or (less than or equal to)).

A commoner resolution is 12 bits. This gives a representation as an integer from 0–4095. With our flow of 0–1800 l/min, one bit would represent just under 0.5 l/min (1800/4095 = 0.44).

This 'coarseness' is not the problem it might at first appear. Although an analog transducer can give any value in its span, it will have inherent errors. Many first line transducers are only 2% accurate. If our flow transducer had 2% accuracy, its measurement could be in error by 36 l/min. Alongside this error, the 7 l/min resolution from an eight bit card is probably quite reasonable.

It is therefore useful to think of the resolution in terms of an error which is to be added to the error from the transducer itself

No of bits	Range	Error
8	0-255	0.5%
10	0-1023	0.1%
12	0-4095	0.025%

Few industrial transducers have an accuracy better than 0.1%, and a 12 bit conversion will add little error in most applications.

The conversion from an analog signal to a digital representation is not instantaneous. Typically signals are read ten times per second. An analog input card thus takes regular 'snapshots' of each analog signal. In Figure 1.3.60(a) this causes no problems, in Figure 1.3.60(b) information is starting to be lost and in Figure 1.3.60(c) a totally false view

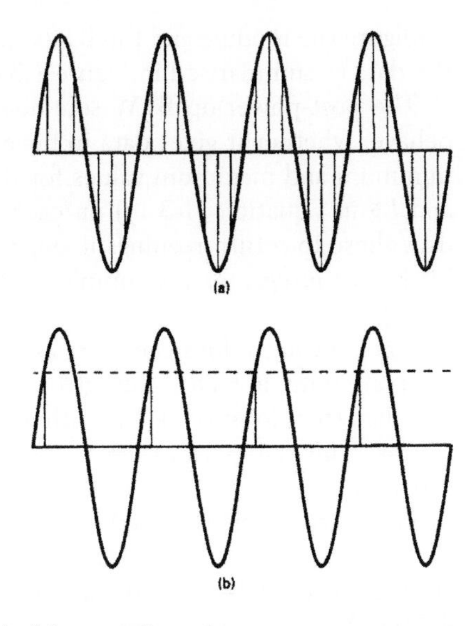

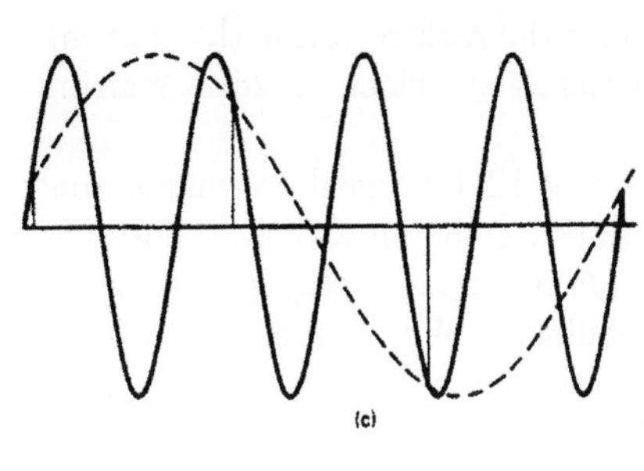

Fig. 1.3.60 The effect of the sampling rate.

of the signal is being given. This latter effect is known as *'aliasing'*.

It is therefore very important to have a sufficiently fast conversion time. Every analog signal will have a maximum frequency at which it can change, and can be represented by a gain/frequency plot as Figure 1.3.61 from which the bandwidth and the critical frequency f_C can be observed. To get a true series of 'snapshots' we must sample the signal at least twice the rate of f_C. If a certain analog signal has a maximum frequency of 2 Hz, we must at least sample it at 4 Hz, or once every 250 ms. This, somewhat simplified, is known as *Shannon's sampling theorem*. In real systems, f_C is rarely known precisely and a scan rate of $4f_C$ to $10f_C$ is normally chosen to give a reasonable safety margin. For our 2 Hz signal, an 8 Hz sampling rate or 12.5 ms conversion time, would be needed. It is good practice to pass the signal through a low pass filter before the ADC to ensure frequencies above f_C are removed. This is known as an *anti-aliasing filter*.

Surprisingly this rarely gives problems. Practical industrial systems, dealing with real plant signals concerned with materials with significant mass, rarely have bandwidths greater than 0.5 Hz, and any frequency higher than this can be considered to be extraneous noise and filtered out. Temperature loops, for example, can often be sampled as slowly as once every few minutes without introducing any errors.

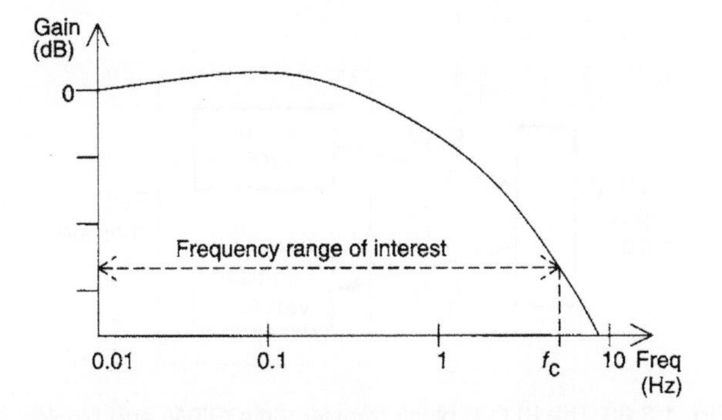

Fig. 1.3.61 Gain/frequency response for an industrial process.

A typical analog input card can read eight 12 bit signals, each ranging from 0 to 4095 in their 'raw' form. Generally, these will need to be accessed via the PLC program and converted to engineering units such as °C, or psi, or l/min.

A common method of handling these signals is shown in Figure 1.3.62. A block of storage locations in the PLC store is directly associated with the analog input card.

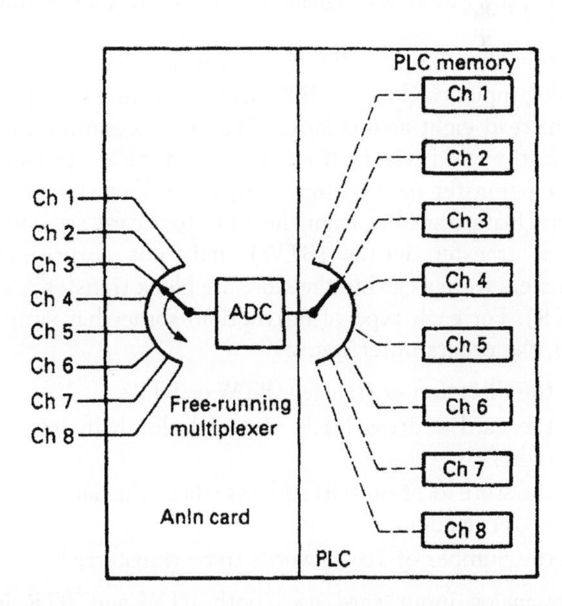

Fig. 1.3.62 Linking channels on an analog input card to a PLC's memory.

The card 'free runs', writing digitised values into the store from where they can be read by the rest of the program. In Siemens PLCs with fixed slot addressing, for example, the store addresses are determined directly by the analog card position in the rack; a card in slot 2 of the first rack will write its values to a block of stores starting at location 192.

Conversion from a raw 12 bit signal to engineering units can have subtle traps for the unwary. In theory the conversion is simple. If N is the raw signal, H_R the high-range signal (corresponding to 4095) and L_R the low-range (corresponding to zero) then the measured value, M_v is simply

$$M_V = \frac{N \times (H_R - L_R)}{4095} + L_R \qquad (1.3.1)$$

If the calculation is done with real (floating point) numbers there should be no problem, and Equation (1.3.1) can be used directly.

If, however, integer numbers have to be used, great care must be taken. If the multiplication $N \times (H_R - L_R)$ is performed first, arithmetic overspill is likely unless 32 bit results can be accommodated. If the division $N/4095$ is performed first, the equation will not work as N is always less than 4095 giving an integer result of zero (and an M_V of L_R). Wherever possible real numbers should be used if Equation (1.3.1) has to be performed.

To avoid this problem, the different manufacturers have devised methods to read analog input signals. In the ABB Master for example, the database for each signal defines H_R, L_R, the sample rate and a name by which the signal will be referred to in the program. There are, obviously, detail differences, so by way of example we will look at the way analog signals are read by an Allen Bradley PLC-5.

The Allen Bradley PLC-5 reads analog signals with an analog input card (1771-IFE) which can in its simplest form read eight analog inputs. The PLC communicates with the card via instructions called block transfers which transfer data to (or from) a block of store locations. Data transfers from the PLC to a card are called block transfer writes (BTW) and, not surprisingly, transfers from a card to the store are block transfer reads (BTR). For each type of instruction, somewhat simplified, the programmer states:

(a) the direction of transfer (BTW or BTR).

(b) the card address (rack, slot and slot half, left or right).

(c) the store location start address where the data are to be received.

(d) the number of 16 bit words to be transferred.

The analog input card uses both BTW and BTR instructions, the BTW being used once, after power up, to configure the module and the BTRs subsequently to read the data as summarised in Figure 1.3.63.

The post-power-up BTW sets how the module is to behave; whether it gives data in binary or BCD and the minimum and maximum values for the input range (H_R and L_R in Equation (1.3.1)) on each channel. The card uses these to return readings in engineering units (in 12 bit binary integer or 2's complement format or 12 bit BCD).

Once set up, values can be read at the required time intervals with a BTR. This gives signal values in the specified store locations along with over-range and similar alarms. The values can then be used elsewhere in the program.

PLCs are often required to provide analog output signals. Like analog inputs, these signals have standard voltage ranges of 1–5 V or 0–10 V or the current range of 4–20 mA.

A typical analog output card, for example, is the Allen Bradley 1771-OFE which has four output channels, each turning a 12 bit (0–095) digital signal into an analog output. Isolation amplifiers are used on the outputs to reduce the effects of noise and allow the signals to connect into external devices fed from different electrical supplies. The digital signals come from storage locations inside the PLC as shown in Figure 1.3.64. This conversion is known as DAC.

For best resolution the PLC should use the full 0–4095 range, but this is frequently impossible. If the PLC, for example, is setting the speed range of a motor from 0–1350 rpm, it will need to convert 0–1350 into the range 4–20 mA. Equation (1.3.1) can be rearranged as

$$X = \frac{4095(N - L_R)}{H_R - L_R} \qquad (1.3.2)$$

where X is the value passed to the DAC (in the range 0–4095). TV is the output number from the PLC in engineering units, and H_R/L_R are the high- and low-range values. As before, great care must be taken to avoid overspill or loss of resolution.

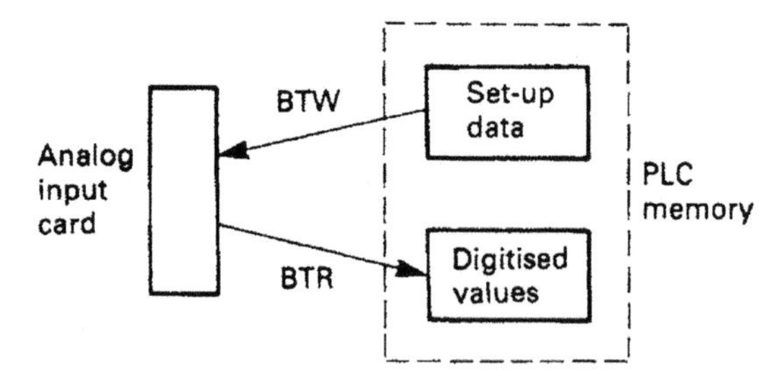

Fig. 1.3.63 The PLC-5 block transfer write (BTW) and block transfer read (BTR) instructions.

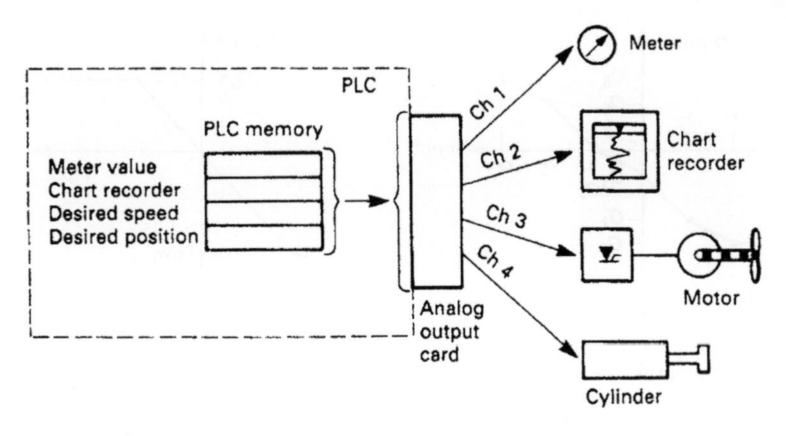

Fig. 1.3.64 Analog output signals.

The PLC-5 communicates with the 1771-OFE with the BTW instruction described previously. The programmer sets up a block of 12 words, the first four of which contain the values, and the balance - set up data such as H_R and L_R. The block of data is then written to the card with a BTW. Figure 1.3.65 shows a typical example where an analog speed reference can be raised or lowered by operator controlled pushbuttons. Note the use of greater than (GTR) and less than (LES) instructions to confine the counter value within the allowed range of 0–1350 rpm.

Ranging as above allows engineering units to be used inside the program; the counter in Figure 1.3.65, for example, holds the required speed directly in rpm, but this is accompanied by a loss of resolution as explained earlier. For the range 0–1350 rpm, we have a resolution of about 0.1%, compared with the theoretical 0.025% resolution available from the card.

There are other operations that can be performed on analog signals. A typical list, for the GEM-80, is

`SQRT` Square root, mainly used with signals from orifice plates.

`LINCON` Performs $X^*(A/B) + C$ with limiting.

`FGEN` Multipoint straight-line function generator used for linearisation as Figure 1.3.66(a).

`LIMIT` Performs limiting of signals as shown in Figure 1.3.66(b).

`RAMP` Rate limiting (with different rise and fall rates).

`DEDBAND` Deadband functions as Figure 1.3.66(c). Useful for preventing 'dither' in closed loop control when PV and SP are close

`ANALAG` First Order lag. Used for filtering.

A simple first-order filter can be produced by any PLC which supports floating point numbers using the procedure shown in Figure 1.3.67(a). This procedure uses just three rungs or three function blocks and is obeyed for one program scan at regular time intervals Δt. V_i is the raw input signal and V_f is the filtered output signal. $V_{f(n-1)}$ is the filtered value obtained on the previous execution Δt seconds ago. The error between V_i and

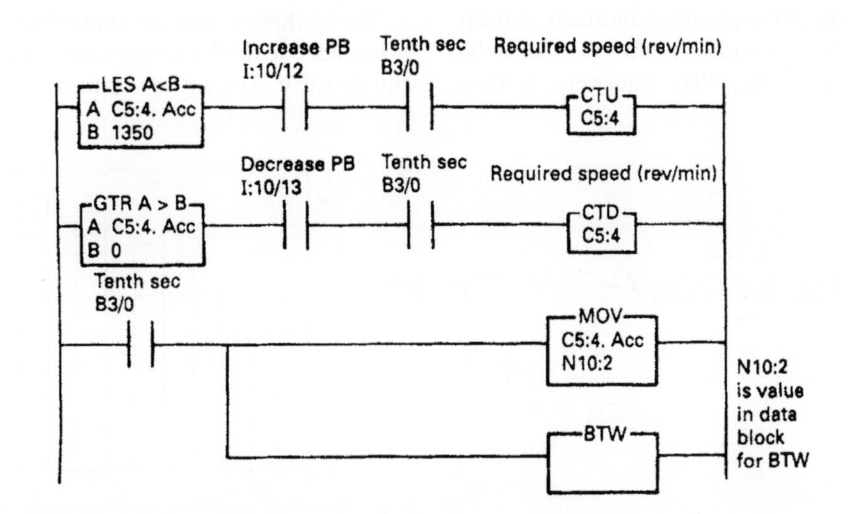

Fig. 1.3.65 Setting the speed for a motor with a counter and an analog output card.

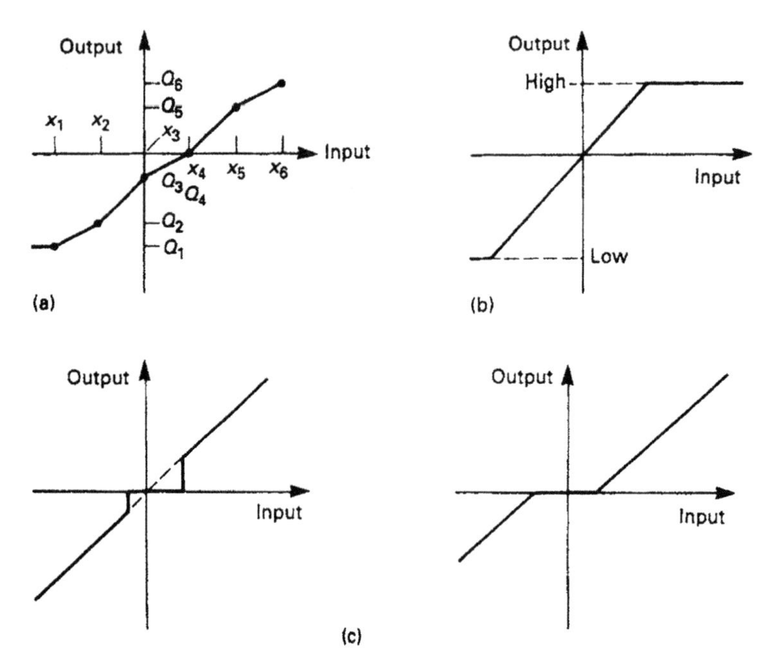

Fig. 1.3.66 GEM-80 special functions for use with analog signals: (a) FGEN with N points at equal intervals X; (b) LIMIT, high and low limits can be different; (c) DEDBAND without and with offset.

$V_{f(n-1)}$ is calculated (V_e) then multiplied by a gain K to give a change V_c. This is added to $V_{f(n-1)}$ to give the new filtered value V_f. Figure 1.3.67(b) shows the response for a step change in V_i with K set at 0.25. At each execution of the routine V_f moves 25% of the difference between V_i and $V_{f(n-1)}$. The gain, K, determines the apparent time constant and must be in the range $0 < K < 1$. The gain K should be set to $\Delta t/T$ where T is the required time constant.

1.3.4.7 Closed loop control

A closed loop system based on PLCs will be similar to Figure 1.3.68. The plant variable, P_V, is read by an analog input card, and the output O_P provided by analog output cards. The setpoint, SP, is provided by the operator or by some program sequence. The PID algorithm is then provided by the program. Chapter 1.2 gives more detail on the theory of closed loop control and an explanation of the PID algorithm.

It is possible to write PID algorithms with four function (+ – */) mathematics, but it needs great care. The program scan time must be known for the integral and derivative routines, and protection against output actuator saturation must be built in to overcome an effect called integral wind-up.

PLCs are becoming increasingly powerful, and most medium-range PLCs now provide a three-term PID function in their instruction set. Figure 1.3.69 shows a ratio temperature control program written for an Allen Bradley PLC-5 processor.

These three rungs are controlling the temperature in a furnace, with the temperature PID block controlling the air valve. The air flow is measured, multiplied by the

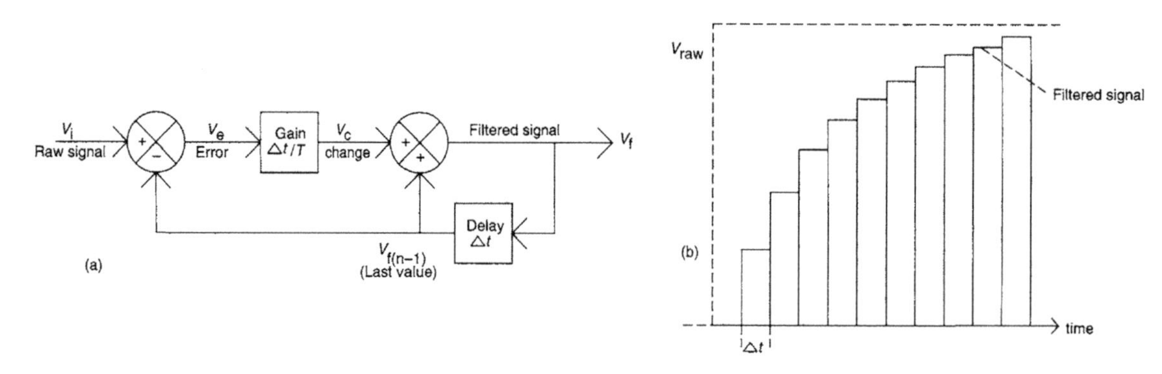

Fig. 1.3.67 Programming a first-order filter: (a) schematic diagram; (b) response to step input.

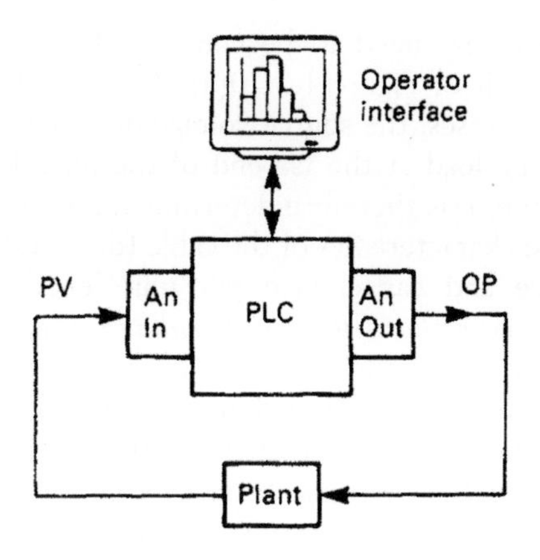

Fig. 1.3.68 Closed loop control with a PLC.

required ratio and used as the setpoint for the gas PID block.

The control blocks in each PID instruction hold the data and working areas for the PID function; things like auto/man status and the sum for the integral action. The setpoint is written directly into the third word of the control block. The process variable is the feedback signal from the variable being controlled, usually obtained from an analog input card. Settings for gain T_i and T_d are also contained in the control block data.

A three-term control algorithm can suffer from integral wind-up in saturation or manual operation. The tieback variable is used to give the current value corresponding to the driven actuator output (possibly after auto/manual changeover) and is used to prevent wind-up and to give bumpless transfer. The control variable is the signal from the PID algorithm, usually sent to an analog output card via auto/manual changeover logic.

The three rungs of Figure 1.3.69 mask, to some extent, the work that must be done elsewhere in the program. Data from the outside world must be obtained with analog input cards, and the controller output(s) must be written to the actuators with analog output cards. The timing of these reads and writes must be regular and linked to the PID instructions.

Auto/manual changeover logic will also be required, linked into the PID instructions with the tieback variable and the auto/man status flag (which makes the integral term track the tieback in manual).

The operator will also require a link to the control, so pushbuttons, displays and alarms must be provided. All of this is in addition to the basic PID control.

1.3.4.8 Intelligent modules

We have so far considered analog input and output modules, which are semi-intelligent (compared to 'dumb' digital input and output cards). These are examples of a more general range of intelligent modules which most manufacturers offer to simplify the designers task.

A typical example is a high-speed counter. We saw earlier in Section 1.3.2.4 that the scan time limits the maximum count rate of a PLC to about 10 Hz. High-speed counter cards are available for use where higher count speeds are needed, or the program scan time introduces an unacceptable random error.

In these, the card contains a bi-directional counter which can be directly driven by a pulse encoder. The counter value can be loaded from the PLC, and read back when needed. The PLC can also download a preset value, allowing the counter card to directly drive outputs according to the relationship between the count and the preset.

Other common intelligent modules are bar code readers (for stock tracking), stepper motor controllers (for position control systems) and vision modules (for quality control applications).

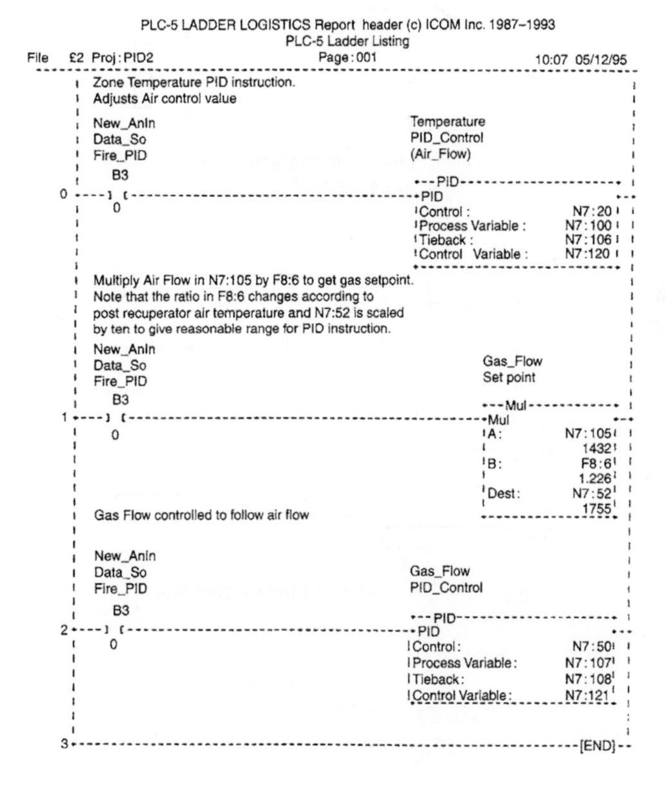

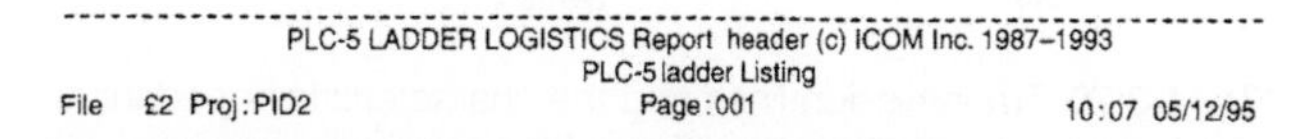

Fig. 1.3.69 PID control on an Allen Bradley PLC-5.

1.3.5 Distributed systems and fieldbus

1.3.5.1 Introduction

For a true distributed control system we need a method where several PLCs or computers can be linked together to allow communication to freely take place between any member of the system.

To achieve this we need to establish a connection topology, some way of sharing the common network that prevents time wasting contention and an address system that allows messages to be sent from one member to another. Such systems are known as *Local Area Networks* (LAN) or *Wide Area Networks* (WAN) depending on the size of the area and the number of stations.

1.3.5.2 Transmission lines

Any network will be based, to some extent, on cable, and at the high speeds used there are aspects of transmission line theory that need to be considered. Consider the simple circuit of Figure 1.3.70(a). At the instance that the switch closes, the source voltage does not know the value of the load at the far end of the line. The initial current step, *i*, is therefore determined not by the load, but by the characteristics of the cable (dependent on the inductance and capacitance per unit length). A line therefore has a *characteristic impedance*, typically 75 Ω or 50 Ω for coax, and 120–150 Ω for biaxial or screened twisted pair. The initial current step will therefore be V/Z where Z is the characteristic impedance.

After a finite time, this current step reaches the load *R*, and produces a voltage step $i \times R$. If *R* is not the same as *Z*, this voltage step will not be the same as *V*, and a reflection will result. Typical results are shown in Figure 1.3.70(b).

This effect occurs on all cables and is normally of no concern as the reflections only persist for a short time. If, however, the propagation delay down the line is similar to the maximum frequency rate of the signal, the reflections can cause problems. It follows that a transmission line should be terminated by a resistance equal to the characteristic impedance of the line. Normally, devices for connecting onto a transmission line have a high input impedance to allow them to tap in anywhere, with terminating resistors being used at the ends of the line.

A side effect of this is that T connections, or spurs, are not allowed (unless the length of the spur is short). In Figure 1.3.70(c) a T has been formed. To the signal, coming from the left, the two legs appear in parallel giving an apparent impedance of $Z/2$ and a reflection.

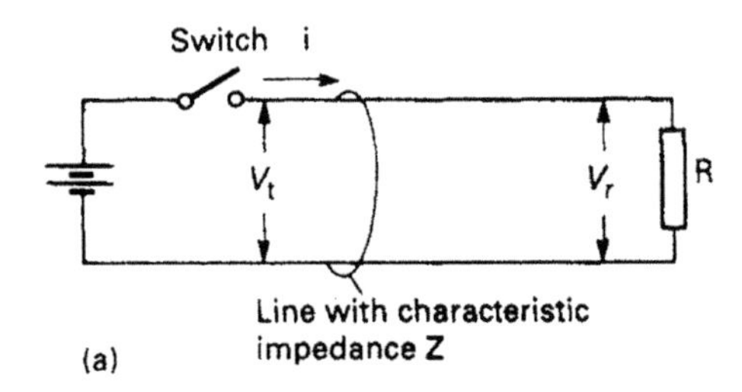

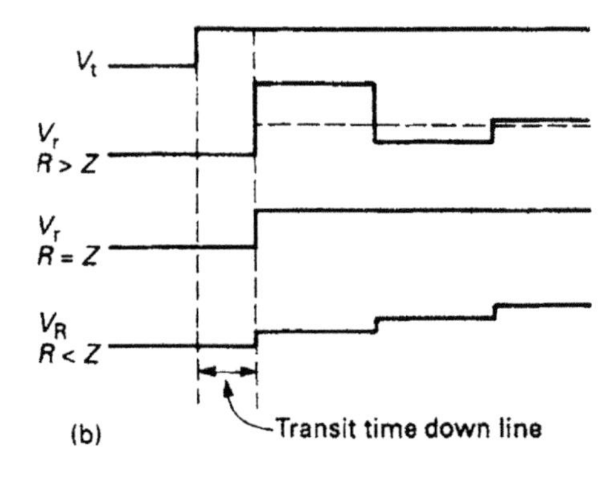

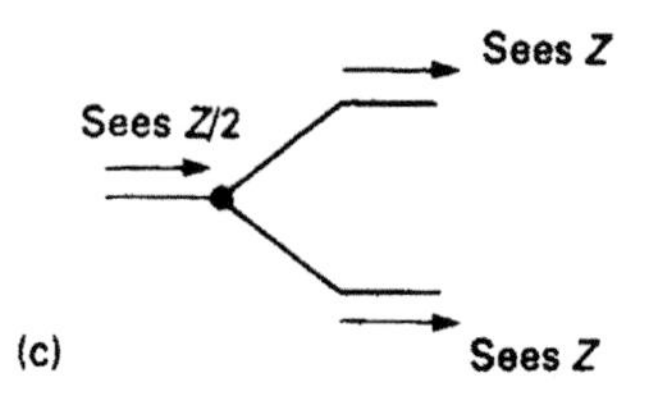

Fig. 1.3.70 Transmission lines and the characteristic impedance: (a) a transmission line; (b) the effect of the terminating resistor; (c) The effect of a 'T' in the line.

1.3.5.3 Network topologies

From the previous section it should be apparent that any network can sensibly only be based on a ring (which needs no terminating resistors) or a line (with a terminating resistor at each end). Figure 1.3.71 is a master/slave system where a common master wishes to receive or send data from/to slave devices, but the slaves never wish to talk to each other. All the slaves have addresses, which allows the master to issue commands such as '*Station 3; give me the value of analogue input 4*' or '*Station 14; your setpoint is 751.2*'. Such systems are often based on RS422 to provide improved noise immunity and allow longer lengths of line.

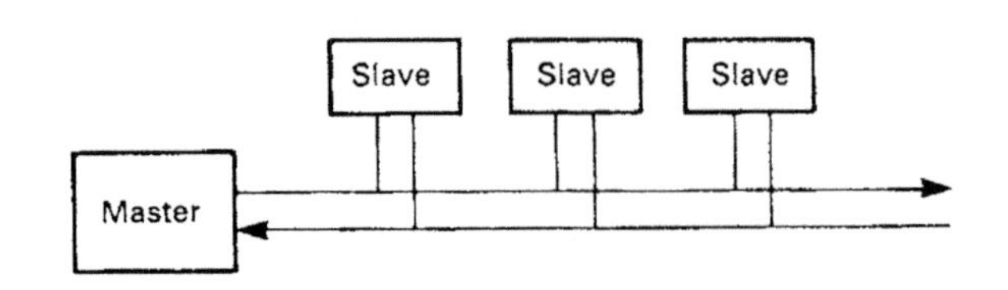

Fig. 1.3.71 A master/slave network.

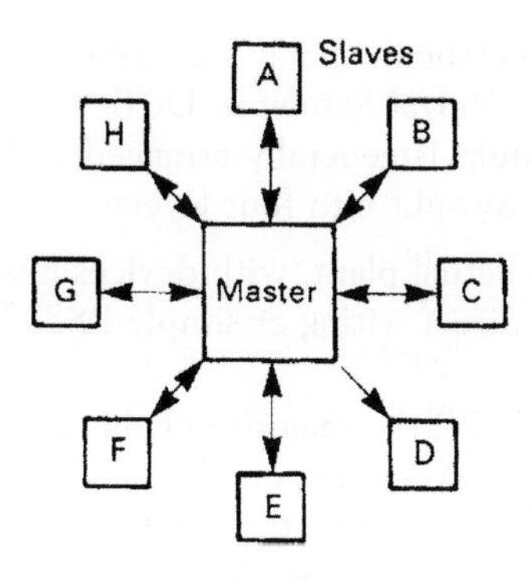

Fig. 1.3.72 A Star network.

The Star network of Figure 1.3.72 is again based on a master with a point to point link to individual stations. This arrangement is commonly used for high-level computer systems. Communication control is performed by the master station. Station to station communication is possible via, and with the cooperation of, the master.

In Figure 1.3.73 all the stations have been connected in a ring. There is no master, and all stations can talk to any other station and all have equal right of access. The term *peer to peer link* is often used for this arrangement. According to Figures 1.3.71 and 1.3.72 control was firmly in the hands of the master. With the ring, some technique is needed to avoid clashes when two stations wish to use the line at the same time. We will discuss this in the following section.

Figure 1.3.74 is probably the most common type of network used by PLCs. It is a single line with terminating resistors and, like the ring, is a peer to peer link where all stations have equal standing.

1.3.5.4 Network sharing

A peer to peer link allows many stations to use the same network. Inevitably two stations will want to communicate at the same time. If no precautions are taken, the result will be chaos. Various methods are used to govern access to the network.

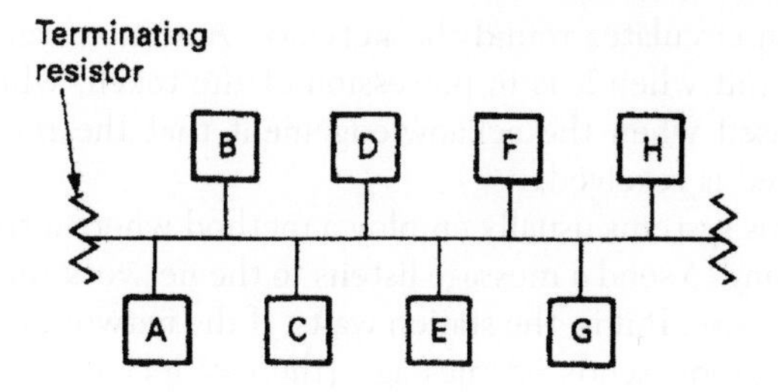

Fig. 1.3.74 A peer to peer link arranged as a single line bus with terminating resistors.

One idea is to allocate time slots into which each station can put its messages. This is known as *Time Division Multiplexing*, or TDM. Whilst it prevents clashes, it can be inefficient as a station will have to wait for its time slot even if no other station has a message to send. To some extent a mismatch between the frequency of messages from different stations can be overcome by giving more slots to hardworking stations. This is sometimes known as *Statistical TDM.*

The *empty time slot* of Figure 1.3.75 uses a packet which continuously circulates around the ring. When a station wishes to send a message it waits for the empty slot to come round, when it adds its message. In Figure 1.3.75, station A wishes to send a message to station D. It waits until the empty packet comes round. Then it puts its message onto the network along with the destination address D. Stations B and C pass the message but ignore it because it is not for their address. Station D matches the address, reads the contents (and appends that it has received the message). Stations E–H ignore it, but pass it on. Station A receives the message back again, sees the acknowledgement and removes its message leaving the empty packet circulating the ring again. A similar idea is a *token passing*, where a 'permit to send'

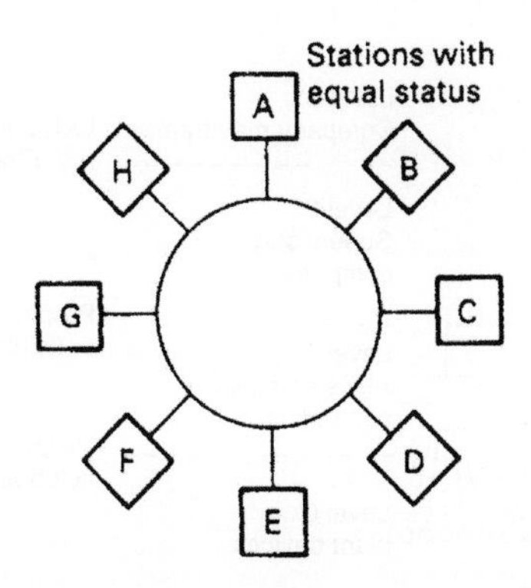

Fig. 1.3.73 A masterless peer to peer or ring network.

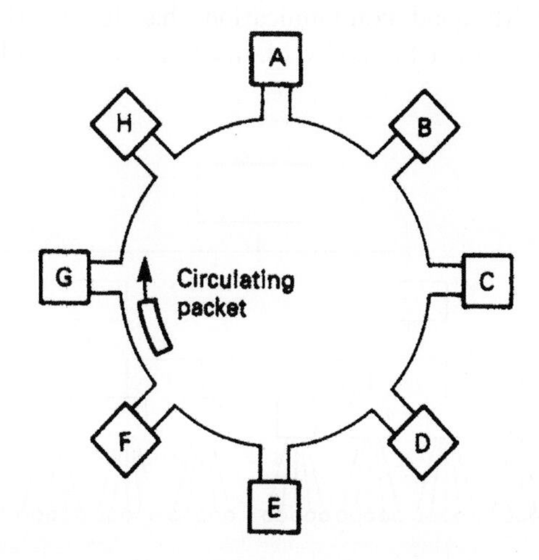

Fig. 1.3.75 Empty slot and token passing network

token circulates round the network. A station can only transmit when it is in possession of the token, which is released when the acknowledgement that the message arrived is received.

Bus systems usually employ a method where a station wishing to send a message listens to the network to see if it is in use. If it is, the station waits. If the network is free, the station sends its message (thereby locking out any other station until the message ends). This is known as *Carrier Sense Multiple Access* (CSMA).

Situations can still arise, however, where two stations simultaneously start to send a message, and a collision (and garbage) results. This situation can easily be detected, and both stations then stop and wait for a random time before trying again. A random time is used to stop the two stations clashing again. This is known as *Carrier Sense Multiple Access with Collision Detection* (or much more simply as CSMA/CD).

There is a fundamental difference between TDM, empty slot, and token passing as one group and CSMA. With the former there is a certain amount of time wasting, but every station is guaranteed access within a specified time. With CSMA there is little time wasting, but a station can, in theory, suffer repeated collisions and never get access at all.

A useful analogy is to consider motor car traffic control. TDM/token passing approximates to traffic lights, CSMA to roundabouts. In heavy traffic the best solution is traffic lights; everyone gets through and the waiting is shared evenly. Roundabouts can 'lock out' one road when the traffic flow is heavy and uneven from one direction. In light traffic, however, roundabouts keep the traffic flowing smoothly.

1.3.5.5 A communication hierarchy

Early process control systems tended to be based on a single large computer or PLC. The advent of cheap PLCs with good communications has led to the development of a hierarchy of machines which split the tasks between them. Such an arrangment is called a *Distributed Control System* or DCS.

Such a system is generally arranged as Figure 1.3.76 with a hierarchy split into four levels.

Level 0 is the actual plant, with devices linking to the next level by direct wiring or simple RS232/422 serial links.
Level 1 is the level the majority of this chapter is concerned with, consisting of PLCs and small computers directly controlling the plant.
Level 2 is supervisory computers for large areas of plants.
Level 3 is the large company mainframe.

Usually the layout is not as clear-cut as this implies. There are also differences between different companies, some number the layers from top to bottom and some ignore level 0.

There are many advantages to distributed systems. The resulting tree is conceptually simple, and as such is easy to design, commission, maintain and modify. A correctly designed system will be, for short periods, fault tolerant and can cope in a limited mode with the failure of individual stations. A distributed system can also bring about an increase in performance as lower-level machines take the work off higher-level machines.

1.3.5.6 Proprietary systems

In this section we will look at a typical proprietary system used to link PLCs from the same manufacturer. Typical examples are the GEM-80s Coronet and ESP, Siemen's Sinec and Modicon's Modbus. For reasons of space we shall consider how machine to machine links are achieved with Allen Bradley PLC-5s which communicate with each other on a peer to peer (no master) token passing highway based on twinaxial cable and operating at 57.6 Kbaud. Their trade name is Data Highway Plus. The PLC stations' addresses are set on switches in each PLC, and up to 64 stations can exist on one line with octal addresses 0–77.

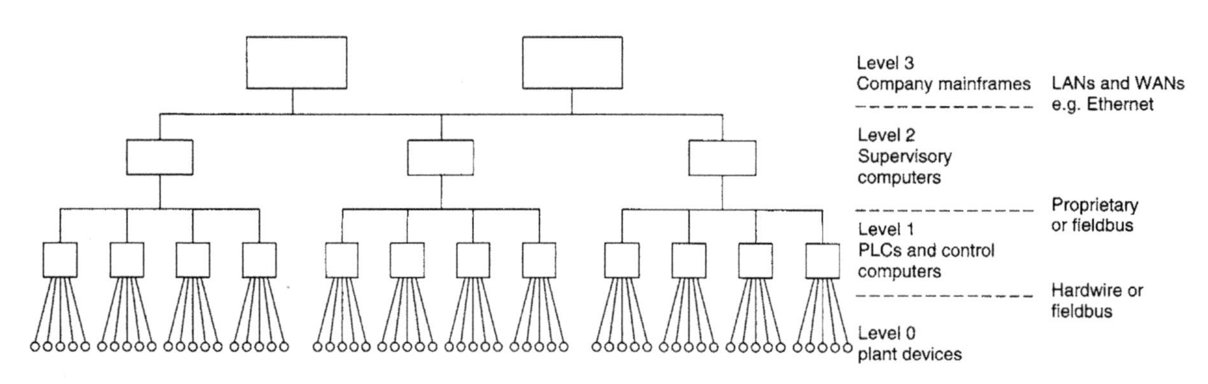

Fig. 1.3.76 A simple communication hierarchy.

Communication is established with a single message (MSG) instruction. This can be set up to read or write a block of data, the programmer specifying:

(1) the start address at the local end;
(2) the start address at the target end;
(3) the length of the block to be transferred (in words); and
(4) the station address at the remote end.

The MSG instruction appears in a program as Figure 1.3.77(a), the transfer being initiated every time the rung goes true. The ENable bit goes true when the transfer is started, and the DoNe bit goes true when it has been successfully completed. The ERRor flag goes true when an error occurs. Common errors are a line fault, a non-existent address at the far end or the PLC at the far end shutdown. The cause of the fault is given in flags set in the message control word. Link statistics (e.g., number of retries) are kept in the processor for diagnostic purposes.

The details of the MSG instruction are set up by the programmer via the screen of Figure 1.3.77(b). These are mostly self-explanatory, with the possible exception of the remote link which is concerned with sending data via a gateway module to a different highway, possibly of a different type.

The data highway is also used by the programming terminal, so a programmer can connect anywhere onto the data highway and link into any machine on the network.

1.3.5.7 Ethernet

Ethernet is a very popular bus-based LAN originated by DEC, Xerox and Intel and is commonly used to link the computers at level 2 in Figure 1.3.76. It uses 50 Ω coaxial cable, with a maximum cable length of 500 m (although this can be extended with repeaters). Up to 1024 stations can be accommodated, although in practical systems the number is far lower. Baseband (i.e. non-modulated) signalling is used with CSMA/CD access control. The raw data rate is 10 Mbaud, giving very fast response at loading levels up to about 20–30% of the theoretical maximum. Beyond this, collisions start to occur.

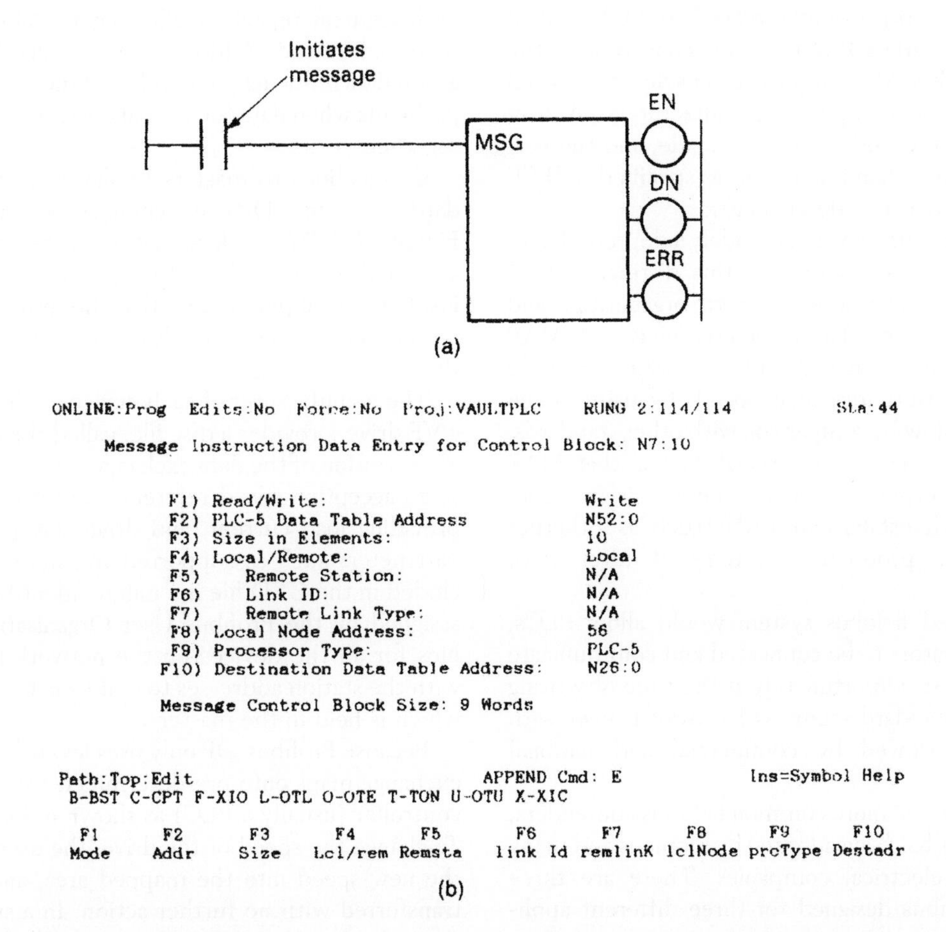

Fig. 1.3.77 The PLC-5 Message (MSG) Instruction: (a) as written in the ladder diagram; (b) as seen in detail on the programming terminal.

Because Ethernet uses CSMA/CD, the successful transmission of a message cannot be guaranteed. It is possible (but unlikely) for a given message to continually suffer from crashes and never get access to the network. In the jargon ethernet is *'nondeterminstic'*. In practice, if the network loading is kept below 30% of its theoretical capacity, this is not a problem. Many PLCs (such as the PLC-5/40E) now can provide direct connection to an Ethernet network.

1.3.5.8 Towards standardisation

We have already discussed the difficulties of linking different equipment. There is normally little problem linking PLC networks to higher-level computers. PLC manufacturers publish their message format and protocols, and interfacing software (called 'drivers') has been written for all common computers and PLCs. The difficulty comes when you want to link two machines from different manufacturers at level 1 in Figure 1.3.76. In many cases, the only economical solution is to do it through the computers and the higher level link.

General Motors (GM) in the USA were faced with this problem and attempted to specify a LAN for industrial control. This was called MAP (Manufacturing Automation Protocol). A similar office based LAN called TOP (Technical Office Protocol) was conceived at the same time. With GM's purchasing muscle, it involved several automation equipment manufacturers. A firm commitment to the OSI model was made, and the network based on broadband token bus as specified in IEEE 802.4 was chosen as it is deterministic.

MAP, however, has not been widely adopted. There appears to be several reasons for this distinct lack of enthusiasm. The first is a bureaucratic organisation and a changing specification. The second reason is cost; MAP links often cost more than the PLC to which they are connected. The third reason is speed; by using token passing MAP is slow by comparison with other standards. The final, and perhaps most crucial, fact is that MAP seems to have settled at a level where it is in direct competition with established LANs such as Ethernet rather than the proprietary systems at level 1 of Figure 1.3.76.

A standardised fieldbus system would allow PLCs, sensors and actuators to be connected and communicate with minimal cost. Unfortunately at the time of writing (early 2002) a standard seems as far away as ever with progress being slowed by commercial and national infighting.

Profibus is one the more common fieldbus contenders, largely because it has been adopted by Siemens and many other German electrical companies. There are three versions of Profibus designed for three different application areas. All use token passing.

The first, called Profibus-DP, for decentralised periphery, is by far the most common and is designed to link intelligent masters (e.g., a PLC), to slave devices such as sensors, drives or actuators. Profibus only uses levels 1 and 2 of the ISO/OSI model. Twisted pair RS485 or fibre optics are used for transmission.

The second, Profibus-FMS, for field message specification, is designed for the higher level with multiple masters, allowing peer to peer communication. Levels 1, 2 and 7 of the ISO/OSI model are used and RS485 or fibre optics for transmission.

Both DP and FMS share the same transmission standards and can consequently work together on the same network.

The final form, designed for process automation in hazardous areas, is Profibus-PA which permits the construction of an intrinsically safe network. Profibus-PA uses slightly different standards to DP and FMS, but can be linked by a segment coupler device.

All are a linear bus system, i.e., a straight line. Transmission speeds from 9.6 kbit/s (up to 1200 m) to 12 Mbit/s (up to 100 m) can be used. Screened twisted pair is used, with terminating resistors at each end of the bus. Up to 32 stations can be used in each segment, each with a unique station address. Segments can be coupled with segment repeaters, allowing a total of 127 stations to be addressed. Addresses are assigned for global or group data reducing the number of messages and time lag problems when data for several devices are to be changed together.

Connections to masters or slaves are made via standard 9 pin D-type connectors, as shown in Figure 1.3.78(a). Terminating resistors are either switched in internally at the end stations or connected inside the final plugs. Note that the terminating resistors require power; this normally comes from the end stations themselves.

The manufacturer of each device on the network, e.g., a VF drive, provides a disc file, called the GSD, which is a description of the data exchange the device can support (e.g., accepting speed reference and run command and providing load current, and drive state) plus operating parameters such as supported transmission speeds. Included in the GSD file is a unique identification number assigned by the Profibus User Organisation. The GSD files for all the devices on the network are used along with the station addresses to build a network description which is held in the master.

Because Profibus-DP only uses levels 1 and 2, the data exchange maps onto pre-determined areas in the master controller (usually a PLC) as shown in Figure 1.3.78(b). To change the speed of the drive, the user simply writes the new speed into the mapped area, and the data are transferred with no further action. In a similar manner, slave data and status is automatically read from the

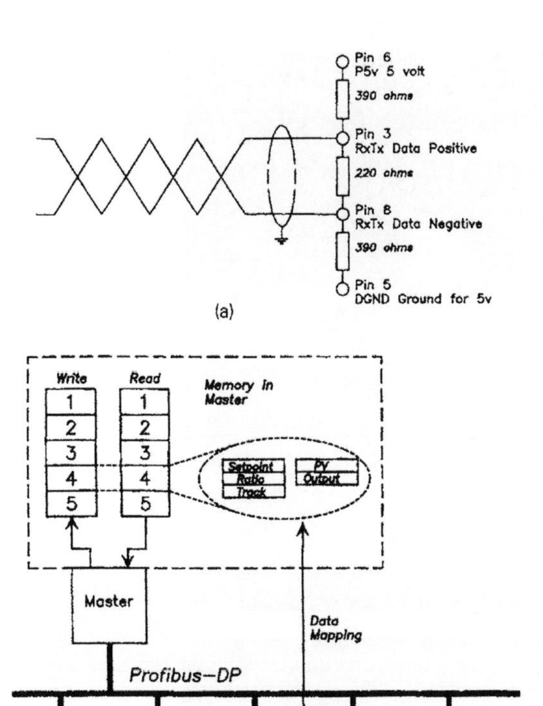

Fig. 1.3.78 Profibus-DP network: (a) connection at a Profibus device. Nonterminating devices use only pins 3 and 8; (b) mapping between a Profibus device and memory in the network master.

mapped area. A Profibus-DP network is thus totally transparent to the user.

A typical example of the problems that any attempts to specify a standardised fieldbus system may encounter is the continual introduction of new ideas. All the communications systems described so far are based on what is called the *source/destination model*. If station A has information for station B, a message is sent with the format:

Source A | DestinationB | Data

If this information is to be sent to several stations, each will need their own message. In applications where multiple setpoints have to be sent to multiple controllers, the delay caused by the time shift between the messages can cause problems, although this can be overcome to some extent by the use of group or global addresses as used by Profibus.

In addition, if station A needs information from station B (the state of an interlock for example), station A must perform a read on each occasion the data are required.

A recent development, called the *producer/consumer model*, uses a different approach. Here data are placed onto the network with no indication as to who it is for. The format is now simply:

Identifier | Data

All stations using these data accept it at the same time, eliminating the need for multiple messages. This significantly reduces the number of messages and hence increases the network speed.

The placement of data onto the network can be done in two ways. The first, and fastest, is *'notify on change'*. Here a station only places information on the network when a new value is different than the old. Stations with an interest in this data assume that the status or value remains the same until notified otherwise. There are obvious dangers in this, and a regular pre-defined 'heartbeat' is included to say a station is active on the network. The second approach updates on a time basis, each data item having its own, or a global, update time.

At the time of writing, Foundation Fieldbus is the only producer/consumer fieldbus network, and Rockwell (Allen Bradley) have also adopted the method for their proprietary ControlNet. The latter is interesting as it combines the ideas of their remote I/O and Data Highway onto one system and allows PLC racks (and their data), to be shared equally amongst several processors and not dedicated to one as before.

1.3.6 Graphics

Operator controls are being increasingly provided by computer graphic screens. These can be a display device designed specifically for a particular range of PLCs (for example the Allen Bradley Panelview and the CEGELEC Imagem) and general purpose graphic display devices (such as ABB/ASEA's excellent Tesselator) or graphics software running on conventional personal computers. Figure 1.3.79 shows some typical examples.

The major advantages are simplicity of installation and flexibility. A graphics terminal has just two connections to the outside world, a serial link connection and a power supply. If it is used to replace a desk full of switches and indicators there are obvious cost savings.

The designer of desks or control stations often has to deal with changes and modifications. Constructing a desk is always a fine balance of time, choosing between waiting until all the requirements are clear, and the minimum time needed to make it. Modifications at the commissioning stage rarely look neat. The displays on a graphical terminal can be modified relatively easily, and, more importantly, the modifications leave no scars. If the

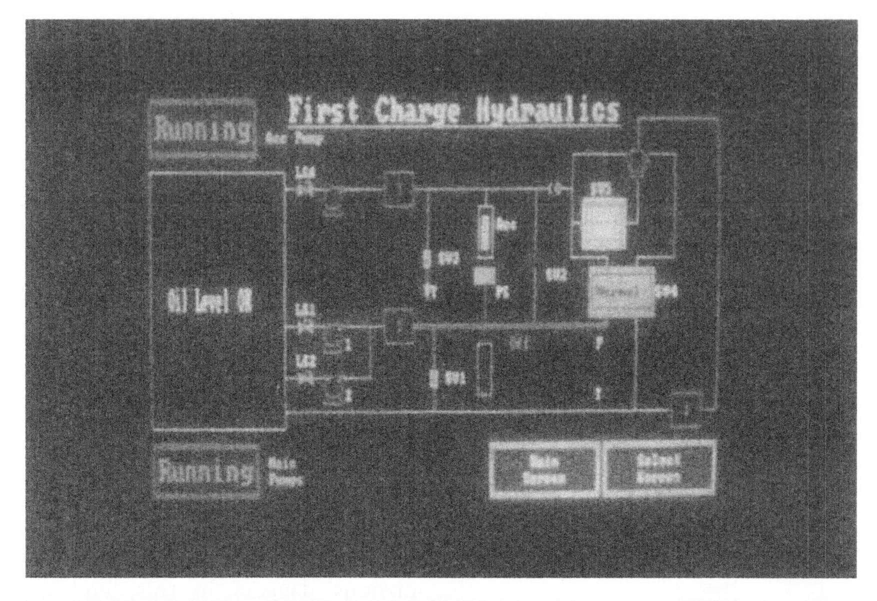

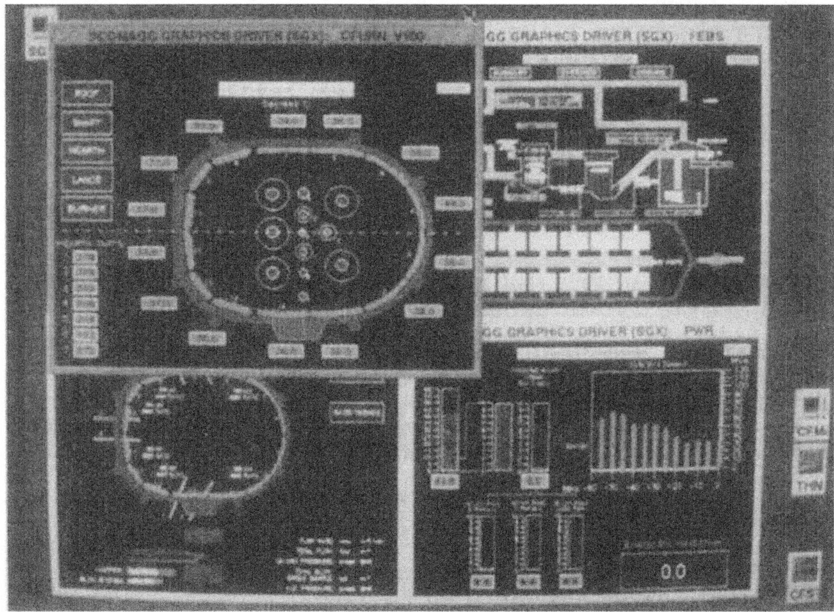

Fig. 1.3.79 Various graphic displays: (a) Allen Bradley touchscreen Panelview using block graphics; (b) high-resolution Scada system.

design of a normal desk can only start when the desk contents are 95% finalised (which is about right) a graphic screen can be started at 75% finalised. This flexibility is of great assistance as no job is ever right first time.

There are disadvantages, though. The most important of these is the limited amount of information that can be displayed on a single screen. It is very easy to overcrowd a screen (giving a screen similar to a page full of text on a word processor) making it difficult for the operator to identify critical items. A useful rule of thumb is not to use more than 25–30% of the screen.

The effect of this is often a need to build up a hierachy of screens: the top screen showing an overview, lower screens showing more and more detail. The problem with this is the time delay needed to shift through the screens. Direct screen to screen movement is possible by calling for a page number (which needs a good human operator memory, or a directory piece of paper, or wasted screen space) or by making all screen changes via an intermediate directory page (with additional delay). These time delays are small (less than a second typically) but the cumulative annoyance is large.

The time taken to update screen data can also be problematical, particularly where a machine to machine link is involved. Again a response time of ~ 1 s is typical, but several seconds is by no means uncommon. The

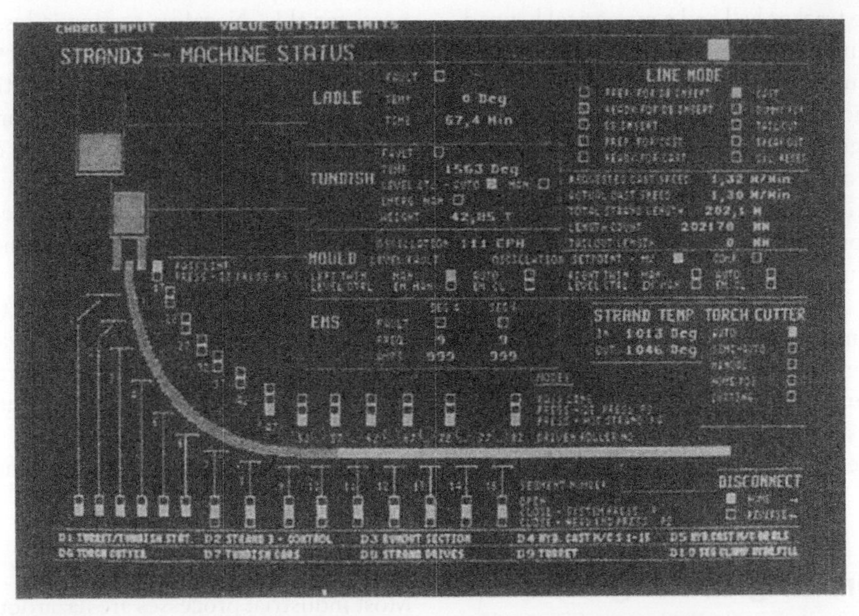

Fig. 1.3.79 (c) the ABB Tesselator. *Photos courtesy of Co-Steel Sheerness, Scomagg and ABB.*

use of a graphic terminal for fault finding on a fast moving plant is not really feasible.

There are generally two types of graphic terminal. The simplest, known as block graphics, has one store location for each character position on the screen and approximates to the old CGA standard on a personal computer.

The second type of display deals not with individual characters, but with individual points on the screen called 'pixels'. A typical medium resolution screen will have 640 (horizontal) by 480 (vertical) pixels, a total of 307,200 points. Each of these can be accessed individually, allowing lines to be drawn at any angle, fill patterns of any type to be used and trend graphs of plant variables to be displayed. Each individual pixel can have its own colour (from over 256 possible colours in some displays) and intensity. The result is an almost photographic resolution. There are additional costs, the most obvious of which is a large store requirement. The system hardware and software is more complex (and hence more expensive) but, perhaps surprisingly, this is not apparent to the designer. Programming for these screens is surprisingly simple with instructions using keywords like

```
DRAW FROM <> to<>.
```

or pick and place functions similar to a good commercial graphics package.

Supervisory Control and Data Acquisition (SCADA) systems are based around graphical objects which are linked to variables in the control systems. The state of the objects on the screen (e.g., size, colour, rotation, etc.) can be changed according to the values of the variables in the control systems giving a very visual image of the operation.

The environment around a display needs to be carefully considered. Most screens are mounted angled up, and are prone to annoying reflections from overhead lights and windows. Bright lighting (and above all direct sunlight) can make a display impossible to read. Displays are also adversely affected by magnetic fields. Close proximity to electric motors, transformers or high current cables will cause a picture to wobble and the colours to change. The effect can be overcome by screening the monitor with a mu-metal cage. Flat screen TFT or LCD displays do not suffer from this effect.

The size and weight of the monitors are often overlooked making them difficult to mount neatly, and even more difficult to change. Access should be made as easy as possible, trying to hold a 25 kg display in place with one hand whilst undoing interminably long mounting screws is not much fun.

Displays fail, and the implication of this needs to be considered in the design. If all the plant control is performed by screens, what will happen during the ten or so minutes it will take to locate a spare and change the faulty unit? Often dual displays are used to overcome this problem.

The operator will obviously need to input data and initiate actions. Keyboards are one approach, but many people are nervous of them and the cable connecting the keyboard always seems prone to damage. In dirty environments keys can become blocked with dirt and membrane keypads with tactile (feel) feedback should be used.

If the operator has to access points anywhere on the screen, a tracker ball is a useful device. Rather like an upside down mouse it controls the movement of a cursor on the screen. All normal actions can be performed with three buttons on the tracker ball and a numerical keypad. Trackerballs work surprisingly well in dirty environments as they are open underneath and dirt seems to fall straight through. Mice perform a similar function but are vulnerable to damage and dirt and seem more suited to an office environment.

A final consideration is security. Most modern graphics systems are based on good-quality personal computers. These have values outside of industry and are vulnerable to theft. Often is it is not the PCs or screens which are stolen, but the internal motherboards and memory cards. Suitable security methods should be used if a PC-based system is to be left unattended for a period of time (e.g., during a Christmas shutdown). Needless to say, backups should not be stored on the same PC as the original system.

1.3.7 Software engineering

Any project goes through six stages during its life. The first of these, analysis, is studying the application to understand what is required. This is by far the most difficult stage as the project requirements are usually unclear. Most projects that come unstuck do so because this first stage has been cut short or overlooked.

Next comes *specification*, which is documenting the analysis so everyone concerned can agree what is to be done and what the end result should be. If you can't produce a specification, how can you sensibly design it? Never say 'we'll sort that out later' because later becomes 3 a.m. as the plant starts up. The final testing procedures must also be defined at the specification stage; again if you don't know how you will test it, how will you know if it's working properly? Defining testing procedures in the cold light of day several months before the final frantic rush to meet a deadline also helps the poor commissioning engineer to resist the pleas for a premature start up.

The importance of these two first stages cannot be overemphasised, too often the users do not know, or do not say, what they want, but once the project is complete they are sure it wasn't that. With these first two difficult stages over, the rest of the project becomes much easier!

The *design* stage can now start (simple with a good specification) followed by *installation*. Next comes *commissioning*. These can also be difficult times, as in any project the control engineer ends up collecting everybody's delays and comes under pressure to 'get the plant away'. It is here that the advantage of the test schedule from the specification stage will be invaluable.

It is not generally understood that commissioning involves both *positive* and *negative* testing. Positive testing is obvious; it is ensuring that when the firkling button is pressed the plant firkles. Practically everyone sees the need for this. Negative testing is less obvious; it ensurs that the control system deals correctly with all the unlikely circumstances and fault conditions. Negative testing takes far longer, because there are many more fault modes than healthy modes. It is very common for people to say '*it works, let's go*' when only the positive testing has been done. Try to resist this pressure; at best it can lead to damaged plant a few years; hence, at worse some safety features could be overlooked.

Finally the plant is handed over to the maintenance department. In commercial software it is generally thought that over 50% of the effort goes into *maintenance* as changes are made to meet new requirements or correct the inevitable bugs. For easy maintenance all the documentation must be complete and up to date.

1.3.8 Safety

Most industrial processes are hazardous, and the safety of all personnel must be of prime importance. This section is a personal view and can only give a simple discussion of safety considerations. The topic of safety is covered by both criminal and civil law. The designer and user of any system must therefore consult the relevant legislation and codes of practice to ensure compliance.

Every single person has a safety responsibility. Employers have a '*duty of care*' for their employees and the public and must ensure that the plant is kept in a safe condition, safe working procedures are devised for all conceivable activities and training in these procedures provided for all relevant employees. Suppliers must ensure that their equipment meets safety criteria, and draw the attention of purchasers to unavoidable hazards (protection and labelling of parts which are live during normal operation for example). Employees must follow safety procedures and not expose themselves (or others) to danger. These responsibilities are covered by the Health and Safety at Work Act 1974 (HASWA) which makes the universal responsibility for safety absolutely clear.

More recently in 1992, a block of EEC Health & Safety Regulations (commonly known as the '*six pack*') introduced the idea of *risk assessment*. This recognises the need to balance the cost and complexity of the safety system against both the likelihood and severity of injury. The procedures outlined use common terms with specific definitions:

`Hazard`: the potential to cause harm.
`Risk`: a function of the likelihood of the hazard occurring and the severity.
`Danger`: the risk of injury.

and outlines procedures to achieve acceptable safety standards. Risk assessment is a legal requirement under most modern legislation, and is covered in detail in standard prEN1050 '*Principles of Risk Assessment*'.

A Health & Safety Executive study of safety in control systems ('*Out of Control*' HSE books 1995 ISBN 0717608476) makes worrisome reading. It suggests that more than 60% of safety related failures are introduced into a system before it is taken into service for the first time. Approximately 44% of safety incidents come from specification errors, 15% from design errors and 6% from poorly thought-out changes during commissioning.

The inclusion of a programmable controller brings additional hazards (and solutions) which must be recognised. A PLC can introduce potentially dangerous situations in different ways. The first (and probably most common) route is via logical errors in the program. These can be the result of oversight, or misunderstanding, on behalf of the original designer who did not appreciate that this set of actions could be dangerous, or by later modifications by people who deliberately (or accidentally) removed some protection to overcome a failure in the middle of the night. '*Midnight programming*' is particularly worrying as usually the only person who knows it has been done is the offending person, and the danger may not be apparent until a considerable time passes and the hazardous condition occurs.

The second possible cause is failure of the input and output modules; in particular, the components connected directly to the plant which will be exposed to high-voltage interference (and possibly direct connected high voltages in the not unlikely event of cable damage). Output modules can also suffer high currents in the (again not unlikely) event of a short circuit. Typical output devices are triacs, thyristors or transistors. The failure mode of these cannot be predicted; all can fail short circuit or open circuit. In these failure conditions the PLC would be unable to control the outputs. Similarly an input signal card can fail in either the 'on' or 'off' state, leaving the PLC misinterpreting a possibly important signal.

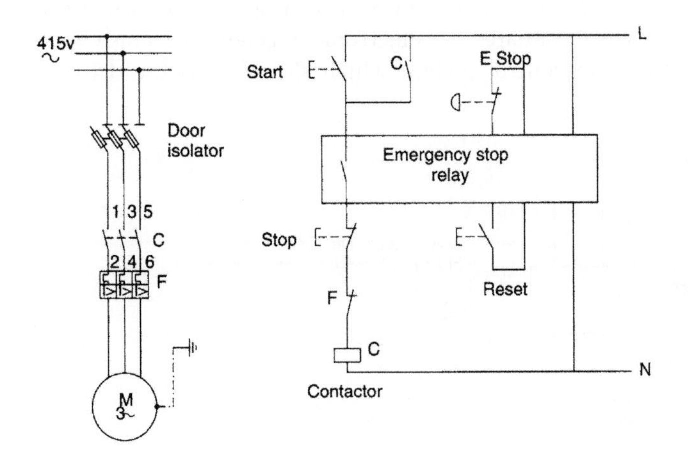

Fig. 1.3.80 A standard hardwired motor starter for a low-risk application. This would normally be considered to be sate. In higher-risk applications there would probably be dual connections on the emergency stop button; dual contactors and the state of the contactors would be monitored by the emergency stop relay.

The next failure mode is the PLC itself. This can be further divided into hardware, software and environmental failures. A hardware failure is concerned with the machine itself; its power supply, its processor, the memory (which contains the 'personality' of the PLC, the user's program, and the data storage). Some of these failures will have predictable effects; a power supply failure will cause all outputs to de-energise, and the PLC supplier will have included memory checks in the design. Environmental effects arise from peculiarities in the installation such as dust, humidity, temperature (and rapid temperature changes) possible water ingress and vibration, and these can result in unexpected operation of output devices.

The final cause is electrical interference (usually called noise). Internally, almost all PLCs work with 5 V signals, but are surrounded by high-voltage, high-current devices. Noise can cause input signals to be misread by the PLC, and, in extreme cases, can corrupt the PLCs internal memory. PLCs generally have internal protection against memory corruption and noise on remote I/O serial lines, so the usual effect of noise is to cause a PLC to stop (and outputs to de-energise). This cannot, however, be relied upon.

Figure 1.3.80 shows a normal motor starter circuit built without a PLC. Safety precautions here are:

- Isolation switch at the MCC removes the supply for maintenance work.
- Normally closed contacts on the stop and emergency stop buttons. A broken wire will look like a stop button being pressed, as will loss of the control supply.
- If the emergency stop is pressed and released, the motor does not restart.
- Isolation and emergency stop have priority over start.

It is still possible to identify dangerous failure modes in this system. The button head of the emergency stop button could unscrew and fall off, or the contacts of the contactor could weld made, or a short could occur between the cores to the stop button but these failure modes are exceedingly rare, and Figure 1.3.80 would be generally accepted as safe for use in normal circumstances.

In Figure 1.3.81(a) the same functions have been provided by an *unsafe* PLC system. To save costs the MCC door isolator has been replaced by a simple switch which makes to say 'Isolate'. Similarly normally open contacts have been used for stop and emergency stop. This is controlled by the unsafe program of Figure 1.3.81(b).

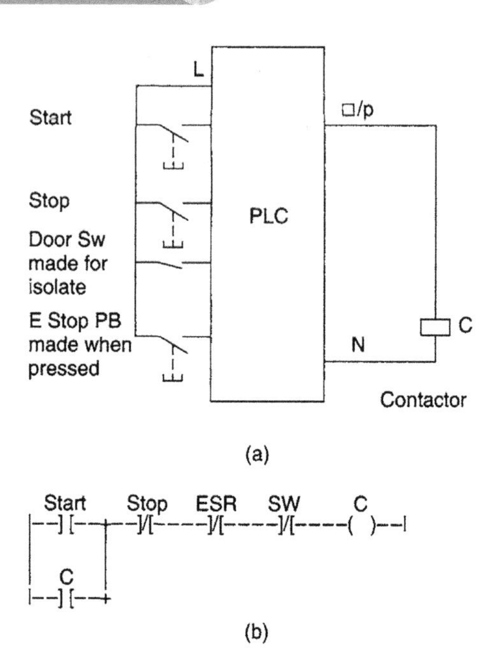

Fig. 1.3.81 An unsafe PLC-based system totally relying on software. The dangers of this system only become apparent when failures occur.

It is important to realise that to the casual user, Figures 1.3.80 and 1.3.81 behave in an identical manner. The differences (and dangers) come in fault, or unusual, conditions. In particular:

- A person using a programming terminal can force inputs or outputs and over-ride the isolation. Although it is unlikely that anyone would do this deliberately, it is easy to confuse similar addresses and swop digits by mistake (forcing 0:23/01 instead of 0:32/01 for example).
- A loss of the input control supply during running will mean the motor cannot be stopped by any means other than totally removing the supply to the system.
- The system is very vulnerable to input and output card faults.

None of these are apparent to the user until an emergency occurs.

A prime rule, therefore, for using PLCs is:

> *'The system should be at least as safe as a conventional system'*

Figure 1.3.82(a) is a revised PLC version of Figure 1.3.80. The isolator has been re-instated with an auxiliary contact as PLC inputs, and normally closed contacts used for the stop and emergency stop buttons. An auxiliary contact has been added to the starter, and this is used to latch the PLC program of Figure 1.3.82(b). The emergency stop is hardwired into the output and is independent of the PLC, and on release the motor will not restart (because the latching auxiliary contact in the program will have been lost). On loss of control supply the program will think the stop button has been pressed, and the motor will stop. Figure 1.3.82 thus behaves in failure as Figure 1.3.80.

Although this example is simple, it illustrates the necessary analysis and considerations that must be applied in more complex systems.

Complex electronic systems can bring increased safety. Consider a thyristor drive controlling the speed of a large d.c. motor. In a typical arrangement there will be an upstream a.c. contactor to enable the drive. Hardwire connection of an emergency stop button into the a.c. contactor will obviously stop the drive, but the inertia of the motor and the load will keep it rotating for several seconds. A thyristor drive, however, can stop the load in less than 1 s by regeneratively braking the motor, but this requires the drive to be alive and functional. The operation of the emergency stop implies a dangerous condition in which the fastest possible stop is required. It is almost certain that at this time the drive controls are functional and there are no 'latent' faults.

Here the emergency stop can operate in two ways. First it initiates an electronic regenerative crash stop via the control system which should stop the drive in

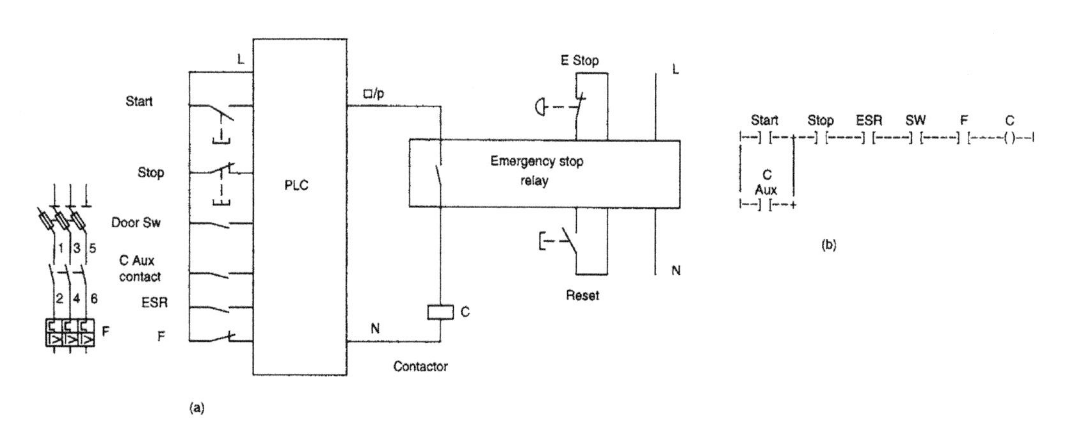

Fig. 1.3.82 A safe PLC system for low-risk applications. As for Figure 1.3.80 more features could be added if the risk was higher.

less than one second. The emergency stop also releases a delay drop-out hardwire relay set for 1.5 s which releases the a.c. contactor. This gives the safest possible reaction to the pressing of the emergency stop button. Safety considerations do not, therefore, explicitly require relay-based, nonelectronic hardware, but the designer must be prepared to justify the design decisions and the methods used.

Where complex control systems are to be used, a common method of improving safety is to duplicate sensors, control systems and actuators. This is known as *redundancy.* A typical application occurs in boilers where feed water is held in a drum. Deviations in water level are dangerous; too low and the boiler will overheat, possibly to the point of melting the boiler tubes; too high and water can be carried over to the downstream turbine with risk of catastrophic blade failure. High- and low-level sensors are therefore usually provided with each being duplicated. The safety system reacts to *any* fault signal, so two sensors have to fail for a dangerous condition to arise. If the probability of a sensor failure in time T is p (where $0 < p < 1$), the probability of both failing is p^2. In a typical case, p will be of the order of 10^{-4} failure per year giving p^2 of 10^{-8}.

There are two disadvantages. The first is that a sensor can fail into a permanently safe signal state, and this failure will be *'latent'*, i.e., hidden from the user with the plant running on one sensor. The second problem is that the plant reliability will go down, since the number of sensors goes up and any sensor failure can result in a shutdown. Both of these effects can be reduced by using *'majority voting'* circuits, taking the vote of two out of three or three out of five signals.

Redundancy can be defeated by *'common mode'* failures. These are failures which affect all the parallel paths simultaneously. Power supplies, electrical interference on cables following the same route and identical components from the same batch from the same supplier are all prone to common mode failure. For true protection, *diverse redundancy* must be used, with differences in components, routes and implementation to reduce the possibility of simultaneous failure.

To give true redundancy it is sensible to provide duplication in the control system as well to protect against hardware and software failures in the system itself. Duplicate control schemes, though, are vulnerable to a form of common mode failure called a *'systematic failure'*. Suppose duplicated temperature sensors are compared, inside a PLC program, with an alarm temperature. Suppose both are identical devices, running the same program containing a bug which inadvertently (but rarely so it does not show up during simple testing) changes the setting for the alarm temperature (from 60 °C say, to 32053 °C). Such an effect could easily occur by a mistype in a MOVE instruction in a totally unrelated part of the program. This error will affect both control systems, and totally remove the redundancy.

If reliance is being made on redundant control systems, therefore, they should be totally different; different machines with different I/O and different programs written by different people with the machines installed running on different power supplies with different types of sensors connected by different cable routes.

The Health and Safety Executive (HSE) became concerned about the safety of direct plant control with computers, and produced an occasional paper OP2 *'Microprocessors in Industry'* in 1981. This was followed in 1987 by two booklets *'Programmable Electronics Systems in Safety Related Applications'*. Book 1 (an Introductory Guide) is a general discussion of the topic, and Book 2 (General Technical Guidelines) goes through the necessary design stages. They suggest a five-stage process:

(i) perform a hazard analysis of the plant or process;

(ii) from this, identify which parts of the control system are concerned with safety and which are concerned purely with efficient production; the latter can be ignored for the rest of the analysis;

(iii) determine the required safety level (based on accepted attainable standards or published material);

(iv) design safety systems to meet or exceed these standards; and

(v) assess the achieved level (by using predicted probability of failure for individual parts of the design); revise the design if the required level has not been achieved.

The books stress the importance of *'Quality'* in the design: quality of components, quality of the suppliers and so on.

The IEC standard *IEC 61508 Functional Safety of Electrical/Electronic/Programmable electronic safety related systems* covers similar grounds to the HSE books. This is based on the ideas of *safety functionality* (what it is designed to protect against and how the protection is achieved e.g. *'open quench valve if temperature rises above 250 °C')* and the *Safety Integrity Level* (or SIL) which, somewhat simplified, is the probability, p, that the safety system will fail to operate on demand. The SIL covers the entire safety system including sensors, control system, and actuators. Four SILs are defined from a basic SIL-1 ($10^{-1} > p > 10^{-2}$) to SIL-4 ($10^{-4} > p > 10^{-5}$). The required SIL is determined from a risk assessment of the system. For continuous protection on a hazardous plant the normal requirement is SIL-3 or SIL-4. Guidelines for architectural constraints (such as keeping the safety system separate from the control system, and using redundancy) are also given. It is probable that IEC 61508 will become a European standard in the near future, and the two HSE books are being re-written to incorporate ideas from IEC 61508.

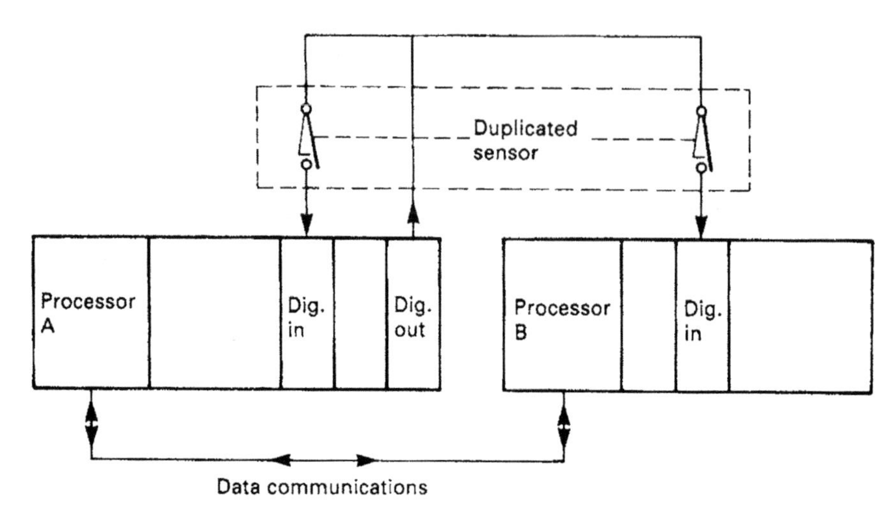

Fig. 1.3.83 Safety critical input with the Siemens 115F PLC.

Surprisingly some fieldbus systems (e.g., specialist versions of Profibus and SafetyBus from Pilz) can achieve SIL-3 which makes a fieldbus safety system attractive in hazardous applications. Extreme care must, of course, be taken.

In America, the Instrument Society of America (ISA) standard S84 follows broadly similar lines to the HSE guidelines and IEC 61508.

Because the HSE books, IEC 61508 and S84 are standards they have the legal status of guidance notes and there is no formal requirement to follow them. In the event of an incident, however, the designers and users must be prepared to justify the actions they have taken and conformance with good practice is a legal defence.

Very high safety levels can be achieved with some PLCs. Siemens market the 115F PLC which has been approved by the German TUV Bayern (Technical Inspectorate of Bavaria) for use in safety critical applications such as transport systems, underground railways, road traffic control and public elevators. The system is based on two 115 PLCs and is a model of diverse redundancy. The two machines run diverse system software and check each other's actions. There is still a responsibility on the user to ensure that no systematic faults exist in the application software.

Inputs are handled as Figure 1.3.83. Diverse (separate) sensors are fed from a pulsed output. A signal is dealt with only if the two processors agree. Obviously, the choice of sense of the signal for safety is important. For an overtravel limit, for example, the sensors should be made for healthy and open for a fault.

Actuators use two outputs (of opposite sense) and two inputs to check the operation as Figure 1.3.84. Each subunit checks the operation of the other by brief pulsing of the outputs allowing the circuit to detect cable damage, faulty output modules and open circuit actuators. If, for

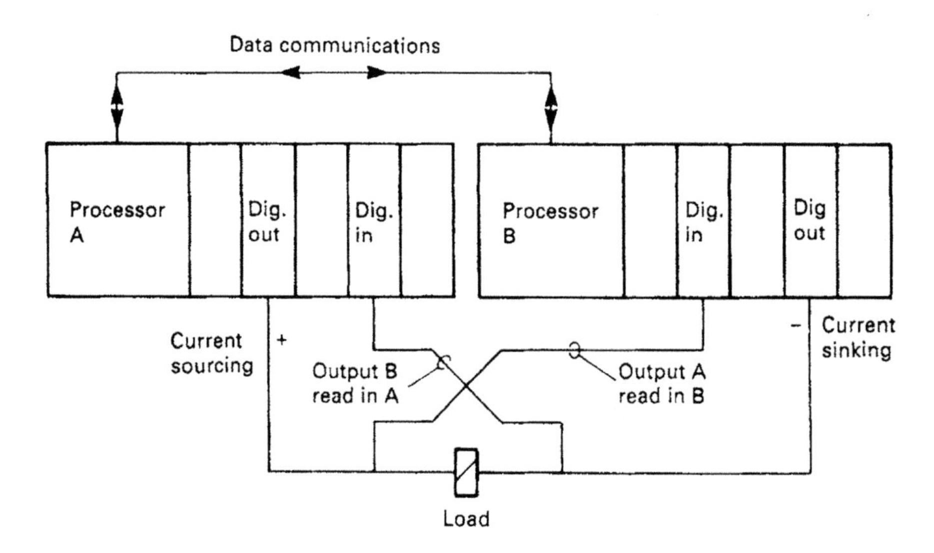

Fig. 1.3.84 Safety critical output with the Siemens 115F PLC.

example, output B fails on, both inputs A and B will go high in the Off state (but the actuator will safely de-energise.)

The operation of Figures 1.3.83 and 1.3.84 is straightforward, but it should not be taken as an immediately acceptable way of providing a fail-safe PLC. The 115F is truly diverse redundant, even the internal integrated circuits are selected from different batches and different manufacturer, and it contains well-tested diverse self-checking internal software. A DIY system would not have these features, and could be prone to common mode or systematic failures.

A PLC system is an electrical system, and is subject to the same legislation as conventional electrical schemes. Apart from the Health and Safety at Work Act, the designer should also observe the Institute of Electrical Engineers Wiring Regulations, and the Electricity at Work Regulations 1990.

Chapter 1.4

1.4

Control systems interfacing

Zhang

1.4.1 Actuator–sensor (as) interface

1.4.1.1 Overview

Industrial automation systems require large amounts of control devices, and the number of binary actuators and sensors on a typical system has increased over the years. Conventional input and output (I/O) methods for wiring include point-to-point connection or bus systems. For example, typical batching valve wiring networks attach each of the I/O to a central location resulting in multiple wire runs for each field device. Large expenditures are needed for cabling conduit, installation, and I/O points. Space for I/O racks and cabling must be accommodated in order to attach only a few field devices. These methods can prove to be too complex for networking simple binary devices and, therefore, too slow for the interactions between the controllers and the controlled devices. Point-to-point wiring is the most common method of wiring in the industry, but large wire bundles take up valuable space, installation is time consuming, and troubleshooting is complex.

Actuator–sensor interface, or AS interface, was developed by a group of sensor manufacturers and introduced into the market in 1994. Since that time, it has become the standard for discrete sensors in process industries throughout the world. AS interface is also a bus system for low-level field applications in industrial automation to communicate with small binary sensors and actuators using the AS interface standard. The AS interface modernizes automation systems effectively and eliminates wire bundles completely, with only one wire cable required, compared to one cable from each device with point-to-point wiring. Junction boxes are also eliminated and the size of the control cabinet needed is significantly reduced. The plug and play wiring supports all typologies. Figure 1.4.1 gives the locations of the AS interface in industrial control networks.

In comparison with conventional I/O wiring methods, the AS interface has many advantages. The most important ones are given below:

(1) *Minimum wiring and cost saving.* AS interface offers a single cable, which uses simple serial connection to the controller, instead of parallel with a multitude of cables.

(2) *Fast and safe installation.* Sensors and actuators are simply installed with modules on the AS interface cable. Contact pins in the modules penetrate the insulation of the cable and establish contact with the copper wire. Incorrect connections are practically impossible because of the design of the cable and the special piercing method.

(3) *Flexible configuration.* Owing to the distributed and modular design, plant sections can be tested in parallel even before the overall solution is finished. This permits flexible modification and expansion.

(4) *Open system.* AS interface is an open system, which means that it is independent of manufacturer and future-proof.

1.4.1.2 Architectures and components

In industrial control networks, as displayed in Figure 1.4.1, there are two types of AS interface architectures.

Industrial Control Technology; ISBN: 9780815515715

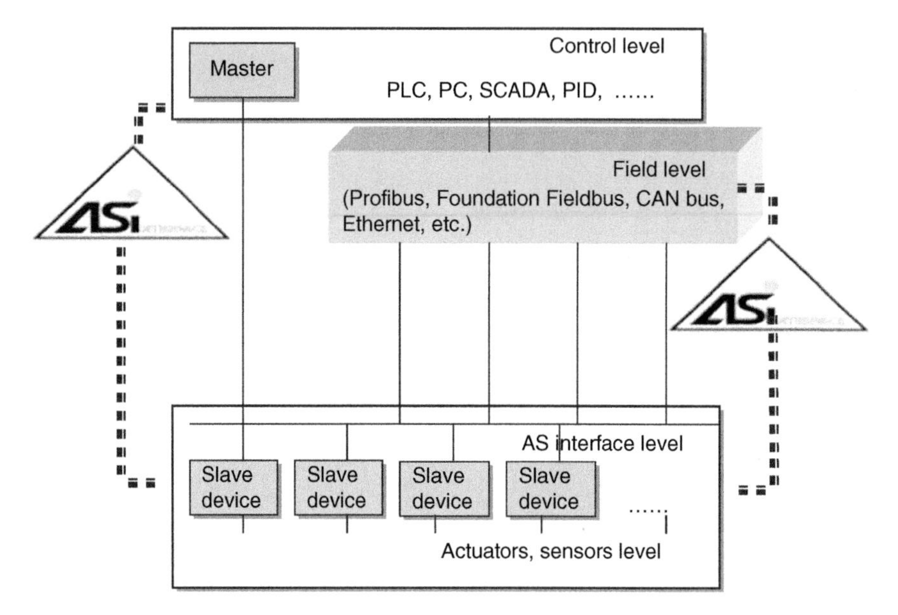

Fig. 1.4.1 The functionality of AS interface in industrial control networks. The AS interface can be at two locations in an industrial control network: between the controllers and the actuators–sensors and between the field level and actuators–sensors.

1.4.1.2.1 AS interface architecture: type 1

In the first type of AS interface architecture, a controller such as programmable logic controller (PLC), supervisory command and data acquisition (SCADA), or personal computer (PC) applies the controls of sensors and actuators via the field level buses, including Fieldbus, PROFIBUS, etc.

As displayed in Figure 1.4.2, the AS interface has gateways directly connected to the field level bus and the I/O module. The I/O module is the device of this AS interface architecture to contact the sensors and the actuators. A field level bus may be able to support several AS interface gateways depending on the system designs, each of which profiles a segment of an industrial control system.

In this type of architecture, AS interface requires the following components:

(1) *Gateways.* Gateways are interface modules between the AS interface and a higher level bus system. They are used when more complex applications are to be solved using standard products. AS interface gateways are the core of the wiring system, which handles the complete data transfer, cyclically polling (master–slave) all participants connected to the wiring system. The AS interface gateway can be placed anywhere in the AS interface segments. One gateway can handle 124 inputs and 124 outputs over 31 addressable I/O modules. For gateways, setup is

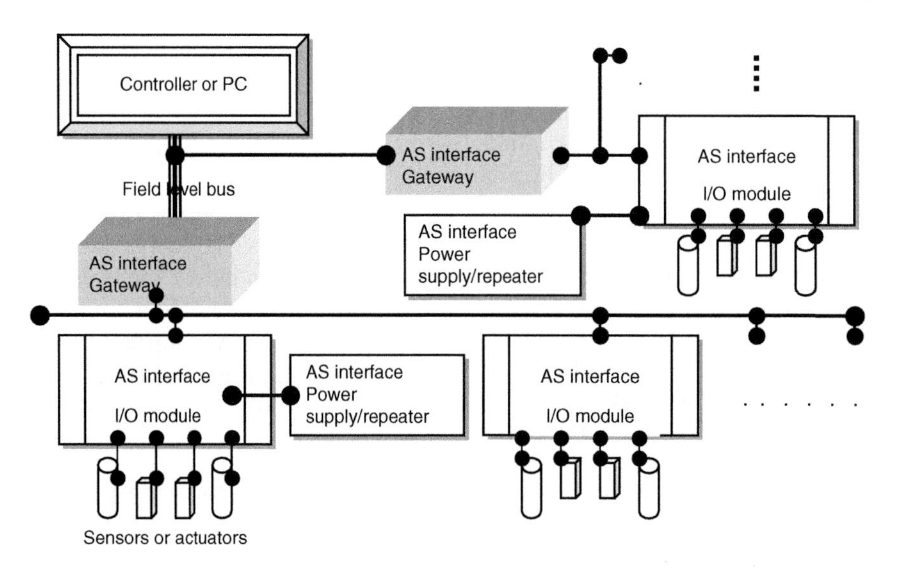

Fig. 1.4.2 AS interface architecture: type 1.

accomplished through the setup tools for the respective system.

(2) *I/O modules.* I/O modules are the interface between standard sensors and actuators and AS interface. I/O modules are available for any kind of application, including flat modules for limited space applications, compact modules for a variety of mounting options, field modules that use cord grips instead of quick disconnects, standard modules that use both the mechanically keyed AS interface cable. and the standard 16 AWG round cable. For enclosures and junction boxes, enclosure modules connect AS interface bus to a power rail system, and junction box modules for use within junction boxes.

(3) *Power supplies and repeaters.* With AS interface, one single cable transmits both power and data. Power supplies contain internal data separation coils so that the capacitive filtering of the supply does not interfere with the data stream. Adding to the high interference immunity of AS interface is the power supply data isolation coil between the voltage transformer and the output so that the data signals are isolated from line noise. Repeaters extend AS interface networks up to 100 additional meters, and by using two in series, an AS interface network can be up to 300 m long. Repeaters do not require a network address and allow I/O modules to be placed anywhere along the network.

(4) *AS interface safety at work.* This extension of AS interface allows for safety equipment to be wired together on a two-wire cable rather than hardwired back to a panel. A maximum of 31 category 4 inputs, for example, E-Stops, can be put on one cable. The parts included in this section are safety slaves, safety monitor, and configuration software.

In many applications safety-relevant functions are to be guaranteed. These are in the form of emergency stop buttons near process lines or the implementation of safe sensors (e.g., safe photo grits and locking of safety related doors) to automatically stop machines. Depending on the safety category, there are requirements of differing severity. Typically, a separate wiring is necessary as well as redundancy or increased protection for the cables.

With the integration of the safety technique into the AS interface line under the terminology "AS interface safety at work," the additional costs can be drastically reduced. The concept designates connection of the safety related switches by safe AS interface module. There is also a safety monitor on the AS interface line permanently observing the communication. The communication happens through a given and predetermined pattern by a dynamic code table with 8 times 4 bits sequence. The safe monitor continuously checks "must" and "actual" values of the communication through comparison. In case of the bit sequence "0000," the safe monitor switches off the safe relay in less than 40 ms. Several safe monitors can be operated in one AS interface line arranged on any position.

Clear benefits of AS interface safety at work include the following: only one AS interface line is required for the communication of safe and nonsafe data; full compatibility with all standard AS interface devices; no specific communication mechanisms required; mixed applications on one and the same AS interface line possible; diagnosis of the safe modules via the standard AS interface master possible.

(5) *AS interface encoders.* In order to be able to meet the real-time requirements of many applications, a "multislave" solution was achieved. The position value, up to 16 bits in length, is transferred to the gateways within a single cycle, via the four integrated AS interface chips used for control purposes. AS interface rotary encoders include 13-bit-Singleturn and 16-bit-Multiturn.

(6) *Accessories.* To make AS interface perfect and to make the installation as easy as possible, various accessories ranging from hand-held addressing devices to mounting bases to simulators for higher level bus systems are offered: sealing for flat cable and adaptor to round cable.

1.4.1.2.2 AS interface architecture: type 2

In the second type of AS interface architecture, as shown in Figure 1.4.3, the AS interface master module resides inside a controller such as PLC, SCADA, or PC. In this type of AS interface architecture, the AS interface master terminal enables the direct connection of AS interface slaves. The AS interface compliant interface supports digital and analog slaves. The AS interface master does not manage the sensors and actuators via the field level buses, but rather via the AS interface slave modules or cables. The slave modules are connected to each other by means of the AS interface cable which can be branched with the cable branch device. Power supply and repeater are used too. A group of slave modules frames a segment of an industrial control network with one interface cable. The AS interface master module is able to support several segments depending on the designed system capabilities.

In this type of architecture, AS interface requires the following components:

(1) *AS interface masters.* The AS interface master automatically controls all communication over the AS interface cable without the need for special software. The master can connect the system to

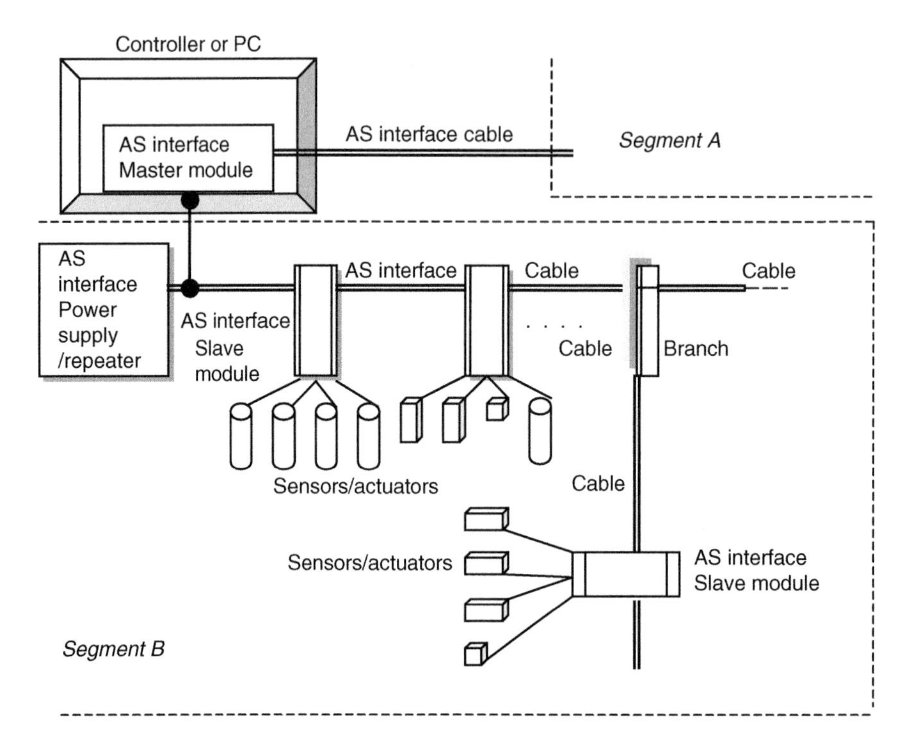

Fig. 1.4.3 AS interface architecture: type 2.

a controller such as PLC, SCADA, or PC, act as a standalone controller, or serve as a gateway to higher level bus systems. There exist the following AS interface masters in the current markets:

(a) *Standard AS interface master.* Up to 31 standard slaves or slaves with the extended addressing mode can be attached to standard AS interface masters.

(b) *Extended AS interface masters.* The extended AS interface masters support 31 addresses that can be used for standard AS interface slaves or AS interface slaves with the extended addressing mode. AS interface slaves with the extended addressing mode can be connected in pairs (programmed as A or B slaves) to an extended AS interface master and can use the same address. This increases the number of addressable AS interface slaves to a maximum of 62. Due to the address expansion, the number of binary outputs is reduced to three per AS interface slave on slaves using the extended addressing mode.

(2) *AS interface slaves.* All the nodes that can be addressed by an AS interface master are defined as AS interface slaves.

(a) *AS interface slave assembly system.* AS interface slaves with the following assembly systems are available:

(i) *AS interface modules.* AS interface modules are AS interface slaves to which up to four conventional sensors and up to four conventional actuators can be connected. The standard coupling module, which is the lower section of a standard device, connects the user module to the yellow AS interface cable. The user module connects the sensors and actuators, while the application modules connect via screw terminals or connectors. Sensors and actuators with a built-in AS interface chip can be directly connected to the AS interface cable.

(ii) *Sensors/actuators with an integrated AS interface connection.* Sensors/actuators with an integrated AS interface connection can be connected directly to the AS interface.

(b) *Addressing mode.* AS interface slaves are available with the following addressing modes:

(i) *Standard slaves.* Standard slaves each occupy one address on the AS interface. Up to 31 standard slaves can be connected to the AS interface.

(ii) *Slaves with an extended addressing mode (A/B slaves).* Slaves with an extended addressing mode can be operated in pairs at the same address with an extended AS interface master. This doubles the number of addressable AS interface slaves to 62.

One of these AS interface slaves must be programmed as an A slave using the addressing unit and the other as a B slave. Due to the address expansion, the number of binary outputs is reduced to three per AS interface slave. Slaves can also be operated with a standard AS interface master. For more detailed information about these functions, refer to the AS interface master discussion in the previous paragraphs.

(c) *Analog slaves.* Analog slaves are special AS interface standard slaves that exchange analog values with the AS interface master. Analog slaves require special program sections in the user program (drivers, function blocks) that execute the sequential transfer of analog data. Analog slaves are intended for operation with extended AS interface masters. The extended AS interface masters handle the exchange of analog data with these slaves automatically. No special drivers or function blocks are required in the user program.

(3) *Further AS interface system components.* The further AS interface components include AS interface cable, AS interface power supply unit, addressing unit, and SCOPE for AS interface.

(a) *AS interface cable.* The trapezoidal AS interface cable is recommended over standard two-wire round cable for quick and simple connection of slaves. The AS interface cable is available in different colors to signify its voltage rating with color assignments as follows:

(i) *Yellow.* The yellow AS interface cable is used for data and control power between the master and its slaves.

(ii) *Black.* It is the external output power cable up to 60 VDC.

(iii) *Red.* It is the external output power cable up to 240 VAC. The AS interface cable, designed as an unshielded two-wire cable, transfers signals and provides the power supply for the sensors and actuators connected using AS interface modules. Networking is not restricted to one type of cable. If necessary, appropriate modules or "T pieces" can be used to change to a simple two-wire cable.

(b) *AS interface power supply unit.* The AS interface power supply unit supplies power to the AS interface nodes connected to the AS interface cable. For actuators with particularly high power requirements, the connection of an additional load power supply may be necessary (e.g., using special application modules). Data and control power are normally transmitted simultaneously via the AS interface cable. Power for the electronics and inputs is supplied by a special AS interface power supply that feeds a symmetrical supply voltage into the AS interface cable via a data-decoupling device.

(c) *Addressing unit.* The addressing unit allows simple programming of AS interface slave addresses.

(d) *SCOPE for AS interface.* SCOPE AS interface is a monitoring program for Windows that can record and evaluate the data exchange in AS interface networks during the commissioning phase and during operation. SCOPE AS interface can be operated on a PC under Windows in conjunction with an AS interface master communications processor.

1.4.1.3 Working principle and mechanism

AS interface utilizes a single, trapezoidal, unshielded two-wire cable, which eliminates the extensive parallel control wiring required with most installations. In a network with AS interface, a simple gateway interfaces the network into the field communication bus. Data and power are transferred over the two-wire network to each of the AS interface compatible field devices. The existing controller sees AS interface as remote I/O; therefore, AS interface connects to the existing network with minimal programming changes. The AS interface system utilizes only one master per network to control the exchange of data. This allows the master to interrogate up to 31 slaves and update all I/O information within 5 ms (10 ms for 62 slaves). For slave connection, an insulated two-wire cable is recommended to prevent reversing polarity. The electrical connection is made using contacts that pierce the insulation of the cable, contacting the two wires, thus eliminating the need to strip the cable and wire to screw terminals. For data exchange to occur, each slave must be programmed with an address that is stored internally in nonvolatile memory and remains even after power is removed.

The tasks and functions of an AS interface master are described below, which is important for understanding the functions, modes, and interfaces available with the AS interface master modules.

1.4.1.3.1 Master–slave principle

The AS interface operates on the master-slave principle. This means that the AS interface master connected to the AS interface cable controls the data exchange with the slaves via the interface to the AS interface cable.

Figure 1.4.4 illustrates the two interfaces of the AS interface master communication processor.

(1) The process data and parameter assignment commands are transferred via the interface between

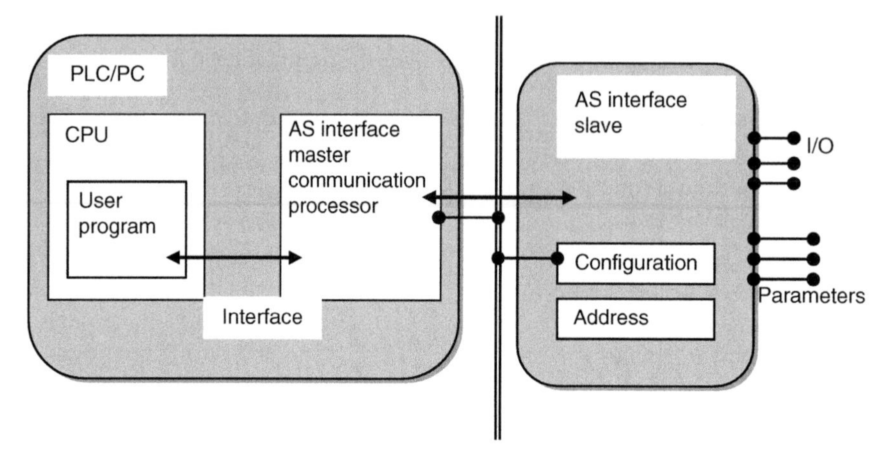

Fig. 1.4.4 AS interface operation.

the master central processing unit (CPU) and the master communication processor. The user programs have suitable function calls and mechanisms available for reading and writing via this interface.

(2) Information is exchanged with the AS interface slaves via the interface between the master communication processor and AS interface cable.

(1) *Tasks and functions of the AS interface master.* The AS interface master specification distinguishes masters with different ranges of functions known as a "profile." For standard AS interface masters and extended AS interface masters, there are three different master classes (M0, M1, M2 for standard masters, and M0e, M1e, M2e for extended masters). The AS interface specification stipulates the functions a master in a particular class must be able to perform.

The profiles have the following practical significance:

(a) *Master profile M0/M0e.* The AS interface master can exchange I/O data with the individual AS interface slaves. The station configuration on the cable, called the "expected configuration," is used to configure the master.

(b) *Master profile M1/M1e.* This profile covers all the functions according to the AS interface master specification.

(c) *Master profile M2/M2e.* The functionality of this profile corresponds to master profile M0/M0e, but in this profile the AS interface master can also assign parameters to the AS interface slaves. The essential difference between extended AS interface masters and standard AS interface masters is that they support the attachment of up to 62 AS interface slaves using the extended addressing mode. Extended AS interface masters also provide particularly simple access for AS interface analog slaves complying with profile specifications.

However, if standard operation (master profile M0) is chosen for use, the following contents can be skipped.

(2) *How an AS interface slave functions*

(a) *Connecting to the AS interface cable.* The AS interface slave has an integrated circuit (AS interface chip) that provides the attachment of an AS interface device (sensor/actuator) to the common bus cable to the AS interface master. The integrated circuit contains these components: four configurable data inputs and outputs; four parameter outputs.

The operating parameters, configuration data with I/O assignment, identification code, and slave address are stored in additional memory (e.g., electrically erasable programmable read-only memory (EEPROM)).

(b) *I/O data.* The useful data for the automation components that were transferred from the AS interface master to the AS interface slave are available at the data outputs. The values at the data inputs are made available to the AS interface master when the AS interface slave is polled.

(c) *Parameter.* Using the parameter outputs of the AS interface slave, the AS interface master can transfer values that are not interpreted as simple data. These parameter values can be used to control and switch over between internal operating modes of the sensors or actuators. It could, for example, be possible to update a calibration value during the various operating phases. This function is possible with slaves with an integrated AS interface connection providing they support the function in question.

(d) *Configuration.* The I/O configuration indicates which data lines of the AS interface slave are used as inputs, outputs, or as bidirectional outputs. The I/O configuration (4 bits) can be found in the description of the AS interface slave. In addition to the I/O configuration, the type of the AS interface slave is described by an identification code; with newer AS interface slaves it is identified by three identification codes (ID code, ID1 code, ID2 code).

For more detailed information on the ID codes, refer to the manufacturer's description.

1.4.1.3.2 Data Transfer

(1) *Information and data structure.* Before introducing the operating phases and the functions during these operating phases, a brief outline of the information structure of the AS interface master–slave system is necessary.

In Figure 1.4.5, the data fields and lists of the system are configured in the system structure diagram as given in Figure 1.4.4.

The following structures are found on the AS interface master:

(a) *Data images.* These contain temporarily stored information:

(i) actual parameters that are an image of the parameters currently on the AS interface slave;

(ii) actual configuration data that contains the I/O configurations and ID codes of all connected AS interface slaves once these data have been read from the AS interface slaves;

(iii) the list of detected AS interface slaves (LDS) that specifies which AS interface slaves were detected on the AS interface bus;

(iv) the list of activated AS interface slaves (LAS) that specify which AS interface slaves were activated by the AS interface master. I/O data are only exchanged with activated AS interface slaves.

(b) *I/O data.* The I/O data are the process input and output data.

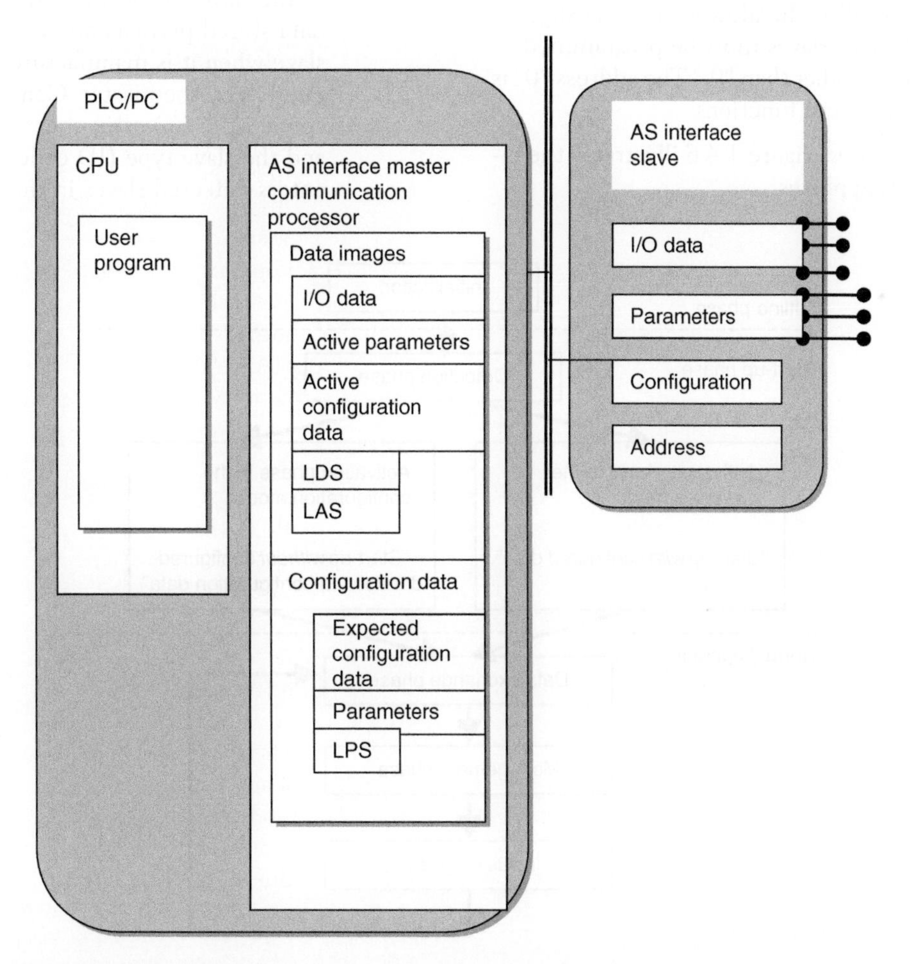

Fig. 1.4.5 Data transfer between the AS interface master and the AS interface slave.

(c) *Configuration data.* These are nonvolatile data (e.g., stored in an EEPROM), which are available unchanged even following a power failure.

(i) Expected configuration data that are selectable comparison values which allow the configuration data of the detected AS interface slaves to be checked.

(ii) List of permanent AS interface slaves (LPS) that specifies the AS interface slaves expected on the AS interface cable by the AS interface master. The AS interface master checks continuously whether all the AS interface slaves specified in the LPS exist and whether their configuration data match the expected configuration data.

The AS interface slave has the following structures:

(a) *I/O data*

(b) *Parameters*

(c) *Actual configuration data.* The configuration data include the I/O configuration and the ID codes of the AS interface slave.

(d) *Address.* The AS interface slaves have address "0" when installed. To allow a data exchange, the AS interface slaves must be programmed with addresses other than "0." The address "0" is reserved for special functions.

(2) *The operating phases.* Figure 1.4.6 illustrates the individual operating phases.

(a) *Initialization mode.* The initialization mode, also known as the offline phase, sets the basic status of the master. The module is initialized after switching on the power supply or following a restart during operation. During the initialization, the images of all the slave inputs and the output data from the point of view of the application are set to the value "0" (inactive).

After switching on the power supply, the configured parameters are copied to the parameter field so that subsequent activation uses the preset parameters. If the AS interface master is reinitialized during operation, the values from the parameters field that may have changed in the meantime are retained.

(b) *Start-up phase.*

(i) *Detection phase.* Detection of AS interface slaves in the start-up phase.

During start-up or after a reset, the AS interface master runs through a start-up phase during which it detects which AS interface slaves are connected to the AS interface cable and what type of slaves these are. The "type" of the slaves is specified by the configuration data stored permanently on the AS interface slave when it is manufactured and can be queried by the master. Configuration files contain the I/O assignment of an AS-I slave and the slave type (ID codes). The master enters detected slaves in the LDS.

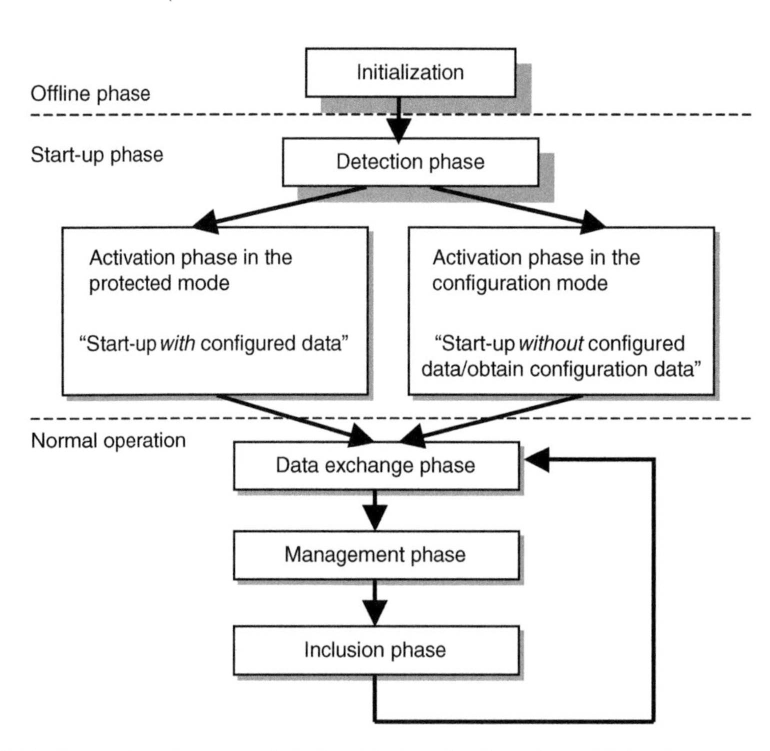

Fig. 1.4.6 How do the individual operating phases work in the data transfer through an AS interface.

(ii) *Activation phase: activating AS interface slaves.* After the AS interface slaves are detected, the master sends a special call which activates them. When activating individual slaves, a distinction is made between two modes on the AS interface master:

Master in the configuration mode: All detected stations (with the exception of the slave with address "0") are activated. In this mode, it is possible to read actual values and to store them for a configuration.

Master in the protected mode: Only the stations corresponding to the expected configuration stored on the AS interface master are activated. If the actual configuration found on the AS interface cable differs from this expected configuration, the AS interface master indicates this.

The master enters activated AS interface slaves in the LAS,

(iii) *Normal mode.* On completion of the start-up phase, the AS interface master switches to the normal mode.

(iv) *Data exchange phase.* In the normal mode, the master sends cyclic data (output data) to the individual AS interface slaves and receives their acknowledgment messages (input data). If an error is detected during the transmission, the master repeats the appropriate poll.

(v) *Management phase.* During this phase, all existing jobs of the control application are processed and sent. Possible jobs are, for example, as follows:

Parameter transfer: Four parameter bits (three parameter bits with AS interface slaves with the extended addressing mode) are transferred to a slave and are used, for example, for a threshold value setting.

Changing slave addresses: This function allows thc addresses of AS interface slaves to be changed by the master if the AS interface slave supports this particular function.

(vi) *Inclusion phase.* In the inclusion phase, newly added AS interface slaves are included in the list of detected AS interface slaves and, providing the configuration mode is selected, they are also activated (with the exception of slaves with address "0"). If the master is in the protected mode, only the slaves stored in the expected configuration of the AS interface master are activated. With this mechanism, slaves that and were temporarily out of service are also included again.

(3) a*Interface functions.* To control the master and slave interaction from the user program, there are various functions available on the interface. The possibilities are explained below. The possible operations and the direction of data flow are illustrated in Figure 1.4.7.

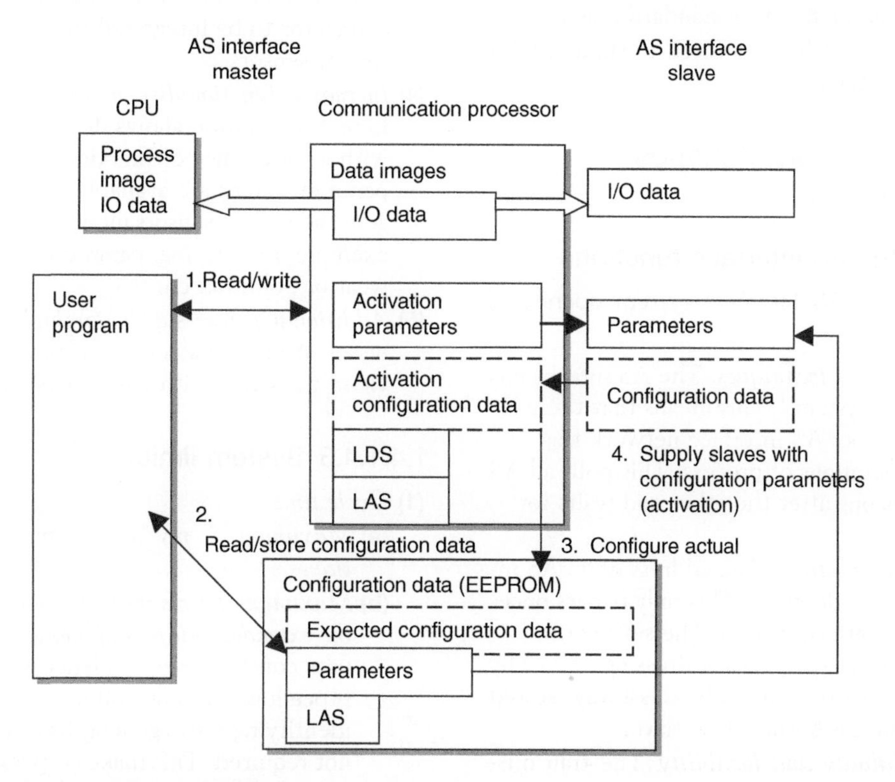

Fig. 1.4.7 How does the AS interface function.

(a) *Read/write.* When writing, parameters are transferred to the slave and the parameter images on the communication processor; when reading, parameters are transferred from the slave or from the communication processor parameter image to the CPU.

(b) *Read and store (configured) configuration data.* Configured parameters or configuration data are read from the nonvolatile memory of the communication processor.

(c) *Configure actual.* When reading, the parameters and configuration data are read from the slave and stored permanently on the communication processor; when writing, the parameters and configuration data are stored permanently on the communication processor.

(d) *Supply slaves with configured parameters.* Configured parameters are transferred from the nonvolatile area of the communication processor to the slaves.

(4) *Operating extended AS interface slaves with standard AS interface masters.* The following information is about operating extended AS interface with standard AS interface masters.

(a) Slaves are connected to standard masters. The most significant slave bit (bit 4) of each A slave must be set to "0." The most significant parameter bit (bit 4) must also be set to "1" (default value). Without these settings, the A slave cannot be operated with a standard master.

(b) B slaves must not be connected to standard AS interface masters.

1.4.1.4 System characteristics and Important Data

1.4.1.4.1 How the AS interface functions

The AS interface or AS interface system operates as outlined below:

(1) *Master–slave access techniques.* The AS interface is a "single master system." This means that there is only one master per AS interface network that controls the operations of process. This polls all AS interface slaves one after the other and waits for a response.

(2) *Electronic address setting.* The address of an AS interface slave is its identifier. This only occurs once within an AS interface system. The setting can either be made using a special addressing unit or by an AS interface master. The address is always stored permanently on the AS interface slave.

(3) *Operating reliability and flexibility.* The transmission technique used (current modulation) guarantees high operating reliability. The master monitors the voltage on the cable and the transferred data. It detects transmission errors and the failure of slaves and sends a message to the controller such as a PLC or PC. The user can then react to this message. Replacing or adding AS interface slaves during normal operation does not affect communication with other AS interface slaves.

1.4.1.4.2 Physical characteristics

The most important physical characteristics of the AS interface and its components are as follows:

(1) *The two-wire cable for data and power supply.* A simple two-wire cable can be used. Shielding or twisting is not necessary. Both the data and the power are transferred on this cable. The power available depends on the AS interface power supply unit used. For optimum wiring, the mechanically coded AS interface cable is available preventing the connections from being reversed and making simple contact with the AS interface application modules using the penetration technique.

(2) *Tree structure network with a cable.* The "tree structure" of the AS interface allows any point on a cable section to be used as the start of a new branch.

(3) *Direct integration.* Practically all the electronics required for a slave have been integrated on a special integrated circuit. This allows the AS interface connector to be integrated directly in binary actuators or sensors.

(4) *Increased functionality, more uses for the customer.* Direct integration allows devices to be equipped with a wide range of functions. Four data and four parameter lines are available. The resulting "intelligent" actuators/sensors increase the possibilities, for example, monitoring, parameter assignment, and wear or pollution checks.

(5) *Additional power supply for higher power requirements.* An external source of power can be provided for slaves with a higher power requirement.

1.4.1.4.3 System limits

(1) *Cycle time*

(a) *Maximum 5 ms with standard AS interface slaves.*

(b) *Maximum 10 ms with AS interface slaves using the extended addressing mode.* AS interface uses constant message lengths. Complicated procedures for controlling transmission and identifying message lengths or data formats are not required. This makes it possible for a master to poll all connected standard slaves within a maximum of 5 ms and to update the data both on the master and slave.

If only one AS interface slave using the extended addressing mode is located at an address, this slave is polled at least every 5 ms. If two extended slaves (A and B slave) share an address, the maximum polling cycle is 10 ms. (B slaves can only be connected to extended masters.)

(2) *Number of connectable AS interface slaves*

(a) *Maximum of 31 standard slaves.*

(b) *Maximum of 62 slaves with the extended addressing mode.* AS interface slaves are the I/O channels of the AS interface system. They are only active when called by the AS interface master. They trigger actions or transmit reactions to the master when commanded. Each AS interface slave is identified by its own address (1–31). A maximum of 62 slaves using the extended addressing mode can be connected to an extended master. Pairs of slaves using the extended addressing mode occupy one address; in other words, the addresses 1–31 can be assigned to two extended slaves.

If standard slaves are connected to an extended master, these occupy a complete address; in other words, a maximum of up to 31 standard slaves can be connected to an extended master.

(3) *Number of inputs and outputs*

(a) *A maximum of 248 binary inputs and outputs with standard modules.*

(b) *A maximum of 248 inputs and 186 outputs with modules using the extended addressing mode.* Each standard AS interface slave can receive 4 bits of data and send 4 bits of data. Special modules allow each of these bits to be used for a binary actuator or a binary sensor. This means that an AS interface cable with standard AS interface slaves can have a maximum of 248 binary attachments (124 inputs and 124 outputs). All typical actuators or sensors can be connected to the AS interface in this way. The modules are used as distributed inputs/outputs.

If modules with the extended addressing mode are used, a maximum of 3 inputs and 3 outputs is available per module; in other words, a maximum of 248 inputs and 186 outputs can be operated with modules using the extended addressing mode.

1.4.1.4.4 Range of functions of the master modules

The functions of the AS interface master modules are stipulated in the AS interface master specification. An overview of these functions can be found in the master module manual provided by the vendor or manufacturer.

The AS interface protocol was created in Germany in 1994 by a consortium of factory automation suppliers. Originally developed to be a low-cost method for addressing discrete sensors in factory automation applications, AS interface has since gained acceptance in process industries due to its high power capability, simplicity of installation and operation, and low-cost adder for devices.

Each AS interface segment can network up to 31 devices. This provides for 124 inputs and 124 outputs, giving a maximum capacity of 248 I/O per network on a v2.0 segment. The AS interface v2.1 specification doubles this to 62 devices per segment, providing 248 inputs and 186 outputs for a total network capacity of 434 I/O points.

Both signal and power are carried on two wires. Up to 8 A at 30 VDC of power are available for field devices such as solenoid valves.

1.4.1.4.5 AS interface in a real-time environment

The system characteristics listed below can offer the AS interface the capability to work in a real-time environment:

(1) Optimized system for binary sensors and actuators and for simple analog elements.

(2) Master–slave principle with cyclic polling.

(3) Tree structure of the network.

(4) Both data and power by means of one unshielded two-wire cable.

(5) Flat cable for contacting by piercing technology.

(6) Modules as remote I/O ports for conventional sensors and actuators.

(7) Integrated slaves with their own AS interface capabilities.

(8) No communication software in the slaves, only firmware in the self-configuring master.

(9) Low costs, simple installation, easy handling, flexible networks, high reliability in an industrial environment, and open and internationally accepted system with many manufacturers and products.

There are three aspects of the AS interface that are of particular importance in real-time applications: connectivity, cycle time, and availability.

(1) *Connectivity.* AS interface has two distinct ways to be connected to the first control level.

The first and most important way is a direct connection as the type 2 of AS interface architecture given in Section 1.4.1.2. In that case, the system's master is part of a controller such as PLC, SCADA, or PC, running at its own cycle time. As the AS interface is an open system, any kind of controller such as PLC, SCADA, or PC manufacturer can build a master for their own system. There are masters available to a lot of systems already, with several more in development.

The second way is to connect AS interface via a coupler to a higher Fieldbus and to use it as a subsystem, which has been given as the type 1 of AS interface architecture in Section 1.4.1.2. In that case, all data from the AS interface network is handled in one node of the Fieldbus and it is connected to the above lying host together with other components of the higher Fieldbus. The application program has to handle all data as usual for the particular Fieldbus. For real-time applications, an analysis of the cycle time and the availability of the combination of the two systems has to be done. AS interface is definitely open to such solutions and offers couplers to most known higher Fieldbus like PROFIBUS, CAN, etc., with others (e.g., LON, Fieldbus Foundation) being in preparation. Together with its tree structure, AS interface thus offers the most flexible networking solution to any application in automation.

(2) *Cycle time.* AS interface is a single-master system with cyclic polling. Thus, any slave is addressed in a definite time.

For a complete net with 31 slaves, the cycle time is 5 ms. It may be shorter with fewer slaves. (With very few slaves the cycle time can be shortened to less than 500 ms.) Analog data with more than 4 bits needs several cycles depending on its length, but without affecting the basic cycle time for binary sensors and actuators.

The cycle time includes all steps from and to the interface to the host system and even includes one repetition. The data exchange with the host happens from here via process I/O images at the end of each cycle stored in, for example, a dual-ported memory at the interface. Therefore no other steps have to be taken into account for a direct connection to the control device.

This is asynchronous coupling, and in real-time applications of the cycle times of both the network and the controller this may present a restriction, but for many systems and applications this is short enough.

(3) *Availability.* Availability in this context means that a system will deliver reliable data and diagnostic values continuously and in time under all specified conditions, especially under severe electromagnetic noise. The answers to three questions are of special importance for real-time applications:

(a) Can electromagnetic noise or other faults disturb the reliability of data?

(b) How much time is necessary for the correction of a faulty transmission?

(c) How often does such a fault happen and can this affect the whole system?

1.4.2 Industrial control system interface devices

In reference to Figure 1.4.1, there exist two kinds of interface between the control level and the actuator/sensor level in industrial control systems, which are AS interface and field level interface. In additional to these two kinds of interface, the interface between controller and either AS interface or field level interface is also of importance in industrial control systems. The interface between controller and either AS interface or field level interface normally resides in the controller's microprocessor unit or chipset to bridge the CPU with exterior environments, which therefore can be defined as controller interface, or simplified as interfaces hereafter. Section 1.4.1 gives a detailed discussion on AS interface. This section concentrates on the field level in Section 1.4.2.1, and interfaces in Section 1.4.2.2.

1.4.2.1 Fieldbus system

In recent years, there have emerged literally hundreds of Fieldbuses developed by different companies and organizations all over the world. The term Fieldbus covers many different industrial control protocols. The following lists some typical Fieldbuses with their applications as shown in Figure 1.4.1.

1.4.2.1.1 Foundation fieldbus

The Foundation Fieldbus can be flexibly used in process automation applications. The specification supports bus-powered field devices as well as allows application in hazardous areas. The Fieldbus Foundation, an independent not-for-profit organization which aims at developing and maintaining an internationally uniform and successful Fieldbus for automation tasks, claimed to establish an international, interoperable Fieldbus standard to replace the expensive, conventional 4–20 mA wiring in the field and enables bidirectional data transmission. The entire communication between the devices and the automation system as well as the process control station takes place over the bus system, and all operating and device data are exclusively transmitted over the Fieldbus. The communication between control station, operating terminals, and field devices simplifies the start-up and parameterization of all components. The communication functions allow diagnostic data, which are provided by up-to-date field devices, to be evaluated. The essential objectives in Fieldbus technology are to reduce installation costs, save time and costs due to simplified planning, as well as improve the operating reliability of the system due to additional performance features. Fieldbus systems are usually implemented in new plants or existing plants that must be extended. To convert an existing plant to Fieldbus technology, the conventional wiring can either be modified into a bus line or be replaced with a shielded bus cable, if required.

(1) *Performance features.* The Foundation Fieldbus provides a broad spectrum of services and functions compared to other Fieldbus systems:
- **(a)** Intrinsic safety for use in hazardous environments.
- **(b)** Bus-powered field devices.
- **(c)** Line or tree topology.
- **(d)** Multimaster capable communication.
- **(e)** Deterministic (predictable) dynamic behavior.
- **(f)** Distributed data transfer (DDT).
- **(g)** Standardized block model for uniform device interfaces
- **(h)** Flexible extension options based on device descriptions.

The characteristic feature of distributed data transfer enables single field devices to execute automation tasks so that they are no longer just sensors or actuators, but contain additional functions. For the description of a device's function(s) and for the definition of a uniform access to the data, the Foundation Fieldbus contains predefined function blocks. The function blocks implemented in a device provide information about the tasks the device can perform. Typical functions provided by sensors include the following: analog input or discrete input (digital input). Control valves usually contain the following function blocks: analog output or discrete output (digital output). The following blocks exist for process control tasks: Proportional and Derivative (PD controller) or Proportional and Integral and Derivative (PID controller). If a device contains such a function block, it can control a process variable independently.

The shift of automation tasks—from the control level down to the field—results in the flexible, distributed processing of control tasks. This reduces the load on the central process control station which can even be replaced entirely in small-scale installations. Therefore, an entire control loop can be implemented as the smallest unit, consisting only of one sensor and one control valve with integrated process controller,

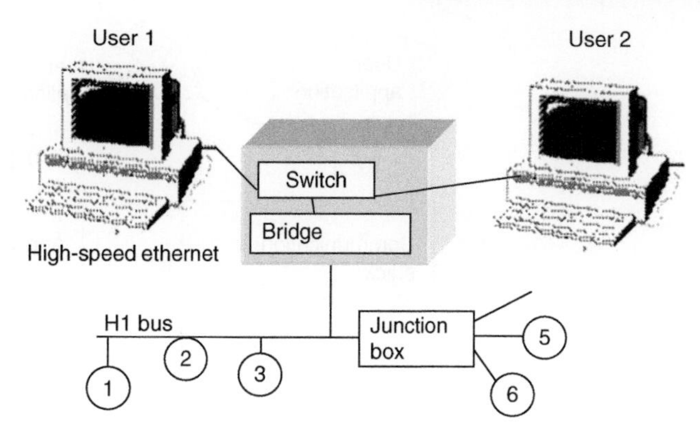

Fig. 1.4.8 Foundation Fieldbus control network.

which communicates over the Foundation Fieldbus (see Figure 1.4.8). The enhanced functionality of the devices leads to higher requirements to be met by the device hardware and comparably complex software implementation and device interfaces.

(2) *Layered communications model.* The Foundation Fieldbus specification is based on the layered communications model and consists of three major functional elements as illustrated in Figure 1.4.9:
- **(a)** Physical layer.
- **(b)** Communication "stack".
- **(c)** User application is made up of function blocks and the device description. It is directly based on the communication stack. Depending on which blocks are implemented in a device, users can access a variety of services. System management utilizes the services and functions of the user application and the application layer to execute its tasks (Figures 1.4.9(b) and (c)). It ensures proper cooperation between the individual bus components as well as synchronizes the measurement and control tasks of all field devices with regard to time.

The Foundation Fieldbus layered communications model is based on the International Standards Organization (ISO)/OSI reference model. As is the case for most Fieldbus systems, and in accordance with an International Electrotechnical Commission (IEC) specification, layers 3–6 are not used. The comparison in Figure 1.4.9 shows that the communication stack covers the tasks of layers 2 and 7 and that Layer 7 consists of the Fieldbus Access Sublayer and the Fieldbus Message Specification.

(3) *Physical layer.* The Foundation Fieldbus model solves pending communication tasks by using two bus systems, the slow, intrinsically safe H1 bus and the fast, higher level H2 bus as given in Figure 1.4.8.

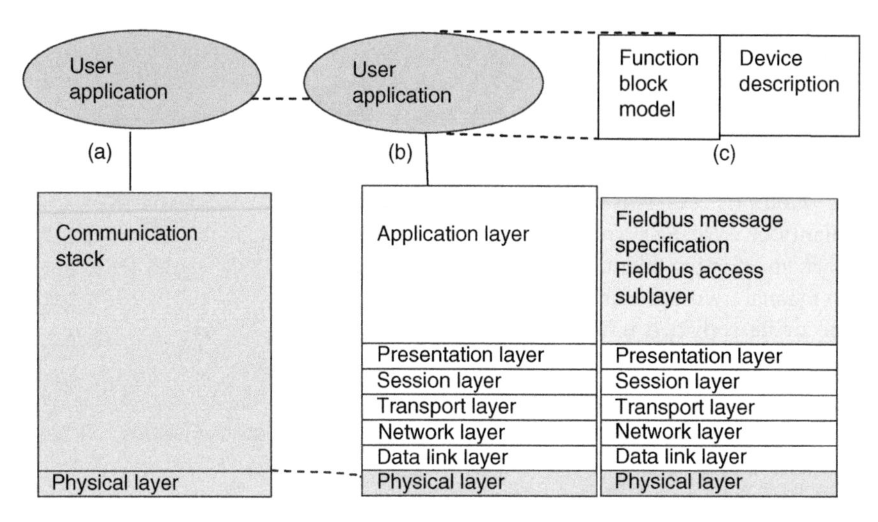

Fig. 1.4.9 Structure and description of the Foundation Fieldbus communication model.

(a) *H1 bus*. The following summary gives a brief overview of the basic values and features of the H1 bus.

(i) Manchester coding is used for data transfer. The data transfer rate is 31.25 kbit/s.

(ii) Proper communication requires that the field devices have enough voltage. Each device should have minimum 9 V. To make sure that this requirement is met, software tools are available which calculate the resulting currents and terminal voltages based on the network topology, the line resistance, and the supply voltage.

(iii) The H1 bus allows the field devices to be powered over the bus. The power supply unit is connected to the bus line in the same way (parallel) as a field device. Field devices powered by supply sources other than the bus must be additionally connected to their own supply sources.

(iv) With the H1 bus it must be ensured that the maximum power consumption of current-consuming devices is lower than the electric power supplied by the power supply unit.

(v) Network topologies used are usually line topology or, when equipped with junction boxes, star, tree, or a combination of topologies. The devices are best connected via short spurs using tee connectors to enable connection/disconnection of the devices without interrupting communication.

(vi) The maximum length of a spur is limited to 120 m and depends on the number of spurs used as well as the number of devices per spur.

(vii) Without repeaters, the maximum length of a H1 segment can be as long as 1900 m. By using up to four repeaters, a maximum of 5×1900 m $= 9500$ m can be achieved. The short spurs from the field device to the bus are included in this total length calculation,

(viii) The number of bus users per bus segment is limited to 32 in intrinsically safe areas. In explosion-hazardous areas, this number is reduced to only a few devices due to power supply limitations.

(ix) Various types of cables are useable for Fieldbus. Type A is recommended as preferred Fieldbus cable, and only this type is specified for the maximum bus length of 1900 m.

(x) Principally, there need to be two terminators per bus segment, one at or near each end of a transmission line.

(xi) It is not imperative that bus cables be shielded, however, it is recommended to prevent possible interferences and for best performance of the system.

The H1 bus can be designed intrinsically safe (Ex-i) to suit applications in hazardous areas. This requires that proper barriers be installed between the safe and the explosion-hazardous area. In addition, only one device, the power supply unit, must supply the Fieldbus with power. All other devices must always, that is, also when transmitting and receiving data, function as current sinks.

Since the capacity of electrical lines is limited in intrinsically safe areas depending on the explosion group—IIB or IIC—the number of devices that can be connected to one segment depends on the effective power consumption of the used devices. Since the

Foundation Fieldbus specification is not based on the Fieldbus intrinsically safe concept (FISCO) model, the plant operator must ensure that intrinsic safety requirements are met when planning and installing the communications network. For instance, the capacitance and inductance of all line segments and devices must be calculated to ensure that the permissible limit values are observed.

(b) *High-speed Ethernet (HSE)*. The HSE is based on standard Ethernet technology. The required components are therefore widely used and are available at low cost. The HSE runs at 100 Mbit/s and cannot be equipped not only with electrical lines, but also with optical fiber cables.

The Ethernet operates by using random (not deterministic) carrier sense multiple access (CSMA) bus access. This method can only be applied to a limited number of automation applications because it requires real-time capability. The extremely high transmission rate enables the bus to respond sufficiently fast when the bus load is low and devices are only few. With respect to process engineering requirements, real-time requirements are met in any case.

If the bus load must be reduced due to the many connected devices, or if several HSE partial networks are to be combined to create a larger network, Ethernet switches must be used (see Figure 1.4.8). A switch reads the target address of the data packets that must be forwarded and then passes the packets on to the associated partial network. This way, the bus load and the resulting bus access time can be controlled to best adapt it to the respective requirements.

(c) *Bridge to H1–HSE coupling*. A communications network that consists of an H1 bus and an HSE network results in a topology as illustrated in Figure 1.4.8. To connect the comparatively slow H1 segments to the HSE network, coupling components, so-called bridges, are required. A bridge is used to connect the individual H1 buses to the fast high-speed Ethernet. The various data transfer rates and data telegrams must be adapted and converted, considering the direction of transmission. This way, powerful and widely branched networks can be installed in larger plants.

(4) *Communication stack*. The field devices used with the Foundation Fieldbus are capable of assuming process control functions. This option is based on distributed communication which ensures that each controlling field device can exchange data with other devices (e.g., reading measuring values, forwarding correction values), all field devices are served in time ("in time" meaning that the processing of the different control loops is not negatively influenced), and two or more devices never access the bus simultaneously.

To meet these requirements, the H1 bus of the Foundation Fieldbus uses a central communication control system.

(a) *Link active scheduler (LAS)*. The LAS controls and schedules the communication on the bus. It controls the bus activities using different commands which it broadcasts to the devices. Since the LAS also continuously polls unassigned device addresses, it is possible to connect devices during operation and to integrate them in the bus communication. Devices that are capable of becoming the LAS are called Link Masters. Basic devices do not have the capability to become LAS. In a redundant system containing multiple Link Masters, one of the Link Masters will become the LAS if the active LAS fails (fail-operational design).

(b) *Communication control*. The communication services of the Foundation Fieldbus (FF) specification utilize scheduled and unscheduled data transmission. Time-critical tasks, such as the control of process variables, are exclusively performed by scheduled services, whereas parameterization and diagnostic functions are carried out using unscheduled communication services.

(i) *Scheduled data transmission*. To solve communication tasks in time and without access conflicts, all time-critical tasks are based on a strict transmission schedule. This schedule is created by the system operator during the configuration of the Foundation Field system. The LAS periodically broadcasts a synchronization signal (TD: Time Distribution) on the Fieldbus so that all devices have exactly the same data link time. In scheduled transmission, the point of time and the sequence are exactly defined. This is why it is called a deterministic system.

(ii) *Unscheduled transmission*. Device parameters and diagnostic data must be transmitted when needed, that is, on request. The transmission of this data is not time critical. For such communication tasks, the Foundation Fieldbus is equipped with the option of unscheduled data transmission.

Unscheduled data transmission is exclusively restricted to the breaks in between scheduled transmission. The LAS grants permission to a device to use the Fieldbus for unscheduled communication tasks if no scheduled data transmission is active.

Permission for a certain device to use the bus is granted by the LAS when it issues a pass token (PT command) to the device. The pass token is sent around to all devices entered in the Live List which is administrated by the LAS. Each device may use the bus as long as required either until it returns the token or until the maximum granted time to use the token has elapsed.

The Live List is continuously updated by the LAS. The LAS sends a special command, the Probe Node (PN), to the addresses not in the Live List, searching for newly added devices. If a device returns a Probe Response (PR) message, the LAS adds the device to the Live List where it receives the pass token for unscheduled communication according to the order submitted for transmission in the Live List. Devices which do not respond to the PT command or return the token after three successive tries are removed from the Live List. Whenever a device is added or removed from the Live List, the LAS broadcasts these changes to all devices. This allows all Link Masters to maintain a current copy of the Live List so that they can become the LAS without the loss of information.

(c) *Communication schedule.* The LAS follows a strict schedule to ensure that unscheduled communication using the token as well as the TD or PN commands do not interfere with the scheduled data transmission.

Before each operation, the LAS refers to the transmission list to check for any scheduled data transmissions. If this is the case, it waits (idle mode) for precisely the scheduled time and then sends a Compel Data (CD) message to activate the operation.

In case there are no scheduled transmissions and sufficient time is available for additional operations, the LAS sends one of the other commands. With PN it searches for new devices, or it broadcasts a TD message for all devices to have exactly the same data link time, or it uses the PT message to pass the token for unscheduled communication. Following this, the sequence starts all over again with the above-mentioned check of the transmission list entries.

It is obvious that this cycle gives scheduled transmission the highest priority and that the scheduled times are strictly observed, regardless of other operations.

(5) *User application layer.* The Fieldbus Access Sublayer (FAS) and Fieldbus Message Specification (FMS) layer form the interface between the data link layer and the user application (see Figure 1.4.9). The services provided by FAS and FMS are invisible for the user. However, the performance and functionality of the communication system considerably depends on these services.

(a) *Fieldbus access sublayer (FAS).* FAS services create Virtual Communication Relationships (VCR) which are used by the higher level FMS layer to execute its tasks (Figure 1.4.10). VCRs describe different types of communication processes and enable the associated activities to be processed more quickly. Foundation Fieldbus communication utilizes three different VCR types as follows:

(i) The Publisher/Subscriber VCR type is used to transmit the I/O data of function blocks. As described above, scheduled data transmission with the CD command is based on this type of VCR. However, the Publisher/Subscriber VCR is also available for unscheduled data transmission; for instance, if a subscriber requests measuring or positioning data from a device.

Client/Server	Report distribution	Publisher/subscriber
Operator communication	Event notification, alarms, trend reports	Data publication
Set point changes Mode and device data changes Upload/download Adjusting alarm values Access display views Remote diagnostics	Send process alarms to operator consoles Send trend reports to data historians	Send actual value of a transmitter to PID block and operator console

Fig. 1.4.10 Virtual communication relationships of the FAS.

(ii) The Client/Server VCR type is used for unscheduled, user-initiated communication based on the PT command. If a device (client) requests data from another device, the requested device (server) only responds when it receives a PT from the LAS. The Client/Server communication is the basis for operator-initiated requests, such as set point changes, tuning parameter access and change, diagnosis, and device upload and download.

(iii) Report distribution communication is used to send alarm or other event notifications to the operator consoles or similar devices. Data transmission is unscheduled when the device receives the PT command together with the report (trend or event notification). Fieldbus devices that are configured to receive the data await and read this data.

(b) *FMS.* FMS provides the services for standardized communication. Data types that are communicated over the Fieldbus are assigned to certain communication services. For a uniform and clear assignment, object descriptions are used.

Object descriptions not only contain definitions of all standard transmission message formats, but also include application-specific data. For each type of object there are special, predefined communication services.

Object descriptions are collected together in a structure called an object dictionary. The object description is identified by its index. (1) Index 0, called the object dictionary header, provides a description of the dictionary itself. (2) Indices between 1 and 255 define standard data types that are used to build more complex object descriptions. (3) The User Application object descriptions can start at any index above 255. The FMS defines Virtual Field Devices (VFD) which are used to make the object descriptions of a field device as well as the associated device data available over the entire network. The VFDs and the object description can be used to remotely access all local field device data from any location by using the associated communication services.

1.4.2.1.2 PROFIBUS

PROFIBUS is the largest Fieldbus in the world with cost-saving solutions in factory automation and process automation plus safety, drives, and motion control coverage. This Fieldbus approach produces significant cost savings in design, installation, and maintenance expenses over the old approach of point-to-point wiring. For many years, PROFIBUS development has continued with undiminished enthusiasm and energy. With more than 13,000,000 nodes installed there are significantly more PROFIBUS nodes installed than of any other Fieldbus. Current PROFIBUS activities are targeted at system integration, PROFIBUS engineering development, and application profiles. Because of these application profiles, PROFIBUS today is the only Fieldbus that provides robust engineering solutions for both factory and process automation.

(1) *Working mechanism.* PROFIBUS is suitable for both fast, time-critical applications and complex communication tasks. PROFIBUS communication is in the international standards IEC 61158 and IEC 61784. The application and engineering aspects are specified in the generally available guidelines of the PROFIBUS User Organization. This fulfills user demand for manufacturer independence and openness and ensures communication between devices of various manufacturers.

PROFIBUS can handle large amounts of data at high speed and can serve the needs of large installations. Based on a real time capable asynchronous token bus principle, PROFIBUS defines multi-master and master–slave communication relations, with cyclic or acyclic access, allowing transfer rates of up to 500 kbit/s. The physical layer (two-wire RS485), the data link layer, and the application layer are all standardized. PROFIBUS distinguishes between confirmed and unconfirmed services, allowing process communication with both broadcast and multitasking protocols. PROFIBUS DP is a master–slave polling network with the ability to upload/download configuration data and precisely synchronized multiple devices on the network. Multiple masters are possible in PROFIBUS, but the outputs of any device can only be assigned to one master. There is no power on the bus.

(2) *Basic types.* PROFIBUS encompasses several Industrial Bus Protocol Specifications, including PROFIBUS-DP, PROFIBUS-PA, PROFIBUS-FMS, PROFInet, PROFIBUS-safe, and PROFIBUS for motion control.

(a) *PROFIBUS-DP.* PROFIBUS-DP is the main emphasis for factory automation; it uses RS485 transmission technology, one of the DP communications protocol versions, and has widespread usage for such items as remote I/O systems, motor control centers, and variable speed drives. PROFIBUS-DP communicates at speeds from 9.6 kbps to 12 Mbps over distances from 100 to 1200 m. PROFIBUS-DP does not natively support intrinsically safe installations. More than 2500 PROFIBUS-compliant products are available from which you can select best-in-class devices to suit your individual needs, with alternative sources usually available.

(b) *PROFIBUS-PA.* PROFIBUS-PA is the main emphasis for process automation, typically with MBP-IS transmission technology, the communications protocol version DP-V1, and the application profile PA devices. PROFIBUS-PA is a full-function Fieldbus that is generally used for process level instrumentation. PROFIBUS-PA communicates at 31.25 kbps and has a maximum distance of 1900 m per segment. PROFIBUS-PA is designed to support intrinsically safe applications. PROFIBUS is also tailored to process automation requirements. It is of modular design and comprises the communication protocol PROFIBUS-DP, different transmission technologies, numerous application profiles, and structured device integration tools. Typical PROFIBUS-PA applications are formed by combining modules suited for or required by the respective applications.

(c) *PROFIBUS-FMS.* PROFIBUS-FMS is designed for communication at the cell level according to Fieldbus message specification. At this level programmable controllers (e.g., PLC and PC) communicate primarily with each other. In this application area a high degree of functionality is more important than fast system reaction times. FMS services are a subset of the services (MMS = Manufacturing Message Specification, ISO 9506) which have been optimized for Fieldbus applications and to which functions for communication object administration and network management have been added. Execution of the FMS services via the bus is described by service sequences consisting of several interactions which are called service primitives. Service primitives describe the interaction between requester and responder.

(d) *PROFInet.* PROFInet is the leading Industrial Ethernet standard for automation that includes plant-wide Fieldbus communication and plant-to-office communication. PROFInet is designed to work from I/O to MES and hence can simultaneously handle standard Ethernet transmissions and real-time transmissions at 1 ms speeds. PROFInet embraces industry standards like TCP/IP, XML, OPC, and ActiveX. Because of the integrated proxy technology it connects other Fieldbuses in addition to PROFIBUS, thus protecting the existing investment in plant equipment and networks (whether that is PROFIBUS or another Fieldbus).

(e) *Motion control with PROFIBUS.* Motion control with PROFIBUS is the main emphasis for drive technology using RS485 transmission technology, the communications protocol version DP V2, and the application profile PROFI drive. The demands of motion control propelled the implementation of functionalities such as clock cycle synchronization or slave-to-slave communication. Decentralized drive applications can be realized economically by means of intelligent drives, since PROFIBUS now also permits the highly dynamic distribution of the technological signals among the drives.

(f) *PROFI-safe.* PROFI-safe is the main emphasis for safety-relevant applications (universal use for almost all industries), using RS485 or MBP-IS transmission technology, one of the available DP versions for communication, and the application profile PROFI-safe. PROFIBUS is the very first Fieldbus in merging standard automation and safety automation in one technology, running on the same bus, using the same communication mechanisms, and thus providing highest efficiency to the user. This supports simple and cost-effective installation and operation.

1.4.2.1.3 Controller area network (CAN bus)

CAN is a serial bus system, which was originally developed for automotive applications in the early 1980s. The CAN protocol was internationally standardized in 1993 as ISO 11898-1 and comprises the data link layer of the seven layer ISO/OSI reference model. CAN bus system can theoretically link up to 2032 devices (assuming one node with one identifier) on a single network. However, due to the practical limitation of the hardware (transceivers), it can only link up to 110 nodes (with 82C250, Philips) on a single network. It offers high-speed communication rate up to 1 Mbit/s thus allowing real-time control. In addition, the error confinement and the error detection feature make it more reliable in noise critical environment.

CAN bus systems provide the following: (1) A multimaster hierarchy, which allows building intelligent and redundant systems. If one network node is defective the network is still able to operate. (2) Broadcast communication. A sender of information transmits to all devices on the bus. All receiving devices read the message and then decide if it is relevant to them. This guarantees data integrity as all devices in the system use the same information. (3) Sophisticated error-detecting mechanisms and retransmission of faulty messages. This also guarantees data integrity.

The CAN serial bus system is used in a broad range of embedded as well as automation control systems. It usually links two or more microcontroller-based physical devices. The original equipment manufacturers (OEM) design embedded control systems; the end user has no or only some knowledge of the embedded network functions and is therefore not responsible for the CAN communication system. However, automation control systems are specified by the end user. The system design including the CAN network services may be implemented by the end users themselves or by a system house.

The main CAN application fields include (1) passenger cars, (2) trucks and buses, (3) off-highway and off-road vehicles, (4) maritime electronics, (5) aircraft and aerospace electronics, (6) factory automation, (7) industrial machine control, (8) lifts and escalators, (9) building automation, (10) medical equipment and devices, (11) nonindustrial control, and (12) non-industrial equipment.

(1) *CAN basic working mechanism.*

(a) *Principles of data exchange.* When data are transmitted by CAN, no stations are addressed, but instead, the content of the message (e.g., rpm or engine temperature) is designated by an identifier that is unique throughout the network. The identifier defines not only the content but also the priority of the message. This is important for bus allocation when several stations are competing for bus access.

If the CPU of a given station wishes to send a message to one or more stations, it passes the data to be transmitted and their identifiers to the assigned CAN chip ("Make ready"). This is all the CPU has to do to initiate data exchange. The message is constructed and transmitted by the CAN chip. As soon as the CAN chip receives the bus allocation ("Send Message") all other stations on the CAN network become receivers of this message ("Receive Message"). Each station in the CAN network, having received the message correctly, performs an acceptance test to determine whether the data received are relevant for that station ("Select"). If the data are of significance for the station concerned they are processed ("Accept"), otherwise they are ignored. Figure 1.4.11 illustrates this scenario.

A high degree of system and configuration flexibility is achieved as a result of the content-oriented addressing scheme. It is very easy to add stations to the existing CAN network without making any hardware or software modifications to the existing stations, provided the new stations are purely receivers. Because the data transmission protocol does not require physical destination addresses for the individual components, it supports the concept of modular electronics and also permits multiple reception (broadcast, multicast) and the synchronization of distributed processes: measurements needed as information by several controllers can be transmitted via the network, in such a way that it is unnecessary for each controller to have its own sensor.

(b) *Nondestructive bitwise arbitration.* For the data to be processed in real time, they must be transmitted rapidly. This not only requires a physical data transfer path with up to 1 Mbit/s but also calls for rapid bus allocation when several stations wish to send messages simultaneously (Figure 1.4.12).

In real-time processing the urgency of messages to be exchanged over the network can

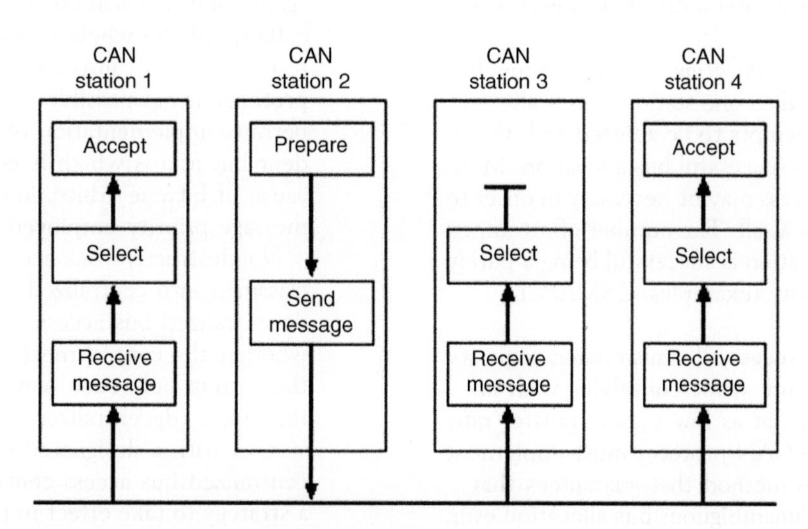

Fig. 1.4.11 Broadcast transmission and acceptance filtering by CAN nodes.

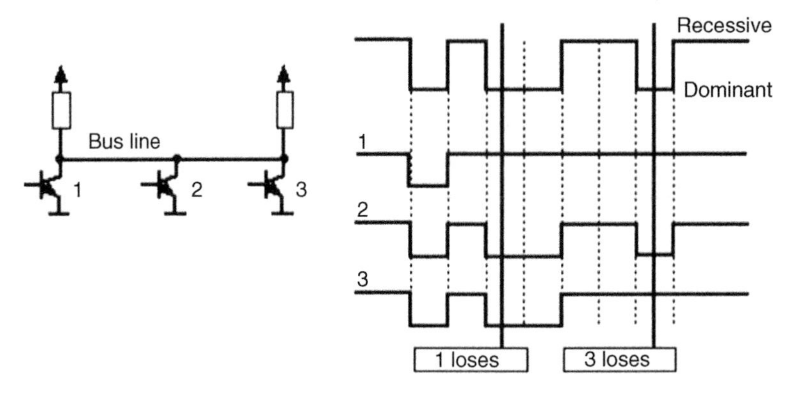

Fig. 1.4.12 Principle of nondestructive bitwise arbitration.

differ greatly: a rapidly changing dimension (e.g., engine load) has to be transmitted more frequently and therefore with fewer delays than other dimensions (e.g., engine temperature) which change relatively slowly.

The priority at which a message is transmitted compared with another less urgent message is specified by the identifier of the message concerned. The priorities are laid down during system design in the form of corresponding binary values and cannot be changed dynamically. The identifier with the lowest binary number has the highest priority. Bus access conflicts are resolved by bitwise arbitration on the identifiers involved by each station observing the bus level bit for bit. In accordance with the "wired" mechanism, by which the dominant state (logical 0) overwrites the recessive state (logical 1), the competition for bus allocation is lost by all those stations with recessive transmission and dominant observation. All "losers" automatically become receivers of the message with the highest priority and do not reattempt transmission until the bus is available again.

(c) *Destructive bus allocation.* Simultaneous bus access by more than one station causes all transmission attempts to be aborted and, therefore, there is no successful bus allocation. More than one bus access may be necessary in order to allocate the bus at all. The number of attempts before bus allocation is successful being a purely statistical quantity (examples: CSMA/CD, Ethernet).

In order to process all transmission requests of a CAN network while complying with latency constraints at as low a data transfer rate as possible, the CAN protocol must implement a bus allocation method that guarantees that there is always unambiguous bus allocation even when there are simultaneous bus accesses from different stations. The method of bitwise arbitration using the identifier of the messages to be transmitted uniquely resolves any collision between a number of stations wanting to transmit, and it does this at the latest within 13 (standard format) or 33 (extended format) bit periods for any bus access period. Unlike the message-wise arbitration employed by the CSMA/CD method, this nondestructive method of conflict resolution ensures that no bus capacity is used without transmitting useful information.

Even in situations where the bus is overloaded, the linkage of the bus access priority to the content of the message proves to be a beneficial system attribute compared with existing CSMA/CD or token protocols: in spite of the insufficient bus transport capacity, all outstanding transmission requests are processed in order of their importance to the overall system (as determined by the message priority). The available transmission capacity is utilized efficiently for the transmission of useful data since "gaps" in bus allocation are kept very small. The collapse of the whole transmission system due to overload, as can occur with the CSMA/CD protocol, is not possible with CAN. Thus, CAN permits implementation of fast, traffic-dependent bus access which is nondestructive because of bitwise arbitration based on the message priority employed.

Nondestructive bus access can be further classified into centralized bus access control or decentralized bus access control depending on whether the control mechanisms are present in the system only once (centralized) or more than once (decentralized). A communication system with a designated station (interalia for centralized bus access control) must provide a strategy to take effect in the event of a failure of the master station. This concept has the disadvantage that the strategy for failure management is difficult and costly to implement and also that the takeover of the central station by a redundant station can be very time consuming. For these reasons and to circumvent the problem of the reliability of the master station (and thus of the whole communication system), the CAN protocol implements decentralized bus control. All major communication mechanisms, including bus access control, are implemented several times in the system because this is the only way to fulfill the high requirements for the availability of the communication system.

In summary it can be said that CAN implements a traffic-dependent bus allocation system that permits, by means of a nondestructive bus access with decentralized bus access control, a high useful data rate at the lowest possible bus data rate in terms of the bus busy rate for all stations. The efficiency of the bus arbitration procedure is increased by the fact that the bus is utilized only by those stations with pending transmission requests. These requests are handled in the order of the importance of the messages for the system as a whole. This proves especially advantageous in overload situations. Since bus access is prioritized on the basis of the messages, it is possible to guarantee low individual latency times in real-time systems.

(d) *Message frame formats.* The CAN protocol supports two message frame formats, the only essential difference being in the length of the identifier (ID). In the standard format the length of the ID is 11 bits, and in the extended format the length is 29 bits. The message frame for transmitting messages on the bus comprises seven main fields (Figure 1.4.13).

A message in the standard format begins with the start bit "start of frame," followed by the "arbitration field," which contains the ID and the remote transmission request (RTR) bit, which indicates whether it is a data frame or a request frame without any data bytes (remote frame). The "control field" contains the IDE (identifier extension) bit, which indicates either standard format or extended format, a bit reserved for future extensions and—in the last 4 bits—a count of the data bytes in the data field. The "data field" ranges from 0 to 8 bytes in length and is followed by the "CRC field," which is used as a frame security check for detecting bit errors. The "ACK field" comprises the ACK slot (1 bit) and the ACK delimiter (1 recessive bit). The bit in the ACK slot is sent as a recessive bit and is overwritten as a dominant bit by those receivers which have at this time received the data correctly (positive acknowledgment). Correct messages are acknowledged by the receivers regardless of the result of the acceptance test. The end of the message is indicated by "end of frame." "Intermission" is the minimum number of bit periods separating consecutive messages. If there is no following bus access by any station, the bus remains idle ("bus idle")

(e) *Detecting and signaling errors.* Unlike other bus systems, the CAN protocol does not use acknowledgment messages but instead signals any errors that occur. For error detection, the CAN protocol implements three mechanisms at the message level:

(i) *Cyclic redundancy check (CRC).* The CRC safeguards the information in the frame by adding redundant check bits at the transmission end. At the receiver end these bits are recomputed and tested against the received bits. If they do not agree, there has been a CRC error.

(ii) *Frame check.* This mechanism verifies the structure of the transmitted frame by checking the bit fields against the fixed format and the frame size. Errors detected by frame checks are designated "format errors."

(iii) *ACK errors.* As mentioned above, frames received are acknowledged by all recipients through positive acknowledgment. If no acknowledgment is received by the transmitter of the message (ACK error) this may mean that there is a transmission error which has been detected only by the recipients, that the ACK field has been corrupted, or that there are no receivers.

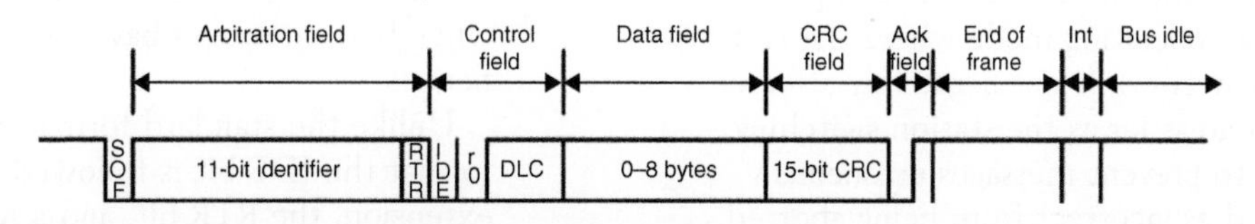

Fig. 1.4.13 Message frame for standard format (CAN Specification 2.0A).

The CAN protocol also implements two mechanisms for error detection at the bit level:

(i) *Monitoring.* The ability of the transmitter to detect errors is based on the monitoring of bus signals: each node which transmits also observes the bus level and thus detects differences between the bit sent and the bit received. This permits reliable detection of all global errors and errors local to the transmitter.

(ii) *Bit stuffing.* The coding of the individual bits is tested at bit level. The bit representation used by CAN is NRZ (nonreturn-to-zero) coding, which guarantees maximum efficiency in bit coding. The synchronization edges are generated by means of bit stuffing, that is, after five consecutive equal bits the sender inserts into the bit stream a stuff bit with the complementary value, which is removed by the receivers. The code check is limited to checking adherence to the stuffing rule.

If one or more errors are discovered by at least one station (any station) using the above mechanisms, the current transmission is aborted by sending an "error flag." This prevents other stations from accepting the message and thus ensures the consistency of data throughout the network.

After transmission of an erroneous message has been aborted, the sender automatically reattempts transmission (automatic repeat request). There may again be competition for bus allocation. As a rule, retransmission will be begun within 23 bit periods after error detection; in special cases the system recovery time is 31 bit periods.

However effective and efficient the method described may be, in the event of a defective station it might lead to all messages (including correct ones) being aborted, thus blocking the bus system if no measures for self-monitoring were taken. The CAN protocol, therefore, provides a mechanism for distinguishing sporadic errors from permanent errors and localizing station failures (fault confinement).

This is done by statistical assessment of station error situations with the aim of recognizing a station's own defects and possibly entering an operating mode where the rest of the CAN network is not negatively affected. This may go as far as the station switching itself off to prevent messages erroneously recognized as incorrect from being aborted.

(f) *Extended format CAN messages.* The SAE "Truck and Bus" subcommittee standardized signals and messages as well as data transmission protocols for various data rates. It became apparent that standardization of this kind is easier to implement when a longer identification field is available.

To support these efforts, the CAN protocol was extended by the introduction of a 29-bit identifier. This identifier is made up of the existing 11-bit identifier (base ID) and an 18-bit extension (ID extension). Thus, the CAN protocol allows the use of two message formats: Standard CAN (Version 2.0A) and Extended CAN (Version 2.0B). As the two formats have to coexist on one bus, it is laid down which message has higher priority on the bus in the case of bus access collisions with differing formats and the same base identifier: the message in standard always has priority over the message in extended format.

CAN controllers that support the messages in extended format can also send and receive messages in standard format. When CAN controllers that only cover the standard format (Version 2.0A) are used in one network, then only messages in standard format can be transmitted on the entire network. Messages in extended format would be misunderstood. However, there are CAN controllers that support only standard format but recognize messages in extended format and ignore them (Version 2.0B passive).

The distinction between standard format and extended format is made using the IDE bit (Identifier Extension Bit) which is transmitted as dominant in the case of a frame in standard format. For frames in extended format it is recessive.

The RTR bit is transmitted dominant or recessive depending on whether data are being transmitted or whether a specific message is being requested from a station. In place of the RTR bit in standard format the substitute remote request (SRR) bit is transmitted for frames with extended ID. The SRR bit is always transmitted as recessive, to ensure that in the case of arbitration the standard frame always has priority bus allocation over an extended frame when both messages have the same base identifier.

Unlike the standard format, in the extended format the IDE bit is followed by the 18-bit ID extension, the RTR bit, and a reserved bit (r1).

All the following fields are identical with standard format. Conformity between the two formats is ensured by the fact that the CAN controllers which support the extended format can also communicate in standard format.

(g) *Implementations of the CAN protocol.* Communication is identical for all implementations of the CAN protocol. There are differences, however, with regard to the extent to which the implementation takes over message transmission from the microcontrollers which follow it in the circuit.

CAN controllers with intermediate buffer (formerly called basicCAN chips) have implemented as hardware the logic necessary to create and verify the bit stream according to protocol. However, the administration of data sets to be sent and received, acceptance filtering in particular, is carried out to only a limited extent by the CAN controller.

Typically, CAN controllers with intermediate buffer have two reception and one transmission buffer. The 8-bit code and mask registers allow a limited acceptance filtering (8 MSB of the identifier). Suitable choice of these register values allows groups of identifiers or in borderline cases all IDs to be selected. If more than the 8 ID-MSBs are necessary to differentiate between messages, then the microcontroller following the CAN controller in the circuit must complement acceptance filtering by software. CAN controllers with intermediate buffer may place a strain on the microcontroller with the acceptance filtering, but they require only a small chip area and can therefore be produced at lower cost. In principle they can accept all objects in a CAN network.

CAN objects consist mainly of three components: identifier, data length code, and the actual useful data. CAN controllers with object storage (formerly called FullCAN) function like CAN controllers with intermediate buffers, but also administer certain objects. Where there are several simultaneous requests they determine, for example, which object is to be transmitted first. They also carry out acceptance filtering for incoming objects. The interface to the following microcontroller corresponds to a RAM. Data to be transmitted are written into the appropriate RAM area, and data received are read out correspondingly. The microcontroller has to administer only a few bits (e.g., transmission request).

CAN controllers with object storage are designed to take as much strain as possible off the local microcontroller. These CAN controllers require a greater chip area, however, and are therefore more expensive. In addition to this, they can only administer a limited number of chips.

CAN controllers are now available which combine both principles of implementation. They have object storage, at least one of which is designed as an intermediate buffer. For this reason there is no longer any point in differentiating between basicCAN and fullCAN.

As well as CAN controllers which support all functions of the CAN protocol, there are also CAN chips which do not require a following microcontroller. These CAN chips are called serial link I/O (SLIO). CAN chips are CAN slaves and have to be administered by a CAN master.

(2) *CAN physical layer.*

(a) *Physical CAN connection.* Data rates (up to 1 Mbit/s) necessitate a sufficiently steep pulse slope, which can be implemented only by using power elements. A number of physical connections are basically possible. However, the users and manufacturers group, CAN in Automation, recommends the use of driver circuits in accordance with ISO 11898. Integrated driver chips in accordance with ISO 11898 are available from several companies (Bosch, Philips, Siliconix, and Texas Instruments). The international users and manufacturers group, CAN in Automation (CiA), also specifies several mechanical connections (cable and connectors) (Figure 1.4.14).

(b) *Physical media.* The basis for transmitting CAN messages and for competing for bus access is the ability to represent a dominant and a recessive bit value. This is possible for electrical and optical media so far. For electrical media the differential output bus voltages are defined in ISO 11898-2 and ISO 11898-3, in SAE J2411, and ISO 11992. With optical media the recessive level is represented by "dark" and the dominant level by "light."

The physical medium most commonly used to implement CAN networks is a differentially driven pair of wires with common return. For vehicle body electronics, single wire bus lines are also used. Some efforts have been made to develop a solution for the transmission of CAN signals on the same line as the power supply.

The parameters of the electrical medium become important when the bus length is increased. Signal propagation, the line resistance, and wire cross-sections are factors when

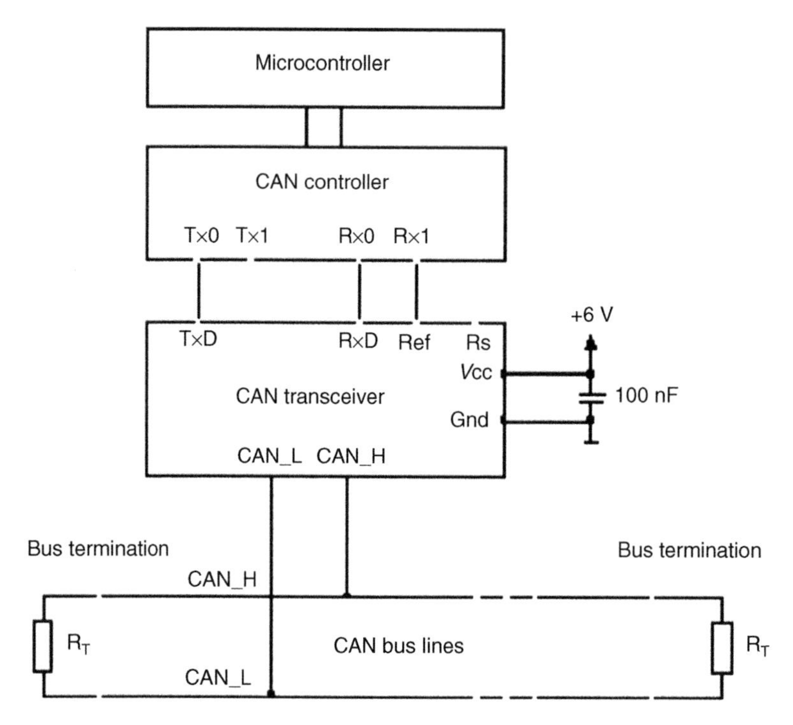

Fig. 1.4.14 Physical CAN connection according to ISO 11898.

dimensioning a network. In order to achieve the highest possible bit rate at a given length, a high signal speed is required. For long bus lines the voltage drops over the length of the bus line. The wire cross-section necessary is calculated by the permissible voltage drop of the signal level between the two nodes farthest apart in the system and the overall input resistance of all connected receivers. The permissible voltage drop must be such that the signal level can be reliably interpreted at any receiving node.

(c) *Network topology.* Electrical signals on the bus are reflected at the ends of the electrical line unless measures against that have been taken. For the node to read the bus level correctly, it is important that signal reflections are avoided. This is done by terminating the bus line with a termination resistor at both ends of the bus and by avoiding unnecessarily long stub lines of the bus. The highest possible product of transmission rate and bus length line is achieved by keeping as close as possible to a single line structure and by terminating both ends of the line. Specific recommendations for this can be found in the applicable standards (i.e., ISO 11898-2 and ISO 11898-3).

It is possible to overcome the limitations of the basic line topology by using repeaters, bridges, or gateways. A repeater transfers an electrical signal from one physical bus segment to another segment. The signal is only refreshed and the repeater can be regarded as a passive component comparable to a cable. The repeater divides a bus into two physically independent segments. This causes an additional signal propagation time. However, it is logically just one bus system.

A bridge connects two logically separated networks on the data link layer (OSI Layer 2). This is so that the CAN identifiers are unique in each of the two bus systems. Bridges implement a storage function and can forward messages or parts thereof in an independent time-delayed transmission. Bridges differ from repeaters since they forward messages, which are not local, whereas repeaters forward all electrical signals including the CAN identifier. A gateway provides the connection of networks with different higher layer protocols. It therefore performs the translation of protocol data between two communication systems. This translation takes place on the application layer (OSI Layer 7).

(d) *Bus access.* For the connection between a CAN controller chip and a two-wire differential bus, a variety of CAN transceiver chips according to different physical layer standards are available (ISO 11898-2 and -3, etc.).

This interface basically consists of a transmitting amplifier and a receiving amplifier transceiver = transmit and receive). Aside from the adaptation of the signal representation between chip and bus medium, the transceiver has to meet a series of additional requirements. As a transmitter it provides sufficient driver output capacity and protects the on-controller-chip driver against overloading. It also reduces electromagnetic radiation. As a receiver the CAN transceiver provides a defined recessive signal level and protects the on-controller-chip input comparator against overvoltages on the bus lines. It also extends the common mode range of the input comparator in the CAN controller and provides sufficient input sensitivity. Furthermore, it detects bus errors such as line breakage, short circuits, shorts to ground, etc. A further function of the transceiver can also be the galvanic isolation of a CAN node and the bus line.

(e) *Physical CAN protocols.* The CAN protocol defines the data link layer and part of the physical layer in the OSI model, which consists of seven layers. The International Standards Organization (ISO) defined a standard, which incorporates the CAN specifications as well as a part of physical layer: the physical signaling, which comprises bit encoding and decoding (Non-Return-to-Zero (NRZ)) as well as bit timing and synchronization.

(i) *Bit encoding.* In the chosen NRZ bit coding the signal level remains constant over the bit time and thus just one time slot is required for the representation of a bit (other methods of bit encoding are, e.g., Manchester or pulse-width modulation). The signal level can remain constant over a longer period of time; therefore, measures must be taken to ensure that the maximum permissible interval between two signal edges is not exceeded. This is important for synchronization purposes. Bit stuffing is applied by inserting a complementary bit after five bits of equal value. Of course the receiver has to unstuff the stuff bits so that the original data content is processed.

(ii) *Bit timing and synchronization.* On the bit level (OSI level one, physical layer), CAN uses synchronous bit transmission. This enhances the transmitting capacity but also means that a sophisticated method of bit synchronization is required. While bit synchronization in a character-oriented transmission (asynchronous) is performed upon the reception of the start bit available with each character, a synchronous transmission protocol is just one start bit available at the beginning of a frame.

To enable the receiver to read the messages correctly, continuous resynchronization is required. Phase buffer segments are, therefore, inserted before and after the nominal sample point within a bit interval.

The CAN protocol regulates bus access by bitwise arbitration. The signal propagation from sender to receiver and back to the sender must be completed within one bit time. For synchronization purposes, a further time segment, the propagation delay segment, is needed in addition to the time reserved for synchronization, the phase buffer segments. The propagation delay segment takes into account the signal propagation on the bus as well as signal delays caused by transmitting and receiving nodes.

Two types of synchronization are distinguished: hard synchronization at the start of a frame and resynchronization within a frame. After a hard synchronization the bit time is restarted at the end of the sync segment. Therefore, the edge, which caused the hard synchronization, lies within the sync segment of the restarted bit time. Resynchronization shortens or lengthens the bit time so that the sample point is shifted according to the detected edge.

(iii) *Interdependency of data rate and bus length.* Depending on the size of the propagation delay segment, the maximum possible bus length at a specific data rate (or the maximum possible data rate at a specific bus length) can be determined. The signal propagation is determined by the two nodes within the system that are farthest apart from each other. It is the time that it takes a signal to travel from one node to the one farthest away (taking into account the delay caused by the transmitting and receiving node), synchronization and the signal from the second node to travel back to the first one. Only then can the first node decide whether its own signal level (recessive in this case) is the actual level on the bus or whether it has been replaced by the dominant level by another node. This fact is important for bus arbitration.

(3) *CAN application layer protocols.* In the CAN world there are different standardized application layer protocols. Some are very specific and related to specific application fields. Examples of CAN-based application layer protocols are given below:

(a) *CANopen.* CANopen is a CAN-based higher layer protocol. It was developed as a standardized embedded network with highly flexible configuration capabilities. CANopen was pre-developed in an Esprit project under the chairmanship of Bosch. In 1995, the CANopen specification was handed over to the CiA international users' and manufacturers' group. Originally, the CANopen communication profile was based on the CAN Application Layer (CAL) protocol. Version 4 of CANopen (CiA DS 301) is standardized as EN 50325-4.

The CANopen specifications cover application layer and communication profile (CiA DS 301), as well as a framework for programmable devices (CiA 302), recommendations for cables and connectors (CiA 303-1), and SI units and prefix representations (CiA 303-2). The application layer as well as the CAN-based profiles is implemented in software.

Standardized profiles (device, interface, and application profiles) developed by CiA members simplify the system design job of integrating a CANopen network system. Off-the-shelf devices, tools, and protocol stacks are widely available at reasonable prices. For system designers, it is very important to reuse application software. This requires not only communication compatibility, but also interoperability and interchangeability of devices. In the CANopen device and interface profiles, defined application objects exist to achieve the interchangeability of CANopen devices. CANopen is flexible and open enough to enable manufacturer-specific functionality in devices, which can be added to the generic functionality described in the profiles.

CANopen unburdens the developer from dealing with CAN-specific details such as bit-timing and implementation-specific functions. It provides standardized communication objects for real-time data (Process Data Objects, PDO), configuration data (Service Data Objects, SDO), and special functions (Time Stamp, Sync message, and Emergency message) as well as network management data (Boot-up message, NMT message, and Error Control).

(b) *CAN Kingdom.* CAN Kingdom unleashes the full power of CAN. It gives system designers maximum freedom to create their own systems, which is not bound to the CSMA/AMP multimaster protocol of CAN but can create systems using virtually any type of bus management and topology. CAN Kingdom opens the possibility for a module designer to design general modules without knowing which system they will finally be integrated into and what type of higher layer CAN protocol it will have. As the system designer can allow only specific modules to be used in the system, the cost advantage of an open system can be combined with the security of a proprietary system!

Since the identifier in a CAN message not only identifies the message but also governs the bus access, a key factor is the enumeration of the messages. Another important factor is to see to it that the data structure in the data field is the same in both the transmitting and receiving modules. By adopting a few simple design rules these factors can be fully controlled and communication optimized for any system. This is done during a short setup phase at the initialization of the system. Including some modules not following the rules of the CAN Kingdom into a CAN Kingdom system is even possible. CAN Kingdom also enforces a conform documentation of modules and systems.

(c) *DeviceNet.* DeviceNet is a low-cost communications link to connect industrial devices (such as limit switches, photoelectric sensors, valve manifolds, motor starters, process sensors, bar code readers, variable frequency drives, panel displays, and operator interfaces) to a network and eliminate expensive hard wiring. The direct connectivity provides improved communication between devices as well as important device-level diagnostics not easily accessible or available through hardwired I/O interfaces. DeviceNet is a simple, networking solution that reduces the cost and time to wire and install factory automation devices, while providing interchangeability of "like" components from multiple vendors. DeviceNet specifications have been developed by the Open DeviceNet Vendor Association (ODVA) and are internationally standardized. Buyers of the DeviceNet Specification receive an unlimited, royalty-free license to develop DeviceNet products.

(d) *J1939-based higher layer protocols.* A J1939 network connects electronic control units (ECU) within a truck and trailer system. The J1939 specification—with its engine, transmission, and brake message definitions—is dedicated to diesel engine applications. It is supposed to replace J1587/J1708 networks. Other industries adopted the general J1939 communication functions, in particular the J1939/21 and J1939/31 protocol definitions—they are required for any J 1939-compatible system. They added other physical layers and defined other application parameters. The ISO standardized the J1939-based truck and trailer communication (ISO 11992) and the J1939-based communication for agriculture and forestry vehicles (ISO 11783). The National Maritime Electronics Association (NMEA) specified the J1939-based communication for navigation systems in marine applications (NMEA 2000). One reason for the incorporation of J1939 specifications into others is the fact that it makes sense to reinvent the basic communication services. An industry-specific document defines the particular combination of layers for that industry.

CiA has developed several CANopen interface profiles for J1939-based networks (CiA DSP 413). Gateways are defined according to ISO 11992-2 and ISO 11992-3. In addition, the CANopen profile family includes a framework for gateways according to SAE J1939/71.

(4) *CAN standards*. The original specification is the Bosch specification. Version 2.0 of this specification is divided into two parts:

(a) Standard CAN (Version 2.0A). Uses 11 bit identifiers.

(b) Extended CAN (Version 2.0B). Uses 29 bit identifiers.

The two parts define different formats of the message frame, with the main difference being the identifier length. There are two ISO standards for CAN. The difference is in the physical layer, where ISO 11898 handles high speed applications up to 1 Mbit/s. ISO 11519 has an upper limit of 125 kbit/s.

(a) *Part A and Part B compatibility*. There are three types of CAN controllers: Part A, Part B passive, and Part B (Table 1.4.1). They are able to handle the different parts of the standard as follows:

Most 2.0A controllers transmit and receive only standard format messages, although some (known as 2.0B passive) will receive extended format messages but then ignore them. 2.0B controllers can send and receive messages in both formats. Note that if 29 bit identifiers are used on a bus that contains part A controllers, the bus will not work!

(b) *CAN bus physical layer*. The physical layer is not part of the Bosch CAN standard. However, in the ISO standards transceiver characteristics are included. CAN transmits signals on the CAN bus which consists of two wires, a CAN-High and CAN-Low. These two wires operate in differential mode, that is, they carry inverted voltages (to decrease noise interference). The voltage levels, as well as other characteristics of the physical layer, depend on which standard is being used,

(i) *ISO 11898*. The voltage levels for a CAN network which follows the ISO 11898 (CAN High Speed) standard are described in Table 1.4.2.

Note that for the recessive state, nominal voltage for the two wires is the same. This decreases the power drawn from the nodes through the termination resistors. These resistors are 120 Ω and are located on each end of the wires. Some people have played with using central termination resistors (i.e., putting them in one place on the bus). This is not recommended since that configuration will not prevent reflection problems.

(ii) *ISO 11519*. The voltage levels for a CAN network which follows the ISO 11519 (CAN Low Speed) standard are described in Table 1.4.3.

ISO 115519 does not require termination resistors. These are not necessary because the limited bit rates (maximum 125 kB/s) make

Table 1.4.1 CAN Part A and Part B compatibility

Message Format\CAN chip type	Part A	Part B passive	Part B
11 bit ID	Ok	Ok	Ok
29 bit ID	Error!	Tolerated on the bus, but ignored	Ok

Table 1.4.2 ISO 11898 parameters for CAN

	Recessive state (V)			Dominant state (V)		
Signal	Min	Nominal	Max	Min	Nominal	Max
CAN-high	2.0	2.5	3.0	2.75	3.5	4.5
CAN-low	2.0	2.5	3.0	0.5	1.5	2.25

Table 1.4.3 ISO 11519 parameters for CAN

	Recessive state (V)			Dominant state (V)		
Signal	Min	Nominal	Max	Min	Nominal	Max
CAN-high	1.6	1.75	1.9	3.85	4.0	5.0
CAN-Low	3.1	3.25	3.4	0	1.0	1.15

Table 1.4.4 CAN Bus Length

Bus length (m)	Maximum bit rate (bit/s)
40	1 Mbit/s
100	500 kbit/s
200	250 kbit/s
500	125 kbit/s
6 km	10 kbit/s

the bus insensitive to reflections. The voltage level on the CAN bus is recessive when the bus is idle.

(iii) *Bus lengths.* The maximum bus length for a CAN network depends on the bit rate used. It is required that the wavefront of the bit signal has time to travel to the most remote node and back again before the bit is sampled. This means that if the bus length is near the maximum for the bit rate used, one should choose the sampling point with utmost care—on the other hand, one should always do that!

Table 1.4.4 gives the different bus lengths and the corresponding maximum bit rates.

(iv) *Cable.* According to the ISO 11898 standard, the impedance of the cable shall be 120 ± 12 Ω. It should be twisted par, shielded or unshielded. Work is in progress on the single-wire standard SAE J2411.

1.4.2.1.4 Interbus

Interbus was one of the very first Fieldbuses to achieve widespread popularity. It continues to be popular because of its versatility, speed, diagnostic and autoaddressing capabilities. Physically, it has the appearance of being a typical line-and-drop-based network, but in reality it is a serial ring shift register. Each slave node has two connectors, one which receives data and one which passes data onto the next slave.

Interbus technology provides an open Fieldbus system, which embraces all the process I/Os required for almost any control system. Interbus is able to fulfill essential requirements of high-performance control concepts, as it is (1) a cost-effective solution with bus systems, which transmits data serially and reduces the amount of parallel cabling required; (2) an open and manufacturer-independent networking system, which can be easily connected with existing control systems; (3) flexible with regard to future modifications or expansions.

With its special features and an extensive product range, Interbus has established itself successfully in all sectors of industry. Its traditional field of application is the automotive industry, but Interbus is also increasingly being used as an automation solution in other areas such as materials handling and conveying, the paper and print industry, the food and beverage industry, building automation, the wood-processing industry, assembly and robotics applications, general mechanical engineering, and, more recently, in process engineering. In addition to standard applications for connecting a large number of sensors and actuators in the field to the higher level control system via a serial bus system, Interbus can also be used to fulfill a variety of special application requirements such as (1) driving synchronically a control loop application in a mill train; and (2) alternative and changing bus configuration in a machining center.

(1) *Operation mechanism.* Interbus works with a master–slave access method, in which the master also establishes the connection to the higher level control or bus system. In terms of topology, Interbus is a ring system with an active connection to communication devices. Starting at the Interbus master, the controller board, all devices are actively connected on the ring system. Each Interbus device (slave) has two separate lines for data transmission: one for forward data transfer and one for return data transfer. This eliminates the need for a return line from the last to the first device, necessary in a simple ring system.

The forward and return lines run in one cable. From the installation point of view, Interbus is similar to bus or linear structures, as only one bus cable connects one device with the next. To enable the structuring of an Interbus system, subring systems (bus segments) can be formed on the main ring, the source of which is the master. These subring systems are connected with bus couplers (also known as bus terminal modules). Figure 1.4.15 illustrates the basic structure of an Interbus system with one main ring and two subring systems.

The remote bus is installed from the controller board. Remote bus devices and bus couplers are connected to the remote bus. Each bus coupler connects the remote bus with a subring system. There are two different types of subring system, which are available in different installation versions:

(a) The local bus (formally known as the I/O bus) is responsible for local management, connects local bus devices, and is typically used to form local I/O compact stations, for example, in the control cabinet. It is also available as a robust version for direct mounting on machines and systems.

(b) The remote bus branch connects remote bus devices and connects distributed devices over large

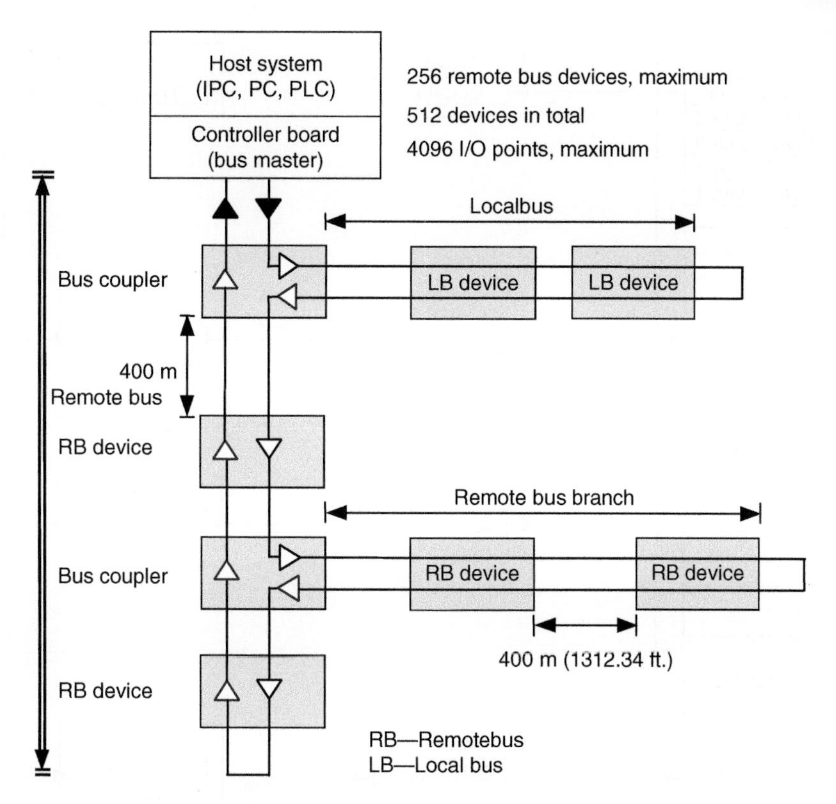

Fig. 1.4.15 Basic structure of an Interbus system.

distances. Remote bus branches can be used to set up complex network topologies, which are ideal for complex technical processes distributed over large distances.

The Interbus remote bus cable forms a recommended standard (RS)-485 connection and, because of the ring structure and the additional need for an equalizing conductor between two remote bus devices, it requires five cables.

Due to the different physical transmission methods, the local bus is available with nine cables and TTL levels for short distances (up to 1.5 m) and as a two-wire cable with a TTY-based current interface for medium distances (up to 10 m).

Due to the integrated amplifier function in each remote bus device, the total expansion of the Interbus system can reach 13 km. To ensure that the system is easy to operate, the number of Interbus devices is limited to a maximum of 512.

Interbus works as a shift register, which is distributed across all bus devices and uses the I/O-based summation frame method for data transmission. Each bus device has data memories, which are combined via the ring connection of the bus system to form a large shift register. Figure 1.4.16 illustrates the data transmission principle. A data packet in the summation frame is made available in the send shift register by the master. The data packet contains all data that is to be transmitted to the bus devices (OUT data). The corresponding data registers in the bus devices contain the data to be transmitted to the master (IN data) (Figure 1.4.16a).

The OUT data is now transferred from the master to the device and the IN data is transferred from the devices to the master in one data cycle. The master starts by sending the loop-back word through the ring. At the end of the data cycle, the master receives the loop-back word. The loop-back word "pulls" the OUT data along behind it while "pushing" the IN data along in front of it. This is called full duplex data transmission (Figure 1.4.16b).

The devices do not have to be addressed explicitly as the physical position of a device in the ring is known and the master can position the information to be transmitted at this point in the summation frame telegram. In the example, the first data word after the loop-back word is addressed to slave 4, for example.

The amount of user data to be pushed through the ring corresponds to the total data length of all bus devices. The bus couplers are integrated into the ring but do not provide any user data. Data widths between 1 bit and 64 bytes per

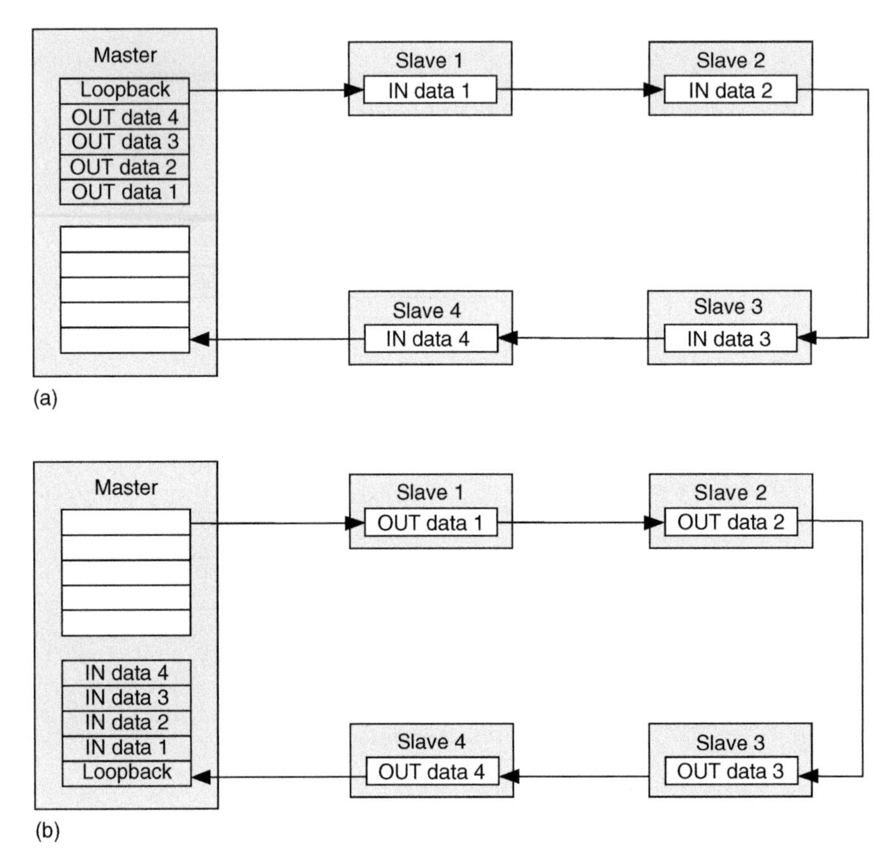

Fig. 1.4.16 Principle of data transmission on Interbus: (a) distribution of data before a data cycle and (b) distribution of data after a data cycle.

data direction are permitted in one Interbus device.

Unlike local bus segments, whose components only really differ in terms of installation technology, Interbus loop (sensor loop, IP 65 local bus) offers a new physical transmission method. The individual devices are connected via a simple two-wire unshielded cable to form a ring. The data and the 24 V power supply for up to 32 sensors are also supplied via the cable. Figure 1.4.17 shows the configuration of an Interbus loop segment.

Data is transmitted as load-independent current signals, which have a higher level of immunity to interference than the voltage signals normally used. The data to be transmitted is modulated using Manchester code on the 24 V supply voltage (Interbus usually uses the NRZ

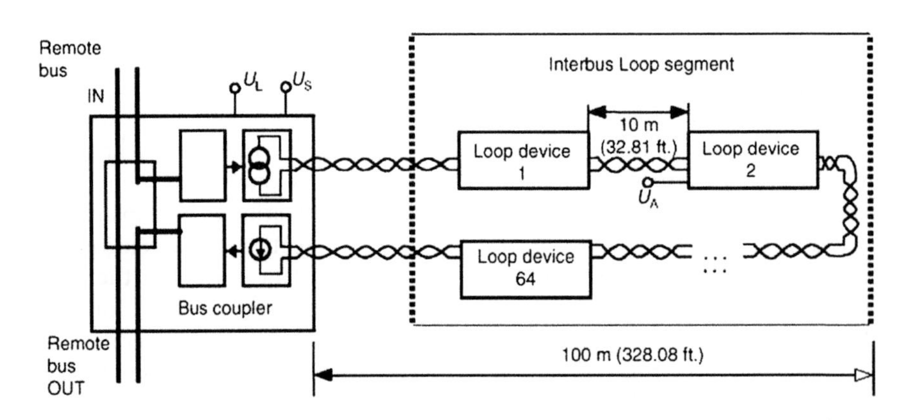

Fig. 1.4.17 Interbus loop segment (U_L, power supply for the bus logic; U_S, power supply for the Interbus loop; U_A, local power supply for actuators).

code). The physical bus characteristics are converted by an appropriate bus terminal module, which can be connected to the Interbus ring at any point in a remote bus segment.

One of the main fields of application of Interbus loop is the connection of individual devices with IP 65 and IP 54 connections directly in the system. An extensive range of functions and devices is available as bus devices. The Interbus protocol is not converted in any way in an Interbus loop, which means that complex gateways are not required and an Interbus loop segment can be used in conjunction with any other type of Interbus device. Data scanning is absolutely synchronous in all parts of the Interbus system. Despite this, the high scanning speed is maintained.

An Interbus system is configured by connecting the bus devices one after the other in a ring. Bus couplers segment the ring according to the application requirements. With Interbus G4 (Generation 4) and later, it is possible to set up complex network topologies, which can be optimized for the structure of the automation system, by integrating bus couplers with an additional bus connection. There are two ways of structuring the configuration of this type of Interbus network:

(i) divide the entire network into various levels;

(ii) assign segment-specific device numbers.

Both configuration methods are explained using the example of an Interbus network configuration with four levels, as illustrated in Figure 1.4.18. The network is split into four different levels starting with the bus master on the main remote bus line as the first level. The branching secondary lines are now assigned a second level. The devices connected to these lines can form additional substructures, etc. In this way, a nesting depth of up to 16 levels can be achieved. The sequence is such that a local bus (formally known as the I/O bus) in a remote bus segment is always assigned to the next level. Segment-specific device numbers are assigned either automatically according to the physical configuration or they can be freely specified by the user. The numbering comprises two components:

< Device number>=< Bus segment number>•<Position number in bus segment>

According to this pattern, the second digit of the device number for all remote devices is zero, for example, 1.0. The second digit is only used by the local bus devices (e.g., I/O modules) connected downstream of the remote device, for example, 1.1.

Bus couplers with an additional remote bus branch appear as two separate remote bus devices with one local bus/remote bus branch, for example, bus coupler 1.0/2.0. When physically assigning this type of remote bus device, the remote bus branch is assigned the next consecutive number, for example, 3.0. Any additional subbranches on this branch are assigned the next

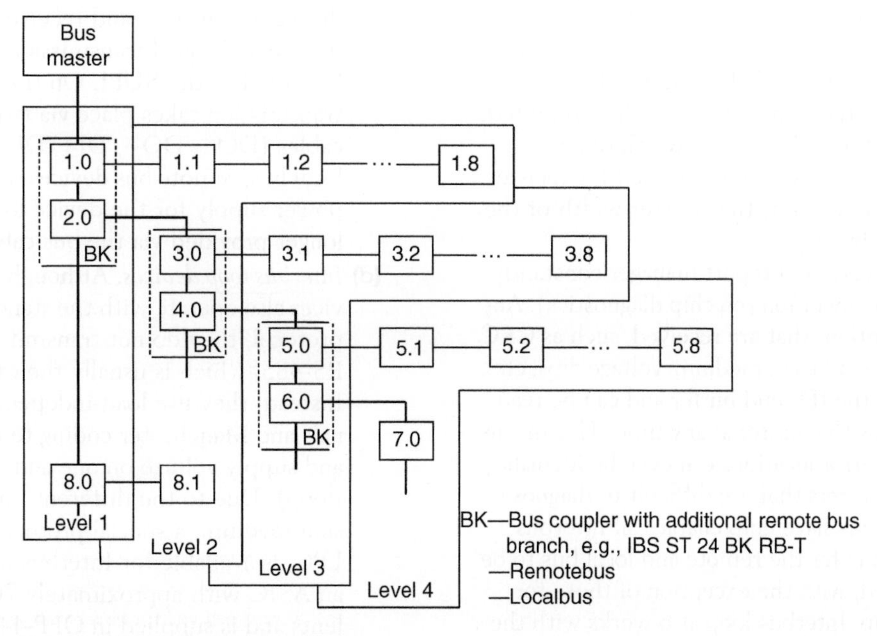

Fig. 1.4.18 Interbus network configuration with four levels.

consecutive number, for example, 4.0, 5.0, etc. The outgoing remote bus from the branch is counted as the last component, for example, 8.0.

Device numbering is a structuring tool and should not be confused with device addressing. Although the device numbers can be used for addressing purposes, this is not absolutely necessary.

(2) *Interbus system devices*

(a) *Protocol chip.* The most important element in the electrical configuration of an Interbus device is the Interbus protocol chip, which manages the complete summation frame protocol and provides the physical interface to the Interbus ring. The bus master and Interbus slave devices use different protocol chips according to their function in the Interbus system. Hardware solutions tailored to meet specific technical requirements are available for both Interbus master and slave solutions.

The parts of the Interbus protocol that correspond to layers 1 and 2 of the OSI reference model are processed entirely in the protocol machine. This means that basic devices require additional software or processing power. The protocol machine also provides physical access to the incoming (IB IN) and outgoing (IB OUT) Interbus interface.

Both shift registers—the ID register and data register— operate as send and receive buffers in the ID and data cycle. The application and/or higher protocol layers can access this buffer via the MPM interface (Multifunction Pin, MFP). The MFP interface can be set according to interface requirements.

The data registers can be expanded with external registers (ToExR, FromExR). The Interbus register chip SRE 1, which, if required, can expand the shift register width of an Interbus device to 64 bytes, is used for register expansion. By default, the register width of the SUPI 3 is 8 bytes.

The diagnostic and report manager constantly monitors the operation (on chip diagnostics). Any error descriptions that are received, such as CRC errors, transient loss of medium, voltage dips, etc., are saved to the ID send buffer and can be read from there by the master at any time. This means that unique error locations can even be identified for sporadic errors that are difficult to diagnose.

The Interbus slave chip enables all Interbus device variants for the remote and local bus to be implemented, with the exception of those for Interbus loop. Interbus loop also works with the Interbus protocol but uses a different physical transmission medium, which requires the protocol chip on the physical interface to be of the same format.

The standard master protocol chip for Interbus masters is the IPMS microcontroller. The IPMS is designed to work with a wide range of different processors. The master chip is often used together with the Motorola CPU 68332. The master firmware, which manages the Interbus functions, is stored in the erasable programmable read-only memory (EPROM).

Only actual bit transmission (Layer 1, parts of Layer 2) takes place via the IPMS. The IPMS is connected to the relevant host system via a shared memory area, which, in its simplest format, is a Dual Port Memory (DPM) or a Multiport Memory (MPM). Interbus masters with IPMS are available in various formats depending on the functions required.

(b) *Local bus devices.* An I/O bus interface for the two-wire protocol with the SUPI 3 is an interface example used to configure an ST local bus. The ST local bus operates with four transmission signals, which, due to the ring format of Interbus, are available twice at the incoming and outgoing local bus interface as IN and OUT signal lines. In addition, one incoming and one outgoing reset line are also available in the local bus. The bus signals can be connected directly to the local bus connectors, as the SUPI 3 meets the Interbus specification for the local bus even without external drivers and receivers.

(c) *Remote bus devices.* If it is used as a remote bus device, the drivers and receivers required for differential signal transmission to RS-485 must be added to the SUPI. On the remote bus, transmission takes place via two twisted pair cables (DO+/DO−, DI+/DI−). Unlike the local bus, remote bus devices require a dedicated power supply for the device logic, as this is no longer provided via the bus cable.

(d) *Interbus loop devices.* Although Interbus loop devices also operate with the standardized Interbus protocol, they do not transmit voltage signals to RS-485, which is usually the case on Interbus. Instead, they use load-independent current signals and Manchester coding to transmit the data and supply voltage on one and the same bus line (loop). Due to the different physical transmission medium, a special protocol chip, the IBS LPC, is available for Interbus loop. This chip is an ASIC with approximately 7000 gate equivalents and is supplied in QFP-44 housing. Special loop diagnostics are integrated into the LPC 2 to extend the familiar diagnostic functions of the SUPI 3.

(3) *Protocol structure.* The Interbus protocol, which has been optimized specifically for the requirements of automation technology, transmits single-bit data from limit switches or to switching devices (process data) and complex programs or data records to intelligent field devices (parameter data) with the same level of efficiency and safety. Process data is transmitted in the fixed and cyclic time slot in real-time conditions, while parameter data comprises the acyclic transmission of larger volumes of data as and when required. The continuity of an Interbus network for very different tasks within an automation system—ensured in essence by the standard protocol—is supported by additional measures: (1) the adaptation of the physical transmission method "downward," making it easy to install and connect individual sensors and actuators; (2) the provision of "upward" interface couplers to connect Interbus networks directly with factory and or company networks (Ethernet networks); (3) the guarantee of easy configuration, project planning, and diagnostics with uniform software tools.

The Interbus protocol is based on the OSI reference model and for reasons of efficiency only takes into account layers 1, 2, and 7 (Figure 1.4.19). Certain functions from layers 3 to 6 have been included in application Layer 7.

The physical layer (Layer 1) defines both the time conditions (such as the baud rate, permissible jitter, etc.) and the formats for encoding information. The data link layer (Layer 2) ensures data integrity and manages cyclic data transfer via the bus using the summation frame protocol. The transmission methods and protocols on layers 1 and 2 can be found in DIN 19 258.

Following on from the data link layer, data access to the Interbus devices takes place in the application layer as required via two different data channels:

(a) The process data channel serves the primary use of Interbus as a sensor and actuator bus. The cyclic exchange of I/O data between the higher level control system and the connected sensors/actuators takes place via this channel.

(b) The parameter channel supplements cyclic data exchange with individual I/O points in connection-oriented message exchange. This type of communication requires additional data packing, as large volumes of information are being exchanged between the individual communication partners. Data is transmitted using communication services based on the client/server model.

Interbus devices almost always have one process data channel. A parameter channel can also be fitted as an optional extra. During operation, an Interbus system requires settings to be made and provides a wide range of diagnostic information. This information is processed by the network management on each layer. More detailed information about readiness for operation, error states, and statistical data can also be accessed and evaluated, and network configuration settings can be made.

The hybrid protocol structure of Interbus for the two different data classes (process data and parameter data) and its independent data transmission via two channels is a decisive factor in the performance of the Interbus protocol. The protocol enables the creation of a seamless network comprising control systems and intelligent field devices right down to individual sensors and actuators.

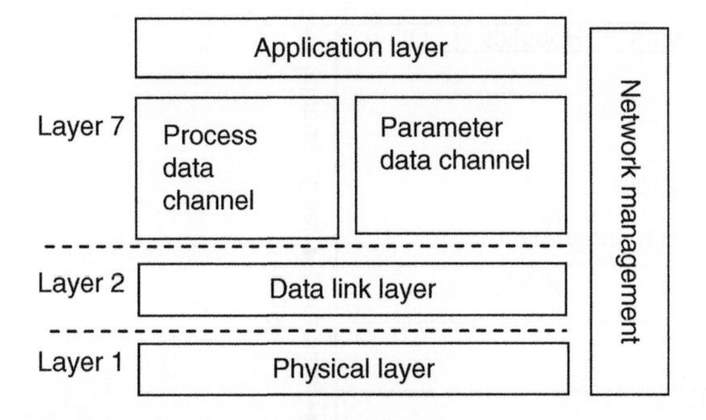

Fig. 1.4.19 Interbus protocol structure.

1.4.2.1.5 Ethernets/hubs

Ethernet is the major local area network (LAN) technology in use today, and is widely used for the LAN-connected PCs and workstations. Ethernet refers to the family of LAN products covered by the IEEE 802.3 standard, and the technology can run over both optical fiber and twisted-pair cables. Over the years, Ethernet has steadily evolved to provide additional performance and network intelligence. More than 300 million switched Ethernet ports have been installed worldwide. Ethernet technology enjoys such wide acceptance because it is easy to understand, deploy, manage, and maintain. Ethernet is low cost and flexible, and supports a variety of network topologies. Although traditional, non-Ethernet-based industrial solutions have a data rate of between 500 kbps and 12 Mbps, Ethernet technology can deliver substantially higher performance. Because it is based on industry standards, it can run and be connected over any Ethernet-compliant device from any vendor.

This continual improvement has made Ethernet an excellent solution for industrial applications. Today, the technology can provide four data rates.

(1) *10BASE-T Ethernet* delivers performance of up to 10 Mbps over twisted-pair copper cable.

(2) *Fast Ethernet* delivers a speed increase of 10 times the 10BASE-T Ethernet specification (100 Mbps) while retaining many of Ethernet's technical specifications. These similarities enable organizations to use 10BASE-T applications and network management tools on Fast Ethernet networks.

(3) *Gigabit Ethernet* extends the Ethernet protocol even further, increasing speed 10-fold over Fast Ethernet to 1000 Mbps, or 1 Gigabit/s. Because it is based upon the current Ethernet standard and compatible with the installed base of Ethernet and Fast Ethernet switches and routers, network managers can support Gigabit Ethernet without needing to retrain or learn a new technology.

(4) *10 Gigabit Ethernet,* ratified as a standard in June 2002, is an even faster version of Ethernet. It uses the IEEE 802.3 Ethernet media access control (MAC) protocol, the IEEE 802.3 Ethernet frame format, and the IEEE 802.3 frame size. Because 10 Gigabit Ethernet is a type of Ethernet, it can support intelligent Ethernet-based network services, interoperate with existing architectures, and minimize users' learning curves. Its high data rate of 10 Gigabits/s makes it a good solution to deliver high bandwidth in wide area networks (WANs) and metropolitan area networks (MANs).

(1) *Industrial Ethernet.* Recognizing that Ethernet is the leading networking solution, many industry organizations are porting the traditional Fieldbus architectures to Industrial Ethernet. Industrial Ethernet applies the Ethernet standards developed for data communication to manufacturing control networks (Figure 1.4.20). Using IEEE standards-based equipment, organizations can migrate all or part of their factory operations to an Ethernet environment at the pace they wish. Instead of using architectures composed of multiple separate networks, Industrial Ethernet can unite a company's administrative, control-level, and device-level networks to run over a single network infrastructure. In an Industrial Ethernet network, Fieldbus-specific information that is used to control I/O devices and other manufacturing components are embedded into Ethernet frames. Because the technology is based on industry standards rather than on custom or proprietary standards, it is more interoperable with other network equipment and networks.

Although Industrial Ethernet is based on the same industry standards as traditional Ethernet technology, the implementation of the two solutions is not always identical. Industrial Ethernet usually requires more robust equipment and a very high level of traffic prioritization when compared with traditional Ethernet networks in a corporate data network.

The primary difference between Industrial Ethernet and traditional Ethernet is the type of hardware used. Industrial Ethernet equipment is designed to operate in harsh environments. It includes industrial-grade components, convection cooling, and relay output signaling. And it is designed to operate at extreme temperatures and under extreme vibration and shock. Power requirements for industrial environments differ from data networks, so the equipment runs using 24 V of DC power. To maximize network availability, it also includes fault-tolerant features such as redundant power supplies.

Industrial Ethernet environments also differ from traditional Ethernet networks in their use of multicasting by hosts for certain applications. Industrial applications often use producer–consumer communication, where information "produced" by one device can be "consumed" by a group of other devices. In a producer–consumer environment, the most important priority for a multicast application

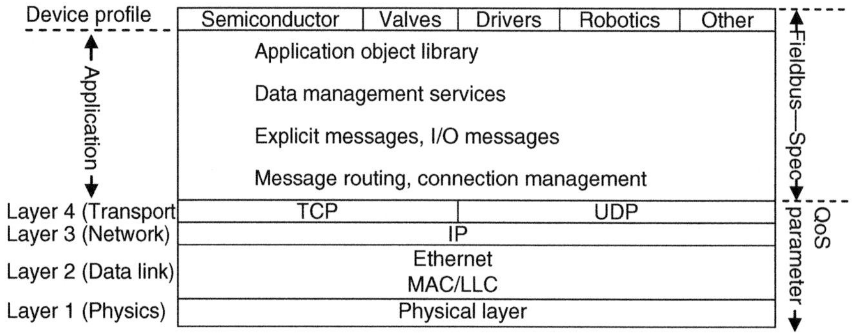

Device profile
Semiconductor | Valves | Drivers | Robotics | Other
Application
Application object library
Data management services
Explicit messages, I/O messages
Message routing, connection management
Layer 4 (Transport) TCP | UDP
Layer 3 (Network) IP
Layer 2 (Data link) Ethernet MAC/LLC
Layer 1 (Physics) Physical layer
Fieldbus—Spec
QoS parameter

Fig. 1.4.20 Using intelligent Ethernet for automation control.

is to guarantee that all hosts receive data at the same time. A traditional Ethernet network, on the other hand, focuses more on the efficient utilization of bandwidth in general, and less on synchronous data access. To help optimize synchronous data access, Industrial Ethernet equipment must include the intelligence and Quality of Services (QoS) features needed to enable organizations to appropriately prioritize multicast transmissions.

Ethernet technology can provide not only excellent performance for manufacturing applications, but also a wide range of network security measures to provide both confidentiality and data integrity. Confidentiality helps ensure that data cannot be accessed by unauthorized users. Data integrity protects data from intentional or accidental alteration.

These network security advantages protect manufacturing devices like PLCs as well as PCs, and apply to both equipment and data security. Manufacturers can use many methods to help ensure network confidentiality and integrity. These network security measures can be grouped into several categories, including access control and authentication, and secure connectivity and management.

Access control is commonly implemented using firewalls or network-based controls protecting access to critical applications, devices, and data so that only legitimate users and information can pass through the network. However, access-control technology is not limited to dedicated firewall devices. Any device that can make decisions to permit or deny network traffic, such as an intelligent switch, is part of an integrated access-control solution. When designing an access-control solution, network administrators can set up filtering decisions based on a variety of criteria, such as an IP address or TCP/UDP port number. Intelligent switches can provide support for this advanced filtering to limit network access to authorized users. At the same time, they can enable organizations to enforce policy decisions based on the IP or MAC address of a laptop or PLC.

Virtual LANs (VLANs) are another access-control solution, providing the ability to create multiple IP subnets within an Ethernet switch. VLANs provide network security and isolation by virtually segmenting factory floor data from other data and users. VLANs can also be used to enhance network performance, separating low-priority end devices from high-priority devices.

Access controls can also include a variety of device or user-authentication services. Authentication services determine who may access a network and what services they are authorized to use. For example, the 802. 1x authentication protocol provides port-based authentication so that only legitimate devices can connect to switch ports. Authentication services are an effective complement to other network security measures in a manufacturing environment. To provide additional protection for manufacturing networks, organizations can take several approaches to authenticate and encrypt network traffic. Using virtual private network (VPN) technology, Secure Sockets Layer (SSL) encryption can be applied to application-layer data in an IP network. Organizations can also use IP Security (IPSec) technology to encrypt and authenticate network packets to thwart network attacks such as sniffing and spoofing.

VPN client software, together with dedicated VPN network hardware, can be used to encrypt device monitoring and programming sessions, and to support strong authentication. Manufacturers can also use Secure Shell (SSH) Protocol encryption for remote terminal logins to network devices. Version 3 of Simple Network Management Protocol (SNMP) also offers support for encryption and authentication of management commands and data.

(2) *Network topology.* Because factory floor applications run in real time, the network must be available to users on a continuous basis, with little or no downtime. Manufacturers can help ensure network reliability using effective network design principles, as well as intelligent networking services. Manufacturers deploying an Ethernet solution should design networks with redundant paths to ensure that a single device outage does not take down the entire network. Two network topologies most often used are ring and hub and spoke. In hub-and-spoke designs (Figure 1.4.21), three layers of switches are usually installed. The first layer is often referred to

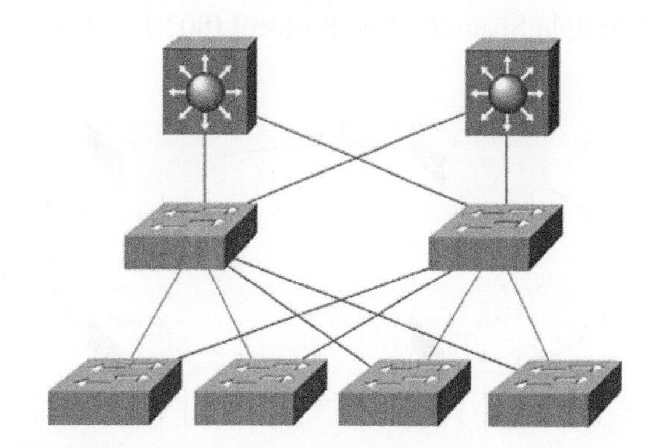

Fig. 1.4.21 Hub-and-spoke network topology.

as the access layer. These switches provide connections for end-point devices like PLCs, robots, and Human–Machine Interfaces (HMIs). A second layer called the distribution layer provides connectivity between the access-layer switches; and a third layer called the core layer provides connectivity to other networks or to the Internet service provider (ISP) via routers. The distribution layer may include switches with routing functions to provide inter-VLAN routing. Access-layer switches, on the other hand, generally provide only Layer 2 (data link) forwarding services. For optimum performance, network equipment at each of these layers must be aware of the information contained within the Layer 2 through Layer 4 packet headers.

In ring topologies (Figure 1.4.22), all devices are connected in a ring. Each device has a neighbor to its left and right. If a connection on one side of the device is broken, network connectivity can still be maintained over the ring via the opposite side of the device. In some situations, manufacturers install dual counterrotating rings to maximize availability. In a ring topology, each switch functions as both an access-layer and a distribution-layer switch.

(3) *Network protocols.* To prevent loops from being formed in the network when devices are interconnected via multiple paths, some organizations use the Spanning Tree Protocol. If a problem occurs on a network node, this protocol enables a redundant alternative link to automatically come back online. The traditional Spanning Tree Protocol has been considered too slow for industrial environments. To address these performance concerns, the IEEE standards committee has ratified anew Rapid Spanning Tree Protocol (802. lw). This protocol provides subsecond convergence times that vary between 200 and 800 ms, depending on network topology. Using 802. lw, organizations can enjoy the benefits of Ethernet networks, with the performance and reliability that manufacturing applications demand. Another spanning-tree option is Multiple Spanning Tree Protocol (802.1s). This enables VLANs to be grouped into spanning-tree instances. Each instance has a spanning-tree topology that is independent from other spanning-tree instances. This architecture provides multiple forwarding paths for data traffic, enables load balancing, and reduces the number of spanning-tree instances needed to support a large number of VLANs.

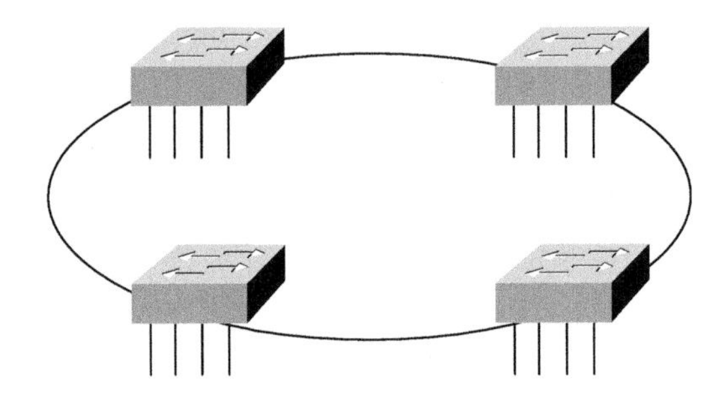

Fig. 1.4.22 Ring topology.

Ethernet switches provide excellent connectivity and performance; however, each switch is another device that must be managed on the factory floor. To make switched Ethernet networks easy to support and maintain, intelligent switches include built-in management capabilities. These intelligent features make it easy to connect manufacturing devices to the network, without creating additional configuration tasks. And they help minimize network downtime if part of the network should fail. One of the most useful intelligent features in a switched Ethernet network is Option 82. In an Ethernet network, Dynamic Host Configuration Protocol (DHCP) lets devices dynamically acquire their IP addresses from a central server. The DHCP server can be configured to give out the same address each time or generate a dynamic one from a pool of available addresses.

Because the interaction of the factory floor devices requires specific addresses, Industrial Ethernet networks usually do not use dynamic address pools. However, static addresses can have drawbacks. Because they are linked to the MAC address of the client, and because the MAC address is often hard-coded in the network interface of the client device, the association is lost when a client device fails and needs to be replaced. Extended fields in the DHCP packet can be filled in by the switch, indicating the location of the device requesting an IP address. The 82nd optional field, called Option 82, carries the specific port number and the MAC address of the switch that received the DHCP request. This modified request is sent on to the DHCP server. If an access server is Option 82-aware, it can use this information to formulate an IP address based on the Option 82 information. Effective use of Option 82 enables manufacturers to minimize administrative demands and maintain maximum network uptime even in the event of the failure of individual devices.

Because manufacturing processes depend on the precise synchronization of processes, network determinism must be optimized to deliver the best possible performance. Data must be prioritized using QoS to ensure that critical information is received first. The multicast applications that are prevalent in manufacturing environments must be well managed using Internet Group Management Protocol (IGMP) snooping.

Many Industrial Ethernet applications depend on IP multicast technology. IP multicast allows a host, or source, to send packets to another group of hosts called receivers anywhere within the IP network using a special form of IP address called the IP multicast group address. While traditional multicast services, such as video or multimedia, tend to scale with the number of streams, Industrial Ethernet multicast applications do not. Industrial Ethernet environments use a producer–consumer model, where devices generate data called "tags" for consumption by other devices. The devices that generate the data are producers and the devices receiving the information are consumers. Multicast is more efficient than unicast, because consumers will often want the same information from a particular producer. Each device on the network can be both a producer and a consumer of traffic.

While most devices generate very little data, networks with a large number of nodes can generate a large amount of multicast traffic, which can overrun end devices in the network. Using mechanisms like QoS and IGMP snooping, organizations can control and manage multicast traffic in manufacturing environments.

Many manufacturing applications depend on multicast traffic, which can introduce performance problems in the network. To address these challenges in an Industrial Ethernet environment, organizations can deploy IGMP snooping. IGMP snooping limits the flooding of multicast traffic by dynamically configuring the interfaces so that multicast traffic is forwarded only to interfaces associated with IP multicast devices. In other words, when a multicast message is sent to the switch, the switch forward the message only to the interfaces that are interested in the traffic.

This is very important because it reduces the load of traffic traversing through the network. It also relieves the hosts from processing frames that are not needed. In a producer–consumer model used by Industrial Ethernet, IGMP snooping can limit unnecessary traffic from the I/O device that is producing, so that it only reaches the device consuming that data. Messages delivered to a given device that were intended for other devices consume resources and slow performance, so networks with many multicasting devices will suffer performance issues if IGMP snooping or other multicast limiting schemes are not implemented.

The IGMP snooping feature allows Ethernet switches to "listen" to the IGMP conversation between hosts. With IGMP snooping, the Ethernet switch examines the IGMP traffic coming to the switch and keeps track of multicast groups and member ports. When the switch receives an "IGMP join" report from a host for a particular multicast group, the switch adds the host port number to the associated multicast forwarding table entry. When it receives an IGMP "leave group" message from a host, it removes the host port from the table entry. After the switch relays the IGMP queries, it deletes entries periodically if it does not receive any IGMP membership reports from the multicast clients. A Layer 3 router normally performs the querying function.

When IGMP snooping is enabled in a network with Layer 3 devices, the multicast router sends out periodic IGMP general queries to all VLANs. The switch responds to the router queries with only one "join" request per MAC multicast group. The switch then creates one entry per VLAN in the Layer 2 forwarding table for each MAC group from which it receives an IGMP join request. All hosts interested in this multicast traffic send "join" requests and are added to the forwarding table entry.

Layer 2 multicast groups learned through IGMP snooping are dynamic. However, in a managed switch, organizations can statically configure MAC multicast groups. This static setting supersedes any automatic manipulation by IGMP snooping. Multicast group membership lists can consist of both user-defined settings and settings learned via IGMP snooping.

(4) *Quality of service (QoS).* An Industrial Ethernet network may transmit many different types of traffic, from routine data to critical control information, or even bandwidth-intensive video or voice. The network must be able to distinguish among and give priority to different types of traffic.

To address these issues, organizations can implement QoS using several techniques. QoS involves three important steps. First, different traffic types in the network need to be identified through classification techniques. Second, advanced buffer-management techniques need to be implemented to prevent high-priority traffic from being dropped during congestion. Finally, scheduling techniques need to be incorporated to transmit high-priority traffic from queues as quickly as possible.

In Layer 2 switches on an Ethernet network, QoS usually prioritizes native, encapsulated Ethernet frames, or frames tagged with 802. 1p class of service (CoS) specifications. More advanced QoS mechanisms take this definition a step further. For example, advanced Ethernet switches can study and interpret the flow of QoS traffic as it is processed through the switch.

A switch can be configured to prioritize frames based on given criteria at different layers of the OSI reference model. For example, traffic could be prioritized according to the source MAC address (in Layer 2) or the destination TCP port (in Layer 4). Any traffic traveling through the interface to which this QoS is applied is classified, and tagged with the appropriate priority. Once a packet has been classified, it is then placed in a holding queue in the switch, and scheduled based on the scheduling algorithm desired.

In an Industrial Ethernet application, real-time I/O control traffic would share network resources with configuration (FTP) and data-collection flows, as well as other traffic, in the upper layers of the OSI reference model. By using QoS to give high priority to real-time UDP control traffic, organizations can prevent delay or jitter from affecting any control functions.

1.4.2.2 Interfaces

Devices connect to the microprocessor using an interface bus. The specification of this bus defines speed between the microprocessor and the connected device that greatly affects the performance of the microprocessor. Peripherals can connect to the controller (or computer) using either an internal or an external interface. This subsection lists main interfaces popularly used in industrial control.

1.4.2.2.1 PCI, ISA, and PCMCIA

(1) *PCI bus.* Intel has developed a standard interface, named the Peripheral Component Interface/Interconnect (PCI) local bus for microprocessors. This technology allows fast memory, disk and video access. The PCI bus is now the main interface bus used in most industrial controllers, and is rapidly replacing the ISA bus for internal interface devices. It is a very adaptable bus and most external buses, such as small computer system interface (SCSI) and universal serial bus (USB) connect to the processor via the PCI bus.

The PCI bus transfers data using the system clock and can operate over a 32- or 64-bit data path. The high transfer rates used in PCI architecture machines limit the number of the PCI bus interfaces to two to three, normally the graphics adapter and hard disk controller. If data is transferred at 64 bits at a rate of 33 MHz, then the maximum transfer rate is 264 MB/s.

(2) *ISA bus.* IBM developed the Industry Standard Architecture or ISA bus for their 80286-based AT (advanced technology) computer. It had the advantage of being able to deal with 16 bits of data at a time. An extra edge connector gives compatibility with the PC bus. This gives an extra 8 data bits and four address lines. Thus, the ISA bus has a 16-bit data and a 24-bit address bus. This gives a maximum of 16 MB of addressable memory and, like the PC bus, it uses a fixed clock rate of 8 MHz. The maximum data rate is thus 2 bytes (16 bits) per clock cycle, giving a maximum throughput of 16 MB/s.

IBM's PC/AT was designed with an expansion bus which not only provided for taking advantage of the new technology, but also remained compatible with the older style 8-bit XT add-in boards. Anticipating that advances in processors would again outpace advances in bus technology, the PC/AT was designed with two separate oscillators. In this way, the microprocessor and expansion bus could be run on different clocks with different speeds. Therefore, a controller or computer running a newer processor with 33 MHz clock speed could also run its expansion bus at an 8 IMHz clock rate. ISA cards are more cumbersome to install than other cards because I/O addresses, interrupts, and clock speed must be set using jumpers and switches on the card itself. The other bus options which use software to set these parameters are called Plug & Play. While there is nothing inferior about using jumpers and switches, it can be more intimidating for novice users. The ISA system, however, does not have a central registry from which to allocate system resources. Consequently, each device behaves as though it has sole access to system resources such as DMA, I/O ports, IRQs, and memory. Obviously, this can cause problems when using multiple add-in boards in a single system.

In practice there is no speed difference between running many serial communication peripherals using a PCI or an ISA bus, though the PCI advantage is obvious for high-speed devices such as video cards. Thus, there is no reason to convert your current ISA serial communication systems to PCI, as ISA will provide equivalent functionality, generally at a lower price. However, if you are starting a new installation using a PC with few or no (as is increasingly the case today) ISA slots, or you prefer using Plug & Play cards, then you should consider using PCI adapters.

Figure 1.4.23 shows a typical connection to the ISA bus. The ALE (sometimes known as BALE) controls the address latch; when active low, it latches the address lines A2–A19 to the ISA bus. The address is latched when ALE goes from a high to a low. The Pentium's data bus is 64 bits wide, whereas the ISA expansion bus is 16 bits wide. It is the bus controller's function to steer data between the processor and the slave device for either 8-bit or 16-bit communications. The following are the descriptions of the ISA signals.

(a) *SA19–SA0*. System Address bits 19:0 are used to address memory and I/O devices within the system. These signals may be used along with LA23–LA17 to address up to 16 MB of memory. Only the lower 16 bits are used during I/O operations to address up to 64 K I/O locations. SA19 is the most significant bit. SA0 is the least significant bit. These signals are gated on the system bus when BALE is high and are latched on the falling edge of BALE. They remain valid throughout a read or write command. These signals are normally driven by the system microprocessor or DMA controller, but may also be driven by a bus master on an ISA board that takes ownership of the bus.

(b) *LA23–LA17*. Unlatched Address bits 23:17 are used to address memory within the system. They are used along with SA19–SA0 to address up to 16 MB of memory. These signals are valid when BALE is high. They are "unlatched" and do not stay valid for the entire bus cycle. Decodes of these signals should be latched on the falling edge of BALE.

(c) *AEN*. Address Enable is used to de-gate the system microprocessor and other devices from the bus during DMA transfers. When this signal is active, the system DMA controller has control of the address, data, and read/write signals. This signal should be included as part of ISA board select decodes to prevent incorrect board selects during DMA cycles.

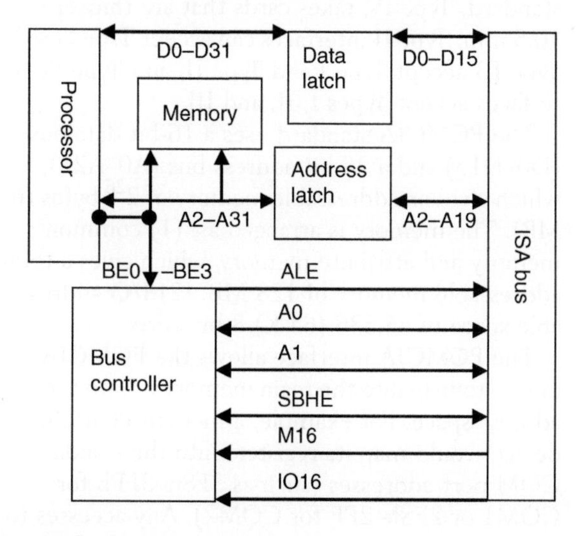

Fig. 1.4.23 ISA bus connections.

(d) *BALE*. Buffered Address Latch Enable is used to latch the LA23–LA17 signals or decodes of these signals. Addresses are latched on the falling edge of BALE. It is forced high during DMA cycles. When used with AEN, it indicates a valid microprocessor or DMA address.

(e) *CLK*. System Clock is a free-running clock typically in the 8–10 MHz range, although its exact frequency is not guaranteed. It is used in some ISA board applications to allow synchronization with the system microprocessor.

(f) *SD15–SD0*. System Data serves as the data bus bits for devices on the ISA bus. SD15 is the most significant bit. SD0 is the least significant bit. SD7–SD0 are used for transfer of data with 8-bit devices. SD15–SD0 are used for transfer of data with 16-bit devices. Sixteen-bit devices transferring data with eight-bit devices convert the transfer into two eight-bit cycles using SD7–SD0.

(g) *–DACK0 to –DACK3 and –DACK5 to –DACK7*. DMA Acknowledge 0 to 3 and 5–7 are used to acknowledge DMA requests on DRQ0–DRQ3 and DRQ5–DRQ7.

(h) *DRQ0 to DRQ3 and DRQ5 to DRQ7*. DMA Requests are used by ISA boards to request service from the system DMA controller or to request ownership of the bus as a bus master device. These signals may be asserted asynchronously. The requesting device must hold the request signal active until the system board asserts the corresponding DACK signal.

(i) *–I/O CH CK*. I/O Channel Check signal may be activated by ISA boards to request that a nonmaskable interrupt (NMI) be generated to the system microprocessor. It is driven active to indicate an incorrect error has been detected.

(j) *I/O CH RDY*. I/O Channel Ready allows slower ISA boards to lengthen I/O or memory cycles by inserting wait states. This signals normal state is active high (ready). ISA boards drive the signal inactive low (not ready) to insert wait states. Devices using this signal to insert wait states should drive it low immediately after detecting a valid address decode and an active read or write command. The signal is released high when the device is ready to complete the cycle.

(k) *–IOR*. I/O Read is driven by the owner of the bus and instructs the selected I/O device to drive read data onto the data bus.

(l) *–IOW*. I/O Write is driven by the owner of the bus and instructs the selected I/O device to capture the write data on the data bus.

(m) *IRQ 3–IRQ 7, IRQ 9–IRQ 12 and IRQ 14–IRQ 15*. Interrupt Requests are used to signal the system microprocessor that an ISA board requires attention. An interrupt request is generated when an IRQ line is raised from low to high. The line must be held high until the microprocessor acknowledges the request through its interrupt service routine. These signals are prioritized with IRQ9–IRQ 12 and IRQ 14–IRQ 15 having the highest priority (IRQ9 is the highest) and IRQ3–IRQ 7 having the lowest priority (IRQ7 is the lowest).

(n) *–SMEMR*. System Memory Read instructs a selected memory device to drive data onto the data bus. It is active only when the memory decode is within the low 1 MB of memory space. SMEMR is derived from MEMR and a decode of the low 1 MB of memory.

(o) *–SMEMW*. System Memory Write instructs a selected memory device to store the data currently on the data bus. It is active only when the memory decode is within the low 1 MB of memory space. SMEMW is derived from MEMW and a decode of the low 1 MB of memory.

(p) *–MEMR*. Memory Read instructs a selected memory device to drive data onto the data bus. It is active on all memory read cycles.

(q) *–MEMW*. Memory Write instructs a selected memory device to store the data currently on the data bus. It is active on all memory write cycles.

(r) *–REFRESH*. Memory Refresh is driven low to indicate a memory refresh operation is in progress.

(s) OSC. Oscillator is a clock with a 70 ns period (14.31818 MHz). This signal is not synchronous with the system clock (CLK).

(t) *RESET DRV*. Reset Drive is driven high to reset or initialize system logic upon power up or subsequent system reset.

(u) *TC*. Terminal Count provides a pulse to signal a terminal count has been reached on a DMA channel operation.

(v) *–MASTER*. Master is used by an ISA board along with a DRQ line to gain ownership of the ISA bus. Upon receiving a –DACK a device can pull –MASTER low which will allow it to control the system address, data, and control lines. After –MASTER is low, the device should wait one CLK period before driving the address and data lines, and two clock periods before issuing a read or write command.

(w) *–MEM CS16*. Memory Chip Select 16 is driven low by a memory slave device to indicate it is capable of performing a 16-bit memory data transfer. This signal is driven from a decode of the LA23–LA17 address lines.

(x) *–I/O CS16*. I/O Chip Select 16 is driven low by an I/O slave device to indicate it is capable of performing a 16-bit I/O data transfer. This signal is driven from a decode of the SA15–SA0 address lines.

(y) *–0WS*. Zero Wait State is driven low by a bus slave device to indicate it is capable of performing a bus cycle without inserting any additional wait states. To perform a 16-bit memory cycle without wait states, –0WS is derived from an address decode.

(z) *–SBHE*. System Byte High Enable is driven low to indicate a transfer of data on the high half of the data bus (D15–D8).

(3) *PCMCIA (PC Card)*. The Personal Computer Memory Card International Association (PCMCIA) interface allows small, thin cards to be plugged into laptop, notebook, or palm computers.

The interface was originally designed for memory cards (Version 1.0), but it has since been adopted for many other types of adapters (Version 2.0), such as fax/modems, sound cards, local area network cards, CD-ROM controllers, digital I/O cards, and so on. Most PCMCIA cards comply with either PCMCIA Type II or Type III. Type I cards are 3.3 mm thick, Type II take cards up to 5 mm thick, and Type III allows cards up to 10.5 mm thick. A new standard, Type IV, takes cards that are thicker than 10.5 mm. Type II interfaces can accept Type I cards, Type III accept Type I and Type II, and Type IV interfaces accept Types I, II, and III.

The PCMCIA standard uses a 16-bit data bus (D0–D15) and a 26-bit address bus (A0–A25), which gives an addressable memory of 2^{26} bytes (64 MB). The memory is arranged as: (1) common memory and attribute memory, which gives a total addressable memory of 128 MB. (2) I/O addressable space of 65,536 (64 K) 8-bit ports.

The PCMCIA interface allows the PCMCIA device to map into the main memory or into the I/O address space. For example, a modem PCMCIA device would map its registers into the standard COM port addresses (such as 3F8h–3FFh for COM1 or 2F8h–2FF for COM2). Any accesses to the mapped memory area will be redirected to the

PCMCIA rather than the main memory or I/O address space. These mapped areas are called windows. A window is defined with a START address and a LAST address.

1.4.2.2.2 IDE

The most popular interface for hard disk drives is the integrated drive electronics (IDE) interface. Its main advantage is that the hard disk controller is built into the disk drive and the interface to the motherboard consists simply of a stripped-down version of the ISA bus. The most common standard is the ANSI-defined ATA-IDE standard. It uses a 40-way ribbon cable to connect to 40-pin header connectors. The standard allows for the connection of two disk drives in a daisy-chain configuration. This can cause problems because both drives have controllers within their drives. The primary drive is assigned as the master, and the secondary driver is the slave. A drive is set as a master or a slave by setting jumpers on the disk drive. They can also be set by software using the cable select pin on the interface. The specifications for the IDE are

(1) Maximum of two devices (hard disks).
(2) Maximum capacity for each disk of 528 MB.
(3) Maximum cable length of 18 in.
(4) Data transfer rates of 3.3, 5.2, and 8.3 MB/s.

A new standard called as enhanced IDE (E-IDE) allows for higher capacities than IDE has

(1) Maximum of four devices (hard disks).
(2) Uses two ports (for master and slaves).
(3) Maximum capacity for each disk of 8.4 GB
(4) Maximum cable length of 18 in.
(5) Data transfer rates of 3.3, 5.2, 8.3, 11.1, and 16.6 MB/s.

The PC is now a highly integrated system. The main elements of modern systems are the processor, the systems controller and the PCI IDE/ISA accelerator, as illustrated in Figure 1.4.24. The system controller provides the main interface between the processor and the level-2 cache, the DRAM, and the PCI bus. It is one of the most important devices in the system and allows data to flow to and from the processor in the correct way. The PCI bus links to interface devices and also the PCI IDE/ISA accelerator (such as PIIX4 device). The PCI IDE/ISA device then interfaces to other buses, such as IDE and ISA. The IDE interface has separate signals for both the primary and secondary IDE drives; these are electrically isolated, which allows drives to be swapped easily without affecting the other port.

The PCI IDE/ISA accelerator is a massively integrated device (the PIIX4 has 324 pins) and provides for an interface to other buses, such as USB and X-Bus. It also handles the interrupts from the PCI bus and ISA bus. It thus has two integrated 82C59 interrupt controllers, which support up to 15 interrupts. The PCI IDE/ISA accelerator also handles DMA transfers (on up to 8 channels), and thus has two integrated 82C37 DMA controllers. Along with this, it has an integrated 82C54, which provides for the system timer, DRAM refresh signal, and the speaker tone output.

The IDE (or AT bus) is the de facto standard for most hard disks in PCs. It has the advantage over older type interfaces that the controller is integrated into the disk drive. Thus, the computer only has to pass high-level commands to the unit and actual control can be achieved with the integrated controller. Several companies developed a standard command set for an AT attachment (ATA). Commands include

(1) read sector buffer—reads contents of the controller's sector buffer;
(2) write sector buffer—writes data to the controller's sector buffer;
(3) check for active;

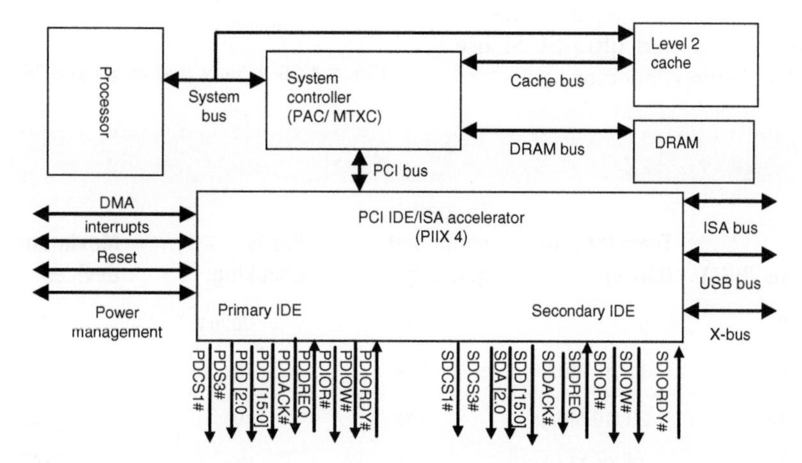

Fig. 1.4.24 IDE system connections.

(4) read multiple sectors;
(5) lock drive door.

The control of the disk is achieved by passing a number of high-level commands through a number of I/O port registers.

1.4.2.2.3 SCSI

Small computer systems interface (SCSI) has an intelligent bus subsystem and can support multiple devices cooperating concurrently. Each device is assigned a priority. The main types of SCSI are

(1) *SCSI-I.* SCSI-I transfers at rate of 5MB/s with an 8-bit data bus and seven devices per controller.
(2) *SCSI-II.* SCSI-II supports for SCSI-I and with one or more of the following:
   (a) Fast SCSI, which uses a synchronous transfer to give 10 MB/s transfer rate. The initiator and target initially negotiate to see if they can both support synchronous transfer. If they can, they then go into a synchronous transfer mode.
   (b) Fast and wide SCSI-2, which doubles the data bus width to 16 bits to give 20 MB/s transfer rate.
   (c) Fifteen devices per master device.
   (d) Tagged command queuing (TCQ), which greatly improves performance and is supported by Windows, NetWare, and OS-2.
   (e) Multiple commands sent to each device.
   (f) Commands executed in whatever sequence will maximize device performance.
(3) *Ultra SCSI (SCSI-III).* Ultra SCSI operates either as 8-bit or 16-bit with either 20 or 40 MB/s transfer rate (Table 1.4.5).

SCSI standard uses a 50-pin header connector and a ribbon cable to connect to up to eight devices. It overcomes the problem existing in IDE, where devices have to be assigned as a master and a slave. SCSI and fast SCSI transfer one byte at a time with a parity check on each byte. SCSI-II, wide SCSI, and ultra SCSI use a 16-bit data transfer and a 68-pin connector.

An SCSI bus is made of an SCSI host adapter connected to a number of SCSI units via SCSI bus. As all units connect to a common bus, only two units can transfer data at a time, either from one SCSI unit to another or from one SCSI unit to the SCSI host. The great advantage of this transfer is that it does not involve the processor.

Each unit on an SCSI unit is assigned a SCSI ID address. In the case of SCSI-I, this ranges from 0 to 7 (where 7 is normally reserved for a tape drive). The host adapter takes one of the addresses; thus a maximum of seven units can connect to the bus. Most systems allow the units to take on any SCSI ID address, but older systems required boot drives to be connected to a specific SCSI address. When the system is initially booted, the host adapter sends out a Start Unit command to each SCSI unit. This allows each of the units to start in an orderly manner (and not overload the local power supply). The host will start with the highest priority address (ID = 7) and finishes with the lowest address (ID = 0). Typically, the ID is set with a rotating switch selector or by three jumpers.

SCSI defines an initiator control and a target control. The initiator requests a function from a target, which then executes the function, as illustrated in Figure 1.4.25, where the initiator effectively takes over the bus for the time to send a command and the target executes the command and then contacts the initiator and transfers any data. The bus will then be free for other transfers. Table 1.4.6 gives the definitions of main SCSI signals. Each of the control signals can be true or false. They can be either OR-tied driven or Non-OR-tied driven. In OR-tied driven, the driver does not drive the signal to the false state. In this case, the bias circuitry of the bus terminators pulls the signal false

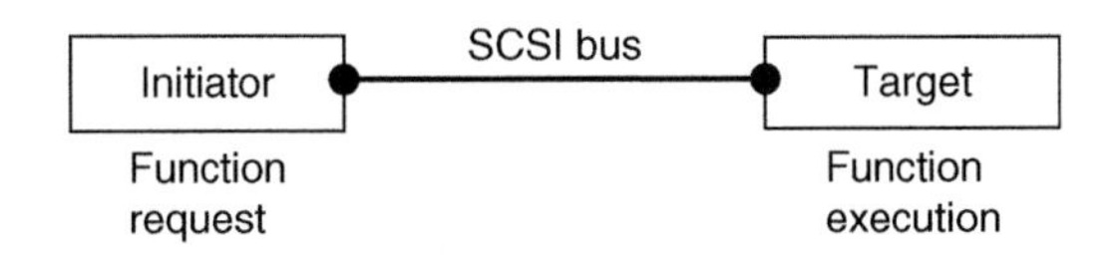

Fig. 1.4.25 Initiator and target in SCSI.

Table 1.4.5 SCSI types

Indices types	Data bus (bits)	Transfer rate (MB/s)	Tagged command queuing	Parity checking	Maximum devices	Pins on cable and connector
SCSI-I	8	5	No	Option	7	50
SCSI-II fast	8	10 (10 MHz)	Yes	Yes	7	50
SCSI-III fast/wide	16	20 (10 MHz)	Yes	Yes	15	68
Ultra SCSI	16	40 (20 MHz)	Yes	Yes	15	68

Table 1.4.6 SCSI main signals

Signals	Definitions
BSY	Indicates that the bus is busy, or not (an OR-tied signal).
ACK	Activated by the initiator to indicate an acknowledgment for an REQ information transfer handshake.
RST	When active (low) resets all the SCSI devices (an OR-tied signal).
ATN	Activated by the initiator to indicate the attention state.
MSG	Activated by the target to indicate the message phase.
SEL	Activated by the initiator; it is used to select a particular target device (an OR-tied signal).
C/D (control/data)	Activated by the target to identify whether there is data or control on the SCSI bus.
REQ	Activated by the target to acknowledge a request for an ACK information transfer handshake.
I/O (input/output)	Activated by the target to show the direction of the data on the data bus. Input defines that data is an input to the initiator, else it is an output.

whenever it is released by the drivers at every SCSI device. If any driver is asserted, then the signal is true. The BSY, SEL, and RST signals are OR-tied. In the ordinary operation of the bus, the BSY and RST signals may be simultaneously driven true by several drivers. However, in non-OR-tied driven, the signal may be actively driven false.

No signals other than BSY, RST, and D(PARITY) are driven simultaneously by two or more drivers.

The SCSI bus allows any unit to talk to any other unit, or the host to talk to any unit. Thus there must be some way of arbitration where units capture the bus. The main phases that the bus goes through are as follows:

(1) *Free bus.* In this state, there are no units that either transfer data or have control of the bus. It is identified by deactivate SEL and BSY (both will be high). Thus, any unit can capture the bus.

(2) *Arbitration.* In this state, a unit can take control of the bus and become an initiator. To do this, it activates the BSY signal and puts its own ID address on the data bus. After a delay, it tests the data bus to determine whether a high-priority unit has put its own address on the bus. If it has, then it will allow the other unit access to the bus. If its address is still on the bus, then it asserts the SEL line. After a delay, it then has control of the bus.

(3) *Selection.* In this state, the initiator selects a target unit and gets the target to carry out a given function, such as reading or writing data. The initiator outputs the OR value of its SCSI-ID and the SCSI-ID of the target onto the data bus (e.g., if the initiator is 2 and the target is 5 then the OR-ed ID on the bus will be 00100100). The target then determines that its ID is on the data bus and sets the BSY line active. If this does not happen within a given time, then the initiator deactivates the SEL signal, and the bus will be free. The target determines that it is selected when the SEL signal and its SCSI ID bit are active and the BSY and I/O signals are false. It then asserts the BSY signal within a selection abort time.

(4) *Reselection.* When the arbitration phase is complete, the wining SCSI device asserts the BSY and SEL signals and has delayed at least a bus clear delay plus a bus settle delay. The wining SCSI device sets the DATA BUS to a value that is the logical OR of its SCSI ID bit and the initiator's CSI ID bit. Sometimes, the target takes some time to reply to the initiator's request. The initiator determines that it is reselected when the SEL and I/O signals and its SCSI ID bit are true and the BSY signal is false. The reselected initiator then asserts the BSY signal within a selection abort time of its most recent detection of being reselected. An initiator does not respond to a reselection phase if other than two SCSI ID bits are on the data bus. After the target detects that the BSY signal is true, it also asserts the BSY signal and waits a given time delay and then releases the SEL signal. The target may then change the I/O signal and the data bus. After the reselected initiator detects the SEL signal is false, it releases the BSY signal. The target continues to assert the BSY signal until it gives up the SCSI bus.

(5) *Command.* The command phase is used by the target to request command information from the initiator. The target asserts the C/D signal and negates the I/O and MSG signals during the REQ/ACK handshake(s) of this phase. The format of the command descriptor block for 6-byte commands is: Byte 0—operation code; Byte 1—logical unit number (MSB, if required); Byte 2—logic block address; Byte 3—logic block address (LSB, if required); Byte 4—transfer length (if required)/parameter list length (if required)/allocation length (if required); Byte 5—control.

(6) *Data.* The data phase covers both the data-in and data-out phases. In the data-in phase, the target requests that data be sent to the initiator from the target. For this purpose, the target asserts the I/O signal and negates the C/D and MSG signals during the REQ/ACK handshake(s) of the phase. In the data-out phase, the target requests that data be sent from the initiator to the target. The target negates the C/D, I/O, and MSG signals during the REQ/ACK handshake(s) of this phase.

(7) *Message.* The message phase covers both the message-out and message-in phases. The first byte transferred in either of these phases can be either a single-byte message or the first byte of a multiple-byte message. Multiple-byte messages are contained completely within a single message phase. The message system allows the initiator and the target to communicate over the interface connection. Each message can be one, two, or more bytes in length. In a single message phase, one or more messages can be transmitted (but a message cannot be split between multiple message phases).

(8) *Status.* The status phase allows the target to request that status information be sent from the target to the initiator. The target shall assert the C/D and I/O signals and negate the MSG signal during the REQ/ACK handshake(s) of this phase. The status phase normally occurs at the end of a command (although in some cases it may occur before transferring the command descriptor block).

1.4.2.2.4 USB and Firewire

(1) *Universal serial bus (USB).* USB is mainly used for the connection of medium bandwidth peripherals such as keyboards, scanner, modem, video, game or graphic controller, and music interface. The great advantage of USB is that it allows for peripherals to be added and deleted from the system without causing any system upsets. The system will also automatically sense the connected device and load the required driver. Basically, USB provides these features: (1) easy to use; (2) self-identifying peripherals with automatic mapping of function to driver and configuration; (3) dynamically attachable and reconfigurable peripherals; and (4) low-speed and medium-speed transfer rate of 1.5 Mbps or 12 Mbps.

USB is a balanced bus architecture that hides the complexity of the operation from the devices connected to the bus. The USB host controller controls system bandwidth. Each device is assigned a default address when the USB device is first powered or reset. Hubs and functions are assigned a unique device address by USB software. All USB devices are attached to the USB via a port on specialized USB devices known as hubs. Hubs indicate the attachment or removal of a USB device in its per port status. Figure 1.4.26 shows an example connection of the USB 2.0 system. In this example, a memory hub is used to provide a fast data transfer (GB/s), while the Firewire connection provides ultra-high-speed connection for video transfers. The USB connection provides low-high and full-speed connections to most of the peripherals that connect to the system. The USB connections can be internal or can connect to an external hub. In the implementation of USB, there are two main ways, as given below.

(a) *Open host controller interface (OHCI).* This method defines the register level interface that enables the USB controller to communicate with the host computer and the operating system. OHCI is an industrial standard hardware interface for operating systems, device drivers, and the BIOS to manage the USB. It optimizes performance of the USB bus while minimizing CPU overhead to control the USB with a scatter/gather bushmaster hardware support. It has efficient isochronous data transfers allowing for high USB bandwidth without slowing down the host CPU. Furthermore, it ensures full compatibility with all USB devices.

(b) *Universal host controller interface (UHCI).* This method defines how the USB controller talks to the host computer and its operating system. It is optimized to minimize host computer design complexity and uses the host CPU to control the USB bus. This method has the advantages of simple design which reduces the transistor count required to implement the USB interface on the host computer, thus reducing the system cost. Furthermore, it can provide full compatibility with all USB devices.

In data transmission, USB supports two types of transfers: stream and message. A stream has no defined structure, whereas a message does. At start-up, one message pipe, control pipe 0, always exits as it provides access to the device's

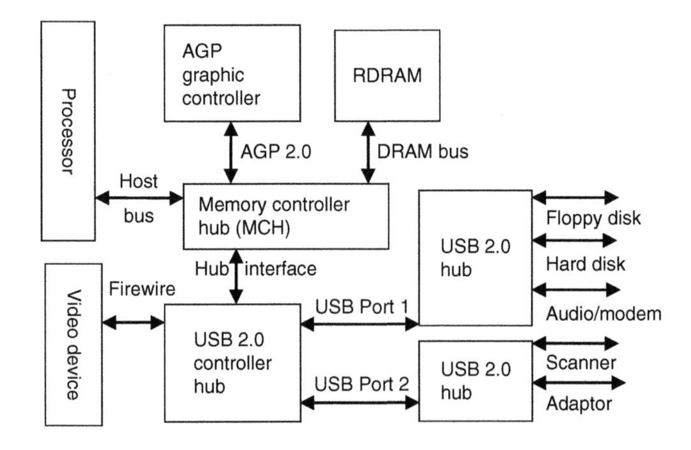

Fig. 1.4.26 An example connection of the USB 2.0 system.

configuration, status, and control information. USB optimizes large data transfers and real-time data transfers. When a pipe is established for an endpoint, most of the pipe's transfer characteristics are determined and remain fixed for the lifetime of the pipe. Each bus transaction of USB involves the transmission of up to three packets which can be (1) token packet transmission, (2) data packet transmission, and (3) handshake packet transmission. With these transfer characteristics, USB defines four transfer types:

(i) *Control transfers.* This is bursty, nonperiodic, host-software-initiated request/response communication typically used for command/status operations.

(ii) *Isochronous transfers.* This is periodic, continuous communication between host and device typically used for time-relevant information. This transfer type also preserves the concept of time encapsulated in the data. This does not imply, however, that the delivery needs of such data are always time-critical.

(iii) *Interrupt transfers.* This is small data, nonperiodic, low frequency, bounded latency, device-initiated communication typically used to notify the host of device service needs.

(iv) *Bulk transfers.* Nonperiodic, large bursty communication typically used for data that can use any available bandwidth and also is delayed until bandwidth is available.

As mentioned earlier, a major advantage of USB is the hot attachment and detachment of devices. USB does this by sensing when a device is attached or detached. When this happens, the host system is notified, and system software interrogates the device. It then determines its capabilities, and automatically configures the devices. All the required drivers are then loaded and applications can immediately make use of the connected device.

(1) *Attachment of USB devices.* All USB devices are addressed using the USB default address when initially connected or after they have been reset. The host determines whether the newly attached USB device is a hub or a function and assigns a unique USB address to the USB device. The host establishes a control pipe for the USB device using assigned USB address and endpoint number zero. If the attached USB device is a hub and USB devices are attached to its ports, then the above procedure is followed for each of the attached USB devices. If the attached USB device is a function, then attachment notifications will be dispatched by USB software to interested host software.

(2) *Removal of USB devices.* When a USB device has been removed from one of its ports, the hub automatically disables the port and provides an indication of device removal to the host. Then the host removes knowledge of the USB device. If the removed USB device is a hub, then the removal process must be performed for all of the USB devices that were previously attached to the hub. If the removed USB device is a function, removal notifications are sent to interested host software.

(2) *Firewire.* The main competitor to USB is the Firewire standard (IEEE 1394–1995 buses), which is a high-speed serial bus typically for video transfers, whereas USB supports low-to-medium-speed peripherals. Firewire supports rates of approximately 100,200, and 400 Mbps, known as S100, S200, and S400, respectively. The future standard promises higher data rates, and ultimately it is envisaged that rates of 3.2 Gbps will be achieved when optical fiber is introduced into the system. It uses point-to-point interconnect with a tree topology: 1000 buses with 64 nodes gives 64,000 nodes. Firewire also can be automatic configuration and hot plugging. In additional to asynchronous transfer, Firewire is able to be isochronous data transfer, where a fixed bandwidth is dedicated to a particular peripheral. However, it has the limit of the maximum cable length as 4.5 m. This should subsequently reduce the costs of production of controller interfaces and peripheral connectors, as well as simplifying the requirements placed on users when setting up their devices. Firewire is a more economical interface bus standard that performs fast and high-bandwidth data transmissions.

There are two bus categories in the Firewire:

(a) *Cable.* This is a bus that connects external devices via a cable. This cable environment is a noncyclic network with finite branches consisting of bus bridges and nodes (cable devices). Noncyclic networks contain no loops and result in a tree topology, with devices daisy-chained and branched (where more than one device branch is connected to a device). Devices on the bus are identified by node IDs. Configuration of the node IDs is performed by the self ID and tree ID processes after every bus reset. This happens every time a device is added to or removed from the bus, and is invisible to the user.

(b) *Backplane.* This type of topology is an internal bus. An internal IEEE-1394 device can be used alone, or incorporated into another backplane bus. Implementation of the backplane specification lags the development of the cable environment, but one could image internal IEEE-1394 hard disks in one computer being directly accessed by another IEEE-1394 connected computer.

One of the key capabilities of IEEE-1394 is isochronous data transfer. Both asynchronous and isochronous are supported, and are useful for different applications. Isochronous transmission transmits data like real-time speech and video, both of which must be delivered uninterrupted, and at the rate expected, whereas asynchronous transmission is used to transfer data that is not tied to a specific transfer time. With IEEE-1394, asynchronous transmission is the conventional transfer method of sending data to an explicit address, and receiving confirmation when it is received. Isochronous, however, is an unacknowledged guaranteed bandwidth transmission method, useful for just-in-time delivery of multimedia type data.

An isochronous "talker" requests an amount of bandwidth and a channel number. Once the bandwidth has been allocated, it can transmit data preceded by a channel ID. The isochronous listeners can then listen for the specified channel ID and accept the data following. If the data is not intended for a node, it will not be set to listen on the specific channel ID. Up to 64 isochronous channels are available, and these must be allocated, along with their respective band-widths, by an isochronous resource manager on the bus.

By comparison, asynchronous transfers are sent to explicit addresses on the 1394 bus. When data is to be sent, it is preceded by a destination address, which each node checks to identify packets for itself. If a node finds a packet addressed to itself, it copies it into its receive buffer. Each node is identified by a 16-bit ID, containing the 10-bit bus ID and 6-bit node or physical ID. The actual packet addressing, however, is 64 bits wide, providing a further 48 bits for addressing a specific offset within a node's memory.

1.4.2.2.5 AGP and Parallel Ports

(1) *AGP.* The accelerated graphic port (AGP) is a major advancement in the connection of three-dimensional graphics applications, and is based on an enhancement of the PCI bus. One of the major improvements is the speed of transfer between the main system memory and the local graphic card. This reduces the need for large areas of memory on the graphics card.

The main gain in moving graphics memory from the display buffer (on the graphic card) to the main memory is the display of text information because (1) it is generally read-only, and does not have to be displayed in any special order. (2) Shifting text does not require a great deal of data transfer and can be easily cached in memory, thus reducing data transfer. A shift in text can be loaded from the cached memory. (3) It is dependent on the graphics quality of the application, rather than the resolution of the display. (4) It is not persistent, as it resides in memory only for the duration that it is required. When it has completed the main memory, it can be assigned to another application. A display buffer, on the other hand, is permanent.

The 440LX was the first AGP set product designed to support the AGP interface. The HOST BRIDGE AGP implementation is compatible with the accelerated graphics port specification 1.0. HOST BRIDGE supports only a synchronous AGP interface, coupling with the host bus frequency. The AGP 1.0 interface can reach a theoretical 528 MB/s transfer rate and AGP 2.0 can achieve a theoretical 1.056 GB/s transfer rate. The actual bandwidth will be limited by the capability of the HOST BRIDGE memory subsystem.

(2) *Parallel port.* The parallel port is hardly the greatest technology. In its standard form, it allows only for simple communications from the PC outwards. However, like the RS-232, the parallel port is a standard port of the PC, and it is cheaper.

All parallel ports use a bidirectional link in either a compatible, nibble, or byte mode. These modes are relatively slow as the software must monitor the handshaking lines (up to 100 kbps). To allow high speeds, the enhanced parallel port and extended capabilities port protocol modes allow high-speed data transfer using automatic hardware handshaking.

1.4.2.2.6 RS-232, RS-422, RS-485, and RS-530

(1) *RS-232.* RS-232 is a TIA/EIA standard for serial transmission between DTE and DCE. Using a 25-pin DB-25 or 9-pin DB-9 connector, its normal cable limitation of 50 ft can be extended to several hundred feet with high-quality cable.

RS-232 defines the purpose and signal timing for each of the 25 lines; however, many applications use less than a dozen. RS-232 transmits positive voltage for a 0 bit, negative voltage for a 1. In 1984, this interface was officially renamed TIA/EIA-232-E standard (E is the current revision, 1991), although most people still call it RS-232. Table 1.4.7 lists the some RE-232 specifications.

(2) *RS-422 and RS-485.* RS-422 is a balanced serial interface for the transmission of digital data. The advantage of a balanced signal is the greater immunity to noise. The EIA describes RS-422 as a DTE to DCE interface for point-to-point connections.

RS-422 was designed for greater distances and higher baud rates than RS-232. In its simplest form, a pair of converters from RS-232 to RS-422 (and back again) can be used to form an "RS-232 extension cord." Data rates of up to 100 kbps and distances up to 4000 ft can be accommodated with RS-422. RS-422 is also specified for multidrop (party-line) applications where only one driver is connected to, and transmits on, a "bus" of up to 10 receivers.

RS-485 is standard for sending serial data. It uses a pair of wires to send a differential signal over distances up to 4000 ft without a repeater. The differential signal makes it very robust; RS-485 is one of the most popular communications methods used in industrial applications where its noise immunity and long-distance capability are a perfect fit. RS-485 is capable of multidrop communications—up to 32 nodes. RS-485 can be configured for their half or full duplex. Half duplex typically uses one pair of wires; full duplex requires two pairs.

Table 1.4.7 RS-232 and RS-422 specifications

Specifications	RS-232	RS-422
Mode of operation	Single-ended	Differential
Total number of drivers and receivers on	1 Driver	1 Driver
one line	1 Receiver	10 Receiver
Maximum cable length (ft)	50	4000
Maximum data rate	20 kbits/s	10 Mbits/s
Maximum driver output voltage (V)	±25	0.25 to +6
Driver output signal level loaded (loaded minimum) (V)	±2.0	±5 to ±15
Driver output signal level unloaded (unloaded maximum) (V)	±6	±25
Driver load impedance (Ω)	3–7 k	100
Maximum driver current in power on high Z state	N/A	N/A
Maximum driver current in power off high Z state	±100 μA	±6 mA @ ±2 V
Slew rate (maximum)	30 V/μs	N/A
Receiver input voltage range (V)	±15	0 to+10
Receiver input sensitivity	±3 V	±200 mV
Receiver input resistance (Ohms)	3–7 k	4 kmin

Both RS-422 and RS-485 use a twisted-pair wire (i.e., 2 wires) for each signal. They both use the same differential drive with identical voltage swings: 0 to +5 V. The main difference between RS-422 and RS-485 is that while RS-422 is strictly for point-to-point communications (and the driver is always enabled), RS-485 can be used for multiple drop systems. Since the basic differential receivers of RS-423-A and RS-422-A are electrically identical, it is possible to interconnect equipment using RS-423-A receivers and generators on one side of the interface with equipment using RS-422-A generators and receivers on the other side of the interface, if the leads of the receivers and generators are properly configured to accommodate such an arrangement and the cable is not terminated.

Table 1.4.7 lists some specifications for RS-422. The data is coded as a differential voltage between the wires. The wires are named A (negative) and B (positive). When B > A then the output is a mark (1 or off) and when A >B then it is counted as a space (0 or on). In general, a mark is +1 VDC for the A line and +4 VDC for the B line. A space is +1 VDC for the B line and +4 VDC for the A line. At the transmitter end the voltage difference should not be less than 1.5 VDC and not exceed 5 VDC. At the receiver end the voltage difference should not be less than 0.2 VDC. The minimum voltage level is −7 VDC and maximum +12 VDC.

(3) *RS-530.* RS-530 employs differential signaling on its send, receive, and clocking signals, as well as its control and handshaking signals. The differential signals for RS-530 are labeled as either "A or B." At both connectors, wire A always connects to A, and B connects to B.

The RS-530 transmitter sends a data 0 (or logic ON) by setting the potential on the A signal to 0.3 V (or more) higher than the voltage on the B signal. The transmitter sends a data 1 (or logic OFF) by setting the potential on the B signal to 0.3 V or more than the voltage on the A signal. The voltage offset (from ground reference) is not to exceed 3 V, however, most receivers can handle much more; check the receiver data sheet for exact limits. This approach is relatively immune to noise when the cable is constructed so that the A and B signal wires are a twisted pair. Shielding the cable is generally not required.

Data 0 = A > B + 0.3 V; Data 1 = B > A + 0.3 V
Example: Data 0: A = 2 V, B = 1 V; Data 1: A = 1 V, B = 2 V.

Most receivers can handle both + and − voltages; again check the data sheet on the part used to be sure. If you have the correct receivers it is possible for the older V.35 (±5 V) signaling to be wired to RS-530 or V11. This is how Cisco and others get many different interfaces on their Smart Serial connectors, and you thought it was magic!

1.4.2.2.7 IEEE-488

Hewlett–Packard originally developed the IEEE-488 bus called the HP-IB to connect and control programmable instruments, and to provide a standard interface for communication between instruments from different sources. The interface quickly gained popularity in the computer industry. Because the interface was so versatile, the IEEE committee renamed it General Purpose Interface Bus (GPIB).

Almost any instrument can be used with the IEEE-488 specification, because it says nothing about the function of the instrument itself, or about the form of the instrument's data. Instead, the specification defines a separate component, the interface, which can be added to the instrument. The signals passing into the interface from the IEEE-488 bus and from the instrument are defined in the standard. The instrument does not have complete control over the interface. Often the bus controller tells the interface what to do. The active controller performs the bus control functions for all the bus instruments.

(1) *IEEE-488 standards.* The IEEE-488.1 standard greatly simplified the interconnection of programmable instruments by clearly defining mechanical, hardware, and electrical protocol specifications. For the first time, instruments from different manufacturers were connected by a standard cable. This standard does not address data formats, status reporting, message exchange protocol, common configuration commands, or device specific commands.

The IEEE-488.2 standard enhances and strengthens the IEEE-488.1 standard by specifying data formats, status reporting, error handling, controller functionality, and common instrument commands. It focuses mainly on the software protocol issues and thus maintains compatibility with the hardware-oriented IEEE-488.1 standard. IEEE-488.2 systems tend to be more compatible and reliable.

The IEEE-488 standard allows up to 15 devices to be interconnected on one bus. Each device is assigned a unique primary address, ranging from 0 to 30, by setting the address switches on the device. A secondary address may also be specified, ranging from 0 to 30. See the device documentation for more information on how to set the device primary and optional secondary address.

The IEEE-488 bus specifies a maximum total cable length of 20 m with no more than 20 devices connected to the bus and at least two-thirds of the devices powered on. A maximum separation of 4 m between devices and an average separation of 2 m over the full bus should be followed. Bus extenders and expanders are available to overcome these system limits. The standard IEEE-488 cable has both a plug and receptacle connector on both ends. Special adapters and nonstandard cables are available for special interconnect applications.

(2) *Interface signals and data lines.* At power-up time, the IEEE-488 interface that is programmed to be the System Controller becomes the Active Controller in charge. The System Controller has several unique capabilities including the ability to send Interface Clear (IFC) and Remote Enable (REN) commands. IFC clears all device interfaces and returns control to the System Controller. REN allows devices to respond to bus data once they are addressed to listen. The System Controller may optionally pass control to another controller, which then becomes Active Controller.

There are three types of devices that can be connected to the IEEE-488 bus (listeners, talkers, and controllers). Some devices include more than one of these functions. The standard allows a maximum of 15 devices to be connected on the same bus. A minimum system consists of one controller and one talker or listener device (i.e., an HP 700 with an IEEE-488 interface and a voltmeter).

It is possible to have several controllers on the bus but only one may be active at any given time. The Active Controller may pass control to another controller which in turn can pass it back or on to another controller. A listener is a device that can receive data from the bus when instructed by the controller and a talker transmits data on to the bus when instructed. The controller can set up a talker and a group of listeners so that it is possible to send data between groups of devices as well.

The IEEE-488 interface system consists of 16 signal lines and 8 ground lines. The 16 signal lines are divided into three groups (8 data lines, 3 handshake lines, and 5 interface management lines).

The lines DIO1 through DIO8 are used to transfer addresses, control information, and data.

The formats for addresses and control bytes are defined by the IEEE 488 standard. Data formats are undefined and may be ASCII (with or without parity) or binary. DIO1 is the Least Significant Bit (note that this will correspond to bit 0 on most computers).

(3) *Handshake lines and handshaking.* The three handshake lines (NRFD, NDAC, DAV) control the transfer of message bytes among the devices and form the method for acknowledging the transfer of data. This handshaking process guarantees that the bytes on the data lines are sent and received without any transmission errors and is one of the unique features of the IEEE-488 bus.

(a) The NRFD (Not Ready for Data) handshake line is identified by a listener to indicate it is not yet ready for the next data or control byte. Note that the controller will not see NRFD released (i.e., ready for data) until all devices have released it.

(b) The NDAC (Not Data Accepted) handshake line is identified by a listener to indicate it has not yet accepted the data or control byte on the data lines. Note that the controller will not see NDAC released (i.e., data accepted) until all devices have released it.

(c) The DAV (Data Valid) handshake line is identified by the talker to indicate that a data or control byte has been placed on the data lines and has had the minimum specified stabilizing time. The byte can now be safely accepted by the devices.

The handshaking process is outlined as follows. When the controller or a talker wishes to transmit data on the bus, it sets the DAV line high (data not valid), and checks to see that the NRFD and NDAC lines are both low, and then it puts the data on the data lines. When all the devices that can receive the data are ready, each releases its NRFD (not ready for data) line. When the last receiver releases NRFD, and it goes high, the controller or talker takes DAV low indicating that valid data is now on the bus. In response each receiver takes NRFD low again to indicate it is busy and releases NDAC (not data accepted) when it has received the data. When the last receiver has accepted the data, NDAC will go high and the controller or talker can set DAV high again to transmit the next byte of data. Note that if after setting the DAV line high, the controller or talker senses that both NRFD and NDAC are high, an error will occur. Also if any device fails to perform its part of the handshake and releases either NDAC or NRFD, data cannot be transmitted over the bus. Eventually a timeout error will be generated. The speed of the data transfer is controlled by the response of the slowest device on the bus; for this reason it is difficult to estimate data transfer rates on the IEEE-488 bus as they are always device dependent.

(4) *Interface management lines.* The five interface management lines Attention (ATN), End or Identify (EOI), Interface Clear (IFC), Remote Enable (REN), and Service Request (SRQ) manage the flow of control and data bytes across the interface.

(a) The ATN (Attention) signal is chosen by the controller to indicate that it is placing an address or control byte on the data bus. ATN is released to allow the assigned talker to place status or data on the data bus. The controller regains control by reasserting ATN; this is normally done synchronously with the handshake to avoid confusion between control and data bytes.

(b) The EOI (End or Identify) signal has two uses. A talker may assert EOI simultaneously with the last byte of data to indicate end-of-data. The controller may assert EOI along with ATN to initiate a parallel poll. Although many devices do not use parallel poll, all devices should use EOI to end transfers (many currently available ones do not).

(c) The IFC (Interface Clear) signal is selected only by the System Controller in order to initialize all device interfaces to a known state. After releasing IFC, the System Controller is the Active Controller.

(d) The REN (Remote Enable) signal is selection only by the System Controller. Its selection does not place devices into remote control mode; REN only enables a device to go into remote mode when addressed to listen. When in remote mode, a device should ignore its local front panel controls.

(e) The SRQ (Service Request) line is like an interrupt: it may be used by any device to request the controller to take some action. The controller must determine which device is calling for the SRQ by conducting a serial poll. The requesting device releases SRQ when it is polled.

1.4.3 HMI in industrial control

1.4.3.1 Overview

The term "user interface" refers to the methods and devices that are used to accommodate interaction between machines and the human beings who use them. The user interface of a mechanical system, a vehicle, or an industrial installation and so on is often referred to as the HMI. In any industrial control system, the HMI can be used to deliver information from machine to people, which allows people to control, monitor, and record the system through devices such as image, keyboard, Ethernet, screen, video, radio, and software. Actually, the HMI can take many forms. Although there are many techniques and methods used in industry, the HMI always accomplishes two fundamental tasks: communicating information from the machine to the user, and communicating information from the user to the machine.

Two industrial applications of the HMI are given below to demonstrate its importance in industry and industrial control. However, in view of the fact that the HMI becomes more and more essential in industry, these two examples are far from representing its applications in industry.

A robot control is a good example that requires working with the HMI. This robot control is based on human speech and gestures instead of keyboard or joystick control. In turn, the robot can also respond to the human control orders by using speech and gestures. When both the operator and the robot understand the environment and the work objects, they can communicate easier and the work task can be completed collaboratively. In most robots, the HMI is a key component in any work cell. The HMI must allow operators to run the equipment and cell in an intuitive manner. It must be configurable to give each level of personnel appropriate access to various layers of functionality. An autonomous service robot operates in the user's own environment, performing independent tasks to reach user goals. Applications include, for instance, delivery agents in hospitals and factories, and cleaning robots in the home or in supermarkets. The latest robot controllers now are beginning to offer built-in HMI functionality, complete with touch-screen interfaces, status indicators, program selection switches, part counters, and various other functions, which can be seen in Figure 1.4.27.

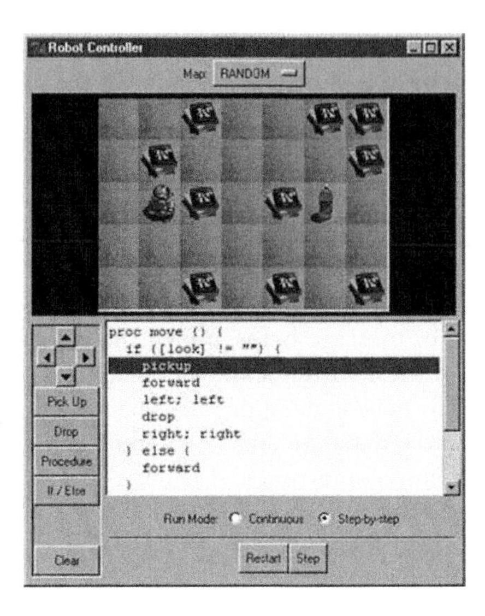

Fig. 1.4.27 A touch-screen shot of the human–machine interface for a robot (courtesy of Siemens).

Another example in this aspect is the SCADA system in which the HMI is an essential component. In industry, the HMI in SCADA was born out of a need for a user-friendly front-end to a control system containing PLC. While a PLC does provide automated, preprogrammed control over a process, they are usually distributed across a plant, making it difficult to gather data from them manually. Additionally, the PLC information is usually in a crude userunfriendly format. The SCADA of HMI gathers information from the PLCs via some form of network, and combines and formats the information. A sophisticated HMI may also be linked to a database, to provide instant trending, diagnostic data, scheduled maintenance procedures, logistic information, detailed schematics for a particular sensor or machine, and expert-system troubleshooting guides. Since about 1998, many companies, especially all major PLC manufacturers, have offered integrated SCADA and HMI systems for its comprehensive range of facilities for industrial automation and process control. These companies recognized the benefits of using the same reliable monitoring and control software throughout their business, from shop floor through to top management. Many of them used open and nonproprietary communications protocols. Numerous specialized third-party SCADA packages with the HMI, offering built-in compatibility with most major PLCs, have also entered the market, allowing mechanical engineers, electrical engineers, and technicians to configure the HMI by themselves, without the need for a custom-made program written by a software developer. Figure 1.4.28 is a diagram including an SCADA system and its control screen in Website.

1.4.3.2 Human–machine interactions

Automated systems have penetrated virtually every area of our private life and our work environments. Development work on technical products is accelerating, and the products themselves are becoming increasingly complex and powerful. Human–machine interaction is already playing a vital role along the entire production process, from planning individual links in the production chain right through to designing the finished product. Innovative technology is made for humans, used and monitored by humans. The optimum

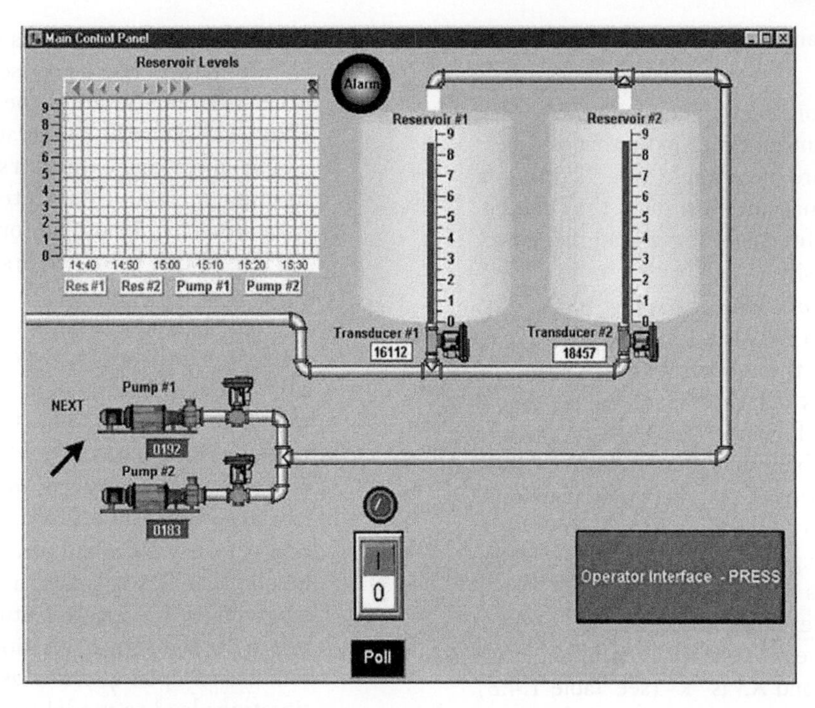

Fig. 1.4.28 A SCADA system and its Website-control screen (courtesy of Siemens).

form for this interaction depends on whether a technical innovation is reliable in operation, is safe, is accepted by personnel, and, last but not least, is cost-effective. This interplay between technology and user, known as human–machine interaction, is hence at the very heart of industrial automation, automated control, and industrial production.

1.4.3.2.1 The models for human–machine interactions

Modeling the human–machine interaction is done simply to depict how human and machine interact in a system. The human–machine interaction model illustrates a typical information flow (or process context) between the "human" and "machine" components of a system. Figure 1.4.29 provides the components involved in each side of the human–machine interaction. The environment side has three components: machine display component, machine CPU component, and machine I/O device component. The human side has another three components: human sensory component, human cognitive component, and human musculoskeletal component.

In modern control systems, a model is a common architecture for grouping several machine configurations under one label. The set of models in a control system corresponds to a set of unique machine behaviors. The operator interacts with the machine by switching among models manually, or monitoring the automatic switching triggered by the machine. However, our understanding of models and their potential contribution to confusion and error is still far from complete. For example, there is widespread disagreement among user interface designers and researchers about what models are, independent of how they affect users. This blurred vision, found not only in the human–machine interaction domain, impedes our ability to develop methods for representing and evaluating human interaction with control systems. This limitation is magnified in high-risk systems such as automated cockpits, for which there is an urgent need to develop methods that will allow designers to identify, early in the design phase, the potential for error. The errors arising from modeling the human–machine

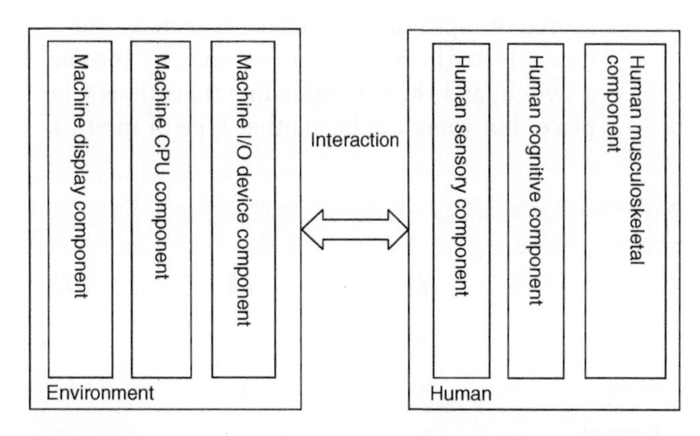

Fig. 1.4.29 The components involved in the human–machine interaction.

interaction are thus an important issue that cannot be ignored.

(1) *Definition and construct.* The constructs of the human–machine interaction model that will be discussed below are measurable aspects of human interaction with machines. As such, they can be used to form the foundation of a systematic and quantitative analysis.

(a) *Models' behaviors.* One of the first treatments of models came in the early 1960s from the science of cybernetics, the comparative study of human control systems and complex machines. The first treatments set forth the construct of a machine with different behaviors. The following is a simplified description of the machine behaviors' construct: A given machine may have several components (e.g., X1, X2, X3). For each component there is a finite set of states. On "Startup," for example, the machine initializes itself so that the active state of component X1 is "a," X2 is "f," and X3 is "k" (see Table 1.4.8). The vector of states (a, f, k) thus defines the machine's configuration on "Startup." Once a built-in test is performed, the machine can move from "Startup" to "Ready." The transition to "Ready" can be defined so that for component X1, state "a" undergoes a transition and becomes state "b"; for X2 state "f" transitions to "g", and for X3, "k" changes to "l". The new configuration (b, g, l) defines the "Ready" model of the machine. Now, there might be a third set of transitions, for example, to "Engaged" (c, h, m), and so on.

The set of configurations labeled "Startup," "Ready" and "Engaged," if embedded in the same physical unit, corresponds to a machine with three different ways of behaving. A real machine whose behavior can be so represented is defined as a machine with "input." The input triggers the machine to change its behavior. One source of input to the machine is manual; the user selects the model, for example, by turning a switch, and the corresponding transitions take place. But there can be another type of input: If some other machine selects the model, the input is "automatic." More precisely, the output of the other machine becomes the input to our machine. For example, a separate machine performs a built-in test and outputs a signal that causes the machine in Table 1.4.8 to transition from "Standby" to "Ready," automatically.

Table 1.4.8 A Machine with different behaviors

	X1	X2	X3
Startup	a	f	k
Ready	b	g	l
Engaged	c	h	m

Here in this book, we first define a model as a machine configuration that corresponds to a unique behavior. This is a very broad definition of the term. Later on, we constrain this definition from our special perspective: user interaction with control systems that employ models.

(b) *Model's error and ambiguity.* Model errors fall into a class of errors that involve forming and carrying out an intention. That is, when a situation is falsely classified, the resulting action may be one that was intended and appropriate for a perceived or expected situation, but inappropriate for the actual situation. However, there remains an open question as to what kind of situations lead to model error. This issue can be addressed by an example of a word processor. Specifically, we looked at situations in which the user's input has different interpretations, depending on model. For example, in one word processing application, the keystroke d can be interpreted as either (1) the literal text "d," or (2) as the command "delete." The interpretation depends on the word processor's active model: either Text or Command.

Model error can be linked to model ambiguity with interjecting the notion of user expectations. In this view, "model ambiguity" will result in model error only when the user has a false expectation about the result of his or her actions. There are two types of ambiguity: one that leads to model error, and one that does not lead to model error. An example of model ambiguity that does lead to model error is timesharing operating systems in which keystrokes are buffered until a "Return" or "Enter" key is pressed. However, when the buffer gets full, all subsequent keystrokes are ignored accordingly. This feature leads to two possible outcomes; all or only a portion of the keystrokes will be processed. The two outcomes depend on the state of buffer which is either "not full" or "full." Since the state of the buffer is unknown to the user, false expectations may occur. The user's action: hitting the "Return" key and seeing only part of what was keyed on the screen is therefore a "model error" because the buffer has already filled up.

An example of model ambiguity that does not lead to model error is a common end-of-line algorithm which determines the number of words in a line. An ambiguity is introduced because the criteria for including the last word on the current line, or wrapping to the next line, are known to the user. Nevertheless, as long as the algorithm works reasonably well, the user will not complain because he or she has not formed any expectation about which word will stay on the line or scroll down, and either outcome is usually acceptable. Therefore, model error will not take place, even though model ambiguity does indeed exist.

(c) *User factors: Task, knowledge, and ability.* One important element that constrains user expectations is the task at hand. If discerning between two or more different machine configurations is not part of the user's task, model error will not occur. Consider, for example, the radiator fan of your car. Do you know what configuration (OFF, ON) it is in? The answer, of course, is no. There is no such indication in most modern cars.

The fan mechanism changes its mode automatically depending on the output of the water temperature sensor. Model ambiguity exists because at any point in time, the fan mechanism can change its model or stay in the current model. The configuration of the fan is completely ambiguous to the driver. But does such model ambiguity lead to model error? The answer is obvious; not at all because monitoring the fan configuration is not part of the driver's task. Therefore, the user's task is an important determinant of which machine configurations must be tracked by the user and which machine configurations need not be tracked.

The second element that is part of the assessment of user expectations is user knowledge about the machine's behaviors. By this, we mean that the user constructs some mental model of the machine's "response map." This mental model allows the user to track the machine's configuration, and most importantly, to anticipate the next configuration of the machine. Specifically, our user must be able to predict what the new configuration will be following a manually triggered event or an automatically triggered event. The problem of reliably anticipating the next configuration of the machine becomes difficult when the number of transitions between configurations is large. Another factor in user knowledge is the number of conditions that must be evaluated as TRUE, before a transition from one model to another takes place. For example, the automated flight control systems of modern aircraft can execute a fully automatic (hands-off) landing. Several conditions (two engaged autopilots, two navigation receivers tuned to the correct frequency, identical course set, and more) must be TRUE before the aircraft will execute automatic landing. Therefore, in order to reliably anticipate the next model configuration of the machine, the user must have a complete and accurate model of the machine's behavior, including its configurations, transitions, and associated conditions. This model, however, does not have to be complete in the sense that it describes every configuration and transition of the machine. Instead, as discussed earlier, the details of the user's model must be merely sufficient for the user's task, which is a much weaker requirement.

The third element in the assessment of user expectations is the user ability to sense the conditions that trigger a transition. Specifically, the user must be able to first sense the events (e.g., a flight director is engaged; aircraft is more than 400 ft above the ground) and then evaluate whether or not the transition to a model (say, a vertical navigation) will take place. These events are usually made known to the user through an interface. Nevertheless, there are more than a few control systems in which the interface does not depict the necessary input events. Such interfaces are said to be incorrect. In large and complex control systems, the user may have to integrate information from several displays in order to evaluate whether the transition will take place or not. For example, one of the conditions for a fully automated landing in a two-engine jetliner is that two separate electrical sources must be online, each one supplying its respective autopilot among the two autopilots. This information is external to the automatic flight control system, in the sense that it involves another aircraft system. The user's job of integrating events, some of which are located in different displays, is not trivial. One important requirement for an efficient design is for the interface to integrate these events and provide the user with a succinct cue.

In summary, we have discussed three elements that help to determine whether a given model ambiguity will or will not lead to false expectations. First is the relationship between model ambiguity and the user's task. If distinguishing between models (e.g., radiator fan is

"ON" or "OFF") is not part of the user's task, no meaningful errors will occur. Second, in a case where model ambiguity is relevant to the user's task, we assess the user's knowledge. If the user has an inaccurate and or incomplete model of the machine's response map, he or she will not be able to anticipate the next configuration and model confusion will occur. Third, we evaluate the user's ability to sense input events that trigger transitions. The interface must provide the user with all the necessary input events. If it does not, no accurate and complete model will help; the user may know what to look for but will never find it. As a result, confusion and model error will occur.

(2) *Classifications and types.* A classification of the human–machine interaction models is proposed here to encompass three types of models in automated control systems: (1) "interface models" that specify the behavior of the interface, (2) "functional models" that specify the behavior of the various functions of a machine, and (3) "supervisory models" that specify the level of user and machine involvement in supervising the process. Before we proceed to discuss this classification, we shall briefly describe a modeling language, "State Charts," that will allow us to represent these models.

Finite State Machine Model is given as a natural medium for describing the behavior of a model-based system. A basic fragment of such a description is a state transition which captures the states, conditions or events, and transitions in a system. The State Chart language is a visual formalism for describing states and transitions in a modular fashion by extending the traditional Finite State Machine to include three unique features: "hierarchy," "concurrency," and "broadcast." The "hierarchy" is represented by substates encapsulated within a superstate. The "concurrency" is shown by means of two or more independent processes working in parallel. The "broadcast" mechanism allows for coupling of components, in the sense that an event in one end of the network can trigger transitions in another. These features of the State Chart are further explained in the following three examples.

(a) *Interface models.* Figure 1.4.30 is a modeling structure of an interface model. It has three concurrently active processes (separated by a broken line): speed knob behavior, speed knob indicator, and speed window display. The behavior of the speed knob (middle process) is either "normal" or "pushed-in." (These two states are depicted, in the State Chart language, by two rounded rectangles.) The initial state of the speed knob is normal (indicated by the small arrow above the state), but when momentarily pushed, the speed knob engages or disengages the Speed Intervene submode of the vertical navigation (VNAV, hereafter) model. The transition between normal and pushed-in is depicted by a solid arrow and the label "push" describes the triggering event. The transition back to normal occurs immediately as the pilot lifts his or her finger (the knob is spring loaded).

The left-most process shown in Figure 1.4.30 is the speed knob indicator. In contrast to many such knobs that have indicators, the Boeing-757 speed knob itself has no indicator and therefore is depicted as a single (blank) state. The right-most process is the speed window display that can be either closed or open. After VNAV is engaged, the speed window display is closed (implying that the source of the speed is from another component, the flight management computer). After VNAV is disengaged, and a semiautomatic model such as vertical speed is active, the speed window display is open, and the

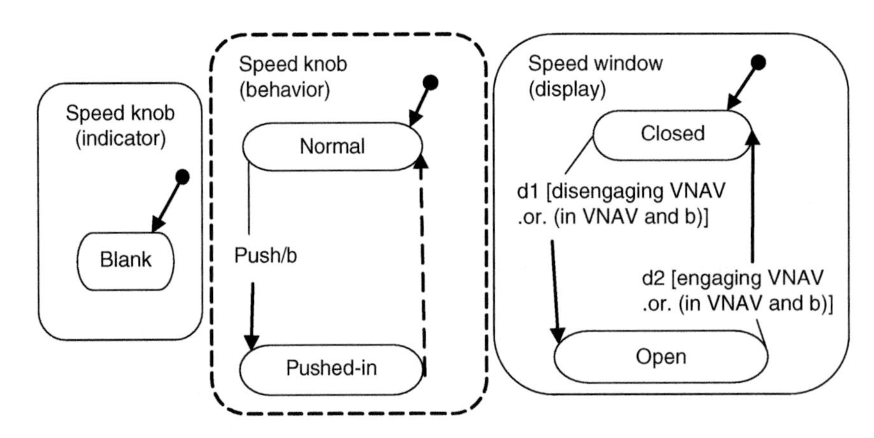

Fig. 1.4.30 An interface model.

pilot can observe the current speed value and enter a new speed value. This logic is depicted in the speed knob indicator process in Figure 1.4.30: transition d1 from closed to open is conditioned by the event "disengaging VNAV," and d2 will take place when the pilot is "engaging VNAV."

When in vertical navigation model, the pilot can engage the Speed Intervene submodel by pushing the speed knob. This event, "push" (which can be seen in speed knob behavior process), triggers event b, which is then broadcast to other processes. Being in VNAV and sensing event b ("in VNAV and b") is another (.OR.) condition on transition d1 from closed to open. Likewise, it is also the condition on transition d2 that takes us back to closed. To this end, the behavior of the speed knob is circular; the pilot can push the knob to close and push it again to open, ad infinitum.

As explained above and seen in Figure 1.4.30, there are two sets of conditions on the transitions between close and open. Of all these conditions, one, namely "disengaging VNAV," is not always directly within the pilot's control; it sometimes takes place automatically (e.g., during a transition from VNAV to the altitude hold model). Manual reengagement of VNAV will cause the speed parameter in the speed window to be replaced by economy speed computed by the flight management computer. If the speed value in the speed window was a restriction required by the American Transport Council, the aircraft will now accelerate/decelerate to the computed speed and the American Transport Council speed restriction will be ignored!

(b) *Functional models.* When we survey the use of models in devices, an additional type of model emerges: the "functional model" which refers to the active function of the machine that produces a distinct behavior. An automatic gearshift mechanism of a car is one example of a machine with different models, each one defining different behaviors.

As we move to discussion of functional models and their uses in machines that control a timed process, we encounter the concept of "dynamics." In dynamic control systems, the configuration and resulting behavior of the machine are a combination of a model and its associated parameter (speed, time, etc.). Referring back to our car example, the active model is the engaged gear that is Drive, and the associated parameter is the speed that corresponds to the angle of the accelerator pedal (say, 65 miles/h). Both model (Drive) and parameter (65 miles/h) define the configuration of the mechanism.

Figure 1.4.31 depicts the structure of a functional model in the dynamic automated control system of a modern airliner. Two concurrent processes are depicted in this modeling structure: (1) models, and (2) parameter sources.

Three models are depicted in the vertical models superstate in Figure 1.4.31: vertical navigation, altitude hold, and vertical speed (the default model). All are functional models related to the vertical aspect of flight. The speed parameter can be obtained from two different sources: the flight management computer or the model control panel. The default source of the speed parameter, indicated by the small arrow in Figure 1.4.31, is the model control panel. As mentioned in the discussion on interface models, engagement of vertical navigation via the model control panel will cause a transition to the flight management computer as the source of speed. This can be seen in Figure 1.4.31 where transition m2 will trigger event rv1, which, in turn, triggers an automatic transition (depicted as a broken line) from "model control panel" to "flight management computer." In many dynamic control mechanisms, some model transitions trigger a parameter source change while others do not. Such independence appears to be a source of confusion to operators.

(c) *Supervisory models.* The third type of model we discuss here is "supervisory models" that

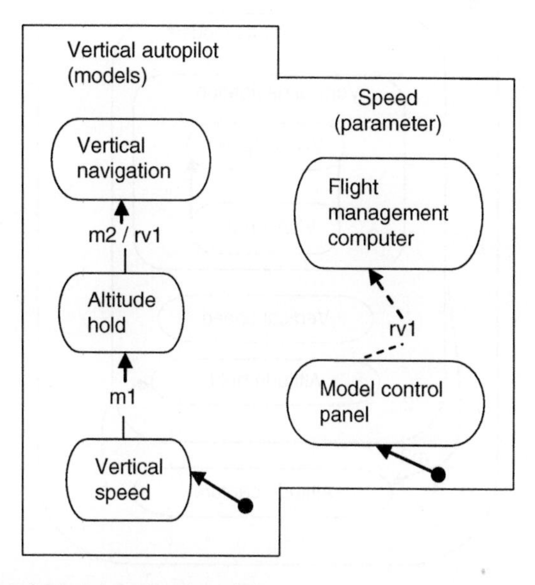

Fig. 1.4.31 A functional model.

sometimes are also referred to as "participatory" or "control" models. Modern automated control mechanisms usually allow the user flexibility in specifying the level of human and machine involvement in controlling the process. That is, the operator may decide to engage a manual model in which he or she is controlling the process; a semiautomatic model in which the operator specifies target values, in real time, and the machine attempts to maintain them; or fully automatic models in which the operator specifies in advance a sequence of target values, that is parameters, and the machine executes these automatically, one after the other.

Figure 1.4.32 is an example of a supervisory model structure that can be found in many control mechanisms, such as the automated flight control system, cruise control of a car, and robots on assembly lines. The modeling structure consists of hierarchical layers of superstates, each with its own set of models. The supervisory models in the Automated Flight Control System are organized hierarchically. Three main levels are described in Figure 1.4.32. The highest level of automation is the vertical navigation model (level 3), depicted as a superstate at the top of the models pyramid. Two submodels are encapsulated in the vertical navigation model; VNAV Speed and VNAV Path; each one exhibiting a somewhat different control behavior. One level below (level 2) are two semiautomatic models: vertical speed and altitude hold.

One model in the Automated Flight Control System, altitude capture, can only be engaged automatically; no direct manual engagement is possible. This model engages automatically when the aircraft is beginning the level-off maneuver to capture the selected altitude. When the aircraft is several hundred feet from the selected altitude, an automatic transition from any climb model to altitude capture takes place (m3). In this aspect, an example can be that a transition from vertical navigation or vertical speed to altitude capture takes place (m3). Finally, when the aircraft reaches the selected altitude, a transition back from altitude capture to altitude hold model also takes place automatically (m4).

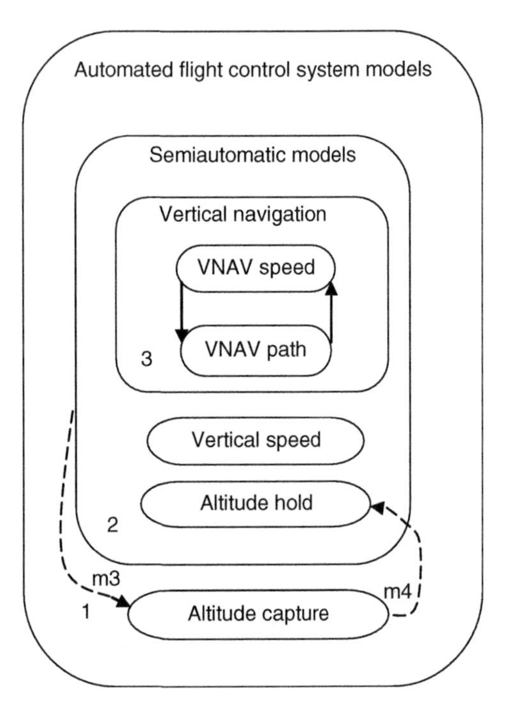

Fig. 1.4.32 A supervisory model.

In summary, we have illustrated a modeling language, Start Charts, for representing human interaction with control systems, and proposed a classification of three different types of models that are employed in computers, devices, and supervisory control systems. The "interface models" change display format. The "functional models" allow for different functions and associated parameters. Last are "supervisory models" that specify the level of supervision (manual, semiautomatic, and fully automatic) in the human–machine system. The three types of models described here are essentially similar, in that they all define the manner in which a certain component of the machine behaves. The component may be the interface only, a function of the machine, or the level of supervision. This commonality brings us back to our general working definition of the term "model," a machine configuration that corresponds to unique behavior.

1.4.3.2.2 Systems of human–machine interactions

There are three architectures of human–machine systems that are currently popular in industrial control: (1) adaptive HMI, (2) supervisory HMI, and (3) distributed human–machine interface.

(1) *Adaptive HMI.* In a complex control system, HMI is attempted to give users the means to perceive and manipulate easily huge quantities of information under resource constraints such as time, cognitive workload, and devices. The intelligence in the HMI makes the control system more flexible and more adaptable. One subset of intelligent user interface is adaptive interfaces. An adaptive interface modifies its behavior according to some defined constraints in order to best satisfy all of them, and varies in the ways and means that are used to achieve adaptation.

In HMI of an industrial control, operators, system, and context are continuously changing and are sometimes in contradiction with one another. Thus, this kind of interface can be viewed as the result of a balance between these three components and their changing relative importance and priority. An adaptive interface aims at assisting the operator in acquiring the most salient information in any context, in most appropriate form, and at the most opportune time. In any industrial control systems, two main factors are considered: the system that generates the information stream and the operator to which this stream is presented. The system and the operator share a common goal: to control the process and to solve any problems that may arise. This common objective makes them cooperate although they may both have their own goals. The role of the interface is to integrate these different goals with different levels of importance and the various constraints that come from the task, the environment, or the interface itself in order to produce an information presentation that best harmonizes the set of all these parameters. Specifically, an industrial process control should react consistently, in a timely fashion without disturbing the operator needlessly in his task. However, the most salient pieces of information should be presented in the most appropriate way.

The two main adaptation triggers used to modify the HMI could be

(a) *The process.* When the process moves from a normal state to a disturbed state, the streams of information may become denser and more numerous. To avoid any cognitive overload problems, the interface thus acts as a filter that channels the streams of information. To this end, it adapts the presentation of the pieces of information in order to help the operator identify and solve the problem more quickly, more easily, and more efficiently.

(b) *The operator.* An operator is much more difficult and trickier to drive the adaptation on the user state. As a matter of fact, the interface has to infer whether the operator reacts incorrectly and needs help based on his actions. Then, it may decide to adapt itself to highlight the problem and suggest solutions to assist the user.

The aim of the adaptation triggered by the process or the operator is to adapt the composition of the streams of information, that is, to adapt the organization and the presentation of the pieces of information on the interface in the best possible way, according to the state of the process and the inferred state of the operator. What is expected is to improve the communication between the system and the user. The means proposed in an adaptive HMI to reach this are the following:

(i) *Highlight relevant pieces of information.* The importance of a piece of information depends on its relevance according to the particular goals and constraints of each of the entities that participates in the communication between the system and the operator.

(ii) *Optimize space usage.* According to the current usage of the resources, it may turn out that it is necessary to reorganize the display space to cope with new constraints and parameters.

(iii) *Select the best representation.* According to the piece of information, its importance, the resources available, and the media currently in usage, the most appropriate media to communicate with the operator should be used.

(iv) *Timeliness of information.* The display of a particular piece of information should be timely regarding the process and the operator. This adaptation should follow the evolution of the process over time, but it should also adapt the timing of the displayed information to the inferred needs of the operator.

(v) *Perspectives.* In traditional interfaces, the operator has to decide what, where, when, and how the information should be presented. Thus, an operator can wonder whether he or she needs an adaptive interface to achieve their task or whether it will constitute more a bother than a help to him or her. This raises at least four questions: (1) from the client's point of view, whether the cost of developing an adaptive HMI is justifiable. (2) From the HMI designer's point of view, whether this kind of interface is usable and how to evaluate its usability. (3) From the developer's point of view, what are the best technical solutions to implement an efficient system within the required time? (4) From the operators' point of view, whether they consider such an adaptive interface as a collaborator or a competitor.

(2) *Supervisory HMI.* Supervisory HMI can be used in such systems and similar ones where there is a considerable distance between the control room and the machine house in a plant. It is from this machine house that the controller such as SCADA or PLC controls the objects which are, for example, pumps, blowers, and purification monitors. To provide the data communication, the supervisory software of the HMI is linked with the controllers over a network such as Ethernet, Control Area Network, and so on. The supervisory software is such that only one person is needed at any one time to monitor the whole plant from a single master device. The generated graphics show a clear representation on screen of the current status of any part of the system. A number of alarms are automatically activated directly on screen if parameters deviate from their tight tolerance band. This ensures extremely rapid updating of the control room screen contents. All the calculations for the controllers are calculated by the control software, using constant feedback from sensors throughout the production process.

According to many applications, the supervisory HMI is indeed an ideal software package for these cases above. Thanks to its interactive configuration and its setting assistants, supervisory HMI is able to straightforwardly get the system up and running and tested out. Supervisory human–machine interface is an open architecture which offers all the functions and options necessary for data collection and graphical representation of data on the operator screen. The system provides comprehensive logging of all measured values with databases. By accessing this database and by using real-time measurements, a wide variety of reports and trend curves can be viewed on the screen or output to a printer.

(3) *Distributed HMI.* The distributed HMI is a component-based approach. In a system of such an approach, the HMI could directly access any controller component, which also means that each controller exposes the HMI. Since all the system components are location transparent the HMI can bind to a component anywhere, be it in-process, local-process, or remote process. The most likely case is remote binding because it would be assumed that the HMI and the controller would reside on different platforms.

In the distributed HMI, the multiple servers are typically used to provide the systems with the flexibility and power of a peer-to-peer architecture. Each controller can have its own HMI server. The assigned server to a controller or a proxy server of a controller is perfect for each controller to manage expansion, frequent system changes, maintenance, and replicated automation lines within or across plants. Instead of a single data server, each controller component provides its own data services through a proxy server.

The primary drawback to decentralized components is the uncertainty of real-time controller performance generally resulting from poorly designed proxyagent use, for example, if the HMI samples controller data at too high a frequency. However, the distributed HMI is ideal for SCADA applications. Its distributed peer-to-peer architecture, reusable components, and remote deployment and maintenance capabilities make supporting SCADA applications remarkably efficient. The software's network services have been optimized for use over slow and intermittent networks, which significantly enhance application deployment and communications.

1.4.3.2.3 Designs of human–machine interactions

The design for a HMI is important, because the HMI of an application will often make or break this application. Although the functionality that an application provides to users is important, the way in which it provides that functionality is of the same importance. An application that is difficult to use will not be used. So, the value of HMI design should not be underestimated.

(1) *Design principles.* The following describes a collection of principles for improving the quality of HMI design.

(a) *The structure principle.* The HMI design should organize the interface purposefully, in meaningful and useful ways based on clear, consistent models that are apparent and recognizable to users, putting related things together and separating unrelated things, differentiating dissimilar things and making similar things resemble one another. The structure principle is concerned with overall interface architecture.

(b) *The simplicity principle.* The HMI design should make simple, common tasks simple to do, communicating clearly and simply in the user's own language, and providing good shortcuts that are meaningfully related to longer procedures.

(c) *The visibility principle.* The HMI design should keep all needed options and materials for a given task visible without distracting the user with extraneous or redundant information. Good designs do not overwhelm users with too many alternatives or confuse them with unneeded information.

(d) *The feedback principle.* The HMI design should keep users informed of actions or interpretations, changes of state or condition, and errors or exceptions that are relevant and of interest to the user through clear, concise, and unambiguous language familiar to users.

(e) *The tolerance principle.* The HMI design should be flexible and tolerant, reducing the cost of mistakes and misuse by allowing undoing and redoing, while also preventing errors wherever possible by tolerating varied inputs and sequences and by interpreting all reasonable actions.

(f) *The reuse principle.* The HMI design should reuse internal and external components and behaviors, maintaining consistency with purpose rather than merely arbitrary consistency, thus reducing the need for users to rethink and remember.

(2) *Design process.*

(a) *Phase one.* The design process begins with a task analysis in which we identify all the stakeholders, examine existing control or production systems and control and production processes, whether they are paper-based or computerized, and identify ways and means to streamline and improve the control or production process. Tasks at this phase are to conduct background research, interview stakeholders, and observe people conducting tasks.

(b) *Phase two.* Once having an agreed-upon objective and set of functional requirements, the next step should go through a design phase to generate a design that meets all the requirements. The goal of the design process is to develop a coherent, easy to understand software front-to-end that makes sense to the eventual users of the system. The design and review cycle should be iterated until we are satisfied with our design.

(c) *Phase three.* The next phase is implementation and test. We are also developing and implementing any performance support aids as necessary, for example, on-line help, paper manuals, etc.

(d) *Phase four.* Once a functional system is complete, we move into the final phase. What constitutes "success" is that people using our system, whether it be an intelligent tutoring system or online decision aid, are able to see solutions that they could not see before and/or better understand the constraints that are in place. We generally conduct formal experiments, comparing performance using our system with perhaps different features turned on and off to make contributions to the literature on decision support and human–machine interaction.

(3) *Design evaluation.* An important aspect of human–machine interaction is the methodology for evaluation of user interface techniques. Precision and recall measures have been widely used for comparing the ranking results of noninteractive systems, but are less appropriate for assessing interactive systems. The standard evaluations emphasize high recall levels. However, in many interactive settings, users require only a few relevant documents and do not care about high recall to evaluate highly interactive information access systems. Useful metrics beyond precision and recall include: time required to learn the system, time required to achieve goals on benchmark tasks, error rates, and retention of the use of the interface over time.

Empirical data involving human users is time consuming to gather and difficult to draw conclusions from. This is due in part to variation in users' characteristics and motivations, and in part to the broad scope of information access activities. Formal psychological studies usually only uncover narrow conclusions within restricted contexts. For example, quantities such as the length of time it takes for a user to select an item from a fixed menu under various conditions have been characterized empirically, but variations in interaction behavior for complex tasks like information access are difficult to account for accurately. A more informal evaluation approach is called a heuristic evaluation in which user interface affordances are assessed in terms of more general properties and without concern about statistically significant results.

1.4.3.3 Interfaces

For all industrial control systems, a high degree of user friendliness at the interface between human and machine is a decisive prerequisite for being accepted by the general public. Ambient Intelligence applications are characterized by multimodal interfaces as well as by the proactive behavior of the controller system. Therefore, various interfaces must be sensibly combined with each other, and the interaction with humans must be perfectly adapted to the individual situation of the human. Specific challenges include, among others, the selection of suitable interfaces for specific applications, the dynamic changes of interfaces based on changes in the state of the human such as "experiences gained" or "accident," as well as the experience-based optimization of such interfaces.

Regarding selection, methods are currently being developed that can suggest suitable interfaces based on a comprehensive characterization of the requirements. This methodology is very comprehensive and complex, since the requirements involve human properties such as their desire for information or personal preferences. Further evaluation and optimization of the methodology are absolutely indispensable. This work urgently requires the collaboration of psychologists. With respect to the dynamic changing of interfaces, this must be supported at least by semiautomatic generation. Experience-based patterns may be a suitable approach for this. Concerning the optimization of interfaces, an increase of acceptance through experience-based optimization can be envisioned. Such assistance systems already exist in vehicles, where, for example, the type of acceleration can be adapted to the style of driving of the respective driver.

1.4.3.3.1 Devices

(1) *Operator interface terminals.* These HMI are operator interface terminals with which users interact in order to control other devices. Some HMIs include knobs, levers, and controls. Others provide programmable function keys or a full keypad. Devices that include a processor or interface to personal computers are also available. Many HMIs include alphanumeric or graphic displays. For ease of use, these displays are often backlit or use standard messages. When selecting HMIs, important considerations include devices supported and devices controlled. Device dimensions, operating temperature, operating humidity, and vibration and shock ratings are other important factors.

Many HMIs include flat panel displays (FPDs) that use liquid crystal display (LCD) or gas plasma technologies. In LCD, an electric current passes through a liquid crystal solution that is trapped between two sheets of polarizing material. The crystals align themselves so that light cannot pass, producing an image on the screen. LCD can be monochrome or color. Color displays can use a passive matrix or an active matrix. Passive matrix displays contain a grid of horizontal and vertical wires with an LCD element at each intersection. In active matrix displays, each pixel has a transistor that is switched directly on or off, improving response times. Unlike LCD, gas plasma displays consist of an array of pixels, each of which contains red, blue, and green subpixels. In the plasma state, gas reacts with the subpixels to display the appropriate color.

These HMIs differ in terms of performance specifications and I/O ports. Performance specifications include processor type, random access memory (RAM), and hard drive capacity, and other drive options. I/O interfaces allow connections to peripherals such as mice, keyboards, and modems. Common I/O interfaces include Ethernet, Fast Ethernet, RS-232, RS-422, RS-485, SCSI, and USB. Ethernet is a LAN protocol that uses a bus or star typology and supports data transfer rates of 10 Mbps. Fast Ethernet is a 100 Mbps specification. RS-232, RS-422, and RS-485 are balanced serial interfaces for the transmission of digital data. SCSI is an intelligent I/O parallel peripheral bus with a standard, device-independent protocol that allows many peripheral devices to be connected to the SCSI port. USB is a four-wire, 12-Mbps serial bus for low-to-medium speed peripheral device connections.

These HMIs are available with a variety of features. For example, some devices are web-enabled or network-able. Others include software drivers, a stylus, and support for a keyboard, mouse, and printer. Devices that provide real-time clock support use a special battery and are not connected to the power supply. Power-over-Ethernet (PoE) equipment eliminates the need for separate power supplies altogether. These, HMI that offer shielding against electromagnetic interference (EMI) and radio frequency interference (RFI) are commonly available. Devices that are designed for harsh environments include enclosures that meet standards from the National Electronics Manufacturers' Association (NEMA).

(2) *Operator interface monitors.* Machine controllers and monitors use electronic numeric control and a monitoring interface for programming and calibrating computerized machinery. This product area includes general-purpose machine controllers, embedded machine controllers, machine monitors, computer numerically controlled (CNC) stepper motors, and CNC router controllers. A machine controller is a programmable, automatic, and CNC device. An embedded machine controller is part of a larger system. A machine monitor is used to collect and display production data from production equipment such as presses. A CNC stepper motor is used to drive a machine tool with power and precision. A CNC router controller is used to cut tool paths. Many other types of machine controllers and monitors are also available.

Machine controllers and monitors consist of many different components. A machine controller uses a microprocessor to perform predetermined control and logical operations. Memory is added to the processor in order to record data from the machine. Often, an input device is used to provide menus or options. Some embedded machine controllers provide 16-axis pulse motion control capabilities. Others include antivibration design mechanisms. Machine monitors track a machine's uptime, downtime, and idle time. They also allow operators to enter a reason for downtime or nonproductive activities. In some cases, a machine monitor can be programmed to require the entry of a reason code after each downtime event. In this way, machine controllers and monitors can be configured to meet the needs of specific machinery and industries.

Machine controllers and monitors are used in many different applications. Some machine control products are used to regulate medical equipment such as respirators. Others are used in aerospace, automotive, or military applications. An embedded machine controller can be used in a printing machine, pipe bending equipment, CNC stepper motor, or CNC router controller. Embedded machine controllers are also used in the manufacture of semiconductors and electronic devices. Machine controllers and monitors with integral software are used in industries where reliability, quality, and cost are important considerations.

(3) *Industrial control pendants.* Industrial control pendants are sophisticated, hand-held terminals that are used to control robot or machine movements from point to point, within a determined space. They consist of a hanging control console furnished with joysticks, pushbuttons, or rotary cam switches. A type of industrial control pendant, teach pendants are the most popular robotics teaching method, and are used widely with all types of robots, in many industries. As the robot moves within this determined space, the various points are recorded into its memory banks, and can be located later on through subsequent playback. There are a number of teach pendant types available, depending on the type of application for which they will be used. If the goal is simply to monitor and control a robotics unit, then a simple control box style is suitable. If additional capabilities such as on the fly programming are required, more sophisticated boxes should be used.

Industrial control pendants are equipped with switches, dials, and pushbuttons through which data is relayed to the robotics unit, and additional monitoring systems if necessary. The relationship between industrial control pendants and their subservient unit is generally established via an interconnected cabling system. However, more advanced wireless devices are also available.

During use, the operator actuates the switches on manual pendants in a specific order. This, in turn, causes the robot, end effectors, or machine, to move to and from the desired points. As the end effector reaches the desired point, the operator uses the record pushbutton to enter the location into the robot, or robot controller's memory banks. This is the most common programming method for playback robots.

The usage of industrial control pendants is common; however, it has a significant disadvantage in that the operator must divert his and her attention away from the movement of the machine during programming in order to locate the appropriate pushbutton to move the robot. The use of a joystick provides a solution to this problem as the movement of the stick in a certain direction propels the robot or machine in that direction. This option is available on more advanced industrial control pendant types.

(4) *SCADA HMI devices.* Distributed control systems (DCSs) and SCADA systems are system architectures for process control applications. A DCS consists of a PLC that is networked both to other controllers and to field devices such as sensors, actuators, and terminals. A DCS may also interface to a workstation. A SCADA system is a process control application that collects data from sensors or other devices on a factory floor or in remote locations. The data is then sent to a central computer for management and process control. SCADA systems provide shop floor data collection and may allow manual input via bar codes and keyboards. Both DCSs and SCADA systems often include integral software for monitoring and reporting.

There are several parts to a supervisory control and data acquisition system or SCADA system. To control SCADA, a SCADA system integrator, SCADA security, and SCADA HMI are required. A SCADA system integrator is used to interface a SCADA system to an external application. SCADA security uses one or more computers at a remote site to monitor and control sensors or shop floor devices. SCADA security includes remote terminal units (RTUs), a communications infrastructure, and a central control room where monitoring devices such as workstations are housed. SCADA HMI is a human machine interface that accounts for human factors in engineering design.

DCSs and SCADA systems are used in a variety of industries. DCSs are used to control traffic lights and manage chemical processing, pharmaceutical, and power generation facilities. SCADA systems are used in warehouses, petrochemical processing, iron and steel production, food processing, and agricultural applications. Providers of DCSs and SCADA systems are located across the United States and around the world.

1.4.3.3.2 Tools

Operator interface mounts and arms are articulating components used to hold and position industrial computer monitors, keyboards, or other operator interfaces. Operator interface mounts and arms are designed to improve the physical and spatial relationships between machines and the humans that operate them. The science of these relationships, called ergonomics, is the study of human–machine interactions. Ergonomically compatible products are designed to maximize productivity and minimize operator fatigue, discomfort, and injury. The goal of using operator interface mounts and arms as part of an ergonomics program is to reduce injuries, illnesses, and musculoskeletal disorders in the workplace.

Several of the most common types of ergonomic operator interface mount and arm products include computer accessories (e.g., keyboard drawer, mouse tray, glare screen, wrist rest, and monitor support arm) and workstation accessories (e.g., instrumentation booms, articulating supports, foot rests, chairs, and document stands). A monitor support arm is a type of operator interface mount that is designed to support computer screens or monitors in workstations, control centers, and operating theaters. A support arm should combine stability and full adjustability to meet the operator's needs. A support arm can be a desk, wall, ceiling, or mobile mounting arm. Articulating supports are movable support arms that a user can readjust for the height or location of monitors and equipment in relation to the user's eyes or hands. Keyboard drawers are used to store unused keyboards. A monitor drawer mounts an LCD and keyboard within a rack frame or enclosure so that a monitor can be folded down and stored when not in use. Instrumentation booms are another type of operator interface mounts and arms. This type of operator interface mount is used to support various types of equipment, including computer or industrial monitors, video equipment, and manufacturing equipment.

The U.S. Occupational Safety and Health Administration (OSHA) has a four-pronged, comprehensive approach to ergonomics, including operator interface mounts and arms, that is designed to quickly and effectively address musculoskeletal disorders in the workplace. The OSHA approach includes industry or task specific guidelines, enforcement actions, outreach and assistance activities, and a national advisory committee.

1.4.3.3.3 Software

The HMI software enables operators to manage industrial and process control machinery via a computer-based graphical user interface (GUI). There are two basic types of HMI software: supervisory level and machine level. The supervisory level is designed for control room environment and used for SCADA, a process control application which collects data from sensors on the shop floor and sends the information to a central computer for processing. The machine level uses embedded, machine-level devices within the production facility itself. Most HMI software is designed for either supervisory level or machine level; however, applications that are suitable for both types of HMI are also available. These software applications are more expensive, but can eliminate redundancies and reduce long-term costs.

Selecting HMI software requires an analysis of product specifications and features. Important considerations include system architectures, standards and platforms; ease of implementation, administration, and use; performance, scalability, and integration; and total costs and pricing. Some HMI software provides data logging, alarms, security, forecasting, operations planning and control (OPC), and ActiveX technologies. Others support data migration from legacy systems. Communication on multiple networks can support up to four channels. Supported networks include ControlNet and DeviceNet. ControlNet is a real-time, control-layer network that provides high-speed transport of both time-critical I/O data and messaging data. DeviceNet is designed to connect industrial devices such as limit switches, photoelectric cells, valve manifolds, motor starters, drives, and operator displays to PLCs and PCs.

Some HMI software runs on Microsoft Windows CE, a version of the Windows operating system that is designed for hand-held devices. Microsoft and Windows are registered trademarks of Microsoft Corporation. Windows CE allows users to deploy the same HMI software on distributed HMI servers, machine-level embedded HMI, diskless open-HMI machines, and portable or pocket-sized HMI devices.

1.4.4 Highway Addressable Remote Transducer (HART) field communications

HART is an acronym for "Highway Addressable Remote Transducer" that represents a two-way digital communication simultaneously with the 4–20 mA analog signaling used by traditional instrumentation equipment in industrial process control. HARTwas developed in the early 1980s by a company named Rosemount Inc. for the host to perform the management of the field devices in industrial systems. In July 1993, the HART Communication Foundation was established to provide worldwide support for application of this technology. HART Specifications continue to be updated to broaden the range of HART applications. A recent HART development, the Device Description Language (DDL), provides a universal software interface to new and existing devices.

1.4.4.1 HART communication

Most industrial control systems include numerous field system functions. The host controller, therefore, should instantly communicate with all the field instruments and devices while control processes are running for (1) device configuration or reconfiguration, (2) device diagnostics, (3) device troubleshooting, (4) reading the values of additional measurements provided by the device, (5) device health and status, and other requirements. A host in the system can be a Distributed Control System, PLC, and Asset Management System, Safety System, or a hand-held device.

By fully using HART communication, industrial control can be benefited in many aspects. Utilizing the full capabilities of HART-enabled devices and systems reduces costs by improving plant operations and increasing efficiency and helps to avoid the high cost of process disruptions and unplanned shutdowns. Properly utilized, the intelligent capabilities of HART-smart devices are a valuable resource for keeping plants operating at maximum efficiency. Real-time HART integration with plant control, safety, and asset management systems unlocks the value of connected devices and extends the capability of systems to detect any problems with the device, its connection to the process, or interference with accurate communication between the device and system.

The world's leading process automation control systems and instrumentation suppliers all support HART Communication in their field device and system products. Most automation system suppliers offer direct HART-enabled I/O and PC-based software applications to leverage the intelligence in HART-smart field devices for continuous device condition monitoring, real-time diagnostics, and multivariable process information.

1.4.4.1.1 HART networks

(1) *Wired HART networks*. There are two kinds of wired HART networks available in the industrial control systems. Figure 1.4.33 displays the architectures of these two wired HART networks; the first (Figure 1.4.33(a)) is a point-to-point HART network, and the second (Figure 1.4.33(b)) is a multiple-dropped HART network. As shown in Figure 1.4.33, wired HART networks include the Host Controller and some field devices that can be transmitters; between the Host and the field devices, this system has an I/O system that can be the system interface with the HART, and a hand-held terminal or hand-held communicator.

The type of network with a single Field Instrument that does both HART network functions and analog signaling is probably the most common type of wired HART network and is called a point-to-point network. In some cases the point-to-point network might have a HART Field Instrument but no permanent HART Master. This might occur, for example, if the user intends primarily analog communication and Field Instrument parameters are set prior to installation. A HART user might also set up this type of network and then later communicate with the Field Instrument using a hand-held communicator (HART Secondary Master). This is a device that clips onto device terminals (or other points in the network) for temporary HART communication with the Field Instrument.

A HART Field Instrument is sometimes configured so that it has no analog signal, only HART function. Several such Field Instruments can be connected together (electrically in parallel) on the same network, as in Figure 1.4.34. These Field Instruments are said to be multiple-dropped. The Master is able to talk to and configure each one, in turn. When Field Instruments are multi-dropped there cannot be any analog signaling. The term "current loop" ceases to have any meaning. Multiple-dropped Field Instruments that are powered from the network draw a small, fixed current (usually 4 mA) so that the number of devices can be maximized. A Field Instrument that has been configured to draw a fixed analog current is said to be "parked." Parking is accomplished by setting the short-form address of the Field Instrument to some number other than 0. A hand-held communicator might also be connected to the network of Figure 1.4.34.

There are few restrictions on building wired HART networks. The topology may be loosely described as a bus, with drop attachments forming secondary busses as desired, which are illustrated in Figure 1.4.35. The whole collection is considered a single network. Except for the intervening lengths of cable, all of the devices are electrically in parallel. The Hand-Held Communicator (HHC) may also be connected virtually anywhere. As a practical matter,

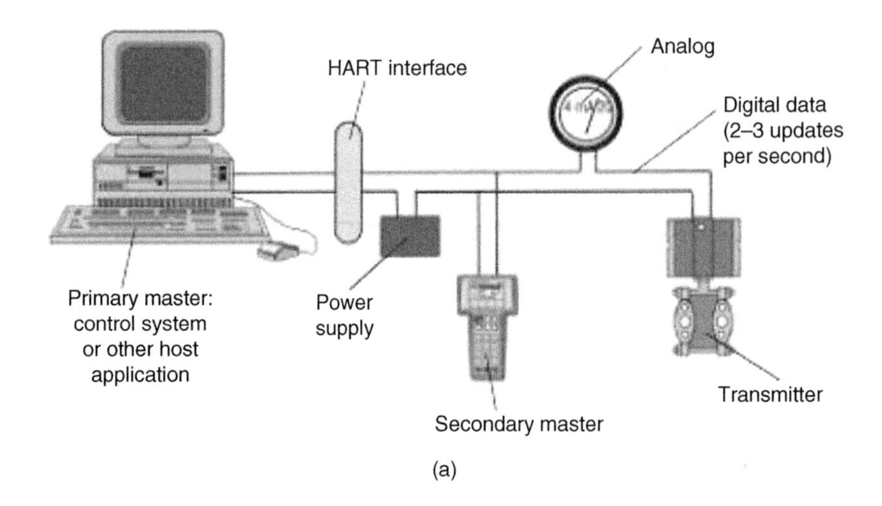

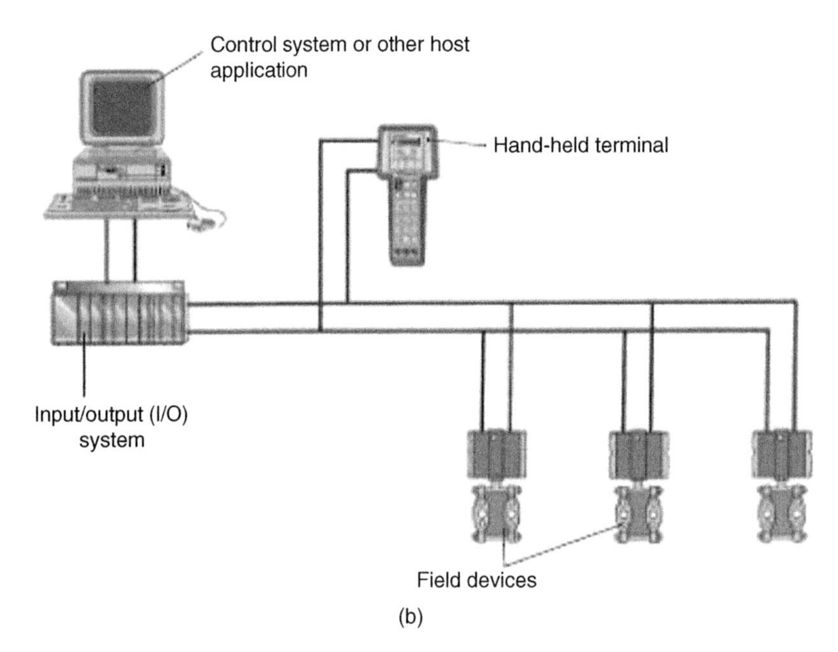

Fig. 1.4.33 The architecture of HART networks: (a) the point-to-point HART network and (b) the multiple-dropped HART network. Note: Instrument power is provided by an internal or external power source that is not shown.

however, most of the cable is inaccessible and the HHC has to be connected at the Field Instrument, in junction boxes, or in controllers or marshaling panels. In intrinsically safe (IS) installations there will likely be an IS barrier separating the control and field areas.

A Field Instrument may be added or removed or wiring changes made while the network is live (powered). This may interrupt an on-going transaction. However, if the network is inadvertently short-circuited, this could reset all devices. The network will recover from the loss of a transaction by retrying a previous communication. If Field Instruments are reset, they will eventually come back to the state they were in prior to the reset. No reprogramming of HART parameters is needed.

Digital signaling brings with it a variety of other possible devices and modes of operation. For example, some Field Instruments are HART only and have no analog signaling. Others draw no power from the network. In still other cases the network may not be powered (no DC). There also exist other types of HART networks that depart from the conventional one described here. These are covered in another section.

(2) *Wireless HART networks*. Wireless HART is the first open and interoperable wireless communication standard designed to address the critical needs of the process industry for reliable, robust, and secure wireless communication in real world industrial plant applications.

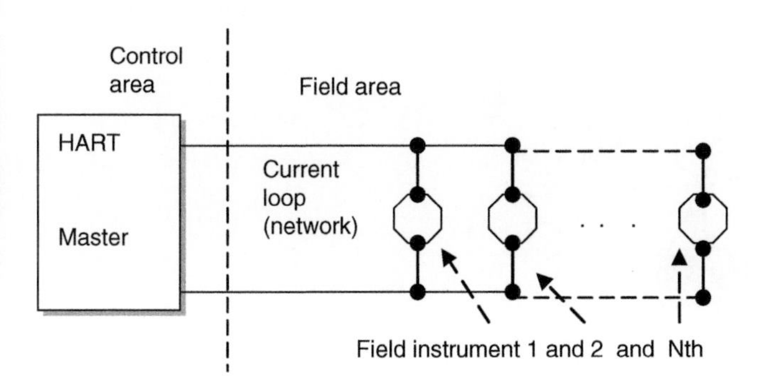

Fig. 1.4.34 HART network with multiple-dropped field instruments.

A Wireless HART Network consists of Wireless HART field devices, at least one Wireless HART gateway, and a Wireless HART network manager. These components are connected into a wireless mesh network supporting bidirectional communication from the HART host to field device and back. Figure 1.4.36 gives the typical Wireless HART network architecture with the principal devices plotted:

(a) *Network Manager.* The Network Manager is an application that manages the mesh network and Network Devices. The Network Manager performs the following functions: (1) forms the mesh network, (2) allows new devices to connect to the network, (3) sets the communication schedule of the devices, (4) establishes the redundant data paths for all communications, and (5) monitors the network.

(b) *Gateway devices.* The gateway device connects the mesh network with a plant automation network, allowing data to flow between the two network devices. The gateway device provides access to the Wireless HART devices by a system or other host application.

(c) *Network devices.* A network device is a node in the mesh network. It can transmit and receive Wireless HART data and perform the basic functions necessary to support network formation and maintenance. Network devices include field devices, router devices, gateway devices, and mesh hand-held devices.

(d) *Field devices.* The field device may be a process connected instrument, a router, or hand-held device. The Wireless HART network connects these devices together.

(e) *Router device.* A router device is used to improve network coverage (to extend a network) so that it is capable of forwarding messages from other network devices.

(f) *Process connected instrument.* Typically a measuring or positioning device used for process monitoring and control, it is also capable of forwarding messages from other network devices.

(g) *Adapter.* An adapter is a device that allows a HART instrument without wireless capability to be connected to a Wireless HART network.

(h) *Hand-held support device.* Hand-held devices are used in the commissioning, monitoring, and maintenance of network devices; they are portable and operated by the plant personnel.

Wireless HART networks can be configured in a number of different topologies to support

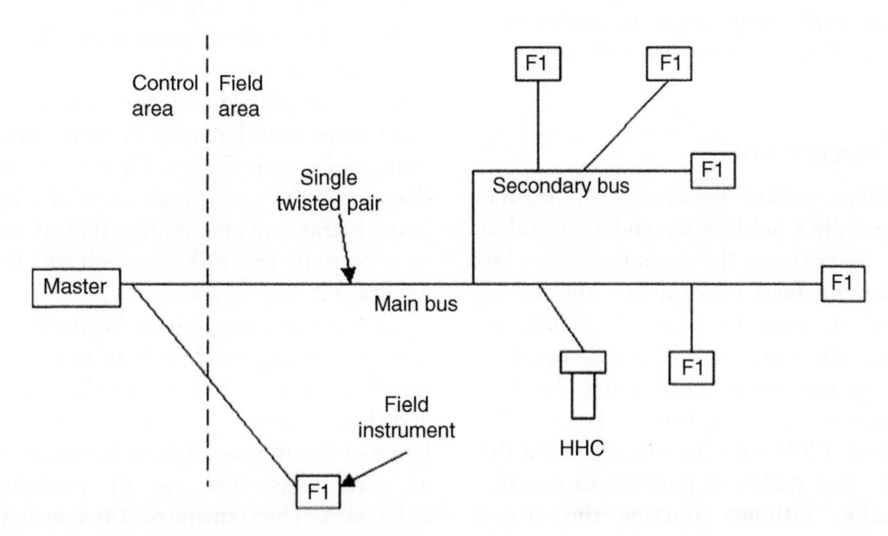

Fig. 1.4.35 HART network showing free arrangement of devices.

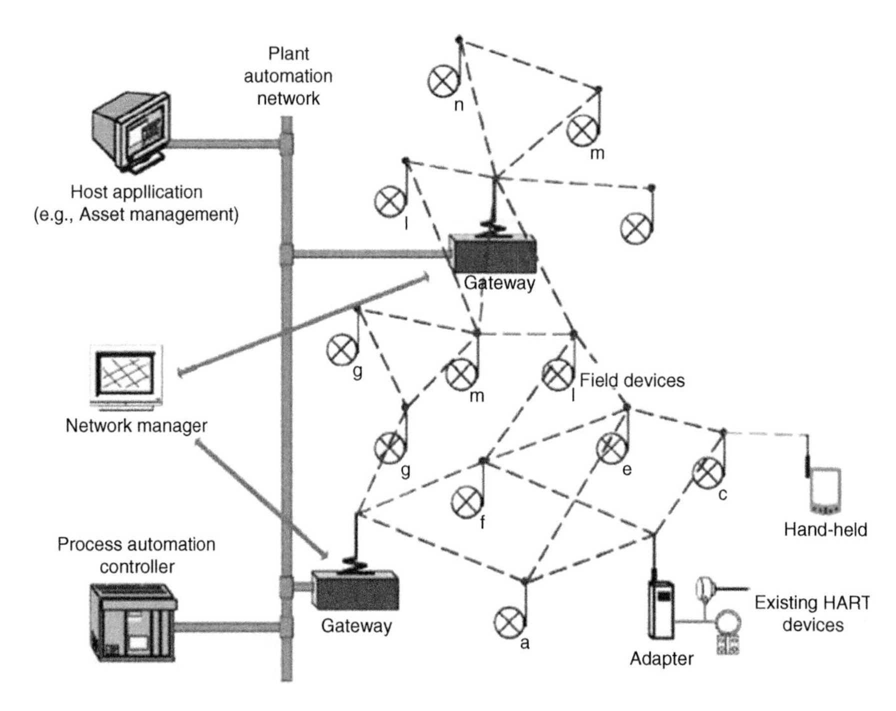

Fig. 1.4.36 Typical wireless HART architecture (courtesy of the HART Communication Foundation).

various application requirements including the following:

(a) *Star network.* Star networks have just one router device that communicates with several end devices. This is one of the simplest network topologies. A star network may be appropriate for small applications.

(b) *Mesh network.* Mesh networks are formed by network devices that are all router devices. Mesh networks provide a robust network with redundant data paths which is able to adapt to changing RF environments.

(c) *Star mesh network.* Star mesh networks are a combination of the star network and mesh network.

1.4.4.1.2 HART mechanism

HART communication occurs between two HART-enabled devices, typically a field device and a control or monitoring system. To perform the communication between the host and the field instruments, the analog measurement signal is used to transmit digital information. For this purpose, an additional signal is modulated to the measurement signal using the Frequency Shift Keying (FSK) process. The two frequencies of the additional signal, 1200 and 2200 Hz, represent the bit values 1 and 0. This makes it possible to transfer additional information without affecting the analog measurement signal. As indicated by Figure 1.4.37, HART provides two simultaneous communication channels: the 4–20 mA analog signal and a digital signal. The 4–20 mA signal communicates the primary measured value (in the case of a field instrument) using the 4–20 mA current loop, the fastest and most reliable industry standard. Additional device information is communicated using a digital signal that is superimposed on the analog signal. The digital signal contains information from the device including device status, diagnostics, and additional measured or calculated values. Together, the two communication channels provide a complete field communication solution that is easy to use and configure, is low cost and is very robust.

The HART signal path from the microprocessor in a sending device to the microprocessor in a receiving device is displayed in Figure 1.4.38. Amplifiers, filters, and the network between these two interfaces have been omitted for simplicity in Figure 1.4.38. At this level the diagram is the same, regardless of whether a Master or Slave is transmitting. Notice that, if the signal starts out as a current, the FSK is a voltage. But if it starts out a voltage it stays a voltage.

The transmitting device begins by turning on its carrier and loading the first byte to be transmitted into its interface circuits. It waits for the byte to be transmitted and then loads the next one. This is repeated until all the bytes of the message (these messages are always defined as commands that are of predefined format) are exhausted. The transmitter then waits for the last byte to be serialized and finally turns off its carrier. With minor

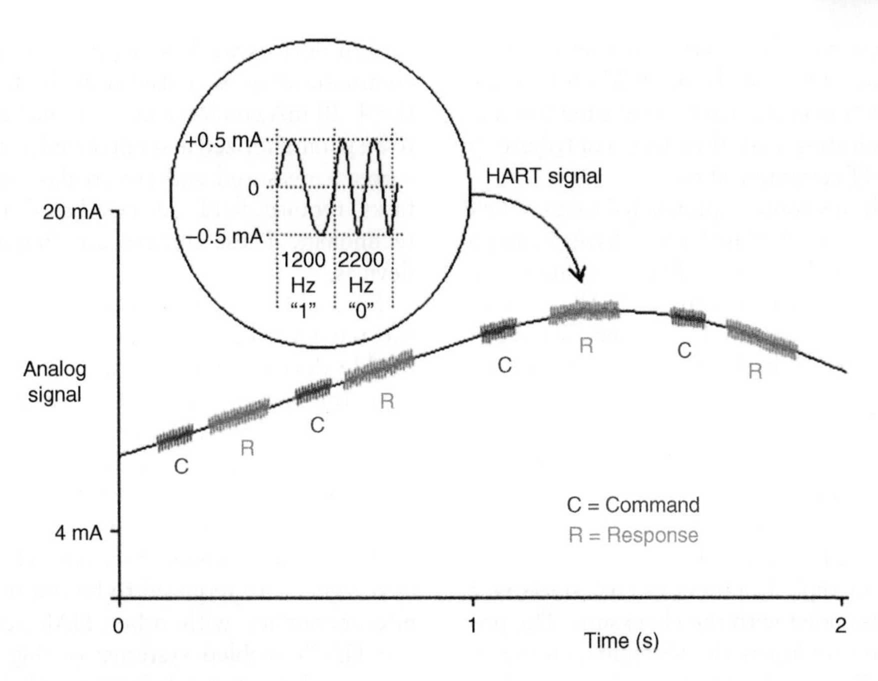

Fig. 1.4.37 HART signaling (digital and analog).

exceptions, the transmitting device does not allow a gap to occur in the serial stream, the start and stop bits are used for synchronization, and the parity bit is part of the HART error detection.

The serial character stream is applied to the modulator of the sending modem. The Modulator operates such that a logic 1 applied to the input produces a 1200 Hz periodic signal at the Modulator output. Logic 0 produces 2200 Hz. The type of modulation used is called Continuous Phase Frequency Shift Keying (CPFSK). "Continuous Phase" means that there is no discontinuity in the modulator output when the frequency changes. When the sender's interface output (modulator input) switches from logic 1 to logic 0, the frequency changes from 1200 to 2200 Hz with just a change in slope of the transmitted waveform. A moment's thought reveals that the phase does not change through this transition. Given the chosen shift frequencies and the bit rate, a transition can occur at any phase.

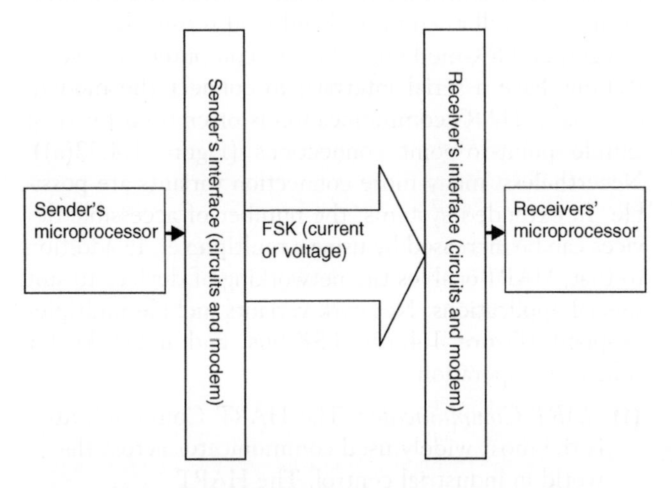

Fig. 1.4.38 HART signaling path.

At the receiving end, the demodulator section of a modem in its interface converts FSK back into a serial bit stream at 1200 bps. Each character is converted back into an 8-bit byte and parity is checked. The receiving microprocessor reads the incoming bytes from its interface and checks parity for each one until there are no more or until parsing of the data stream indicates that this is the last byte of the message. The receiving processor accepts the incoming message only if its amplitude is high enough to cause carrier detect to be asserted. In some cases, the receiving processor will have to test an I/O line to make this determination. In others, the carrier detect signal gates the receive data so that nothing (no transitions) reaches the receiving interface unless carrier detect is asserted.

HART protocol puts most of the responsibility (such as timing and arbitration) into the Masters. This eases the Field Instrument software development and puts the complexity into the device that is more suited to deal with it. A Master typically sends a command and then expects a reply. A Slave waits for a command and then sends a reply. The command and associated reply are called a transaction. There are typically periods of silence (no device is allowed communicating) between transactions. A Slave accesses the network as quickly as possible in response to a Master. Network access by Masters requires arbitration. Masters arbitrate by observing who sent the last transmission (a Slave or the other Master) and by using timers to delay their own transmissions. Thus, a Master allows time for the other Master to start a transmission. The timers constitute dead time when no device is communicating and therefore contribute to "overhead" in HART communication.

Each HART field instrument (in normal cases, a field instrument plays a role of Slave) must have a unique address. Each command sent by a Master contains the address of the desired Field Instrument. All Field Instruments examine the command. The one that recognizes its own address sends back a response. This address is incorporated into the command message sent by a Master and is echoed back in the reply by the Slave. Addresses are either 4 bits or 38 bits and are called short and long or "short frame" and "long frame" addresses, respectively. A Slave can also be addressed through its tag (an identifier assigned by the user).

Each command or reply is a message that starts with the preamble and is ended with the checksum. The preamble is allowed to vary in length, depending on the requirements in the Slave end. Different Slaves can have different preamble length requirements, so that a Master might need to maintain a table of these values. A Master will use the longest possible preamble when talking to a Slave for the first time. Once the Master reads the Slave's preamble, it first checks the length requirement (a stored HART parameter), then will subsequently use this new length when talking to that Slave. The checksum at the end of the message is used for error control. It is the exclusive-OR of all of the preceding bytes, starting with the start delimiter. The checksum, along with the parity bit in each character, creates a message matrix having so-called vertical and longitudinal parity. If a message is in error, this usually necessitates a retry.

One more feature, available in some Field Instruments, is burst mode. A Field Instrument that is burst-mode capable can repeatedly send a HART reply without a repeated command. This is useful in getting the fastest possible updates (about 2–3 times per second) of process variables. If burst mode is to be used, there can be only one bursting Field Instrument on the network. A Field Instrument remembers its mode of operation during power down and returns to this mode on power up. Thus, a Field Instrument that has been parked will remain so through power down. Similarly, a Field Instrument in burst-mode will begin bursting again on power up.

1.4.4.2 HART system

HART communication in industrial control comprises two folders: HART connection system and HART protocol. This section focuses on the HART system, and the next section will be on the HART protocol. HART system devices work to support the HART protocol by communicating their data over the transmission lines of the 4–20 mA connections. This enables the field devices to be parameterized and initialized in a flexible manner or to read measured and stored data (records). All these tasks require field devices based on microprocessor technology. These devices are frequently called smart devices.

For building and maintaining a HART-enabled system, the technical kernel will be choosing the HART-compatible devices, installing the system's devices, configuring the system's devices, and calibrating the system's devices. The key is to make sure that the engineers or designers are requesting or specifying devices or systems that are fully compliant with the HART protocol specification and are tested and registered with the HART Communication Foundation (the HCF). The engineers or designers are required to be assured of these aspects: interoperability with other HART-compatible devices and HART-enabled systems; getting a device that will provide the powerful features of HART technology; specifying a product that will fully integrate into your HART-enabled applications.

1.4.4.2.1 HART system devices

The devices constructing a HART connection system have several features that significantly reduce the time required to fully commission a HART-enabled network. When less time is required for commissioning, substantial cost savings are hence achieved. Devices which support the HART protocol are grouped into master (host) and slave (field) devices. Master devices include communicator or hand-held terminals as well as PC-based workplaces that stay in a control room. HART slave devices, on the other hand, include sensors, transmitters, and various actuators. The variety ranges from two-wire and four-wire devices to intrinsically safe versions for use in hazardous environments. Field devices and communicators as well as compact hand-held terminals have an integrated FSK-modem, whereas computers or workstations have a serial interface to connect the modem externally. HART communication is often used for such simple point-to-point connections (Figure 1.4.33(a)). Nevertheless, many more connection variants are possible. In extended systems, the number of accessible devices can be increased by using a multiplexer. In addition to that, HART enables the networking of devices to suit special applications. Network variants include multiple-dropped (Figure 1.4.39), FSK-bus, and networks for split-range operation.

(1) *HART Communicator.* The HART Communicator is the most widely used communicator across the world in industrial control. The HART

Communicators are portable devices; their weights have been evenly distributed for comfortable one-handed operation in the field. The result is the universal, user upgradeable, intrinsically safe, rugged and reliable Field Communicator. In HART-enabled systems, the HART Communicators are often defined by engineers as the Second Master (Figure 1.4.33(a)) or Hand-held Terminals (Figure 1.4.33(b)).

With a memory and a microprocessor or some application-specific integrated circuits, the HART Communicator provides a complete solution for configuring and monitoring all HART devices and all Fieldbus devices of an industrial system.

It is comprised of three main components plus accessories. These parts consist of the hand-held; the HART interface hardware, and the application software Suite. The Communicator runs on a robust, real-time, operating system. This trio of hardware and software comprises a complete HART field communicator that can be powerful, multi-faceted, and portable all in one.

The hardware for the HART Communicators primarily includes the HART interface and the pinch connectors. The HART interface is designed to interface to the multiple connectors located on the bottom of the hand-held, allowing communication between the Palm and the HART network. The pinch connectors easily connect to any HART network for instant communication. Most of the HART interface requires no batteries, running solely off the hand-held's internal power supply. Its compact size and low power consumption makes the interface an ideal solution for portability.

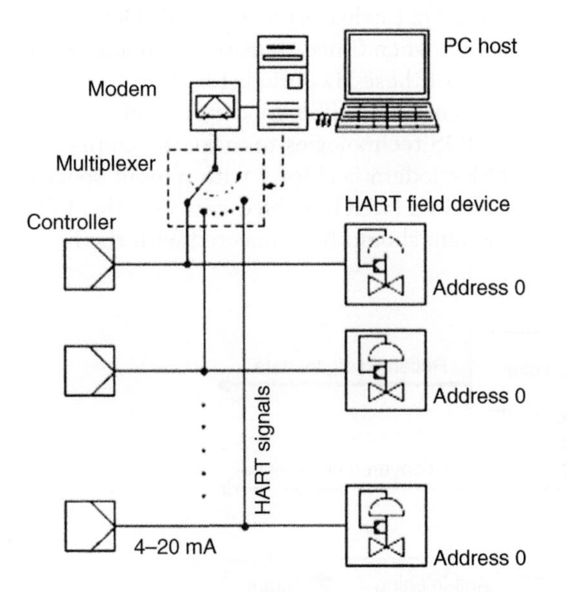

Fig. 1.4.39 A HART connecting system including the FSK-modem and HART-multiplexer, and HART-buses (courtesy of the SAMSON, Inc.).

The software suite for the HART Communicators includes some distinct applications. Each of these applications is preloaded onto the hand-held and designed for a particular function. The main application of the suite allows communication, monitoring, and configuration of HART-compatible devices. The software is based upon manufacturer device description files (DDL) and thus allows access to all menus and parameters as designed by the manufacturer. This software application allows for the logging of device variable values over time. A wide range of variables can be logged automatically at a user selectable sample time, or manually one by one. These logs can be saved and transferred to a PC for further analysis.

This graphing application allows device variables to be trended over time in an easy to view graphical format. Device parameters can be simultaneously graphed in various colors for easy identification. The display makes it easy to read in both bright sunlight and in normal lighting. To make sure all conditions are covered, a multilevel backlight is added, allowing the display to be viewed in those areas of your plant with dim light. The touch-sensitive display and large physical navigation buttons provide for efficient use both on the bench and in the field.

User upgradeable HART and Fieldbus devices, as well as functional updates to existing devices are introduced continually by device vendors. Keeping up to date with the required Device Description (DD) drivers for all the devices in plant can be a real challenge. Nowadays, with the Easy Upgrade option, keeping communicators updated with the most current DDs is an easy job.

(2) *FSK-modem.* There are often two kinds of modem required in the HART-enabled industrial control networks: USB Modem and FSK modem. These two kinds of modem are all connected with the host PC in the HART-enabled networks. This USB-modem is just an ordinary PC modem used for computer networks, without special design for HART functions. However, the FSK-modem should be particularly designed for HART functions. The following will focus on the FSK-modem.

The FSK-modem is designed to provide HART communication capabilities for the implementation of Frequency Shift Keying (FSK) techniques to transfer data. The FSK-modem is also required to conform to the HART network's physical layer. For this purpose, most FSK-modems operate at the Bell 202 standard and are made into a chipset containing some integrated circuits. As shown in

Figure 1.4.37, the FSK is the frequency modulation of a carrier of digital capability. For Simplex or Half Duplex operation, the FSK-modem uses a single carrier in which the communication can only be transmitted in one direction at a time. For Full Duplex, the FSK-modem uses multiple carriers so that data communication can be simultaneous in both directions.

The basic block diagram for the FSK-modem chipset is depicted in Figure 1.4.40, which illustrates the mechanism of a data modulation and demodulation. This chip is divided into three main parts: receive, transmit, and clock recovery. Both receive and transmit blocks are separated and data can be processed in each direction independently.

(a) *Modulator for transmitting data.* The data transmit part in this chipset performs modulation, in which the scrambler is to make a nearly flat spectrum of output signal. The output of the scrambler is connected to a long digital FIR filter. This FIR filter compensates distortion of transmission line and removes sharp edges rising from High–Low or Low–High logic transitions. The FIR filter makes the transmitted signal spectrum narrow to fit bandwidth and compensates the signal for the receiver's side. The transmit wave's shapes are stored in an EPROM. In this way the transmitted waveform is synthesized not only from the present bit's state, but also the four that preceded it and four to come. Data burnt into EPROM represents filter response for each of 256 combinations.

(b) *Demodulator for receiving data.* At the receiving side, the audio signal going from the discriminator of the transceiver is passed through a low-pass filter to eliminate pertinent higher frequencies, and remove out of band spurious noise and residue. The signal is then limited and detected by sampling at the correct instant. At this point, the detected data, still randomized, are passed through a descrambler, where the original data are recovered and the result goes off to terminal. A descrambler, like a scrambler, is simply to provide the invert function of the scrambler and perforce requires some number of bits to synchronize.

(c) *Clock recovery.* The heart of the receiver is a digital phase locked network (DPLL), which must extract a clock from the received audio stream. It is needed to time the receiver functions, including the all-important data detector. Each waveform has a phase shift of 360/256° from another and is made up of 16 samples. The received audio signal is limited, and a zero crossing detector circuit generates one cycle of 9600 Hz for each zero-crossing (a proto-clock). This is compared with a locally generated clock in a phase detector based on an up/down counter. The counter increments if one clock is early, decrements otherwise. This count then addresses an EPROM mentioned above. In this way, the local clock slips rapidly into phase with that of the incoming data. Local clock signal is derived from the output of EPROM. Output data are converted to sine voltage with maximum amplitude.

Some FSK-modems have one more function block in their chipsets; that is Carrier Detect. The carrier detect function in the FSK-modem is responsible for checking whether or not the modulated or demodulated waves fall in some range of frequencies. The carrier detect output is active low whenever a valid carrier tone between some Hz (inclusive) is detected. Detection occurs when timed transitions remain within the band of these Hz periods for 10 ns to 1 bit time.

Some of the FSK-modem manufacturers use CMOS technologies to make the chipset. The FSK-modem is chipset with pin-out specification. In a HART-enabled network, this FSK-modem should be connected with the host

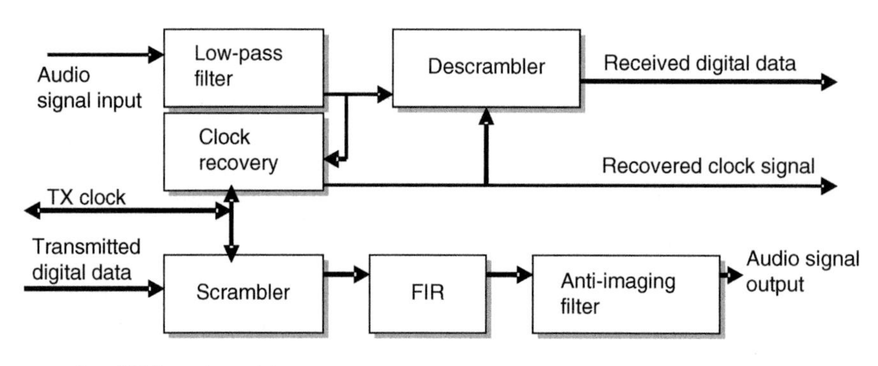

Fig. 1.4.40 Block diagram of an FSK-modem chipset.

microprocessor or the CPU. Figure 1.4.41 illustrates how this modem's pins connect with the CPU in an FSK-modem.

(3) *HART multiplexer.* In the HART-enabled industrial networks, the HART multiplexer acts as a gateway between the network management computer and the HART-compatible field devices. Many field devices are distributed over a wide area in industrial process systems, and must be monitored and adapted to the changes in the processing environment. The process system with the HART multiplexer enables on-line communication between an asset management computer or workstation and those intelligent field devices that support the HART protocol so as to simultaneously fit into the changes in the process system. All actions on the field device are parallel to the transmission of the 4–20 mA measurement signals and have no influence on the measurement value processing through the process system.

Each HART multiplexer, regardless of whether it is a slave or a master, provides a connection to a specified number of field instruments. Up to thousands of field units can communicate and exchange data with a computer or workstation. Working with a hand-held terminal (HART-communicator) is also possible since the HART protocol accepts two masters (computer and hand-held terminal) in one system. Systems can be easily expanded and the advantages of the HART communications can be exploited. At the present technical level, the system consists of a maximum of 31 HART Multiplexer masters which are linked to the computer with an RS485 interface. Each HART Multiplexer master controls up to 15 HART Multiplexer slaves.

At present, the HART Multiplexers are specified as the following three types:

(a) *HART Multiplexer Master.* This is a HART Multiplexer that can operate up to 256 analog field instruments. The built-in slave unit operates the first 16 loops. If more than 16 loops are required, additional slave units can be connected. The slave units are connected to the master with a 14-pin flat cable. The connector for the ribbon cable is found on the same housing side as the connectors for the interface and the power supply.

The analog signals are separately linked to a termination board with a 26-pin cable for each unit. Sixteen leads are reserved for the HART signal of the analog measurement circuits. The remaining 10 leads are sent to ground. This unit is designed with removable terminals and can be connected to a Power Rail.

(b) *HART Multiplexer Slave.* This is a HART Multiplexer that can operate up to 16 analog field instruments at the present. In this case, the slave can only be operated with the HART Multiplexer Master and is powered by the master across a 14-pin flat cable connection. Up to 15 slaves can be connected to the master. The slave address is set with a 16 position rotary switch (addresses 1–16). If only one slave is connected to the master, then the slave address should be 1. If multiple slaves are connected, slaves are to be assigned addresses in ascending order. The analog signals are fed into the slave by means of a 26-pin ribbon cable. Sixteen leads are reserved for the HART signal of the analog measurement circuits. The remaining 10 leads are assigned to ground.

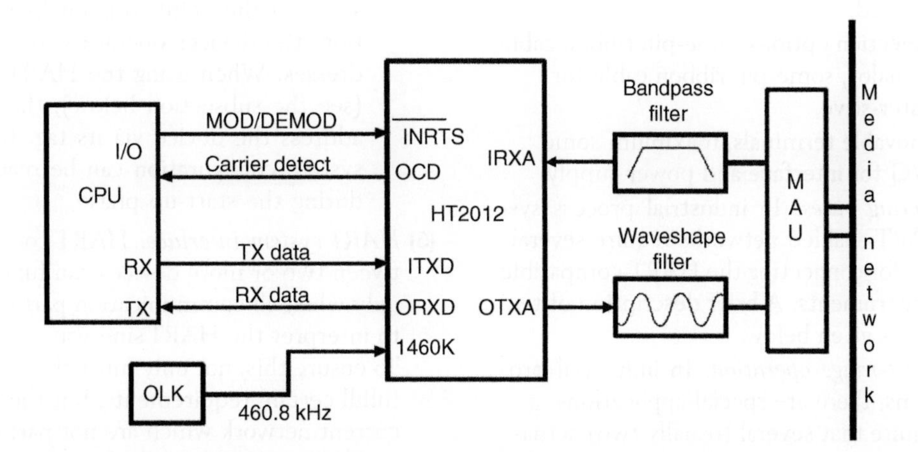

Fig. 1.4.41 Typical hardware design of an FSK-modem.

(c) *HART flexible interface.* This is a flexible interface board with a HART pick-up connector. This flexible interface board has 16 terminal blocks to connect up to 16 smart field devices. This board can be used for general purpose applications or in conjunction with intrinsic safety barriers for hazardous area applications.

The specification of a HART Multiplexer should include these important technical data:

(i) *HART signal channels*:

(1) leakage current (A at some temperature range),

(2) output termination External (measured by Ω),

(3) output voltage (mVpp),

(4) output impedance (measured by Ω, capacitively linked),

(5) input impedance per HART conventions,

(6) input voltage range (mV-Vpp),

(7) input voltage,

(ii) *Power supply:*

(1) nominal voltage (VDC),

(2) power consumption (less and equal to some W).

(iii) *Interface:*

(1) type is RS-xxx (Some number of wire multidrops),

(2) transmission speed 9600, 19,200, 38,400 baud,

(3) address selection (32) possible RS-xxx addresses,

(4) the transmission speed (unit is "baud") at the "ON" and "OFF" state for every switch.

(iv) *Mechanical data:*

(1) mounting some mm DIN rail or wall mounted,

(2) connection options some-pin ribbon cable for analog; some-pin ribbon cable for master-slave,

(3) removable terminals, maximum some AWG for interface and power supply.

(4) *HART connecting buses.* In industrial process systems, the HART-enabled networks require several kinds of buses for connecting the HART-compatible devices and instruments. A brief description of two of these buses is given below.

(a) *Bus for split-range operation.* In industrial process systems, there are special applications which require that several (usually two) actuators receive the same control signal. A typical example is the split-range operation of control valves. One valve operates in the nominal current range from 4 to 12 mA, while another valve uses the current range from 12 to 20 mA. The split-range operation technique is a solution for this case. In split-range operation, the control valves are connected in series in the current network. When both valves have a HART interface, the HART host device must be able to distinguish with which valve it must communicate. To achieve this, the HART protocol revision 6 (anticipated for autumn 1999) and later will be extended by one more network variant. As is the case for multiple-dropped mode, each device is assigned to an address from 1 to 15. The analog 4–20 mA signals preserve its device-specific function, which is, for control valves, the selection of the required travel.

(b) *FSK-bus.* The HART protocol can be extended by the FSK-bus. Similar to a device bus, the FSK-bus can connect approximately 100 HART-compatible devices and address these devices with the technical level at the present. This requires special assembly-type isolating amplifiers (e.g., TET 128). The only reason for the limited number of participants is that each additional participant increases the signal noise. The signal quality is therefore no longer sufficient to properly evaluate the telegram. The HART devices are connected to their analog current signal and the common FSK-bus line with the isolating amplifier (Figure 1.4.42). From the FSK-bus viewpoint, the isolating amplifiers act as impedance converters. This enables devices with high load to be integrated in the communication network. To address these devices, a special, long form of addressing is used. During the configuration phase, the bus address and the tag number of each device are set with the point-to-point line. During operation, the devices operate with the long addresses. When using the HART command 11 (see the subsection below), the host can also address the device via its tag. In this way, the system configuration can be read and checked during the start-up phase.

(5) *HART system interface.* HART communication between two or more devices can function properly only when all communication participants are able to interpret the HART sine-wave signals correctly. To ensure this, not only must the transmission lines fulfill certain requirements, but the devices in the current network which are not part of the HART communication can impede or even prevent the

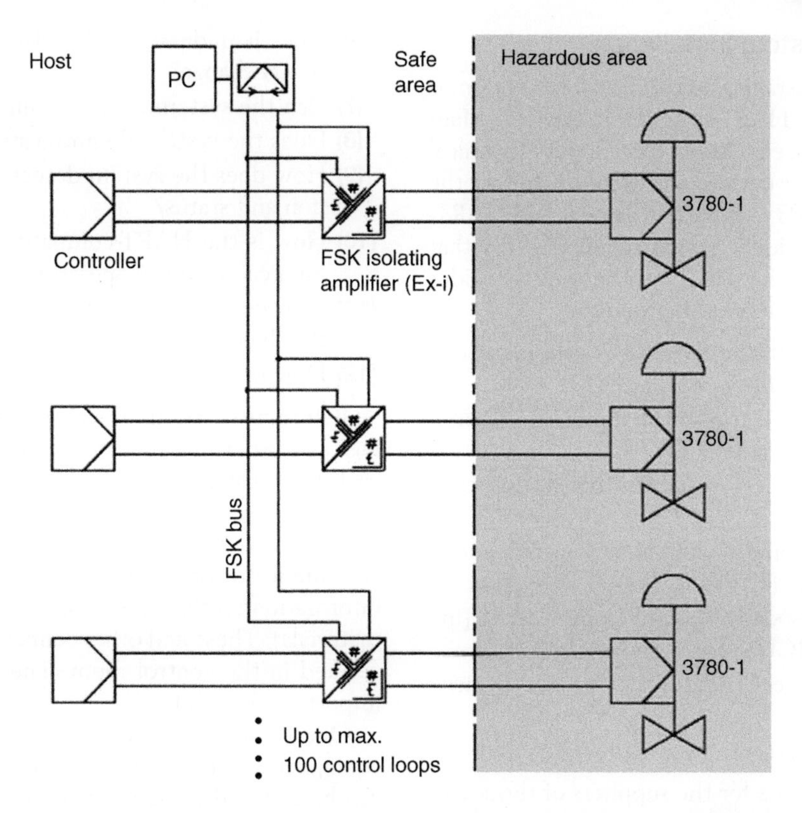

Fig. 1.4.42 The connection architecture of an HART network with the FSK-bus (courtesy of the SAMSON, Inc.).

transmission of the data. The reason is that the inputs and outputs of these devices are specified only for the 4–20 mA technology. Because the I/O resistances change with the signal frequency, such devices are likely to short-circuit the higher frequency HART signals (1200–2200 Hz).

Where a HART communication system is connected with other kinds of communication systems, gateways could be the best interface devices to convert the HART protocol into the protocols of the networks to be coupled. In most cases, when complex communications must be performed, Fieldbus systems would be the preferred choice. Even though there is no complex protocol conversion, the HART-enabled system is capable of communicating over long distances. Furthermore, the HART data signals can be transmitted over telephone lines using HART/ CCITT converters, in which the Field devices directly connect to dedicated lines owned by the telephone company, being able to communicate with the centralized host located many kilometers away. However, as already mentioned, within a HART-enabled system, the HART-compatible field devices also require an appropriate communication interface that could be, for example, an integrated FSK-modem or a HART-Multiplexer.

As mentioned earlier, the HART signals are imposed on the conventional analog current signal. Whether the devices in the networks are designed in four-wire technique including an additional power supply or in two-wire technique, HART communication can be used for both cases. However, it is important to note that the maximum permissible load of a HART device is fixed. The load of a HART device is limited by the HART specification. Another limitation is due to the process controller. The output of the process controller must be able to provide the power for the connected two-wire device.

The higher the power consumption of a two-wire device is, the higher its load is. The additional functions of a HART-communicating device increase its power consumption, and hence the load, compared to non-HART devices. When retrofitting HART devices into an already existing installation, the process controller must be checked for its ability to provide the power required by the HART-compatible device. The process controller must be able to provide at least the load impedance of the HART device at 20 mA.

1.4.4.2.2 HART system installation

The first task before installing a HART-enabled system is checking to verify the HART-compatible devices. Manufacturers have different levels of HART technical implementation in their devices and systems. In fact the capabilities of the HART-compatible devices and HART-enabled system vary widely, which requires when that engineers, when specifying HART technology, consider such factors and parameters as the following:

(1) Registered device at the HART Communication Foundation (the HCF).
(2) Registered Device-Description at the HART Communication Foundation (the HCF).
(3) What is the number of variables this device can measure?
(4) Does this device comply with HART specification (in reference of the HCF)?
(5) Does the device respond to HART Command 48 (in reference of the HCF' specifications)?
(6) What unique or special features does this device support?
(7) What diagnostic features does the device contain?

These questions below are for the suppliers of those I/O interface devices:

(1) How much HARTcapability is embedded into the I/O and how smart is it?
(2) Can the I/O validate and secure the 4–20 mA signal?
(3) Is there one HART modem per channel, or is the I/O multiplexed? How fast can it update the HART digital values?
(4) In what ways does the system support access to multivariable HART data from multivariable devices?
(5) Can you merely "push a button" on the I/O to calibrate the network current or check the range?
(6) Does the I/O support multiple-dropped networks?
(7) Does the I/O automatically scan and monitor the HART-compatible field devices or is the scanning only possible using "pass through"?

These questions below should be asked of the control system suppliers:

(1) Does your host use a "native" device description or does it require a different file type?
(2) Does the system make it easy to use all HART capabilities?
(3) How much training is required to learn how to get and use HART data?
(4) Review the configuration of a HART-compatible device using the control system.
(5) Can the system use secondary digital process variables?
(6) Does it understand the HART-compatible device status change?
(7) Can the system detect configuration changes?
(8) Does the system do notification by exception?
(9) How does the system detect changes in configuration and status?
(10) How is the HART-compatible device status communicated to the operators?
(11) How do you perform tests when there is an error in the device?
(12) How open is the system to third party software?

Before installation, it is also necessary to enter device tags and other identification and configuration data into each field instrument. After installation, the instrument identification (tag and descriptor) can be verified in the control room using a configuration tool such as hand-held communicator or computer. Some field devices provide information on their physical configuration (e.g., wetted materials). These and other configuration data can also be verified in the control room. The verification process is important for safety.

Once a field instrument has been identified and its configuration data confirmed, the analog network integrity can be checked using the network test feature, which is supported by many HART-compatible devices. The network test feature enables the analog signal from a HART transmitter to be fixed at a specific value to verify network integrity and ensure proper connection to support devices such as indicators, recorders, and DCS displays. Use the HART protocol network test feature to check analog network integrity and ensure a proper physical connection among all network devices. Additional integrity can be achieved if the analog value is compared to the digital value being reported in a device. For example, someone might have provided an offset to the 4–20 mA analog value that has not been accounted for in the control system. By comparing the digital value of the Primary Variable to the analog value, the network integrity can be verified.

There are some ways to integrate HART data and leverage the intelligence in smart field devices. Several simple and cost-effective integration strategies are listed below in order to get more from currently installed HART-compatible devices and instruments (Figure 1.4.43).

(1) *Point-to-point integration.* This is the most common way to use HART. The communication capability of HART-compatible devices allows them to be configured and set up for specific applications, reducing costs and saving time in commissioning and maintenance. With connection to the 4–20 mA wires, a device can be integrated from remote locations by connecting anywhere on the current network to obtain device status and diagnostic information.

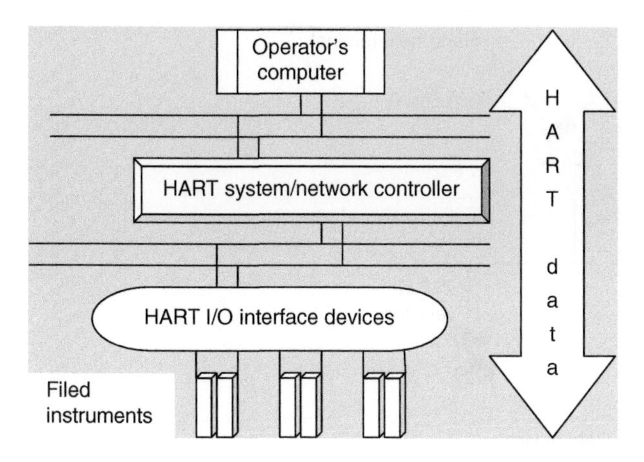

Fig. 1.4.43 The data flow diagram for the integration of the HART data.

(2) *HART-to-analog integration.* Signal extractors communicate with HART-compatible devices in real-time (simultaneously) to convert the intelligent information in these devices into 4–20 mA signals for input into an existing analog control system. Add this capability one device at a time to get more from the intelligent HART-compatible devices.

(3) *HART-plus-analog integration.* New HART-multiplexer packaging solutions make it easy to communicate with HART-compatible devices by replacing the existing I/O termination panels. The analog control signal continues on to the control system as it does today but the HART data is sent to a device asset management system providing valuable diagnostics information 24/7. Although the control system is not aware of the HART data, this solution provides better access to device diagnostics for asset management improvements.

(4) *Full HART integration.* Upgrading a Field or Remote I/O system provides an integrated path to continuously put HART data directly into your control system. Most new control systems are HART-capable and many suppliers offer software and I/O solutions to make upgrades simple and cost-effective. Continuous communication between the field device and control system enables problems with the devices, its connection to the process, or inaccuracies in the 4–20 mA control signal to be detected automatically so that corrective action can be taken before there is negative impact to the process operation.

1.4.4.2.3 HART system configuration

The purpose of the Device Configuration is for accessing its HART Data. There are several methods of accessing the intelligent information in the HART-compatible device on a temporary or a permanent basis. The configuration of a HART-compatible device can be achieved by using the software and hardware tools.

To configure a single device on a temporary basis, a universal hand-held configuration tool is needed, with a power supply, a load resister, and a HART-compatible device. Or, configuration can be achieved by using a computer which is capable of running a device configuration application and using a HART-modem.

(1) *Universal hand-held communicators.* HART hand-held communicators are available from major instrumentation suppliers across the world and are supported by the member companies of the HCF. Using DD files, the communicator can fully configure any HART-compatible device for which it has a DD installed. If the communicator does not have the DD for a specific device, it will still communicate and configure the device using the HART Universal and Common Practice commands.

There are 35–40 standard data items in every registered HART-compatible device. The data can be accessed by any approved configuration tool such as a communicator. These items do not require the use of a DD and typically include the basic functionality for all devices. These are the Universal and Common Practice commands required of every registered HART device. To access the device specific data, a current DD is required and provides the communicator with the information it needs to fully access all the device specific capabilities.

A HART hand-held communicator, if equipped, can also facilitate record keeping of device configurations. After a device is installed, its configuration data can be stored in memory or on a disk for later archiving or printing. There are many types of hand-held communicators available today; their features and ability should be compared to find those that meet your specific requirements.

(2) *Computer-based device configuration and management tools.* A HART-compatible device can be configured with a desktop or laptop computer (or other portable models) by using a computer-based software application and a HART interface modem (Figure 1.4.44). The advantages of using a computer include an improved screen display and support for more DDs and Device Configurations due to additional computer-based memory storage capacity. Due to the critical nature of device configurations in the plant environment, these computers can also be used as backup storage for data from hand-held communicators.

Software applications are available from many suppliers. It is important to review their features to

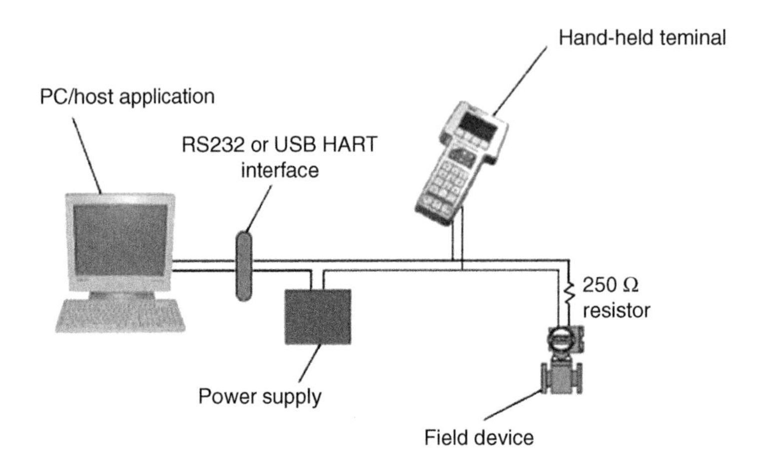

Fig. 1.4.44 The connection of a computer with a HART-compatible device for configuration (courtesy of the HCF).

determine ease of use, ability to add or download the DDs, and general functionality.

1.4.4.2.4 HART system calibration

In order to take advantage of the digital capabilities of HART-compatible devices and instruments, especially for precisely reporting the data values of process control, it is essential that these devices and instruments should be calibrated correctly. Like the calibration procedure of other devices, a calibration procedure for HART-compatible devices and instruments consists of a verification test, an adjustment to within acceptable tolerance if necessary, and a final verification test if an adjustment has been made. Furthermore, data from the calibration is collected and used to complete a report of calibration, documenting instrument performance over time.

(1) *Functional parts of HART devices.* For a HART-compatible device, a multiple-point test between input and output does not provide an accurate representation of its operation. Just like a conventional device, the measurement process begins with a technology that converts a physical quantity into an electrical signal. However, the similarity to a conventional device ends here. Instead of a purely mechanical or electrical path between the input and the resulting 4–20 mA output signal, a HART-compatible device has a microprocessor that manipulates the input data. As shown in Figure 1.4.45, there are typically three calculation sections involved, and each of these sections may be individually tested and adjusted in the calibration procedure.

Prior to the first box in Figure 1.4.45, the microprocessor of this device measures some electrical property that is affected by the process variable of interest. The measured value may be voltage, capacitance, reluctance, inductance, frequency, or some other property. However, before it can be used by the microprocessor, it must be transformed to a digital count by an analog to digital (A/D) converter.

In the first box, the microprocessor of this device must rely upon some form of equation or table to relate the raw count value of the electrical measurement to the actual property (PV) of interest

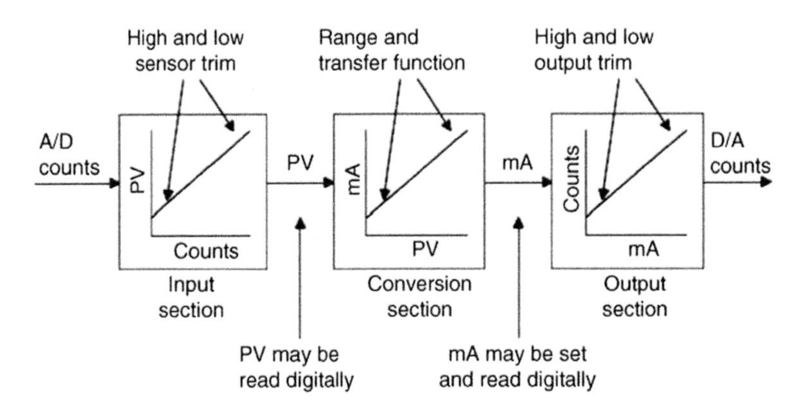

Fig. 1.4.45 A functional block diagram of HART-compatible devices.

such as temperature, pressure, or flow. The principal form of this table is usually established by the manufacturer, but most HART-compatible devices and instruments include commands to perform field adjustments. This is often referred to as a sensor trim. The output of the first box is a digital representation of the process variable. When engineers read the process variable using a communicator, this is the value that they can see.

The second box in Figure 1.4.45 is strictly a mathematical conversion from the process variable to the equivalent milliamp representation. The range values of the instrument (related to the zero and span values) are used in conjunction with the transfer function to calculate this value. Although a linear transfer function is the most common, pressure transmitters often have a square-root option. Other special instruments may implement common mathematical transformations or user defined break point tables. The output of the second block is a digital representation of the desired instrument output. When engineers read the network current using a HART-communicator, this is the value that they see. Many HART-compatible instruments support a command which puts the instrument into a fixed output test mode. This overrides the normal output of the second block and substitutes a specified output value.

The third box in Figure 1.4.45 is the output section where the calculated output value is converted to a count value that can be loaded into a digital to analog converter. This produces the actual analog electrical signal. Once again the microprocessor must rely on some internal calibration factors to get the output correct. Adjusting these factors is often referred to as a current loop trim or 4–20 mA trim.

(2) *Basic steps of HART calibration.* This analysis in (1) above tells us why a proper calibration procedure for a HART-compatible instrument is significantly different from that for a conventional instrument. The specific calibration requirements depend upon the application. If the application uses the digital representation of the process variable for monitoring or control, then the sensor input section (the first box in Figure 1.4.45) must be explicitly tested and adjusted. Please note that this reading is completely independent of the milliamp output, and has nothing to do with the zero or span settings. The PV as read with HART communication continues to be accurate even when it is outside the assigned output range.

If the current network output is not used (i.e., the instrument is used as a digital only device), then the input section calibration is all that is required. If the application uses the milliamp output, then the output section must be explicitly tested and calibrated. Please note that this calibration is independent of the input section, and again, has nothing to do with the zero and span settings.

The same basic multiple point test and adjust technique are employed, but with a new definition for output. To run a test, use a calibrator to measure the applied input, but read the associated output (PV) with a communicator. Error calculations are simpler because there is always a linear relationship between the input and output, and both are recorded in the same engineering units. In general, the desired accuracy for this test will be the manufacturer's accuracy specification. If the test does not pass, then follow the procedure recommended by the manufacturer for trimming the input section. This may be called a sensor trim and typically involves one or two trim points. Pressure transmitters also often have a zero trim, where the input calculation is adjusted to read exactly zero (not low range). Do not confuse a trim with any form of reranging or any procedure that involves using zero and span buttons.

The same basic multiple point test and adjust technique is employed again, but with a new definition for input. To run a test, use a communicator to put the transmitter into a fixed current output mode. The input value for the test is the mA value. The output value is obtained using a calibrator to measure the resulting current. This test also implies a linear relationship between the input and output, and both are recorded in the same engineering units (milliamps). The desired accuracy for this test should also reflect the manufacturer's accuracy specification. If the test does not pass, then follow the procedure recommended by the manufacturer for trimming the output section. This may be called a 4–20 mA trim, a current loop trim, or a D/A trim. The trim procedure should require two trim points close to or just outside of 4 and 20 mA. Do not confuse this with any form of reranging or any procedure that involves using zero and span buttons.

After calibrating both the Input and Output sections, a HART-compatible device or instrument should operate correctly. The middle block in Figure 1.4.45 only involves computations. That is why the range, units, and transfer function can be changed without necessarily affecting the calibration. Notice also that even if the instrument has an unusual transfer function, it only operates in the conversion of the input value to a milliamp output value, and therefore is not involved in the testing or calibration of either the input or output sections.

(3) *Performance verification of HART calibration.* If the goal of this calibration is to validate the overall performance of a HART-compatible device or instrument, it needs just to run a zero and span test like that applied to a conventional instrument. However, passing this test does not definitely indicate that the transmitter is operating correctly, which is due to the following reasons.

Many HART-compatible instruments support a parameter called damping. If this is not set to zero, it can have an adverse effect on tests and adjustments. Damping induces a delay between a change in the instrument input and the detection of that change in the digital value for the instrument input reading and the corresponding instrument output value. This damping induced delay may exceed the settling time used in the test or calibration. The settling time is the amount of time the test or calibration waits between setting the input and reading the resulting output. It is advisable to adjust the instrument damping value to zero prior to performing tests or adjustments. After calibration, be sure the damping constant is returned to its required value.

There is a common misconception that changing the range of a HART-compatible instrument by using a communicator somehow calibrates the instrument. Remember that a true calibration requires a reference standard, usually in the form of one or more pieces of calibration equipment to provide an input and measure the resulting output. Therefore, since a range change does not reference any external calibration standards, it is really a configuration change, not a calibration. Please note that in the block diagram of HART-compatible devices (Figure 1.4.45), changing the range only affects the second box. It has no effect on the digital process variable as read by a communicator.

Using only the zero and span adjustments to calibrate a HART-compatible instrument (the standard practice associated with conventional instruments) often corrupts the internal digital readings. As shown in Figure 1.4.45, there is more than one output to consider. The digital PV and milliamp values read by a communicator are also outputs, just like the analog current network.

The proper way to correct a zero drift condition is to use a zero trim. This adjusts the instrument input block so that the digital PV agrees with the calibration standard. If intending to use the digital process values for trending, statistical calculations, or maintenance tracking, then it should disable the external zero and span buttons and avoid using them entirely.

1.4.4.3 HART protocol

HART protocol is widely recognized as the industry standard for digitally enhanced 4–20 mA field instrument communication in process control. In industrial process control, the HART protocol provides a uniquely backward compatible solution for filed instrument communication as both 4–20 mA analog and digital signals are transmitted simultaneously on the same wiring.

1.4.4.3.1 HART protocol model

HART communication uses a master–slave protocol which means that a field device as slave speaks only when it is spoken to by a device as master. In every communication, a master device sends a "command" message first; while receiving this command message, the slave device processes it and then sends back a "response" message to that master device (Figure 1.4.46). Both the "command" and "response" messages include the HART-data and must be formatted in accordance with the relevant HCF specifications.

The HART protocol can be used in various modes for communicating information to and from smart field instruments and central control or monitoring equipment, which includes analog plus digital signals, and digital-only signals. Digital master–slave communication simultaneous with the 4–20 mA analog signals is the most common. This mode, depicted in Figure 1.4.46, allows digital information from the slave device to be updated twice per second in the master. The 4–20 mA analog signals are continuous and can still carry the primary variable for control. Please note that when a communication between the master and the slave is engaging, "interrupt" is definitely not allowed.

"Burst" is an optional communication mode (Figure 1.4.46) which allows a single slave device to continuously broadcast a standard HART response message. This mode

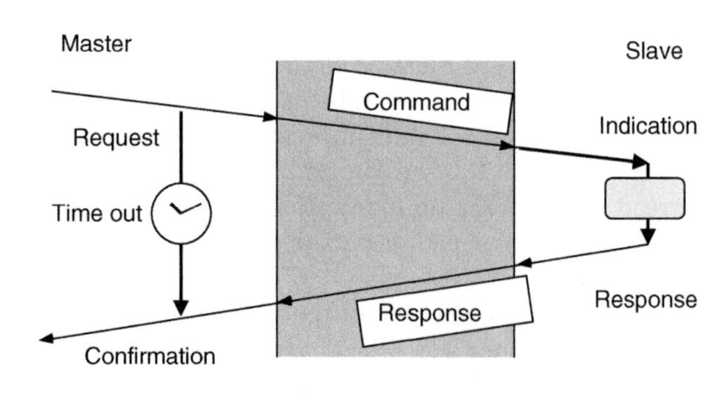

Fig. 1.4.46 The HART master–slave protocol model.

frees the master from having to send repeated command requests to get updated process variable information. The same HART response message (PV or other, see Figure 1.4.45) is continuously broadcast by the slave until the master instructs the slave to do otherwise. Data update rates of 3–4 per second are typical with "burst" mode communication and will vary with the chosen command. Please note that the "Burst" mode should be used only in single slave device networks.

Two masters (primary and secondary) can communicate with slave devices in a HART network. Secondary masters, such as hand-held communicators, can be connected almost anywhere on the network and communicate with field devices without disturbing communication with the primary master. A primary master is typically a PLC, or computer-based central control or monitoring system. A typical installation with two masters is shown in Figures 1.4.33 and 1.4.44. From an installation perspective, the same wiring used for conventional 4–20 mA analog instruments carries the HART communication signals. Allowable cable run lengths will vary with the type of cable and the devices connected, but in general up to 3000 m for a single twisted pair cable with shield and 1500 m for multiple twisted pair cables with a common shield. Unshielded cables can be used for short distances. Intrinsic safety barriers and isolators which pass the HART signals are readily available for use in hazardous areas.

The HART protocol also has the capability to connect multiple field devices on the same pair of wires in a multiple-dropped network configuration as shown in Figure 1.4.33(b). In multiple-dropped networks, communication is limited to master-slave digital only. The current through each slave device is fixed at a minimum value to power the device (typically 4 mA) and no longer has any meaning relative to the process.

The HART protocol utilizes the OSI 7-layer reference model. As is the case for most of the communication systems on the field level, the HART protocol implements only the layers 1, 2, and 7 of the OSI model. The layers 3–6 remain empty since their services are either not required or provided by the Application Layer 7 (see Figure 1.4.47).

The Application Layer defines the commands, responses, data types, and status reporting supported by the protocol. In addition, there are certain conventions in HART (e.g., how to trim the network current) that are also considered part of the Application Layer. While

OSI layer	Function	HART layer	
Application layer	Provides the user with network capable applications	Command oriented. Predefined data types and application procedures	
Presentation layer	Converts application data between network and local machine formats		
Session layer	Connection management service for applications		
Transport layer	Provides network independent, transport message transfer	Autosegmented transfer of large datasets, reliable stream transport, negotiated segment sizes.	
Network layer	End to end routing of packets. Resolving network addresses		Power-optimized, redundant path, self-healing wireless mesh network
Data Link layer	Establishes data packet structure, framing, error detection, bus arbitration.	A binary, byte-oriented, token passing, master–slave protocol	Secure and reliable, time synched TDMA/CSMA, frequency agile with ARQ
Physical layer	Mechanical/electrical connection. Transmits raw bit stream	Simultaneous analog and digital signaling, normal 4–20 mA copper wiring.	2.4 Hz wireless, 802.15.4 based radios, 10 dBm T×power
		Wired HART	Wireless HART

Fig. 1.4.47 HART protocol implementing the OSI 7-layer model.

the Command Summary, Common Tables, and Command Response Code Specifications all establish mandatory Application Layer practices (including data types, common definitions of data items, and procedures), the Universal Commands specify the minimum Application Layer content of all HART-compatible devices.

1.4.4.3.2 HART protocol commands

In the communication routines of the application layer of the HART protocol, the master devices and operating programs are based on HART commands to give instructions or send messages plus data to a field device. Once receiving the command message, the field devices immediately process it and then respond by sending back a response message which can contain requested status reports and/or the data of the field device. Table 1.4.9 provides the classes of the HART commands, and Table 1.4.10 gives a summary of the HART commands.

Figure 1.4.48 is the standard format of both the HART command and response messages. In accordance with the HART command specification, this format includes the following:

(1) First, the preamble, of between 5 and 20 bytes of hex FF (all 1s), helps the receiver to synchronize to the character stream.

(2) The start character may have one of several values, indicating the type of message: master to slave, slave to master, or burst message from slave; also the address format: short frame or long frame.

(3) The address field includes both the master address (a single bit: 1 for a primary master, 0 for a secondary master) and the slave address. In the short frame format, the slave address is 4 bits containing the "polling address" (0–15). In the long frame format, it is 38 bits containing a "unique identifier" for that particular device. (One bit is also used to indicate if a slave is in burst mode.)

(4) The command byte contains the HART command for this message. Universal commands are in the range 0–30; common-practice commands are in the range 32–126; device-specific commands are in the range from 128 to 253.

(5) The byte count byte contains the number of bytes to follow in the status and data bytes. The receiver uses this to know when the message is complete. (There is no special "end of message" character.)

(6) The status field (also known as the "response code") is two bytes, only present in the response message from a slave. It contains information about communication errors in the outgoing message, the status of the received command, and the status of the device itself.

(7) The data field may or may not be present, depending on the particular command. A maximum length of 25 bytes is recommended, to keep the overall message duration reasonable. (But some devices have device-specific commands using longer data fields.) See also the HART data field.

(8) Finally, the checksum byte contains an "exclusive-OR" or "longitudinal parity" of all previous bytes (from the start character onward). Together with the parity bit attached to each byte, this is used to detect communication errors.

1.4.4.3.3 HART protocol data

(1) *HART data.* There are several types of data or information that can be communicated from a HART-compatible device. This includes:

Table 1.4.9 HART commands classes

Universal commands	Command practice commands	Device-specific commands
All devices using the HART protocol must recognize and support the universal commands. Universal commands provide access to information useful in normal operations. For example, read primary variable and units, read manufacturer and device type, read current output and percentage of range, and read sensor serial number and limits	Common practice commands provide functions implemented by many, but not necessarily all, HART communication devices. The HART specifications recommend devices to support these commands when applicable. Examples of common practice commands are read a selection of up to four dynamic variables, write damping time constant, write transmitter range, set fixed output current and perform self-test	Device-specific commands represent functions that are unique to each field device. These commands access setup and calibration information as well as information about the construction of the device. Information on device-specific commands is available from device manufacturers or in the Field Device Specification document. Examples of device-specific commands are read or write sensor type; start, stop, or clear totalizer; read or write alarm relay set point; etc.

Table 1.4.10 HART commands summary

Universal commands	Command practice commands	Device-specific commands (example)
Read manufacturer and device type	Read selection of up to four dynamic variables	Read or write low-flow cut-off
Read primary variable (PV) and units	Write damping time constant	Start, stop, or clear totalizer
Read current output and percentage of range	Write device range values	Read or write density calibration factor
Read up to four pre-defined dynamic variables	Calibrate (set zero, set span)	Choose PV (mass, flow, or density)
Read or write 8-character tag, 16-character descriptor, date	Set fixed output current	Read or write materials or construction information
Read or write 32-character message	Perform self-test	Trim sensor calibration
Read device range values, units, and damping time constant	Perform master reset	PID enable
Read or write final assembly number	Trim PV zero	Write PID set point
Write polling address	Write PV unit	Valve characterization
	Trim DAC zero and gain	Valve set point
	Write transfer function (square root/linear)	Travel limits
	Write sensor serial number	User units
	Read or write dynamic variable assignments	Local display information

(a) device data
(b) supplier data
(c) measurement data
(d) calibration data.

The following is a summary of these data items available for communication between HART-compatible devices and a Host.

(a) *Process variable values.*

(i) Primary process variable (analog): 4–20 mA current signals continuously transmitted to host.

(ii) Primary process variable (digital): Digital value in engineering units, IEEE floating point, up to 24-bit resolution.

(iii) Percent range: Primary process variable expressed as percent of calibrated range.

(iv) Loop current: loop current value in milliamps.

(v) Secondary process variable 1: Digital value in engineering units available from multivariable devices.

(vi) Secondary process variable 2: digital value in engineering units available from multivariable devices

(vii) Secondary process variable 3: digital value in engineering units available from multivariable devices

(b) *Commands from host to device:*

(i) set primary variable units

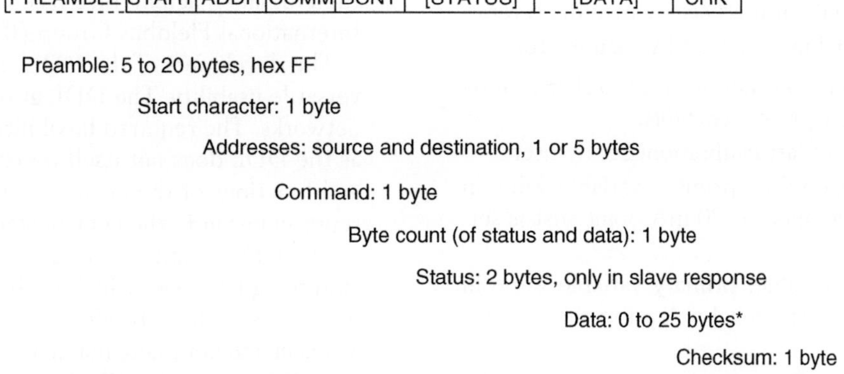

PREAMBLE	START	ADDR	COMM	BCNT	[STATUS]	[DATA]	CHK

Preamble: 5 to 20 bytes, hex FF
Start character: 1 byte
Addresses: source and destination, 1 or 5 bytes
Command: 1 byte
Byte count (of status and data): 1 byte
Status: 2 bytes, only in slave response
Data: 0 to 25 bytes*
Checksum: 1 byte

*25 bytes is a recommended maximum data length
The maximum number of data bytes is not defined by the protocol specifications.

Fig. 1.4.48 The standard format of the HART protocol command and response frames.

(ii) set upper range
(iii) set lower range
(iv) set damping value
(v) set message
(vi) set tag
(vii) set date
(viii) set descriptor
(ix) perform loop test: force loop current to specific value
(x) initiate self-test: start device self-test
(xi) get more status available information.

(c) *Status and diagnostic alerts:*

(i) Device malfunction: Indicates device self-diagnostic has detected a problem in device operation.
(ii) Configuration changed: Indicates device configuration has been changed.
(iii) Cold start: Indicates device has gone through power cycle.
(iv) More status available: Indicates additional device status data available.
(v) Primary variable analog output fixed: Indicates device in fixed current mode.
(vi) Primary variable analog output saturated: Indicates 4–20 mA signal is saturated.
(vii) Secondary variable out of limits: Indicates secondary variable value outside the sensor limits.
(viii) Primary variable out of limits: Indicates primary variable value outside the sensor limits.

(d) *Device identification:*

(i) Instrument tag: User defined, up to 8 characters.
(ii) Descriptor: User defined, up to 16 characters.
(iii) Manufacturer name (code): code established by HCF and set by manufacturer.
(iv) Device type and revision: Set by manufacturer.
(v) Device serial number: set by manufacturer.
(vi) Sensor serial number: set by manufacturer.

(e) *Calibration information for 4–20 mA transmission of primary process variable.*

(i) Date: date of last calibration, set by user.
(ii) Upper range value: primary variable value in engineering units for 20 mA point that is set by user.
(iii) Lower range value: primary variable value in engineering units for 4 mA point that is set by user.
(iv) Upper sensor limit: set by manufacturer.
(v) Lower sensor limit: set by manufacturer.
(vi) Sensor minimum span: set by manufacturer,
(vii) PV damping: primary process variable damping factor, set by user.
(viii) Message: scratch pad message area (32 characters), set by user.
(ix) Loop current transfer function: relationship between primary variable digital value and 4–20 mA current signal.
(x) Loop current alarm action: loop current action on device failure (upscale/downscale).
(xi) Write protect status: device write-protect indicator.

(2) *DDL device description.* The HART commands in the application layer are based on the services of the lower layers and enable an open communication between the master and the field devices. All the HART-compatible devices, no matter their manufacturers, are capable for this openness and intercommunications as long as the field devices operate exclusively with the universal and common-practice commands. In a HART-enabled system, the user does not need more than the simple HART standard notation for the status and fault messages.

When the user wants the message to contain further device-related information or that special properties of a field device are also used, the common-practice and universal commands are not sufficient. Using and interpreting the data requires that the user know their meaning. However, this knowledge is not available in further extending systems which can integrate new components with additional options. To eliminate the adaptation of the master device's software whenever an additional status message is included or a new component is installed, the device description language (DDL) was accordingly developed.

The DDL is not limited to the HART applications. It was developed and specified for all the Fieldbus, independent of the HART protocol, by the Human–machine Interface workshop of the International Fieldbus Group (IFG).

The developers of the DDL aimed at achieving versatile usability. The DDL also finds use in field networks. The required flexibility is ensured insofar as the DDL does not itself determine the number and functions of the device interfaces and their representation in the control stations.

The DDL is simply a language, similar to a programming language, which enables the device manufacturers to describe all communication options in an exact and complete manner. The DDL allows the manufacturer to describe:

(a) attributes and additional information on communication data elements,

(b) all operating states of the device,
(c) all device commands and parameters,
(d) the menu structure, thus providing a clear representation of all operating and functional features of the device.

Having the device description of a field device and being able to interpret it, a master device is equipped with all necessary information to make use of the complete performance features of the field device.

Device-specific and manufacturer-specific commands can also be executed and the user is provided with a universally applicable and uniform user interface, enabling him or her to clearly represent and perform all device functions. Thanks to this additional information, clear, exact and, hence, safer operation and monitoring of a process is made possible.

The master device does not read the device description as readable text in DDL syntax, but as short, binary-coded DD data record specially generated by the DDL encoder (or DDL compiler). For devices with sufficient storage capacity, this short form opens up the possibility of storing the device description already in the firmware of the field device. During the parameterization phase, it can be read by the corresponding master device.

1.4.4.4 HART integration

1.4.4.4.1 Basic industrial field networks

There are many different field networks available in industrial control for instrument engineers today. Understanding these networks' classifications should allow us to choose the right tools for the control applications. Some common industrial communication protocols will be introduced below by focusing on the native applications and placing them in three basic categories:

(1) Sensor networks are those protocols initially designed to support discrete I/O.
(2) Device networks are those protocols originally focused on process instrumentation.
(3) Control networks are those protocols typically used to connect controllers and I/O systems.

(1) *Sensor networks*. Sensor level protocols, with a principal focus on supporting the digital communications for those discrete sensors and actuators. Sensor level protocols tend to have very fast cycle times and, since they are often promoted as an alternative to PLC discrete I/O, the cost of a network node should be relatively low. These protocols listed below are the simplest forms of sensor networks available today.

(a) *AS-i*. AS-i acts as a network-based replacement for a discrete I/O card. Consequently, AS-i offers perhaps the simplest network around consisting of up to 31 slave devices with 248 I/O bits and the following functionality: (1) the master polls each slave, (2) the master message contains four output bits, (3) the slave answers immediately with four input bits, 4) diagnostics are included in each message, and (5) a worst-case scan time of less than 5 ms.

(b) *CAN*. The CAN defines only basic, low level signaling and medium access specifications which are both simple and unique. Even though CAN medium access is technically CSMA/CD, this classification can provide simple, highly reliable, prioritized communication between intelligent devices, sensors, and actuators in automotive applications. Of these advantages, reliability is paramount. Network errors while driving a car on a busy interstate highway are unacceptable. Today, CAN is used in a vast number of vehicles and in a variety of other applications.

As a result: (1) a large number of different chips and vendors support CAN, (2) the total chip volume is huge, and (3) the parts cost is small.

(c) *DeviceNet*. DeviceNet is a well-established machine and manufacturing automation network supported by a substantial number of products and vendors in the semiconductor industry. DeviceNet specifies physical (connectors, network terminators, power distribution, and wiring) and application layer operation based on the CAN standard. While all the popular application layer hierarchies are supported (client and server, master and slave, peer-to-peer, publisher and subscriber), in practice most devices support only master–slave operation which results in significantly lower costs. In comparison with the CAN, DeviceNet provides configurable structure to the operation of the network, allowing for the selection of polled, cyclic, or event driven network operation.

(d) *Interbus*. Interbus is a popular industrial network that uses a ring topology in which each slave has an input and an output connector. Interbus is one of the few protocols that are full duplex-data transmitted and received at the same time.

Interbus communication is cyclical, efficient, fast, and deterministic (e.g., 4096 digital inputs and outputs scanned in 14 ms). Of the ring topology, Interbus commissioning offers us these advantages: (1) Node addresses are not required because the master can automatically identify the nodes on the network. (2) Slaves provide identification information that allows the master to determine the quantity of the data provided by the slave. (3) Using this data, the master explicitly maps the data to and from the slave into the bit stream as it shifts through the network.

Interbus works as a large network-based shift register in which a bus cycle begins with the network master transmitting a bit stream. As the first slave receives the bits, they are echoed passing the data on to the next slave in the ring. This process is simultaneous with the data being shifted from the master to the first slave. In turn, the data from the first slave is then being shifted into the master.

(2) *Device networks.* Device network protocols support process automation that is fundamentally continuous and analog, more complex transmitters, and valve-actuators. Transmitters typically include pressure, level, flow, and temperature. The valve-actuators can include those intelligent controllers, motorized valves, and pneumatic positioners. Three device networks are prevalent in the process automation industry: Foundation Fieldbus H1, HART, and PROFIBUS-PA.

(a) *Foundation Fieldbus H1.* In August 1996, the Foundation Fieldbus released its H1 Specifications that focus on "the network is the control system." Of the fundamental differences from the controller and I/O approach used in traditional systems, Foundation Fieldbus specifies not only a communication network but also control functions. As a result, the purpose of the communications is to pass data to facilitate proper operation of the distributed control application. Its success relies on synchronized cyclical communication and on a well-defined applications layer.

In the Foundation Fieldbus H1, communications occur within framed intervals of fixed time duration (a fixed repetition rate) and are divided into two phases and scheduled cyclical data exchange and acyclic (e.g., configuration and diagnostics) communication. The communication is controlled by polling the network and thereby prompting the process data to be placed on the bus. This is done by passing a special token to grant the bus to the appropriate device, resulting in the cyclic data being generated at regular intervals.

In the Foundation Fieldbus H1 networks, the application layer defines function blocks, which include analog in, analog out, transducer, and the blocks. The data on the network is the transfer from one function block to another in the network-based control system. The data is complex and includes the digital value, engineering units, and status on the data (to indicate the PID is manual or the measured value is suspect).

(b) *HART.* HART is unique among device networks because it is fundamentally an analog communications protocol. All the other protocols use digital signaling, while HART uses modulated communications because the "HART digital communications" modulate analog signals centered in a frequency band separated from the 4–20 mA signaling. HART enhances smart 4–20 mA field devices by providing two-way communication that is backward compatible with existing installations, which allows HART to support two communications channels simultaneously: a one-way channel carrying a single process value (the 4–20 mA signal) and a bidirectional channel to communicate digital process values, status, and diagnostics. Consequently, the HART protocol can be used in traditional 4–20 mA applications, and allows the benefits of digital communication to be realized in existing plant installations.

HART is a simple, easy-to-use protocol because of these facts: (1) Two masters are supported using token passing to provide bus arbitration. (2) Allows a field device to publish process data ("burst mode"). (3) Cyclical process data includes floating-point digital value, engineering units, and status. (4) Operating procedures standardized (for current loop reranging, loop test, and transducer calibration). (5) Standardized identification and diagnostics provided.

(c) *PROFIBUS-PA.* PROFIBUS-PA (process automation) was introduced to extend PROFIBUS-DP (decentralized peripherals) in order to support process automation. PROFIBUS-PA, which operates over the same H1 physical layer as Foundation Fieldbus H1, is essentially an LAN for communication with process instruments.

PROFIBUS-PA networks are fundamentally master–slave model, so sophisticated bus arbitration is not necessary. PROFIBUS-PA also defines profiles for common process instruments, including both mandatory and optional propertied (data items). When a field device supports a profile some configuration of the device should be possible without being device-specific.

(3) *Control networks.* Control networks are focused on providing a communication backbone that allows integration of controllers, I/O, and subnetworks. Control networks stand at the crossroads between the growing capabilities of industrial networks and the penetration of enterprise networks into the control system. As such, the control networks are able to move huge chunks of heterogeneous data and operate at high data rates. A brief overview of these four control networks is given below; they are ControlNet, Industrial Ethernet, Ethernet/IP, and PROFIBUS-DP.

(a) *ControlNet.* ControlNet was developed as a high performance network suitable for both manufacturing and process automation. ControlNet uses Time Division Multiple Access (TDMA) to control the access to the network, which means a network cycle is assigned a fixed repetition rate. Within the bus cycle, data items are assigned a fixed time division for transmission by the corresponding device. Data objects are placed on ControlNet within a designated time slot and at precise levels. Once the data to be published is identified along with its time slot, any device on the network can be configured to use the data. In the second half of a bus cycle, acyclic communications occur. ControlNet is more efficient than polled or token passing protocols because its data transmission is very deterministic.

(b) *Industrial Ethernet.* Many industrial communication protocols specify mechanisms to embed their protocols in Ethernet. Ethernet addresses only the lower layers of communications networks, but does not address the meaning of the data it transports. Even though Ethernet effectively communicates many protocols simultaneously over the same wire, it provides no guarantees that the data can be exchanged between different protocols.

In Industrial Ethernet, the protocol being adopted is TCP/IP and not Ethernet at all. TCP/IP is using two approaches to support the session, presentation, and application layers of the corresponding industrial protocol. First, the industrial protocol is simply encapsulated in the TCP/IP, allowing the shortest development time for defining industrial protocol transportation over TCP/IP. The second approach actually maps the industrial protocol to TCP and UDP services. While this strategy takes more time and effort to develop, it results in a more complete implementation of the industrial protocol on top of the TCP/IP. UDP is a connectionless, unreliable communication service that works well for broadcasts to multiple recipients and fast, low-level signaling. UDP is used by several industrial protocols (for time synchronization). TCP is a connection oriented data stream that can be mapped to the data and I/O functions in some industrial protocols.

(c) *Ethernet/IP.* Ethernet/IP is a mapping of the "Control and Information Protocol (CIP)" used in both ControlNet and DeviceNet to TCP/IP (not Ethernet). While all the basic functionality of ControlNet is supported, the hard real-time determinism that ControlNet offers is not present. CIP is being promoted as a common, object-oriented mechanism for supporting both manufacturing and process automation functions.

Ethernet/IP is a good contribution to the growing discussion of Industrial Ethernet. However, Ethernet/IP is a recent development and is basically the application of an existing industrial network's application layer to the TCP/IP.

(d) *PROFIBUS-DP.* PROFIBUS-DP (Decentralized Peripherals) is a master-slave protocol used primarily to access remote I/O. Each node is polled cyclically updating the status and data associated with the node. Operation is relatively simple and fast. PROFIBUS-DP also supports Fieldbus Message Specification (FMS), which is a more complex protocol for demanding applications that includes support for multiple masters and peer-to-peer communication.

1.4.4.4.2 Choosing the right field networks

Before choosing the best suitable network for an industrial control application, it is necessary to review all the open network types available today. By carefully reviewing each of these types of open communication networks, their respective strengths and weaknesses, in particular some important technical factors, ought to be understood. Please note that the specified application will be far more than the technology used. The specific, measurable benefits must drive this selection process.

The first factor that should be considered is the cost of each network. The second is the network connectivity in which the bottom line is that you want your data. It is very important to remember that all communications networks cause changes. Although the actual changes can be very difficult to foretell accurately, the effect of these changes must be realistically considered.

In some instances, we have to make a decision on choosing between the HART and other types of field networks. Before making such a decision, it is important to understand the differences between the HART and other field networks. For example, if all the choices are

Table 1.4.11 A comparison between HART and Foundation Fieldbus

Feature	HART	Foundation Fieldbus (FF)
Technology acceptance	Well proven as Large Installed base Will continue to be sold as a replacement unit Simple for technicians' competence	Proven as Growing Installed base Training required New investment to occur increasingly in the FF
Power limitation Advances in silicon power consumption same for HART/FF; thus FF will always have capability for more functionality	35 mW to 4 mA available for the HART signal Cannot "mirror" Fieldbus, however may provide an 80% solution	FF minimum power requirement of 8 mA No spec limit; Ultimate FF segment power budget FF devices order of more magnitude than that powering an Event for IS
Communication performance	100 bits/s Additional burden on host	FF H1 communicates at 31250 bits/s
Transmitter diagnostics	Device only Includes predictive No knowledge of other devices	Device + other devices
Advanced diagnostics	Does not have the processing power	For example, Statistical Process Monitoring and Machinery Health monitoring
Push or poll	Polled for HART status periodically Status can be missed	Events are latched/time stamped in the device Sent by the device There is no chance of missing field problems with FF
Two-way communication to other devices	No	Yes
Multiple-dropped	Very limited Theory 15 devices: real around 3 slow series loop	True multiple-dropped: physically 32 devices realistically 12–16
Use in safety instrumented systems	All devices wired individually	Devices 2007 offers the holding back technology
Control in the field and advanced applications	Does not support the function block model	Function block model supports interoperable control in the field where blocks can reside in the field device
Multiple variables	In digital mode only It is limited	Yes
Footprint and hardware reduction	No	Renders obsolete all separate signal conditioners, isolation amplifier cards, output cards, CPU cards, I/P converters, etc.
Future proof devices—typical upgrade capability In the field	No	Ability to download new version of firmware over H1 link None disconnected device from the H1 segment Communicate with this devices conformant with Fieldbus specification
Full specifications in the devices	No	Embedded at the factory Travels with the instrument Upload directly to "Smart Instrument" software Reduces commissioning time Reduces time to perform diagnostics
Commissioning speed	Hours for individually wired devices	The networking capability of FF allows the user to commission a device in 10 s

constrained between the HART and the Foundation Fieldbus, comparisons of these two types in every category are necessary. Table 1.4.11 is a list of the differences in elemental technical features between the HART and the Foundation Fieldbus, which should be a help in choosing between the HART and the Foundation Fieldbus.

1.4.4.4.3 Integrating the HART with other field networks

The integration of the HART network with the Foundation Fieldbus network is taken as an example here to demonstrate the strategies and techniques of integrating

HART with other field networks. As mentioned above, one fundamental difference between HART and Fieldbus devices is that with Fieldbus devices, you can implement control strategies inside the field devices themselves; however, even if you can implement the same control strategies with HART devices, the execution of actual control algorithms would go on in the control system computer or PLC. A well-designed control system will allow an integrated control strategy to use devices independent of communication protocol.

HART and Fieldbus devices have the capability of providing a wide range of diagnostics data about the device's safety. In fact, the diagnostic capabilities of HART and Fieldbus devices are nearly the same. The types of device diagnostics change widely, depending on the type of device. Measurement transmitters will have diagnostics related to the status of the transducer and measurement logic in the device. Control devices such as valves will provide a lot of information about the mechanical condition of the device. Both transmitters and valves will provide diagnostic information about the communication electronics in their respective devices.

Although the diagnostic data provided by HART and Fieldbus is very similar, the way they get to the control system and the way they get to the operator or technician to see it can be quite different. This has to do with the speed and characteristics of the communication technology used by these two protocols. Fieldbus basically uses point-to-point communication technology. This means when a Fieldbus device detects a diagnostics condition it wants to report, it can send an event out on the bus with the related information. The control system picks up the event and immediately displays or annunciates it on the console. HART devices, on the other hand, have to continually undergo polling to see if there is anything to report. Because the polling occurs at 1200 bps with HART, there are limitations on how many devices it can poll for alters in a specific time frame. An operator can poll a small number of critical devices for alters within seconds or a large number of devices within minutes. However, it is possible to implement an effective diagnostic alert system with HART as long as you understand the restrictions on response and device count.

Once the operator or maintenance engineer is aware of a problem in a filed device either through an alert or some other means, the actual display of the status information from HART and Fieldbus devices is very similar. Usually, a record of this status event will automatically log into the control system. The logging of HART and Fieldbus device problems should normally look the same on a well-integrated system.

The type of portable maintenance tools required in systems of HART and Fieldbus devices is also an important factor for consideration. Portable tools currently fall into two general categories. The first is intrinsically safe hand-held devices. Several are available for HART only devices. A combined HART and Fieldbus intrinsically hand-held device has become available, too. This integrated tool allows the user to configure and diagnose HART and Fieldbus devices while in the field. The laptop computer is the second type of portable tool. However, these types of computers cannot go in hazardous areas of a plant. As far as small hand-held computers go, how practical these will be in a plant environment remains an open question.

References

Alan R. Dewey in EMERSON. 2005. HART, Fieldbus Work Together in Integrated Environment. http://www.emersonprocess.com/home/library/articles/protocol/ protocol0507_teamwork.pdf. Accessed date: October 2007.

Analog Services, Inc. (http://www.analogservices.com). 2006. HART Book. http:// www.analogservices.com/about_part0.htm. Accessed date: October 2007.

AS-INTERFACE (http://www.as-interface.net). 2007a. AS-Interface System. http://www.as-interface.net/System/. Accessed date: May.

AS-INTERFACE (http://www.as-interface.net). 2007b. AS-Interface Products. http://www.as-interface.net/Products/. Accessed date: May.

CiA (http://www.can-cia.org). 2005a. Registered Free Download; CAN Physical Layer Specification Version 2.0. http://www.can-cia.org/downloads/ciaspecifications/?557. Accessed date: July.

CiA (http://www.can-cia.org). 2005b. Registered Free Download; CAN Application Layer Specification Version 1.1. http://www.can-cia.org/downloads/ciaspecifications/?1169. Accessed date: July.

CiA (http://www.can-cia.org). 2005c. Registered Free Download; CANopen Specification Version 1.3. http://www.can-cia.org/downloads/ciaspecifications/?1136. Accessed date: July.

Commfront (http://www.commfront.com). 2007. RS232, 485,422,530 Buses. http://www.commfront.com/CommFront-Home.htm. Accessed date: July.

Cyber (http://cyber.felk.cvut.cz). 2006. Supervisory Human Operation. http://cyber.felk.cvut.cz/gerstner/biolab/bio_web/projects/iga2002/index.html. Accessed date: October.

David Belohrad and Miroslav Kasal. 1999. FSK Modem with GALs. http://www.isibrno.cz/ ~ belohrad/radioelektronika99-fskmodem.pdf. Accesseddate: October 2007.

Degani, Asaf, Shafto, Michael, Kirlik, Alex. 2006. Modes in Human–Machine Systems: Review, Classification, and Application. http://ic www.arc.nasa.gov/people/asaf/interface_design/pdf/Modes%20in%20Human–;machine%20Systems.pdf. Accessed date: October.

Degani, Asaf. 2006. Modeling Human–Machine Systems: On Modes, Error, and Patterns of Interaction. http://ase.arc.nasa.gov/people/asaf/hai/pdf/Degani_Thesis.pdf. Accessed date: October.

ESD-Electronics (http://www.esd-electronics.com). 2005. Controller Area Network. http://www.esd-electronics.com/german/PDF-file/CAN/Englisch/intro-e.pdf. Accessed date: July.

Fieldbus (http://www.fieldbus.org). 2005a. FOUNDATION Technology. http://www.fieldbus.org/index.php?option=com_content&task=view&id=45&Itemid=195. Accessed date: July.

Fieldbus (http://www.fieldbus.org). 2005b. Profibus Technology. http://www.pepperl-fuchs.com/pa/interbtob/profibus/default_e.html. Accessed date: July.

Fieldbus Centre (http://www.knowthebus.org). 2005a. FOUNDATION Fieldbus. http://www.knowthebus.org/fieldbus/foundation.asp. Accessed date: July.

Fuji Electric (http://web1.fujielectric.co.jp). 2007. Fuji AS-I Technologies. http://www.fujielectric.co.jp/fcs/eng/as-interface/as_i/index.html. Accessed date: May.

Grid Connect (http://www.industrialethernet.com). 2005. Industrial Ethernet. http://www.industrialethernet.com/etad.html. Accessed date: July.

Groover Mikell P 2001 Automation, Production Systems, and Computer-Integrated Manufacturing, Second Edition. New Jersey: Prentice Hall.

H. Kirrmann in ABB Research Center of Switzerland. 2006. The HART Protocol. AI_41 1_HART.ppt. Accessed date: October 2007.

Hardware Secrets (http://www.hardwaresecrets.com). 2007a. PCI Bus Tutorial. http://www.hardwaresecrets.com/article/190. Accessed date: May.

Hardware Secrets (http://www.hardwaresecrets.com). 2007b. AGP Bus Tutorial. http://www.hardwaresecrets.com/article/155. Accessed date: July.

Harris, Don. 2006. Human–Machine Interaction. http://www.cranfield.ac.uk/soe/postgraduate/hf_module9.htm. Accessed date: October.

HCF (HART Communication Foundation). 2007. HCF—Main Pages. http://www.hartcomm2.org/index.html. Accessed date: October.

HIT (http://www.hit.bme.hu). 2007. GPIB Tutorial. http://www.hit.bme.hu/~papay/edu/GPIB/tutor.htm. Accessed date: July.

HMS Industrial Networking (http://www.anybus.com). 2005a. AS-Interface Technologies. http://www.anybus.com/technologies/asi.shtml. Accessed date: July.

HMS Industrial Networking (http://www.anybus.com). 2005b. AS-Interface Products. http://www.anybus.com/products/asinterface.shtml. Accessed date: July.

HMS Industrial Networking (http://www.anybus.com). 2005c. Interbus Connectivity: http://www.anybus.com/products/interbus.shtml?gclid=CLXtoK7UjYwCFT4GQgod3T_GBw. Accessed date: July.

Honey Well (http://hpsweb.honeywell.com). 2005a. http://hpsweb.honeywell.com/Cultures/en-US/Products/Systems/ExperionPKS/FoundationFieldbusIntegration/default.htm. Accessed date: July.

Honey Well (http://hpsweb.honeywell.com). 2007. Wireless HART. http://hpsweb.honeywell.com/Cultures/en-US/Products/wireless/SecondGenerationWireless/default.htm. Accessed date: October.

IBM (http://www.ibm.com/us). 2005. IDE Subsystem. http://publib.boulder.ibm.com/infocenter/pseries/v5r3/index.jsp?topic=/com.ibm.aix.kernelext/doc/kernextc/ide_subsys.htm. Accessed date: May.

Interbus Club (http://www.interbusclub.com). 2005a. Interbus Technology. http://www.interbusclub.com/en/index.html. Accessed date: July.

Interbus Training in Web (http://pl.et.fh-duesseldorf.de/prak/prake/index.asp). 2007.

Interbus Basic. http://pl.et.fh-duesseldorf.de/prak/prake/download/IBS_grundlagen.pdf. Accessed date: May.

Interface Bus (http://www.interfacebus.com). 2005a. PCI Bus Pins. http://www.interfacebus.com/Design_PCI_Pinout.html. Accessed date: May.

Interface Bus (http://www.interfacebus.com). 2005b. PCMCIA 16 bits Bus. http://www.interfacebus.com/Design_Connector_PCMCIA.html. Accessed date: May.

Interface Bus (http://www.interfacebus.com). 2005c. RS-485 Bus. http://www.interfacebus.com/Design_Connector_RS485.html. Accessed date: May.

IO Tech (http://www.iotech.com). 2007. IEEE-488 Standard. http://www.iotech.com/an06.html. Accessed date: July.

Israel, Johann Habakuk and Anja Naumann. 2006. http://useworld.net/ausgaben/4-2007/0 1-Israel_Naumann.pdf. Accessed date: October.

IXXAT (http://www.ixxat.com). 2005. CAN Application Layer. http://www.ixxat.com/can_application_layer_introduction_en, 7524, 5873.html. Accessed date: July.

Jaffe, David. 2006. Enhancing Human–Machine Interaction. http://ability.stanford.edu/Press/rehabman.pdf. Accessed date: October.

Jim Russell. 2007. HART v Foundation Fieldbus. http://www.iceweb.com.au/Instrument/FieldbusPapers/HART%20v%20FF%20PAPERfinal.pdf. Accessed date: October.

Kenneth L. Holladay, P. E. in Southwest Research Institute of the USA. 1991. Calibrating HART Transmitters. http://www.transcat.com/PDF/Hart_Transmitter_Calibration.pdf. Accessed date: October 2007.

Kvaser (http://www.kvaser.com). 2005. Controller Area Network. http://www.kvaser.com/can/. Accessed date: July.

Microchip (http://ww1.microchip.com). 2005a. CAN Basics. http://ww1.microchip.com/downloads/en/AppNotes/00713a.pdf. Accessed date: July.

Microchip (http://ww1.microchip.com). 2005b. CAN Physical Layer. http://ww1.microchip.com/downloads/en/AppNotes/00228a.pdf. Accessed date: July.

MOXA (http://www.moxa.com). 2005a. Industrial Ethernet Technologies. http://www.moxa.com/Zones/Industrial_Ethernet/Tutorial.htm. Accessed date: July.

MOXA (http://www.moxa.com). 2005b. Industrial Ethernet Products. http://www.moxa.com/product/Industrial_Ethernet_Switches.htm. Accessed date: July.

Murray, Steven A. 2006. Human–Machine Interaction with Multiple Autonomous Sensors. http://www.spawar.navy.mil/robots/research/hmi/ifac.html. Accessed date: October.

PC Guide (http://www.pcguide.com). 2005a. IDE Guide. http://www.pcguide.com/ref/hdd/if/ide/unstdIDE-c.html. Accessed date: May.

PC Guide (http://www.pcguide.com). 2005b. SCSI Guide. http://www.pcguide.com/ref/hdd/if/scsi/. Accessed date: May.

PEPPERL + FUCHS (http://www.am.pepperl-fuchs.com). 2007. HART Multiplexers. http://www.am.

pepperl-fuchs.com/products/productsubfamily.jsp?division=PA&productsubfamily_id=1343. Accessed date: October.

PEPPERL+FUCHS (http://www.am.pepperl-fuchs.com). 2005a. AS-Interface. http://www.pepperl-fuchs.com/cgi-bin/site_search.pl. Accessed date: July.

PEPPERL+FUCHS (http://www.am.pepperl-fuchs.com). 2005b. Fieldbus Technology; Foundation Fieldbus: AS-Interface. http://www.pepperl-fuchs.com/pa/interbtob/communication/default_e.html. Accessed date: July.

PHM (http://www.phm.lu). 2007. IDE Pins Out. http://www.phm.lu/Documentation/Connectors/IDE.asp. Accessed date: May.

PROFIBUS (http://www.profibus.com). 2005a. Profibus Technical Description. http://www.profibus.com/pb/technology/description/. Accessed date: July.

PROFIBUS (http://www.profibus.com). 2005b. Profibus Specification. http://www.profibus.com/pall/meta/downloads/. Accessed date: July.

QSI (http://www.qsicorp.com). 2006. Human–Machine Interface Devices. http://www.qsicorp.com/product/industrial/?gclid=CK-apKS1l4wCFSQHEgod6iqP5w. Accessed date: October.

Quatech (http://www.quatech.com). 2005. ISA Bus Overviews. http://www.quatech.com/support/comm-over-isa.php. Accessed date: July.

Samson (http://www.samson.de). 2005. FOUNDATION Fieldbus Technical Information. http://www.samson.de/pdf_en/l454en.pdf. Accessed date: July.

SAMSON. 2005. SAMSON Technical Information—HART Communications. http://www.samson.de/pdf_en/l452en.pdf. Accessed date: October 2007.

Schneider-Electric (http://www.automation.schneider-electric.com). 2006. Human–Machine Interface. http://www.automation.schneider-electric.com/as-guide/EN/pdf_files/asg-8-Human–Machine-interface.pdf. Accessed date: October.

SCSI Library (http://www.scsilibrary.com). 2005. SCSI. http://www.scsilibrary.com. Accessed date: May.

Semiconductors (http://www.semiconductors.bosch.de).2005.CAN Specifications. http://www.semiconductors.bosch.de/pdf/can2spec.pdf. Accessed date: July.

SIEMENS (http://www.automation.siemens.com). 2005a. Siemens AS-Interface Technologies. http://www.automation.siemens.com/cd/as-interface/html_76/asisafe.htm. Accessed date: July.

SIEMENS (http://www.automation.siemens.com). 2005b. Siemens AS-Interface Products. http://www.automation.siemens.com/infocenter/order_form.aspx?tab=3&nodekey=key_1994569&lang=en. Accessed date: July.

SIEMENS (http://www.automation.siemens.com). 2005c. Siemens Profibus: http://www.automation.siemens.com/net/h tml_76/produkte/020_produkte.htm. Accessed date: July.

SIEMENS (http://www.automation.siemens.com). 2005d. Industrial Ethernet Technologies and Products. http://www.automation.siemens.com/net/html_7 6/produkte/040_produkte.htm. Accessed date: July.

Simons, C.L. and Parmee, L.C. 2006. Human–Machine Interaction Software Design. http://www.ip-cc.org.uk/INTREP-COINT-SIMONS-2006.pdf. Accessed date: October.

SMAR International Corporation (http://www.smar.com). 2006. HART Tutorial. http://www.smar.com/PDFs/Catalogues/Hart_Tutorial.pdf. Accessed date: October 2007.

Softing (http://www.softing.com). 2005a. FOUNDATION Fieldbus. http://www.softing.com/home/en/industrial-automation/products/foundation-fieldbus/index.php. Accessed date: July.

Softing (http://www.softing.com). 2005b. Profibus. http://www.softing.com/home/en/industrial-automation/products/profibus-dp/index.php?navanchor=3010004. Accessed date: July.

Tech Soft (http://www.techsoft.de). 2007. IEEEE-488 Tutorial. http://www.techsoft.de/htbasic/tutgpibm.htm?tutgpib.htm. Accessed date: July.

Techfest (http://www.techfest.com). 2005. ISA Bus Technology. http://www.techfest.com/hardware/bus/isa.htm. Accessed date: July.

Texas Instruments (http://sparc.feri.uni-mb.si). 2005. CAN Introduction. http://sparc.feri.uni-mb.si/Sistemidaljvodenja/Vaje/pdf/Inroduction%20to%20CAN.pdf. Accessed date: July.

The UK AS-i Expert Alliance (http://www.as-interface.com). 2007a. AS-Interface Technologies. http://www.as-interface.com/asitech.asp. Accessed date: May.

The UK AS-i Expert Alliance (http://www.as-interface.com). 2007b. AS-Interface Products. http://www.as-interface.com/asi_literature.asp. Accessed date: May.

Vector Germany (http://www.can-solutions.com). 2005. CAN Mechanism. http://www.can-solutions.com/?gclid=CLWxnai6jIwCFQrlQgodnx3gBg. Accessed date: July.

YOKOGAWA (http://www.yokogawa.com). 2007. HART Communicator. http://www.yokogawa.com/us/mi/MetersandInstruments/us-ykgw-yhcypc.htm. Accessed date: October.

Section Two

Motors and drives

Chapter 2.1

2.1

Electromechanical systems

Crowder

In the design of any complex system, all the relevant design details must be considered to ensure the development of a successful product. In the development of motion systems, problems in the design process are most likely to occur in the actuator or motor–drive system. When designing any actuation system, mechanical designers work with electrical and electronic systems engineers, and if care is not taken, confusion will result. Here we discuss some of the electric motor–drive systems in common use, and to identify the issues that arise in the selection of the correct components and systems for specific applications.

A key step in the selection of any element of a drive system is a clear understanding of the process being undertaken. Section 2.1.1 provides an overview to the principles of industrial automation, and Sections 2.1.2 and 2.1.3 consider machine tools and industrial robotics, respectively. Section 2.1.4 considers a number of other applications domains.

2.1.1 Principles of automation

Within manufacturing, automation is defined as the technology which is concerned with the application of mechanical, electrical, and computer systems in the operation and control of manufacturing processes. In general, an automated production process can be classified into one of three groups: fixed, programmable, or flexible.

- *Fixed automation* is typically employed for products with a very high production rate; the high initial cost of fixed-automation plant can therefore be spread over a very large number of units. Fixed-automation systems are used to manufacture products as diverse as cigarettes and steel nails. The significant feature of fixed automation is that the sequence of the manufacturing operations is fixed by the design of the production machinery, and therefore the sequence cannot easily be modified at a later stage of a product's life cycle.
- *Programmable automation* can be considered to exist where the production equipment is designed to allow a range of similar products to be produced. The production sequence is controlled by a stored program, but to achieve a product change-over, considerable reprogramming and tooling changes will be required. In any case, the process machine is a stand-alone item, operating independently of any other machine in the factory; this principle of automation can be found in most manufacturing processes and it leads to the concept of islands of automation. The concept of programmable automation has its roots in the Jacquard looms of the nineteenth century, where weaving patterns were stored on a punched-card system.
- *Flexible automation* can be considered to be an enhancement of programmable automation in which a computer-based manufacturing system has the capability to change the manufacturing program and the physical configuration of the machine tool or cell with a minimal loss in production time. In many systems the machining programs are prepared at a location remote from the machine, and they are then transmitted as required over a computer-based local-area communication network.

Electric Drives and Electromechanical Systems; ISBN: 9780750667401

The basic design of machine tools and other systems used in manufacturing processes changed little from the eighteenth century to the late 1940s. There was a gradual improvement during this period as the metal cutting changed from an art to a science; in particular, there was an increased understanding of the materials used in cutting tools. However, the most significant change to machine-tool technology was the introduction of *numerical-control* (NC) and *computer-numerical-control* (CNC) systems.

To an operator, the differences between these two technologies are small: both operate from a stored program, which was originally on punched tape, but more recently computer media such as magnetic tapes and discs are used. The stored program in a NC machine is directly read and used to control the machine; the logic within the controller is dedicated to that particular task. A CNC machine tool incorporates a dedicated computer to execute the program. The use of the computer gives a considerable number of other features, including data collection and communication with other machine tools or computers over a computer network. In addition to the possibility of changing the operating program of a CNC system, the executive software of the computer can be changed, which allows the performance of the system to be modified at minimum cost. The application of NC and CNC technology permitted a complete revolution of the machine tool industry and the manufacturing industries it supported. The introduction of electronic systems into conventional machine tools was initially undertaken in the late 1940s by the United States Air Force to increase the quality and productivity of machined aircraft parts. The rapid advances of electronics and computing systems during the 1960s and 1970s permitted the complete automation of machine tools and the parallel development of industrial robots. This was followed during the 1980s by the integration of robots, machine tools, and material handling systems into computer-controlled factory environments. The logical conclusion of this trend is that individual product quality is no longer controlled by direct intervention of an operator. Since the machining parameters are stored either within the machine or at a remote location for direct downloading via a network a capability exists for the complete repeatability of a product, both by mass production and in limited batches (which can be as small as single components). This flexibility has permitted the introduction of management techniques, such as just-in-time production, which would not have been possible otherwise.

A typical CNC machine tool, robot, or multi-axis system, whatever its function, consists of a number of common elements (see Figure 2.1.1). The axis position, or the speed controllers, and the machining process

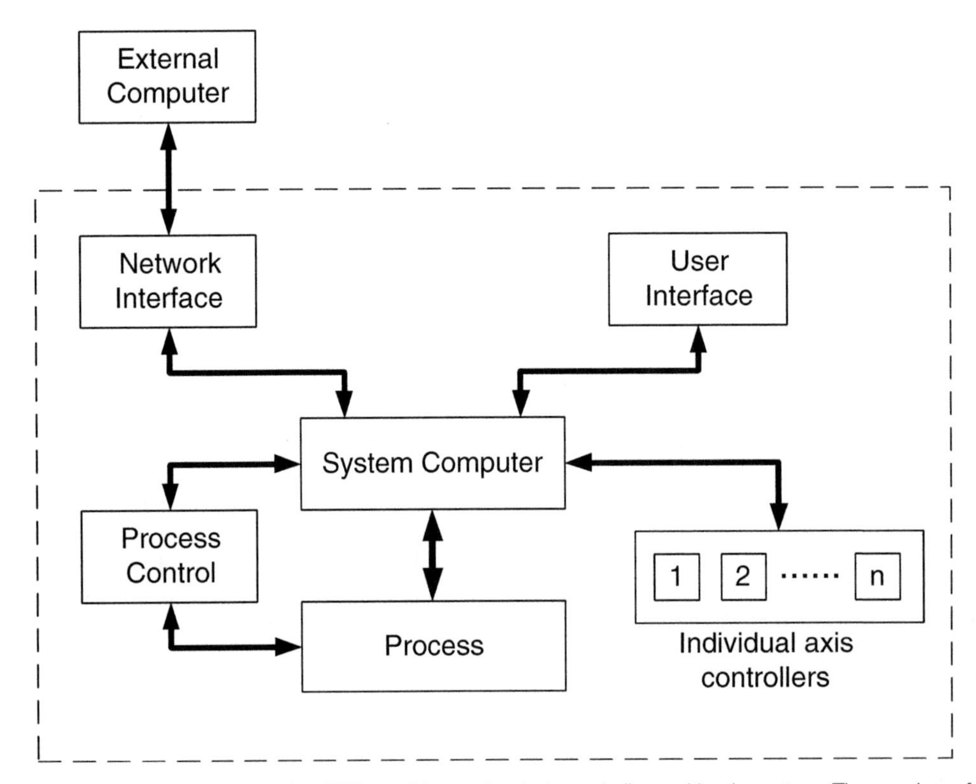

Fig. 2.1.1 The outline of the control structure for CNC machine tool, robot, or similar multi-axis system. The number of individual motion axes, and the interface to the process are determined by the system's functionality.

controller are configured to form a hierarchical control structure centred on the main system computer. The overall control of the system is vested in the system computer, which, apart from sequencing the operation of the overall system, handles the communication between the operator and the factory's local-area network. It should be noted that industrial robots, which are considered to be an important element of an automated factory, can be considered to be just another form of machine tool. In a machine tool or industrial robot or related manufacturing systems, controlled motion (position and speed) of the axes is necessary; this requires the provision of actuators, either linear or rotary, associated power controllers to produce motion, and appropriate sensors to measure the variables.

2.1.2 Machine tools

Despite advances in technology, the basic stages in manufacturing have not changed over the centuries: material has to be moved, machined, and processed. When considering current advanced manufacturing facilities it should be remembered that they are but the latest step in a continuing process that started during the Industrial Revolution in the second half of the eighteenth century. The machine-tool industry developed during the Industrial Revolution in response to the demands of the manufacturers of steam engines for industrial, marine, and railway applications. During this period, the basic principles of accurate manufacturing and quality were developed by, amongst others, James Nasmyth and Joseph Whit-worth. These engineers developed machine tools to make good the deficiencies of the rural workers and others drawn into the manufacturing towns of Victorian England, and to solve production problems which could not be solved by the existing techniques. Increased accuracy led to advantages from the interchangeability of parts in complex assemblies. This led, in turn, to mass production, which was first realised in North America with products (such as sewing machines and typewriters) whose commercial viability could not be realised except by high-volume manufacturing. The demands of the market place for cost reductions and the requirement for increased product quality has led to dramatic changes in all aspects of manufacturing industry, on an international scale, since 1970. These changes, together with the introduction of new management techniques in manufacturing, have necessitated a considerable improvement in performance and costs at all stages of the manufacturing process. The response has been a considerable investment in automated systems by manufacturing and process industries.

Machining is the manufacturing process in which the geometry of a component is modified by the removal of material. Machining is considered to be the most versatile of production processes since it can produce a wide variety of shapes and surface finishes. To fully understand the requirements in controlling a machine tool, the machining process must be considered in some detail. Machining can be classified as either *conventional machining*, where material is removed by direct physical contact between the tool and the workpiece, or *non-conventional machining*, where there is no physical contact between the tool and the workpiece.

2.1.2.1 Conventional machining processes

In a conventional machining operation, material is removed by the relative motion between the tool and the workpiece in one of five basic processes: turning, milling, drilling, shaping, or grinding. In all machining operations, a number of process parameters must be controlled, particularly those determining the rate of material removal; and the more accurately these parameters are controlled the higher is the quality of the finished product. In sizing the drives of the axes in any machine tool, the torques and speed drives that are required in the machining process must be considered in detail. Figure 2.1.2 illustrates a turning operation where the tool is moved relative to the workplace. The power required by the turning operation is of most concern during the roughing cut (that is, when the cutting depth is at its maximum), when it is essential to ensure that the drive system will produce sufficient power for the operation. The main parameters are the tangential cutting force, F_t, and the cutting speed, V_c. The cutting speed is defined as

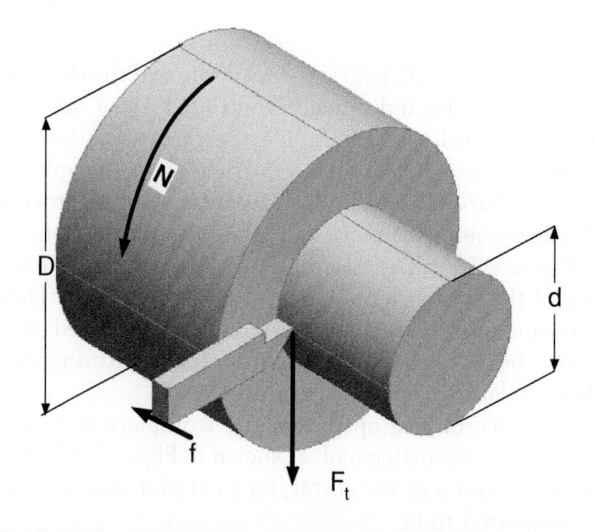

Fig. 2.1.2 The turning process, where a workpiece of initial diameter D is being reduced to d; F_t is the tangential cutting force, N is the spindle speed, and f the feed rate. In the diagram the depth of the cut is exaggerated.

Table 2.1.1 Machining data

Material	Cutting speed, V_c	Specific cutting force, K	Material removal rate, R_p
Low carbon steel	90–150	2200	25
Cast iron	60–90	1300	35
Aluminium	230–730	900	80

the relative velocity between the tool and the surface of the workpiece (m min^{-1}). The allowable range depends on the material being cut and the tool: typical values are given in Table 2.1.1. In a turning operation, the cutting speed is directly related to the spindle speed, N (rev min^{-1}), by

$$V_c = D\pi N \qquad (2.1.1)$$

The tangential force experienced by the cutter can be determined from knowledge of the process. The specific cutting force, K, is determined by the manufacturer of the cutting tool, and is a function of the materials involved, and of a number of other parameters, for example, the cutting angles. The tangential cutting force is given by

$$F_t = \frac{Kf(D-d)}{2} \qquad (2.1.2)$$

Knowledge of the tangential forces allows the power requirement of the spindle drive to be estimated as

$$\text{Power} = \frac{V_c F_t}{60} \qquad (2.1.3)$$

In modern CNC lathes, the feed rate and the depth of the cut will be individually controlled using separate motion-control systems. While the forces will be considerably smaller than those experienced by the spindle, they still have to be quantified during any design process. The locations of the radial and axial forces are also shown in Figure 2.1.2; their magnitudes are, in practice, a function of the approach and cutting angles of the tool. Their determination of these magnitudes are discussed elsewhere, but it can be found in texts or manufacturers' data sheets relating to machining processes.

In a face-milling operation, the workpiece is moved relative to the cutting tool, as shown in Figure 2.1.3. The power required by the cutter, for a cut of depth, W_c, can be estimated to be

$$\text{Power} = \frac{dfW_c}{R_p} \qquad (2.1.4)$$

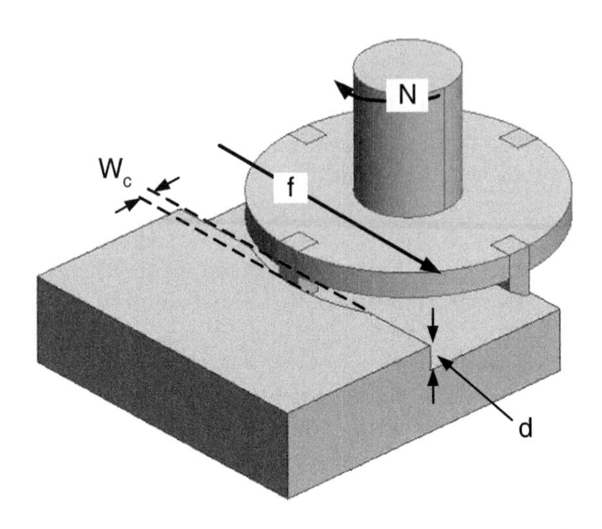

Fig. 2.1.3 The face-milling process where the workpiece is being reduced by d: f is the feed rate of the cutter across the workpiece, W_c is the depth of the cut, and N is the rotary speed of the cutting head.

where R_p is the quantity of material removed in m^3 min^{-1} kW^{-1} and the other variables are defined in Figure 2.1.3. A number of typical values for R_p are given in Table 2.1.1. The determination of the cutting forces is discussed elsewhere, because the resolution of the forces along the primary axes is a function of the angle of entry and of the path of the cutter relative to the material being milled. A value for the sum of all the tangential forces can, however, be estimated from the cutting power; if V_c is the cutting speed, as determined by Equation (2.1.1), then

$$\sum F_t = \frac{60,000 \times \text{Power}}{V_c} \qquad (2.1.5)$$

The forces and powers required in the drilling, planing, and grinding processes can be determined in a similar manner. The sizes of the drives for the controlled axes in all types of conventional machine tools must be carefully determined to ensure that the required accuracy is maintained under all load conditions. In addition, a lack of spindle or axis drive power will cause a reduction in the surface quality, or, in extreme cases, damage to the machine tool or to the workpiece.

2.1.2.2 Non-conventional machining

Non-conventional processes are widely used to produce products whose materials cannot be machined by conventional processes, for example, because of the workpiece's extreme hardness or the required operation cannot be achieved by normal machine processes (for example, if there are exceptionally small holes or complex profiles). A range of non-conventional processes are now available, including

- laser cutting and electron beam machining,
- electrochemical machining (ECM),
- electrodischarge machining (EDM),
- water jet machining.

In laser cutting (see Figure 2.1.4(a)), a focused high-energy laser beam is moved over the material to be cut. With suitable optical and laser systems, a spot size with a diameter of 250 μm and a power level of 10^7 W mm^{-2} can be achieved. As in conventional machining the feed speed has to be accurately controlled to achieve the required quality of finish: the laser will not penetrate the material if the feed is too fast, or it will remove too much material if it is too slow. Laser cutting has a low efficiency, but it has a wide range of applications, from the production of cooling holes in aerospace components to the cutting of cloth in garment manufacture. It is normal practice, because of the size and delicate nature of laser optics, for the laser to be fixed and for the workpiece to be manoeuvred using a multi-axis table. The rigidity of the structure is critical to the quality of the spot, since any vibration will cause the spot to change to an ellipse, with an increase in the cutting time and a reduction in the accuracy. It is common practice to build small-hole laser drills on artificial granite bed-plates since the high density of the structure damps vibration.

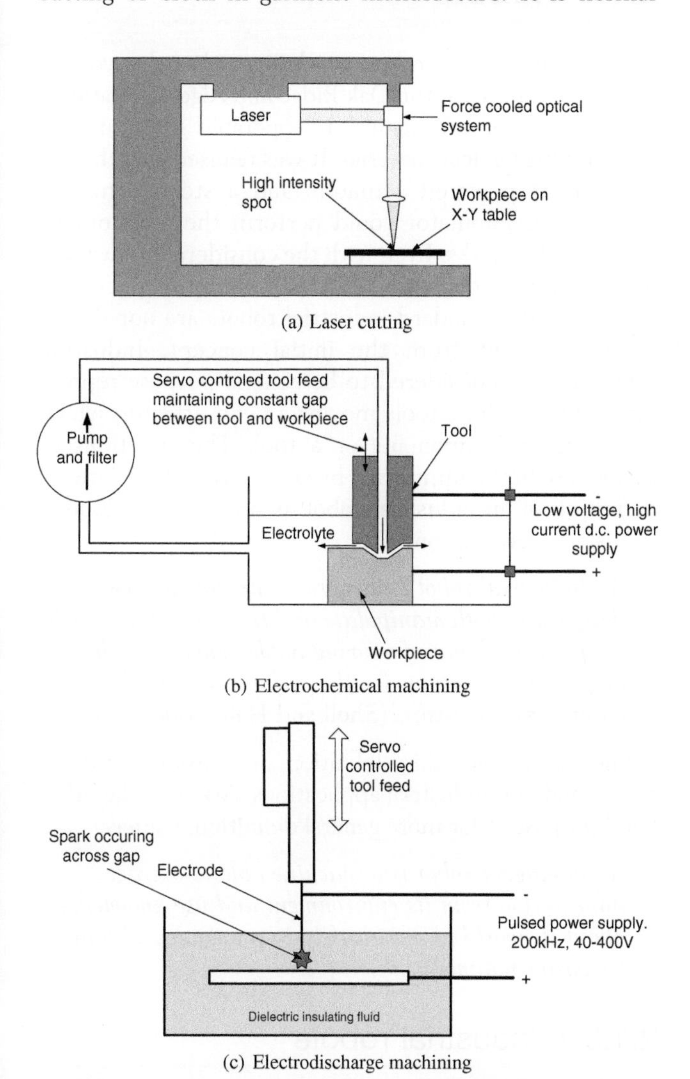

Fig. 2.1.4 The principles of the main unconventional machining processes.

In electron beam machining, a focused beam of electrons is used in a similar fashion to a laser; however, the beam is generated and accelerated by a cathode–anode arrangement. As the beam consists of electrons it can be steered by the application of a magnetic field. The beam can be focused to 10–200 μm and a density of 6500 GW mm^{-2}. At this power a 125 μm diameter hole in a steel sheet 1.25 mm thick can be cut almost instantly. As in the case of a laser, the beam source is stationary and the workpiece is moved on an *X–Y* table. The process is complicated by the fact that it is undertaken in a vacuum due to the nature of the electron beam. This requires the use of drives and tables that can operate in a vacuum, and do not contaminate the environment.

ECM can be considered to be the reverse of electroplating. Metal is removed from the workpiece, which takes up the exact shape of the tool. This technique has the advantage of producing very accurate copies of the tool, with no tool wear, and it is widely used in the manufacture of moulds for the plastics industry and aerospace components. The principal features of the process are shown in Figure 2.1.4(b). A voltage is applied between the tool and the workpiece, and material is removed from the workpiece in the presence of an electrolyte. With a high level of electrolyte flow, which is normally supplied via small holes in the tooling, the waste product is flushed from the gap and held in solution prior to being filtered out in the electrolyte-supply plant. While the voltage between the tool and the workpiece is in the range 8–20 V, the currents will be high. A metal removal rate of 1600 mm^3 min^{-1} per 1000 A is a typical value in industry. In order to achieve satisfactory machining, the gap between the tool and the workpiece has to be kept in the range 0.1–0.2 mm. While no direct machining force is required, the feed drive has to overcome the forces due to the high electrolyte pressure in the gap. Due to the high currents involved, considerable damage would occur if the feed rate was higher than the required value, and the die and the blank tool collided. To ensure this does not occur, the voltage across the gap is closely monitored, and is used to modify the predefined feed rates, and, in the event of a collision, to remove the machining power.

In EDM (see Figure 2.1.4(c)), a controlled spark is generated using a special-purpose power supply between the workpiece and the electrode. As a result of the high temperature (10,000 °C) small pieces of the workpiece and the tool are vaporised; the blast caused by the spark removes the waste so that it can be flushed away by the electrolyte. The choice of the electrode (for example, copper, carbon) and the dielectric (for example, mineral oil, paraffin, or deionised water) is determined by the material being machined and the quality of the finish required. As material from the workpiece is removed, the electrode is advanced to achieve a constant discharge voltage.

Due to the nature of the process, the electrode position tends to oscillate at the pulse frequency, and this requires a drive with a high dynamic response; a hydraulic drive is normally used, even if the rest of the machine tool has electric drives. A number of different configuration can be used, including wire machining, small-hole drilling, and die sinking. In electrodischarge wire machining, the electrode is a moving wire, which can be moved relative to the workpiece in up to five axes; this allows the production of complex shapes that could not be easily produced by any other means.

Water jet machining involves the use of a very high pressure of water directed at the material being cut. The water is typically pressurised between 1300 and 4000 bar. This is forced through a hole, typically 0.18–0.4 mm diameter, giving a nozzle velocity of over 800 m s^{-1}. With a suitable feed rate, the water will cleanly cut through a wide range of materials, including paper, wood, and fibreglass. If an abrasive powder, such as silicon carbide, is added to the water a substantial increase in performance is possible though at a cost of increased nozzle ware. With the addition of an abrasive powder, steel plate over 50 mm thick can easily be cut. The key advantages of this process include very low side forces, which allows the user to machine a part with walls as thin as 0.5 mm without damage, allowing for close nesting of parts and maximum material usage. In addition, the process does not generate heat hence it is possible to machine without hardening the material, generating poisonous fumes, recasting, or distortion. With the addition of a suitable motion platform, three-dimensional machining is possible, similar to electrodischarge wire cutting.

2.1.2.3 Machining centres

The introduction of CNC systems has had a significant effect on the design of machine tools. The increased cost of machine tools requires higher utilisation; for example, instead of a manual machine running for a single shift, a CNC machine may be required to run continually for an extended period of time. The penalty for this is that the machine's own components must be designed to withstand the extra wear and tear. It is possible for CNC machines to reduce the non-productive time in the operating cycle by the application of automation, such as the loading and unloading of parts and tool changing. Under automatic tool changing a number of tools are stored in a magazine; the tools are selected, when they are required, by a program and they are loaded into the machining head, and as this occurs the system will be updated with changes in the cutting parameters and tool offsets. Inspection probes can also be stored, allowing in-machine inspection. In a machining centre fitted with automatic part changing, parts can be presented to the machine on pallets, allowing for work to be removed from an area of the machine without stopping the machining cycle. This will give a far better usage of the machine, including unmanned operation. It has been estimated that seventy per cent of all manufacturing is carried out in batches of fifty or less. With manual operation (or even with programmable automation) batches of these sizes were uneconomical; however, with the recent introduction of advanced machining centres, the economic-batch size is equal to one.

2.1.3 Robots

The development of robots can be traced to the work in the United States at the Oak Ridge and Algonne National Laboratories of mechanical teleoperated manipulators for handling nuclear material. It was realised that, by the addition of powered actuators and a stored program system, a manipulator could perform the autonomous and repetitive tasks. Even with the considerable advances in sensing systems, control strategies, and artificial intelligence, the standard industrial robots are not significantly different from the initial concept. Industrial robots can be considered to be general-purpose reprogrammable machine tools moving an end effector, which holds either components or a tool. The functions of a robot are best summarised by considering the following definition of an industrial robot as used by the Robotic Industries Association:

> *An industrial robot is a reprogrammable device designed to both manipulate and transport parts, tools, or specialised manufacturing implements through programmed motions for the performance of specific manufacturing tasks.* (Shell and Hall, 2000, p. 499)

While acceptable, this definition does exclude mobile robots and non-industrial applications. Arkin on the other hand proposes a far more general definition, namely:

> *An intelligent robot is a machine able to extract information from its environment and use knowledge about its world to move safely in a meaningful and purposive manner.*

2.1.3.1 Industrial robots

Depending on the type of robot and the application, the mechanical structure of a conventional robot can be divided into two parts: the main manipulator and a wrist assembly. The manipulator will position the end effector while the wrist will control its orientation. The structure of the robot consists of a number of links and joints; a joint allows relative motion between two links. Two types of joints are used: a revolute joint to produce rotation, and a linear or prismatic joint. A minimum of six joints are required to achieve complete control of the end effector's position and orientation. Even though a large number of robot configurations are possible, only five configurations are commonly used in industrial robotics:

- *Polar* (Figure 2.1.5(a)). This configuration has a linear extending arm (Joint 3) which is capable of being rotated around the horizontal (Joint 2) and vertical axes (Joint 3). This configuration is widely used in the automotive industry due to its good reach capability.
- *Cylindrical* (Figure 2.1.5(b)). This comprises a linear extending arm (Joint 3) which can be moved vertically up and down (Joint 2) around a rotating column (Joint 1). This is a simple configuration to control, but it has limited reach and obstacle-avoidance capabilities.
- *Cartesian and gantry* (Figure 2.1.5(c)). This robot comprises three orthogonal linear joints (Joints 1–3). Gantry robots are far more rigid than the basic Cartesian configuration; they have considerable reach capabilities, and they require a minimum floor area for the robot itself.
- *Jointed arm* (Figure 2.1.5(d)). These robots consists of three joints (Joints 1–3) arranged in an anthropomorphic configuration. This is the most widely used configuration in general manufacturing applications.
- *Selective-compliance-assembly robotic arm (SCARA)* (Figure 2.1.5(e)). A SCARA robot consists of two rotary axes (Joints 1 and 2) and a linear joint (Joint 3). The arm is very rigid in the vertical direction, but is compliant in the horizontal direction. These attributes make it suitable for assembly tasks.

A conventional robotic arm has three joints; this allows the tool at the end of the arm to be positioned anywhere in the robot's working envelope. To orientate the tools, three additional joints are required; these are normally mounted at the end of the arm in a *wrist* assembly (Figure 2.1.5(f)). The arm and the wrist give the robot the required six degrees of freedom which permit the tool to be positioned and orientated as required by the task.

The selection of a robot is a significant problem for a design engineer, and the choice depends on the task to be performed. One of the earliest applications of robotics was within a foundry; such environments were considered to be hazardous to human operators because of the noise, heat, and fumes from the process. This is a classic application of a robot being used to replace workers because of environmental hazards. Other tasks which suggest the use of robots include repetitive work cycles, the moving of difficult or hazardous materials, and requirements for multishift operation. Robots that have been installed in manufacturing industry are normally employed in one of four application groups: materials handling, process operations, assembly, or inspection. The control of a robot in the performance of a task necessitates that all the joints can be accurately controlled. A basic robot controller is configured as a hierarchial structure, similar to that of a CNC machine tool; each

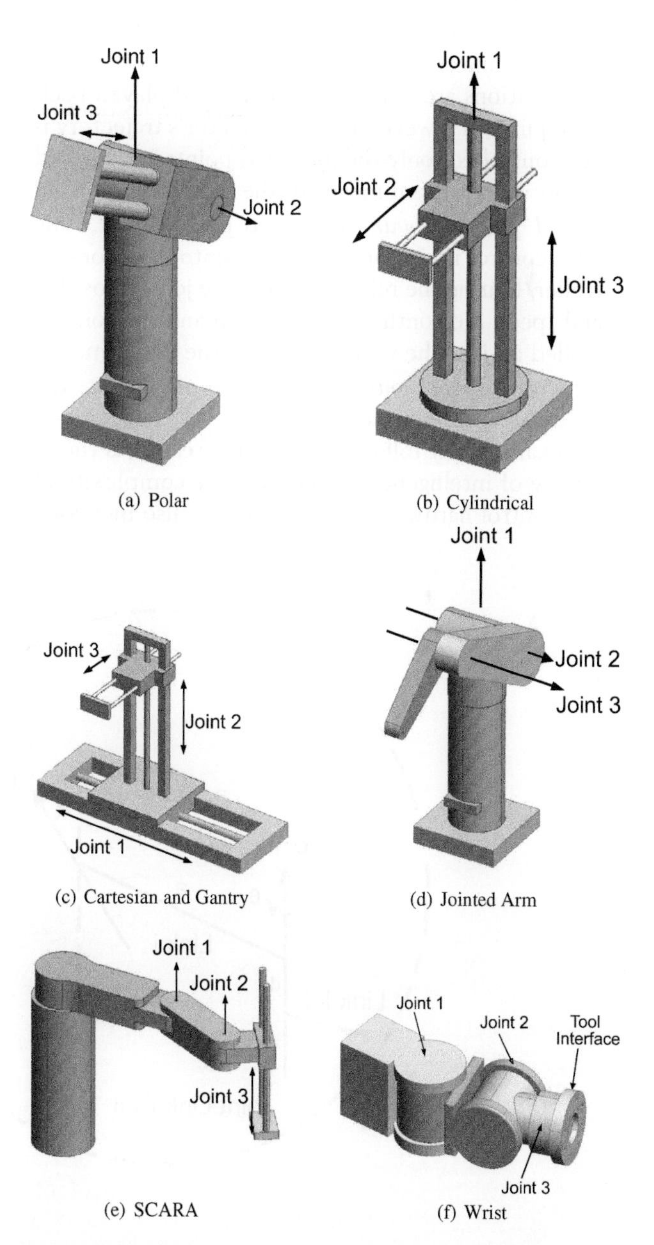

Fig. 2.1.5 The standard configurations of joints as found in industrial robots, together with the wrist.

joint actuator has a local motion controller, with a main supervisory controller which coordinates the motion of each joint to achieve the end effector trajectory that is required by the task. As robot control theory has developed so the sophistication of the controller and its algorithms has increased. Controllers can be broadly classified into one of four groups:

- *Limited-sequence control.* This is used on low-cost robots which are typically designed for pick-and-place operation. Control is usually achieved by the use of mechanical stops on the robot's joint which control the end positions of each movement. A step-by-step sequential controller is used to sequence the joints and hence to produce the correct cycle.
- *Stored program with point-to-point control.* Instead of the mechanical stops of the limited-sequence robot, the locations are stored in memory and played back as required. However, the end effector's trajectory is not controlled; only the joint end points are verified before the program moves to the next step.
- *Stored program with continuous-path control.* The path control is similar to a CNC contouring controller. During the robot's motion the joint's position and speed are continually measured and are controlled against the values stored in the program.
- *Intelligent-robot control.* By the use of sensors, the robot is capable of interacting with its environment for example, by following a welding seam. As the degree of intelligence is increased the complexity of the control hardware and its software also increase.

The function of the robot is to move the end effector from an initial position to a final position. To achieve this, the robot's control system has to plan and execute a motion trajectory; this trajectory is a sequence of individual joint positions, velocities, and accelerations that will ensure that the robot's end effector moves through the correct sequence of positions. It should be recognised that even though robotic manipulators are being considered, there is no difference between their control and the control of the positioning axes of a CNC machine tool.

The trajectory that the end effector, and hence each joint, has to follow can be generated from a knowledge of the robot's kinematics, which defines the relationships between the individual joints. Robotic kinematics is based on the use of homogeneous transformations. A transformation of a space H is represented by a 4×4 matrix which defines rotation and translation; given a point u, its transform V can be represented by the matrix product

$$V = Hu \tag{2.1.6}$$

Following an identical argument, the end of a robot arm can be directly related to another point on the robot or anywhere else in space. Since a robot consists of a number of links and joints, it is convenient to use the homogeneous matrix, ${}^{0}T_{i}$ (see Figure 2.1.6). This relationship specifies that the location of the ith coordinate frame with respect to the base coordinate system is the chain product of successive coordinate transformation

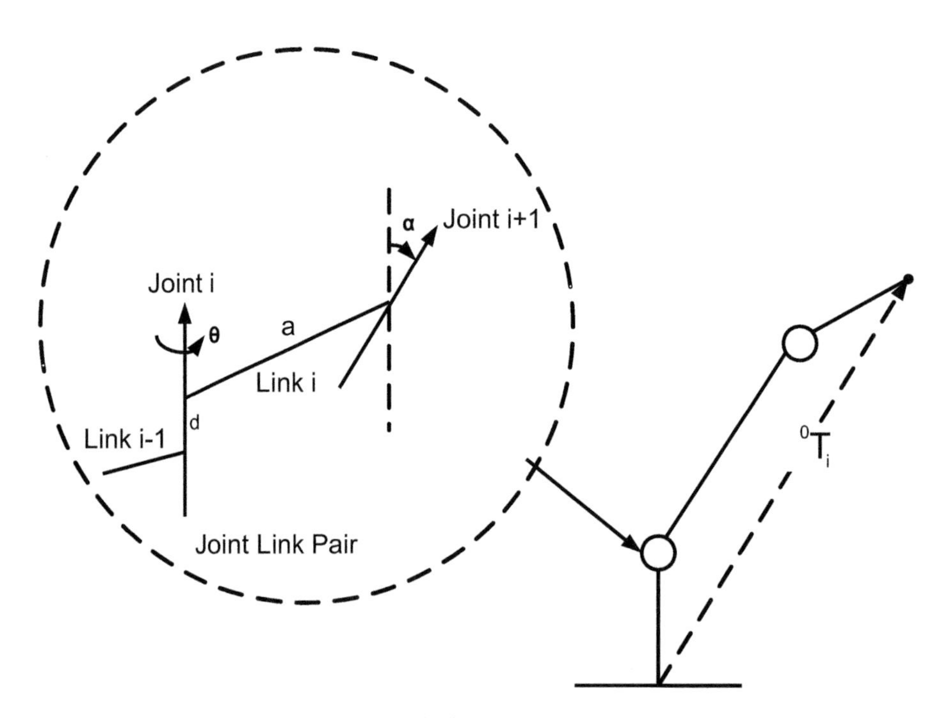

Fig. 2.1.6 Joint-transformation relationships for a robotic manipulator.

matrices for each individual joint-link pair, ${}^{i-1}A_i$, which can be expressed as

$$ {}^0T_i = {}^0A_i{}^1A_2{}^2A_3.......{}^{i-1}A_i \tag{2.1.7}$$

$$ {}^0T_i = \begin{bmatrix} x_i & y_i & z_i & p_i \\ 0 & 0 & 0 & 1 \end{bmatrix} \tag{2.1.8}$$

where $[x_i\ y_i\ z_i]$ is the orientation matrix of the ith co-ordinate with respect to the base coordinate system, and $[p_i]$ is the position vector which points from the origin of the base coordinate system of the ith coordinate frame. Each ${}^{i-1}A_i$ transformation contains information for a single joint-link pair, and it consists of the four parameters: d, the distance between two links along the i – 1th joint axis; θ, the joint angle; a, the distance between two joint axes; and α, the angle between two joint axes.

In any joint-link pair only one parameter can be a variable; θ in a rotary joint, and d for a prismatic or linear joint. The general transformation for a joint-link pair is given by

$$ {}^0A_i = \begin{bmatrix} \cos\theta_i & -\cos\alpha_i\sin\theta_i & \sin\alpha_i\cos\theta_i & a_i\cos\theta_i \\ \sin\theta_i & \cos\alpha_i\sin\theta_i & -\sin\alpha_i\cos\theta_i & a_i\cos\theta_i \\ 0 & \sin\alpha_i & \cos\alpha_i & d_i \\ 0 & 0 & 0 & 1 \end{bmatrix} \tag{2.1.9}$$

The solution of the end effector position from the joint variables is termed *forward kinematics*, while the determination of the joint variables from the robot's position in termed *inverse kinematics*. To move the joints to the required position the actuators need to be driven under closed-loop control to a required position, within actuator space. The mapping between joint, Cartesian and actuator space is shown in Figure 2.1.7. The inverse kinematic is essentially non-linear, as we are given ${}^0T_i$ and are required to find values for $\theta_1, \ldots, \theta_n$. If we consider a six-axis robot ${}^0T_6$ has 16 variables, of which four are trivial, from which we are required to solve for six joints. In principle we have 12 equations with 6 unknowns. However, within the rotational element of the matrix, Equation (2.1.7), only three variables are independent, reducing the number of equations to six. These equations are highly non-linear, transcendental equations, that are difficult to solve. As with any set of non-linear equations their are multiple solutions that need to be considered, the approaches used can either be analytic, or more recently approaches based on neural networks have been investigated.

In order to determine the change of joint position required to change the end effectors' position, use is made on inverse kinematics. Consider the case of a six-axis manipulator that is required to move an object, where the manipulator is positioned with respect to the base frame by a transform O (see Figure 2.1.8). The position and orientation of the tool interface of the six-axis manipulator is described by ${}^0T_6$, and the position of the end effector relative to the tool interface is given by E. The object to be moved is positioned at G, relative to the origin, and the location of the end effector relative to

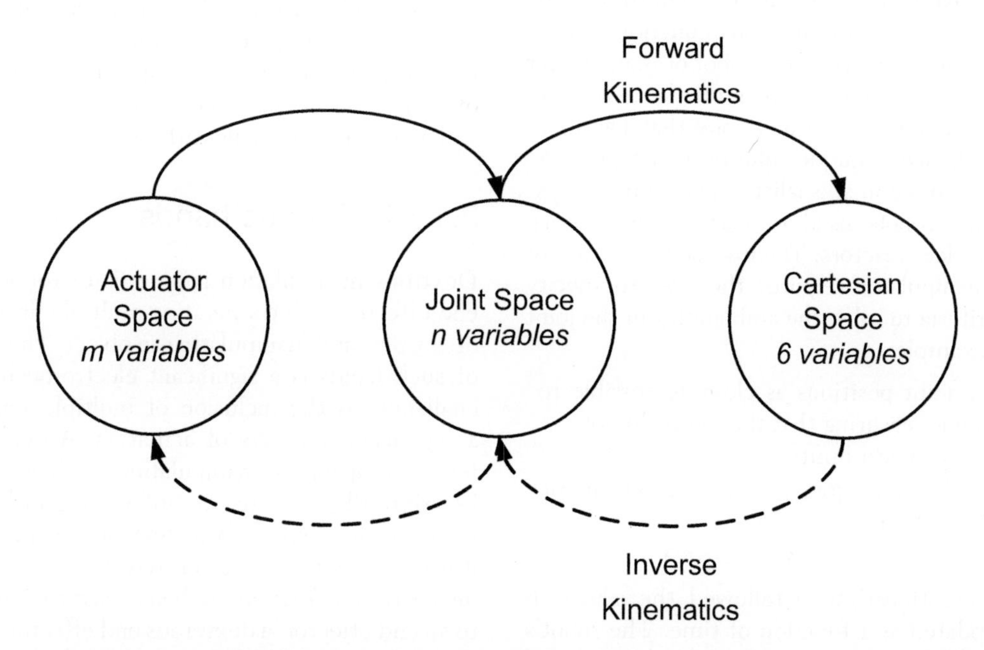

Fig. 2.1.7 The mapping between the kinematic descriptions. The number of variables in Cartesian space is 6, while the number of variables in joint and actuator space is determined by the manipulators's design.

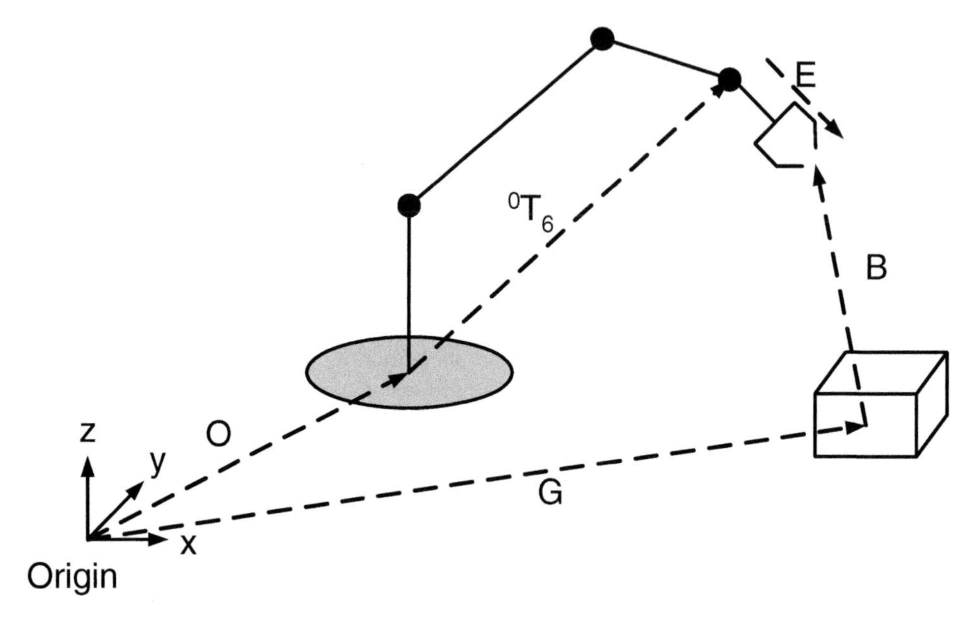

Fig. 2.1.8 The transformations that need to be considered when controlling the position and trajectory of a six-axis manipulator.

the object is B. Hence, it is possible to equate the position of the end effector by two routes, firstly via the manipulator and secondly via the object, giving

$$O^0T_6E = BG \tag{2.1.10}$$

and hence

$$^0T_6 = O^{-1}BGE^{-1} = {}^0A_i{}^1A_2{}^2A_3 \ldots\ldots{}^5A_6 \tag{2.1.11}$$

As $^0T_6$ is limited to six variables in Cartesian space, the six individual joint positions are determined by solving the resultant six simultaneous equations. However, problems will occur when the robot or manipulator is considered to be kinematically redundant. A kinematically redundant manipulator is one that has more than six joints, hence a unique solution is not possible. They are widely found in specialist applications, for example snake-like robots used to inspect the internal structures of nuclear reactors. The inverse kinematics of a redundant manipulator requires the user to specify a number of criteria to solve the ambiguities in the joint positions, for example:

- Maintain the joint positions as close as possible to a specified value, ensuring that the joints do not reach their mechanical limits.
- Minimise the joint torque, potential or kinematic energy of the system.
- Avoid obstacles in the robot's workspace.

In order for a smooth path to be followed, the value of B needs to be updated as a function of time. The robot's positional information is used to generate the required joint position by the inverse kinematic solution of the $^0T_6$ matrix. In practice, the algorithms required to obtain these solutions are complicated by the occurrence of multiple solutions and singularities, which are resolved by defining the path, and is solution prior to moving the robot. Usually, it is desirable that the motion of the robot is smooth; hence, the first derivative (that is, the speed) must be continuous. The actual generation of the path required by the application can be expressed as a cubic or higher-order polynomial, see Section 2.2.4. As the robot moves, the dynamics of the robot changes. If the position loops are individually closed, a poor end effector response results, with a slow speed and unnecessary vibration. To improve the robot's performance, and increasingly that of CNC machine tools, considerable use of is made of the real-time solution of dynamic equations and adaptive control algorithms.

2.1.3.2 Robotic hands

Dextrous manipulation is an area of robotics where an end effector with cooperating multiple fingers is capable of grasping and manipulating an object. The development of such hands is a significant electromechanical design challenge, as the inclusion of multiple fingers requires a significant numbers of actuators. A dextrous end effector is capable of manipulating an object so that it can be arbitrarily relocated to complete a task. One of the main characteristics of the dextrous manipulation is that it is object and not task centred. It should be noted that *dexterity* and *dextrous* are being used to define attributes to an end effector: a dexterous end effector may not have the ability to undertake a task that a human considers as dextrous. As dextrous manipulation is quintessentially a human activity, a majority of the dextrous robotic end effectors developed to date have significant anthropomorphic characteristics. In view of the importance of this research area a considerable body of research literature on the analysis of the grasp quality and its control is currently available.

As a dextrous end effector needs to replicate some or all the functionality of the human hand, an understanding of human hand functionality is required in the design process. It is recognised that there are five functions attributed to the hand: manipulation, sensation and touch, stabilisation as a means of support, protection, and expression and communication. In robotic systems only the first three need to be considered. The hand can function either dynamically or statically. Its function as a whole is the sum of many sub-movements; these movements may be used to explore an object, involve actions such as grasping and carrying as well as provide dexterity and maintain stability.

The hand may be used in a multitude of postures and movements, which in most cases involve both the thumb and other digits. There are two basic postures of the human hand: the power grasp and the precision grasp. The power grasp (Figure 2.1.9(a)), used where full strength is needed, is where the object is held in a clamp formed by the partly flexed fingers and often a wide area of the palm. The hand conforms to the size and shape of the object. All four fingers are more or less flexed, with each finger accommodating a position so that force can be applied and the force applied to the object to perform a task or resist motion. In a precision grasp (Figure 2.1.9(b)) there is a greater control of the finger and thumb position than in the power grasp. The precision grasp is carried out between the tip of the thumb and that of one or more of the fingers. The object is held relatively lightly and manipulated between the thumb and related finger or fingers.

The human hand consists of a palm, four fingers and a thumb. The internal structure consists of 19 major bones, 20 muscles within the hand, a number of tendons from forearm muscles, and a considerable number of ligaments. The muscles in the body of the hand are smaller and less powerful than the forearm muscles and are used more for the precise movements rather than the power grasps. A hand is covered with skin that contains sensors and provides a protective compliant covering.

The classification of movements of the hand in which work is involved can be placed in two main areas: prehensile and non-prehensile. A prehensile movement is a controlled action in which an object is held in a grasp or pinching action partly or wholly in the working envelope of the hand, while a non-prehensile movement is one, which may involve the whole hand, fingers, or a finger but in which no object is grasped or held. The movement may be a pushing one such as pushing an object, or a finger-lifting action such as playing the piano.

The dynamic specification of the hand can be summarised as:

- Typical forces in the range 285–534 N during a power grasp.
- Typical forces in the range 55–133 N during a precision grasp.
- Maximum joint velocity 600 s^{-1}.
- Maximum repetitive motion frequency, 5 Hz.

2.1.3.2.1 End effector technologies

The development of dextrous hands or end effectors has been of considerable importance to the academic robotic research community for many years, and while in no way exhaustive does however present some of the thinking that has gone into dextrous robotic systems.

- *University of Southampton.* A significant robotic end effector designs was the Whole Arm Manipulator. This manipulator was developed at for insertion into a human sized rubber glove, for use in a conventional glove box (Figure 2.1.10). Due to this design requirement, the manipulator has an anthropomorphic end effector with four adaptive fingers and a prehensile

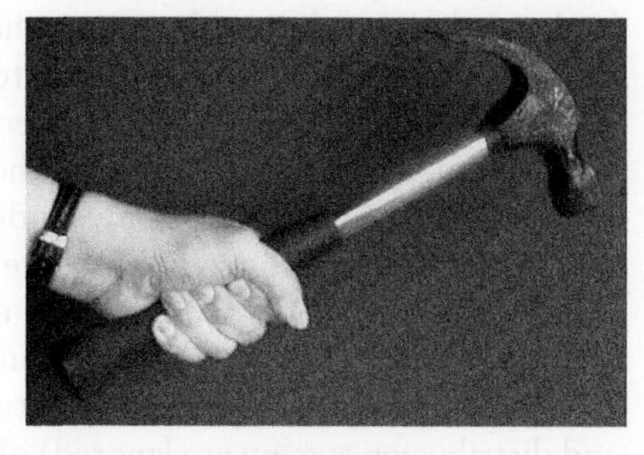

(a) Power grasp

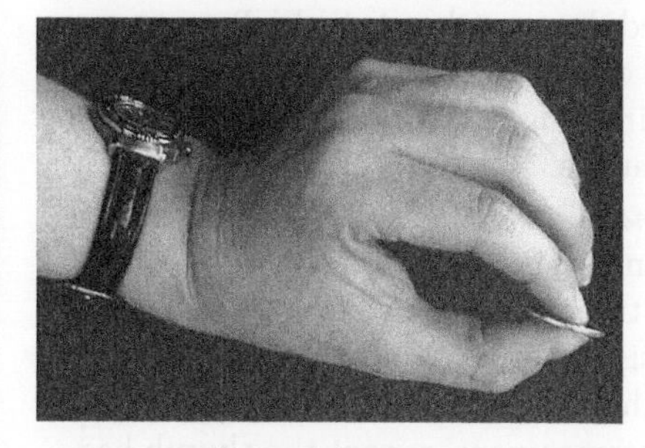

(b) Precision grasp

Fig. 2.1.9 The power and precision grasp of the human hand.

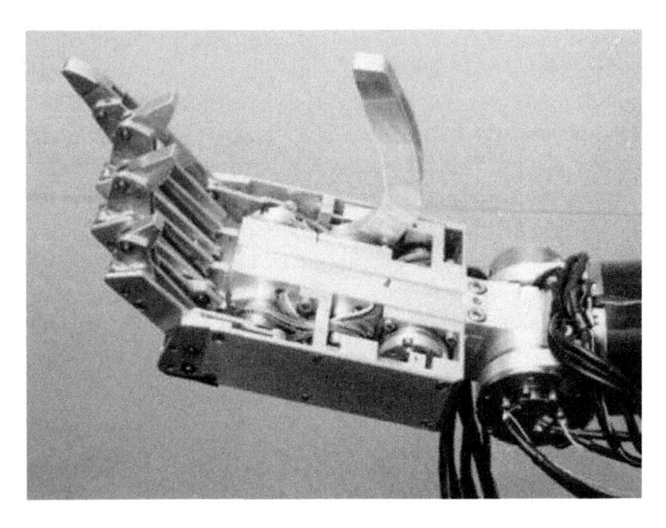

Fig. 2.1.10 The whole arm manipulator's anthropomorphic hand.

thumb. Due to size constraints the degrees of freedom within the hand were limited to three, Figure 4.1.2.

- *Stanford/JPL hand.* The Stanford/JPL hand (some times termed the Salisbury hand) was designed as a research tool in the control and design of articulated hands. In order to minimise the weight and volume of the hand the motors are located on the forearm of the serving manipulator and use Teflon-coated cables in flexible sleeves to transmit forces to the finger joints. To reduce coupling and to make the finger systems modular, the designers used four cables for each three degree of freedom finger making each finger identical and independently controllable.
- *Universities of Southampton and Oxford.* The work at the University of Southampton on prosthetic hands has continued both at Southampton and at Oxford. The mechanics of the hand are very similar to the Whole Arm Manipulator with solid linkages and multiple motors; however, the flexibility and power capabilities are closly tailored to prosthetic applications as opposed to industrial handling.
- *BarrettHand.* One of the most widely cited commercial multifingered dextrous hands is the BarrettHand, this hand combines a high degree of dexterity with robust engineering and is suitable for light engineering applications.
- *UTAH–MIT hand.* The Utah–MIT Dextrous hand is an example of an advanced dextrous system. The hand comprises three fingers and an opposed thumb. Each finger consists of a base roll joint, and three consecutive pitch joints. The thumb and fingers have the same basic arrangement, except the thumb has a lower yaw joint in place of the roll joint. The hand is tendon driven from external actuators.
- *Robonaut Hand.* The Robonaut Hand is one of the first systems being specifically developed for use in outer space; its size and capability is close to that of a suited astronaut's hand. Each Robonaut Hand has a total of fourteen degrees of freedom, consisting of a forearm which houses the motors and drive electronics, a two degree of freedom wrist, and a five finger, twelve degree of freedom hand.

The design of the fingers and their operation are the key to the satisfactory operation of a dextrous hand. It is clear that two constraints exist. The work by Salisbury indicated that the individual fingers should be multi-jointed, with a minimum of three joints and segments per finger. In addition a power grasp takes place in the lower part of the finger, while during a precision grasp it is the position and forces applied at the fingertip that is of the prime importance. It is normal practice for the precision and power grasp not occurring at the same time.

In the design of robotic dextrous end effectors, the main limitation is the actuation technology; it is recognised that an under-actuated approach may be required, where the number of actuators used is less than the actual number of degrees of freedom in the hand. Under-actuation is achieved by linking one or more finger segments or fingers together; this approach was used in Southampton's Whole Arm Manipulator. The location and method of transmission of power is crucial to the successful operation of any end effector, the main being that the end effector size should be compact and consistent with the manipulator.

2.1.3.2.2 Actuation

Both fully and under-actuated dextrous artificial hands have been developed using electric, pneumatic, or hydraulic actuators. The use of electrically powered actuators have, however, been the most widely used, due to both its convenience and its simplicity compared to the other approaches. The use of electrically powered actuator systems ensures that the joint has good stiffness and bandwidth. One drawback with this approach is the relatively low power-to-weight/volume ratio which can lead to a bulky solution; however, the developments in magnetic materials and advanced motor design have (and will continue to) reduced this problem. In many designs the actuators are mounted outside the hand with power transmission being achieved by tendons. On the other hand, pneumatic actuators exhibit relatively low actuation bandwidth and stiffness and as a consequence, continuous control is complex. Actuation solutions developed on the basis of pneumatic actuators (if the pump and distribution system are ignored) offer low weight and compact actuators that provide considerable force. Hydraulic actuators can be classified somewhere in between pneumatics and electrically powered actuators. With hydraulics stiffness is good due to the low compressibility of the fluid. While pneumatic actuators can be used with gas pressures up to 5–10 MPa, hydraulic actuators will work with up to 300 MPa. One approach that is being considered at present is the development of artificial muscles. Klute and co-workers provide a detailed overview of the biomechanics approach to aspects of muscles and joint actuation. In addition the paper presents details of a range of muscle designs, including those based on pneumatic design which are capable of providing 2000 N of force. This force equates to that provided by the human's triceps. The design consists of a inflatable bladder sheathed double helical weave so that the actuator contracts lengthwise when it expands radially. Other approaches to the design for artificial mussels have been based on technologies including shape-memory alloy (see Section 9.5), electroresistive gels, and stepper motors connected to ball screws.

When considering conventional technologies the resultant design may be bulky and therefore the actuators have to be placed somewhere behind the wrist to reduce system inertia. In these systems power is always transmitted to the fingers by using tendons or cables. Tendon transmission systems provide a low-inertia and low-friction approach for low-power systems. As the force transmitted increases considerable problems can be experienced with cable wear, friction and side loads in the pulleys. One of the main difficulties in controlling tendon systems is the that force is unidirectional – a tendon cannot work in compression. The alternative approach to joint actuation is to use a solid link which has a bi-directional force characteristic, thus it can both push and pull a finger segment. The use of a solid link reduced the number of connections to an individual finger segment. The disadvantage of this approach is a slower non-linear dynamic response, and that ball screw or crank arrangement is required close to the point of actuation. Irrespective of the detailed design of the individual fingers, they are required to be mounted on a supporting structure.

2.1.3.3 Mobile robotics

In recent years there has been a considerable increase in the types and capabilities of mobile robots, and in general three classes can be identified: UAV (unmanned aerial vehicles), UGV (unmanned ground vehicles), and UUV (unmanned underwater vehicles). In certain cases the design and control theory for a mobile robot has drawn heavily on biological systems, leading to a further class, biologically inspired robotics. An early example of this type of robot was the *Machina Specu-latrix* developed by W. Grey Walter (Holland, 2003), which captured a number of principles including simplicity, exploration, attraction, aversion, and discernment. Since this original work, a considerable number of robots have been developed including both wheeled and legged. The applications for mobile robots are wide-ranging and include:

- *Manufacturings systems.* Mobile robots are widely used to move material around factories. The mobile robot is guided through the factory by the use of underfloor wiring or visual guidelines. In most systems the robots follow a fixed path under the control of the plant's controller, hence they are able to move product on-demand.
- *Security systems.* The use of a mobile robot is considered to be a cost-effective approach to patrolling large warehouses or other buildings. Equipped with sensors they are able to detect intruders and fires.
- *Ordinance and explosive disposal.* Large number of mobile robots have been developed to assist with searching and disposal of explosives, one example being the British Army's Wheelbarrow robots that have been extensively used in Northern Ireland. The goal of these robots is to allow the inspection and destruction of a suspect device from a distance without risking the life of a bomb disposal officer.
- *Planetary exploration.* Figure 2.1.11 shows an artist's impression of one of the two Mars rovers that were landed during January 2004. *Spirit* and *Opportunity* have considerably exceeded their primary objective of exploring Mars for 90 days. At the time of writing, both rovers have been on Mars for over a year and have travelled approximately 3 km. During this time,

Fig. 2.1.11 An artist's impression of the rover *Spirit* on the surface of Mars. The robotic arm used to position scientific instruments is clearly visible. Image reproduced courtesy of NASA/JPL-Caltech.

sending back to Earth over 15 GB of data, which included over 12,000 images. Of particular interest is that to achieve this performance each rover incorporated 39 d.c. brushed ironless rotor motors. The motors were of standard designs with a number of minor variation, particularly as the motors have to endure extreme conditions, such as variations in temperature which can range from −120 °C to +25 °C.

2.1.3.4 Legged robots

While the majority of mobile robots are wheeled, there is increasing interest in legged systems, partly due to increased research activity in the field of biologically inspired robotics. One example is shown in Figure 2.1.12. Many legged designs have been realised, ranging from military logistic carriers to small replicas of insects. These robots, termed biometric robots, mimic the structure and movement of humans and animals. Of particular interest to the research community is the construction and control of dynamically stable legged robots. In the design of these systems the following constraints exist:

- The robot must be self-supporting. This puts severe limits on the force/mass and power/mass ratio of the actuators.
- The actuators of the robot must not be damaged during impact steps or falls and must maintain stability following an impact.
- The actuators need to be force controllable because the algorithms used for robot locomotion are force based.

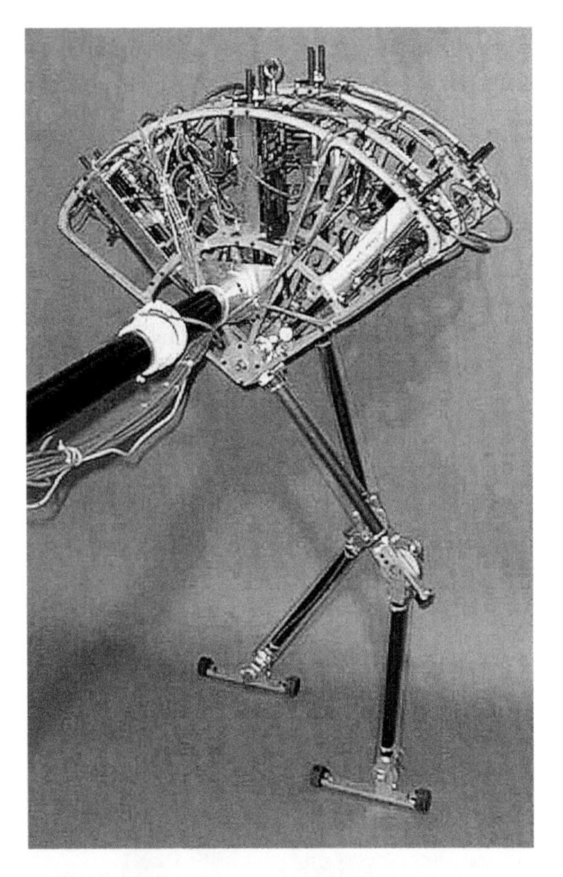

Fig. 2.1.12 Spring Flamingo, a six degree of freedom planar biped robot, was developed at the MIT Leg Laboratory. Spring Flamingo is capable of human-like walking at speeds of up to 1.25 m s^{-1}. Picture reproduce with permission from Jerry Pratt, Yobotics, Cincinnati, OH. (a) Block diagram of the actuator's control loop. (b) The key features of a series elastic actuator. The output carriage is connected to the ball screw nut solely by the four springs; the required support bearings have been omitted for clarity.

One of the most successful sets of legged robots has been based on a series elastic actuator, which has a spring in series with the transmission and the actuator output. The spring reduces the actuators' bandwidth somewhat, but for a low bandwidth application, such as walking, this is unimportant. In exchange, series elastic actuators are low motion, high force/mass, high power/mass actuators with good force control as well as impact tolerance. In addition they have low impedance and friction, and thus can achieve high-quality force control.

Figure 2.1.13(a) shows the architecture of a series elastic actuator. It should be recognised that the series elastic actuators is similar to any motion actuator with a load sensor and closed-loop control system.

The series elastic actuator uses active force sensing and closed-loop control to reduce friction and inertia. By measuring the compression of the compliant element, the force on the load can be calculated using Hooke's law. A feedback controller calculates the error between the actual force and the desired force; applying appropriate current to the motor will correct any force errors. The actuator's design introduces compliance between the actuator's output and the load, allowing for greatly increased control gains.

In practice the series elastic actuator consists of two subassemblies: a drive train subassembly and an output carriage subassembly, Figure 2.1.13(b). When assembled, the output carriage is coupled to the drive train through springs. During operation, the servomotor directly drives the ball screw, the ball nut direction of travel depending on the direction of motor rotation. The rotary motion of the motor is converted to linear motion of a ball nut which pushes on the compression springs that transmit forces to the load. The force on the load is calculated by measuring the compression of the springs using position transducers, such as a linear potentiometer or linear variable differential transformer.

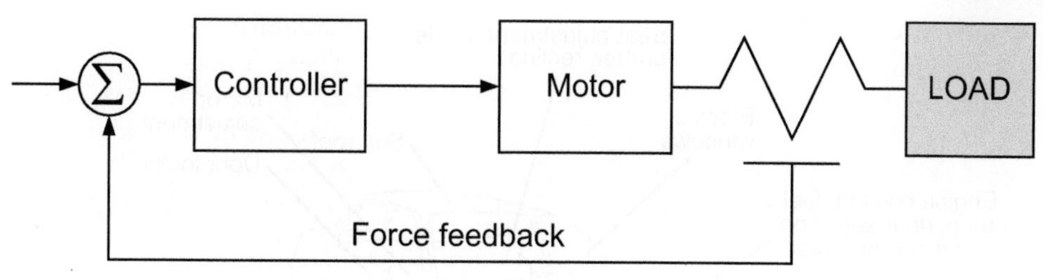

(a) Block diagram of the actuator's control loop

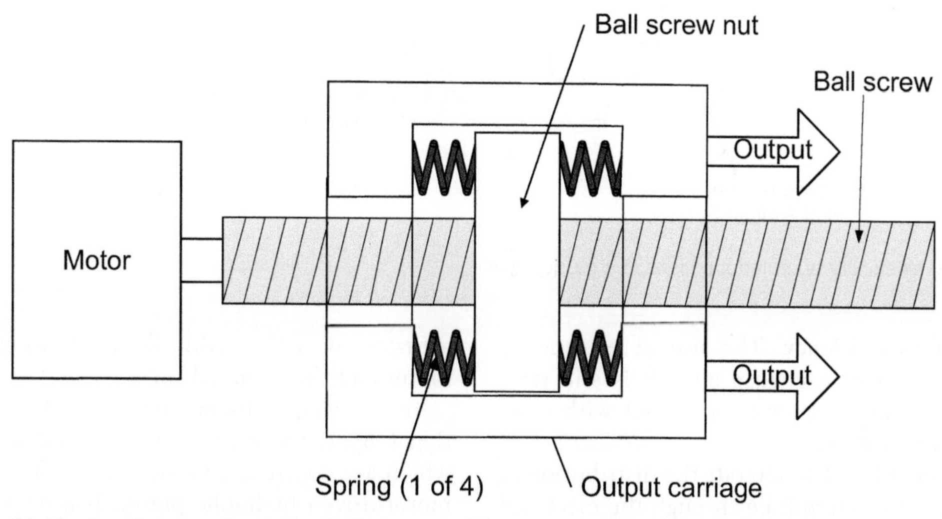

(b) The key features of a series elastic actuator. The output carriage is connected to the ball screw nut solely by the four springs; the required support bearings have been omitted for clarity.

Fig. 2.1.13 The operation of the series elastic actuator.

2.1.4 Other applications

2.1.4.1 Automotive applications

In a modern car, small electric motors undertake functions that were formerly considered the domain of mechanical linkages or to increase driver comfort or safety. The conventional brushed d.c. motors, can be found in body and convenience areas, for example windscreen wipers and electric windows. Increasingly brushless motors are also being used in open-loop pump drives and air conditioning applications. Figure 2.1.14 shows a number of electrically operated functions in a modern car. Many top of the range models currently, or will shortly incorporate systems such as intelligent brake-control, throttle-by-wire and steer-by-wire that require a sensor, a control unit and an electric motor. It has been estimated that the electrical load in a car will increase from to around 2.5 kW, with a peak value of over 12 kW. This implies that the electrical system will have to be redesigned from the current 12 V d.c. technology to distribution and utilisation at higher voltages. One of the possible options is a multivoltage system with some functions remaining at 12 V, and others operating from voltages as high as 48 V.

2.1.4.2 Aerospace applications

The flying surfaces of civil aircraft are conventionally powered through three independent and segregated hydraulic systems. In general, these systems are complex to install and costly to maintain. The concept of replacing the hydraulic system with electric actuation, coupled with changes to the electric generation technology and flight control systems, is commonly termed either the *more-electric-aircraft* or the *all-electric-aircraft* depending on the amount of electrical system incorporated, Figure 2.1.15. Electrically powered flight systems are not new; a number of aircraft developed in the 1950s and 1960s incorporated electrically actuated control functions, however they were exceptions to the general design philosophy of the time. Recently, there has been increasing interest in electrically powered actuators due to the increased reliability of power electronics, and the continual drive for the reduction of operating cost of the aircraft though weight reduction

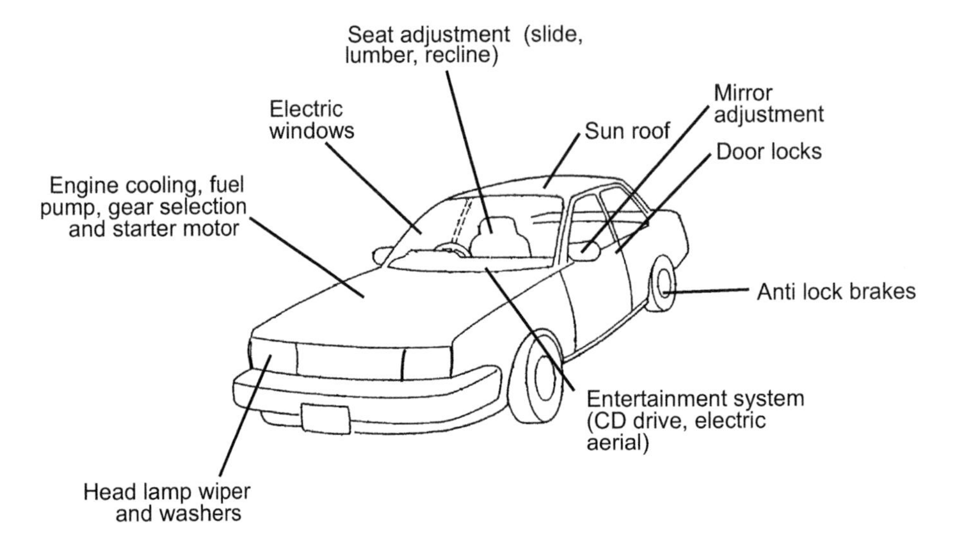

Fig. 2.1.14 Typical electrically operated functions in a modern car.

and increased fuel efficiency. The use of electrically powered flying surfaces in civilian aircraft is still rare, but a number of military aircraft are fitted with electrically powered actuators.

In the more- or all-electric-aircraft the distribution of power for flight actuation will be through the electrical system, as opposed to the currently used bulk hydraulic system. It has been estimated that the all-electric-aircraft could have a weight reduction of over 5000 kg over existing designs, which could be converted into an increase in range or a reduction in fuel costs. In order to implement power-by-wire, high-performance electrically powered actuators and related systems are required. Electrically powered flight actuators can take one of two principal configurations, the electromechanical actuator with mechanical gearing, and the electrohydrostatic actuator, or EHA, with fluidic gearing, between the motor and the actuated surface.

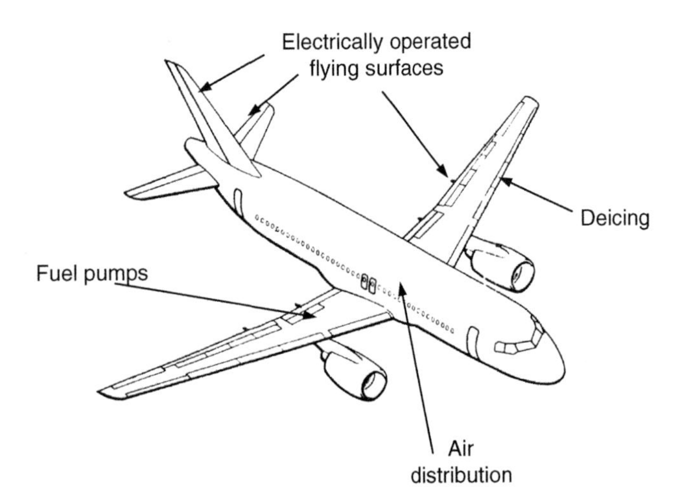

Fig. 2.1.15 The all-electric-aircraft concept, show the possible location of electrically powered actuators, or drives within a future civil aircraft.

In an EHA, hydraulic fluid is used to move a conventional hydraulic actuator, the speed and direction of which are controlled by the fluid flow from an electric motor-driven hydraulic pump. If a displacement pump (see Figure 2.1.16) is used, where the piston's diameter is d_p and the pitch diameter is d_{pp}; the flow rate $Q(t)$ as a function of the pump speed, $\omega_p(t)$ can be determined to be

$$Q(t) = D\omega_p(t) \tag{2.1.12}$$

where the pump constant, D, is given by

$$D = \frac{\pi d_p^2 d_{pp} \tan \alpha}{4} \tag{2.1.13}$$

Hence the flow rate in a variable-displacement pump unit can be controlled by adjustment of the swash plate angle, α, and hence piston displacement. In this approach two motor-drives are required, a fixed-speed drive for the pump, and a small variable-speed drive for positioning the swash plate. A different approach is just to control the rotational speed of a displacement pump, ω_p, where α is fixed this design only requires the use of a single variable-speed motor drive.

Figure 2.1.17 shows a possible concept for an electrohydrostatic actuator suitable for medium power surfaces, such as the ailerons. In most future designs the fixed pump option will be used for the rudder, which requires a far higher power output. In the actuator the basic hydraulic system consists of the pump, actuator, and accumulator. A valve ensures that the low pressure sides of the pump and actuator are maintained at the accumulator's pressure, therefore ensuring that

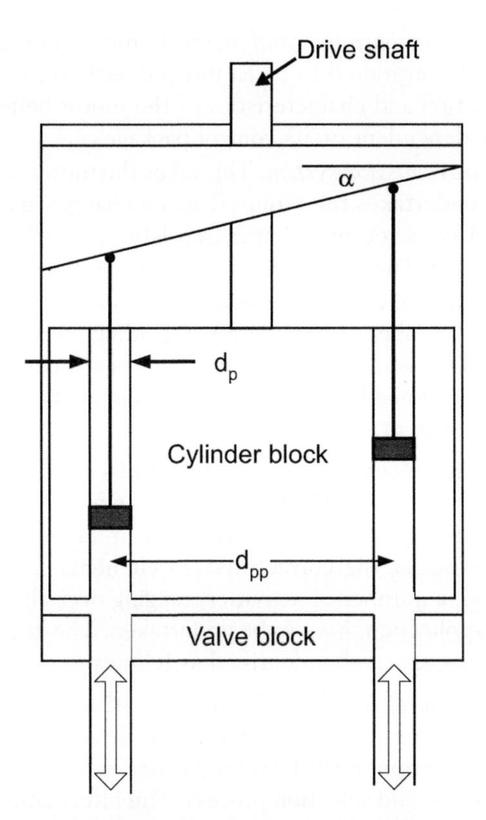

Fig. 2.1.16 The displacement hydraulic pump used in an EHA. Driving the valve cylinder causes the pistons to operate, the amount of stroke is determined either by the rotation speed, or the swash plate angle, α. The clearances between the cylinder block and, the valve block and casing have been exaggerated.

cavitation does not occur in the system. As envisaged in this application, the motor can have a number of special features, in particular a flooded 'air gap', allowing the motor to be cooled by the hydraulic fluid. The cooling oil is taken from the high-pressure side of the pump, and returned to the accumulator via an additional valve. The accumulator has a number of functions: maintaining the low pressure in the system to an acceptable value, acting as the hydraulic fluid's thermal radiator, and making up any fluid loss. It is envisaged that the unit is sealed at manufacture, and then the complete actuator considered to be a line replacement unit.

The flow of hydraulic fluid, and hence the actuator's displacement, is determined by the pump's velocity. To obtain the specified slew rate, the motor speed of approximately 10,000 rpm will be required, depending on the pump and actuators size. It should be noted that when the actuator is stationary, low-speed operation (typically 100 rpm) is normally required, because of the leakage flow across the actuator and pump. The peak pressure differential within a typical system is typically 200 bar.

The motor used in this application can be a sinusoidally wound permanent-magnet synchronous motor, the speed controller with vector control to achieve good low speed performance. An outer digital servo loop maintains the demanded actuator position, with an LVDT measuring position. The controller determines the motor, and hence pump velocity. Power conversion is undertaken using a conventional three phase IGBT bridge. In an aircraft application the power will be directly

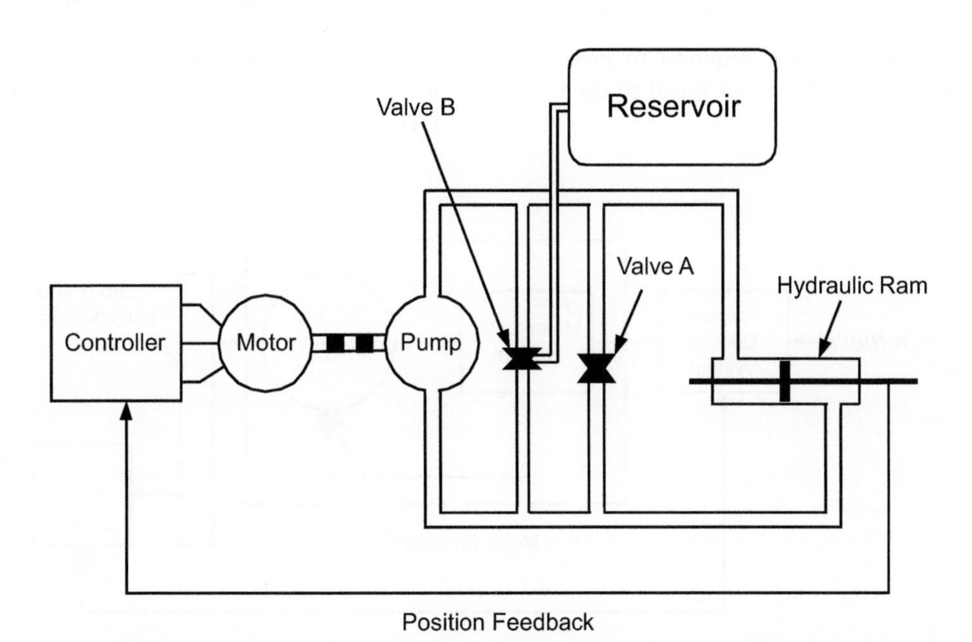

Fig. 2.1.17 Concept of an electrohydrostatic actuator for use in an aircraft. Value A is a bypass valve that can allow the ram to move under external forces in the case of a failure, while valve B ensures that the pressure of the input side of the pump does not go below that of the reservoir.

supplied from the aircraft's bus, in the all-electric-aircraft this is expected to be at 270 V d.c., as opposed to the current 110 V, 400 Hz ac systems. To prevent excessive bus voltages when the motor drive is regenerating under certain aerodynamic conditions, particularly when the aerodynamic loading back drives the actuator, a bus voltage regulator to dissipate excess power is required.

2.1.5 Motion-control systems

In this brief review of the motion requirements of machine tools, robotics and related systems, it is clear that the satisfactory control of the axes, either individually or as a coordinate group, is paramount to the implementation of a successful system. In order to achieve this control, the relationship between the mechanical aspects of the complete system and its actuators needs to be fully understood. Even with the best control system and algorithms available, it will not perform to specification if the load cannot be accelerated or decelerated to the correct speed within the required time and if that speed cannot be held to the required accuracy.

A motion-control system consists of a number of elements (see Figure 2.1.18) whose characteristics must be carefully determined in order to optimise the performance of the complete system. A motion-control system consists of five elements:

- *The controller,* which implements the main control algorithms (normally either speed or position control) and provides the interface between the motion-control system and the main control system and/or the user.
- *The encoders and transducers,* required to provide feedback of the load's position and speed to the controller.
- *The motor controller, and motor.* In most cases, these can be considered to be an integral package, as the operation and characteristics of the motor being totally dependent on its control package.
- *The transmission system.* This takes the motor output and undertakes the required speed changes and, if required, a rotary-to-linear translation.
- *The load.* The driven elements greatly influences the operation of the complete system. It should be noted that a number of parameters, including inertia, external loads, and friction, may vary as a function of time, and need to be fully determined at the start of the design process.

The key to successful implementation of a drive system is full identification of the applications needs and hence its requirements; these are summarised in Table 2.1.2. In order to select the correct system elements for an application, a number of activities, ranging over all aspects of the application, have to be undertaken. The key stages of the process can be identified as follows:

- *Collection of the data.* The key to satisfactory selection and commissioning of a motor–drive system is the collection of all the relevant data before starting the sizing and selection process. The information obtained will mostly relate to the system's operation, but may also include commercial considerations.
- *Sizing of the system.* The size of the various drive components can be determined on the basis of the data collected earlier.
- *Identification of the system to be used.* Once the size of the various elements and the application requirements are known, the identity of the various elements can be indicated. At this stage, the types of motor, feedback transducer, and controller can finalised.

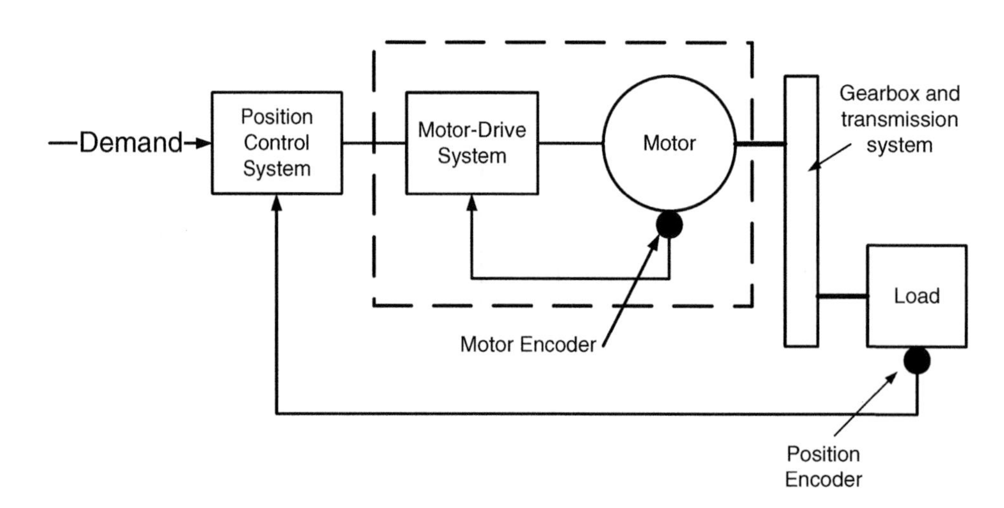

Fig. 2.1.18 Block diagram of an advanced electric motion-control system.

Table 2.1.2 Requirements to be considered in the selection of a motor–drive system

Load	Maximum speed
	Acceleration and deceleration
	Motion profile
	Dynamic response
	External forces
Environmental factors	Safety and risk
	Electromagnetic compatibility
	Climatic and humidity ranges
	Electrical supply specifications and compatibility
Life-cycle costs	Initial costs
	Operational costs
	Maintenance costs
	Disposal costs
System integration	Mechanical fittings
	Bearing and couplings
	Cooling
	Compatibility with existing systems

- *Selection of the components.* Using the acquired knowledge, the selection process can be started. If the items cannot be supplied and need to be re-specified, or the specification of a component is changed, the effect on the complete system must be considered.
- *Verification.* Prior to procuring the components, a complete check must be made to ensure that the system fits together in the space allocated by the other members of the design team.
- *Testing.* Theoretically, if all the above steps have been correctly followed, there should be no problems. But this is not always the case in the real world, commissioning modification may be required. If this is required care must be taken to ensure that the performance of the system is not degraded.

One of the main design decisions that has to be taken is the selection of the correct motor technology. With the rapid development in this field, number of options are available; each option will have benefits and disadvantages. In the consideration of the complete system the motor determines the characteristic of the drive, and it also determines the power converter and control requirements. A wide range of possibilities exist, however, only a limited number of combinations will have the broad characteristics which are necessary for machine-tool and robotic applications, namely:

- a high speed-to-torque ratio,
- four-quadrant capability,
- the ability to produce torque at standstill,
- a high power-to-weight ratio,
- high system reliability.

The following motor–drive systems satisfy these criteria, and are widely used in machine tool, robotic, and other high-performance applications:

- brushed, permanent-magnet, d.c. motors with a pulse width modulated or linear drive systems;
- brushless, d.c., permanent-magnet motors, either with trapezoidal or sinusoidal windings;
- vector, or flux-controlled induction motors;
- stepper motors.

With the exception of brushed, permanent-magnet d.c. motors, all the other machines are totally dependent on their power controller, and they will be treated as integrated drives. The list above covers most widely used motors; however, recent development have allowed the introduction of other motors, ranging from the piezo-electric motor to the switched reluctance motor.

2.1.6 Summary

This chapter has briefly reviewed a number of typical application areas where high-performance servo drives are required. It has been clearly demonstrated that the satisfactory performance of the overall system is dependent on all the components in the motor–drive system and its associated controllers; in particular, it is dependent on its ability to provide the required speed and torque performance. The determination of the characteristics that are required is a crucial step in the specification of such systems.

Chapter 2.2

2.2

Analysing a drive system

Crowder

To achieve satisfactory operation of any motion-control system, all the components within the system must be carefully selected. If an incorrect selection is made, either in the type or the size of the motor and/or drive for any axis, the performance of the whole system will be compromised. It should be realised that oversizing a system is as bad as undersizing; the system may not physically fit and will certainly cost more. In the broadest sense, the selection of a motor–drive can be considered to require the systematic collection of data regarding the axis, and its subsequent analysis.

In Chapter 2.1 an overview of a number of applications were presented, and their broad application requirements identified. This chapter considers a number of broader issues, including the dynamic of both rotary and linear systems as applied to drive, motion profiles, and aspects related to the integration of a drive system into a full application. With the increasing concerns regarding system safety in operation the risks presented to and by a drive are considered, together with possible approaches to their mitigation.

2.2.1 Rotary systems

2.2.1.1 Fundamental relationships

In general a motor drives a load through some form of transmission system in a drive system and although the motor always rotates the load or loads may either rotate or undergo a translational motion. The complete package will probably also include a speed-changing system, such as a gearbox or belt drive. It is convenient to represent such systems by an equivalent system (see Figure 2.2.1);

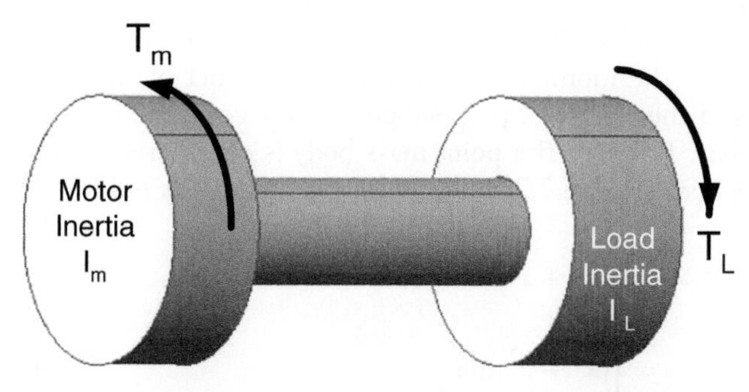

Fig. 2.2.1 The equivalent rotational elements of a motor–drive system.

the fundamental relationship that describes such a system is

$$T_m = T_L + I_{tot}\frac{d\omega_m}{dt} + B\omega_m \qquad (2.2.1)$$

where I_{tot} is the system's total moment of inertia, that is, the sum of the inertias of the transmission system and load referred to the motor shaft, and the inertia of the motor's rotor (in kg m^2); B is the damping constant (in N rad^{-1} s); ω_m is the angular velocity of the motor shaft (in rad s^{-1}); T_L is the torque required to drive the load referred to the motor shaft (in Nm), including the external load torque, and frictional loads (for example, those caused by the bearings and by system inefficiencies); T_m is the torque developed by the motor (in Nm).

When the torque required to drive the load (that is, $TL + B\omega_m$) is equal to the supplied torque, the system is in balance and the speed will be constant. The load accelerates or decelerates depending on whether the supplied torque is greater or lower than the required driving torque. Therefore, during acceleration, the motor has to

Electric Drives and Electromechanical Systems; ISBN: 9780750667401

supply not only the output torque but also the torque which is required to accelerate the inertia of the rotating system. In addition, when the angular speed of the load changes, for example from ω_1 to ω_2, there is a change in the system's kinetic energy, E_k, given by

$$\Delta E_k = \frac{I_{tot}(\omega_2^2 - \omega_1^2)}{2} \quad (2.2.2)$$

I_{tot} is the total moment of inertia that is subjected to the speed change. The direction of the energy flow will depend on whether the load is being accelerated or decelerated. If the application has a high inertia and if it is subjected to rapid changes of speed, the energy flow in the drive must be considered in some detail, since it will place restrictions on the size of the motor and its drive, particularly if excess energy has to be dissipated.

2.2.1.1.1 Moment of inertia

In the consideration of a rotary system, the body's *moment of inertia* needs to be considered, which is the rotational analog of mass for linear motion. For a point mass the moment of inertia is the product of the mass and the square of perpendicular distance to the rotation axis, $I = mr^2$. If a point mass body is considered within a body, Figure 2.2.2, the following definitions hold:

$$I_{xx} = \int (y^2 + z^2)\mathrm{d}m \quad (2.2.3)$$

$$I_{yy} = \int (x^2 + z^2)\mathrm{d}m \quad (2.2.4)$$

$$I_{yy} = \int (x^2 + y^2)\mathrm{d}m \quad (2.2.5)$$

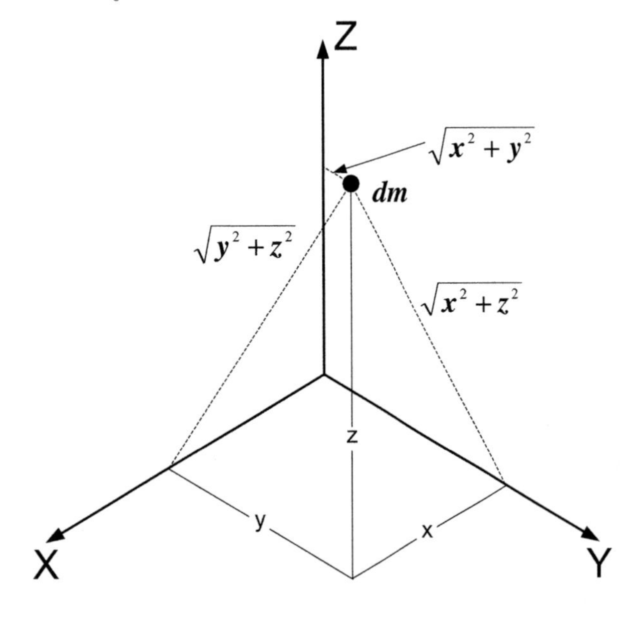

Fig. 2.2.2 Calculation of the moment of inertia for a solid body, the elemental mass, together the values of *r* for all three axes.

Table 2.2.1 Moment of inertia for a number of bodies with uniform density

Body	I_{xx}	I_{yy}	I_{zz}
Slender bar	–	$\frac{ml^2}{12}$	$\frac{ml^2}{12}$
Cuboid	$\frac{m}{12}(a^2 + c^2)$	$\frac{m}{12}(a^2 + c^2)$	$\frac{m}{12}(b^2 + c^2)$
Disc	$\frac{mR^2}{4}$	$\frac{mR^2}{4}$	$\frac{mR^2}{2}$
Cylinder	$\frac{m}{12}(3R^2 + h^2)$	$\frac{m}{12}(3R^2 + h^2)$	$\frac{mR^2}{2}$
Sphere	$\frac{2}{5}mR^2$	$\frac{2}{5}mR^2$	$\frac{2}{5}mR^2$

For a number of basic components, the moments of inertia is given in Table 2.2.1. From this table it is possible to calculate the moment of inertia around one axis, and then compute the moment of inertia, I, around a second parallel axis, using the *parallel axis theorem*, where

$$I = I_G + Md^2 \quad (2.2.6)$$

where I_G is the moment of inertia of the body, M is the mass, and d is the distance between the new axis of rotation and the original axis of rotation.

2.2.1.2 Torque considerations

The torque which must be overcome in order to permit the load to be accelerated can be considered to have the following components:

- *Friction torque*, T_f, results from relative motion between surfaces, and it is found in bearings, lead screws, gearboxes, slideways, etc. A linear friction model that can be applied to a rotary system is given in Section 2.2.3.
- *Windage torque*, T_w, is caused by the rotating components setting up air (or other fluid) movement, and is proportional to the square of the speed.
- *Load torque*, T_L, required by the application, the identification of which has been discussed in part within Chapter 2.1. The load torque is also required to drive the power train, which will be discussed in Chapter 4.1.

2.2.1.3 Gear ratios

In a perfect speed-changing system (see Figure 2.2.3), the input power will be equal to the output power, and the following relationships will apply:

$$T_o = \pm n T_i \tag{2.2.7a}$$

$$\omega_o = \pm \frac{\omega_i}{n} \tag{2.2.7b}$$

$$I_{eff} = \frac{I_L}{n^2} \tag{2.2.7c}$$

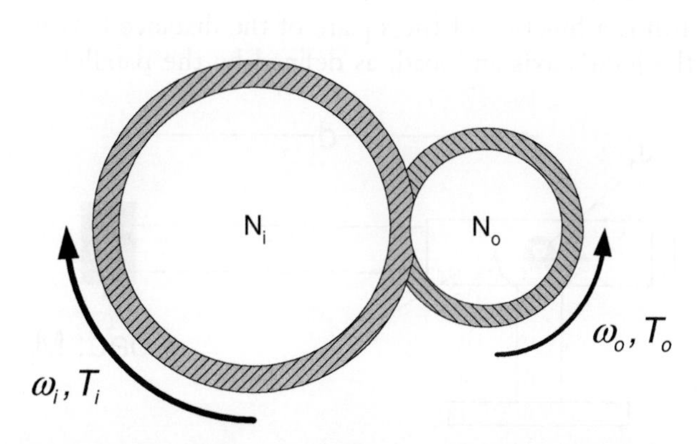

Fig. 2.2.3 The relationship between the input and output of a single-stage gear. The gear ratio is calculated from the ratio of teeth on each gearwheel, N_o:N_i or n:1.

The '±' is determined by the design of the gear train and the number of reduction stages this is discussed more fully in Section 4.1.

If a drive system incorporating a gearbox is considered, Figure 2.2.4, the dynamics of the system can be written in terms of the load variables, giving

$$T_m - \frac{T_L}{n} = T_{diff} = \alpha_L(I_L + I_m n^2) + \omega_L(B_L + B_m n^2) \tag{2.2.8}$$

where I_L is the load inertia, I_m is the motor's inertia, B_L is the load's damping term, B_m is the motor's damping term, α_L is the load's acceleration, and ω_L is the load's speed. Whether the load accelerates or decelerates depends on the difference between the torque generated by the motor and the load torque reflected through the gear train, T_{diff}. In Equation (2.2.8), the first bracketed term is the effective inertia and the second the effective damping. It should be noted that in determining the effective value, all the rotating components need to be considered. Therefore, the inertia of the shafts, couplings, and of the output stage of the gearbox need to be added to the actual load inertia to determine the effective inertia. It should also be noted that if $n \gg 1$ then the motor's inertia will be a significant part of the effective inertia.

As noted in Chapter 2.1, the drives of robots and machine tools must continually change speed to generate the required motion profile. The selection of the gear ratio and its relationship to the torque generated from the motor must be fully considered. If the load is required to operate at constant speed, or torque, the optimum gear ratio, n^*, can be determined. In practice, a number of cases need to be considered, including acceleration with and without an externally applied load torque and the effects of variable load inertias.

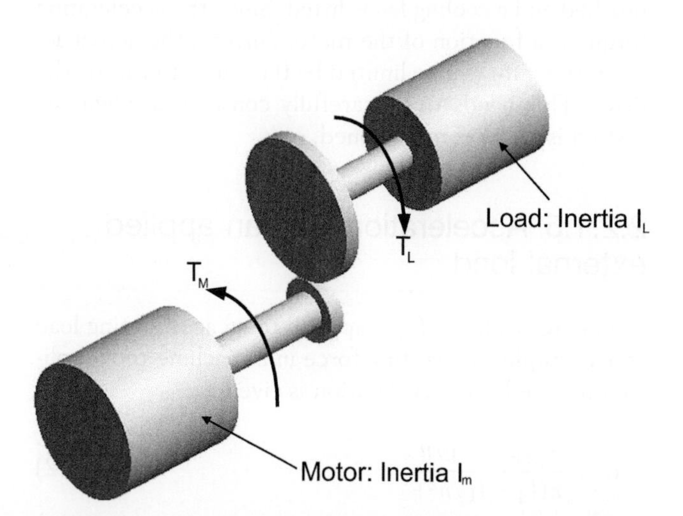

Fig. 2.2.4 A motor connected through gearing to an inertial load.

2.2.1.4 Acceleration without an external load

If a motor is capable of supplying a peak torque of T_{peak} with a suitable drive, the acceleration, α, of a load with an inertia I_L, through a gear train of ratio n:1 is given by

$$\alpha = \frac{T_{peak}}{n(I_d + I_L/n^2)} \quad (2.2.9)$$

The term in parentheses is the total inertia referred to the motor; I_L includes the inertia of the load, and the sum of inertias of the gears, shafts, and couplings rotating at the system's output speed; I_d includes the inertias of the motor's rotor, connecting shafts, gears, and couplings rotating at the motor's output speed. In the case of a belt drive, the inertias of the belt, pulleys, and idlers must be included and referred to the correct side of the speed changer. The optimum gear ratio, n^*, can be determined from Equation (2.2.9), by equating $dT_{peak}/dn = 0$, to give

$$n^* = \sqrt{\frac{I_L}{I_d}} \quad (2.2.10)$$

Therefore, the inertia on the input side of the gearing has to be equal to the reflected inertia of the output side to give a maximum acceleration of the load of

$$\alpha_{peak} = \frac{T_{peak}}{2I_d n^*} \quad (2.2.11)$$

The value of α_{peak} is the load acceleration; the acceleration of the motor will be n^* times greater. The acceleration parameters of a motor should be considered during its selection. In practice, this will be limited by the motor's construction, particularly if the motor is brushed and a cooling fan is fitted. Since the acceleration torque is a function of the motor current, the actual acceleration rate will be limited by the current limit on the drive. This needs to be carefully considered when the system is being commissioned.

2.2.1.5 Acceleration with an applied external load

If an external load, T_L, is applied to an accelerating load (for example, the cutting force in a machine tool application), the load's acceleration is given by

$$\alpha = \frac{T_{peak} - T_L/n}{n(I_d + I_L/n^2)} \quad (2.2.12)$$

This value is lower than that given by Equation (2.2.9) for an identical system.

The optimum gear ratio for an application, where the load is accelerating with a constant applied load, can be determined from this equation, in an identical manner to that described above, giving:

$$n^* = \sqrt{\frac{I_L\alpha - T_L}{I_d\alpha}} \quad (2.2.13)$$

The peak acceleration for such a system will be,

$$\alpha_{peak} = \frac{T_{peak} - T_L/n^*}{2I_m n^*} \quad (2.2.14)$$

The use of this value of the optimal gear ratio given by Equation (2.2.10), results in a lower acceleration capability; this must be compensated by an increase in the size of the motor–drive torque rating. In sizing a continuous torque or speed application, the optimal value of the gear ratio will normally be selected by comparing the drives' continuous rating with that of the load. As noted above, the calculation of the optimal gear ratio for acceleration is dependent on the drive's peak-torque capability. In most cases, the required ratios obtained will be different, and hence in practice either the acceleration or the constant-speed gear ratio will not be at their optimum value. In most industrial applications, a compromise will have to be made.

2.2.1.6 Accelerating loads with variable inertias

As has been shown, the optimal gear ratio is a function of the load inertia: if the gear ratio is the optimum value, the power transfer between the motor and load is optimised. However, in a large number of applications, the load inertia is not constant, either due to the addition of extra mass to the load, or a change in load dimension.

Consider polar robot shown in Figure 2.2.5; the inertia that joint, J_1, has to overcome to accelerate the robot's arm is a function of the square of the distance between the joint's axis and load, as defined by the parallel axis

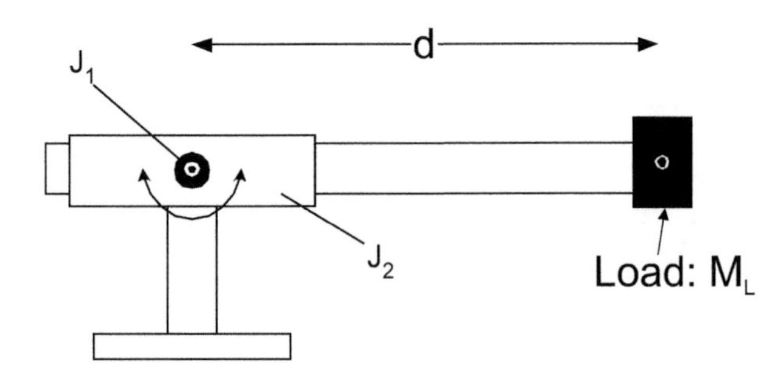

Fig. 2.2.5 The effective load inertia of a rotary joint, J_1, changes as the linear joint, J_2 of polar robot extends or retracts (the *YY* axis of the joint and the load point out of the page.

theorem. The parallel axis theorem states that the inertia of the load in Figure 2.2.5 is given by

$$I_{load} = d^2 M_L + I_{YY} \quad (2.2.15)$$

where d is the distance from the joint axis to the parallel axis of the load – in this case *YY*, I_{YY} is the inertia of the load about this axis, and M_L is the mass of the load.

If a constant peak value in the acceleration is required for all conditions, the gear ratio will have to be optimised for the maximum value of the load inertia. At lower values of the inertia, the optimum conditions will not be met, although the load can still be accelerated at the required value.

Example 2.2.1

Consider the system shown Figure 2.2.5, where the rotary axis is required to be accelerated at $\alpha_{max} = 10$ rad s^{-1}, irrespective of the load inertia. A motor with inertia $I_m = 2 \times 10^{-3}$ kg m^2 is connected to the load through a conventional gearbox. As the arm extends the effective load inertia increases from $I_{min} = 0.9$ kg m^2 to $I_{max} = 1.2$ kg m^2.

The optimum gear ratio, n^*, can be calculated, using Equation (2.2.10). The gear ratio has limiting values of 6.7 and 31.7, given the range of the inertia. To maintain performance at the maximum inertia the larger gear ratio is selected, hence the required motor torque is

$$T = 31.7\alpha_{max}\left(I_m + \frac{I_{max}}{31.7^2}\right) = 1.3 \text{ Nm}$$

If the lower gear ratio is selected, the motor torque required to maintain the same acceleration is 3 Nm, hence the system is grossly overpowered.

2.2.2 Linear systems

From the viewpoint of a drive system, a linear system is normally simpler to analyse than a rotary system. In such systems a constant acceleration occurs when a constant force acts on a body of constant mass:

$$\ddot{x} = \frac{F}{m} \quad (2.2.16)$$

where $\ddot{x}$ is the linear acceleration, F the applied force, and m the mass of the object being accelerated. As with the rotary system, a similar relationship to Equation (2.2.1) exists:

$$F_m = F_L + m_{tot}\ddot{x} + B_L\dot{x} \quad (2.2.17)$$

where m_{tot} is the system's total mass; B_L is the damping constant (in N m^{-1} s); $\dot{x}$ is the linear velocity (in m s^{-1}); F_L is the force required to drive the load (in N), including the external load forces and frictional loads (for example, those caused by any bearings or other system inefficiencies); F_m is the force (in N) developed by a linear motor or a rotary-to-linear actuator.

The kinetic energy change for a linear system can be calculated from

$$\Delta E_k = \frac{m_{tot}(\dot{x}_2^2 - \dot{x}_1^2)}{2} \quad (2.2.18)$$

for a speed change from $\dot{x}_2$ to $\dot{x}_1$.

2.2.3 Friction

In the determination of the force required within a drive system it is important to accurately determine the frictional forces this is of particular importance when a retrofit is being undertaken, when parameters may be difficult to obtain, and the system has undergone significant amounts of wear and tear. Friction occurs when two load-bearing surfaces are in relative motion. The fundamental source of friction is easily appreciated when it is noted that even the smoothest surface consists of microscopic peaks and troughs. Therefore, in practice, only a few points of contact bear the actual load, leading to virtual welding, and hence a force is required to shear these contact points. The force required to overcome the surface friction, F_f, for a normally applied load, N, is given by the standard friction model

$$F_f = \mu N \quad (2.2.19)$$

where μ is the coefficient of friction; typical values are given in Table 2.2.2. In order to minimise frictional forces, lubrication or bearings are used, as discussed in Section 4.1.

Table 2.2.2 Typical values for the coefficient of friction, μ, between two materials

Materials	Coefficient of friction
Aluminum and aluminum	1.05–1.35
Aluminum and mild steel	0.61
Mild steel and brass	0.51
Mild steel and mild steel	0 74
Tool steel on brass	0.24
Tool steel on PTFE	0.05–0.3
Tool steel on stainless steel	0.53
Tool steel on polyethylene	0.65
Tungsten carbide and mild steel	0.4–0.6

This basic model is satisfactory for slow-moving, or very large loads. However, in the case of high speed servo application the variation of the Coulomb friction with speed as shown in Figure 2.2.6(a), may need to be considered. The Coulomb friction at a standstill is higher than its value just above a standstill; this is termed the stiction (or static friction). The static frictional forces is the result of the interlocking of the irregularities of two surfaces that will increase to prevent any relative motion. The stiction has to be overcome before the load will move. An additional component to the overall friction is the viscous friction which increases with the speed; if this is combined with Coulomb friction and stiction, the resultant characteristic (known as the *general kinetic friction model)* is shown in Figure 2.2.6(b). This curve can be defined as

$$F_f = \begin{cases} F_f(\dot{x}), & \dot{x} \neq 0 \\ F_e, & \dot{x} = 0,\ \ddot{x} = 0,\ |F_e| < F_s \\ F_s \mathrm{sgn}(F_e), & \dot{x} = 0,\ \ddot{x} = 0,\ |F_e| > F_s \end{cases} \quad (2.2.20)$$

where the classical friction model is given by

$$F_f(\dot{x}) = F_c \mathrm{sgn}(\dot{x}) + B\dot{x} \quad (2.2.21)$$

where F_c is the Coulomb friction level and B the viscous friction coefficient. The 'sgn' function is defined as

$$\mathrm{sgn}(\dot{x}) = \begin{cases} +1, & \dot{x} > 0 \\ 0, & \dot{x} = 0 \\ -1, & \dot{x} < 0 \end{cases}$$

In the analysis F_e is the externally applied force, and F_s is the breakaway force, which is defined as the limit between static friction (or stiction) and the kinetic friction.

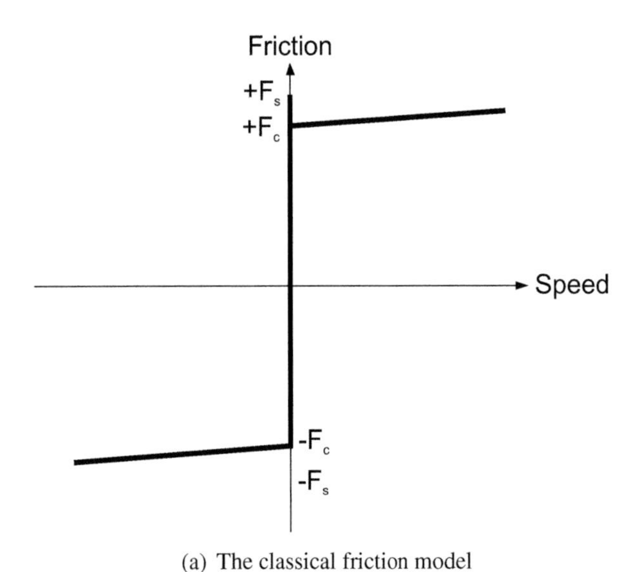

(a) The classical friction model

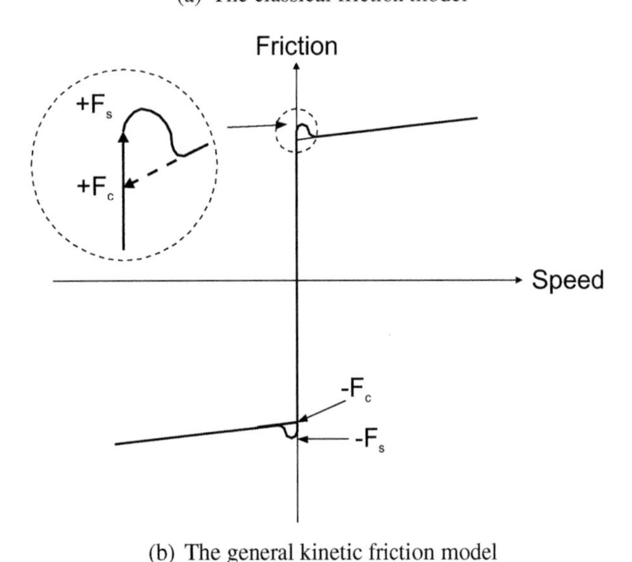

(b) The general kinetic friction model

Fig. 2.2.6 The friction between two surfaces as a function of speed, using the classical or general kinematic model. F_S is the *breakaway* or *stiction* frictional force. F_C is the coulomb frictional force.

2.2.4 Motion profiles

In this section, the methods of computing the trajectory or motion profile that describes the design motion of the system under consideration, are considered. The motion profile refers to the time history of position velocity and acceleration for each degree of freedom. One significant problem is how to specify the problem – the user does not want to write down a detailed function, but rather specify a simple description of the move required. Examples of this approach include:

- In *a point-to-point* system the objective is to move the tool from one predetermined location to another, such as in a numerically controlled drill, where actual tool path is not important. Once the tool is at the correct location, the machining operation is initiated. The control variables in a point-to-point system are the X- and Y-coordinates of the paths' starting and finishing points (Figure 2.2.7(a)).
- A *straight-cut* control system is capable of moving the cutting tool parallel to one of the major axes at a fixed rate which is suitable for machining, the control variable being the feed speed of the single axis (Figure 2.2.7(b)).
- *Contouring* is the most flexible, where the relative motion between the work-piece and the tool is continuously controlled to generate the required geometry. The number of axes controlled at any one time can be as high as five (three linear and two rotational motions); this gives the ability to produce with relative ease, for example, plane surfaces at any

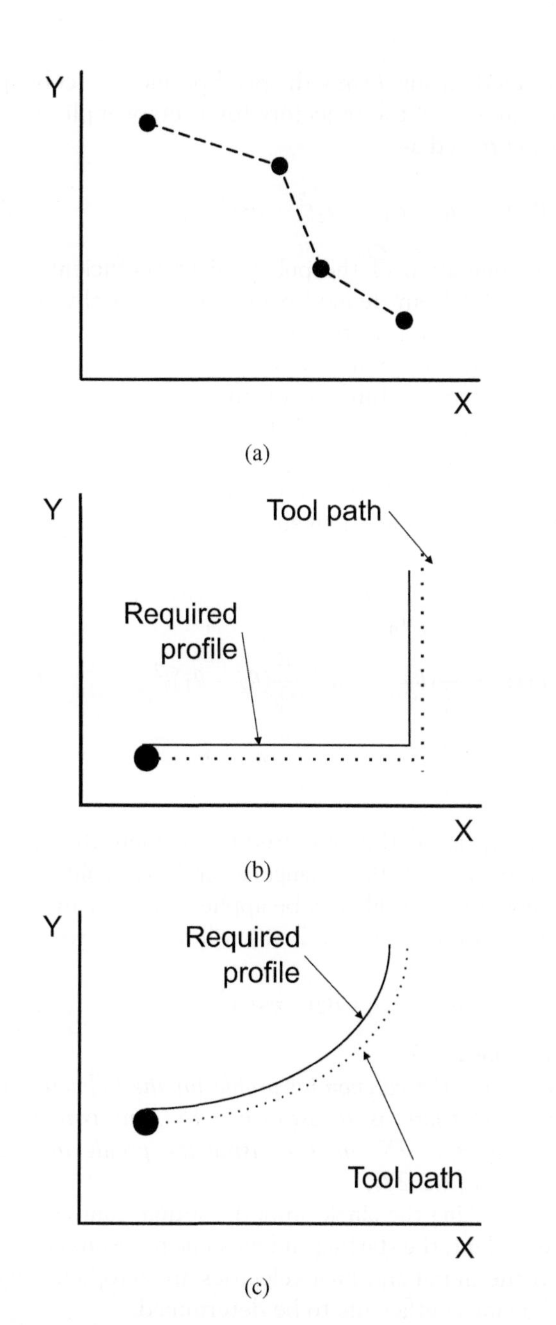

Fig. 2.2.7 Tool paths: (a) point to point, (b) straight cut, and (c) contouring. In cases (b) and (c) the tool path is offset by the radius of the cutter or, in the case of a robot application the size of the robot's end effector.

orientation, or circular and conical paths. The control variable in a contouring system is the relationship between the speed of all the axes under control. If a smooth curved path is to be generated (Figure 2.2.7(c)), there will be a constant relationship between the speeds of the X- and Y-axes. To generate a curve, the speeds, and hence the acceleration of the individual axes will vary as a function of the position; this is identical to the function of a robot's controller. The path followed by the cutter in a contouring machine tool is generated by the controller, on the basis of knowledge of the profile required and the size of the cutter. The cutter has to follow a path that will produce the required profile; this requires careful design of the profile to ensure that the cutter will not have to follow corners or radii that would be physically impossible to cut.

The paths described in Figure 2.2.7 are defined in terms of the machine's *X* and *Y* coordinates the actual profile or trajectory of the individual joint axes have to be generated at run time, the rate at which the profile is generated is termed the path-update rate, and for a typical system lies between 60 Hz and 2000 Hz.

In an indexing, or point-to-point, application, either a triangular or a trapezoidal motion profile can be used, the trapezoidal profile being the most energy-efficient route between any two points. If a specific distance must be moved within a specific time, t_m (seconds), the peak speed, and acceleration, can be determined; for a triangular profile in a rotary system, Figure 2.2.8(a), the required peak speed and acceleration can be determined as

$$N_{max} = \frac{2\theta_m}{t_m} \tag{2.2.22}$$

$$\alpha = \frac{2N_{max}}{t_m} \tag{2.2.23}$$

where θ_m is the distance moved in revolutions, N_{max}, the maximum speed required, and α is the acceleration. For a trapezoidal motion profile, and if the time spent on acceleration, deceleration and at constant velocity are equal, Figure 2.2.8(b), the peak speed and acceleration are given by

$$N_{max} = \frac{3\theta_m}{2t_m} \tag{2.2.24}$$

$$\alpha = \frac{3N_{max}}{t_m} \tag{2.2.25}$$

In order to determine the power requirements of the drive system, the acceleration/deceleration duty cycle, *d*, needs to be determined. If the dwell time between each move is t_d, then the duty cycle for a triangular profile is

$$d = \frac{t_m}{t_m + t_d} \tag{2.2.26}$$

and for the trapezoidal profile

$$d = \frac{2t_m}{3(t_m + t_d)} \tag{2.2.27}$$

The general form of the above equations can also be used for linear motions, though care must be taken to ensure consistency of the units.

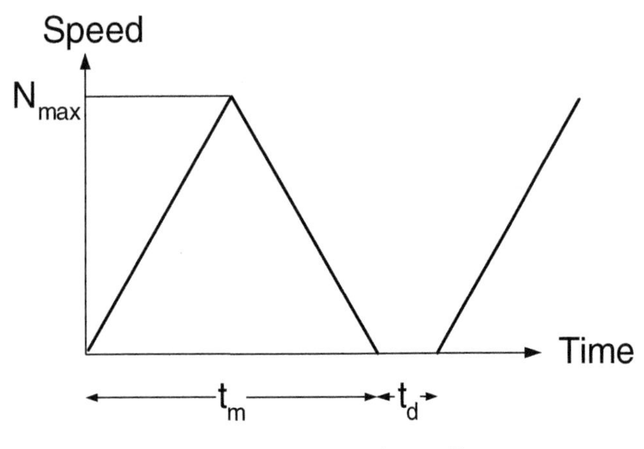

(a) Triangular motion profile

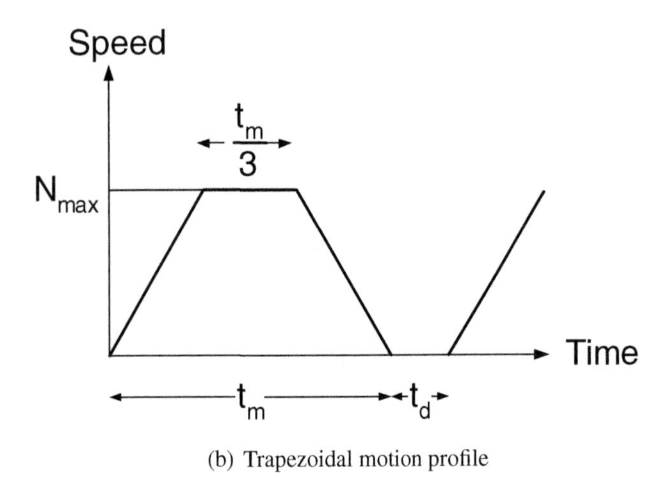

(b) Trapezoidal motion profile

Fig. 2.2.8 Motion profiles, the total distance covered and time is identical in both cases.

Example 2.2.2

Compare the triangular and trapezoidal motion profiles, when a disc is required to move 90° in 1.0 s.

1. Using a triangular profile, the peak speed of the table will be $N_{max} = 30$ rev min^{-1} $= 3.142$ rad s^{-1} and the acceleration $\alpha = 3600$ rev min^{-2} $= 6.282$ rad s^{-2}.
2. Using a trapezoidal profile, with one-third of the move at constant speed, the peak speed will be N_{max} $= 22.5$ rev min^{-1} $= 2.365$ rad s^{-1} and the acceleration $\alpha = 4050$ rev min^{-2} $= 7.1$ rad s^{-2}.

It can be noted that the peak speed is higher for a triangular motion. Even though the acceleration is higher for the trapezoidal profile, it is applied for a shorter time period, hence the energy dissipated in the motor will be lower than for triangular motion profile.

The motion profiles defined above, while satisfactory for many applications, result in rapid changes of speed. In order to overcome this the motion trajectory can be defined as a continuous polynomial, the load will be accelerating and decelerating continually to follow the path specified, giving a smooth speed profile. If a cubic polynomial is used the trajectory for a rotary application can be expressed as

$$\theta(t) = a_0 + a_1 t + a_2 t^2 + a_3 t^3 \tag{2.2.28}$$

The generation of the polynomial's coefficients can be calculated from defined parameters, typically the positions and speeds at the start and end of the move. This will allow the joint's velocity and accelerations to be determined as a function of time.

If a path is required that moves a load from θ_1 to θ_2, and the speeds at both ends of of the motion path are zero, it is possible to determine the speeds and accelerations required are defined by the following equations:

$$\theta(t) = \theta_1 + \frac{3}{t_m^2}(\theta_2 - \theta_1)t^2 + \frac{2}{t_m^3}(\theta_2 - \theta_1)t^3 \tag{2.2.29}$$

$$\dot{\theta}(t) = \frac{6}{t_m^2}(\theta_2 - \theta_1)t + \frac{6}{t_m^3}(\theta_2 - \theta_1)t^2 \tag{2.2.30}$$

$$\ddot{\theta}(t) = \frac{6}{t_m^2}(\theta_2 - \theta_1) + \frac{12}{t_m^3}(\theta_2 - \theta_1)t \tag{2.2.31}$$

where t_m is the time required to complete the move. As in the case of the triangular and trapizoidal profile, a polynomial profile can be applied to linear motions, in which case Equation (2.2.28) will be expressed as

$$x(t) = a_0 + a_1 t + a_2 t^2 + a_3 t^3 \tag{2.2.32}$$

Example 2.2.3

Determine the polynomial profile for the following application. If a joint is at rest at $\theta = 15°$, and is required to move to $\theta = 75°$ in 3 s, using the profile defined by Equation (2.2.28).

In making the single smooth motion, four constraints are evident, the starting and finishing positions are known and the initial and final velocities are zero, allowing the following coefficients to be determined:

$a_0 = 15.0$
$a_1 = 0$
$a_2 = 20.0$
$a_3 = -4.44$

when substituted in the Equation (2.2.28), we obtain:

$$\theta = 15.0 + 20.0t^2 - 4.44t^3$$

$$\dot{\theta} = 40.0t + 13.33t^2$$

$$\ddot{\theta} = 40.0 - 26.66t$$

Figure 2.2.9 shows the position, velocity and acceleration functions for the required profile. It should be noted that

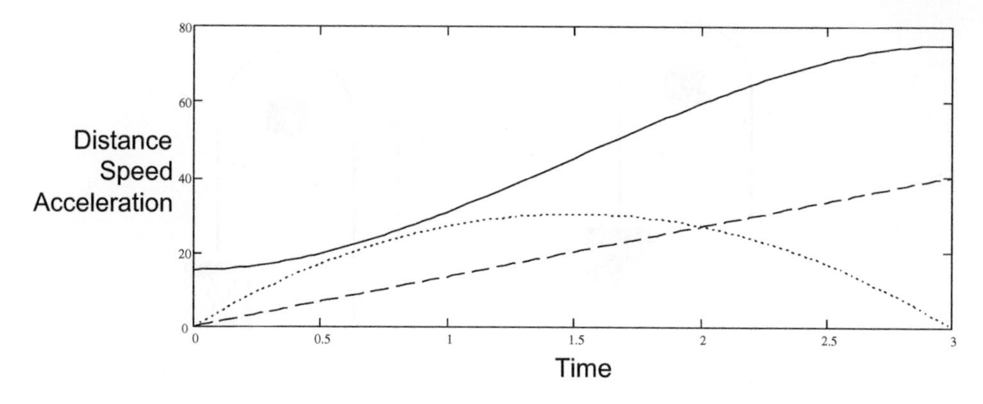

Fig. 2.2.9 Position, velocity and acceleration profiles for a single cubic segment where $\theta_1 = 15°$, $\theta_2 = 75°$, the move time is given by $t_m = 3$ s, at the start and end positions the system is at rest. The distance moved is shown as a solid line, the speed as dotted, and the acceleration as dashed.

the velocity profile of a movement where the distance moved as a function of time is a cubic polynomial, is a parabola, and the acceleration is linear.

In many case the path is required to pass through a number of intermediate or via points without changing speed. The via points can be determined as a set of positions. In order to determine the path, Equation (2.2.28) is applied for each segment, however the velocity at the start and/or end points of cubic segments are non-zero, and can be specified by a one of the following approaches:

- The user specifies the velocities at the via point.
- The system computes a velocity based on a function of the joint position.
- The system computes the velocity at the via point to maintain constant acceleration through the via point.

For certain applications higher-order polynomials can be used to specify paths, for example for a quintic polynomial, Equation (2.2.33) can be used. However this needs all three parameters at the start and finish to be specified, which gives a linear set of six equations that require solving, to determine the coefficients of the profile.

$$\theta(t) = a_0 + a_1 t + a_2 t^2 + a_3 t^3 + a_4 t^4 + a_5 t^5 \tag{2.2.33}$$

2.2.5 Assessment of a motor–drive system

The first step to the successful sizing and selection of a system is the collating of the information about the system and its application. Apart from the electrical and mechanical aspects this must also include details of the operating environment; for if these details are not considered at an early stage, the system which is selected may not be suitable for the application.

2.2.5.1 Mechanical compatibility

The mechanical requirements of the motor must be identified at an early stage in the sizing and selection procedure. Items that are frequently overlooked include any restrictions in dimensions and orientation resulting from the mechanical design. If such restrictions can be identified at an early stage unsatisfactory performance of the equipment after it is installed may be prevented. In particular, if the motor is mounted in the vertical position, special shimming or bearing preloads may be required. Apart from ensuring that all the mechanical parts fit together, the problems of assembly and maintenance should be considered at this stage; there is no more frustrating (or costly) problem than a junction box that cannot be reached, or having to dismantle half a machine to replace a motor or position encoder.

The correct identification and determination of the load and any externally applied forces often is the most critical aspect of sizing a motor–drive combination. The worst-case masses, forces, and speeds need to be accurately determined if the motor–drive is to be correctly sized. Even if the exact mass of a component is unknown, it can be determined from the component's volume and density.

In an application where the linear-motion axis is horizontal, these values can normally be determined without significant problems. However, when the axis of motion is vertical, the total weight of the load, if it is not counterbalanced, will appear as a constant load on the motor, which must be taken into account. When the load is counterbalanced, the design must be carefully analysed. Two possible counterbalancing schemes can be used, either using a second mass, or a pneumatic system. If the former strategy is used, Figure 2.2.10(a), the effective load inertia must include the counterbalance when the load's operating requirement is calculated. If a pneumatic system is used, Figure 2.2.10(b), to support the load, the percentage of the load which is supported must be

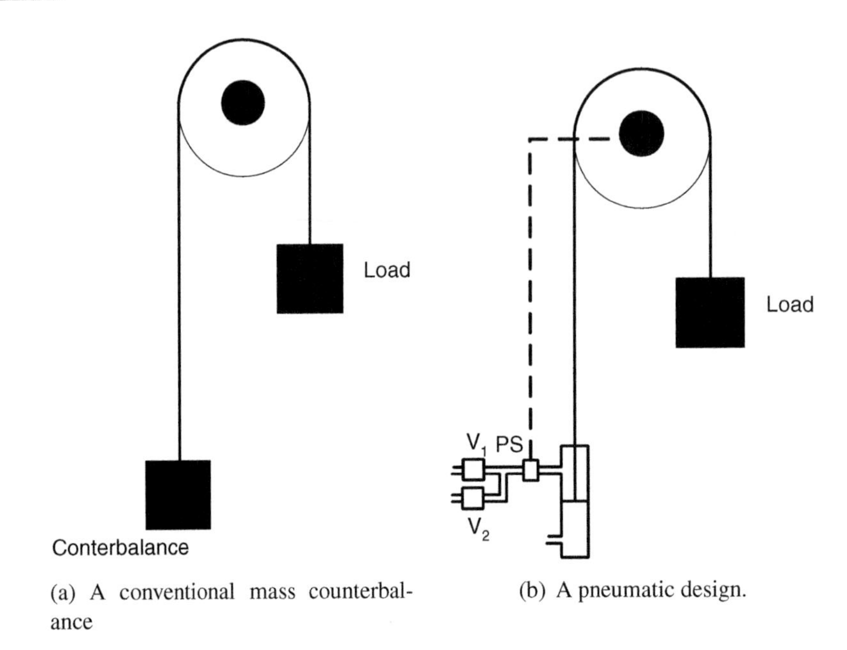

(a) A conventional mass counterbalance

(b) A pneumatic design.

Fig. 2.2.10 The use of a counterbalance to reduce the motor's continuous loading in a vertical drive application. In the pneumatic design the valves V_1 and V_2 ensure that the pressure difference in the cylinder remains constant; if the air pressure is lost a low pressure switch activates a brake.

determined, and the motor must again handle the unsupported load as a constant-thrust loading. Systems of this type should be carefully designed to ensure there is no possibility of damage because of a failure of the system. This may require the fitting of brakes, overspeed detectors, and, in the case of the hydraulic or pneumatic systems, pressure switches.

The area that most often prevents successful motor–drive sizing is poor estimation of the frictional forces. Modern antifriction devices such as air bearings, hydrostatic ways, linear-roller or ball-thrust bearings have done much to minimise this problem, as discussed in Section 4.1.4. In most cases the frictional forces can be safely estimated for horizontal motion using Equation (2.2.19). Again in a vertical application, the forces on the slideways need to be carefully resolved. If the drive is being used as a retrofit on an older design of machine tool which incorporates dovetail slides or similar bearings, the actual coefficients of friction can be many times the nominal value for the given material-to-material contact. In practice, the actual friction should be carefully determined by practical measurement or an adequate safety margin should be used in the sizing of the motor–drive in order to reflect the degree of uncertainty which is present.

2.2.5.2 Electromagnetic compatibility

Electromagnetic interference (EMI) can affect all types of electrical and electronic equipment to varying degrees; such interference has increased in importance because of recent European legislation, which recognises the importance of removing these potential problem areas at the design stage. In an assessment of equipment for compatibility, the emissions and the susceptibility of the equipment must be considered over a very broad frequency range, normally d.c. (0 Hz) to 110 MHz. In consequence, electromagnetic compatibility has a considerable influence over the design and application of a system.

The most obvious sources of electromagnetic radiation are the power converters which are used in motor controllers and any associated switch gear. With the increasing use of microprocessor-based controllers, any interference can have serious safety implications. In general, the main route of interference into or out of a piece of equipment is via the cabling. A cable which is longer than one metre should be considered to be a problem area. The use of the grounding and shielding of cables must be a high priority, as must careful design of the panel layout. While these measures cost little to implement, failure to do so will be costly. Figure 2.2.11 shows the precautions which should be taken in a typical drive system.

Radio-frequency interference (RFI) is electromagnetic emission in the range 150 kHz to 110 MHz which are of sufficient strength to be capable of interfering with any form of telecommunications; the main sources of RFI are power electronics, relays, and motor commutators. Recent legislation has placed strict limits on RFI emissions, and these limits should be met before a piece of equipment can be placed in service.

Due to problems of interference between systems, a range of standards (including the British Standards BS 800, BS 6667 and BS EN 60529) specify the amount of

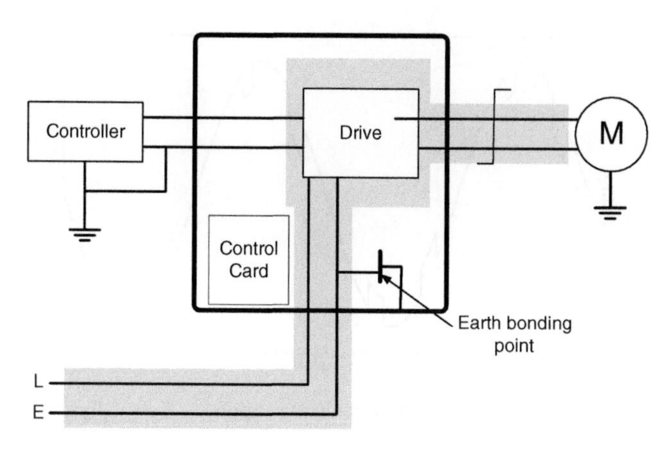

Fig. 2.2.11 A possible wiring layout for a drive system in a normal industrial environment. The motor cables are twisted to reduce interference, the cabinet is provided with a bonding point. While the controller is separately earthed, care needs to be taken to prevent earth loops. In order to prevent problems, any sensitive electronics should be placed a distance away from various elements as shown by the shaded areas.

RFI which is permitted in the supply. However, the European Community has produced a directive, EEC82/499, that covers the maximum RFI levels in the supply and the output; this has been in force throughout Europe since 1992. The suppression of RFI requires considerable thought and can only be achieved by the use of filters. The screening of cables will not affect the RFI being transmitted around a system. It should be noted that directive EEC82/499 states that screened cables are not an acceptable form of RFI suppression. The only acceptable form of suppression is by the use of in-line filters in all the supply lines, including the earth return. The design of suitable filtering should be carried out in partnership with the system user, its supplier, and the electricity supply authority.

2.2.5.3 Wiring considerations

While the installation of motor power cables needs to be considered for any EMI and RFI effects, other factors must also be considered if the separation of the motor and drive is excessive. The pulses from a pulse width modulated drive will act as an impulse on the motor cables, and reflection will occur due to the mismatch in impedance between the motor and the cable. If the length of the cable is excessive, the reflected pulse will combine with other pulses to increase the motor voltage to over twice the nominal supply voltage. This reflected-wave theory is well understood in terms of transmission-line theory. The net effect of this high-voltage spike is the possibility that the motor's insulation will break down. In the selection of a motor, the following points should be considered:

- If the supply voltage is over 440 V (such as the supply of 460–480 V in the USA), the voltage spikes will be in excess of those experienced in European applications.
- If the drive is to be retrofitted to a motor with unknown insulation specifications, this problem can only be resolved by consultation with the supplier of the original motor, and it may require replacement of the motor by a motor with enhanced insulation capabilities.

In the design of the electrical supply system to a drive system, it is important to ensure that the system is fully and correctly earthed. A good earthing system is required:

- To ensure safety of operators and other personnel by limiting touch voltages to safe values, by providing a low-resistance path for fault current so that the circuit protective devices operate rapidly to disconnect the supply. The resistance of the earth path must be low enough so that the potential rise on the earth terminal and any metalwork connected to it is not hazardous.
- To limit EMI and RFI as discussed in Section 2.2.5.2, by providing a noise-free ground.
- To ensure correct operation of the electricity supply network and ensure good power quality.

The actual design of a complete earthing system is complex and reference should be made to the relevant national standards, within the UK reference should be made to BS7671:2001 (*Requirements for electrical installations. IEE Wiring Regulations. Sixteenth edition*) and BS7430:1998 (*Code of practice for earthing*).

In the construction of a drive system, bonding is applied to all accessible metalwork – whether associated with the electrical installation (known as exposed-metalwork) or not (extraneous-metalwork) – is connected the system earth. The bonding must be installed so that the removal of a bond for maintenance of equipment does not break the connection to any other bond.

As noted above the provision of a good earth is fundamental to the prevention of EMI and RFI problems. It is common practice to use a single point or star earthing system to avoid the problems of common mode impedance coupling. However, care needs to be exercised when shielded cables are used, as loops may inadvertently be formed, which will provide a path for any noise current.

2.2.5.4 Supply considerations

While the quality of public-utility supplies in Western Europe is normally controlled within tight specifications, considerable voltage fluctuations may have to be accommodated in a particular application. In cases where the drive system is used on sites with local generation (for example, on oil rigs and ships), considerable care needs to be taken in the specification of the voltage limits. Since the peak speed of a motor is dependent on the supply voltage, consideration needs to be given to what happens during a period of low voltage. As a guideline, drives are normally sized so that they can run at peak speed at eighty per cent of the nominal supply voltage. If a system is fed from a vulnerable supply, considerable care will have to be taken to ensure that the drive, its controller, and the load are all protected from damage; this problem is particularly acute with the introduction of microprocessor systems, which may lock-up or reset without warning if they are not properly configured, leading to a possibly catastrophic situation.

In practice, the supply voltage can deviate from a perfect sinewave due to the following disturbances.

- *Overvoltage*. The voltage magnitude is substantially higher than its nominal value for a significant number of cycles. This can be caused sudden decreases in the system load, thus causing the supply to rise rapidly.
- *Undervoltage or brownout*. The voltage is substantially lower than its nominal value for a significant number of cycles. Undervoltages can be caused by a sudden increase in load, for example a machine tool or induction motor starting.
- *Blackout or outage*. The supply collapses to zero for a period of time that can range from a few cycles to an extended period of time.
- *Voltage spikes*. These are superimposed on the normal supply waveform, and are non-repetitive. A spike can be either differential-mode or a common mode. Occasional large voltage spikes can be caused by rapid switching of power factor correction capacitors, power lines or motors in the vicinity.
- *Chopped voltage waveform*. This refers to a repetitive chopping of the waveform and associated ringing. Chopping of the voltage can be caused by ac-to-dc line frequency thyristor converters, Figure 2.2.12(a).
- *Harmonics*. A distorted voltage waveform contains harmonic voltage components at harmonic frequencies (typically low-order multiples of the line frequency). These harmonics exist on a sustained basis. Harmonics can be produced by a variety of sources including magnetic saturation of transformers or harmonic currents injected by power electronic loads, Figure 2.2.12(b).
- *Electromagnetic interference*. This refers to high-frequency noise, which may be conducted on the power line or radiated from its source, see Section 2.2.5.2.

The effect of power line disturbances on drive systems depends on a number of variables including the type and magnitude of the disturbance, the type of equipment and how well it is designed and constructed, and finally the power conditioning equipment fitted to the system or the individual drive.

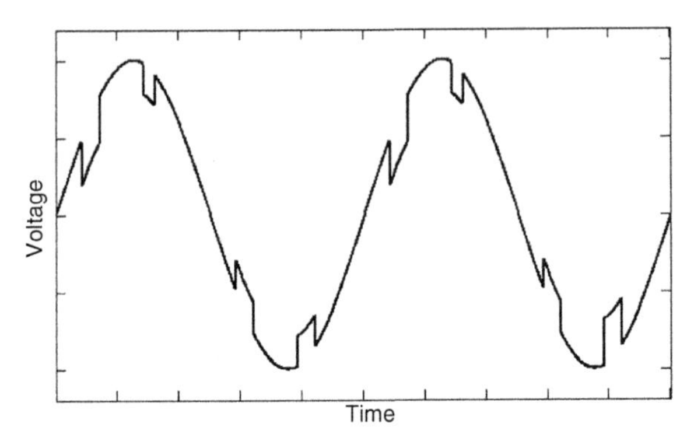

(a) Distortion caused by chopping applied to the supply.

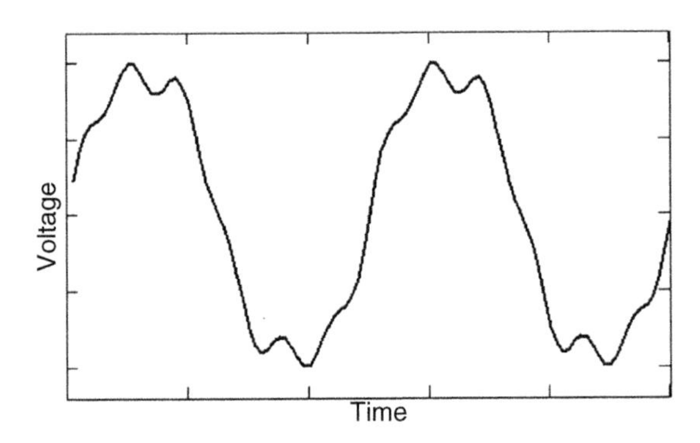

(b) Distortion caused by the addition of externally generated harmonics to the supply

Fig. 2.2.12 Distortion in the supply input to a drive system due to externally generated harmonics or chopping.

Sustained under- and over-voltages will cause equipment to trip out, which is both highly undesirable and with a high degree of risk in certain applications. Large voltage spikes may cause a hardware (particularly in power semiconductors) failure in the equipment. Manufacturers of critical equipment often provide a certain degree of protection by including surge arrestors or snubbers in their designs. However, spikes of very large magnitude in combination with a higher frequency may result in a stress-related hardware failure, even if normal protection standards are maintained. A chopped voltage and harmonics have the potential to interfere with a drive system if it is not designed to be immune from such effects.

2.2.5.4.1 Power conditioning

Power conditioning provides an effective way of suppressing some or all of the electrical disturbances other than the power outages and frequency. Typical methods of providing power conditioning include:

- metal-oxide varistors, which provide protection against line-mode voltage spikes,

- electromagnetic interference filters, which help to prevent the effect of the chopped waveforms on the equipment as well as to prevent the equipment from conducting high-frequency noise into supply,
- isolation transformers with electrostatic shields, which not only provide galvanic isolation, but also provide protection against voltage spikes,
- ferroresonant transformers, which provide voltage regulation as well as line spike filtering.

2.2.5.4.2 Interface with the utility supply

All power electronic converters (including those used to protect critical loads) can add to the supply line disturbances by distorting the supply waveform. To illustrate the problems due to current harmonies in the input current of a power electronic load, consider the block diagram of Figure 2.2.13. Due to the finite internal impedance of the supply source, the voltage waveform at the point of common coupling to other loads will become distorted, which may cause additional malfunctions. In addition to the waveform distortion, other problems due to the harmonic currents include: additional heating and over-voltages (due to resonance conditions) in the utility distribution and equipment, errors in metering and malfunction of utility relays and interference with communications and control signal.

One approach to minimise this impact is to filter the harmonic currents and the EMI produced by the power electronic loads. An alternative, in spite of a small increase in the initial cost, is to design the power electronic equipment such that the harmonic currents and the EMI are prevented or minimised from being generated in the first place.

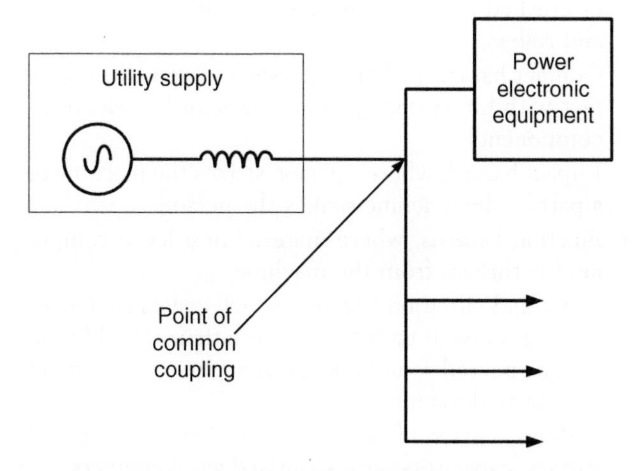

Fig. 2.2.13 The utility interface, showing the *point of common coupling,* where the supply distortion caused by each individual load is combined, due to the finite impedance of the supply – here represented by a simple inductance.

2.2.5.4.3 Standards

In view of the increased amount of power electronic equipment connected to the utility systems, various national and international agencies have been considering limits to the amount of harmonic current injection to maintain good power supply quality. As a consequence a number of standards have been developed, including:

- EN 60555–1:1987, *The Limitation of Disturbances in Electricity Supply Networks caused by Domestic and Similar Appliances Equipped with Electronic Device*, and EN 61000-3–2:1995 *Electromagnetic compatibility (EMC) – Part 3–2: Limits – Limits for harmonic current emissions (equipment input current up to and including 16A per phase)*. Both these European Standards are prepared by CENELEC (European Committee for Electrotechnical Standardisation).
- IEC Norm 555-3, prepared by the International Electrotechnical Commission.
- *IEEE Recommended Practices and Requirements for Harmonic Control in Electrical Power Systems*, IEEE Standard 519-1992.

The CENELEC and IEC standards specify the limits on harmonics within the supply, while the IEEE standards contain recommended practices and requirement for harmonic control in electric power system as well as specifying requirements on the user as well as on the utility.

2.2.5.5 Protection from the environment

A significant proportion of drive systems have to operate in relatively poor environments. The first line in this protection is the provision of a suitable protective enclosure. Two basic classes exist for non-hazardous areas and hazardous areas. An enclosure for non-hazardous areas is classified by the use of an IP code number specified in IEC publication 60529, which indicates the degree of protection from ingress of solid objects including personnel contact, dust, and liquid. In the United States the National Electrical Manufacturers Association (NEMA) classification should be referred to.

A brief definition of the IP classifications are given in Table 2.2.3. If a drive has to operate in a hazardous environment, where an explosive gas/air mixture is present, the formal United Kingdom definition is contained in BS 6345; careful consideration has to be given to the design of the enclosure and all external connections. It is recommended that the designers of systems for this type of environment consult the relevant specialist agencies.

These general application problems can never be solved by one specific formula; rather, the requirements of the various equipment must be recognised, and an

Table 2.2.3 The IP rating system allows designers to specify a motor, or an enclosure, or other components, with a specified degree of protection from dust, water, and impact. The two numerals can be used to specify the protection afforded to a component. In a number of cases a third numeral can be attached this defines the protection against impact

First number Protection against solid objects	Second number Protection against liquids
0 No protection	No protection
1 Objects up to 50.00 mm	Protection against vertically falling drops of water
2 Objects up to 12.00 mm	Direct sprays up 15° from the vertical
3 Objects up to 2.50 mm	Direct sprays up 60° from the vertical
4 Objects up to 1.00 mm	Water sprayed in all direction; limited ingress is permitted
5 Protection against dust; limited ingress is permitted, but no harmful deposits	Protection from low-pressure jets of water in all directions; limited ingress is permitted
6 Totally protected against dust	
7	Protection from immersion in water up to a depth between 15 cm and 1 m
8	Immersion under pressure

optimum system should be selected by careful attention to detail. For example a system protected to IP54 is dust protected, and also protected against splashing water.

The NEMA system takes a different approach, by classifying individual cubicles or systems for a specific application, for example, a NEMA-3 system is defined as being for outdoor use and providing a degree of protection against windblown dust, rain and sleet, and will be undamaged by the formation of ice on the enclosure – this equates to IP64 protection.

2.2.5.6 Drive hazards and risk

It is a legal requirement, placed on both the supplier and user, that the equipment should be designed, manufactured, installed, operated, and maintained to avoid dangerous situations. Within the United Kingdom these requirements are embodied in the relevant Acts of Parliament, and they are enforced by the Health and Safety Executive, which issues a range of notes for guidance for the designers of equipment. Regulations in other countries will be covered by national legislation, and this needs to be considered during the design process. In understanding risks it is worth considering the concepts of hazards, risk, and danger – and how they can be determined and designed out of a system.

A *hazard* is any condition with the potential to cause an accident, and the exposure to such a hazard is known as the corresponding *danger*. As part of the design process an estimate of the damage that may result if an accident occurs, together with the likelihood that such damage will occur, is termed the *risk* associated with the hazard.

2.2.5.6.1 Principles of risk management

Some hazards are inherent within a design; for example, the spindle of a lathe is hazardous by its very nature. Other hazards are contingent upon some set of conditions, such as improper maintenance, unsafe design, or inadequate operating instructions. Several distinct types of hazards can be associated with machine tools and similar systems:

- Entrapment and entanglement hazards, where part or all of a person's body or clothes may be pinched or crushed as parts move together, including gears and rollers.
- Contact hazards, where a person can come into contact with hot surfaces, sharp edges, or live electric components.
- Impact hazard, where a person strikes the machine or a part of the machine strikes the person.
- Ejection hazards, where material or a loose component is thrown from the machine.
- Noise and vibration hazards, which can cause loss of hearing, a loss of tactile sense, or fatigue. In addition, an unexpected sound may cause a person to respond in a startled manner.
- Sudden release of stored energy from mechanical springs, capacitors, or pressurised gas containers.
- Environmental and biological hazards associated with a design, its manufacture, operation, repair, and disposal.

Within any form of risk assessment, the first step is to identify the hazards, namely those with the potential for causing harm. It should be noted that some physical hazards might be present for the complete life cycle of the system whilst others may exist only during the installation, or during maintenance. The second step is to identify the possible accidents or failure modes associated with each hazard, or combinations of hazards, that could lead to the release of the hazard potential and then to determine the times in the life cycle at which such events could occur. To be successful in finding the majority of these events requires the use of a systematic approach, such as a hazard and operability study.

Accidents, however, do not just happen and the third step is to study the possible range of triggering mechanisms, or conditions, which can give rise to each failure or accident. For some events a combination or sequence of triggering conditions will be needed, in other cases only one. The underlying causes, or the conditions which initiate the trigger, often relate back to earlier phases of the project, for example to the design or planning stages. Risk assessment is the estimation of the probabilities or likelihoods that the necessary sequence of triggering events will occur for each particular hazard potential to be released, and an estimation of the consequences of each accident or failure. The latter may involve fatalities, serious injuries, long-term health problems, environmental pollution, and financial losses. Risk management is an extension of risk assessment and typically it involves the steps described above, together with the introduction of preventative measures. The measures may be designed to reduce or eliminate the hazards themselves, the triggering conditions, or on the magnitude of the potential consequences.

2.2.5.6.2 A risk assessment methodology

This section describes the development of a practical risk assessment methodology, as part of risk management of engineering systems; in particular how the process is undertaken. It is clear that the risk assessment methodology should satisfy a number of basic requirements, as shown in Figure 2.2.14. The approach should be capable of:

- Identifying significant hazards at various stages in the equipment's life.
- Identifying the failure mechanisms that could lead to a release of each hazard's potential and the associated triggering conditions.
- Assessing the nature and severity of the consequences of each type of physical failure and other undesired events.
- Enabling estimates to be made of the likelihood of each type of physical failure and other undesired events.
- Assessing the resulting risks.
- Determining the control measures that could reduce the likelihood of undesired events and mitigate their consequences.

The following five step methodology for dealing effectively with hazards has been found to be effective:

1. *Review existing standards*. These will include those provided by the British Standards Institute (BSI), Institution of Electrical Engineers, American

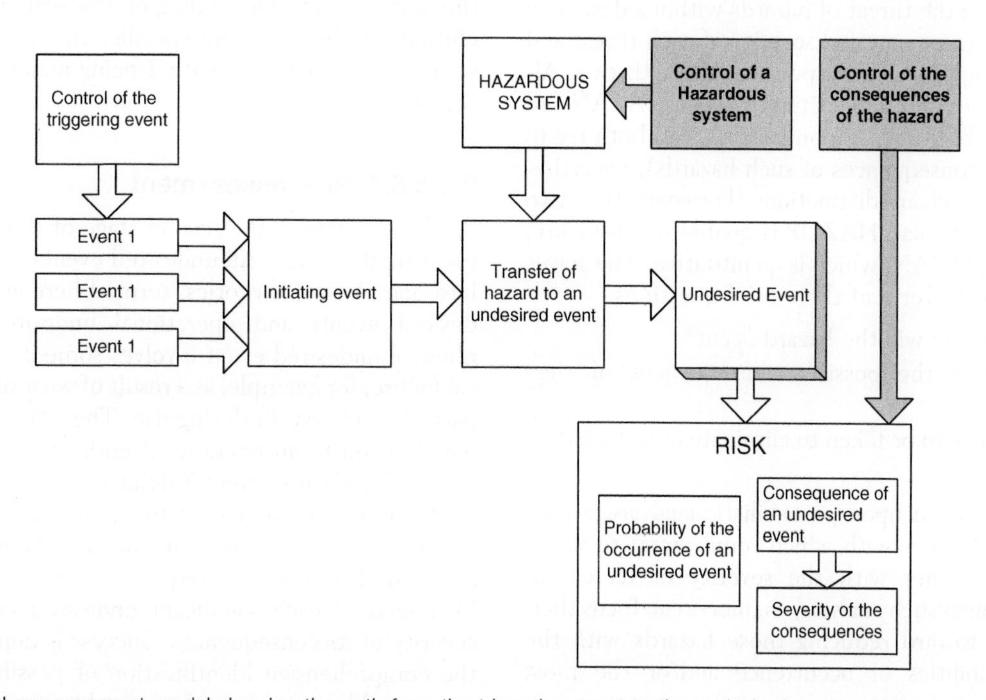

Fig. 2.2.14 Risk management model showing the path from the triggering event to the undesired event and the subsequent risk.

National Standards Institute, Underwriters Laboratory, and Institution of Electronic and Electrical Engineers. This review will determine if standards and requirements exist for the product or system being considered.

2. *Identify known hazards.* Studying recognised standards should make it possible to identify the hazards usually associated with a system.
3. *Identify unknown hazards.* These hazards include those identified in standards that must be eliminated. The design team must follow a systematic approach to identify these undiscovered hazards lurking within the design and in its use or misuse by the operator. Several techniques can be used to identify the unknown hazards, including hazard and operability studies (HAZOP), hazards analysis (HAZAN), fault tree analysis (FTA), and failure modes and effects analysis (FMEA).
4. *Determine the characteristics of hazards.* This stage attempts to determine the frequency, relative severity, and charcterictics of each hazard. By doing so, the designer can focus initially upon those hazards that can result in the most damage and/or those that have the greatest risk associated with them.
5. *Elimination and reduction of the hazard.* Following identification of the hazard, they can be ranked in order of severity and occurrence; the designer can concentrate on their elimination.

2.2.5.6.3 Hazards analysis

HAZAN seeks to identify the most effective way in which to reduce the threat of hazards within a design by estimating the frequency and severity of each threat and developing an appropriate response to these threats. Although there are some similarities between HAZAN and HAZOP (e.g., both focus upon hazards, and both try to anticipate the consequences of such hazards), nevertheless there are clear distinctions between the two methods. In particular, HAZOP is qualitative in nature, in contrast to HAZAN, which is quantitative. The stages of HAZAN in the form of three brief questions:

- How frequently will the hazard occur?
- How large is the possible consequences of the hazard?
- What action is to be taken to eliminate or reduce the hazard?

HAZAN is based upon probabilistic analysis in estimating the frequency with which some threat to safety may occur, together with the severity of its consequences. Through such analysis, engineers can focus their initial efforts toward reducing those hazards with the highest probabilities of occurrence and/or the most severe consequences.

Failure modes and effects analysis: When using failure modes and effects analysis (FMEA) to troubleshoot a design, one begins by focusing upon each basic component one at a time and tries to determine every way in which that component might fail. All components of a design should be included in the analysis, including such elements as warning labels, operation manuals, and packaging. One then tracks the possible consequences of such failures and develops appropriate corrective actions. As part of an FMEA exercise an analysis of all the system's components are produced. A format can be used through which all components or parts can be listed, together with the following information:

- Failure models, identifying all ways in which the part can fail to perform its intended function should be identified.
- Failure causes, identifies the underlying reasons leading to a particular failure.
- Identifying how that a particular failure mode has occurred.
- Details of the protective measures that have been incorporated to prevent any failure.
- A weighted value of the severity, occurrence, and detection of the event.

Example 2.2.4

Consider the risks associated with the tachogenerator within a motor–drive system.

An illustration of the FEMA format, which takes a bottom up approach is shown in Table 2.2.4.

The rating is a subjective measures of the consequence of an undesirable event upon the operators, company, and the system itself. Depending on the scale used the resolution can be company specific. In this example the scale runs from 1 to 7, with 2 being major repairs being required.

2.2.5.6.4 Risk assessment

Risk assessment is the second stage of the risk management methodology. All undesired events can be grouped into one of two categories, termed here as physical undesired events and operational undesired events. A physical undesired event involves some degree of physical failure, for example, as a result of wear or corrosion of part of a subsystem during use. The latter may, or may not, lead on to an operational undesired event. An operational undesired event is defined as an event leading to death or injury, or a near miss, in which there is no physical failure of any part of the equipment being assessed. The next step requires a determination of the likelihood of each significant undesired event and the severity of its consequences. Success is dependent upon the comprehensive identification of possible undesired events and knowing how these can be related back to the

Table 2.2.4 FMEA risk assessment for a tachogenerator as fitted to a motor–drive system. *P* is the probability, *S* the seriousness of the fault, *D* the likelihood that the fault will reach the customer and $R = P \times S \times D$) is the priority measure. *P*, *S*, and *D* are measure on a scale of 1–5.

Failure mode	Cause	Hazard	*P*	*S*	*D*	*R*	Corrective action
Plug failure	Using as a step	Overspeed	1	4	1	4	Safety cover and warning label
Incorrect wiring	Assembly fault	Overspeed	2	4	1	8	QA documentation
Broken coupling	Metal fatigue due to misalignment	Overspeed	2	4	2	16	QA and inspection
Wiring failure	Fatigue caused by vibration	Overspeed	1	4	2	8	Design and use of cable restraints

initiating events, which caused them. The process involves determining the likelihood that the initiating event will be detected, before serious damage can occur; determining the corresponding likelihood of recovery from, or correction of, the initiating event; assessing the likelihood that the initiating event will escalate to give rise to an undesired event; and finally, determining the consequences of the undesired event, and their severity. Since data on the frequencies of these types of event are unlikely to be available in most situations, the various likelihoods are obtained by expert judgement using specially selected teams of experts, for example as convened in the process industries for HAZOP studies or historical data. As part of the process the consequences of all significant undesired events also need to be assessed. The severity of the consequences can be expressed in financial terms for the physical damage that may occur and in terms of injury/harm to people for the operational undesired events.

Risk assessment of software: Many engineering systems now incorporate computer-based control systems, which must be incorporated into any risk assessment process. It has been estimated that for every million lines of code, there are 20,000 bugs. Of these, 90% are found during the testing phase, of the remaining 2000, 200 are found in the first year of use. The problem is such that the remainder will remain dormant until a set of trigger conditions occur. The risks of using software with any system can be minimised by techniques such as protective redundancy or *N* modular redundancy. *N* modular redundancy depends on the assumption that a programme can be completely, constantly, and unambiguously specified, and that the independently developed programmes will fail independently. To fully implement this a number of versions of the programme must be developed using different languages or compilers, and running of hardware supplied by different manufactures.

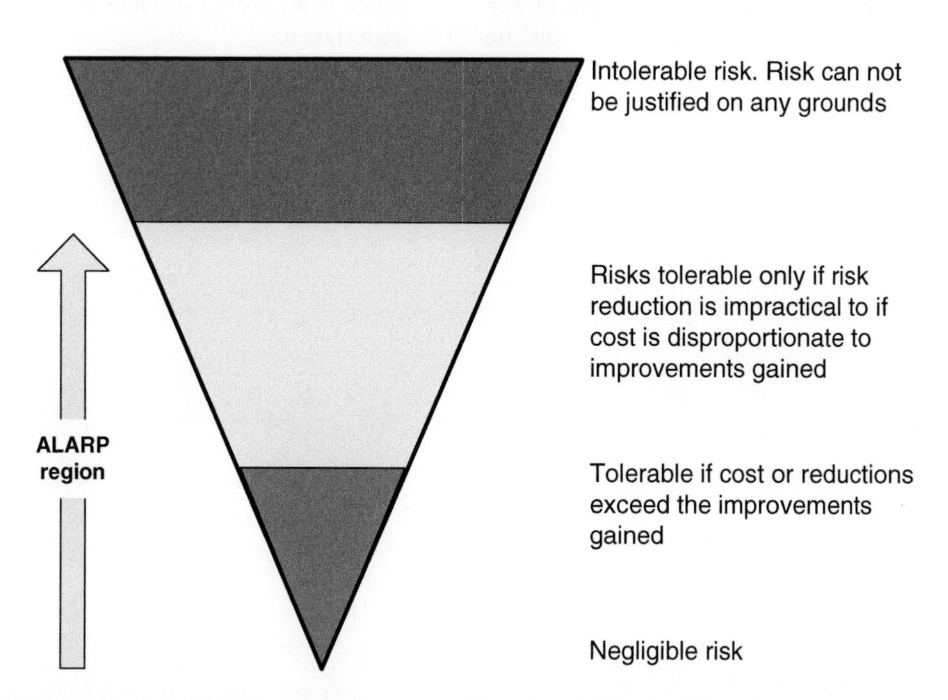
Intolerable risk. Risk can not be justified on any grounds
Risks tolerable only if risk reduction is impractical to if cost is disproportionate to improvements gained
Tolerable if cost or reductions exceed the improvements gained
Negligible risk
ALARP region

Fig. 2.2.15 A diagrammatic representation of the ALARP principle. Within the ALARP region the costs of removing a risk have to be justified against the benefits.

Preventative measures: When the most significant risks have been determined, the next stage of the methodology requires that the underlying causes should be targeted for control, as shown in Figure 2.2.14. In general each possible triggering condition leading to an undesired event can be attributed to a stage of design, assembly, or maintenance and this indicates whether it is the manufacturer or user of the system who is failing in their responsibilities. Finally, remedial action should be specified, indicating how the undesired events and their effects can be controlled. This involves a hierarchy of control measures, with these being applied not only to the cause of the initiating event and triggering conditions, but also to the hazards and the consequences of failure. For example, one might use interlock systems, and other safety features to protect or distance the operator from the hazard, thereby reducing her or his exposure to the danger. Finally, if a hazard cannot be eliminated or further reduced, one should prepare an appropriate set of warnings and instructions for the operator so that he or she can take precautions to avoid the danger.

ALARP: Within the scope of considering preventative assessment the concept of As Low As Reasonably Practical, or ALARP, needs to be discussed. The ALARP principle is fundamental to the regulation of health and safety in the UK and requires that risks should be weighed against the costs of implementing the control measures. These measures must then be taken to reduce or eliminate the risks unless the cost of doing so is obviously unreasonable compared with the risk. The principles are summarised in Figure 2.2.15. Within the ALARP process, a value has to be put on the hazards, its consequence and control. As can be appreciated this can be a very problematic area, particularly when human life has to be assigned a monetary value. In addition the level of risk needs to be quantified, and while this in engineering terms can be rigorously achieved, the public's perceived level of risk is far more difficult to quantify. This can be illustrated by comparing the public perception of risk between public and private transport.

2.2.6 Summary

In this chapter the criteria and parameters that affect the selection and sizing of a motor–drive system were considered. In this process, particular emphasis has to be placed on the following:

- The load torque and the speed reflected back to the motor through the power-transmission chain have to be correctly determined. A range of power-transmission elements were considered, and the selection for a particular application will depend on the speed and the precision which are required.
- The drive and the motor must be suited to the application. Points that should be considered include the physical environment, the electrical compatibility, and the emission requirements. In certain cases, the optimum system will need to be modified in because of the customer's commercial requirements.
- Drives by their very nature can pose a significant hazard to the user or its immediate environment, hence in all cases a suitable risk analysis should be undertaken.

Chapter 2.3

2.3

Motors, motor control and drives

Drury

2.3.1 Introduction

Electric motors can be found in applications from computer disk drives, domestic appliances, automobiles to industrial process lines, liquid pumping, conveyors, weaving machines and many more. Modern building services are heavily reliant upon motors and drives, which can be found at the heart of air handling and elevator systems. Even in the world of theatre and film, the electric motor is at the centre of the action, moving scenery and allowing actors to perform death-defying feats in complete safety. If it moves, it is reasonable to expect to find that an electric motor is somehow responsible.

The flexibility of power transmission that was introduced by the electric motor has been harnessed and controlled by the application of torque, speed and position controllers, which all fall under the generic term of drives or *variable speed drives* (*VSD*). This precision of control has been central to all aspects of industrial automation and has opened up new and demanding applications such as automatic production and sectional process lines, machine tool axis control, glass engraving, embroidery machines and precision polishing machines. It has also facilitated considerable reductions in energy consumption by regulation of flow through speed control in fan and pump type loads, where power consumption is proportional to the cube of the speed. This ability to reduce energy consumption continues to be a major stimulus to growth in the variable speed drives market.

Many alternating current (ac) motors operate at fixed speed, being connected directly to the fixed-frequency supply system, but of the 70 per cent of all electrical power which flows through electric motors, a significant proportion passes through semiconductor conversion in the form of drives. The importance of this technology is self-evident, though the selection of appropriate equipment is often less clear. It would be a difficult, if not impossible, task to detail every motor type and associated power conversion circuit available, and the focus here will be on those of greatest practical importance in the broad base of industries.

All electric motors have a stationary component, the stator, and a moving component, the rotor. In conventional motors, the rotor turns within the stator, but special motors are available for applications such as material handling conveyors where the rotor rotates outside the stator. In linear motors, the rotor moves along the path of the stator. In all cases, the rotor and stator are separated by an air gap. The stator and rotor usually have a laminated steel core to reduce the losses arising from time-varying magnetic fields.

Almost all electric motors adhere to common fundamental principles of operation. This so-called unified theory of electric machines is important in the design and analysis of motor performance, but is not always helpful in understanding the principles of operation, and is therefore not included here. Only the characteristics relevant to operation and control of the practically important motors are presented.

2.3.2 The direct current motor

History will recognize the vital role played by direct current (dc) motors in the development of industrial power transmission systems, the dc machine being the

Newnes Electrical Power Engineer's Handbook; ISBN: 9780750662680

first practical device to convert electrical power into mechanical power. Inherently straightforward operating characteristics, flexible performance and high efficiency have encouraged the widespread use of dc motors in many types of industrial drive applications. The basic construction of a dc motor is shown in Figure 2.3.1.

Standard dc motors are readily available in one of two main forms:

- *wound-field*, where the magnetic flux in the motor is controlled by the current flowing in a field or excitation winding, usually located on the stator.
- *permanent magnet*, where the magnetic flux in the motor is created by permanent magnets which have a curved face to create a constant air gap to the conventional armature, located on the rotor. These are commonly used at powers up to approximately 3 kW.

Torque in a dc motor is produced by the product of the magnetic field created by the field winding or magnets and the current flowing in the armature winding. The action of a mechanical commutator switches the armature current from one winding to another to maintain the relative position of the current to the field, thereby producing torque independent of rotor position.

The circuit of a shunt-wound dc motor (Figure 2.3.2) shows the armature M, the armature resistance R_a and the field winding. The armature supply voltage V_a is supplied typically from a controlled thyristor system and the field voltage V_f from a separate bridge rectifier.

As the armature rotates, an *electromotive force* (emf) E_a is induced in the armature circuit and is called the *back-emf* since it opposes the applied voltage V_a (according to Lenz's law). The E_a is related to armature speed and main field flux by

$$E_a = k_1 n \phi \tag{2.3.1}$$

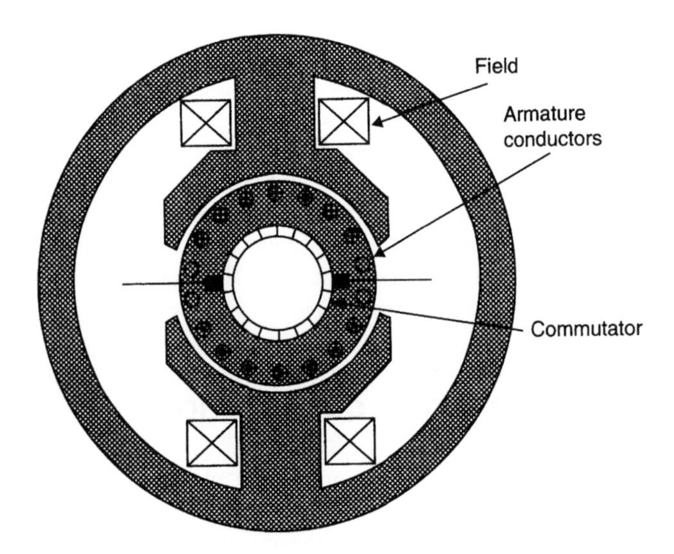

Fig. 2.3.1 DC motor in schematic form.

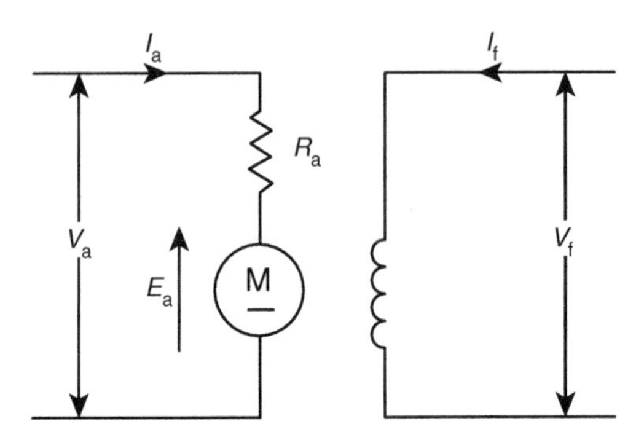

Fig. 2.3.2 Shunt wound dc motor.

where n is the speed of rotation, ϕ is the field flux and k_1 is a motor constant. From Figure 2.3.1 it is seen that the terminal armature voltage V_a is given by

$$V_a = E_a + I_a R_a \tag{2.3.2}$$

Multiplying each side of Equation (2.3.2) by I_a gives:

$$V_a I_a = E_a I_a + I_a{}^2 R_a \tag{2.3.3}$$

(or total power supplied = power output + armature losses). The interaction of the field flux and armature flux produces an armature torque as given in the equation:

$$\text{Torque } M = k_2 I_f I_a \tag{2.3.4}$$

where k_2 is a motor constant and I_f is the field current. This confirms the straightforward and linear characteristic of the dc motor and consideration of these simple equations will show its controllability and inherent stability. The speed characteristic of a motor is generally represented by curves of speed against input current or torque and its shape can be derived from Equations (2.3.1) and (2.3.2):

$$k_1 n \phi = V_a - (I_a R_a) \tag{2.3.5}$$

If the flux is held constant by holding the field current constant in a properly compensated motor then

$$n = k_2 [V_a - (I_a R_a)] \tag{2.3.6}$$

From Equations (2.3.4) and (2.3.6), it follows that full control of the dc motor can be achieved through control of the field current and the armature current. In the dc shunt wound motor shown in Figure 2.3.2, these currents can be controlled independently. Most industrial dc motor controllers or drives are voltage fed; that is to say that a voltage is applied, and the current is controlled by measuring the current and adjusting the voltage to give the desired current. This basic arrangement is shown in Figure 2.3.3.

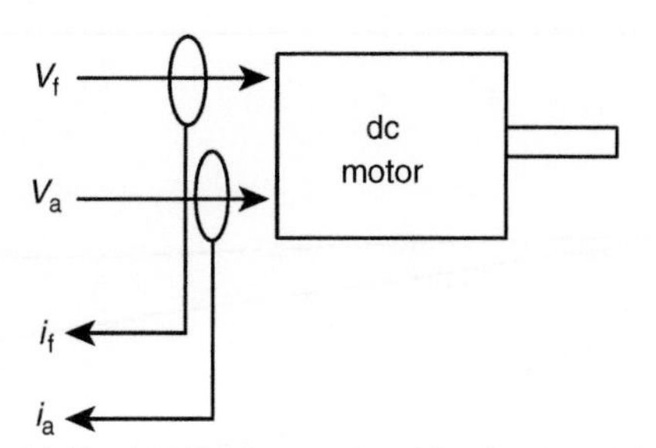

(a) Fundamental control and feedback variables

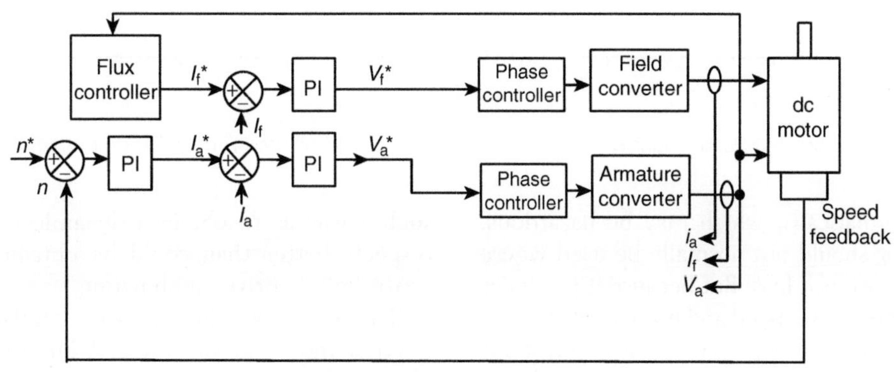

(b) Typical drive control scheme

Fig. 2.3.3 Control structure for a shunt wound dc motor.

DC motors exist in other formats. The series dc motor shown in Figure 2.3.4 has the field and armature windings connected in series. In this case the field current and armature current are equal and show characteristically different performance results, though still defined by Equations (2.3.4) and (2.3.6).

In the shunt motor the field flux ϕ is only slightly affected by armature current, and the value of $I_a R_a$ at full load rarely exceeds 5 per cent of V_a, giving a torque–speed curve shown typically as 'a' in Figure 2.3.6, where speed remains sensibly constant over a wide range of load torque.

The compound-wound dc motor shown in Figure 2.3.5 combines both shunt and series characteristics.

The shape of the torque–speed characteristic is determined by the resistance values of the shunt and series fields. The slightly drooping characteristic (curve 'b' in Figure 2.3.6) has the advantage in many applications of reducing the mechanical effects of shock loading.

The series dc motor curve, ('c' in Figure 2.3.6) shows that the initial flux increases in proportion to current, falling away due to magnetic saturation. In addition the armature circuit includes the resistance of the field winding and the speed becomes roughly inversely proportional to the current. If the load falls to a low value the

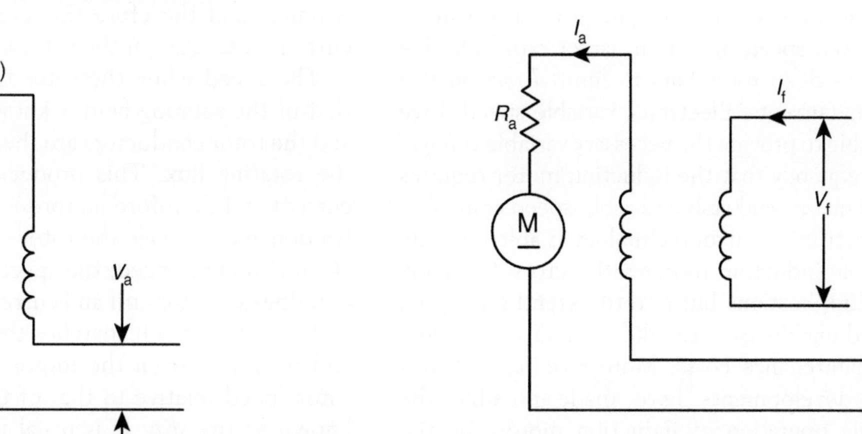

Fig. 2.3.4 Schematic of series dc motor.

Fig. 2.3.5 Compound dc motor.

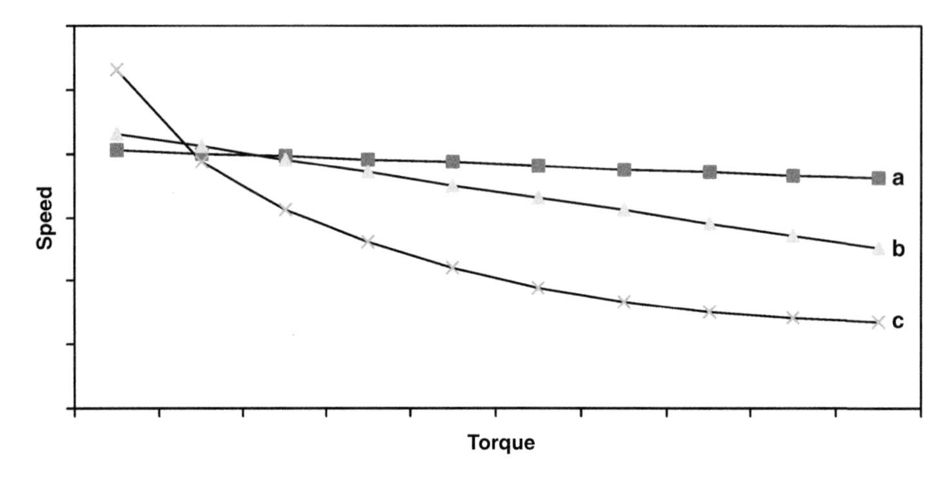

Fig. 2.3.6 Torque–speed characteristic (a, shunt wound dc motor; b, compound dc motor; c, series dc motor).

speed increases dramatically, which may be hazardous, so, the series motor should not normally be used where there is a possibility of load loss. But because it produces high values of torque at low speed and its characteristic is falling speed with load increase, it is useful in applications such as traction and hoisting, and some mixing duties where initial stiction is dominant.

Under semiconductor converter control with speed feedback from a tachogenerator, the shape of the speed–load curve is largely determined within the controller. It has become standard to use a shunt dc motor with converter control even though the speed–load curve, when under open-loop control is often slightly drooping.

The power-speed limit for the dc motor is approximately 3×10^6 kW rev/min, due to restrictions imposed by the commutator.

2.3.3 The cage induction motor

This simplest form of ac induction motor or asynchronous motor is the basic, universal workhorse of industry. Its general construction is shown in Figure 2.3.7. It is usually designed for fixed-speed operation, larger ratings having such features as deep rotor bars to limit *direct on line (DOL)* starting currents. Electronic variable speed drive technology is able to provide the necessary variable voltage, current and frequency that the induction motor requires for efficient, dynamic and stable variable-speed control.

Modern electronic control technology is able not only to render the ac induction motor satisfactory for many modern drive applications but also to extend greatly its application and enable users to take advantage of its low capital and maintenance costs. More striking still, microelectronic developments have made possible the highly dynamic operation of induction motors by the application of flux vector control. The practical effect is that it is now possible to drive an ac induction motor in such a way as to obtain a dynamic performance in all respects better than could be obtained with a phase-controlled dc drive combination.

The stator winding of the standard industrial induction motor in the integral kilowatt range is three phase and is sinusoidally distributed. With a symmetrical three-phase supply connected to these windings, the resulting currents set up, in the air gap between the stator and the rotor, a travelling wave magnetic field of constant magnitude and moving at synchronous speed. The rotational speed of this field is f/p revolutions per second, where f is the supply frequency (hertz) and p is the number of pole pairs (a four-pole motor, for instance, having two pole pairs). It is more usual to express speed in revolutions per minute, as $60\ f/p$ (rpm).

The emf generated in a rotor conductor is at a maximum in the region of maximum flux density and the emf generated in each single rotor conductor produces a current, the consequence being a force exerted on the rotor which tends to turn it in the direction of the flux rotation. The higher the speed of the rotor, the lower the speed of the rotating stator flux field relative to the rotor winding, and therefore the smaller is the emf and the current generated in the rotor cage or winding.

The speed when the rotor turns at the same rate as that of the rotating field is known as synchronous speed and the rotor conductors are then stationary in relation to the rotating flux. This produces no emf and no rotor current and therefore no torque on the rotor. Because of friction and windage the rotor cannot continue to rotate at synchronous speed; the speed must therefore fall and as it does so, rotor emf and current, and therefore torque, will increase until it matches that required by the losses and by any load on the motor shaft. The difference in rotor speed relative to that of the rotating stator flux is known as the *slip*. It is usual to express slip as a percentage of the synchronous speed. Slip is closely proportional to torque from zero to full load.

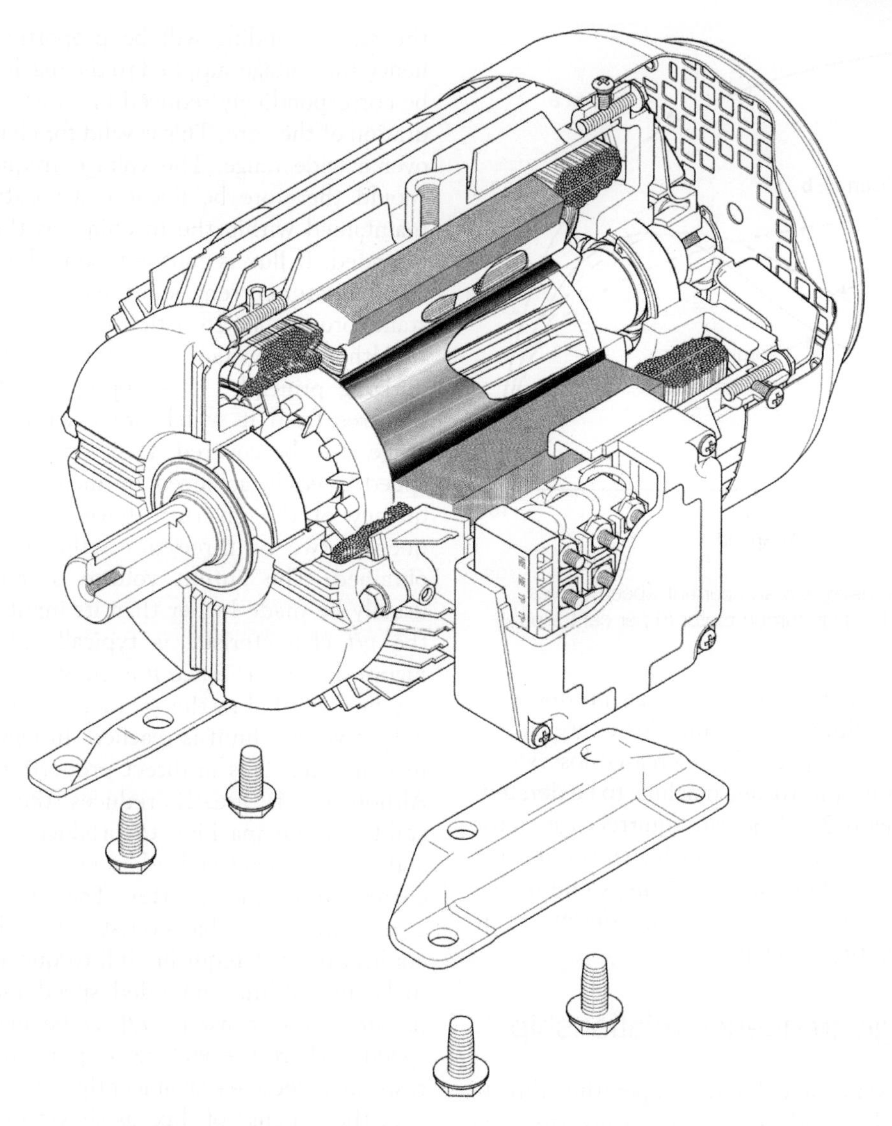

Fig. 2.3.7 Sectional view of a totally enclosed induction motor (courtesy of Brook Crompton).

The most popular squirrel cage induction motor is of a 4-pole design. Its synchronous speed with a 50 Hz supply is therefore 60 f/p, or 1500 rpm. For a full-load operating slip of 3 per cent, the speed will then be $(1 - s)60\ f/p$, or 1455 rpm.

2.3.3.1 Torque characteristics

A disadvantage of the squirrel cage machine is its fixed rotor characteristic. The starting torque is directly related to the rotor circuit impedance, as is the percentage slip when running at load and speed. Ideally, a relatively high rotor impedance is required for good starting performance (torque against current) and a low rotor impedance provides low full-load speed slip and high efficiency.

This problem can be overcome to a useful extent for DOL application by designing the rotor bars with special cross sections as shown in Figure 2.3.8 so that rotor eddy currents increase the impedance at starting when the rotor flux (slip) frequency is high. Alternatively, for special high starting torque motors, two or even three concentric sets of rotor bars are used. Relatively costly in construction but capable of a substantial improvement in

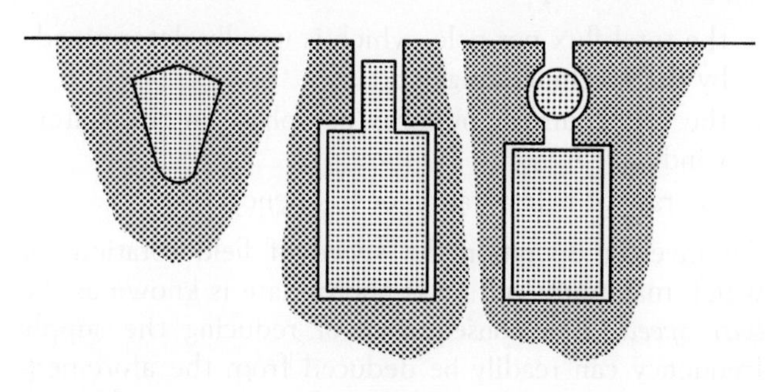

Fig. 2.3.8 Typical rotor bar profiles.

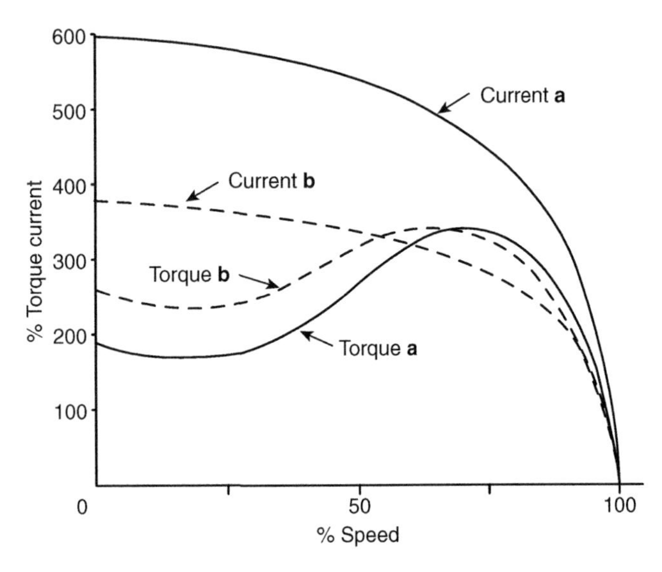

Fig. 2.3.9 Typical torque–speed and current–speed curves (a – standard motor, b – high torque motor (6 per cent slip)).

starting performance, this form of design produces an increase in full load slip. Since machine losses are closely proportional to working speed slip, increased losses may require such a high starting torque machine to be derated.

The curves in Figure 2.3.9 indicate squirrel cage motor characteristics. In the general case, the higher the starting torque the greater the full load slip. This is one of the important parameters of squirrel cage design as it influences the operating efficiency.

2.3.3.2 Voltage–frequency relationship

To convert a constant speed motor operating direct-on-line to a variable speed drive using an inverter it is necessary to consider the effect of frequency on flux and torque. The magnitude of the field created by the stator winding is controlled broadly by the voltage impressed upon the winding by the supply. This is because the resistance of the winding results in only a small voltage drop, even at full load current, and therefore in the steady state the supply voltage must be balanced by the emf induced by the rotating field. This induced emf depends on the product of three factors:

- the total flux per pole, which is usually determined by the machine designer;
- the total number of turns per phase in the stator winding;
- the rate of field rotation or frequency.

For inverter operation the speed of field rotation for which maximum voltage is appropriate is known as the *base speed*. The consequence of reducing the supply frequency can readily be deduced from the aforementioned relationship. For the same flux the induced emf in the stator winding will be proportional to frequency, hence the voltage supplied to the machine windings must be correspondingly reduced in order to avoid heavy saturation of the core. This is valid for changes in frequency over a wide range. The voltage–frequency relationship should therefore be linear if a constant flux is to be maintained within the machine, as the motor designer intended. If flux is constant, so is the motor torque for a given stator current, and the drive therefore has a constant-torque characteristic.

Although constant v/f control is an important underlying principle, it is appropriate to point out departures from it which are essential if a wide speed range is to be covered. First, operation above the base speed is easily achieved by increasing the output frequency of the inverter above the normal mains frequency; two or three times the base speed is easily obtained. The output voltage of an inverter cannot usually be made higher than its input voltage therefore the v/f characteristic is typically like that shown in Figure 2.3.10(a). Since v is constant above base speed, the flux will fall as the frequency is increased after the output voltage limit is reached. In Figure 2.3.10(b), the machine flux falls in direct proportion to the v/f ratio. Although this greatly reduces the core losses, the ability of the machine to produce torque is impaired and less mechanical load is needed to draw full load current from the inverter. The drive is said to have a constant-power characteristic above base speed. Many applications not requiring full torque at high speeds can make use of this extended speed range. Second, departure from constant v/f is beneficial at very low speeds, where the voltage drop arising from the stator resistance becomes significantly large. This voltage drop is at the expense of flux, as shown in Figure 2.3.10(b). To maintain a truly constant flux within the machine the terminal voltage must be increased above the constant v/f value to compensate for the stator resistance effect. Indeed, as output frequency approaches zero, the optimum voltage becomes the voltage equal to the stator IR drop. Compensation for stator resistance is normally referred to as *voltage boost* and almost all inverters offer some form of adjustment so that the degree of voltage boost can be matched to the actual winding resistance. It is normal for the boost to be gradually tapered to zero as the frequency progresses towards base speed. Figure 2.3.10(c) shows a typical scheme for *tapered boost*. It is important to appreciate that the level of voltage boost should increase if a high starting torque is required, since in this case the IR drop will be greater by virtue of the increased stator current. In this case automatic load-dependent boost control is useful in obtaining the desired low speed characteristics. Such a strategy is referred to as *constant v/f control* and is a feature of

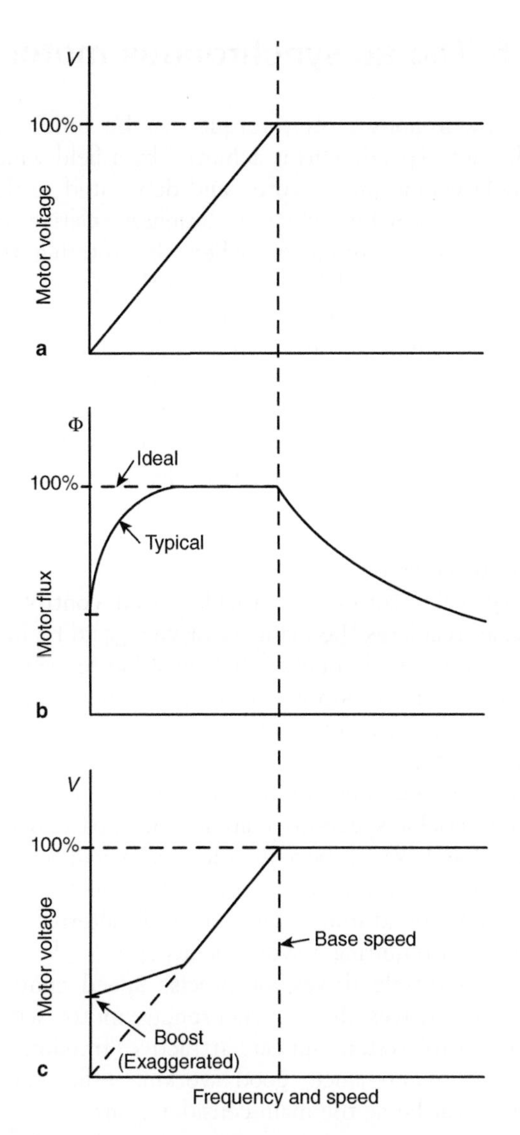

Fig. 2.3.10 Voltage–frequency characteristics (a, linear *v/f* below base speed; b, typical motor flux with linear *v/f* (showing fall in flux at low frequency as well as above base speed); c, modified *v/f* characteristic with low frequency boost (to compensate for stator resistance effects in steady state)).

most commercially available ac drives, though more advanced open-loop strategies are now becoming more available.

So far the techniques described have been based on achieving constant flux within the air gap of the machine or, if that is not possible, then the maximum flux. Constant flux is the ideal condition if the highest torque is required because the load cannot be predicted with certainty, or if the most rapid possible acceleration time is desired. A large number of inverters are used however for variable air volume applications where control of airflow is obtained by variable speed fans. The torque required by a fan follows a square law characteristic with respect to speed and reducing the speed of a fan by 50 per cent will reduce its torque requirement to 25 per cent. As the load is entirely predictable there is no need for full torque capability and hence flux to be maintained, and higher motor efficiency can be obtained by operating at a reduced flux level. A further benefit is that acoustic noise, a major concern in air conditioning equipment, is significantly reduced. It is therefore common for inverters to have an alternative square law *v/f* characteristic or, ideally, a self-optimizing economy feature so that rapid acceleration to meet a new speed demand is followed by settling to a highly efficient operating point. A loss of stability may result from such underfluxing with some control strategies.

2.3.3.3 Flux vector control

High-performance operation of ac motors can be achieved by the use of *flux vector control*. A detailed explanation is beyond the scope of this book, but because this subject is prone to considerable misrepresentation, a few words of explanation may be helpful.

The purpose of flux vector control is to decouple the flux-producing current from the torque-producing current, giving a control structure equivalent to that of the dc motor shown in Figure 2.3.3(a). This is shown in Figure 2.3.11.

Here, v_{sx} is the demand to the voltage source for flux-producing current and v_{sy} is the demand to the voltage source for torque-producing current. These demands are fed through polar to rectangular (or quadrature to magnitude and phase) conversion using a calculated phase angle θ_{ref}, of the rotor position with respect to the field. Unlike the dc motor controller it is not possible to measure directly the flux-producing and torque-producing currents and in flux vector control, these quantities have to be calculated from the measured motor currents, known applied voltages and motor parameters. The performance and quality of this form of control system is critically dependent upon the accuracy

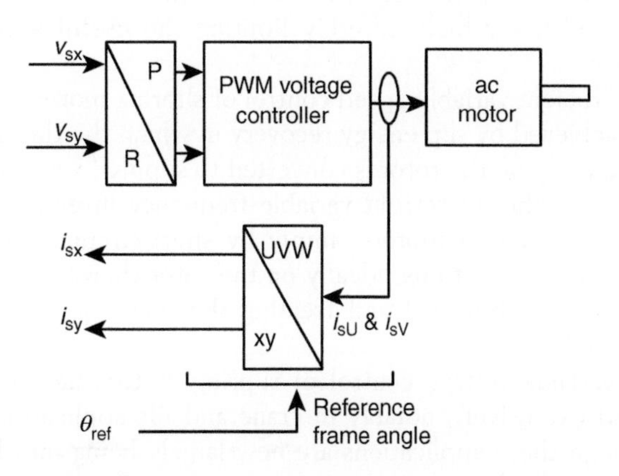

Fig. 2.3.11 Control structure for an ac motor.

of the calculation of the phase angle, θ_{ref}. If the rotor position is measured, as is the case in *closed-loop flux vector* systems, then the phase angle can be calculated very precisely, and excellent flux and torque control is possible at frequencies including 0 Hz. Key performance measures include dynamics, where the performance of dc motors is easily exceeded, and torque linearity, which can be controlled within 2 per cent over the entire operating range in motoring and braking operation. In open-loop control systems, where there is no rotor speed or position measurement, the quality of the control system is critically dependent upon the performance of the phase angle estimator. Modern industrial systems have to balance performance with robustness in operation, and robustness when a standard controller is applied to different industrial motors.

2.3.4 The slipring induction motor

The wound rotor or slipring ac machine addresses some of the disadvantages of the cage induction motor, but with the handicap of extra cost and the complexity of brushes and insulated rotor windings.

With the correct value of (usually) resistance inserted in the rotor circuit, a near-unity relationship between torque and supply current at starting can be achieved, such as 100 per cent full load torque (FLT), with 100 per cent full load current (FLC) and 200 per cent FLT with 200 per cent FLC. This is comparable with the starting capability of the dc machine. Not only high starting efficiency but also smoothly controlled acceleration historically gave the slipring motor a great popularity for lift, hoist and crane applications. It has had a similar popularity with fan engineers, providing a limited range of air volume control (either 2:1 or 3:1 reduction) at constant load by the use of speed-regulating variable resistances in the rotor circuit. Although a fan presents a square-law torque-speed characteristic, so that motor currents fall considerably with speed, losses in the rotor regulator at lower motor speeds are still relatively high, severely limiting the useful speed range.

Efficient variable-speed control of slipring motors can be achieved by slip energy recovery in which the the slip frequency on the rotor is converted to supply frequency. It is possible to retrofit variable frequency inverters to existing slipring motors simply by short-circuiting the slipring terminations (ideally on the rotor thereby eliminating the brushes) and treating the motor as a cage machine.

Variable voltage control of slipring motors has been used extensively, notably in crane and lift applications, though these applications are now largely being met by flux vector drives.

2.3.5 The ac synchronous motor

In a synchronous motor, torque can be produced at synchronous speed. This is achieved by a field winding, generally wound on the rotor, and dc excited so that it produces a rotor flux which is stationary relative to the rotor. Torque is produced when the rotating three-phase field produced by currents in the stator winding and the rotor field are stationary relative to each other, hence there must be physical rotation of the rotor at *synchronous speed* n_s in order that its field travels in step with the stator field axis. At any other speed a rotor pole would approach alternately a stator 'north' pole field, then a 'south' pole field, changing the resulting torque from a positive to a negative value at a frequency related to the speed difference, the mean torque being zero.

A typical inverter for variable speed control automatically regulates the main stator voltage to be in proportion to motor frequency. It is possible to arrange an excitation control loop which monitors the main stator voltage and increases the excitation field voltage proportionately.

The ac synchronous motor has attractive features for inverter variable speed drive applications, particularly at ratings of 40 kW and above. Not least is overall cost when compared with a cage induction motor and inverter, or a dc shunt wound motor with converter alternatives. In applications requiring a synchronous speed relationship between multiple drives, or precise speed control of single large drives the ac synchronous motor with inverter control system appears attractive, freedom from brushgear maintenance, good working efficiency and power factor being the main considerations.

2.3.6 The brushless servomotor

A synchronous machine with permanent magnets on the rotor is the heart of the modern brushless servomotor drive. The motor stays in synchronism with the frequency of supply, though there is a limit to the maximum torque which can be developed before the rotor is forced out of synchronism, *pull-out torque* being typically between 1.5 and 4 times the continuously rated torque. The torque–speed curve is therefore simply a vertical line.

The industrial application of brushless servomotors has grown significantly for the following reasons:

- reduction of price of power conversion products;
- establishment of advanced control of pulse-width modulation (PWD) inverters;
- development of new, more powerful and easier to use permanent magnet materials;

- the developing need for highly accurate position control;
- the manufacture of all these components in a very compact form.

They are, in principle, easy to control because the torque is generated in proportion to the current. In addition, they have high efficiency, and high dynamic responses can be achieved.

Brushless servomotors are often called brushless dc servomotors because their structure is different from that of dc servomotors. They rectify current by means of transistor switching within the associated drive or amplifier, instead of a commutator as used in dc servomotors. Confusingly, they are also called ac servomotors because brushless servomotors of the synchronous type (with a permanent magnet rotor) detect the position of the rotational magnetic field to control the three-phase current of the armature. It is now widely recognized that *brushless ac* refers to a motor with a sinusoidal stator winding distribution which is designed for use on a sinusoidal or PWM inverter supply voltage. *Brushless dc* refers to a motor with a trapezoidal stator winding distribution which is designed for use on a square wave or block commutation inverter supply voltage.

The brushless servomotor lacks the commutator of the dc motor, and has a device (the drive, sometimes referred to as the amplifier) for making the current flow according to the rotor position. In the dc motor, increasing the number of commutator segments reduces torque variation. In the brushless motor, torque variation is reduced by making the coil three phase and, in the steady state, by controlling the current of each phase into a sine wave.

2.3.6.1 Stationary torque characteristics

The simple equivalent circuit of Figure 2.3.12 represents a motor which uses permanent magnets to supply the field flux. This is a series circuit comprising the armature resistance R_a, and back emf E. If the voltage drop across the transistors is ignored, the equation for the voltage is

$$V = R_a I_a + K_e n \quad (2.3.7)$$

K_e is known as the *back emf constant* of the motor, and n is the speed. Therefore, the torque T is given by

$$T = K_t I_a = (K_t/R_a)(V - K_e n) \quad (2.3.8)$$

K_t is known as the *torque constant* of the motor.

Figure 2.3.13 shows the relation between T and n at two different voltages. The torque decreases linearly as the speed increases and the slope is $K_t K_e/R_a$ which is independent of the terminal voltage and the speed. Such characteristics make speed or position control relatively easy.

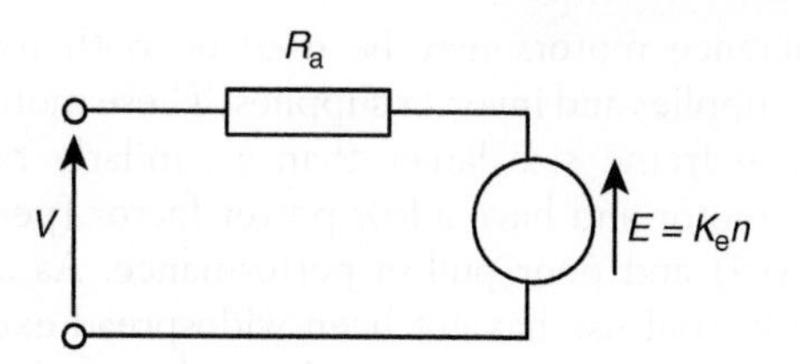

Fig. 2.3.12 Simplified equivalent circuit.

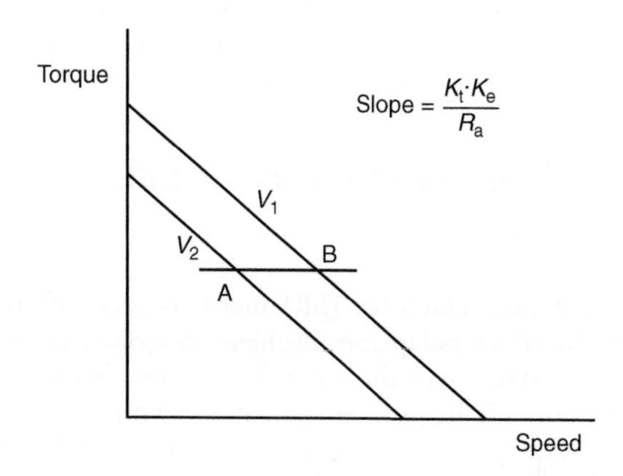

Fig. 2.3.13 Torque–speed characteristics.

The starting torque and the speed are given by

$$T_s = K_t V/R_a \quad (2.3.9)$$

$$n_o = V/K_e \quad (2.3.10)$$

In modern drive systems, a flux vector controller is applied to many brushless servomotors. The control is almost exactly as for the induction motor except zero flux-producing current is demanded for operation in the normal frequency range.

2.3.7 The reluctance motor

The reluctance motor is arguably the simplest synchronous motor of all, the rotor consisting of a set of iron laminations shaped so that it tends to align itself with the field produced by the stator.

The stator winding is identical to that of a three-phase induction motor. The rotor is different, containing saliency which provides a preferred path for the flux. This is the feature which tends to align the rotor with the rotating magnetic field, making it a form of synchronous machine. In order to start the motor a form of cage needs to be incorporated into the rotor design, and the motor can then start as an induction motor. Once a higher speed is reached, the reluctance torque 'pulls in' the rotor to run synchronously in much the same way as a permanent magnet rotor.

Reluctance motors may be used on both fixed frequency supplies and inverter supplies. These motors tend to be one frame size larger than a similarly rated induction motor and have a low power factor (perhaps as low as 0.4) and poor pull-in performance. As a result, their industrial use has not been widespread except for special applications such as textile machines where large numbers of reluctance motors may be connected to a single 'bulk' inverter and may maintain synchronism. Even in this application, as the cost of inverters has reduced, bulk inverters are infrequently used and the reluctance motor is now rarely seen.

2.3.8 The switched reluctance motor

The switched reluctance (SR) motor is very different from the other polyphase machines described because both the stator and the rotor have salient poles. The motor can only be used in conjunction with its specific power converter and control, and consequently only overall characteristics are relevant.

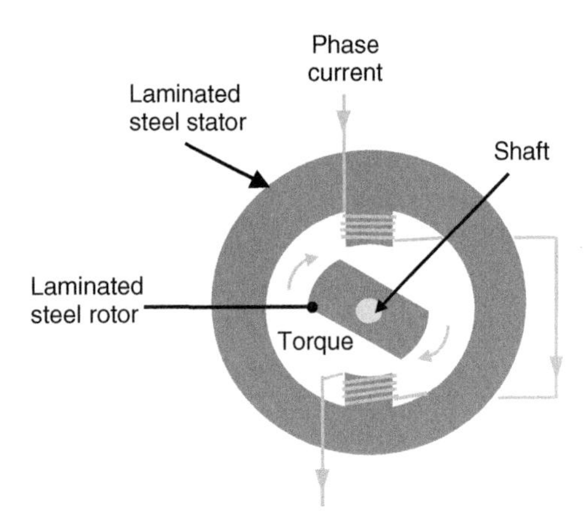

Fig. 2.3.14 Simple reluctance motor.

The SR motor produces torque through the magnetic attraction which occurs between stator electromagnets and a corresponding set of salient poles formed on a simple rotor made only of ferromagnetic material. The intuitively straightforward principle of torque production is easily visualized in the very simple reluctance motor illustrated, in cross section, in Figure 2.3.14.

If current is applied to the winding the rotor will turn until it reaches a position where it is aligned with the coils, at which point the inductance of the magnetic circuit is a minimum. The characteristic variation of inductance of a SR motor is shown in Figure 2.3.15.

If the machine is lightly magnetically loaded and a modest level of torque is produced, then the steel from which the rotor and stator are made of will behave magnetically in an approximately linear fashion. That is, for a given number of turns on the windings, the magnetic flux of a phase varies approximately in proportion to the phase current. If linearity is assumed, then it can be shown that the torque produced as a function of angle θ is

$$T = [i^2(dL/d\theta)]/2 \tag{2.3.11}$$

Equation (2.3.11) shows that the torque is not dependent upon the direction of the current, but depends upon whether the current is applied when the inductance L is rising or falling with angular position.

The phase currents are always switched synchronously with the mechanical position of the rotor. At low speeds, the phases are energized over the entire region of rising inductance, and active current limiting is required from the controller. Torque is controlled by adjusting the magnitude of the phase current. As speed increases, the rise and fall times (especially the latter) of the phase current occupy significant rotor angle, and it is usual to advance the turn-on and turn-off angles with respect to rotor position. The torque is now controlled by both the current limit level and by the switching angles, though current is usually used as the primary control variable. At high speeds, the rise and fall times occupy still greater

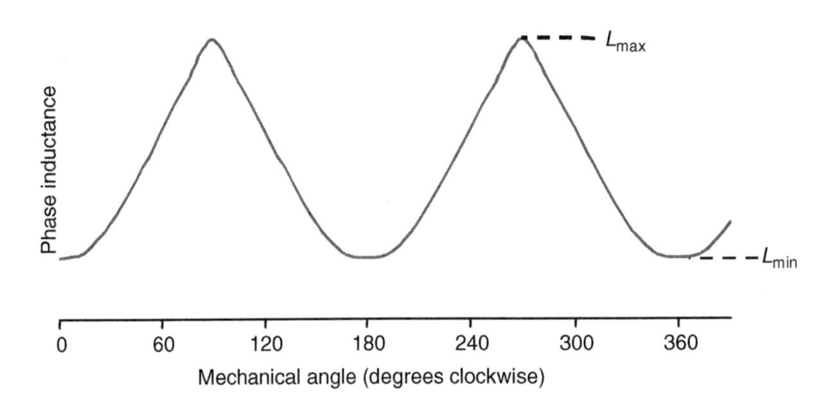

Fig. 2.3.15 Variation of inductance with rotor angle.

Fig. 2.3.16 Cross section of three-phase 6–4 SR motor.

rotor angles. The current naturally self-limits and it is usual to control the torque using only the switching angles. The shape of the current waveform is greatly influenced by the high rate of change of inductance with respect to time.

By choosing appropriate switching angles and current levels, together with an appropriate electromagnetic design, the torque–speed characteristic of the switched reluctance drive can be tailored to suit the application. Furthermore, simply by changing the control parameter selection with torque and speed, a given machine design can be made to offer a choice of different characteristics.

The simple single-phase machine in Figure 2.3.14 is capable of producing torque over only half of its electrical cycle. More demanding applications use higher pole numbers on the rotor and stator, with the stator poles wound and connected into multiple identical phases. Figure 2.3.16 illustrates the cross section of a three-phase 6–4 machine, diametrically opposed coils being connected together to form three-phase circuits denoted A, B and C.

The excitation of the phases is interleaved equally throughout the electrical period of the machine. This means that torque of the desired polarity can be produced continuously. The number of phases can in theory be increased without limit, but one to four phases are most common for commercial and industrial applications. Many different combinations of pole count are possible. It is sometimes beneficial to use more than one stator pole pair per phase, so that, for example, the 12–8 pole structure is commonly used for three-phase applications. Each phase circuit then comprises four stator coils connected and energized together. Increasing the phase number brings the advantage of smoother torque. Self-starting in either direction requires at least three phases.

These SR drives are finding application in high-volume appliances and some industrial applications which can take good advantage of their characteristics, notably high starting torque and where less importance is placed on the smoothness of rotation. Considerable advances have been made in improving the noise characteristics of this drive, but this can still be a limiting factor where a broad operating speed range is required.

2.3.9 Mechanical and duty cycle considerations

2.3.9.1 Mounting of the motor

Internationally agreed coding applies to a range of standard mountings for electric motors, dc and ac, which covers all the commonly required commercial arrangements. Within IEC 60034-7 (EN 60034-7) there are examples of all practical methods of mounting motors. The NEMA publishes alternative standards within NEMA Standards publication No. MG1: Motors and Generators.

There are many standard mounting arrangements described in IEC 60034-7. The most usual types of construction for small- and medium-sized motors are shown in Figure 2.3.17.

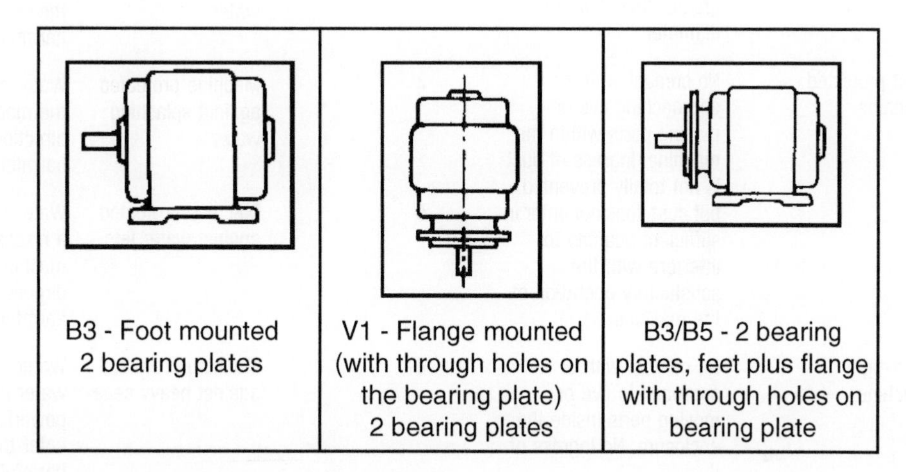

Fig. 2.3.17 Common mounting arrangements for motors.

2.3.9.2 Degree of protection

All types of electric motors are classified in accordance with a standard coding to indicate the degree of protection afforded by any design against mechanical contact and against various degrees of ambient contamination.

The designation defined in IEC 60034-5 (EN 60034-5) consists of the alphabets IP followed by two numerals signifying conformance with specific conditions. Additional information may be included by a supplementary alphabet following the second numeral. This system is contained within NEMA MG 1 but is not universally adopted by the industry in the United States. The first characteristic numeral indicates the degree of protection provided by the enclosure with respect to persons and also to the parts of the machine inside the enclosure. The commonly used numbers are given in Table 2.3.1.

Table 2.3.1 IP designation: first characteristic numeral

First characteristic numeral	Brief description	Definition
0	Non-protected machine	No special protection
2	Machine protected against solid objects greater than 12 mm in diameter	No contact by fingers or similar objects not exceeding 80 mm in length with or approaching live or moving parts inside the enclosure. Ingress of solid objects exceeding 12 mm in diameter.
4	Machine protected against solid objects greater than 1 mm in diameter	No contact with or approaching live or moving parts inside the enclosure by wires or strips of thickness greater than 1 mm in diameter.
5	Dust protected machine	No contact with or approaching live or moving parts within the machine. Ingress of dust is not totally prevented but dust does not enter in sufficient quantity to interfere with the satisfactory operation of the machine.
6	Dust-tight machine	No contact with or approach to live or moving parts inside the enclosure. No ingress of dust.

The second characteristic numeral indicates the degree of protection provided by the enclosure with respect to harmful effects due to the ingress of water. The commonly used numbers are given in Table 2.3.2.

Motors for hazardous areas have to meet very special requirements and flameproof motors in particular have special enclosures.

2.3.9.3 Duty cycles

The capacity of an electrical machine is often temperature dependent, and the duty cycle of the application may significantly affect the rating. IEC 60034-1 defines a number of specific duty cycles with the designation S1 through to S10. For example, S3 describes the duty of an application with an intermittent duty shown in Figure 2.3.18.

This form of duty rating refers to a sequence of identical duty cycles, each cycle consisting of an on-load and off-load period, the motor coming to rest during the latter. Starting and braking are not taken into account; it is being assumed that the times taken up by these are short in comparison with the on-load period, and do not appreciably affect the heating of the motor.

Table 2.3.2 IP designation: second characteristic numeral

Second characteristic numeral	Brief description	Definition
0	Non-protected machine	No special protection
1	Machine protected against dripping water	Dripping water (vertically falling drops) shall have no harmful effects
3	Machine protected against spraying water	Water falling as a spray at an angle up to 60° from the vertical shall have no harmful effect
4	Machine protected against splashing water	Water splashing against the machine from any direction shall have no harmful effect
5	Machine protected against water jets	Water projected by a nozzle against the machine from any direction shall have no harmful effect
6	Machine protected against heavy seas	Water from heavy seas or water projected in powerful jets shall not enter the machine in harmful quantities

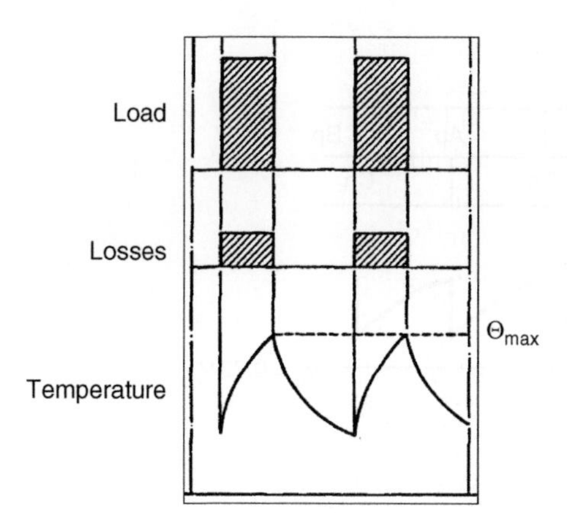

Fig. 2.3.18 Load, losses and temperature for duty type S3.

When stating the motor power for this form of duty, it is necessary to state the *cyclic duration factor,* which is the on-time as a percentage of the cycle time. The duration for one cycle must be shorter than 10 minutes and preferred cyclic duration factors (15%, 25%, 40% and 60%) are specified.

2.3.10 Drive power circuits

The control of a dc motor is based upon the precise control of the voltage or current supplied to the armature and field windings. AC motor control is based upon the precise control of the supply in terms of frequency, the magnitude of the voltage and, in some cases, the phase angle of the supply in relation to rotor position.

A large number of power semiconductor-based converter circuits exist to form voltage or current sources. The principles of many of these circuits are reviewed in Chapter 3.1 and only the most important practical types of circuit for motor drives are discussed in the following sections.

2.3.10.1 DC motor drive systems

In Section 2.3.2 it is shown that complete control of a dc machine can be achieved by controlling the armature voltage V_a and the field current I_f. Two power converters are employed for this purpose in most variable speed drives which employ the separately excited dc machine. (In referring to the number of converters in a drive, it is common to ignore the field converter; this nomenclature will be adopted here.) It is relatively common in simple drives for the field converter to be a single-phase uncontrolled bridge thereby applying fixed field voltage.

Where the variation in motor resistance with temperature, or a poorly regulated supply, results in unacceptable variations in field current, a controlled power converter is employed with current control. Such field

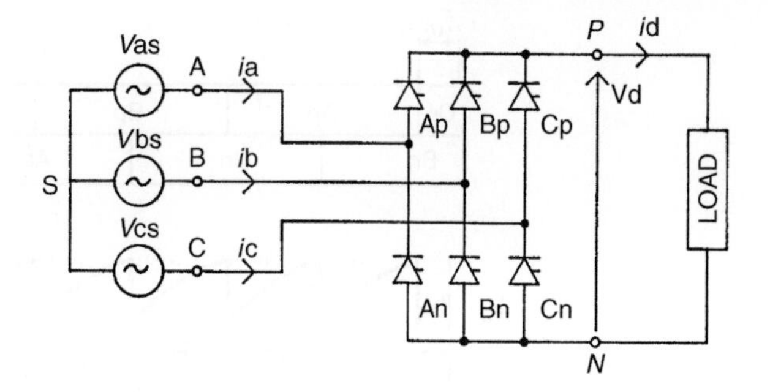

Fig. 2.3.19 Three-phase fully controlled bridge.

controllers are further discussed later as applied to field-weakening control.

2.3.10.1.1 AC to dc power conversion

The three-phase fully controlled converter dominates all but the lowest powers where single-phase converters are used and Figure 2.3.19 shows the power circuit together with associated ac/dc relationships. Figure 2.3.20 shows how the dc voltage can be varied by adjusting the firing delay angle α. The *pulse number p*, of this bridge is 6 and energy flow can be from ac to dc or dc to ac.

The salient 'ideal' characteristics of the three-phase fully controlled bridge are shown in Table 2.3.3.

The above characteristics are based upon idealized conditions of negligible ac inductance and constant dc current, but these are not often found in practice. It is not possible to consider all practical effects here, but the effect of dc link current ripple on ac supply harmonics is of great practical industrial importance mainly in relation to three-phase bridges. Practical experience has led to the adoption by many of $I_5 = 0.25I_1$ (the ideal being $0.2I_1$), $I_7 = 0.13I_1$ (the ideal being $0.14I_1$), $I_{11} = 0.09I_1$ (the ideal being $0.11I_1$) and $I_{13} = 0.07I_1$ (the ideal being $0.08I_1$). In general, the amplitudes of higher harmonics are rarely of significance with regard to supply distortion. Under conditions of very high dc current ripple, the fifth harmonic can assume a considerably higher value than that quoted here. A practical example would be an application with a very capacitive dc load such as a voltage source inverter; in such a case where no smoothing choke is used, I_5 could be as high as $0.5I_1$.

2.3.10.1.2 Single-converter drives

Figure 2.3.21 shows a single-converter dc drive. In its most basic form the motor will drive the load in one direction only without braking or reverse running. It is said to be a *single-quadrant drive,* only operating in one quadrant of the torque–speed characteristic. Such drives have wide application from simple machine tools to fans, pumps, extruders, agitators and printing machines.

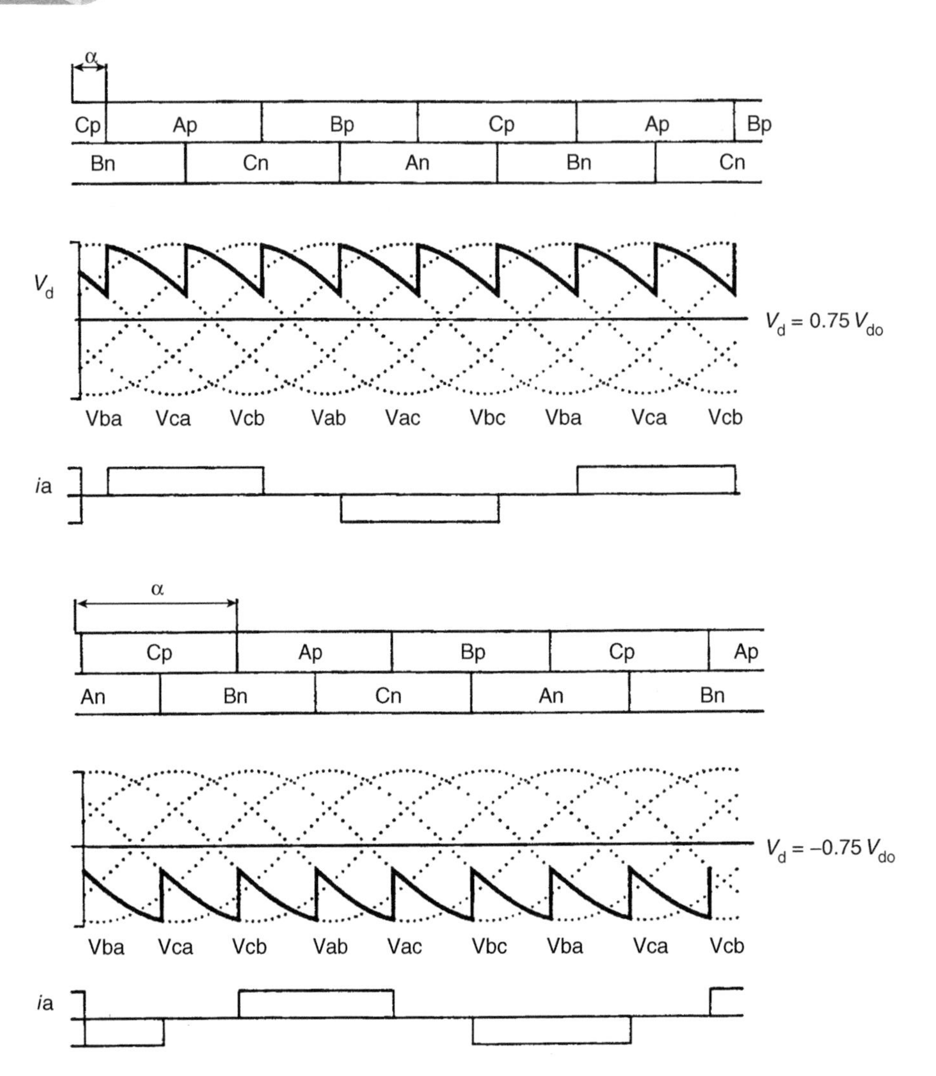

Fig. 2.3.20 Three-phase fully controlled bridge: dc voltage control.

For full *four-quadrant* operation, a drive is required to operate in both the forward and reverse directions and to provide regenerative braking. This is illustrated in Figure 2.3.22. A single fully controlled converter can still be used but some means of reversing, either the field or armature connections, as shown in Figure 2.3.23, must be added.

Table 2.3.3 Ideal characteristics of a three-phase fully controlled bridge	
Firing angle	α
V_{do}	$\frac{3\sqrt{2}}{\pi}V_s$
V_d/V_{do}	$\cos\alpha$
I_s/I_d	$\sqrt{(3/2)}$
Overall power factor	$(3/\pi)\cos\alpha$
Supply current nth harmonic/I_d	0 for $n = 3, 6, 9, \ldots, 0$ for n even $\sqrt{6}/n\pi$ for n odd
Phase of supply current harmonics	$n\alpha$

Reversal of the armature current can involve bulky (high current) reversing switches, but due to the low inductance of the armature circuit it can be completed in typically 0.2 s. Field current reversal takes longer, typically in the order of 1 s, but lower cost reversing switches

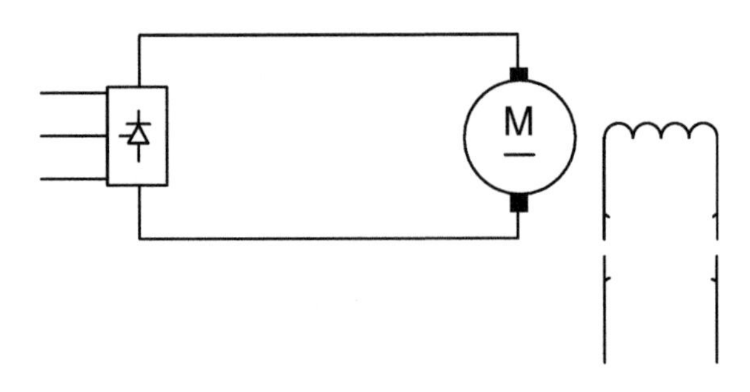

Fig. 2.3.21 Single-phase converter dc drive.

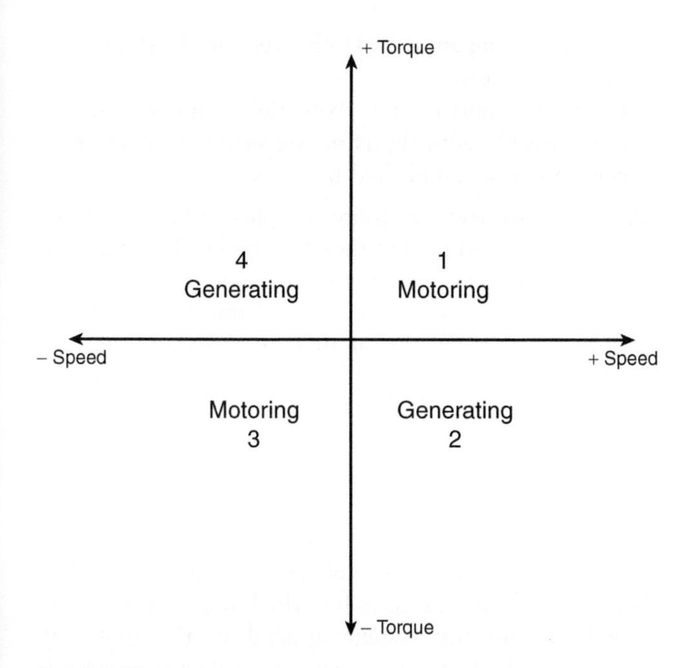

Fig. 2.3.22 Quadrants of operation.

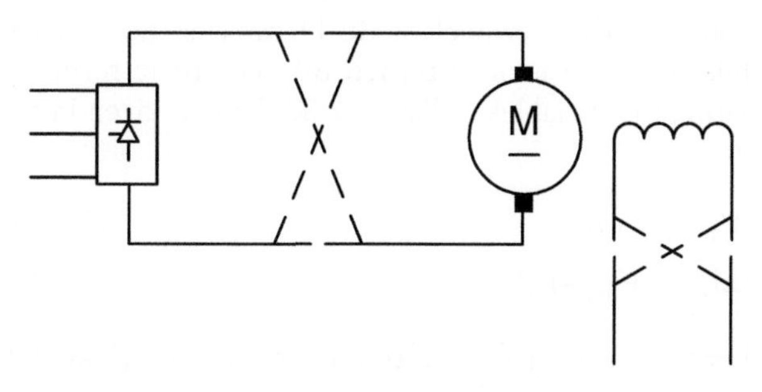

Fig. 2.3.23 Single-phase converter reversing/regenerative dc drive.

may be used. The field reversal time can be reduced by using higher voltage field converters to force the current. Forcing voltages up to 4 per unit are used but care must be taken not to over-stress the machine. This increased voltage cannot be applied continuously and either a switched ac supply or a controlled field converter is required. Armature and field reversal techniques are used where torque reversals are infrequent such as hoists, presses, lathes and centrifuges.

2.3.10.1.3 Dual-converter drives

When a four-quadrant drive is required to change the direction of torque rapidly, the delays associated with reversing switches described earlier may be unacceptable and a dual converter comprising two fully controlled power converters connected in inverse-parallel can be used as shown in Figure 2.3.24. Bridge 1 conducts when the armature current I_a is required to be positive and bridge 2 when it is required to be negative.

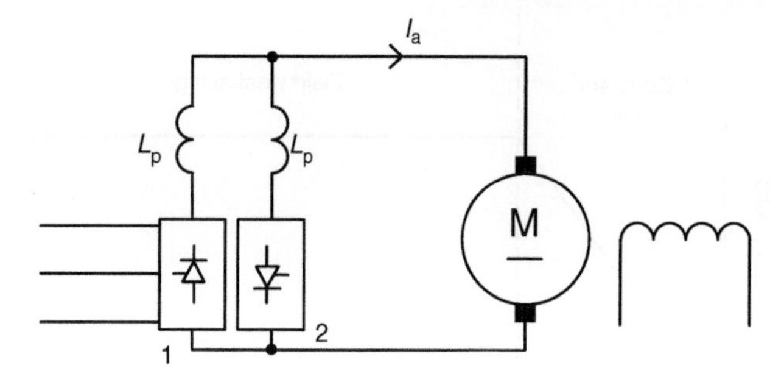

Fig. 2.3.24 Single-phase dual-converter dc drive.

There are two common forms of dual converter. In the first, both bridges are controlled simultaneously to give the same mean output voltage. The instantaneous voltages from the rectifying and inverting bridges cannot be identical, and reactors L_p are included to limit the current circulating between them. The principal advantage of this system is that when the motor torque, and hence current, is required to change direction, there need be no delay between the conduction of one bridge and the other. This is the *dual converter bridge with circulating current*.

In the second, the *circulating current-free dual converter*, only one bridge at a time is allowed to conduct. The cost and losses associated with the L_p reactors can then be eliminated, and economies can also be made in the drive control circuits. The penalty is a time delay of typically 10 ms as the current passes through zero, while it is ensured that the thyristors in one bridge have safely turned off before those in the second are fired. This circulating current-free dual converter is the most common industrial four-quadrant drive and is used in many demanding applications. Paper, plastics and textile machines where rapid control of tension is required are good examples.

2.3.10.1.4 Field control

By weakening the field as speed increases, a constant power characteristic can be achieved. The field can be controlled by a three-phase (or often a single-phase) fully controlled bridge. By including this converter in a speed-control loop, it can be arranged that as speed increases, the armature voltage rises to the point where it matches the pre-set reference in the field controller. Above that speed, an error signal is produced by the voltage loop, which causes the field controller to weaken the motor field current and thereby restore armature voltage to the set-point level. The resulting characteristics are shown in Figure 2.3.25.

2.3.10.1.5 DC to dc power conversion

The principles of dc to dc power converters are explained in Section 3.1.4.1. They provide the means to change one

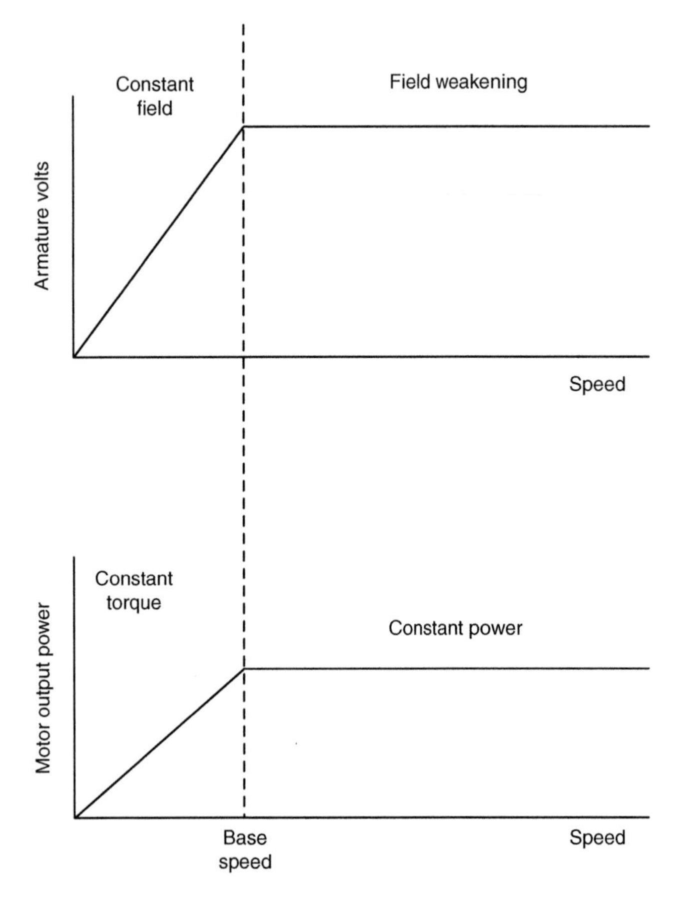

Fig. 2.3.25 Constant power operation using a field controller.

dc voltage to another; step-up (or boost) converters are available and have potential for the future, but step-down (or buck) converters are of greatest commercial interest and will be concentrated upon here. These converters are usually supplied from an uncontrolled ac to dc converter or a battery supply; their output can be used to control a dc machine as in the case of the controlled ac to dc converters.

Several important limitations of ac to dc converters are overcome by the dc–dc converter, as follows:

- The dc ripple frequency is determined by the ac and is, for a 50 Hz supply frequency, 100 Hz for single-phase and 300 Hz for three-phase fully controlled bridges. This means that additional smoothing components are often required when using high-speed machines, permanent magnet motors or other special motors with low armature inductance.
- As a result of the delay inherent in thyristor switching (3.3 ms in a 50 Hz three-phase converter) the current control loop band width of the converter is limited to approximately 100 Hz, which is too low for many servo drive applications.
- Thyristor controlled ac to dc converters have an inherently poor input power factor at low output voltages. The near-unity power factor can be achieved using an uncontrolled rectifier feeding a dc to dc converter.
- Electronic short-circuit protection is not economically possible with thyristor converters. Protection is normally accomplished by fuses.

DC to dc converters are more complex and less efficient than ac–dc converters, but they find application mainly in dc servo drives, rail traction drives and small fractional drives employing permanent magnet motors. In the examples which follow, the circuits are illustrated showing bipolar transistors, but MOSFETs, IGBTs and, at higher powers, GTOs are widely used.

2.3.10.1.5.1 Single-quadrant step-down dc to dc converter

The most basic dc to dc converter is shown in Figure 2.3.26. The output voltage is changed by *PWM*, that is, by varying the time for which the transistor T is turned on and the voltage applied to the motor is therefore in the form of a square wave of varying period. The principle of PWM is explained in Section 3.1.4.4. Because the motor is inductive the current waveform is smoothed, the flywheel diode D carrying the current whilst the transistor is turned off. The basic formulae relating the variables in Figure 2.3.26 are as given in the equations:

$$V_a = V_{dc}tf \tag{2.3.12}$$

$$\Delta I_a = V_{dc}/4L_af \tag{2.3.13}$$

where f is the frequency of transistor 'on pulse' (Hz), ΔI_a is the maximum deviation of armature current and t is the on-pulse duration.

Applications for this circuit are normally limited to drives below 5 kW and simple variable speed applications.

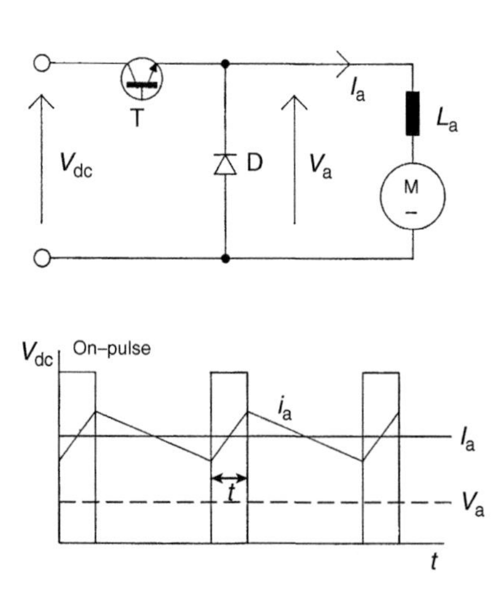

Fig. 2.3.26 Single quadrant dc to dc converter.

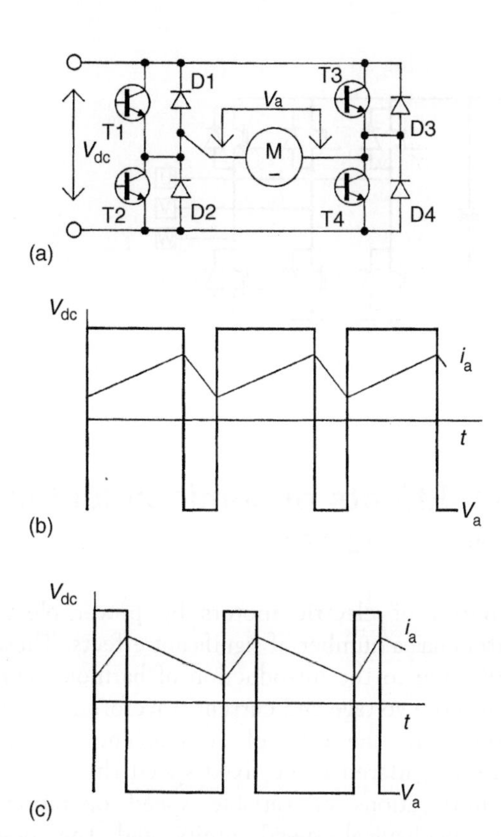

Fig. 2.3.27 Four-quadrant dc to dc converter (a) circuit, (b) forward motoring, (c) reverse braking.

2.3.10.1.5.2 Four-quadrant dc to dc converter (the H-bridge)

Figure 2.3.27 shows a basic four-quadrant converter. During motoring, positive output transistors T1 and T4 are switched on during the on-period, whilst diodes D2 and D4 conduct during the off-period. When D2 and D4 conduct, the motor supply is reversed and consequently the voltage is reduced to zero at 50 per cent duty cycle. Any reduction of duty cycle below 50 per cent will cause the output voltage to reverse but with current in the same direction; hence the speed is reversed and the drive is regenerating. With transistors T2 and T3 conducting, the current is reversed and hence the full four-quadrant operation is obtained. These converters are widely used in high-performance dc drives such as servos.

2.3.10.2 AC motor drive systems

2.3.10.2.1 AC to ac power converters with intermediate dc link

This category of ac drive commonly termed the *variable frequency inverter* is by far the most important in respect of the majority of industrial applications. It is considered here as a complete converter, although the input stages have been considered earlier in isolation, and their individual characteristics are, of course, applicable. Alternative input stages to some of the drives are also applicable.

The concept of these inverter drives is well understood, being rectification of fixed frequency, smoothing and then inverting to give variable frequency, variable voltage to supply an ac machine. Within this broad concept two major categories exist. First, in the *Voltage Source Inverter*, the converter impresses a voltage on the motor, and the impedance of the machine determines the current. Second, in the *Current Source Inverter*, the converter impresses a current on the motor, and the impedance of the machine determines the voltage. For most industrial applications the PWM voltage source inverter is applied, and only this converter will be considered here.

2.3.10.2.2 General characteristics of a voltage source inverter

Voltage source inverters can be considered as a voltage source behind an impedance, and consequently they are very flexible in their application. Major inherent features include the following:

- Multi-motor loads can be supplied. This can be very economical in applications such as roller table drives and spinning machines.
- Inverter operation is not dependent upon the motor characteristics. Various machines (induction, synchronous or even reluctance) can be used, provided the current drawn is within the current rating of the inverter although care should be taken where a low power factor motor is used to ensure that the inverter can provide the required reactive power.
- Open-circuit protection is inherent. This is useful in applications where the cables between the inverter and the motor are insecure or subject to damage, etc.
- The facility to ride through mains dips can easily be provided by buffering the dc voltage link with capacitance or, where necessary, a battery.
- Motoring operation only in both directions is possible without the addition of resistive dumps for braking energy or expensive regenerative converters to feed energy back to the supply.

In the PWM inverter drive shown in Figure 2.3.28, the dc link voltage is uncontrolled and derived from a simple diode bridge. The output voltage can be controlled electronically within the inverter by use of the PWM techniques explained in Section 3.1.4.4. In this method, the transistors are switched on and off many times within a half cycle to generate a variable voltage output which is normally low in harmonic content. A PWM waveform is illustrated in Figure 2.3.29.

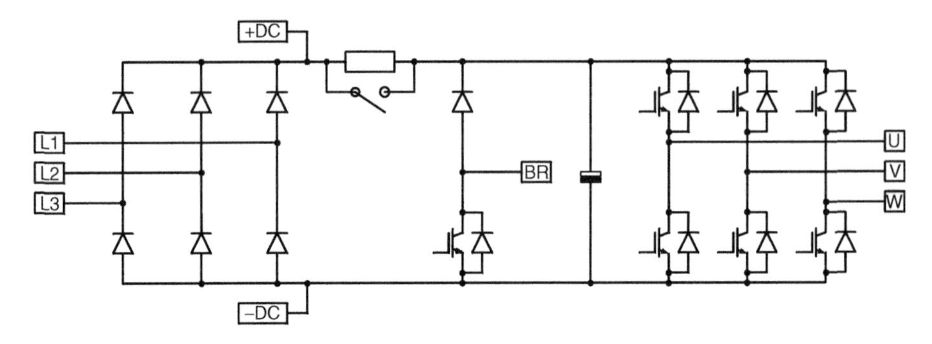

Fig. 2.3.28 Typical power circuit of a PWM voltage-fed inverter.

A large number of PWM techniques exist, each having different performance notably in respect to the stability and audible noise of the driven motor. The use of PWM virtually eliminates low-speed torque pulsations since negligible low-order harmonics are present so this is an ideal solution where a drive system is to be used across a wide speed range.

Since voltage and frequency are both controlled with the PWM, quick response to changes in demand can be achieved. Furthermore, with a diode rectifier as the input circuit a high power factor, approaching unity, is offered to the incoming ac supply over the entire speed and load range.

The PWM inverter drive efficiency typically approaches 98 per cent but this figure is heavily affected by the choice of switching frequency, the losses being greater where a higher switching frequency is used. In practice the maximum fundamental output frequency is usually restricted to about 1 kHz for a transistor-based system. The upper frequency limit may be improved by making a transition to a less sophisticated PWM waveform with a lower switching frequency and ultimately to a square wave if the application demands it. However, with the introduction of faster switching power semiconductors, these restrictions to switching frequency and minimum pulse-width have been eased.

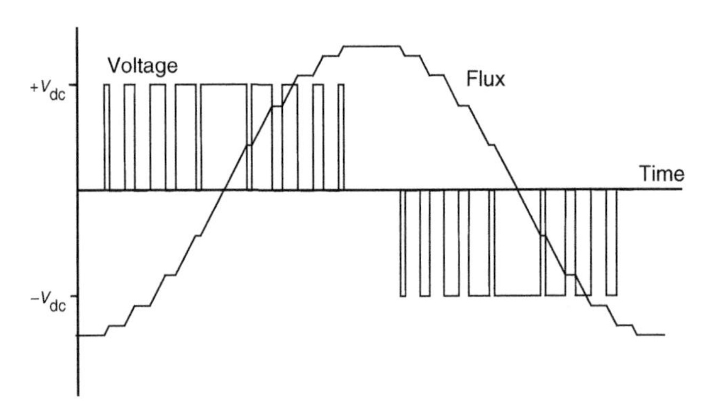

Fig. 2.3.29 PWM inverter output voltage and flux waveforms.

2.3.11 Effects of semiconductor power converters

The control of electric motors by power electronic converters has a number of significant effects. These are primarily due to the introduction of harmonic components into the voltage and current waveforms applied to the motor. In the case of ac machines which are normally considered to be fixed speed there are additional implications of variable speed operation, including mechanical speed limits and the possible presence of critical speeds within the operating speed range.

2.3.11.1 Effects upon dc machines

The effects due to deviation from a smooth dc supply are, in general, well understood by drive and motor manufacturers. The impact of ripple in the dc current increases the rms current, which leads to increased losses and hence reduced torque capacity. The harmonics associated with the current ripple lead to the now universal practice of using laminated magnetic circuits, which are designed to minimize eddy currents. With the chopper converters used in servo amplifiers and traction drives, frequencies in excess of 2 kHz can be impressed on the motor and special care is needed to select a motor with sufficiently thin laminations.

The ripple content of the dc currents significantly affects commutation within a dc machine. The provision of a smoothing choke can be extremely important, and a recommendation should be made by the motor manufacturer depending upon the supply converter used.

Besides the thermal and commutation impacts, the ripple current also results in pulsating torque, which can cause resonance in the drive train. Laminating the stator field poles not only improves the thermal characteristic of the motor but also its dynamic behaviour by decreasing the motor time constant.

2.3.11.2 Effects upon ac machines

It is often stated that standard ac motors can be used without problem on modern PWM inverters and while such claims may be largely justified, switching converters do have an impact and limitations do exist. The NEMA MG1:1987, Part 17A gives guidance on the operation of constant-speed cage induction motors on 'a sinusoidal bus with harmonic content' and general purpose motors used with variable-voltage or variable-frequency controls or both.

2.3.11.2.1 Machine rating: thermal effects

The operation of ac machines on a non-sinusoidal supply inevitably results in additional losses in the machine, which fall into three main categories:

- *Stator copper loss.* This is proportional to the square of the rms current, but additional losses in the winding conductors due to skin effect must also be considered.
- *Rotor 'copper' loss.* The rotor resistance is different for each harmonic current in the rotor. This is due to skin effect and is pronounced in deep bar rotors. The rotor 'copper' loss must be calculated independently for each harmonic and the increase caused by harmonic currents can be a significant component of the total losses, particularly with PWM inverters having higher harmonics for which slip and rotor resistance are high.
- *Iron loss.* This is increased by the harmonic components in the supply voltage. The increase in iron loss due to the main field is usually negligible, but there is a significant increase in loss due to end winding leakage and slew leakage fluxes at the harmonic frequencies.

The total increase in losses does result in increased temperatures within the motor, but these cannot be readily represented by a simple de-rating factor since the harmonic losses are not evenly distributed throughout the machine and the distribution will vary according to the design of the motor. This has special implications for machines operating in a hazardous atmosphere.

Many fixed-speed motors have shaft-mounted cooling fans and operation away from the rated speed of the machine results in reduced or increased cooling. This needs to be taken into account when specifying a motor for variable-speed duty.

2.3.11.2.2 Machine insulation

The fast-rising voltage created by a PWM drive can result in a transiently uneven voltage distribution through a winding. For supply voltages up to 500 V, the voltage imposed by a correctly designed inverter is well within the capability of a standard motor of reputable manufacture, but for higher supply voltages an improved winding insulation system is generally required to ensure that the intended working life of the motor is achieved.

There can also be short-duration voltage over-shoots because of reflection effects in the motor cable, which is a system effect caused by the combined behaviour of the drive, cable and motor. The length of the motor cable can increase the peak motor voltage, but in applications with cables of 10 m or less, no special considerations are generally required. Output inductors (chokes) or output filters are sometimes used with drives for reasons such as long-cable driving capability or radio frequency suppression. In such cases no further precautions are required because these devices also reduce the peak motor voltage and increase its rise-time.

The IEC 60034-17 gives a profile for the withstand capability of a minimum standard motor, in the form of a graph of peak terminal voltage against voltage rise-time. The standard is based on research on the behaviour of motors constructed with the minimum acceptable level of insulation within the IEC motor standard family. Tests show that standard PWM drives with cable lengths of 20 m or more produce voltages outside the IEC 60034-17 profile. However most motor manufacturers produce, as a standard, machines with a capability substantially exceeding the requirements of IEC 60034-17.

2.3.11.2.3 Bearing currents

The sum of the three stator currents in an ac motor is ideally zero and there is no further path of current flow outside the motor, but in practice there are conditions which result in currents flowing through the bearings. These conditions include:

- *Magnetic asymmetry.* An asymmetric flux distribution within an electrical machine can result in an induced voltage from one end of the rotor shaft to the other. If the bearing breakover voltage is exceeded, a current flows through both the bearings. In some large machines, it is a common practice to fit an insulated bearing, usually at the non-drive end, to stop such currents.
- *Supply asymmetry.* With PWM inverter supplies, it is impossible to achieve perfect balance between the phases instantaneously, when pulses of different widths are produced. The resulting neutral voltage is not zero with respect to earth, and its presence equates to that of a common mode voltage source. This is sometimes referred to as a *zero sequence voltage.* It is proportional in magnitude to the dc link voltage in the inverter (itself proportional to the supply voltage), and has a frequency equal to the switching frequency of the inverter.

The risk of bearing currents can be minimized by adopting a grounding strategy which keeps all system components grounded at the same potential. This needs to be achieved for all frequencies and high inductance paths must be avoided, for instance keeping cable runs as short as possible. In addition, a low-impedance path should be defined for the common-mode currents to return to the inverter. As the common-mode current flows through the three phase conductors in the supply cable, the best return path is through a shield around that cable. This could be in the form of a screen. Such measures are well defined by most reputable manufacturers in their EMC guidance.

2.3.12 The commercial drive

The block schematic shown in Figure 2.3.30 illustrates the normal arrangement of an inner torque or current loop with an outer speed loop. Alternatively, where torque control is required in, for example, a tensioning application, then the outer speed loop can be removed, or made subservient to a torque loop with an external torque demand.

However, the modern industrial drive comprises much more than a speed and torque controller. Recent reviews of industrial drive specifications and marketing literature from a broad range of suppliers confirm that the ability to turn the shaft of a motor could be considered a very small part of the feature set of a modern drive. Figure 2.3.31 is an illustration of the additional functionality frequently found; this generalization changes from supplier to supplier and by the sector of the drive market being considered.

It is rare that an industrial drive stands alone in an application. In the majority of cases, drives are part of a system and it is necessary for the parts of the system to communicate with one another, transmitting commands and data. This communication can be in many forms from traditional analog signals through to wireless communication systems. The drives industry has been working to produce lower cost, higher performance drives, with good flexible and dynamic interfaces to other industrial products such as PLCs and HMIs. Other suppliers have taken a more holistic view of the needs of their customers, moving from a component supply situation to a solution provider. The ability to interface, efficiently and with appropriate dynamics, with other areas of a machine or factory automation system is of increasing importance in the design of industrial drives. As

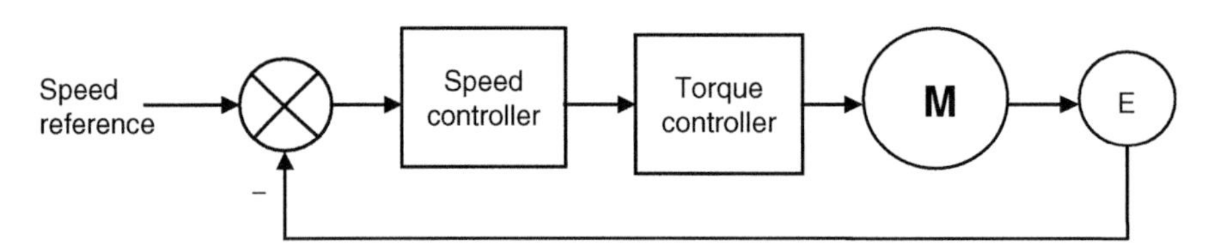

Fig. 2.3.30 Basic schematic diagram of a variable speed drive.

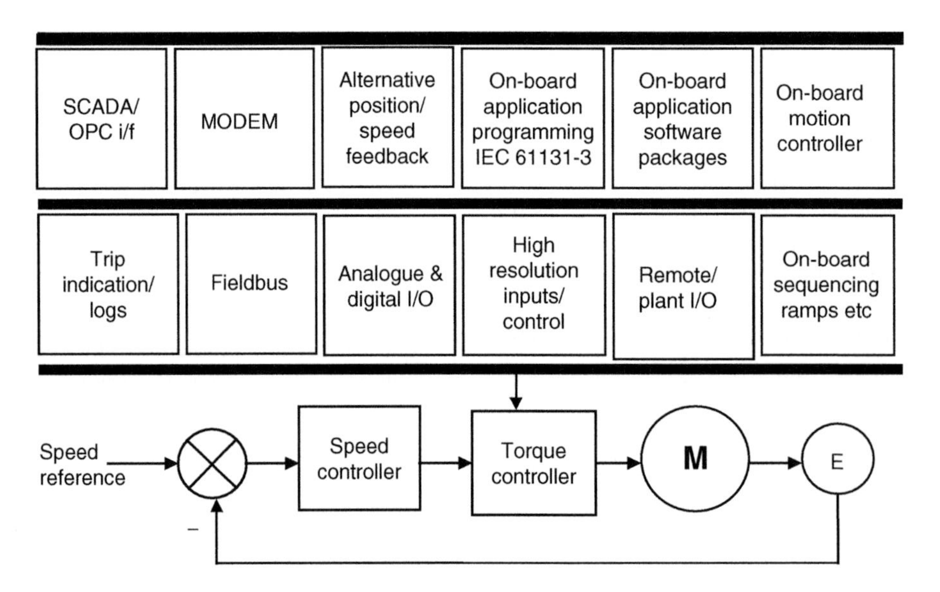

Fig. 2.3.31 Elements of a modern drive.

Table 2.3.4 National and international standards relating to motors and drives

IEC	Subject of standard	UL/NEMA
60034-1	Rotating electrical machines – Pt 1: rating and performance	
60034-2	Rotating electrical machines – Pt 2: methods for determining losses and efficiency of rotating electrical machinery from tests	
60034-5	Rotating electrical machines – Pt 5: classification of degrees of protection provided by enclosures of electrical machines (IP code)	
60035-6	Rotating electrical machines – Pt 6: methods of cooling (IC code)	
60035-7	Rotating electrical machines – Pt 7: classification of types of construction, mounting arrangements and terminal box position (IM code)	
60034-8	Rotating electrical machines – Pt 8: terminal markings and direction of rotation	
60034-9	Rotating electrical machines – Pt 9: noise limits	
60034-11	Rotating electrical machines – Pt 11: thermal protection	
60034-12	Rotating electrical machines – Pt 12: starting performance of single-speed three-phase cage induction motors	
60034-14	Rotating electrical machines – Pt 14: mechanical vibration of certain machines with shaft heights 56 mm and higher	
60034-15	Rotating electrical machines – Pt 15: impulse voltage withstand levels of rotating ac machines with form-wound stator coils	
60034-17	Rotating electrical machines – Pt 17: cage induction motors when fed from converters	
60034-18	Rotating electrical machines – Pt 18: functional evaluation of insulation systems	
60034-19	Rotating electrical machines – Pt 19: specific test methods for dc machines on conventional and rectifier-fed supplies	
60034-20-1	Rotating electrical machines – Pt 20-1: control motors – stepping motors	
60034-25	Rotating electrical machines – Pt 25: guide for the design and performance of cage induction motors specifically designed for converter supply	
60146-1	Semiconductor converters – general requirements and line commutated converters	
60146-2	Semiconductor converters – Pt 2: self-commutating semiconductor converters including direct dc converters	
60146-6	Semiconductor converters – Pt 6: application guide for the protection of semiconductor converters against overcurrent by fuses	
61800-1	Adjustable speed electrical power drive systems – Pt 1: rating specifications for low voltage adjustable speed dc power drive systems	
61800-2	Adjustable speed electrical power drive systems – Pt 2: rating specifications for low voltage adjustable speed ac power drive systems	
61800-3	Adjustable speed electrical power drive systems – Pt 3: EMC requirements and specific test methods	
61800-4	Adjustable speed electrical power drive systems – Pt 4: rating specifications for ac power drive systems above 1000 V ac and not exceeding 35 kV	
61800-5	Adjustable speed electrical power drive systems – Pt 5: safety requirements	
61800-6	Adjustable speed electrical power drive systems – Pt 6: guide for determination of types of load duty and corresponding current ratings	
61800-7	Adjustable speed electrical power drive systems – Pt 7: generic interface and use of profiles for power drive systems	
	Standard for power conversion equipment	UL 508C
	Motors and generators	NEMA MG 1

machine and factory control moves towards more distributed control structures, the drive will grow in importance as the hub a system controller. To fulfil this role, more flexibility and configurability in drives will result.

2.3.13 Standards

Key IEC and North American standards relating to electrical machines and drives are given in Table 2.3.4.

References

Drury, W. (ed), *The Control Techniques Drives and Controls Handbook*, The Institution of Electrical Engineers, 2001, ISBN 0 85296 793 4. (A handbook detailing all important practical aspects of industrial variable speed drives and their successful application)

Chalmers, B.J. Electric Motor Handbook, Butterworths, 1988, ISBN 0-408-00707-9. (A practical reference book covering many aspects of characteristics, specification, design, selection, commissioning and maintenance)

Vas, P. Sensorless Vector and Direct Torque Control, Oxford University Press, 1998, ISBN 0198564651. (General background to the theory of vector control of motors)

Chapter 2.4

2.4

Variable speed drives

Barnes

2.4.1 The need for variable speed drives

There are many and diverse reasons for using variable speed drives. Some applications, such as paper making machines, cannot run without them while others, such as centrifugal pumps, can benefit from energy savings. In general, variable speed drives are used to:

- match the speed of a drive to the process requirements,
- match the torque of a drive to the process requirements,
- save energy and improve efficiency.

The needs for speed and torque control are usually fairly obvious. Modern electrical VSDs can be used to accurately maintain the speed of a driven machine to within ±0.1%, independent of load, compared to the speed regulation possible with a conventional fixed speed squirrel cage induction motor (SCIM), where the speed can vary by as much as 3% from no load to full load.

The benefits of energy savings are not always fully appreciated by many users. These savings are particularly apparent with centrifugal pumps and fans, where load torque increases as the square of the speed and power consumption as the cube of the speed. Substantial cost savings can be achieved in some applications.

An everyday example, which illustrates the benefits of variable speed control, is the motorcar. It has become such an integral part of our lives that we seldom think about the technology that it represents or that it is simply a variable speed platform. It is used here to illustrate how variable speed drives are used to improve the speed, torque and energy performance of a machine.

It is intuitively obvious that the speed of a motorcar must continuously be controlled by the driver (the operator) to match the traffic conditions on the road (the process). In a city, it is necessary to obey speed limits, avoid collisions and to start, accelerate, decelerate and stop when required. On the open road, the main objective is to get to a destination safely in the shortest time without exceeding the speed limit. The two main controls that are used to control the speed are the accelerator, which controls the driving torque, and the brake, which adjusts the load torque. A motorcar could not be safely operated in city traffic or on the open road without these two controls. The driver must continuously adjust the fuel input to the engine (the drive) to maintain a constant speed in spite of the changes in the load, such as an uphill, downhill or strong wind conditions. On other occasions he may have to use the brake to adjust the load and slow the vehicle down to standstill.

Another important issue for most drivers is the cost of fuel or the cost of energy consumption. The speed is controlled via the accelerator that controls the fuel input to the engine. By adjusting the accelerator position, the energy consumption is kept to a minimum and is matched to the speed and load conditions. Imagine the high fuel consumption of a vehicle using a fixed accelerator setting and controlling the speed by means of the brake position.

2.4.2 Fundamental principles

The following is a review of some of the fundamental principles associated with variable speed drive applications.

Practical Variable Speed Drives and Power Electronics; ISBN: 9780750658089

- *Forward direction:* This refers to motion in one particular direction, which is chosen by the user or designer as being the forward direction. The forward direction is designated as being positive (+ve). For example, the forward direction for a motorcar is intuitively obvious from the design of the vehicle. Conveyor belts and pumps also usually have a clearly identifiable forward direction.
- *Reverse direction:* This refers to motion in the opposite direction. The reverse direction is designated as being negative (−ve). For example, the reverse direction for a motorcar is occasionally used for special situations such as parking or un-parking the vehicle.
- *Force:* Motion is the result of applying one or more forces to an object. Motion takes place in the direction in which the resultant force is applied. So force is a combination of both magnitude and direction. A force can be +ve or −ve depending on the direction in which it is applied. A force is said to be +ve if it is applied in the forward direction and −ve if it is applied in the reverse direction. In SI units, force is measured in *newtons.*
- *Linear velocity (v) or speed (n):* Linear velocity is the measure of the linear distance that a moving object covers in a unit of time. It is the result of a linear force being applied to the object. In SI units, this is usually measured in *meters per second (m/s). Kilometers per hour (km/h)* is also a common unit of measurement. For motion in the forward direction, velocity is designated positive (+ve). For motion in the reverse direction, velocity is designated negative (−ve).
- *Angular velocity (ω) or rotational speed (n):* Although a force is directional and results in linear motion, many industrial applications are based on rotary motion. The rotational force associated with rotating equipment is known as torque. Angular velocity is the result of the application of torque and is the angular rotation that a moving object covers in a unit of time. In SI units, this is usually measured in radians per second (rad/s) or revolutions per second (rev/s). When working with rotating machines, these units are usually too small for practical use, so it is common to measure rotational speed in revolutions per minute (rev/min).
- *Torque:* This is the product of the tangential force *F*, at the circumference of the wheel, and the radius *r* to the center of the wheel. In SI units, torque is measured in newton-meters (Nm). A torque can be +ve or −ve depending on the direction in which it is applied. A torque is said to be +ve if it is applied in the forward direction of rotation and −ve if it is applied in the reverse direction of rotation.

Using the motorcar as an example, Figure 2.4.1 illustrates the relationship between direction, force, torque, linear speed and rotational speed. The petrol engine develops rotational torque and transfers this via the transmission and axles to the driving wheels, which convert torque (*T*) into a tangential force (*F*). No horizontal motion would take place unless a resultant force is exerted horizontally along the surface of the road to propel the vehicle in the forward direction. The higher the magnitude of this force, the faster the car accelerates. In this example, the motion is designated as being forward, so torque, speed, acceleration are all +ve.

- *Linear acceleration (a):* It is the rate of change of linear velocity, usually in m/s^2.

$$\text{Linear acceleration,} \quad a = \frac{dv}{dt} \ \text{m/s}^2$$

  - Linear acceleration is the increase in velocity in either direction.
  - Linear deceleration or braking is the decrease in velocity in either direction.

- *Rotational acceleration (a):* It is the rate of change of rotational velocity, usually in rad/s^2.

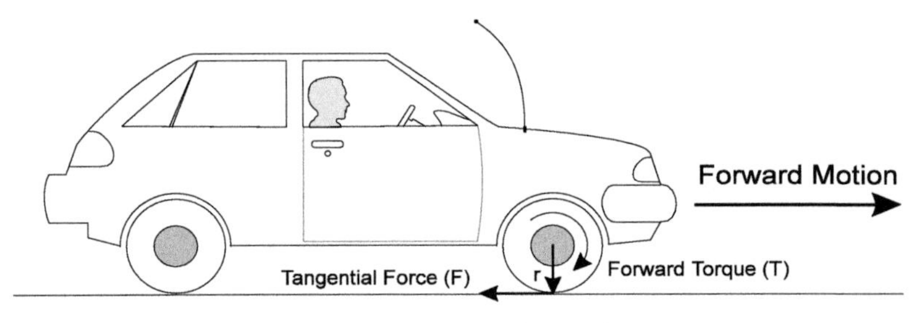

Torque (Nm) = Tangential Force (N) × Radius (m)

Fig. 2.4.1 The relationship between torque, force and radius.

Rotational acceleration, $a = \frac{d\omega}{dt}\ \text{rad/s}^2$

- Rotational acceleration is the increase in velocity in either direction.
- Rotational deceleration or Braking is the decrease in *velocity* in either direction.

In the example in Figure 2.4.2, a motorcar sets off from standstill and accelerates in the forward direction up to a velocity of 90 km/h (25 m/s) in a period of 10 s.

In variable speed drive applications, this acceleration time is often called the *ramp-up time.* After traveling at 90 km/h for a while, the brakes are applied and the car decelerates down to a velocity of 60 km/h (16.7 m/s) in 5 s. In variable speed drive applications, this deceleration time is often called the *ramp-down time*.

From the example outlined in Figure 2.4.3, the acceleration time (ramp-up time) to 20 km/h in the reverse direction is 5 s. The braking period (ramp-down time) back to standstill is 2 s.

There are some additional terms and formulae that are commonly used in association with variable speed drives and rotational motion.

- *Power*: The rate at which work is being done by a machine. In SI units, it is measured in watts. In practice, power is measured in kilowatts (kW) or megawatts (MW) because watts are such a small unit of measurement.

In rotating machines, power can be calculated as the product of torque and speed. Consequently, when a rotating machine such as a motorcar is at standstill, the output power is zero. This does not mean that input power is zero! Even at standstill with the engine running, there are a number of power losses that manifest themselves as heat energy.

Using SI units, power and torque are related by the following very useful formula, which is used extensively in VSD applications:

$$\text{Power (kW)} = \frac{\text{Torque (Nm)} \times \text{Speed (rev/min)}}{9550}$$

Alternatively,

$$\text{Torque (Nm)} = \frac{9550 \times \text{Power (kW)}}{\text{Speed (rev/min)}}$$

- *Energy*: The product of power and time and represents the rate at which work is done over a period of time. In SI units it is usually measured as kilowatt-hours (kWh). In the example of the motorcar, the fuel consumed over a period of time represents the energy consumed.

$$\text{Energy (kWh)} = \text{Power (kW)} \times \text{Time (h)}$$

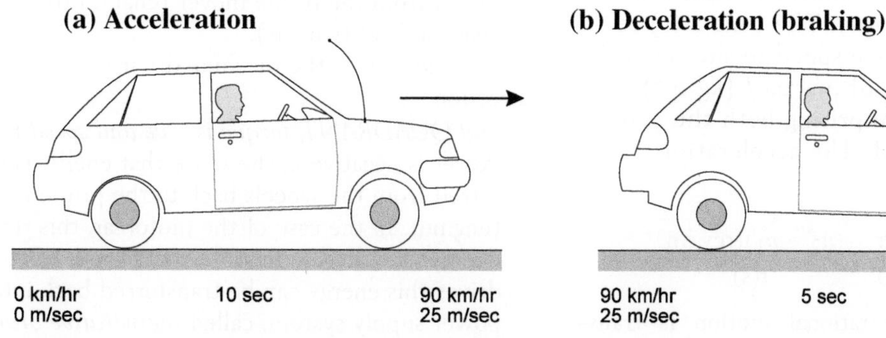

$$Acceleration = \frac{v_2 - v_1}{t} = \frac{25-0}{10} = +2.5$$

$$Deceleration = \frac{v_2 - v_1}{t} = \frac{16.7-25}{5} = -1.67\ \text{m/sec}^2$$

Fig. 2.4.2 Acceleration and deceleration (braking) in the forward direction.

REVERSE DIRECTION

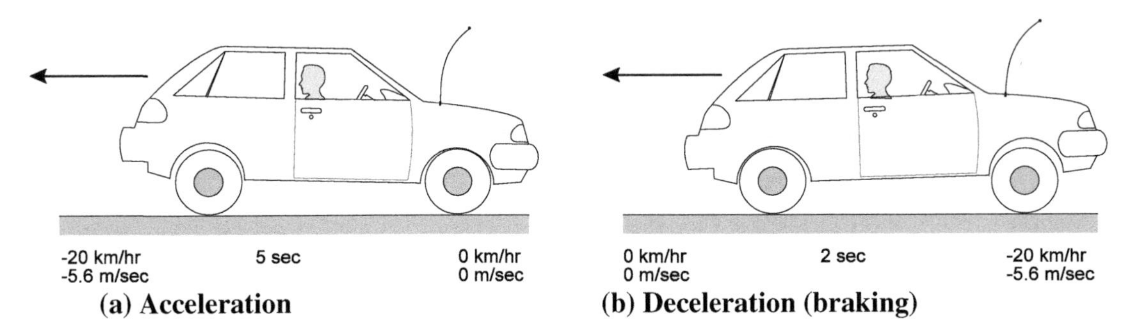

$$Acceleration = \frac{v_2 - v_1}{t} = \frac{-5.6 - 0}{5} = -1.12 \text{ m/sec}^2$$

$$Deceleration = \frac{v_2 - v_1}{t} = \frac{0 - (-5.6)}{2} = +2.8 \text{ m/sec}^2$$

Fig. 2.4.3 Acceleration and deceleration (braking) in the reverse direction.

- *Moment of inertia*: The property of a rotating object that resists change in rotational speed, either acceleration or deceleration. In SI units, moment of inertia is measured in kg m^2.

  This means that, to accelerate a rotating object from speed n_1 (rev/min) to speed n_2 (rev/min), an acceleration torque T_A (Nm) must be provided by the prime mover in addition to the mechanical load torque. The time t (s) required to change from one speed to another will depend on the moment of inertia J (kg m^2) of the rotating system, comprising both the drive and the mechanical load. The acceleration torque will be

$$T_A(\text{Nm}) = J(\text{kg m}^2) \times \frac{2\pi}{60} \times \frac{(n_2 - n_1)(\text{rev/m})}{t(\text{s})}$$

In applications where rotational motion is transformed into linear motion, for example on a crane or a conveyor, the rotational speed (n) can be converted to linear velocity (v) using the diameter (d) of the rotating drum as follows:

$$v(\text{m/s}) = \pi \, d \, n(\text{rev/s}) = \frac{\pi \, dn(\text{rev/min})}{60}$$

therefore

$$T_A(\text{Nm}) = J(\text{kg m}^2) \times \frac{2}{d} \times \frac{(v_2 - v_1)(\text{m/s})}{t(\text{s})}$$

From the above power, torque and energy formulae, there are four possible combinations of acceleration/braking in either the forward/reverse directions that can be applied to this type of linear motion. Therefore, the following conclusions can be drawn:

- *1st QUADRANT, torque is +ve and speed is +ve*: Power is positive in the sense that energy is transferred from the prime mover (engine) to the mechanical load (wheels).

This is the case of the machine driving in the forward direction.

- *2nd QUADRANT, torque is −ve and speed is +ve*: Power is negative in the sense that energy is transferred from the wheels back to the prime mover (engine). In the case of the motorcar, this returned energy is wasted as heat. In some types of electrical drives this energy can be transferred back into the power supply system, called *regenerative braking*.

This is the case of the machine braking in the forward direction.

- *3rd QUADRANT, torque is −ve and speed is −ve*: Power is positive in the sense that energy is transferred from the prime mover (engine) to the mechanical load (wheels).

This is the case of the machine driving in the reverse direction.

- *4th QUADRANT, If torque is +ve and speed is −ve*: Power is negative in the sense that energy is transferred from the wheels back to the prime mover (engine). As above, in some types of electrical drives this power can be transferred back into the power supply system, called regenerative braking.

This is the case of the machine braking in the reverse direction.

These four quadrants are summarized in Figure 2.4.4.

2.4.3 Torque–speed curves for variable speed drives

In most variable speed drive applications torque, power, and speed are the most important parameters. Curves, which plot torque against speed on a graph, are often used to illustrate the performance of the VSD. The speed variable is usually plotted along one axis and the torque variable along the other axis. Sometimes, power is also plotted along the same axis as the torque. Since energy consumption is directly proportional to power, energy depends on the product of torque and speed. For example, in a motorcar, depressing the accelerator produces more torque that provides acceleration and results in more speed, but more energy is required and more fuel is consumed.

Again using the motorcar as an example of a variable speed drive, torque–speed curves can be used to compare two alternative methods of speed control and to illustrate the differences in energy consumption between the two strategies:

- *Speed controlled by using drive control*: adjusting the torque of the prime mover. In practice, this is done by adjusting the fuel supplied to the engine, using the accelerator for control, without using the brake. This is analogous to using an electric variable speed drive to control the flow of water through a centrifugal pump.
- *Speed controlled by using load control*: adjusting the overall torque of the load. In practice, this could be done by keeping a fixed accelerator setting and using the brakes for speed control. This is analogous to controlling the water flow through a centrifugal pump by throttling the fluid upstream of the pump to increase the head.

Using the motorcar as an example, the two solid curves in Figure 2.4.5 represent the drive torque output of the engine over the speed range for two fuel control conditions:

- High fuel position – accelerator full down.
- Lower fuel position – accelerator partially down.

The two dashed curves in Figure 2.4.5 represents the load torque changes over the speed range for two

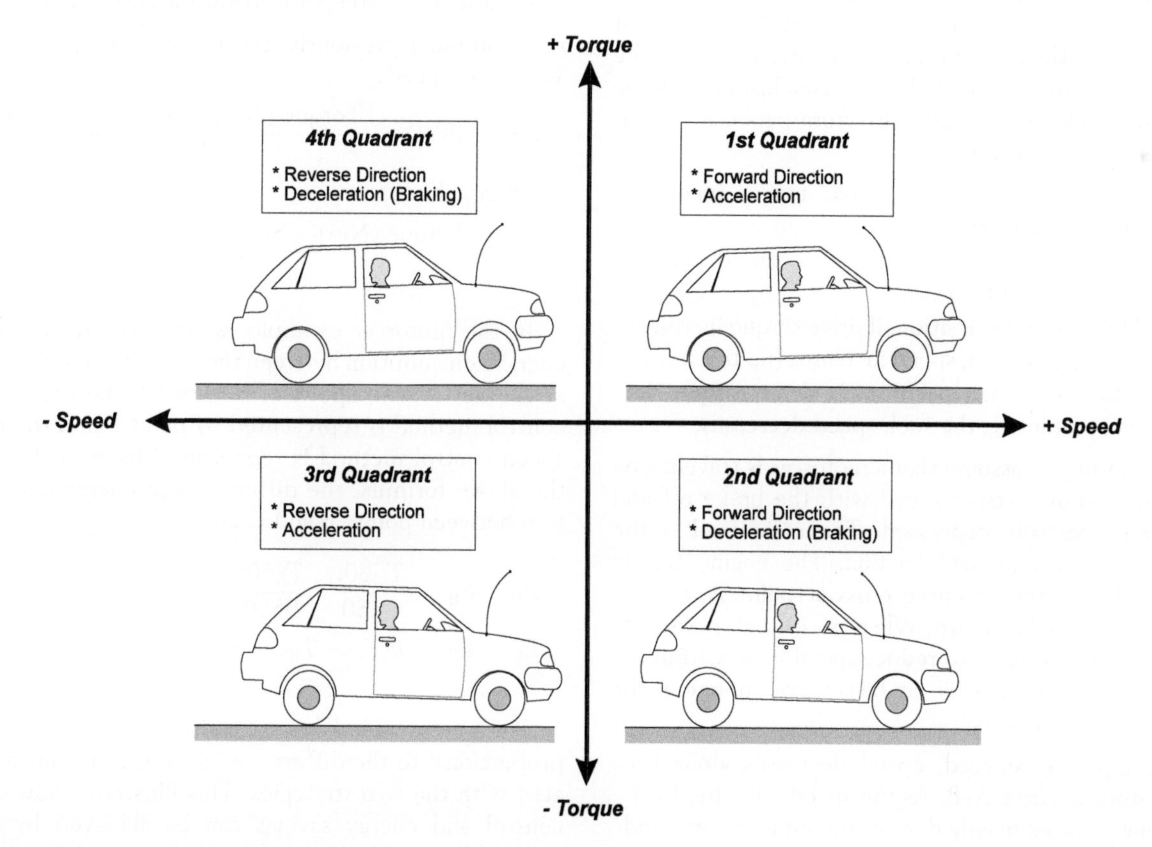

Fig. 2.4.4 The four quadrants of the torque-speed diagram for a motorcar.

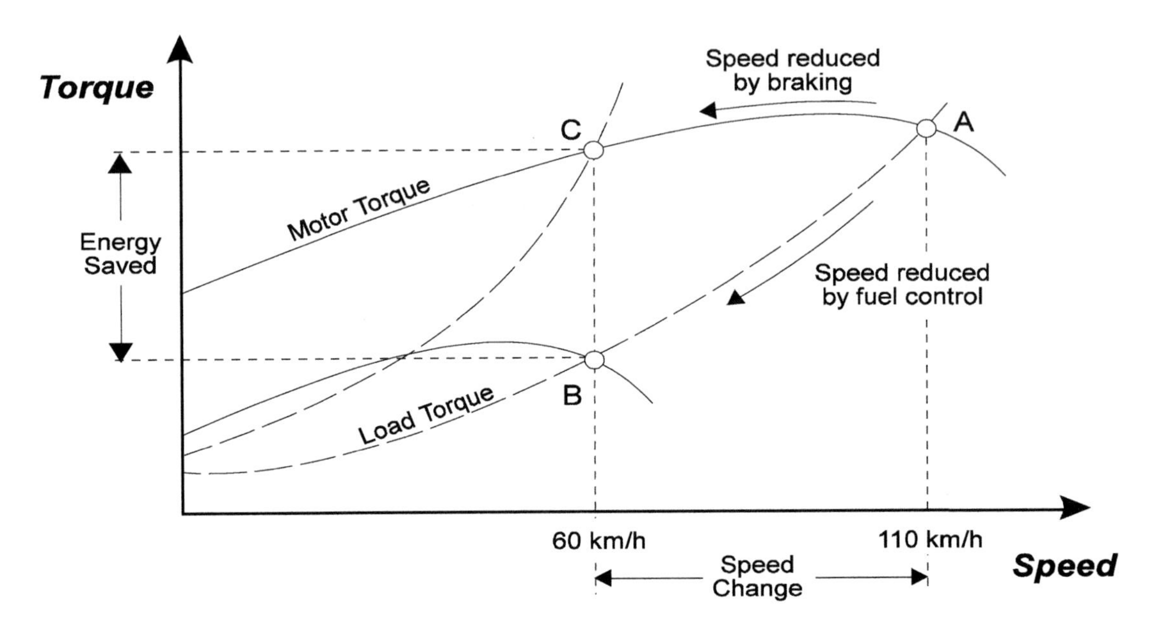

Fig. 2.4.5 Torque–speed curves for a motorcar.

mechanical load conditions. The mechanical load is mainly due to the wind resistance and road friction, with the restraining torque of the brakes added.

- Wind & friction plus brake ON – high load torque.
- Wind & friction plus brake OFF – low load torque.

As with any drive application, a *stable speed* is achieved when the drive torque is equal to the load torque, where the drive torque curve intersects with the load torque curve. The following conclusions can be drawn from Figure 2.4.5 and also from personal experience driving a motorcar:

- Fixed accelerator position, if load torque increases (uphill), speed drops.
- Fixed accelerator position, if load torque decreases (downhill), speed increases.
- Fixed load or brake position, if drive torque increases by increasing the fuel, speed increases (up to a limit).
- Fixed load or brake position, if drive torque decreases by reducing the fuel, speed decreases.

As an example, assume that a motorcar is traveling on an open road at a stable speed with the brake off and accelerator partially depressed. The main load is the wind resistance and road friction. The engine torque curve and load torque curve cross at point A, to give a stable speed of 110 km/h. When the car enters the city limits, the driver needs to reduce speed to be within the 60 km/h speed limit. This can be achieved in one of the two ways listed above:

- Fuel input is reduced, speed decreases along the load-torque curve A–B. As the speed falls, the load torque reduces mainly due to the reduction of wind resistance. A new stable speed of 60 km/h is reached at a new intersection of the load-torque curve and the engine-torque curve at point B.
- The brake is applied with a fixed fuel input setting, speed decreases along the drive-torque curve A–C due to the increase in the load torque. A new stable speed is reached when the drive-torque curve intersects with the steeper load-torque curve at 60 km/h.

As mentioned previously, the power is proportional to Torque × Speed:

$$\text{Power (kW)} = \frac{\text{Torque (Nm)} \times \text{Speed (rev/min)}}{9550}$$

$$\text{Energy (kWh)} = \frac{\text{Torque (Nm)} \times \text{Speed (rev/min)} \times \text{Time (h)}}{9550}$$

In the motorcar example, what is the difference in energy consumption between the two different strategies at the new stable speed of 60 km/h? The drive speed control method is represented by point B and the brake speed control method is represented by point C. From the above formula, the differences in energy consumption between points B and C are

$$E_C - E_B = \frac{T_C 60t}{9550} - \frac{T_B 60t}{9550}$$

$$E_C - E_B = k(T_C - T_B)$$

The energy saved by using drive control is directly proportional to the difference in the load torque associated with the two strategies. This illustrates how speed control and energy savings can be achieved by using a variable speed drive, such as a petrol engine, in a motorcar. The added advantages of a variable speed drive strategy are the reduced wear on the transmission, brakes and other components.

The same basic principles apply to industrial variable speed drives, where the control of the speed of the prime mover can be used to match the process conditions. The control can be achieved manually by an operator. With the introduction of automation, speed control can be achieved automatically, by using a feedback controller which can be used to maintain a process variable at a preset level. Again referring to the motorcar example, automatic speed control can be achieved using the 'auto-cruise' controller to maintain a constant speed on the open road.

Another very common application of VSDs for energy savings is the speed control of a centrifugal pump to control fluid flow. Flow control is necessary in many industrial applications to meet the changing demands of a process. In pumping applications, *Q–H* curves are commonly used instead of torque-speed curves for selecting suitable pumping characteristics and they have many similarities. Figure 2.4.6 shows a typical set of *Q–H* curves. Q represents the flow, usually measured in m^3/h and *H* represents the head, usually measured in meters. These show that when the pressure head increases on a centrifugal pump, the flow decreases and vice versa. In a similar way to the motorcar example above, fluid flow through the pump can be controlled either by controlling the speed of the motor driving the pump or alternatively by closing an upstream control valve (throttling). Throttling increases the effective head on the pump that, from the *Q–H* curve, reduces the flow.

From Figure 2.4.6, the reduction of flow from Q_2 to Q_1 can be achieved by using one of the following two alternative strategies:

- Drive speed control, flow decreases along the curve A–B and to a point on another *Q–H* curve. As the speed falls, the pressure/head reduces mainly due to the reduction of friction in the pipes. A new stable flow of Q_1 m^3/h is reached at point B and results in a head of H_2.
- Throttle control, an upstream valve is partially closed to restrict the flow. As the pressure/head is increased by the valve, the flow decreases along the curve A–C. The new stable flow of Q_1 m^3/h is reached at point C and results in a head of H_1.

From the well-known pump formula, the power consumed by the pump is

$$\text{Pump power (kW)} = k \times \text{Flow (m}^3\text{/h)} \times \text{Head (m)}$$
$$\text{Pump power (kW)} = k \times Q \times H$$
$$\text{Absorbed energy (kWh)} = k \times Q \times H \times t$$
$$E_C - E_B = (kQ_1H_1t) - (kQ_1H_2t)$$
$$E_C - E_B = kQ_1\,(H_1 - H_2)t$$
$$E_C - E_B = K(H_1 - H_2)$$

With flow constant at Q_1, the energy saved by using drive speed control instead of throttle control is directly proportional to the difference in the head associated with the two strategies. The energy savings are therefore

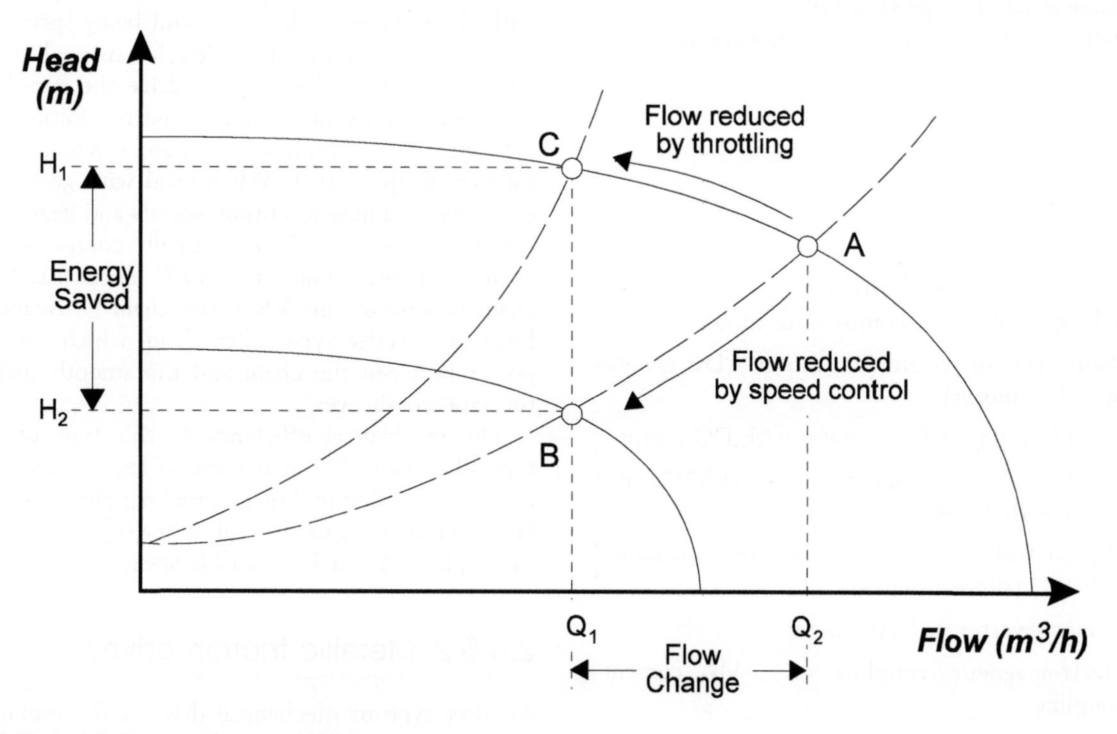

Fig. 2.4.6 Typical *Q–H* curves for a centrifugal pump.

a function of the difference in the head between the point B and point C. The energy savings on large pumps can be quite substantial and these can readily be calculated from the data for the pump used in the application.

There are other advantages in using variable speed control for pump applications:

- Smooth starting, smooth acceleration/deceleration to reduce mechanical wear and water hammer.
- No current surges in the power supply system.
- Energy savings are possible. These are most significant with centrifugal loads such as pumps and fans because power/energy consumption increases/decreases with the cube of the speed.
- Speed can be controlled to match the needs of the application. This means that speed, flow or pressure can be accurately controlled in response to changes in process demand.
- Automatic control of the process variable is possible, for example to maintain a constant flow, constant pressure, etc. The speed control device can be linked to a process control computer such as a PLC or dcS.

2.4.4 Types of variable speed drives

The most common types of variable speed drives used today are shown in Figure 2.4.7.

Variable speed drives can be classified into three main categories each with their own advantages and disadvantages:

- Mechanical variable speed drives
  - Belt and chain drives with adjustable diameter sheaves
  - Metallic friction drives
- Hydraulic variable speed drives
  - Hydrodynamic types
  - Hydrostatic types
- Electrical variable speed drives
  - Schrage motor (AC commutator motor)
  - Ward-Leonard system (AC motor – DC generator – DC motor)
  - Variable voltage DC converter with DC motor
  - Variable voltage, variable frequency (VVVF) converter with AC motor
  - Slip control with wound rotor induction motor (WRIM) (slipring motor)
  - Cycloconverter with AC motor
  - Electromagnetic coupling or 'eddy current' coupling
  - Positioning drives (servo and stepper motors)

2.4.5 Mechanical variable speed drive methods

Historically, electrical VSDs, even DC drives, were complex and expensive and were only used for the most important or difficult applications. So mechanical devices were developed for insertion between a fixed speed electric drive motor and the shaft of the driven machine.

Mechanical variable speed drives are still favored by many engineers (mainly mechanical engineers!) for some applications mainly because of simplicity and low cost.

As listed above, there are basically two types of mechanical construction.

2.4.5.1 Belt and chain drives with adjustable diameter sheaves

The basic concept behind adjustable sheave drives is very similar to the gear changing arrangement used on many modern bicycles. The speed is varied by adjusting the ratio of the diameter of the drive pulley to the driven pulley.

For industrial applications, an example of a continuously adjustable ratio between the drive shaft and the driven shaft is shown in Figure 2.4.8. One or both pulleys can have an adjustable diameter. As the diameter of one pulley increases, the other decreases thus maintaining a nearly constant belt length. Using a V-type drive belt, this can be done by adjusting the distance between the tapered sheaves at the drive end, with the sheaves at the other end being spring loaded. A hand-wheel can be provided for manual control or a servo-motor can be fitted to drive the speed control screw for remote or automatic control. Ratios between 2:1 and 6:1 are common, with some low power units capable of up to 16:1. When used with gear reducers, an extensive range of output speeds and gear ratios are possible. This type of drive usually comes as a totally enclosed modular unit with an AC motor fitted. On the chain version of this VSD, the chain is usually in the form of a wedge type roller chain, which can transfer power between the chain and the smooth surfaces of the tapered sheaves.

The mechanical efficiency of this type of VSD is typically about 90% at maximum load. They are often used for machine tool or material handling applications. However, they are increasingly being superseded by small single phase AC or DC variable speed drives.

2.4.5.2 Metallic friction drives

Another type of mechanical drive is the metallic friction VSD unit, which can transmit power through the

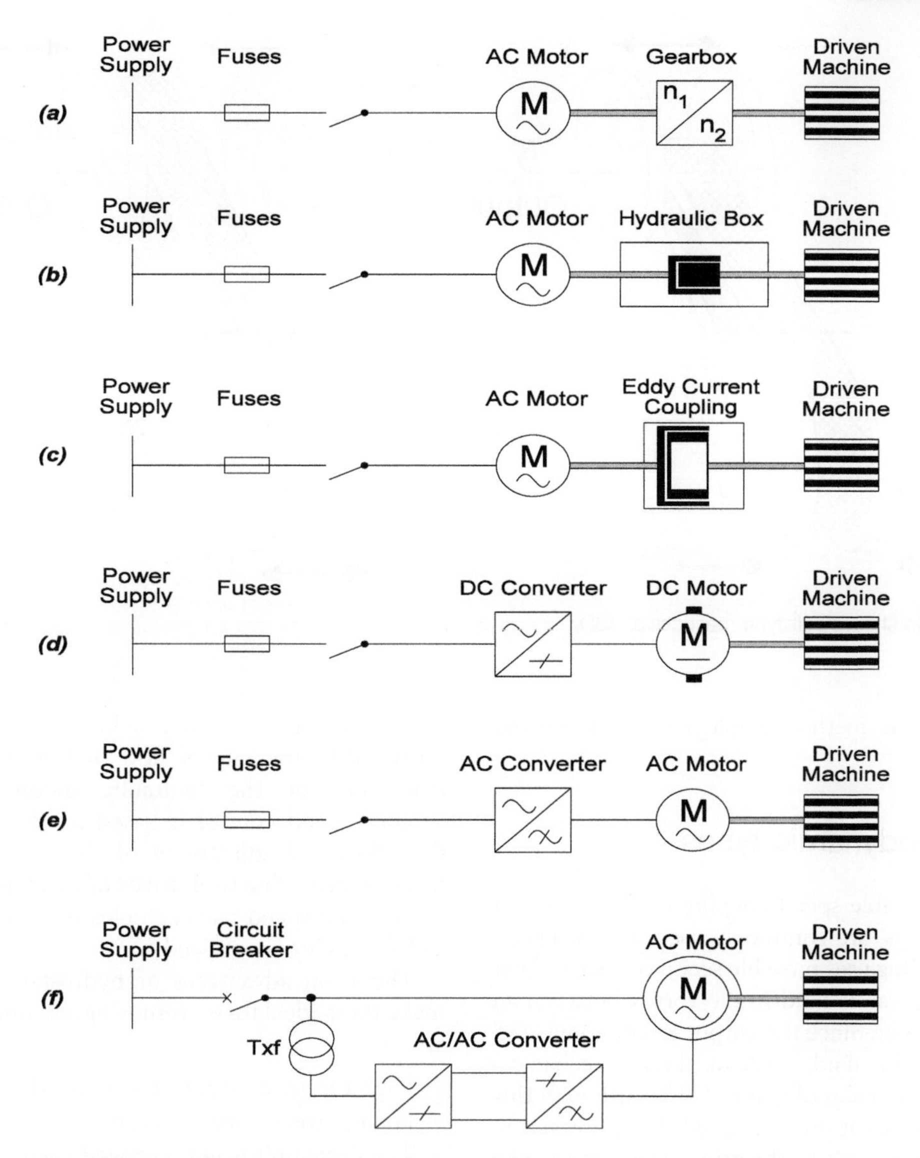

Fig. 2.4.7 Main types of variable speed drive for industrial applications. (a) Typical mechanical VSD with an AC motor as the prime mover. (b) Typical hydraulic VSD with an AC motor as the prime mover. (c) Typical electromagnetic coupling or eddy current coupling. (d) Typical electrical VSD with a DC motor and DC voltage converter. (e) Typical electrical VSD with an AC motor and AC frequency converter. (f) Typical slip energy recovery system or static Kramer system.

friction at the point of contact between two shaped or tapered wheels. Speed is adjusted by moving the line of contact relative to the rotation centers. Friction between the parts determines the transmission power and depends on the force at the contact point.

The most common type of friction VSD uses two rotating steel balls, where the speed is adjusted by tilting the axes of the balls. These can achieve quite high capacities of up to 100 kW and they have excellent speed repeatability. Speed ratios of 5:1 up to 25:1 are common.

To extend the life of the wearing parts, friction drives require a special lubricant that hardens under pressure. This reduces metal-to-metal contact, as the hardened lubricant is used to transmit the torque from one rotating part to the other.

2.4.6 Hydraulic variable speed drive methods

Hydraulic VSDs are often favored for conveyor drive applications because of the inherently *soft-start* capability of the hydraulic unit. They are also frequently used in all types of transportation and earthmoving equipment because of their inherently high starting torque. Both of the two common types work on the same basic principle where the prime mover, such as a fixed speed electric motor or a diesel/petrol engine, drives a hydraulic pump to transfer fluid to a hydraulic motor. The output speed can be adjusted by controlling the fluid flow rate or pressure. The two different types outlined below are

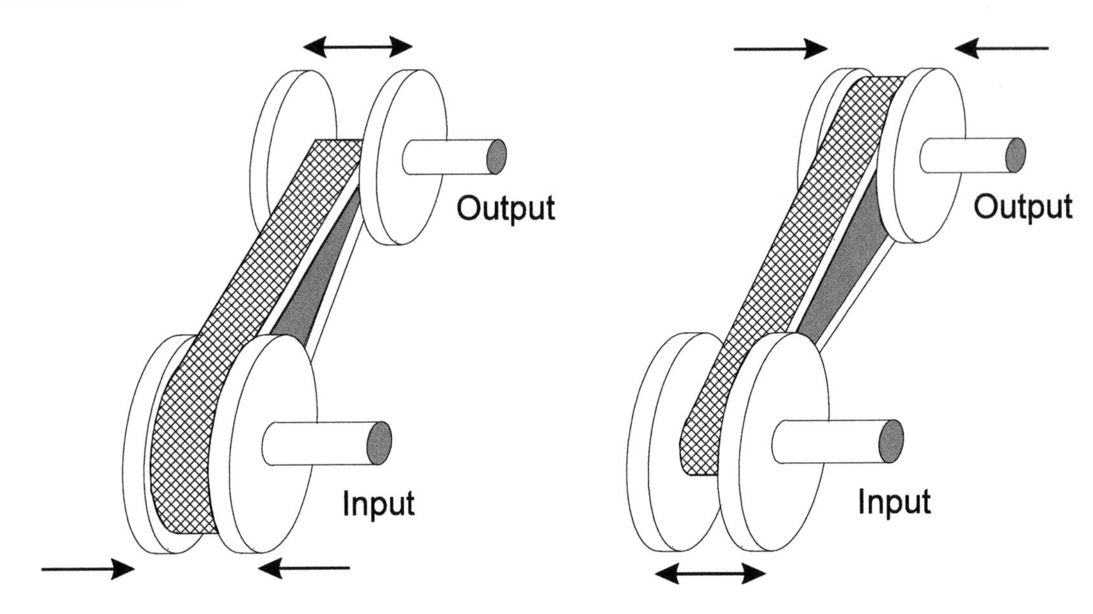

Fig. 2.4.8 Adjustable sheave belt-type mechanical VSD.

characterized by the method employed to achieve the speed control.

2.4.6.1 Hydrodynamic types

Hydrodynamic variable speed couplings, often referred to as fluid couplings, are commonly used on conveyors. This type of coupling uses movable scoop tubes to adjust the amount of hydraulic fluid in the vortex between an impeller and a runner. Since the output is only connected to the input by the fluid, without direct mechanical connection, there is a slip of about 2–4%. Although this slip reduces efficiency, it provides good shock protection or soft-start characteristic to the driven equipment. The torque converters in the automatic transmissions of motorcars are hydrodynamic fluid couplings.

The output speed can be controlled by the amount of oil being removed by the scoop tube, which can be controlled by manual or automatic control systems. Operating speed ranges of up to 8:1 are common. A constant speed pump provides oil to the rotating elements.

2.4.6.2 Hydrostatic type

This type of hydraulic VSD is most commonly used in mobile equipment such as transportation, earthmoving and mining machinery. A hydraulic pump is driven by the prime mover, usually at a fixed speed, and transfers the hydraulic fluid to a hydraulic motor. The hydraulic pump and motor are usually housed in the same casing that allows closed circuit circulation of the hydraulic fluid from the pump to the motor and back.

The speed of the hydraulic motor is directly proportional to the rate of flow of the fluid and the displacement of the hydraulic motor. Consequently, variable speed control is based on the control of both fluid flow and adjustment of the pump and/or motor displacement. Practical drives of this type are capable of a very wide speed range, steplessly adjustable from zero to full synchronous speed.

The main advantages of hydrostatics VSDs, which make them ideal for earthmoving and mining equipment, are:

- High torque available at low speed.
- High power-to-weight ratio.
- The drive unit is not damaged even if it stalls at full load.
- Hydrostatics VSDs are normally bi-directional.

Output speed can be varied smoothly from about 40 rev/min to 1450 rev/min up to a power rating of about 25 kW. Speed adjustment can be done manually from a hand-wheel or remotely using a servo-motor. The main disadvantage is the poor speed holding capability. Speed may drop by up to 35 rev/min between 0% and 100% load.

Hydrostatic VSDs fall into four categories, depending on the types of pumps and motors:

- *Fixed displacement pump – fixed displacement motor*: The displacement volume of both the pump and the motor is not adjustable. The output speed and power are controlled by adjusting a flow control valve located between the hydraulic pump and motor. This is the cheapest solution, but efficiency is low, particularly at low speeds. So these are applied only where small speed variations are required.

- *Variable displacement pump – fixed displacement motor*: The output speed is adjusted by controlling the pump displacement. Output torque is roughly constant relative to speed if pressure is constant. Thus power is proportional to speed. Typical applications include winches, hoists, printing machinery, machine tools and process machinery.
- *Fixed displacement pump – variable displacement motor*: The output speed is adjusted by controlling the motor displacement. Output torque is inversely proportional to speed, giving a relatively constant power characteristic. This type of characteristic is suitable for machinery such as rewinders.
- *Variable displacement pump – variable displacement motor*: The output speed is adjusted by controlling the displacement of the pump, motor or both. Output torque and power are both controllable across the entire speed range in both directions.

2.4.7 Electromagnetic or 'eddy current' coupling

The electromagnetic or 'eddy current' coupling is one of the oldest and simplest of the electrically controlled variable speed drives and has been used in industrial applications for over 50 years. In a similar arrangement to hydraulic couplings, eddy current couplings are usually mounted directly onto the flange of a standard SCIM between the motor and the driven load as shown in Figure 2.4.9.

Using the principles of electromagnetic induction, torque is transferred from a rotating drum, mounted onto the shaft of a fixed speed electric motor, across the air gap to an output drum and shaft, which is coupled to the driven load. The speed of the output shaft depends on the slip between the input and output drums, which is controlled by the magnetic field strength. The field winding is supplied with DC from a separate variable voltage source, which was traditionally a variac but is now usually a small single-phase thyristor converter.

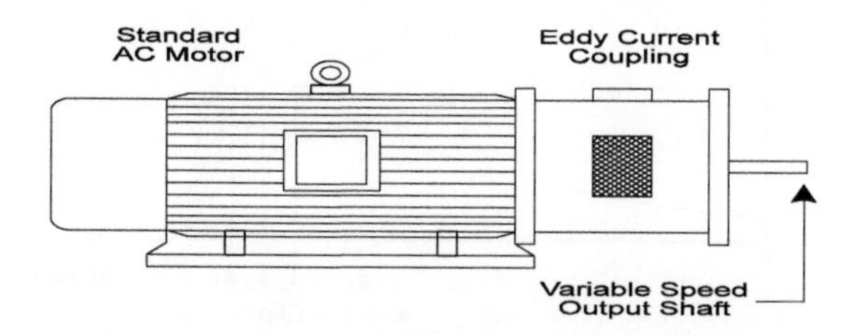

Fig. 2.4.9 Eddy current coupling mounted onto SCIM.

There are several slightly different configurations using the electromagnetic induction principle, but the most common two constructions are shown in Figure 2.4.10. It comprises a cylindrical input drum and a cylindrical output drum with a small air gap between them. The output drum, which is connected to the output shaft, is capable of rotating freely relative to the input drum. A primary electromagnetic field is provided by a set of field coils that are connected to an external supply.

In configuration Figure 2.4.10(a), the field coils are mounted directly onto the rotating output drum, which then requires sliprings to transfer the excitation current to the field coils. On larger couplings, this arrangement can be difficult to implement and also sliprings create additional maintenance problems. In configuration Figure 2.4.10(b), the field coils are supported on the frame with the output drum closely surrounding it. This configuration avoids the use of sliprings.

The operating principle is based on the following:

- When a conducting material moves through the flux lines of a magnetic field, eddy currents are induced in the surface of the material, which flow in circular paths.
- The magnitude of the eddy currents is determined by the primary flux density and the rate at which the rotating part cuts these primary flux lines, i.e. the magnitude of the eddy currents depends on the magnetic field strength and the relative speed between the input and output shafts.
- These eddy currents collectively establish their own magnetic field which interacts with the primary magnetic flux in such a way as to resist the relative motion between them, thus providing a magnetic coupling between input and output drums.
- Consequently, torque can be transferred from a fixed speed prime mover to the output shaft, with some slip between them.
- The output torque and the slip are dependent on the strength of the electromagnetic field, which can be controlled from an external voltage source.

In the practical implementation, the input and output drums are made from a ferromagnetic material, such as iron, with a small air gap between them to minimize the leakage flux. The field coils, usually made of insulated copper windings, are mounted on the static part of the frame and are connected to a DC voltage source via a terminal box on the frame. Variable speed is obtained by controlling the field excitation current, by adjusting the voltage output of a small power electronic converter and control circuit. Speed adjustments can be made either manually from a potentiometer or remotely via

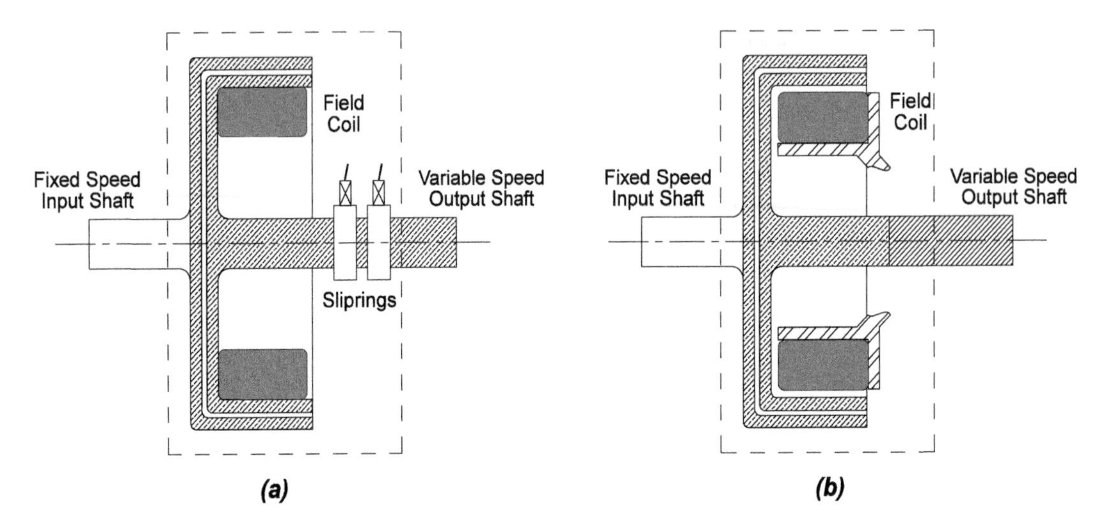

Fig. 2.4.10 Cross section of the eddy current couplings. (a) Field coils mounted onto the output drum. (b) Field coils mounted onto the fixed frame.

a 4–20 mA control loop. An important feature of the eddy current coupling is the very low power rating of the field controller, which is typically 2% of the rated drive power.

When this type of drive is started by switching on the AC motor, the motor quickly accelerates to its full speed. With no voltage applied to the field coils, there are no lines of flux and no coupling, so the output shaft will initially be stationary. When an excitation current is applied to the field coils, the resulting lines of flux cut the rotating input drum at the maximum rate and produce the maximum eddy current effect for that field strength.

The interaction between the primary flux and the secondary field produced by the eddy currents establishes an output torque, which accelerates the output shaft and the driven load. As the output drum accelerates, the relative speed between the two drums decreases and reduces the rate at which the lines of flux cut the rotating drum. The magnitude of the eddy currents and secondary magnetic field falls and, consequently, reduces the torque between them.

With a constant field excitation current, the output shaft will accelerate until the output torque comes into equilibrium with that of the driven machine. The output speed can be increased (reduce the slip) by increasing the field excitation current to increase the primary magnetic field strength. The output speed can be reduced (increase slip) by reducing the field excitation current.

To transfer torque through the interaction of two magnetic fields, eddy currents must exist to set up the secondary magnetic field. Consequently, there must always be a difference in speed, called the *slip*, between the input drum and the output drum. This behavior is very similar to that of the AC SCIM and indeed the same principles apply. The eddy current coupling produces a torque–speed curve quite similar to a SCIM as shown in Figure 2.4.11.

Theoretically, the eddy current coupling should be able to provide a full range of output speeds and torques from zero up to just below the rated speed and torque of the motor, allowing of course for slip. In practice, this is limited by the amount of torque that can be transferred continuously through the coupling without generating excessive heat.

When stability is reached between the motor and the driven load connected by an *eddy current coupling*, the output torque on the shaft is equal to the input torque from the AC motor. However, the speeds of the input and output shafts will be different due to the slip. Since power is a product of torque and speed, the difference between the input and output power, the losses, appears as heat in the coupling. These losses are dissipated through cooling fins on the rotating drums.

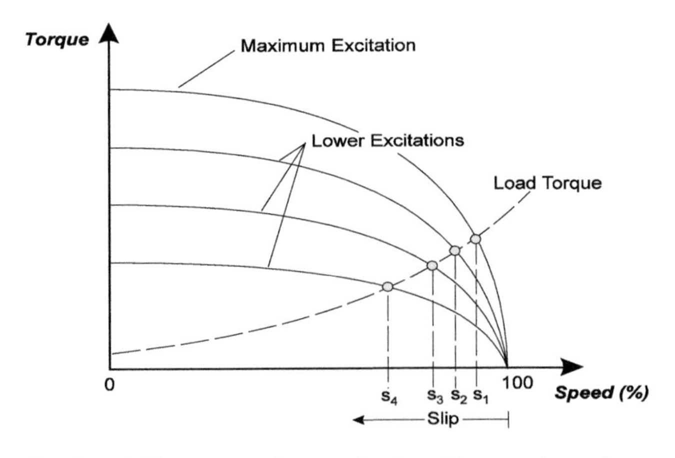

Fig. 2.4.11 Torque–speed curves for the eddy current coupling.

These losses may be calculated as follows:

$$\text{Losses} = (P_I - P_O)\ \text{kW}$$

$$\text{Losses} = \frac{T(n_1 - n_2)}{9550}\ \text{kW}$$

The worst case occurs at starting, with the full rated torque of the motor applied to the driven load at zero output speed, the losses in the coupling are the full rated power of the motor. Because of the difficulty of dissipating this amount of energy, in practice it is necessary to limit the continuous torque at low speeds.

Alternatively, some additional cooling may be necessary for the coupling, but this results in additional capital costs and low energy efficiency. In these cases, other types of VSDs may be more suitable. Consequently, an eddy current coupling is most suited to those types of driven load, which have a low torque at low speed, such as centrifugal pumps and fans. The practical loadability of the eddy current coupling is shown in Figure 2.4.12.

A major drawback of the eddy current coupling is its poor dynamic response. Its ability to respond to step changes in the load or the speed setpoint depends on the time constants associated with the highly inductive field coil, the eddy currents in the ferro-magnetic drums and the type of control system used. The field coil time constant is the most significant factor and there is very little that can be done to improve it, except possibly to use a larger coupling. Closed loop speed control with tachometer feedback can also be used to improve its performance. But there are many applications where the dynamic response or output speed accuracy are not important issues and the eddy current coupling has been proven to be a cost-effective and reliable solution for these applications.

2.4.8 Electrical variable speed drive methods

In contrast to the mechanical and hydraulic variable speed control methods, electrical variable speed drives are those in which the speed of the electric motor itself, rather than an intermediary device, is controlled. Variable speed drives that control the speed of DC motors are loosely called *DC variable speed drives* or simply *DC drives* and those that control the speed of AC motors are called *AC variable speed drives* or simply *AC drives.* Almost all electrical VSDs are designed for operation from the standard three-phase AC power supply system.

Historically, two of the best-known electrical VSDs were the schrage motor and the Ward-Leonard system. Although these were both designed for operation from a three-phase AC power supply system, the former is an AC commutator motor while the latter uses a DC generator and motor to effect speed control.

2.4.8.1 AC commutator motor – schrage motor

The schrage motor is an AC commutator motor having its primary winding on the rotor. The speed was changed by controlling the position of the movable brushes by means

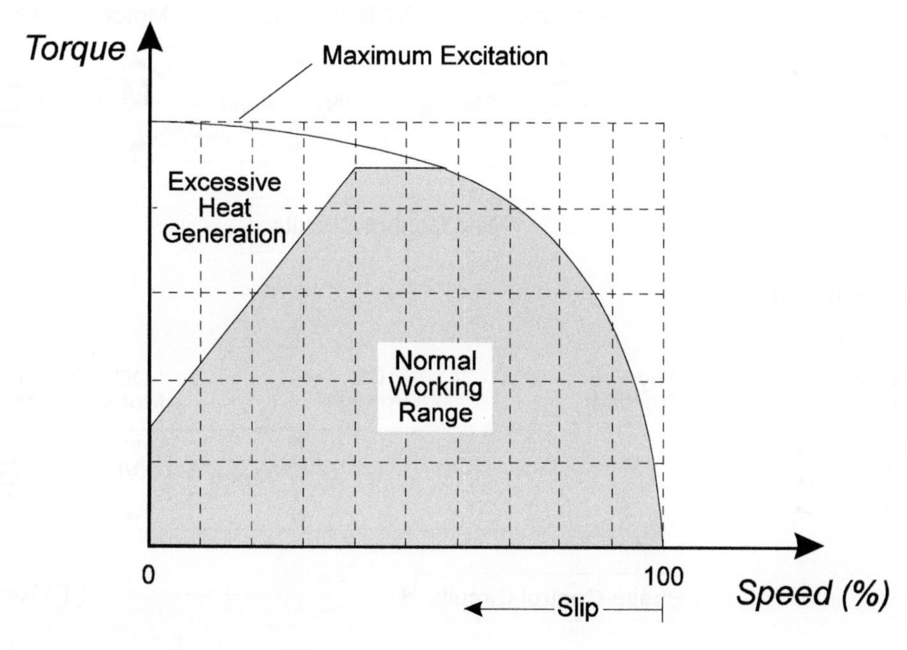

Fig. 2.4.12 Loadability of the eddy current coupling.

of a hand-wheel or a servo-motor. Although it was very popular in its time, this type of motor is now too expensive to manufacture and maintain and is now seldom used.

2.4.8.2 Ward-Leonard system

The Ward-Leonard system comprises a fixed speed three-phase AC induction motor driving a separately excited DC generator that, in turn, feeds a variable voltage to a shunt wound DC motor. So this is essentially a DC variable speed drive.

DC drives have been used for variable speed applications for many decades and historically were the first choice for speed control applications requiring accurate speed control, controllable torque, reliability and simplicity. The basic principle of a DC variable speed drive is that the speed of a separately excited DC motor is directly proportional to the voltage applied to the armature of the DC motor. The main changes over the years have been concerned with the different methods of generating the variable DC voltage from the three-phase AC supply.

In the case of the Ward-Leonard system, the output voltage of the DC generator, which is adjusted by controlling the field voltage, is used to control the speed of the DC motor as shown in Figure 2.4.13. This type of variable speed drive had good speed and torque characteristics and could achieve a speed range of 25:1. It was commonly used for winder drives where torque control was important. It is no longer commonly used because of the high cost of the three separate rotating machines. In addition, the system requires considerable maintenance to keep the brushes and commutators of the two DC machines in good condition.

In modern DC drives, the motor-generator set has been replaced by a thyristor converter. The output DC voltage is controlled by adjusting the *firing angle* of the thyristors connected in a bridge configuration connected directly to the AC power supply.

2.4.8.3 Electrical variable speed drives for DC motors (DC drives)

Since the 1970s, the controlled DC voltage required for DC motor speed control has been more easily produced from the three-phase AC supply using a *static* power electronic AC/DC converter, or sometimes called a *controlled rectifier.* Because of its low cost and low maintenance, this type of system has completely superseded the Ward-Leonard system. There are several different configurations of the AC/DC converter, which may contain a full-wave 12-pulse bridge, a full-wave 6-pulse bridge or a half-wave 3-pulse bridge. On larger DC drive systems, 12-pulse bridges are often used.

The most common type of AC/DC converter, which meets the steady state and dynamic performance requirements of most VSD applications, comprises a 6-pulse thyristor bridge, electronic control circuit and a DC motor as shown Figure 2.4.14. The 6-pulse bridge produces less distortion on the DC side than the 3-pulse

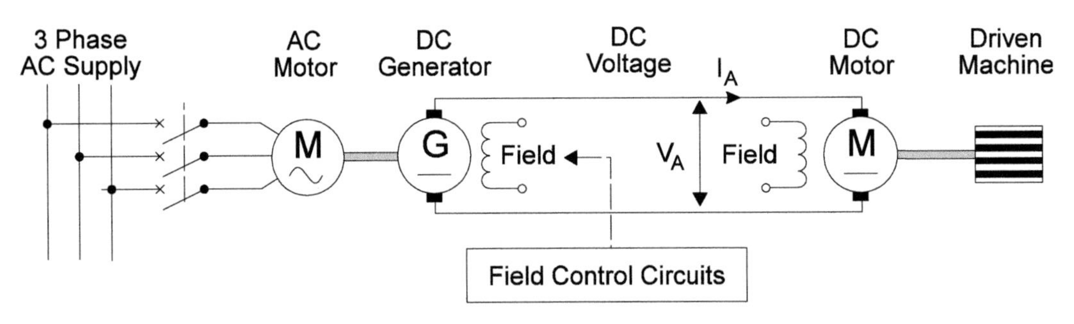

Fig. 2.4.13 The Ward-Leonard system.

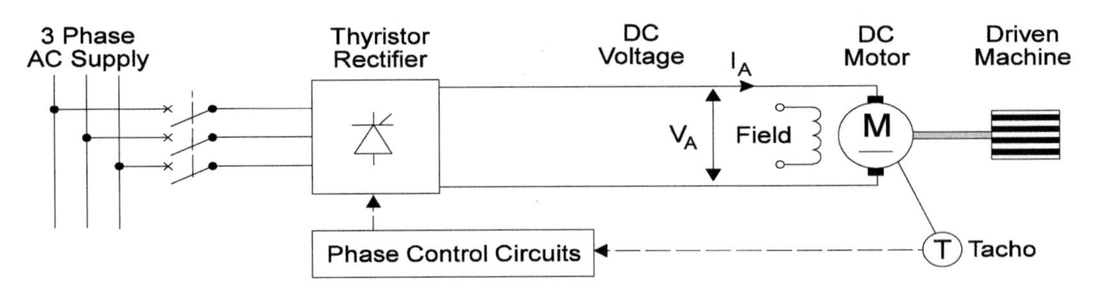

Fig. 2.4.14 Basic construction of a 6-pulse DC variable speed drive.

bridge and also results in lower losses in the DC motor. On larger DC drive systems, 12-pulse bridges are often used to reduce the harmonics in the AC power supply system.

The efficiency of an AC/DC converter is high, usually in excess of 98%. The overall efficiency of the DC drive, including the motor, is lower and is typically about 90% at full load depending on the size of the motor. The design and performance of power electronic converters is described in detail in Chapter 3.2.

AC/DC converters of this type are relatively simple and robust and can be built for VSDs of up to several megawatts with good control and performance characteristics. Since the DC motor is relatively complex and expensive, the main disadvantage of this type of VSD in comparison to an AC VSD, is the reliability of the DC motor. Although the maintenance requirements of a DC motor are inherently higher than an AC induction motor, provided that the correct brush grade is used for the speed and current rating, the life of the commutator and brushgear can be quite long and maintenance minimal.

The fundamental principles of a DC variable speed drive (Figure 2.4.15), with a shunt wound DC motor, are relatively easy to understand and are covered by a few simple equations as follows:

- The armature voltage V_A is the sum of the internal armature EMF V_E and the volt drop due to the armature current I_A flow through the armature resistance R_A:

Armature voltage, $V_A = V_E + I_A R_A$

- The DC motor speed is directly proportional to the armature back EMF V_E and indirectly proportional to the field flux Φ, which in turn depends on the field excitation current I_E. Thus, the rotational speed of the motor can be controlled by adjusting either the armature voltage, which controls V_E, or the field current, which controls the Φ:

Motor speed, $n \propto \dfrac{V_E}{\Phi}$

- The output torque T of the motor is proportional to the product of the armature current and the field flux:

Output torque, $T \propto I_A \Phi$

- The direction of the torque and direction of rotation of the DC motor can be reversed either by changing the polarity of Φ, called field reversal, or by changing the polarity of I_A, called armature current reversal. These can be achieved by reversing the supply voltage connections to the field or to the armature.
- The output power of the motor is proportional to the product of torque and speed:

Output power, $P \propto Tn$

From these equations, the following can be deduced about a DC motor drive:

- The speed of a DC motor can be controlled by adjusting either the armature voltage or the field flux or both. Usually the field flux is kept constant, so the motor speed is increased by increasing the armature voltage.
- When the armature voltage V_A has reached the maximum output of the converter, additional increases in speed can be achieved by reducing the field flux. This is known as the *field weakening* range. In the field

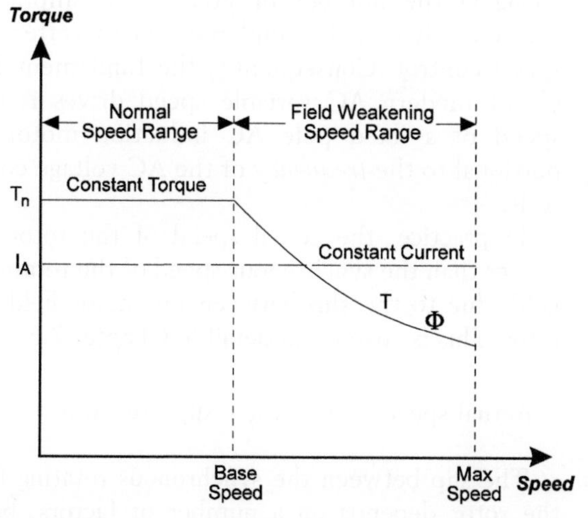

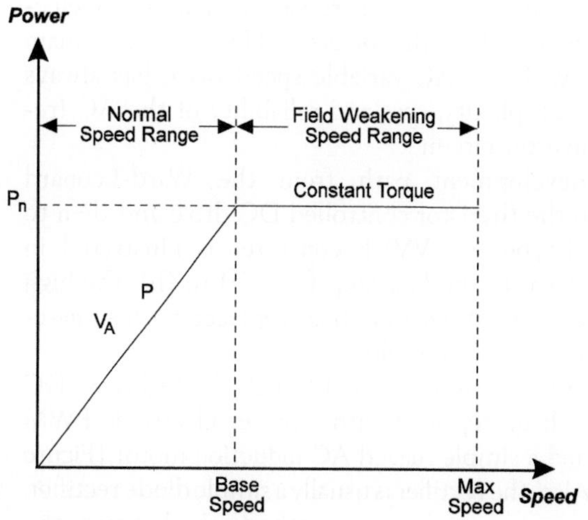

Fig. 2.4.15 Torque and power of a DC drive over the speed range.

weakening range, the speed range is usually limited to about 3:1, mainly to ensure stability and continued good commutation.
- The motor is able to develop its full torque over the normal speed range. Since torque is not dependent on V_A, the full-load torque output is possible over the normal speed range, even at standstill (zero speed).
- The output power is zero at zero speed. In the normal speed range and at constant torque, the output power increases in proportion to the speed.
- In the field weakening range, the motor torque falls in proportion to the speed. Consequently, the output power of the DC motor remains constant.

Although a DC machine is well suited for adjustable speed drive applications, there are some limitations due to the mechanical commutator and brushes, which

- impose restrictions on the ambient conditions, such as temperature and humidity;
- are subject to wear and require periodic maintenance;
- limit the maximum power and speed of machines that can be built.

2.4.8.4 Electrical variable speed drives for AC motors (AC drives)

One of the lingering problems with thyristor controlled DC drives is the high maintenance requirement of the DC motor. Since the 1980s, the popularity of AC variable speed drives has grown rapidly, mainly due to advances in power electronics and digital control technology affecting both the cost and performance of this type of VSD. The main attraction of the AC VSDs is the rugged reliability and low cost of the *squirrel cage AC induction motor* compared to the DC motor.

In the AC VSD, the mechanical commutation system of the DC motor, has been replaced by a power electronic circuit called the *inverter.* However, the main difficulty with the AC variable speed drive has always been the complexity, cost and reliability of the AC frequency inverter circuit.

The development path from the Ward-Leonard system to the thyristor controlled DC drive and then to the PWM-type AC VVVF converter is illustrated in Figure 2.4.16. In the first step from (a) to (b), the high cost motor-generator set has been replaced with a phase-controlled thyristor rectifier.

In the second step from (b) to (d), the high cost DC motor has been replaced with a power electronic PWM inverter and a simple rugged AC induction motor (Figure 2.4.17). Also, the rectifier is usually a simple diode rectifier.

Frequency control, as a method of changing the speed of AC motors, has been a well-known technique for decades, but it has only recently become a technically viable and economical method of variable speed drive control. In the past, DC motors were used in most variable speed drive applications in spite of the complexity, high cost and high maintenance requirements of the DC motors. Even today, DC drives are still often used for the more demanding variable speed drive applications. Examples of this are the sectional drives for paper machines, which require fast dynamic response and separate control of speed and torque.

Developments in power electronics over the last 10–15 years has made it possible to control not only the speed of AC induction motors but also the torque. Modern AC variable speed drives, with *flux-vector control*, can now meet all the performance requirements of even the most demanding applications.

In comparison to DC drives, AC drives have become a more cost-effective method of speed control for most variable speed drive applications up to 1000 kW. It is also the technically preferred solution for many industrial environments where reliability and low maintenance associated with the AC SCIM are important.

The fundamental principles of an AC variable speed drive are relatively easy to understand and are covered by a few simple equations as follows:

- The speed (n) of the motor can be controlled either by adjusting the supply frequency (f) or the number of poles (p). In an AC induction motor, the synchronous speed, which is the speed at which the stator field rotates, is governed by the simple formula:

$$\text{Synchronous speed,} \quad n_S = \frac{120f}{p} \text{ rev/min}$$

Although there are special designs of induction motors, whose speed can be changed in one or more steps by changing the number of poles, it is impractical to continuously vary the number of poles to effect smooth speed control. Consequently, the fundamental principle of modern AC variable speed drives is that the speed of a fixed pole AC induction motor is proportional to the *frequency* of the AC voltage connected to it.

In practice, the actual speed of the rotor shaft is slower than the synchronous speed of the rotating stator field, due to the **slip** between the stator field and the rotor. This is covered in detail in Chapter 2.5.

$$\text{Actual speed,} \quad n = (n_S - \text{slip}) \text{ rev/min}$$

The slip between the synchronous rotating field and the rotor depends on a number of factors, being the stator voltage, the rotor current and the mechanical load

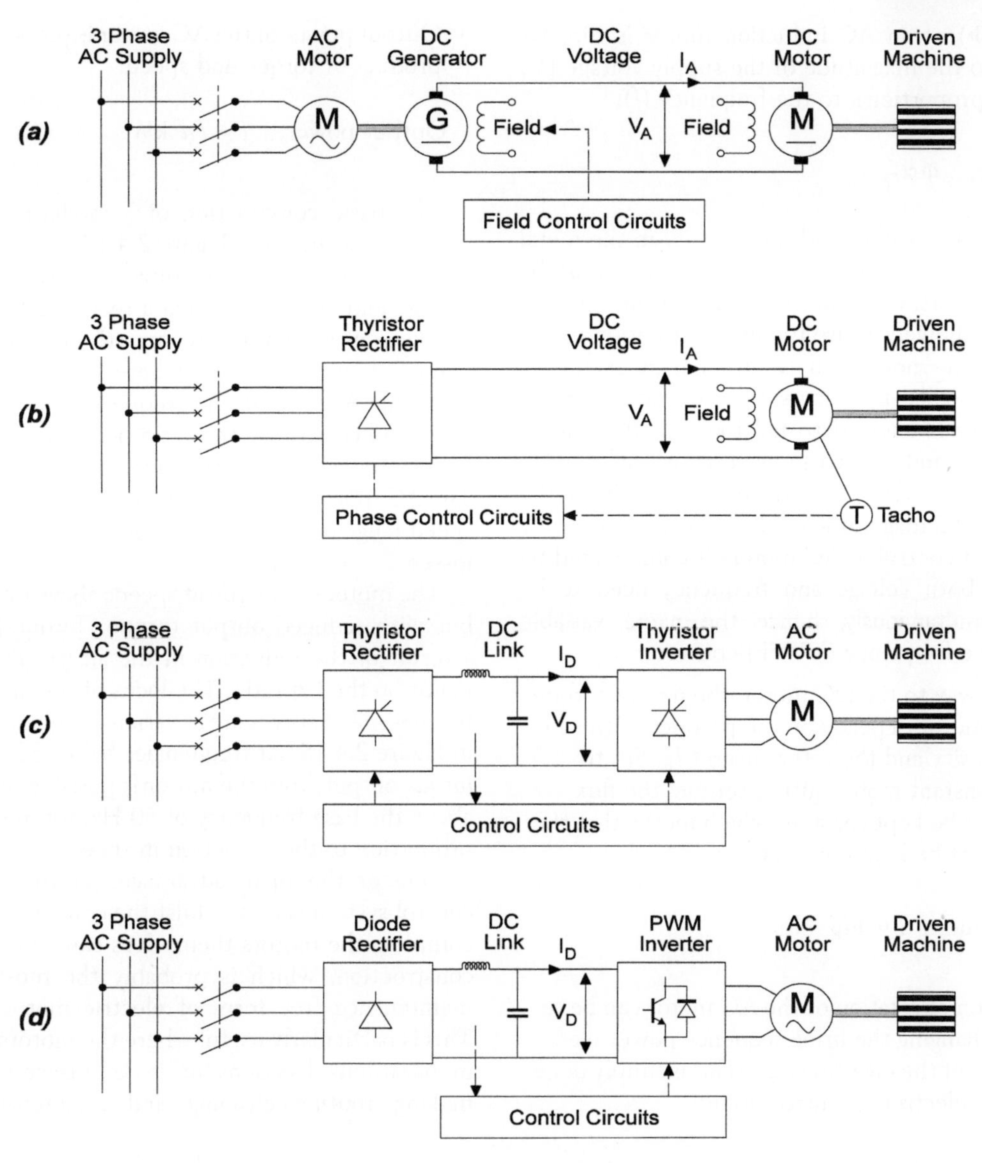

Fig. 2.4.16 Main components of various types of variable speed drive: (a) Ward-Leonard system; (b) thyristor controlled DC drive; (c) voltage source inverter (PAM) AC drive; (d) PWM voltage source (PWM) AC drive.

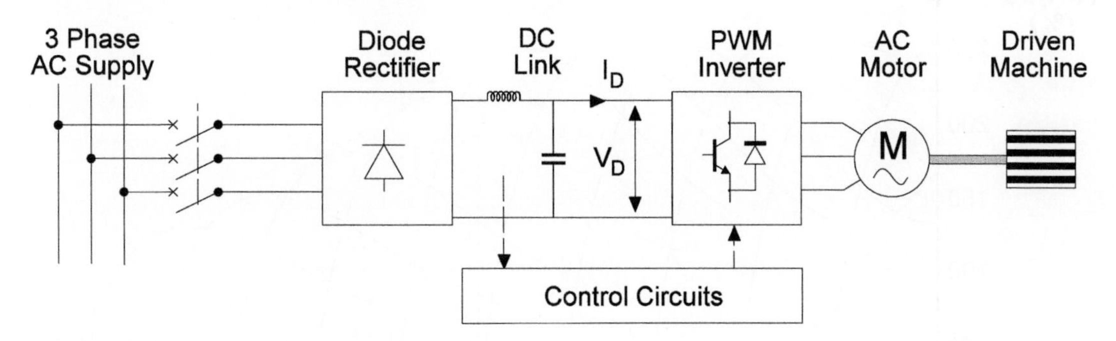

Fig. 2.4.17 Main components of a typical PWM-type AC drive.

on the shaft. Consequently, the speed of an AC induction motor can also be adjusted by controlling the slip of the rotor relative to the stator field. Slip control is discussed in Section 2.4.8.5.

Unlike a shunt wound DC motor, the stator field flux in an induction motor is also derived from the supply voltage and the flux density in the air gap will be affected by changes in the frequency of the supply voltage. The air-gap flux (Φ) of an AC induction motor is directly proportional to the magnitude of the supply voltage (V) and inversely proportional to the frequency (f):

$$\text{Air-gap flux,} \quad \Phi \propto \frac{V}{f}$$

To maintain a constant field flux density in the metal parts during speed control, the stator voltage must be adjusted in proportion to the frequency. If not and the flux density is allowed to rise too high, saturation of the iron parts of the motor will result in high excitation currents, which will cause excessive losses and heating. If the flux density is allowed to fall too low, the output torque will drop and affect the performance of the AC drive. Air-gap flux density is dependent on both the frequency and the magnitude of the supply voltage.

So the speed control of AC motors is complicated by the fact that both voltage and frequency need to be controlled simultaneously, hence the name variable voltage, variable frequency (VVVF) converter.

- In a similar way to the DC motor, the output torque of the AC motor depends on the product of the air-gap flux density and the rotor current I_R. So, to maintain constant motor output torque, the flux density must be kept constant which means that the ratio V/f must be kept constant.

$$\text{Output torque,} \quad T \propto \Phi I_R \text{ Nm}$$

- The direction of rotation of the AC motor can be reversed by changing the firing sequence power electronic valves of the inverter stage. This is simply done through the electronic control circuit.
- Output power of the AC motor is proportional to the product of torque and speed:

$$\text{Output power,} \quad P \propto Tn \text{ kW}$$

The basic construction of a modern AC frequency converter is shown in Figure 2.4.17.

The mains AC supply voltage is converted into a DC voltage and current through a rectifier. The DC voltage and current are filtered to smooth out the peaks before being fed into an inverter, where they are converted into a variable AC voltage and frequency. The output voltage is controlled so that the ratio between voltage and frequency remains constant to avoid over-fluxing the motor. The AC motor is able to provide its rated torque over the speed range up to 50 Hz without a significant increase in losses.

The motor can be run at speeds above rated frequency, but with reduced output torque. Torque is reduced as a result of the reduction in the air-gap flux, which depends on the V/f ratio. The locus of the induction motor torque–speed curves are at various frequencies are shown in Figure 2.4.18. At frequencies below 50 Hz, a *constant torque* output from the motor is possible. At frequencies above the base frequency of 50 Hz, torque is reduced in proportion to the reduction in speed.

One of the main advantages of this VVVF speed control system is that, whilst the controls are necessarily complex, the motors themselves can be of squirrel cage construction, which is probably the most robust, and maintenance free form of electric motor yet devised. This is particularly useful where the motors are mounted in hazardous locations or in an inaccessible position, making routine cleaning and maintenance difficult.

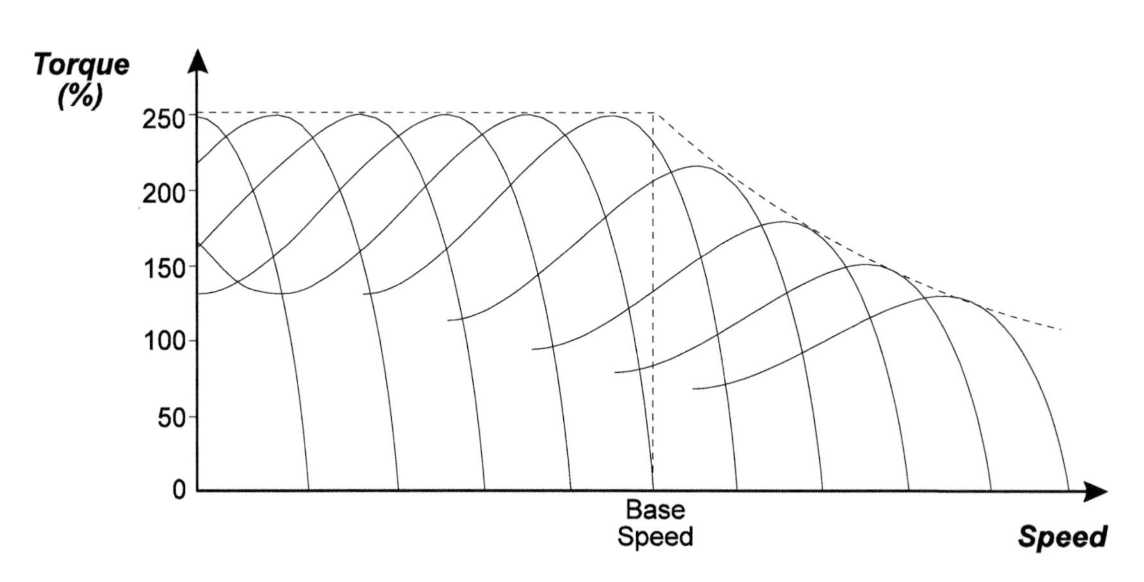

Fig. 2.4.18 Locus of the motor torque–speed curves at various frequencies.

Where a machine needs to be built into a flameproof, or even waterproof enclosure, this can be done more cheaply with a squirrel cage AC induction motor than for a DC motor.

On the other hand, an additional problem with standard AC squirrel cage motors, when used for variable speed applications, is that they are cooled by means of a shaft mounted fan. At low speeds, cooling is reduced, which affects the *loadability* of the drive. The continuous output torque of the drive must be derated for lower speeds, unless a separately powered auxiliary fan is used to cool the motor. This is similar to the cooling requirements of DC motors, which require a separately powered auxiliary cooling fan. From the equations above, the following deductions can be made about an AC drive (Figure 2.4.19):

- The speed of an AC induction motor can be controlled by adjusting the frequency and magnitude of the stator voltage. Motor speed is proportional to frequency, but the voltage must be simultaneously adjusted to avoid over-fluxing the motor.
- The AC motor is able to develop its full torque over the normal speed range, provided that the flux is held constant (V/f ratio kept constant). A standard AC motor reaches its rated speed, when the frequency has been increased to rated frequency (50 Hz) and stator voltage V has reached its rated magnitude.
- The speed of an AC induction motor can be increased above its nominal 50 Hz rating, but the V/f ratio will fall because the stator voltage cannot be increased any further. This results in a fall of the air-gap flux and a reduction in output torque. As with the DC motor, this is known as the *field weakening* range. The performance of the AC motor in the field weakening range is similar to that of the DC motor and is characterized by constant power, reduced torque.
- The output power is zero at zero speed. In the normal speed range and at constant torque, the output power increases in proportion to the speed.
- In the field weakening range, the motor torque falls in proportion to the speed and the output power of the AC motor remains constant.

2.4.8.5 Slip control AC variable speed drives

When an AC induction motor is started direct-on-line (DOL), the electrical power supply system experiences a current surge which can be anywhere between 4 and 10 times the rated current of the motor. The level of inrush current depends on the design of the motor and is independent of the mechanical load connected to the motor. A standard SCIM has an inrush current typically of 6 times the rated current of the motor. The starting torque, associated with the inrush current, is typically between 1.5 and 2.5 times the rated torque of the motor. When the rotor is stationary, the slip is 100% and the speed is zero. As the motor accelerates, the slip decreases and the speed eventually stabilizes at the point where the motor output torque equals the mechanical load torque, as illustrated in Figures 2.4.20 and 2.4.22.

The basic design of an SCIM and a WRIM are very similar, the main difference being the design and construction of the rotor. The design and performance of AC induction motors is described in considerable detail in Chapter 2.5. In AC induction motors, the slip between

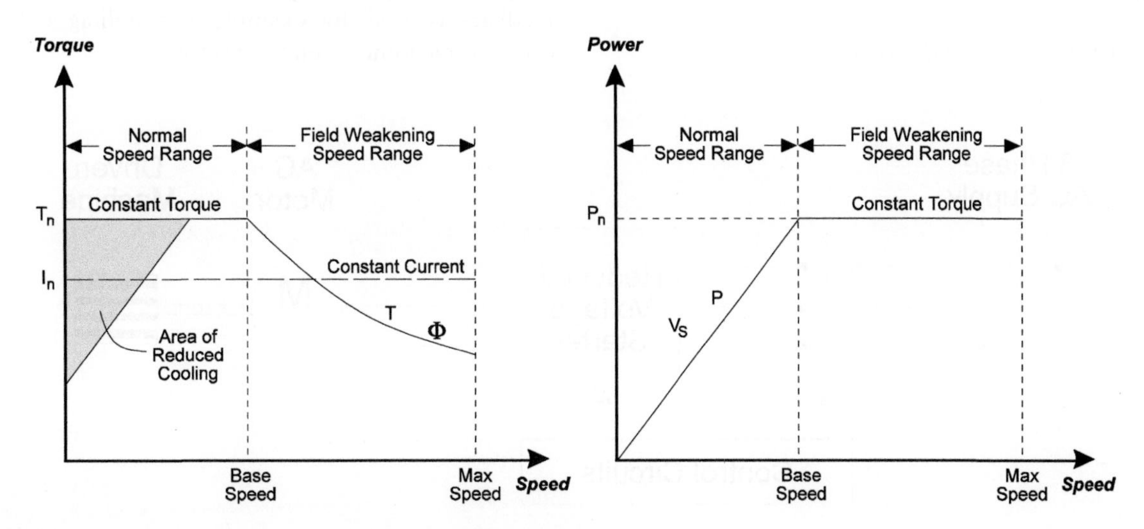

Fig. 2.4.19 Torque and power of an AC drive over the speed range.

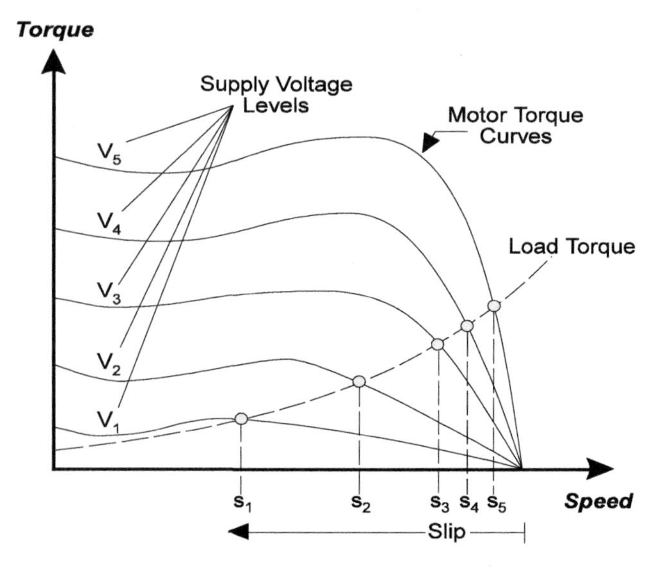

Fig. 2.4.20 Torque–speed curves of an induction motor with reduced supply voltage. V_1, low level of supply voltage; V_4, high level of supply voltage; V_5, full rated supply voltage.

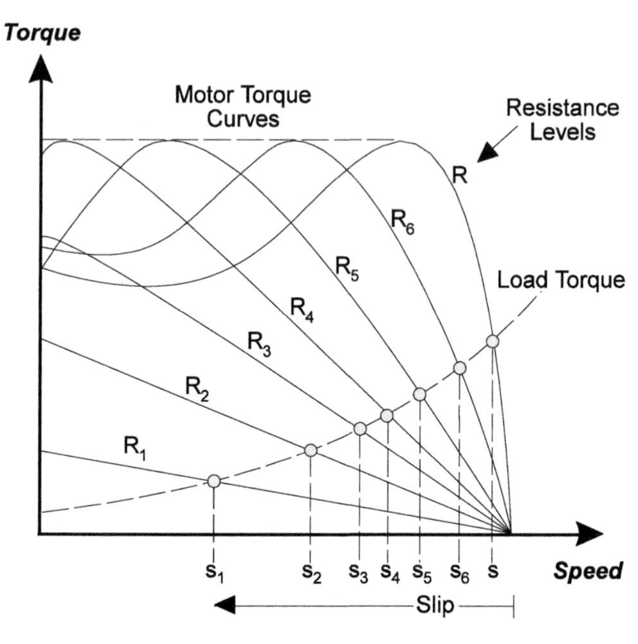

Fig. 2.4.22 Torque–speed curves of a WRIM with external rotor resistance. R_1, high external rotor resistance connected; R_6, low external rotor resistance connected; R, normal rotor resistance.

the synchronous rotating stator field and the rotor is mainly dependent on the following two factors, either of which can be used to control the motor speed:

- *Stator voltage*: affects both the flux and the rotor current.
- *Rotor current*: for an SCIM, this depends on the rotor design. For a WRIM, this depends on the external rotor connections

2.4.8.5.1 Stator voltage control

The reduction of the AC supply voltage to an induction motor has the effect of reducing both the air-gap flux (Φ) and the rotor current (I_R). The output torque of the motor behaves in accordance with the following formula:

Output torque, $T \propto \Phi I_R$ Nm

Since both Φ and I_R decrease with the voltage, the output torque of the motor falls roughly as the square of the voltage reduction. So when voltage is reduced, torque decreases, slip increases and speed decreases. The characteristic curves in Figure 2.4.20 show the relationship between torque and speed for various values of the supply voltage.

From this figure, the speed stabilizes at the point where the motor torque curve, for that voltage, intersects with the load-torque curve. The application of this technique for speed control is very limited because the resulting speed is dependent on the mechanical load torque. Consequently, speed holding is poor unless speed feedback is used, for example by installing a shaft encoder or tachometer on the motor.

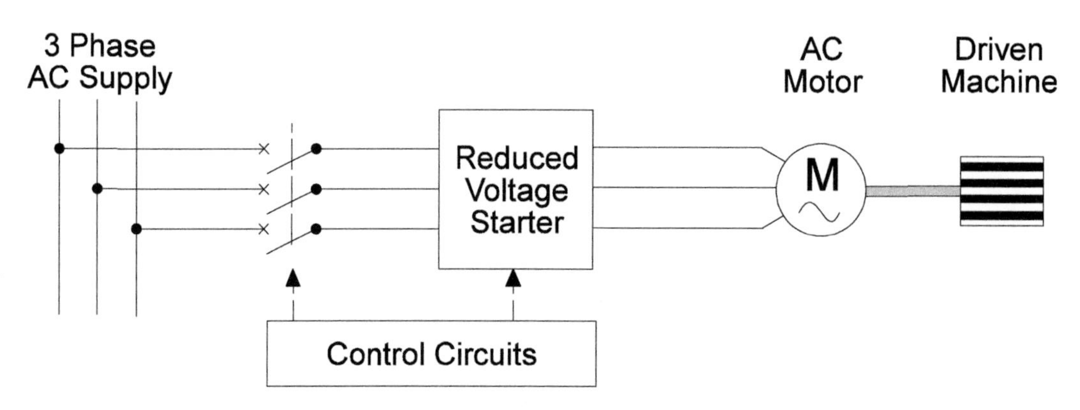

Fig. 2.4.21 Typical connections of a reduced voltage starter with an SCIM.

Reduced voltage control is not usually for speed control in industry, but for motor torque control, mainly for *soft starting* SCIMs. Reduced voltage (reduced torque) soft starting has the following main advantages:

- Reduces mechanical shock on the driven machinery, hence the name soft starting.
- Reduces the starting current surge in the electrical power supply system.
- Reduces water hammer during starting and stopping in pumping systems.

The following devices are commonly used in industry for reduced voltage starting and are typically connected as shown in Figure 2.4.21:

- Auto-transformers in series with stator: reactive volt drop.
- Reactors in series with stator: reactive volt drop.
- Resistors in series with stator: resistive volt drop.
- Thyristor bridge with electronic control: chopped voltage waveform.

Referring to the above, the characteristics of stator voltage control are as follows:

- Starting current inrush decreases as the square of the reduction in supply voltage.
- Motor output torque decreases as the square of the reduction in supply voltage.
- For reduced stator voltage, starting torque is always lower than DOL starting torque.
- Reduced voltage starting is not suitable for applications that require a high break-away torque.
- Stator voltage control is not really suitable for speed control because of poor speed holding capability.

2.4.8.5.2 Rotor current control

Rotor current control is another effective method of slip control that has successfully been used with induction motors for many decades. With full supply voltage on the stator, giving a constant flux Φ, the rotor current I_R can be controlled by adjusting the effective rotor resistance R_R.

For this method of control, it is necessary to have access to the three-phase rotor windings. A special type of induction motor, known as a *wound rotor induction motor (WRIM)*, sometimes also called a *slipring motor*, is used for these applications. In a WRIM, the connections to the rotor windings are brought out to terminals via three slip-rings and brushes, usually mounted at the non-drive end of the shaft. By connecting external resistance banks to the rotor windings, the rotor current can be controlled. Since output torque is proportional to the product of Φ and I_R, with a constant field flux, the rotor current affects the torque–speed characteristic of the motor as shown in Figure 2.4.22.

Increasing the rotor resistance reduces the rotor current and consequently the output torque. With lower output torque, the slip increases and speed decreases.

As with stator voltage control, this method of speed control has a number of limitations. The speed holding capability, for changes in mechanical load, is poor. Again, this method of control is more often used to control starting torque rather than for speed control. In contrast to stator voltage control, which has low starting torque, *rotor current control* can provide a high starting torque with the added advantage of soft starting.

The following devices are commonly used in industry for rotor current control:

- Air-cooled resistor banks with bypass contactors.
- Oil-cooled resistor banks with bypass contactors.
- Liquid resistor starters with controlled depth electrodes.
- Thyristor converters for rotor current control and slip energy recovery (SER).

Some of the characteristics of rotor current control are as follows:

- Starting current inrush is reduced in direct proportion to the rotor resistance.
- Starting torque, for certain values of rotor resistance, is higher than DOL starting torque and can be as high as the breakdown torque.
- Starting with a high external rotor resistance, as resistance is decreased in steps, the starting torque is progressively increased from a low value up to the breakdown torque.
- This type of starting is ideal for applications that require a high pull-away torque with a soft start, such as conveyors, crushers, ball mills, etc.
- Rotor current control is not ideal for speed control because of poor speed holding capability. However, it can be used for limited speed control, provided the speed range is small, typically 70–100% of motor rated speed. Motor speed holding is improved with the use of a shaft encoder or tachometer.

When the rotor current is controlled by external rotor resistors, a considerable amount of heat, known as the slip energy, needs to be dissipated in the resistor banks. In practice, rotor resistors are used for starting large induction motors and to accelerate heavy mechanical loads up to full speed. At full speed, the resistors are bypassed by means of contactors and the motor runs with a shorted rotor. Consequently, these losses occur for relatively short periods of time and are not considered to be of major significance.

However, when rotor resistors are used for speed control over an extended period of time, the energy losses can be high and the overall efficiency of the drive low. At constant output torque, the energy losses in the resistors are directly proportional to the slip. So as the speed is decreased, the efficiency decreases in direct proportion. For example, a WRIM running at 70% of rated speed at full load will need to dissipate roughly 30% of its rated power in the rotor resistors (Figure 2.4.23).

2.4.8.5.3 SER system

The SER system is a further development of rotor current control, which uses power electronic devices, instead of resistors, for controlling the rotor current. The main components of the SER system are shown in Figure 2.4.24. The rotor current is controlled by adjusting the firing angle of the rectifier bridge. With the rectifier bridge turned off, the rotor current is zero and with the thyristor bridge full on, the rotor current approaches rated current. The rectifier bridge can be controlled to provide any current between these outer limits. Instead of dumping the slip energy into a resistor, it is smoothed through a large choke and converted back into three-phase AC currents, which are pumped back into the mains at 50 Hz through a matching transformer. The thyristors of the rectifier bridge are commutated by the rotor voltage, while the thyristors of the inverter bridge are commutated by the supply voltage. The DC link allows the two sides of the converter to run at different frequencies. The tacho is used for speed feedback to improve the speed holding capability of this variable speed drive.

Using SER technique, the slip energy losses can be recovered and returned to the power supply system, thus improving the efficiency of the drive.

Some interesting aspects of the SER system are as follows:

- The rotor connected SER converter need only be rated for the slip energy, which depends on the required speed range. For example, for a speed range of 80–100% of rated motor speed, the SER converter should be rated at roughly 20% of motor power rating. If the speed range needs to be broadened to 70–100%, the rating of the SER converter needs to be increased

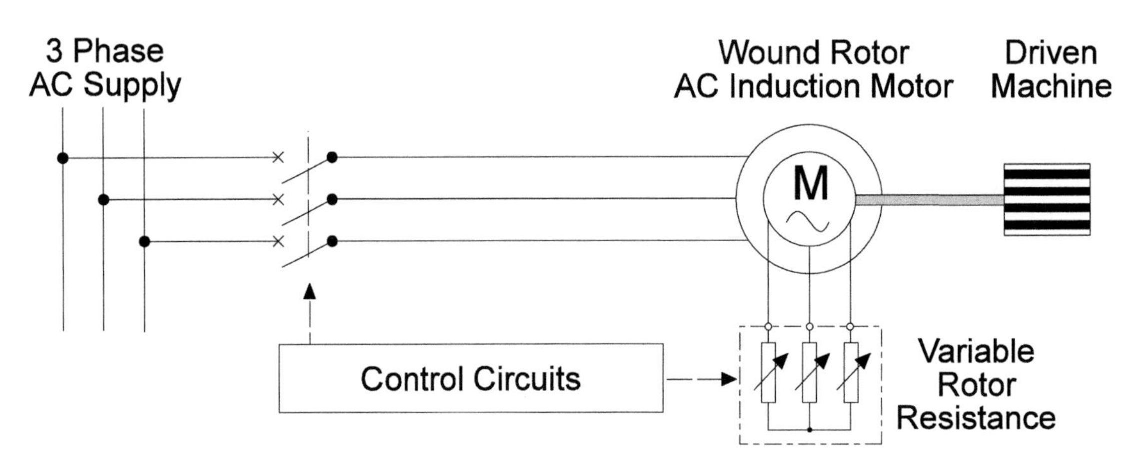

Fig. 2.4.23 Typical connections of a WRIM with rotor resistance starter.

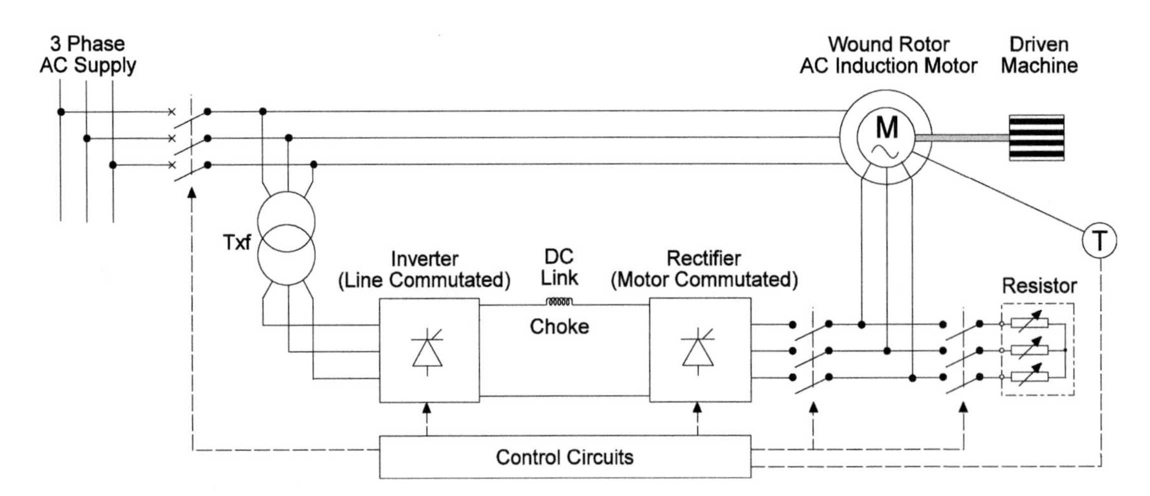

Fig. 2.4.24 The main components of a SER system.

to roughly 30% of motor power rating. In contrast, stator connected VVVF converters, commonly used for the speed control of SCIMs, need to be rated for >100% of the motor power rating.

- Because the SER converter rating is lower than motor rating, the slip power at starting would exceed the rating of the converter. It has become common practice to use an additional rotor resistance starter, selected by contactors from the control circuit, for the starting period from standstill. These resistors can be air-cooled, oil-cooled or the liquid type. Once the WRIM motor has been accelerated up to the variable speed range, the SER converter is connected and the resistors disconnected. These resistors have the added advantage of providing a standby solution in the event of a SER converter failure, when the motor can be started and run at fixed speed without the SER system.
- For additional flexibility, a bypass contactor is usually provided to short circuit the rotor windings and allow the motor to run at fixed speed.

The SER system is most often used by large water supply authorities for soft starting and limited speed control of large centrifugal pumps, typically 1–10 MW. In these applications, they are a more cost-effective solution than the equivalent stator connected AC or DC drives. Another increasingly common application is the starting and limited speed control (70–100%) of large SAG mills in mineral processing plants, typically 1–5 MW.

2.4.8.6 Cycloconverters

A cycloconverter is a converter that synthesizes a three-phase AC variable frequency output directly from a fixed frequency three-phase AC supply, without going via a DC link. The cycloconverter is not new and the idea was developed over 50 years ago using mercury arc rectifiers (Figure 2.4.25).

The low frequency AC waveform is produced using two back-to-back thyristors per phase, which are allowed to conduct alternatively. By suitable phase angle control, the output voltage and load current can be made to change in magnitude and polarity in cyclic fashion. The main limitation of the cycloconverter is that it cannot generate frequencies higher than the AC supply frequency. In fact, a frequency of about 30% of the supply frequency is the highest practically possible with reasonable waveforms. The lower the frequency, the better the waveform. The system is inherently capable of regeneration back into the mains.

The cycloconverter requires a large number of thyristors, and the control circuitry is relatively complex but, with the advent of microprocessors and digital electronics, the implementation of the control circuits has become more manageable.

Because of the low frequency output, cycloconverters are suited mainly for large slow speed drives, where it is used to drive either a large induction motor or a synchronous motor. Typical applications are SAG or ball mills, rotary cement kilns, large crushers, mine-winders, etc.

2.4.8.7 Servo-drives

Servo-drives are used in those drive applications which require a high level of precision, usually at relatively low powers. This often includes rapid stop–start cycles, very

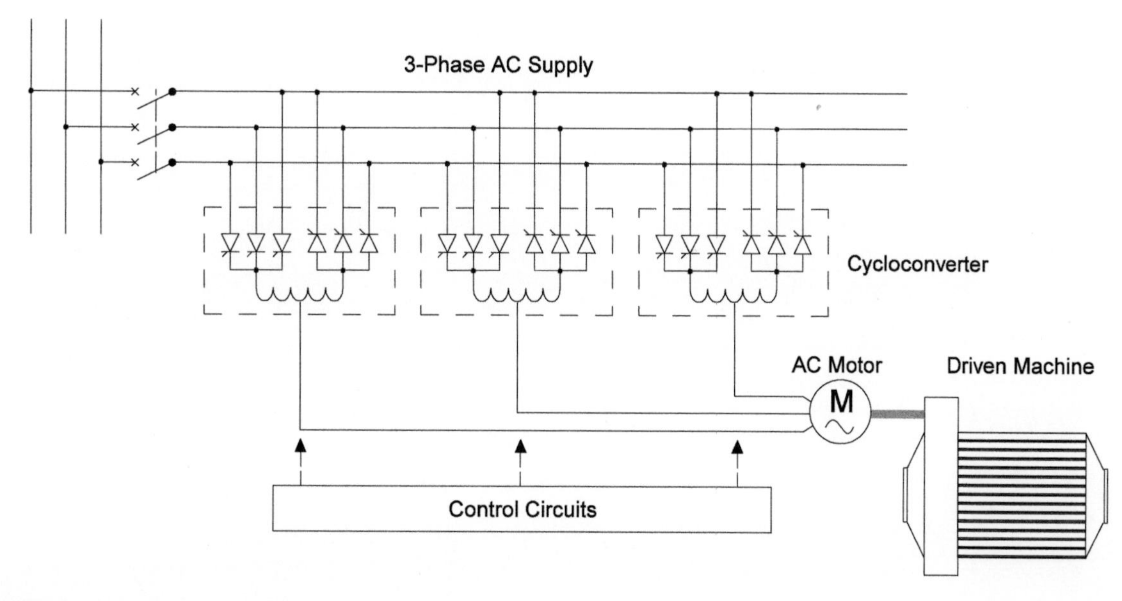

Fig. 2.4.25 The main components of a cycloconverter.

high acceleration torques, accurate positioning with controllable velocity and torque profiles.

The use of servo-drives for industrial manufacturing and materials handling has also become far more common, particularly for accurate positioning systems. This type of drive differs from a normal open loop VVVF drive in the following respects:

- Accuracy and precision of the motor speed and torque output are far in excess of what is normally possible with AC induction motors.
- A servo-motor is usually designed to operate with a specific servo-converter.
- Response of the servo-drive system to speed change demand is extremely fast.
- Servo-drives provide full torque holding at zero speed.
- Servo-drive inertia is usually very low to provide rapid response rates.

Servo-drives are beyond the scope of this book and will not be covered here.

Chapter 2.5

2.5

Induction motors

Barnes

2.5.1 Introduction

For industrial and mining applications, three-phase alternating current (AC) induction motors are the prime movers for the vast majority of machines. These motors can be operated either directly from the mains or from adjustable frequency drives. In modern industrialized countries, more than half the total electrical energy used is converted to mechanical energy through *AC induction motors*. The applications for these motors cover almost every stage of manufacturing and processing. Applications also extend to commercial buildings and the domestic environment. They are used to drive pumps, fans, compressors, mixers, agitators, mills, conveyors, crushers, machine tools, cranes, etc. It is not surprising to find that this type of electric motor is so popular, when one considers its simplicity, reliability and low cost.

In the last decade, it has become increasingly common practice to use three-phase squirrel cage AC induction motors with *variable voltage variable frequency (VVVF) converters* for variable speed drive (VSD) applications. To clearly understand how the VSD system works, it is necessary to understand the principles of operation of this type of motor.

Although the basic design of induction motors has not changed very much in the last 50 years, modern insulation materials, computer-based design optimization techniques and automated manufacturing methods have resulted in motors of smaller physical size and lower cost per kW. International standardization of physical dimensions and frame sizes means that motors from most manufacturers are physically interchangeable and they have similar performance characteristics.

The reliability of squirrel cage AC induction motors, compared to DC motors, is high. The only parts of the squirrel cage motor that can wear are the bearings. Sliprings and brushes are not required for this type of construction. Improvements in modern pre-lubricated bearing design have extended the life of these motors.

Although *single-phase* AC induction motors are quite popular and common for low-power applications up to approx 2.2 kW, these are seldom used in industrial and mining applications. Single-phase motors are more often used for domestic applications.

The information in this chapter applies mainly to three-phase squirrel cage AC induction motors, which is the type most commonly used with VVVF converters.

2.5.2 Basic construction

The AC induction motor comprises two electromagnetic parts:

- Stationary part called the *stator*.
- Rotating part called the *rotor*, supported at each end on bearings.

The stator and the rotor (Figure 2.5.1) are each made up of:

- An *electric circuit*, usually made of insulated copper or aluminum, to carry current.
- A *magnetic circuit*, usually made from laminated steel, to carry magnetic flux.

Practical Variable Speed Drives and Power Electronics; ISBN: 9780750658089

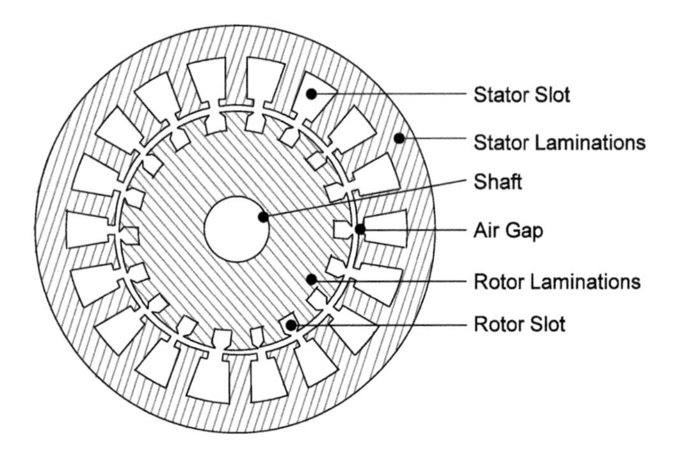

Fig. 2.5.1 Stator and rotor laminations.

2.5.2.1 The stator

The *stator* is the outer stationary part of the motor, which consists of:

- The *outer cylindrical frame* of the motor, which is made either of welded sheet steel, cast iron or cast aluminum alloy. This may include feet or a flange for mounting.
- The *magnetic path*, which comprises a set of slotted steel laminations pressed into the cylindrical space inside the outer frame. The magnetic path is laminated to reduce eddy currents, lower losses and lower heating.
- A set of *insulated electrical windings*, which are placed inside the slots of the laminated magnetic path. The cross-sectional area of these windings must be large enough for the power rating of the motor. For a three-phase motor, three sets of windings are required, one for each phase.

2.5.2.2 The rotor

This is the rotating part of the motor. As with the stator above, the rotor consists of a set of slotted steel laminations pressed together in the form of a cylindrical magnetic path and the electrical circuit. The electrical circuit of the rotor can be either:

- Wound-rotor type, which comprises three sets of insulated windings with connections brought out to three sliprings mounted on the shaft. The external connections to the rotating part are made via brushes onto the sliprings. Consequently, this type of motor is often referred to as a *slipring motor*.
- Squirrel cage rotor type, which comprises a set of copper or aluminum bars installed into the slots, which are connected to an end-ring at each end of the rotor. The construction of these rotor windings resembles a '*squirrel cage*'. Aluminum rotor bars are usually die-cast into the rotor slots, which results in a very rugged construction. Even though the aluminum rotor bars are in direct contact with the steel laminations, practically all the rotor current flows through the aluminum bars and not in the laminations.

2.5.2.3 The other parts

The other parts, which are required to complete the induction motor (Figure 2.5.2) are:

- *Two end-flanges* to support the two bearings, one at the drive-end (DE) and the other at the non-drive-end (NDE).
- *Two bearings* to support the rotating shaft, at DE and NDE.
- *Steel shaft* for transmitting the torque to the load.

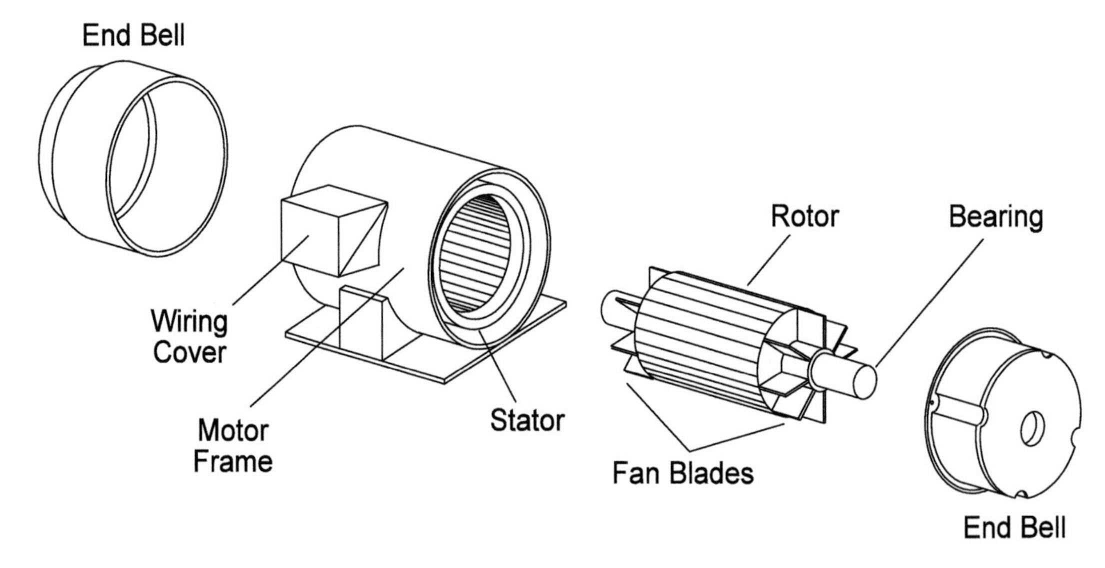

Fig. 2.5.2 Assembly details of a typical AC induction motor.

- *Cooling fan* located at the NDE to provide forced cooling for the stator and rotor.
- *Terminal box* on top or either side to receive the external electrical connections.

2.5.3 Principles of operation

When a three-phase AC power supply is connected to the stator terminals of an induction motor, three-phase AC flow in the stator windings. These currents set up a changing magnetic field (flux pattern), which rotates around the inside of the stator. The speed of rotation is in synchronism with the electric power frequency and is called the *synchronous speed*.

In the simplest type of three-phase induction motor, the rotating field is produced by three fixed stator windings, spaced 120° apart around the perimeter of the stator (Figure 2.5.3). When the three stator windings are connected to the three-phased power supply, the flux completes one rotation for every *cycle* of the supply voltage. On a 50 Hz power supply, the stator flux rotates at a speed of 50 revolutions per second, or 50 × 60 = 3000 rev/min.

A motor with only one set of stator electrical windings per phase, as described above, is called a *two pole motor (2p)* because the rotating magnetic field comprises two rotating poles, one North pole and one South pole. In some countries, motors with two rotating poles are also sometimes called a *two pole-pair motor.*

If there was a permanent magnet inside the rotor, it would follow in synchronism with the rotating magnetic field. The rotor magnetic field interacts with the rotating stator flux to produce a rotational force. A permanent

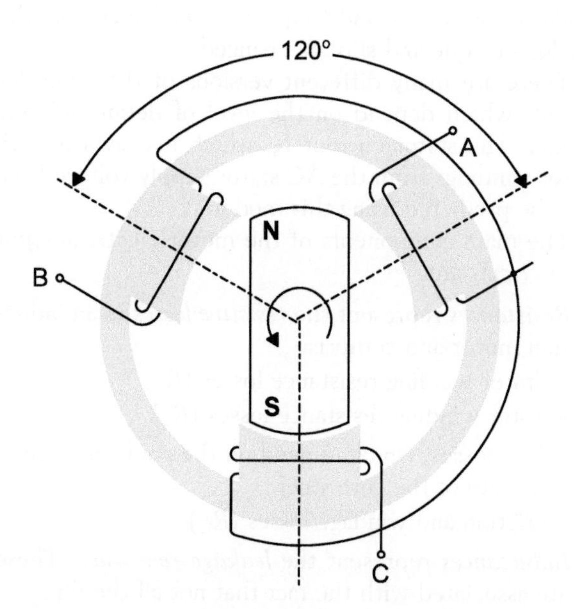

Fig. 2.5.3 Basic (simplified) principle of a two pole motor.

magnet is only being mentioned because the principle of operation is easy to understand. The magnetic field in a normal induction motor is induced across the rotor air gap as described below.

If the three windings of the stator were re-arranged to fit into half of the stator slots, there would be space for another three windings in the other half of the stator. The resulting rotating magnetic field would then have four poles (two North and two South), called a *four pole motor.* Since the rotating field only passes three stator windings for each power supply cycle, it will rotate at half the speed of the above example, at 1500 rev/min.

Consequently, induction motors can be designed and manufactured with the number of stator windings to suit the base speed required for different applications:

- two pole motors, stator flux rotates at 3000 rev/min;
- four pole motors, stator flux rotates at 1500 rev/min;
- six pole motors, stator flux rotates at 1000 rev/min;
- eight pole motors, stator flux rotates at 750 rev/min;
- etc.

The speed at which the stator flux rotates is called the *synchronous speed* and, as shown above, depends on the number of poles of the motor and the power supply frequency:

$$n_o = \frac{f \times 60}{\text{pole–pairs}} = \frac{f \times 60}{p/2} \text{ rev/min}$$

$$n_o = \frac{f \times 120}{p} \text{ rev/min}$$

where n_o is the synchronous rotational speed in rev/min; f the power supply frequency in Hz; p the number of motor poles.

To establish a current flow in the rotor, there must first be a voltage present across the rotor bars. This voltage is supplied by the magnetic field created by the stator current. The rotating stator magnetic flux, which rotates at *synchronous speed,* passes from the stator iron path, across the air gap between the stator and rotor and penetrates the rotor iron path as shown in Figure 2.5.4. As the magnetic field rotates, the lines of flux cut across the rotor conductors. In accordance with *Faraday's Law,* this induces a voltage in the rotor windings, which is dependent on the rate of change of flux.

Since the rotor bars are short circuited by the end-rings, current flows in these bars will set up its own magnetic field. This field interacts with the rotating stator flux to produce the rotational force. In accordance with *Lenz's law,* the direction of the force is that which tends to reduce the changes in flux field, which means that the rotor will accelerate to follow the direction of the rotating flux.

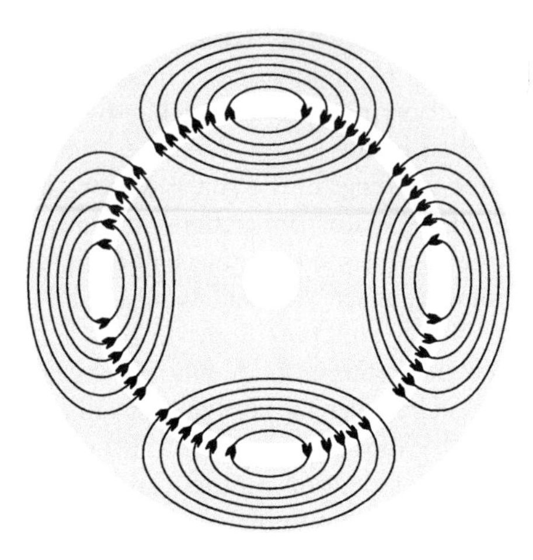

Fig. 2.5.4 Flux distribution in a four pole machine at any one moment.

At starting, while the rotor is stationary, the magnetic flux cuts the rotor at *synchronous speed* and induces the highest rotor voltage and, consequently, the highest rotor current. Once the rotor starts to accelerate in the direction of the rotating field, the rate at which the magnetic flux cuts the rotor windings reduces and the induced rotor voltage decreases proportionately. The frequency of the rotor voltage and current also reduces.

When the speed of the rotor approaches synchronous speed at no-load, both the magnitude and frequency of the rotor voltage become small. If the rotor reached synchronous speed, the rotor windings would be moving at the same speed as the rotating flux, and the induced voltage (and current) in the rotor would be zero. Without rotor current, there would be no rotor field and consequently no rotor torque. To produce torque, the rotor must rotate at a speed slower (or faster) than the synchronous speed. Consequently, the rotor settles at a speed slightly less than the rotating flux, which provides enough torque to overcome bearing friction and windage. The actual speed of the rotor is called the slip speed and the difference in speed is called the slip. Consequently, induction motors are often referred to as asynchronous motors because the rotor speed is not quite in synchronism with the rotating stator flux. The amount of slip is determined by the load torque, which is the torque required to turn the rotor shaft.

For example, on a four pole motor, with the rotor running at 1490 rev/min on no-load, the rotor frequency is 10/1500 of 50 Hz and the induced voltage is approximately 10/1500 of its value at starting. At no-load, the rotor torque associated with this voltage is required to overcome the frictional and windage losses of the motor.

As shaft load torque increases, the slip increases and more flux lines cut the rotor windings, which in turn increases rotor current, which increases the rotor magnetic field and consequently the rotor torque. Typically, the slip varies between about 1% of synchronous speed at no-load to about 6% of synchronous speed at full load:

$$\text{Slip, } s = \frac{(n_o - n)}{n_o} \text{ per unit}$$

and actual rotational speed is

$$n = n_o(1 - s) \text{ rev/min}^2$$

where n_o is the synchronous rotational speed in rev/min; n the actual rotational speed in rev/min; s the slip in per unit.

The direction of the rotating stator flux depends on the phase sequence of the power supply connected to the stator windings. The phase sequence is the sequence in which the voltage in the three-phases rises and reaches a peak. Usually, the phase sequence is designated A–B–C, L1–L2–L3 or R–W–B (Red–White–Blue). In Europe, this is often designated as U–V–W and many IEC style motors use this terminal designation. If two supply connections are changed, the phase sequence A–C–B would result in a reversal of the direction of the rotating stator flux and the direction of the rotor.

2.5.4 The equivalent circuit

To understand the performance of an AC induction motor operating from a VVVF converter, it is useful to electrically represent the motor by an *equivalent circuit* (Figure 2.5.5). This clarifies what happens in the motor when stator voltage and frequency are changed or when the load torque and slip are changed.

There are many different versions of the equivalent circuit, which depend on the level of detail and complexity. The stator current I_S, which is drawn into the stator windings from the AC stator supply voltage V, can then be predicted using this model.

The main components of the motor electrical *equivalent circuit* are:

- *Resistances* represent the *resistive losses* in an induction motor and comprise
  - Stator winding resistance losses (R_S)
  - Rotor winding resistance losses (R_R)
  - Iron losses, which depend on the grade and flux density of the core steel
  - Friction and windage losses (R_C).
- *Inductances* represent the *leakage reactance*. These are associated with the fact that not all the flux produced by the stator windings cross the air gap to

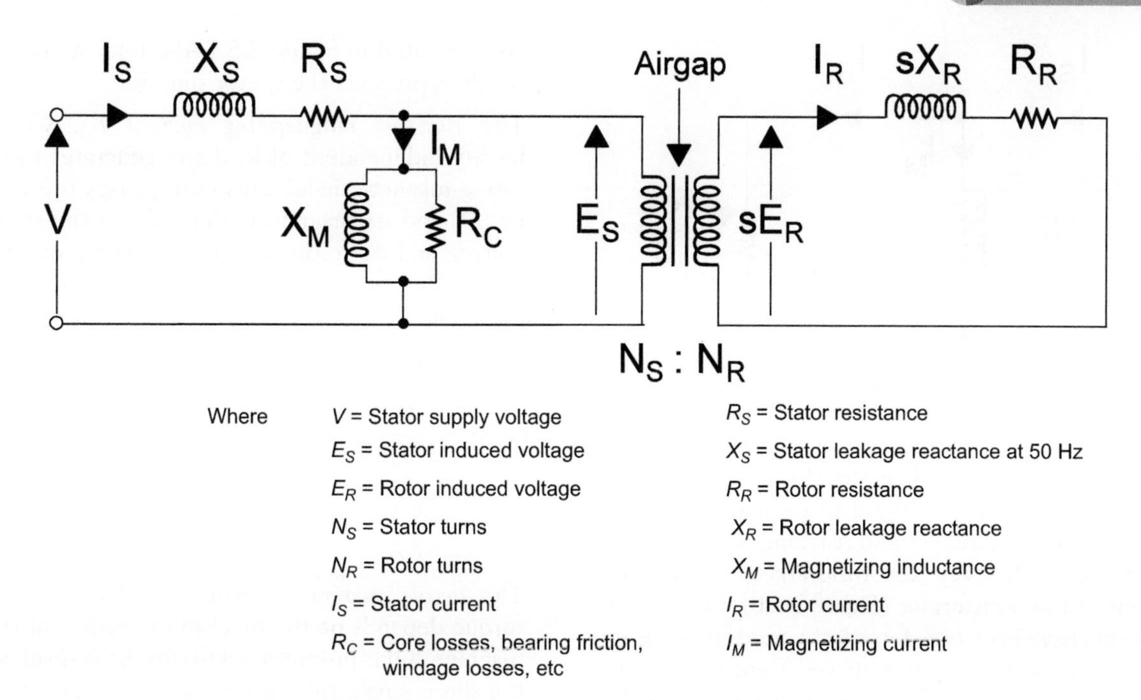

Fig. 2.5.5 The equivalent circuit of an AC induction motor.

link with the rotor windings and not all of the rotor flux enters the air gap to produce torque.

- Stator leakage reactance (X_S shown in Figure 2.5.6)
- Rotor leakage reactance (X_R shown in Figure 2.5.6).
- Magnetizing inductance (X_M, which produces the magnetic field flux).

In contrast with a DC motor, the AC induction motor does not have separate field windings. As shown in the equivalent circuit, the stator current therefore serves a double purpose:

- It carries the current (I_M), which provides the rotating magnetic field.
- It carries the current (I_R), which is transferred to the rotor to provide shaft torque.

The stator voltage E_S is the *theoretical* stator voltage that differs from the supply voltage by the volt drop across X_S and R_S. X_M represents the magnetizing inductance of the core and R_C represents the energy lost in the core losses, bearing friction and windage losses. The rotor part of the equivalent circuit consists of the induced voltage, E_R, which, as discussed earlier, is proportional to the slip and the rotor reactance, X_R, which depends on frequency and is consequently also dependent on slip.

This equivalent circuit is quite complex to analyze because the transformer, between the stator and rotor,

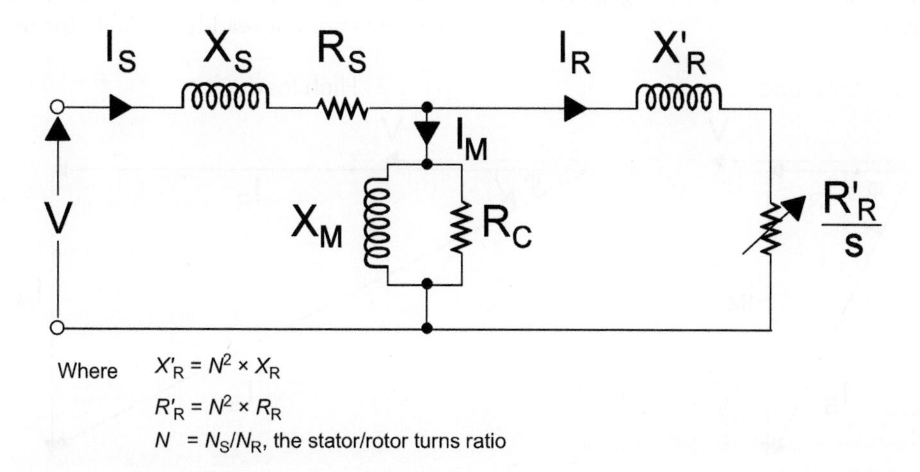

Fig. 2.5.6 The simplified equivalent circuit of an AC induction motor.

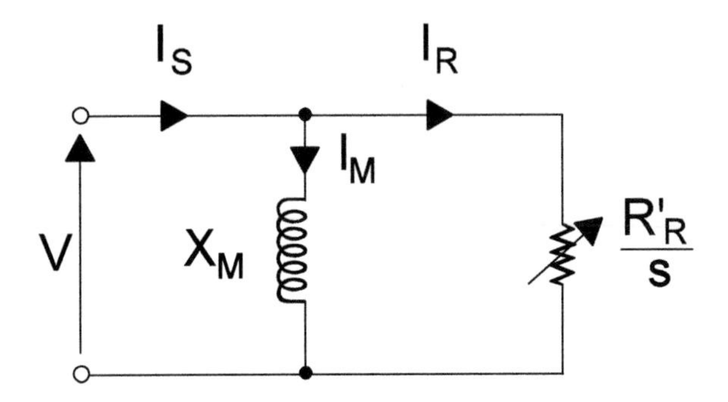

Fig. 2.5.7 The very simplified equivalent circuit of an AC induction motor.

has a ratio that changes when the slip changes. Fortunately, the circuit can be simplified by mathematically adjusting the rotor resistance and reactance values by the turns ratio $N^2 = (N_S/N_R)^2$, i.e. '*transferring*' them to the stator side of the transformer (Figure 2.5.6). Once these components have been transferred, the transformer is no longer relevant and it can be removed from the circuit. This mathematical manipulation must also adjust for the variable rotor voltage, which depends on slip. The equivalent circuit can be re-arranged and simplified as shown in Figure 2.5.6.

In this modified equivalent circuit, the rotor resistance is represented by an element that is dependent on the slip s. This represents the fact that the induced rotor voltage and consequently current depends on the slip. Consequently, when the induction motor is supplied from a power source of constant voltage and frequency, the current I_S drawn by the motor depends primarily on the slip.

The equivalent circuit can be simplified even further to represent only the most significant components (Figure 2.5.7), which are:

- Magnetizing inductance (X_M)
- Variable rotor resistance (R'_R/s)

All other components are assumed to be negligibly small and have been left out.

As illustrated in Figure 2.5.7, the total stator current I_S largely represents the vector sum of:

- The *reactive magnetizing current* I_M, which is largely independent of load and generates the rotating magnetic field. This current lags the voltage by 90° and its magnitude depends on the stator voltage and its frequency. To maintain a constant flux in the motor, the V/f ratio should be kept constant:

$$X_M = j\omega L_M = j(2\pi f)L_M$$

$$I_M = \frac{V}{j(2\pi f)L_M}$$

$$I_M = k\left(\frac{V}{f}\right), \quad \text{where } k = \text{constant}$$

- The *active current* I_R, which produces the rotor torque depends on the mechanical loading of the machine and is proportional to slip. At no-load, when the slip is small, this current is small. As load increases and slip increases, this current increases in proportion. This current is largely in phase with the stator voltage.

Figure 2.5.8 shows the current vectors for low-load and high-load conditions.

2.5.5 Electrical and mechanical performance

The angle between the two main stator components of voltage V and current I_S is known as the power factor angle represented by the angle ϕ and can be measured at the stator terminals. As shown, the stator current is the vector sum of the magnetizing current I_M, which is in quadrature to the voltage, and the torque producing current I_R, which is in phase with the voltage. These two currents are not readily available for measurement.

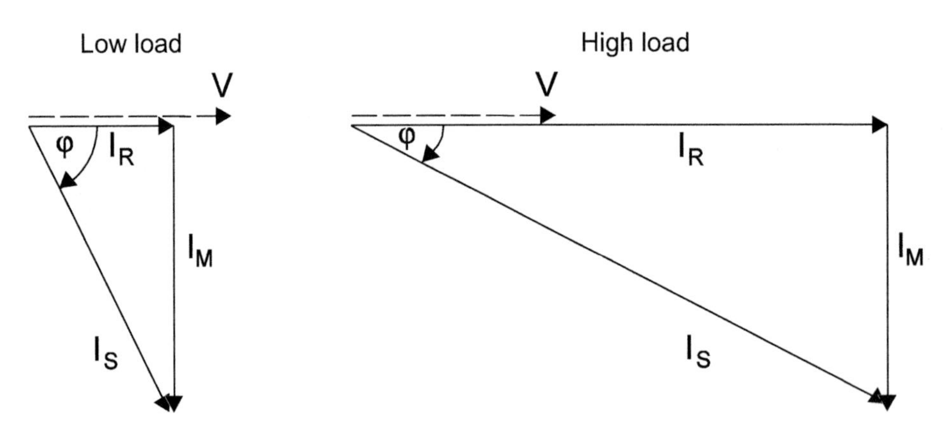

Fig. 2.5.8 Stator current for low-load and high-load conditions.

Consequently, the total apparent motor power S also comprises two components, which are in quadrature to one another,

$$S = P + jQ \text{ kVA}$$

- *Active power* P can be calculated by

$$P = \sqrt{3} \times V \times I_R \text{ kW}$$

or

$$P = \sqrt{3} \times V \times I_S \times \cos\phi \text{ kW}$$

- *Reactive power* Q, can be calculated by

$$Q = \sqrt{3} \times V \times I_M \text{ kVAr}$$

or

$$Q = \sqrt{3} \times V \times I_S \times \sin\phi \text{ kVAr}^3$$

where S is the total apparent power of the motor in kVA; P the active power of the motor in kW; Q the reactive power of the motor in kVAr; V the phase–phase voltage of the power supply in kV; I_S the stator current of the motor in amps; ϕ the phase angle between V and I_S (power factor = cos ϕ).

Not all the electrical input power P_I emerges as mechanical output power P_M. A small portion of this power is lost in the stator resistance ($3I^2R_S$) and the core losses ($3I_M^2R_C$) and the rest crosses the air gap to do work on the rotor. An additional small portion is lost in the rotor ($3I^2R'_R$). The balance is the mechanical output power P_M of the rotor.

Another issue to note is that the magnetizing path of the equivalent circuit is mainly inductive. At no-load, when the slip is small (slip $s \rightarrow 0$), the equivalent circuit shows that the effective rotor resistance $R'_R/s \rightarrow$ infinity. Therefore, the motor will draw only no-load magnetizing current. As the shaft becomes loaded and the slip increases, the magnitude of R'_R/s decreases and the current rises sharply as the output torque and power increases.

This affects the *phase relationship* between the stator voltage and current and the power factor cos ϕ. At no-load, the power factor is low, which reflects the high component of magnetizing current. As mechanical load grows and slip increases, the effective rotor resistance falls, active current increases and power factor improves.

When matching motors to mechanical loads, the two most important considerations are the torque and speed. The *torque–speed curve*, which is the basis of illustrating how the torque changes over a speed range, can be derived from the equivalent circuit and the equations above. By reference to any standard textbook on three-phase AC induction motors, the *output torque* of the motor can be expressed in terms of the speed as follows:

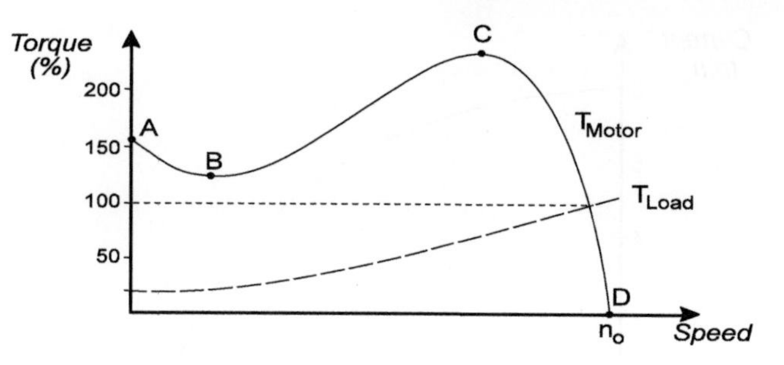

Fig. 2.5.9 Torque–speed curve for a three-phase AC induction motor. A is called the breakaway starting torque. B is called the pull-up torque. C is called the pull-out torque (or breakdown torque or maximum torque). D is the synchronous speed (zero torque).

$$T_M = \frac{3 \times s \times V^2 \times R'_R}{[(R_s + R'_R)^2 + s(X_s + X'_R)^2]n_o}$$

This equation and the curve in Figure 2.5.9 shows how the motor output torque T_M varies when the motor runs from standstill to full speed under a constant supply voltage and frequency. The torque requirements of the mechanical load are shown as a dashed line.

At starting, the motor will not pull away unless the starting torque exceeds the load breakaway torque. Thereafter, the motor accelerates if the motor torque always exceeds the load torque. As the speed increases, the motor torque will increase to a maximum T_{Max} at point C.

On the torque–speed curve, the final drive speed (and slip) stabilizes at the point where the *load torque* exactly equals the *motor output torque*. If the load torque increases, the motor speed drops slightly, slip increases, stator current increases, and the motor torque increases to match the load requirements.

The range CD on the torque–speed curve is the stable operating range for the motor. If the load torque increased to a point beyond T_{Max}, the motor would stall because, once the speed drops sufficiently back to the unstable portion ABC of the curve, any increase in load torque requirements T_L and any further reduction in drive speed, result in a lower motor output torque.

The relationship between stator current I_S and speed in an induction motor, at its rated voltage and frequency, is shown in Figure 2.5.10. When an induction motor is started direct-on-line (DOL) from its rated voltage supply, the stator current at starting can be as high as 6–8 times the rated current of the motor. As the motor approaches its rated speed, the current falls to a value determined by the mechanical load on the motor shaft (Figure 2.5.10).

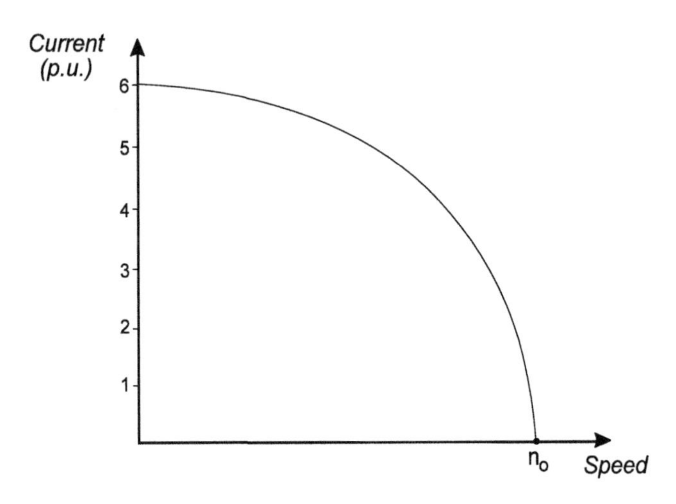

Fig. 2.5.10 Current–speed characteristic of a three-phase AC induction motor.

Some interesting observations about the AC induction motor that can be deduced from the above equations are:

- Motor output torque is proportional to the square of the voltage

$T_M \propto V^2$

Consequently, starting an induction motor with a reduced voltage starter, such as soft-starters, star-delta starters, auto-transformer starters, etc., means that motor-starting torque is reduced by the square of the reduced voltage.

- The efficiency of an induction motor is approximately proportional to $(1 - s)$ i.e. as speed drops and slip increases, efficiency drops:

$\text{Eff} \propto (1 - s)$

The induction motor operates as a slipping clutch with the slip power being dissipated as heat from the rotor as 'copper losses'. On speed control systems that rely on slip, such as wound-rotor motors with variable resistors, slip-recovery systems, etc., speed variation is obtained at the cost of motor efficiency.

Efficient use of an induction motor means that slip should be kept as small as possible. This implies that, from an efficiency point of view, the ideal way to control the speed of an induction motor is the stepless control of frequency.

Three-phase AC induction motors typically have slip values, at full load, of:

- 3–6% for small motors;
- 2–4% for larger motors.

This means that the speed droop from no-load to full load is small and therefore this type of motor has an almost constant speed characteristic.

One of the most fundamental and useful formulae for rotating machines is the one that relates the *mechanical output power* P_M of the motor to torque and speed,

$$P_M = \frac{(T_M \times n)}{9550} \text{ kW}$$

where P_M is the motor output power in kW; T_M, actual rotation speed in Nm; N, actual rotational speed in rev/min.

2.5.6 Motor acceleration

An important aspect of correctly matching a motor to a load is the calculation of the acceleration time of the motor from standstill to full running speed. Acceleration time is important to avoid over-heating the motor due to the high starting currents. So it is often necessary to know how long the machine will take to reach full rated speed. Manufacturers of electric motors usually specify a maximum starting time, during which acceleration can safely take place. This can be a problem during the acceleration of a high inertia load, such as a fan.

Figure 2.5.9 shows the motor torque curve and the load torque curve plotted on the same graph for a speed range from standstill to full speed. Assuming DOL starting, the time taken to accelerate a mechanical load to full speed depends on:

- *Acceleration torque* (T_A), which is the difference between the motor torque (T_M) and the load torque (T_L), $T_A = (T_M - T_L)$
- *Total moment of inertia* (J_{Tot}) of the rotating parts which is the sum of
  - moment of inertia of the rotor;
  - referred value of the moment of inertia of the load.

For acceleration to occur, the output torque of the motor must exceed the mechanical load torque. The bigger the *acceleration torque*, the shorter the acceleration time and vice versa. When the motor torque is less than the load torque, the motor will stall. Figure 2.5.11 shows an example of the *acceleration torque* of a motor, started DOL, driving a centrifugal pump load, whose torque requirement is low at starting and increases as the square of the speed.

The *acceleration torque* at starting is roughly equal to the rated motor torque, increases as the pump drive accelerates and then falls to zero as the motor reaches its rated speed. A steady state speed is reached when *motor torque* T_M matches the *load torque* T_L. The time taken

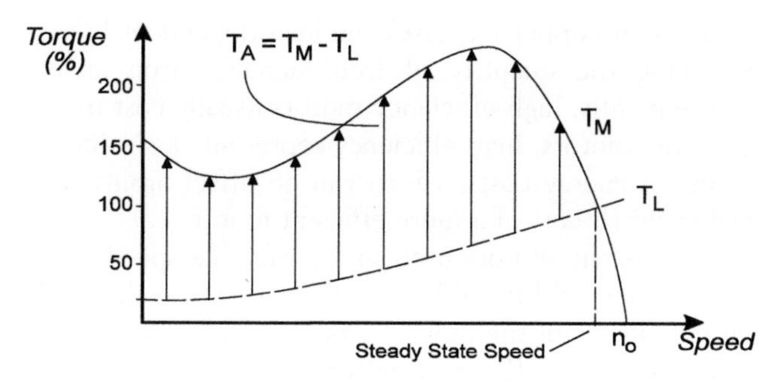

Fig. 2.5.11 Acceleration torque during the starting of an AC induction motor.

from standstill to reach this stable speed is called the *acceleration time.*

The rate of acceleration of the drive system also depends on the *moment of inertia (J)* of the rotating object. The higher its value, the longer it takes to increase speed:

$$T_A = J\frac{d\omega}{dt}\text{ Nm}$$

where J is the inertia of the drive system in kg m^2, ω the rotational speed in rad/s.

If this formula is adjusted to convert the rotational speed from rad/s to rev/min:

$$T_A = J\frac{2\pi}{60}\frac{dn}{dt}\text{ Nm}$$

where n is the rotational speed in rev/min.

Re-arranging

$$\frac{dt}{dn} = J\frac{2\pi}{60}\frac{1}{T_A}$$

This is integrated with respect to speed, from starting speed (n_1) to final speed (n_2). The total acceleration time t_d is given by

$$t_d = J\frac{2\pi}{60}\int_{n1}^{n2}\frac{1}{T_A}\,d\,n\text{ s}$$

If the acceleration torque were constant over the acceleration period, this formula would simplify to

$$t_d = J\frac{2\pi}{60}\frac{(n_2 - n_1)}{T_A}\text{ s}$$

Inertia can be calculated using the formula:

$$J = \frac{G \times D^2}{4}\text{ kg m}^2$$

where J is the moment of inertia of the rotating object (in kg m^2); G the mass (in kg) D the diameter of gyration (m).

It is not usually necessary to calculate the value of J from first principles because this may be obtained from the manufacturer of the motor as well as the driven machine.

2.5.7 AC induction generator performance

The performance of the three-phase AC induction motor has been described for the speed range from zero up to its rated speed at 50 Hz, where it behaves as a motor. A motor converts electrical energy to mechanical energy. The induction motor will always run at a speed lower than synchronous speed because, even at no-load, a small slip is required to ensure that there is sufficient torque to overcome friction and windage losses.

If, by some external means, the rotor speed was increased to the point that there was no slip, the induced voltage and current in the rotor fall to zero and torque output is zero.

If the rotor speed is, by some external means increased above this, the rotor will run faster than the rotating stator field and the rotor conductors again start to cut the lines of magnetic flux. Induced voltage reappears in the rotor, but in the opposite direction. From Lenz's Law, this results in currents that oppose this change and the power flows in the opposite direction from the driven rotor to the stator. Power flows from the mechanical prime mover, through the induction machine into the electrical supply connected to the stator. When the speed of the machine exceeds the synchronous speed n_o, it then operates as an *induction generator.*

This situation can often occur in the case of cranes, hoists, inclined conveyors, etc., where the load 'overruns' the motor.

The torque–speed curve can be extended to cover the induction generator region as well (Figure 2.5.12). The shape of the curve in the generator region is identical to the motor region because exactly the same equivalent circuit applies. The only difference is that the slip is negative and active power is transferred back into the mains.

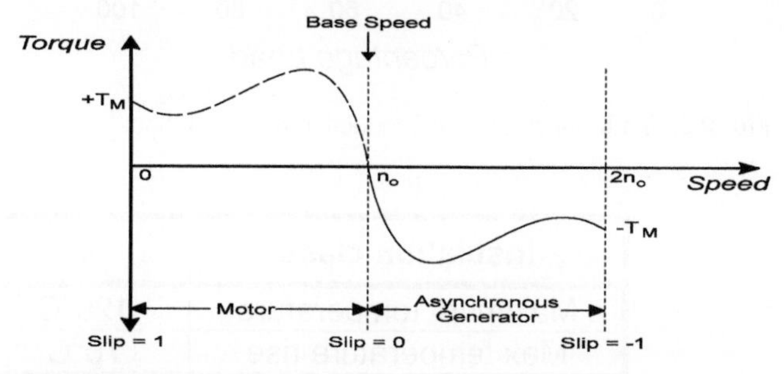

Fig. 2.5.12 Transition from induction motor to induction generator.

2.5.8 Efficiency of electric motors

The efficiency of a machine is a measure of how well it converts the input electrical energy into mechanical output energy. It is directly related to the losses in the motor, which depend on the design of the machine. Referring to the equivalent circuit of an induction motor, the losses comprise the following:

- *Load-dependent losses*: These are mainly the copper losses due to the load current flowing through the resistance of the stator and rotor windings and shown in the equivalent circuit as roughly $I_S^2(R_S + R'_R)$. These losses are proportional to the square of the stator current.
- *Constant losses*: These losses are mainly due to the friction, windage and iron losses and are almost independent of load. They are represented in the equivalent circuit as $I_M^2 R_C$.

Since the constant losses are essentially independent of load, while the stator and rotor losses depend on the square of the load current, the overall efficiency of an AC induction motor drops significantly at low-load levels, as shown in Figure 2.5.13.

Because of price competition, AC motor manufacturers are under pressure to economize on the quality and quantity of materials used in the motor. Reducing the quantity of copper increases the load-dependent losses. Reducing the quantity of iron increases iron losses. Consequently, high-efficiency motors usually cost more. On large motors, high efficiency represents a significant saving in energy costs, which can be offset against the higher initial cost of a more efficient motor.

For electric motors used in AC variable speed drive applications, additional harmonic currents result in additional losses in the motor, making it even more desirable to use high-efficiency motors.

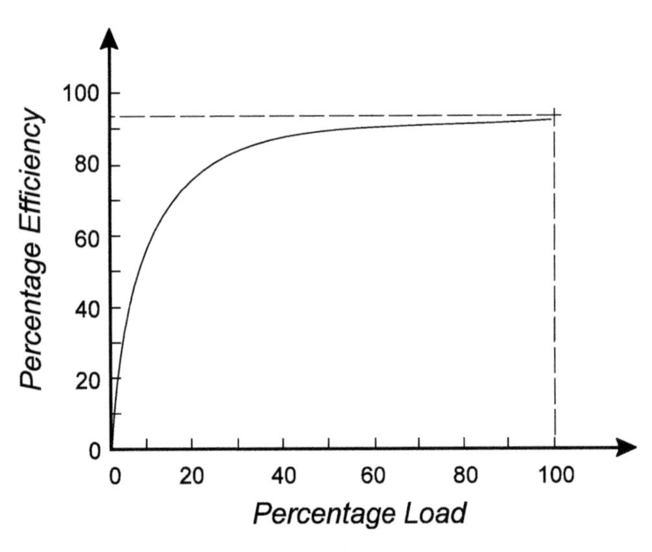

Fig. 2.5.13 Efficiency of an AC induction motor vs load.

2.5.9 Rating of AC induction motors

AC induction motors should be designed or selected to match the load requirements of any particular application. Some mechanical loads require the motor to run continuously at a particular load torque. Other loads may be cyclical or with numerous stops and starts.

The key consideration in matching a motor to a load is to ensure that the temperature inside the motor windings does not rise, as a result of the load cycle, to a level that exceeds the *critical temperature*. This *critical temperature* is that level which the stator and rotor winding insulation can withstand without permanent damage. Insulation damage can shorten the useful life of the motor and eventually results in electrical faults.

The temperature rise limits of insulation materials are classified by standards organizations, such as IEC 34.1 and AS 1359.32. These standards specify the maximum permissible temperatures that the various classes of insulation materials should be able to withstand. A safe temperature is the sum of the maximum specified ambient temperature and the permitted temperature rise due to the mechanical load.

For purposes of motor design, most motor specifications, such as IEC, AS/NZS, specify a maximum ambient temperature of 40°C. The temperature rise of the induction machine is the permissible increase in temperature, above this maximum ambient, to allow for the losses in the motor when running at full load. The maximum critical temperatures for each insulation class and the temperature rise figures, which are specified by IEC 34.1 and AS 1359.32 for rotating electrical machines, are shown in Figure 2.5.14.

From these tables, note that electrical rotating machines are designed for an overall temperature rise to

Insulation class	E	B	F	H
Maximum temperature	120°C	130°C	155°C	180°C
Max temperature rise	70°C	80°C	100°C	125°C

Fig. 2.5.14 Maximum temperature ratings for insulation materials.

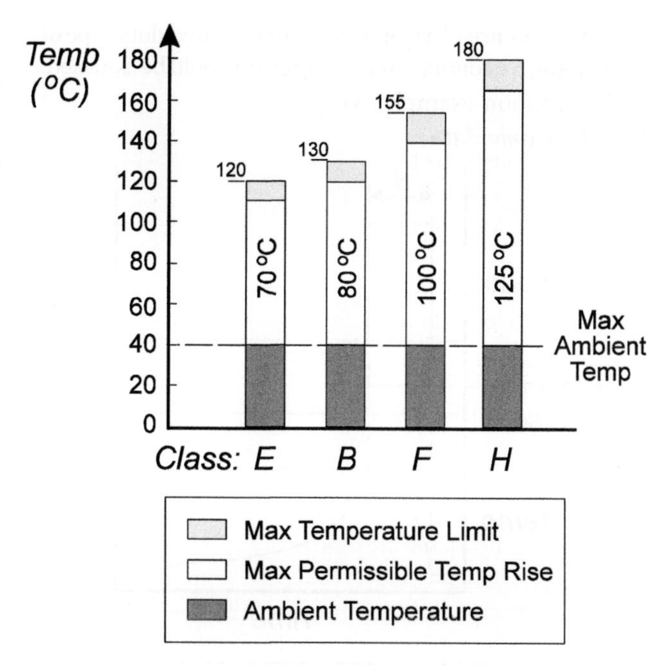

Fig. 2.5.15 Summary of temperature rise for classes of insulation materials according to IEC 34.1.

a level that is below the maximum specified for the insulation materials.

For example, using class-F insulation,

$$\text{Max ambient} + \text{Max temperature rise}$$
$$= 40\ ^\circ\text{C} + 100\ ^\circ\text{C} = 140\ ^\circ\text{C}$$

which gives a thermal reserve of 15 °C. The larger the thermal reserve, the longer the life expectancy of the insulation material.

When operating continuously at the maximum rated temperature of its class, the life expectancy of the insulation is about 10 years. Most motors do not operate at such extreme conditions because an additional safety margin is usually allowed between the calculated load torque requirements and the actual size of the motor chosen for the application. So life expectancy of a motor, which is correctly matched to its load and with suitable safety margins, can reasonably be taken as between 15 and 20 years.

If additional thermal reserve is required, the motor can be designed for an even lower temperature. It is common practice for the better quality manufacturers to design their motors for class-B temperature rise but to actually use class-F insulating materials (Figure 2.5.15). This provides an extra 20 °C thermal reserve that will extend the life expectancy to more than 20 years. This also means that the motor could be used at higher ambient temperatures of up to 50 °C or more, theoretically up to 65 °C.

In manufacturer's catalogues, the rating of three-phase AC induction motors are usually classified in terms of the following:

- Rated output power, in kW
- Rated speed, depends on the number of poles
- Rated for a continuous duty cycle S1 (see below)
- Rated at an ambient temperature not exceeding 40 °C
- Rated at an altitude not exceeding 1000 m above sea-level, which implies an atmospheric pressure of above 1050 mbar.
- Rated for a relative humidity of less than 90%.

AC induction motors often need to operate in environmental conditions where the *ambient temperature* and/or the *altitude* exceed the basis for the standard IEC or AS rating. Where the ambient temperature is excessively high, temperature de-rating tables are available from the motor manufacturers. An example of one manufacturer's de-rating table, for both temperature and altitude, is shown in Figure 2.5.16. As pointed out earlier, better quality AC induction motors have a built-in *thermal reserve*. In some cases, where the ambient temperature is only marginally higher than 40 °C, this reserve may be used with no additional de-rating for temperature.

Ambient temperature	Permissible output % of rated output	Altitude above Sea Level	Permissible output % of rated output
30°C	107 %	1000m	100 %
40°C	100 %	1500m	96 %
45°C	96 %	2000m	92 %
50°C	92 %	2500m	88 %
55°C	87 %	3000m	84 %
60°C	82 %	3500m	80 %
70°C	65 %	4000m	76 %

Fig. 2.5.16 Motor de-rating for temperature and altitude.

In motor mounting positions that are exposed to continuous direct sunshine, motors should be provided with a protective cover.

At *high altitudes*, where there is a reduced atmospheric pressure, the cooling of electrical equipment is degraded by the reduced ability of the air to remove the heat from the cooling surfaces of the motor. When the air pressure falls with increased altitude, the density of the air falls and, consequently, its thermal capacity is reduced. In accordance with the standards, AC induction motors are rated for altitudes up to 1000 m above sea-level. Rated power and torque output should be de-rated for altitudes above that.

When a motor needs to be de-rated for both temperature and altitude, the de-rating factors given in the table above should be multiplied together. For example, for a motor operating at above 2500 m in an ambient temperature of 50 °C, the overall de-rating factor should be (0.92 × 0.88) × 100%, or 81%.

2.5.10 Electric motor duty cycles

The rated output of an AC induction motor given in manufacturer's catalogues is based on some assumptions about the proposed application and duty cycle of the motor. It is common practice to base the motor rating on the *continuous running duty cycle* S1. When a motor is to be used for an application duty cycle other than the S1 continuous running duty, some precautions need to be taken in selecting a motor and the standard motors may be re-rated for the application. The duty cycles are normally calculated so that the average load over a period of time is lower than the continuous load rating S1.

In the standards, several different duty cycles are defined. In IEC 34.1 and AS 1359.30, eight different duty types are defined by the symbols S1–S8 as follows:

S1: Continuous running duty

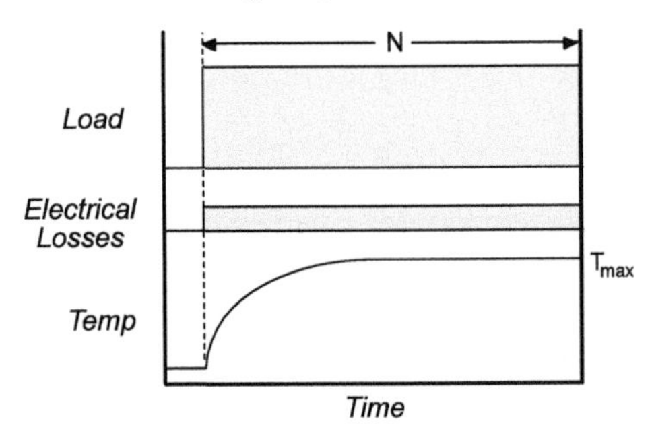

N = Operation at Constant Load
T_{max} = Max Temp Attained

- Operation at constant mechanical load for a period of sufficient duration for thermal equilibrium to be reached.
- In the absence of any indication of the rated duty type of a motor, S1 continuous running duty should be assumed.
- Designation example: S1.

S2: Short-time duty

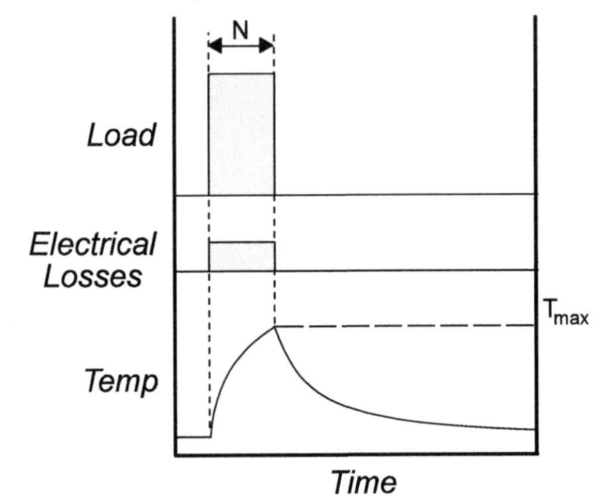

N = Operation at Constant Load
T_{max} = Max Temp Attained

- Operation at constant load, for a period of time which is less than that required to reach thermal equilibrium, followed by a rest and motor de-energized period of sufficient duration for the machine to re-establish temperatures to within 2 °C of the ambient or the coolant temperature.
- The values 10, 30, 60 and 90 are recommended periods for the rated duration of the duty cycle.
- Designation example: S2 – 60 min.

S3: Intermittent periodic duty not affected by the starting process.

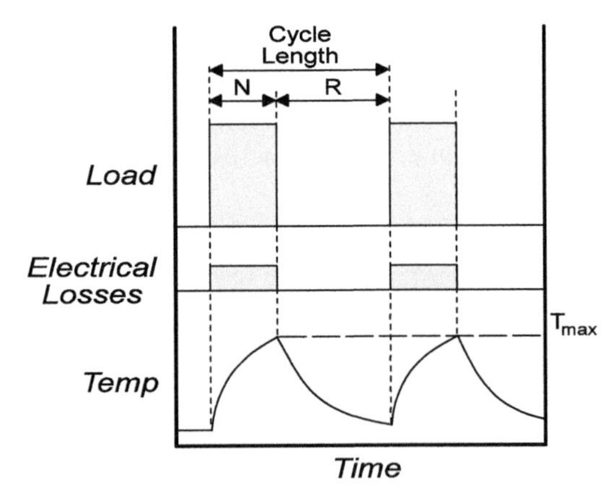

N = Operation at Constant Load
R = Rest
T_{max} = Max Temp Attained

$$\text{Operating Factor} = \frac{N}{N+R} \times 100$$

- A sequence of identical duty cycles, each comprising a period of operation at constant load and a period of rest when the motor is de-energized.
- The period of the duty cycle is too short for thermal equilibrium to be reached.
- Assumed that the starting current does not significantly affect the temperature rise.
- The duration of one duty cycle is 10 min.
- The following items should also be specified for this duty cycle:
  - The *cyclic duration factor*, which represents the percentage duration of the loaded period as a percentage of the total cycle.
  - Recommended values for cyclic duration factor are 15%, 25%, 40%, 60%.
- Designation example: S3 – 25%.

S4: Intermittent periodic duty affected by the starting process.

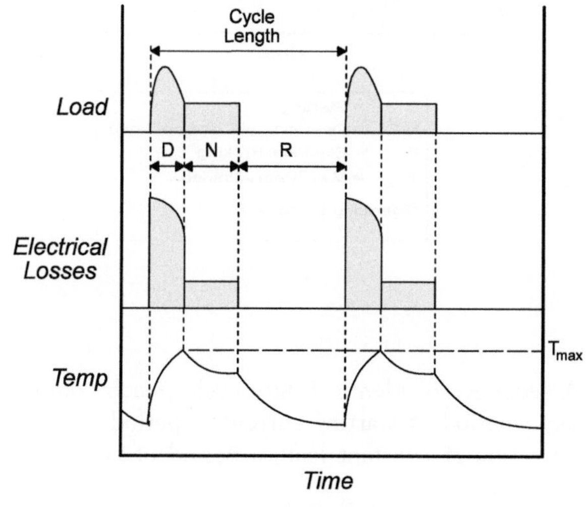

- A sequence of identical duty cycles, each comprising a period of significant starting current, a period of operation at constant load and a period of rest when the motor is de-energized.
- The period of the duty cycle is too short for thermal equilibrium to be reached.
- Assumed that the starting current is significant.
- The motor is brought to rest by the load or by mechanical braking, where the motor is not thermally loaded.
- The following items should also be specified for this duty cycle.
  - The *cyclic duration factor*, which represents the percentage duration of the loaded period as a percentage of the total cycle.
  - The number of *load cycles per hour* (c/h).
  - The *inertia factor FI*, which is the ratio of the *total* moment of inertia to the moment of inertia of the motor rotor.
  - The *moment of inertia* of the motor rotor (J_M).
  - The *average moment of resistance* T_V, during the change of speed given with rated load torque.
- Designation example: S4 – 25% – 120 c/h – (F_I = 2) – ($J_M = 0.1$ kg m^2) – ($T_V = 0.5T_N$)

S5: Intermittent periodic duty affected by the starting process and also by electric braking

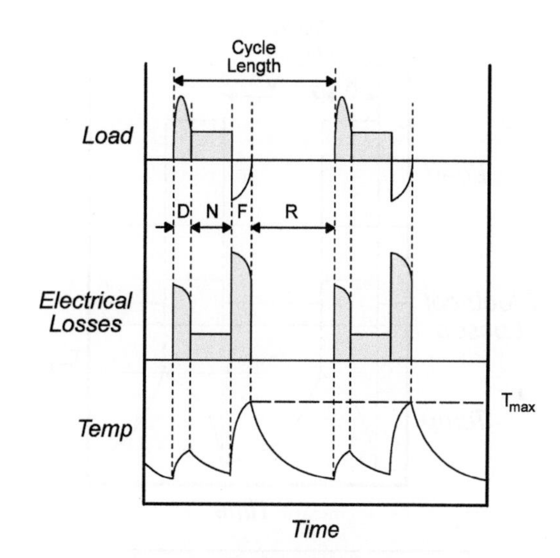

- A sequence of identical duty cycles, each comprising a period of significant starting current, a period of operation at constant load, a period of rapid electric braking and a period of rest when the motor is de-energized.
- The period of the duty cycle is too short for thermal equilibrium to be obtained.

- The following items should also be specified for this duty cycle.
  - The *cyclic duration factor,* which represents the duration of the loaded period as a percentage of the total cycle.
  - The number of *load cycles per hour* (c/h).
  - The *inertia factor* F_I, which is the ratio of the total moment of inertia to the moment of inertia of the motor rotor.
  - The *moment of inertia* of the motor rotor (J_M).
  - The *permissible average moment of resistance* T_V, during the change of speed given with rated load torque.
- Designation example: S5 – 40% – 120 c/h – (F_I = 3) – (J_M = 1.3 kg m^2) – (T_V = 0.3T_N)

S6: Continuous operation, periodic duty with intermittent load

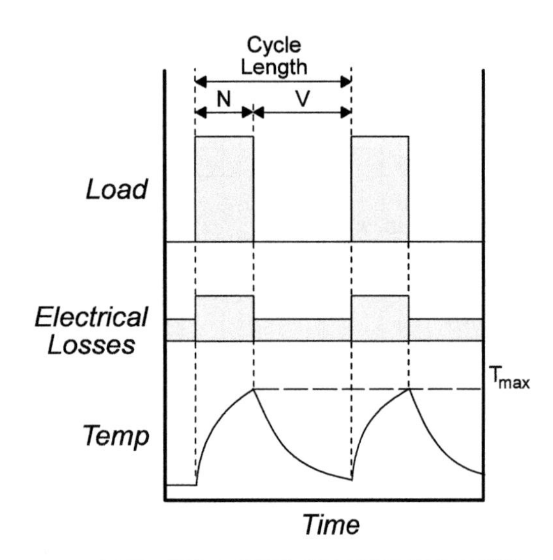

- A sequence of identical duty cycles, where each cycle consists of a period at constant load and a period of operation at no-load (no-load current only), but with no period of de-energization.
- The period of the duty cycle is too short for thermal equilibrium to be obtained.
- Recommended values for the cyclic duration factor are 15%, 25%, 40% and 60%.
- The duration of the duty cycle is 10 min.
- Designation example: S6 – 40%.

S7: Uninterrupted periodic duty, affected by the starting process and also by electric braking

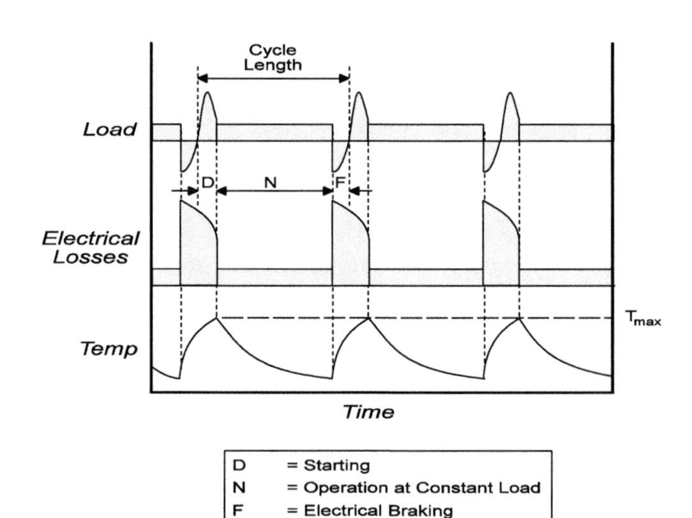

- A sequence of identical duty cycles, each comprising a period of starting current, a period of operation at constant load, a period of electric braking.
- The braking method is too short for thermal equilibrium to be obtained.
- The following items should also be specified for this duty cycle.
  - The number of load cycles per hour (c/h).
  - The inertia factor F_I which is the ratio of the total moment of inertia to the moment of inertia of the motor rotor.
  - The moment of inertia of the motor rotor (J_M).
  - The permissible average moment of resistance T_V, during the change of speed given with rated load torque.
- Designation example: S7 – 500 c/h – (F_I = 2) – (J_M = 0.08 kg m^2) – (T_V = 0.3T_N). *S8: Uninterrupted periodic duty with recurring speed and load changes*

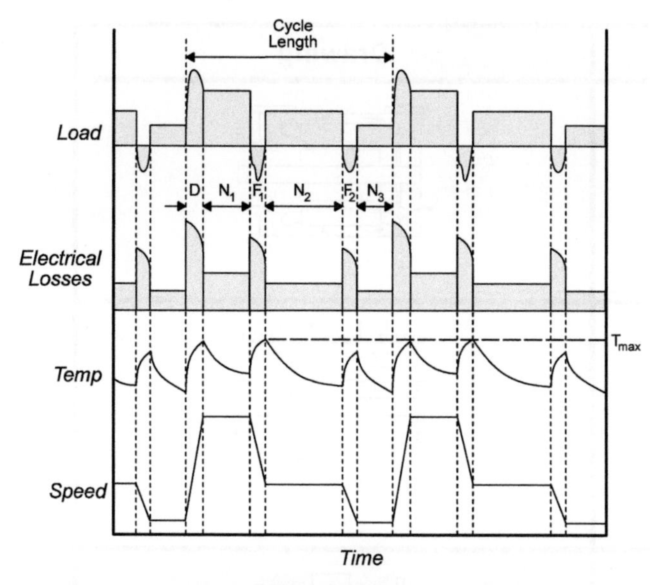

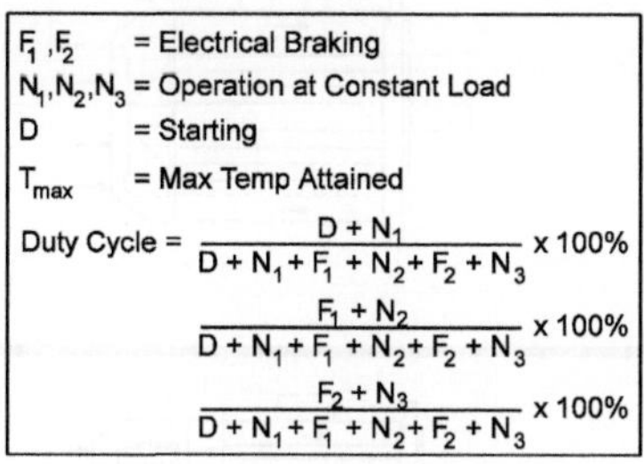

F1, F2 = Electrical Braking
N1, N2, N3 = Operation at Constant Load
D = Starting
Tmax = Max Temp Attained
Duty Cycle = (D + N1) / (D + N1 + F1 + N2 + F2 + N3) x 100%
(F1 + N2) / (D + N1 + F1 + N2 + F2 + N3) x 100%
(F2 + N3) / (D + N1 + F1 + N2 + F2 + N3) x 100%

- A sequence of identical duty cycles, each comprising a period of operation at constant load corresponding to a predetermined speed of rotation, followed by one or more periods of operation at other constant loads corresponding to different speeds of rotation.
- The period of the duty cycle is too short for thermal equilibrium to be obtained.
- This type of duty cycle is used for pole changing motors.
- The following items should also be specified for this duty cycle:
  - The number of load cycles per hour (c/h).
  - The inertia factor F_I which is the ratio of the total moment of inertia to the moment of inertia of the motor rotor.
  - The permissible average moment of resistance T_V, during the change of speed given with rated load torque.
  - The cyclic duration factor for each speed of rotation.
  - The moment of inertia of the motor rotor (J_M).
  - The combinations of the load and the speed of rotation are listed in the order in which they occur in use.
- Designation examples:
  - S8 – 30 c/h – ($F_I = 30$) – $T_V = 0.5T_N$ – 24 kW – 740 rev/min – 30%
  - S8 – 30 c/h – ($F_I = 30$) – $T_V = 0.5T_N$ – 60 kW – 1460 rev/min – 30%
  - S8 – 30 c/h – ($F_I = 30$) – $T_V = 0.5T_N$ – 45 kW – 980 rev/min – 40% – ($J_M = 2.2$ kg m^2).

2.5.11 Cooling and ventilation of electric motors (IC)

All rotating electrical machines generate heat as a result of the electrical and mechanical losses inside the machine. Losses are high during starting or dynamic braking. Also, losses usually increase with increased loading. Cooling is necessary to continuously transfer the heat to a cooling medium, such as the air. The different methods of cooling rotating machines are classified in the standards IEC 34.6 and AS 1359.21 (Figure 2.5.17).

For AC induction motors, cooling air is usually circulated internally and externally by one or more fans mounted on the rotor shaft. To allow for operation of the machine in either direction of rotation, fans are usually of the bi-directional type and made of a strong plastic material, aluminum, or steel. In addition, the external frames of the motor are usually provided with cooling ribs to increase the surface area for heat radiation.

The most common type of AC motor is the *totally enclosed fan cooled* (TEFC) motor, which is provided with an external forced cooling fan mounted on the NDE of the shaft, with cooling ribs running axially along the outer surface of the motor frame. These are designed to keep the air flow close to the surface of the motor along its entire length, thus improving the cooling and self-cleaning of the ribs. An air gap is usually left between the ribs and the fan cover for this purpose.

Internally, on smaller TEFC motors, the rotor end-rings are usually constructed with ribs to provide additional agitation of the internal air for even distribution of temperature and to allow the radiation of heat from the end shields and frame.

Special precautions need to be taken when standard TEFC induction motors are used with AC variable speed drives, powered by VVVF converters. For operation at speeds below the rated frequency of 50 Hz, the shaft-mounted fan cooling efficiency is lost. For constant torque loads, it is sometimes necessary to install a separately powered forced cooling fan (IC 43) to maintain

Code	Description	Drawing
IC 01	- Open machine - Fan mounted on shaft - Often called 'drip-proof' motor	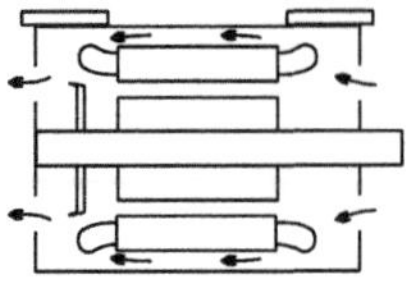
IC 40 (New : IC 410)	- Enclosed machine - Surface cooled by natural convection and radiation - No external fan	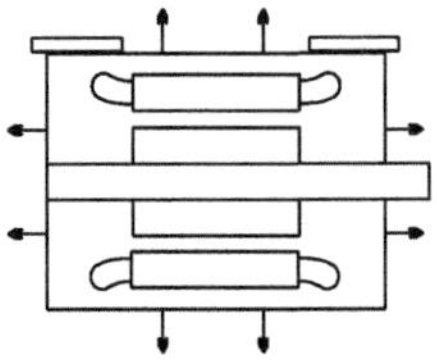
IC 41 (New : IC 411)	- Enclosed machine - Smooth or finned casing - External shaft-mounted fan - Often called TEFC motor	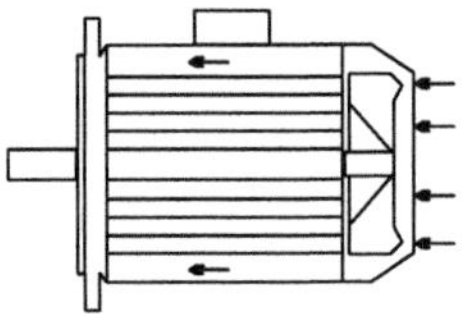
IC 43 A (New : IC 416A)	- Enclosed machine - Smooth or finned casing - External motorized Axial fan supplied with machine	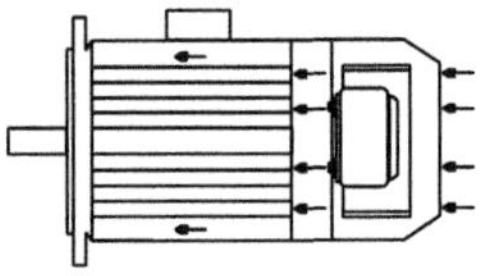
IC 43 R (New : IC 416R)	- Enclosed machine - Smooth or finned casing - External motorized Radial fan supplied with machine	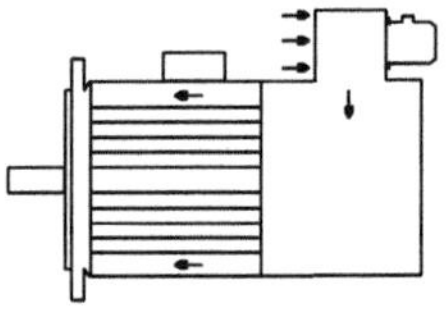
IC 61 (New : IC 610)	- Enclosed machine - Heat Exchanger fitted - Two separate air circuits - Shaft-mounted Fans - Often called CacA motor	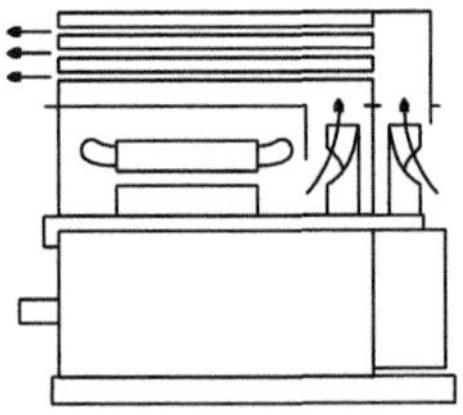

Fig. 2.5.17 Designation of the most common methods of cooling.

adequate cooling at low speeds. On the other hand, for prolonged operation at high speeds above 50 Hz, the shaft mounted fan works well but may make excessive noise. Again, it may be advisable to fit a separately powered cooling fan.

Larger rotating machines can have more elaborate cooling systems with heat exchangers.

The system used to describe the method of cooling is currently being changed by IEC, but the designation system currently in use is as follows:

- A prefix comprising the letters IC (index of cooling).
- A letter designating the cooling medium, this is omitted if only air is used.
- Two numerals which represent:
  1. The cooling circuit layout.
  2. The way in which the power is supplied to the circulation of the cooling fluid, fan, no fan, separate forced ventilation, etc.

2.5.12 ADegree of protection of motor enclosures (IP)

The degree of protection (also called *index of protection – IP*) which is provided by the enclosure of the motor, is classified in the standards IEC 34.5 and AS 1359.20 (Figure 2.5.18).

The system used to describe the index of protection is as follows:

- A prefix comprising the letters IP (index of protection).
- Three numerals which represent:
  1. The protection against contact and ingress of solid objects, such as dust.
  2. The protection against ingress of liquids, such as water.
  3. The mechanical protection and its resistance to impact.

This third numeral is often not used in practice.

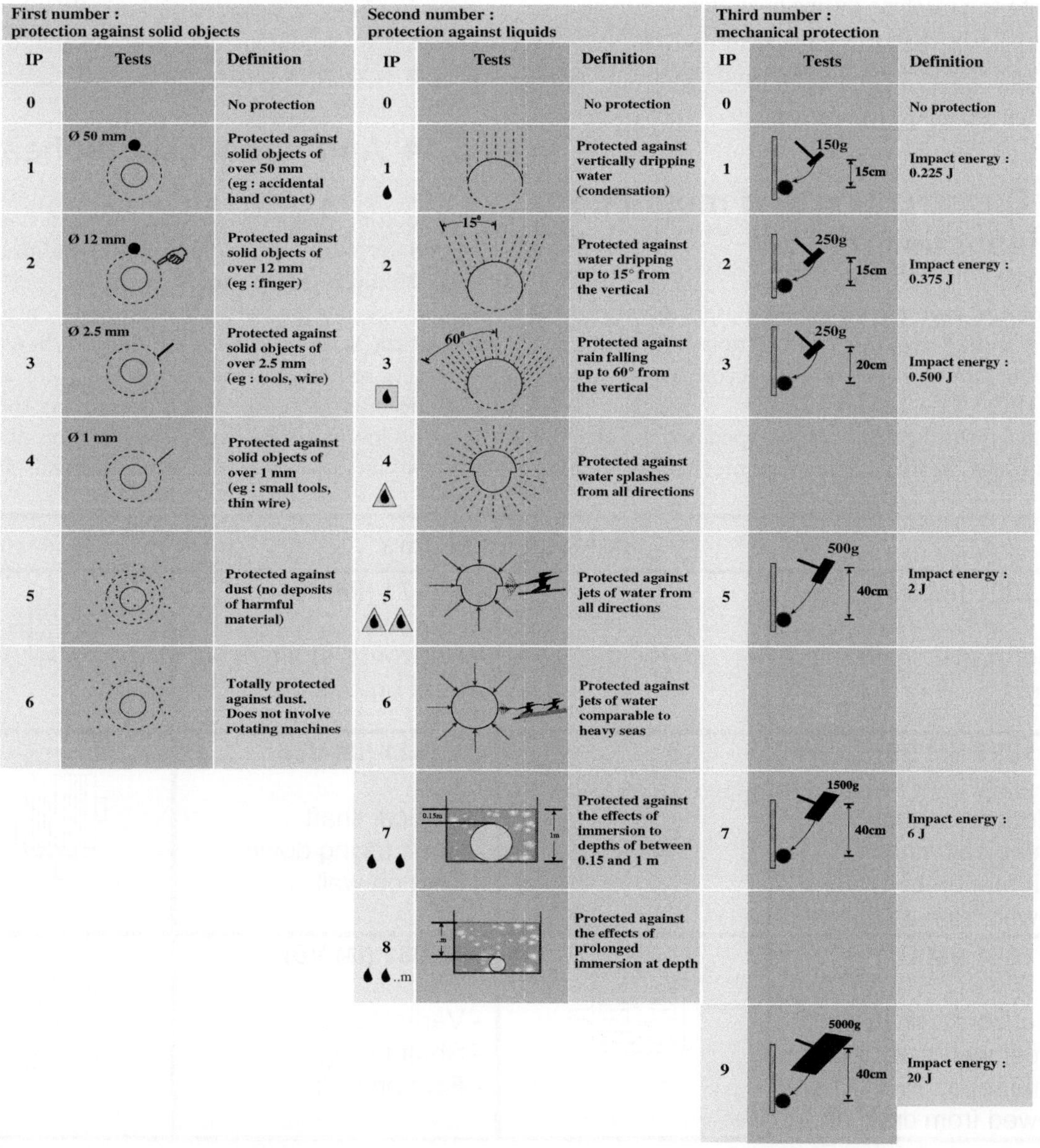

First number : protection against solid objects			Second number : protection against liquids			Third number : mechanical protection		
IP	Tests	Definition	IP	Tests	Definition	IP	Tests	Definition
0		No protection	0		No protection	0		No protection
1	Ø 50 mm	Protected against solid objects of over 50 mm (eg : accidental hand contact)	1		Protected against vertically dripping water (condensation)	1	150g 15cm	Impact energy : 0.225 J
2	Ø 12 mm	Protected against solid objects of over 12 mm (eg : finger)	2	15°	Protected against water dripping up to 15° from the vertical	2	250g 15cm	Impact energy : 0.375 J
3	Ø 2.5 mm	Protected against solid objects of over 2.5 mm (eg : tools, wire)	3	60°	Protected against rain falling up to 60° from the vertical	3	250g 20cm	Impact energy : 0.500 J
4	Ø 1 mm	Protected against solid objects of over 1 mm (eg : small tools, thin wire)	4		Protected against water splashes from all directions			
5		Protected against dust (no deposits of harmful material)	5		Protected against jets of water from all directions	5	500g 40cm	Impact energy : 2 J
6		Totally protected against dust. Does not involve rotating machines	6		Protected against jets of water comparable to heavy seas			
			7	0.15m 1m	Protected against the effects of immersion to depths of between 0.15 and 1 m	7	1500g 40cm	Impact energy : 6 J
			8 ..m	m	Protected against the effects of prolonged immersion at depth			
						9	5000g 40cm	Impact energy : 20 J

Fig. 2.5.18 Summary of the index of protection.

This system of degrees of protection does not relate to protection against corrosion. For example, a machine with an index of protection of *IP557*, is protected as follows:

4. Machine is protected against accidental personal contact of moving parts, such as the fan, and against the ingress of dust.

   Test result: No risk of direct contact with rotating parts (test finger)
   No risk that dust could enter the machine in harmful quantities.

5. Machine is protected against jets of water from all directions from hoses 3 m away and with a flow rate less than 12.5 l/s at 0.3 bar.

   Test result: No damage from water projected onto the machine during operation.

6. Machine is resistant to impacts of up to 6 J.

   Test result: Damage caused by impacts does not affect running of the machine.

2.5.13 Construction and mounting of AC induction motors

Modern squirrel cage AC induction motors are available in several standard types of construction and mounting arrangements. These are classified in accordance with the standards IEC 34.7 and AS 1359.22.

Mounting position needs to be specified to ensure that drain plugs, bearings and other mechanical details are correctly located and dimensioned during assembly.

The system used to describe the mounting arrangements is as follows:

- A prefix comprising the letters IM (index of mounting)
- Four numerals which represent:
  1. Type of construction
  2. Type of construction
  3. Mounting position
  4. Mounting position.

A summary of the mounting designations is shown in Figure 2.5.19.

A previous system of designation used letters *B* (horizontal mounting) and *V* (vertical mounting). This system has been superseded in both IEC 34.7 and AS 1359.22. The old designations are shown in the table in brackets (Figures 2.5.19–2.5.21).

2.5.14 Anti-condensation heaters

When rotating electrical machines need to stand idle for long periods of time in severe climatic conditions, such as a high humidity environment, moisture can be drawn into the machine and absorbed into and onto the insulation of the stator and rotor windings. When a machine is de-energized after it has been running for a period of time, the internal temperature is high. As the machine cools, the low pressure inside the machine draws external moist air into the machine via the seals around the

Foot Mounted Motors			
IM 1001 (IM B3) - Horizontal shaft - Feet on floor	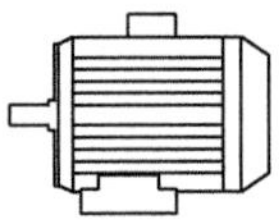	**IM 1071 (IM B8)** - Horizontal shaft - Feet on ceiling	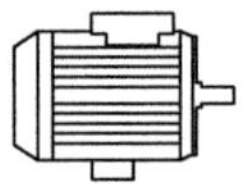
IM 1051 (IM B6) - Horizontal shaft - Feet wall mounted with feet on LHS when viewed from drive end	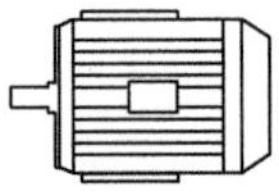	**IM 1011 (IM V5)** - Vertical shaft - Shaft facing down - Feet on wall	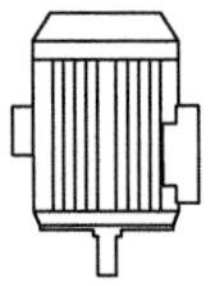
IM 1061 (IM B7) - Horizontal shaft - Feet wall mounted with feet on RHS when viewed from drive end	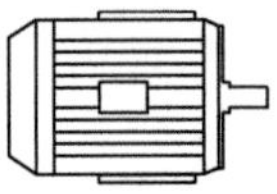	**IM 1031 (IM V6)** - Vertical shaft - Shaft facing up - Feet on wall	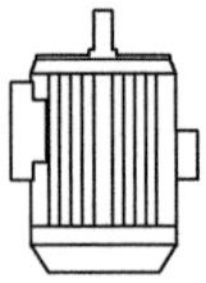

Fig. 2.5.19 Mounting designations for foot mounted motors.

Flange Mounted Motors			
IM 3001 (IM B5) - Horizontal shaft	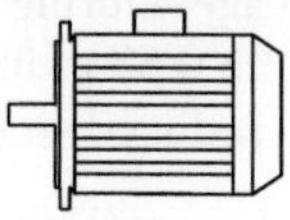	**IM 2001 (IM B35)** - Horizontal shaft - Feet on floor	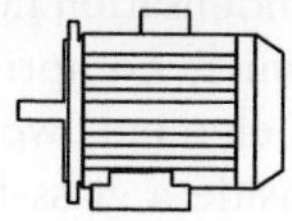
IM 3011 (IM V1) - Vertical shaft - Shaft facing down	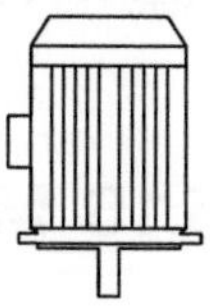	**IM 2011 (IM V15)** - Vertical shaft - Shaft facing down - Feet on wall	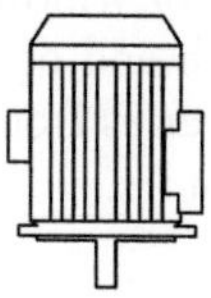
IM 3031 (IM V3) - Vertical shaft - Shaft facing up	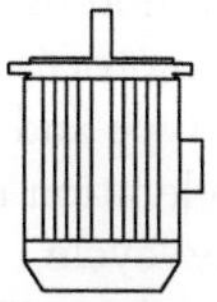	**IM 2031 (IM V36)** - Vertical shaft - Shaft facing up - Feet on wall	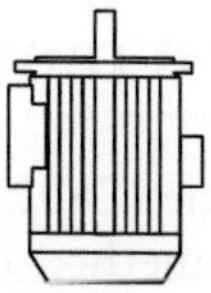

Fig. 2.5.20 Mounting designations for flange mounted motors.

shaft. The moisture degrades the performance of the insulation materials by providing a partially conductive path between the windings and the frame of the machine. When the machine is energized, electrical breakdown of the insulation can occur. Standby motors or generators, which have not been used for some time, can fail to operate when they are needed.

Under these conditions, where a motor is expected to stand idle for long periods in an environment of high humidity, it may be necessary to specify additional winding impregnation treatment and consideration should also be given to anti-condensation heaters. These are fitted inside the motor and their connections brought out to terminals. The heaters are energized from a 240 V

Face Mounted Motors			
IM 3601 (IM B14) - Horizontal shaft	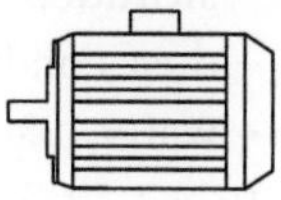	**IM 2101 (IM B34)** - Horizontal shaft - Feet on floor	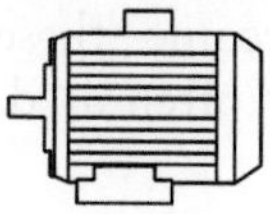
IM 3611 (IM V18) - Vertical shaft - Shaft facing down	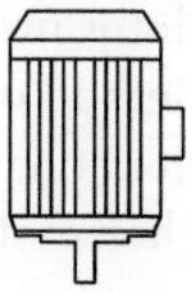	**IM 2111 (IM V58)** - Vertical shaft - Shaft facing down - Feet on wall	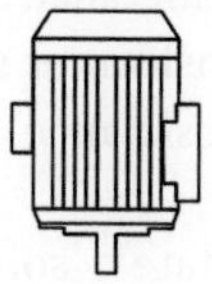
IM 3631 (IM V19) - Vertical shaft - Shaft facing up	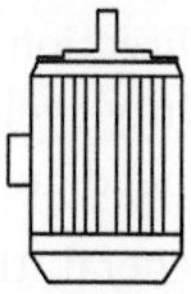	**IM 2131 (IM V69)** - Vertical shaft - Shaft facing up - Feet on wall	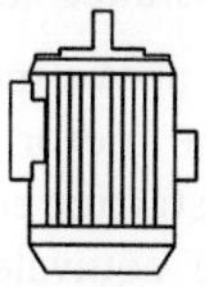

Fig. 2.5.21 Mounting designations for face mounted motors.

supply when the motor is not in use to prevent condensation forming inside the windings.

Anti-condensation heaters are normally in the form of a tape, which comprises a flat glass-fiber tape with a heating element woven into it. This tape is then inserted inside a glass-fiber sleeve and wrapped around the stator winding overhang, braced and impregnated with the stator winding. One heater element is normally fitted to each end of the stator winding. A typical rating of a heater varies from 25 W, for small motors, to 200 W for large motors.

2.5.15 Methods of starting AC induction motors

Direct-on-line (DOL) starting is the simplest and most economical method of starting an AC squirrel cage induction motor. A suitably rated contactor is used to connect the stator windings of the motor directly to the three-phase power supply. While this method is simple and produces a reasonable level of starting torque, there are a number of disadvantages:

- The starting current is very high, between 3 and 8 times the full load current. Depending on the size of the motor, this can result in voltage sags in the power system.
- The full torque is applied instantly at starting and the mechanical shock can eventually damage the drive system, particularly with materials handling equipment, such as conveyors.
- In spite of the high starting current, for some applications the starting torque may be relatively low, only 1.0–2.5 times full load torque.

To overcome these problems, other methods of starting are often used. Some common examples are as follows:

- Star-delta starting
- Series inductance starting (e.g. series chokes)
- Auto-transformer starting
- Series resistance starting (e.g. liquid resistance starter).
- Solid state soft-starting (e.g. smart motor controller).
- Rotor resistance starting, requires a slipring motor.

Most of the above motor-starting techniques *reduce the voltage* at the motor stator terminals, which effectively reduces the starting current as well as the starting torque.

From the equivalent circuits and formulae for AC induction motors, covered earlier in this chapter, the following conclusions can be drawn about *reduced voltage starting*:

- Both the stator current and output torque during starting are proportional to the square of the voltage. During star-delta starting, the voltage is reduced to 0.58 of its rated value. The current and torque are reduced to 0.33 of prospective value.

$$I_{Start} \propto (\text{Voltage})^2$$
$$T_{Start} \propto (\text{Voltage})^2$$

2.5.16 Motor selection

The correct selection of an AC induction motor is based on a thorough understanding of the application for which the motor is to be used. This requires knowledge about the type and size of the mechanical load, its starting and acceleration requirements, running speed requirements, duty cycle, stopping requirements, and the environmental conditions. The following checklist and reference to the preceding sections provides a guide to the selection procedure.

When selecting an electric motor, the following factors should be considered:

- Type and torque requirements of the mechanical load
- Method of starting
- Acceleration time
- Type of construction of AC induction motor:
  - squirrel cage rotor,
  - wound-rotor with sliprings,
  - foot mounted,
  - flange mounted.
- Environmental conditions:
  - ambient temperature,
  - altitude,
  - dust conditions,
  - water.
- Required degree of protection of the enclosure
- Insulation class
- Motor protection
- Method of cooling
- Mounting arrangement:
  - horizontal,
  - vertical.
- Cable connections
- Direction of rotation
- Duty cycle
- Speed control (if required).

In general, the selection of the motor is dictated by the type of load and the environment in which it will operate. The selection of a cage motor or slipring motor is closely related to the size of the machine, the acceleration time required (determined by load) and the method of starting (determined by the electrical supply limitations).

From the point of view of price, reliability, and maintenance, the cage motor is usually the first choice. In general, slipring motors are required when:

- The load has a high starting torque requirement, but the supply dictates a low starting current.
- The acceleration time is long due to high-load inertia, such as in a fan.
- Where duty dictates frequent starting, inching or plugging.

These are general comments because cage motors can be successfully used in all the above situations.

Slipring motors are sometimes used for limited speed control. The slip can be controlled by controlling the external rotor resistance. As demonstrated earlier, the overall efficiency of this method is poor, so this method can only be used if the speed does not deviate too far from the rated speed. The slip power is dissipated as heat in the external rotor resistors.

Section Three

Electronic drive control

Chapter 3.1

3.1

Electronic circuits and devices

Forsyth

3.1.1 Introduction

The use of solid-state techniques for the control and conversion of electrical power is now commonplace, with applications ranging from small dc power supplies for electronic devices, through actuator and motor drive systems, to large active filters and static power compensators for power transmission and distribution systems.

The aim here is to review the main classes of circuits and devices that are currently used, emphasizing the basic operating principles and characteristics. The opening sections cover the traditional diode-based ac–dc converters, starting with the three-pulse circuit and leading to the twelve-pulse configuration, and including the supply current characteristics. Thyristor phase control techniques are also briefly described. The following sections introduce the principal active devices that are in common use, namely the MOSFET and the IGBT (insulated gate bipolar transistor), then, starting with simple dc–dc converter switching cells, switching and inverter circuits are explained, leading to sinusoidal pulse width modulation (PWM) methods. Finally, there is a short review of high-frequency power supplies.

3.1.2 Diode converters

3.1.2.1 Three-pulse rectifier

The three-pulse rectifier in Figure 3.1.1 supplies a resistive load in series with a filter inductor, the inductor being large enough to ensure that the load current is continuous and ripple-free. The circuit is a half-wave rectifier, with each supply line, A, B and C, being connected through a single diode to the top of the load, the neutral wire forming the return path. The circuit waveforms follow directly from the assumption of a continuous and smooth load current:

- One of the three diodes must always be in conduction to provide a path for the load current.
- Only one diode may conduct at once, since otherwise two of the supply lines would have the same voltage, and this only occurs momentarily as the voltages cross each other.
- The diode that conducts at any instant is determined by the supply line with the largest positive voltage, since this voltage will forward bias the diode in that line and reverse bias the other two diodes.

Therefore the diodes conduct in sequence for $2\pi/3$ radians or 120°, and the rectifier output voltage v_D is the maximum value of the three line-to-neutral voltages. Since the v_D waveform repeats every $2\pi/3$ radians, or three times each utility cycle, it is known as a three-pulse rectifier, and the output voltage ripple frequency is three times the supply frequency.

As the repetition period of v_D is $2\pi/3$ radians, the average value of the waveform may be calculated by integrating over this interval:

$$\text{average } v_D = \frac{1}{2\pi/3}\int_{\pi/6}^{5\pi/6} V_{LN}\sin\theta\, d\theta = \frac{3\sqrt{3}\,V_{LN}}{2\pi} = \frac{3V_{LL}}{2\pi} \quad (3.1.1)$$

Newnes Electrical Power Engineer's Handbook; ISBN: 9780750662680

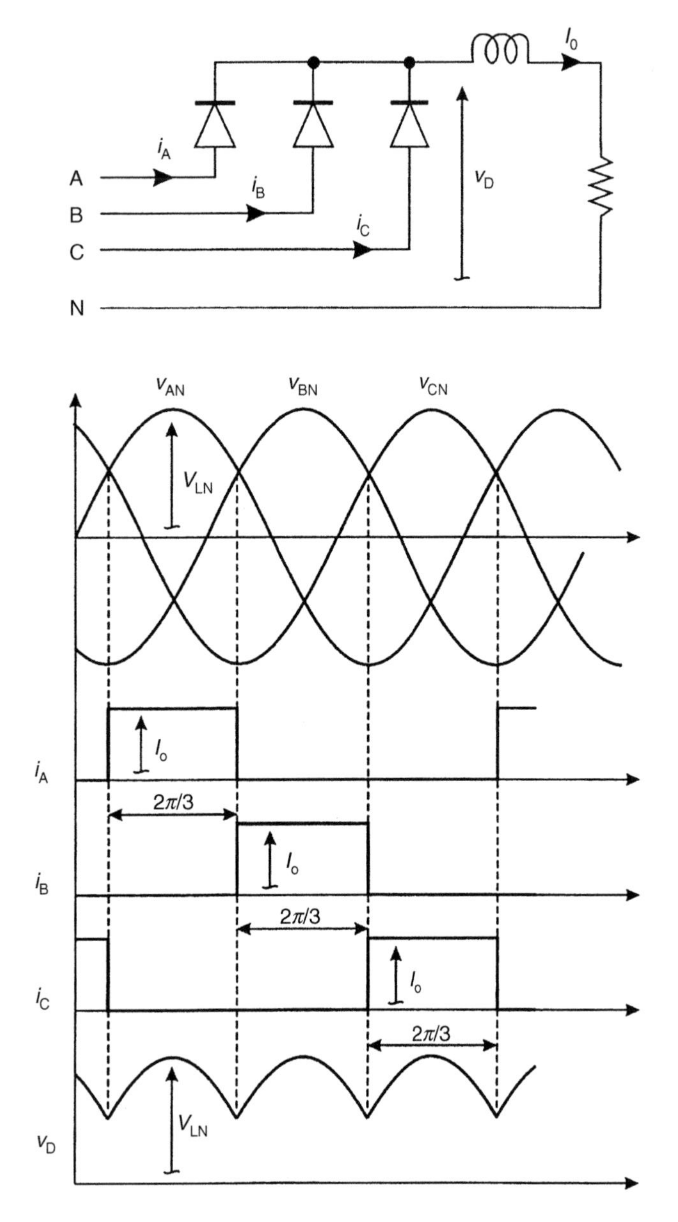

Fig. 3.1.1 Three-pulse rectifier.

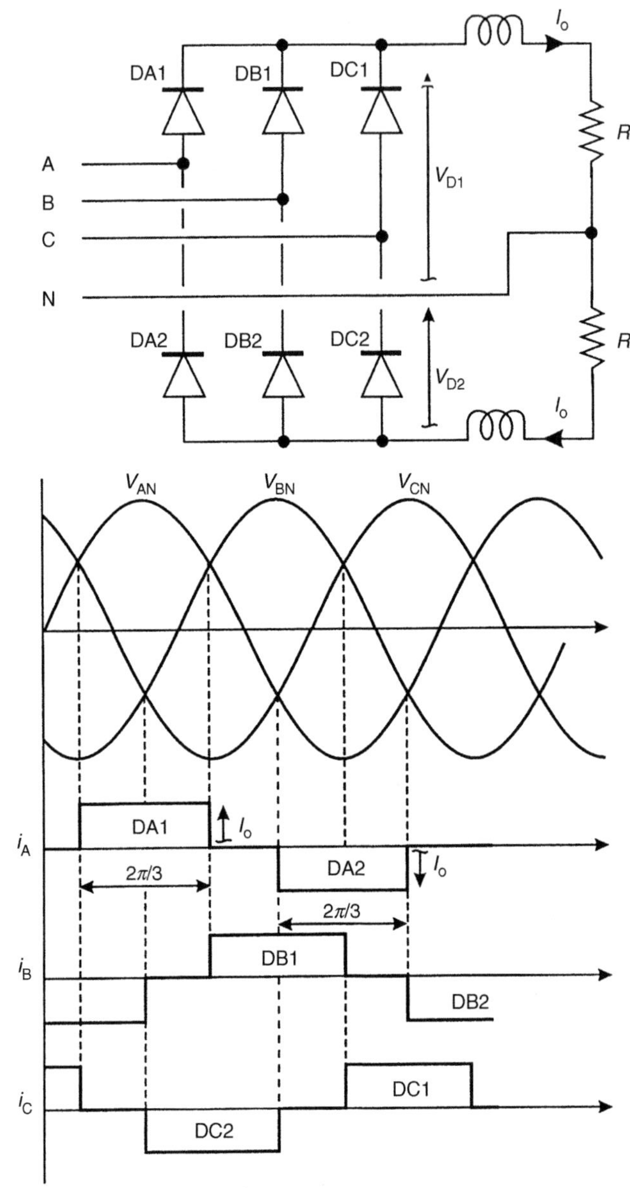

Fig. 3.1.2 Six-pulse rectifier as two back-to-back three-pulse circuits.

where V_{LL} is the amplitude or peak value of the line-to-line supply voltage.

Since the circuit only draws a unidirectional current from each supply line, and also uses the neutral wire for the return current, it is rarely used in practice, but it is simple to understand, and is helpful in analyzing more complex circuits.

3.1.2.2 Six-pulse rectifier

Connecting two three-pulse circuits, one operating on the positive half cycle and the other on the negative half cycle, in a back-to-back configuration forms the six-pulse rectifier, as shown in Figure 3.1.2. The upper three-pulse rectifier here is identical to that described in Section 3.1.2.1, whilst the lower circuit is essentially formed from the first by reversing the diodes, enabling it to operate on the negative parts of the supply voltage waveform.

Since the average values of v_{D1} and v_{D2} are the same, the currents in the two load resistors will be the same, provided that these have equal values R. Under these conditions, the line currents will consist of 120° pulses of $+I_o$ as DA1, DB1 and DC1 conduct, and 120° pulses of $-I_o$ as DA2, DB2 and DC2 conduct. The negative pulses will flow between the positive pulses, making the line currents quasi-squarewaves.

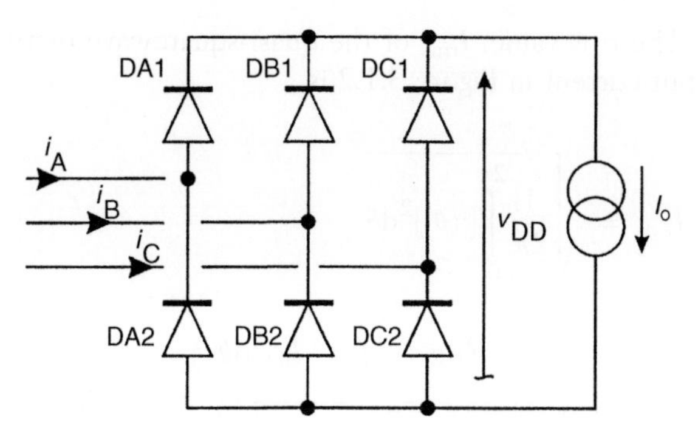

Fig. 3.1.3 Six-pulse rectifier.

Furthermore, since the two resistors carry equal currents, the neutral current will be zero, and the neutral connection may be removed without affecting the circuit operation. This leaves the standard three-wire, six-pulse rectifier of Figure 3.1.3, into which the load is simply drawn as a constant current element, representing its highly inductive nature.

Figure 3.1.4 shows the detail of the two three-pulse output voltages v_{D1} and v_{D2} and the total output voltage $v_{DD} = v_{D1} + v_{D2}$. Since the ripple components in v_{D1} and v_{D2} are phase-shifted by $\pi/3$ radians, the resultant ripple in v_{DD} is reduced in amplitude and has a repetition interval of $\pi/3$ radians, giving six pulses per period of the input waveform. The average of v_{DD} is simply twice the average value of the three-pulse waveforms and from Equation (3.1.1):

$$\text{average } V_{DD} = 2\frac{3V_{LL}}{2\pi} = \frac{3V_{LL}}{\pi} \quad (3.1.2)$$

The six-pulse rectifier is widely used for the conversion of ac to dc, for example as the input stage of a variable speed induction motor drive system, described in Chapter 2.3.

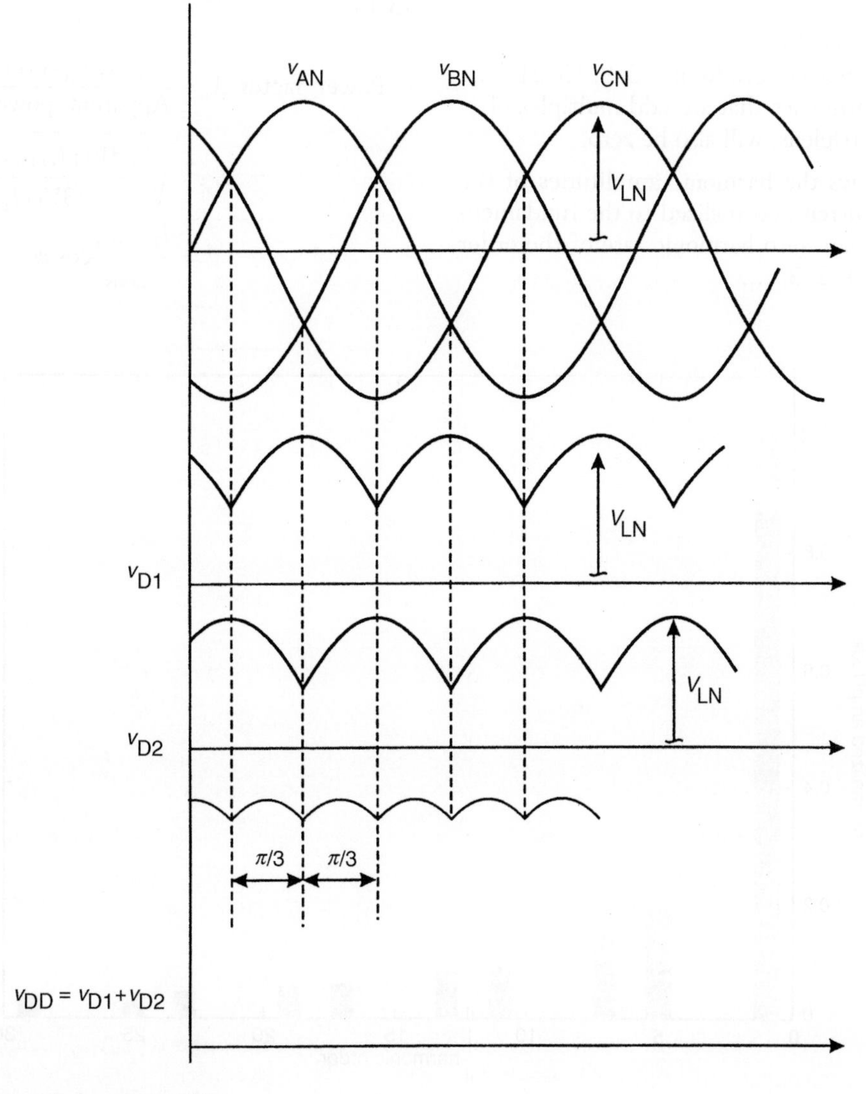

Fig. 3.1.4 Output voltage of six-pulse rectifier.

3.1.2.3 Characteristics of the six-pulse input current

Fourier analysis of the quasi-squarewave currents drawn by the six-pulse rectifier shows that the harmonic amplitudes of the currents are given by the equation:

$$\text{harmonic amplitudes} = \frac{4I_o}{n\pi}\sin\left[\frac{n\pi}{3}\right] \text{for odd values of } n \tag{3.1.3}$$

where n is the harmonic number. The fundamental component of each line current is in phase with the respective line-to-neutral voltage, as is evident from Figure 3.1.2.

The following conclusions may be drawn from Equation (3.1.3):

- the fundamental current has an amplitude of $2\sqrt{3}I_o/\pi$;
- there are no even harmonics;
- the sine function will be zero for $n = 3, 9, 15, 21, \ldots$, therefore, the harmonics that are odd multiples of three, known as triplens, will also be zero.

Figure 3.1.5 shows the harmonic amplitudes of the quasi-squarewave current, normalized to the fundamental amplitude. The non-zero harmonics are of the order $6k \pm 1$, where $k = 1, 2, 3, \ldots$.

The rms value, I_{rms} of the quasi-squarewave rectifier input current in Figure 3.1.2 is

$$I_{rms} = \sqrt{\frac{1}{2\pi}\int_0^{2\pi}[i(\theta)]^2 d\theta} = \sqrt{\frac{\int_0^{2\pi/3} I_o^2 d\theta + \int_{\pi}^{5\pi/3}(-I_o)^2 d\theta}{2\pi}} = \sqrt{2/3}I_o \tag{3.1.4}$$

The active power transfer to loads that draw non-sinusoidal currents from a sinusoidal supply is solely due to the fundamental component of the current, and the six-pulse rectifier therefore has a power factor of less than unity even though the fundamental component of current is in phase with the voltage. The power factor of the six-pulse rectifier is given by Equations (3.1.5) and (3.1.6).

$$\text{Power factor } \lambda = \frac{\text{Active power}}{\text{Apparent power}} = \sqrt{\frac{\sqrt{3}V_{LL}I_{L(1)rms}\cos\phi_1}{\sqrt{3}V_{LL}I_{Lrms}}} = \frac{I_{L(1)rms}}{I_{Lrms}}\cos\phi_1 \tag{3.1.5}$$

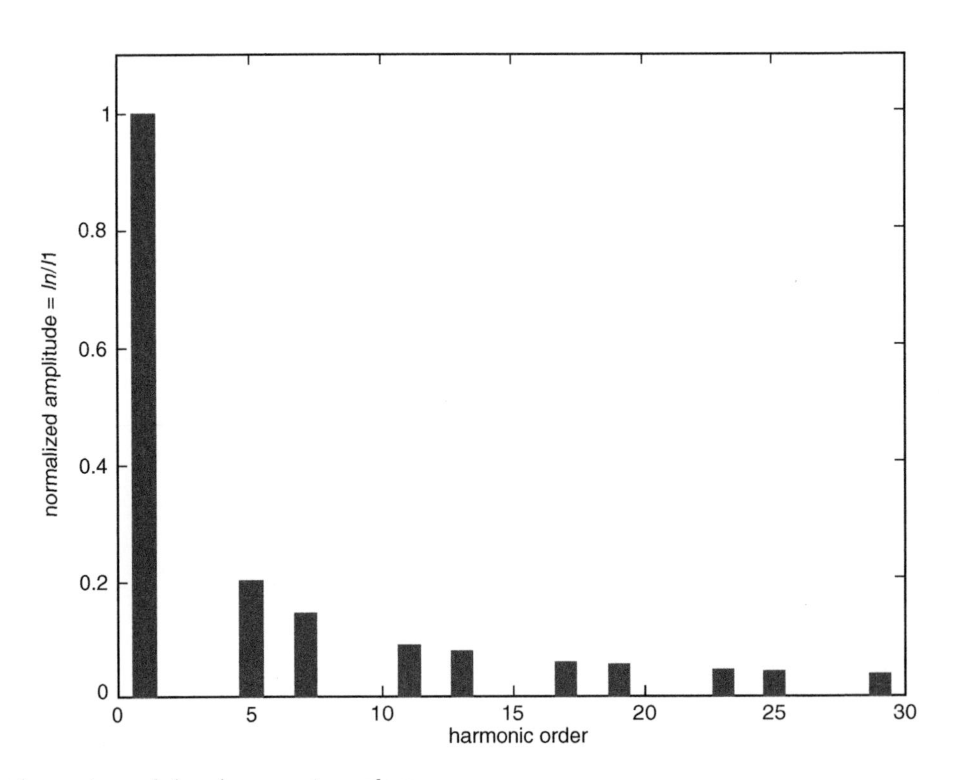

Fig. 3.1.5 Harmonic spectrum of six-pulse current waveform.

where $I_{L(1)rms}$ is the rms fundamental component of line current, I_{Lrms} is the rms line current and ϕ_1 is the phase angle of the fundamental current with respect to the voltage. Since $\phi_1 = 0$, the power factor is

$$\text{Power factor } \lambda = \frac{I_{L(1)rms}}{I_{Lrms}} = \frac{(2\sqrt{3}I_o/\pi)/\sqrt{2}}{\sqrt{2/3}I_o} = \frac{3}{\pi} = 0.955 \tag{3.1.6}$$

The practical operation of the rectifier differs slightly from this idealized representation because of the presence of source inductance, which slows down the edges of the quasi-squarewave input currents and gives rise to an output voltage regulation effect. Ripple current in the dc inductor also modifies the shape of the input currents.

3.1.2.4 Twelve-pulse rectifier with delta–star transformer

To enhance the characteristics of the six-pulse rectifier, reducing the input current harmonics and the output voltage ripple, multiple rectifiers may be combined with a phase-shifting device, a transformer or an autotransformer.

As an example, Figure 3.1.6 shows one of the simplest ways of combining two six-pulse rectifiers to produce a twelve-pulse circuit.

The first rectifier in Figure 3.1.6 is supplied through a delta–star transformer, whilst the second operates through a delta–delta transformer. The turns ratio of the delta–star transformer is $1 : 1/\sqrt{3}$, so the secondary line-to-line voltages have the same amplitude as the primary voltages, but are advanced in phase by 30°. The delta–delta transformer has a 1:1 turns ratio and its primary and secondary voltages and currents are identical, neglecting magnetizing current; it does not provide any phase shift. The two sets of transformer primaries are connected in parallel, whilst the rectifier outputs are connected in series across a common load. This results in the two six-pulse rectifiers delivering the same load current. The two transformers could be replaced by a single device having two sets of secondaries; one connected in star, the other in delta.

Figure 3.1.7 shows the current waveforms for the first supply line in the twelve-pulse rectifier. The diagram shows that in the two six-pulse rectifiers, each draw quasi-squarewave currents of value $\pm I_o$, i_{X1} and i_{X2} being the currents in the first input line to each rectifier. A phase shift of $\pi/6$ radians or 30° is shown

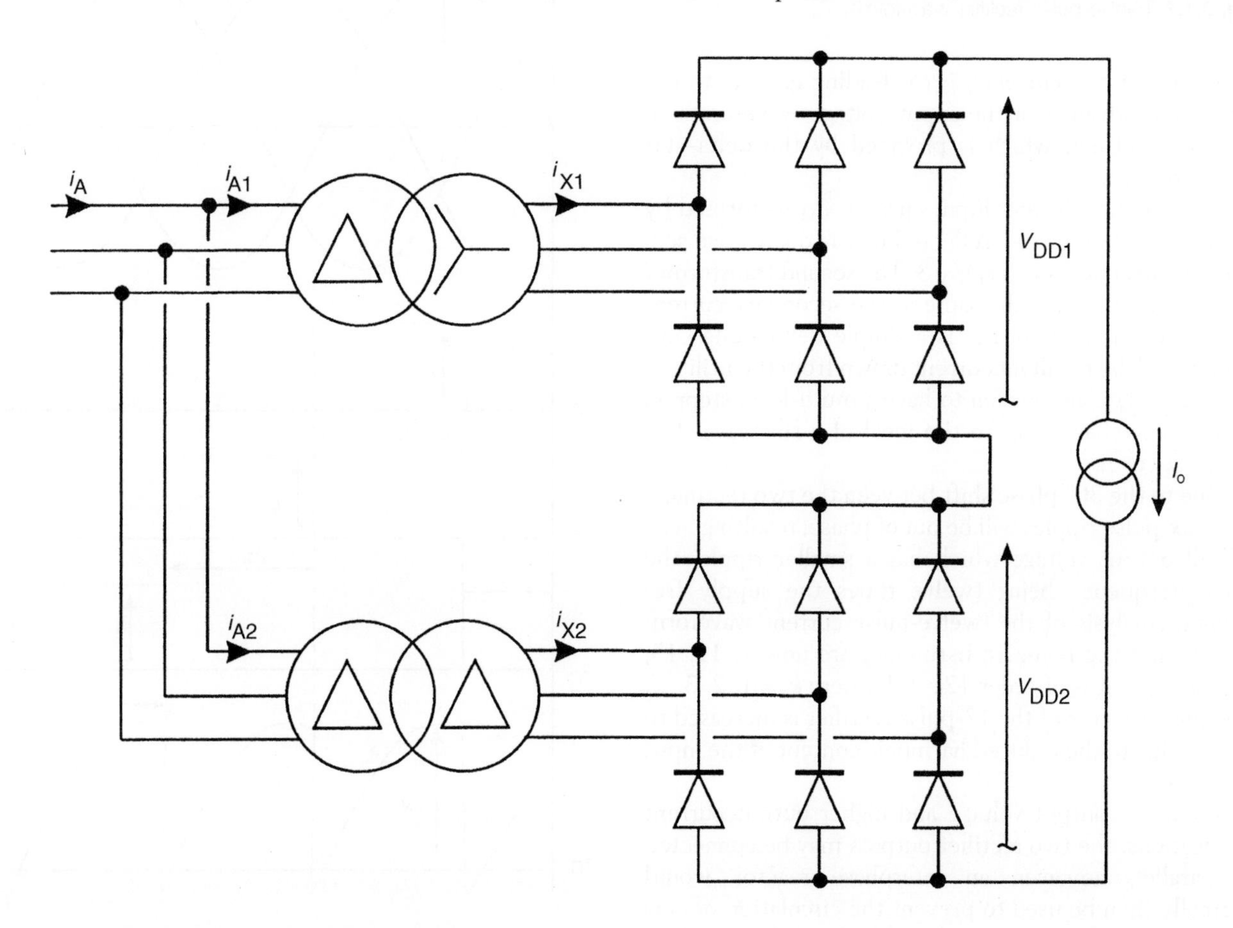

Fig. 3.1.6 Twelve-pulse rectifier – series connection.

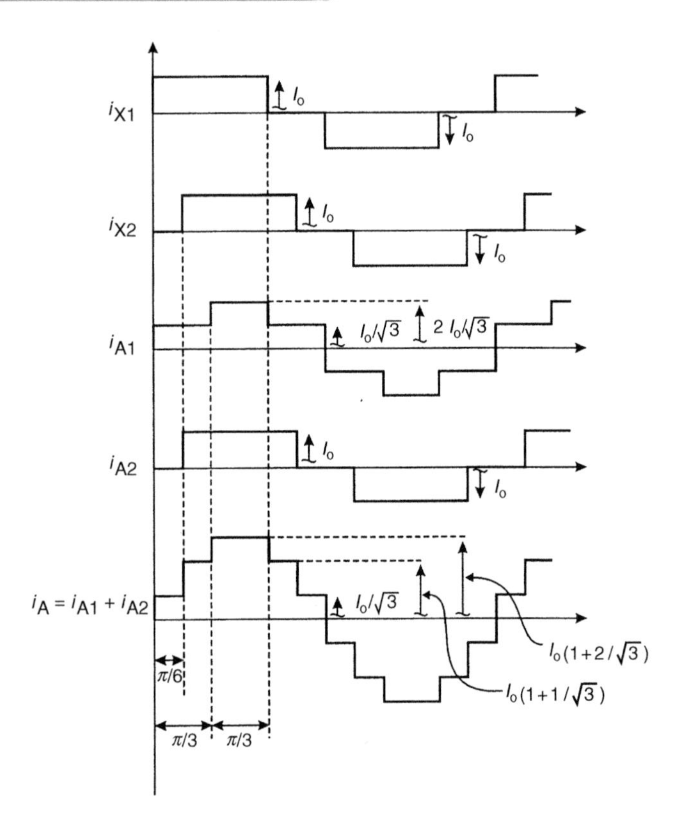

Fig. 3.1.7 Twelve-pulse rectifier waveforms.

between the two currents, i_{X1} is leading i_{X2} due to the 30° phase advance in the input voltage waveforms to the first rectifier, which is provided by the delta–star transformer.

The first transformer input current, i_{A1}, is formed by the combination of two reflected quasi-squarewave secondary currents, $(i_{X1} - i_{Z1})/\sqrt{3}$. The second transformer input current, i_{A2}, is the same as the secondary current i_{X2}, since this transformer is a simple delta–delta configuration. The resultant current drawn from the utility is $i_A = i_{A1} + i_{A2}$, and is seen to have a multi-level stepped shape somewhat closer to the ideal of a sinewave than either i_{A1} or i_{A2}.

Due to the 30° phase shift between the two rectifiers, their six-pulse ripples will be out of phase, resulting in an overall output voltage which has a smaller ripple, the ripple frequency being twelve times the supply frequency. Analysis of the twelve-pulse current waveform reveals that the non-zero harmonics are now 1, 11, 13, 23, 25,..., that is of order $12k \pm 1$ where $k = 1, 2, 3, \ldots$. The power factor of the 12-pulse rectifier is increased to 0.9886 due to the reduced harmonic content of the input currents.

For lower output voltage and higher output current applications, the two rectifier outputs may be connected in parallel, however, an interphase reactor would normally then be used to prevent the circulation of currents between the two rectifiers. Twelve-pulse rectifiers would normally be used in higher power applications, or in environments having very stringent power quality specifications, such as an aircraft.

3.2.2.5 Controlled rectifiers

By replacing the diodes in the circuits described in the previous sections by thyristor devices, the output voltage of the rectifiers may be controlled, and a reversal of power flow is also possible if a source of energy is present in the dc circuit. The *thyristor* is a comparatively old power device that has limited control characteristics. A gate pulse must be applied to the control terminal to switch the device into its conducting state, however, once the device is in conduction it cannot be turned off and will only return to the off state when the circuit current naturally falls to zero.

The operating principle of *thyristor-controlled rectifiers* is illustrated using the three-pulse circuit in Figure 3.1.1. The waveforms with thyristor control are shown in Figure 3.1.8. As before, each device conducts

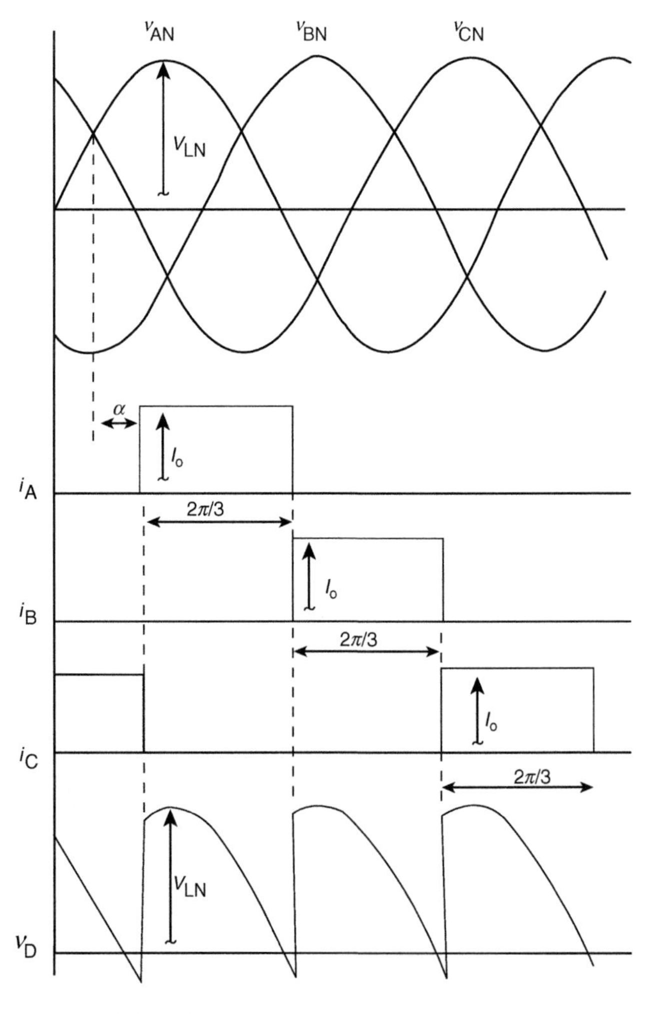

Fig. 3.1.8 Waveforms for phase-controlled three-pulse rectifier.

the dc load current in sequence for one-third of the supply cycle, however, the conduction of each device is delayed by the firing delay angle α. This results in the output voltage v_D being modified. The average value of v_D is given by the equation:

$$\text{average } v_D = \frac{3V_{LL}}{2\pi}\cos\alpha \qquad (3.1.7)$$

The average output voltage falls to zero as α is increased to 90°. If a source of emf is present in the dc circuit, orientated to maintain the current flow, then α may be increased beyond 90° and the circuit enters the inversion mode of operation. The average value of V_D becomes negative, but the direction of current in the dc circuit is unchanged, and power flows from the energy source in the dc circuit into the ac system. To reverse the dc circuit current and obtain four-quadrant operation of the output, two rectifier circuits must be used in a back-to-back configuration.

The output voltage of the six- and twelve-pulse rectifiers may be controlled in a similar manner. However, a disadvantage of this mode of operation is that it results in a phase shift of the input currents, reducing the power factor. Thyristor-controlled rectifiers are widely used for the control of dc motor drives, as described in Section 2.3.10.1.

3.1.3 Active devices

The *MOSFET* and *IGBT* (*insulated gate bipolar transistor*) are the predominant active power devices in use today, covering virtually all mainstream applications, the principal exception being the *GTO* (*gate turn off thyristor*), which is found in specialized high-power systems. The device symbols are shown in Figure 3.1.9.

The MOSFET and IGBT both have an insulated gate control terminal and enhancement type control characteristics; they are normally off and the gate voltage must be increased beyond the threshold voltage (typically around 4 V) to bring the devices into conduction. Drive voltages of 12 or 15 V are normally used to ensure that the devices are fully switched on. Since the gate drive circuit must only charge and discharge the input capacitance of the device at the switching instants, the power consumption of the drive circuit is low, but pulse currents of several amperes are required to ensure rapid switching.

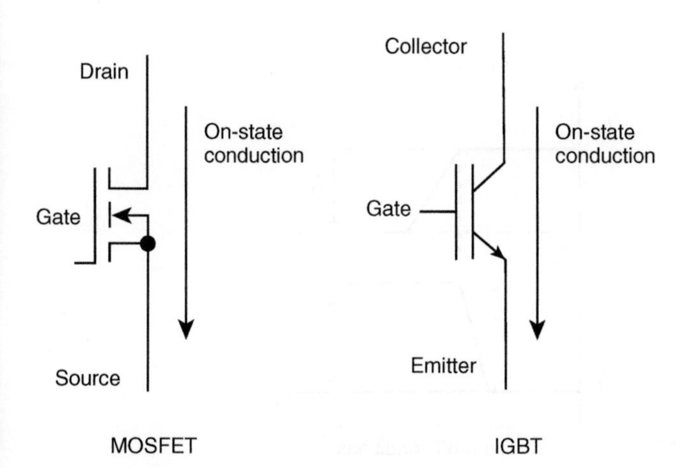

Fig. 3.1.9 MOSFET and IGBT symbols.

The MOSFET is a majority carrier device and is characterized by a constant on-state resistance, so the rms current must be used to estimate conduction losses. The on-state resistance has a positive temperature coefficient and typically rises by a factor of 1.5–2.0 for a 100 °C temperature rise. The switching speed of the MOSFET is very high, current rise and fall times of tens of nanoseconds being achievable. The MOSFETs rated at a few hundred volts are available with current carrying capabilities of up to a hundred amperes, whilst devices with voltage ratings approaching 1000 V tend to have current ratings of just a few amperes, the values of on-state resistance being correspondingly higher. Very few MOSFETs are available with voltage ratings in excess of 1000 V. The MOSFET is therefore used in lower power applications with switching frequencies of up to a few hundred kilohertz. The applications include high-frequency power supplies, dc–dc converters and small servo drives.

The IGBT is a minority carrier device, and through the use of conductivity modulation it is able to operate with much higher current densities than the MOSFET. The device is characterized by a constant on-state voltage, typically in the region of 2.0–3.0 V, requiring the use of the average forward current in estimates of on-state losses. The on-state voltage of the IGBT usually has a positive temperature coefficient, rising by approximately 20 per cent for a 100 °C temperature rise. Due to the recombination time of the stored carriers within the device, the IGBT exhibits a tail current characteristic at turn off; the current rapidly falls to around 10 per cent of its on-state level then decays down to zero comparatively slowly, the overall switching time being a significant fraction of a microsecond. This effect limits the maximum operating frequency of the device to a few tens of kilohertz. The IGBTs are available with current ratings of up to several hundred amperes and with off-state voltages of up to several kilovolts. The devices are widely used in three-phase inverters and converters and have been used in small high-voltage dc power transmission systems.

3.1.3.1 Inductive switching waveforms

In the majority of power electronic circuits the operation of an active device results in the commutation of an inductive current to or from the device and a freewheel diode path. The simple equivalent circuit in Figure 3.1.10 illustrates this basic switching process. The inductive current path is represented by a constant current element.

Assuming that the voltages and currents change linearly at the switching instants, then the turn-on and turn-off waveforms are as shown in Figure 3.1.10. At the turn-on instant, the transistor current must rise beyond the full load current level by an amount I_{rr}, the *peak reverse recovery current* of the diode, before the diode can support reverse voltage and the transistor voltage can collapse to its on-state level. During the switching transient, the transistor experiences high instantaneous power dissipation and a significant energy loss. A similar effect is seen to occur at turn-off, where the transistor voltage must rise to the off-state level, forward biasing the diode, before the transistor current can fall to zero.

The average power loss in a transistor due to the inductive switching waveforms increases with operating frequency and this is a limit to the maximum operating frequency of a device. *Snubber circuits* have been used to shape the switching waveforms and limit the power losses in the devices but since the MOSFET and IGBT are much more robust than the bipolar power transistors that they replaced, these are now less common.

3.1.4 Principles of switching circuits

3.1.4.1 DC–dc converters

The simplest and most common dc–dc converter is the *step-down chopper* or *buck converter*, and is shown in Figure 3.1.11 along with idealized waveforms. The circuit operates from a dc source, V_{in}, and supplies power at a lower voltage, V_o, to a load element, which is shown here as a resistor in parallel with a smoothing capacitor, but could instead be the armature of a dc motor, as described in Section 2.3.10.1. The inductor acts as an energy storage buffer between the input and the output.

When the transistor is switched on the voltage V_D is equal to V_{in}, the freewheel diode is reverse biased, and a voltage of $V_{in}-V_o$ drives a linearly increasing current, I_L, through the inductor. When the transistor turns off, the inductor current diverts to the freewheel diode, voltage

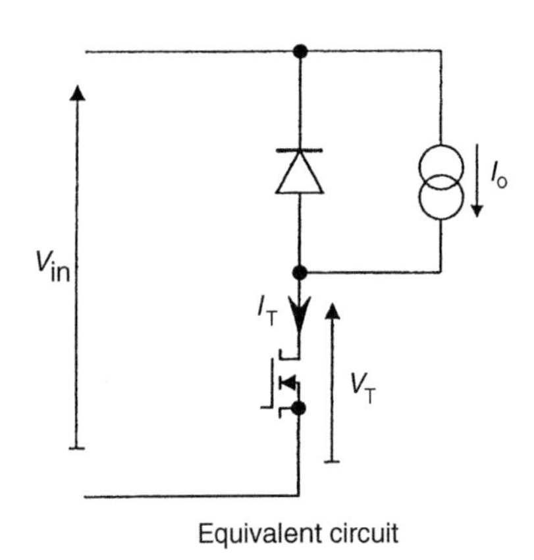

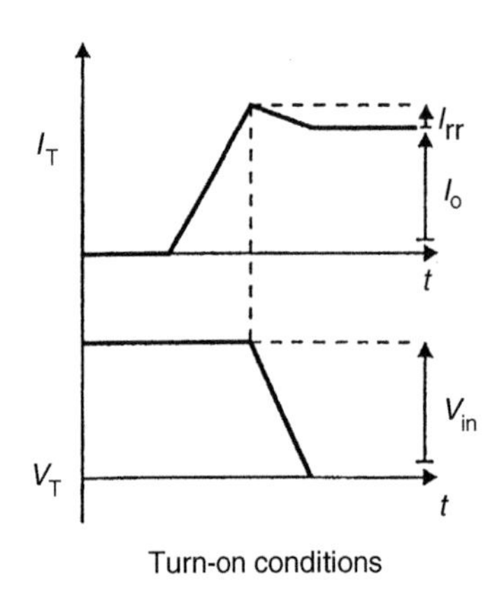

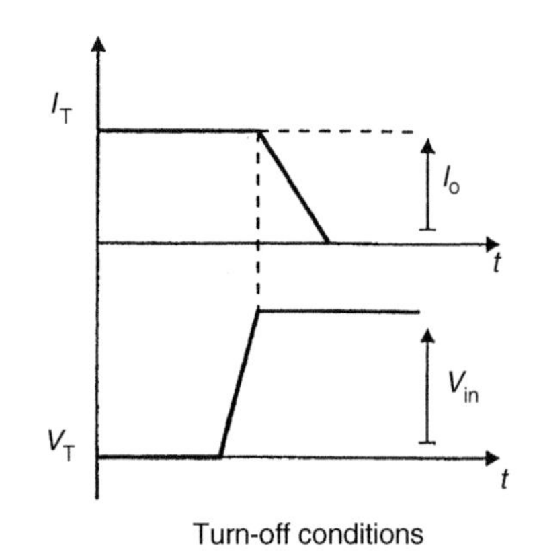

Fig. 3.1.10 Inductive switching waveforms.

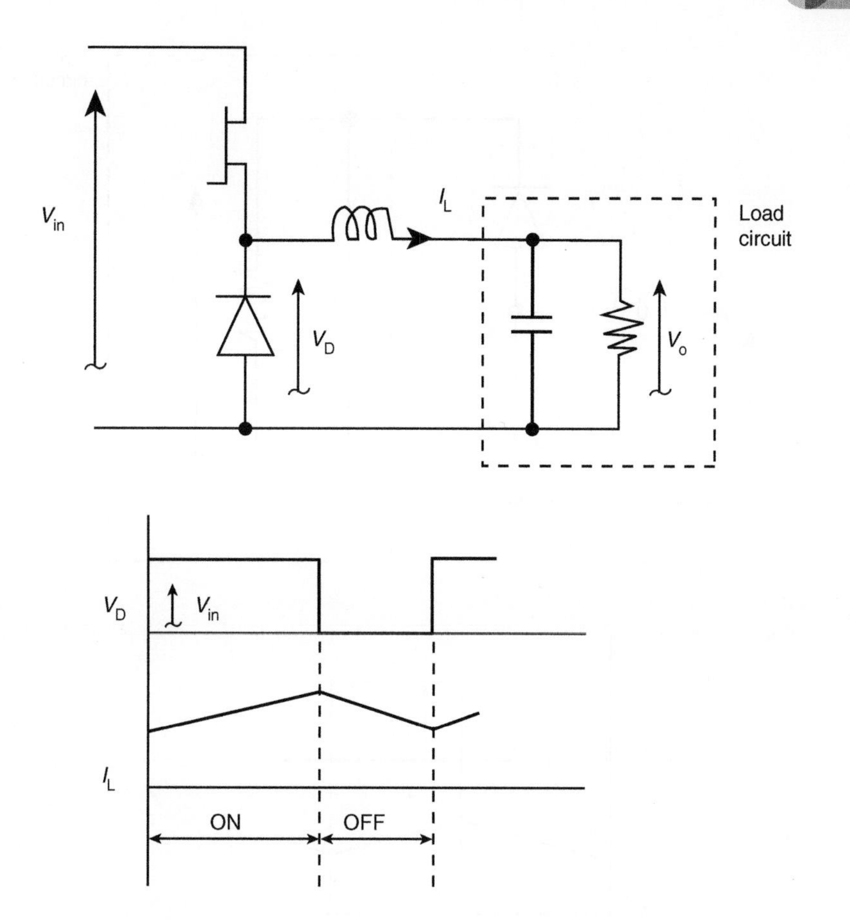

Fig. 3.1.11 Step-down chopper or buck converter.

V_D is zero and there is a reverse voltage of V_o across the inductor resulting in a linear fall in current. The load voltage V_o is equal to the average of the V_D waveform since no average or dc voltage is dropped across an ideal inductor, and is given by the equation:

$$V_o = \text{average of } V_D = DV_{in} \tag{3.1.8}$$

where D is the *duty ratio* or *on time-to-period ratio* of the transistor, and $0 \le D \le 1$.

The *step-up chopper* or *boost converter* (Figure 3.1.12) is essentially a circuit dual of the buck converter. The transistor and diode again conduct the inductor current alternately. The inductor current rises linearly when the transistor is turned on, the inductor voltage being equal to V_{in}, and the current falls linearly when the transistor is turned off since the inductor voltage is then reversed with a value of $V_o - V_{in}$.

The average of the V_D waveform is now equal to the input voltage, and since the average of V_D is $V_o(1 - D)$, the output voltage may be expressed in the equation:

$$V_o = \frac{V_{in}}{1 - D} \tag{3.1.9}$$

which confirms the voltage step-up operating characteristics of the circuit.

3.1.4.2 Two- and four-quadrant converters

By re-drawing the boost converter circuit with the input on the right, the buck and boost circuits may be combined to form the bidirectional converter shown in Figure 3.1.3. To ensure proper circuit operation the transistors must be switched in anti-phase. With a dc source connected to the left-hand terminals of the bidirectional converter and a load element on the right, the circuit operates as a buck converter and the inductor current flows to the right. If the dc source is connected to the right-hand terminals and the load connected on the left, the inductor current will be reversed and the circuit will operate as a boost converter.

The circuit therefore provides two-quadrant operation with unidirectional voltage at both terminals, but bidirectional currents. The circuit could be used to control the charging and discharging of a battery, or to provide two-quadrant operation of a dc machine, with unidirectional voltage and therefore speed, but with bidirectional current and torque. The bidirectional converter in Figure 3.1.13 is commonly referred to as an *inverter leg* since it forms the building block of dc–ac inverter circuits.

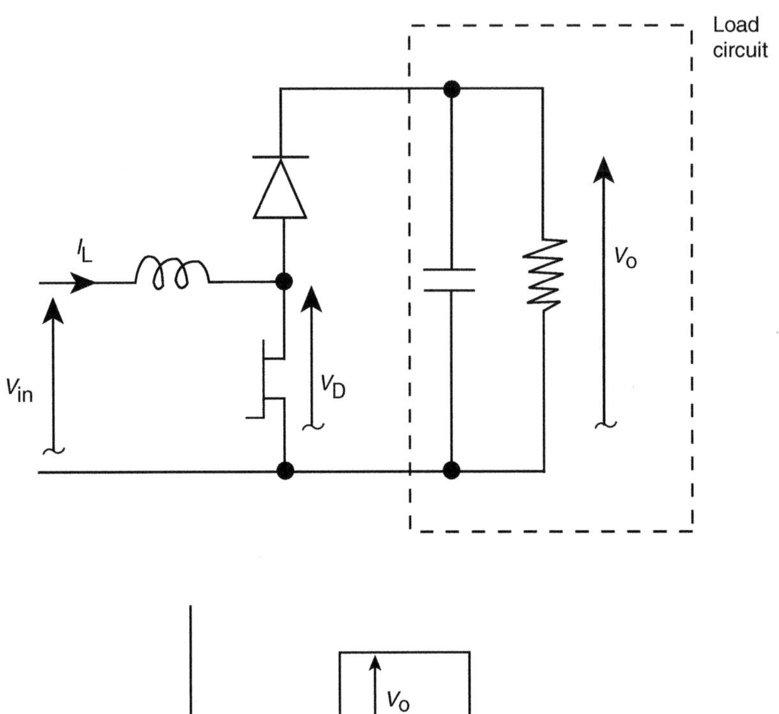

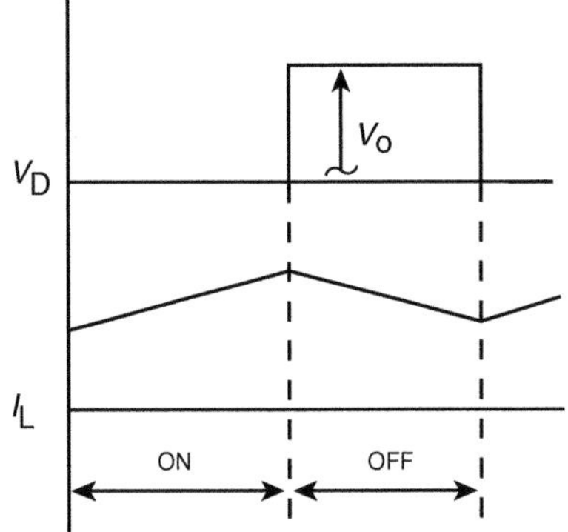

Fig. 3.1.12 Step-up chopper or boost converter.

To achieve a four-quadrant output, two of the bidirectional circuits operating from a common dc source may be combined as shown in Figure 3.1.14. The inductor is no longer shown but it is implied and it might typically be formed from the load impedance, such as the inductance of a motor.

The four-quadrant output is formed by the difference of the two bidirectional converter outputs. The resultant circuit is sometimes known as an *H-bridge* or a *single-phase inverter*, and it has a wide variety of applications, requiring different operating patterns for the transistors. The transistors in each leg of the circuit would normally be operated in anti-phase to ensure that the leg output voltages are always defined by the state of the transistors. Typical applications include:

- *four-quadrant drive for a dc machine*, enabling motoring and generating operation with both

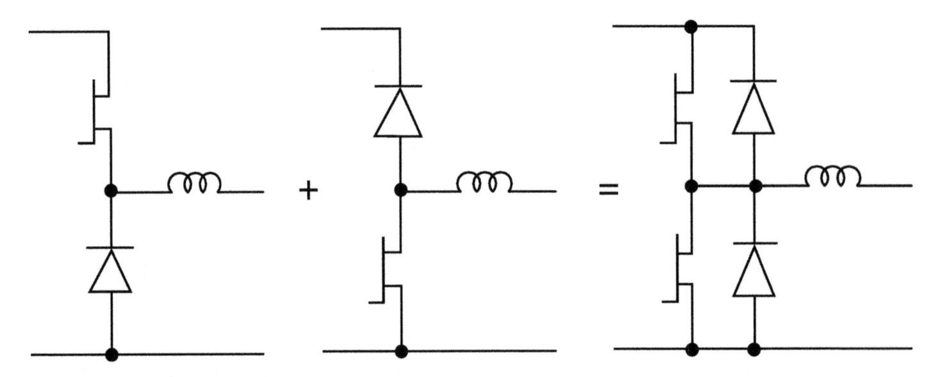

Fig. 3.1.13 Combination of the buck and boost converters.

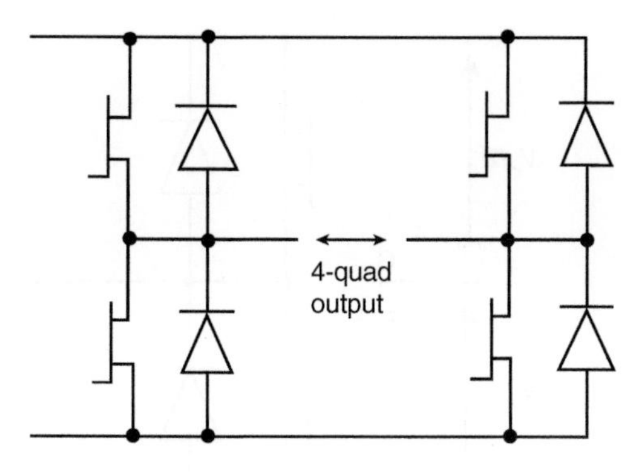

Fig. 3.1.14 Single-phase inverter.

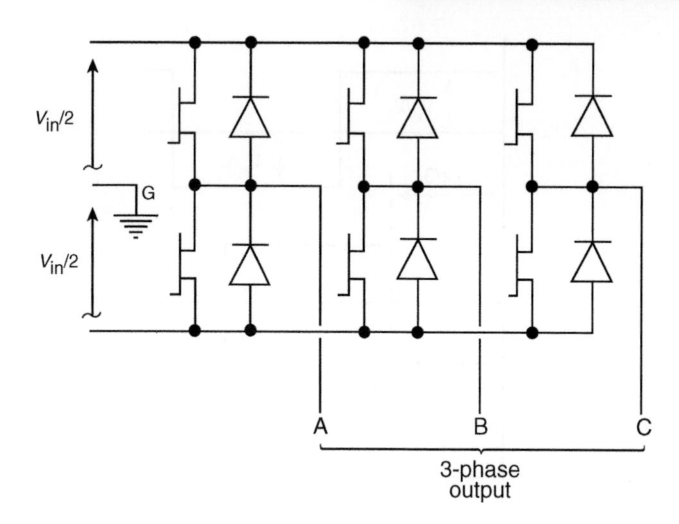

Fig. 3.1.15 Three-phase inverter.

forward and reverse rotation, as described in Section 2.3.10.1. Diagonally opposite transistors could be operated in synchronism with variable duty ratio. With a duty ratio of 0.5 the average machine voltage would be zero, and increasing or decreasing the duty ratio could produce positive or negative average voltages.

- *high-frequency ac output* to drive a high-frequency transformer in a switched-mode power supply. Here the two legs could each be operated with 0.5 duty ratios, and a variable phase shift introduced between the legs to produce a controllable quasi-squarewave output voltage.
- *synthesizing an ac output from a dc source*, for example to supply ac equipment in the absence of a utility connection, or to interface a dc energy source such as a photovoltaic array with utility. In this case, diagonally opposite devices could be operated in synchronism with a 0.5 duty ratio, producing a squarewave ac output voltage, however, a filter circuit may be required to remove the unwanted harmonics from the squarewave. Alternatively, the devices could be operated at a much higher frequency than the utility using sinusoidal PWM techniques, discussed in Section 3.1.4.4.

Since the single-phase inverter produces a four-quadrant output, it cannot only be used to synthesize an ac output from a dc power source, but can also be used to control the flow of power from an ac source to a dc load. If sinusoidal PWM techniques are used, then it is known as a PWM rectifier.

3.1.4.3 Three-phase inverter

To synthesize a set of three-phase voltages from a dc source, three inverter legs are connected together as shown in Figure 3.1.15. The circuit has two operating modes, either the transistors operate at the same frequency as the ac output waveforms, known as quasi-squarewave or six-step operation, or alternatively the devices operate at a much higher frequency than the ac output using a form of sinusoidal PWM.

Figure 3.1.16 illustrates the simpler *quasi-squarewave* operation of the three-phase inverter. The transistors in each leg operate in anti-phase with duty ratios of 0.5, the leg output voltages, measured with respect to a notional ground at the mid-point of the dc input, are therefore symmetrical squarewaves of $\pm V_{in}/2$. By displacing the switching actions in the three legs by $2\pi/3$ radians or 120° as shown, the resultant line-to-line output voltages form a set of mutually displaced quasi-squarewaves. The V_{AB} line-to-line voltage is shown as an example and consists of 120° intervals of positive and negative voltage separated by 60° intervals of zero voltage. Since the line-to-line output voltages are identical in form to the ideal input currents drawn by the six-pulse rectifier (Figure 3.1.2) the waveforms will also have an identical frequency spectrum (Figure 3.1.5), the non-zero harmonics being of order $6k \pm 1$. From Equation (3.1.3) the amplitude of the fundamental component of the line-to-line waveforms is $2\sqrt{3}V_{in}/\pi$.

One of the largest application areas for the three-phase inverter is in variable speed drive systems for ac motors, as described in Section 2.3.10.2.

3.1.4.4 Sinusoidal PWM

The most common operating mode for inverter circuits is with the transistors switching at a much higher frequency than the ac output waveform that is being synthesized, typically twenty times greater or more, which implies switching frequencies in the region of 1–20 kHz for a mains frequency inverter. The duty ratio of the transistors is varied throughout the mains cycle to shape the required sinusoidal output voltage. The main advantage

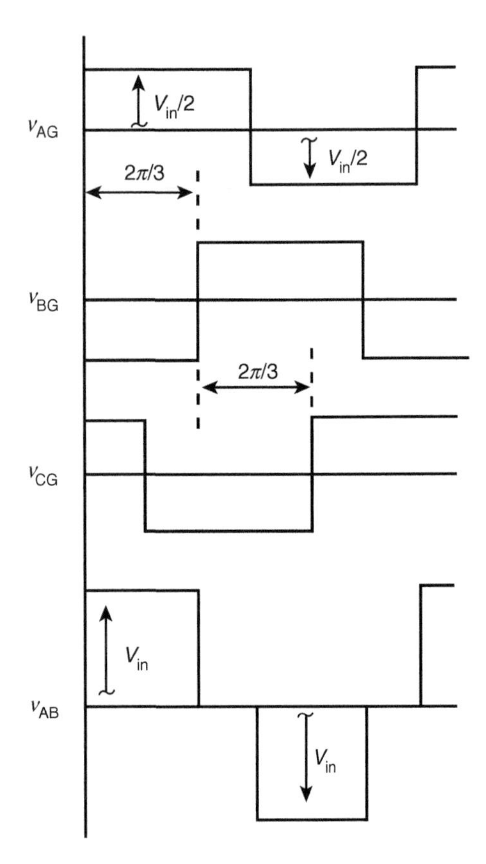

Fig. 3.1.16 Quasi-squarewave operation of the three-phase inverter.

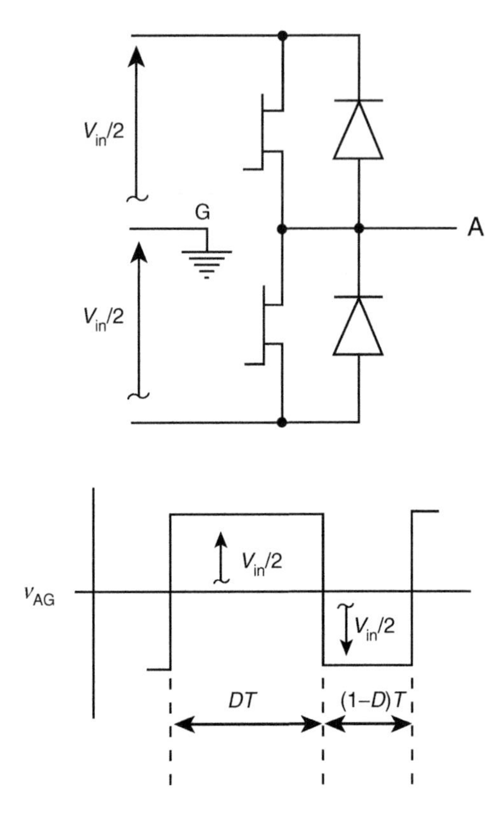

Fig. 3.1.17 Inverter leg and output voltage for one PWM cycle.

of *PWM* methods for inverter control is the harmonic purity of the output waveform. Apart from the required fundamental component, the frequency spectrum of the output contains only switching frequency-related harmonics, and these are at high frequency, typically clustered around integer multiples of the switching frequency. These frequency components may be removed by a small high-frequency filter if necessary, however, the load impedance itself often acts as a low pass filter, making additional filtering unnecessary.

The basic principle of PWM control is explained with reference to the single inverter leg in Figure 3.1.17. The leg output voltage v_{AG} is shown with respect to a notional ground at the mid-point of the dc supply. The transistors operate in anti-phase with switching period T, the upper device having a duty ratio or on time-to-period ratio D, and the lower device therefore having a duty ratio of $(1 - D)$. The local average, $\overline{v}_{AG}$, of the v_{AG} waveform, taken across the arbitrary switching cycle in Figure 3.1.17 is given by the equation:

$$\overline{v}_{AG} = \frac{V_{in}}{2}(2D - 1) \tag{3.1.10}$$

By varying D from one cycle to the next, the local average of v_{AG} may be forced to follow a required waveform and to synthesize a sinusoidal output waveform, the duty ratio should be varied according to the equation,

$$D = 0.5(1 + M \sin \omega t) \tag{3.1.11}$$

where ω is the angular frequency of the waveform to be synthesized, and the parameter M is known as the *depth of modulation* or the *modulation index*. With a modulation index of zero the duty ratio will be constant, whilst a modulation index of unity will result in the duty ratio varying over the full range of $0 \leq D \leq 1$. Substituting Equation (3.1.11) into Equation (3.1.10) gives the local average output voltage in the equation,

$$\overline{v}_{AG} = \frac{V_{in}}{2} M \sin \omega t \tag{3.1.12}$$

which has the required sinusoidal form, the amplitude being $(V_{in}/2)M$. The modulation index therefore controls the amplitude of the synthesized output.

Equation (3.1.12) shows that with a modulation index of unity the amplitude of the output waveform is limited to $V_{in}/2$. A modulation index of greater than unity, known as *over-modulation*, allows the amplitude of the fundamental component of the output to be increased, although Equation (3.1.12) breaks down in this region. However, over-modulation has the disadvantage of introducing low-order harmonics in the output spectrum. Alternatively, the technique of third harmonic injection may be used to increase the amplitude of the fundamental component of the output by around 15 per cent without degrading the frequency spectrum. This is achieved by adding a small amount of third harmonic to the modulating waveform. The addition of the third harmonic does not affect the line-to-line output waveforms of a three-phase inverter since the third harmonic is a common mode component in three-phase systems.

A variety of methods have been developed for the practical implementation of sinusoidal PWM and the determination of the required duty ratio variation. Naturally sampled PWM is intuitively the most simple, and involves comparing the sinusoidal modulating waveform with a triangular carrier signal at the switching frequency. The analog implementation of this technique is fraught with offset and drift problems, and, being analog in nature, the method does not readily lend itself to digital implementations. Instead, methods such as regularly sampled PWM have been developed, which use sample and hold techniques to simplify the calculations within a processor. Most recently, the method of space vector PWM has become common for three-phase inverters. This uses a single rotating vector to represent the required three-phase output waveforms, and the transistor switching patterns for each of the three legs are calculated simultaneously, resulting in a very efficient implementation, furthermore the third harmonic injection to increase the maximum fundamental output is inherently provided.

3.1.5 High-frequency power supplies

High-frequency power supplies or *switched-mode power supplies* are widely used to derive low-voltage dc supplies from the ac utility, and operate by first rectifying the ac input to provide high-voltage dc, which is then converted into high-frequency ac, typically at around 100 kHz, and then passed through a physically small high-frequency transformer, before finally being rectified and smoothed to form the output. Power levels range from a few watts to a few kilowatts.

Traditionally, at these low power levels the input rectifier consists of a single-phase bridge and large smoothing capacitor, however in recent years this arrangement has become unacceptable for all but the very lowest power levels due to the high harmonic content of the currents drawn by such a rectifier circuit. The input current to the rectifier consists of narrow pulses at the peaks of the ac voltage waveform. Instead of adding large low-frequency filters at the rectifier input, active high-power factor rectifiers are increasingly being used to provide high-quality, near-sinusoidal input currents by using high-frequency converter techniques. A common example, based on the boost converter, is illustrated in Figure 3.1.18.

A full-wave rectified voltage waveform is presented as the input to the boost converter, which, through the use of an input current feedback loop, is controlled to offer a constant input resistance, that is, to draw an input current that follows the full-wave rectified voltage. As a result the input power factor is almost unity. However,

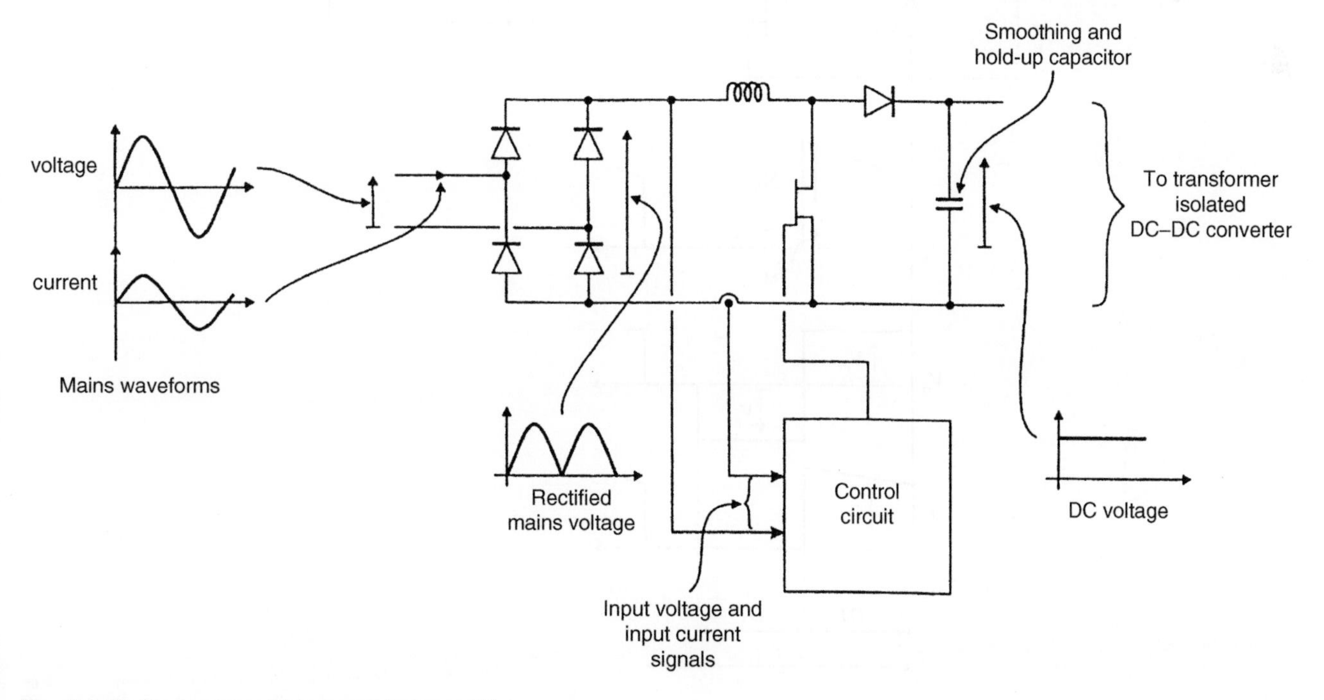

Fig. 3.1.18 Single-phase high-power factor rectifier.

a small high-frequency filter may be required at the input to provide a path for the high-frequency inductor ripple current. To ensure proper operation, the boost converter output voltage must be maintained above the peak of the ac input, which is achieved by sizing the output capacitor to accommodate the twice mains frequency pulsation in power flow that occurs through the boost converter, and also by modulating the amplitude of the boost converter input current to compensate for changes in the power drawn from the power supply output.

High-frequency coupled dc–dc converter circuits that are commonly used in switched mode power supplies are the *forward converter* and the *flyback converter*, and these are briefly reviewed in the following sections.

3.1.5.1 Forward converter

This is a transformer-isolated derivative of the buck converter, and the two circuits operate in a very similar manner. In the two-transistor version of the circuit, shown in Figure 3.1.19, the devices operate in synchronism with a duty ratio D. The waveforms show a steady-state switching cycle, the transistors being turned on at the origin. The inductor current is assumed to be continuous with a small triangular ripple component. The waveforms show the primary voltage and current and the secondary current.

When the transistors turn on, the dc input voltage is applied to the primary winding. The voltage is reflected across to the secondary winding multiplied by the turns ratio and acts to forward bias the series diode and reverse bias the freewheel diode; the inductor voltage is therefore given by $NV_{in} - V_o$, causing a linear rise in the inductor current during the transistor on-time. The inductor current flows through the transformer secondary and is reflected through the turns ratio into the primary, transferring energy from the source. In addition, the transformer magnetizing current rises linearly from zero.

When the transistors turn off, the inductor current diverts to the freewheel diode, the inductor voltage is equal to $-V_o$ and the inductor current falls linearly. By equating the positive and negative inductor volt-seconds, the expression for the voltage conversion ratio is given in the equation:

$$\frac{V_o}{V_{in}} = ND \quad (3.1.13)$$

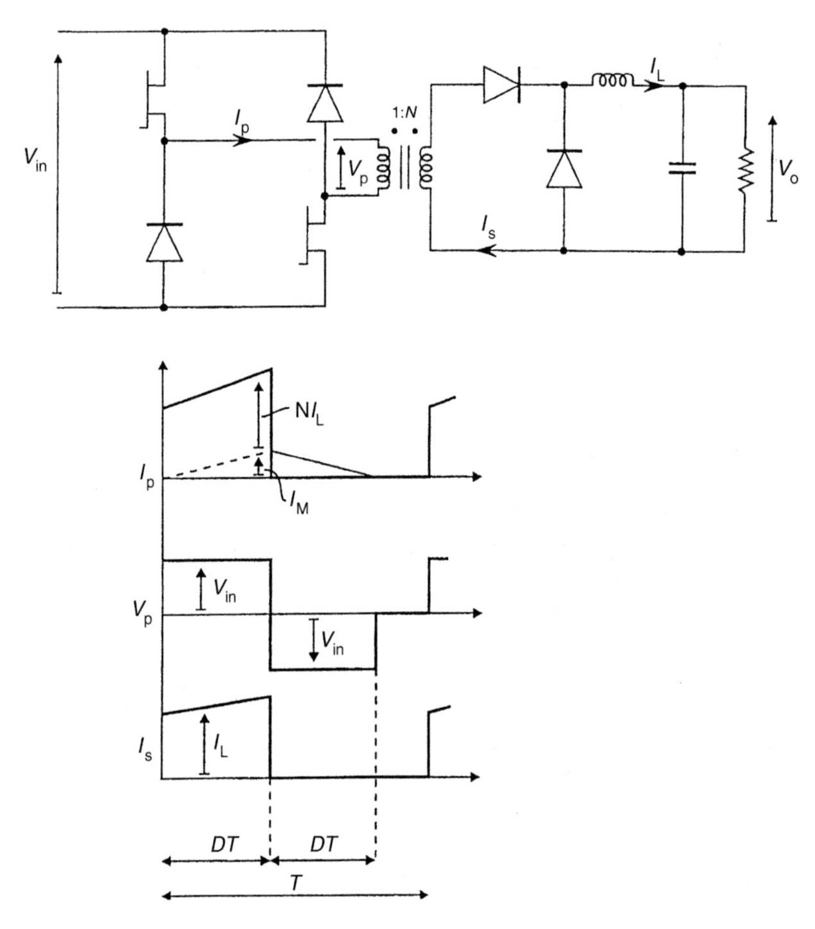

Fig. 3.1.19 Two-transistor forward converter.

The conversion ratio expression is similar to that of the buck converter, but is multiplied by the transformer turns ratio.

When the transistors turn off, the stored energy in the transformer core acts to maintain the flow of magnetizing current, the conduction path being through the two diodes in the primary circuit and the dc source. As a result, the primary voltage reverses, the magnetizing current falls linearly, returning the stored magnetic energy to the dc source. Since the reverse voltage applied to the primary is equal in magnitude to the forward voltage, the current decays to zero in a time equal to the transistors on time. Therefore, for the magnetizing current to fall to zero before the start of the next cycle, the transistor duty ratio must be limited to a maximum of 0.5. The limitation on the duty ratio results in poor utilization of the transformer, power flowing through it only for up to 50 per cent of the time.

For applications of a few hundred watts, a single transistor circuit is sometimes used with a third winding to allow transformer reset, but this circuit has the disadvantage that the transistor off-state voltage becomes $2V_{in}$. For higher powers, around 1 kW and above, a four-transistor full-bridge circuit (single-phase inverter) is used to drive the transformer. This allows a true bi-directional voltage waveform to be impressed on the primary and enables power to flow through the transformer for almost 100 per cent of the cycle.

3.1.5.2 Flyback converter

This is often preferred at power levels below 100 W, where its simple circuit topology with only one magnetic component makes it the most economic solution. At higher power levels the circuit simplicity is out-weighed by the disadvantages of high ripple current in the output filter capacitor, high reverse voltages across the devices and poor utilization of the wound component.

The circuit and waveforms are shown in Figure 3.1.20. Here, V_p is the primary winding voltage, I_p the primary winding current, I_s the secondary winding current and I_c is the filter capacitor current. The currents in the magnetic component are assumed to be zero at the start of the cycle.

The turn-on instant is shown at the origin and when the transistor is conducting, the input voltage is impressed across the primary winding, causing a linear increase in current as energy is stored in the core and the flux increases. While the transistor is on, the secondary winding cannot conduct since the diode is reverse biased by a voltage $NV_{in} + V_o$.

The primary current is rapidly reduced to zero as the transistor turns off, and to maintain the core flux a current starts to flow in the secondary winding. The secondary current then falls linearly to zero as the stored

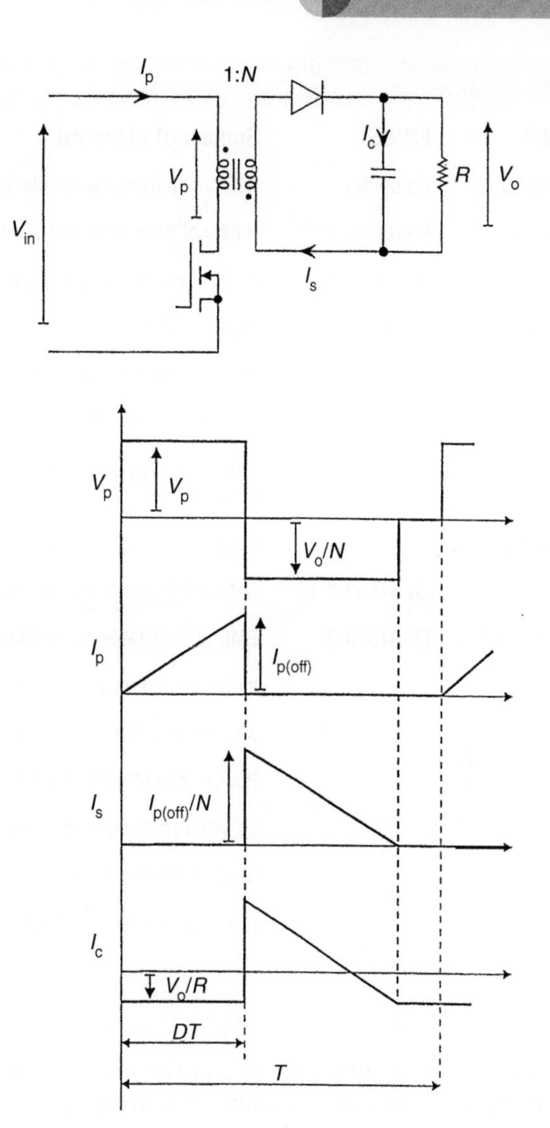

Fig. 3.1.20 Flyback converter.

magnetic energy is transferred to the load. It is important to note that the alternate conduction of the primary and secondary is quite different from the normal operation of a transformer, and results in the entire energy that passes through the circuit each cycle being stored momentarily in the core. The component is therefore known as a coupled inductor rather than a transformer, and due to the energy storage requirement, the coupled inductor tends to be significantly larger than a transformer designed for the same power and frequency.

The voltage conversion ratio for the converter may be obtained by equating the input and output energy per cycle as in the equation,

Energy through put per cycle

$$= \frac{L_p I^2_{p(off)}}{2} = \frac{L_p}{2}\left[\frac{V_{in}DT}{L_p}\right]^2 = \frac{V_o^2 T}{R} \tag{3.1.14}$$

Table 3.1.1 International and national standards relating to power electronics

IEC	EN/BS	Subject of standard	N. American
60065	EN 60065	Safety requirements for electronic apparatus for household use	
60335-2	EN 60335-2	Safety of household and similar apparatus	
60601-1	EN 60601-1	Medical electrical equipment – Pt 1: general requirements for safety	
60950-1	EN 60950-1	Safety of IT equipment	CSA 22.2 no 950-95 UL1950
	EN 61000	General emission standard (section 14.3.1)	
	EN 61000-3-2	Emissions: mains harmonic current	
61010-1	EN 61010-1	Safety requirements for electrical equipment for measurement, control and laboratory use – Pt 1: general requirements	
61204		LV power supply devices, dc output – performance and safety	
	EN 61558-2-7	Safety of power transformers and power supply units	
	EN 61800-3	EMC of adjustable speed drives	
		Industrial control equipment	UL 508
		Medical and dental equipment	UL 544
		Power units other than class 2	UL 1012
		LV video products without CRT displays	UL 1409
		Medical electrical equipment – Pt 1: general requirements for safety	UL 2601-1
		Electrical equipment for laboratory use – Pt 1: general requirements	UL 3101-1

where L_p is the self-inductance of the primary winding and D is the transistor duty ratio. Re-arranging gives the equation:

$$\frac{V_o}{V_{in}} = \frac{D}{\sqrt{2L_p/(RT)}} \quad (3.1.15)$$

The conversion ratio is a linear function of D, but also depends upon load, operating frequency and the primary inductance. Surprisingly, the turns ratio N does not appear in Equation (3.1.15); this is because the converter operates by transferring a fixed packet of energy to the output each cycle, which is independent of N.

3.1.6 Standards

There is a wide range of standards covering the performance of power electronic equipment, particularly with regard to electromagnetic compatibility and safety. Some standards are generic whilst others apply to specific classes of equipment. Examples of the most common standards are listed in Table 3.1.1.

References

Mohan, Undeland and Robbins: *Power Electronics: Converters, Applications and Design*, Third Edition, John Wiley, 2003

Chapter 3.2

3.2

Electronic converters

Barnes

3.2.1 Introduction

This chapter deals with the active components (e.g., diodes, thyristors, transistors) and passive components (e.g., resistors, chokes, capacitors) used in *power electronic* circuits and converters. *Power electronics* is that field of electronics which covers the conversion of electrical energy from one form to another for high power applications. It applies to circuits in the following power ranges:

- Power ratings up to the MVA range
- Frequency ratings up to about 100 kHz

Power electronics is a rapidly expanding field in electrical engineering and the scope of the technology covers a wide spectrum. Therefore, the emphasis will be on the components used in converters used for the speed control of electric motors. Components used for other applications such as power supplies, high-frequency generators, etc. will not be covered in great detail.

3.2.2 Definitions

The following are the common terms used in the field of power electronics.

- **Power electronic components**, are those semiconductor devices, such as diodes, thyristors, transistors, etc. that are used in the power circuit of a converter. In power electronics, they are used in the non-linear switching mode (ON/OFF mode) and not as linear amplifiers.
- **Power electronic converter** or 'converter' for short, is an assembly of power electronic components that converts one or more of the characteristics of an electric power system. For example, a converter can be used to change
  - AC to DC
  - DC to AC
  - Frequency
  - Voltage level
  - Current level
  - Number of phases

The following graphic symbols are used to designate the different types of converter.

- **Rectifier** is that special type of converter that converts AC to DC

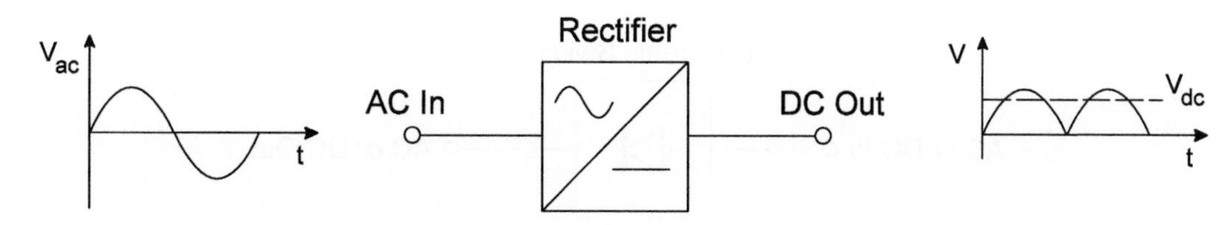

Practical Variable Speed Drives and Power Electronics; ISBN: 9780750658089

- **Inverter** is that special type of converter that converts DC to AC.

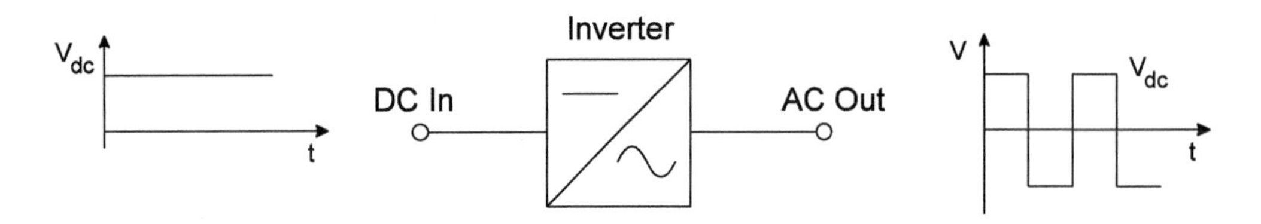

- **AC converter** is that special type of converter that converts AC, of one voltage and frequency, to AC of another voltage and frequency, which are often variable.

  An *AC frequency converter* is a special type of AC converter.

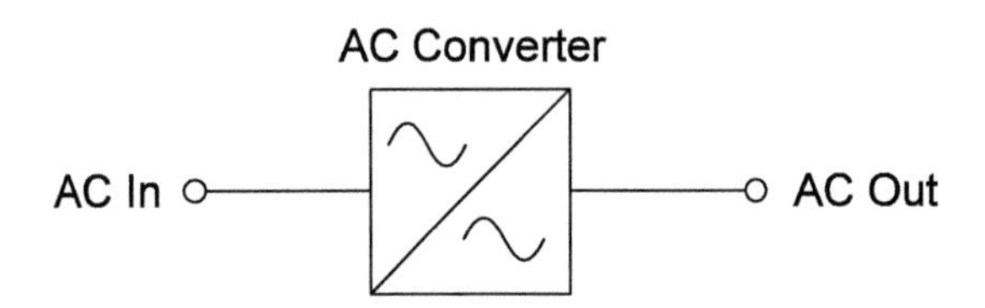

In a power electronic AC converter, it is common to use an intermediary DC link with some form of smoothing.

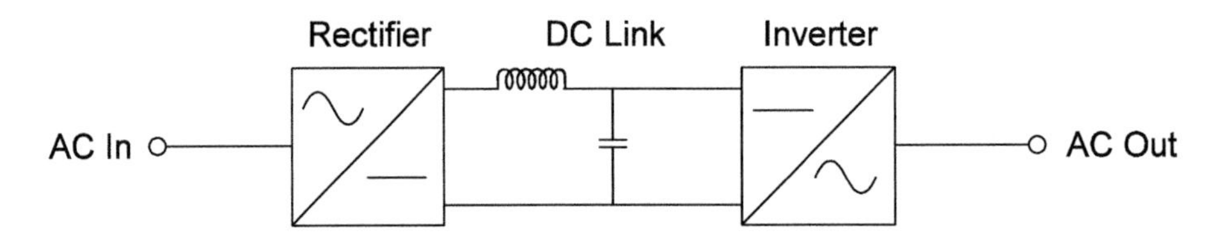

- **DC converter** is one that converts DC of one voltage to DC of another voltage.

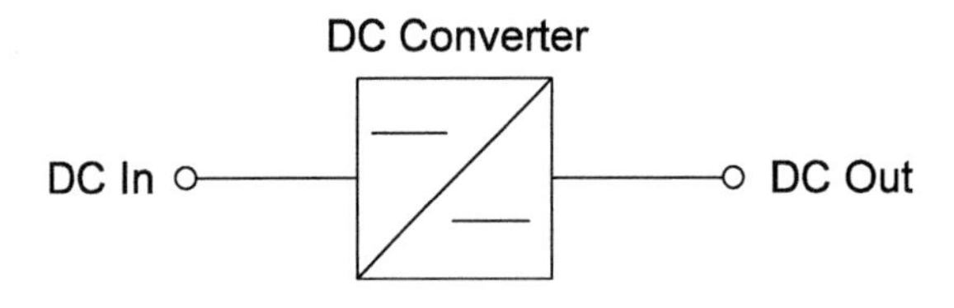

In a DC converter, it is common to use an intermediary AC link, usually with galvanic isolation via a transformer.

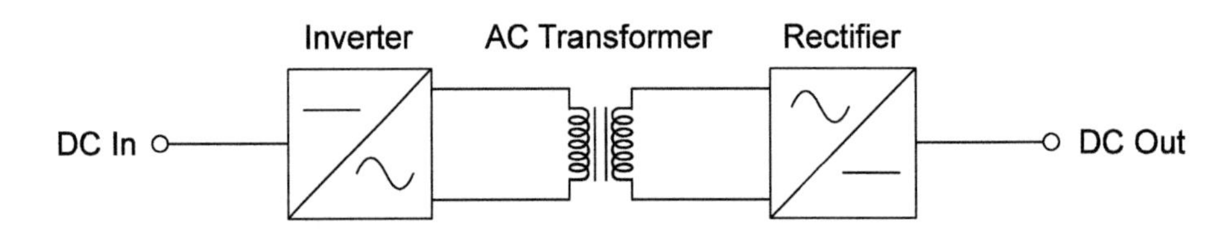

- **Electronic switch** is one that electronically connects or disconnects an AC or DC circuit and can usually be switched ON and/or OFF. Conduction is usually permitted in **one direction only**.

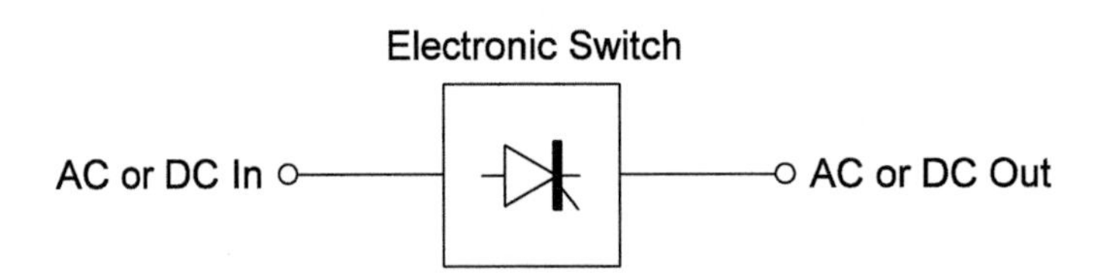

The following components are those devices that are most commonly used as electronic switches in *power electronic converters*. Developments in semiconductor technology have made these power electronic components smaller, more reliable, more efficient (lower losses), cheaper and able to operate at much higher voltages, currents and frequencies. The idealised operating principles of these components can be described in terms of simple mathematical expressions.

- Power diodes
- Power thyristors
- Gate turn-off (GTO) thyristors
- MOS controlled thyristors (MCTs)
- Power bipolar junction transistors (BJTs)
- Field effect transistors (FETs, MOSFET)
- Integrated gate bipolar transistors (IGBT)
- Resistors (provide resistance)
- Reactors or chokes (provide inductance)
- Capacitors (provide capacitance)

In power electronic circuits, semiconductor devices are usually operated in the bi-stable mode, which means that they are operated in either one of two stable conditions:

- Blocking mode: fully switched OFF
  - Voltage across the component is **high**
  - Current through the component is **low** (only leakage current)
- Conducting mode: fully switched ON
  - Voltage across the component is **low**
  - Current through the component is **high**

Diodes and thyristors are inherently bi-stable but transistors are not. Transistors must be biased fully ON to behave like bi-stable devices.

3.2.3 Power diodes

Power diode is a two-terminal semiconductor device with a relatively large single P–N junction. It consists of a two-layer silicon wafer attached to a substantial copper base. The base acts as a heat-sink, a support for the enclosure and also one of the electrical terminals of the diode (Figure 3.2.1). The other surface of the wafer is connected to the other electrical terminal. The enclosure seals the silicon wafer from the atmosphere and provides adequate insulation between the two terminals of the diode. The two terminals of a diode are called the *anode (A)* and the *cathode (K)*. These names are derived from the days when *Valves* were commonly used.

SYMBOL:

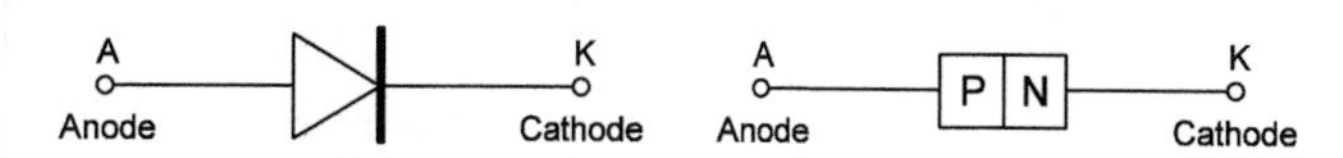

IDEAL:

Forward conduction: Resistanceless
Reverse blocking: Lossless
Switch ON/OFF time: Instantaneous

Many different mechanical designs are commonly used for diodes, some of which are shown below. Power diodes rated from a few amperes are usually stud mounted but it is increasingly common (more economical) to have several diodes encapsulated into an insulated module. Examples are full wave rectifiers, 6-pulse diode bridges, etc.

The base of this type of diode module is usually not electrically active, so it can be mounted directly onto the heat-sink of a converter. Larger units for high current ratings are usually of the disc type, which provides a larger area of contact between the case and the heat-sink for better cooling.

When the anode is positive relative to the cathode, it is said to *be forward biased* and the diode conducts current. When the anode is negative relative to the cathode the diode is said to be *reverse biased* and the flow of current is blocked. The typical characteristic of a power diode is shown in Figure 3.2.2.

Unfortunately, power diodes have several limitations:

- In the *conduction mode*, when the diode *is forward biased*
  - Real diodes are not resistanceless and there is *a forward volt drop* of between 0.5 and 1.0 V during conduction
  - As a result, there is a limit to how much current can continuously flow without overheating. This is the maximum rated current of the diode.
- In the blocking mode, when the diode is *reverse biased*
  - there is a small leakage current
  - there is a limit to how much voltage it can withstand before reverse breakdown and current can

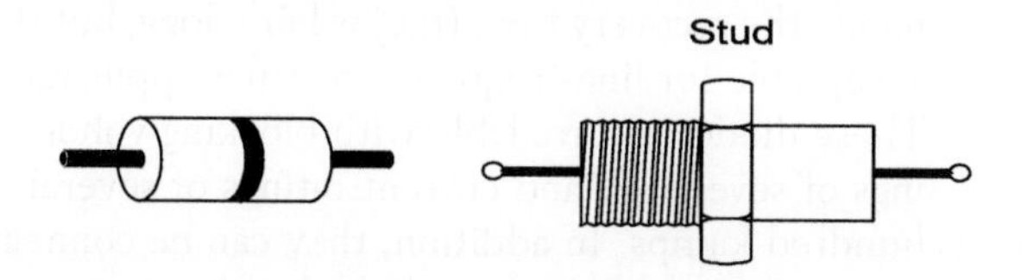

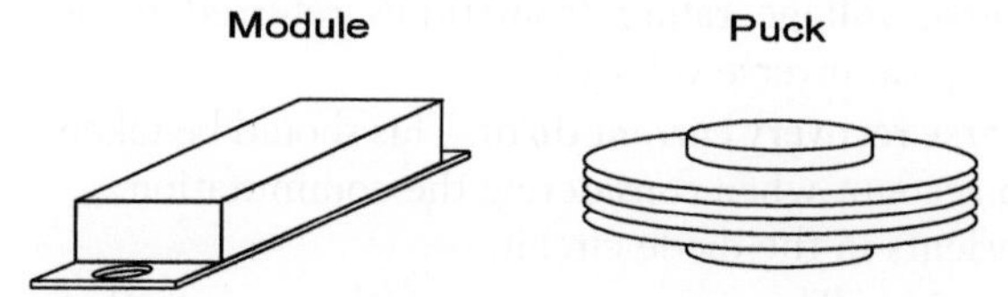

Fig. 3.2.1 Typical mechanical construction of diodes.

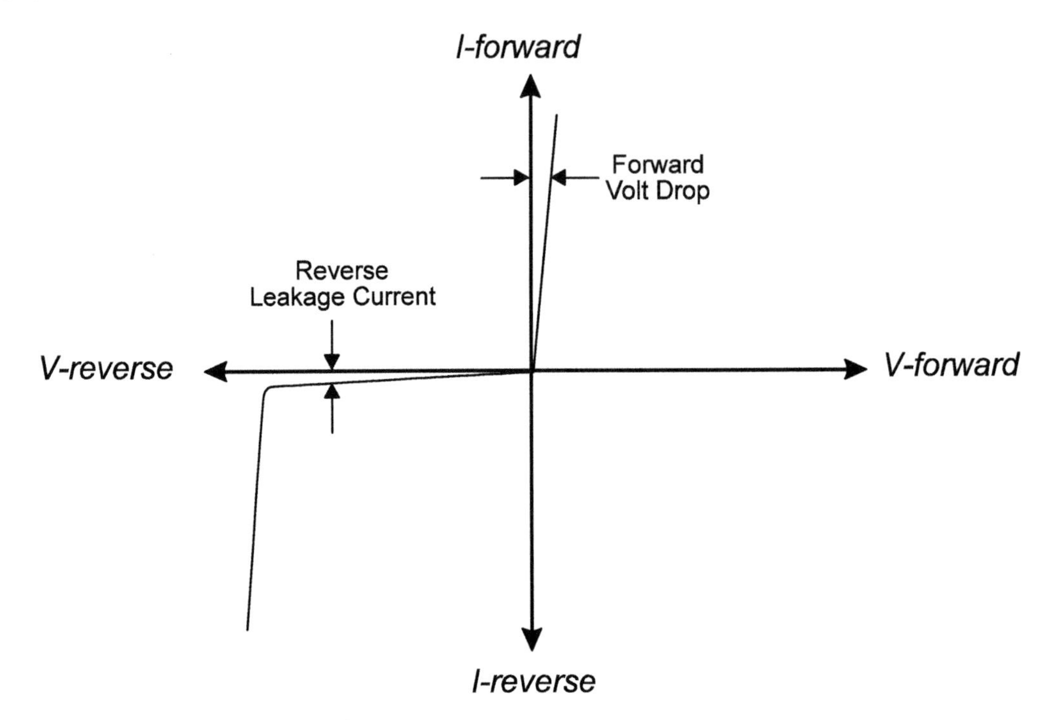

Fig. 3.2.2 Typical characteristic of a power diode.

start to flow in the reverse direction. It is a sound common practice to select diodes with a reverse voltage limit of at least twice the value that will practically occur.

- The *commutation time* from the blocking mode to the conduction mode and vice versa takes a finite time.

A power diode must be rated for the electrical environment in which it is to be used. The following are the most important factors that must be considered when choosing a power diode for a converter application:

- **Forward current rating**. The current rating is based on a certain wave shape and should be taken as a guide only. The real selection should be based on the total power losses in the diode taking into account the actual wave shape, load cycle and cooling conditions.
- **Forward voltage drop**. This has an effect on current sharing between parallel circuits that include diodes.
- **Forward surge current** capability (rate of rise of current di/dt)
- **Reverse voltage rating** (sometimes referred to as PIV - peak inverse voltage)
- **Reverse recovery current di/dt**. This should be taken into account when considering the commutation transients in the diode circuit.
- **I^2t rating**. This is a measure of the energy that a diode can handle in the case of a short circuit without permanent damage. It gives a guide to the correct choice of high speed fuses to protect the diode. Briefly, a protection fuse must be chosen with an I^2t rating lower than the diode.

Depending on the application requirements, various types of diode are available:

- Schottky diodes
- These diodes are used where a low forward voltage drop, typically 0.4 V, is needed for low output voltage circuits. These diodes have a limited blocking voltage capability of 50–100 V.
- Fast recovery diodes
- These diodes are designed for use in circuits where fast recovery times are needed, for example in combination with controllable switches in high frequency circuits. Such diodes have a recovery time (t_{RR}) of less than a few microsecs.
- Line frequency diodes
- The ON-state voltage of these diodes is designed to be as low as possible to ensure that they switch ON quickly in rectifier bridge applications. Unfortunately the recovery time (t_{RR}) is fairly long, but this is acceptable for line-frequency rectifier applications. These diodes are available with blocking voltage ratings of several kV and current ratings of several hundred kamps. In addition, they can be connected in series or parallel to satisfy high voltage or current requirements.

3.2.4 Power thyristors

Thyristors are often referred to as silicon controlled rectifiers (SCRs). This was the name originally given to the device when it was invented by General Electric (USA) in about 1957. This name has never been universally accepted and used. The name accepted by both the IEC and ANSI/IEEE is *reverse blocking triode thyristor* or simply *thyristor.* The name *thyristor* is a generic term that is applied to a family of semiconductor devices that have the regenerative switching characteristics. There are many devices in the thyristor family including the power thyristor, the GTO, the FCT, the triac, etc.

A thyristor consists of a four-layer silicon wafer with three P–N junctions. It has two power terminals, called the *anode (A)* and *cathode (K)*, and a third control terminal called the *gate (G)*. High voltage, high power thyristors sometimes also have a fourth terminal, called an auxiliary cathode and used for connection to the triggering circuit. This prevents the main circuit from interfering with the gate circuit.

A thyristor is very similar to a power diode in both physical appearance and construction, except for the gate terminal required to trigger the thyristor into the conduction mode (Figure 3.2.3).

SYMBOL:

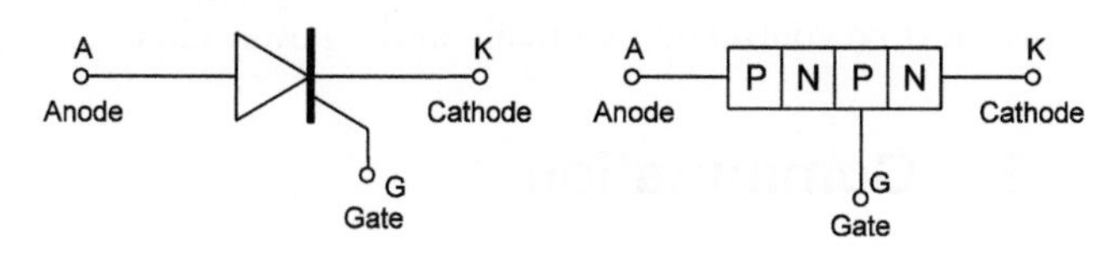

IDEAL:

Forward conduction: Resistanceless
Forward blocking: Lossless (no leakage current)
Reverse blocking: Lossless (no leakage current)
Switch ON/OFF time: Instantaneous

As with power diodes, smaller units are usually of the stud type but it is also increasingly common to have two or more thyristors assembled into a thyristor module. The base of this type of pack is not electrically active, so it can be mounted directly onto the heat-sink of a converter. Large thyristor units are usually of the disc type for better cooling.

Most converters for the speed control of motors are air-cooled, the smaller units using natural convectional cooling over the heat-sink and the larger units using a fan for forced cooling.

A *thyristor is a controllable device,* which can be switched from a blocking state (high voltage, low current) to a conducting state (low voltage, high current) by a suitable gate pulse. Forward conduction is blocked until an external positive pulse is applied to the gate terminal. A thyristor cannot be turned off from the gate. During forward conduction, its behavior resembles that of a power diode and it also exhibits a forward voltage drop of between 1 and 3 V. Like the diode, conduction is blocked in the reverse biased direction. A typical characteristic of the thyristor is shown in Figure 3.2.4.

There are several ways in which a thyristor can be turned ON or brought into *forward conduction.*

- **Positive current gate pulse**. This is the normal way that a thyristor is brought into conduction. The gate pulse must be of a suitable amplitude and duration, depending on the size of the thyristor.
- **High forward voltage**. An excessively high forward voltage between the anode and the cathode can cause enough leakage current to flow to trigger the turn on process.
- **High rate of rise of forward voltage**, dV/dt. A high dV/dt can produce enough leakage current to trigger the turn ON process.
- **Excessive temperature**. The leakage current increases with temperature, so high temperature can aggravate the above two problems.

A thyristor must be suitable for the electrical environment in which it is used. The following are some of the more important factors which must be considered when choosing a thyristor for a converter application:

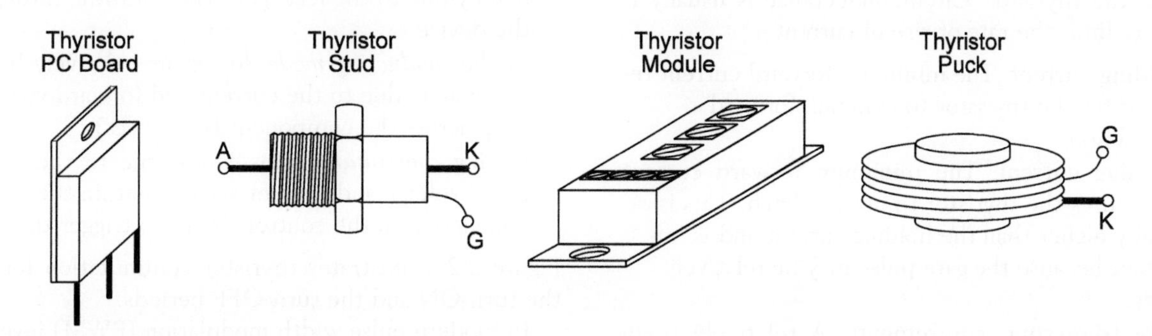

Fig. 3.2.3 Typical mechanical construction of thyristors.

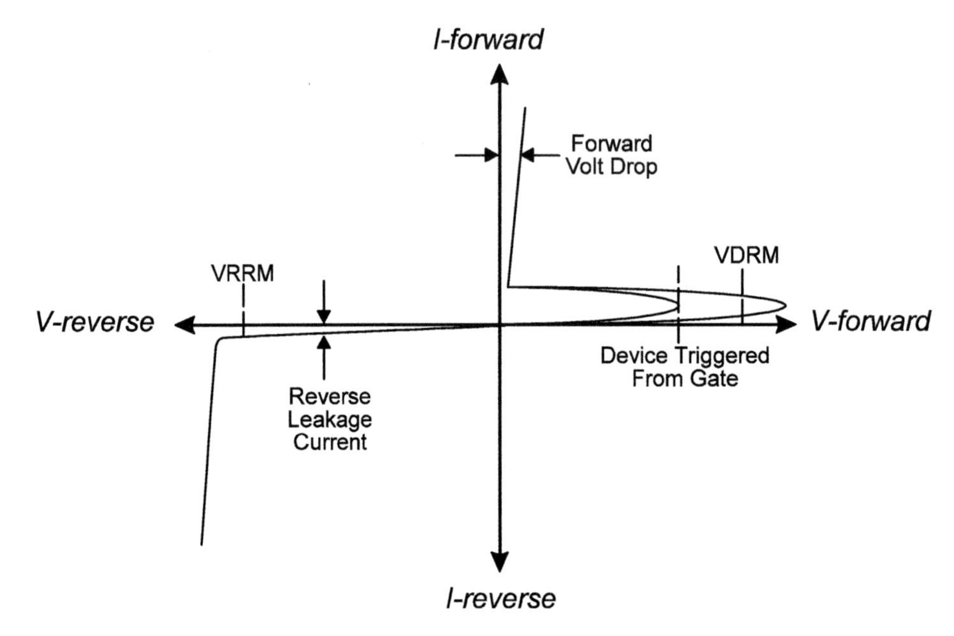

Fig. 3.2.4 Typical characteristic of a thyristor.

- Same factors outlined above for diodes.
- The **power losses** in the thyristor comprise the conduction losses, switching losses (turn ON and turn OFF), gate power losses, forward off state losses and reverse blocking losses. The data sheet usually provides curves for estimating power losses for various wave shapes.
- **Peak forward voltage** (PFV). This is the forward anode voltage that the device must withstand without switching on and without damage.
- **Rate of rise of forward voltage** dV/dt should not be too high; typically it should be less than about 200 V/μsec. A parallel RC snubber circuit is usually required to protect the thyristor.
- **Rate of rise of anode current** di/dt should not be too high, typically it should be less than about 100 A/μsec. The current is initially concentrated around the gate and takes a finite time to spread over the conducting area.

If the rate of rise is too high, local overheating could damage the thyristor. Circuit inductance is usually required to limit the rate of rise of current.

- **Holding current**. The minimum forward current required for the thyristor to maintain forward conduction.
- **Latching current**. The minimum forward current that causes the thyristor to initially latch. This is usually higher than the holding current and is important because the gate pulse may be relatively short.
- **Gate triggering** requirements. A relatively small gate pulse will turn the thyristor on. Typically, a value of 100 mA for 10 μsec is the threshold. In practice, a much higher value should be used for optimum thyristor operation. Also, the turn on time is affected by the magnitude of the gate pulse.

The thyristor is turned off when it becomes reverse biased and/or the forward current falls below the holding current. This must be controlled externally in the power circuit.

3.2.5 Commutation

The transitional period from blocking to conducting, and vice versa, is called *commutation* and the period during which a component turns ON/OFF, is called the *commutation period.* During commutation, the component comes under electrical stress due to changes in the circuit conditions and the thermal stress due to losses. These losses produce heat in the component and also stress the insulation and current paths.

- In the *blocking mode,* losses are usually small and mainly due to the leakage current flowing through the device
- In the *conducting mode,* losses are relatively higher and mainly due to the current and forward volt drop across the component (I^2R losses)
- During *commutation,* losses are due to the transitional voltage and current activity within the component and in the control circuit to trigger the gate.

Figure 3.2.5 illustrates thyristor commutation for both the turn-ON and the turn-OFF periods.

In modern pulse width modulation (PWM) inverters, there is a tendency to use electronic switches operating at

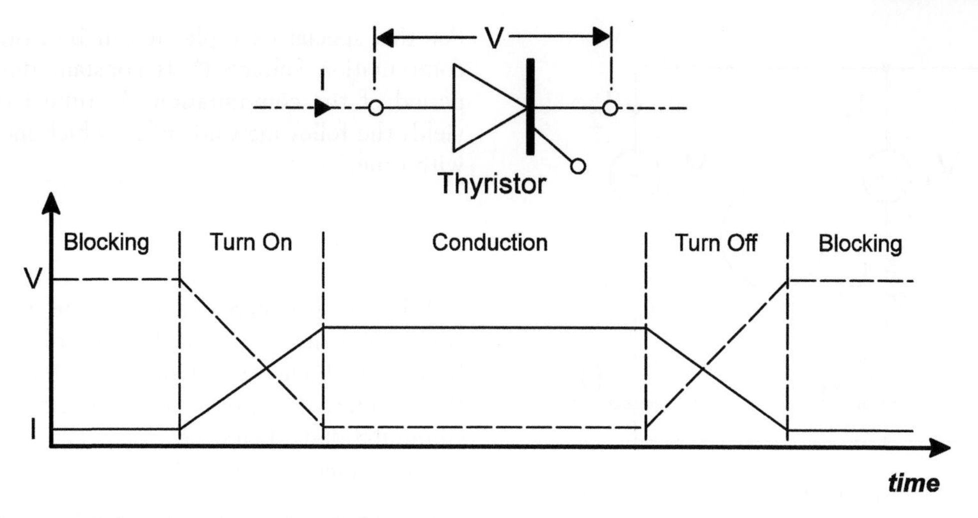

Fig. 3.2.5 Simple commutation of an electronic switch.

high switching frequencies to achieve faster responses or better output wave shapes. Unfortunately, the increased number of commutations results in higher losses both in the triggering circuits as well as the power circuits of the components.

Losses may be reduced by using devices that have the following characteristics:

- Low leakage current during blocking
- Low forward volt drop during conduction
- High switching speed, short commutation period
- Low triggering losses in the control circuit

3.2.6 Power electronic rectifiers (AC/DC converters)

The first stage of an AC frequency converter is the conversion of a three-phase AC power supply to a smooth DC voltage and current. Simple bi-stable devices, such as the diode and thyristor, can effectively be used for this purpose.

Initially, when analyzing power electronic circuits, it will be assumed that the bi-stable semiconductor devices, such as the diodes and thyristors, are ideal switches with no losses and minimal forward voltage drop. It will also be assumed that the reactors, capacitors, resistors and other components of the circuits have ideal linear characteristics with no losses. Once the operation of a circuit is understood, the imperfections associated with the practical components can be introduced to modify the performance of the power electronic circuit.

In power electronics, the operation of any converter is dependent on the *switches* being turned ON and OFF in a sequence. Current passes through a switch when it is ON and is blocked when it is OFF. As mentioned above, the word *commutation* is used to describe the transfer of current from one switch turning OFF to another turning ON.

In a diode rectifier circuit, a diode turns ON and starts to conduct current when there is a forward voltage across it, i.e., the forward voltage across it becomes positive. This process usually results in the forward voltage across another diode becoming negative, which then turns off which stops conducting current. In a thyristor rectifier circuit, the switches additionally need a gate signal to turn them ON and OFF.

The factors affecting commutation may be illustrated in the idealized diode circuit in Figure 3.2.6, which shows two circuit branches, each with its own variable DC voltage source and circuit inductance. Assume, initially, that a current I is flowing through the circuit and that the magnitude of the voltage V_1 is larger than V_2. Since $V_1 > V_2$, diode $D1$ has a positive forward voltage across it and it conducts a current I_1 through its circuit inductance L_1. Diode D_2 has a negative forward voltage across it and is blocking and carries no current.

Consequently, at time t_1

$$I_1 = I$$

$$I_2 = 0$$

Suppose that voltage V_2 is increased to a value larger than V_1, the forward voltage across diode D_2 becomes positive and it then starts to turn on. However, the circuit inductance L_1 prevents the current I_1 from changing instantaneously and diode D_1 will not immediately turn OFF. So, both diodes D_1 and D_2 remain ON for an overlap period called the commutation time t_c.

With both diodes turned ON, a closed circuit is established which involves both branches. The effective circuit voltage $V_C = (V_2 - V_1)$, called the **commutation voltage**, then drives a circulating current $\boldsymbol{i}_c$, called the

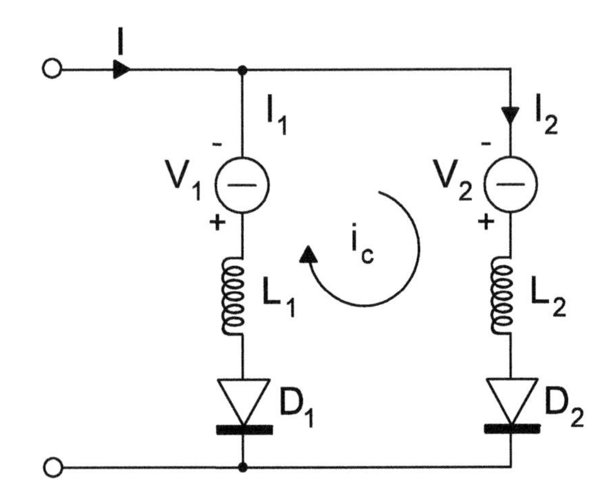

Fig. 3.2.6 Simple circuit to illustrate commutation from diode D1 to D2.

commutation current, through the two branches which have a total circuit inductance of $L_c = (L_1 + L_2)$.

In this idealized circuit, the volt drop across the diodes and the circuit resistance have been ignored. From basic electrical theory of inductive circuits, the current i_c increases with time at a rate dependent on the circuit inductance. The magnitude of the commutation current may be calculated from the following equations:

$$(V_2 - V_1) = (L_1 + L_2)\frac{di_c}{dt}$$

$$V_c = L_c\frac{di_c}{dt}$$

$$\frac{di_c}{dt} = \frac{V_c}{L_c}$$

If the commutation starts at a time t_1 and finishes at a time t_2, the magnitude of the commutation current I_c at any time t, during the commutation period, may be calculated by integrating the above equation from time t_1 to t.

$$I_c = \frac{1}{L_c}\int V_c \, dt$$

During the commutation period:

- It is assumed that the overall current through the circuit remains constant.
  $I = (I_1 + I_2)$ *constant*

As the circulating commutation current increases:

- Current (I_2) through the diode that is turning ON **increases** in value
  $I_2 = I_c$ *increasing*
- Current (I_1) through the diode that is turning OFF **decreases** in value
  $I_1 = I - I_c$ *decreasing*

For this special example, it can be assumed that the commutation voltage V_c is constant during the short period of the commutation. At time t the integration yields the following value of I_c, which increases linearly with time.

$$I_c = \frac{V_c}{L_c}(t - t_1)$$

When I_C has increased to a value equal to the load current I at time t_2, then all the current has been transferred from branch 1 to branch 2 and the current through the switch that is turning off has decreased to zero. The commutation is then over.

Consequently, at time t_2

$$I_1 = 0$$

$$I_2 = I_c = I$$

At the end of commutation when $t = t_2$, putting I_c equal to I in the above equation, the time taken to transfer the current from one circuit branch to the other (commutation time), may be calculated.

$$I = \frac{V_c(t_2 - t_1)}{L_c}$$

$$I = \frac{V_c t_c}{L_c}$$

$$t_c = \frac{I\,L_c}{V_c}$$

It is clear from this equation that the commutation time t_c depends on the overall circuit inductance ($L_1 + L_2$) and the commutation voltage.

From this we can conclude that:

- A large circuit inductance will result in a long commutation time.
- A large commutation voltage will result in a short commutation time.

In practice, a number of deviations from this idealized situation occur.

- The *diodes* are not ideal and do not turn off immediately when the forward voltage becomes negative. When a *diode* has been conducting and is then presented with a reverse voltage, some reverse current can still flow for a few microseconds as indicated in Figure 3.2.7. The current I_1 continues to decrease beyond zero to a negative value before returning to zero. This is due to the free charges that must be removed from the P–N junction before blocking is achieved.
- Even if the commutation time is very short, the commutation voltage of an AC fed rectifier bridge does not remain constant but changes slightly during the

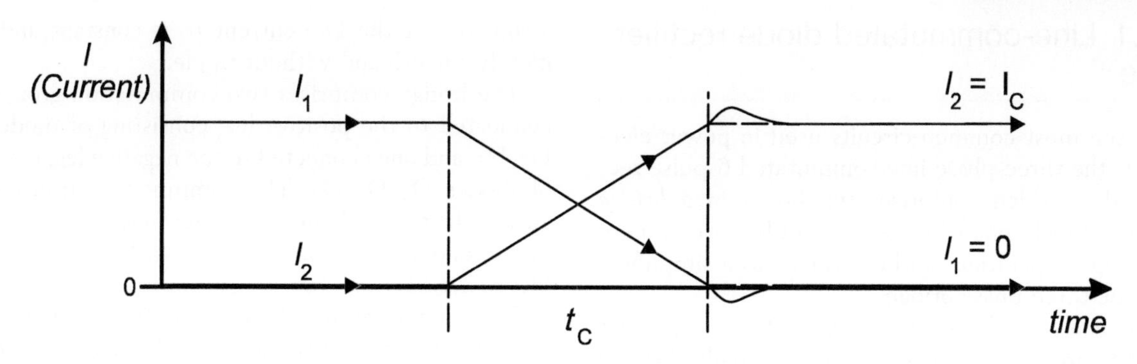

Fig. 3.2.7 The currents in each branch during commutation.

commutation period. An increasing commutation voltage will tend to reduce the commutation time.

In practical power electronic converter circuits, commutation follows the same basic sequence outlined above. The figure below shows a typical 6-pulse rectifier bridge circuit to convert three-phase AC currents I_A, I_B and I_C, to a DC current I_D.

This type of circuit is relatively simple to analyze because only two of the six diodes conduct current at any one time. The idealized commutation circuit can easily be identified. In this example, commutation is assumed to be taking place from diode D_1 to D_3 in the positive group, while D_2 conducts in the negative group.

In power electronic bridge circuits, it is conventional to number the diodes D_1 to D_6 in the sequence in which they turned ON and OFF. When V_A is the highest voltage and V_C the lowest, D1 and D_2 are conducting.

In a similar way to the idealized circuit in Figure 3.2.6, when V_B rises to exceed V_A, D_3 turns on and commutation transfers the current from diode D_1 to D_3. As before, the commutation time is dependent on the circuit inductance (L) and the commutation voltage (V_B–V_A).

As can be seen from the 6-pulse diode rectifier bridge example above in Figure 3.2.8, commutation is usually initiated by external changes. In this case, commutation is controlled by the three-phase supply line voltages. In other applications, commutation can also be initiated or controlled by other factors, depending on the type of converter and the application. Therefore, converters are often classified in accordance with the source of the external changes that initiate commutation.

- In the above example, the converter is said to be **line commutated** because the source of the commutation voltage is on the mains supply line.
- A converter is said to be **self-commutated** if the source of the commutation voltage comes from within the converter itself. Gate-commutated converters are typical examples of this.

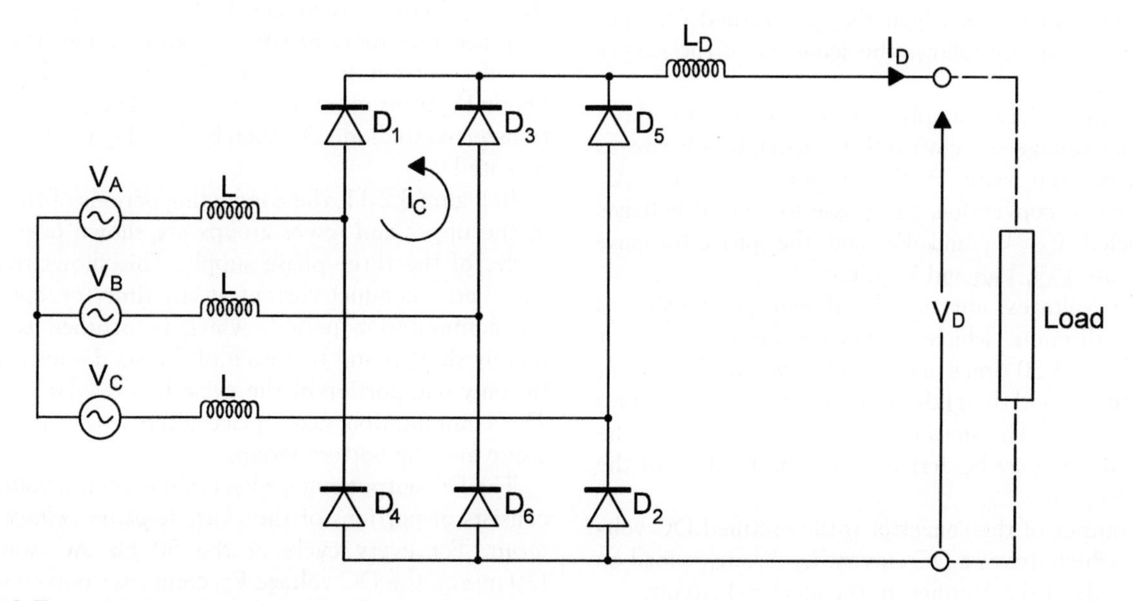

Fig. 3.2.8 Three-phase commutation with a 6-pulse diode bridge.

3.2.6.1 Line-commutated diode rectifier bridge

One of the most common circuits used in power electronics is the three-phase line-commutated 6-pulse rectifier bridge, which comprises six diodes in a *bridge connection*. Single-phase bridges will not be covered here because their operation can be deduced as a simplification of the three-phase bridge.

In the analysis of the various types of converter that follow, the procedure will be to assume initially that the conditions and components are ideal. Once the principles have been established, any deviations from the ideal will be discussed. The following *ideal* assumptions are made:

- The supply voltages are '*stiff*' and completely sinusoidal
- Commutations are instantaneous and have no recovery problems
- Load currents are completely smooth
- Transformers and other line components are linear and ideal
- There is no volt drop in power electronic switches

These assumptions are made to gain an understanding of the circuits and to make estimates of currents, voltages, commutation times, etc. Thereafter, the limiting conditions that affect the performance of the practical converters and their deviation from the ideal conditions will be examined to bridge the gap from the ideal to the practical.

In the diode bridge, the diodes are not controlled from an external control circuit. Instead, commutation is initiated externally by the changes that take place in the supply line voltages, hence the name **line-commutated rectifier**.

According to convention, the diodes are labeled D1 to D_6 in the sequence in which they are turned ON and OFF. This sequence follows the sequence of the supply line voltages.

The three-phase supply voltages comprise three sinusoidal voltage waveforms 120° apart which rise to their maximum value in the sequence A – B – C. According to convention, the phase-to-neutral voltages are labeled V_A, V_B and V_C and the phase-to-phase voltages are V_{AB}, V_{BC} and V_{CA}, etc.

These voltages are usually shown graphically as a vector diagram, which rotates counter-clockwise at a frequency of 50 times per second. A vector diagram of these voltages and their relative positions and magnitudes is shown below. The sinusoidal voltage waveforms of the supply voltage may be derived from the rotation of the vector diagram.

The output of the converter is the rectified DC voltage V_D, which drives a DC current I_D through a load on the DC side of the rectifier. In the idealized circuit, it is assumed that the DC current I_D is constant and completely smooth and without ripple.

The bridge comprises two commutation groups, one connected to the positive leg, consisting of diodes D_1–D_3–D_5, and one connected to the negative leg, consisting of diodes D_4–D_6–D_2. The commutation transfers the current from one diode to another in sequence and each diode conducts current for 120° of each cycle as shown in Figure 3.2.11.

In the upper group, the positive DC terminal follows the highest voltage in the sequence V_A–V_B–V_C via diodes D_1–D_3–D_5. When V_A is near its positive peak, diode D1 conducts and the voltage of the +DC terminal follows V_A. The DC current flows through the load and returns via one of the lower group diodes. With the passage of time, V_A reaches its sinusoidal peak and starts to decline. At the same time, V_B is rising and eventually reaches a point when it becomes equal to and starts to exceed V_A. At this point, the forward voltage across diode D_3 becomes positive and it starts to turn on. The commutating voltage in this circuit, V_B–V_A starts to drive an increasing commutation current though the circuit inductances and the current through D_3 starts to increase as the current in D_1 decreases. In a sequence of events similar to that described above, commutation takes place and the current is transferred from diode D_1 to diode D_3. At the end of the commutation period, diode D_1 is blocking and the +DC terminal follows V_B until the next commutation takes place to transfer the current to diode D_5. After diode D_5, the commutation transfers the current back to D1 and the cycle is repeated.

In the lower group, a very similar sequence of events takes place, but with negative voltages and the current flowing from the load back to the mains. Initially, D_2 is assumed to be conducting when V_C is more negative than V_A. As time progresses, V_A becomes equal to V_C and then becomes more negative. Commutation takes place and the current is transferred from diode D_2 to D_4. Diode D_2 turns off and D_4 turns on. The current is later transferred to diode D_6, then back to D_2 and the cycle is repeated.

In Figure 3.2.11, the conducting periods of the diodes in the upper and lower groups are shown over several cycles of the three-phase supply. This shows that only two diodes conduct current at any time (except during the commutation period, which is assumed to be infinitely short!!) and that each of the six diodes conducts for only one portion of the cycle in a regular sequence. The commutation takes place alternatively in the top group and the bottom group.

The DC output voltage V_D is not a smooth voltage and consists of portions of the *phase-to-phase voltage waveforms*. For every cycle of the 50 Hz AC waveform (20 msec), the DC voltage V_D comprises portions of the six voltage pulses, V_{AB}, Vac, V_{BC}, V_{BA}, V_{CA}, V_{CB}, etc, hence the name 6-pulse rectifier bridge.

The average magnitude of the DC voltage may be calculated from the voltage waveform shown above. The average value is obtained by integrating the voltage over one of the repeating 120° portions of the DC voltage curve. This integration yields an average magnitude of the voltage V_D as follows.

$V_D = 1.35 \times (RMS - Phase\ Voltage)$
$V_D = 1.35 \times V_{RMS}$
For example, if $V_{RMS} = 415$ V,
$V_D = 560$ V DC

If there is sufficient inductance in the DC circuit, then the DC current I_D will be fairly steady and the AC supply current will comprise segments of DC current from each diode in sequence. As an example, the current in the A-phase is shown in Figure 3.2.9. The non-sinusoidal current that flows in each phase of the supply mains can affect the performance of other AC equipment connected to the supply line that are designed to operate with sinusoidal waveforms.

In practice, to ensure that the diode reverse blocking voltage capability is properly specified, it is necessary to know the magnitude of the reverse blocking voltage which appears across each of the diodes. Theoretically, the maximum reverse voltage across a diode is equal to the peak of the phase–phase voltage. For example, the reverse voltage V_{CA} and V_{CB} appears across diode D_5 during the blocking period. In practice, a factor of safety of 2.5 is commonly used for specifying the reverse blocking capability of diodes and other power electronic switches. On a rectifier bridge fed from a 415 V power supply, the reverse blocking voltage V_{bb} of the diode must be higher than 2.5 × 440 V = 1100 V. Therefore, it is common practice to use diodes with a reverse blocking voltage of 1200 V (Figure 3.2.10).

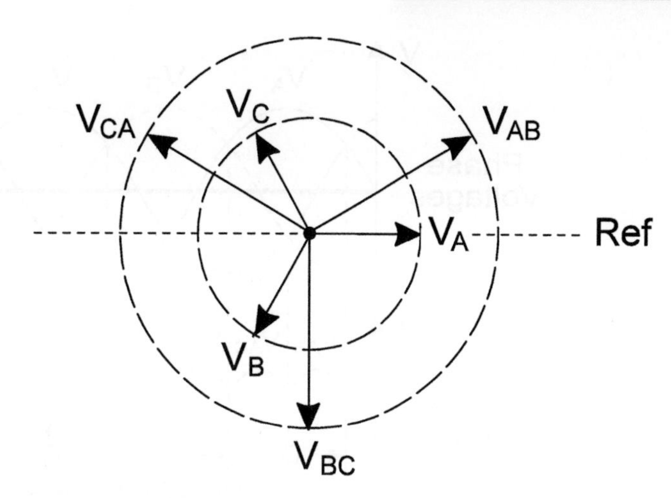

Fig. 3.2.10 Vector diagram of the three-phase mains supply voltages.

3.2.6.2 The line-commutated thyristor rectifier bridge

The output DC voltage and operating sequence of the diode rectifier above is dependent on the continuous changes in the supply line voltages and is not dependent on any control circuit. This type of converter is called an *uncontrolled diode rectifier bridge* because the

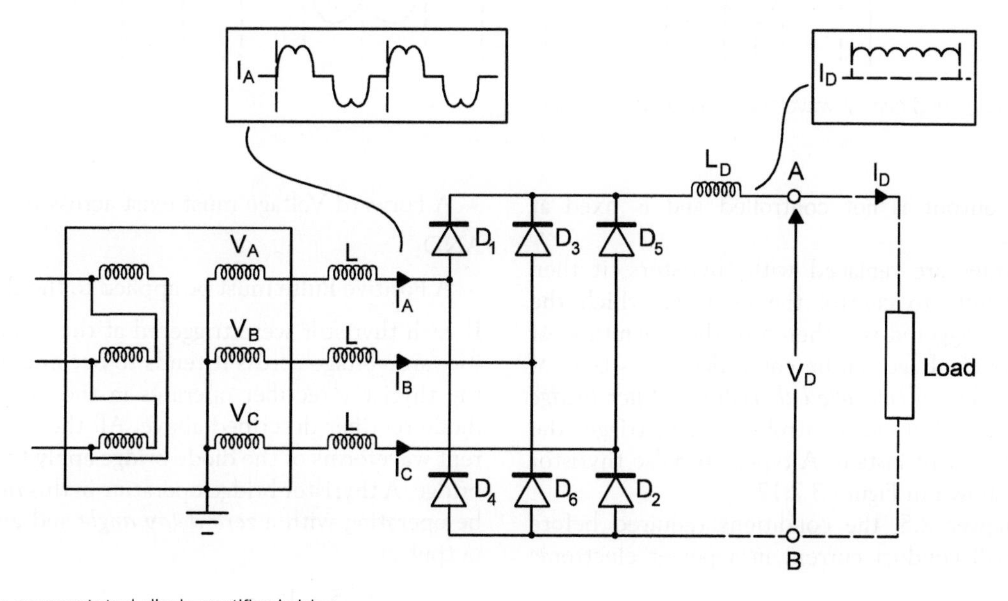

Fig. 3.2.9 Line-commutated diode rectifier bridge.

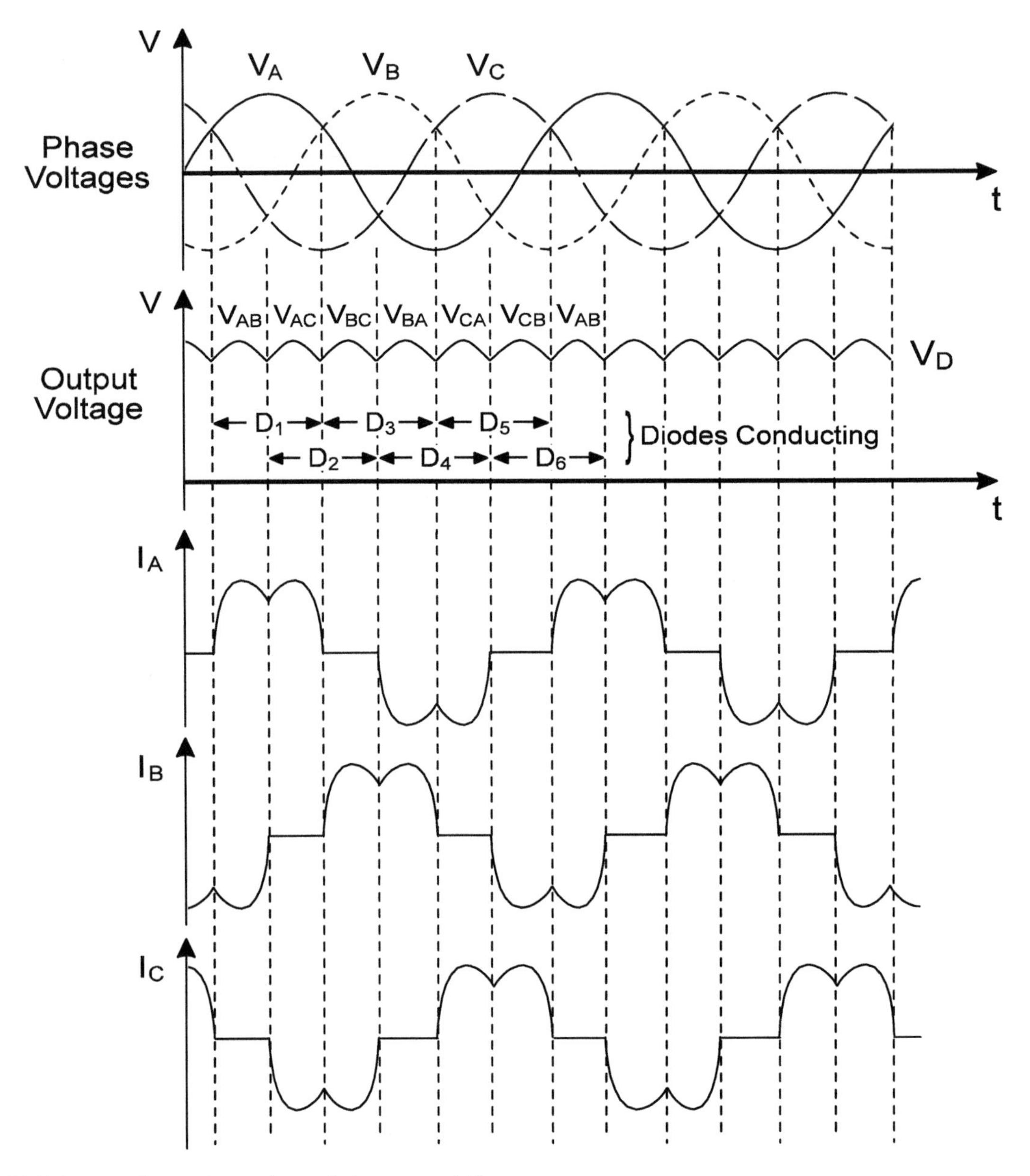

Fig. 3.2.11 Voltage and current waveforms during commutation.

DC voltage output is not controlled and is fixed at $1.35 \times V_{RMS}$.

If the diodes are replaced with thyristors, it then becomes possible to control the point at which the thyristors are triggered and therefore the magnitude of the DC output voltage can be controlled. This type of converter is called a *controlled thyristor rectifier bridge* and requires an additional control circuit to trigger the thyristor at the right instant. A typical 6-pulse thyristor converter is shown in Figure 3.2.12.

From Chapter 2.5, the conditions required before a thyristor will conduct current in a power electronic circuit are:

- A Forward Voltage must exist across the thyristor

AND

- A Positive Pulse must be applied to the thyristor gate

If each thyristor were triggered at the instant when the forward voltage across it tends to become positive, then the thyristor rectifier operates in the same way as the diode rectifier described above. All the voltage and current waveforms of the diode bridge apply to the thyristor bridge. A thyristor bridge operating in this mode is said to be operating with a *zero delay angle* and gives a voltage output of:

$$V_D = 1.35 \times V_{RMS}$$

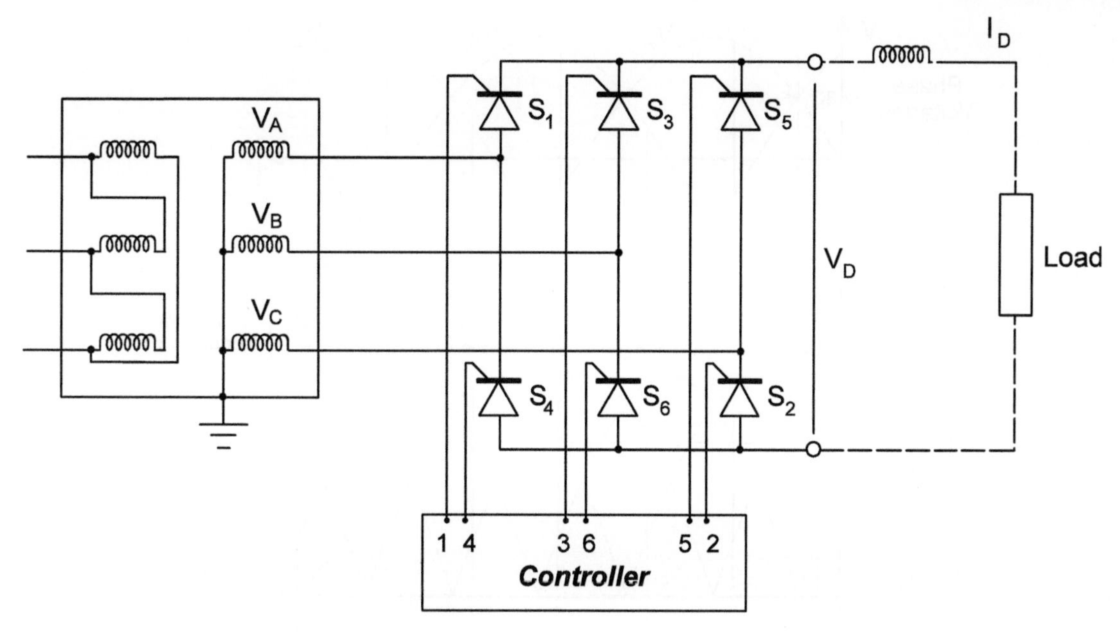

Fig. 3.2.12 6-pulse controlled thyristor rectifier bridge.

The output of the rectifier bridge can be controlled by delaying the instant at which the thyristor receives a triggering pulse. This delay is usually measured in degrees from the point at which the switch CAN turn on, due to the forward voltage becoming positive.

The angle of delay is called the *delay angle*, or sometimes *the firing angle*, and is designated by the symbol α. The reference point, for the angle of delay, is the point where a phase voltage wave crosses the voltage of the previous phase and becomes positive relative to it. A diode rectifier can be thought of as a converter with a *delay angle of* $\alpha = 0°$.

The main purpose of controlling a converter is to control the magnitude of the DC output voltage. In general, the bigger the delay angle, the lower the average magnitude of the DC voltage. Under steady-state operation of a controlled thyristor converter, the delay angle for each switch is the same. Figure 3.2.13 shows the voltage waveforms where the triggering of the switches has been delayed by an angle of α degrees.

In the positive switch group, the positive DC terminal follows the voltage associated with the switch that is in conduction in the sequence V_A–V_B–V_C. Assume, initially, that thyristor S_1 associated with voltage V_A is conducting and S_3 is not yet triggered. The voltage on the + **bus** on the DC side follows the declining voltage V_A because, in the absence of S_3 conduction, there is still a forward voltage across S_1 and it will continue to conduct. When S_3 is triggered after a delay angle = α, the voltage on + **bus** jumps to V_B, whose value it then starts to follow. At this instant, with both S_1 and S_3 conducting, a negative commutation voltage equal to V_B–V_A appears across the switch S_1 for the commutation period, which then starts to turn off. With the passage of time, V_B reaches its sinusoidal peak and starts to decline, followed by + DC terminal. At the same time, V_C is rising and when S_5 is triggered, the same sequence of events is repeated and the current is commutated to S_5.

As with the diode rectifier, the average magnitude of the DC voltage V_D can be calculated by integrating the voltage waveform over a 120° period representing a repeating portion of the DC voltage. At a delay angle α, the DC voltage is given by:

$$V_D = 1.35 \times (\textit{RMS} - \textit{Phase voltage}) \times \cos\alpha$$

$$V_D = 1.35 \times V_{RMS} \times \cos\alpha$$

This formula shows that the theoretical DC voltage output of the thyristor rectifier with a firing angle $\alpha = 0$ is the same as that for a diode rectifier (Figure 3.2.14). It also shows that the average value of the DC voltage will decrease as the delay angle is increased and is dependent on the cosine of the delay angle. When $\alpha = 90°$, then $\cos\alpha = 0$ and $V_D = 0$, which means that the average value of the DC voltage is zero. The instantaneous value of the DC voltage is a *saw-tooth* voltage as shown in the figure below.

If the delay angle is increased further, the average value of the DC voltage becomes negative. In this mode of operation, the converter operates as an *inverter*. It is interesting to note that the direction of the DC current remains unchanged because the current can only flow through the switches in one direction. However, with a negative DC voltage, the direction of the power flow is reversed and the power flows from the DC side to the

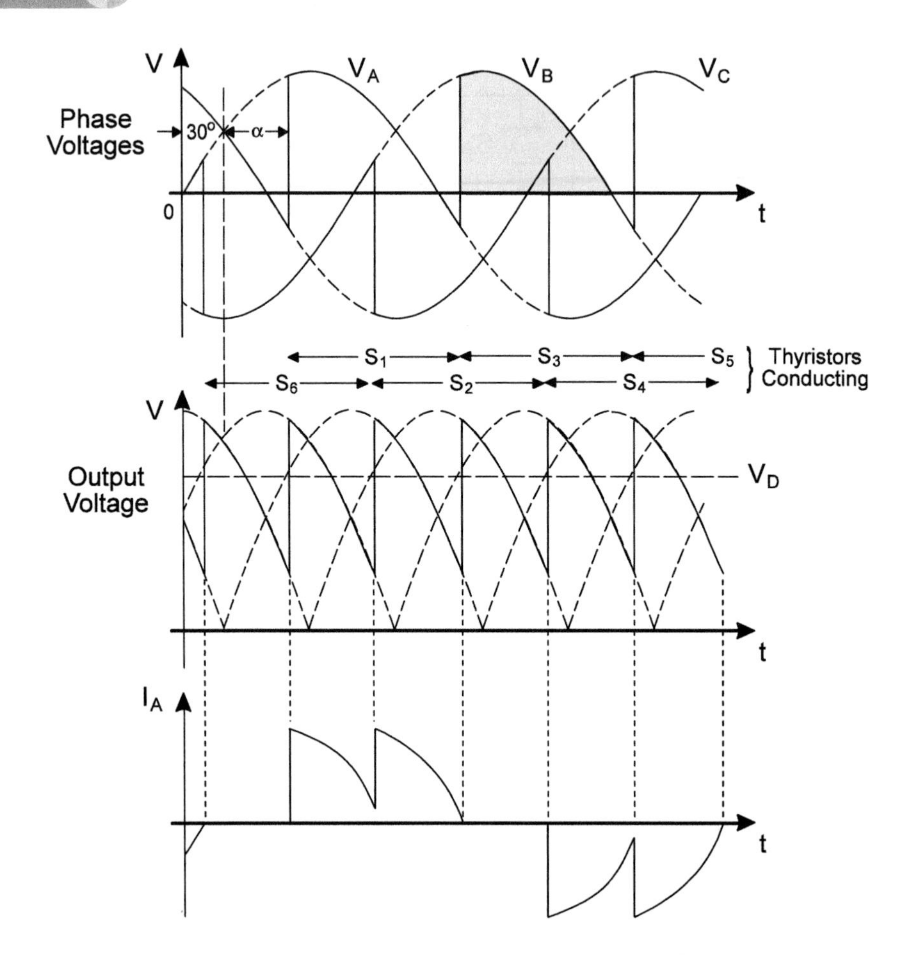

Fig. 3.2.13 Voltage waveforms of a controlled rectifier.

AC side. Steady-state operation in this mode is only possible if there is a voltage source on the DC side. The instantaneous value of the DC voltage for $\alpha > 90°$ is shown in Figure 3.2.15.

In practice, the commutation is not instantaneous and lasts for a period dependent on the circuit inductance and the magnitude of the commutation voltage. As in the idealized case, it is possible to estimate the commutation time from the commutation circuit inductance and an estimate of the average commutation voltage.

As in the diode rectifier, the steady DC current I_D comprises segments of current from each of the three phases on the AC side. On the AC side, the current in each phase comprises non-sinusoidal blocks, similar to those associated with the diode rectifier and with similar harmonic consequences. In the case of the diode bridge, with a delay angle of $\alpha = 0$, the angle between the phase current and the corresponding phase voltage on the AC side is roughly zero. Consequently, the power factor is roughly unity and converter behaves something like a resistive load.

For the controlled rectifier, with a delay angle of α, the angle between the phase current and the corresponding phase voltage is also roughly α, but normally called the power factor angle Ø. This angle should be called the *displacement factor* because it does not really represent power factor (see later). Consequently, when the delay angle of the thyristor rectifier is changed to reduce the DC voltage, the angle between the phase current and voltage also changes by the same amount. The converter then behaves like a resistive-inductive load with a displacement factor of cosØ. It is well known that the *power factor* associated with a controlled rectifier falls when the DC output voltage is reduced.

Delay angle	Converter behavior
$\alpha = 0°$	Behaves like a resistive load
$0° < \alpha < 90°$	Behaves like a resistive/inductive load and absorbs active power
$\alpha = 90°$	Behaves like an inductive load with no active power drawn
$\alpha > 90°$	Behaves like an inductive load but is also a source of active power

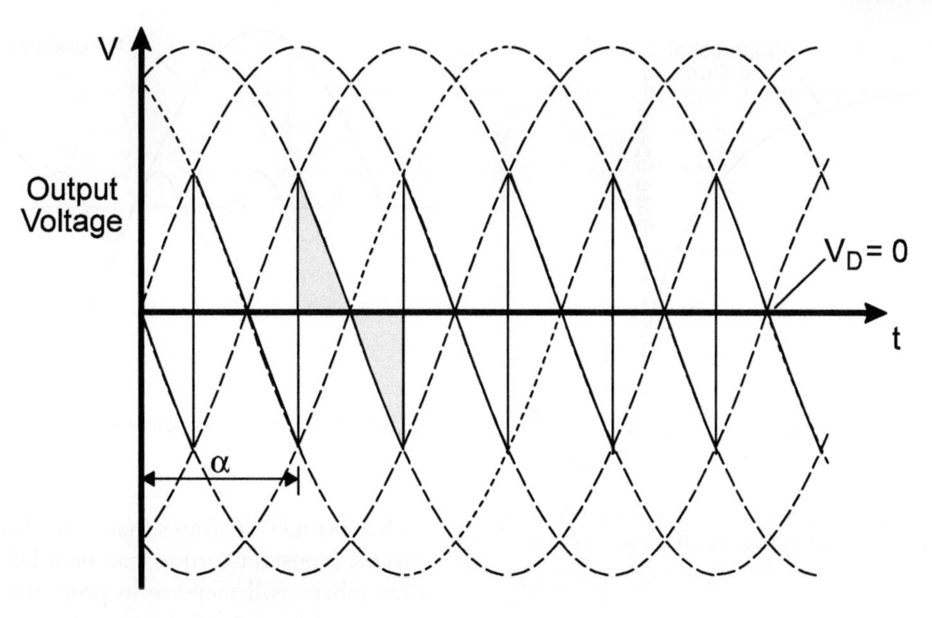

Fig. 3.2.14 DC output voltage for delay angle $\alpha = 90°$.

A common example of this is a DC motor drive controlled from a thyristor converter. As the DC voltage is reduced to reduce the DC motor speed at constant torque, the power factor drops and more reactive power is required at the supply line to the converter (Figure 3.2.16).

Figure 3.2.17 summarizes the possible vector relationships between the phase voltage and the fundamental component of the phase current in the supply line for the various values of delay angle α.

The phase current on the AC side is, fundamentally, a non-sinusoidal *square wave*. By applying the principles of harmonic analysis, using the Fourier transform, this non-sinusoidal wave can be resolved into a fundamental (50 Hz) sinusoidal wave plus a number of sinusoidal harmonics. The fundamental waveform has the highest amplitude and therefore the most influence on the power supply system. In a 6-pulse rectifier bridge, the fifth harmonic has the highest magnitude, theoretically 20% of the fundamental current (Figure 3.2.18).

The RMS value of the fundamental current can be calculated from the following formula, which is derived from fundamental principles:

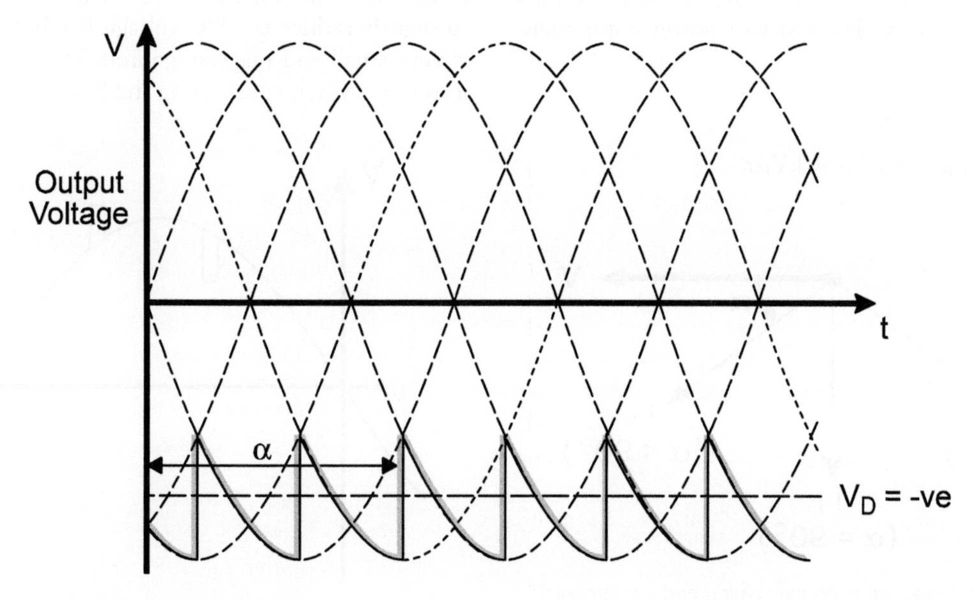

Fig. 3.2.15 DC voltage when the delay angle $\alpha > 90°$.

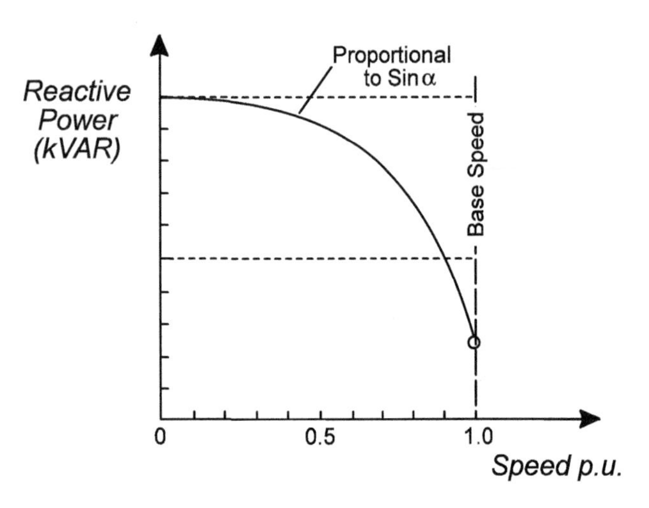

Fig. 3.2.16 Reactive power requirements of a DC motor drive with a constant torque load fed from a line-commutated converter.

$$I_1 = \sqrt{3}\frac{\sqrt{2}}{\pi}I_D = 0.78\, I_D \text{ A}$$

The corresponding apparent power S_1 kVA is given by:

$$S_1 = \sqrt{3}V_{RMS}I_1 \text{ kVA}$$
$$S_1 = \sqrt{3}V_{RMS}\, 0.78\, I_D \text{ kVA}$$
$$S_1 = 1.35\, V_{RMS}\, I_D \text{ kVA}$$

The active power component is given by:

$$P_1 = S_1 \text{ Cos } \varphi \text{ kW}$$
$$P_1 = 1.35\, V_{RMS}\, I_D \text{ kW}$$

This confirms that the active power calculated on the AC side is identical to the power calculated for the DC side ($V_D.I_D$), since from the previous formula $V_D = 1.35\ V_{RMS}\cos\alpha$. The reactive power component is given by:

$$Q_1 = S_1 \text{ Sin } \varphi \text{ kVAr}$$
$$Q_1 = 1.35\, V_{RMS}\, I_D \text{ Sin } \varphi \text{ kVAr}$$

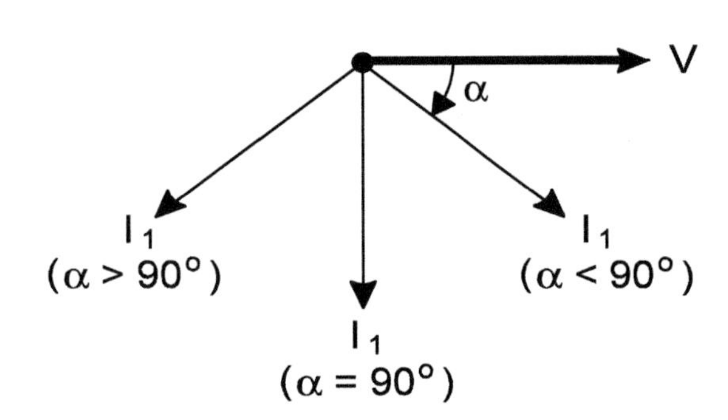

Fig. 3.2.17 Vector diagram of phase voltage and fundamental current for a controlled thyristor rectifier bridge.

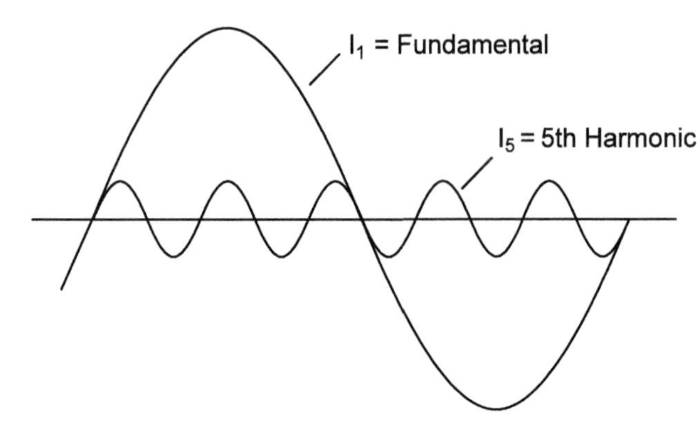

Fig. 3.2.18 The fundamental current and the fifth harmonic current.

This formula illustrates that, if the load current is held constant (constant torque load on a DC motor), the reactive power will increase in proportion to sinα as the triggering *delay angle* is increased.

Looking at the rectifier from the three-phase supply, an effective phase-to-phase short circuit occurs across the associated supply lines during commutation, when the two sequential switches are conducting. For example, when switch S_3 is triggered and switch S_1 continues to conduct, the voltage of V_A and V_B must be equal at switches themselves (except for the small volt drop across the switches). The commutation voltage V_B–V_A drives a circulating current through S_1 and S_3 and the circuit inductance 2L. Depending on the *delay angle*, the commutation voltage can be quite large. At the voltage source, the magnitude of the voltages V_A and V_B are depressed during this period by an amount dependent on the circulating current and circuit inductance. This additional non-desirable effect in the supply line is called *voltage notching* (Figure 3.2.19). The effect of notching is to slightly reduce the DC voltage V_D, but this reduction is very small and may be ignored. However, notching is important when considering the losses in the converter.

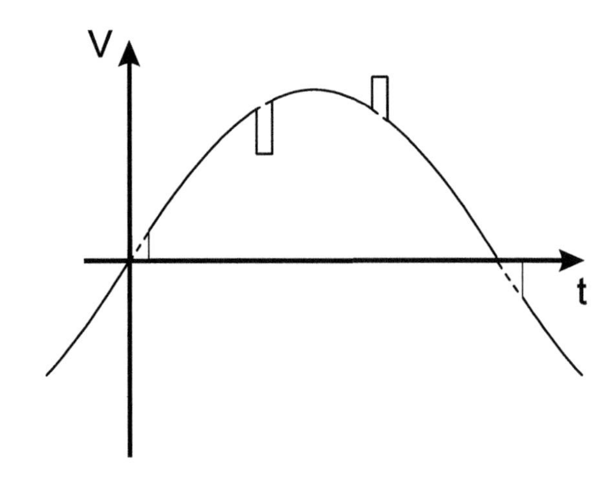

Fig. 3.2.19 Voltage notching in the supply line.

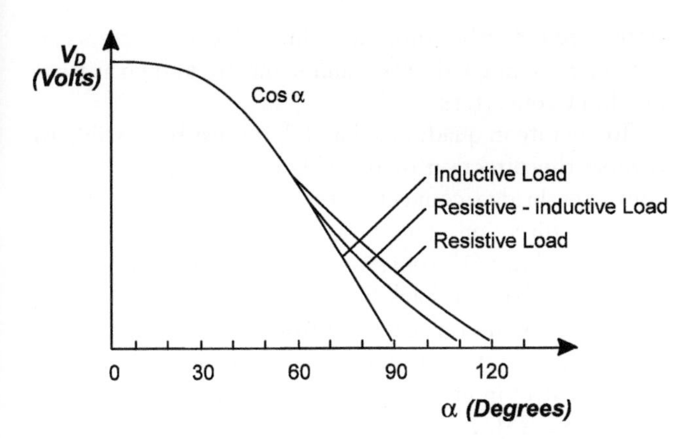

Fig. 3.2.20 Deviation of DC voltage from theoretical vs delay angle.

3.2.6.3 Practical limitations of line-commutated converters

The above analysis covers the theoretical aspects of uncontrolled and controlled converters. In practice, the components are not ideal and the commutations are not instantaneous. This results in certain deviations from the theoretical performance (Figure 3.2.20).

One of the most important deviations is that the DC load current is never completely smooth. The reason for this is fairly obvious. Accepting that the instantaneous DC voltage V_D can never be completely smooth, if the load is purely *resistive*, the DC load current cannot be completely smooth because it will linearly follow the DC voltage. Also, at delay angles $\alpha > 60°$, the DC output voltage becomes discontinuous and, consequently, so would the DC current. In an effort to maintain a smooth DC current, practical converters usually have some inductance L_D in series with the load on the DC side. For complete smoothing, the value of L_D should theoretically be infinite, which is not really practical.

The practical consequence of this is that the theoretical formula for the calculated value of DC voltage ($V_D = 1.35\ V_{RMS} \cos\alpha$) is not completely true for all values of delay angle α. Practical measurements confirm that it only hold true for delay angles up to about 75°, but this depends on the type of load and, in particular, the DC load inductance. Experience shows that for a particular delay angle $\alpha > 60°$, the average DC voltage will be higher than the theoretical value as shown in the figure below.

3.2.6.4 Applications for line-commutated rectifiers

An important application of the line-commutated converter is the DC motor drive. Figure 3.2.21 shows a single *controlled* line-commutated converter connected to the armature of a DC motor. The converter provides a variable DC voltage V_A to the armature of the motor, controlled from the control circuit of the converter.

When the delay angle is less than 90°, the DC voltage is positive and a positive current I_A flows into the armature of the DC motor to deliver active power to the load. The drive system is said to be operating in the *first quadrant* where the motor is running in the forward direction with active power being transferred from the supply to the motor and its mechanical load.

The motor field winding is usually separately excited from a simple diode rectifier and carries a field magnetizing current I_F. For a fixed field current, the speed of the motor is proportional to the DC voltage at the armature. The speed can be controlled by varying the delay angle of the converter and its output armature voltage V_A.

If the delay angle of the converter is increased to an angle greater than 90°, the voltage V_D will become negative and the motor will slow to a standstill. The

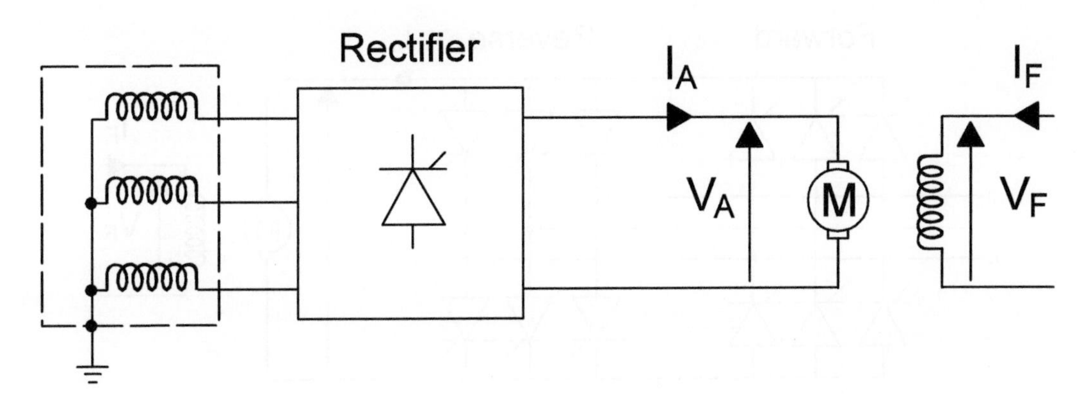

Fig. 3.2.21 Converter fed DC motor drive.

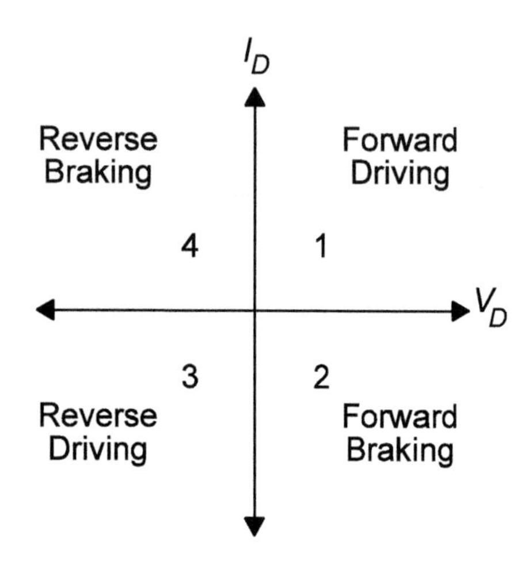

Fig. 3.2.22 Operating quadrants for variable speed drives.

current I_D also reduces to zero and the supply line can be disconnected from the motor without breaking any current.

Consequently, to stop a DC motor, the delay angle must be increased to value sufficiently larger than 90° to ensure that the voltage V_D becomes negative. With V_D negative and I_D still positive, the converter transiently behaves like an inverter and transfers active power from the motor to the supply line. This also acts as a brake to slow the motor and its load quickly to standstill. In this situation, the drive system is said to be operating in the *second quadrant* where the motor is running in the forward direction but the active power is being transferred back from the motor to the supply line (Figure 3.2.22).

The concept of the four operating quadrants has been covered in Chapter 2.4, (Figure 3.2.23). It illustrates the four possible operating states of any drive system and also shows the directions of V_D and I_D for the DC motor drive application.

The converters discussed so far have been *single converters*, which are only able to operate with positive DC current ($I_D = +\text{ve}$), which means that the motor can only run in the forward direction but active power can be transferred in either direction. Single DC converters can only operate in quadrants 1 and 4 and are known as two-quadrant converters.

To operate in quadrants 3 and 2, it must be possible to reverse the direction of I_D. This requires an additional converter bridge connected for current to flow in the opposite direction. This type of converter is known as a *four-quadrant DC converter*, and sometimes also called a double or back-to-back 6-pulse rectifier.

With a DC motor drive fed from a four-quadrant DC converter, operation in all four quadrants is possible with speed control in either the forward or reverse direction (Figure 3.2.23). A change of direction of the motor can quickly be achieved. Converter-1 is used as a controlled rectifier for speed control in the forward direction of rotation, while converter-2 is blocked, and vice versa in the reverse direction.

Assume, initially, that the motor is running in the forward direction under the control of converter-1 with a *delay angle* of <90°. Converter-2 is blocked. The changeover sequence from running in the forward direction to the reverse direction is as follows:

- Converter-1 delay angle increased to $\alpha > 90°$. This means that DC voltage $V_D < 0$ and DC current I_D is decreasing.
- When $I_D = 0$, converter-1 is blocked and thyristor firing is terminated.
- After small delay, converter-2 is unblocked and starts in the inverter mode with a firing angle greater than 90°.
- If the motor is still turning in the forward direction, converter-2 DC current I_D starts to increase in the negative direction and the DC machine acts as a generator and is braked to standstill, returning energy to the supply line.
- As the firing angle is reduced $\alpha < 90°$, converter-2 changes from the inverter to rectifier mode and, as voltage V_D increases, the motor starts to rotate in the opposite direction.

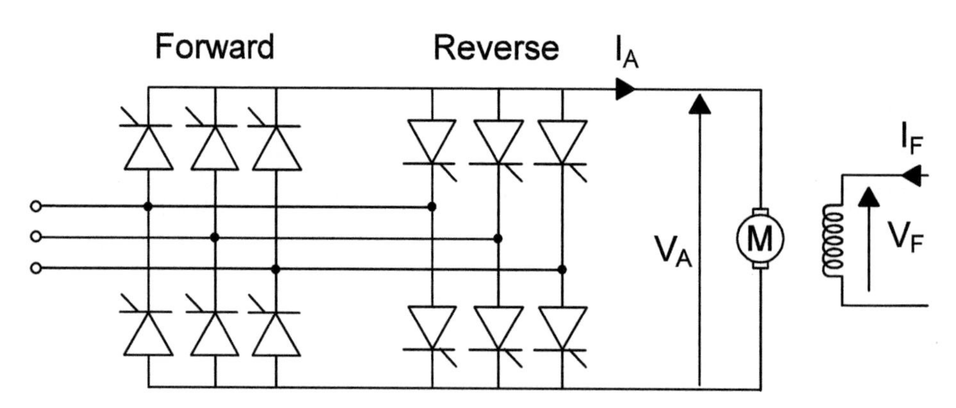

Fig. 3.2.23 Four-quadrant line-commutated rectifier.

In a DC motor drive, reversal of the direction of rotation can also be achieved by using a single converter and changing the direction of the excitation current. This method can only be used where there are no special drive requirements for changing over from forward to reverse operation. In this case, the changeover is done mechanically using switches in the field circuit during a period at standstill. Considerable time delays are required during standstill to remagnetize the field in the reverse direction.

There are many practical applications for both uncontrolled and controlled line-commutated rectifiers. Some of the more common applications include the following:

- DC motor drives with variable speed control
- DC supply for variable voltage variable frequency inverters
- Slip-energy recovery converters for wound rotor induction motors
- DC excitation supply for machines
- High voltage DC converters
- Electrochemical processes

3.2.7 Gate-commutated inverters (DC/AC converters)

Most modern AC variable speed drives in the 1–500 kW range are based on *gate-commutated devices* such as the GTO, MOSFET, BJTand IGBT, which can be turned ON and OFF by low power control circuits connected to their control gates.

The difficulties experienced with thyristor commutation in the early days of PWM inverters have largely been overcome by new developments in power electronic technology. Diodes and thyristors are still used extensively in line-commutated rectifiers.

Starting with a DC supply and using these semiconductor *power electronic switches,* it is not possible to obtain a pure sinusoidal voltage at the load. On the other hand, it may be possible to generate a near-sinusoidal current. Consequently, the objective is to control these switches in such a way that the current through the inductive circuit should approximate a sinusoidal current as closely as possible.

3.2.7.1 Single-phase square wave inverter

To establish the principles of gate-controlled inverter circuits, Figure 3.2.24 shows four semiconductor power switches feeding an inductive load from a single-phase supply.

This circuit can be considered to be an electronic reversing switch, which allows the input DC voltage V_D to be connected to the inductive load in any one of the following ways:

(1) S1 = on, S_4 = on, giving $+V_D$ at the load

(2) S_2 = on, S_3 = on, giving $-V_D$ at the load

(3) S1 = on, S_2 = on, giving zero volts at the load
S_3 = on, S_4 = on, giving zero volts at the load

(4) S1 = on, S_3 = on, giving a short circuit fault
S_2 = on, S_4 = on, giving a short circuit fault

However, these four switches can be controlled to give a square waveform across the inductive load as shown in Figure 3.2.24. This makes use of switch configuration (1) and (2), but not switch configuration (3) or (4). Clearly, for continued safe operation, option (4) should always be avoided. In the case of a purely inductive load, the current waveform is a triangular waveform as shown in Figure 3.2.25.

In the first part of the cycle, the current is negative although only switches S_1 and S_4 are ON. Since most power electronic devices cannot conduct negatively, to avoid damage to the switches, this negative current would have to be diverted around them. Consequently, diodes are usually provided in anti-parallel with the switches to allow the current flow to continue. These diodes are sometimes called *reactive* or *free-wheeling diodes* and conduct whenever the voltage and current polarities are opposite. This occurs whenever there is a reverse power flow back to the DC supply.

The frequency of the periodic square wave output is called the fundamental frequency. Using Fourier analysis, any repetitive waveform can be resolved into a number of sinusoidal waveforms, comprising one sinusoid at fundamental frequency plus a number of sinusoidal harmonics at higher frequencies, which are multiples of the fundamental frequency. The harmonic spectrum for a single-phase square wave output is

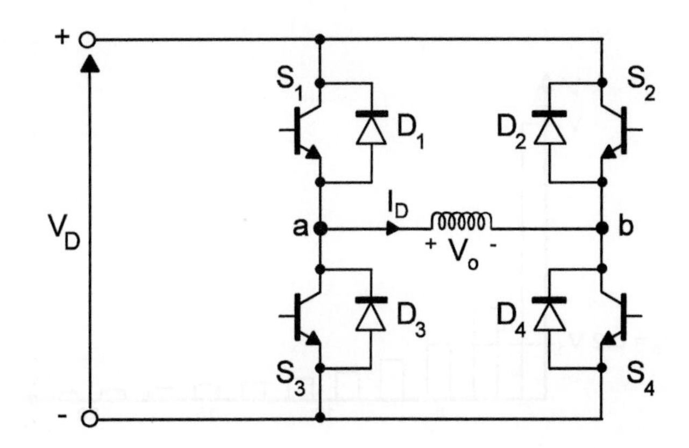

Fig. 3.2.24 Single-phase DC to AC inverter.

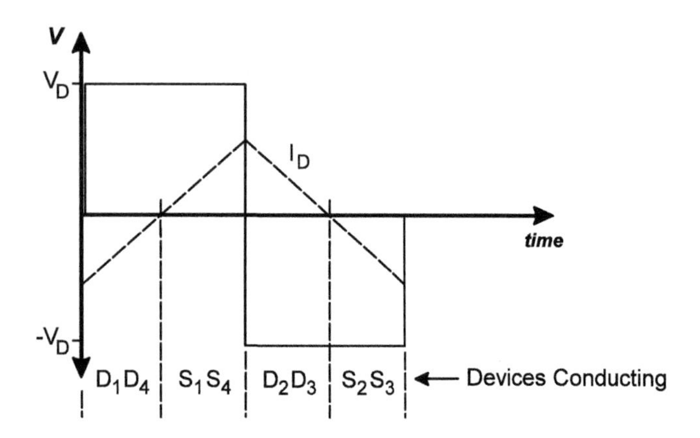

Fig. 3.2.25 Square wave modulation waveforms.

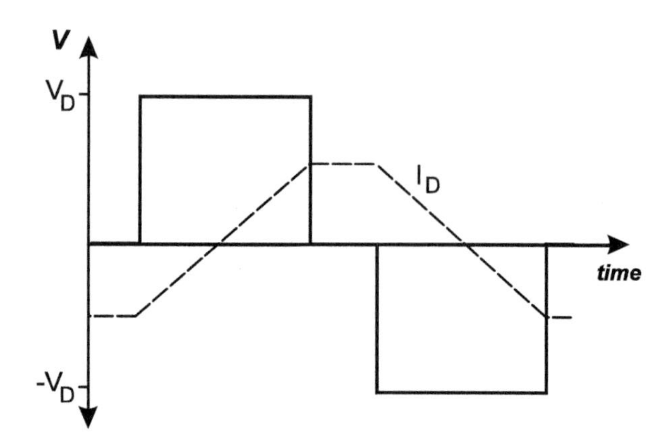

Fig. 3.2.27 Square wave modulation with reduced voltage pulse width.

shown in Figure 3.2.26. The amplitude of the higher order harmonics voltages falls off rapidly with increasing frequency.

The RMS value of the fundamental sinusoidal voltage component is:

$$V_1 = 2\frac{\sqrt{2}}{\pi}V_d \text{ V}$$

The RMS value of the nth harmonic voltage:

$$V_n = \frac{V_1}{n} \text{ V}$$

This illustrates that the square wave output voltage has several unwanted components of reasonably large magnitude at frequencies close to the fundamental. The current flowing in the load as a result of the output voltage is distorted, as demonstrated by the non-sinusoidal current wave-shape. In this example, the current has a triangular shape.

If the square-wave voltage were presented to a single-phase induction motor, the motor would run at the frequency of the square-wave but, being a linear device

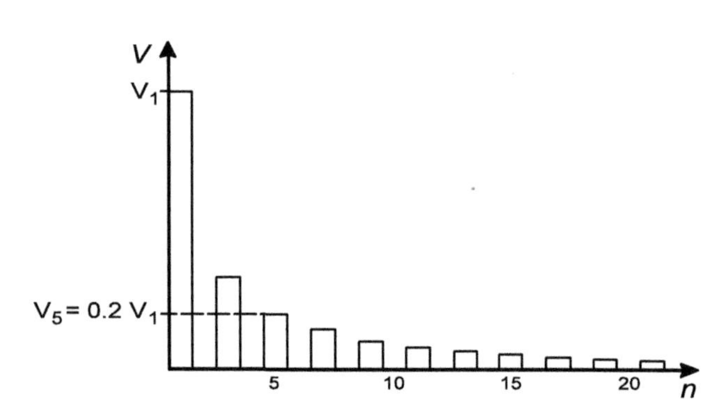

Fig. 3.2.26 Square-wave harmonic spectrum.

(inductive/ resistive load), it would draw non-sinusoidal currents and would suffer additional heating due to the harmonic currents. These currents may also produce pulsating torques.

To change the speed of the motor, the fundamental frequency of the inverter output can be changed by adjusting the speed of the switching. To increase frequency, switching speed can be increased and to decrease frequency, switching speed can be decreased.

If it is required to also control the magnitude of the output voltage, the average inverter output voltage can be reduced by inserting periods of zero voltage, using switch configuration (3) as shown in Figure 3.2.24. Each half-cycle then consists of a square pulse which is only a portion of a half-period.

The process of changing the width of the pulse to reduce the average RMS value of a waveform is called pulse width modulation (PWM). In the single-phase example in Figure 3.2.27, PWM it possible to control the RMS value of the output voltage. The fundamental sinusoidal component of voltage is continuously variable in the following range:

$$zero\ -\ 2\frac{\sqrt{2}}{\pi}V_D \text{ V}$$

The harmonic spectrum of this modified waveform depends on the fraction that the pulse is of the full square wave, but is broadly similar to the waveform shown in Figure 3.2.26.

3.2.7.2 Single-phase PWM inverter

The fact that the voltage supply to the stator of an AC induction motor is a square wave and is distorted is not in itself a problem for the motor. The main problem comes from the distortion of the current waveform, which results in extra copper losses and shaft torque pulsations. The ideal inverter output is one that results in a current waveform of low harmonic distortion.

Since an AC induction motor is predominantly inductive, with a reactance that depends on the frequency ($X_L = j2\pi fL$), it is beneficial if the **voltage harmonic distortion** can be pushed into the high frequencies, where the motor impedance is high and not much distorted current will flow.

One technique for achieving this is *sine-coded pulse width modulation (sine-PWM)*. This requires the power devices to be switched at frequencies much greater than that of the fundamental frequency producing a number of pulses for each period of the desired output period. The frequency of the pulses is called the *modulation frequency*. The width of the pulses is varied throughout the cycle in a sinusoidal manner giving a voltage waveform as shown in Figure 3.2.28. The figure also shows the current waveform for an inductive load showing the improvement in the waveform.

The improvement in the current waveform can be explained by the harmonic spectrum shown in Figure 3.2.29. It can be seen that, although the voltage waveform still has many distortion components, they now occur at higher harmonic frequencies, where the high load impedance of the motor is effective in reducing these currents.

Increasing the *modulation frequency* will improve the current waveform, but at the expense of increased losses in the switching devices of the inverter. The choice of modulation frequency depends on the type of switching device and its frequency. With the force-commutated thyristor inverter (10 years ago), a modulation frequency of up to 1 kHz was possible. With the introduction of GTOs and BJTs, this could be pushed up to around 5 kHz. With IGBTs, the modulation frequency could be as high as 20 kHz. In practice, a maximum modulation frequency of up to 12 kHz is common with IGBT inverters up to about the 22 kW motor size and 8 kHz for motors up to about 500 kW. The choice of modulation frequency is a trade-off between the losses in the motor and in the inverter. At low modulation frequencies, the losses in the inverter are low and those in the motor are high. At high modulation frequencies, the losses in the inverter increase, while those in the motor decrease.

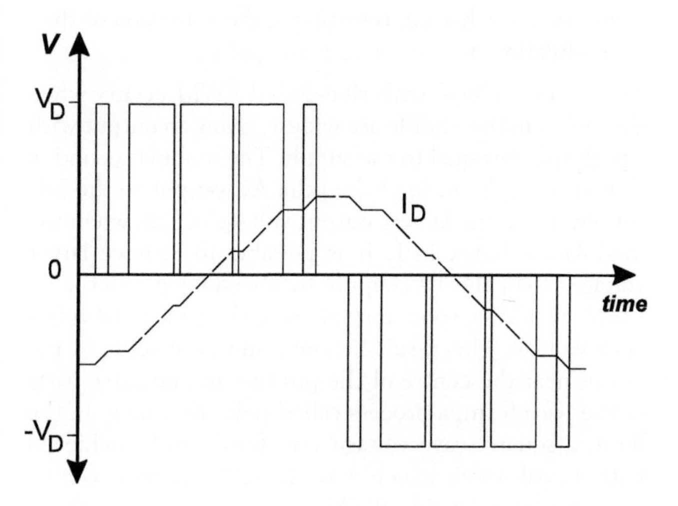

Fig. 3.2.28 Sine-coded pulse width modulated voltage and current.

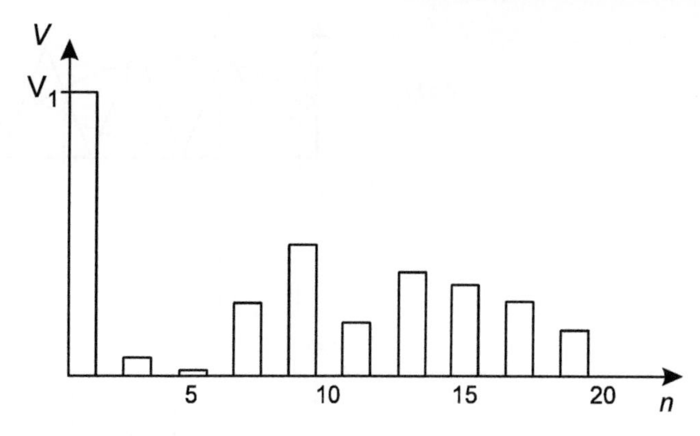

Fig. 3.2.29 Harmonic spectrum for a PWM inverter.

One of the most common techniques for achieving sine-coded PWM in practical inverters is the sine-triangle intersection method shown in Figure 3.2.30.

A triangular *saw-tooth* waveform is produced in the control circuit at the desired inverter switching frequency. This is compared in a comparator with a sinusoidal reference signal, which is equal in frequency and proportional in magnitude to that of the desired sinusoidal output voltage. The voltage V_{AN} (Figure 3.2.30(b)) is switched high whenever the reference waveform is greater than the triangle waveform. The voltage V_{BN} (Figure 3.2.30(c)) is controlled by the same triangle waveform but with a reference waveform shifted by 180°.

The actual phase-to-phase output voltage is then V_{AB} (Figure 3.2.30(d)), which is the difference between V_{AN} and V_{BN}, which consists of a series of pulses each of whose width is related to the value of the reference sinewave at that time. The number of pulses in the output voltage V_{AB} is double that in the inverter leg voltage V_{AN}. For example, an inverter switching at 5 kHz should produce switching distortion at 10 kHz in the output phase-to-phase voltage. The polarity of the voltage is alternatively positive and negative at the desired output frequency.

It can also be seen that the reference sine-wave in Figure 3.2.30 is given a DC component so that the pulse produced by this technique has a positive width. This puts a DC bias on the voltage of each leg as shown

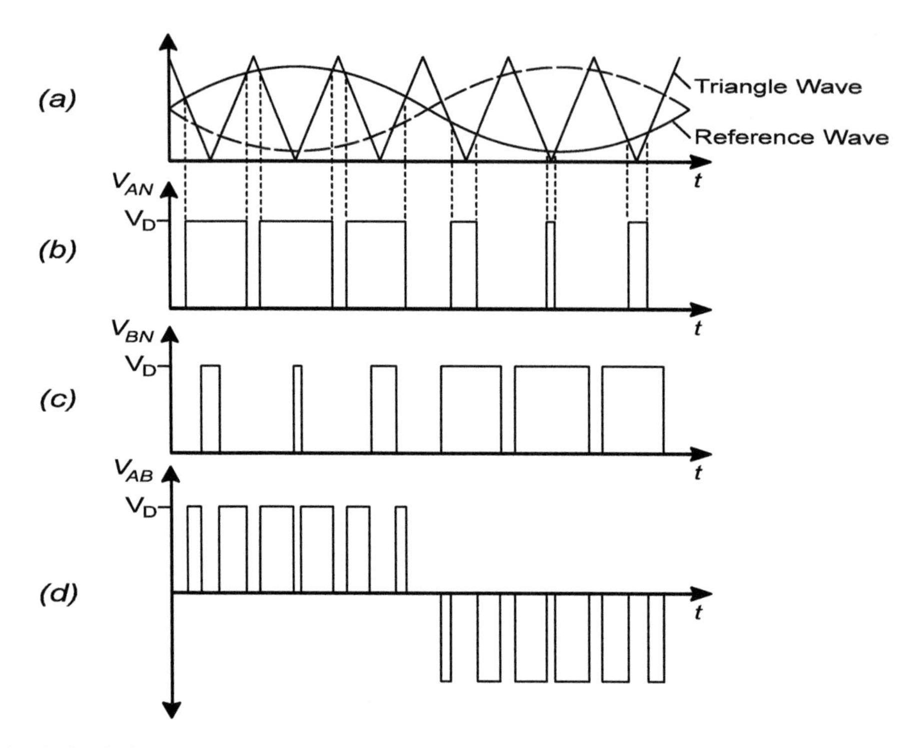

Fig. 3.2.30 Principle of triangle intersection PWM.

in Figures 3.2.30 (b) and (c). However, each leg has the same DC offset which disappears from the load voltage.

The technique using sine-triangle intersection is particularly suited for use with the older analog control circuits, where the two reference waveforms were fed into a comparator and the output of the comparator was used to trigger the inverter switches.

Modern *digital techniques* operate on the basis of a switching algorithm, for example, by producing triggering pulses proportional to the area under a part of the sine wave. In recent times, manufacturers have developed a number of different algorithms that optimise the performance of the output waveforms for AC induction motors. These techniques result in PWM output waveforms which are similar to those shown in Figure 3.2.30.

The sine-coded PWM voltage waveform is a composite of a high frequency square wave at the pulse frequency (the switching carrier) and the sinusoidal variation of its width (the modulating waveform). It has been found that, for lowest harmonic distortion, the modulating waveform should be synchronised with the carrier frequency, so that it should contain an integral number of carrier periods. This requirement becomes less important with high carrier frequencies of more than about 20 times the modulating frequency.

The voltage and frequency of a sinusoidal PWM waveform are varied by changing the reference waveform of Figure 3.2.30(a) giving outputs as shown in Figure 3.2.31.

- Figure 3.2.31(a) shows a base case, with the rated *V/f* ratio
- Figure 3.2.31(b) shows the case where the voltage reference is halved, resulting in the halving of each pulse
- Figure 3.2.31(c) shows the case where the reference frequency is halved, resulting in the extension of the modulation over twice as many pulses

The largest voltage with sine-coded PWM occurs when the pulses in the middle are widest, giving an output with a peak voltage equal to the supply. The modulation index is defined as the ratio of the peak AC output to the DC supply. Thus, the largest output voltage occurs when the modulation index is 1. It is possible to achieve larger voltages than the DC supply by abandoning strict sine-PWM by adding some distortion to the sinusoidal reference voltage. This results in the removal of some of the pulses near the centre of the positive and negative parts of the waveform, a process called pulse dropping. In the limit, a square wave voltage waveform can be achieved with a peak value which is up to 127% of what can be achieved by strict sine-PWM.

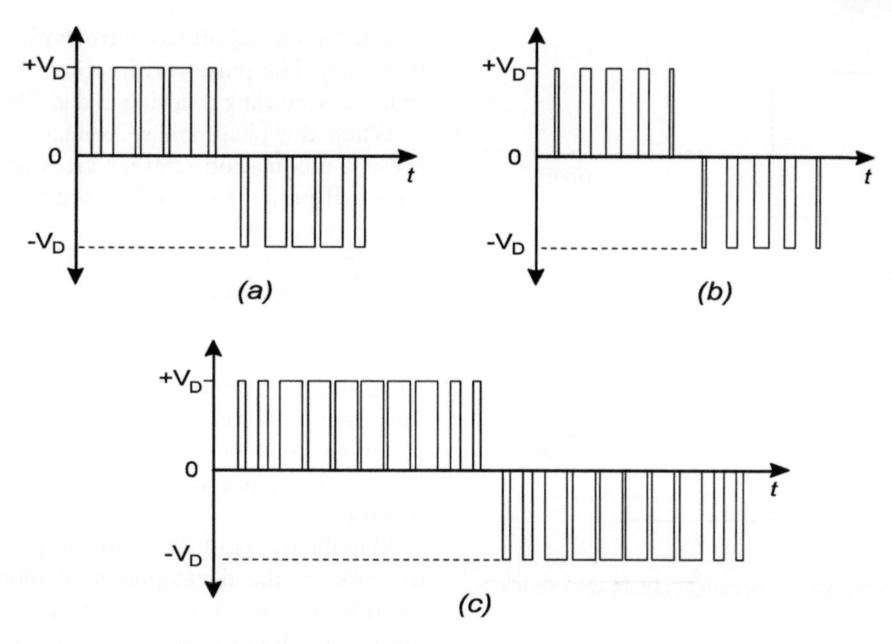

Fig. 3.2.31 Variation of frequency and voltage with sinusoidal PWM.

3.2.7.3 Three-phase inverter

A three-phase inverter could be constructed from three inverters of the type shown in Figure 3.2.24. However, it is more economical to use a 6-pulse (three-leg) bridge inverter as shown in Figure 3.2.32.

In its simplest form, a square output voltage waveform can be obtained by switching each leg high for one half-period and low for the next half-period, at the same time ensuring that each phase is shifted one-third of a period (120°) as shown in Figure 3.2.33. The resulting phase-to-phase voltage waveform comprises a series of square pulses whose widths are two-thirds of the period of the switch in each phase. The resulting voltage waveform is called a *quasi-square wave (QSW)* voltage. This simple technique was used in early voltage source inverters (VSI) which used forced-commutated thyristors in the inverter bridge. To maintain a constant *V/f* ratio, the magnitude of the DC bus voltage was controlled by the rectifier bridge to keep a fixed ratio to the output

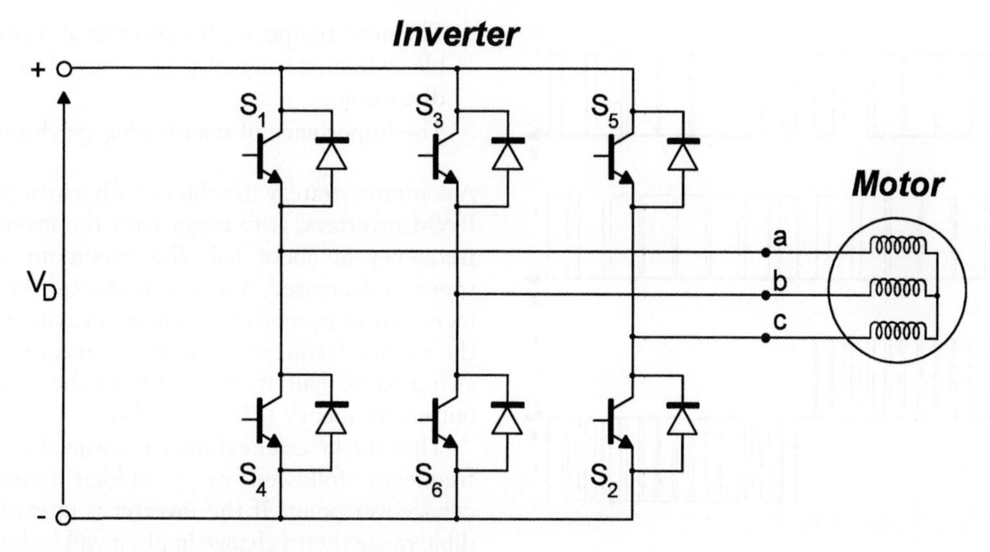

Fig. 3.2.32 Three-phase inverter using gate controlled switches.

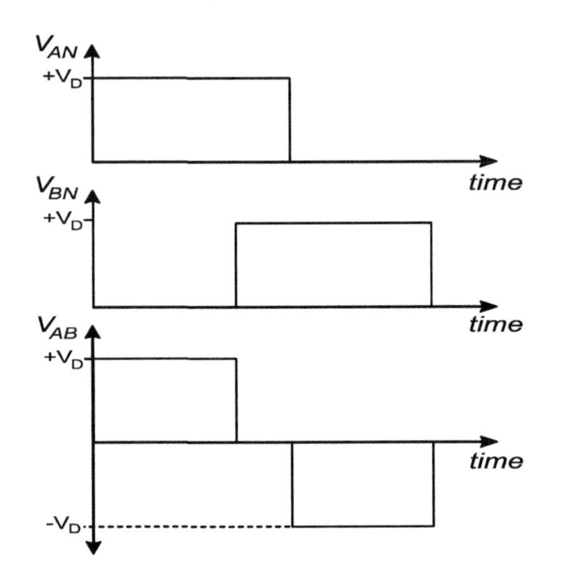

Fig. 3.2.33 Quasi-square wave modulation output waveforms.

frequency, which was controlled by the inverter bridge. This technique was sometimes also called *pulse amplitude modulation (PAM)*.

The output voltage of a three-phase converter has a harmonic spectrum very similar to the single-phase square wave, except that the triplen harmonics (harmonics whose frequency is a multiple of three times the fundamental frequency) have been eliminated. In an inverter with a three-phase output, this means that the 3rd, 9th, 15th, 21st, etc. harmonics are eliminated. To develop a three-phase variable voltage AC output of a particular frequency, the voltages V_{AN}, V_{BN}, V_{CN} on the three output terminals a, b, and c in Figure 3.2.32 can be

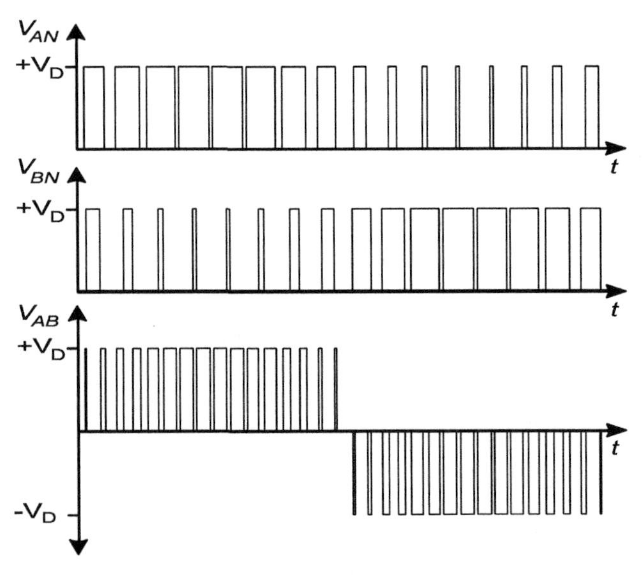

Fig. 3.2.34 Output voltage waveform of a three-phase sine-coded PWM.

modulated on and off to control both the voltage and the frequency. The pulse-width ratio over the period can be changed according to a sine-coded PWM algorithm.

When the phase–phase voltage V_{AB} is formed, the present modulation strategy gives only positive pulses for a half-period followed by negative pulses for a half-period, a condition known as *consistent pulse polarity.* It can be shown that *consistent pulse polarity* guarantees lowest harmonic distortion with most of the distortion being at twice the inverter chopping frequency. The presence of both positive and negative pulses throughout the whole period of the phase–phase voltage (*inconsistent pulse polarity*) gives distortion at the inverter chopping frequency, where it will have more effect on current distortion and is a sign of a poor modulation scheme.

Manufacturers of AC frequency converters continue to work on the development of more efficient PWM algorithms in an attempt to improve the current wave form. The ultimate objective is a completely sinusoidal current, which produces no harmonic losses in the motor. These more advanced PWM algorithms have become possible as a result of the increased speed and power of microprocessors. Most reputable PWM inverters can operate at modulation frequencies between 2 kHz and 16 kHz and produce a current waveform, which is sufficiently sinusoidal to overcome the problem of motor derating for harmonic losses. However, as a result of the high PWM frequencies, a new problem has emerged, the high frequency leakage current due to the motor cable capacitance.

In practical inverters, there are two conflicting requirements which need to be met when it is required to accelerate a motor from standstill to rated speed with constant *V/f* ratio.

- The need to operate the inverter at its highest possible switching frequency to achieve low current distortion
- The importance of maintaining synchronization

A common strategy to achieve both, particularly for older PWM inverters, is to begin with the inverter switching frequency at about half the maximum value. As the speed is increased, the saw-tooth carrier frequency is increased in proportion to maintain synchronism. When the carrier frequency reaches its maximum, it is then switched to half its value for further increase in the output frequency (Figure 3.2.34).

Thus the inverter exhibits a continual ramp increase in frequency followed by a sudden reduction at the changeover point. If the inverter is operating in the audible range then a change in pitch will be heard similar to the sound of a car engine as the car accelerates through the gears, hence the term '*gear-changing*'.

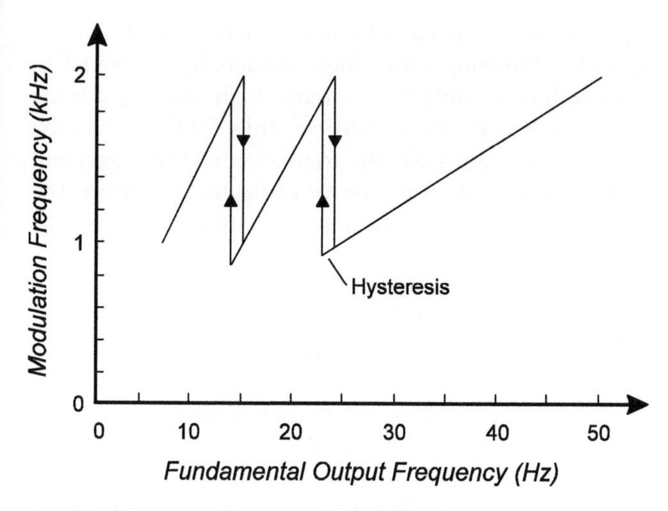

Fig. 3.2.35 : Changing the modulation frequency in steps (gear-changing).

When the motor speed is reduced from maximum to zero, there is a similar change in carrier frequency with output frequency. However, the changeover points must be different, otherwise an inverter sitting at one of the changeover frequencies might continually oscillate between the upper and lower carrier frequency. This is avoided by introducing hysteresis in the control scheme as shown in Figure 3.2.35.

3.2.8 Gate-controlled power electronic devices

A number of gate-controlled devices have become available in the past decade, which are suitable for use as bi-stable switches on power inverters for AC variable speed drives. These can be divided into two main groups of components:

- Those based on thyristor technology such as GTO and FCT
- Those based on transistor technology such as the BJT, FET and the IGBT

3.2.8.1 GTO thyristor

A *GTO thyristor* is another member of the thyristor family and is very similar in appearance and performance to a *normal thyristor*, with the important additional feature that it can be turned off by applying a negative current pulse to the gate. *GTO thyristors* have high current and voltage capability and are commonly used for larger converters, especially when self-commutation is required.

SYMBOL:

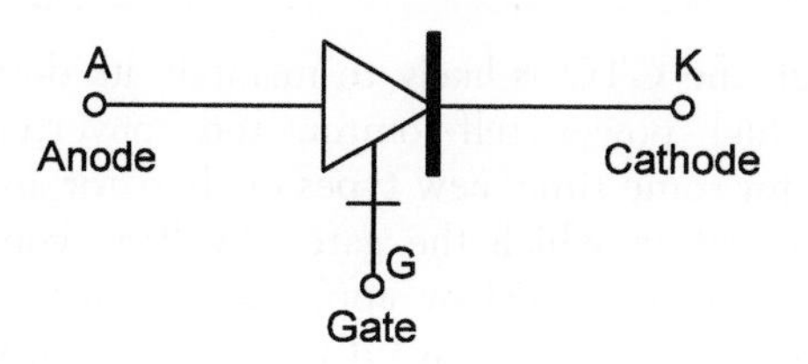

IDEAL:

Forward conduction: Resistance (less)
Forward blocking: Loss (less) (no leakage current)
Reverse blocking: Loss (less) (no leakage current)
Switch ON/OFF time: Instantaneous

The performance of a GTO is similar to a normal thyristor. Forward conduction is blocked until a positive pulse is applied to the gate terminal. When the GTO has been turned on, it behaves like a thyristor and continues to conduct even after the gate pulse is removed, provided that the current is higher than the holding current. The GTO has a higher forward voltage drop of typically 3–5 V. Latching and holding currents are also slightly higher.

The important difference is that the GTO may be turned off by a negative current pulse applied to the gate terminal. This important feature permits the GTO to be used in self-commutated inverter circuits. The magnitude of the off pulse is large and depends on the magnitude of the current in the power circuit. Typically, the gate current must be 20% of the anode current. Consequently, the triggering circuit must be quite large and this results in additional commutation losses. Like a thyristor, conduction is blocked in the reverse-biased direction or if the holding current falls below a certain level.

Since the GTO is a special type of thyristor, most of the other characteristics of a thyristor covered above also apply to the GTO and will not be repeated here. The mechanical construction of a GTO is very similar to a normal thyristor with stud types common for smaller units and disc types common for larger units.

GTO thyristors are usually used for high voltage and current applications and are more robust and tolerant to over-current and over-voltages than power transistors. GTOs are available for ratings up to 2500 A and 4500 V. The main disadvantages are the high gate current required to turn the GTO off and the high forward volt drop.

Power electronic converters of all types are usually controlled by an electronic control circuit which controls the ON/OFF state of the power electronic devices and provides the interface for the external controls. Until recently, all control circuits were of the analog type using operational amplifiers (Op-Amps). Modern control circuits are usually of the **digital** type using microprocessors.

3.2.8.2 Field controlled thyristors

Although the GTO is likely to maintain its dominance for the high power, self-commutated converter applications for some time, new types of thyristor are under development in which the gate is voltage controlled. **Turn ON** is controlled by applying a **positive voltage** signal to the gate and **turn OFF** by a **negative voltage**. Such a device is called *an FCT* and the name highlights the similarity to the FET. The FCT is expected to eventually supersede the GTO because it has a much simpler control circuit in which both the cost and the loss may be substantially reduced. Small FCTs have become available and it is expected that larger devices will come into use in the next few years. Development of a practical cost-effective device has been a bit slower than expected.

3.2.8.3 Power BJTs

Transistors have traditionally been used as amplification devices, where control of the base current is used to make the transistor conductive to a greater or lesser degree. Until recently, they were not widely used for power electronic applications. The main reasons were that the control and protective circuits were considerably more complicated and expensive and transistors were not available for high power applications. They also lacked the overload capacity of a thyristor and it is not feasible to protect transistors with fuses.

In the mid-1980s, the NPN transistor known as a *BJT* has become a cost-effective device for use in power electronic converters. Modern BJTs are usually supplied in an encapsulated module and each BJT has two **power terminals**, called the *collector (C)* and *emitter (E)* and a third **control terminal** called the *base (B)*.

SYMBOL:

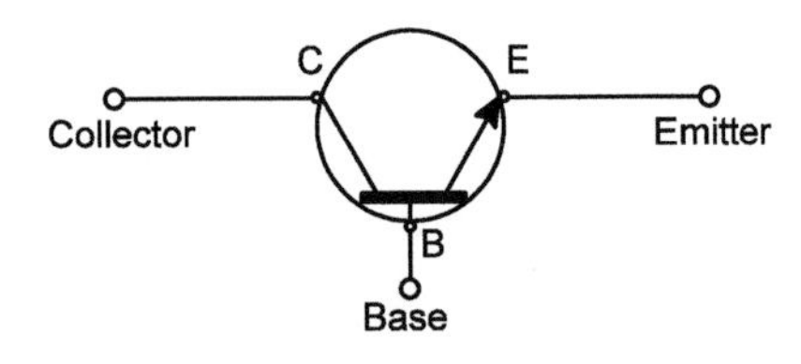

IDEAL:

Forward conduction: Resistance (less)
Forward blocking: Loss (less) (no leakage current)
Reverse blocking: Loss (less) (no leakage current)
Switch ON/OFF time: Instantaneous

A transistor is not inherently a *bi-stable (ON/OFF) device*. To make a transistor suitable for the conditions in a power electronic circuit where it is required to switch from the **blocking state (high voltage, low current)** to the **conducting state (low voltage, high current)** it must be used in its extreme conditions, fully OFF to fully ON. This potentially stresses the transistor and the trigger and protective circuits must be co-ordinated to ensure the transistor is not permitted to operate outside its *safe operating area*.

Suitable control and protective circuits have been developed to protect the transistor against over-current when it is turned ON and against over-voltage when it is turned OFF.

When turned ON, the control circuit must ensure that the transistor does not come out of saturation; otherwise it will be required to dissipate high power. In practice, the control system has proved to be cost effective, efficient, and reliable in service.

Transistors do not tolerate reverse voltages. When BJTs are used in inverter bridges, they must be protected against high reverse voltages by means of a reverse diode in series or in parallel. For the same reason, transistors are not used in rectifier bridges, which have to be able to withstand reverse voltages.

In general, transistors were considered to be less robust and less tolerant of overloads and 'spikes' than thyristors. *GTO thyristors* were often preferred for converters. In spite of the earlier problems experienced with transistors, AC converters have used power transistors at power ratings up to about 150 kW at 415 V.

The main advantage of transistors is that they can be turned ON and OFF from the *base* terminal, which makes them suitable for *self-commutated inverter circuits*. This results in power and control circuits which are simpler than those required for thyristors (Figure 3.2.36).

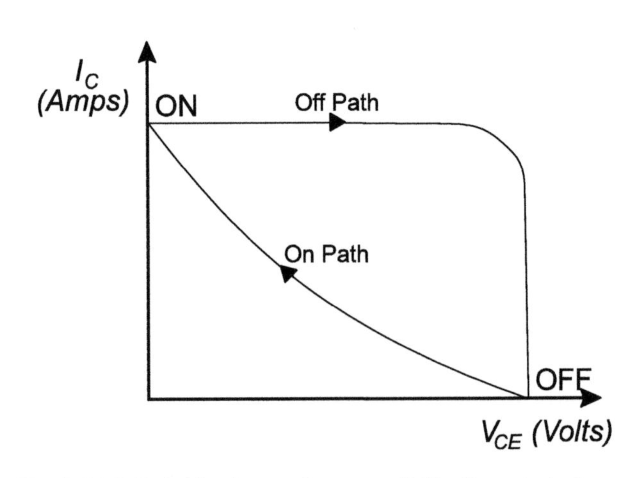

Fig. 3.2.36 Switching locus of a power BJT with an inductive load.

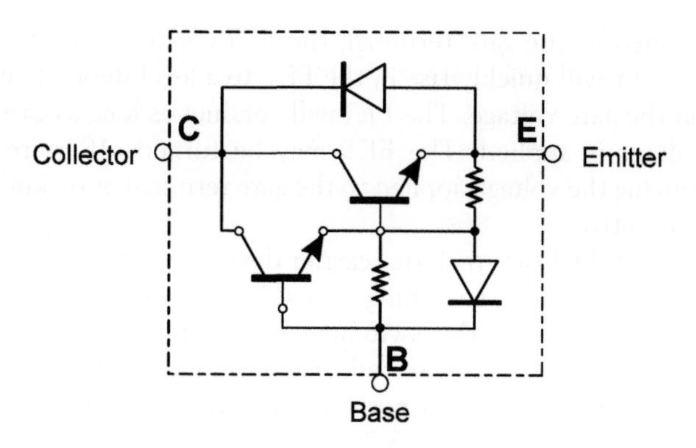

Fig. 3.2.37 Power Darlington transistor.

Unfortunately, the base amplification factor of a transistor is fairly low (usually 5 to 10 times); so, the trigger circuit of the transistor must be driven by an auxiliary transistor to reduce the magnitude of the base trigger current required from the control circuit. The emitter current from the auxiliary transistor drives the base of the main transistor using the Darlington connection. Figure 3.2.37 shows a double Darlington connection, but for high power applications, two auxiliary transistors (*triple Darlington*) may be used in cascade to achieve the required amplification factor. The overall amplification factor is approximately the product of the amplification factors of the two (or three) transistors.

Transistors, used in VSD applications, are usually manufactured as an integrated circuit and encapsulated into a three-terminal module, complete with the other necessary components, such as the resistors and anti-parallel protection diode. The module has an insulated base suitable for direct mounting onto the heat-sink. This type of module is sometimes called a *power Darlington transistor module.*

As shown in Figure 3.2.37, the anti-parallel diode protects the transistors from reverse biasing. In practice, this diode in the integrated construction is slow and may not be fast enough for inverter applications. Consequently, converter manufacturers sometimes use an external fast diode to protect the transistors.

Figure 3.2.38 shows the saturation characteristic of Toshiba MG160 S1UK1 triple Darlington power transistor rated at 1400 V, 160 A with a built-in free-wheeling diode.

Although the control circuits are completely different, the power circuit performance of a BJT is similar to a GTO thyristor. Forward conduction is blocked until a positive current is applied to the gate terminal and will conduct as long as the voltage is applied.

During forward conduction, it also exhibits a forward voltage drop which causes losses in the power circuit. The BJT may be turned OFF by applying a negative current to the gate. The main advantages of the *BJT* are:

- Good power-handling capabilities
- Low forward conduction voltage drop

The main disadvantages of *BJTs* are:

- Relatively slow switching times
- Inferior safe operating area
- Has complex **current controlled** gate driver requirements

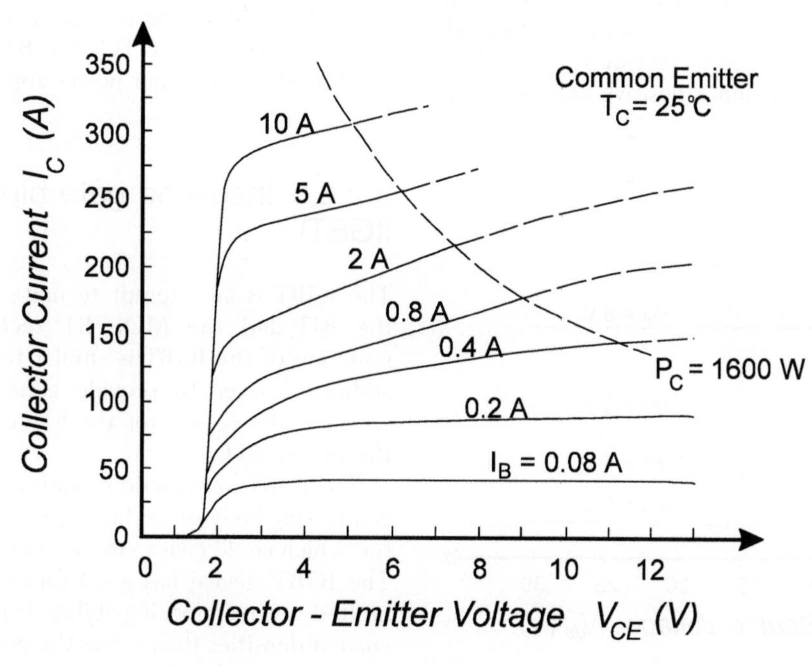

Fig. 3.2.38 Characteristics of a 160 A bipolar junction transistor (BJT).

Power BJT are available for ratings up to a maximum of about 300 A and 1400 V. For VSDs requiring a higher power rating, GTOs are usually used in the inverter circuit.

3.2.8.4 Field effect transistor (FET)

An FET is a special type of transistor that is particularly suitable for high speed switching applications (Figure 3.2.39). Its main advantage is that its *gate* is **voltage controlled** rather than **current controlled**. It behaves like a voltage controlled resistance with the capacity for high frequency performance.

FETs are available in a special construction known as the MOSFET. MOS stands for metal oxide silicon. The MOSFET is a three-terminal device with terminals called the *source (S)*, *drain (D)* and the *gate (G)*, corresponding to the emitter, collector and gate of the NPN transistor.

SYMBOL:

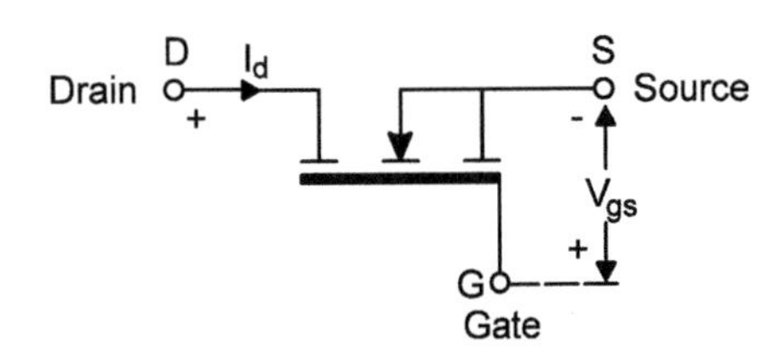

IDEAL:

Forward conduction: Resistance (less)
Forward blocking: Loss (less) (no leakage current)
Reverse blocking: Loss (less) (no leakage current)
Switch ON/OFF time: Instantaneous

The overall performance of an FET is similar to a power transistor, except that the gate is **voltage controlled**. Forward conduction is blocked if the gate voltage is low, typically less than 2 V. When a positive voltage V_{gs} is applied to the gate terminal, the FET conducts and the current will quickly rise in the FET to a level dependent on the gate voltage. The FET will conduct as long as gate voltage is applied. The FET may be turned off by removing the voltage applied to the gate terminal or making it negative.

MOSFETs are majority carrier devices, so they do not suffer from long switching times. With their very short switching times, the switching losses are low. Consequently, they are best suited to high frequency switching applications. A typical performance characteristic of an FET is shown in Figure 3.2.39.

Initially, high speed switching was not an important requirement for AC converter applications. With the development of *PWM* inverters, high frequency switching has become a desirable feature to provide a smooth output current waveform. Consequently, power FETs were not widely used until recently.

At present, FETs are only used for small PWM frequency converters. Ratings are available from about 100 A at 50 V to 5 A at 1000 V, but for VSD applications MOSFETs need to be in the 300–600 V range. The advantages and disadvantages of MOSFETs are almost exactly the opposite of *BJTs*.

The main advantages of a power MOSFET are

- High speed switching capability (10 nsec to 100 nsec)
- Relatively simple protection circuits
- Relatively simple voltage controlled gate driver with low gate current

The main disadvantages of a power MOSFET are

- Relatively low power handling capabilities
- Relatively high forward voltage drop, which results in higher losses than GTOs and BJTs, limits the use of MOSFETs for higher power applications

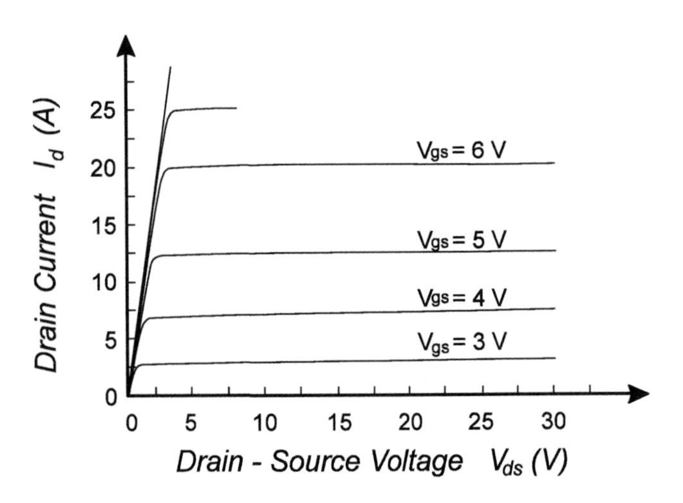

Fig. 3.2.39 Typical characteristic of a field effect transistor.

3.2.8.5 Insulated gate bipolar transistor (IGBT)

The IGBT is an attempt to unite the best features of the BJT and the MOSFET technologies. The construction of the IGBT is similar to a MOSFET with an additional layer to provide conductivity modulation, which is the reason for the low conduction voltage of the power BJT.

The IGBT construction avoids the MOSFET's reverse conducting body diode but introduces a parasitic thyristor, which could give spurious operation in early devices. The IGBT device has good forward blocking but very limited reverse blocking ability. It can operate at higher current densities than either the power BJT or MOSFET allowing a smaller chip size.

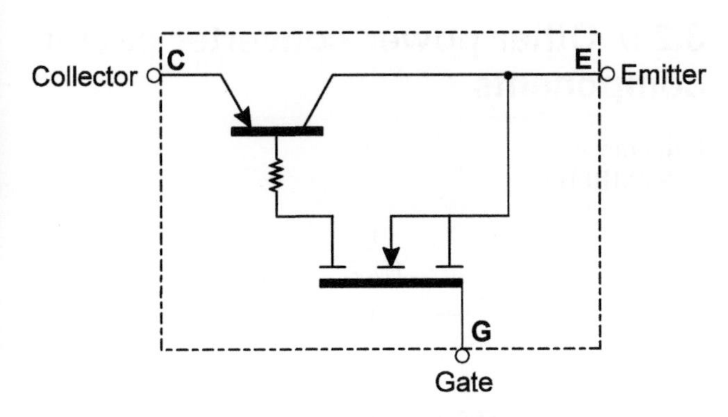

Fig. 3.2.40 The equivalent circuit of an IGBT.

The IGBT is a three-terminal device. The power terminals are called the *emitter (E)* and *collector (C)*, using the BJT terminology, while the control terminal is called the *gate (G)*, using the MOSFET terminology.

SYMBOL:

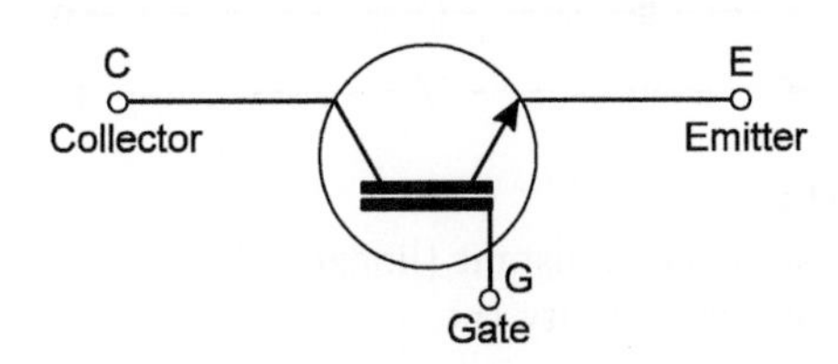

IDEAL:

Forward conduction: Resistance (less)
Forward blocking: Loss (less) (no leakage current)
Reverse blocking: Loss (less) (no leakage current)
Switch ON/OFF time: Instantaneous

The electrical equivalent circuit of the IGBT, (Figure 3.2.40), shows that the IGBT can be considered to be a hybrid device, similar to a darlington transistor configuration, with a MOSFET driver and a power bipolar PNP transistor. Although the circuit symbol above suggests that the device is related to an NPN transistor, this should not be taken too literally.

The gate input characteristics and gate drive requirements are very similar to those of a power MOSFET. The threshold voltage is typically 4 V. Turn-ON requires 10–15 V and takes about 1 µs. Turn-OFF takes about 2 µs and can be obtained by applying zero volts to the gate terminal. Turn-OFF time can be accelerated, when necessary, by using a negative drive voltage. IGBT devices can be produced with faster switching times at the expense of increased forward voltage drop.

An example of a practical IGBT driver circuit is shown in Figure 3.2.41. This circuit can drive two IGBTs, connected to a 1000 V supply, at a switching frequency of 10 kHz with propagation times of no more than 1 µs.

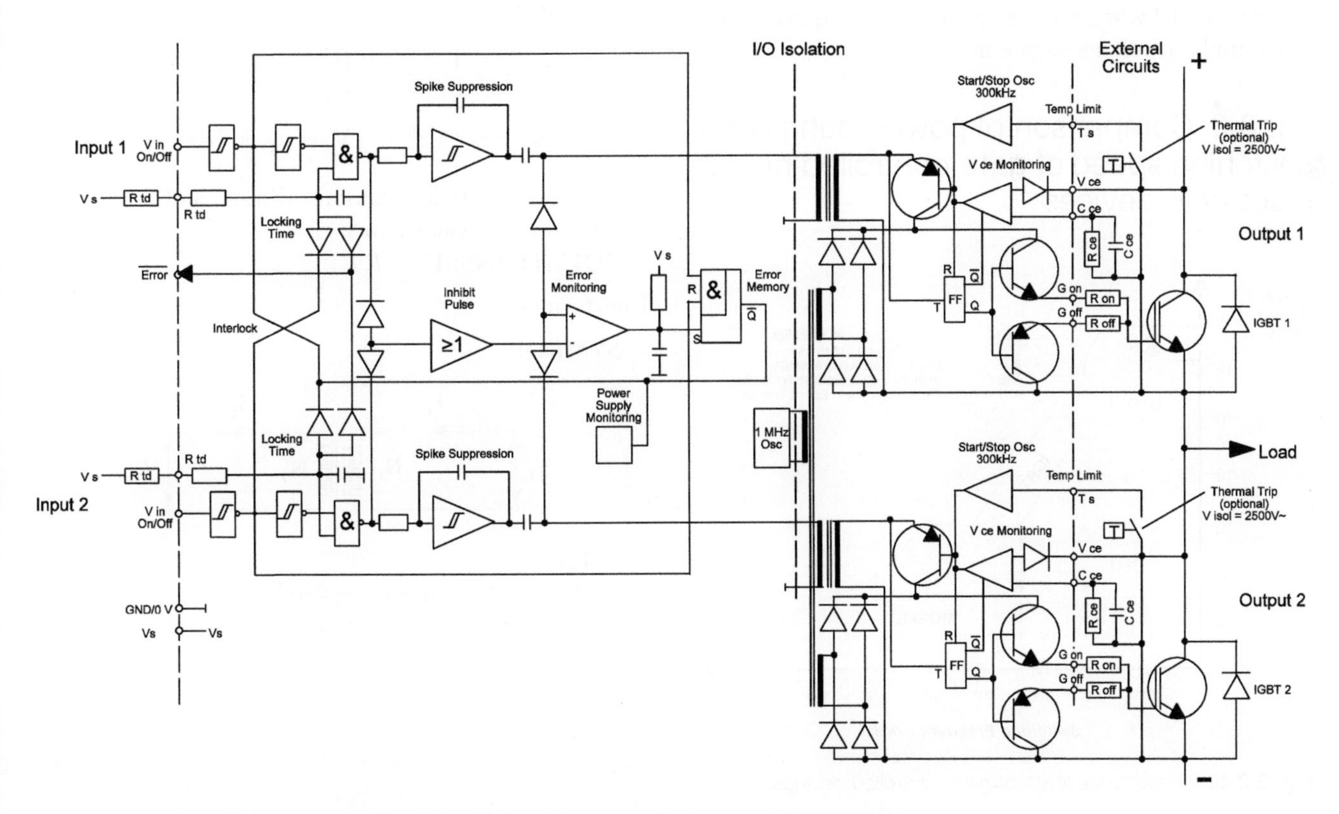

Fig. 3.2.41 Circuit diagram of semikron SKHI 20 hybrid double IGBTor double MOSFET driver.

IGBTs are currently available in ratings from a few amperes up to around 500 A at 1500 V, which are suitable for three-phase AC VSDs rated up to about 500 kW at 380 V/415 V/480 V. They can be used at switching frequencies up to 100 kHz. BJTs have now largely been replaced by IGBTs for AC variable speed drives.

The main advantages of the IGBT are:

- Good power handling capabilities
- Low forward conduction voltage drop of 2–3 V, which is higher than for a BJT but lower than for a MOSFET of similar rating
- This voltage increases with temperature making the device easy to operate in parallel without danger of thermal instability
- High speed switching capability
- Relatively simple voltage controlled gate driver
- Low gate current

Some other important features of the IGBT are (Figure 3.2.42):

- There is no secondary breakdown with the IGBT, giving a good *safe operating area* and low switching losses
- Only small snubbers are required
- The inter-electrode capacitances are not as relatively important as in a MOSFET, thus reducing miller feedback
- There is no body diode in the IGBT, as with the MOSFET, and a separate diode must be added in anti-parallel when reverse conduction is required, for example in voltage source inverters

3.2.8.6 Comparison of power ratings and switching speed of gate-controlled power electronic devices

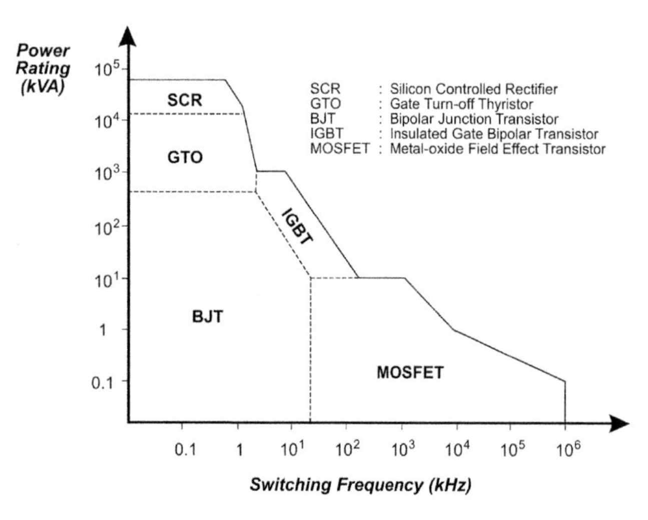

Fig. 3.2.42 Performance limits of gate-controlled devices.

3.2.9 Other power converter circuit components

Inductance

SYMBOL:

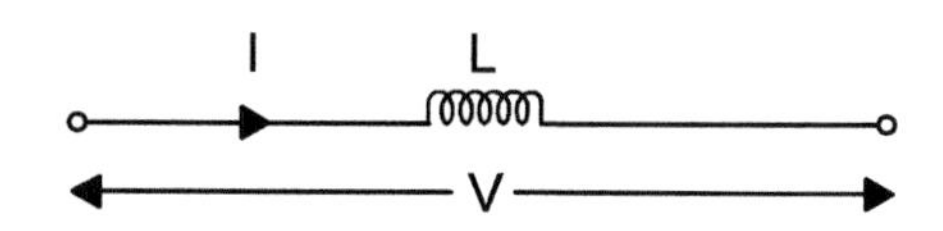

IDEAL:

Inductance: constant (linear)

Resistance: zero (no losses)

EQUATIONS: $V = L\frac{dI}{dt}$ $X_L = j\,2\pi f\,L$

Capacitance

SYMBOL:

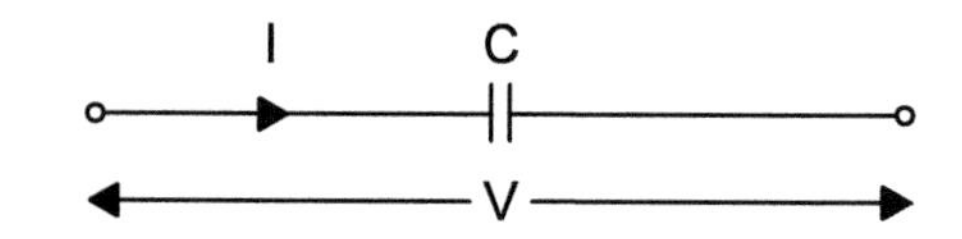

IDEAL:

Capacitance: constant (linear)

Resistance: infinity (no losses)

EQUATIONS: $I = C\frac{dV}{dt}$ $X_c = \frac{1}{j\,2\pi f\,C}$

Resistance

SYMBOL:

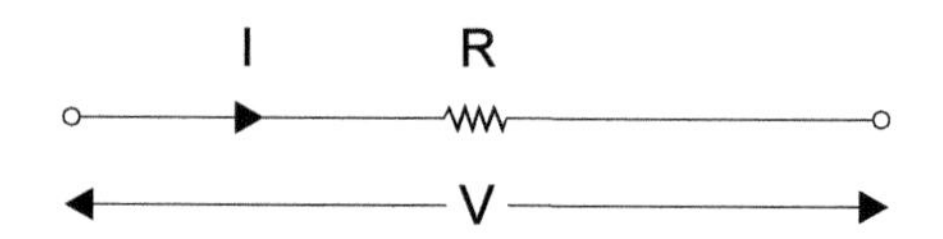

IDEAL: Resistance: constant (linear) and free of inductance and capacitance

EQUATIONS: $V = R\,I$

Transformer

SYMBOL:

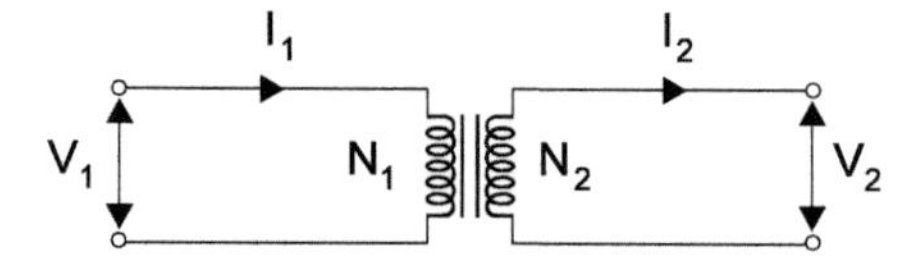

IDEAL:

Magnetizing current negligible

Free of losses and capacitance

EQUATIONS: $I_1 \times N_1 = I_2 \times N_2$ $\frac{V_1}{N_1} = \frac{V_2}{N_2}$

Section Four

Power transmission

Chapter 4.1

4.1

Power transmission

Crowder

While Chapters 2.1 and 2.2 have considered the analysis of a proposed motor–drive system and obtaining the application requirements, it must be recognised that the system comprises a large number of mechanical component. Each of these components, for example couplings, gearboxes and lead screws, will have their own inertias and frictional forces, which all need to be considered as part of the sizing process. This chapter considers power transmission components found in applications, and discusses their impact on overall system performance, and concludes with the process required to determine the detailed specifications of the motor and the drive.

The design parameters of the mechanical transmission system of the actuator must be identified at the earliest possible stage. However, it must be realised that the system will, in all probability, be subjected to detailed design changes as development proceeds. It should also be appreciated that the selection of a motor and its associated drive, together with their integration into a mechanical system, is by necessity an iterative process; any solution is a compromise. For this reason, this chapter can only give a broad outline of the procedures to be followed; the detail is determined by the engineer's insight into the problem, particularly for constraints of a non-engineering nature, such as a company's or a customer's policy, which may dictate that only a certain range of components or suppliers can be used.

In general, once the overall application, and the speed and torque (or in the case of a linear motor, speed and force) requirements of the total system have been clearly identified, various broad combinations of motors and drives can be reviewed. The principles governing the sizing of a motor–drive are largely independent of the type of motor being considered. In brief, adequate sizing involves determining the motor's speed range, and determining the continuous and intermittent peak torque or force which are required to allow the overall system to perform to its specification. Once these factors have been determined, an iterative process using the manufacturer's specifications and data sheets will lead to as close an optimum solution as is possible.

4.1.1 Gearboxes

As discussed in Section 2.2.1.3 a conventional gear train is made up of two or more gears. There will be a change in the angular velocity and torque between an input and output shaft; the fundamental speed relationship is given by

$$n = \pm\frac{\omega_i}{\omega_o} = \pm\frac{N_o}{N_i} \tag{4.1.1}$$

where N_i and ω_i are the number of teeth on, and the angular velocity of, the input gear, and N_o and ω_o are the number of teeth on, and the angular velocity of, the output gear. In Equation (4.1.1) a negative sign is used when two external gears are meshing, Figure 4.1.1(a), or a positive sign indicates that system where an internal gear is meshing with an internal gear, Figure 4.1.1(b).

In the case where an idler gear is included, the gear ratio can be calculated in an identical fashion, hence for an external gear train, Figures 4.1.1(c) and (d),

$$n = \frac{\omega_{in}}{\omega_{out}} = \left(-\frac{N_2}{N_1}\right)\left(-\frac{N_3}{N_2}\right) = \frac{N_3}{N_1} \tag{4.1.2}$$

The direction of the output shaft is reversed for an internal gear train, Figure 4.1.1(d). In practice the actual gear train can consist of either a spur, or helical gear

Electric Drives and Electromechanical Systems; ISBN: 9780750667401

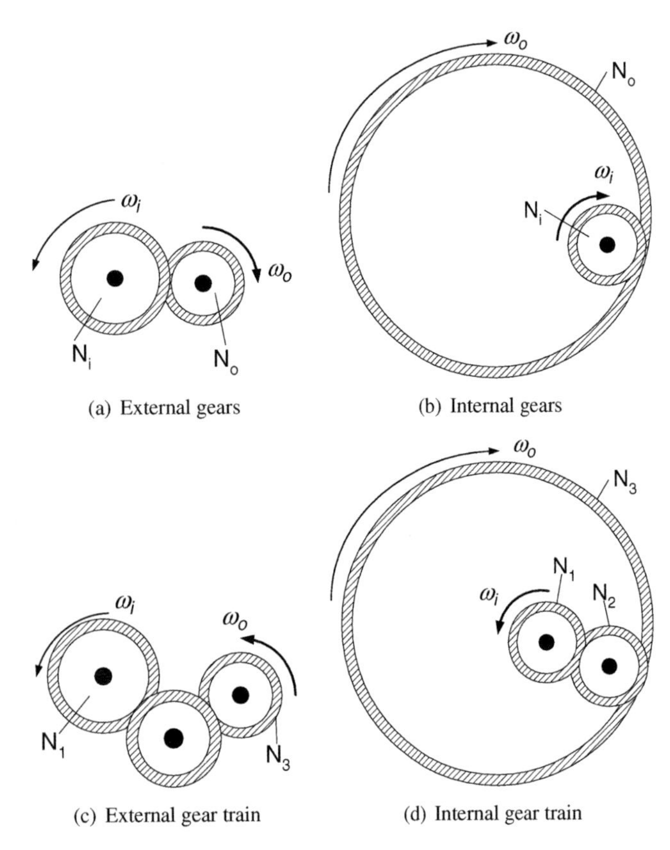

Fig. 4.1.1 Examples of the dependency of direction and velocity of the output shaft on the type of gearing.

wheels. A spur gear (see Figure 4.1.2(a)) is normally employed within conventional gear trains, and has the advantage of producing minimal axial forces which reduce problems connected with motion of the gear bearings. Helical gears (see Figure 4.1.2(b)) are widely used in robotic systems since they give a higher contact ratio than spur gears for the same ratio; the penalty is axial gear load. The limiting factors in gear transmission are the stiffness of the gear teeth, which can be maximised by selecting the largest-diameter gear wheel which is practical for the application, and backlash or lost motion between individual gears. The net result of these problems is a loss in accuracy through the gear train, which can have an adverse affect on the overall accuracy of a controlled axis.

In many applications conventional gear trains can be replaced by complete gearboxes (in particular, those of a planetary, harmonic or cycloid design) to produce compact drives with high reduction ratios.

4.1.1.1 Planetary gearbox

A Planetary gearbox is coaxial and is particularly suitable for high torque, low-speed applications. It is extremely price-competitive against other gear systems and offers high efficiency with minimum dimensions. For similar output torques the planetary gear system is the most compact gearbox on the market. The internal details of a planetary gearbox are shown in Figure 4.1.3; a typical planetary gearbox consists of the following:

- A sun gear, which may or may not be fixed.
- A number of planetary gears.
- Planet gear carrier.
- An internal gear ring, which may not be used on all systems.

This design results in relatively low speeds between the individual gear wheels and this results in a highly efficient design. One particular advantage is that the gearbox has no bending moments generated by the transmitted torque; consequently, the stiffness is considerably higher than in comparable configuration. Also, they can be assembled coaxially with the motor, leading to a more compact overall design. The relationship for a planetary gearbox can be shown to be

$$\frac{\omega_{\text{sun}} - \omega_{\text{carrier}}}{\omega_{\text{ring}} - \omega_{\text{carrier}}} = -\frac{N_{\text{ring}}}{N_{\text{sun}}} \tag{4.1.3}$$

(a) Spur gears.

(b) Helical gears.

Fig. 4.1.2 Conventional gears.

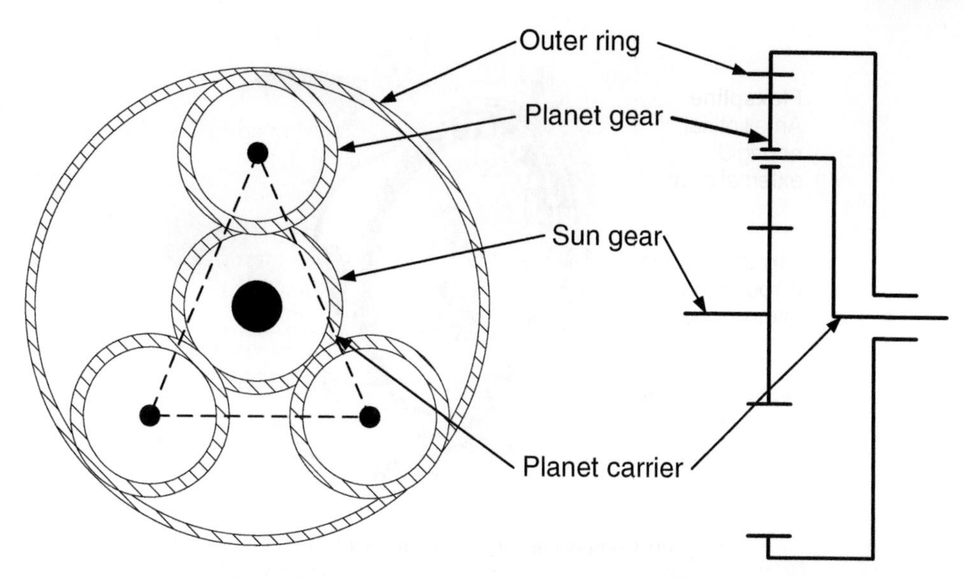

Fig. 4.1.3 A planetary gearbox; the output from the gearbox is from the three planet gears via the planet carrier, while the sun gear is driven. In this case the outer ring is fixed, the input is via the sun, and the output via the planet carrier.

where ω_{sun}, $\omega_{carrier}$ and ω_{ring} are the angular speeds of the sun gear, planet carrier and ring with reference to ground. N_{ring}, and N_{sun} are the number of teeth on the sun and ring respectively. Given any two angular velocities, the third can be calculated – normally the ring if fixed hence $\omega_{ring} = 0$. In addition, it is important to define the direction of rotation; normally clockwise is positive, and counterclockwise is negative.

Example 4.1.1

A planetary gearbox has 200 teeth on its ring, and 40 teeth on its sun gear. The input to the sun gear is 100 rev min^{-1} clockwise. Determine the output speed if the ring is fixed, or rotating at 5 rev min^{-1} either clockwise or counterclockwise.

Rearranging Equation (4.1.3), gives

$$\omega_{carrier} = \frac{N_{sun}\omega_{sun} - N_{ring}\omega_{ring}}{N_{sun} - N_{ring}}$$

- When the ring is rotated at 5 rev min^{-1} clockwise, the output speed is 18.75 rev min^{-1} conterclockwise.
- When the ring is fixed, the output speed is 25 rev min^{-1} conterclockwise.
- When the ring is rotated at 5 rev min^{-1} counterclockwise, the output speed is 31.25 rev min^{-1} counterclockwise.

This simple example demonstrates that the output speed can be modified by changing the angular velocity of the ring, and that the direction of the ring adds or subtracts angular velocity to the output.

4.1.1.2 Harmonic gearbox

A harmonic gearbox will provide a very high gear ratio with minimal backlash within a compact unit. As shown in Figure 4.1.4(a), a harmonic drive is made up of three main parts: the circular spline, the wave generator and the flexible flexspline. The design of these components depends on the type of gearbox, in this example the flexispline forms a cup. The operation of an harmonic gearbox can be appreciated by considering the circular spline to be fixed, with the teeth of the flexspline to engage on the circular spline. The key to the operation is the difference of two teeth (see Figure 4.1.4(b)) between the flexspline and the circular spline. The bearings on the elliptical-wave generator support the flexspline, while the wave generator causes it to flex. Only a small percentage of the flexspline's teeth are engaged at the ends of the oval shape assumed by the flexspline while it is rotating, so there is freedom for the flexspline to rotate by the equivalent of two teeth relative to the circular spline during rotation of the wave generator. Because of the large number of teeth which are in mesh at any one time, harmonic drives have a high torque capability; in addition the backlash is very small, being typically less that 30″ of arc.

In practice, any two of the three components that make up the gearbox can be used as the input to, and the output from, the gearbox, giving the designer considerable flexibility. The robotic hand shown in figure incorporates three harmonic gearboxes of a pancake design where the flexispline is a cylinder equal in width to the wave generator.

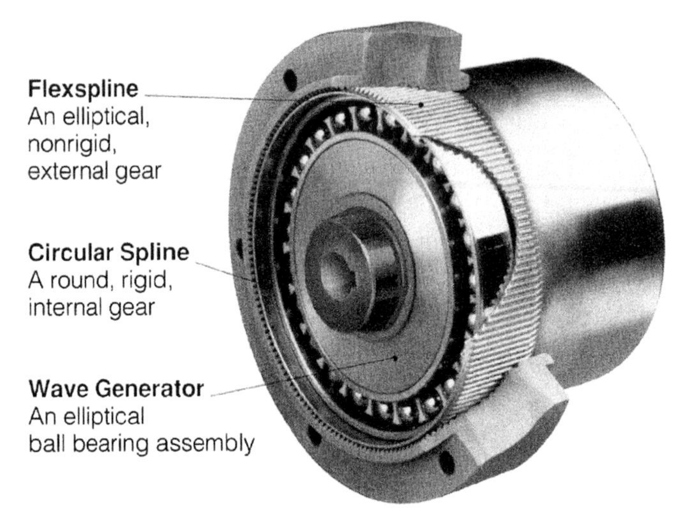

(a) Components of a harmonic gearbox

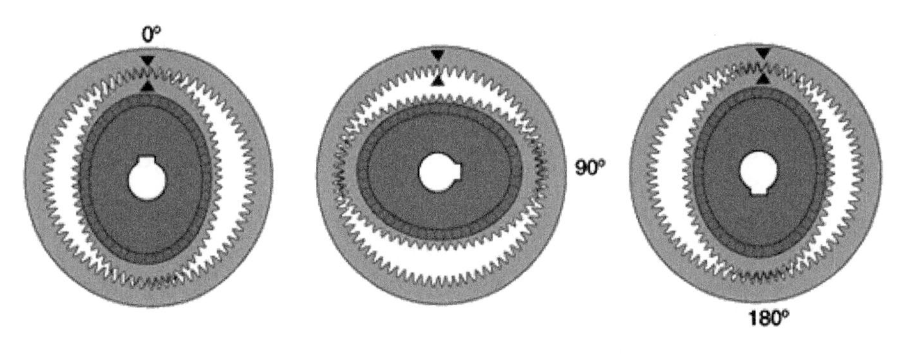

(b) Operation of a harmonic gear box, for each 360° rotation of the wave generator the flexspline moves 2 teeth. The deflection of the flexspline has been exaggerated.

Fig. 4.1.4 Construction and operation of an HDC harmonic gearbox. Reproduced with permission from Harmonic Drive Technologies, Nabtesco Inc, Peabody, MA.

4.1.1.3 Cycloid gearbox

The cycloid gearbox is of a coaxial design and offers high reduction ratios in a single stage, and is noted for its high stiffness and low backlash. The gearbox is suitable for heavy duty applications, since it has a very high shock load capability of up to 500%. Commercially cycloid gearboxes are available in a range of sizes with ratios between 6:1 and 120:1 and with a power transmission capability of up to approximately 100 kW. The gearbox design, which is both highly reliable and efficient, undertakes the speed conversion by using rolling actions, with the power being transmitted by cycloid discs driven by an eccentric bearing.

The significant features of this type of gearbox are shown in Figure 4.1.5. The gearbox consists of four main components:

- A high-speed shaft with an eccentric bearing.
- Cycloid disc(s).

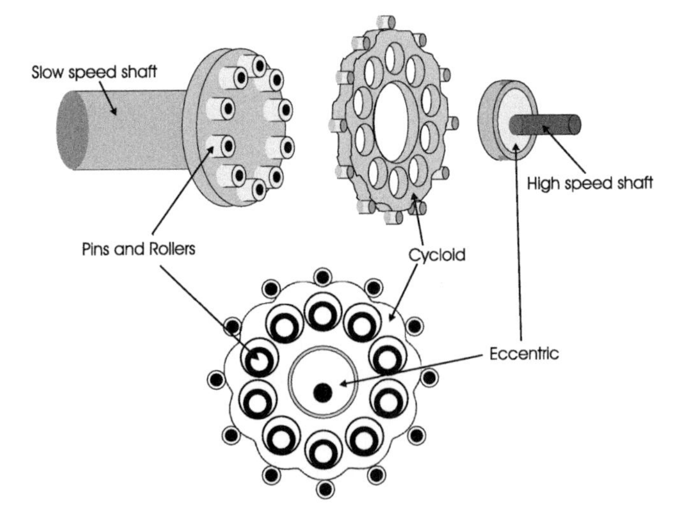

Fig. 4.1.5 A schematic diagram of a cycloid speed reducer. The relationship between the eccentric, cylcoid and slow-speed output shaft is clearly visible. It should be noted that in the diagram only one cycloid disc is shown, commercial systems typically have a number of discs, to improve power handelling.

- Ring gear housing with pins and rollers.
- Slow-speed shaft with pins and rollers.

As the eccentric rotates, it rolls the cycloid disc around the inner circumference of the ring gear housing. The resultant action is similar to that of a disc rolling around the inside of a ring. As the cycloid disc travels clockwise around the gear ring, the disc turns counter-clockwise on its axis. The teeth of the cycloid discs engage successively with the pins on the fixed gear ring, thus providing the reduction in angular velocity. The cycloid disc drives the low-speed output shaft. The reduction ratio is determined by the number of 'teeth' on the cycloid disc, which has one less 'tooth' than there are rollers on the gear ring. The number of teeth on the cycloid disc equals the reduction ratio, as one revolution of the high-speed shaft, causes the cycloid disc to move in the opposite direction by one 'tooth'.

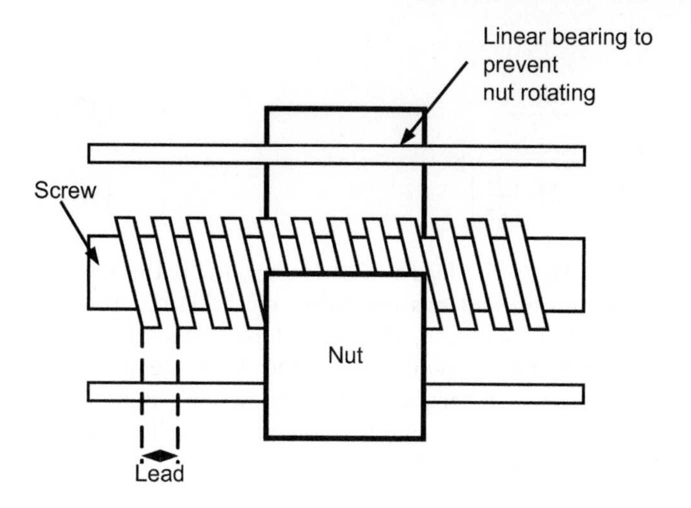

Fig. 4.1.6 The construction of a lead screw. The screw illustrated is single start with an ACME thread.

4.1.2 Lead and ball screws

The general arrangement of a lead screw is shown in Figure 4.1.6. As the screw is rotated, the nut, which is constrained from rotating, moves along the thread. The linear speed of the load is determined by the rotational speed of the screw and the screw's lead. The distance moved by one turn of the lead screw is termed the lead: this should not be confused with the pitch, which is the distance between the threads. In the case of a single start thread, the lead is equal to the pitch; however, the pitch is smaller than the lead on a multi-start thread. In a lead screw there is direct contact between the screw and the nut, and this leads to relatively high friction and hence an inefficient drive. For precision applications, ball screws are used due to their low friction and hence their good dynamic response. A ball screw is identical in principle to a lead screw, but the power is transmitted to the nut via ball bearings located in the thread on the nut (see Figure 4.1.7).

The relationship between the rotational and linear speed for both the lead and ball screw is given by

$$N_L = \frac{V_L}{L} \tag{4.1.4}$$

where N_L is the rotational speed (in rev min^{-1}), V_L is the linear speed in m min^{-1} and L is the lead (in metres). The inertia of the complete system is the sum of the screw inertia J_s and the reflected inertia of the load J_L

$$I_{tot} = I_s + I_L \tag{4.1.5}$$

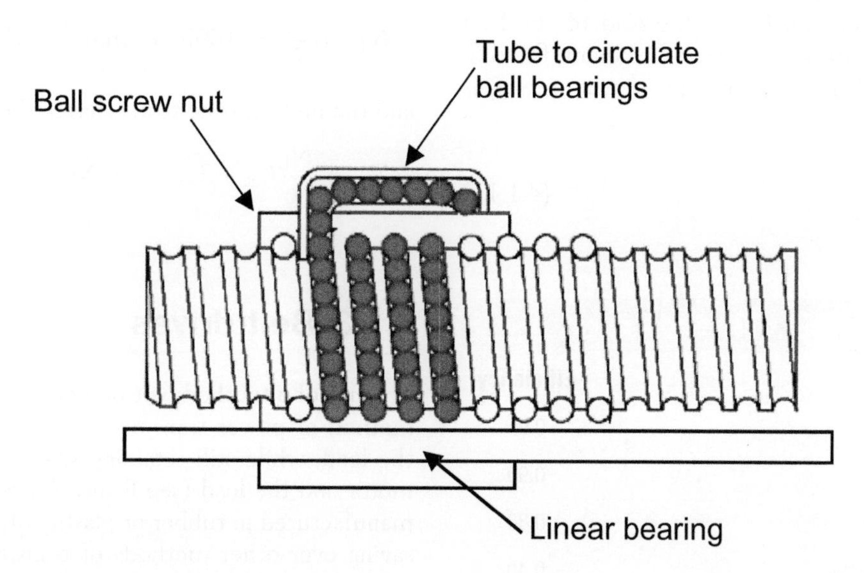

Fig. 4.1.7 The cross section of a high-performance ball screw, the circulating balls are clearly visible.

where

$$J_s = \frac{M_s r^2}{2} \quad (4.1.6)$$

$$J_L = M_L \left[\frac{L}{2\pi}\right]^2 \quad (4.1.7)$$

where M_L is the load's mass (in kg), M_s is the screw's mass (in kg) and r is the radius of the lead screw (in metres). In addition, the static forces, both frictional and the forces required by the load, need to be converted to a torque at the lead screw's input. The torque caused by external forces, F_L, will result in a torque requirement of

$$T_L = \frac{LF_L}{2\pi} \quad (4.1.8)$$

and a possible torque resulting from slideway friction of

$$T_f = \frac{LM_L g \cos\theta\mu}{2\pi} \quad (4.1.9)$$

where θ is the inclination of the slideway. It has been assumed so far that the efficiency of the lead screw is one hundred per cent. In practice, losses will occur and the torques will need to be divided by the lead-screw efficiency, ε, see Table 4.1.1, hence

$$T_{required} = \frac{(T_f + T_L)}{\varepsilon} \quad (4.1.10)$$

A number of linear digital actuators are based on stepper-motor technology, where the rotor has been modified to form the nut of the lead screw. Energisation of the windings will cause the lead screw to move a defined distance, which is typically in the range 0.025–0.1 mm depending on the step angle and the lead of the lead screw. For a motor with a step angle of θ radians, fitted to a lead screw of lead L, the incremental linear step, S, is given by

$$S = \frac{\theta L}{2\pi} \quad (4.1.11)$$

Table 4.1.1 Typical efficiencies for lead and ball screws

System type	Efficiency
Ball screw	0.95
Lead screw	0.90
Rolled-ball lead screw	0.80
ACME threaded lead screw	0.40

Example 4.1.2

Determine the speed and torque requirements for the following lead-screw application:

- *The length (L_s) of a lead screw is 1 m, its radius (R_s) is 20 mm and is manufactured from steel ($\rho = 7850$ kg m^{-3}). The lead (L) is 6 mm rev^{-1}. The efficiency (ε) of the lead screw is 0.85.*
- *The total linear mass (M_L) to be moved is 150 kg. The coefficient of friction (μ) between the mass and its slipway is 0.5. A 50 N linear force (F_L) is being applied to the mass.*
- *The maximum speed of the load (V_L) has to be 6 m min^{-1} and the time (t) the system is required to reach this speed in 1 s.*

The mass of the lead screw and its inertia are calculated first:

$$M_s = \rho\pi R_s^2 L_s = 9.97 \text{ kg} \quad \text{and}$$

$$J_s = \frac{M_s R_s^2}{2} = 1.97 \times 10^{-3} \text{ kg m}^{-2}$$

The total inertia can be calculated by adding the reflected inertia from the load to the lead screw's inertia:

$$J_{tot} = J_s + M_L\left(\frac{L}{2\pi}\right) = 2.11 \text{ kg m}^{-2}$$

The torque required to drive the load against the external and frictional forces, allowing for the efficiency of the lead screw, is given by

$$T_{ext} = \frac{1}{\varepsilon}\left(\frac{LF_L}{2\pi} + \frac{LM_L g\mu}{2\pi}\right) = 0.79 \text{ Nm}$$

The input speed required is given by

$$N_L = \frac{V_L}{L} = 1000 \text{ rev min}^{-1} = 104.7 \text{ rad s}^{-1}$$

and the input torque to accelerate the system is given by

$$T_{in} = \frac{N_L}{t} J_{tot} + T_{ext} = 1 \text{ Nm}$$

4.1.3 Belt drives

The use of a toothed belt or a chain drive is an effective method of power transmission between the motor and the load, while still retaining synchronism between the motor and the load (see Figure 4.1.8). The use of belts, manufactured in rubber or plastic, offers a potential cost saving over other methods of transmission. Typical applications that incorporate belt drives include printers,

Fig. 4.1.8 Synchronous belts and pulleys suitable for servo-drive applications.

ticketing machines, robotics and scanners. In the selection of the a belt drive, careful consideration has to be given to ensuring that positional accuracy is not compromised by selection of an incorrect component. A belt drive can be used in one of two ways, either as a linear drive system (for example, positioning a printer head) or as speed changer.

In a linear drive application, the rotational input speed is given by

$$N_i = \frac{V_L}{\pi D} \tag{4.1.12}$$

where D is the diameter of the driving pulley (in metres), and V_L is the required linear speed (in m s^{-1}). The inertia of the transmission system, J_{tot}, must include the contributions from all the rotational elements, including the idler pulleys, any rotating load and the belt:

$$I_{tot} = I_p + I_L \tag{4.1.13}$$

where I_p is the sum of the inertias of all the rotating elements. The load and belt inertia is given by

$$I_L = \frac{MD^2}{4} \tag{4.1.14}$$

where the mass, M, is the sum of the linear load (if present) and the transmission-belt masses. An external linear force applied to the belt will result in a torque at the input drive shaft of

$$T_{in} = \frac{DF}{2} \tag{4.1.15}$$

In a linear application, the frictional force, F_f, must be carefully determined as it will result in an additional torque

$$T_f = \frac{DF_f}{2} \tag{4.1.16}$$

If a belt drive is used as a speed changer, the output speed is a ratio of the pulley diameters

$$n = \frac{D}{d} \tag{4.1.17}$$

and the input torque which is required to drive the load torque, T_i, is given by

$$T_i = \frac{T_{out}}{n} \tag{4.1.18}$$

The inertia seen at the input to the belt drive is the sum of the inertias of the pulleys, the belt, the idlers and the load, taking into account the effects of the gearing ratio; that is

$$I_{tot} = I_{p1} + I_{belt} + \frac{I_L + I_{p2}}{n^2} \tag{4.1.19}$$

Where the inertia of the belt can be calculated from Equation (4.1.14) and J_{p2} is the inertia of the driven pulley modified by the gear ratio. The drive torque which is required can then be computed; the losses can be taken into account by using Equation (4.1.10).

The main selection criteria for a belt or chain is the distance, or pitch, between the belt's teeth (this must be identical to the value for the pulleys) and the drive characteristics. The belt pitch and the sizes of the pulleys will directly determine the number of teeth which are in mesh at a particular time, and hence the power that can be transmitted. The power that has to be transmitted can be determined by the input torque and speed. The greater the number of teeth in mesh, the greater is the power that can be transmitted; the number of teeth in mesh on the smaller pulley, which is the system's limiting value, and can be determined from

$$\text{Teeth in mesh} = \left[\pi - 2\sin^{-1}\frac{(D-d)}{C}\right] \times \frac{\text{Teeth on the small pulley}}{2\pi} \tag{4.1.20}$$

The selection of the correct belt requires detailed knowledge of the belt material, together with the load and drive characteristics. In the manufacturer's data sheets, belts and chains are normally classified by their power-transmission capabilities. In order to calculate the effect that the load and the drive have on the belt, use is made of an application factor, which is determined by the load and/or drive. Typical values of the application factors are given in Table 4.1.2, which are used to determine the belt's power rating, P_{belt}, using

$$P_{belt} = \text{Power requirements} \times \text{application factor} \tag{4.1.21}$$

Table 4.1.2 Typical application factors for belt drives

	Drive characteristic		
Load	**Smooth running**	**Slight shocks**	**Moderate shocks**
Smooth	1.0	1.1	1.3
Moderate shocks	1.4	1.5	1.7
Heavy shocks	1.8	1.9	2.0

Example 4.1.3
Determine the speed and torque requirements for the following belt drive:

- *A belt drive is required to position a 100 g load. The drive consists of two aluminium pullies ($\rho = 2770$ kg m^{-3}), 50 mm in diameter and 12 mm thick driving a belt weighting 20 g. The efficiency (ε) of the drive is 0.95.*
- *The maximum speed of the load (V_L) is 2 m min^{-1} and the acceleration time (t) is 0.1 s.*

Calculate the moment of inertia of the pulley

$$M_p = \rho \pi R_p^2 t_p = 0.065 \text{ kg} \quad \text{hence}$$

$$I_p = \frac{M_p R_p^2}{2} = 2 \times 10^{-5} \text{ kg m}^2$$

The reflected inertia of the belt and load is given by

$$I_L = \frac{MD^2}{4} = 7.5 \times 10^{-5} \text{ kg m}^2$$

The total driven inertia can now be calculated

$$I_{tot} = 2J_p + I_L = 11.5 \times 10^{-5} \text{ kg m}^2$$

The required peak input speed is

$$N_i = \frac{V_L}{piD} = 763 \text{ rev min}^{-1}$$

and hence the torque requirement can be determined

$$T_{in} = \frac{1}{\varepsilon}\left(\frac{IN_i}{t}\right) = 0.098 \text{ Nm}$$

4.1.4 Bearings

In the case of a rotating shaft, the most widely used method of support is by using one or a number bearing. A considerable number of different types of bearing are commonly available. The system selected is a function of the loads and speeds experienced by the system; for very high speed application air or magnetic bearings are used instead of the conventional metal-on-metal, rolling contacts. When considering the dynamics of a system, the friction and inertia of individual bearings, though small, must need to be taken into account.

4.1.4.1 Conventional bearings

The bearing arrangement of a rotating component, e.g. a shaft, generally requires two bearings to support and locate the component radially and axially relative to the stationary part of the machine. Depending on the application, load, running accuracy and cost the following approaches can be considered:

- Locating and non-locating bearing arrangements.
- Adjusted bearing arrangements.
- Floating bearing arrangements.

4.1.4.1.1 Locating and non-locating bearing arrangements

The locating bearing at one end of the shaft provides radial support and at the same time locates the shaft axially in both directions. It must, therefore, be fixed in position both on the shaft and in the housing. Suitable bearings are radial bearings which can accommodate combined loads, e.g. deep groove ball bearings. The second bearing then provides axial location in both directions but must be mounted with radial freedom (i.e. have a clearance fit) in its housing. The deep groove ball bearing and a cylindrical roller bearing, shown in Figure 4.1.9(a), illustrate this concept.

4.1.4.1.2 Adjusted bearing arrangements

In an adjusted bearing arrangements the shaft is axially located in one direction by the one bearing and in the opposite direction by the other bearing. This type of arrangement is referred to as *cross located* and is generally used on short shafts. Suitable bearings include all types of radial bearings that can accommodate axial loads in at least one direction, for example the taper roller bearings shown in Figure 4.1.9(b).

4.1.4.1.3 Floating bearing arrangements

Floating bearing arrangements are also cross located and are suitable where demands regarding axial location are moderate or where other components on the shaft serve to locate it axially. Deep groove ball bearings will satisfy this arrangement, Figure 4.1.9(c).

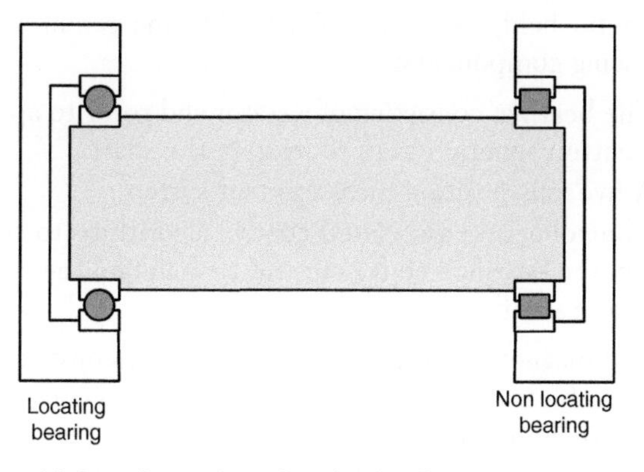

(a) Locating and non-locating bearing arrangement

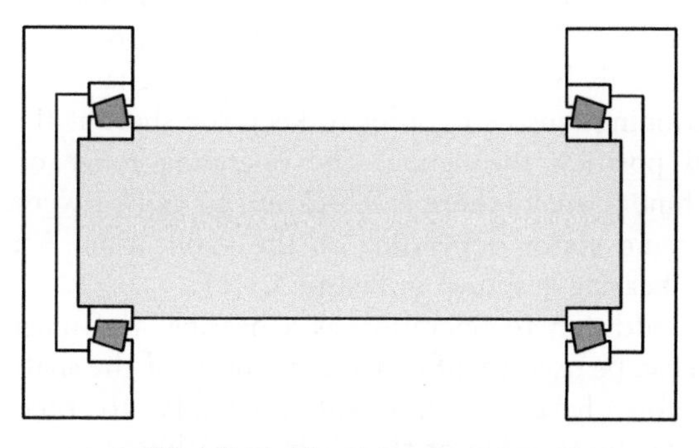

(b) Adjusted bearing arrangement

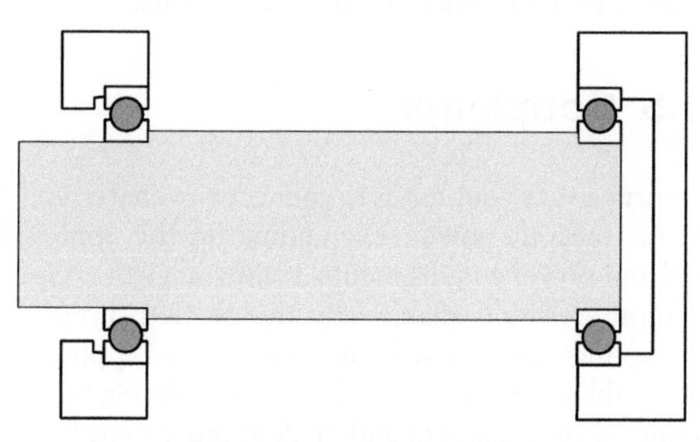

(c) Floating bearing arrangements

Fig. 4.1.9 Three approaches to supporting a rotating shaft.

4.1.4.1.4 Bearing friction

Friction within a bearing is made up of the rolling and sliding friction in the rolling contacts, in the contact areas between rolling elements and cage, as well as in the guiding surfaces for the rolling elements or the cage, the properties of the lubricant and the sliding friction of contact seals when applicable.

The friction in these bearing is either caused by the metal-to-metal contact of the balls or rollers on the bearing cage, or by the presence of lubrication within the bearing. The manufacturer will be able to supply complete data, but, as an indication, the friction torque, T_b, for a roller bearing can be determined using the following generally accepted relationship

$$T_b = 0.5 B_l d \mu_b \qquad (4.1.22)$$

where d is the shaft diameter and B_l is the bearing load computed from the radial load, F_r, and the axial load, F_a, in the bearings, given by

$$B_l = \sqrt{F_r^2 + F_a^2} \qquad (4.1.23)$$

The value of the coefficient of friction for the bearing, μ_b, will be supplied by the manufacturer; some typical values are given in Table 4.1.3.

Table 4.1.3 Typical coefficients of friction for roller bearings

Bearing types	Coefficient of friction, μ_b
Deep grove	0.0015–0.003
Self-aligning	0.001–0.003
Needle	0.002
Cylindrical, thrust	0.004

The friction due to the lubrication depends on the amount of the lubricant, its viscosity, and on the speed of the shaft. At low speeds the friction is small, but it increases as the speed increases. If a high-viscosity grease is used rather than an oil, the lubrication friction will be higher and this can, in extreme cases, give rise to overheating problems. The contribution of the lubricant to the total bearing friction can be computed using standard equations.

4.1.4.2 Air bearings

Air bearings can either be of an aerostatic or an aerodynamic design. In practice aerodynamic bearings are used in turbomachinery, where speeds of up to 36,000 rev min^{-1} in high-temperature environments are typically found. In an aerostatic air bearing, Figure 4.1.10, the two bearing surfaces are separated by a thin film of pressurised air. The compressed air is supplied by a number of nozzles in the bearing housing. The distance between the bearing surfaces is about 5–30 μm. As the object is supported by a thin layer of air, the friction between the shaft and its housing can be considered to be virtually zero.

The use of an air bearing gives the system designer a number of advantages including:

- High rotational accuracy typically greater than 5×10^{-8} m is achievable and will remain constant over time as there is no wear due to the absence of contact between the rotating shaft and the housing.

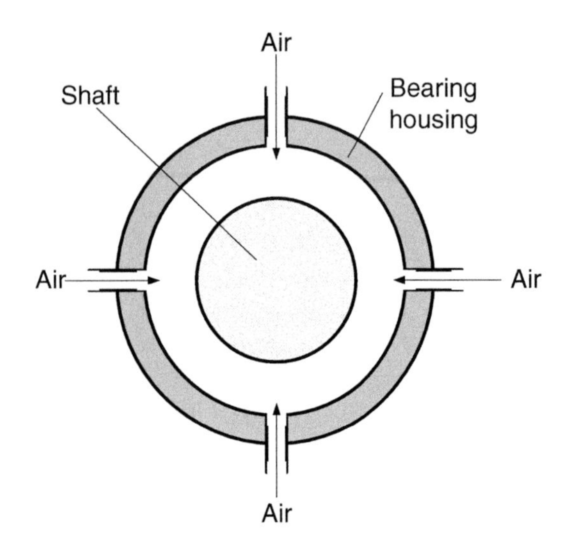

Fig. 4.1.10 Cross section of an air bearing: the dimension of the airgap have been greatly exaggerated.

- Low frictional drag, allow high rotational speeds; shaft speed of up to 200,000 rev min^{-1} with suitable bearings can be achieved.
- Unlimited life due to the absence of metal-to-metal contact, provided that the air supply is clean.
- High stiffness which is enhanced at speed due to a lift effect.

In machine tool applications, the lack of vibration and high rotational accuracy of an air bearing will allow surface finishes of up to 0.012 microns to be achieved.

4.1.4.3 Magnetic bearings

In a magnetic bearing the rotating shaft is supporting in a powerful magnetic field, and as with the air bearing gives a number of significant advantages:

- No contact, hence no wear, between the rotating and stationary parts. As particle generation due to wear is eliminated, magnetic bearings are suited to clean room applications.
- Operating through a wide temperature range, typically −250 to 220 °C: for this reason magnetic bearings are widely used in superconducting machines.
- A non-magnetic sheath between the stationary and rotating parts allows operation in corrosive environments.
- The bearing can be submerged in process fluid under pressure or operated in a vacuum without the need for seals.
- The frictional drag on the shaft is minimal, allowing exceptionally high speeds.

To maintain clearance, the shaft's position is under closed loop control by controlling the strength of the magnetic field, hence a magnetic bearing requires the following components:

- The bearing, consisting of a stator and rotor to apply electromagnetic forces to levitate the shaft.
- A five axis position measurement system.
- Controller and associated control algorithms to control the bearing's stator current to maintain the shaft at a pre-defined position.

The magnetic bearing stator has a similar construction to a brushless direct current (d.c.) motor and consists of a stack of laminations wound to form a series of north and south poles. The current is supplied to each winding will produce an attractive force that levitates the shaft inside the bearing. The controller controls the current applied to the coils by monitoring the position signal from the positioning sensors in order to keep the shaft at the desired position throughout the operating range of the machine. Usually there is 0.5–2 mm air gap between the rotor and stator depending on the application. A magnetic bearing is shown in Figure 4.1.11.

In addition to operation as a bearing, the magnetic field can be used to influence the motion of the shaft and therefore have the inherent capability to precisely control the position of the shaft to within microns and additionally to virtually eliminate vibrations.

4.1.5 Couplings

The purpose of a coupling is to connect two shafts, end-to-end, to transmit power. Depending on the application speed and power requirements a wide range of couplings are commercially available, and this section summarises the couplings commonly found in servo type applications.

A flexible coupling is capable of compensating for minor amounts of misalignment and random movement between the two shafts. Such compensation is vital because perfect

Fig. 4.1.11 A Radial magnetic bearing, manufactured by SKF Magnetic Bearings, Calgary, Canada.

alignment of two shafts is extremely difficult and rarely attained. The coupling will, to varying degrees, minimise the effect of misaligned shafts. If not properly compensated a minor shaft misalignment can result in unnecessary wear and premature replacement of other system components.

In certain cases, flexible couplings are selected for other functions. One significant application is to provide a break point between driving and driven shafts that will act as a *mechanical fuse* if a severe torque overload occurs. This assures that the coupling will fail before something more costly breaks elsewhere along the drive train. Another application is to use the coupling to dampen the torsional vibration that occurs naturally in the driving and/or driven system.

Currently there are a large number of flexible couplings due to the wide range of applications. However, in general flexible couplings fall into one of two broad categories, *elastomeric* or *metallic*. The key advantages and limitations of the designs are briefly summarised in

Table 4.1.4 Summary of the key characteristics of elastomeric couplings

Advantages	Limitations
No lubrication required	Difficult to balance as an assembly
Good vibrational damping and shock absorption	Not torsionally stiff
Field replaceable elastomers elements	Larger than a metallic coupling of the same torque capacity
Capable of accommodating more misalignment than a metallic bellow coupling	Poor overload torque capacity

Table 4.1.5 Summary of the key characteristics of metallic couplings

Advantages	Limitations
Torsionally stiff	Fatigue or wear plays a major role in failure
High-temperature capability	May need lubrication
Good chemical resistance possible	Complex assembly may be required
Low cost per unit torque transmitted	Require very careful alignment
High speed and large shaft size capability	Cannot damp vibration or absorb shock
Zero backlash	High electrical conductivity

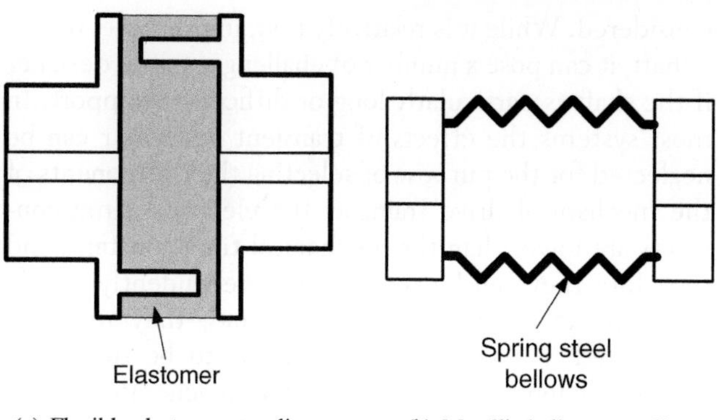

(a) Flexible elastomer coupling. (b) Metallic bellows coupling.

Fig. 4.1.12 Cross sections of commonly used couplings.

Tables 4.1.4 and 4.1.5 to allow the user to select the match the correct coupling to the application.

Elastomeric couplings use a non-metallic element within the coupling, through which the power is transmitted, Figure 4.1.12(a). The element is manufactured from a compliant medium (for example rubber or plastic) and can be in compression or shear. Compression flexible couplings designs, include those based on jaw, pin and bushing, and doughnut designs while shear couplings include tyre and sleeve moulded elements.

In practice there are two basic failure modes for elastomeric couplings. Firstly, break down can be due to fatigue from cyclic loading when hysteresis that results in internal heat build up if the elastomer exceeds its design limits. This type of failure can occur from either misalignment or torque beyond its capacity. Secondly, the compliant component can break down from environmental factors such as high ambient temperatures, ultraviolet light or chemical contamination. It should be noted that all elastomers have a limited shelf life and will in practice require replacement as part of maintenance programme, even if these failure conditions are not present.

Metallic couplings transmit the torque through designs where loose fitting parts are allowed to roll or slide against one another (for example in designs based on gear, grid, chain) or through the flexing/bending of a membrane (typically designed as a disc, diaphragm, beam or bellows), Figure 4.1.12(b). Those with moving parts generally are less expensive, but need to be lubricated and maintained. Their primary cause of failure in a flexible metallic couplings is wear, so overloads generally shorten the couplings life through increased wear rather than sudden failure.

4.1.6 Shafts

A linear rotating shaft supported on bearings can be considered to be the simplest element in a drive system: their static and dynamic characteristics need to be considered. While it is relatively easy, in principle, to size a shaft, it can pose a number of challenges to the designer if the shaft is particularly long or difficult to support. In most systems the effects of transient behaviour can be neglected for the purpose of selecting the components of the mechanical drive train, as the electrical time constants are lower than the mechanical time constant, and therefore they can be considered independently. While such effects are not commonly found, they must be considered if a large-inertia load has to be driven by a relatively long shaft, where excitation generated either by the load (for example, by compressors) or by the drive's power electronics needs to be considered.

4.1.6.1 Static behaviour of shafts

In any shaft, torque is transmitted by the distribution of shear stress over its cross section, where the following relationship, commonly termed the *Torsion Formula*, holds

$$\frac{T}{I_o} = \frac{G\theta}{L} = \frac{\tau}{r} \quad (4.1.24)$$

where T is the applied torque, I_o is the polar moment of area, G is the shear modulus of the material, θ is the angle of twist, L is the length of the shaft, τ is the shear stress and r the radius of the shaft.

In addition, we can use the torsion equation to determine the stiffness of a circular shaft

$$K = \frac{T}{\theta} = \frac{G\pi r^4}{2L} \quad (4.1.25)$$

where the polar moment of area of a circular shaft is given by

$$I_o = \frac{\pi r^4}{2} \quad (4.1.26)$$

Example 4.1.4

Determine the diameter of a steel shaft required to transmit 3000 Nm, without exceed the shear stress of 50 MNm^{-1} *or a twist of 0.1 rad m*$^{-1}$*. The shear modulus for steel is approximately 80 GNm*$^{-2}$*.*

Using Equations (4.1.24) and (4.1.26) it is possible to calculate the minimum radius for both the stress and twist conditions

$$r_{stress} = \frac{\tau_{max} I_o}{T} = \sqrt[3]{\frac{2T}{\pi\tau_{max}}} = 33.7 \text{ mm}$$

$$r_{twist} = \sqrt[4]{\frac{2TL}{G\theta\pi}} = 22.1 \text{ mm}$$

To satisfy both constraints the shaft should not have a radius of less than 33.7 mm.

4.1.6.2 Transient behaviour of shafts

In most systems the effects of transient behaviour can be neglected for the purpose of selecting the components of the mechanical drive train, because, in practice, the electrical time constants are normally smaller than the mechanical time constant. However, it is worth examining the effects of torque pulsations on a shaft within a system. These can be generated either by the load (such as a compressor) or by the drive's power electronics. While these problems are not commonly found, they must be considered if a large-inertia load has to be driven by a relatively long shaft.

The effect can be understood by considering Figure 4.1.13; as the torque is transmitted to the load, the shaft will twist and carry the load. The twist at the motor end, θ_m, will be greater than the twist at the load end, θ_L, because of the flexibility of the shaft; the transmitted torque will be proportional to this difference. If K is the shaft stiffness (Nm rad^{-1}), and B is the damping constant (Nm rad^{-1} s) then for the motor end of the shaft

$$T_m = I_m s^2\theta_m + Bs(\theta_m - \theta_L) + K(\theta_m - \theta_L) \quad (4.1.27)$$

and at the load end the torque will turn the load in the same direction as the motor, hence

$$Bs(\theta_m - \theta_L) + K(\theta_m - \theta_L) = I_m s^2\theta_m + T_L \quad (4.1.28)$$

where s is the differential operator, d/dt. If these equations are solved it can be shown that the undamped natural frequency of the system is given by

$$\omega_o = \sqrt{\frac{K}{I_m} + \frac{K}{I_L}} \quad (4.1.29)$$

$$\omega_o = \sqrt{1 - \varsigma^2} \quad (4.1.30)$$

$$\varsigma^2 = \sqrt{\frac{B}{2\sqrt{K}}\left(\frac{1}{I_m} + \frac{1}{I_L}\right)} \quad (4.1.31)$$

and the damped oscillation frequency is given by

$$\omega_n = \sqrt{1 + \frac{B^2}{4K}\left(\frac{1}{I_m} + \frac{1}{I_L}\right)} \quad (4.1.32)$$

In order to produce a stable system, the damped oscillation frequency must be significantly different to any torque pulsation frequencies produced by the system.

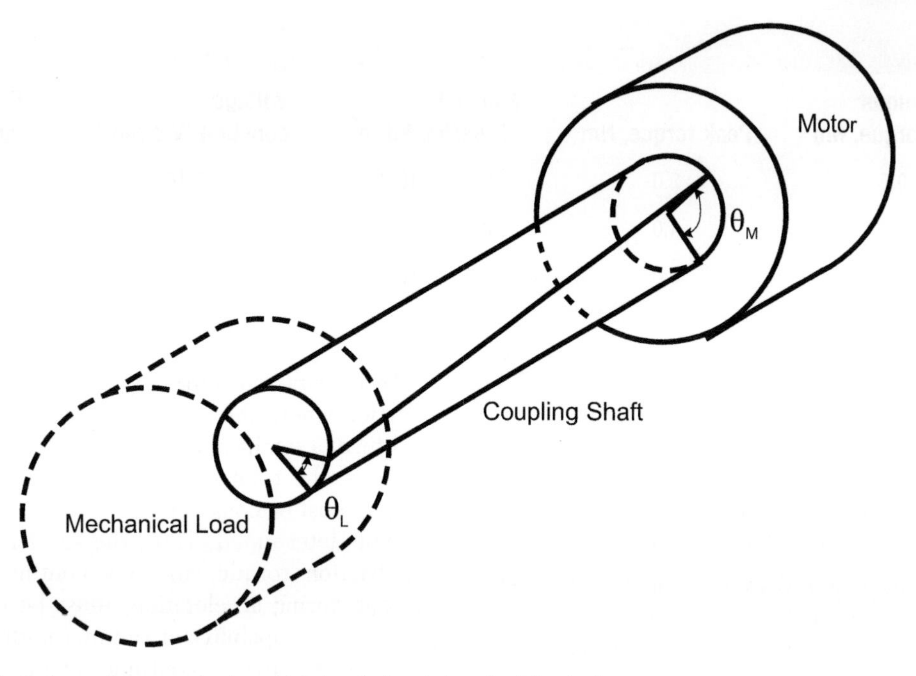

Fig. 4.1.13 The effect of coupling a motor to a high-inertia load via a flexible shaft.

4.1.7 Linear drives

For many high-performance linear applications, including robotic or similar high-performance applications, the use of lead screws, timing belts or rack and pinions driven by rotary motors, are not acceptable due to constrains imposed by backlash and limited acceleration. The use of a linear three phase brushless motor or the piezoelectric motor, provides a highly satisfactory solution to many motion control problems. If the required application requires only a small high-speed displacement, the voice coil can be considered.

The following advantages are apparent when a linear actuator is compared to conventional system based on a driving a belt or lead screw:

- When compared to a belt and pulley system, a linear motor removes the problems associated with the compliance in the belt. The compliance will cause vibration when the load comes to rest, and this limits the speed and acceleration of a belt drive. It should be noted that a high-performance belt drive can have a repeatability error in excess of 50 μm.
- As there are no moving parts in contact, a linear motor has significant advantages over ball and lead screw drives due to the removal of errors caused by wear on the nut and screw and due to friction, which is common if the drive has a high duty cycle. Even with the use of a high-performance ball screw the wear may become significant for certain application over time.
- As the length of a lead screw or ball screw is increased, so its maximum operating speed is limited, due to the flexibility of the shaft leading to vibration, particularly if a resonant frequency is hit – this is magnified as the length of the shaft increases. While the speed of the shaft can be decreased, by increasing the pitch, the system's resolution is compromised.

While the linear motor does provide a suitable solution for many applications, it is not inherently suitable for vertical operation, largely due to the problems associated with providing a fail-safe brake. In addition, it is more difficult to seal against environmental problems compared with a rotary system, leading to restrictions when the environment is particularly hostile, for example when there is excessive abrasive dust or liquid present. Even with these issues, linear motors are widely used is many applications, including high-speed robotics and other high-performance positioning systems.

4.1.8 Review of motor–drive sizing

This chapter has so far discussed the power transmission elements of a drive system, while Chapter 2.2 has looked at issues related to the determination of a drive's requirements. This concluding section provides an overview of how this information is brought together, and the size of the motor and its associated drive are identified. The objective of the sizing procedure is to determine the required output speed and torque of the motor and

Table 4.1.6 Typical d.c. brushed motor data. All motors are rated for a maximum speed of 5000 rev min^{-1} at terminal voltage of 95 V

Type	Continuous stall torque, Nm	Peak torque, Nm	Moment of inertia, kg m^2	Voltage constant, V s rad^{-1}	Current constant, Nm A^{-1}
M1	0.5	2.0	1.7×10^{-4}	0.18	0.18
M2	0.7	4.0	2.8×10^{-4}	0.18	0.18
M3	1.2	8.0	6.0×10^{-4}	0.18	0.18

hence to allow a required system to be selected. The process is normally started once the mechanical transmission system has been fully identified and quantified.

The main constraints that have to be considered during the sizing procedure when a conventional motor is being used can be summarised as follows:

- The peak torque required by the application must be less than both the peak stall torque of the motor and the motor's peak torque using the selected drive.
- The root-mean-square (r.m.s.) torque required by the application must be less than both the continuous torque rating of the motor and the continuous torque which can be delivered by the motor with the specified drive system.
- The maximum speed required by the application must be no greater than approximately eighty per cent of the maximum no-load speed of the motor-drive combination; this allows for voltage fluctuations in the supply to the drive system.
- The motor's speed–torque characteristics must not be violated; in addition with a d.c. brushed motor, the commutation characteristics of the motor must not be exceeded.

It should be noted that if a linear motor is used in an application the same set of constraints need to be considered; however, force is considered to be the main driver as opposed to torque in the sizing process.

The operating regimes of the motor and its associated controller must be considered; two types of duty can be identified. The main determining factor is a comparison of the time spent accelerating and decelerating the load with the time spent at constant speed. In a continuous-duty application the time spent accelerating and decelerating is not critical to the application, hence the maximum required torque (the external load torque plus the drive-train's friction torque) needs to be provided on a continuous basis; the peak torque and the r.m.s. torque requirements are not significantly different to that of the continuous torque. The motor and the controller are therefore selected primarily by considering the maximum speed and continuous torque requirements.

An intermittent-duty application is defined as an application where the acceleration and deceleration of the load form a significant part of the motor's duty cycle. In this case the total system inertia, including the motor inertia, must be considered when the acceleration torque is being determined. Thus, the acceleration torque plus the friction torque, and any continuous load torque present during acceleration, must be exceeded by the peak torque capability of the motor–drive package. Additionally, the drive's continuous torque capability must exceed the required r.m.s. torque resulting from the worst-case positioning move.

The difference between these two application regimes can be illustrated by considering a lathe, shown in Figure 2.1.2. The spindle drive of a lathe can be considered to be a continuous-duty application since it runs at a constant speed under a constant load, while the axis drives are intermittent-duty applications because the acceleration and deceleration required to follow the tool path are critical selection factors.

The confirmation of suitable motor–drive combinations can be undertaken by the inspection of the supplier's motor–drive performance data, which provides information on the maximum no-load speed and on the continuous torque capability, together with the torque sensitivity of various motor frame sizes and windings. Tables 4.1.6 and 4.1.7 contain information extracted from typical manufacturer's data sheets relating to d.c. brushed motors and drives. In the sizing process it is normal to initially consider only a small number of the key electrical and mechanical parameters.

Table 4.1.7 Typical current data (in amps) supplied by manufacturers, for drives capable of driving d.c. brushed motor. All the drives are capable of supplying the 95 V required for the motors detailed in Table 4.1.6

Type	Continuous current	Peak current
D1	5	10
D2	10	20
D3	14	20

If significant problems with motor and drive selection are experienced, a detailed discussion with the suppliers will normally resolve the problem.

As discussed above, two operating regimes can be identified:

- In a *continuous-duty application* the acceleration and deceleration requirements are not considered critical; the motor and the controller can be satisfactorily selected by considering the maximum speed and continuous torque requirements.
- An *intermittent-duty application* is defined as an application where the acceleration and deceleration of the load form a significant part of the motor's duty cycle, and need to be considered during the sizing process.

4.1.8.1 Continuous duty

For continuous duty, where the acceleration performance is not of critical importance, the following approach can be used:

- Knowledge of the required speed range of the load, and an initial estimation of the gear ratios required, will permit the peak motor speed to be estimated. In order to prevent the motor from not reaching its required speed, due to fluctuations of the supply voltage, the maximum required speed should be increased by a factor of 1.2. It should be noted that this factor is satisfactory for most industrial applications, but it may be refined for special applications, for example, when the system has to operate from a restricted supply as is found in aircraft and offshore-oil platforms.
- Using the drive and the motor manufacturers' data sheet, it will be possible to locate a range of motors that meets the speed requirement when the drive operates at the specified supply voltage. If the speed range is not achievable, the gear ratio should be revised.
- From the motor's data, it will normally be possible to locate a motor–drive that meets the torque requirement; this will also allow the current rating of the drive to be determined. A check should then be undertaken to ensure that the selected system can accelerate the load to its required speed in an acceptable time.

Example 4.1.5

Determine the motor's speed and torque requirement for the system detailed below, and hence identify a suitable motor and associated drive:

- *The maximum load speed 300 rev min*$^{-1}$*, a non-optimal gearbox with a ratio of 10:1 has been selected. The gearbox's moment of inertia referred to its input shaft is* 3×10^{-4} *kg m*^{2}*.*
- *The load's moment of inertia has been determined to* 5×10^2 *kg m*^{2}
- *The maximum load torque is 8 Nm.*

Based on this information the minimum motor speed required can be determined including an allowance for voltage fluctuations

$$\text{Motor speed} = 300 \times \text{gear ratio} \times 1.2$$
$$= 3600 \text{ rev min}^{-1}$$

$$\text{Continuous torque} = \frac{8}{10} = 0.8 \text{ Nm}$$

Consideration of the motor data given in Table 4.1.6 indicates that motor M3 is capable of meeting the requirements. The required speed is below the peak motor speed of 5000 rev min^{-1} ± 10%, and the required torque is below the motor's continuous torque rating. At the continuously torque demand the motor requires 4.5 A, hence the most suitable drive from those detailed in Table 4.1.7, will be drive D1.

To ensure that the motor–drive combination is acceptable, the acceleration can be determined for the drives peak output of 10 A. At this current the torque generated by the motor is 1.8 Nm, well within the motor rating. Using Equation (2.2.12), and noting that the gearbox's moment of inertia is added to that of the motor to give:

$$\alpha = \frac{T_{\text{peak}} - T_{\text{L}}/n}{n(I_{\text{d}} + I_{\text{L}}/n^2)} = \frac{1.8 - 8/10}{10(9 \times 10^{-4} + 5 \times 10^{-2}/10^2)}$$
$$= 71.4 \text{ rad s}^{-1}$$

Hence the load will be accelerated to a peak speed of 300 rev min^{-1}, within 0.5 s, which is satisfactory. In practice the acceleration rate would be controlled, so that the system, in particular the gear teeth, would not experience significant shock loads.

4.1.8.2 Intermittent duty

When the acceleration performance is all important, the motor inertia must be considered, and the torque which is necessary to accelerate the total inertia must be determined early in the sizing process. A suitable algorithm is as follows:

- Using the application requirements and the required speed profile determine the required speeds and acceleration.
- Estimate the minimum motor torque for the application using Equation (2.2.1).

- Select a motor–drive combination with a peak torque capability of at least 1.5–2 times the minimum motor-torque requirement to ensure a sufficient torque capability.
- Recalculate the acceleration torque required, this time including the inertia of the motor which has been selected.
- The peak torque of the motor–drive combination must exceed, by a safe margin of at least fifteen per cent, the sum of the estimated friction torque and the acceleration torque and any continuous torque loading which is present during acceleration. If this is not achievable, a different motor or gear ratio will be required.
- The motor's r.m.s. torque requirement can then be calculated as a weighted time average, using;

$$T_{rms} \leq T_{cm} + \sqrt{T_f^2 + dT_a^2} \tag{4.1.33}$$

  where T_{cm} is the continuous motor-torque requirement, T_f is the friction torque at the motor, T_a is the acceleration torque and d is the duty cycle.
- The selected motor–drive combination is evaluated for maximum speed and continuous torque capabilities as in Section 4.1.8.1.
- If no motor of a given size can meet all the constraints, then a different, usually larger, frame must be considered, and the procedure must be repeated.

In practice, it is usual for one or two iterations to be undertaken in order to find an acceptable motor–drive combination. The approximate r.m.s.-torque equation used above not only simplifies computation, but it also allows an easy examination of the effects of varying the acceleration/deceleration duty cycle. For example, the effects of changes in the dwell time on the value of r.m.s. torque can be immediately identified. Should no cost-effective motor–drive be identified, the effects of varying the speed-reduction ratio and inertias can easily be studied by trying alternative values and sizing the reconfigured system.

Sometimes, repeated selections of motors and drives will not yield a satisfactory result; in particular, no combination is able to simultaneously deliver the speed and the continuous torque which is required by the application, or to simultaneously deliver the peak torque and the r.m.s. torque required. In certain cases motor-drive combinations can be identified, but the size or cost of the equipment may appear to be too high for the application, and changes will again be required.

Example 4.1.6

Identify a suitable motor and its associated drive for the application detailed below:

- *The load is a rotary disc which has a moment of inertia of 1.34 kg m^2. The estimated frictional torque referred to the table's drive input is 5 Nm, and the external load torque is 8 Nm.*
- *The table is driven through a 20:1 gear box, which has a moment of inertia of 3×10^{-4} kg m^2 referred to its input shaft.*
- *The table is required to index 90.0° (θ) in 1 s (t_m), and then dwell for a further 2 s. A polynomial speed profile is required.*

The selection process starts with the determination of the peak load speed and acceleration, using Equation (2.2.28), the maximum speed occurs at $t = 0.5$ s and maximum acceleration occurs at $t = 0$ s:

$$\dot{\theta}(0.5) = \frac{6\theta t}{t_m^2} + \frac{6\theta t^2}{t_m^3} = 7.1 \text{ rad s}^{-1}$$

$$\ddot{\theta}(0) = \frac{6\theta}{t_m^2} = 9.4 \text{ rad s}^{-2}$$

The peak torque can now be calculated, at the input to the table. The torque is determined by the peak acceleration, and the load and friction torques, giving

$$T_{peak} = 8 + 5 + 1.34 \times 9.4 = 25.6 \text{ Nm}$$

This equates to 1.28 Nm peak torque from the motor. Using the motors defined in Table 4.1.6, it appears that motor M1 is a suitable candidate as it is capable of supplying over four times the required torque. If the motor's moment of inertia is now included in the calculation, the peak torque requirement is

$$\begin{aligned} T_{peak} &= \frac{25.6}{20} + (3 \times 10^{-4} + 1.7 \times 10^{-4}) \\ &\quad \times (9.4 \times 20) \\ &= 1.37 \text{ Nm} \end{aligned}$$

which is well within the capabilities of the motor and drive D1, detailed in Table 4.1.7. The required peak current is 7 amps. The r.m.s. torque can now be calculated using Equation (4.1.33):

$$T_{rms} \leq T_{cm} + \sqrt{T_f^2 + dT_a^2} = 0.67 \text{ Nm}$$

This figure is in excess of the continuous torque rating of the motor M1, and in certain applications could lead to the motor overheating. In addition, while the current is below the peak rating it is greater than the continuous rating: in practice this could result in the drive cutting-out due to motor overheating. Thus a case can be made to change the motor and drive – in practice this decision would be made after careful consideration of the application.

If motor M2 is selected, and the above calculations are repeated, the motor's torque requirement becomes 1.4 Nm, and the r.m.s torque becomes 0.69 Nm. While marginal, M2 can be used along as the friction or load torque do not increase, if the drive is also changed to D2 there is no possibility of any overheating problems in the system.

As a final check the motor's peak speed is determined to be 1350 rev min^{-1}; this is well within the specification of the selected motor and drive.

This short example illustrates how a motor and drive can be selected; however, the final decision needs a full understanding of the drive and its application. If the application only requires a few indexing moves, the selection of motor M1 could be justified; however, if a considerable number of indexes are required, motor M2 could be the better selection. This example has only considered the information given above; in practice the final decision will be influenced on the technical requirements of the complete process, and commercial requirements. While this example has been undertaken for a d.c. brushed motor, the same procedure is followed for any other type of drive – the only differences being the interpretation of the motor and drive specifications.

4.1.8.3 Inability to meet both the speed and the torque requirements

In the selection of motors, the limitations of both the motor and the drive forces a trade-off between the speed and the torque capabilities. Thus, it is usually advantageous to examine whether some alteration in the mechanical elements may improve the overall cost effectiveness of the application. Usually the speed-reduction ratios used in the application are the simplest mechanical parameter which can be investigated. If the speed required of the motor is high, but the torque seems manageable, a reduction in the gear ratio may solve the problem. If the torque required seems high but additional speed is obtainable, then the gear ratio should be increased. The goal is to use the smallest motor–drive combination that exceeds both the speed and torque requirement by a minimum of ten to twenty per cent. Sometimes the simple changing of a gear or pulley size may enable a suitable system to be selected.

A further problem may be the inability to select a drive that meets both the peak and the continuous torque requirements. This is particularly common in intermittent-motion applications. Often, the peak torque is achievable but the drive is unable to supply the continuous current required by the motor. As has been shown earlier, while optimum power transfer occurs when the motor's rotor and the reflected load inertia are equal this may not give the optimum performance for an intermittent drive, hence the gear ratios in the system need to be modified and the sizing process repeated.

Example 4.1.7

Consider Example 4.1.6 above, and consider the impact of performance due to a change in the reduction ratio.

In Example 4.1.6 the speed and torque requirements, using motor M1, were calculated to be 1.35 Nm and 1350 rev min^{-1}. The speed requirement is well within the motor specifications. If the gear ratio was changed to 40:1, the motor's peak speed requirement increases to 2700 rev min^{-1} and the r.m.s. torque drops to 0.35 Nm, and a peak torque of 0.818 Nm. These figures are well within the specification of motor M1 and drive D1.

This example illustrates a different approach to resolving the sizing problem encountered earlier. The change in gear ratio can easily be achieved at the design state, and in all possibility be cheaper that going to a larger motor and drive system.

4.1.8.4 Linear motor sizing

So far in this section we have considered the sizing of conventional rotary motors. We will now consider the sizing of a linear motor. Due to the simplicity of a linear drive, the process is straightforward compared to combining a lead screw, ball screw or belt drive with a conventional motor. As with all other sizing exercises, the initial process is to identify the key parameters, before undertaking the detailed sizing process. A suitable algorithm is as follows:

- Using the application requirements and the required speed profile determine the required speed and acceleration.
- Estimate the minimum motor force for the application using Equation (2.2.2).
- Select a motor–drive combination with a peak force capability of at least 1.5–2 times the minimum force requirement to ensure a sufficient capability.
- Recalculate the acceleration force required, this time including the mass of the moving part of the selected motor.
- The peak force of the motor–drive combination must exceed, by a safe margin of at least fifteen per cent, the sum of the estimated friction force and the acceleration force and any continuous force which is present during acceleration. If this is not achievable, a different motor will be required.
- The motor's r.m.s. torque requirement can then be calculated as a weighted time average; in addition, this will allow the motor's temperature to be estimated.

Example 4.1.8

Determine the size of the linear motor, and drive required to move a mass of $M_L = 40$ kg, a distance of $d = 750$ mm in time of $t_m = 400$ ms.

- *The system has a dwell time of $t_d = 300$ ms, before the cycle repeats.*
- *Assume that the speed profile is triangular, and equal times are spent accelerating, decelerating and at constant speed.*
- *Assume the frictional force, $F_f = 3$ N.*
- *The motors's parameters are: force constant is $K_F = 40\ \mathrm{N\,A^{-1}}$, back emf constant $K_{emf} = 50\ \mathrm{V\,m^{-1}\,s}$, winding resistance, $Rt_w = 2\ \Omega$ and thermal resistance of the coil assembly, $Rt_{c-a} = 0.15\ ^{\circ}\mathrm{C W^{-1}}$.*

The acceleration and peak speed can be determined using the process determined in Section 2.2.4, hence

$$\dot{x} = \frac{3d}{2t_m} = 2.4\ \mathrm{m\,s^{-1}}$$

and

$$\ddot{x} = \frac{3\dot{x}}{t_m} = 28.8\ \mathrm{m\,s^{-2}}$$

The acceleration force required is given by

$$F_a = M_L\ddot{x} + F_f = 1155\ \mathrm{N}$$

This now allows the calculation of the r.m.s. force requirement

$$F_{rms} = \sqrt{\frac{2t_mF_a^2 + t_mF_f^2}{t_m + t_d}} = 635.5\ \mathrm{N}$$

The drives current and voltage requirements can therefore be calculated

$$V_{drive} = \dot{x}K_{emf} + I_{peak}R_w = 178\ \mathrm{V}$$

$$I_{peak} = \frac{F_a}{K_F} = 28.8\ \mathrm{A}$$

$$I_{continuous} = \frac{F_{rms}}{K_F} = 15.9\ \mathrm{A}$$

As linear motors are normally restricted to temperature rises of less than 100 °C, the temperature rise over ambient needs to be calculated

$$T_{rise} = I^2_{continuous}R_wRt_{c\text{-}a} = 76\ ^{\circ}\mathrm{C}$$

4.1.9 Summary

This chapter has reviewed the characteristics of the main mechanical power transmission components commonly used in the construction of a drive system, together with their impact on the selection of the overall drive package. The chapter concluded by discussing the approach to sizing drives. One of the key points to be noted is that the motor–drive package must be able to supply torques and speed which ensure that the required motion profile can be followed. To assist with the determination of the required values, a sizing procedure was presented. It should be remembered over-sizing a drive is as undersizing.

Chapter 4.2

4.2

Gears and gearboxes

Mobley

A gear is a disc or wheel with teeth around its periphery – either on the inside edge (i.e., internal gear) or on the outside edge (i.e., external gear). A gear is used to provide a positive means of power transmission, which is effected by the teeth on one gear meshing with the teeth on another gear or rack (i.e., straight-line gear).

Gear drives are packaged units used for a wide range of power-transmission applications. They are used to transmit power to a driven piece of machinery and to change or modify the power that is transmitted. Modifications include reducing speed and increasing output torque, increasing speed, changing the direction of shaft rotation, or changing the angle of shaft operation.

4.2.1 Configuration

There are several different types of gears used in industry. Many are complex in design and manufacture and several have evolved directly from the spur gear, which is referred to as the basic gear. Types of gears are: spur, helical, bevel, and worm. Table 4.2.1 summarizes the characteristics of each gear type.

4.2.1.1 Spur gears

The spur gear is the least expensive of all gears to manufacture and is the most commonly used. It can be manufactured to close tolerances and is used to connect parallel shafts that rotate in opposite directions. It gives excellent results at moderate peripheral speeds and the tooth load produces no axial thrust. Because contact is simultaneous across the entire width of the meshing teeth, it tends to be noisy at high speeds. However, noise and wear can be minimized with proper lubrication.

There are three main classes of spur gears: external tooth, internal tooth, and rack-and-pinion. The external tooth variety shown in Figure 4.2.1 is the most common. Figure 4.2.2 illustrates an internal gear and Figure 4.2.3 shows a rack or straight-line spur gear.

The spur gear is cylindrical and has straight teeth cut parallel to its rotational axis. The tooth size of spur gears is established by the diametrical pitch. Spur gear design accommodates mostly rolling, rather than sliding, contact of the tooth surfaces and tooth contact occurs along a line parallel to the axis. Such rolling contact produces less heat and yields high mechanical efficiency, often up to 99 per cent.

An internal spur gear, in combination with a standard spur-gear pinion, provides a compact drive mechanism for transmitting motion between parallel shafts that rotate in the same direction. The internal gear is a wheel that has teeth cut on the inside of its rim and the pinion is housed inside the wheel. The driving and driven members rotate in the same direction at relative speeds inversely proportional to the number of teeth.

4.2.1.2 Helical gears

Helical gears, which are shown in Figure 4.2.4, are formed by cutters that produce an angle that allows several teeth to mesh simultaneously. Helical gears are superior to spur gears in their load-carrying capacity and quietness and smoothness of operation, which results from the sliding contact of the meshing teeth. A disadvantage, however, is higher friction and wear that accompanies this sliding action.

Plant Engineer's Handbook; ISBN: 9780750673280

Table 4.2.1 Gear characteristics overview

Gear type	Characteristics: Attributes/positives	Characteristics: Negatives
Spur, external	Connects parallel shafts that rotate in opposite directions, inexpensive to manufacture to close tolerances, moderate peripheral speeds, no axial thrust, high mechanical efficiency	Noisy at high speeds
Spur, internal	Compact drive mechanism for parallel shafts rotating in same direction	
Helical, external	Connects parallel and non-parallel shafts; superior to spur gears in load-carrying capacity, quietness, and smoothness; high efficiency	Higher friction than spur gears, high-end thrust
Helical, double (also referred to as herringbone)	Connects parallel shafts, overcomes high-end thrust present in single-helical gears, compact, quiet and smooth operation at higher speeds (1000–12,000 fpm or higher), high efficiencies	
Helical, cross	Light loads with low power transmission demands	Narrow range of applications, requires extensive lubrication
Bevel	Connects angular or intersecting shafts	Gears overhang supporting shafts resulting in shaft deflection and gear misalignment
Bevel, straight	Peripheral speeds up to 1000 fpm in applications where quietness and maximum smoothness not important, high efficiency	Thrust load causes gear pair to separate
Bevel, zerol	Same ratings as straight-bevel gears and use same mountings, permits slight errors in assembly, permits some displacement due to deflection under load, highly accurate, hardened due to grinding	Limited to speeds less than 1000 fpm due to noise
Bevel, spiral	Smoother and quieter than straight-bevel gears at speeds greater than 1000 fpm or 1000 rpm, evenly distributed tooth loads, carry more load without surface fatigue, high efficiency, reduces size of installation for large reduction ratios, speed-reducing and speed-increasing drive	High tooth pressure, thrust loading depends on rotation and spiral angle
Bevel, miter	Same number of teeth in both gears, operate on shafts at 90°	
Bevel, hypoid	Connects non-intersecting shafts, high pinion strength, allows the use of compact straddle mounting on the gear and pinion, recommended when maximum smoothness required, compact system even with large reduction ratios, speed-reducing and speed-increasing drive	Lower efficiency, difficult to lubricate due to high tooth-contact pressures, materials of construction (steel) require use of extreme-pressure lubricants
Planetary or epicyclic	Compact transmission with driving and driven shafts in line, large speed reduction when required	
Worm, cylindrical	Provide high-ratio speed reduction over wide range of speed ratios (60:1 and higher from a single reduction, can go as high as 500:1), quiet transmission of power between shafts at 90°, reversible unit available, low wear, can be self-locking	Lower efficiency; heat removal difficult, which restricts use to low-speed applications
Worm, double-enveloping	Increased load capacity	Lower efficiencies

Source: the Plant Performance Group.

Fig. 4.2.1 Example of a spur gear.

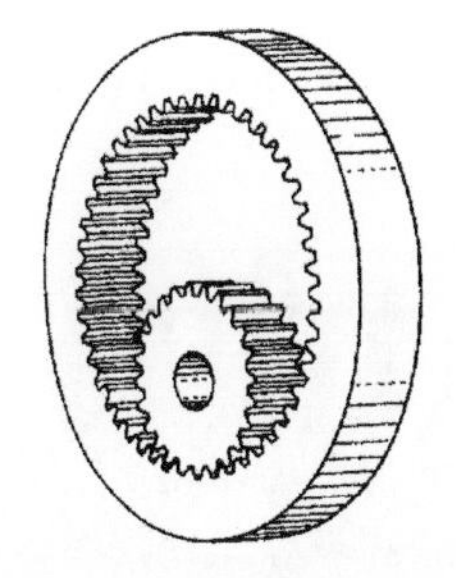

Fig. 4.2.2 Example of an internal spur gear.

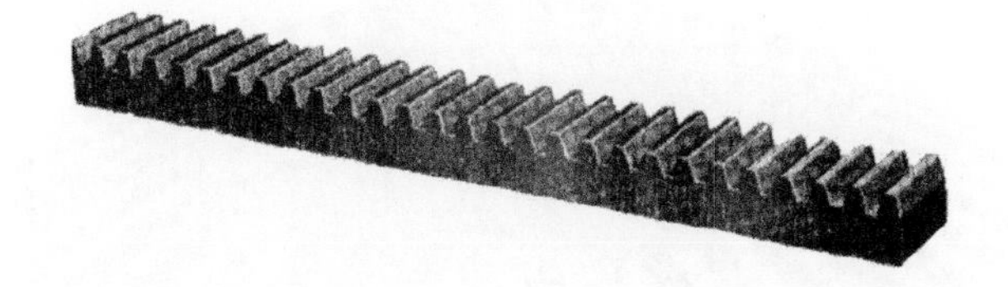

Fig. 4.2.3 Rack or straight-line gear.

Fig. 4.2.4 Typical set of helical gears.

Single-helical gears are manufactured with the same equipment as spur gears, but the teeth are cut at an angle to the axis of the gear and follow a spiral path. The angle at which the gear teeth are cut is called the helix angle, which is illustrated in Figure 4.2.5. This angle causes the position of tooth contact with the mating gear to vary at each section. Figure 4.2.6 shows the parts of a helical gear.

It is very important to note that the helix angle may be on either side of the gear's centerline. Or, if compared to the helix angle of a thread, it may be either a 'right-hand' or a 'left-hand' helix. Figure 4.2.7 illustrates a helical gear

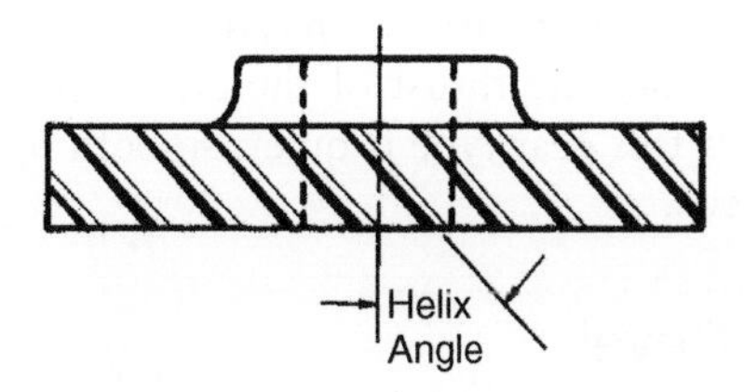

Fig. 4.2.5 The angle at which the teeth are cut.

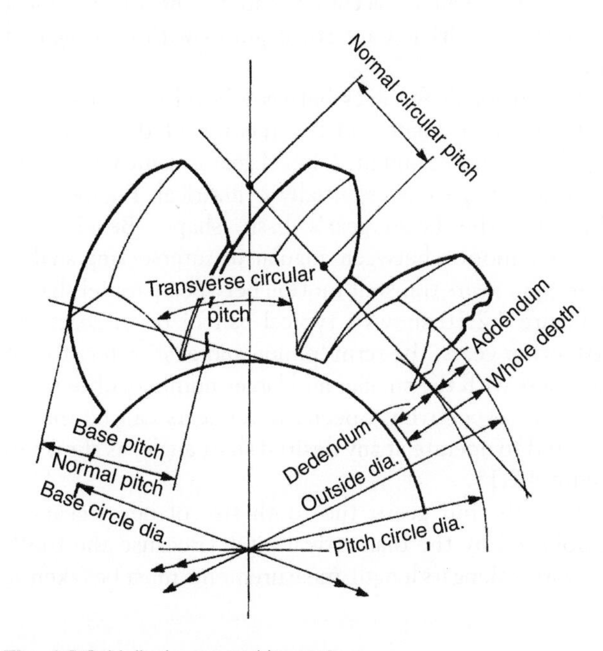

Fig. 4.2.6 Helical gear and its parts.

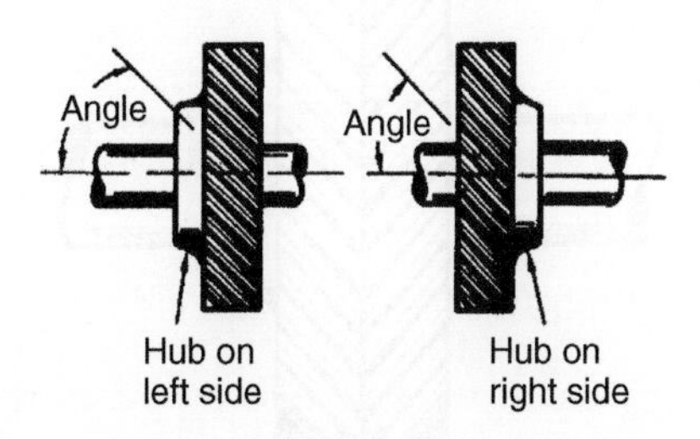

Fig. 4.2.7 Helix angle of the teeth – the same no matter from which side the gear is viewed.

as viewed from opposite sides. A pair of helical gears must have the same pitch and helix angle, but must be of opposite hand (one right hand and one left hand).

4.2.1.3 Herringbone

The double-helical gear, also referred to as the herringbone gear (Figure 4.2.8), is used for transmitting power between parallel shafts. It was developed to overcome the disadvantage of the high-end thrust that is present with single-helical gears.

The herringbone gear consists of two sets of gear teeth on the same gear, one right hand and one left hand. Having both hands of gear teeth causes the thrust of one set to cancel out the thrust of the other. Thus, another advantage of this gear type is quiet, smooth operation at higher speeds.

4.2.1.4 Bevel

Bevel gears are used most frequently for 90° drives, but other angles can be accommodated. The most typical application is driving a vertical pump with a horizontal driver.

Two major differences between bevel gears and spur gears are their shape and the relation of the shafts on which they are mounted. A bevel gear is conical in shape, while a spur gear is essentially cylindrical. Figure 4.2.9 illustrates the bevel gear's basic shape. Bevel gears transmit motion between angular or intersecting shafts, while spur gears transmit motion between parallel shafts.

Figure 4.2.10 shows a typical pair of bevel gears. As with other gears, the term 'pinion and gear' refers to the members with the smaller and larger numbers of teeth in the pair, respectively. Special bevel gears can be manufactured to operate at any desired shaft angle, as shown in Figure 4.2.11.

As with spur gears, the tooth size of bevel gears is established by the diametrical pitch. Because the tooth size varies along its length, measurements must be taken at

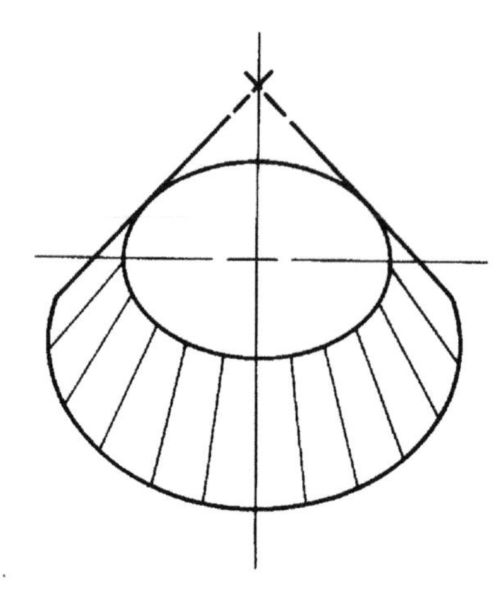

Fig. 4.2.9 Basic cone shape of bevel gears.

Fig. 4.2.10 Typical set of bevel gears.

Fig. 4.2.8 Herringbone gear.

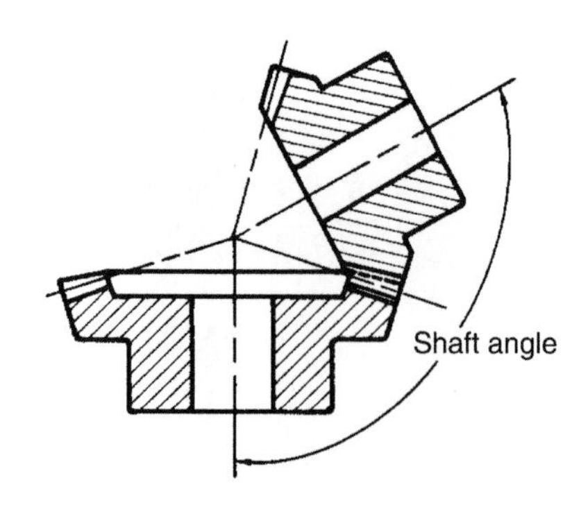

Fig. 4.2.11 Shaft angle, which can be at any degree.

a specific point. Note that, because each gear in a bevel-gear set must have the same pressure angle, tooth length, and diametrical pitch, they are manufactured and distributed only as mated pairs. Like spur gears, bevel gears are available in pressure angles of 14.5° and 20°.

Because there generally is no room to support bevel gears at both ends due to the intersecting shafts, one or both gears overhang their supporting shafts. This is referred to as an overhung load. It may result in shaft deflection and gear misalignment, causing poor tooth contact and accelerated wear.

4.2.1.5 Straight or plain

Straight-bevel gears, also known as plain bevels, are the most commonly used and simplest type of bevel gear (Figure 4.2.12). They have teeth cut straight across the face of the gear. These gears are recommended for peripheral speeds up to 1000 feet per minute in cases where quietness and maximum smoothness are not crucial. This gear type produces thrust loads in a direction that tends to cause the pair to separate.

4.2.1.6 Zerol

Zerol-bevel gears are similar to straight-bevel gears, carry the same ratings, and can be used in the same mountings. These gears, which should be considered as spiral-bevel gears having a spiral angle of zero, have curved teeth that lie in the same general direction as straight-bevel gears. This type of gear permits slight errors in assembly and some displacement due to deflection under load. Zerol gears should be used at speeds less than 1000 feet per minute because of excessive noise at higher speeds.

4.2.1.7 Spiral

Spiral-bevel gears (Figure 4.2.13) have curved oblique teeth that contact each other gradually and smoothly from one end of the tooth to the other, meshing with a rolling contact similar to helical gears. Spiral-bevel gears are smoother and quieter in operation than straight-bevel gears, primarily due to a design that incorporates two or more contacting teeth. Their design, however, results in high tooth pressure.

Fig. 4.2.12 Straight or plain bevel gear.

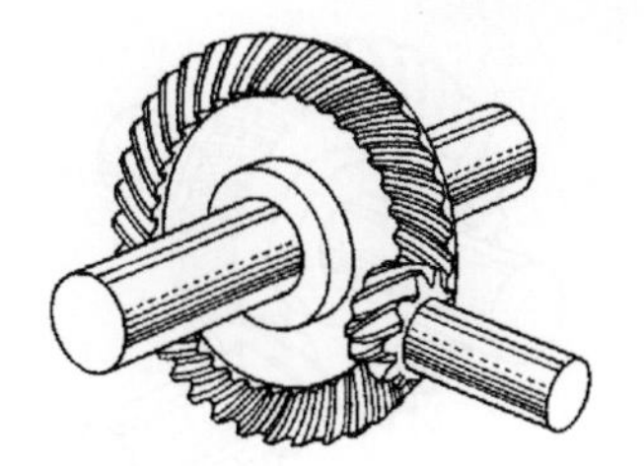

Fig. 4.2.13 Spiral-bevel gear.

This type of gear is beginning to supersede straight-bevel gears in many applications. They have the advantage of ensuring evenly distributed tooth loads and carry more load without surface fatigue. Thrust loading depends on the direction of rotation and whether the spiral angle of the teeth is positive or negative.

4.2.1.8 Miter

Miter gears are bevel gears with the same number of teeth in both gears, operating on shafts at 90°, as shown in Figure 4.2.14. Their primary use is to change direction in a mechanical drive assembly. Since both the pinion and gear have the same number of teeth, there is no mechanical advantage generated by this type of gear.

4.2.1.9 Hypoid

Hypoid-bevel gears are a cross between a spiral-bevel gear and a worm gear (Figure 4.2.15). The axes of a pair of hypoid-bevel gears are non-intersecting and the distance between the axes is referred to as the 'offset'. This configuration allows both shafts to be supported at both ends and provides high strength and rigidity.

Although stronger and more rigid than most other types of gears, they are less efficient and extremely difficult to lubricate because of high tooth-contact

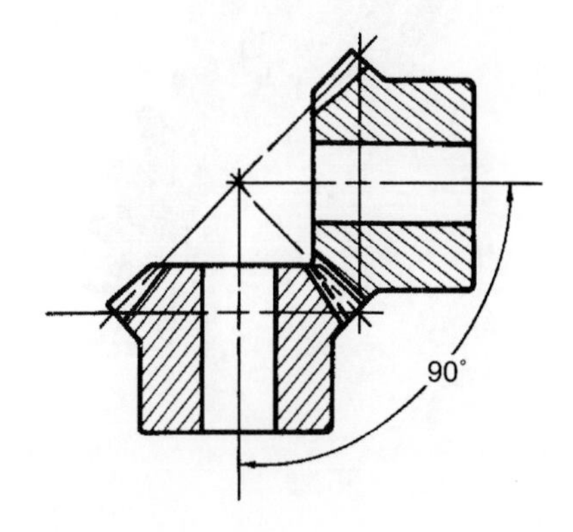

Fig. 4.2.14 Miter gear shaft angle.

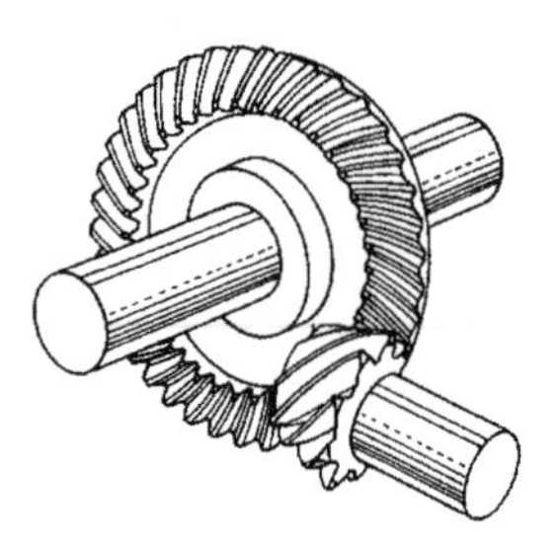

Fig. 4.2.15 Hypoid-bevel gear.

pressures. Further increasing the demands on the lubricant is the material of construction as both the driven and driving gears are made of steel. This requires the use of special extreme-pressure lubricants that have both oiliness and anti-weld properties that can withstand the high contact pressures and rubbing speeds.

Despite its demand for special lubrication, this gear type is in widespread use in industrial and automotive applications. It is used extensively in rear axles of automobiles having rear-wheel drives and is increasingly being used in industrial machinery.

4.2.1.10 Worm

The worm and gear, which are illustrated in Figure 4.2.16, are used to transmit motion and power when a high-ratio speed reduction is required. They accommodate a wide range of speed ratios (60:1 and higher can be obtained from a single reduction and can go as high as 500:1). In most worm gear sets, the worm is most often the driver and the gear the driven member. They provide a steady, quiet transmission of power between shafts at right angles and can be self-locking. Thus, a torque on the gear will not cause the worm to rotate.

Fig. 4.2.16 Worm gear.

The contact surface of the screw on the worm slides along the gear teeth. Because of the high level of rubbing between the worm and wheel teeth, however, slightly less efficiency is obtained than with precision spur gears. Note that large helix angles on the gear teeth produce higher efficiencies. Another problem with this gear type is heat removal, a limitation that restricts their use to low-speed applications.

One of the major advantages of the worm gear is low wear, which is due mostly to a full-fluid lubricant film. In addition, friction can be further reduced using metals having low coefficients of friction. For example, the wheel is typically made of bronze and the worm of highly finished hardened steel.

Most worms are cylindrical in shape with a uniform pitch diameter. However, a variable pitch diameter is used in the double-enveloping worm. This configuration is used when increased load capacity is required.

4.2.2 Performance

With few exceptions, gears are one-directional power transmission devices. Unless a special bi-directional gear set is specified, gears have a specific direction of rotation and will not provide smooth, trouble-free power transmission when the direction is reversed. The reason for this one-directional limitation is that gear manufacturers do not finish the non-power side of the tooth profile. This is primarily a cost-savings issue and should not affect gear operation.

The primary performance criteria for gear sets include: efficiency, brake horsepower, speed transients, startup, backlash, and ratios.

4.2.2.1 Efficiency

Gear efficiency varies with the type of gear used and the specific application. Table 4.2.2 provides a comparison of the approximate efficiency range of various gear types. The table assumes normal operation where torsional loads are within the gear set's designed horsepower range. It also assumes that startup and speed change torques are acceptable.

4.2.2.2 Brake horsepower

All gear sets have a recommended and maximum horsepower rating. The rating varies with the type of gear set, but must be carefully considered when evaluating a gearbox problem. The maximum installed motor horsepower should never exceed the maximum recommended horsepower of the gearbox. This is especially true of worm gear sets. The soft material used for these

Table 4.2.2 Gear efficiencies

Gear type	Efficiency range, %
Bevel gear, hypoid	90–98
Bevel gear, miter	Not available
Bevel gear, spiral	97–99
Bevel gear, straight	97–99
Bevel gear, zerol	Not available
Helical gear, external	97–99
Helical gear–double, external (herringbone)	97–99
Spur gear, external	97–99
Worm, cylindrical	50–99
Worm, double-enveloping	50–98

Source: Adapted by Integrated Systems, Inc. from 'Gears and Gear Drives,' 1996 Power. Transmission Design, Penton Publishing Inc., Ohio, pp. A199–A211.

gears is easily damaged when excess torsional load is applied.

The procurement specifications or the vendor's engineering catalog will provide all of the recommended horsepower ratings needed for an analysis. These recommendations assume normal operation and must be adjusted for the actual operating conditions in a specific application.

4.2.2.3 Speed transients

Applications that require frequent speed changes can have a severe, negative impact on gearbox reliability. The change in torsional load caused by acceleration and deceleration of a gearbox may exceed its maximum allowable horsepower rating. This problem can be minimized by decreasing the ramp speed and amount of braking that is applied to the gear set. The vendor's O&M manual and/or technical specifications should provide detailed recommendations that define the limits to use in speed change applications.

4.2.2.4 Startup

Start–stop operation of a gearbox can accelerate both gear and bearing wear and may cause reliability problems. In applications like the bottom discharge of storage silos, where a gear set drives a chain or screw conveyor system and startup torque is excessive, care must be taken to prevent overloading the gear set.

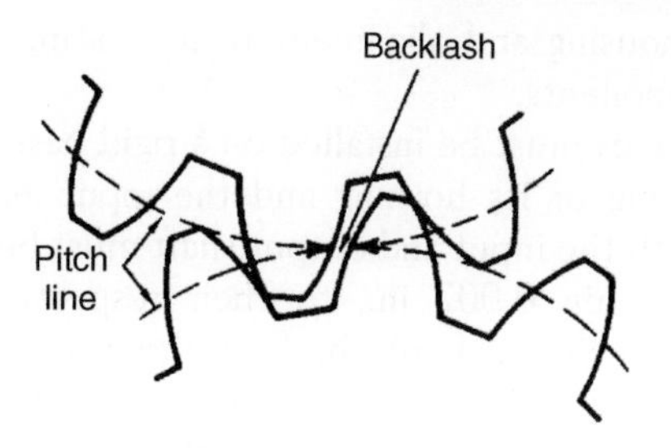

Fig. 4.2.17 Backlash.

4.2.2.5 Backlash

Gear backlash is the play between teeth measured at the pitch circle. It is the distance between the involutes of the mating gear teeth, as illustrated in Figure 4.2.17.

Backlash is necessary to provide the running clearance needed to prevent binding of the mating gears, which can result in heat generation, noise, abnormal wear, overload, and/or failure of the drive. In addition to the need to prevent binding, some backlash occurs in gear systems because of the dimensional tolerances needed for cost-effective manufacturing.

During the gear-manufacturing process, backlash is achieved by cutting each gear tooth thinner by an amount equal to one-half of the backlash dimension required for the application. When two gears made in this manner are run together (i.e., mate), their allowances combine to provide the full amount of backlash.

The increase in backlash that results from tooth wear does not adversely affect operation with non-reversing drives, or drives with continuous load in one direction. However, for reversing drives and drives where timing is critical, excessive backlash that results from wear usually cannot be tolerated.

4.2.2.6 Ratios

Ratios used in defining and specifying gears are gear tooth ratio, contact ratio, and hunting ratio. The gear tooth ratio is the ratio of the larger to the smaller number of teeth in a pair of gears. The contact ratio is a measure of overlapping tooth action which is necessary to assure smooth, continuous action. For example, as one pair of teeth passes out of action, a succeeding pair of teeth must have already started action. The hunting ratio is the ratio of the number of gear and pinion teeth. It is a means of ensuring that each tooth in the pinion contacts every tooth in the gear before it contacts any gear tooth a second time.

4.2.3 Installation

Installation guidelines provided in the vendor's O&M manual should be followed for proper installation of the gearbox housing and alignment to its mating machine-train components.

Gearboxes must be installed on a rigid base that prevents flexing of its housing and the input and output shafts. Both the input and output shaft must be properly aligned, within 0.002 in., to their respective mating shafts. Both shafts should be free of any induced axial forces that may be generated by the driver or driven units.

Internal alignment is also important. Internal alignment and clearances of new gearboxes should be within the vendor's acceptable limits, but there is no guarantee that this will be true. All internal clearances (e.g., backlash and center-to-center distances) and the parallel relationship of the pinion and gear shafts should be verified for any gearbox that is being investigated.

4.2.4 Operating methods

Two primary operating parameters govern effective operation of gear sets or gearboxes: maximum torsional power rating and transitional torsional requirements.

Each gear set has a specific maximum horsepower rating. This is the maximum torsional power that the gear set can generate without excessive wear or gear damage. Operating procedures should ensure that the maximum horsepower is not exceeded throughout the entire operating envelope. If the gear set was properly designed for the application, its maximum horsepower rating should be suitable for steady-state operation at any point within the design-operating envelope. As a result, it should be able to provide sufficient torsional power at any set point within the envelope.

Two factors may cause overload on a gear set: excessive load or speed transients. Many processes are subjected to radical changes in the process or production loads. These changes can have a serious effect on gear-set performance and reliability.

Operating procedures should establish boundaries that limit the maximum load variations that can be used in normal operation. These limits should be well within the acceptable load rating of the gear set.

The second factor, speed transients, is a leading cause of gear-reliability problems. The momentary change in torsional load created by rapid changes in speed can have a dramatic, negative impact on gear sets. These transients often exceed the maximum horsepower rating of the gears and may result in failure. Operating procedures should ensure that torsional power requirements during startup, process-speed changes, and shutdown do not exceed the recommended horsepower rating of the gear set.

Section Five

Hydraulic and pneumatic systems

Chapter 5.1

5.1

Hydraulic fundamentals

Mobley

5.1.1 Introduction

The study of hydraulics deals with the use and characteristics of liquids and gases. Since the beginning of time, man has used fluids to ease his burden. Earliest recorded history shows that devices such as pumps and water wheels were used to generate useable mechanical power.

Fluid power encompasses most applications that use liquids or gases to transmit power in the form of mechanical work, pressure and/or volume in a system. This definition includes all systems that rely on pumps or compressors to transmit specific volumes and pressures of liquids or gases within a closed system. The complexity of these systems range from a simple centrifugal pump used to remove casual water from a basement to complex airplane control systems that rely on high-pressure hydraulic systems.

Fluid power systems have been developing rapidly over the past thirty-five years. Fluid power filled a need during World War II for an energy transmission system with muscle, which could easily be adapted to automated machinery. Today, fluid power technology is seen in every phase of man's activities. Fluid power is found in areas of manufacturing such as metal forming, plastics, basic metals, and material handling. Fluid power is evident in transportation as power and control systems of ships, airplanes, and automobiles. The environment is another place fluid power is hard at work compacting waste materials and controlling floodgates of hydroelectric dams. Food processing, construction equipment, and medical technology are a few more areas of fluid power involvement. Fluid power applications are only limited by imagination.

There are alternatives to fluid power systems. Each system, regardless of the type, has its own advantages and disadvantages. Each has applications where it is best suited to do the job. This is probably the reason you won't find a fluid power wristwatch or hoses carrying fluid power replacing electrical power lines.

5.1.1.1 Advantages of fluid power

If a fluid power system is properly designed and used, it will provide smooth, flexible, uniform action without vibration, and is unaffected by variation of load. In case of an overload, an automatic release of pressure can be guaranteed, so that the system is protected against breakdown or excessive strain. Fluid power systems can provide widely variable motions in both rotary and linear transmission of power and the need for manual control can be minimized. In addition, fluid power systems are economical to operate.

Fluid power includes hydraulic, hydro-pneumatic and pneumatic systems. Why are hydraulics used in some applications, pneumatics in others, or combination systems, in still others? Both the user and the manufacturer must consider many factors when determining which type of system should be used in a specific application.

In general, pneumatic systems are less expensive to manufacture and operate, but there are factors that prohibit their universal application. The compressibility of air, as any gas, limits the operation of pneumatic systems. For example, a pneumatic cylinder cannot maintain the position of a suspended load without a constant supply of air pressure. The load will force the air trapped within the cylinder to compress and allow the suspended

Plant Engineer's Handbook; ISBN: 9780750673280

load to creep. This compressibility also limits the motion of pneumatic actuators when under load.

Pneumatic systems can be used for applications that require low to medium pressure and only accurate control. Applications that require medium pressure, more accurate force transmission, and moderate motion control can use a combination of hydraulics and pneumatics, or hydro-pneumatics. Hydraulics systems must be used for applications that require high pressure and/or extremely accurate force and motion control.

The flexibility of fluid power, of both hydraulic and pneumatic elements, presents a number of problems. Since fluids and gases have no shape of their owns; they must be positively confined throughout the entire system. This is especially true in hydraulics where leakage of hydraulic oil can result in safety or environmental concerns. Special consideration must be given to the structural integrity of the parts of a hydraulic system. Strong pipes, tubing, and hoses, as well as the containers must be provided. Leaks must be prevented. This is a serious problem with the high pressure obtained in many hydraulic system applications.

5.1.1.2 Fluid power systems vs. mechanical systems

Some desirable characteristics of fluid power systems when compared with mechanical systems are mentioned below. A fluid power system is often a simpler means of transmitting energy. There are fewer mechanical parts than in an ordinary industrial system. Since there are fewer mechanical parts, a fluid power system is more efficient and more dependable. In the common industrial system, there is no need to worry about hundreds of moving parts failing, with fluid or gas as the transmission medium.

With fluid or gas as the transmission medium, various components of a system can be located at convenient places on the machine. Fluid power can be transmitted and controlled quickly and efficiently up, down, and around corners with few controlling elements.

Since fluid power is efficiently transmitted and controlled, it gives freedom in designing a machine. The need for gear, cam, and lever systems is eliminated. Fluid power systems can provide infinitely variable speed, force, and direction control with simple, reliable elements.

5.1.1.3 Fluid power vs. electrical systems

Mechanical force can be more easily controlled using fluid power. The simple use of valves and rotary or linear actuators control speed, direction, and force. The simplicity of hydraulic and pneumatic components greatly increases their reliability. In addition, components and overall system size are typically much smaller than comparable electrical transmission devices.

5.1.1.4 Special problems

The operation of the system involves constant movement of the hydraulic fluid within its lines and components. This movement causes friction within the fluid itself and against the containing surfaces. Excessive friction can lead to serious losses in efficiency or damage to system components. Foreign matter must not be allowed to accumulate in the system, where it will clog small passages or score closely fitted parts. Chemical action may cause corrosion. Anyone working with hydraulic systems must know how a fluid power system and its components operate, both in terms of the general principles common to all physical mechanisms and of the peculiarities of the specific arrangement at hand.

The word hydraulics is based on the Greek word for water, the first used form of hydraulic power transmission. Initially, hydraulics covered the study of the physical behavior of water at rest and in motion. It has been expanded to include the behavior of all liquids, although it is primarily limited to the motion or kinetics of liquids.

5.1.1.5 Hazards

Any use of a pressurized medium, such as hydraulic fluid, can be dangerous. Hydraulic systems carry all the hazards of pressurized systems and special hazards related directly to the composition of the fluid used.

When using oil as a fluid in a high-pressure hydraulic system, the possibility of fire or an explosion exists. A severe fire hazard is generated when a break in the high-pressure piping occurs and the oil is vaporized into the atmosphere. Extra precautions against fire should be practiced in these areas.

If oil is pressurized by compressed air, an explosive hazard exists. If high-pressure air encounters the oil, it may create a diesel effect, which may result in an explosion. A carefully followed preventive maintenance plan is the best precaution against explosions.

5.1.2 Basic hydraulics

Fluid power systems have developed rapidly over the past thirty-five years. Today, fluid power technology is used in every phase of human existence. The extensive use of hydraulics to transmit power is due to the fact that properly constructed fluid power systems possess a number of favorable characteristics. They eliminate the need for complicated systems of gears, cams, and levers. Motion can be transmitted without the slack or mechanical looseness inherent in the use of solid machine parts. The fluids used are not subject to breakage as are mechanical parts, and the mechanisms are not subjected to great wear.

The operation of a typical fluid power system is illustrated in Figure 5.1.1. Oil from a tank or reservoir flows through a pipe into a pump. An electric motor, air motor, gas or steam turbine, or an internal combustion engine can drive the pump. The pump increases the pressure of the oil. The actual pressure developed depends on the design of the system.

The high-pressure oil flows in piping through a control valve. The control valve changes the direction of oil flow. A relief valve, set at a desired, safe operating pressure, protects the system from an over-pressure condition. The oil that enters the cylinder acts on the piston, with the pressure acting over the area of the piston, developing a force on the piston rod. The force on the piston rod enables the movement of a load or device.

5.1.2.1 States of matter

The material that makes up the universe is known as matter. Matter is defined as any substance that occupies space and has weight. Matter exists in three states: solid, liquid, and gas. Each has distinguishing characteristics. Solids have a defined volume and a definite shape. Liquids have a definite volume, but take the shape of their containing vessels. Gases have neither a definite shape nor volume. Gases not only take the shape of the containing vessel, but also expand to fill the vessel, regardless of its volume. Examples of the states of matter are iron, water, and air.

Matter can change from one state to another. Water is a good example. At high temperatures, above 212 °F, it is in a gaseous state known as steam. At moderate temperatures, it is liquid, and at low temperatures, below 32 °F, it becomes ice, a solid. In this example, the temperature is the dominant factor in determining the state that the substance assumes.

Pressure is another important factor that affects changes in the state of matter. At pressures higher than atmospheric, 14.7 psi, water boils and thus changes to steam at temperatures below 212 °F. Pressure is also a critical factor in changing some gases to liquids or solids. Normally, when pressure and chilling are both applied to a gas, the gas assumes a liquid state. Liquid air, which is a mixture of oxygen and nitrogen, is produced in this manner.

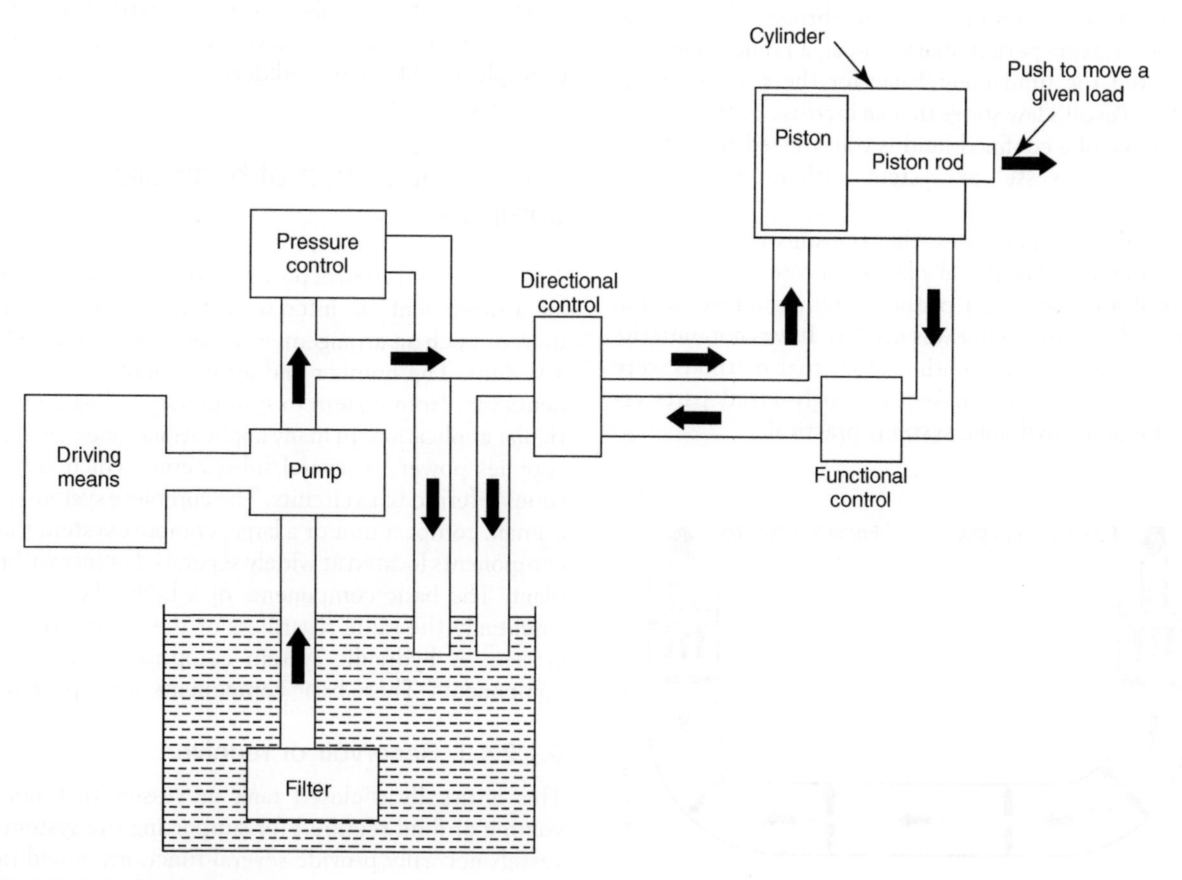

Fig. 5.1.1 Basic hydraulic system.

In the study of fluid power, we are concerned primarily with the properties and characteristics of liquids and gases. However, one should keep in mind that the properties of solids also affect the characteristics of liquids and gases. The lines and components, which are solids, enclose and control the liquid or gas in their respective systems.

5.1.2.2 Development of hydraulics

The use of hydraulics is not new. The Egyptians and people of ancient Persia, India, and China conveyed water along channels for irrigation and other domestic purposes. They used dams and sluice gates to control the flow and waterways to direct the water to where it was needed. The ancient Cretins had elaborate plumbing systems. Archimedes studied the laws of floating and submerged bodies. The Romans constructed aqueducts to carry water to their cities.

After the breakup of the ancient world, there were few new developments for many centuries. Then, over a comparatively short period, beginning near the end of the seventeenth century, Italian physicist, Evangelista Torricelle, French physicist, Edme Mariotte, and later, Daniel Bernoulli conducted experiments to study the force generated by the discharge of water through small openings in the sides of tanks and through short pipes. During the same period, Blaise Pascal, a French scientist, discovered the fundamental law for the science of hydraulics. Pascal's law states that an increase in pressure on the surface of a confined fluid is transmitted throughout the confining vessel or system without any loss of pressure.

Figure 5.1.2 illustrates the transmission of forces through liquids. For Pascal's law to become effective for practical applications, a piston or ram confined within a close tolerance cylinder was needed. It was not until the latter part of the eighteenth century that methods were developed that could make the snugly fitted parts required making hydraulic systems practical.

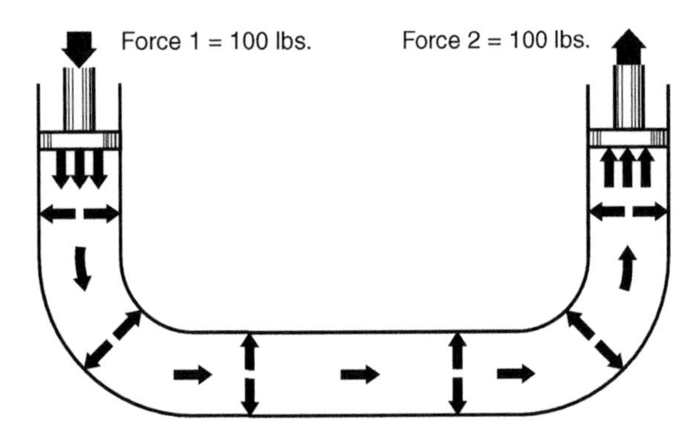

Fig. 5.1.2 Transmission of forces.

This was accomplished by the invention of machines that were used to cut and shape the necessary closely fitted parts and, particularly, by the development of gaskets and packing. Since that time, components such as valves, pumps, actuating cylinders, and motors have been developed and refined to make hydraulics one of the leading methods of transmitting power.

5.1.2.3 Use of hydraulics

The hydraulic press, invented by an Englishman, John Brahmah, was one of the first workable machines that used hydraulics in its operation. It consisted of a plunger pump piped to a large cylinder and a ram. This press found wide use in England because it provided a more effective and economical means of applying large, uniform forces in industrial uses.

Today, hydraulic power is used to operate many different tools and mechanisms. In a garage, a mechanic raises the end of an automobile with a hydraulic jack. Dentists and barbers use hydraulic power to lift and position their chairs. Hydraulic doorstops keep heavy doors from slamming. Hydraulic brakes have been standard equipment on automobiles since the 1930s. Most automobiles are equipped with automatic transmissions that are hydraulically operated. Power steering is another application of hydraulic power. Construction workers depend upon hydraulic power for their equipment. For example, the blade of a bulldozer is normally operated by hydraulic power.

5.1.2.4 Operation of hydraulic components

To transmit and control power through pressurized fluids, an arrangement of interconnected components is required. Such an arrangement is commonly referred to as a system. The number and arrangement of the components vary from system to system, depending on the particular application. In many applications, one main system supplies power to several subsystems, which are sometimes referred to as circuits. The complete system may be a small, compact unit or a large, complex system that has components located at widely separated points within the plant. The basic components of a hydraulic system are essentially the same regardless of the complexity of the system. A hydraulic system comprises of seven basic components. These basic components are explain below.

5.1.2.4.1 Reservoir or receiver

This is usually a closed tank or vessels that hold the volume of fluid required for supporting the system. The vessels normally provide several functions in addition to holding fluid reserves. The major functions include filtration of the fluid, heat dissipation, and water separation.

5.1.2.4.2 Hydraulic pump

This is the energy source for hydraulic systems. It converts electrical energy into dynamic, hydraulic pressure. In almost all cases, hydraulic systems utilize positive displacement pumps as their primary power source. These are broken down into two primary sub-classifications: constant-volume or variable-volume. In the former, the pumps are designed to deliver a fixed output (i.e. both volume and pressure) of hydraulic fluid. In the latter, the pump delivers only the volume or pressure required for specific functions of the system or its components.

5.1.2.4.3 Control valves

The energy generated by the hydraulic pump must be directed and controlled so that the energy can be used. A variety of directional and functional control valves are designed to provide a wide range of control functions.

5.1.2.4.4 Actuating devices

The energy within a hydraulic system is of no value until it is converted into work. Typically, this is accomplished by using an actuating device of some type. This actuating device may be a cylinder, which converts the hydraulic energy into linear mechanical force; a hydraulic motor, that converts energy into rotational force; or a variety of other actuators designed to provide specific work functions.

5.1.2.4.5 Relief valves

Most hydraulic systems use a positive displacement pump to generate energy within the system. Unless the pressure is controlled, these pumps will generate excessive pressure that can cause catastrophic failure of system component. A relief valve is always installed downstream of the hydraulic pump to prevent excessive pressure and to provide a positive relief, should a problem develop within the system. The relief valve is designed to open at a preset system pressure. When the valve opens, it diverts flow to the receiver tank or reservoir.

5.1.2.4.6 Lines (pipe, tubing, or flexible hoses)

All systems require some means to transmit hydraulic fluid from one component to another. The material of the connecting lines will vary from system to system or within the system.

5.1.2.4.7 Hydraulic fluid

The fluid provides the vehicle that transmits input power, such as from a hydraulic pump to the actuator device or devices that perform work.

5.1.3 Forces in liquids

The study of liquids is divided into two main parts: liquids at rest, hydrostatics; and liquids in motion, hydraulics. The effect of liquids at rest can often be expressed by simple formulas. The effects of liquids in motion are more difficult to express due to frictional and other factors whose actions cannot be expressed by simple mathematics.

Liquids are almost incompressible. For example, if a pressure of 100 pounds per square inch, psi, is applied to a given volume of water that is at atmospheric pressure, the volume will decrease by only 0.03 per cent. It would take a force of approximately 32 tons to reduce its volume by 10 per cent; however, when this force is removed, the water immediately returns to its original volume. Other liquids behave in about the same manner as water.

Another characteristic of a liquid is the tendency to keep its free surface level. If the surface is not level, the liquid will flow in the direction which tends to make the surface level.

5.1.3.1 Liquids at rest (hydrostatics)

In the study of fluids at rest, we are concerned with the transmission of force and the factors which affect the forces in liquids. Additionally, pressure in and on liquids and factors affecting pressure are of great importance.

5.1.3.1.1 Pressure and force

The terms force and pressure are used extensively in the study of fluid power. It is essential that we distinguish between these terms. Force is the total pressure applied to or generated by a system. It is the total pressure exerted against the total area of a particular surface and is expressed in pounds or grams.

Pressure is the amount of force applied to each unit area of a surface and is expressed in $\mathrm{lb/in.^2}$ (psi) or $\mathrm{g/cm^2}$. Pressure may be exerted in one direction, in several directions, or in all directions.

A formula is used in computing force, pressure, and area in fluid power systems. In this formula, P refers to pressure; F indicates force, and A represents area. Force equals pressure times area. Thus, the formula is written:

$$F = P \times A$$

Pressure equals force divided by area. By rearranging the formula, this statement may be condensed to

$$P = \frac{F}{A}$$

Since area equals force divided by pressure, the formula is written:

$$A = \frac{F}{P}$$

5.1.3.1.2 Atmospheric pressure

The atmosphere is the entire mass of air that surrounds the earth. While it extends upward for about 500 miles, the section of primary interest is the portion that rests on the earth's surface and extends upward for about 7(1/2) miles. This layer is called the troposphere.

If a column of air 1 in.^{2} extended to the 'top' of the atmosphere could be weighed, this column of air would weigh approximately 14.7 pounds at sea level. Thus, atmospheric pressure, at sea level, is approximately 14.7 lb/in.^{2} or psi.

Atmospheric pressure decreases by approximately 1.0 psi for every 2343 ft of elevation. Elevations below sea level, such as in excavations and depressions, atmospheric pressure increases. Pressures under water differ from those under air only because the weight of the water must be added to the pressure of the air.

Atmospheric pressure can be measured by any of several methods. The common laboratory method uses a mercury column barometer. The height of the mercury column serves as an indicator of atmospheric pressure. At sea level and at a temperature of 0 °C, the height of the mercury column is approximately 30 in., or 76 cm. This represents a pressure of approximately 14.7 psi. The 30-in. column is used as a reference standard.

Atmospheric pressure does not vary uniformly with altitude. It changes more rapidly at lower altitudes because of the compressibility of air, which causes the air layers close to the earth's surface to be compressed by the air masses above them. This effect, however, is partially counteracted by the contraction of the upper layers due to cooling. The cooling tends to increase the density of the air.

Atmospheric pressures are quite large, but in most instances, practically the same pressure is present on all sides of objects so that no single surface is subjected to a greater load. Atmospheric pressure acting on the surface of a liquid, Figure 5.1.3 view (a), is transmitted equally throughout the liquid to the walls of the container, but is balanced by the same atmospheric pressure acting on the outer walls of the container. In view (b), atmospheric pressure acting on the surface of one piston is balanced by the same pressure acting on the surface of the other piston. The different areas of the two surfaces make no difference, since for a unit of area, pressures are balanced.

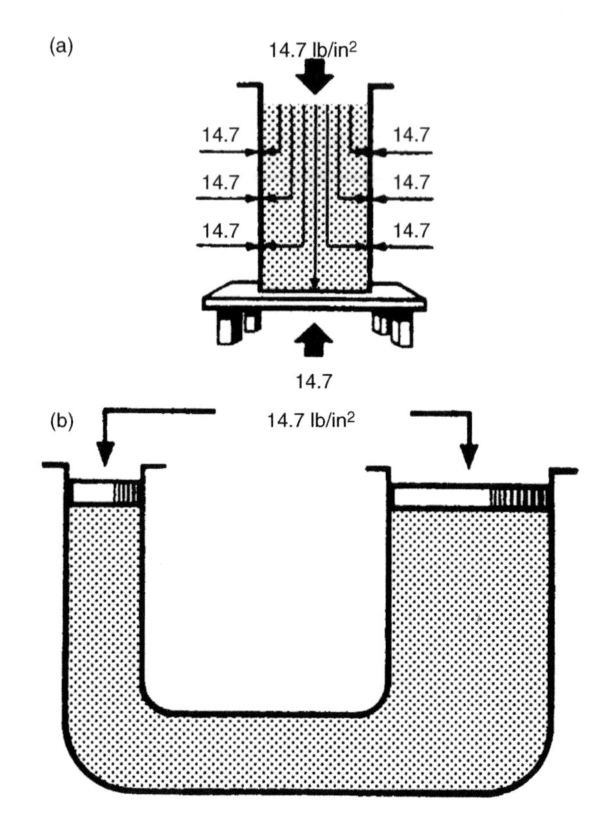

Fig. 5.1.3 Effects of atmospheric pressure.

$$= \frac{62.4 \text{ lbs}}{1 \text{ ft}^2}$$

5.1.3.1.3 Pascal's law

The foundation of modern hydraulics was established with Pascal's discovery that pressure in a fluid acts equally in all directions. This pressure acts at right angles to the containing surfaces. If some type of pressure gauge, with an exposed face, is placed beneath the surface of a liquid, Figure 5.1.4, at a specific depth and pointed in different directions, the pressure will read the same. Thus, we can say that pressure in a liquid is independent of direction.

Pressure due to weight of a liquid, at any level, depends on the depth of the fluid from the surface. If the exposed face of the pressure gauge, Figure 5.1.4, is moved closer to the surface of the liquid, the indicated pressure will be less.

When the depth is doubled, the indicated pressure is also doubled. Thus, the pressure in a liquid is directly proportional to the depth. Consider a container with vertical sides, Figure 5.1.5, that is 1 ft high and 1 ft wide.

Let it be filled with water 1 foot deep, thus providing 1 ft^3 of water. We know that 1 ft^3 of water weighs 62.4 pounds.

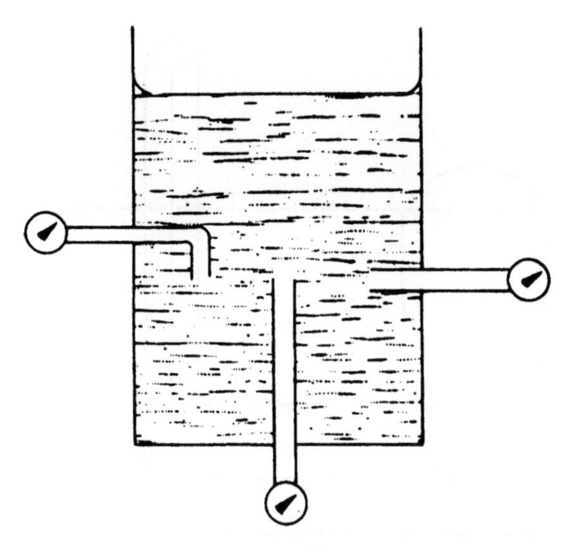

Fig. 5.1.4 Pressure of a liquid is independent of direction.

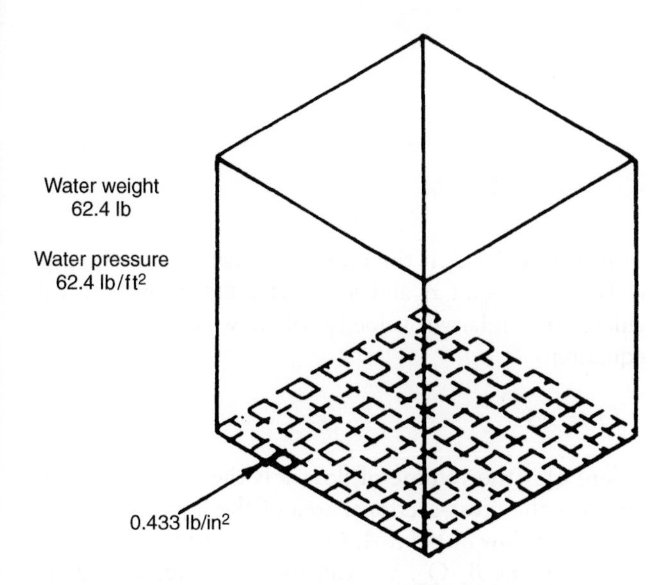

Fig. 5.1.5 Water pressure in a 1-cubic foot container.

Using this information and the equation for pressure, we can calculate the pressure on the bottom of the container:

$$P = \frac{F}{A}$$

$$P = \frac{62.4}{144} = 0.433\ \text{lbs/in.}^2$$

Since there are 144 in.^{2} in 1 ft^2, this can be stated as follows: the weight of a column of water 1 ft high, having a cross-sectional area of 1 in.^{2}, is 0.433 pounds. If the depth of the column is tripled, the weight of the column will be 3×0.433 or 1.299 pounds and the pressure at the bottom will be 1.299 lb/in.^{2} (psi), since the pressure is equal to the force divided by the area. Thus, the pressure at any depth in a liquid is equal to the weight of the column of liquid at the depth divided by the cross-sectional area of the column at that depth. The volume of a liquid that produces the pressure is referred to as the fluid head of the liquid. The pressure of a liquid due to its fluid head is also dependent on the density of the liquid.

If we let A equal any cross-sectional area of a liquid column and h equal the depth of the column, the volume becomes Ah. Using the equation $D = W/V$, the weight of the liquid above area A is equal to AhD, or

$$D = \frac{W}{Ah} W = Ah \times D$$

Since pressure is equal to the force per unit area, set A equal to 1. Then the formula for pressure becomes:

$$P = hD$$

It is essential that h and D be expressed in similar units. That is, if D is expressed in pounds per cubic foot, the value of h must be expressed in feet. If the desired pressure is to be expressed in pounds per square inch, the pressure formula becomes:

$$P = \frac{hD}{144}$$

Pascal was also the first to prove by experiment that the shape and volume of a container in no way alters pressure. Thus in Figure 5.1.6, if the pressure due to the weight of the liquid at a point on horizontal line H is 8 psi, the pressure is 8 psi everywhere at level H in the system.

5.1.3.2 Liquids in motion (hydraulics)

In the operation of fluid power systems, there must be flow of fluid. The amount of flow will vary from system to system. To understand fluid power systems, it is necessary to understand some of the characteristics of liquids in motion.

Liquids in motion have characteristics different from liquids at rest. Frictional resistances within the fluid, viscosity, and inertia contribute to these differences. Inertia, which means the resistance a mass offers to being set in motion, will be discussed later in this section. Other relationships of liquids in motion you must be familiar with. Among these are volume and velocity of flow; flow rate and speed; laminar and turbulent flow; and more importantly, the force and energy changes which occur in flow.

5.1.3.2.1 Volume and velocity of flow

The volume of a liquid passing a point in a given time is known as its volume of flow or flow rate. The volume of

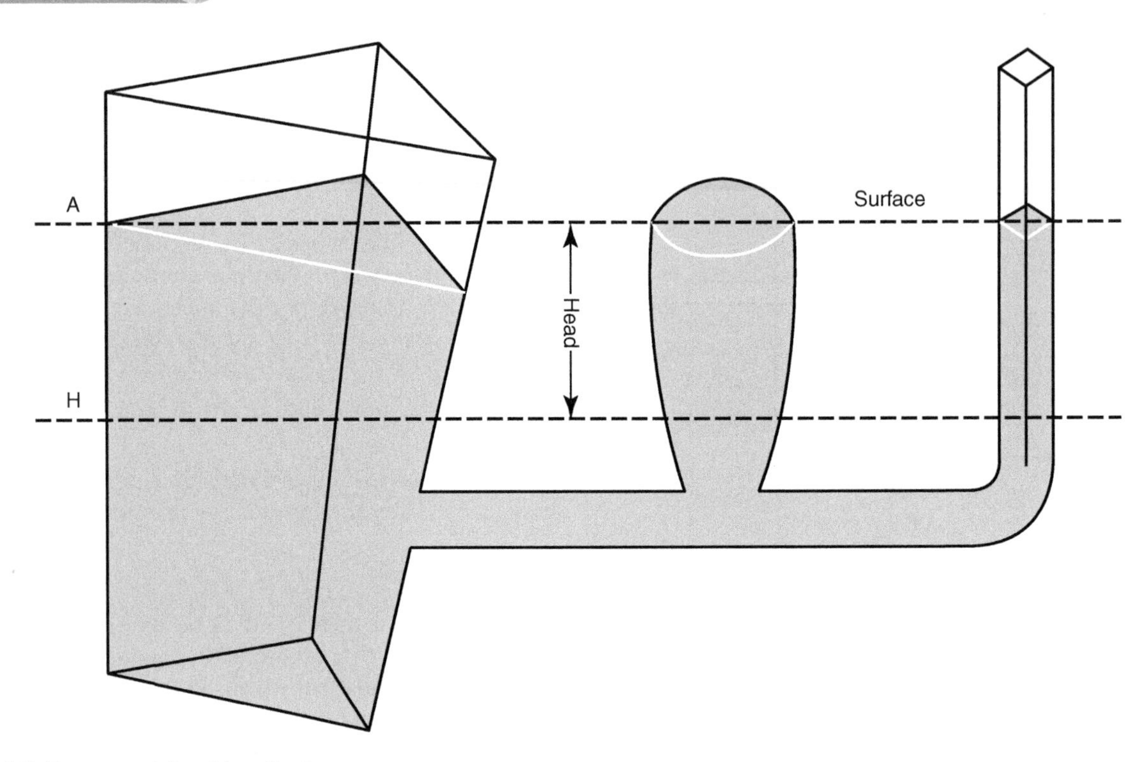

Fig. 5.1.6 Pressure relationship with shape.

flow is usually expressed in gallons per minute (gpm) and is associated with the relative pressures of the liquid, such as 5 gpm at 40 psig. The velocity of flow or velocity of the fluid is defined as the average speed at which the fluid moves past a given point. It is usually expressed in feet per minute (fpm) or inches per second (ips). Velocity of flow is an important consideration in sizing the hydraulic piping and other system components.

Volume and velocity of flow must be considered together. With other conditions unaltered, the velocity of flow increases as the cross-section or size of the pipe decreases, and the velocity of flow decreases as the cross-section or pipe size increases. For example, the velocity of flow is slow at wide parts, yet the volume of liquid passing each part of the stream is the same.

In Figure 5.1.7, if the cross-sectional area of the pipe is 16 in.^{2} at point A and 4 in.^{2} at point B, we can calculate the relative velocity of flow using the flow equation:

$$Q = vA$$

where Q is the volume of flow, v is the velocity of flow, and A is the cross-sectional area of the liquid. Since the volume of flow at point A, Q_1, is equal to the volume of flow at point B, Q_2, we can use this equation to determine the ratio of the velocity of flow at point A, v_1, to the velocity of flow at point B, v_2. Since

$$Q_1 = Q_2, \quad \text{then } A_1 v_1 = A_2 v_2$$

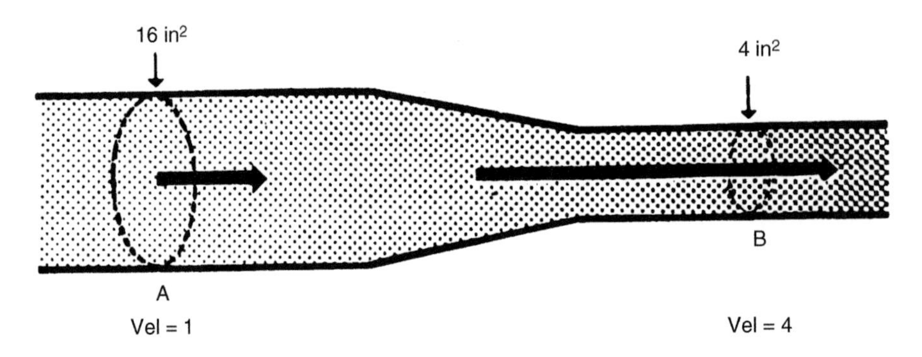

Fig. 5.1.7 Volume and velocity of flow.

From Figure 5.1.7:

$$A_1 = 16 \text{ in.}^2, \quad A_2 = 4 \text{ in.}^2$$

Substituting:

$$16v_1 = 4v_2 \text{ or } v_2 = 4v_1$$

Therefore, the velocity of flow at point B is four times greater than the velocity of flow at point A.

5.1.3.2.2 Volume of flow and speed

When you consider the cylinder volume that must be filled and the distance that the piston must travel, you can relate the volume of flow to the speed of the piston. The volume of the cylinder is found by multiplying the piston area by the length the piston must travel. This length is known as stroke.

Suppose you have determined that two cylinders have the same volume and that one cylinder is twice as long as the other. In this case, the cross-sectional area of the longer tube will be half of the cross-sectional area of the other tube. If fluid is pumped into each cylinder at the same rate, both pistons will reach their full travel at the same time. However, the piston in the smaller cylinder must travel twice as fast because it has twice as far to go. There are two ways of controlling the speed of the piston, (1) by varying the size of the cylinder and (2) by varying the volume of flow (gpm) to the cylinders.

5.1.3.2.3 Streamline and turbulent flow

At low velocities or in tubes of small diameter, flow is streamlined. This means that a given particle of fluid moves straight forward without bumping into other particles and without crossing their paths. Streamline flow is often referred to as laminar flow, which is defined as a flow situation in which fluid moves in parallel lamina or layers. An example of streamline flow is an open stream flowing at a slow, uniform rate with logs floating on its surface. The logs represent particles of fluid. As long as the stream flows at a slow, uniform rate, each log floats downstream in its own path, without crossing or bumping into the others.

If the stream narrows and the volume of flow remains the same, the velocity will increase. If the velocity increases sufficiently, the water becomes turbulent. Swirls, eddies, and cross-motions are set up in the water. As this happens, the logs are thrown against each other and against the banks of the stream, and the paths followed by different logs will cross and re-cross.

Particles of fluid flowing in pipes act in the same manner. The flow is streamlined if the fluid flows slowly enough, and remains streamlined at greater velocities if the diameter of the pipe is small. If the velocity of flow or size of pipe is increased sufficiently, the flow becomes turbulent.

While a high velocity of flow will produce turbulence in any pipe, other factors contribute to turbulence. Among these are the roughness of the inside of the pipe, obstructions, the degree of curvature of bends, and the number of bends in the pipe. In setting up or maintaining fluid power systems, care should be taken to eliminate or minimize as many causes of turbulence as possible, since energy consumed by turbulence is wasted.

While designers of fluid power equipment do what they can to minimize turbulence, it cannot be avoided. For example, in a 4-in. pipe at 68 °F, flow becomes turbulent at velocities over approximately 6 in. per second (ips) or about 3 ips in a 6-in. pipe. These velocities are far below those commonly encountered in fluid power systems, where velocities of 5 ft per second (fps) and above are common. In laminar flow, losses due to friction increase directly with velocity. With turbulent flow, these losses increase much more rapidly.

5.1.3.2.4 Factors involved in flow

An understanding of the behavior of fluids in motion, or solids for that matter, requires an understanding of the term inertia. Inertia is the term used by scientists to describe the property possessed by all forms of matter that make it resist being moved when it is at rest and to resist any change in its rate or motion when it is moving.

The basic statement covering inertia is Newton's first law of motion. His first law states: A body at rest tends to remain at rest, and a body in motion tends to remain in motion at the same speed and direction, unless acted on by some unbalanced force. This simply says what you have learned by experience – that you must push an object to start it moving and push it in the opposite direction to stop it again.

A familiar illustration is the effort a pitcher must exert to make a fast pitch and the opposition the catcher must put forth to stop the ball. Similarly, the engine to make an automobile begin to roll must perform considerable work; although, after it has attained a certain velocity, it will roll along the road at uniform speed if just enough effort is expended to overcome friction, while brakes are necessary to stop its motion. Inertia also explains the kick or recoil of guns and the tremendous striking force of projectiles.

5.1.3.2.5 Inertia and force

To overcome the tendency of an object to resist any change in its state of rest or motion, some force that is not otherwise canceled or balanced must act on the object. Some unbalanced force must be applied whenever fluids are set in motion or increased in velocity;

while conversely, forces are made to do work elsewhere whenever fluids in motion are retarded or stopped.

There is a direct relationship between the magnitude of the force exerted and the inertia against which it acts. This force depends on two factors: (1) the mass of the object and (2) the rate at which the velocity of the object is changed. The rule is that the force, in pounds, required to overcome inertia is equal to the weight of the object multiplied by the change in velocity, measured in feet per second (fps) and divided by 32 times the time, in seconds, required to accomplish the change. Thus, the rate of change in velocity of an object is proportional to the force applied. The number 32 appears because it is the conversion factor between weight and mass.

Five physical factors act on a fluid to affect its behavior. All of the physical actions of fluids in all systems are determined by the relationship of these five factors to each other. Summarizing, these five factors are:

1. Gravity, which acts at all times on all bodies, regardless of other forces.
2. Atmospheric pressure, which acts on any part of a system exposed to the open air.
3. Specific applied forces, which may or may not be present, but which are entirely independent of the presence or absence of motion.
4. Inertia, which comes into play whenever there is a change from rest to motion, or the opposite, or whenever there is a change in direction or in rate of motion.
5. Whenever there is motion, friction is always present.

5.1.3.2.6 Kinetic energy

An external force must be applied to an object in order to give it a velocity or to increase the velocity it already has. Whether the force begins or changes velocity, it acts over a certain distance. Force acting over a certain distance is called work. Work and all forms into which it can be changed are classified as energy. Obviously then, energy is required to give an object velocity. The greater the energy used, the greater the velocity will be.

Disregarding friction, for an object to be brought to rest or for its motion to be slowed down, a force opposed to its motion must be applied to it. This force also acts over some distance. In this way, energy is given up by the object and delivered in some form to whatever opposed its continuous motion. The moving object is therefore a means of receiving energy at one place and delivering it to another point. While it is in motion, it is said to contain this energy, as energy of motion or kinetics energy.

Since energy can never be destroyed, it follows that if friction is disregarded the energy delivered to stop the object will exactly equal the energy that was required to increase its speed. At all times, the amount of kinetic energy possessed by an object depends on its weight and the velocity at which it is moving.

The mathematical relationship for kinetic energy is stated in the rule. Kinetic energy, in foot-pounds, is equal to the force, in pounds, which created it, multiplied by the distance through which it was applied, or to the weight of the moving object, in pounds, multiplied by the square of its velocity, in feet per second, and divided by 64.

The relationship between inertia forces, velocity, and kinetic energy can be illustrated by analyzing what happens when a gun fires a projectile against the armor of an enemy ship. The explosive force of the powder in the breach pushes the projectile out of the gun, giving it a high velocity. Because of its inertia, the projectile offers opposition to this sudden velocity and a reaction is set up that pushes the gun backward. The force of the explosion acts on the projectile throughout its movement in the gun. This is force acting through a distance producing work. This work appears as kinetic energy in the speeding projectile. The resistance of the air produces friction, which uses some of the energy and slows down the projectile. When the projectile hits the target, it tries to continue moving. The target, being relatively stationary, tends to remain stationary because of inertia. The result is that a tremendous force is set up that either leads to the penetration of the armor or the shattering of the projectile. The projectile is simply a means to transfer energy from the gun to the enemy ship. This energy is transmitted in the form of energy in motion or kinetic energy.

A similar action takes place in a fluid power system in which the fluid takes the place of the projectile. For example, the pump in a hydraulic system imparts energy to the fluid, which overcomes the inertia of the fluid at rest and causes it to flow through the lines. The fluid flows against some type of actuator that is at rest. The fluid tends to continue flowing, overcomes the inertia of the actuator, and moves the actuator to do work. Friction uses up a portion of the energy as the fluid flows through the lines and components.

5.1.3.2.7 Relationship of force, pressure, and head

In dealing with fluids, forces are usually considered in relation to the areas over which they are applied. As previously discussed, a force acting over a unit area is a pressure, and pressure can alternately be stated in pounds per square inch or in terms of head, which is the vertical height of the column of fluid whose weight would produce that pressure.

All five of the factors that control the actions of fluids can be expressed either as force or in terms of equivalent pressures or head. In either situation, the different factors are referred to in the same terms.

5.1.3.2.8 Static and dynamic factors

Gravity, applied forces, and atmospheric pressure are examples of static factors that apply equally to fluids at rest or in motion. Inertia and friction are dynamic forces that apply only to fluids in motion. The mathematical sum of gravity, applied forces, and atmospheric pressure is the static pressure obtained at any one point in a fluid system at a given point in time. Static pressure exists in addition to any dynamic factors that may also be present at the same time.

Remember that Pascal's law states that a pressure set up in a fluid acts equally in all directions and at right angles to the containing surfaces. This covers the situation only for fluids at rest. It is true only for the factors making up static head. Obviously, when velocity becomes a factor, it must have a direction of flow. The same is true of the force created by velocity. Pascal's law alone does not apply to the dynamic factors of fluid power systems.

The dynamic factors of inertia and friction are related to the static factors. Velocity head and friction head are obtained at the expense of static head. However, a portion of the velocity head can always be reconverted to static head. Force, which can be produced by pressure or head when dealing with fluids, is necessary to start a body moving if it is at rest, and is present in some form when the motion of the body is arrested. Therefore, whenever a fluid is given velocity, some part of its original static head is used to impart this velocity, which then exists as velocity head.

5.1.3.2.9 Bernoulli's principle

Review the system illustrated in Figure 5.1.8. Chamber A is under pressure and is connected by a tube to chamber B, which is also under pressure. The pressure in chamber A is static pressure of 100 psi. The pressure at some point (X) along the connecting tube consists of a velocity pressure of 10 psi exerted in a direction parallel to the line of flow, plus the unused static pressure of 90 psi. The static pressure (90 psi) follows Pascal's law

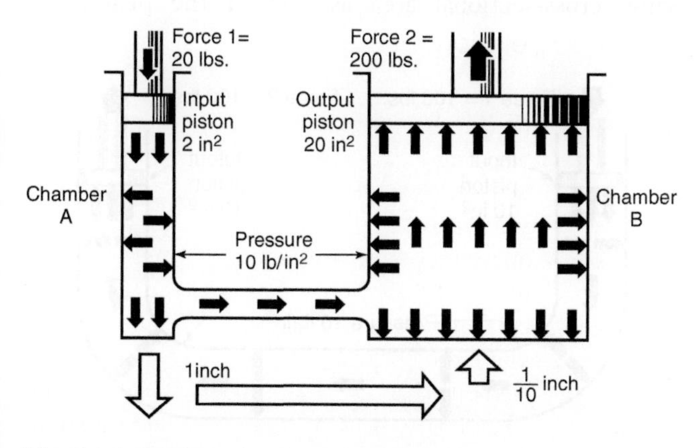

Fig. 5.1.8 Relation of static and dynamic factors.

and exerts equal pressure in all directions. As the fluid enters chamber B, it slows down and its velocity is reduced. As a volume of liquid moves from a small, confined space into a larger area, the fluid will expand to fill the greater volume. The result of this expansion is a reduction of velocity and a momentary reduction in pressure.

In the example, the force required to absorb the fluid's inertia equals the force required to start the fluid moving originally, so that the static pressure in chamber B is equal to that in chamber A.

This example disregards friction. Therefore, it would not be encountered in actual practice. Force or head is also required to overcome friction. Unlike inertia, this force cannot be recovered. Even though the energy required overcoming friction still exists, it has been converted to heat. In an actual system, the pressure in chamber B would be less than that in chamber A. The difference would be the amount of pressure used to overcome friction within the system.

At all points in a system, the static pressure is always equal to the original static pressure less any velocity head at a specific point in the system and less the friction head required to reach that point. Since both the velocity head and friction head represent energy and energy cannot be destroyed, the sum of the static head, the velocity head, and the friction head at any point in the system must add up to the original static head. This is known as Bernoulli's principle, which states: For the horizontal flow of fluids through a tube, the sum of the pressure and the kinetic energy per unit volume of the fluid is constant. This principle governs the relationship of the static and dynamic factors in hydraulic systems.

5.1.3.2.10 Minimizing friction

Fluid power equipment is designed to reduce friction as much as possible. Since energy cannot be destroyed, some of the energy created by both static pressure and velocity is converted to heat energy as the fluid flows through the piping and components within a hydraulic system. As friction increases, so does the amount of dynamic and static energy that is converted into heat.

To minimize the loss of useable energy lost to its conversion to heat energy, care must be taken in the design, installation, and operation of hydraulic system. As a minimum the following factors must be considered:

Proper fluid must be chosen and used in the system. It must have the best viscosity, operating temperature range, and other characteristics that are conducive to proper operation of the system and lowest possible friction component.

Fluid flow is also critical for proper operation of a hydraulic system. Turbulent flow should be avoided as much as possible. Clean, smooth pipe or tubing should be used to provide laminar flow and the lowest friction possible within the system. Sharp, close radius bends and sudden changes in cross-sectional area are avoided.

System components, such as pumps, valves, and gauges, create both turbulent flow and high-friction components. Pressure drop, or a loss of pressure, is created by a combination of turbulent flow and friction as the fluid flows through the unit. System components that are designed to provide minimum interruption of flow and pressure should be selected for the system.

5.1.3.3 Transmission of force through liquids

When the end of a solid bar is struck, the main force of the blow is carried straight through the bar to the other end (Figure 5.1.9, view A). This happens because the bar is rigid. The direction of the blow almost entirely determines the direction of the transmitted force. The more rigid the bar, the less force is lost inside the bar or transmitted outward at right angles to the direction of the blow.

When a force is applied to the end of a column of confined liquid (Figure 5.1.9, view B), it is transmitted straight through to the other end. It is also equal and undiminished in every direction throughout the column – forward, backward, and sideways – so that the containing vessel is literally filled with the added pressure.

So far, we have explained the effects of atmospheric pressure on liquids and how external forces are distributed through liquids. Let us now focus our attention on forces generated by the weight of liquids themselves. To do this, we must first discuss density, specific gravity, and Pascal's law.

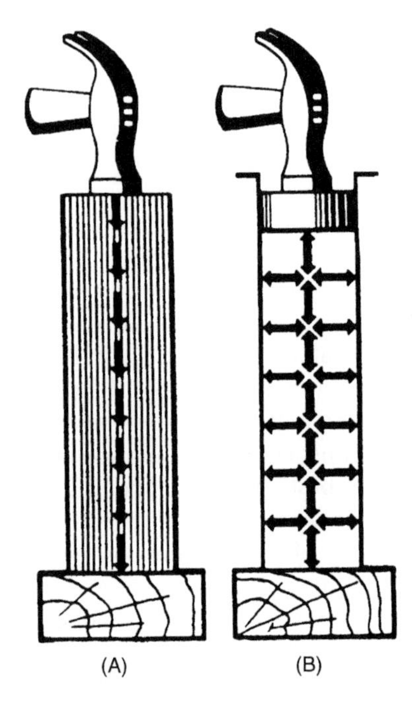

Fig. 5.1.9 Transmission of force: (A) solid; (B) fluid.

5.1.3.4 Pressure and force in hydraulic systems

According to Pascal's law, any force applied to a confined fluid is transmitted uniformly in all directions throughout the fluid regardless of the shape of the container. Consider the effect of this in the system shown in Figure 5.1.10.

If there is a resistance on the output piston and the input piston is pushed downward, a pressure is created through the fluid which acts equally at right angles to surfaces in all parts of the container. If force 1 is 100 pounds and the area of the input piston is 10 in.^{2}, then the pressure in the fluid is 10 psi:

$$\frac{100 \text{ lbs}}{10 \text{ in.}^2}$$

Note: Fluid pressure cannot be created without resistance to flow. In this case, the equipment to which the output piston is attached provides resistance. The force of resistance acts against the top of the output piston. The pressure is created in the system, by the input piston pushing on the underside of the output piston with a force of 10 psi.

In this case, the fluid column has a uniform cross section, so the area of the output piston is the same as the area of the input piston, or 10 in.^{2}. Therefore, the upward force on the output piston is 100 pounds and is equal to the force applied to the input piston. All that was accomplished in this system was to transmit the 100 pounds of force around the bend. However, this principle underlies practically all mechanical applications of hydraulics or fluid power.

At this point, you should note that since Pascal's law is independent of the shape of the container, it is not necessary that the tube connecting the two pistons has the same cross-sectional area as that of the pistons. A

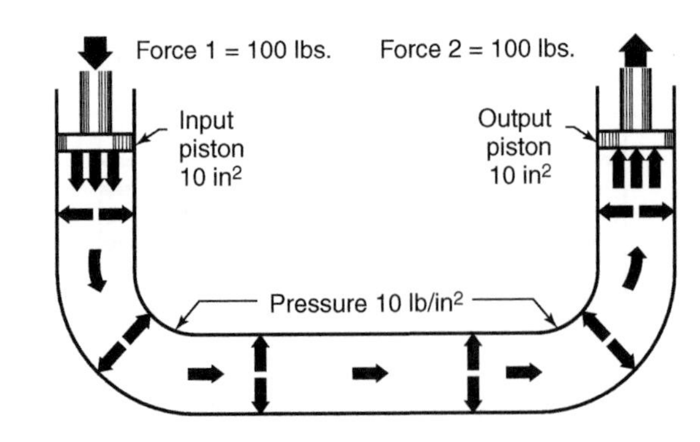

Fig. 5.1.10 Force transmitted through fluid.

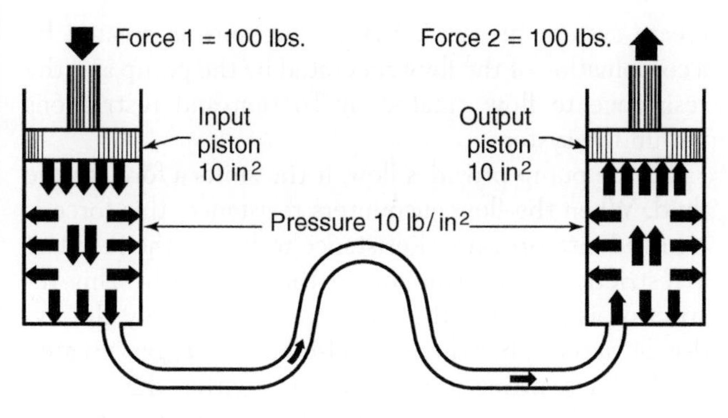

Fig. 5.1.11 Transmitting force through small pipe.

connection of any size, shape or length will do, as long as an unobstructed passage is provided. Therefore, the system shown in Figure 5.1.11, with a relatively small, bent pipe connecting two cylinders will act the same as the system shown in Figure 5.1.10.

5.1.3.5 Multiplication of force

Unlike the preceding discussion, hydraulic systems can provide mechanical advantage or a multiplication of input force. Figure 5.1.12 illustrates an example of an increase in output force. Assume that the area of the input piston is 2 in.^{2}. With a resistant force on the output piston, a downward force of 20 pounds acting on the input piston will create a pressure of 20/2 or 10 psi in the fluid. Although this force is much smaller than the force applied in Figures 5.1.10 and 5.1.11, the pressure is the same. This is because the force is applied to a smaller area.

This pressure of 10 psi acts on all parts of the fluid container, including the bottom of the output piston. The upward force on the output piston is 200 pounds (10 psi $\times$ piston area). In this case, the original force has been multiplied tenfold while using the same pressure in the fluid as before. In any system with these dimensions, the ratio of output force to input force is always 10:1, regardless of the applied force. For example, if the applied force of the input piston is 50 pounds, the pressure in the system will be 25 psi. This will support a resistant force of 500 pounds on the output piston. The system works the same in reverse.

If we change the applied force and place a 200-pound force on the output piston, Figure 5.1.12, making it the input piston, the output force on the input piston will be one-tenth the input force, or 20 pounds. Therefore, if two pistons are used in a fluid power system, the force acting on each piston is directly proportional to its area, and the magnitude of each force is the product of the pressure and the area of each piston.

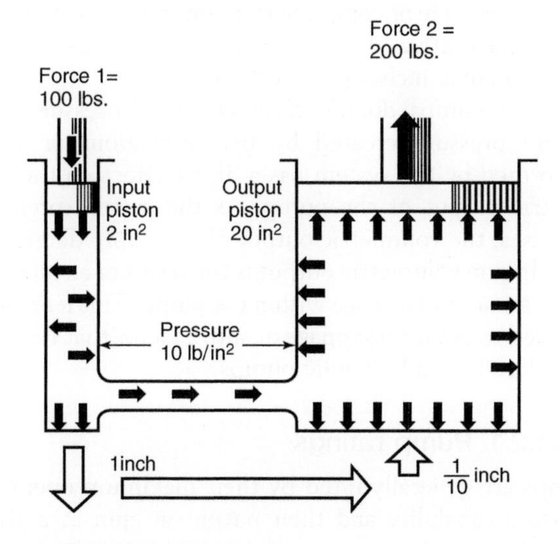

Fig. 5.1.12 Multiplication of forces.

5.1.3.5.1 Differential areas

Figure 5.1.13 is a simple example of differential pressure. The figure illustrates a single piston, with a surface area of 6 in.^{2}, attached to a piston rod, with an area of 2 in.^{2}. Without any external force applied to the end of the piston rod, an equal force, 20 psig, is applied to both sides of the piston, and will cause the piston to move to the right. This motion is the result of differential forces. Even though the input force, 20 psig, is applied to both sides of the piston the difference in area, 6 in.^{2} on the left face and 2 in.^{2} on the right face, will cause the piston to move. The opposed faces of the piston behave like two pistons acting against each other. The area of one face is the full cross-sectional area of the cylinder or 6 in.^{2}, while the area of the opposing face is the area of the piston minus the area of the piston rod, or 2 in.^{2}. This leaves an effective area of 4 in.^{2} on the piston rod side of the piston.

The force acting on the left side of the piston is equal to 20 psi $\times$ 6 in.^{2} or 120 pounds. The opposing force generated by the right side of the piston is 20 psi $\times$ 4 in.^{2}

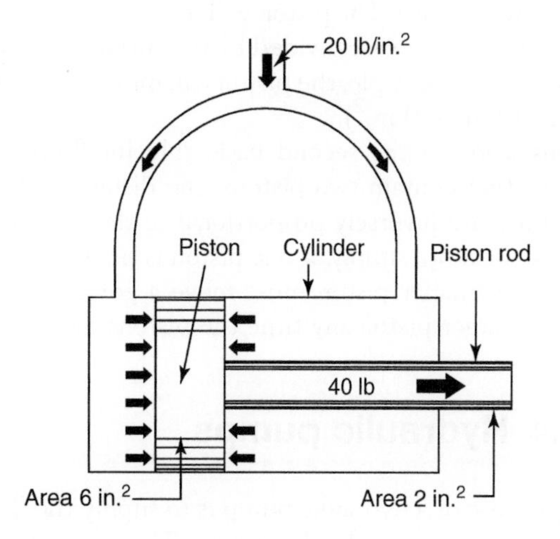

or 80 pounds. Therefore, there is a net unbalanced force of 40 pounds (120–80) acting at the right, and the piston will move in that direction.

5.1.3.5.2 Volume and distance factors

You have learned that if a force is applied to a system and the cross-sectional areas of the input and output are equal, the force on the input piston will support an equal resistant force on the output piston. The pressure of the liquid at this point is equal to the force applied to the input piston divided by the piston's area. Let us now look at what happens when a force greater than the resistance is applied to the input piston.

In the system illustrated in Figure 5.1.10, assume that the resistant force on the output piston is 100 pounds. If a force slightly greater than 100 pounds is applied to the input piston, the pressure in the system will be slightly greater than 10 psi. This increase in pressure will overcome the resistant force on the output piston. If the input piston is forced downward 1 in., the movement displaces 10 in.^{3} of fluid. The fluid must go somewhere. Since the system is closed and the fluid is practically incompressible, the fluid will move the right side of the system. Because the output piston also has a cross-sectional area of 10 in.^{2}, it will move upward 1 in. to accommodate the 10 in.^{3} of fluid. You may generalize this by saying that if two pistons in a closed system have equal cross-sectional areas and one piston is pushed and moved, the other piston will move the same distance in the opposite direction. This is because a decrease in volume in one part of the system is balanced by an equal increase in volume in another part of the system.

Apply this reasoning to the system is Figure 5.1.11. If the input piston is pushed down a distance of 1 in., the volume in the left cylinder will decrease by 2 in.^{3}. At the same time, the volume in the right cylinder will increase by 2 in.^{2}. Since the diameter of the right cylinder cannot change, the piston must move upward to allow the volume to increase. The piston will move a distance equal to the volume increase divided by the surface area of the piston. In this example, the piston will move one-tenth of an inch (2 in.^{3}/20 in.^{2}).

This leads to the second basic rule for fluid power systems that contain two pistons: the distances the pistons move are inversely proportional to the areas of the pistons. Or more simply, if one piston is smaller than the other, the smaller piston must move a greater distance than the larger piston any time the pistons move.

5.1.4 Hydraulic pumps

The purpose of a hydraulic pump is to supply the flow of fluid required by a hydraulic system. The pump does not create system pressure. System pressure is created by a combination of the flow generated by the pump and the resistance to flow created by friction and restrictions within the system.

As the pump provides flow, it transmits a force to the fluid. When the flow encounters resistance, this force is changed into pressure. Resistance to flow is the result of a restriction or obstruction in the flow path. This restriction is normally the work accomplished by the hydraulic system, but there can also be restrictions created by the lines, fittings, or components within the system. Thus, the load imposed on the system or the action of a pressure-regulating valve controls the system pressure.

5.1.4.1 Operation

A pump must have a continuous supply of fluid available to its inlet port before it can supply fluid to the system. As the pump forces fluid through the outlet port, a partial vacuum or low-pressure area is created at the inlet port. When the pressure at the inlet port of the pump is lower than the atmospheric pressure, the atmospheric pressure acting on the fluid in the reservoir must force the fluid into the pump's inlet. This is called a suction lift condition.

5.1.4.2 Performance

Pumps are normally rated by their volumetric output and discharge pressure. Volumetric output is the amount of fluid a pump can deliver to its outlet port in a certain period of time and at a given speed. Volumetric output is usually expressed in gpm.

Since changes in pump speed affect volumetric output, some pumps are rated by their displacement. Pump displacement is the amount of fluid the pump can deliver per cycle or complete rotation. Since most pumps use a rotary drive, displacement is usually expressed in terms of cubic inches per revolution.

While pumps do not directly create pressure, the system pressure created by the restrictions or work performed by the system has a direct affect on the volumetric output of the pump. As the system pressure increases, the volumetric output of the pump decreases. This drop in volumetric output is the result of an increase in the amount of leakage within the pump. This leakage is referred to as pump slippage or slip, a factor that must be considered in all hydraulic pumps.

5.1.4.2.1 Pump ratings

Pumps are generally rated by their maximum operating pressure capability and their output in gpm at a given operating speed.

5.1.4.2.2 Pressure

The manufacturer, based on reasonable service life expectancy under specified operating conditions, determines the pressure rating of a pump. It is important to note that there is no standard industry-wide safety factor in this rating. Operating at higher pressure may result in reduced pump life or damage that is more serious.

5.1.4.2.3 Displacement

The flow capacity of a pump can be expressed as its displacement per revolution or by its output in gpm. Displacement is the volume of liquid transferred in one complete cycle of pump operation. It is equal to the volume of one pumping chamber multiplied by the number of chambers that pass the outlet during one complete revolution or cycle. Displacement is expressed in cubic inches per revolution.

Most pumps that are used in hydraulic applications have a fixed displacement which cannot be changed except by replacing certain components. However, in some, it is possible to vary the size of the pumping chamber and thereby the displacement by means of external controls. Some unbalanced vane pumps and many piston units can be varied from maximum to zero delivery or even to reverse flow without modification to the pump's internal configuration.

5.1.4.2.4 Volumetric efficiency

In theory, a pump delivers an amount of fluid equal to its displacement each cycle or revolution. In reality, the actual output is reduced because of internal leakage or slippage. As pressure increases, the leakage from the outlet to the inlet or to the drain also increases and the volumetric efficiency decreases.

Volumetric efficiency is equal to the actual output divided by the theoretical output. It is expressed as a percentage:

$$\text{Efficiency} = \frac{\text{Actual output}}{\text{Theoretical output}} \times 100$$

For example, if a pump theoretically should deliver 10 gpm by delivers only 9 gpm at 1000 psig, its volumetric efficiency at that pressure is 90 per cent.

$$\text{Efficiency} = \frac{9\ \text{gpm}}{10\ \text{gpm}} \times 100 = 90\%$$

If the discharge pressure is increased, the amount of slippage will increase. If we increase the pressure in the above example, to 1500 psig, the actual output may drop to 8 gpm. Therefore, the volumetric efficiency will decrease to 80 per cent at 1500 psig.

5.1.5 Hydraulic fluids

Selection and care of the hydraulic fluid for a machine will have an important effect on how it performs and on the life of the hydraulic components. During the design of equipment that requires fluid power, many factors are considered in selecting the type of system to be used-hydraulic, pneumatic, or a combination of the two. Some of the factors required are speed and accuracy of operation, surrounding atmospheric conditions, economic conditions, availability of replacement fluid, required pressure level, operating temperature range, contamination possibilities, cost of transmission lines, limitations of the equipment, lubricity, safety to the operators, and expected service life of the equipment.

After the type of system has been selected, many of these same factors must be considered in selecting the fluid for the system. This section deals with the properties and characteristics desired of hydraulic fluids; types of hydraulic fluids; hazards and safety precautions for working with, handling, and disposing of hydraulic liquids; types and control of contamination; and sampling.

5.1.5.1 Purpose of the fluid

As a power transmission medium, the fluid must flow easily through lines and component passages. Too much resistance to flow creates considerable power loss. The fluid also must be as incompressible as possible so that action is instantaneous when the pump is started or a valve shifts.

5.1.5.1.1 Lubrication

In most hydraulic components, the hydraulic fluid provides internal lubrication. Pump elements and other wear parts slide against each other on a film of fluid, Figure 5.1.14.

For long component life, the oil must contain the necessary additives to ensure high anti-wear characteristics. Not all hydraulic oils contain these additives.

5.1.5.1.2 Sealing

In many applications, hydraulic fluid is the only seal against pressure inside system components. In Figure 5.1.14, there is no seal ring between the valve spool and body to prevent leakage from the high-pressure passage to the low-pressure passage. The close mechanical fit and viscosity of the hydraulic fluid determine leakage rate.

5.1.5.1.3 Cooling

Circulation of the hydraulic oil through lines, heat exchangers, and the walls of the reservoir, Figure 5.1.15,

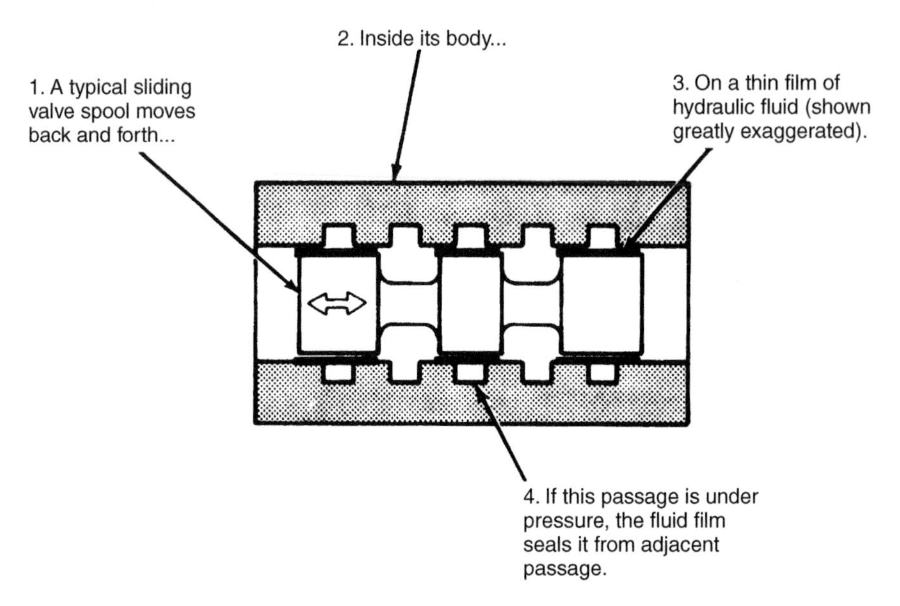
2. Inside its body...
1. A typical sliding valve spool moves back and forth...
3. On a thin film of hydraulic fluid (shown greatly exaggerated).
4. If this passage is under pressure, the fluid film seals it from adjacent passage.

Fig. 5.1.14 Fluid lubricates working parts.

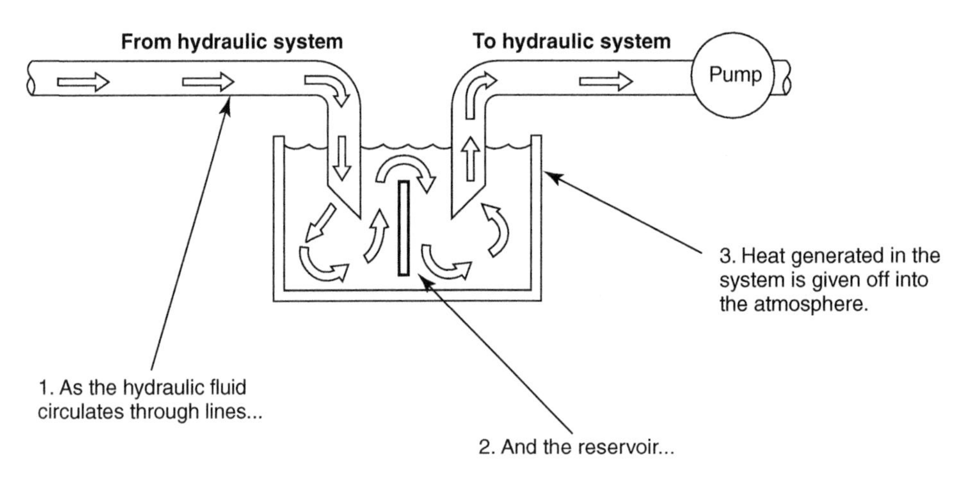
From hydraulic system
To hydraulic system
Pump
3. Heat generated in the system is given off into the atmosphere.
1. As the hydraulic fluid circulates through lines...
2. And the reservoir...

Fig. 5.1.15 Circulating oil cools system.

gives up heat that is generated within the system. Without this cooling, the heat generated by the hydraulic pump and mechanical work performed by system actuators would build up to a point that damage of system components could cause premature failure of the system.

5.1.5.2 Properties

If fluidity, the physical property of a substance that enables it to flow and incompressibility were the only properties required, any liquid that is not too thick might be used in a hydraulic system. However, a satisfactory liquid for a particular system must possess a number of other properties. The most important properties and some characteristics are discussed in the following paragraphs.

5.1.5.2.1 Density and specific gravity

The density of a substance is its weight per unit of volume. The unit of volume in the English system of measurement is 1 cubic foot or 1 ft^3. To find the density of a substance, you must know its weight and volume. You then divide its weight by its volume to find the weight per unit volume.

In equation form, this is written as

$$D = \frac{W}{V}$$

Example: The liquid that fills a certain container weighs 1496.6 pounds. The container is 4 ft long, 3 ft wide, and

2 ft deep. Its volume is 24 ft^3 (4 ft × 3 ft × 2 ft). If 24 ft^3 of this liquid weighs 1497.6 pounds, then 1 ft^3 weighs:

$$D = \frac{1497.6}{24}$$

or 62.4 pounds. Therefore, the density of the liquid is 62.4 pounds per cubic foot or 62.4 lbs/ft^3. This is the density of water at 40 °C and is usually used as the standard for comparing densities of other substances. The temperature of 40 °C was selected because water has its maximum density at this temperature. This standard temperature is used whenever the density of liquids and solids is measured. Changes in temperature will not change the weight of a substance but will change the volume of the substance by expansion or contraction, thus changing the weight per unit volume, density.

In physics, the word specific implies a ratio. Weight is the measure of the earth's attraction for a body, which is called gravity. Thus, the ratio of the weight of a unit volume of some substance to the weight of an equal volume of a standard substance, measured under standard pressure and temperature, is called specific gravity. The terms specific weight and specific density are also sometimes used to express this ratio:

$$\text{Specific gravity, Sp.Gr.} = \frac{\text{Weight of the substance}}{\text{Weight of an equal volume of water}}$$

or, the specific gravity of water is 1.0 or a density of 62.4 lbs/ft^3. If a cubic foot of a liquid weighs 68.64 pounds, then its specific gravity is 1.1 or

$$\frac{68.64}{62.4}$$

$$\text{Sp.Gr.} = \frac{\text{Density of the substance}}{\text{Density of water}}$$

Thus, the specific gravity of the liquid is the ratio of its density to the density of water. If the specific gravity of a liquid or solid is known, the density can be obtained by multiplying its specific gravity by the density of water. For example, if a hydraulic fluid has a specific gravity of 0.8, 1 ft^3 of the fluid weighs 0.8 times the weight of water (62.4 pounds) or 49.92 pounds.

Specific gravity and density are independent of the size of the sample and depend only on the substance of which it is made.

5.1.5.2.2 Viscosity

Viscosity is one of the most important properties of hydraulic fluids. It is a measure of a fluid's resistance to flow. A liquid such as gasoline which flows easily has a low viscosity, and a liquid such as tar which flows slowly has a high viscosity. The viscosity of a liquid is affected by changes in temperature and pressure. As the temperature of liquid increases, its viscosity decreases. That is, a liquid flows more easily when it is hot than when it is cold. The viscosity of a liquid will increase as the pressure on the liquid increases.

A satisfactory liquid for a hydraulic system must be thick enough to give a good seal at pumps, motors, valves, and so on. These components depend on close fits for creating and maintaining pressure. Any internal leakage through these clearances results in loss of pressure, instantaneous control, and pump efficiency. Leakage losses are greater with thinner liquids (low viscosity). A liquid that is too thin will also allow rapid wearing of moving parts, or of parts that operate under heavy loads. On the other hand, if the liquid is too thick, viscosity too high, the internal friction of the liquid will cause an increase in the liquid's flow resistance through clearances of closely fitted parts, lines, and internal passages. This results in pressure drops throughout the system, sluggish operation of the equipment, and an increase in power consumption.

5.1.5.2.2.1 Measurement of viscosity

Viscosity is normally determined by measuring the time required for a fixed volume of a fluid, at a given temperature, to flow through a calibrated orifice or capillary tube. The instruments used to measure the viscosity of a liquid are known as viscosimeters.

In decreasing order of exactness, methods of defining viscosity include absolute (poise) viscosity; kinematic viscosity in centistokes; relative viscosity in Saybolt universal seconds (SUS); and Society of Automotive Engineers (SAE) numbers.

5.1.5.2.2.2 Absolute viscosity

The resistance when moving one layer of liquid over another is the basis for the laboratory method of measuring *absolute viscosity. Poise viscosity* is defined as the force (pounds) per unit of area, in square inches, required to move one parallel surface at a speed of 1 cm-per-second past another parallel surface when the two surfaces are separated by a fluid film 1 cm thick, Figure 5.1.16. In the metric system, force is expressed in *dynes* and area in square centimeters. Poise is also the ratio between the shearing stress and the rate of shear of the fluid:

$$\text{Absolute viscosity} = \frac{\text{Shear stress}}{\text{Rate of shear}}$$

$$1 \text{ poise} = 1 \times \left(\frac{\text{Dyne second}}{\text{Square centimeter}}\right)$$

A smaller unit of absolute viscosity is the centipoise, which is one-hundredth of a poise or 1 centipoise = 0.01 poise.

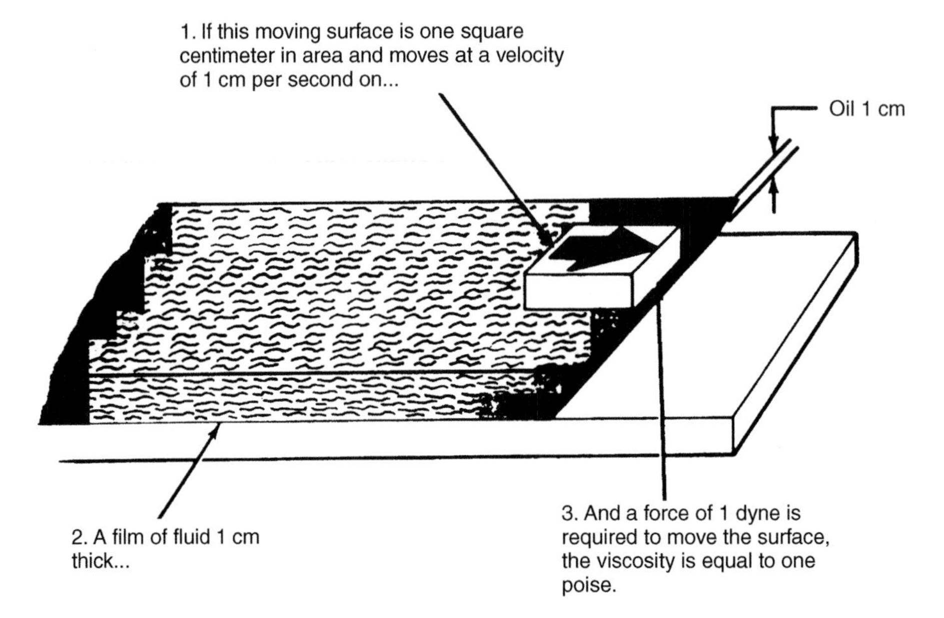
1. If this moving surface is one square centimeter in area and moves at a velocity of 1 cm per second on...
Oil 1 cm
2. A film of fluid 1 cm thick...
3. And a force of 1 dyne is required to move the surface, the viscosity is equal to one poise.

Fig. 5.1.16 Measuring absolute viscosity.

5.1.5.2.2.3 Kinematic viscosity

The concept of kinematic viscosity is the outgrowth of the use of a head of liquid to produce a flow through a capillary tube. The coefficient of absolute viscosity, when divided by the density of the liquid is called the kinematic viscosity. In the metric system, the unit of viscosity is called the *stoke* and it has the units of centimeters squared per second. One one-hundredth of a stoke is a centistoke. The relationship between absolute and kinematic viscosity can be stated as

$$\text{Centipoise} = \text{Centistoke} \times \text{Density}$$

or

$$\text{Centistoke} = \frac{\text{Centipoise}}{\text{Density}}$$

5.1.5.2.2.4 SUS viscosity

For most practical purposes, it will serve to know the relative viscosity of the fluid. Relative viscosity is determined by timing the flow of a given quantity of the hydraulic fluid through a standard orifice at a given temperature. There are several methods in use. The most acceptable method in the United States is the *Saybolt viscosimeter*, Figure 5.1.17.

The time it takes for the measured quantity of liquid to flow through the orifice is measured with a stopwatch. The viscosity in SUS equals the elapsed time.

Obviously, a thick liquid will flow slowly, and the SUS viscosity will be higher than for a thin liquid, which flows faster. Since oil becomes thicker at low temperatures and thins when warmed, the viscosity must be expressed as a specific SUS at a given temperature. Tests are usually conducted at either 100 °F or 210 °F.

For industrial applications, hydraulic oil viscosity is typically approximately 150 SUS at 100 °F. It is a general rule that the viscosity should never go below 45 SUS or above 4000 SUS, regardless of temperature. Where temperature extremes are encountered, the fluid should have a high viscosity index.

5.1.5.2.2.5 SAE number

SAE numbers have been established by the Society of Automotive Engineers to specify ranges of SUS viscosities of oils at SAE test temperatures. Winter numbers (5W, 10W, 20W) are determined by tests at 0 °F. Summer numbers (20W, 30W, etc.) designate the SUS range at 210 °F. Table 5.1.1 is a chart of the temperature ranges.

The following formulas may be used to convert centistokes (cSt units) to approximate SUS units. For SUS values between 32 and 100:

$$\text{cST} = 0.226 \times \text{SSU} - \frac{195}{\text{SSU}}$$

For SUS values greater than 100

$$\text{cST} = 0.226 \times \text{SSU} - \frac{135}{\text{SSU}}$$

Although the viscometers discussed above are used in laboratories, there are other viscometers in the supply system that are available for local use. These viscometers can be used to test the viscosity of hydraulic fluids

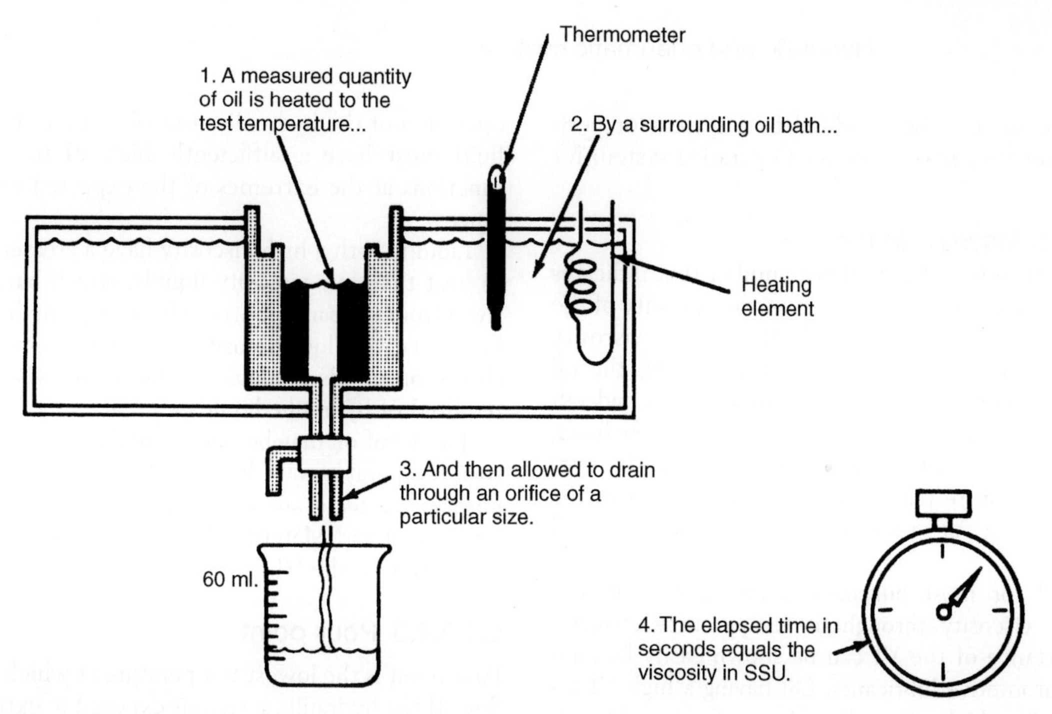

Fig. 5.1.17 Saybolt viscosimeter.

Table 5.1.1 SAE viscosity numbers for crankcase oils

		Viscosity range[b]			
		At 0 °F		At 210 °F	
SAE number	Viscosity units[a]	Minimum	Maximum	Minimum	Maximum
5W			Less than		
	Centipoise	–	1200	–	–
	Centistokes	–	1300	–	–
	SUS	–	6000	–	–
10W			Less than		
	Centipoise	1200[c]	2400	–	–
	Centistokes	1300	2600	–	–
	SUS	6000	12,000	–	–
20W			Less than		
	Centipoise	2400[d]	9600	–	–
	Centistokes	2600	10,500	–	–
	SUS	12,000	48,000	–	–
20	Centistokes	–	–	5.7	Less than 9.6
	SUS	–	–	45	58
30	Centistokes	–	–	9.6	Less than 12.9
	SUS	–	–	58	70
40	Centistokes	–	–	12.9	Less than 16.8
	SUS	–	–	70	85
50	Centistokes	–	–	16.8	Less than 22.7
	SUS	–	–	85	110

[a] The official values in this classification are based upon 210 °F viscosity in centistokes (ASTM D 445) and 0 °F viscosities in centipoise (ASTM D260-2). Approximate values in other units of viscosity are given for information only. The approximate values at 0 °F were calculated using an assumed oil density of 0.9 gm/cc at that temperature.
[b] The viscosity of all oils included in this classification shall not be less than 3.0 centistokes at 210 °F (39 SUS).
[c] Minimum viscosity at 0 °F may be waived provided viscosity at 210 °F is not below 4.2 centistokes (40 SUS).
[d] Minimum viscosity at 0 °F may be waived provided viscosity at 210 °F is not below 5.7 centistokes (45 SUS).

either prior to their being added to a system or periodically after they have been in an operating system for a while.

5.1.5.2.2.6 Viscosity index

The viscosity index, VI, of oil is a number that indicates the effect of temperature changes on the viscosity of the oil. A low VI signifies a relatively large change of viscosity with changes of temperature. In other words, the oil becomes extremely thin at high temperatures and extremely thick at low temperatures. On the other hand, a high VI signifies relatively little change in viscosity over a wide temperature range. Figure 5.1.18 illustrates the relative change of viscosity with changes in oil temperature.

Ideal oil for most purposes is one that maintains a constant viscosity throughout temperature changes. The importance of the VI can be shown easily by considering automotive lubricants. Oil having a high VI resists excessive thickening when the engine is cold and, consequently, promotes rapid starting and prompt circulation; it resists excessive thinning when the motor is hot and thus provides full lubrication and prevents excessive oil consumption.

Another example of the importance of the VI is the need for a high viscosity index hydraulic oil for military aircraft, since hydraulic control systems may be exposed to temperatures ranging from below −65 °F at high altitudes to over 100 °F on the ground. For the proper operation of the hydraulic control system, the hydraulic fluid must have a sufficiently high VI to perform its functions at the extremes of the expected temperature range.

Liquids with a high viscosity have a greater resistance to heat than low viscosity liquids, which have been derived from the same source. The average hydraulic liquid has a relatively low viscosity. Fortunately, there is a wide choice of liquids available for use in the viscosity range required of hydraulic liquids.

The VI of oil may be determined if its viscosity at any two temperatures is known. Tables based on a large number of tests are issued by the American Society for Testing and Materials (ASTM). These tables permit calculation of the VI from known viscosity.

5.1.5.2.3 Pour point

Pour point is the lowest temperature at which a fluid will flow. If the hydraulic system is exposed to extremely low temperatures, it is a very important specification. For a rule of thumb, the pour point should be 20 °F below the lowest temperature that may be encountered.

5.1.5.2.4 Lubricating power

If motion takes place between surfaces in contact, friction tends to oppose the motion. When pressure forces the liquid of a hydraulic system between the surfaces of moving parts, the liquid spreads out into a thin film

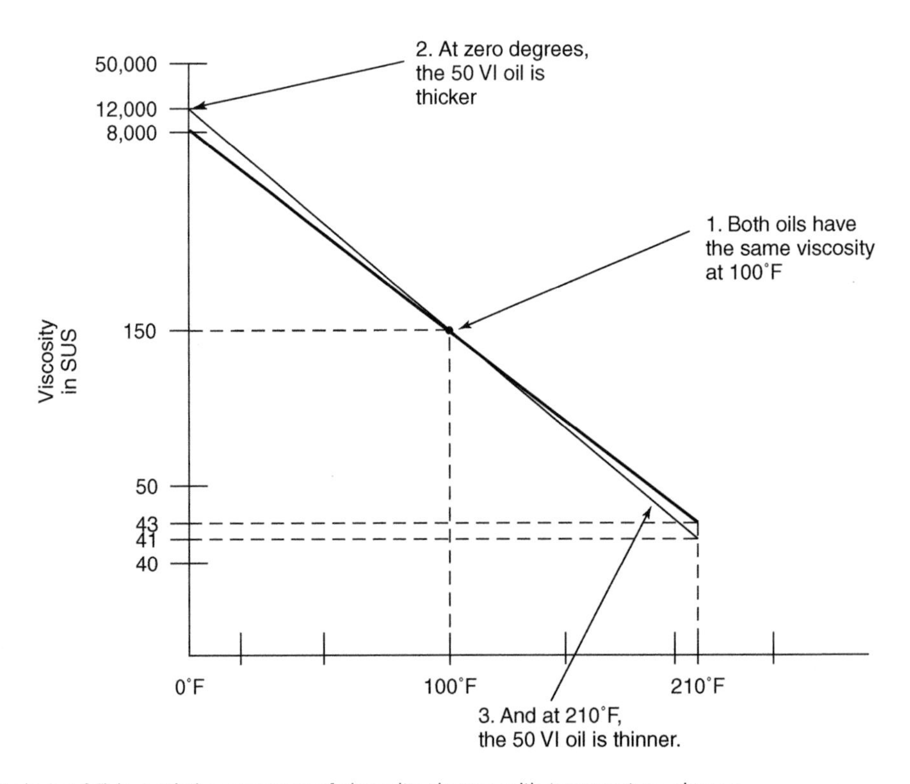

Fig. 5.1.18 Viscosity Index (VI) is a relative measure of viscosity change with temperature change.

which enables the parts to move more freely. Different liquids, including oils, vary greatly not only in their lubricating ability but also in film strength. Film strength is the capability of a liquid to resist being wiped or squeezed out from between the surfaces when spread out in an extremely thin layer. A liquid will no longer lubricate if the film breaks down, since the motion of part against part wipes the metal clean of liquid.

Lubricating power varies with temperature changes; therefore the climatic and working conditions must enter into the determination of the lubricating qualities of a liquid. Unlike viscosity, which is a physical property, the lubricating power and film strength of a liquid is directly related to its chemical nature. Lubricating qualities and film strength can be improved by the addition of certain chemical agents.

5.1.5.2.5 Chemical stability

Chemical stability is another property which is exceedingly important in the selection of a hydraulic liquid. It is defined as the liquid's ability to resist oxidation and deterioration for long periods. All liquids tend to undergo unfavorable changes under severe operating conditions. This is the case, for example, when a system operates for a considerable period at high temperatures.

Excessive temperatures, especially extremely high temperatures, have a great effect on the life of a liquid. The temperature of the liquid in the reservoir of an operating hydraulic system does not always indicate the operating conditions throughout the system. Localized hot spots occur on bearings, gear teeth, or at other points where the liquid under pressure is forced through small orifices. Continuous passage of the liquid through these points may produce local temperatures high enough to carbonize the liquid or turn it into sludge; yet the liquid in the reservoir may not indicate an excessively high temperature.

Liquids may break down if exposed to air, water, salt, or other impurities, especially if they are in constant motion or subjected to heat. Some metals, such as zinc, lead, brass, and copper, have undesirable chemical reactions with certain liquids.

These chemical reactions result in the formation of sludge, gums, carbon, or other deposits, which clog openings, cause valves and pistons to stick or leak, and give poor lubrication to moving parts. Once a small amount of sludge or other deposits is formed, the rate of formation generally increases more rapidly. As these deposits are formed, certain changes in the physical and chemical properties of the liquid take place. The liquid usually becomes darker, the viscosity increases, and damaging acids are formed.

The extent to which changes occur in different liquids depends on the type of liquid, type of refining, and whether it has been treated to provide further resistance to oxidation. The stability of liquids can be improved by the addition of oxidation inhibitors. Inhibitors selected to improve stability must be compatible with the other required properties of the liquid.

5.1.5.2.6 Freedom from acidity

An ideal hydraulic liquid should be free from acids that cause corrosion of the metals in the system. Most liquids cannot be expected to remain completely non-corrosive under severe operating conditions. When new, the degree of acidity of a liquid may be satisfactory; but after use, the liquid may tend to become corrosive as it begins to deteriorate.

Many systems are idle for long periods after operating at high temperatures. This permits moisture to condense in the system, resulting in rust formation. Certain corrosion- and rust-preventive additives are added to hydraulic liquids. Some of these additives are effective only for a limited period. Therefore, the best procedure is to use the liquid specified for the system for the time specified by the system manufacturer and to protect the liquid and the system as much as possible from contamination by foreign matter, from abnormal temperatures, and from misuse.

5.1.5.2.7 Flashpoint

Flashpoint is the temperature at which a liquid gives off vapor in sufficient quantity to ignite momentarily or flash when a flame is applied. A high flashpoint is desirable for hydraulic liquids because it provides good resistance to combustion and a low degree of evaporation at normal temperatures. Required flashpoint minimums vary from 300 °F for the lightest oils to 510 °F for the heaviest oils.

5.1.5.2.8 Firepoint

Fire point is the temperature at which a substance gives off vapor in sufficient quantity to ignite and continue to burn when exposed to a spark or flame. Like flashpoint, a high fire point is required of desirable hydraulic liquids.

5.1.5.2.9 Minimum toxicity

Toxicity is defined as the quality, state, or degree of being toxic or poisonous. Some liquids contain chemicals that are a serious toxic hazard. These toxic or poisonous chemicals may enter the body through inhalation, by absorption through the skin, or through the eyes or the mouth. The result is sickness and, in some cases, death. Manufacturers of hydraulic liquids strive to produce suitable liquids that contain no toxic chemicals and, as a result, most hydraulic liquids are free of harmful chemicals. Some fire-resistant liquids are toxic, and suitable protection and care in handling must be provided.

5.1.5.2.10 Density and compressibility

A fluid with a specific gravity of less than 1.0 is desired when weight is critical, although with proper system design, a fluid with a specific gravity greater than 1.0 can be tolerated. Where avoidance of detection by military units is desired, a fluid that sinks rather than rises to the surface of the water is desirable. Fluids having a specific gravity greater than 1.0 are desired, as leaking fluid will sink, allowing the vessel with the leak to remain undetected.

Under extreme pressure, a fluid may be compressed up to 7 per cent of its original volume. Highly compressible fluids produce sluggish system operation. This does not present a serious problem in small, low-speed operations, but it must be considered in the operating instructions.

5.1.5.2.11 Foaming tendencies

Foam is an emulsion of gas bubbles in the fluid. In a hydraulic system, foam results from compressed gases in the hydraulic fluid. A fluid under high pressure can contain a large volume of air bubbles. When this fluid is depressurized, as when it reaches the reservoir, the gas bubbles in the fluid expand and produce foam. Any amount of foaming may cause pump cavitation and produce poor system response and sponge control. Therefore, defoaming agents are often added to fluids to prevent foaming. Minimizing air in fluid systems is discussed later in this chapter.

5.1.5.2.12 Cleanliness

Cleanliness in hydraulic systems has received considerable attention recently. Some hydraulic systems, such as aerospace hydraulic systems, are extremely sensitive to contamination. Fluid cleanliness is of primary importance because contaminants can cause component malfunction, prevent proper valve seating, cause wear in components, and may increase the response time of servo valves. Fluid contaminants are discussed later in this chapter.

The inside of a hydraulic system can only be kept as clean as the fluid added to it. Initial fluid cleanliness can be achieved by observing stringent cleanliness requirements (discussed later in this chapter) or by filtering all fluid added to the system.

5.1.5.3 Contamination

Hydraulic fluid contamination may be described as any foreign material or substance whose presence in the fluid is capable of adversely affecting system performance or reliability. It may assume many different forms, including liquids, gases, and solid matter of various composition, sizes, and shapes. Solid matter is the type most often found in hydraulic systems and is generally referred to as particulate contamination. Contamination is always present to some degree, even in new, unused fluid, but must be kept below a level that will adversely affect system operation. Hydraulic contamination control consists of requirements, techniques, and practices necessary to minimize and control fluid contamination.

5.1.5.3.1 Classification

There are many types of contaminants which are harmful to hydraulic systems and liquids. These contaminants may be divided into two different classes – particulate and fluid.

Particulate contamination: This class of contaminants includes organic, metallic solid, and inorganic solid contaminants. These contaminants are discussed in the following paragraphs.

Organic: Wear, oxidation, or polymerization produces organic solids or semisolids found in hydraulic systems. Minute particles of O-rings, seals, gaskets, and hoses are present, due to wear or chemical reactions. Synthetic products, such as neoprene, silicones, and hypalon, though resistant to chemical reaction with hydraulic fluids, produce small wear particles. Oxidation of hydraulic fluids increases with pressure and temperature, although antioxidants are blended into hydraulic fluids to minimize such oxidation. The ability of a hydraulic fluid to resist oxidation or polymerization in service is defined as its oxidation stability. Oxidation products appear as organic acids, asphaltics, gums, and varnishes. These products combine with particles in the hydraulic fluid to form sludge. Some oxidation products are oil soluble and cause the hydraulic fluid to increase in viscosity; other oxidation products are not oil soluble and form sediment.

Metallic solids: Metallic contaminants are usually present in a hydraulic system and may range in size from microscopic particles to particles readily visible to the naked eye. These particles are the result of wearing and scoring of bare metal parts and plating materials, such as silver and chromium. Although practically all metals commonly used for part fabrication and plating may be found in hydraulic fluids, the major metallic materials found are ferrous, aluminum, and chromium particles. Because of their continuous high-speed internal movement, hydraulic pumps usually contribute most of the metallic particulate contamination present in hydraulic systems. Metal particles are also produced by other hydraulic system components, such as valves and actuators, due to body wear and the chipping and wearing away of small pieces of metal plating materials.

Inorganic solids: This contaminant group includes dust, paint particles, dirt, and silicates. Glass particles from glass bead peening and blasting may also be found as contaminants. Glass particles are very undesirable contaminants due to their abrasive effect on synthetic rubber seals and the very fine surfaces of critical moving parts. Atmospheric dust, dirt, paint particles, and other materials are often drawn into hydraulic systems from external sources. For example, the wet piston shaft of a hydraulic actuator may draw some of these foreign materials into the cylinder past the wiper and dynamic seals, and the contaminant materials are then dispersed in the hydraulic fluid. Contaminants may also enter the hydraulic fluid during maintenance when tubing, hoses, fittings, and components are disconnected or replaced. It is therefore important that all exposed fluid ports be sealed with approved protective closures to minimize such contamination.

5.1.5.3.2 Fluid contamination

Air, water, solvent, and other foreign fluids are in the class of fluid contaminants.

Air: Hydraulic fluids are adversely affected by dissolved, entrained, or free air. Air may be introduced through improper maintenance or because of system design. Any maintenance operation that involves breaking into the hydraulic system, such as disconnecting or removing a line or component, will invariably result in some air being introduced into the system. This source of air can and must be minimized by pre-filling replacement components with new filtered fluid prior to their installation. Failing to pre-fill a filter-element bowl with fluid is a good example of how air can be introduced into the system. Although pre-filling will minimize introduction of air, it is still important to vent the system where venting is possible.

Most hydraulic systems have built-in sources of air. Leaky seals in gas-pressurized accumulators and reservoirs can feed gas into a system faster than it can be removed, even with the best of maintenance. Another lesser-known but major source of air is air that is sucked into the system past actuator piston rod seals. This occurs when the piston rod that is stroked by some external means while the actuator itself is not pressurized.

Water: Water is a serious contaminant of hydraulic systems. Hydraulic fluids are adversely affected by dissolved, emulsified, or free water. Water contamination may result in the formation of ice, which impedes the operation of valves, actuators, and other moving parts. Water can also cause the formation of oxidation products and corrosion of metallic surfaces.

Solvents: Solvent contamination is a special form of foreign fluid contamination in which the original contaminating substance is a chlorinated solvent. Chlorinated solvents or their residues may, when introduced into a hydraulic system, react with any water present to form highly corrosive acids.

Chlorinated solvents, when allowed to combine with minute amounts of water often found in operating hydraulic systems, change chemically into hydrochloric acids. These acids then attack internal metallic surfaces in the system, particularly those that are ferrous, and produce a severe rust-like corrosion.

Foreign Fluids: Foreign fluids other than water and chlorinated solvents can seriously contaminate hydraulic systems. This type of contamination is generally a result of lube oil, engine fuel, or incorrect hydraulic fluid being introduced inadvertently into the system during servicing. The effects of such contamination depend on the contaminant, the amount in the system, and how long it has been present.

Note: It is extremely important that the different types of hydraulic fluids are not mixed in one system. If different types of hydraulic fluids are mixed, the characteristics of the fluid required for a specific purpose are lost. Mixing the different types of fluids usually results in a heavy, gummy deposit that will clog passages and require a major cleaning. In addition, seals and packing installed for use with one fluid are usually not compatible with other fluids and damage to the seals will result.

5.1.5.4 Contamination control

Maintaining hydraulic fluid within allowable contamination limits for both water and particulate matter is crucial to the care and protection of hydraulic equipment. Filters will provide adequate control of the particular contamination problem during all normal hydraulic system operations if the filtration system is installed properly and filter maintenance is performed properly. Filter maintenance includes changing elements at proper intervals.

Control of the size and amount of contamination entering the system from any other source is the responsibility of the personnel who service and maintain the equipment. During installation, maintenance, and repair of hydraulic equipment, the retention of cleanliness of the system is of paramount importance for subsequent satisfactory performance.

The following maintenance and servicing procedures should be adhered to at all times to provide proper contamination control:

1. All tools and the work area (workbenches and test equipment) should be kept in a clean, dirt-free condition.
2. A suitable container should always be provided to receive the hydraulic liquid that is spilled during component removal or disassembly.

   Note: The reuse of drained hydraulic liquid is prohibited in most hydraulic systems. In some large-capacity systems the reuse of fluid is permitted.

When liquid is drained from these systems for reuse, it must be stored in a clean and suitable container. The liquid must be strained and/or filtered when it is returned to the system reservoir.

3. Before hydraulic lines or fittings are disconnected, the affected area should be cleaned with an approved dry-cleaning solvent.
4. All hydraulic lines and fittings should be capped or plugged immediately after disconnection.
5. Before any hydraulic components are assembled, their parts should be washed with an approved dry-cleaning solvent.
6. After the parts have been cleaned in dry-cleaning solvent, they should be dried thoroughly with clean, low-lint cloths and lubricated with the recommended preservative or hydraulic liquid before assembly.

   Note: Only clean, low-lint type II or I cloths as appropriate should be used to wipe or dry component parts.
7. All packing and gaskets should be replaced during the assembly procedures.
8. All parts should be connected with care to avoid stripping metal slivers from threaded areas. All fittings and lines should be installed and torqued according to applicable technical instructions.
9. All hydraulic servicing equipment should be kept clean and in good operating condition.

Some hydraulic fluid specifications contain particle contamination limits that are so low that the products are packaged under clean room conditions. Very slight amounts of dirt, rust, and metal particles will cause them to fail the specification limit for contamination. Since these fluids are usually all packaged in hermetically sealed containers, the act of opening a container may allow more contaminants into the fluid than the specification allows. Therefore, extreme care should be taken in the handling of these fluids. In opening the container for use, observation, or tests, it is extremely important that the container be opened and handled in a clean environment. The area of the container to be opened should be flushed with filtered solvent (petroleum ether or isopropyl alcohol), and the device used for opening the container should be thoroughly rinsed with filtered solvent. After the container is opened, a small amount of the material should be poured from the container and disposed of prior to pouring the sample for analysis. Once a container is opened, if the contents are not totally used, the unused portion should be discarded. Since the level of contamination of a system containing these fluids must be kept low, maintenance on the system's components must be performed in a clean environment commonly known as a controlled-environment work center.

5.1.5.5 Hydraulic fluid sampling

The condition of a hydraulic system, as well as its probable future performance, can best be determined by analyzing the operating fluid. Of particular interest are any changes in the physical and chemical properties of the fluid and excessive particulate or water contamination, either of which indicates impending trouble.

Excessive particulate contamination of the fluid indicates that the filters are not keeping the system clean. This can result from improper filter maintenance, inadequate filters, or excessive ongoing corrosion and wear.

1. All samples should be taken from circulating systems, or immediately upon shutdown, while the hydraulic fluid is within 5 °C (9 °F) of normal system operating temperature. Systems not up to temperature may provide non-representative samples of system dirt and water content, and such samples should either be avoided or so indicated on the analysis report. The first oil coming from the sampling point should be discarded, since it can be very dirty and does not represent the system. As a rule, a volume of oil equivalent to one to two times the volume of oil contained in the sampling line and valve should be drained before the sample is taken.
2. Ideally, the sample should be taken from a valve installed specifically for sampling. When sampling valves are not installed the taking of samples from locations where sediment or water can collect, such as dead ends of piping, tank drains, and low points of large pipes and filter bowls, should be avoided if possible. If samples are taken from pipe drains, sufficient fluid should be drained before the sample is taken to ensure that the sample actually represents the system. Samples are not to be taken from the tops of reservoirs or other locations where the contamination levels are normally low.
3. Unless otherwise specified, a minimum of one sample should be taken for each system located wholly within one compartment. For ship's systems extending into two or more compartments, a second sample is required. An exception to this requirement is submarine external hydraulic systems, which require only one sample. Original sample points should be labeled and the same sample points used for successive sampling. If possible, the following sampling locations should be selected:
   - (a) A location that provides a sample representative of fluid being supplied to system components;
   - (b) A return line as close to the supply tank as practical but upstream of any return line filter;
   - (c) For systems requiring a second sample, a location as far from the pump as practical.

Operation of the sampling point should not introduce any significant amount of external contaminants into the collected fluid.

5.1.6 Reservoirs, strainers, filters, and accumulators

Fluid power systems must have a sufficient and continuous supply of uncontaminated fluid to operate efficiently. This section covers: hydraulic reservoirs; various types of strainers and filters; and accumulators that are typically installed in fluid power systems.

5.1.6.1 Reservoirs

A hydraulic system must have a reserve of fluid in addition to that contained in the pumps, actuators, pipes and other components of the system. This reserve fluid must be readily available to make up losses of fluid from the system, to make up for compression of fluid under pressure, and to compensate for the loss of volume as the fluid cools. This extra fluid is contained in a tank usually called a reservoir. A reservoir may sometimes be referred to as a sump tank, service tank, operating tank, supply tank, or base tank.

In addition to providing storage for the reserve fluid needed for the system, the reservoir acts as a radiator for dissipating heat from the fluid. It also acts as a settling tank where heavy particles of contamination may settle out of the fluid and remain harmlessly on the bottom until removed by cleaning or flushing the reservoir. Also, the reservoir allows entrained air to separate from the fluid.

Most reservoirs have a capped opening for filling, an air vent, an oil level indicator or dipstick, a return line connection, a pump inlet or suction line connection, a drain line connection, and a drain plug (see Figure 5.1.19).

The inside of the reservoir generally will have baffles to prevent excessive sloshing of the fluid and to put a partition between the fluid return line and the pump suction or inlet line. The partition forces the returning fluid to travel farther around the tank before being drawn back into the active system through the pump inlet line. This aids in settling the contamination and separating air entrained in the fluid.

Large reservoirs are desirable for cooling. A large reservoir also reduces re-circulation, which helps settle contamination, and separates entrained air. As a rule of

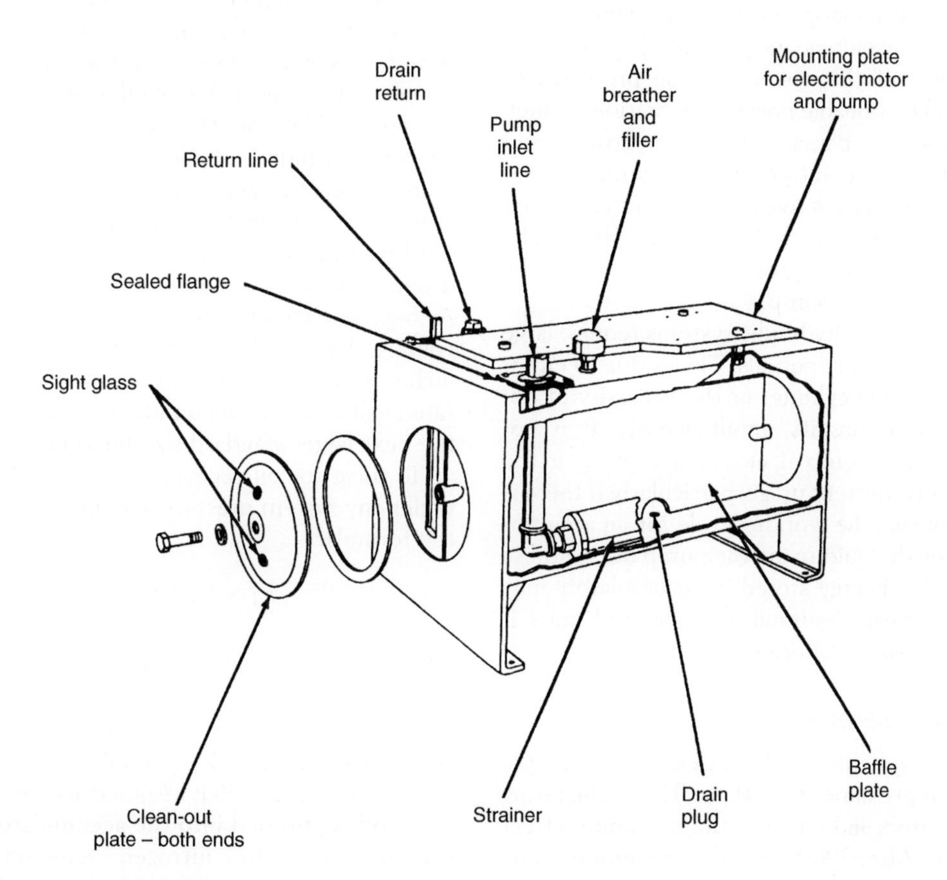

Fig. 5.1.19 Non-pressurized reservoir.

thumb, the ideal reservoir should be two to three times the pump outlet per minute. However, due to space limitations in mobile and aerospace systems, the benefits of a large reservoir may have to be sacrificed. But they must be large enough to accommodate thermal expansion of the fluid and changes in fluid level due to system operation.

5.1.6.2 Accumulators

An accumulator is a pressure storage reservoir in which hydraulic fluid is stored under pressure from an external source. The storage of fluid under pressure serves several purposes in hydraulic systems.

In some hydraulic systems, it is necessary to maintain the system pressure within a specific pressure range for long periods. It is very difficult to maintain a closed system without some leakage, either external or internal. Even a small leak can cause a decrease in pressure. By using an accumulator, leakage can be compensated for and the system pressure can be maintained within acceptable range for extended periods. Accumulators also compensate for thermal expansion and contraction of the liquid due to variations in temperature or generated heat.

A liquid flowing at a high velocity in a pipe will create a backward surge when stopped suddenly. This sudden stoppage causes an instantaneous pressure 2–3 times the operating pressure of the system. These pressures or shocks produce objectionable noise and vibrations, which can cause considerable damage to piping, fittings, and components. The incorporation of an accumulator enables such shocks and surges to be absorbed or cushioned by the entrapped gas, thereby reducing their effects. The accumulator also dampens pressure surges caused by pulsing delivery from the pump.

There are times when hydraulic systems require large volumes of liquid for short periods. This is due to either the operation of a large cylinder or the necessity of operating two or more circuits simultaneously. It is not economical to install a pump of such large capacity in the system for only intermittent usage, particularly if there is sufficient time during the working cycle for an accumulator to store enough liquid to aid the pump during these peak demands. The energy stored in accumulators may be also used to actuate hydraulically operated units if normal hydraulic system failure occurs.

5.1.6.2.1 Accumulator sizing

Most accumulator systems should be designed to operate at a maximum oil pressure of 3000 psi. This is the rating of most accumulators and will give the maximum effect for the least cost. Also, 3000 psi is the maximum rating for most hydraulic valves.

A rule of thumb for the nitrogen pre-charged level is one-half the maximum oil pressure. This is acceptable for most applications. The pre-charge should be replenished when it falls to one-third the maximum hydraulic oil pressure. On a 3000-psi hydraulic system, initial pre-charge should be 1500 psi and replenishment level of 1000 psi. Most applications will tolerate a wide variation in pre-charge pressure.

Accumulators are catalog-rated by gas volume when all oil is discharged and usually rated in quarts or gallons (i.e. 1 US gallon = 231 in.^{3}). The amount of oil which can be stored is approximately half the gas volume. Only a part of the stored oil can be used each cycle because the oil pressure decreases as oil is discharged. The problem in selecting accumulator size is to have sufficient capacity so system pressure, at the end of the discharge, does not fall below a value which will do the job.

For illustration, we are using an application on which accumulator oil will be used on the extension stroke of a cylinder to supplement the oil delivery from a pump, to increase speed. Retraction will be by pump volume alone. A fully charged accumulator system pressure of 3000 psi is assumed.

First, select the cylinder bore for sufficient force not only at 3000 psi but also at some selected lower pressure to which it will be allowed to fall during discharge. Next, calculate the number of cubic inches of oil required to fill the cylinder cavity during its extension stroke. Using the time, in seconds, allowed for the full extension stroke, calculate the cubic inches of oil which can be obtained for the pump alone. Subtract the calculated pump volume from the cylinder volume to find the volume of oil required from the accumulator. Use Table 5.1.2 to find how many cubic inches of oil would be supplied from a 1-gallon size accumulator before its terminal pressure dropped below the minimum acceptable pressure level. A 5-gallon accumulator would supply 5 times this volume. Finally, divide this figure into the total cubic inches needed for the application. This is the minimum rated gallon size of accumulator capacity. Select at least the next larger standard size for your application.

To solve for oil recovery from any size accumulator, under any system pressure and any pre-charge level, use the formula:

$$D = (0.95 \times P_1 \times V_1 \div P_2) - (0.95 \times P_1 \times V_1 \div P_3)$$

where D is cubic inches of oil discharge; P_1 is pre-charge pressure in psi; P_2 is system pressure after volume D has been discharged; P_3 is maximum system pressure at full accumulator charge; V_1 is catalog rated gas volume, in cubic inches; and 0.95 is assumed accumulator efficiency.

As oil is pumped into the accumulator, compressing the nitrogen, the nitrogen temperature increases (Charles' law). Therefore, the amount of oil stored will

Table 5.1.2 Accumulator selection table

Minimum acceptable system PSI	Cubic inch discharge
2700	12
2600	17
2500	22
2400	27
2300	33
2200	40
2100	46
2000	55
1900	63
1800	73
1700	84
1600	96
1500	109

not be quite as much as calculated using Boyle's law unless sufficient time is allowed for the accumulator to cool to atmospheric temperature. Likewise, when oil is discharged, the expanding nitrogen is cooled. So, the discharge volume will not be quite as high as that calculated using Boyle's law. In Table 5.1.2 and the formula above, an allowance of 5 per cent has been included as a safety factor. After making a size calculation from the table, allow enough extra capacity for contingencies.

5.1.6.3 Filtration

Clean hydraulic fluid is essential for proper operation and acceptable component life in all hydraulic systems. While every effort must be made to prevent contaminants from entering the system, contaminants that do find their way into the system must be removed. Filtration devices are installed at key points in fluid power systems to remove the contaminants that enter the system along with those that are generated during normal operations of the system.

The filtering devices used in hydraulic systems are commonly referred to as strainers and filters. Since they share a common function, the terms are often used interchangeably. Generally, devices used to remove large particles of foreign matter from hydraulic systems are referred to as strainers, while those used to remove the smallest particles are called filters.

5.1.6.3.1 Strainers

Strainers are used primarily to catch only very large particles and will be found in applications where this type of protection is required. Most hydraulic systems have a strainer in the reservoir at the inlet to the suction line of the pump. A strainer is used in lieu of a filter to reduce its chance of being clogged and starving the pump. However, since this strainer is located in the reservoir, its maintenance is frequently neglected. When heavy dirt and sludge accumulate on the suction strainer, the pump soon begins to cavitate. Pump failure follows quickly.

5.1.6.3.2 Filters

The most common device installed in hydraulic systems to prevent foreign particles and contaminations from remaining in the system are referred to as filters. They may be located in the reservoir, in the return line, in the pressure line, or in any other location in the system where the designer of the system decides they are needed to safeguard the system against impurities.

Filters are classified as full-flow or proportional flow. In full-flow types of filters, all of the fluid that enters the filter passes through the filtering element, while in proportional types only a portion of the fluid passes through the element.

5.1.6.3.3 Filter elements

Filter elements may be divided into two classes: surface and depth. Surface filters are made of closely woven fabric or treated paper with a uniform pore size. Fluid flows through the pores of the filter material and contaminants are stropped on the filter's surface. This type of filter element is designed to prevent the passage of a high percentage of solids of a specific size.

Depth filters on the other hand are composed of layers of fabric or fibers, which provide many tortuous paths for the fluid to flow through. The pores or passages must be larger than the rated size of the filter if particles are to be retained in the depth of the media rather than on the surface.

Filter elements may be of the 5-micron, woven mesh, micronic, porous metal, or magnetic type. The micronic and 5-micron elements have non-cleanable filter media and are disposed of when they are removed. Porous metal, woven mesh, and magnetic filter elements are usually designed to be cleaned and reused.

5.1.6.4 Heat exchangers

The conversion of hydraulic force to mechanical work generates excessive heat. This heat must be removed from the hydraulic fluid to prevent degradation of the fluid and possible damage to system components.

Table 5.1.3 Heat and power equivalents

1 Horsepower (HP)	2545 BTU per hour
1 Horsepower (HP)	42.4 BTU per minute
1 British thermal unit per hour (BTU/hr)	0.000393 HP or 0.293 watts
1 British thermal unit per minute (BTU/min)	0.0167 BTU per hour or 17.6 watts

Heat goes into the hydraulic oil at every point in the system where there is a pressure loss due to oil flow without mechanical work being produced. Examples are pressure relief and reducing valves, flow control valves, and flow resistance in plumbing lines and through components. Hydraulic pumps and motors also produce heat at about 15 per cent of their working horsepower (Table 5.1.3). Power loss and heat generation due to the above causes can be calculated using one of the following formulas:

$$\text{Horsepower heat} = \frac{\text{PSI} \times \text{GPM}}{1714}$$

$$\text{or HP} = \text{PSI} \times \text{GPM} \times .000583$$

$$\text{BTU/Hr of heat generation} = 1.5 \times \text{PSI} \times \text{GPM}$$

The pressure (PSI) for calculating heat generation in a flow control valve, for example, is the inlet minus the outlet pressure, or the pressure drop across the valve.

Sometimes power loss and heat generation occur intermittently and to find the average amount of heat that will go into the oil, the *average* power loss should be calculated. Usually, taking an average over a 1-hour period should be sufficient. Therefore, hydraulic systems should include a positive means of heat removal. Normally, a heat exchanger is used for this purpose.

The exact size of a heat exchanger needed on a new system cannot be accurately calculated because of too many unknown factors. On existing systems, by making tank measurements and measuring air and oil temperatures, a rather accurate calculation can be made of the heat exchanger capacity needed to reduce the maximum oil temperature.

The theoretical maximum cooling capacity of a heat exchanger for a hydraulic system will never have to be greater than the input horsepower to the system. Usually its capacity can be considerably less, based on the calculated input horsepower. A rule of thumb is to provide a heat exchanger removal capacity of about 25 per cent of the input horsepower. Rarely, even on inefficient systems, would a capacity of more than 50 per cent be required.

When ordering a heat exchanger the information furnished to your supplier must include the maximum rate of oil flow, in GPM, through the heat exchanger, and the horsepower or BTU per hour of heat to be removed. On water-cooled models, state the maximum rate of water flow that will be available. For best water usage, the water flow should be approximately one-half of the oil flow. Specify the temperature of the cooling water.

5.1.6.4.1 Heat load

Heat load on the waterside of a shell and tube heat exchanger can be calculated as

$$\text{BTU per hour} = \text{GPM} \times 500 \times \text{Temperature differential}$$

The temperature differential is the difference between the inlet and outlet oil temperature in degrees Fahrenheit.

Heat load on the shell side of the exchanger can also be calculated by

$$\text{BTU per hour} = \text{GPM} \times 210 \times \text{Temperature differential}$$

5.1.7 Actuators

One of the outstanding features of fluid power systems is that force, generated by the power supply, controlled and directed by suitable valving, and transported by lines, can be converted with ease to almost any kind of mechanical motion. Either linear or rotary motion can be obtained by using a suitable actuating device.

An actuator is a device that converts fluid power into mechanical force and motion. Cylinders, hydraulic motors, and turbines are the most common types of actuating devices used in fluid power systems. This section describes various types of actuating devices and their applications.

5.1.7.1 Hydraulic cylinders

An actuating cylinder is a device that converts fluid power into linear, or straight-line, force and motion. Since linear motion is back-and-forth motion along a straight line, this type of actuator is sometimes referred to as a reciprocating, or linear motor. The cylinder consists of a ram, or piston, operating within a cylindrical bore. Actuating cylinders may be installed so that the cylinder is anchored to a stationary structure and the ram or piston is attached to the mechanism to be operated, or the piston can be anchored and the cylinder attached to the movable mechanism.

5.1.7.1.1 Piston rod column strength

Long, slim piston rods may buckle if subjected to too heavy a push load. Table 5.1.4 suggests the minimum diameter piston rod to use under various conditions of load and unsupported rod length. It should be used in accordance with the instructions in the next paragraph. There must be no side load or bending stress at any point along the piston rod.

Exposed rod length is shown along the top of the table. This is usually somewhat longer than the actual stroke of the cylinder. The vertical scale, column 1, shows the load on the cylinder, and is expressed in English tons, i.e. 1 ton equals 2000 pounds. If both the end of the rod and the *front* end of the cylinder barrel are rigidly supported a smaller rod may have sufficient column strength and you may use as *exposed* length of piston rod that is one-half of the actual total rod length. For example, if the actual rod length is 80 in., and if the cylinder barrel and rod end are supported as described, you could enter the table in the column marked 40. On the other hand, if hinge mounting is used on both cylinder and rod, you may not be safe in using actual exposed rod length. Instead, you should use about twice the actual rod length. For example, if the actual rod length is 20 in., you should enter the table in the 40-in. column.

When mounted horizontally or at any angle other than vertical, hinge-mounted cylinders create a bending stress on the rod when extended. In part, this bending stress is created by the cylinder's weight. On large bore and/or long stroke hinge-mounted cylinders, a trunnion mount should be used in instead of tang or clevis mounts. In addition, the trunnion should be positioned so that the cylinder's weight is balanced when the rod is fully extended.

Table 5.1.4 Minimum recommended piston rod diameter

Load in tons	Exposed length of piston rod, inches							
	10	20	40	60	70	80	100	120
$\frac{1}{2}$			$\frac{3}{4}$	1				
$\frac{3}{4}$			$\frac{13}{16}$	$1\frac{1}{16}$				
1		$\frac{5}{8}$	$\frac{7}{8}$	$1\frac{1}{8}$	$1\frac{1}{4}$	$1\frac{3}{8}$		
$1\frac{1}{2}$		$\frac{11}{16}$	$\frac{15}{16}$	$1\frac{3}{16}$	$1\frac{3}{8}$	$1\frac{1}{2}$		
2		$\frac{3}{4}$	1	$1\frac{3}{16}$	$1\frac{7}{16}$	$1\frac{9}{16}$	$1\frac{13}{16}$	
3	$\frac{13}{16}$	$\frac{7}{8}$	$1\frac{1}{8}$	$1\frac{3}{8}$	$1\frac{9}{16}$	$1\frac{5}{8}$	$1\frac{7}{8}$	
4	$\frac{15}{16}$	1	$1\frac{3}{16}$	$1\frac{1}{2}$	$1\frac{5}{8}$	$1\frac{3}{4}$	2	$2\frac{1}{4}$
5	1	$1\frac{1}{8}$	$1\frac{5}{16}$	$1\frac{9}{16}$	$1\frac{3}{4}$	$1\frac{7}{8}$	$2\frac{1}{8}$	$2\frac{3}{8}$
$7\frac{1}{2}$	$1\frac{3}{16}$	$1\frac{1}{4}$	$1\frac{7}{16}$	$1\frac{3}{4}$	$1\frac{7}{8}$	2	$2\frac{1}{4}$	$2\frac{1}{2}$
10	$1\frac{3}{8}$	$1\frac{7}{16}$	$1\frac{5}{8}$	$1\frac{7}{8}$	2	$2\frac{1}{8}$	$2\frac{7}{16}$	$2\frac{3}{4}$
15	$1\frac{11}{16}$	$1\frac{3}{4}$	$1\frac{7}{8}$	$2\frac{1}{8}$	$2\frac{1}{4}$	$2\frac{3}{8}$	$2\frac{11}{16}$	3
20	2	2	$2\frac{1}{8}$	$2\frac{3}{8}$	$2\frac{1}{2}$	$2\frac{5}{8}$	$2\frac{7}{8}$	$3\frac{1}{4}$
30	$2\frac{3}{8}$	$2\frac{7}{16}$	$2\frac{1}{2}$	$2\frac{3}{4}$	$2\frac{3}{4}$	$2\frac{7}{8}$	$3\frac{1}{4}$	$3\frac{1}{2}$
40	$2\frac{3}{4}$	$2\frac{3}{4}$	$2\frac{7}{8}$	3	3	$3\frac{1}{4}$	$3\frac{1}{2}$	$3\frac{3}{4}$
50	$3\frac{1}{8}$	$3\frac{1}{8}$	$3\frac{1}{4}$	$3\frac{3}{8}$	$3\frac{1}{2}$	$3\frac{1}{2}$	$3\frac{3}{4}$	4
75	$3\frac{3}{4}$	$3\frac{3}{4}$	$3\frac{7}{8}$	4	4	$4\frac{1}{8}$	$4\frac{3}{8}$	$4\frac{1}{2}$
100	$4\frac{3}{8}$	$4\frac{3}{8}$	$4\frac{3}{8}$	$4\frac{1}{2}$	$4\frac{3}{4}$	$4\frac{3}{4}$	$4\frac{7}{8}$	5
150	$5\frac{3}{8}$	$5\frac{3}{8}$	$5\frac{3}{8}$	$5\frac{1}{2}$	$5\frac{1}{2}$	$5\frac{1}{2}$	$5\frac{3}{4}$	6

5.1.7.1.2 Hydraulic cylinder forces

Tables 5.1.5 and 5.1.6 provide the mechanical forces, both extension and retraction, that can be generated by hydraulic cylinders. The tables are divided into the two principal operating pressure ranges associated with hydraulic applications.

Values in bold type show the extension forces, using the full piston area. Values in italic type are for the retraction force for various piston rod diameters. Remember that force values are *theoretical*, derived by calculation. Experience has shown that probably 5 per cent but certainly no more than 10 per cent additional pressure will be required to make up cylinder losses.

For pressures not shown, the effective piston areas in the third column can be used as power factors. Multiply effective area by pressure to obtain cylinder force produced. Values in two or more columns can be added for a pressure not listed, or, force values can be obtained by interpolating between the next higher and the next lower pressure columns.

Pressure values along the top of each table are differential pressures across the two cylinder ports. This is the pressure to just balance the load and not the pressure that must be produced by the system pump. There will be circuit flow losses in pressure and return lines due to oil flow and these will require additional pressure. When designing a system, be sure to allow sufficient pump pressure, about 25–30 per cent, both to supply the cylinder and to satisfy system flow losses.

5.1.7.1.3 Hydraulic motors

A fluid power motor is a device that converts fluid power energy into rotary motion and force. The function of a motor is opposite that of a pump. However, the design and operation of fluid power motors are very similar to pumps. Therefore a thorough knowledge of pumps will help one understand the operation of fluid power motors.

Motors have many uses in fluid power systems. In hydraulic power drives, pumps and motors are

Table 5.1.5 Hydraulic cylinder force, low pressure range 500–1500 psi

Bore diameter (in.)	Rod diameter (in.)	Effective area (in.^{2})	Pressure differential across cylinder ports				
			500 psi	750 psi	1000 psi	1250 psi	1500 psi
$1\frac{1}{2}$	None	1.77	884	1325	1767	2209	2651
	$\frac{5}{8}$	1.46	730	1095	1460	1825	2190
	1	0.982	491	736	982	1227	1473
2	None	3.14	1571	2356	3142	3927	4712
	1	2.36	1178	1767	2356	2945	3534
	$1\frac{3}{8}$	1.66	828	1243	1657	2071	2485
$2\frac{1}{2}$	None	4.91	2454	3682	4909	6136	7363
	1	4.12	2062	3092	4123	5154	6188
	$1\frac{3}{8}$	3.42	1712	2568	3424	4280	5136
	$1\frac{3}{4}$	2.50	1252	1878	2503	3129	3755
3	None	7.07	3534	5301	7069	8836	10,603
	1	6.28	3142	4712	6283	7854	9425
	$1\frac{3}{8}$	5.58	2792	4188	5584	6980	8376
	$1\frac{3}{4}$	4.66	2332	3497	4663	5829	6995
$3\frac{1}{4}$	None	8.30	4148	6222	8298	10,370	12,444
	$1\frac{3}{8}$	6.81	3405	5108	6811	8514	10,216
	$1\frac{3}{4}$	5.89	2945	4418	5891	7363	8836
	2	5.15	2577	3866	5154	6443	7731
4	None	12.57	6284	9425	12,567	15,709	18,851
	$1\frac{3}{4}$	10.16	5081	7621	10,162	12,702	15,243
	2	9.43	4713	7069	9425	11,782	14,138
	$2\frac{1}{2}$	7.66	3829	5744	7658	9573	11,487
5	None	19.64	9818	14,726	19,635	24,544	29,453
	2	16.49	8247	12,370	16,493	20,617	24,740
	$2\frac{1}{2}$	14.73	7363	11,045	14,726	18,408	22,089
	3	12.57	6283	9425	12,566	15,708	18,850
	$3\frac{1}{2}$	10.01	5007	7510	10,014	12,517	15,021
6	None	28.27	14,137	21,206	28,274	35,343	42,411
	$2\frac{1}{2}$	23.37	11,683	17,524	23,365	29,207	35,048
	3	21.21	10,603	15,094	21,205	26,507	31,808
	$3\frac{1}{2}$	18.65	9326	13,990	18,653	23,316	27,979
	4	15.71	7854	11,781	15,708	19,635	23,562
7	None	38.49	19,243	28,864	38,485	48,106	57,728
	3	31.42	15,708	23,562	31,416	39,271	47,125
	$3\frac{1}{2}$	28.87	14,432	21,648	28,864	36,080	43,296
	4	25.92	12,960	19,439	25,910	32,399	38,879
	$4\frac{1}{2}$	22.58	11,291	16,936	22,581	28,226	33,872
	5	18.85	9425	14,138	18,850	23,563	28,275
8	None	50.27	25,133	37,699	50,265	62,831	75,398
	$3\frac{1}{2}$	40.64	20,322	30,483	40,644	50,805	60,966
	4	37.70	18,850	28,274	37,699	47,124	56,549
	$4\frac{1}{2}$	34.36	17,181	25,771	34,361	42,951	51,542
	5	30.63	15,315	22,973	30,630	38,288	44,945
	$5\frac{1}{2}$	26.51	13,254	19,880	26,507	33,134	39,761
10	None	78.54	39,270	58,905	78,540	98,175	117,810
	$4\frac{1}{2}$	62.64	31,318	46,977	62,636	78,295	93,954
	5	58.91	29,453	44,179	58,905	73,631	88,358
	$5\frac{1}{2}$	54.78	27,391	41,087	54,782	68,478	82,172
	7	40.06	20,028	30,041	40,055	50,069	60,083

Table 5.1.5 Hydraulic cylinder force, low pressure range 500–1500 psi (*cont'd*)

Bore diameter (in.)	Rod diameter (in.)	Effective area (in.^{2})	Pressure differential across cylinder ports				
			500 psi	750 psi	1000 psi	1250 psi	1500 psi
12	None	113.1	56,550	84,825	113,100	141,375	169,650
	$5\frac{1}{2}$	89.34	44,671	67,007	89,342	111,678	134,013
14	None	153.9	76,970	115,455	153,940	192,425	230,910
	7	115.5	57,728	86,591	115,455	144,319	173,183

Table 5.1.6 Hydraulic cylinder force, high pressure range 2000–5000 psi

Bore diameter (in.)	Rod diameter (in.)	Effective area (in.^{2})	Pressure differential across cylinder ports				
			2000 psi	2500 psi	3000 psi	4000 psi	5000 psi
$1\frac{1}{2}$	None	1.77	3534	4418	5301	7068	8836
	$\frac{5}{8}$	1.46	2921	3651	4381	5841	7302
	1	0.982	1963	2454	2945	3927	4909
2	None	3.14	6283	7854	9425	12,566	15,708
	1	2.36	4712	5890	7069	9425	11,781
	$1\frac{3}{8}$	1.66	3313	4142	4970	6627	8283
$2\frac{1}{2}$	None	4.91	9817	12,271	14,726	19,635	24,544
	1	4.12	8247	10,308	12,370	16,493	20,617
	$1\frac{3}{8}$	3.42	6848	8560	10,271	13,695	17,119
	$1\frac{3}{4}$	2.50	5007	6259	7510	10,014	12,517
3	None	7.07	14,137	17,672	21,206	28,274	35,343
	1	6.28	12,567	15,708	18,850	25,133	31,416
	$1\frac{3}{8}$	5.58	11,167	13,959	16,751	22,335	27,919
	$1\frac{3}{4}$	4.66	9327	11,658	13,990	18,653	23,317
$3\frac{1}{4}$	None	8.30	16,592	20,740	24,837	33,183	41,479
	$1\frac{3}{8}$	6.81	13,622	17,027	20,433	27,244	34,055
	$1\frac{3}{4}$	5.89	11,781	14,726	17,672	23,562	29,453
	2	5.15	10,308	12,886	15,463	20,617	25,771
4	None	12.57	25,134	31,418	37,701	50,268	62,835
	$1\frac{3}{4}$	10.16	20,323	25,404	30,485	40,647	50,809
	2	9.43	18,851	23,564	28,276	37,702	47,127
	$2\frac{1}{2}$	7.66	15,317	19,146	22,975	30,633	38,292
5	None	19.64	39,270	49,088	58,905	78,540	98,175
	2	16.49	32,987	41,234	49,480	65,974	82,467
	$2\frac{1}{2}$	14.73	29,453	36,816	44,179	58,905	73,632
	3	12.57	25,133	31,416	37,699	50,266	62,832
	$3\frac{1}{2}$	10.01	20,028	25,035	30,042	40,056	50,070
6	None	28.27	56,548	70,685	84,822	113,090	141,370
	$2\frac{1}{2}$	23.37	46,731	58,413	70,096	93,461	116,827
	3	21.21	42,411	53,014	63,616	84,822	106,027
	$3\frac{1}{2}$	18.65	37,306	46,632	55,959	74,612	93,265
	4	15.71	31,416	39,270	47,124	62,832	78,540
7	None	38.49	76,970	96,213	115,455	153,940	192,425
	3	31.42	62,833	78,541	94,249	125,666	157,082
	$3\frac{1}{2}$	28.87	57,728	72,160	86,592	115,456	144,320
	4	25.92	51,838	64,798	77,757	103,676	129,595
	$4\frac{1}{2}$	22.58	45,162	56,453	67,743	90,324	112,905
	5	18.85	37,700	47,125	56,550	75,400	94,250

(*Continued*)

Table 5.1.6 Hydraulic cylinder force, high pressure range 2000–5000 psi (*cont'd*)

Bore diameter (in.)	Rod diameter (in.)	Effective area (in.^{2})	Pressure differential across cylinder ports				
			2000 psi	2500 psi	3000 psi	4000 psi	5000 psi
8	None	50.27	100,530	125,663	150,795	201,060	251,325
	$3\frac{1}{2}$	*40.64*	*81,288*	*101,610*	*121,932*	*162,576*	*203,220*
	4	*37.70*	*75,398*	*94,248*	*131,097*	*150,796*	*188,495*
	$4\frac{1}{2}$	*34.36*	*68,722*	*85,903*	*103,083*	*137,444*	*171,805*
	5	*30.63*	*61,260*	*76,575*	*91,890*	*125,520*	*153,150*
	$5\frac{1}{2}$	*26.51*	*53,014*	*66,268*	*79,521*	*106,028*	*132,535*
10	None	78.54	157,080	196,350	235,620	314,160	392,700
	$4\frac{1}{2}$	*62.64*	*125,272*	*156,590*	*187,908*	*250,544*	*313,180*
	5	*58.91*	*117,810*	*147,263*	*176,715*	*235,620*	*294,525*
	$5\frac{1}{2}$	*54.78*	*109,564*	*136,955*	*164,346*	*219,128*	*273,910*
	7	*40.06*	*80,110*	*100,138*	*120,165*	*160,220*	*200,275*
12	None	113.1	226,200	282,750	339,300	452,400	565,500
	$5\frac{1}{2}$	*89.34*	*178,684*	*223,355*	*268,026*	*357,368*	*446,710*
14	None	153.9	307,880	384,850	461,820	615,760	769,700
	7	*115.5*	*230,910*	*288,638*	*346,365*	*461,820*	*577,275*

combined with suitable lines and valves to form hydraulic transmissions. The pump commonly referred to as the A-end is driven by some outside source, such as an electric motor. The pump delivers pressurized fluid to the hydraulic motor, referred to as the B-end. The hydraulic motor is actuated by this flow and through mechanical linkage conveys rotary motion and force to do work.

Fluid motors may be either fixed or variable displacement. Fixed-displacement motors provide constant torque and variable speed. Controlling the amount of input flow varies the speed. Variable-displacement motors are constructed so that the working relationship of the internal parts can be varied to change displacement. The majority of the motors used in fluid power systems are the fixed-displacement type.

Although most fluid power motors are capable of providing rotary motion in either direction, some applications require rotation in only one direction. In these applications, one port of the motor is connected to the system pressure line and the other port to the return line. The flow of fluid to the motor is controlled by a flow control valve, a two-way directional control valve or by starting and stopping the power supply. Varying the rate of fluid flow to the motor may control the speed of the motor.

In most fluid power systems, the motor is required to provide actuating power in either direction. In these applications, the ports are referred to as working ports, alternating as inlet and outlet ports. Either a four-way directional control valve or a variable-displacement pump usually controls the flow to the motor.

Fluid motors are usually classified according to the type of internal element which is directly actuated by the pressurized flow. The most common types of elements are gears, vanes and pistons. All three of these types are adaptable for hydraulic systems, but only the vane type is used on pneumatic systems.

5.1.8 Control valves

It is impossible to design a practical fluid power system without some means of controlling the volume and pressure of the fluid, and directing that flow to the proper operating units. This is accomplished by the inclusion of control valves in the hydraulic circuit.

A valve is defined as any device by which the flow of fluid may be started, stopped, regulated, or directed by a movable part that opens or obstructs passage of the fluid. Valves must be able to accurately control fluid flow and system pressure and to sequence the operation of all actuators within a hydraulic system.

Hydraulic control values can utilize a variety of actuators that activate their function. Normally these actuators use manual, electrical, mechanical, or pneumatic power sources.

5.1.8.1 Valve classification

Valves are classified by their intended use: flow control, pressure control, and direction control. Some valves have multiple functions that fall into more than one classification.

5.1.8.1.1 Flow control valves

Flow control valves are used to regulate the flow of fluids. Control of flow in hydraulic systems is critical because the rate of movement of fluid-powered machines or actuators depends on the rate of flow of the pressurized fluid.

5.1.8.1.2 Pressure control

The safe and efficient operation of fluid power systems, system components, and related equipment requires a means to control pressure within the system. There are many types of automatic pressure control valves. Some of them merely provide an escape for excess pressures; some only reduce the pressure; and some keep the pressure within a pre-set range.

Some fluid power systems, even when operated normally, may temporarily develop excessive pressure. For example, when an unusually strong work resistance is encountered, system pressure may exceed design limits. Relief valves are used to control this excess pressure.

Relief valves are automatic valves, used on system lines and equipment to prevent over-pressurization. Most relief valves simply open at a preset pressure and shut when the pressure returns to normal limits. They do not maintain flow or pressure at a given amount, but prevent pressure from rising above a specified level.

Main system relief valves are generally installed between the pump or pressure source and the first system isolation valve. The valve must be large enough to allow the full output of the hydraulic pump to be delivered back to the reservoir. This design feature, called a full-flow bypass, is essential for all hydraulic systems. The location of the valve is also critical. If the valve were installed downstream from the system isolator valve, the pump could be deadheaded when the system was shutdown.

Smaller relief valves are often used in isolated parts of the system where a check valve or directional control valve prevents pressure from being relieved through the main system relief valve or where pressures must be relieved at a specific set point lower than the main system pressure. These small relief valves are also used to relieve pressures caused by thermal expansion of fluids.

Figure 5.1.20 shows a typical relief valve. System pressure simply acts under the valve disk at the inlet of the valve. When the system pressure exceeds the preload force exerted by the valve spring, the valve disk will lift off of its seat. This will allow some of the system fluid to escape through the valve outlet. Flow will continue until the system pressure is reduced to a level below the spring force.

All relief valves have an adjustment for increasing or decreasing the set relief pressure. Some relief valves are equipped with an adjusting screw for this purpose. This adjusting screw is usually covered with a cap, which must be removed before an adjustment can be made.

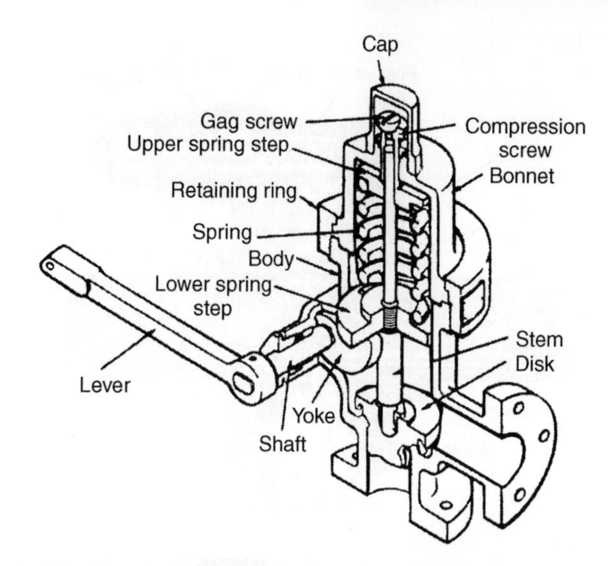

Fig. 5.1.20 Cutaway of relief valve.

5.1.8.1.3 Pressure regulators

Pressure regulators, often referred to as unloading valves, are used in fluid power systems to regulate pressure. In hydraulic systems, the pressure regulator is used to unload the pump and to maintain or regulate system pressure at the desired values.

All hydraulic systems do not require pressure regulators. The open-center system does not require a pressure regulator. Many systems are equipped with variable-displacement pumps, which contain a pressure-regulating device.

Pressure regulators are made in a variety of types. However, the basic operating principles of all regulators are similar to the one illustrated in Figure 5.1.21.

A regulator is open when it is directing fluid under pressure into the system, Figure 5.1.21, view (a). In the closed position, Figure 5.1.21, view (b), the fluid in the part of the system beyond the regulator is trapped at the desired pressure and the fluid from the pump is bypassed into the return line and back to the reservoir. To prevent constant opening and closing (chatter), the regulator is designed to open at pressure somewhat lower than the closing pressure. This difference is known as differential or operating range. For example, assume that a pressure regulator is set to open when the system pressure drops below 600 psi and close when the pressure rises above 800 psi. The differential or operating range is 200 psi.

Referring to Figure 5.1.21, assume that the piston has an area of 1 in.^{2}, the pilot valve has a cross-sectional area of 1 in.^{2}, and the piston spring provides 600 pounds of force that pushes the piston against its seat. When the system pressure is less than 600 psi, fluid from the pump will enter the inlet port, flow to the top of the regulator, and then to the pilot valve. When the system pressure at the valve inlet increases to the point where the force it

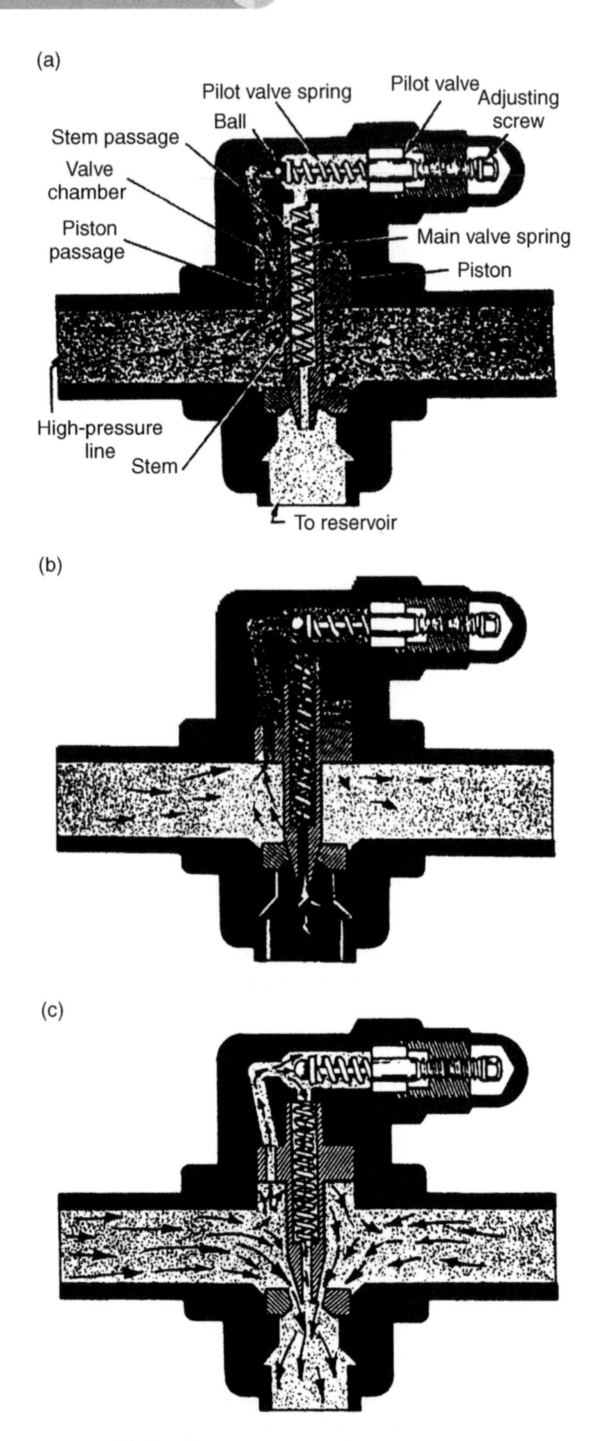

Fig. 5.1.21 Hydraulic pressure regulator.

creates against the front of the check valve exceeds the force created against the back of the check valve, the check valve opens. This allows fluid to flow into the system and to the bottom of the regulator against the piston. When the system force exceeds the force exerted by the spring, the piston moves up, causing the pilot valve to unseat. Since the fluid will take the path of least resistance, it will pass through the regulator and back to the reservoir through the bypass line.

When the fluid from the pump is suddenly allowed a free path to return, the pressure on the input side of the check valve drops and the check valve closes. The fluid in the system is then pressurized until a power unit is actuated or until pressure is slowly lost through normal internal leakage within the system.

When the system pressure decreases to a point slightly below 600 psi, the spring forces the piston down and closes the pilot valve. When the pilot valve is closed, the fluid cannot flow directly to the return line. This causes the pressure to increase in the line between the pump and the regulator. This pressure opens the check valve, causing fluid to enter the system.

In summary, when the system pressure decreases, the pressure regulator will open, sending fluid to the system. When the system pressure increases, the regulator will close, allowing the fluid from the pump to flow through the regulator and back to the reservoir. The pressure regulator takes the load off of the pump and regulates system pressure.

5.1.8.1.4 Sequence valves

Sequence valves control the sequence of operation between two branches in a hydraulic circuit. In other words, they enable one component within the system to automatically set another component into motion. An example of the use of a sequence valve is in an aircraft landing gear actuating system.

In a landing gear actuating system, the landing gear doors must open before the landing gear starts to extend. Conversely, the landing gear must be completely retracted before the doors close. A sequence valve installed in each landing gear actuating line performs this function.

A sequence valve is somewhat similar to a relief valve except that, after the set pressure has been reached, the sequence valve diverts the fluid to a second actuator or motor to do work in another part of the system. Figure 5.1.22 shows an installation of two sequence valves that control the sequence of operation of three actuating cylinders. Fluid is free to flow into cylinder A. The first sequence valve (1) blocks the passage of fluid until the piston in cylinder A moves to the end of its stroke. At this time, sequence valve 1 opens, allowing fluid to enter cylinder B. This action continues until all three pistons complete their strokes.

5.1.8.1.5 Pressure-reducing valves

Pressure-reducing valves provide a steady pressure into a part of the system that operates at a pressure lower that normal system pressure. A reducing valve can normally be set for any desired downstream pressure within its design limits. Once the valve is set, the reduced pressure

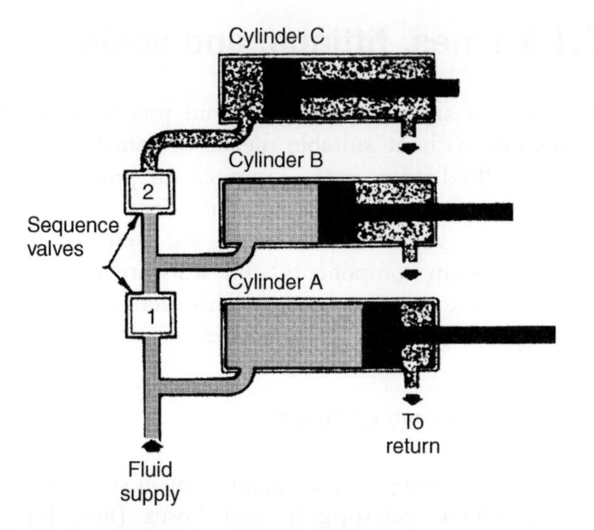

Fig. 5.1.22 Installation of sequence valve.

will be maintained regardless of changes in the supply pressure and system load variations.

There are various designs and types of pressure-regulating valves. The spring-loaded reducer and the pilot-controlled valve are the most common.

5.1.8.1.6 Directional control valves

Directional control valves are designed to direct the flow of fluid, at the desired time, to the point in a fluid power system where it will do work. The driving of a ram back and forth in its cylinder is an example of when a directional control valve is used. Various other terms are used to identify directional valves, such as selector valve, transfer valve, and control valve. Here, we shall use the term directional control valve to identify these valves.

Directional control valves for hydraulic and pneumatic systems are similar in design and operation. However, there is one major difference. The return port of a hydraulic valve is ported through a return line to the reservoir. Any other differences are pointed out in the discussion of these valves.

Directional control valves may be operated by differences in pressure acting on opposite sides of the valve elements or may be positioned manually, mechanically, or electrically. Often two or more methods of operating the same valve will be used in different phases of its action.

Directional control valves may be classified in several ways. Some of the different ways are by the type of control, the number of ports in the valve housing, and the specific function that the valve performs. The most common method is by the type of valving element used in the construction of the valve. The most common types of valving elements are the ball, cone, sleeve, poppet, rotary spool, and sliding spool. The basic operating principles of the poppet, rotary spool, and sliding spool types are discussed in this text.

The poppet fits into the center bore of the seat, Figure 5.1.23. The seating surfaces of the poppet and the seat are lapped or closely machined so that the center bore will be sealed when the poppet is seated (shut). The action of the poppet is similar to that of the valves in an automobile engine. In most valves, the poppet is held in the seated position by a spring.

The valve consists primarily of a movable poppet, which closes against the valve seat. In the closed position, fluid pressure on the inlet side tends to hold the valve tightly closed. A small amount of movement from a force applied to the top of the poppet stem opens the poppet and allows fluid to flow through the valve.

The rotary spool directional control valve, Figure 5.1.24, has a round core with one or more passages or recesses in it. The core is mounted within a stationary sleeve. As the core is rotated within the stationary sleeve, the passages or recesses connect or block the ports in the sleeve. The ports in the sleeve are connected to the appropriate lines of the fluid system.

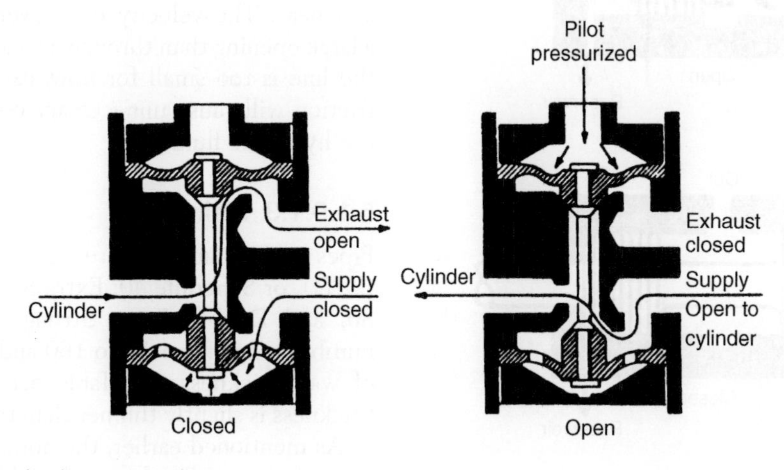

Fig. 5.1.23 Operation of a simple poppet valve.

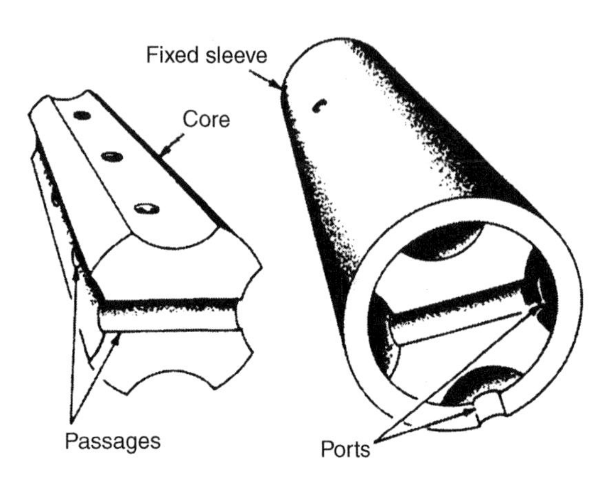

Fig. 5.1.24 Parts of a rotary spool valve.

The operation of a simple sliding spool directional control valve is shown in Figure 5.1.25. The valve is so named because of the shape of the valving element that slides back and forth to block or open ports in the valve housing. The sliding element is referred to as the spool or piston. The inner piston areas, or lands, are equal. Thus, fluid under pressure, which enters the valve from the inlet ports, acts equally on both inner piston areas, regardless of the position of the spool. Sealing is usually accomplished by a very closely machined fit between the spool and the valve body or sleeve. For valves with more ports, the spool is designed with more pistons or lands on a common shaft. The sliding spool is the most common type of directional control valve.

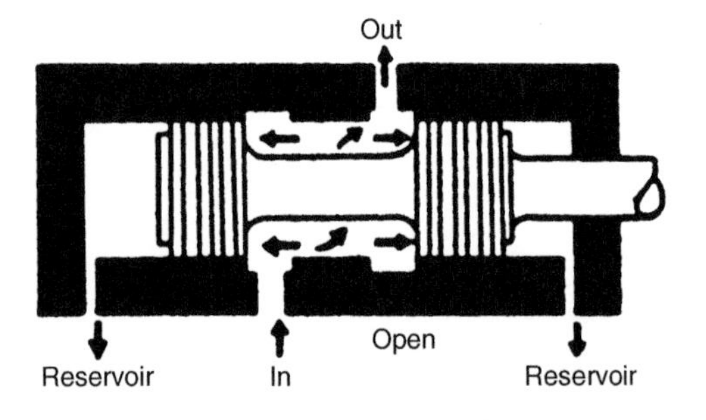

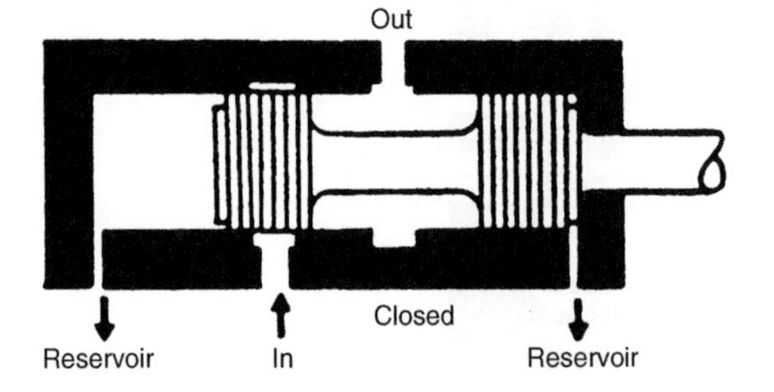

Fig. 5.1.25 Two-way sliding spool valve.

5.1.9 Lines, fittings, and seals

The control and application of fluid power would be impossible without suitable means of transferring the hydraulic fluid between the reservoir, the power source, and the points of application. Fluid lines are used to transfer the hydraulic fluid, fittings are used to connect lines to system components, and seals are used in all components to prevent leakage. This section is devoted to these critical system components.

5.1.9.1 Types of lines

Three types of lines are used in fluid power systems: pipe (rigid), tubing (semi-rigid), and hoses (flexible). A number of factors are considered when the type of line is selected for a particular application. These factors include the type of fluid, the required system pressure, and the location of the system. For example, heavy pipe might be used for a large, stationary systems, but comparatively lightweight tubing must be used in mobile applications. Flexible hose is required in installations where units must be free to move relative to each other.

5.1.9.1.1 Pipe and tubing

There are three important dimensions of any tubular product: outside diameter (OD), inside diameter (ID) and wall thickness. Sizes of pipe are listed by the nominal, or approximate, ID and the wall thickness. The actual OD and the wall thickness list sizes of tubing.

5.1.9.1.2 Selection

The material, ID, and wall thickness are the three primary considerations in the selection of lines for a particular fluid power system. The ID of the line is important because it determines how much fluid can pass through the line without loss of power due to excessive friction and heat. The velocity of a given flow is less through a large opening than through a small opening. If the ID of the line is too small for flow, excessive turbulence and friction will cause unnecessary power loss and overheat the hydraulic fluid.

5.1.9.1.3 Sizing

Pipes are available in three different weights: Standard (STD), or Schedule 40; Extra Strong (XS), or Schedule 80; and Double Extra Strong (XXS). The schedule number range from 10 to 160 and cover 10 distinct sets of wall thickness, see Table 5.1.7. Schedule 160 wall thickness is slightly thinner than the double extra strong.

As mentioned earlier, the nominal ID determines the size of pipes. For example, the ID for a (1/4)-in.

Table 5.1.7 Wall thickness schedule designation for pipe

		Inside diameter (ID)		
Nominal size	**Pipe OD**	**Schedule 40**	**Schedule 80**	**Schedule 160**
$\frac{1}{8}$	0.405	0.269	0.215	
$\frac{1}{4}$	0.540	0.364	0.302	
$\frac{3}{8}$	0.675	0.493	0.423	
$\frac{1}{2}$	0.840	0.622	0.546	0.466
$\frac{3}{4}$	1.050	0.824	0.742	0.815
1	1.315	1.049	0.957	0.815
$1\frac{1}{4}$	1.660	1.380	1.278	1.160
$1\frac{1}{2}$	1.900	1.610	1.500	1.338
2	2.375	2.067	1.939	1.689

Schedule 40 pipe is 0.364 in., and the ID for a (1/2)-in. Schedule 40 pipe is 0.622 in.

It is important to note that the IDs of all pipes of the same nominal size are not equal. This is because the OD remains constant and the wall thickness increases as the schedule number increases. For example, a nominal 1-in. Schedule 40 pipe has a 1.049-in. ID. The same size Schedule 80 pipe has a 0.957-in. ID, while Schedule 160 pipe has 0.815-in. ID. In each case, the OD is 1.315-in. and the wall thickness varies. The actual wall thickness is the difference between the OD and ID divided by 2.

Tubing differs from pipe in its size classification. Its actual OD designates tubing. Thus, (5/8)-in. tubing has an OD of (5/8)-in. As indicated in Table 5.1.7, tubing is available in a variety of wall thickness. The diameter of tubing is often measured and indicated in (1/16)ths. Thus, No. 6 tubing is 6/16- or 3/8-in. OD, No. 8 tubing is 8/16 or 1/2-in., and so forth.

The wall thickness, material used, and ID determines the bursting pressure of a line or fitting. The greater the wall thickness in relation to the ID and the stronger the metal, the higher the bursting pressure. However, the greater the ID for a given wall thickness, the lower the bursting pressure. This is because force is the product of area and pressure.

5.1.9.1.4 Materials

The pipe and tubing used in fluid power systems are commonly made from steel, copper, brass, aluminum, and stainless steel. Each of these metals has its distinct advantages and disadvantages in certain applications.

Steel pipe and tubing are relatively inexpensive and are used in many hydraulic and pneumatic applications. Steel is used because of its strength, suitability for bending and flanging and adaptability to high pressures and temperatures. Its chief disadvantage is its comparatively low resistance to corrosion.

Copper pipe and tubing are sometimes used for fluid power lines. Copper has high resistance to corrosion and is easily drawn or bent. However, it is unsatisfactory for high temperatures and has a tendency to harden and break due to stress and vibration.

Aluminum has many of the characteristics and qualities required for fluid power lines. Is has high resistance to corrosion and is easily drawn or bent. In addition, it has the outstanding characteristic of being lightweight. Since weight elimination is a vital factor in the design of aircraft, aluminum alloy tubing is used in the majority of aircraft fluid power systems.

An improperly piped system can lead to serious power loss and possible harmful fluid contamination. Therefore, in maintenance and repair of fluid power system lines, the basic design requirements must be kept in mind. Two primary requirements are as follows:

1. The lines must have the correct ID to provide the required volume and velocity of flow with the least amount of turbulence during all demands on the system.
2. The lines must be made of the proper material and have the wall thickness to provide sufficient strength to both contain the fluid at the required pressure and withstand the surges of pressure that may develop in the system.

5.1.9.1.5 Preparation

Fluid power systems are designed as compactly as possible, to keep the connecting lines short. Every section of line should be anchored securely in one or more places so that neither the weight of the line nor the effects of vibration are carried on the joints. The aim is to minimize stress throughout the system.

Lines should normally be kept as short and free of bends as possible. However, tubing should not be assembled in a straight line, because a bend tends to eliminate strain by absorbing vibration and compensates for thermal expansion and contraction. Bends are preferred to elbows, because bends cause less loss of power. Some of the correct and incorrect methods of installing tubing are illustrated in Figure 5.1.26.

Bends are described by their radius measurements. The ideal bend radius is 2(1/2)–3 times the ID, as shown in Figure 5.1.27. For example, if the ID of a line is 2 in., the radius of the bend should be between 5 and 6 in.

While friction increases markedly for sharper curves than this, it also tends to increase up to a certain point for gentler curves. The increases in friction in a bend with a radius of more than 3 pipe diameters results from

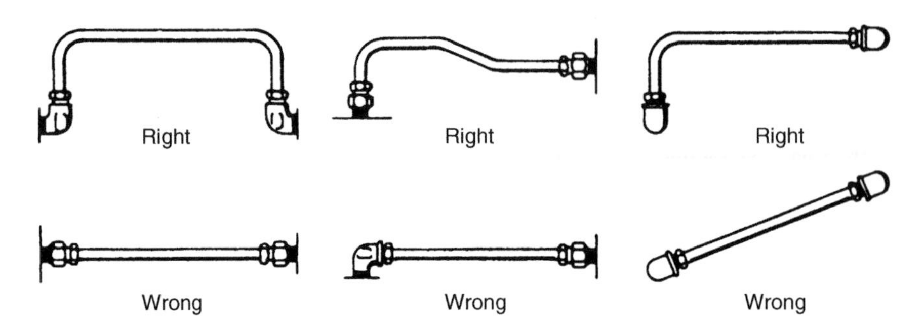

Fig. 5.1.26 Correct and incorrect methods of installation.

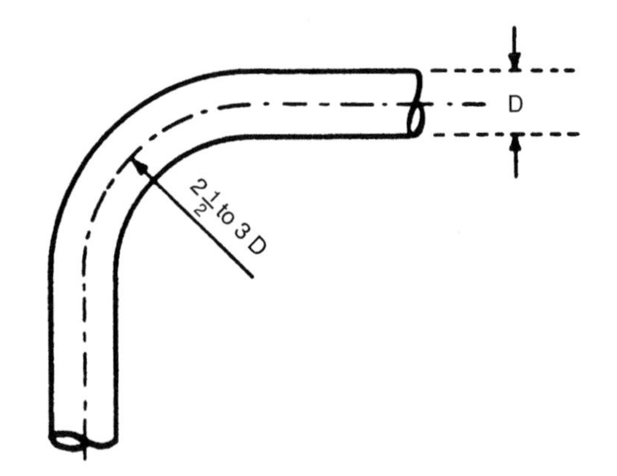

Fig. 5.1.27 Ideal bend radius.

increased turbulence near the outside edges of the flow. Particles of fluid must travel a longer distance in making the change in direction. When the radius of the bend is less than 2(1/2) pipe diameters, the increased pressure loss is due to the abrupt change in the direction of flow, especially for particles near the inside edge of the flow.

5.1.9.1.6 Tube cutting and deburring

The objective of cutting tubing is to produce a square end that is free from burrs. Tubing may be cut using a standard tube cutter, Figure 5.1.28, or a fine-toothed hacksaw. When you use the standard tube cutter, place the tube in the cutter with the cutting wheel at the point where the cut is to be made. Apply light pressure on the tube by tightening the adjusting knob. Too much pressure applied to the cutting wheel at one time may deform the tubing or cause excessive burrs. Rotate the cutter, adjust the tightening knob after each complete rotation.

After the tubing is cut, remove all burrs and sharp edges from the inside and outside of the tube with a deburring tool, Figure 5.1.29. Clean out the tubing. Make sure no foreign particles remain.

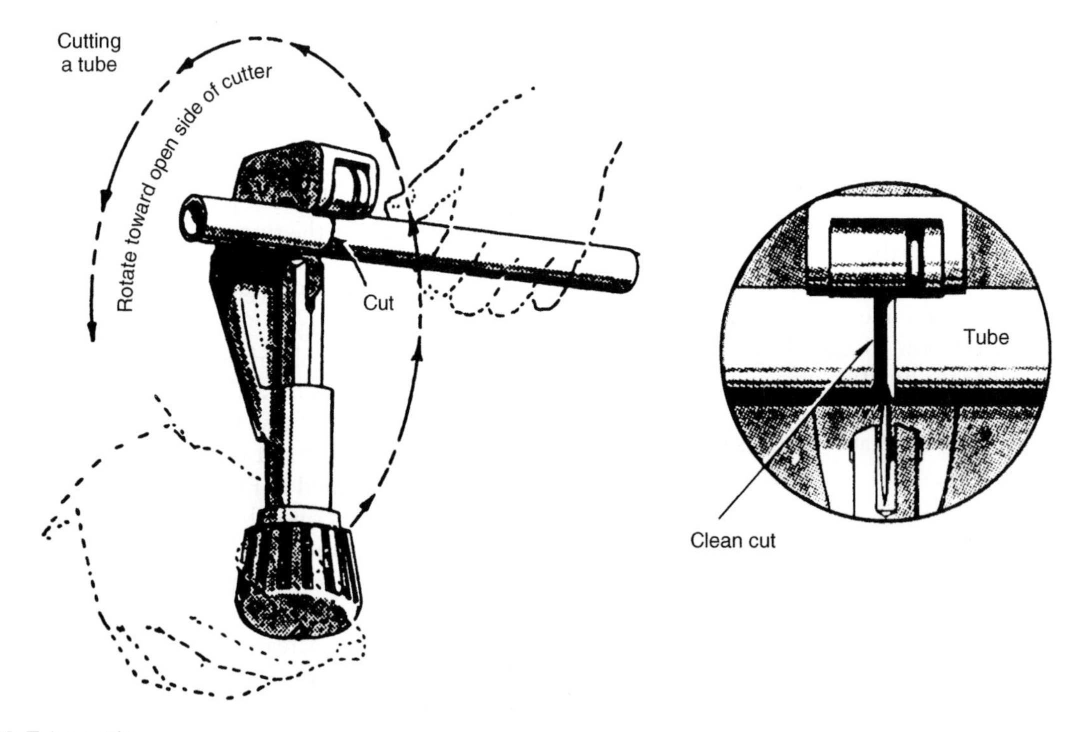

Fig. 5.1.28 Tube cutting.

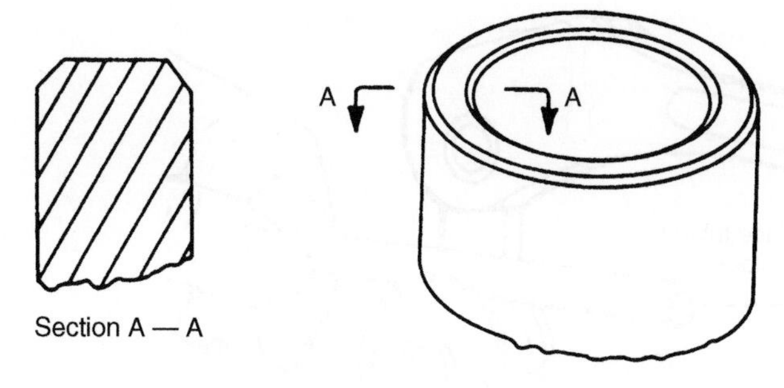

Fig. 5.1.29 Properly burred tubing.

5.1.9.1.7 Tube bending

The objective in tube bending is to obtain a smooth bend without flattening the tube. Tube bending is usually done with either a hand tube bender or a mechanically operated bender.

The hand tube bender shown in Figure 5.1.30, consists of a handle, a radius block, a clip, and a slide bar. The handle and slide bars are used as levers to provide the mechanical advantage necessary to bend the tubing. The radius block is marked in degrees or bend ranging from 0° to 180°. The slide bar has a mark which is lined up with the zero mark on the radius block. The tube is inserted in the tube bender and after the marks are lined up, the slide bar is moved around until the mark on the slide bar reaches the desired degree of bend on the radius block.

5.1.9.1.8 Tube flaring

Tube flaring is a method of forming the end of a tube into a funnel shape so a threaded fitting can hold it. When a flared tube is prepared, a flare nut is slipped onto the tube and the end of the tube is flared. During tube installation, the flare is seated to a fitting with the inside of the flare against the cone-shaped end of the fitting, pulling the inside of the flare against the seating surface of the fitting.

Either of two flaring tools in Figure 5.1.31 may be used. One gives a single flare and the other gives a double flare. The flaring tool consists of a split die block that has holes for various sizes of tubing. It also has a clamp to lock the end of the tubing inside the die block and a yoke with a compressor screw and cone that slips over the die block and forms the 45° flare on the end of the tube. A double flaring tube has adapters that turn in the edge of the tube before a regular 45° double flare is made.

To use the single flaring tool, first check to see that the end of the tubing has been cut off squarely and has had the burrs removed for both the inside and outside. Slip the flare nut onto the tubing before you make the flare. Then, open the die block. Insert the end of the tubing into the hole corresponding to the OD of the tubing so that the end protrudes slightly above the top face of the die blocks. The amount by which the tubing extends above the blocks determines the finished diameter of the flare. The flare must be large enough to seat properly against the fitting, but small enough that the threads of the flare nut will slide over it. Close the die block and secure the tool with the wing nut. Use the handle of the yoke to tighten the wing nut. Then place the yoke over the end of the tubing and tighten the handle to force the cone into the end of the tubing. The completed flare should be slightly visible above the face of the die blocks.

5.1.9.1.9 Flexible hose

Shock-resistant, flexible hose (Figure 5.1.32) assemblies are required to absorb the movements of mounted equipment under both normal operating conditions and extreme conditions. They are also used for their noise-attenuating properties and to connect moving parts of certain equipment. The two basic hose types are synthetic rubber and polytetrafluoroethylene (PTFE), such as DuPont's Teflon fluorocarbon resin.

Rubber hoses are designed for specific fluid, temperature, and pressure ranges and are provided in various specifications. Rubber hoses consist of a minimum of three layers: a seamless synthetic rubber tube reinforced with one or more layers of braided or spiraled cotton, wire, or synthetic fiber; and an outer cover. The inner tube is designed to withstand the attack of the fluid that passes through it. The braided or spiraled layers determine the strength of the hose. The greater the number of layers, the greater is the pressure rating. Hoses are provided in three pressure ranges: low, medium, and high. The outer cover is designed to withstand external abuse and contains identification markings.

5.1.9.1.10 Sizing

The size of a flexible hose is identified by the dash (–) number, which is the ID of the hose expressed in 16ths of an inch. For example, the ID of a –64 hose is 4 in. For a few hose styles this is the nominal and not the true ID.

5.1.9.1.11 Cure date

Synthetic rubber hoses deteriorate due to aging. A cure date is used to ensure that they do not deteriorate beyond material and performance specifications. The cure date is the quarter and year the hose was manufactured. For example, 1Q89 or 1/89 means the hose was made during the first quarter of 1989. The cure date limits the length of time a rubber hose can be safely used in fluid power applications. The normal shelf life of rubber hose is 4 years.

5.1.9.1.12 Application

As mentioned earlier, flexible hose is available in three pressure ranges: low, medium, and high. When replacing

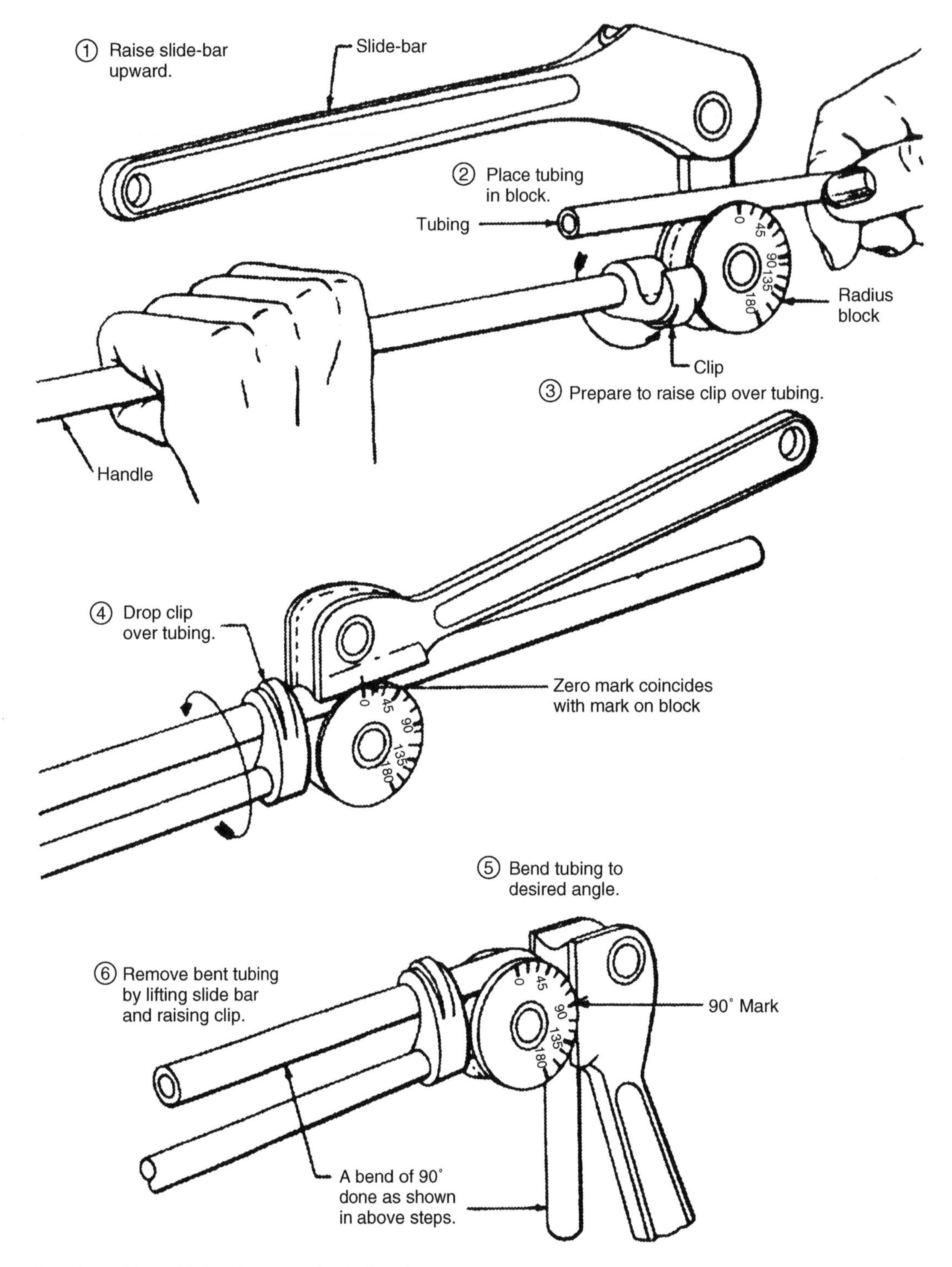
① Raise slide-bar upward.
Slide-bar
② Place tubing in block.
Tubing
Radius block
Clip
③ Prepare to raise clip over tubing.
Handle
④ Drop clip over tubing.
Zero mark coincides with mark on block
⑤ Bend tubing to desired angle.
⑥ Remove bent tubing by lifting slide bar and raising clip.
90° Mark
A bend of 90° done as shown in above steps.

Fig. 5.1.30 Bending tubing with hand-operated tube bender.

hoses, it is important to ensure that the replacement hose is a duplicate of the one removed in length, OD, material, type and contour. In selecting hose, several precautions must be observed. The selected hose must:

1. Be compatible with the system fluid.
2. Have a rated pressure greater than the design pressure of the system.
3. Be designed to give adequate performance and service for infrequent transient pressure peaks up to 150 per cent of the working pressure of the hose. And,
4. Have a safety factor with a burst pressure at a minimum of 4 times the rated working pressure.

There are temperature restrictions applied to the use of hoses. Rubber hose must not be used where the operating

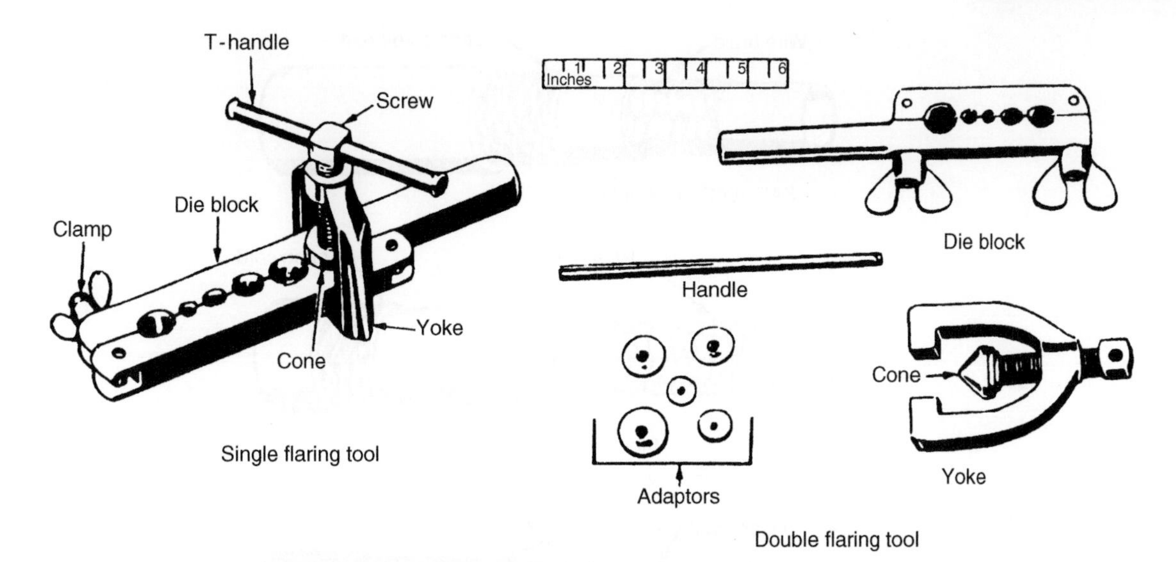

Fig. 5.1.31 Flaring tools.

temperature exceeds 200 °F. PTFE hoses in high-pressure air systems must not be used where the temperature exceeds 350 °F.

5.1.9.1.13 Installation

Flexible hose must not be twisted during installation. This will reduce the life of the hose and may cause the fittings to loosen. You can determine whether a hose is twisted by looking at the layline that runs along the length of the hose. If the layline does not spiral around the hose, the hose is not twisted. If the layline does spiral around the hose, the hose is twisted and must be untwisted. Flexible hose should be protected from chafing by using a chafe-resistant covering wherever necessary.

The minimum bend radius for flexible hose varies according to the size and construction of the hose and the pressure under which the system operates. Current applicable technical publications contain tables and graphs showing the minimum bend radii for the different types of installations. Bends that are too sharp will reduce the bursting pressure of flexible hose considerably below its rated valve.

Flexible hose should be installed so that it T is subjected to a minimum of flexing during operation. Support clamps are not necessary with short installations; but for hose of considerable length (48 in. for example), clamps should be placed not more than 24 in. apart. Closer supports are desirable and in some cases may be required.

A flexible hose must never be stretched tightly between two fittings. About 5–8 per cent of the total length must be allowed as slack to provide freedom of movement under pressure. When under pressure, flexible hose contracts in length and expands in diameter. Examples of correct and incorrect installations of flexible hose are illustrated in Figure 5.1.33.

5.1.9.2 Types of fittings and connectors

Some type of connector or fitting must be provided to attach the lines to the components of the system and to connect sections of line to each other. There are many different types of connectors and fittings provided for this purpose. The type of connector or fitting required for a specific system depends on several factors. One determining factor is the type of fluid line (pipe, tubing, or flexible hose) used in the system. Other determining factors are the type of fluid medium and maximum operating pressure of the system. Some of the most common types of fittings and connectors are described in the sections that follow.

5.1.9.2.1 Pipes and tubing

High-pressure pipe or tubing can also be used for hydraulic circuits. In these applications, special threading or fittings are required to connect circuit components.

Threaded connectors: There are several different types of threaded connectors. In the type discussed in this section, both the connector and the end of the fluid line are threaded. These connectors are used in some low-pressure fluid power systems and are usually made of steel, copper, and brass is available in a variety of designs.

Threaded connectors (Figure 5.1.34) are made with standard pipe threads cut on the inside surface (female). The end of the pipe is threaded with outside threads (male). Standard pipe threads are tapered slightly to ensure tight connections. The amount of taper is approximately 3/4-inch in diameter per foot of thread.

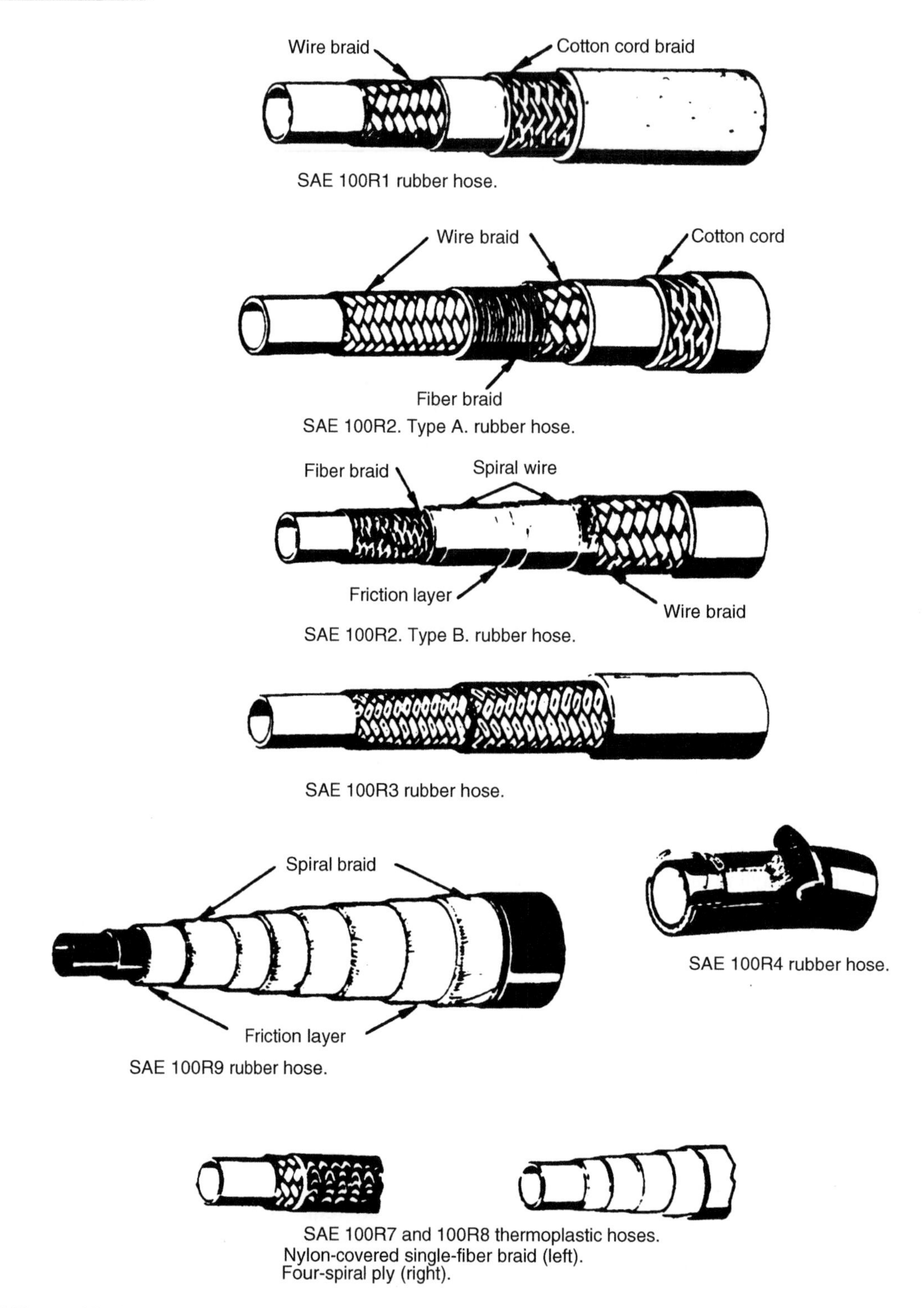

Fig. 5.1.32 Types of flexible hose.

Metal is removed when pipe is threaded, thinning the pipe and exposing new and rough surfaces. Corrosion agents work more quickly at such points than elsewhere. If pipes are assembled with no protective compounds on the threads, corrosion sets in at once and the two sections stick together so that the threads seize when disassembly is attempted. The result is damaged threads and pipes. To prevent seizing, a suitable pipe thread compound is sometimes applied to the threads. The two end threads must be kept free of compound so that it will not contaminate the hydraulic fluid. Pipe compound, when improperly applied, may get inside the lines and damage

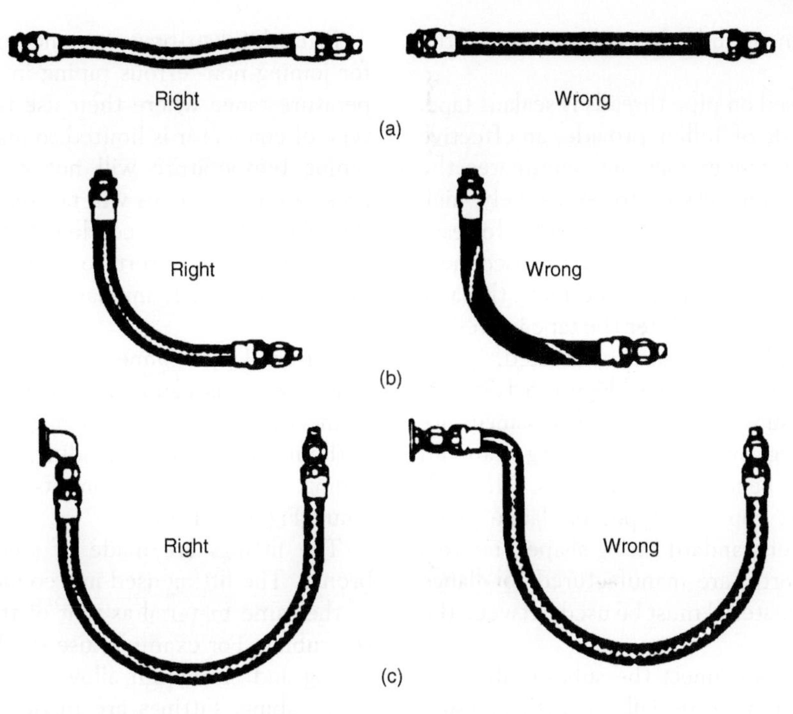
Right
Wrong
(a)
Right
Wrong
(b)
Right
Wrong
(c)

Fig. 5.1.33 Correct and incorrect installation of flexible hose.

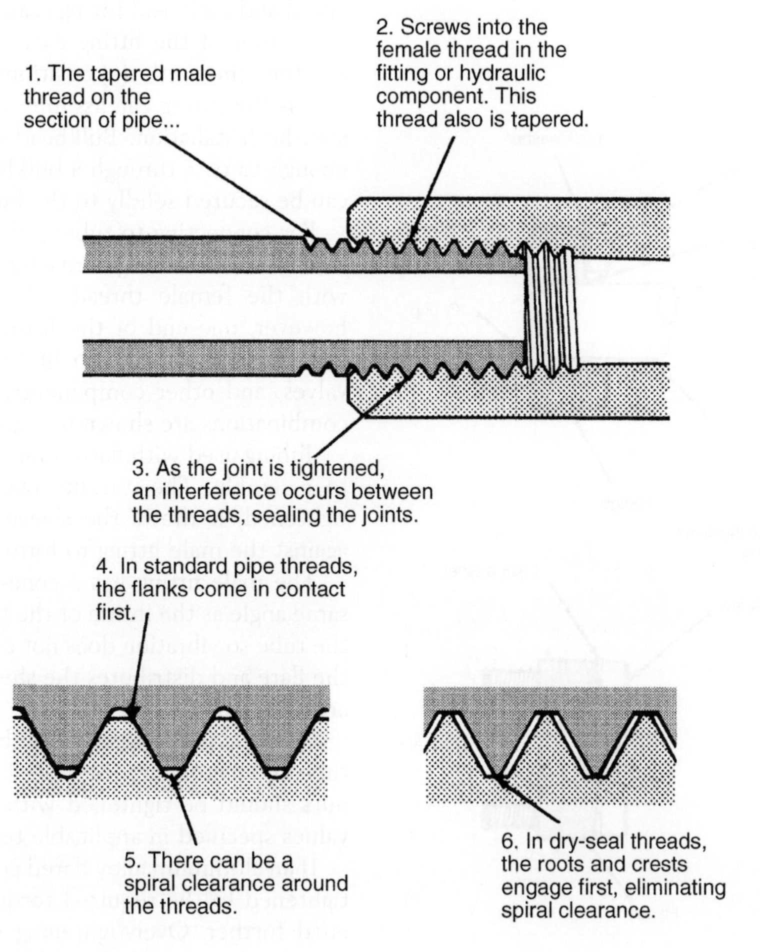
1. The tapered male thread on the section of pipe...
2. Screws into the female thread in the fitting or hydraulic component. This thread also is tapered.
3. As the joint is tightened, an interference occurs between the threads, sealing the joints.
4. In standard pipe threads, the flanks come in contact first.
5. There can be a spiral clearance around the threads.
6. In dry-seal threads, the roots and crests engage first, eliminating spiral clearance.

Fig. 5.1.34 Threaded connectors.

pumps, control equipment and other components of the system.

Another material used on pipe threads is sealant tape. This tape, which is made of Teflon, provides an effective means of sealing pipe connections and eliminates the necessity of torquing connections to excessively high values in order to prevent leaks. It also provides for ease of maintenance whenever it is necessary to disconnect pipe joints. The tape is applied over the male threads, leaving the first thread exposed. After the tape is pressed firmly against the threads, the joint is connected.

Flanged: Bolted flange connectors, Figure 5.1.35, are suitable for most pressures now in use. The flanges are attached to the piping by welding, brazing, tapered threads, or rolling and bending into recesses. Those illustrated are the most common types of flange joints used. The same types of standard fitting shapes, i.e. tee, cross, elbow, and so forth, are manufactured for flange joints. Suitable gasket material must be used between the flanges.

Welded: Welded joints connect the subassemblies of some fluid power systems, especially in high-pressure systems that use pipe for fluid lines. The welding is done according to standard specifications that define the materials and techniques.

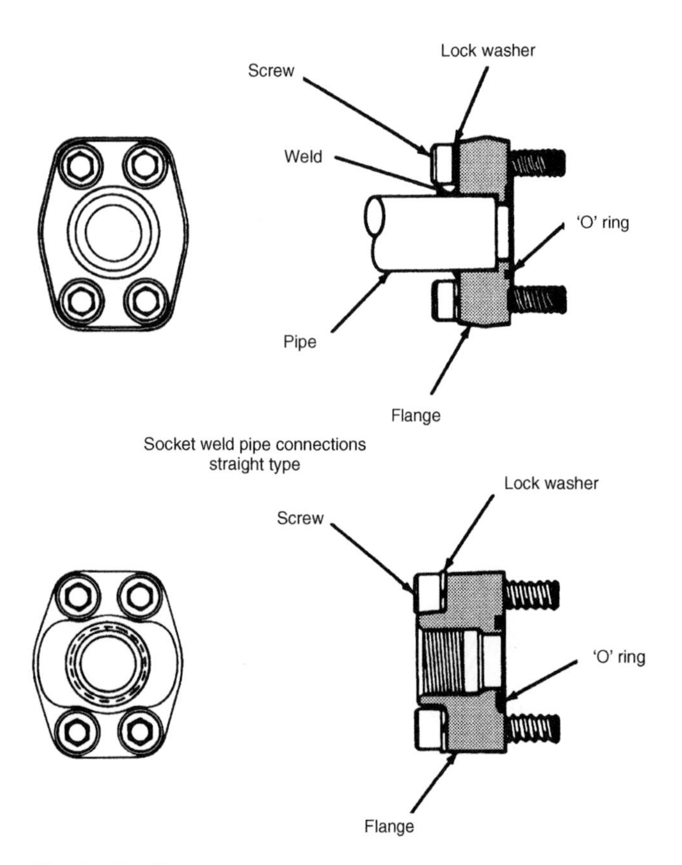

Fig. 5.1.35 Four types of bolted flange connectors.

Brazed: Silver-brazed connectors are commonly used for joining non-ferrous piping in the pressure and temperature range where their use is practical. Use of this type of connector is limited to installations in which the piping temperature will not exceed 425 °F and the pressure in cold lines will not exceed 3000 psi. Heating the joint with an oxyacetylene torch melts the alloy. This causes the alloy insert to melt and fill the few thousandths of an inch annular space between the pipe and fitting.

Flared: These connectors are commonly used in fluid power systems containing lines made of tubing. These connectors provide safe, strong, dependable connections without the need for threading, welding, or soldering the tubing. The connector consists of a fitting, a sleeve, and a nut, Figure 5.1.36.

The fittings are made of steel, aluminum alloy, or bronze. The fitting used in a connection should be made of the same material as that of the sleeve, the nut, and the tubing. For example, use steel connectors with steel tubing and aluminum alloy connectors with aluminum alloy tubing. Fittings are made in union, 45° and 90° elbows, tee, and various other shapes, Figure 5.1.37.

Tees, crosses, and elbows are self-explanatory. Universal and bulkhead fittings can be mounted solidly with one outlet of the fitting extending through a bulkhead and the other outlet(s) positioned at any angle. Universal means the fitting can assume the angle required for the specific installation. Bulkhead means the fitting is long enough to pass through a bulkhead and is designed so it can be secured solidly to the bulkhead.

For connecting to tubing, the ends of the fittings are threaded with straight machine threads to correspond with the female threads of the nut. In some cases, however, one end of the fitting may be threaded with tapered pipe threads to fit threaded ports in pumps, valves, and other components. Several of these thread combinations are shown in Figure 5.1.37.

Tubing used with flare connectors must be flared prior to assembly. The nut fits over the sleeve and, when tightened, it draws the sleeve and tubing flare tightly against the male fitting to form a positive seal.

The male fitting has a cone-shaped surface with the same angle as the inside of the flare. The sleeve supports the tube so vibration does not concentrate at the edge of the flare and distributes the shearing action over a wider area for added strength.

Correct and incorrect methods of installing flared-tube connectors are illustrated in Figure 5.1.38. Tubing nuts should be tightened with a torque wrench to the values specified in applicable technical publications.

If an aluminum alloy flared connector leaks after being tightened to the required torque, it must not be tightened further. Over-tightening may severely damage or completely cut off the tubing flare or may result in

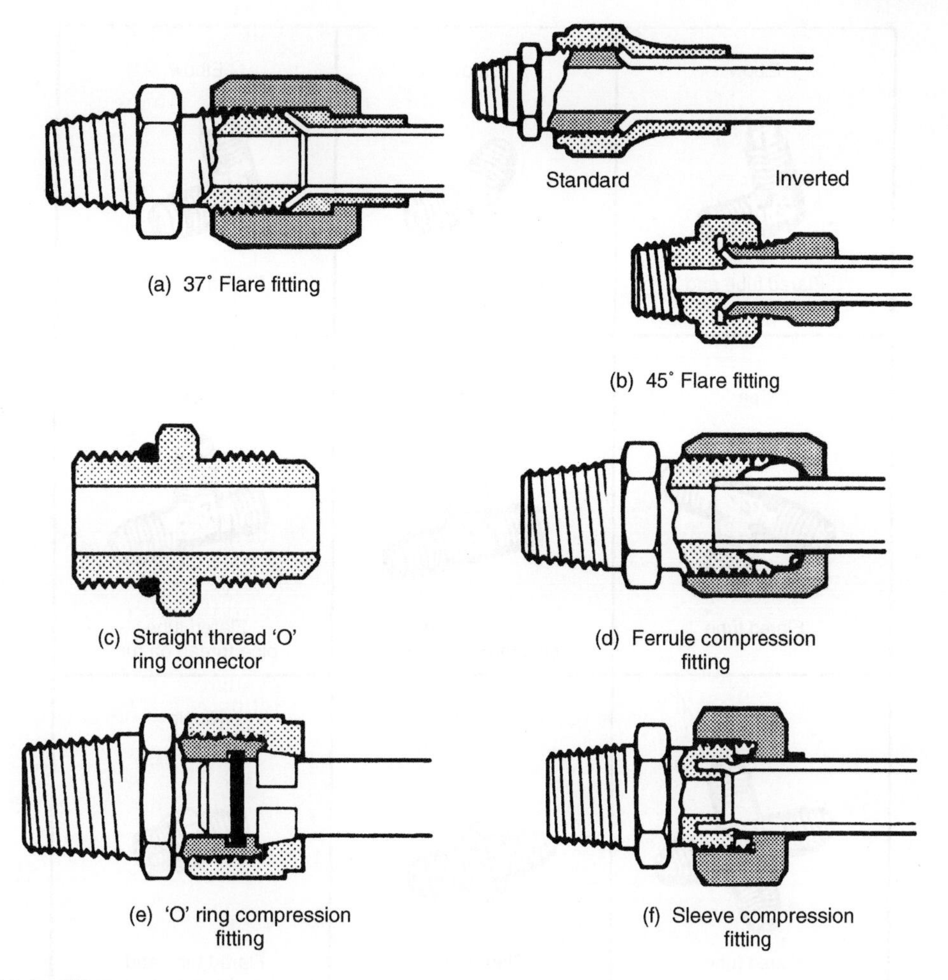

Fig. 5.1.36 Flared tube fitting.

damage to the sleeve or nut. The leaking connection must be disassembled and the fault corrected.

If steel tube connection leaks, it may be tightened 1/16 turn beyond the specified torque in an attempt to stop the leakage. If the connection continues to leak, it must be disassembled and the problem corrected.

5.1.9.2.2 Connectors for flexible hose

There are various types of end fittings for both piping connection side and hose connection side of hose fittings. Figure 5.1.39 shows commonly used fittings.

5.1.9.2.3 Quick-disconnect couplings

Self-sealing, quick-disconnect couplings, Figure 5.1.40, are used at various points in many fluid power systems. These couplings are installed at locations where frequent uncoupling of the lines is required for inspection, test, and maintenance. Quick-disconnect couplings are also commonly used in pneumatic systems to connect sections of air hose and to connect tools to the air pressure lines. This provides a convenient method of attaching and detaching tools and sections of lines without losing pressure.

Quick-disconnect couplings provide a means for quickly disconnecting a line without the loss of fluid from the system or the entrance of foreign matter into the system. Several types of quick-disconnect couplings have been designed for use in fluid power systems. Figure 5.1.40 illustrates a coupling that is used with portable pneumatic tools. The male section is connected to the tool or to the line leading from the tool. The female section, which contains the shutoff valve, is installed in the line leading from the pressure source. These connectors can be separated or connected by very little effort on the part of the operator.

The most common quick-disconnect coupling for hydraulic systems consists of two parts, held together by a union nut. Each part contains a valve which is held open when the coupling is connected, allowing fluid to flow in either direction. When the coupling is disconnected, a spring in each part closes the valves, preventing the loss of fluid and entrance of foreign matter.

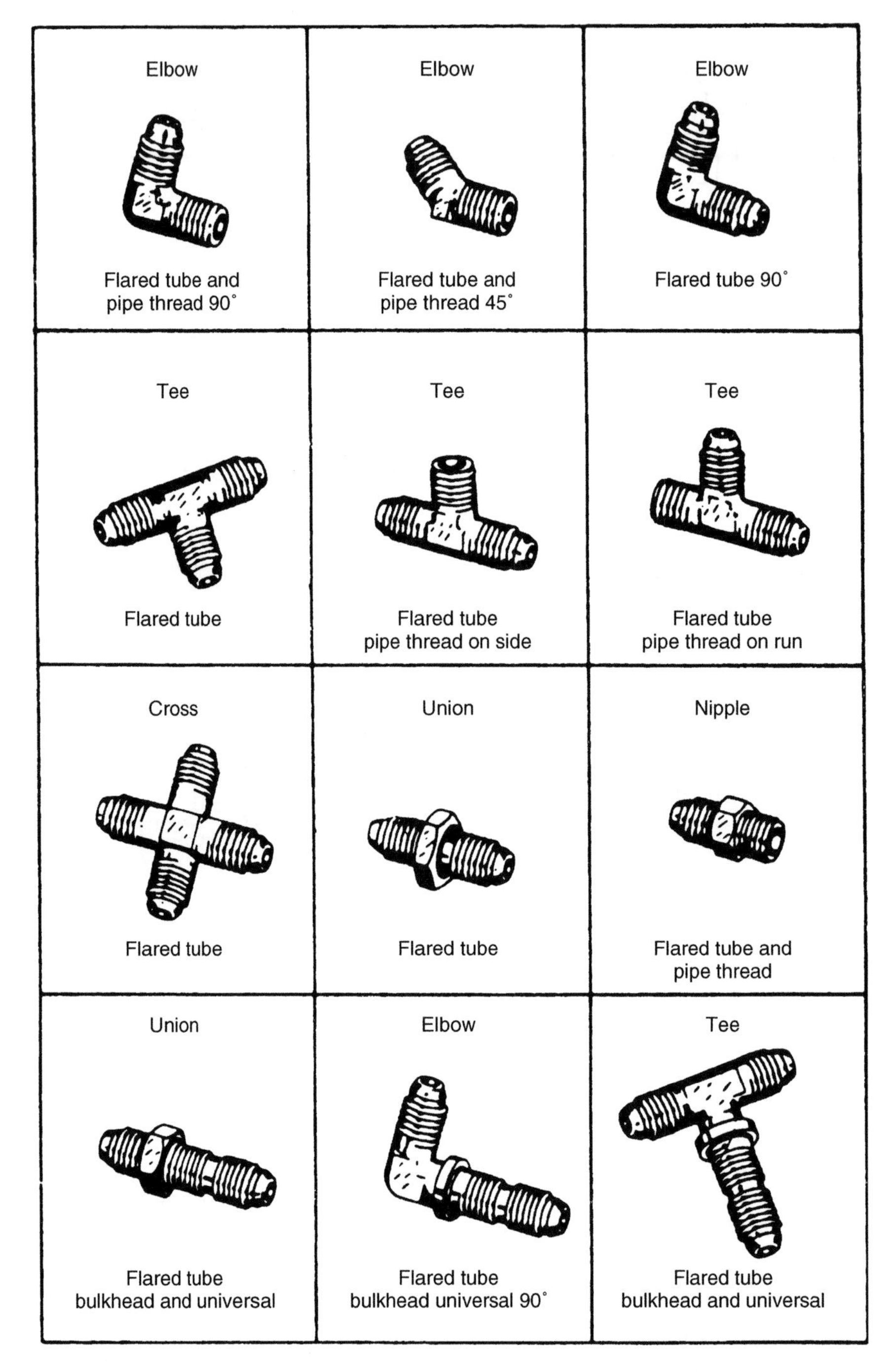

Fig. 5.1.37 Flared tube fittings.

5.1.9.2.4 Manifolds

Some fluid power systems are equipped with manifolds in the pressure supply and/or return lines. A manifold is a fluid conductor that provides multiple connection ports. Manifolds eliminate piping, reduce joints, which are often a source of leakage, and conserve space. For example, manifolds may be used in systems that contain several subsystems. One common line connects the pump to the manifold.

There are outlet ports in the manifold to provide connections to each subsystem. A similar manifold may be used in the return system. Lines from the control valves of the subsystem connect to the inlet ports of the manifold, where the fluid combines into

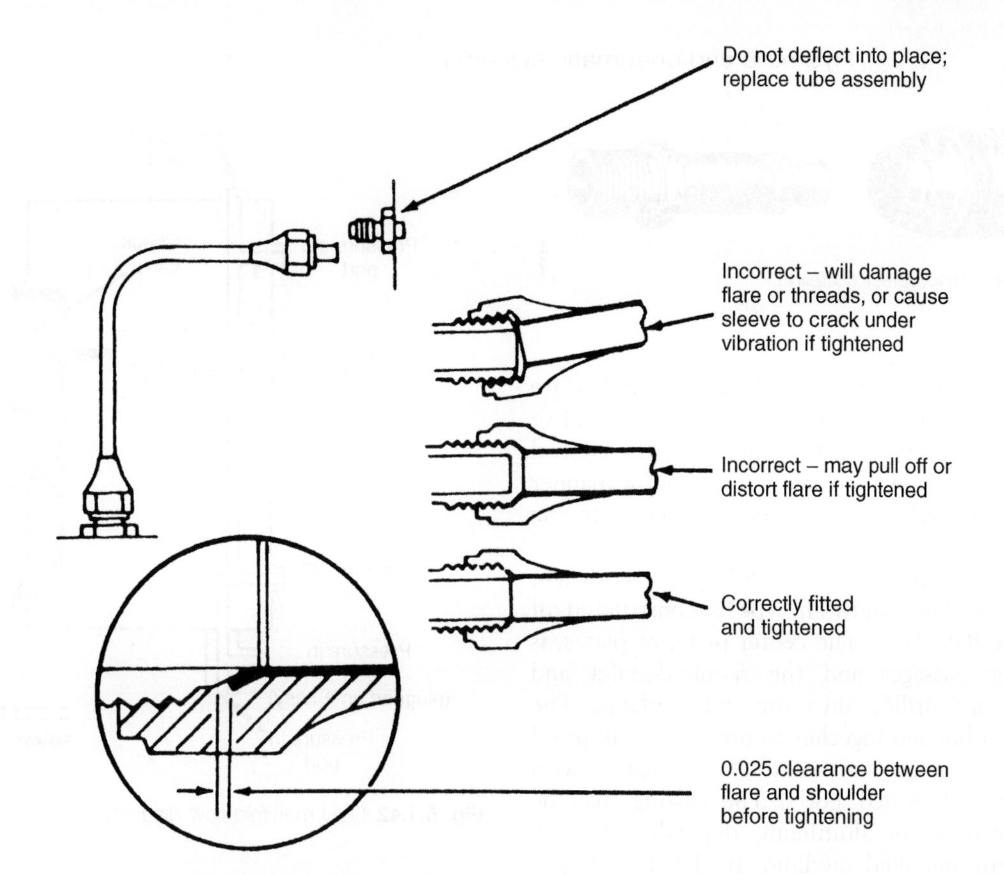

Fig. 5.1.38 Correct and incorrect method of installing flared fittings.

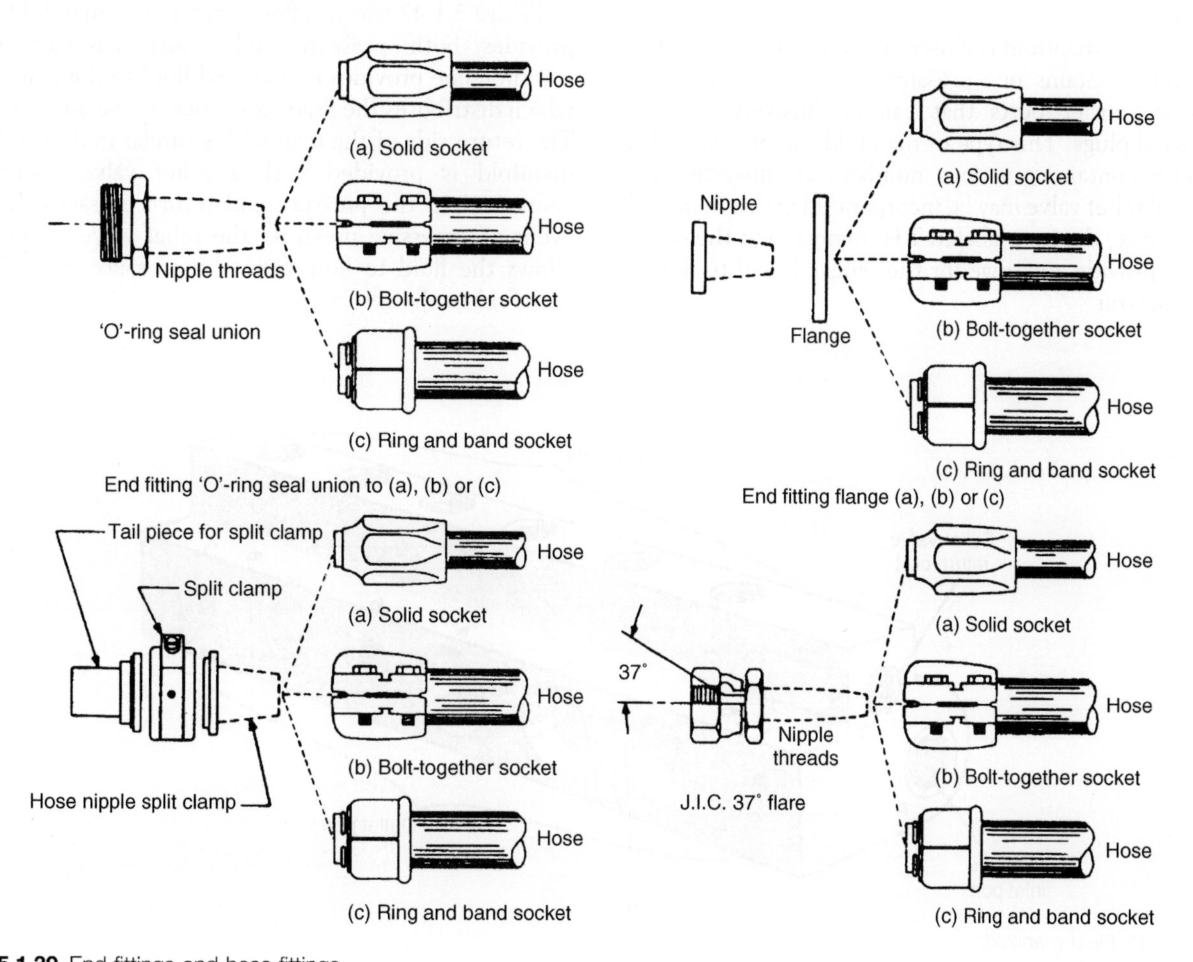

Fig. 5.1.39 End fittings and hose fittings.

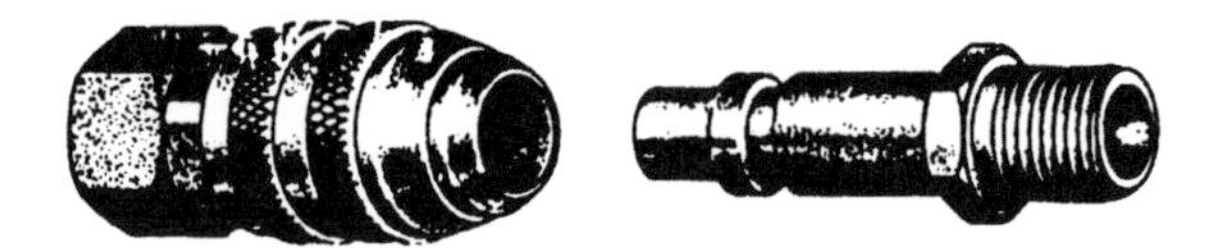

Fig. 5.1.40 Quick-disconnect coupling.

one outlet line to the reservoir. Some manifolds are equipped with check valves, relief valves, filters, and so on, required for the system. In some cases, the control valves are mounted on the manifold in such a manner that ports of the valves are connected directly to the manifold.

Manifolds are usually one of three types: sandwich, cast, or drilled. The sandwich type is constructed of three or more flat plates. The center plate, or plates, is machined for passages and the required inlet and outlet ports are drilled into the outer plates. The plates are then bonded together to provide a leak-proof assembly. The cast type of manifold is designed with cast passages and drilled ports. The casting may be iron, steel, bronze, or aluminum, depending on the type of system and fluid medium. In the drilled type manifold, all ports and passages are drilled in a block of metal.

A simple manifold is illustrated in Figure 5.1.41. This manifold contains one pressure inlet port and several pressure outlet ports that can be blocked off with threaded plugs. This type of manifold can be adapted to systems containing various numbers of subsystems. A thermal relief valve may be incorporated in this manifold. In this case, the port labeled T is connected to the return line to provide a passage for the relieved fluid to flow to the reservoir.

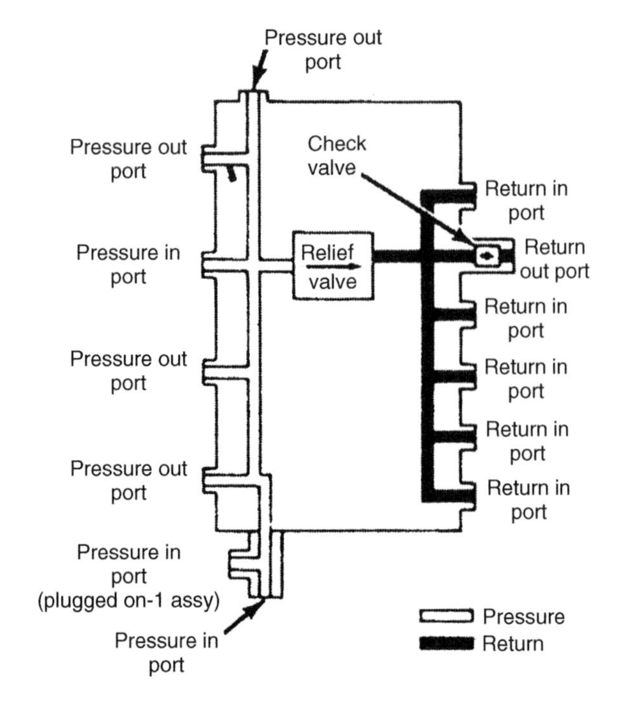

Fig. 5.1.42 Fluid manifold flow diagram.

Figure 5.1.42 shows a flow diagram in a manifold that provides both pressure and return passages. One common line provides pressurized fluid to the manifold, which distributes the fluid to any one of five outlet ports. The return side of the manifold is similar in design. This manifold is provided with a relief valve, which is connected to the pressure and return passages. In the event of excessive pressure, the relief valve opens and allows the fluid to flow from the pressure side of the manifold to the return side.

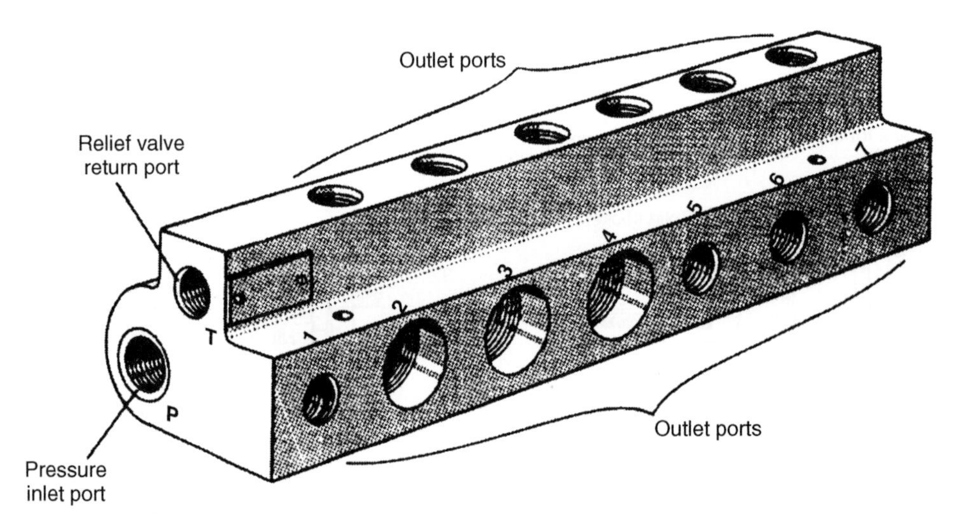

Fig. 5.1.41 Fluid manifold.

Chapter 5.2

5.2

Pneumatic fundamentals

Mobley

5.2.1 Introduction

The purpose of pneumatics is to do work in a controlled manner. The control of pneumatic power is accomplished using valves and other control devices that are connected together in an organized circuit. The starting point in this organized circuit is the air compressor, where the air is pressurized.

The pressurized air goes to an air receiver for storage and then is processed for use by passing through filters, dryers, and, in some cases, lubricators. This pressurized air is normally classified as *instrument air* when it is used in control systems. This air must be moist and oil free to prevent the control devices from clogging up.

The law states that any pressurized air system must be fitted with a pressure relief valve. This valve prevents the system from being over pressurized and becoming a hazard to personnel or damaging equipment. A pressure switch is an electro/pneumatic control device that is installed on the air receiver to regulate the output of the air compressor. When the air pressure reaches its maximum set point, the regulator is activated and transmits a signal to a solenoid valve on the air compressor. This solenoid valve opens to direct lubricating oil to hydraulically keep the suction valves shut on both the low- and high-pressure cylinders on the air compressor. The air compressor will remain in this mode until the pressure drops to the lower set point and deactivates the pressure regulator. This, in turn, de-energizes the solenoid valve on the air compressor causing it to release the lubricating oil pressure on the low- and high-pressure suction valves. The air compressor returns to normal operation pumping air into the receiver until the maximum pressure set point is reached; when this happens, the control cycle starts again. Using a pressure switch prevents the air compressor from running continuously.

Figure 5.2.1 shows a typical compressed air supply system. The following describes the functions of the system components:

- Compressor – compresses the air.
- Pressure switch – turns the air compressor on and off.
- Pressure relief valve – relieves air pressure at 110 per cent of the operating maximum pressure. This device is fitted to the receiver by law.
- Check valve – permits the compressed air to flow away from the compressor and will not allow any air to return to the compressor.
- Air receiver – the receiver stores the pressurized air. By law, it must have the following fitting installed on it:

  (a) Pressure relief valve.
  (b) Pressure gage.
  (c) Access hand-hole.
  (d) Drain valve.
- Pressure regulator – controls the system pressure to the manifold.
- Pressure gage – indicates internal pressure of the system. Must be fitted by law.
- Filter – this device cleans the air of dirt and contaminants.
- Lubricator – this device adds a small amount of oil to the air in order to lubricate equipment.
  Note: this is only installed when needed. Most systems are oil-free.

Plant Engineer's Handbook; ISBN: 9780750673280

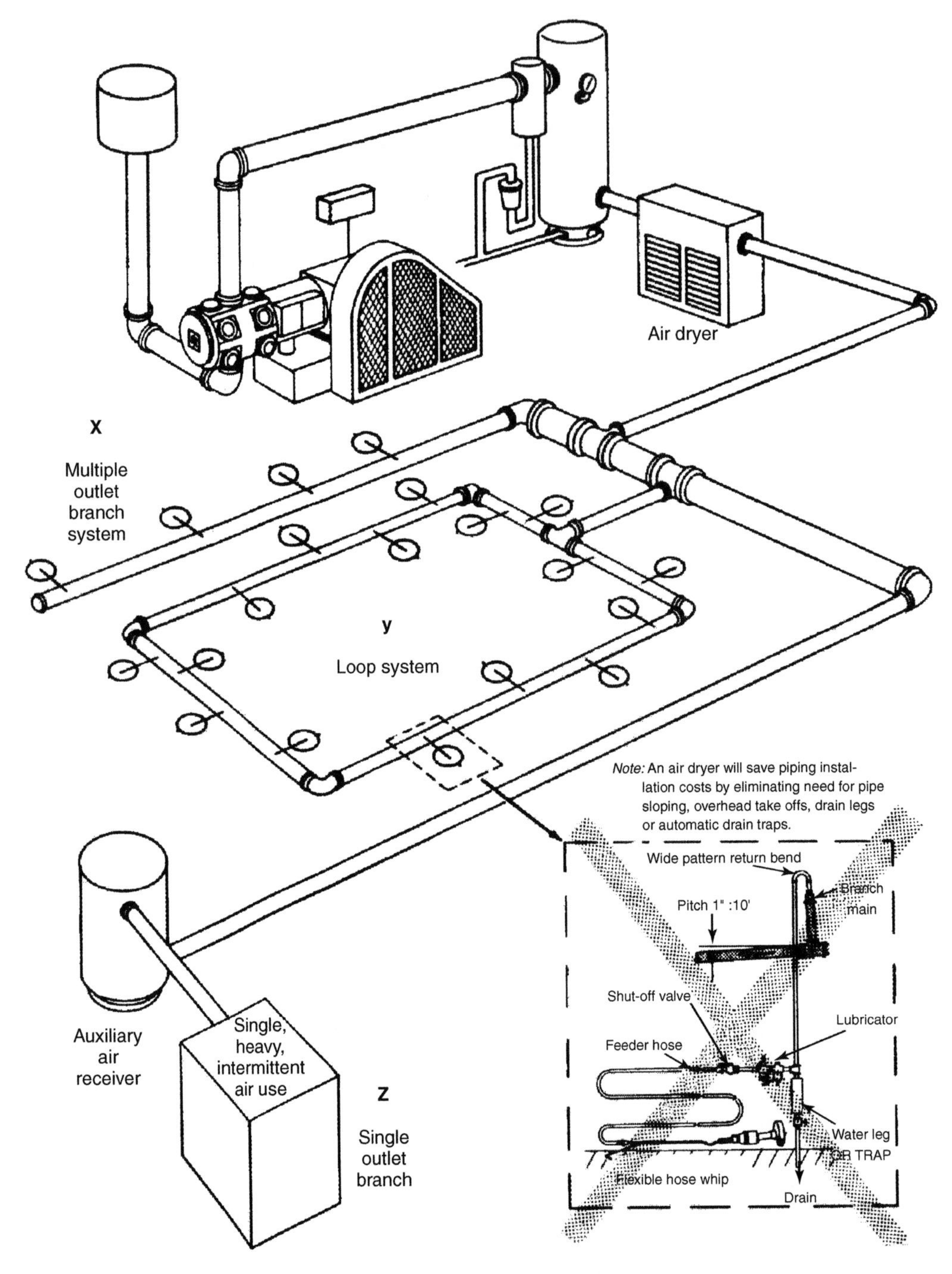

Fig. 5.2.1 Typical compressed air supply.

- Pressure manifold – this distributes the air to the various pressure ports.
- Needle valves – control the airflow to the various systems that are to be operated.

5.2.1.1 Hazards of compressed air

People often lack respect for the power in compressed air because air is so common, and it is viewed as harmless. At sufficient pressures, compressed air can cause damage if an accident occurred. To minimize the hazards of working with compressed air, all safety precautions should be followed closely. Reasons for general precautions follow.

Small leaks or breaks in the compressed air system can cause minute particles to be blown at surprisingly high speeds. Always wear safety glasses when working in the vicinity of any compressed air system. Goggles in place of glasses are recommended if contact lenses are worn.

Compressors can make an exceptional amount of noise while running. The noise of the compressor, in addition to the drain valves lifting, creates noise to require hearing protection. The area around compressors should always be posted as a hearing protection zone.

Pressurized air can do the same type of damage as pressurized water. Treat all operations on compressed air systems with the same care taken on liquid systems. Closed valves should be slowly cracked open and both sides allowed to equalize prior to opening the valve further.

5.2.2 Characteristics of compressed air

Pascal's law states that the pressure of a gas or liquid exerts force equally in all directions against the walls of its container. The force is measured in terms of force per unit area (pounds per square inch - psi). This law is for liquids and gases at rest and neglects the weight of the gas or liquid. It should be noted that the field of fluid power is divided into two parts: pneumatics and hydraulics. These two have many characteristics in common. The difference is hydraulic systems use liquids, and pneumatics use gases, usually air. Liquids are only slightly compressible and, in hydraulic systems, this property can often be neglected. Gases, however, are very compressible.

Three properties of gases must be well understood in order to gain an understanding of pneumatic power systems. These are its temperature, pressure, and volume. Physical laws that define their efficiency and system dynamics govern compressed air systems and compressors. These laws include the following.

5.2.2.1 Thermodynamics

Both the first and second laws of thermodynamics apply to all compressors and compressed air systems.

5.2.2.1.1 First law

This law states that energy cannot be created or destroyed during a process, such as compression and delivery of air or gas, although it may change from one form of energy to another. In other words, whenever a quantity of one kind of energy disappears, an exactly equivalent total of other kinds of energy must be produced.

5.2.2.1.2 Second law

This is more abstract and can be stated in several ways:

- Heat cannot, by itself, pass from a colder to a hotter body.
- Heat can be made to go from a body at lower temperature to one at higher temperature only if external work is done.
- The available energy of the isolated system decreases in all real processes.
- Heat or energy, like water, by itself, will flow only downhill.

These statements say that energy exists at various levels and is available for use only if it can move from a higher to a lower level. In thermodynamics, a measure of the unavailability of energy has been devised and is known as *entropy*. It is defined by the differential equation:

$$dS = \frac{dQ}{T}$$

Entropy, as a measure of unavailability, increases as a system loses heat, but remains constant when there is no gain or loss of heat, as in an adiabatic process.

5.2.2.1.3 Boyle's law

If a fixed amount of gas is placed in a container of variable volume (such as a cylinder fitted with a piston), the gas will fill completely the entire volume, however large it may be. If the volume is changed, the pressure exerted by the gas will also change. As the volume decreases, the pressure increases. This property is called Boyle's law and can be written as:

$$P_1 \times V_1 = P_2 \times V_2$$

where P_1, initial absolute pressure; V_1, initial volume of air or gas; P_2, final pressure (psia); V_2, final volume of air or gas.

According to Boyle's law, pressure of a gas is inversely proportional to the volume, if the temperature is held constant. For example, 2 ft^3 at 4 psi would exert only 1 psi if allowed to expand 8 ft^3.

$$4 \text{ psia} \times 2 \text{ cubic feet} = P_2 \times 8 \text{ ft}^3$$

$$P_2 = \frac{4 \text{ psia} \times 2 \text{ ft}^3}{8 \text{ ft}^3}$$

$$P_2 = 1 \text{ psia}$$

In calculations that involve gas pressure and volume, *absolute pressure* or pounds per square inch absolute (psia) must be used.

5.2.2.1.4 Charles' law

If a fixed quantity of gas is held at a constant pressure and heated or cooled, its volume will change. According to Charles' law, the volume of a gas at constant pressure is directly proportional to the absolute temperature. This is shown by the following equation:

$$\frac{V_1}{V_2} = \frac{T_1}{T_2}$$

It is important to remember that *absolute temperature* must be considered, not temperature, according to normal Fahrenheit or Centigrade scales. The absolute Fahrenheit scale is called Rankine and the absolute Centigrade is called Kelvin. For conversion:

0°fahrenheit = 460°rankine
0°centigrade = 273 kelvin

Thus gas at 700°F would be 530° on Rankine scale.

5.2.2.2 Combined effect of pressure, volume, and temperature

Pressure, temperature, and volume are properties of gases that are completely interrelated. Boyle's law and Charles' law may be combined into the following ideal gas law, which is true for any gas:

$$\frac{P_1 V_1}{T_1} = \frac{P_2 V_2}{T_2}$$

According to this law, if the three conditions of a gas are known in one situation, then if any condition is changed and the effect on the others may be predicted.

5.2.2.2.1 Dalton's law

This states that the total pressure of a mixture of ideal gases is equal to the sum of the partial pressures of the constituent gases. The partial pressure is defined as the pressure each gas would exert if it alone occupied the volume of the mixture at the mixture's temperature.

Dalton's law has been proved experimentally to be somewhat inaccurate, the total pressure often being higher than the sum of the partial pressures. This is especially true during transitions as pressure is increased. However, for engineering purposes it is the best rule available and the error is minor.

When all contributing gases are at the same volume and temperature, Dalton's law can be expressed as

$$p = p_a + p_b + p_c + \cdots$$

5.2.2.2.2 Amagat's law

This is similar to Dalton's law, but states that the volume of a mixture of ideal gases is equal to the sum of the partial volumes that the constituent gases would occupy if each existed alone at the *total* pressure and temperature of the mixture. As a formula this becomes:

$$V = V_a + V_b + V_c + \cdots$$

5.3.2.2.3 Perfect gas formula

Starting with Charles' and Boyle's laws, it is possible to develop the formula for a given weight of gas:

$$pV = WR_1T$$

where W is weight and R_1 is a specific constant for the gas involved. This is the perfect gas equation. Going one step further, by making W, in pounds, equal to the molecular weight of the gas (one mole), the formula becomes:

$$pV = R_0T$$

In this very useful form, R_0 is known as the *universal gas constant*, has a value of 1545, and is the same for all gases. The *specific* gas constant (R_1) for any gas can be obtained by dividing 1545 by the molecular weight. R_0 is only equal to 1545 when gas pressure (p) is in PSIA; volume (V) is expressed as cubic feet per pound mole; and temperature (T) is in Rankine or absolute, i.e., °F + 460.

5.2.2.2.4 Avogadro's law

Avogadro states that equal volumes of all gases, under the same conditions of pressure and temperature, contain the same number of molecules. This law is very important and is applied in many compressor calculations.

The *mole* is particularly useful when working with gas mixtures. It is based on Avogadro's law that equal volumes of gases at given pressure and temperature (pT) conditions contain equal number of molecules. Since this is so, then the *weight* of these equal volumes will be proportional to their molecular weights. The volume of one *mole* at any desired condition can be found by the use of the perfect gas law.

$$pV = R_0T \text{ or } pV = 1545\ T$$

5.2.2.3 Gas and vapor

By definition, *a gas* is that fluid form of a substance in which it can expand indefinitely and completely fill its container. A *vapor* is a gasified liquid or solid or a substance in gaseous form. These definitions are in general use today.

All gases can be liquefied under suitable pressure and temperature conditions and therefore could be called vapors. The term gas is most generally used when conditions are such that a return to the liquid state, i.e., condensation, would be difficult within the scope of the operations being conducted. However, a gas under such conditions is actually a superheated vapor.

5.2.2.4 Changes of state

Any given pure substance may exist in three states: as a solid, as liquid, or as vapor. Under certain conditions, it may exist as a combination of any two phases and changes in conditions may alter the proportions of the two phases. There is also a condition where all three phases may exist at the same time. This is known as the *triple point*. Water has a triple point at near 32 °F and 14.696 psia. Carbon dioxide may exist as a vapor, a liquid, and solid simultaneously at about minus 69.6 °F and 75 psia. Substances under proper conditions may pass directly from a solid to a vapor phase. This is known as *sublimation*.

5.2.2.5 Changes of state and vapor pressure

As liquid physically changes into a gas, their molecules travel with greater velocity and some break out of the liquid to form a vapor above the liquid. These molecules create a *vapor pressure* that, at a specified temperature, is the only pressure at which a pure liquid and its vapor can exist in equilibrium. If in a closed liquid–vapor system, the volume is reduced at constant temperature, the pressure will increase imperceptibly until condensation of part of the vapor into liquid has lowered the pressure to the original vapor pressure corresponding to the temperature. Conversely, increasing the volume at constant temperature will reduce the pressure imperceptibly and molecules will move from the liquid phase to the vapor phase until the original vapor pressure has been restored. For every substance, there is a definite vapor pressure corresponding to each temperature.

The temperature corresponding to any given vapor pressure is obviously the *boiling point* of the liquid and also the dew point of the vapor. Addition of heat will cause the liquid to boil and removal of heat will start condensation. The three terms, saturation temperature, boiling point, and dew point, all indicate the same physical temperature at a given vapor pressure. Their use depends on the context in which they appear.

5.2.2.6 Critical gas conditions

There is one temperature above which a gas will not liquefy due to pressure increase. This point is called the *critical temperature*. The pressure required to compress and condense a gas at this critical temperature is called the *critical pressure*.

5.2.2.6.1 Relative humidity

Relative humidity is a term frequently used to represent the quantity of moisture or water vapor present in a mixture although it uses partial pressures in so doing. It is expressed as:

$$\mathrm{RH}(\%) = \frac{\begin{array}{c}\text{Actual partial}\\ \text{vapor pressure} \times 100\end{array}}{\begin{array}{c}\text{Saturated vapor pressure at}\\ \text{existing mixture temperature}\end{array}} = \frac{p_v \times 100}{p_\lambda}$$

Relative humidity is usually considered only in connection with atmospheric air, but since it is unconcerned with the nature of any other components or the total mixture pressure, the term is applicable to vapor content in any problem. The saturated water vapor pressure at a given temperature is always known from steam tables or charts. It is the existing partial vapor pressure, which is desired and therefore calculable when the relative humidity is stated.

5.2.2.6.2 Specific humidity

Specific humidity used in calculations on certain types of compressors is a totally different term. It is the ratio of the weight of water vapor to the weight of *dry air* and is usually expressed as pounds, or grains, of moisture per pound of dry air. Where *pa* is the partial air pressure, specific humidity can be calculated as

$$\mathrm{SH} = \frac{W_v}{W_a}$$

$$\mathrm{SH} = \frac{0.622 p_v}{p p_v} = \frac{0.622 p_v}{p_a}$$

5.2.2.6.3 Degree of saturation

The degree of saturation denotes the actual relationship between the weight of moisture existing in a space and the weight that would exist if the space were saturated.

$$\text{Degree of saturation}(\%) = \frac{\mathrm{SH}_{\text{actual}} \times 100}{\mathrm{SH}_{\text{saturated}}}$$

A great many dynamic compressors handle air. Their performance is sensitive to density of the air, which varies with moisture content. The practical application of partial pressures in compression problems centers to a large degree on the determination of mixture volumes or weights to be handled at the intake of each stage of compression, the determination of mixture molecular weight, specific gravity, and the proportional or actual weight of components.

5.2.2.6.4 Psychrometry

Psychrometry has to do with the properties of the air–water vapor mixtures found in the atmosphere. Psychrometry tables, published by the US Weather Bureau, give detailed data about vapor pressure, relative humidity, and dew point at the sea-level barometer of 30 in Hg, and at certain other barometric pressures. These tables are based on relative readings of dry bulb and wet bulb atmospheric temperatures as determined simultaneously by a sling psychrometer. The dry bulb reads ambient temperature while the wet bulb reads a lower temperature influenced by evaporation from a wetted wick surrounding the bulb of a parallel thermometer.

5.2.2.6.5 Compressibility

All gases deviate from the perfect or ideal gas laws to some degree. In some cases the deviation is rather extreme. It is necessary that these deviations be taken into account in many compressor calculations to prevent compressor and driver sizes being greatly in error.

Compressibility is experimentally derived from data about the actual behavior of a particular gas under pVT changes. The compressibility factor, Z, is a multiplier in the basic formula. It is the ratio of the actual volume at a given pT condition to ideal volume at the same pT condition. The ideal gas equation is therefore modified to

$$pV = ZR_0T \quad \text{or} \quad Z = \frac{pV}{R_0T}$$

In the above equation, R_0 is 1545 and p is pounds per square foot.

5.2.3 Generation of pressure

Keeping with the subject of pressure, the basic concepts will be treated in the working sequence: pressure generation, transmission, storage, and utilization in a pneumatic system.

Pumping quantities of atmospheric air into a tank or other pressure vessel produces pressure. Pressure is *increased* by progressively increasing the amount of air in a confined space. The effects of pressure exerted by a confined gas result from the average of forces acting on container walls caused by the rapid and repeated bombardment from an enormous number of molecules present in a given quantity of air. This is accomplished in a controlled manner by *compression*, a decrease in the space between the molecules. Less volume means that each particle has a shorter distance to travel, thus proportionately more collisions occur in a given span of time, resulting in a higher pressure. Air compressors are designed to generate particular pressures to meet individual application requirements.

Basic concepts discussed here are atmospheric pressure; vacuum; gage pressure; absolute pressure; Boyle's law or pressure/volume relationship; Charles' law or temperature/volume relationship; combined effects of pressure, temperature, and volume; and generation of pressure or compression.

5.2.3.1 Atmospheric pressure

In the physical sciences, pressure is usually defined as the perpendicular force per unit area, or the stress at a point within a confined fluid. This force per unit area acting on a surface is usually expressed in pounds per square inch.

The weight of the earth's atmosphere pushing down on each unit of surface constitutes atmospheric pressure, which is 14.7 psi at sea level. This amount of pressure is called *one atmosphere*. Because the atmosphere is not evenly distributed about earth, atmospheric pressure can vary, depending upon geographic location. Also, obviously, atmospheric pressure decreases with higher altitude. A barometer using the height of a column of mercury or other suitable liquid measures atmospheric pressure.

5.2.3.2 Vacuum

It is helpful to understand the relationship of vacuum to the other pressure measurements. Vacuums can range from atmospheric pressure down to 'zero absolute pressure', representing a 'perfect' vacuum (a theoretical condition involving the total removal of all gas molecules from a given volume). The amount of vacuum is measured with a device called a vacuum gage.

Vacuum is a type of pressure. A gas is said to be under vacuum when its pressure is below atmospheric pressure, i.e., 14.7 psig at sea level. There are two methods of stating this pressure, but only one is accurate in itself.

A differential gage that shows the difference in the system and the atmospheric pressure surrounding the system usually measures vacuum. This measurement is expressed as

Millimeters of mercury – Vacuum (mm Hg Vac)
Inches of mercury – Vacuum (in Hg Vac)
Inches of water – Vacuum (in H_2O Vac)
Pounds per square inch – Vacuum (psi Vac)

Unless the barometric or atmospheric pressure is also given, these expressions do not give an accurate specification of pressure. Subtracting the vacuum reading from the atmospheric pressure will give an absolute pressure which is accurate. This may be expressed as

Inches of mercury – Absolute (in Hg Abs)
Millimeters of mercury – Absolute (mm Hg Abs)
Pounds per square inch – Absolute (psia)

The word *absolute* should never be omitted, otherwise one is never sure whether a vacuum is expressed in differential or absolute terms.

5.2.3.3 Perfect vacuum

A perfect vacuum is space devoid of matter. It is absolute emptiness. The space is at zero pressure *absolute*. A perfect vacuum cannot be obtained by any known means, but can be closely approached in certain applications.

5.2.3.4 Gage pressure

Gage pressure is the most often used method of measuring pneumatic pressure. It is the relative pressure of the compressed air within a system. Gage pressure can be either positive or negative, depending upon whether its level is above or below the atmospheric pressure reference. Atmospheric pressure serves as the *reference level* for the most significant types of pressure measurements. For example, if we inflate a tire to 30 psi, an ordinary tire-pressure gage will express this pressure as the value in excess of atmospheric pressure, or 30 psig ('g' indicates gage pressure). This reading shows the numerical value of the *difference* between atmospheric pressure and the air pressure in the tire.

5.2.3.5 Absolute pressure

A different reference level, absolute pressure, is used to obtain the total pressure value. Absolute pressure is the *total* pressure, i.e., gage and atmospheric, and is expressed as psia or pounds per square inch absolute. To obtain absolute pressure, simply add the value of atmospheric pressure (14.7 psi at sea level) to the gage pressure reading.

Absolute pressure (psia) values must be used when computing the pressure changes in a volume and when pressure is given as one of the conditions defining the amount of gas contained within a sample.

5.2.4 Compressors

A compressor must operate within a system that is designed to acquire and compress a gas. These systems must include the following components regardless of compressor type.

5.2.4.1 Lubrication system

The lubrication system has two basic functions: to lubricate the compressor's moving components and to cool the system by removing heat from the compressor's moving parts. While all compressors must have a lubrication system, the actual design and function of these systems will vary depending on compressor type.

The lubricating system for centrifugal or dynamic compressors is designed to provide bearing lubrication. In smaller compressors, the lubrication systems may consist of individual oil baths located at each of the main shaft bearings. In larger compressors, such as a bull-gear design, a positive system is provided to inject oil into the internal, tilting-pad bearings located at each of the pinion shafts inside the main compressor housing.

In positive lubrication systems, a gear-type pump is normally used to provide positive circulation of clean oil within the compressor. In some cases, the main compressor shaft directly drives this pump. In others, a separate motor-driven pump is used.

Positive displacement compressors use their lubrication system to provide additional functions. The lubrication system must inject sufficient quantities of clean fluid to provide lubrication for the compressor's internal parts, such as pistons and lobes, and to provide a positive seal between moving and stationary parts.

The main components of a positive displacement compressor's lubrication system consist of an oil pump, filter, and heat exchanger. The crankcase of the compressor acts as the oil sump. A lockable drain cock is installed at the lowest end of the crankcase to permit removal of any water accumulation that has resulted from sweating of the crankcase walls. The oil passes through a strainer into the pump. It then flows through the heat exchanger, where it is cooled. After the heat exchanger, the cooled oil flows directly to the moving parts of the compressor before returning to the crankcase sump. A small portion is diverted to the oil injector if one is installed.

The oil that is injected into the cylinder seals the space between the cylinder wall and the piston rings. This prevents compressed air from leaking past the pistons, and thus improves the compressor's overall efficiency.

5.2.4.1.1 Lube pump

The oil pump is usually gear driven from the crankshaft so that it will start pumping oil immediately on start-up of the compressor. In compressors that work in an oil-free system, oil injectors are not used. Oil separators are installed on the discharge side after leaving the aftercooler.

5.2.4.1.2 Oil separator

The basic purpose of an oil separator (Figure 5.2.2) is to clean the pressurized air of any oil contamination, which is highly detrimental to pneumatically controlled instrumentation. A separator consists of an inlet, a series of internal baffle plates, a wire mesh screen, a sump, and an outlet. The pressurized air enters the separator and immediately passes through the baffle plates. As the air impinges on the baffle plates it is forced into making sharp directional changes as it passes through each baffle section. As a result, the oil droplets separate from the air and collect on the baffles before dropping into the separator's sump.

After the air clears the baffle section, it then passes through the wire mesh screen where any remaining oil is trapped. The relatively oil-free air continues to the air reservoir for storage. The air reservoir acts as a final separator where moisture and oil is eventually removed. The air reservoir has drain traps installed at its lowest point where any accumulated moisture/oil is automatically discharged.

As a part of any routine maintenance procedure, these discharge traps should periodically be manually bypassed to ensure that the trap is functioning, and no excessive water accumulation is evident.

5.2.4.2 Compressor selection

Air power compressors generally operate at pressures of 500 psig or lower, with the majority in the range of 125 psig or less. All major types of compressors (i.e., reciprocating, vane, helical lobe, and dynamic) are used for this type of service. Choice is limited somewhat by capacity at 100 psig of about 10,500 ft^3 per minute but can be built to approximately 28,000 cfm. The vane-type rotary has an upper listed size of 3700 cfm as a twin unit and the helical lobe rotary can be used to nearly 20,000 cfm. The centrifugal can be built to very large sizes. It is currently offered in the proven, moderate speed designs starting at a minimum of about 5000 cfm.

5.2.4.2.1 Selection criteria

The following guidelines should be used for the selection process. While the criteria listed are not all inclusive, they will provide definition of the major considerations that should be used to select the best compressor for a specific application.

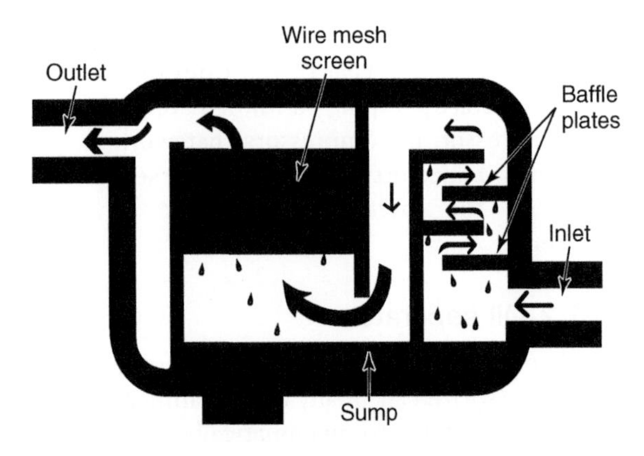

Fig. 5.2.2 Basic oil separator system.

5.2.4.2.2 Application

The mode of operation of a specific application should be the first consideration. The inherent design of each type of compressor defines the acceptable operating envelope or mode of operation that it can perform with reasonable reliability and life cycle costs. For example, a bull-gear-type centrifugal compressor is not suitable for load-following applications but will prove exceptional service in constant-load and volume applications.

Load factor is the ratio of actual compressed air output, while the compressor is operating, to the rated full-load output during the same period. It should never be 100 per cent, a good rule being to select an installation for from 50 to 80 per cent load factor, depending on the size, type and number of compressors involved. Proper use of load factor results in: more uniform pressure, a cooling-off period, less maintenance, and ability to increase use of air without additional compressors.

Load factor is particularly important with air-cooled machines where sustained full-load operation results in an early build-up of deposits on valves and other parts. This build-up increases the frequency of maintenance required to maintain compressor reliability. Intermittent operation is always recommended for these units. The frequency and duration of *unloaded* operation depends on the type, size, operating pressure of the compressor. Air-cooled compressors for higher than 200-psig pressure application are usually rated by a rule that states that the *compressing time* shall not exceed 30 min or less than 10 min. Shutdown or unloaded time should be at least equal to compression time or 50 per cent.

Rotary screw compressors are exceptions to this 50 per cent rule. Each time a rotary screw compressor unloads, both the male and female rotor instantaneously shifts axially. These units are equipped with a balance piston or heavy-duty thrust bearing that is designed to absorb the tremendous axial forces that result from this instantaneous movement, but they are not able to fully protect the compressor or its components. The compressor's design accepted the impact loading that results from this unload shifting and incorporated enough axial strength to absorb a normal unloading cycle. If this type of compressor is subjected to constant or frequent unloading, as in a load-following application, the cycle frequency is substantially increased and the useful life of the compressor is proportionally reduced. There have been documented cases where either the male or female rotor actually broke through the compressor's casing as a direct result of this failure mode.

The only compressor that is ideally suited for load-following applications is the reciprocating type. These units have an absolute ability to absorb the variations in pressure and demand without any impact on either reliability or life cycle cost. The major negative of the reciprocating compressor is the pulsing or constant variation in pressure that is produced by the reciprocating compression cycle. Properly sized accumulators and receiver tanks will resolve most of the pulsing.

5.2.4.2.3 Life cycle costs

All capital equipment decisions should be based on the true or life cycle cost of the system. Life cycle cost includes all costs that will be incurred beginning with specification development before procurement to final decommissioning cost at the end of the compressor's useful life. In many cases, the only consideration is the actual procurement and installation cost of the compressor. While these costs are important, they represent less than 20 per cent of the life cycle cost of the compressor.

The cost evaluation must include the recurring costs, such as power consumption, maintenance, etc. that are an integral part of day-to-day operation. Other costs that should be considered include training of operators and maintenance personnel who must maintain the compressor.

5.2.5 Air dryers

Air entering the first stage of any air compressor carries with it a certain amount of native moisture. This is unavoidable, although the quantity carried will vary widely with the ambient temperature and relative humidity. Figure 5.2.3 shows the effect of ambient temperature and relative humidity on the quantity of moisture in atmospheric air entering a compressor at 14.7 psia. Under any given condition, the amount of water vapor entering the compressor per 1000 ft^3 of mixture may be approximated from these curves.

In any air–vapor mixture, each component has its own partial pressure and the air and the vapor are each indifferent to the existence of the other. It follows that the conditions of either component may be studied without reference to the other. In a certain volume of mixture, each component fills the full volume at its own partial pressure. The water vapor may saturate this space or it may be superheated.

As this vapor is compressed, its volume is reduced while at the same time the temperature automatically increases. As a result, the vapor becomes superheated. More pounds of vapor are now contained in one cubic foot than when originally entering the compressor.

Under the laws of vapor, the maximum quantity of a particular vapor a given space can contain is dependent solely upon the vapor temperature. As the compressed water vapor is cooled, it will eventually reach the temperature at which the space becomes saturated, now containing the maximum it can hold. Any further cooling will force part of the vapor to condense into its liquid form – water.

The curves contained in Figure 5.2.4 show what happens over a wide range of pressures and temperatures. However, these are saturated vapor curves based on starting with 1000 ft^3 of *saturated* air. If the air is not saturated at the compressor's inlet, and it usually is not, use Figure 5.2.5 to obtain the initial water vapor weight entering the system per 1000 ft^3 of compressed air. By reading left in Figure 5.2.6 from the juncture of the final pressure and final temperature, obtain the maximum weight of vapor that this same 1000 ft^3 can hold after compression and cooling to saturation. If the latter is less than the former, the difference will be condensed. If the latter is higher, there will be no condensation. It is evident that the lower the temperature and the greater the pressure of compressed air, the greater will be the amount of vapor condensed.

5.2.5.1 Problems caused by water in compressed air

Few plant operators need to be told of the problems caused by water in compressed air. They are most apparent to those who operate pneumatic tools, rock drills, automatic pneumatic powered machinery, paint and other sprays, sandblasting equipment, and pneumatic controls. However, almost all applications, particularly of 100-psig power, could benefit from the elimination of water carryover. The principal problems might be summarized as:

1. Washing away of required lubrication.
2. Increase in wear and maintenance.
3. Sluggish and inconsistent operation of automatic valves and cylinders.
4. Malfunctioning and high maintenance of control instruments.
5. Spoilage of product by spotting in paint and other types of spraying.
6. Rusting of parts that have been sandblasted.
7. Freezing in exposed lines during cold weather.
8. Further condensation and possible freezing of moisture in the exhaust of those more efficient tools which expand the air considerably.

A fact to remember is that water vapor, *as vapor*, does no harm in most pneumatic systems. It is only when the vapor condenses and remains in the system as a liquid

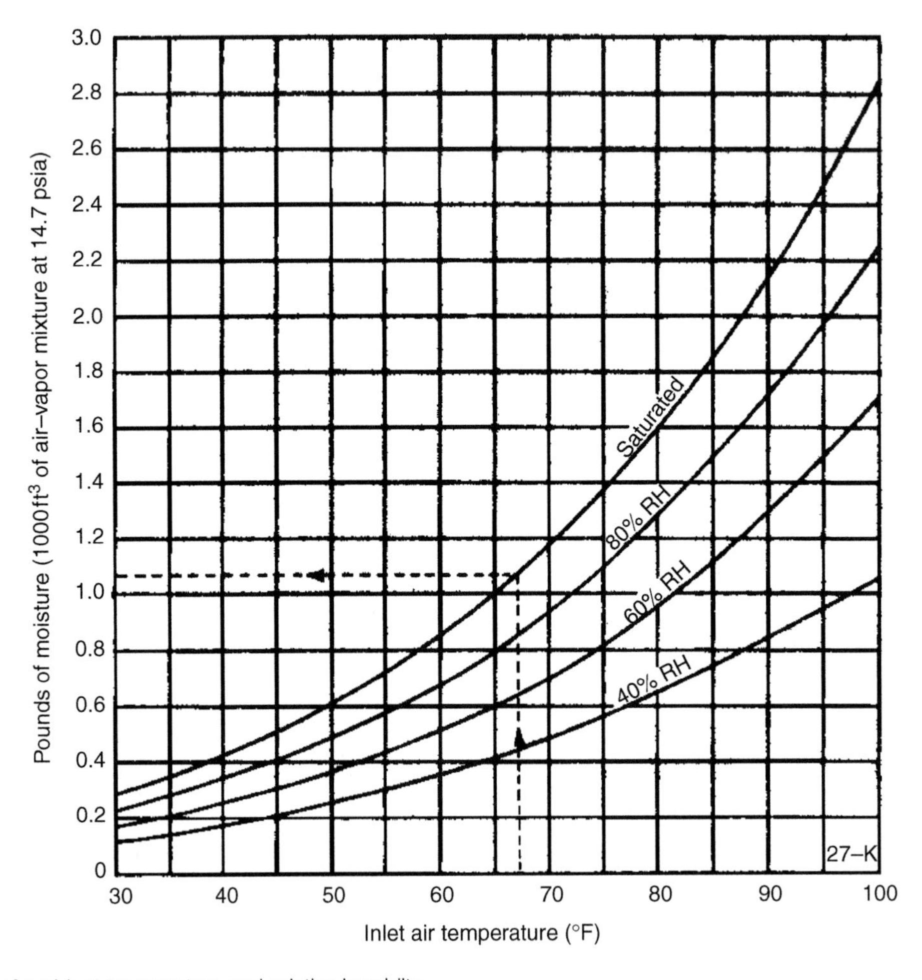

Fig. 5.2.3 Effects of ambient temperature and relative humidity.

that problems exist. The goal, therefore, is to condense and remove as much of the vapor as is economically possible.

In conventional compressed air systems, vapor and liquid removal is limited. Most two-stage compressors will include an intercooler between stages. On air-cooled units for 100–200 psig service, the air between stages is not cooled sufficiently to cause substantial liquid drop out and provision is not usually made for its removal. Water-cooled intercoolers used on larger compressors will usually cool sufficiently to condense considerable moisture at cooler pressure. Drainage facilities must always be provided and used. Automatic drain traps are normally included to drain condensed water vapor.

All compressed air systems should always include a water-cooled aftercooler between the compressor and receiver tank. Properly designed and maintained after-coolers, in normal summer conditions, condense at 100 psig, up to 70 per cent of the vapor entering the system. Most of this condensation will collect in the aftercooler or the receiver tank. Therefore, both must be constantly drained.

The problem with a conventional system that relies on heat exchangers (i.e., aftercoolers) for moisture removal is temperature. The aftercooler will remove only liquids that have condensed at a temperature between the compressed air and cooling water temperature. In most cases, this differential will be about 20–50° lower than the compressed air temperature or around 70–90 °F. As long as the compressed air remains at or above this temperature range, any remaining vapor that it contains will remain in a vapor or gaseous state. However, when the air temperature drops below this range, additional vapor will condense into water.

5.2.6 Dried air systems

This system involves processing the compressed air or gas after the aftercooler and receiver to further reduce moisture content. This requires special equipment, a higher first cost, and a higher operating cost. These costs must be balanced against the gains obtained. They may show up as less wear and maintenance of tools and air-

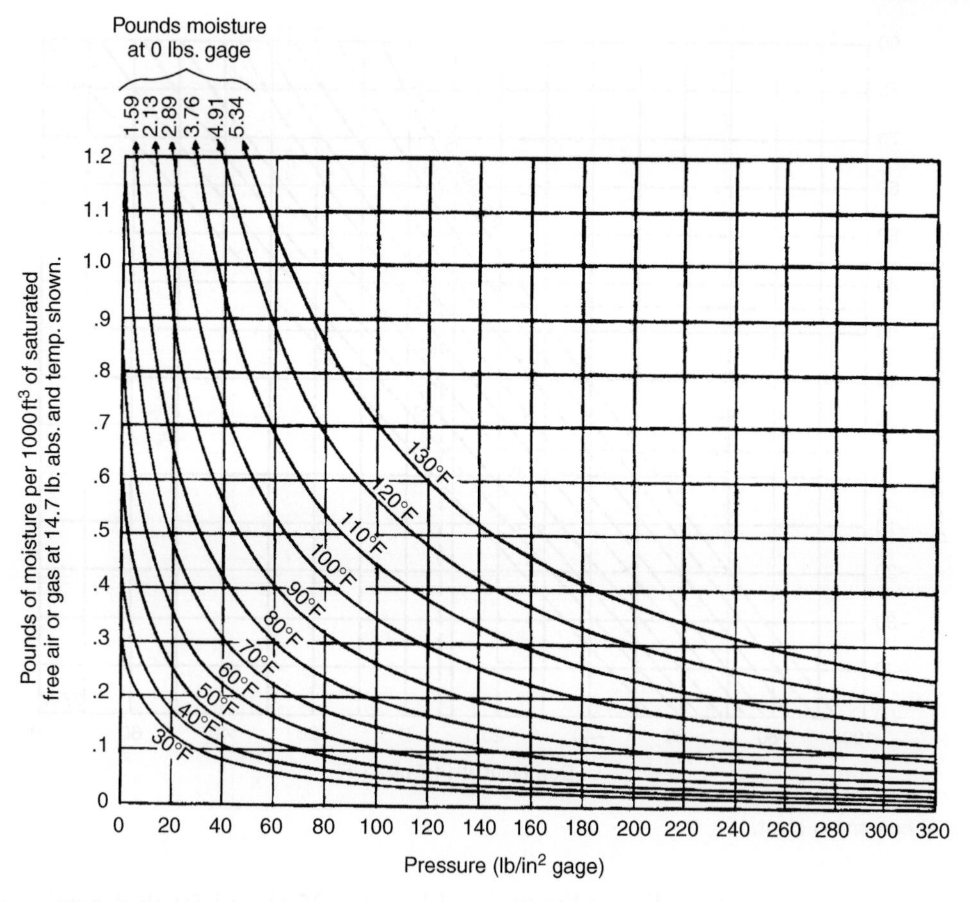

Fig. 5.2.4 Moisture remaining in saturated air or gas when compressed isothermally to pressure shown.

operated devices, greater reliability of devices and controls, and greater production as a result of fewer outages for repairs. In many cases, reduction or elimination of product spoilage or a better product quality may also result.

The degree of drying desired will vary with the pneumatic equipment and application involved. The aim is to eliminate further condensation in the airlines and pneumatic tools or devices. Prevailing atmospheric conditions also have an influence on the approach that is most effective. In many 100-psig installations, a dew point at line pressure of from 500 °F to 350 °F is adequate. Other applications, such as instrument air systems, will require dew points of minus 500 °F.

Terminology involves drier outlet dew point at the *line* pressure or the pneumatic circuit. This is the saturation temperature of the remaining moisture contained in the compressed air or gas. If the compressed gas temperature is never reduced below the outlet dew point beyond the drying equipment, there will be no further condensation.

Another value sometimes involved when the gas pressure is reduced before it is used is the dew point at that lower pressure condition. A major example is the use of 100 psig (or higher) gas reduced to 15 psig for use in pneumatic instruments and controls. This dew point will be lower because the volume involved increases as the pressure is decreased. The dew point at atmospheric pressure is often used as a reference point for measurement of drying efficiency. This is of little interest when handling *compressed* air or gas.

Figure 5.2.7 enables one to determine dew point at reduced pressure. The left scale shows the dew point at the elevated pressure. Drop from the intersection of this value and the elevated pressure line to the reduced pressure line and then back to the left to read the dew point at the reduced pressure.

Figure 5.2.8 shows graphically the amount of moisture remaining in the vapor form when the air–vapor mixture is conditioned to a certain dew point. This curve is based on a volume of 1000 ft^3 or an air–vapor mixture *at its total pressure*. For example, 1000 ft^3 at 100-psig air at 50 °F and 1000 ft^3 of 15 psig air at 50 °F will hold the same vapor at the dew point. However, 1000 ft^3 at 100 psig and 50 °F reduced to 15 psig will become 3860 ft^3 at 50 °F. As a result, it now capable of holding 3.86 times as much vapor and the dew point will not be

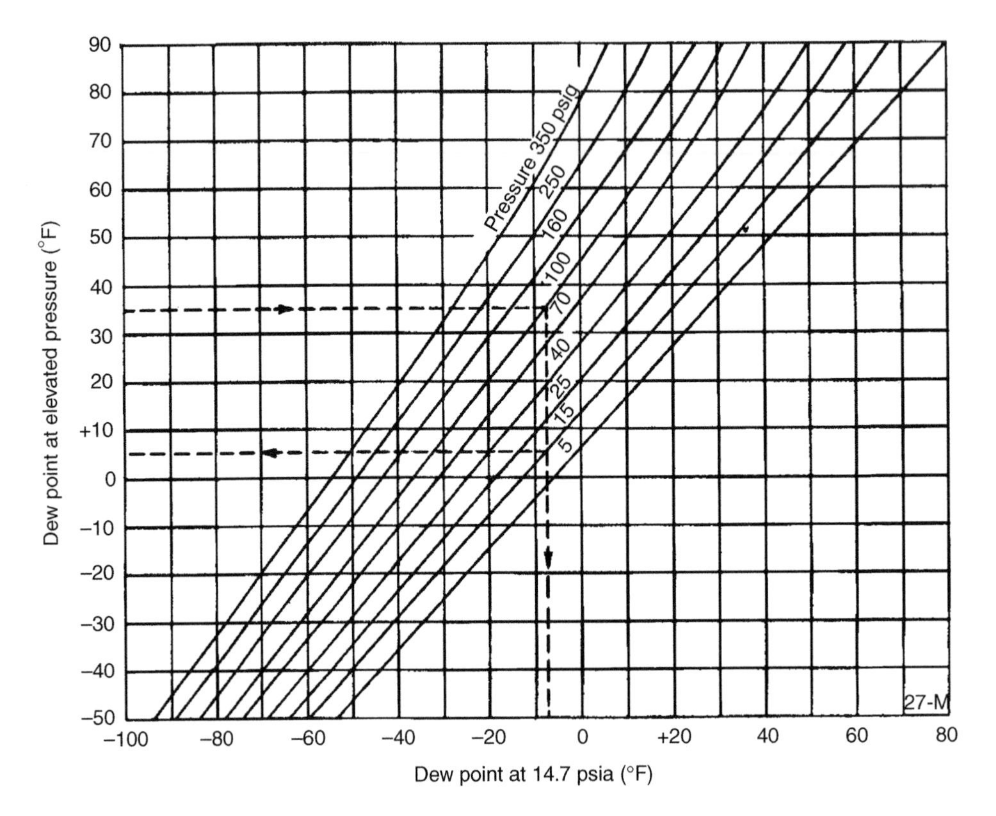

Fig. 5.2.5 Dew point conversion chart.

reached until the mixture temperature is lowered to its saturation temperature.

5.2.6.1 Drying methods

There are three general methods of drying compressed air: chemical, adsorption, and refrigeration. In all cases, aftercooling and adequate condensation removal must be done ahead of this drying equipment. The initial and operating costs and the results obtained vary considerably.

These methods are primarily for water vapor removal. Removal of lubricating oil is secondary, although all drying systems will reduce its carryover. It should be understood that complete elimination of lubricating oil, particularly in the vapor form, is very difficult and that when absolutely oil-free air is required, some form of non-lubricated compressor is the best guaranteed method.

5.2.6.1.1 Chemical dryers

Chemical dryers are materials which combine with or absorb moisture from air when brought into close contact. There are two general types. One, using deliquescent material in the form of pellets or beads, is reputed to obtain a dew point, with 700 °F inlet air to the dryer, of between 35 °F and 50 °F depending on the specific type of deliquescent material. The material turns into a liquid as the water vapor is absorbed. This liquid must be drained off and the pellets or beads replaced periodically. Entering air above 900 °F is not generally recommended.

The second type of chemical dryer utilizes an ethylene glycol liquid to absorb the moisture. Standard dew point reduction claimed is 400 °F, but greater reductions are said to be possible with special equipment. The glycol is regenerated (i.e., dried) in a still using fuel gas or steam as a heating agent. The released moisture is vented to atmosphere. The regenerated glycol is re-circulated by a pump through a water-cooled heat exchanger that lowers the glycol temperature before returning to the dryer vessel.

5.2.6.1.2 Adsorption

Adsorption is the property of certain extremely porous materials to hold vapors in the pores until the desiccant is either heated or exposed to a drier gas. The material is a solid at all times and operates alternately through drying and reactivation cycles with no change in composition. Adsorbing materials in principal use are activated alumina and silica gel. Molecular sieves are also used. Atmospheric dew points of −1000 °F are readily obtained using adsorption.

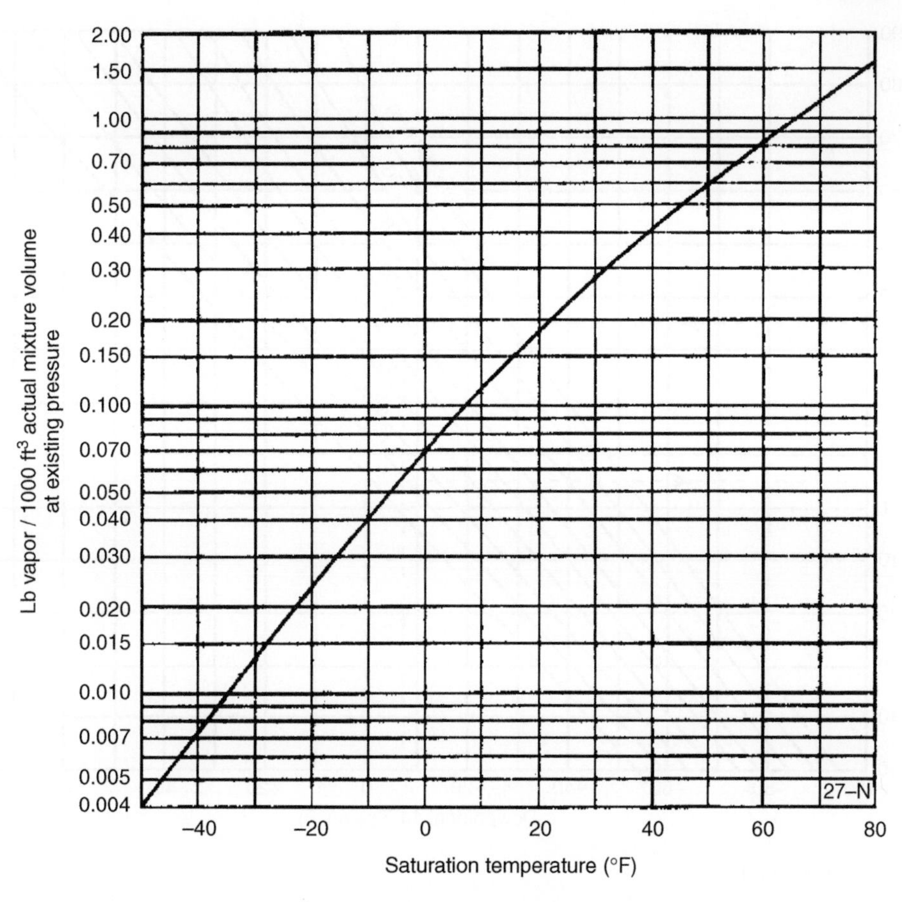

Fig. 5.2.6 Moisture in air at any pressure for 100 ft^3 of actual volume at existing pressures.

Reactivation or regeneration is usually obtained by diverting a portion of the already dried air through a reducing valve or orifice, reducing its pressure to atmospheric, and passing it through the wet desiccant bed. This air, with the moisture it has picked up from the saturated desiccant bed, is vented to atmosphere. The diverted air may vary from 7 to 17 per cent of the mainstream flow, depending upon the final dew point desired from the dryer. Heating the activating air prior to its passing through the desiccant bed, or heating the bed itself, is often done to improve the efficiency of the regeneration process. This requires less diverted air since each cubic foot of diverted air will carry much more moisture out of the system. Other modifications are also available to reduce or even eliminate the diverted air quantity.

5.2.6.1.3 Refrigeration

Refrigeration for drying compressed air is growing rapidly. It has been applied widely to small installations, sections of larger plants, and even to entire manufacturing plant systems. Refrigerated air dryers have been applied to the air system both before and after compression. In the *before compression* system, the air must be cooled to a lower temperature for a given *final line pressure dew point*. This takes more refrigeration power for the same end result. Partially offsetting this is a saving in air compressor power per 1000 ft^3 per minute (cfm) of *atmospheric* air compressed due to the reduction in volume at the compressor inlet caused by the cooling and the removal of moisture. There is also a reduction in discharge temperature on single-stage compressors that may at time have some value. As atmospheric (inlet) dew point of 350 °F is claimed.

When air is refrigerated *following compression*, two systems have been used. Flow of air through directly refrigerated coils is used predominately in the smaller and moderate-sized systems (Figure 5.2.9). These are generally standardized for cooling to 350 °F, which is the dew point obtained at line pressure.

The larger systems chill water that is circulated through coils to cool the air. A dew point at line pressure of about 500 °F is obtainable with this method. Figure 5.2.10 illustrates a typical system of a *chiller–dryer* unit. The designs shown are regenerative since the incoming air is partially cooled by the outgoing air stream. This reduces the size and first cost of the refrigeration compressor and exchanger. It also reduces

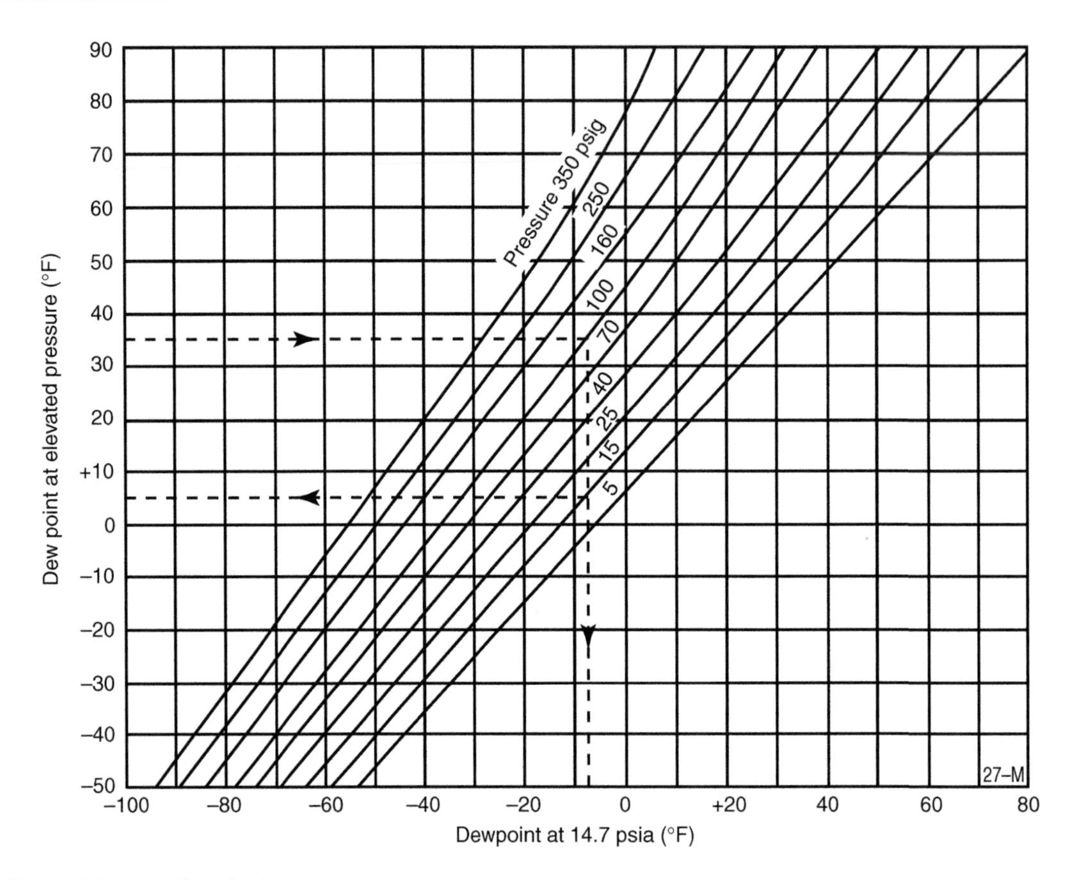

Fig. 5.2.7 Dew point conversion chart.

power cost and reheats the air returning to the line. Reheating of the air after it is dried has several advantages: the air volume is increased and less *free air* is required; chance of line condensation is still further reduced; and sweating of the cold pipe leaving the dryer is eliminated. Reheating dryers seldom need further reheating.

5.2.6.1.4 Combination systems

The use of a combination dryer should be investigated when a very low dew point is necessary. Placing a refrigeration system ahead of an adsorption dryer will allow the more economical refrigeration unit to remove most of the vapor and reduce the load on the desiccant.

5.2.6.2 Dry air distribution system

Most plants are highly dependent upon their compressed air supply and it should be assured that the air is in at least a reasonable condition at all times, even if the drying system is out of use for maintenance or repair. It is possible that the line condensation would be so bad that some air applications would be handicapped or even shutdown if there were no protection. A vital part of the entire endeavor to separate water in the conventional compressed air system is also the trapping of dirt, pipe scale, and other contaminates. This is still necessary with a dried air system. As a minimum, all branch lines should be taken off the top of the main and all feeder lines off the top of branch lines.

Absolute prevention of line freezing can be obtained only when the dew point of the line air is below any temperature to which it may be exposed. Freezing is always possible if there is line condensation. For example, when airlines are run outdoors in winter weather or pass through cold storage rooms, the ambient temperatures will change the dew point and cause any moisture in the air to condense and freeze.

The dryer unit has an air inlet, an air outlet, a waste air outlet, two heater coils, and two 4-way reversing valves. Although the illustration shows the two tanks in a functioning mode, they are universal in operation. By this, we mean that this changes the tanks over when the active tank becomes totally saturated, and the tank that was on the regeneration cycle then becomes the active unit.

With this process, there will always be one tank in active service, and one on regeneration or stand-by. A timer governs the changeover process. The timer changes the position of the four-way, or reversing, valves. This operation reverses the airflow through the tanks; the tank

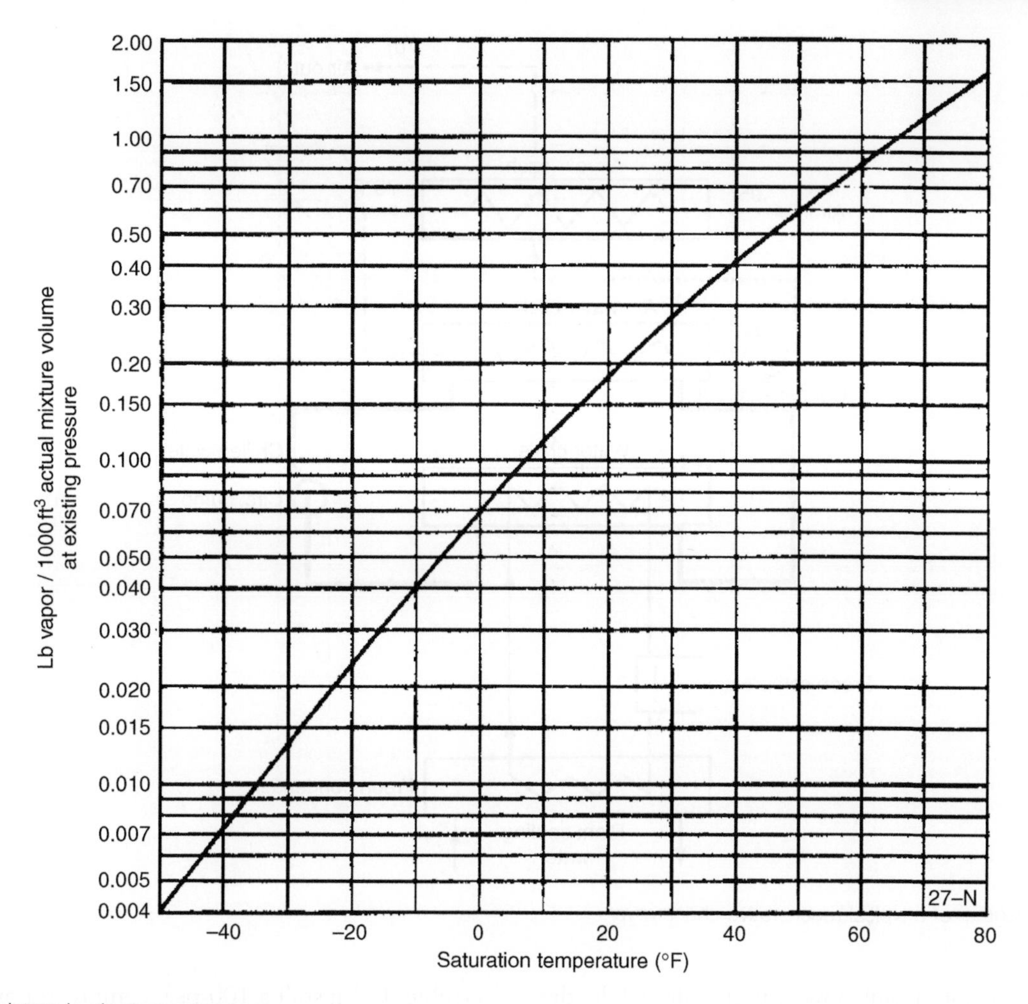

Fig. 5.2.8 Moisture in air at any pressure.

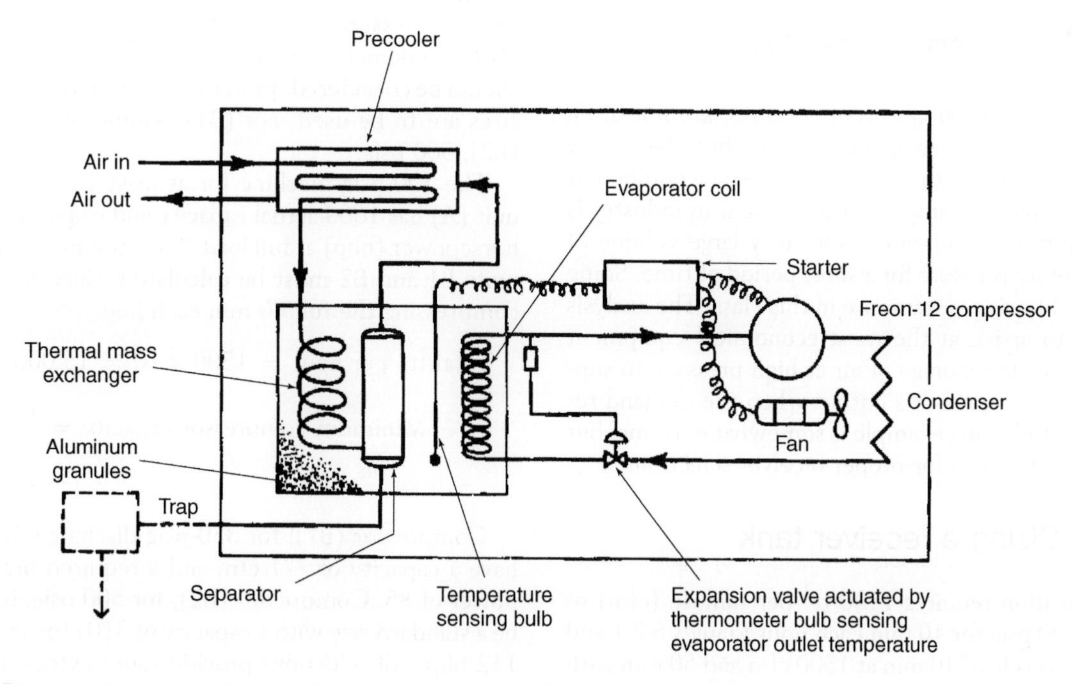

Fig. 5.2.9 Diagram of equipment furnished in small self-contained direct-refrigeration dryer.

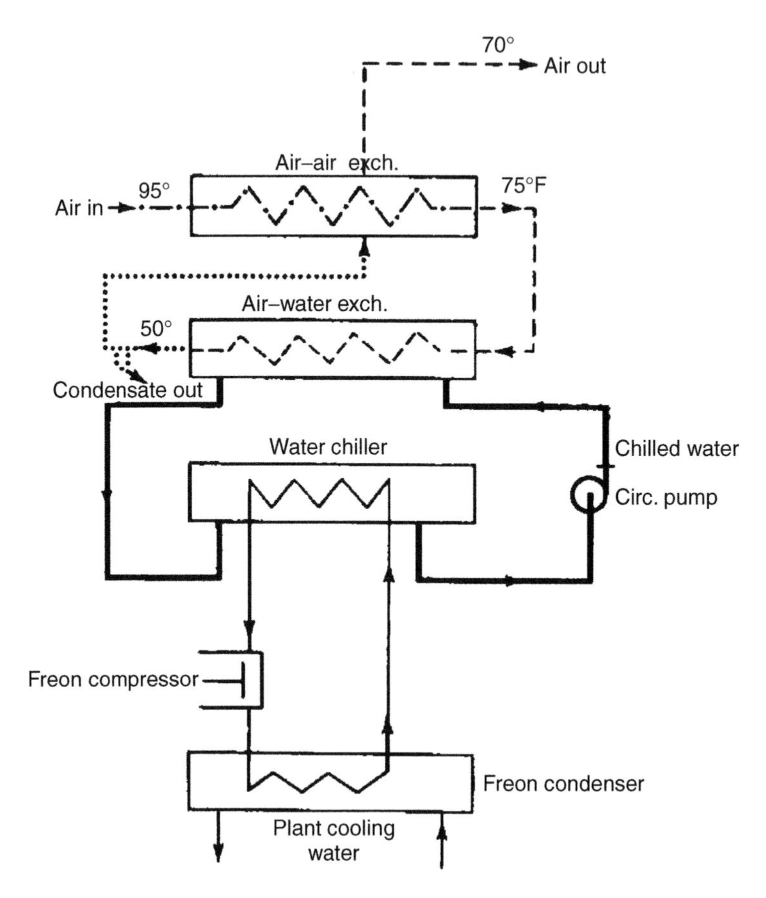

Fig. 5.2.10 Diagram of a typical regenerative chiller–dryer.

that was regenerated will now dry the air, while the saturated tank will be regenerated.

5.2.7 Air reservoir (receivers)

Air receivers, being simple volume tanks, are not often thought of as highly engineered items, but the use of simple engineering with receivers can reduce equipment costs. A pertinent example not infrequent in industry is the intermittent requirement for fairly large volume of air at moderate pressure for a short period of time. Some boiler soot blowing systems are in this class. The analysis necessary to arrive at the most economical equipment often involves the storage of air at high pressure to supplement the compressor's output when the demand requires. The following example is somewhat extreme, but emphasizes the need for proper receiver selection.

5.2.7.1 Sizing a receiver tank

This application requires 1500 ft^3 per minute (cfm) of free air at 90 psig for 10 min each hour (Tables 5.2.1 and 5.2.2). This cycle of 10 min at 1500 cfm and 50 min with zero demand repeats hourly. Alternates obviously are possible; (A) install a 100-psig compressor and standard accessories large enough for the maximum demand requirements; or, (B) install a smaller compressor, but for a higher pressure, and store the air in receivers during the off or no demand period. At least two storage pressures should be considered. In all cases commercial compressor sizes are to be used. For (B1) assume 350 psig and for (B2), 500 psig.

The 100-psig machine for at least 1500-cfm output, unit (A) has 1660 actual capacity and requires 309 brake horsepower (bhp) at full load. The minimum capacity for units B1 and B2 must be calculated, since these can be compressing the full 60 min each hour.

$$\text{Total ft}^3 \text{ per hour} = 1500 \times 10 = 15{,}000$$

$$\text{Minimum compressor capacity} = \frac{15{,}000}{60} = 250 \text{ cfm}$$

Compressor (B1), for 350-psig discharge, is found to have a capacity of 271 cfm and a required brake horsepower of 85. Compressor (B2), for 500 psig, is found to be a standard size with a capacity of 310 cfm and requires 112 bhp. All selections provide some extra capacity for emergency and losses are economical two-stage designs.

Air compressed to 100 psig, to be used at 90 psig, provides no possibility of storage. Storage is practical at the other pressures selected. Compressor units B and C will operate at full load at their rated or lower pressures. Since the receiver pressure will fall during the 10 min when air is used faster than the compressor can replenish it, the unit will be operating at full capacity and will supply some of the demand. The full demand need not be stored.

$$\text{Cfm to be stored} = \text{Total ft}^3/\text{hour} - \text{Compressor cfm} \times \text{Minutes of demand}$$

The cubic feet of air to be stored represents the free air, at 14.7 psia, that must be packaged into the receiver above the minimum pressure required by the demand. In this case, the demand pressure is 90 psig, but an allowance for line losses and the necessary reducing valve pressure drop would prevent the use of any air stored below 110 psig. The receiver has a volume of (V) expressed in cubic feet.

$$\text{Useful free air stored} = \frac{V \times \text{Pressure drop}}{14.7}$$

$$V = \frac{\text{Useful free air stored} \times 14.7}{\text{Pressure drop}}$$

This receiver volume may be in one or several tanks, the most economical number being chosen.

The final selection can be made only after consideration of the first cost of the compressor, motor, starter, aftercooler, and receiver. The cost of installation, including foundations, piping, and wiring, as well as the operating power cost must also be considered. In the example, Table 5.2.3 provides a basic comparison of the three options.

Not all problems of this nature will result in the selection of the intermediate storage pressure. Although few will favor the 100-psig level, many will be more economical at the 500-psig level. Experience indicates that there is seldom any gain in using a higher storage pressure than 500 psig since larger receivers required become very expensive above this level and power cost increases.

Air reservoirs are classified as pressure vessels and have to conform to the ASME Pressure Vessel Codes. As such, the following attachments must be fitted:

- Safety valves;
- Pressure gages;
- Isolation valves;
- Manhole or inspection ports;
- Fusible plug.

Air reservoirs are designed to receive and store pressurized air. Pressure regulating devices are installed to maintain the pressure within operational limits. When the air reservoir is pressurized to the maximum pressure set point, the pressure regulator causes the air compressor to off-load compression by initiating an electrical solenoid valve to use lubricating oil to hydraulically hold open the low-pressure suction valve on the compressor.

As the compressed air is used, the pressure drops in the reservoir until the low-pressure set point is reached. At this point, the pressure regulated solenoid valve is de-energized. This causes the hydraulic force to drop off on the low-pressure suction valve, restoring it to the full compression cycle.

This cycling process causes drastic variations in noise levels. These noises should not be regarded as problems, unless accompanied by severe knocking or squealing

Table 5.2.1 Comparison of receiver options

Unit	B1	B2
Capacity (cfm)	271	310
Demand period (min)	10	10
Total cubic feet required	15,000	15,000
Delivered during demand period	2710	3100
Cubic feet to be stored	12,290	11,900

Table 5.2.2 Receiver volume required

Unit	B1	B2
Storage pressure (psig)	350	500
Minimum pressure (psig)	110	110
Pressure drop (psi)	240	390
Free air to be stored (ft^3)	12,290	11,900
Receiver volume (V)	752	448

Table 5.2.3 Economic comparison

Unit	A	B1	B2
Pressure (psig)	100	350	500
Installed cost	100%	52%	63%
Fixed charges	100%	52%	63%
Power cost	100%	64%	93%
Oil, water, attendance	100%	100%	100%

noises. Figure 5.2.11 shows a typical hydraulic unloader and its location on the compressor.

5.2.8 Safety valves

All compressed air systems that use a positive displacement compressor must be fitted with a pressure relief or safety valve (Figure 5.2.12) that will limit the discharge or inter-stage pressures to a safe maximum limit. Most dynamic compressors must have similar protection due to restrictions placed on casing pressure, power input, and/or keeping out of surge range.

Two types of pressure relief devices are available, *safety valves* and *relief valves*. Although these terms are often used interchangeably, there is a difference between the two. Safety valves are used with gases. The disk overhangs the seat to offer additional thrust area after the initial opening. This fully opens the valve immediately, giving maximum relief capacity. These are often called *pop-off safety valves*.

With relief valves, the disk area exposed to overpressure is the same whether the valve is open or closed. There is a gradual opening, the amount depending upon the degree of overpressure. Relief valves are used with liquids where a relatively small opening will provide pressure relief.

Positive displacement machines use *safety valves*. There are ASME standards of materials, sizing, and only ASME stamped valves should be used. The relieving capacity of a given size of safety valve varies materially with the make and design. Care must be taken to assure proper selection.

An approved safety valve is usually of the 'huddling chamber' design. In this valve the static pressure acting on the disk area causes initial opening. As the valve pops, the air space within the huddling chamber between seat and the blowdown ring fills with pressurized air and builds up more pressure on the roof of the disk holder.

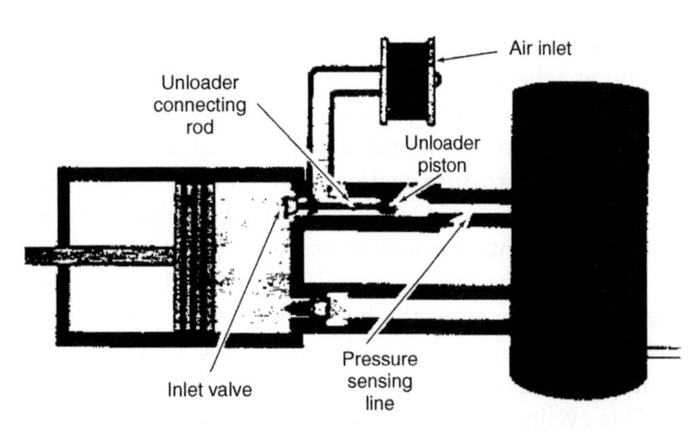

Fig. 5.2.11 Typical hydraulic unloader and its location on the compressor.

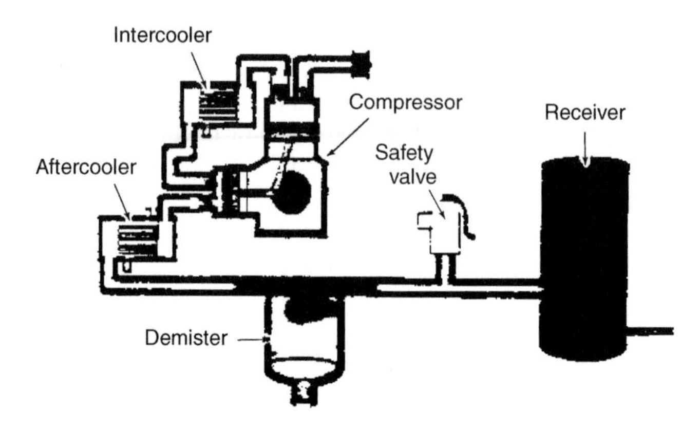

Fig. 5.2.12 Illustrates how a safety valve functions.

This temporary pressure increases the upward thrust against the spring, causing the disk and its holder to lift to full pop opening.

After a predetermined pressure drop, which is referred to as blowdown, the valve closes with a positive action by trapping pressurized air on top of the disk holder. The pressure drop is adjusted by raising or lowering the blowdown ring. Raising the ring increases the pressure drop while lowering it decreases the drop.

Most state laws and safe practice require a safety relief valve ahead of the first stop valve in every positive displacement compressed air system. It is set to release at 1.25 times the normal discharge pressure of the compressor or at the maximum working pressure of the system, whichever is lower. The relief valve piping system sometimes includes a manual vent valve and/or a bypass valve to the suction to facilitate start-up and shutdown operations. Quick line sizing equations are (1) line connection, $d/1.75$; (2) bypass, $d/4.5$; (3) vent, $d/6.3$; and (4) relief valve port, $d/9$.

The safety valve is normally situated atop the air reservoir. There must be no restriction on all blow-off points. Compressors can be hazardous to work around because they do have moving parts. Ensure that clothing is kept away from belt drives, couplings, and exposed shafts.

In addition, high-temperature surfaces around cylinders and discharge piping are exposed. Compressors are notoriously noisy. For this reason, ear protection should be worn. When working around high-pressurized air systems, wear safety glasses and do not search for leaks with bare hands. High-pressure leaks can cause severe friction burns.

5.2.9 Coolers

The amount of moisture that air can hold is inversely proportional to the pressure of the air. As the pressure of

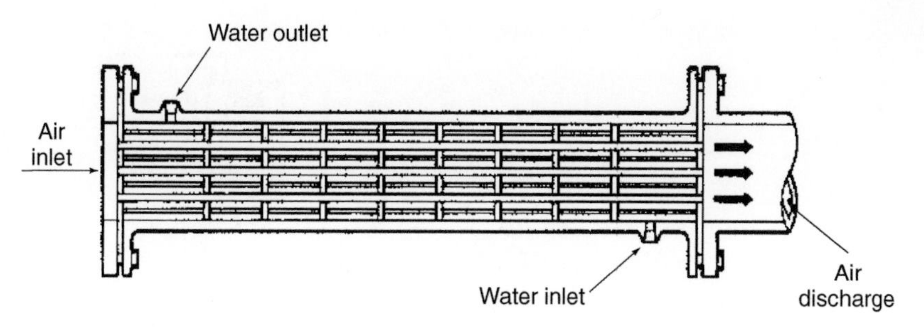

Fig. 5.2.13 Compressor air cooler.

the air increases, the amount of moisture that air can hold decreases. The amount of moisture air can hold is also proportional to the temperature of the air. As the temperature of the air decreases, the amount of moisture it can hold decreases. The pressure change of compressed air is larger than the temperature change of the compressed air. This causes the moisture in the air to condense out of the compressed air. The moisture in compressed air systems can cause corrosion, water hammers, and freeze damage. Therefore, it is important to avoid moisture in compressed air systems. Coolers are used to address the problems by moisture in compressed air systems.

Coolers are frequently used on the discharge of a compressor. These are called aftercoolers, and their purpose is to remove the heat generated during the compression of the air. The decrease in temperature promotes the condensing of any moisture present in the compressed air. This moisture is collected in condensate traps that are either automatically or manually drained.

If the compressor is of the multi-stage type, there may be an intercooler, which is located after the first-stage discharge and second-stage suction. The principle of the intercooler is the same as the principle of the aftercoolers, and the result is drier, cooler compressed air. The structure of the individual cooler depends on the pressure and volume of the air it cools. Figure 5.2.13 illustrates a typical compressor air cooler.

This combinations of drier compressed air (which helps prevent corrosion) and cooler compressed air (which allows more air to be compressed for a set volume) is the reason the air coolers are worth the investment.

Chapter 5.3

5.3

Pump types and characteristics

Girdhar, Moniz and Mackay

The transfer of liquids against gravity existed from time immemorial. A pump is one such device that expends energy to raise, transport, or compress liquids. The earliest known pump devices go back a few thousand years. One such early pump device was called 'Noria', similar to the Persian and the Roman water wheels. Noria was used for irrigating fields (Figure 5.3.1).

The ancient Egyptians invented water wheels with buckets mounted on them to transfer water for irrigation. More than 2000 years ago, a Greek inventor, *Ctesibius*, made a similar type of pump for pumping water (Figure 5.3.2).

During the same period, Archimedes, a Greek mathematician, invented what is now known as the 'Archimedes' screw' – a pump designed like a screw rotating within a cylinder (Figure 5.3.3). The spiraled tube was set at an incline and was hand operated. This type of pump was used to drain and irrigate the Nile valley.

Fig. 5.3.2 Model of a piston pump made by Ctesbius.

Fig. 5.3.1 Noria water wheel (from the Ripley's believe it not).

In the fourth-century Rome, Archimedes' screw was used for the Roman water supply systems, highly advanced for that time. The Romans also used screw pumps for irrigation and drainage work.

Screw pumps can also be traced to the ore mines of Spain. These early units were all driven by either man or animal power.

The mining operations of the Middle Ages led to the development of the suction (piston) pump, types of which are described by Georgius Agricola in De re

Practical Centrifugal Pumps: Design, Operation and Maintenance; ISBN: 9780750662734

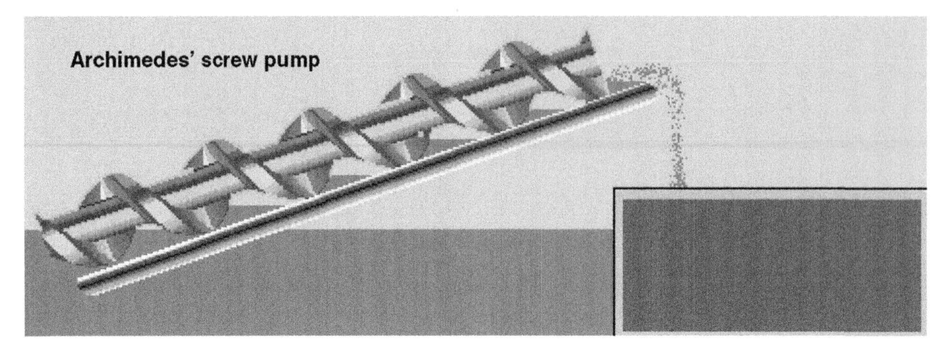

Fig. 5.3.3 Archimedes' screw pump.

metallica (1556). Force pumps, utilizing a piston-and-cylinder combination, were used in Greece to raise water from wells (Figure 5.3.4).

Adopting a similar principle, air pumps operated spectacular musical devices in Greek temples and amphitheaters, such as the water organ.

5.3.1 Applications

Times have changed, but pumps still operate on the same fundamental principle – expend energy to raise, transport, or compress liquids. Over time, the application of pumps in the agricultural domain has expanded to cover other domains as well. The following are a few main domains that use pumps extensively:

- *Water supply*: to supply water to inhabited areas.
- *Drainage*: to control the level of water in a protected area.
- *Sewage*: to collect and treat sewage.
- *Irrigation*: to make dry lands agriculturally productive.
- *Chemical industry*: to transport fluids to and from various sites in the chemical plant.
- *Petroleum industry*: used in every phase of petroleum production, transportation, and refinery.

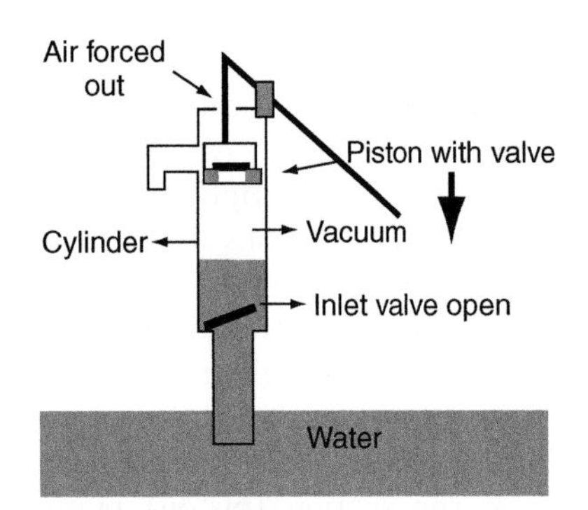

Fig. 5.3.4 Reciprocating hand pump in suction stroke.

- *Pharmaceutical and medical field*: to transfer of chemicals in drug manufacture; pump fluids in and out of the body.
- *Steel mills*: to transport cooling water.
- *Construction*: bypass pumping, well-point dewatering, remediation, and general site pumping applications.
- *Mining*: heavy-duty construction, wash water, dust control fines and tailings pumping, site dewatering, groundwater control, and water runoff.

Pumps are also used for diverse applications like in transfer of potatoes, to peel the skin of hazelnuts in chocolate manufacture, and to cut metal sheets in areas that are too hazardous to allow cutting by a gas flame torch. The artificial heart is also a mechanical pump. The smallest pump ever made is no bigger than the tip of a finger. It moves between 10 and 30 nl of liquid in one cycle (10- to 30-thousandths of a drop of water). It was not found to have any practical use so maybe it was created just for the records!

5.3.2 Pump types

Pumps can be classified on various bases.

Pumps based on their principle of operation are primarily classified into:

- positive displacement pumps (reciprocating, rotary pumps),
- roto-dynamic pumps (centrifugal pumps),
- others.

5.3.2.1 Positive displacement pumps

Positive displacement pumps, which lift a given volume for each cycle of operation, can be divided into two main classes: reciprocating and rotary.

Reciprocating pumps include piston, plunger, and diaphragm types. The rotary pumps include gear, lobe, screw, vane, regenerative (peripheral), and progressive cavity pumps.

5.3.2.2 Roto-dynamic pumps

Roto-dynamic pumps raise the pressure of the liquid by first imparting velocity energy to it and then converting this to pressure energy. These are also called centrifugal pumps. Centrifugal pumps include radial, axial, and mixed flow units.

A radial flow pump is commonly referred to as a straight centrifugal pump; the most common type is the volute pump. Fluid enters the pump through the eye of impeller, which rotates at high speed. The fluid is accelerated *radially* outward from the pump casing. A partial vacuum is created that continuously draws more fluid into the pump if properly primed.

In the axial flow centrifugal pumps, the rotor is a propeller. Fluid flows parallel to the axis of the shaft. The mixed flow, the direction of liquid from the impeller acts as an in-between that of the radial and axial flow pumps.

5.3.2.3 Other types

The other types include electromagnetic pumps, jet pumps, gas lift pumps, and hydraulic ram pumps.

5.3.3 Reciprocating pumps

Reciprocating pumps are positive displacement pumps and are based on the principle of the 2000-year-old pump made by the Greek inventor, Ctesibius.

5.3.3.1 Plunger pumps

Plunger pumps comprise of a cylinder with a reciprocating plunger in it (Figure 5.3.5). The head of the cylinder houses the suction and the discharge valves.

In the suction stroke, as the plunger retracts, the suction valve opens causing suction of the liquid within the cylinder.

In the forward stroke, the plunger then pushes the liquid out into the discharge header. The pressure built in the cylinder is marginally over the pressure in the discharge.

The gland packings help to contain the pressurized fluid within the cylinder. The plungers are operated using the slider-crank mechanism. Usually, two or three cylinders are placed alongside and their plungers reciprocate from the same crankshaft. These are called as duplex or triplex plunger pumps.

5.3.3.2 Diaphragm pumps

Diaphragm pumps are inherently plunger pumps. The plunger, however, pressurizes the hydraulic oil and this pressurized oil is used to flex the diaphragm and cause the pumping of the process liquid.

Diaphragm pumps are primarily used when the liquids to be pumped are hazardous or toxic. Thus, these pumps are often provided with diaphragm rupture indicators.

Diaphragm pumps that are designed to pump hazardous fluids usually have a double diaphragm which is separated by a thin film of water (for example, see Figure 5.3.6). A pressure sensor senses the pressure of this water. In a normal condition, the pressure on the process and oil sides of the diaphragms is always the same and the pressure between the diaphragms is zero.

However, no sooner does one of them ruptures than the pressure sensor records a maximum of process discharge pressure. The rising of this pressure is an indicator of the diaphragm rupture (Figure 5.3.7).

Even with the rupture of just one diaphragm, the process liquid does not come into contact with the atmosphere.

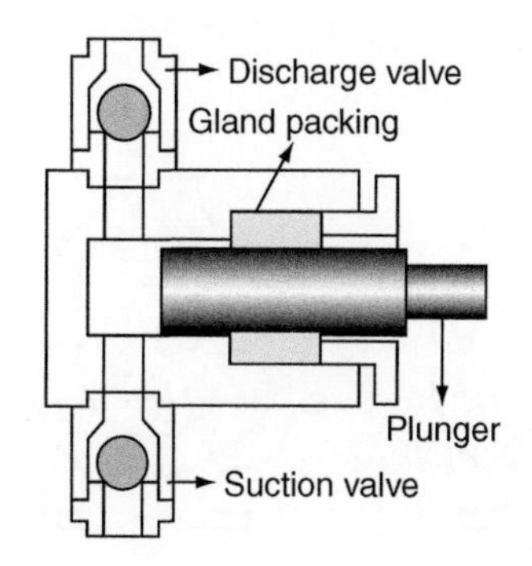

Fig. 5.3.5 Plunger pump.

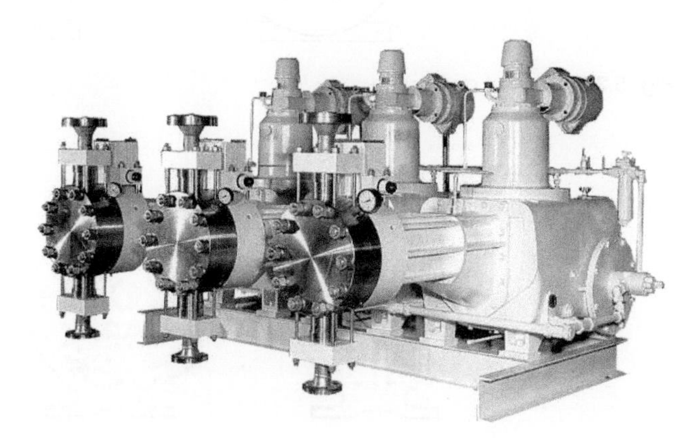

Fig. 5.3.6 Double diaphragm pumps (Lewa pumps).

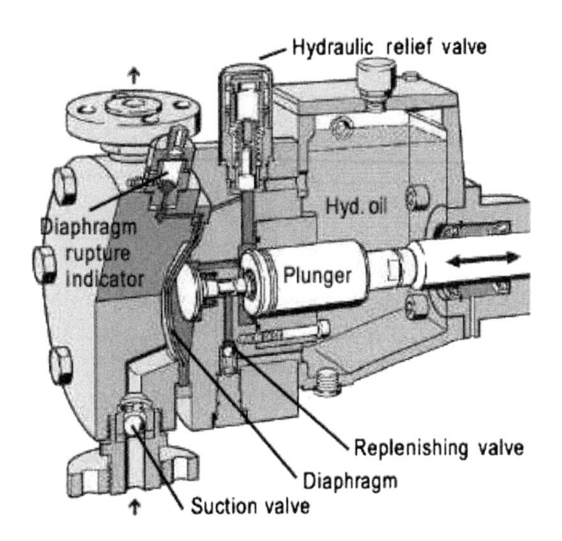

Fig. 5.3.7 Diaphragm pump.

5.3.4 Rotary pumps

5.3.4.1 Gear pump

Gear pumps are of two types:

1. external gear pump,
2. internal gear pump.

5.3.4.1.1 External gear pump

In external gear pumps, two identical gears rotate against each other. The motor provides the drive for one gear. This gear in turn drives the other gear. A separate shaft supports each gear, which contains bearings on both of its sides (Figure 5.3.8).

As the gears come out of the mesh, they create expanding volume on the inlet side of the pump. Liquid flows into the cavity and is trapped by the gear teeth while they rotate.

Liquid travels around the interior of the casing in the pockets between the teeth and the casing. The fine side clearances between the gear and the casing allow recirculation of the liquid between the gears.

Finally, the meshing of the gears forces liquid through the outlet port under pressure. As the gears are supported on both sides, the noise levels of these pumps are lower and are typically used for high-pressure applications such as the hydraulic applications.

5.3.4.1.2 Internal gear pump

Internal gear pumps have only two moving parts (Figure 5.3.9). They can operate in either direction, which allows for maximum utility with a variety of application requirements.

In these pumps, liquid enters the suction port between the large exterior gears, rotor, and the smaller interior gear teeth, idler. The arrows indicate the direction of the pump and the liquid.

Liquid travels through the pump between the teeth of the 'gear-within-a-gear' principle. The crescent shape divides the liquid and acts as a seal between the suction and the discharge ports.

The pump head is now nearly flooded as it forces the liquid out of the discharge port.

Fig. 5.3.8 External gear pump.

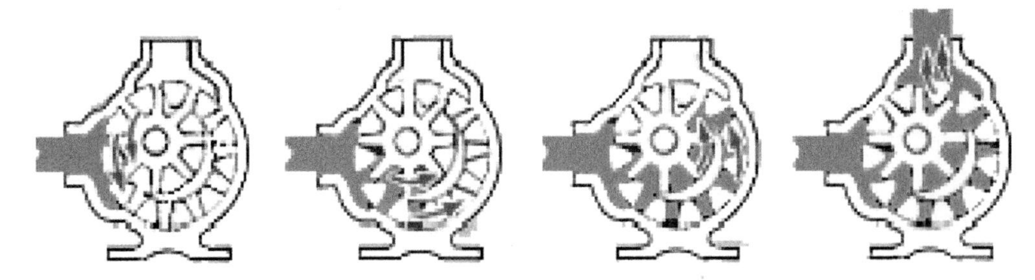

Fig. 5.3.9 Internal gear pump.

Rotor and idler teeth mesh completely to form a seal equidistant from the discharge and suction ports. This seal forces the liquid out of the discharge port.

The internal gear pumps are capable of handling liquid from very low to very high viscosities. In addition to superior high-viscosity handling capabilities, internal gear pumps offer a smooth, nonpulsating flow. Internal gear pumps are self-priming and can run dry.

5.3.4.2 Lobe pump

The operation of the lobe pumps is similar to the operation of the external gear pumps (Figure 5.3.10). Here, each of the lobes is driven by external timing gears. As a result, the lobes do not make contact.

Pump shaft support bearings are located in the gearbox, and since the bearings are not within the pumped liquid, pressure is limited by the location of the bearing and shaft deflection.

As the lobes come out of mesh, they create expanding volume on the inlet side of the pump. The liquid then flows into the cavity and is trapped by the lobes as they rotate.

The liquid travels around the interior of the casing in the pockets between the lobes and the casing and it does not pass between the lobes.

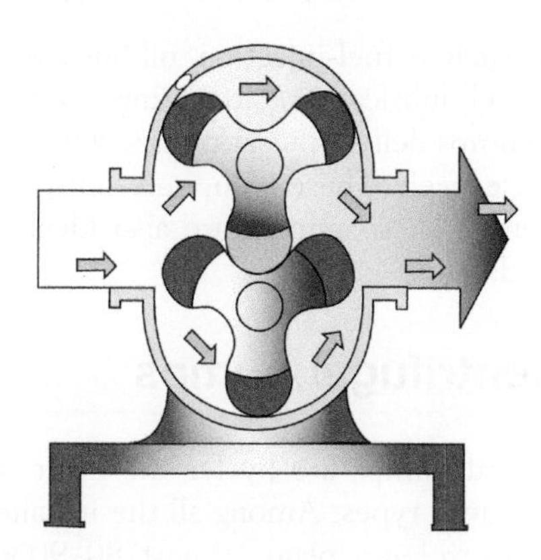

Fig. 5.3.10 Lobe pump.

Finally, the meshing of the lobes forces the liquid through the outlet port under pressure. Lobe pumps are frequently used in food applications because they can handle solids without damaging the product. The particle size pumped can be much larger in lobe pumps than in any other of the PD types.

5.3.4.3 Vane pump

A vane pump too traps the liquid by forming a compartment comprising of vanes and the casing (Figure 5.3.11). As the rotor turns, the trapped liquid is traversed from the suction port to the discharge port.

A slotted rotor or impeller is eccentrically supported in a cycloidal cam. The rotor is located close to the wall of the cam so a crescent-shaped cavity is formed. The rotor is sealed in the cam by two side plates. Vanes or blades fit within the slots of the impeller. As the impeller rotates and fluid enters the pump, centrifugal force, hydraulic pressure, and/or pushrods push the vanes to the walls of the housing. The tight seal among the vanes, rotor, cam, and side plate is the key to the good suction characteristics common to the vane pumping principle.

The housing and cam force fluid into the pumping chamber through the holes in the cam. Fluid enters the pockets created by the vanes, rotor, cam, and side plate.

As the impeller continues around, the vanes sweep the fluid to the opposite side of the crescent where it is squeezed through the discharge holes of the cam as the vane approaches the point of the crescent. Fluid then exits the discharge port.

Vane pumps are ideally suited for low-viscosity, nonlubricating liquids.

5.3.4.4 Progressive cavity pump

A progressive cavity pump consists of only one basic moving part, which is the driven metal rotor rotating within an elastomer-lined (elastic) stator (Figure 5.3.12).

As the rotor turns, chambers are formed between the rotor and stator. These chambers progress axially from

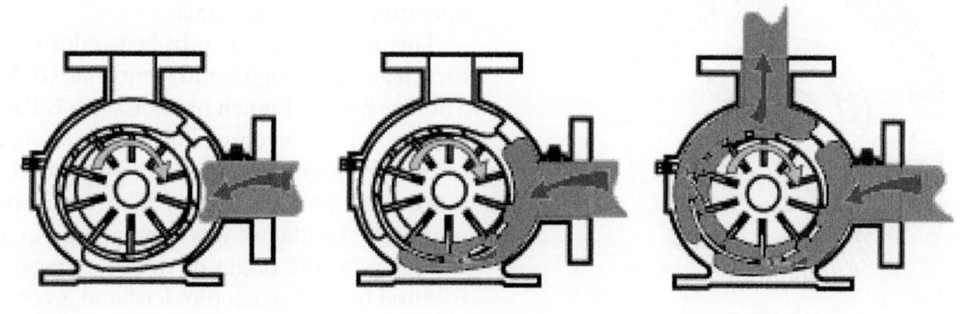

Fig. 5.3.11 Vane pump.

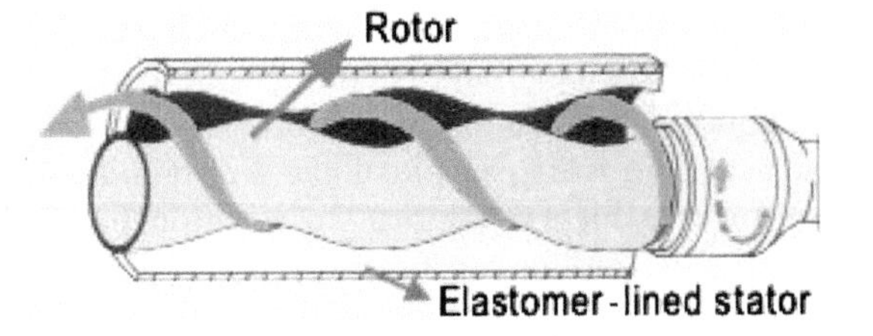

Fig. 5.3.12 Vane pump progressive cavity pump.

the suction to the discharge end, moving the fluid. By increasing the pitch of the rotor and stator, additional chambers or stages are formed.

The Vane pumps are solutions to the special pumping problems of municipal and industrial wastewater and waste processing operations. Industries, such as chemical, petrochemical, food, paper and pulp, construction, mining, cosmetic, and industrial finishing, find these pumps are ideally suited for pumping fluids with non-abrasive material inclusion.

5.3.4.5 Peripheral pump

As shown in Figure 5.3.13, the impeller has a large number of small radial vanes on both of its sides. The impeller runs in a concentric circular casing. Interaction between the casing and the vanes creates a vortex in the spaces between the vanes and the casing, and the mechanical energy is transmitted to the pumped liquid.

Peripheral pumps are relatively inefficient and have poor self-priming capability. They can handle large amounts of entrained gas. They are suitable to low flow and high-pressure applications with clean liquids.

5.3.4.6 Screw pump

In addition to the previously described pumps based on the Archimedes' screw, there are pumps fitted with two or three spindles crews housed in a casing.

Three-spindle screw pumps, as shown in Figure 5.3.14, are ideally suited for a variety of marine and offshore

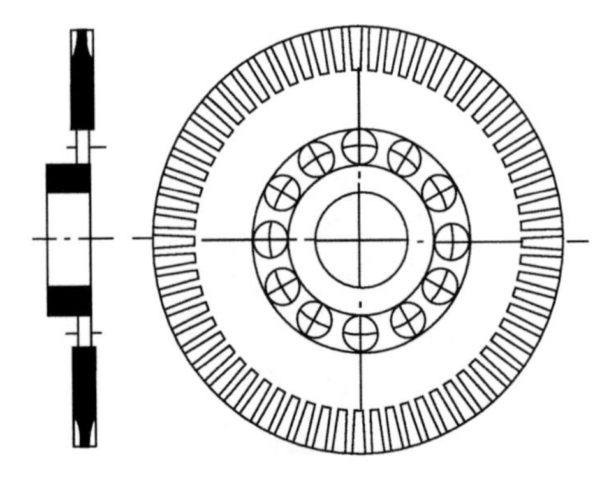

Fig. 5.3.13 Peripheral pump impeller.

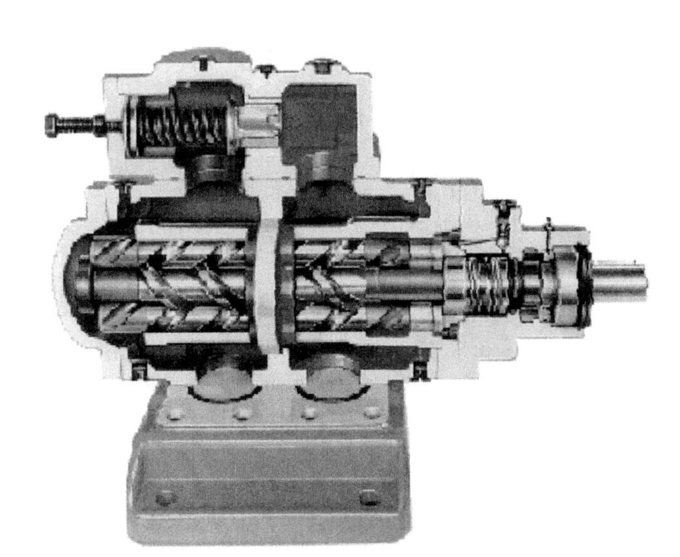

Fig. 5.3.14 Three-spindle screw pump – Alweiller pumps.

applications such as fuel-injection, oil burners, boosting, hydraulics, fuel, lubrication, circulating, feed, and many more. The pumps deliver pulsation free flow and operate with low noise levels. These pumps are self-priming with good efficiency. These pumps are also ideal for highly viscous liquids.

5.3.5 Centrifugal pumps

The centrifugal pumps are by far the most commonly used of the pump types. Among all the installed pumps in a typical petroleum plant, almost 80–90% are centrifugal pumps. Centrifugal pumps are widely used because of their design simplicity, high efficiency, wide range of capacity, head, smooth flow rate, and ease of operation and maintenance.

The 'modern' era pumps began during the late seventeenth and early eighteenth centuries AD. British engineer Thomas Savery, French physicist Denis Papin, and British blacksmith and inventor Thomas Newcomen contributed to the development of a water pump that used steam to power the pump's piston. The steam-powered water pump's first wide use was in pumping water out of mines.

However, the origin of the centrifugal impeller is attributed to the French physicist and inventor Denis Papin in 1689 (Figure 5.3.15).

Fig. 5.3.15 Denis Papin.

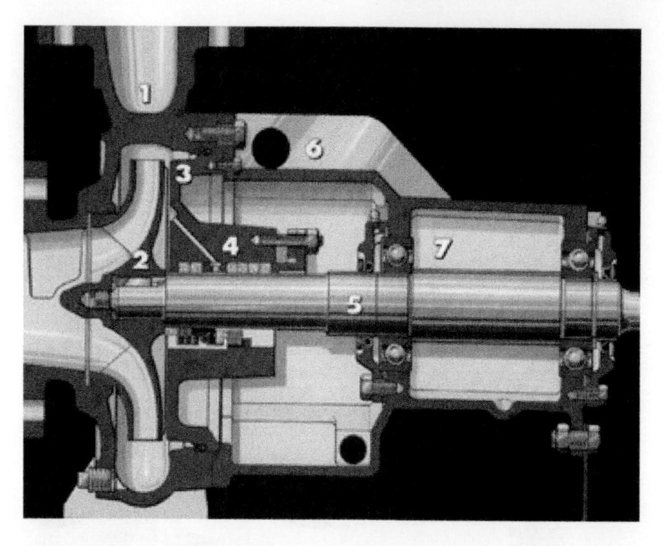

Fig. 5.3.16 Centrifugal pump – basic construction.

Papin's contribution lies in his understanding of the concept of creating a forced vortex within a circular or spiral casing by means of vanes. The pump made by him had straight vanes.

Following Papin's theory, Combs presented a paper in 1838 on curved vanes and the effect of curvature, which subsequently proved to be an important factor in the development of the centrifugal impeller. In 1839, W.H. Andrews introduced the proper volute casing and in 1846, he used a fully shrouded impeller.

In addition, in 1846, W.H. Johnson constructed the first three-stage centrifugal pump, and in 1849, James S. Gwynne constructed a multistage centrifugal pump and began the first systematic examination of these pumps.

Around the same time, British inventor, John Appold conducted an exhaustive series of empirically directed experiments to determine the best shape of the impeller, which culminated in his discovery that efficiency depends on blade curvature. Appold's pump of 1851 with curved blades showed an efficiency of 68%, thus improving pump efficiency three-fold.

The subsequent development of centrifugal pumps was very rapid due to its relatively inexpensive manufacturing and its ability to handle voluminous amounts of fluid. However, it has to be noted that the popularity of the centrifugal pumps has been made possible by major developments in the fields of electric motors, steam turbines, and internal combustion (IC) engines. Prior to this, the positive displacement type pumps were more widely used.

The centrifugal pump has a simple construction, essentially comprising a volute (1) and an impeller (2) (refer to Figure 5.3.16). The impeller is mounted on a shaft (5), which is supported by bearings (7) assembled in a bearing housing (6). A drive coupling is mounted on the free end of the shaft.

The prime mover, which is usually an electrical motor, steam turbine, or an IC engine, transmits the torque through the coupling.

As the impeller rotates, accelerates, and displaces the fluid within itself, more fluid is drawn into the impeller to take its place; if the pump is properly primed. The impeller thus, impacts kinetic or velocity energy to the fluid through mechanical action. This velocity energy is then converted to pressure energy by the volute. The pressure of the fluid formed in the casing has to be contained and this is achieved by an appropriate sealing arrangement (4). The seals are installed in the seal housing (3).

The normal operating speed of pumps is 1500 rpm (1800 rpm) and 3000 rpm (3600 rpm). However, there are certain designs of pumps that can operate at speeds in the range of 5000–25,000 rpm.

5.3.5.1 Types of centrifugal pumps

Centrifugal pumps can be categorized in various ways. Some of the main types are on the following basis.

5.3.5.1.1 Orientation of the pump shaft axis

This refers to the plane on which the shaft axis of the pump is placed. It is either horizontal or vertical as shown in Figure 5.3.17.

5.3.5.1.2 Number of stages

This refers to the number of sets of impellers and diffusers in a pump. A set forms a stage and it is usually single, dual, or multiple (more than two) stages (Figure 5.3.18).

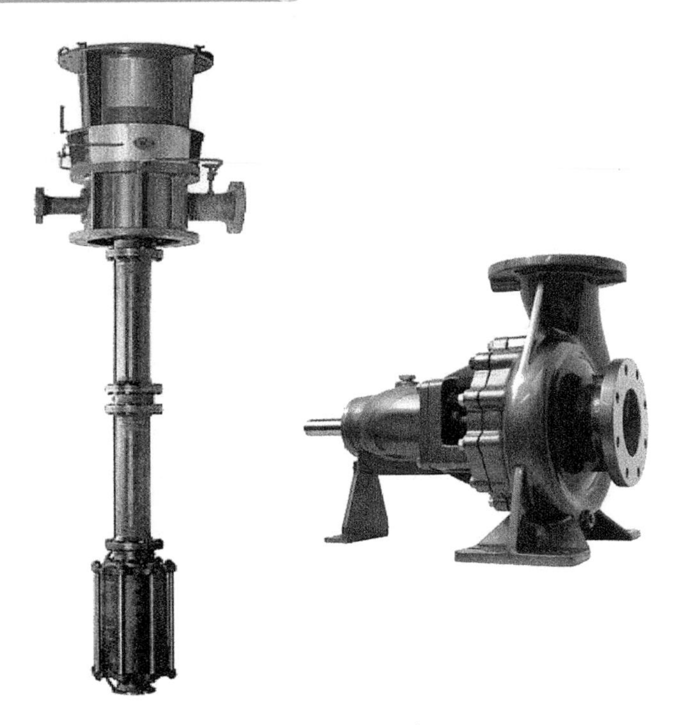

Fig. 5.3.17 Vertical pump and horizontal pump.

5.3.5.1.3 Suction flange orientation

This is based on the orientation of the pump suction flange. This orientation could be horizontal (also known as End) or vertical (also known as Top) (Figure 5.3.19).

5.3.5.1.4 Casing split

This classification is based on the casing split. It is either radial (perpendicular to shaft axis) or axial (plane of the shaft axis) (Figure 5.3.20).

5.3.5.1.5 Bearing support

This is judged based on the location of the bearings supporting the rotor. If the rotor is supported in the form of a cantilever (Figure 5.3.22), it is called as an Overhung type of pump. When the impellers on the rotor are supported with bearings on either side, the pump is called as an in-between bearings pump.

5.3.5.1.6 Pump support

This refers to how the pump is supported on the base frame. It could be a center-line (Figure 5.3.21a) support or foot-mounted support (Figure 5.3.21b).

5.3.5.1.7 Shaft connection

The closed coupled pumps are characterized by the absence of a coupling between the motor and the pump. The motor shaft has an extended length and the impeller is mounted on one end (Figure 5.3.22).

The vertical monobloc pumps have the suction and discharge flanges along one axis and can be mounted between pipelines. They are also termed as 'in-line pumps'.

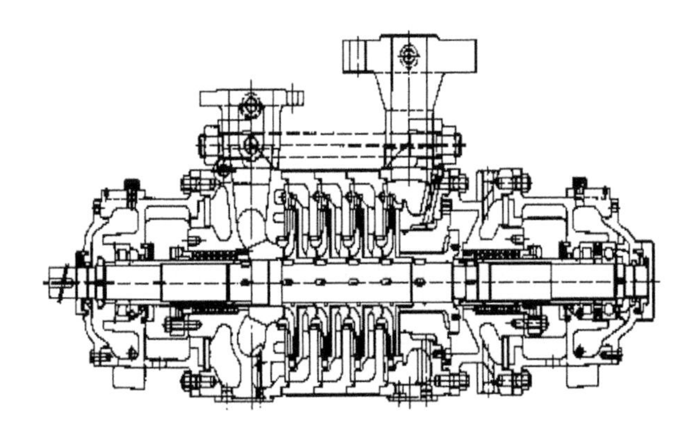

Fig. 5.3.18 Multistage pump.

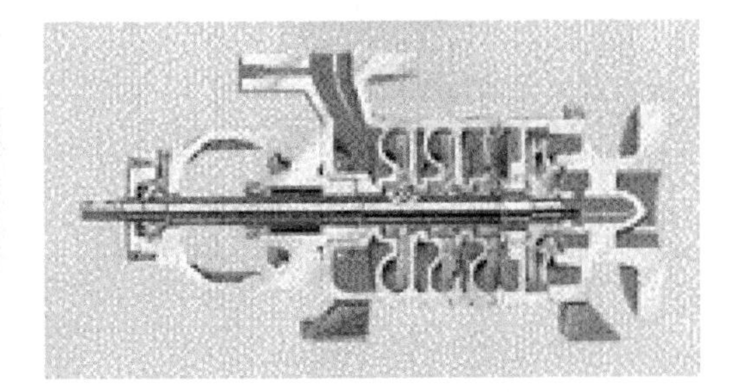

Fig. 5.3.19 Multistage pump with end suction.

Fig. 5.3.20 Axial split casing.

Fig. 5.3.22 Closed coupled monobloc pumps with end suction.

5.3.5.1.8 Sealless pumps

Pumps are used to build the pressure in a liquid and if necessary to contain it within the casing. At the interface of the rotating shaft and the pump casing, mechanical seals are installed to do the job of product containment. However, seals are prone to leakages and this maybe unacceptable in certain critical applications. To address this issue, sealless pumps have been designed and manufactured.

These are of two types – canned and magnetic drive pumps:

1. *Canned pumps*: In the construction of this second type of sealless pump, the rotor comprises of an impeller, shaft, and the rotor of the motor. These are housed within the pump casing and a containment shell (Figure 5.3.23). The hazardous or the toxic liquid is confined within this shell and casing.

   The rotating flux generated by the stator passes through the containment shell and drives the rotor and the impeller.

2. *Magnetic drive pumps*: In magnetic drive pumps, the rotor comprises of an impeller, shaft, and driven magnets. These housed within the pump casing and the containment shell ensures that the usually hazardous/toxic liquid is contained within a metal shell (Figure 5.3.24).

   The driven magnets take their drive from the rotating drive magnets, which are assembled on a different shaft that is coupled to the prime mover.

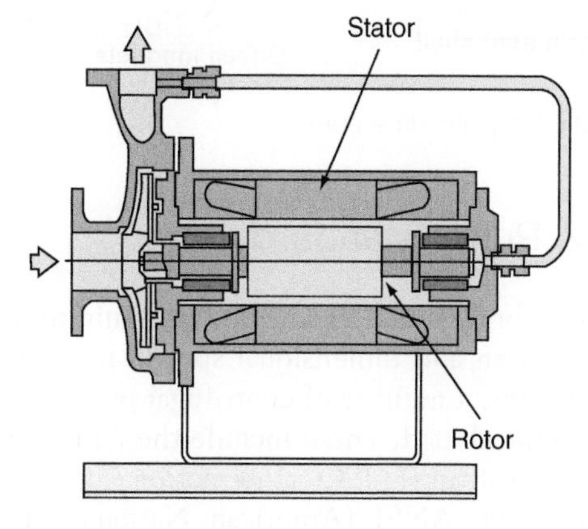

Fig. 5.3.23 Canned pump.

(a)

(b)

Fig. 5.3.21 Models of pump supports.

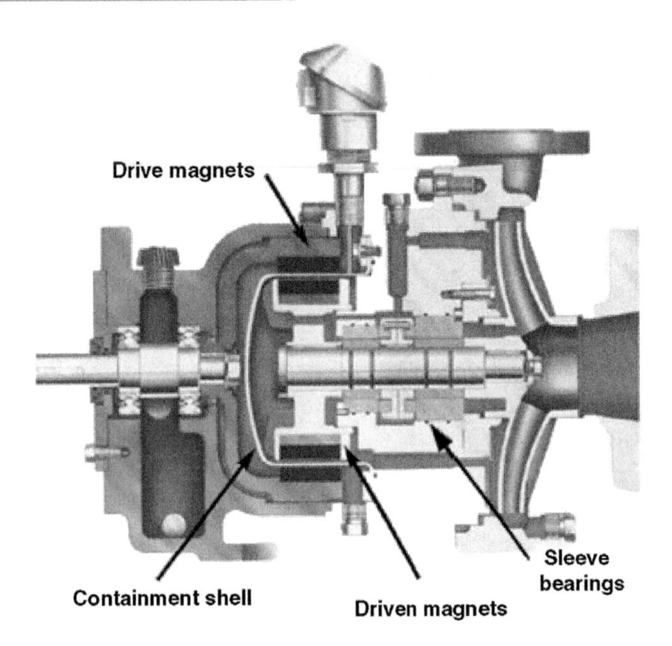

Fig. 5.3.24 Magnetic drive pump.

5.3.5.2 Pump standards

In order to bring about uniformity and minimum standards of design and dimensional specifications for centrifugal pumps, a number of centrifugal pump standards have been developed. These include the API (American Petroleum Institute), ISO (International Organization for Standards), ANSI (American National Standards Institute), DIN (German), NFPA (Nation Fire Protection Agency), and AS-NZ (Australia–New Zealand).

Some of the famous standards, which are used in the development and manufacture of centrifugal pumps are API 610, ISO 5199, 2858, ANSI B73.1, DIN 24256, NFPA-21.

In addition to the above, there are many National Standards. Some of these are:

- *France*: NF E 44.121
- *United Kingdom*: BS 5257
- *German*: DIN 24256
- *Australia & New Zealand*: AS 2417-2001, grades 1 and 2.

Usually, the service criticality or application of the pump forms the deciding factor for a choice of standard. A critical refinery pump handling hazardous hydrocarbons would be in all probability built as per standard API 610 (Figure 5.3.25).

However, ordinary applications do not require the entire API-specified features and so the premium that comes with an API pump is not justified. Such pumps can be purchased built to lesser demanding standards like the ANSI B73.1. One big advantage of ANSI pumps is the outline dimensional interchangeability of same size

Fig. 5.3.25 Pump built to API 610 standard.

pumps regardless of brand or manufacturer, something that is not available in the API pumps.

In a similar way, pumps meant for firewater applications are usually built to the design specifications laid out in NFPA-21.

There are some standards like the ISO 2858, which are primarily meant as dimensional standards. This does not provide any requirement for the pump's construction. The standard from ISO that addresses the design aspects of pumps is ISO 5199.

For a good comparative study of the API, ANSI, and ISO standards, it is recommended to read the technical paper called, 'ISO-5199 Standard Addresses Today's Reliability Requirements For Chemical Process Pumps', by Pierre H. Fabeck, Product Manager, Durco Europe, Brussels, Belgium and R. Barry Erickson, Manager of Engineering, The Duriron Company, Incorporated, Dayton, Ohio. This paper was presented at the 7th (1990) Pumps Symposium at the Texas A&M University.

5.3.5.3 Pump applications

The classification of pumps in the above sections is based on the construction of the pump and its components. However, on the basis of the applications for which they are designed, pumps tend to be built differently.

Some of the applications where typical pumps can be found are:

- petroleum and chemical process pumps;
- electric, nuclear power pumps;
- waste/wastewater, cooling tower pumps;
- pulp and paper;
- slurry;
- pipeline, water-flood (injection) pumps;
- high-speed pumps.

As this needs an introduction to the components/construction of the pump, these are covered in detail in subsequent topics.

Chapter 5.4

5.4

Centrifugal pump design

Girdhar, Moniz and Mackay

In Chapter 5.3, we have studied different types of centrifugal and positive displacement pumps and their various distinguishing external features. In this chapter, we will learn about various types of internal components of centrifugal pumps.

The diversity among pumps does not only limit itself to the external features of the machines but also extends to its internal components. This is especially true in the case of centrifugal pumps. The basic components are essentially the same in almost every design but depending on the design and its applications, the construction features of the internal components differ to meet various requirements.

5.4.1 Impellers

The impeller of the centrifugal pump converts the mechanical rotation to the velocity of the liquid. The impeller acts as the spinning wheel in the pump.

It has an inlet eye through which the liquid suction occurs. The liquid is then guided from the inlet to the outlet of the impeller by vanes. The angle and shape of the vanes are designed based on flow rate. The guide vanes are usually cast with a back plate, termed *shroud* or *back cover*, and a front plate, termed *front cover*.

Impellers are generally made in castings and very rarely do come across fabricated and welded impellers.

Impellers can have many features on them like balancing holes and back vanes. These help in reducing the axial thrust generated by the hydraulic pressure.

In order to reduce recirculation losses and to enhance the volumetric efficiency of the impellers, they are provided with wearing rings. These rings maybe either on the front side or on both the front and backsides of the impeller. It is also possible to have an impeller without any wearing rings.

The casting process, as mentioned above, is the primary method of impeller manufacture. Smaller size impellers for clean water maybe cast in brass or bronze due to small section thickness of shrouds and blades. Recently, plastic has also been introduced as casting material.

For larger impellers and in most of the applications, cast iron is the first choice of the material. The grade used is ASTM A-48-40 (minimum tensile strength is 40,000 psi or 2720 kgs/cm^2).

This is used for a maximum peripheral speed of 55 m/s and a maximum temperature of 200 °C. When the temperature exceeds 200 °C, carbon steel castings of the grade A-216 WCA/WCC are recommended.

The adequacy of cast steel is dependent on its usage in handling of abrasives like ash, sand, or clinker. In such cases, the impellers could also be cast in 12% Cr steels (A-743 CA15). Stainless steel castings (A-744 CF8M) are used for their high corrosion resistance and for low-temperature applications. In case of low-temperature applications (not lower than 100 °C), ferritic steel castings containing 3.5% nickel can be used (A-352 LC3). For temperatures until 200 °C, A-276-Type 304 castings are used.

Marine applications may demand castings made from aluminum bronze (B-148 – Alloy C 95,800). Copper bronze casting grade adopted is B-150-Alloy 63,200. Caustic acid solutions and other corrosive liquids may demand special materials. For example, sulfuric acid (concentration 67% and at a temperature of 60–70 °C) needs hi-silicon cast iron (15% Si).

Practical Centrifugal Pumps: Design, Operation and Maintanance; ISBN: 9780750662734

During the casting process, it is important to keep the liquid contact surfaces of the impeller as smooth as possible. Thus, the composition of the core sand mixture and the finish of the core play an important part in the casting process. Largely, the relative smoothness of the liquid path determines the efficiency of the pump.

In a closed impeller design, the contact surface area of the metal with the liquid is higher which results in high friction losses. When the impeller's diameter is large, the problem becomes more acute and so there is a higher demand for smoother surface. Friction losses are related to fifth power of the diameter.

Subsequent to the casting and surface finishing operations, the impellers are dynamically balanced. The limits of residual unbalance are generally specified in ISO 1940, or even in API, which has a stricter limit. The balance of impellers alone is insufficient. Once the pump rotor components are ready, these should be mounted assembly wise on the balancing machine and balanced to stated limits.

5.4.1.1 Construction of impellers

There are three types of construction seen in an impeller. These are based on the presence or absence of the impeller covers and shrouds.

The three types (Figure 5.4.1) are:

1. closed,
2. semi-open,
3. open.

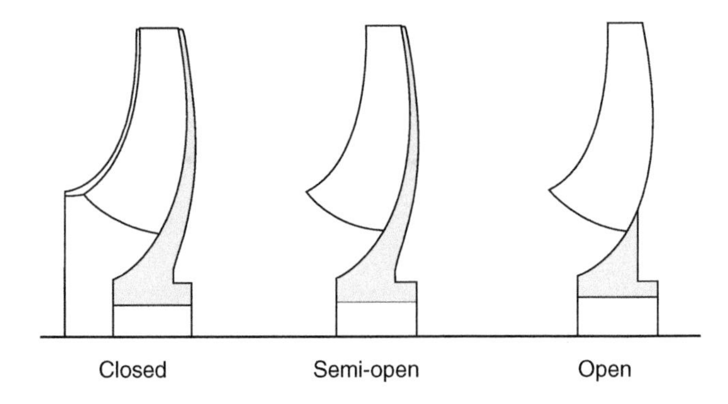

Fig. 5.4.1 Types of impellers.

5.4.1.1.1 Closed impellers

The closed impeller consists of radial vanes (typically 3–7 in number), which are enclosed from both sides by two disks termed 'shrouds'. These have a wear ring on the suction eye and may or may not have one on the back shroud. Impellers that do not have a wear ring at the back typically have back vanes. Pumps with closed type impellers and wear rings on both sides have a higher efficiency.

5.4.1.1.2 Semi-open impellers

The semi-open type impellers are more efficient due to the elimination of disk friction from the front shroud and are preferred when the liquid used may contain suspended particles or fibers. The axial thrust generated in semi-open impellers is usually higher than closed impellers.

5.4.1.1.3 Open impellers

There are three types of back shroud configurations. The first one is a fully scalloped open impeller as shown in Figure 5.4.2.

The back shroud is almost taken out and thus the axial thrust caused by the hydraulic pressure is almost eliminated.

The second type is known as the partially scalloped open type of impeller as shown in Figure 5.4.3. It experiences a greater axial thrust than the fully scalloped open impeller. However, this has higher efficiency and head characteristics.

The third type is known as the fully back shroud open impeller (Figure 5.4.4) where there is an open impeller with a full back shroud. It normally has almost 5% higher efficiency than a fully scalloped impeller, though it has diminished head generation capabilities.

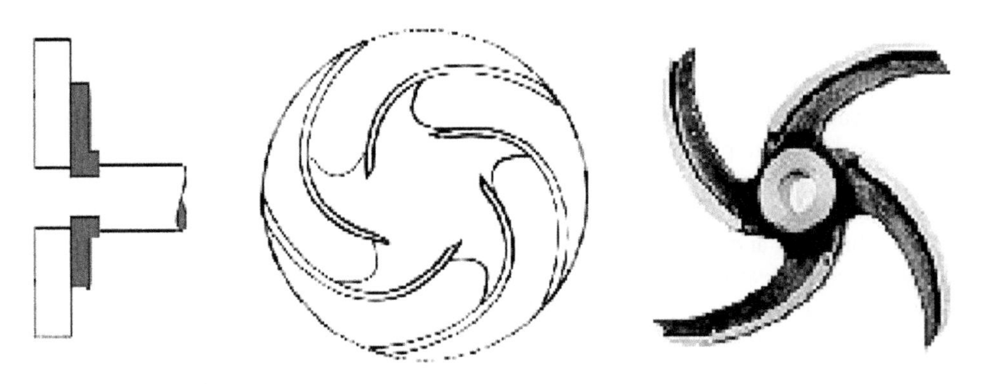

Fig. 5.4.2 Fully scalloped open impeller.

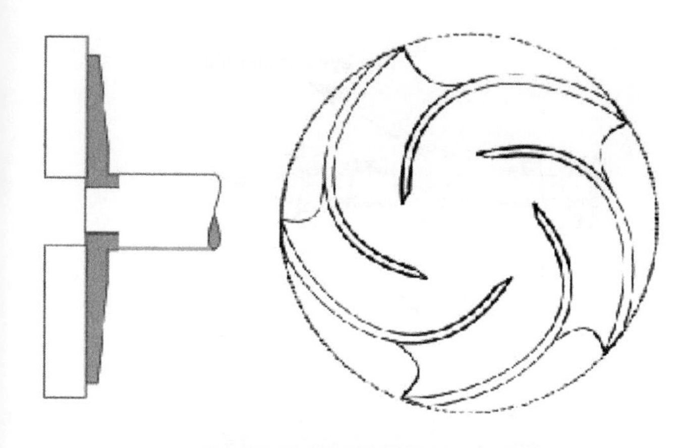

Fig. 5.4.3 Partially scalloped open impeller.

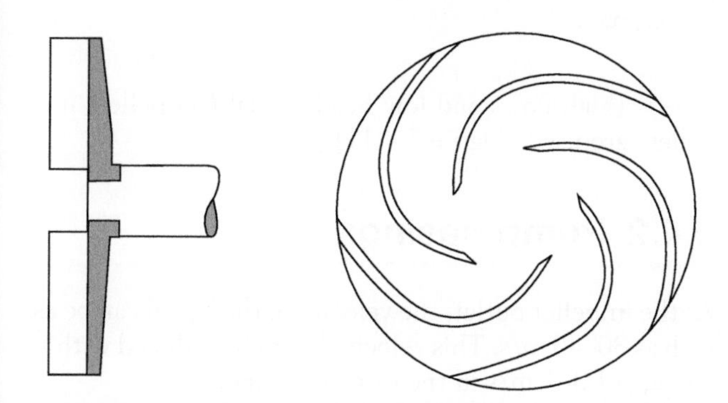

Fig. 5.4.4 Fully back shroud open impeller.

The fully shrouded open impellers experience the maximum axial thrust among the open impeller types. To reduce this effect, back vanes are provided to relieve the hydraulic pressure that generates the axial thrust.

The vortex or non-clog impellers (Figure 5.4.5) are the fully shrouded open type of impellers. These are used in applications where the suspended solid's size maybe large or the solid's maybe of crystals and fibers type. The vortex impeller does not impart energy directly to the liquid. Instead it creates a whirlpool, best described as a vortex. The vortex in turn imparts energy to the liquid or pumpage. The location of the impeller is usually above the volute, so it experiences hardly any radial forces. This

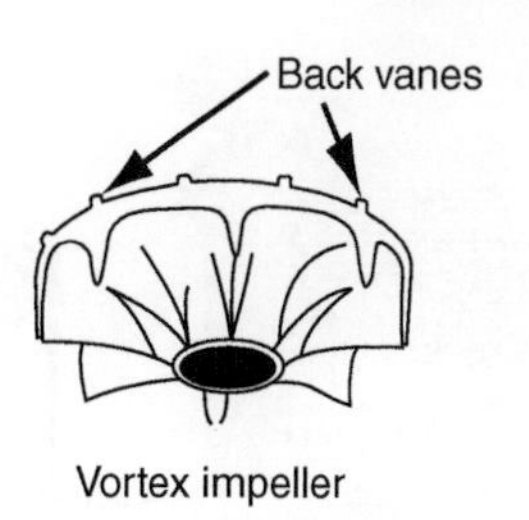

Fig. 5.4.5 Vortex impeller.

allows extended operation of the pump even at closed discharge conditions.

Some of the other non-clogging designs of impellers in the closed and semi-open types are shown in Figures 5.4.6 and 5.4.7.

In general, most of the open impellers are of the partially scalloped and fully shroud types. Fully open impellers are rarely used because of its lower efficiency and the bending load on the vanes.

5.4.1.2 Impeller suction

In general, an impeller has one eye or a single opening through which liquid suction occurs. Such impellers are called as single-suction impellers. Pumps with a single-suction impeller (impeller having suction cavity on one side only) are of a simple design but the impeller is subjected to higher axial thrust imbalance due to the flow on one side of the impeller only.

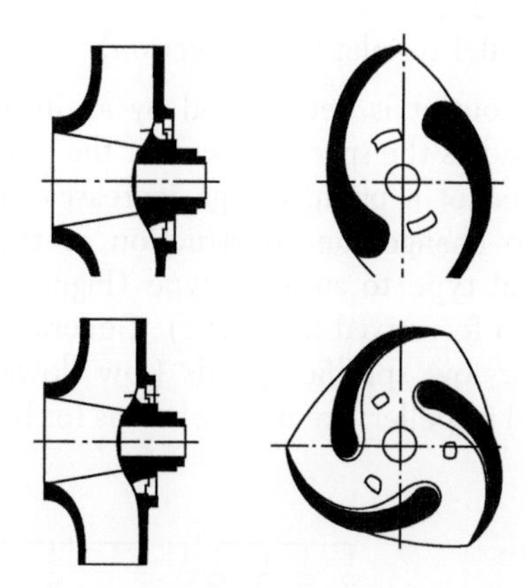

Fig. 5.4.6 2- & 3-Passage closed non-clog impellers.

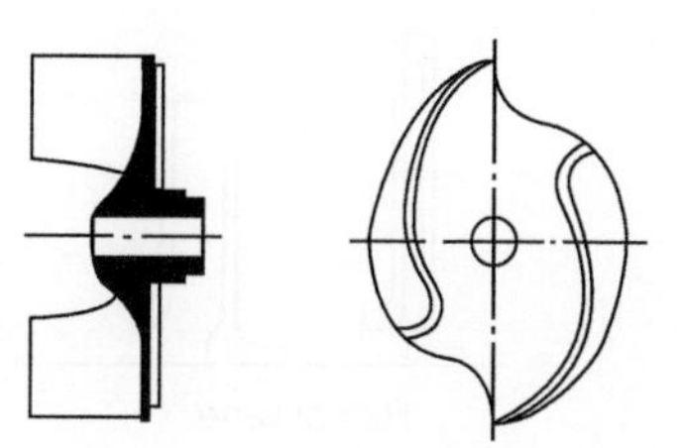

Fig. 5.4.7 Semi-open 2-passage non-clog S-shaped impeller.

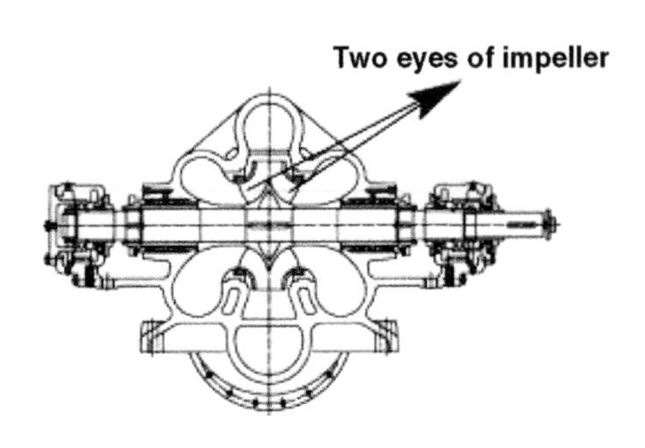

Fig. 5.4.8 Pump with double suction impeller.

In certain pumps, the flow rate is quite high. This can be managed by having one impeller with two suction eyes. Pumps with double-suction impeller (impeller having suction cavities on both sides) has lower NPSH-r than single-suction impeller. Such a pump is considered hydraulically balanced but is susceptible to an uneven flow on both sides if the suction piping is improper.

Generally, flows that are more than 550 m^3/h (or 153 l/s) may necessitate a double suction impeller (Figure 5.4.8).

5.4.1.3 Flow outlet from impeller

The flow direction of the liquid at the outlet of the impeller can be:

- radial (perpendicular to inlet flow direction),
- mixed,
- axial (parallel to inlet flow direction).

The flow outlet is determined by an important parameter called as the specific speed of the pump. As the specific speed of a pump design increases, it becomes necessary to change the construction of the impeller from a radial type to an axial type (Figure 5.4.9, and Figure 5.4.10 for mixed flow type). Generally, it can be said that for low specific speeds (low flows and high heads) radial impellers are used whereas for high specific speeds (high flows and low heads) axial (propeller) impellers are used (Figure 5.4.11).

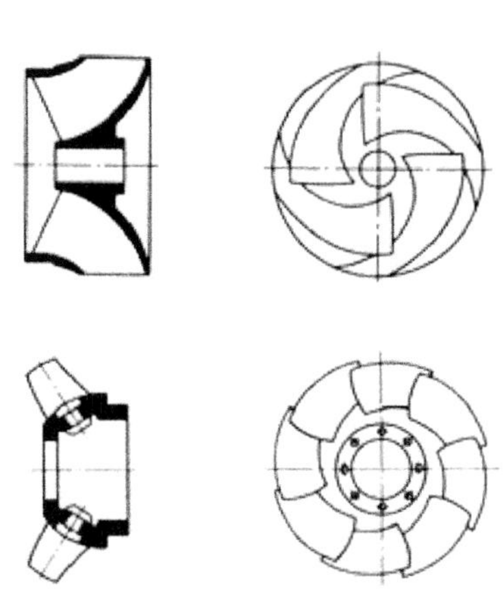

Fig. 5.4.10 Mixed flow impeller and propeller vaned mixed flow type impeller.

5.4.2 Pump casings

At the impeller outlet, the velocity of the liquid can be as high as 30–40 m/s. This velocity has to be reduced within a range of 3–7 m/s in the discharge pipe.

Velocity reduction is carried out in the pump casing by recuperators. The kinetic energy in the liquid at

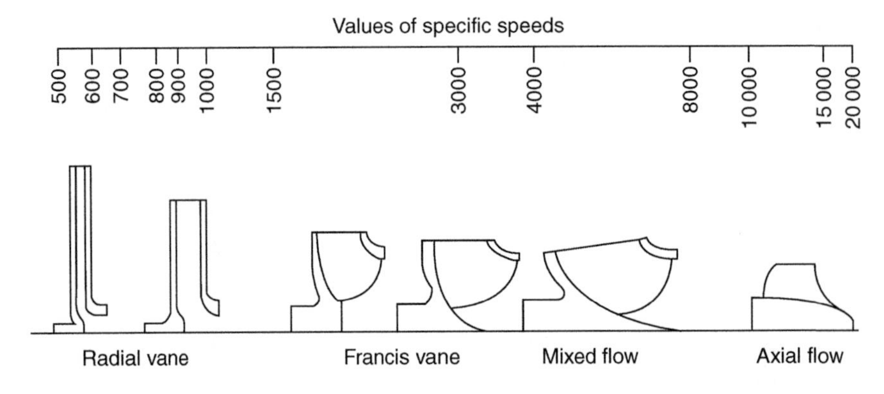

Fig. 5.4.9 Shapes of impellers according to their specific speeds.

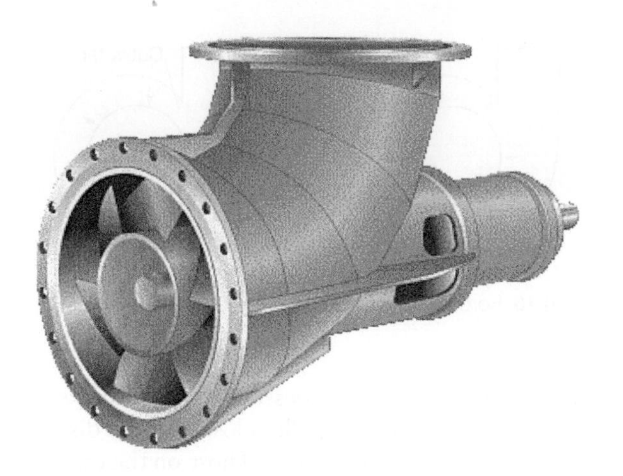

Fig. 5.4.11 Pump with axial flow impeller.

the outlet is converted to pressure energy by the recuperators.

Here, energy conversion has to be undertaken with a minimal loss to have an insignificant effect on pump efficiency.

Some of the recuperators are:

- vaneless guide ring,
- concentric casing,
- volute casing,
- diffuser ring vanes,
- diagonal diffuser vanes,
- axial diffuser vanes.

5.4.2.1 Vaneless guide ring

A vaneless guide ring consists of two smooth disks (Figure 5.4.12). The distance between the two guide rings is either constant or is increased toward the outlet.

It follows that the conversion of kinetic energy of the liquid to pressure energy is entirely proportional to the ratio of the outlet diameter (D_o) of the ring to the inlet diameter (D_i).

The breadth of the ring has little role in the generation of liquid head, though it is observed that rings of constant breadth are more efficient than those with higher breadths at the outlet diameter (B_o).

Due to the above, the vaneless guide ring is used in pumps where liquid velocities are lower. It is thus found in pumps developing low heads. For larger heads, the outlet diameter of the ring would become larger and this maybe unpractical.

Vaneless guide rings are usually used in mixed flow impeller pumps of higher specific speeds along with an annular delivery passage of constant cross section. These may also be found in lower specific speed pumps handling liquid with solid matter.

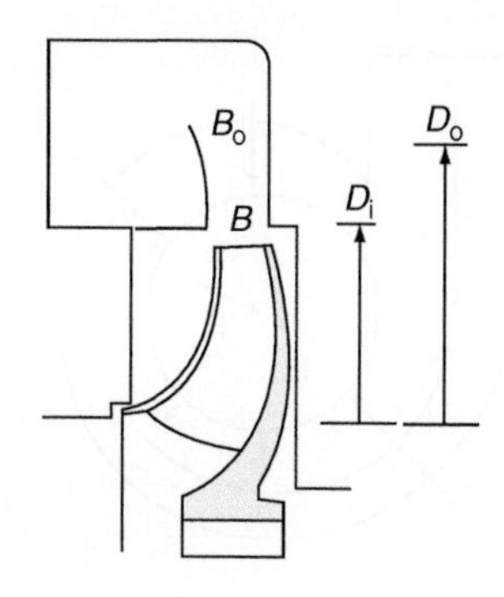

Fig. 5.4.12 Vaneless guide ring.

5.4.2.2 Concentric casing

Concentric casings are usually found in single-stage centrifugal pumps and in the last stage of multistage pumps (Figure 5.4.13).

In some of the earlier designs of a single-stage centrifugal pump for larger heads, an annular delivery passage is used in conjunction with a *diffuser ring*. The liquid outlet is through a conical diffuser.

The ratio of the impeller diameter to the diameter of the casing is not less than 1.15 and not more than a ratio of 1.2.

The volute width is designed to accommodate the maximum width of the impeller. The capacity at the most efficient point of operation is controlled by the volute diameter (d).

To minimize the recirculation in the volute, a cutwater tongue is used. In addition, this helps in significantly reducing the radial loads on the shaft.

In pumps with a specific speed of less than 600 (US-gpm, feet, rpm), the concentric casing provides higher efficiency than a conventional volute casing. Above the specific speed, N_s of 600, the efficiency progressively drops.

The concentric casings are used:

- for less flow and higher head; low specific speeds N_s is in the range of 500–600;

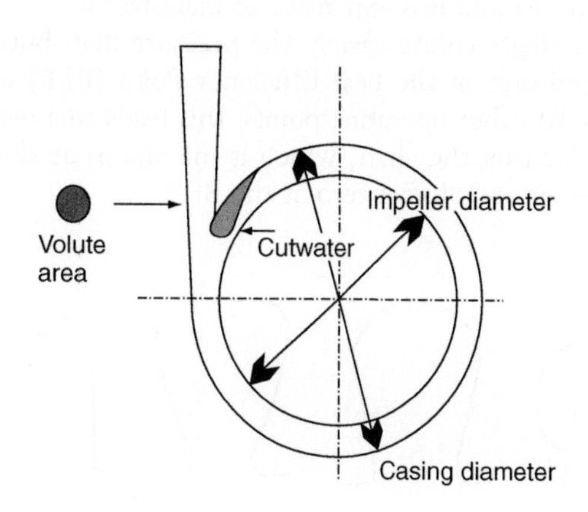

Fig. 5.4.13 Concentric casing pump.

- where the pump casing has to accommodate several impeller sizes;
- where pump has to use a fabricated casing;
- where volute passage has to be machined from a casting;
- where foundry limitations result in higher impeller width.

5.4.2.3 Volute casing

Volute casings when manufactured with smooth surfaces offer insignificant hydraulic losses. In pumps with volute casings, it is possible to trim down impeller vanes and shrouds with minimal effect on efficiency.

In volute casings, the kinetic energy is converted into pressure only in the diffusion chamber immediately after the volute throat. The divergence angle is between 7° and 13°.

The volutes encountered can be of various cross sections and these are shown in Figure 5.4.14.

The first two profiles are of circular cross section; the third is called as the trapezoidal cross section, which is typically found in single-stage pumps. The last profile is the rectangular cross section.

The rectangular section is used in small single-stage pumps and in multistage pumps. It is economical to manufacture due to its low pattern cost and production time. The hydraulic losses are minimal in the specific speed range of less than 1100.

Volute casings are manufactured in various designs and these are:

- single volute casing,
- double volute casing.

5.4.2.3.1 Single volute casing

Single volute designs are the most commonly found designs and those designed on the basis of constant velocity are the most efficient among all types. They are easy to cast and less expensive to manufacture.

In a single volute casing, the pressure distribution is balanced only at the Best Efficiency Point (BEP) of the pump. At other operating points, this leads to a residual radial load on the shaft, which is maximum at shut-off conditions and almost zero at the BEP.

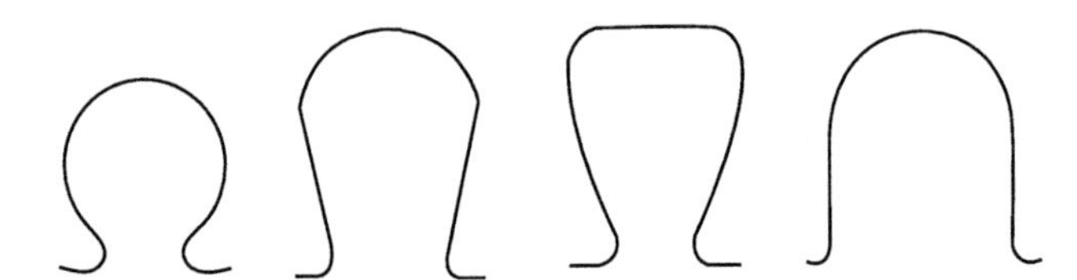

Fig. 5.4.14 Different volute cross-section shapes.

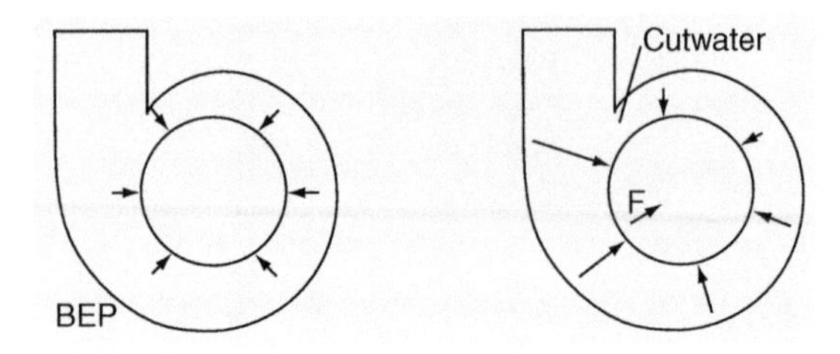

Fig. 5.4.15 Forces as generated in a single volute.

At low flow rates, the pressure distribution is such that the surfaces of the impeller closest to the discharge are acted upon by high pressures. Those on the other side of the cutwater are acted upon by comparatively low pressures (Figure 5.4.15).

The resulting unbalanced forces can be assumed to be acting at a point 240° from the cutwater and acting in a direction which points to the center of the impeller.

Theoretically, these casings can be used over the entire range of specific speed pumps; however, these are used mainly on low capacity, low specific speed pumps. They can also be used in pumps handling slurries and solids.

5.4.2.3.2 Double volute casing

A double volute casing design is actually two single volute designs combined in an opposed arrangement (Figure 5.4.16). The total throat area of the two volutes is identical to that which would be used on a comparable single volute design.

Single volute designs inherently generate a radial load on the shaft. The double volute designs limit this radial force to a greater extent.

In this design, the volute is symmetrical about its centerline; however, the two passages carrying the liquid to the discharge flange are not symmetrical. As a result, the pressure forces around the impeller periphery do not cancel and this leads to some radial force.

The hydraulic performance of the double volute is on a par with the single volute design. At the BEP, the

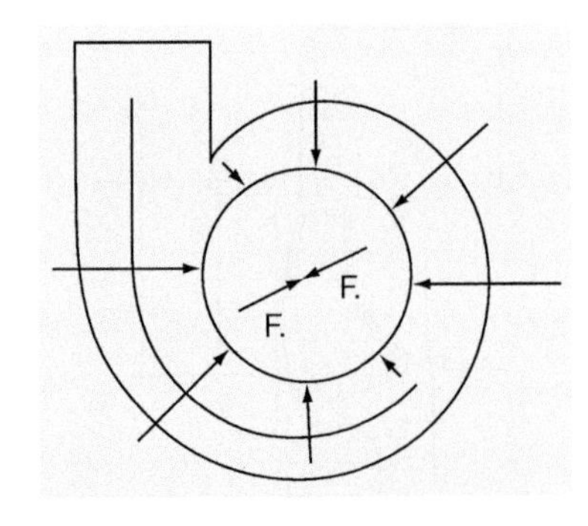

Fig. 5.4.16 Balance of forces in double volute.

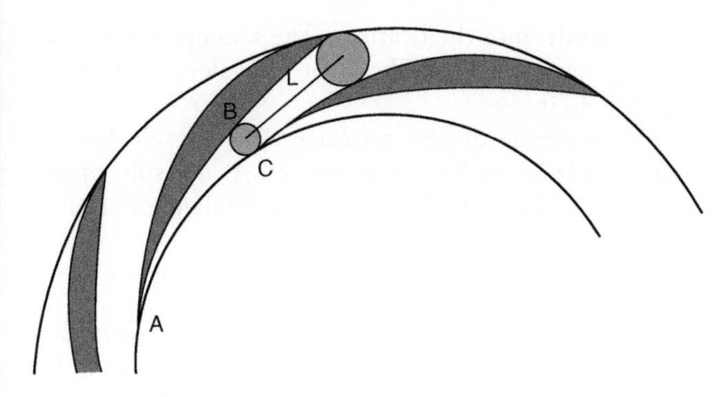

Fig. 5.4.17 Vaned diffuser.

efficiency is marginally lower but is higher at operating points; lower and higher than BEP. Thus, for flows over the entire range, the double volute design is preferred.

Therefore, flow rate is the basic criterion that determines the selection of one design over another. For flows under 125 m^3/h, double volute designs are not used since it becomes difficult to manufacture and clean them in smaller casing. In larger pumps, double volutes are invariably used.

5.4.2.4 Vaned diffuser ring

The vaned diffuser ring has a series of symmetrically placed vanes forming gradually widening passages (Figure 5.4.17). This ring comprises of a series of vanes set around the impeller. The flow from the vaned diffuser is collected in a volute or circular casing and is discharged through the discharge pipe.

In these passages, the velocity head is converted to pressure energy. The distance BC shown in Figure 5.4.17 is called as the throat.

The design of the vaned diffuser is similar to the volute except that there are many throats in a vaned diffuser compared to just one continuous expanding section in the volute.

From the throat onward, the area of the vane channel increases progressively so that, further, a slight increase in pressure takes place. The centerline of the vane channel after the throat maybe straight or curved. The straight diffusing channel is slightly more efficient but results in a larger casing.

The vane surface from the vane inlet to the outlet can be shaped like a volute but even a circular arc works fine.

The number of diffuser vanes is usually one more than the impeller vanes, as it is found that the number of diffuser vanes should not be much larger than the number of impeller vanes.

With just one vane more than the impeller, it insures that one impeller passage does not extend over several diffuser passages.

5.4.2.5 Diagonal diffuser vanes

Diagonal diffuser vanes are recuperators for the mixed flow impeller pumps.

The functions of the diagonal diffuser vanes are:

- To change the direction of flow of the liquid leaving the impeller and direct it along the axis of the pump.
- To reduce the velocity of liquid and convert it to pressure.

The vanes are disposed in the axial direction forming channels with no sudden changes in cross section. They make it possible to use impellers of different diameters and breadths so as to extend the range of application of the given model of diffuser.

As the specific speed increases, the profiles of impellers and diffusers change and approximate to the shapes of impellers and diffusers of propeller pumps.

5.4.2.6 Axial diffuser vanes

Axial diffuser vanes are vanes placed behind the impeller of an axial flow pump (Figure 5.4.18). The functions of these vanes are similar to those of a mixed flow pump.

The vanes usually number 5–8. The lower number is found in pumps with a lower specific speed (diffuser type-1).

The efficiency is influenced to a certain extent by the shape of the diffuser passage. This depends on the number of vanes and their axial length and the distance between the impeller blades and the diffuser vanes.

Shorter and higher number of vanes (diffuser type-2) for the same flow and head give better efficiency.

When specific speeds are higher, these vanes are superfluous and a simple conical diffuser is constructed in their place.

5.4.3 Wearing rings

The impeller is a rotating component and it is housed within the pump casing. To prevent frictional contact, a gap between these two parts is essential.

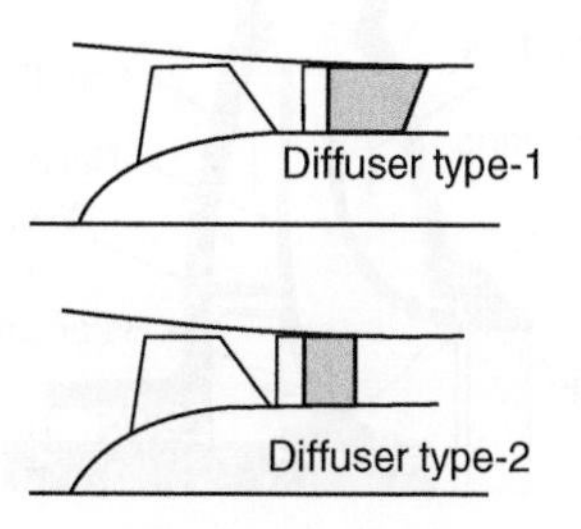

Fig. 5.4.18 Axial flow pumps with diffuser behind the propeller.

So there exists a gap between the periphery of an impeller intake and the pump casing. In addition, there is a pressure difference between them, which results in the recirculation of the pumped liquid. This leakage reduces the efficiency of the pump.

The other advantages of lower clearance is that reduced leakage prevents erosion due to suction recirculation and also provides a much better rotor dynamic stability to the pump. As a result, the vibration of the pump operates with lesser vibrations.

Thus, it is essential to keep this gap or clearance at an optimum value. When this clearance is kept at a lower value, the efficiency improves but there is always a risk of contact of the impeller with the casing.

Such a frictional contact may render the impeller or the casing useless which would be a loss since these are expensive parts. Therefore, in the areas of the impeller intake, metallic rings are fitted on the impeller eye as well as on the pump casing.

Accordingly, the wearing ring on the impeller is called as impeller wearing rings and the one fitted on the casing is called as the case wearing ring (Figure 5.4.19).

The cross section of wearing rings shown in Figure 5.4.20 is fitted with an impeller eye and is called as the front wearing ring. However, in some cases, wearing rings are installed even at the back shroud of the impeller.

Usually, these are required when impellers are provided with balancing holes in order to minimize the axial thrust coming onto the pump impeller and

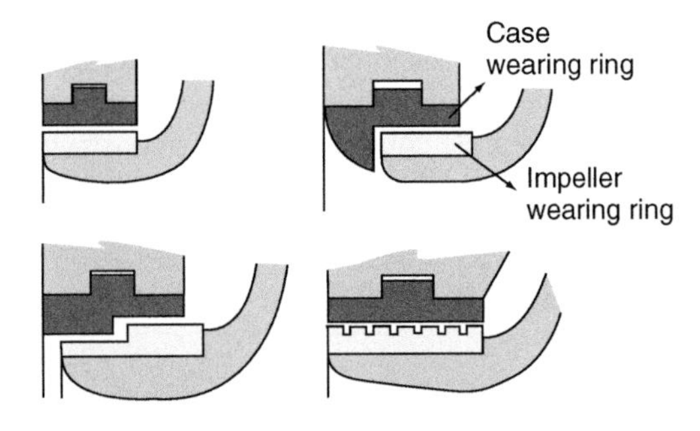

Fig. 5.4.19 Wearing rings of different types.

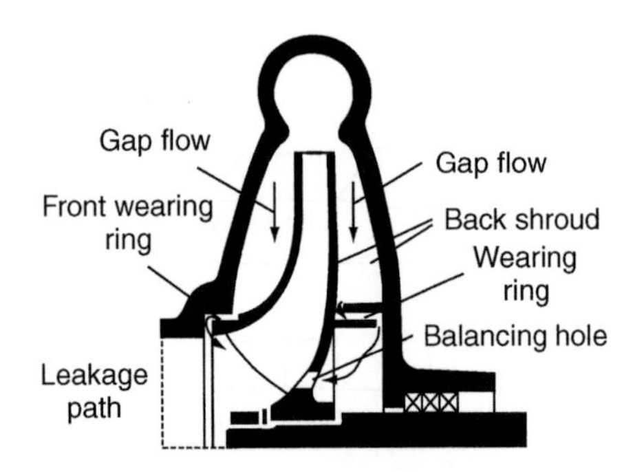

Fig. 5.4.20 Locations of wearing rings.

consequently onto the bearings. The arrangement of the wearing rings on the back of the impeller is shown in Figure 5.4.20.

The material of the wearing rings is selected to prevent seizure on frictional contact. As a result, materials like SS-316 which have galling tendencies are not considered for this application.

The other materials considered favorably are:

- Austenitic gray iron castings – ASTM A-436, Type-1.
- Austenitic ductile iron castings – ASTM A-439, Type-D2.
- 12% Chrome steels – AISI 420 (hardenable).
- 18 Cr–8 Ni steel castings – AISI 304.
- Copper alloy sand castings (bronze) – B-584, Alloy C 90,500.
- Aluminum bronze sand castings – B-148, Alloy C 95,800
- Monel – K 500
- Nickel 200.

The hardness range of the case wearing ring is in the region of 225–275 BHN, whereas the corresponding impeller wearing ring is kept harder by about 50–100 BHN. The range of hardness varies from 325 to 375 BHN.

API 610 standard for centrifugal pumps provides guidelines on the minimum recommended wearing ring clearances for metallic wearing rings. However, these clearances have to be in line with the pumping temperatures, thermal expansion, and galling tendencies of the ring material and the efficiency of the pump.

For materials that have galling tendencies and pumps operating at temperatures above 260 °C diameters are provided with an additional clearance of 5 mils (0.127 mm) over and above those recommended in Table 5.4.1.

For 26 in. and above, the diametrical clearance is recommended to be 0.037 in. plus 0.001 in. clearance for every additional inch of impeller diameter.

For example, a 30 in. impeller wearing ring diameter will have a minimum recommended clearance of 0.037 in. + 0.004 in. = 0.041 in.

API is also quite particular in the way the rings need to be fitted to the impellers. API 610 does not recommend tack welding of rings to impellers. They should be pressed with locking pins or threaded dowels, in the radial or axial direction.

Thermoplastic composite materials are also now being considered as ideal wearing ring materials. They can be applied to the stationary wear part or with the mating component remaining in steel. The use of thermoplastic composite material provides for greater hardness differential between wear parts, the thermoplastic serving as a sacrificial component.

Table 5.4.1 Minimum diametrical clearance

Diameter of rotating member at clearance inches		Minimum diametrical clearance	
From	To	Inches	mm
<2		0.01	0.254
2.000	2.499	0.011	0.28
2.500	2.999	0.012	0.30
3.000	3.499	0.014	0.36
3.500	3.999	0.016	0.41
4.000	4.499	0.016	0.41
4.500	4.999	0.016	0.41
5.000	5.999	0.017	0.43
6.000	6.999	0.018	0.46
7.000	7.999	0.019	0.48
8.000	8.999	0.02	0.51
9.000	9.999	0.021	0.53
10.000	10.999	0.022	0.56
11.000	11.999	0.023	0.58
12.000	12.999	0.024	0.61
13.000	13.999	0.025	0.64
14.000	14.999	0.026	0.66
15.000	15.999	0.027	0.69
16.000	16.999	0.028	0.71
17.000	17.999	0.029	0.74
18.000	18.999	0.03	0.76
19.000	19.999	0.031	0.79
20.000	20.999	0.032	0.81
21.000	21.999	0.033	0.84
22.000	22.999	0.034	0.86
23.000	23.999	0.035	0.89
24.000	24.999	0.036	0.91
25.000	25.999	0.037	0.94

Thermoplastics too have their limitations, however in some cases, they provide the best alternative.

Thermoplastic composite materials are non-galling and have a lower coefficient of friction. They demonstrate excellent wear resistance in clean liquids. Some of these plastics contain reinforced carbon fibers, which greatly enhance the mechanical properties of these plastics. As a result, they can be a direct replacement of the metal wearing rings.

Due to the reduced friction and low galling tendencies, it is possible to almost have half of the clearances that would be considered as optimum with metal wearing rings.

This possibility allows improving pump efficiency especially in low specific speed pumps.

However, the limitations of such materials are that:

- maximum life is obtained in clean fluids;
- they do not have a wide compatibility with various chemicals.

5.4.4 Shaft

The pump rotor assembly comprises of the shaft, impeller, sleeves, seals (rotating element), bearings or bearing surfaces, and coupling halves. The shaft, however, is the key element of the rotor.

The prime mover drives the impeller and displaces the fluid in the impeller and pump casing through the shaft.

The pump shaft is a stressed member for during operation it can be in tension, compression, bending, and torsion. As these loads are cyclic in nature, the shaft failure is likely due to fatigue.

The shaft design depends on the evaluations of either the torsion shear stress at the smallest diameter of the shaft or a comprehensive fatigue evaluation taking into consideration the combined loads, the number of cycles, and the stress concentration factors. The design at all times involves sophisticated finite-element computer evaluations.

The shaft design is limited not only to the stress evaluation but is also dependent on other factors such as:

- shaft deflection,
- key stresses,
- mounted components,
- critical speeds (rotordynamics).

The most common pump shaft material is plain carbon steel, typically BS-970-En 8.

Higher grades include BS-970-En19 or AISI 4140, ASTMA-322, Grade-4140 (quenched and tempered).

Austenitic steel shafts may also be used of grade ASTM A-276, Type 316 and AISI 304. Some applications like sour water with pH less than 7, drain water or slightly acidic non-aerated liquids, and hydrocarbons containing corrosive aqueous phase may demand shafts made from aluminum bronze material. The recommended grade is

B-150-Alloy C 63 200. Special applications may call for Monel or even Hastalloy C shafts.

The mechanical seals or gland packing, in contact with the shaft, can cause excessive wear due to frictional contact or fretting corrosion. As a sacrificial component, shaft sleeves are used. These are fitted closely onto the shaft; and seals and gland packing are exposed to the sleeve rather than the shaft. It is far less expensive to replace a sleeve than the complete shaft.

The material of construction of the pump sleeves is similar to that of the shaft but the standardization favors the use of SS-316. The portion of the sleeve that is exposed to the secondary seal of the mechanical seals such as an O-Ring or a Teflon wedge is hard coated. The plasma sprayed, hard coating can be of Chrome-oxide, tungsten carbide, or alumina. This offers hardness around 70–72 Rc. The surface is then provided with a ground finish.

5.4.5 Stuffing boxes

The stuffing box is a chamber or a housing that serves to seal the shaft where it passes through the pump casing (Figure 5.4.21).

In a stuffing box, 4–6 suitable packing rings are placed and a gland (end plate) for squeezing and pressing them down the shaft.

The narrow passage, between the shaft and the packing housed in the stuffing box, provides a restrictive path to the liquid, which is at a high pressure within the pump casing.

The restrictive path causes a pressure drop, prevents leakage resulting in considerable friction between the shaft and the packing, and causes the former to heat up. It is thus good practice to tighten the gland just enough to allow for a minimal leak through the packing. This slight leakage of the liquid acts as a lubricant as well as a coolant. Obviously, this cannot be allowed for hazardous and toxic liquids, but then gland packings are also not used in such applications.

When pumps are handling dirty or high pressure liquid, lantern rings are used. These are rings with holes drilled along its circumference.

A lantern ring substitutes one of the packing rings in the stuffing box and is situated at the pump end or midway between the packings.

In applications where the discharge pressure of the pump is higher, a restrictive bush is placed at the throat of the stuffing box.

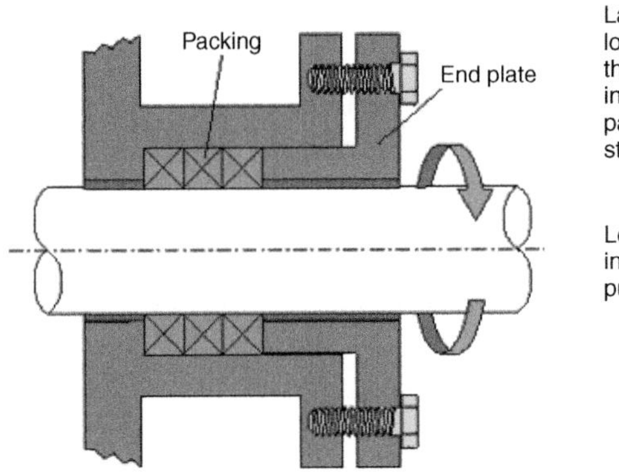

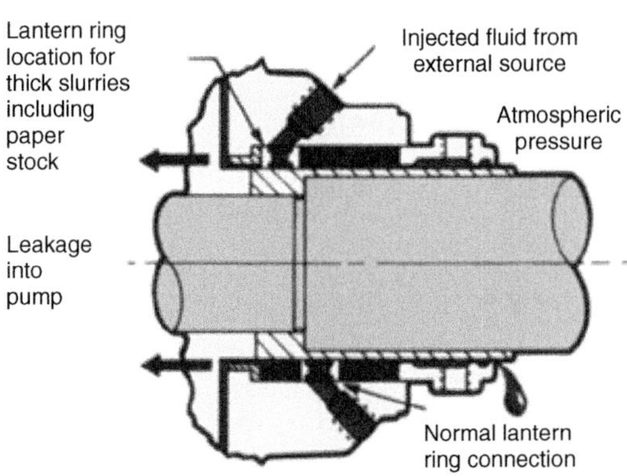

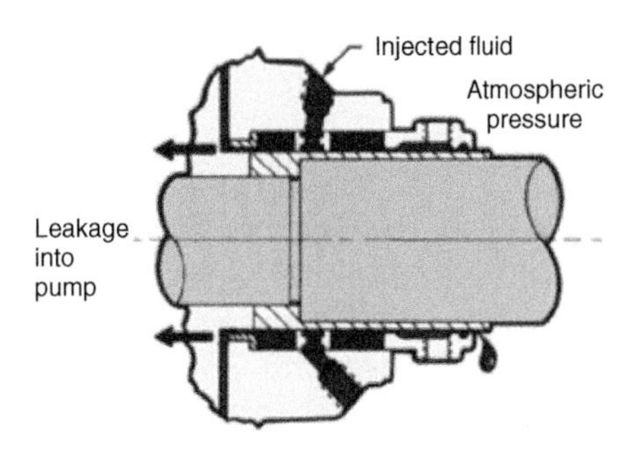

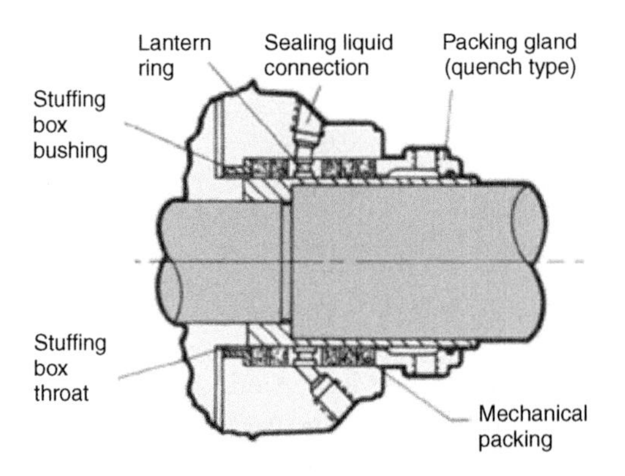

Fig. 5.4.21 Stuffing box.

When the liquid being handled is at a higher temperature (above 120 °C), the stuffing box has an integrally cast water jacket housing. This allows for water circulation and keeps the packings at a lower temperature.

When toxic or corrosive liquids are handled, it is necessary to insure complete sealing of the stuffing box. Leakage of such liquids is a hazard to the plant personnel and can also be detrimental to the outer surface of the pump and foundation. It can also result in the loss of a valuable product.

This is achieved by first reducing the pressure in front of the packings. The reduction in pressure is brought about by having radial blades at the back shroud of an auxiliary impeller (see Figure 5.4.22). This auxiliary impeller is also called as a repeller.

As the repeller rotates with a shaft, it throws the liquid outwards thus reducing the pressure in front of the packings. The pressure generated by the repeller is dependent on the length of the blades and its clearance with the casing.

Another common design is to cast back vanes of the main impeller itself. The back vanes help reduce the pressure acting on the packing.

The magnitude of the work done due to the friction between the shaft and the packings is influenced by:

- kind of packing quality,
- length of the gland,
- diameter of the shaft,
- speed of rotation,
- pressure acting on gland,
- volume of liquid passing through the packing.

In a properly operating stuffing box, the friction losses are usually of the order of 1% of the total pump power. This is independent of the size and kind of pump.

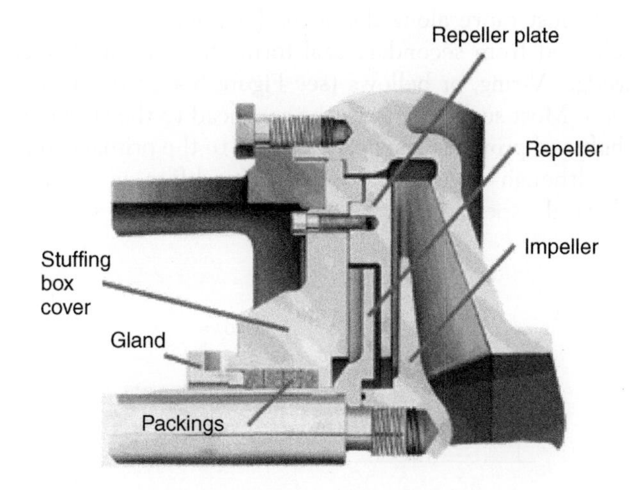

Fig. 5.4.22 Stuffing box for a corrosive liquid.

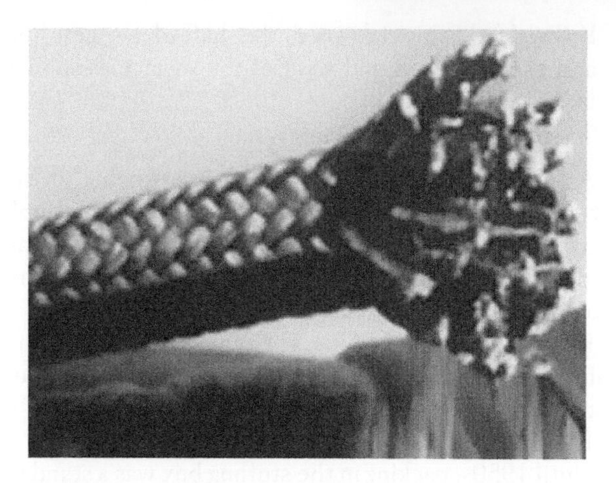

Fig. 5.4.23 A high-performance filament packing (EGK®** Filament Packing – Style 2070).

The present day packing used in pumps are predominantly made of PTFE (teflon)/graphite filaments. These are braided and formed into square shapes. They offer heat dissipation and low friction qualities. For example, see Figure 5.4.23 for a high-performance filament packing.

Packing properties can be enhanced by including other special materials during braiding of the packings, to handle contaminants, acids, alkalis, temperature, speed, and other factors. For example, a material called as Aramid can enhance the mechanical properties of the packing. This helps in prevention of extrusion of packing and withstands slurries and abrasives in liquids.

5.4.6 Mechanical seals and seal housings

The stuffing boxes described above have many disadvantages and these include:

- A persistent leakage and loss of product if the shaft surface is not smooth.

Fig. 5.4.24 Wear on shaft/sleeve due to tight packing.

- If the gland is too tightened, the shaft/sleeve gets hot and there can be rapid wear of the surface as shown in Figure 5.4.24.
- They require constant supervision.

As a result, the use of gland packing is being phased out but is still used in noncritical and low-power applications. In most of the applications, mechanical seals are used. Most of the disadvantages of packing are eliminated by the use of mechanical seals.

From its origins in 1930s, the technology of mechanical seals continues to evolve at a rapid pace. This is, especially, in regard to the enhancement of the reliability of seals.

Until 1950s, packing in the stuffing box was a standard method of shaft sealing. As operating conditions became more demanding and pumps were used on a greater variety of fluids, mechanical seals were designed to handle these changing conditions.

Mechanical seals comprise of two perfectly lapped mating faces. One face is stationary and the other is rotating. The leakage resistance in gland packing is along the axis of the shaft but in seals, it is orthogonal.

The seal faces cannot run mating with each other without any lubricant (Figure 5.4.25). This can lead to an early wear and seal damage results in leakage. Usually, the sealant fluid is injected in the seal housing at a specified pressure, which lubricates and cools the faces.

The fluid between the faces can escape into the atmosphere and this is called as fugitive emissions. In some applications, fugitive emissions are unacceptable and in such cases, multiple seal arrangements are used.

However, due to its precise design, mechanical seal demands careful attention to precision during pump assembly.

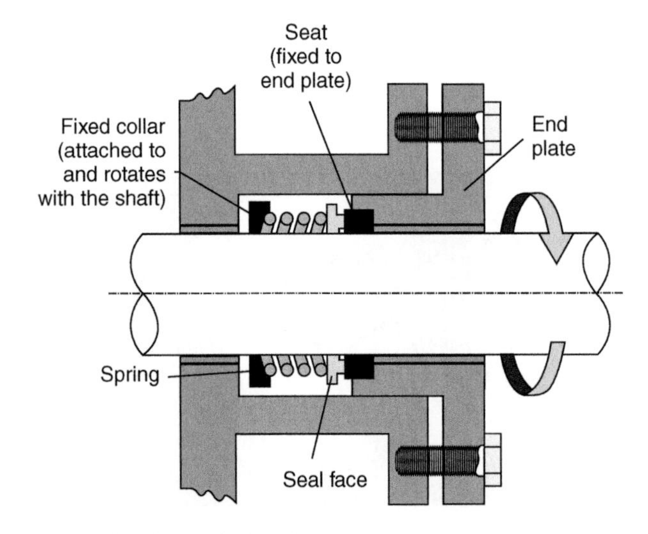

Fig. 5.4.25 Mechanical seal.

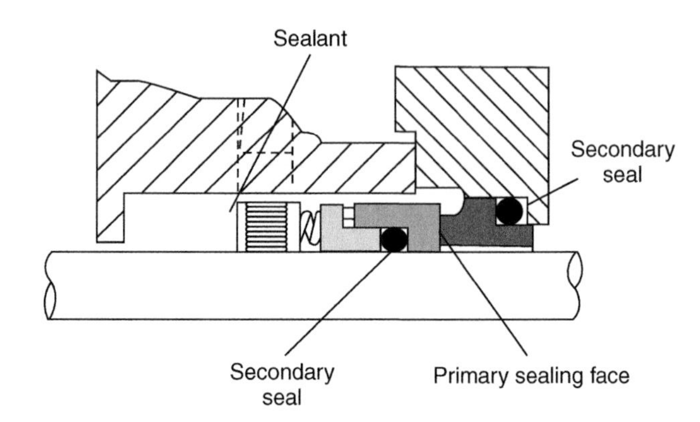

Fig. 5.4.26 Points of sealing in a mechanical seal (pusher type unbalanced seal).

There are three points of sealing as shown in Figure 5.4.26, common to all mechanical seal installations:

1. at the mating surfaces of the primary and mating rings;
2. between the rotating component and the shaft or sleeve;
3. between the stationary component and the gland plate.

When a seal is installed on a sleeve, there is an additional point of sealing between the shaft and sleeve. Certain mating ring designs may also require an additional seal between the gland plate and stuffing box.

Normally, the mating surfaces of the seal faces are made of dissimilar materials and held in contact with a spring. Preload from the spring pressure holds the primary and mating rings together during shutdown or when there is a lack of liquid pressure.

The secondary seal between the shaft and sleeve must be partially dynamic. As the seal faces wear, the primary ring must move slightly forward. Because of vibration from the machinery, shaft run out, and due to thermal expansion of the shaft to the pump casing, the secondary seal must move along the shaft. Flexibility in sealing is achieved from secondary seal forms such as an O-ring, wedge, V-ring, or bellows (see Figure 5.4.27 for bellow seal). Most seal designs fix the seal head to the sleeve or shaft and provide for a positive drive to the primary ring.

Although mechanical seals may differ in various physical aspects, they are fundamentally the same in

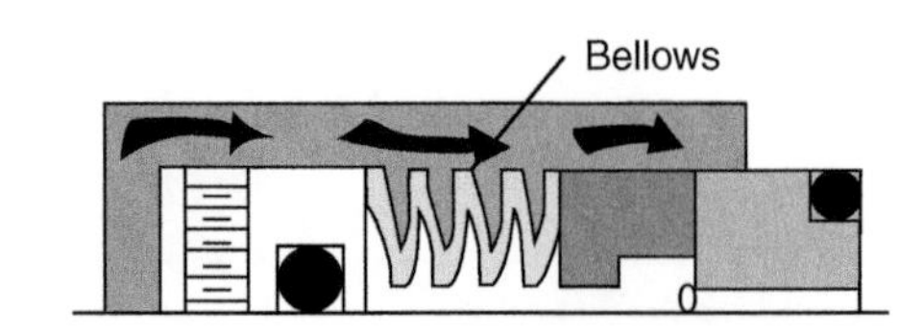

Fig. 5.4.27 Bellow seal.

principle. The wide variation in design is the result of the many methods used to provide flexibility, ease of installation, and economy.

A seal arrangement is used to describe the design of a particular seal system and the number of seals used on a pump.

The most common sealing arrangements maybe defined as:

- single seal installations,
- internally mounted,
- externally mounted,
- dual seal installation,
- tandem seals,
- double seals,
- externally pressurized,
- internally pressurized.

A single seal mounted inside the seal chamber represents at least 75% of all installations. It is the most economical sealing system available to the industry.

Just as gland packings are housed in stuffing boxes, the mechanical seals are housed in seal housings. Research has indicated that providing an enlarged bore can provide distinctive advantages depending on the applications. Some of the seal housings suggested are given in Figure 5.4.28.

Studies demonstrate that the fluid flow within the enlarged chamber is increased. This aids in the removal of seal generated heat.

Optimum selection of enlarged chambers can help deal with:

- gases in seal housing under start stop conditions,
- light hydrocarbons with low boiling points,
- liquids with solid particles.

These designs help in improving the reliability of the seals.

5.4.7 Bearing housing/bearing isolators

5.4.7.1 Cantilevers or overhung impeller pumps

Overhung impeller pumps usually employ anti-friction bearings only. In a typical bearing housing arrangement, the radial ball or cylindrical roller bearing is located

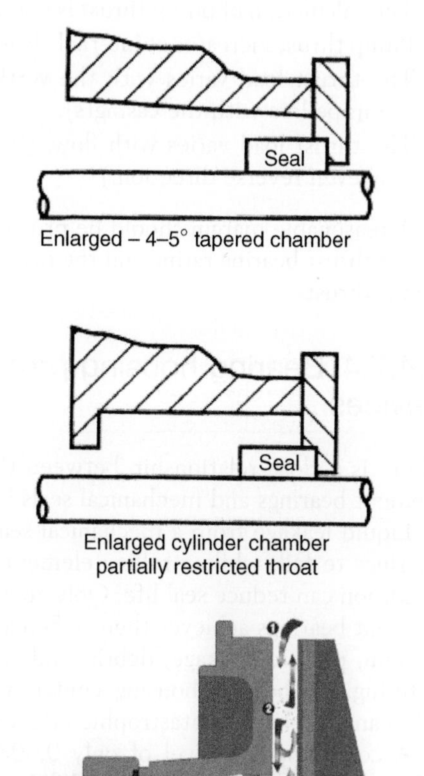

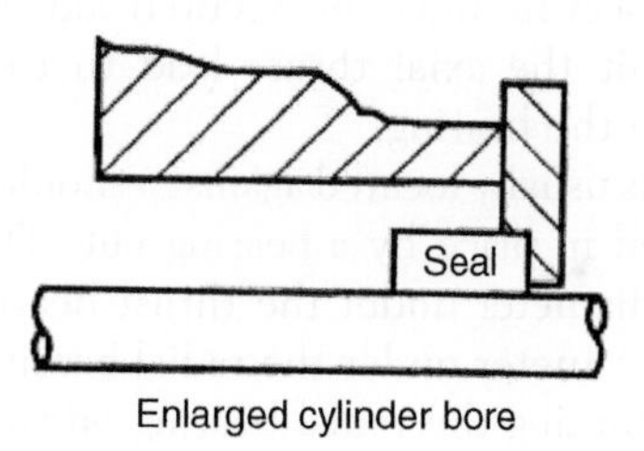

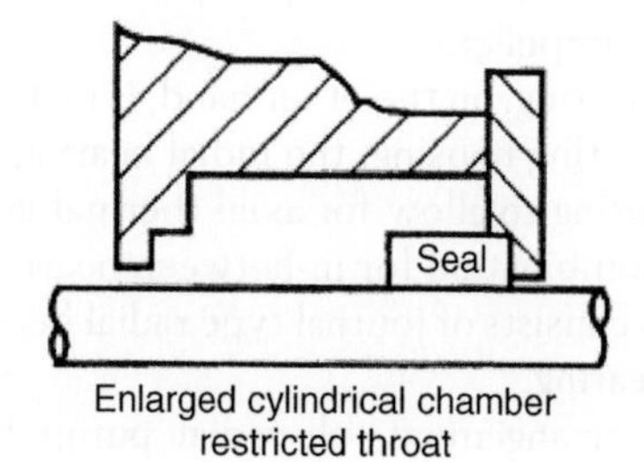

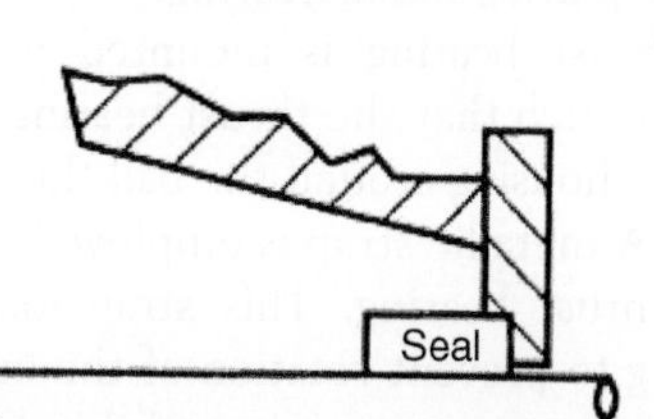

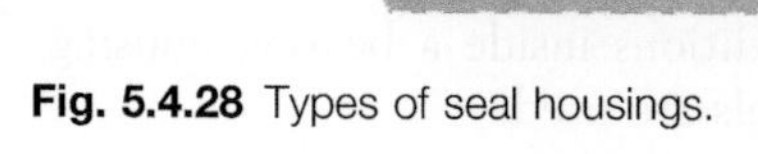

Fig. 5.4.28 Types of seal housings.

adjacent to the impeller or inboard position. It is arranged to take only radial loads.

The thrust bearing is located closest to the coupling and usually consists of a duplex pair of angular contact bearings or double row angular contact bearings.

Typically, when a ball bearing is mounted on the inboard side, the coupling side or the outboard side is provided with a duplex angular contact bearing in a face-to-face arrangement or a double row angular contact bearing.

With a cylindrical roller bearing on the inboard side, the outboard bearings are mounted back-to-back so that the axial thrust load can be carried in either direction. This duplex bearing pair carries both the unbalanced axial thrust loading and the radial load.

5.4.7.2 In-between bearing or fully supported shaft pumps

In-between bearing pumps, the ball radial bearing and the ball thrust bearing combination have individual bearing housings.

The radial bearing is normally located at the coupling end of the pump. The ball thrust bearing is located at the outboard pump end.

The thrust bearing must be secured axially on the shaft to transmit the axial thrust load to the bearing housing through the bearing.

The bearing is usually located against a shoulder on the shaft and locked in place by a bearing nut. This means that the shaft diameter under the thrust bearing is less than the shaft diameter under the radial bearing.

Thus, by mounting the radial bearing on the inboard (or coupling) end of the pump shaft, a larger shaft diameter is available to transmit pump torque from the coupling to the impeller.

The thrust bearing, on the other hand, is locked axially in the thrust bearing housing; the radial bearing is axially loose in its housing to allow for axial thermal growth.

A popular combination for in-between bearing double suction pumps consists of journal type radial bearings and a ball thrust bearing.

In such an arrangement, all radial pump loads are handled by the journal radial bearing.

The ball thrust bearing is mounted in the thrust bearing housing such that the thrust bearing carries only axial loads. The housing around the ball thrust bearing is radially loose. A metallic strap is employed on the outer rings of the thrust bearing. This strap locks into the bearing housing to prevent rotation of the outer rings.

Such a bearing arrangement is useful in higher horsepower and higher speed applications where ball radial bearings would be impractical due to speed, load, and lubrication limitations.

Due to the location of the ball thrust bearing on the outboard end of the shaft, the shaft diameter under the ball thrust bearing can be relatively small since no torque is transmitted from this end of the shaft.

5.4.7.3 Vertical pumps

In vertical pumps, the unbalanced axial hydraulic forces as well as the static weight of the rotating element (i.e. pump shaft and impeller(s)) is taken up by the thrust bearing, which by design, maybe located within the driver or normally at the head end within the pump casing.

These bearings could be ordinary ball bearings, angular contact bearings, split inner race angular contact bearings, and even the spherical roller thrust bearings in larger pumps like the vertical deep well pump.

Typically, these bearings are rated to handle at least thrice the maximum thrust load. This is due to number of varying factors in the determination of the thrust loads generated by such pumps. The most severe thrust loads are generated at the time of shutting down the pump and or from impact loading, as a result of water hammer.

Some of these are mentioned below:

- The calculation of pump thrust is not highly accurate.
- Pump thrust increases as internal clearances increase.
- The thrust load varies with the vertical position of the impellers with the casing(s).
- The thrust load varies with flow. (In some cases, it may even reverse direction.)

A reasonable margin should be provided between the driver thrust bearing rating and the maximum calculated pump thrust.

5.4.7.4 Bearing housing protection devices

There is a close relationship between the life of rolling element bearings and mechanical seals in pumps.

Liquid leakage from a mechanical seal may cause the bearings to fail, while a rolling element bearing in poor condition can reduce seal life. Only about 10% of rolling element bearings achieves their 3–5-year design life.

Rain, product leakage, debris, and wash-down water entering the bearing housing contaminate the bearing lubricant and have a catastrophic effect on bearing life.

A contamination level of only 0.002% water in the lubricating oil can reduce bearing life by as much as 48%. A level of 0.10% water will reduce bearing life by as much as 90%.

To improve the conditions inside a bearing housing, various types of end seals are used.

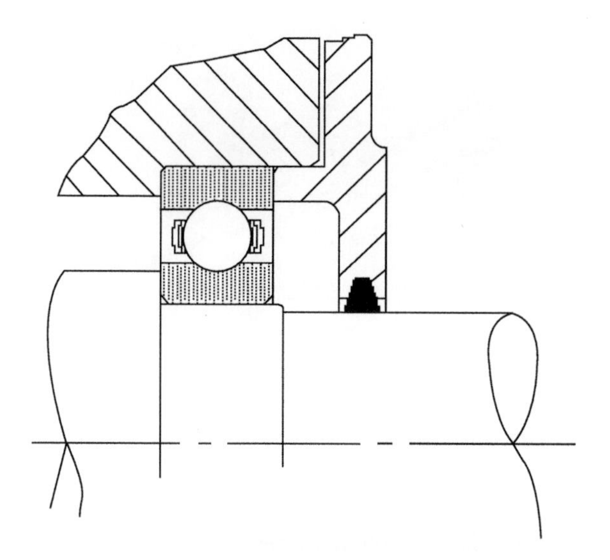

Fig. 5.4.29 Felt seal.

In almost every case, the normal operating life and quality of the end seal is not nearly as good as that of the rolling element bearings. Improving the quality of the end seals will increase the life of rolling element bearings.

5.4.7.4.1 Felt and lip seals

One of the earliest, bearing housing isolators was the 'felt' (Figure 5.4.29). The bearing covers are provided with a groove in which a felt strip is cut and inserted. The felt acts as a barrier for oil and dust from the atmosphere.

The lip or the oil seals (Figure 5.4.30) have low initial cost, availability, and are common. New lip seals provide protection in both static and dynamic modes. Their major disadvantage is short protection life due to wear of the elastomer.

Life expectancy of a common single lip seal can be as low as 3000 h, or 3–4 months. Thus, while a bearing is designed to last from 3 to 5 years of continuous operation, the lip seal will provide protection for only a few months.

The temperature limits of lip seals are –40 to 400 °F (–42 to 203 °C) for Viton.

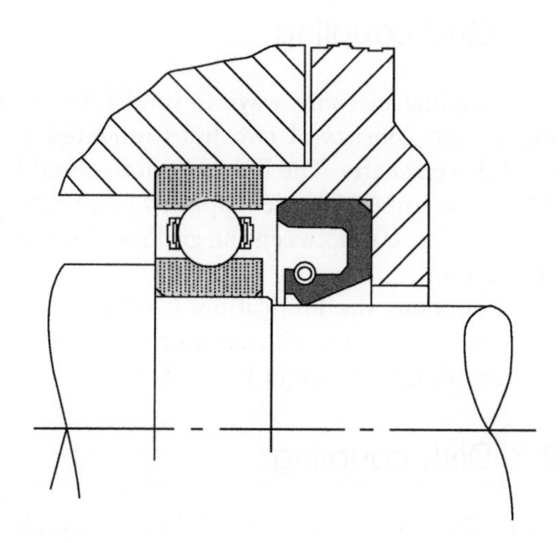

Fig. 5.4.30 Lip seal.

5.4.7.4.2 Labyrinths

Labyrinths are devices that contain a tortuous path, making it difficult for contaminants to enter the bearing housing (Figure 5.4.31). Labyrinths are devices that contain a tortuous path, which in turn discourages and hence minimizes leakage of fluid without there being any physical contact between the stationary and moving elements that make up the seal. Labyrinth seal design may vary hence selection must be based on its suitability for the application and purpose.

The advantages of labyrinths are their non-wearing and self-venting features. With no contacting parts to wear out, a labyrinth can be reused for a number of equipment rebuilds. Because the labyrinth provides an open, however difficult, path to the atmosphere, the bearing housing vent can be removed and the tapped hole can be plugged with a temperature gage.

The disadvantages of labyrinths include a higher initial cost than lip seals and the existence of an open path to the atmosphere, which can enable contamination of the lubricant by atmospheric condensate as the housing chamber 'breathes' during temperature fluctuations in humid environments. Also, they do not work as well in a static mode as in a dynamic, rotating mode.

The temperature limits of labyrinths are determined by the elastomer driving the rotor and holding the stator in place, the same as for the lip seal.

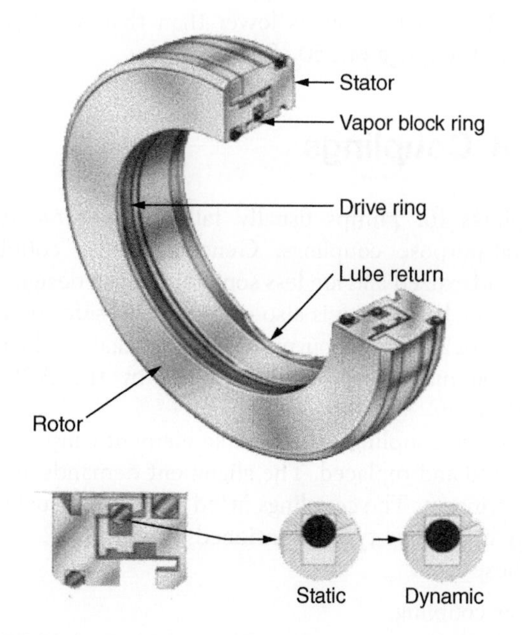

Fig. 5.4.31 A special type of labyrinth seal (Inpro seals).

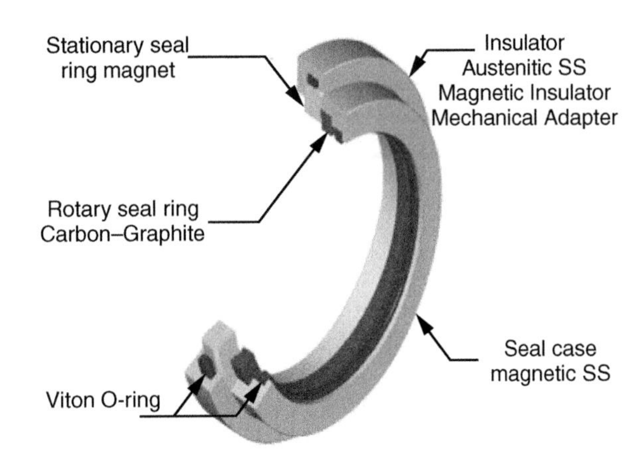

Fig. 5.4.32 Magnetic seal.

5.4.7.4.3 Magnetic seals

Magnetic seals use a two-piece end face mechanical seal with optically flat seal faces held together by magnetic attraction (Figure 5.4.32).

They have a design life equivalent to mechanical seals and rolling element bearings and can be repaired.

The major advantage of magnetic seals is the hermetic seal they provide for the bearing housing. Because of the positive seal, other arrangements must be made to allow for the 'breathing' that results from expansion and contraction of the air pocket above the lubricant during normal temperature changes.

Disadvantages of magnetic seals include higher initial cost and a shorter life than the almost infinite life of a labyrinth.

Magnetic seals are generally not recommended with dry sump, oil mist lubricated bearing housings, or grease-lubricated bearings. The upper operating temperature limit of magnetic seals is lower than that of labyrinth seals, in the range of 250 °F (121 °C).

5.4.8 Couplings

Couplings for pumps usually fall in the category of general-purpose couplings. General-purpose couplings are standardized and are less sophisticated in design. The cost of such coupling is also on the lower side. In addition, there are special purpose couplings that can be used on turbo machines and are covered by the API 671 specification.

In these couplings, the flexible element can be easily inspected and replaced. The alignment demands are not very stringent. The couplings fitted on pumps usually fall in any of the five types mentioned below.

These are:

1. gear coupling,
2. grid coupling,

Fig. 5.4.33 Gear coupling.

3. disk coupling,
4. elastomeric compression type,
5. elastomeric shear type.

5.4.8.1 Gear coupling

Gear couplings comprise of two hubs with external teeth that engage the internal teeth on a two or one piece flanged shroud or sleeve (Figure 5.4.33).

One hub with shroud is mounted on the pump shaft and the other on the shaft of the prime mover. The flanges are bolted after packing grease between the meshing gears.

To contain the grease within an enclosed space, the flange faces and the shroud are provided with suitable static seals (like O-rings).

Some of the couplings may have a spacer between the flanged shrouds. In case of any misalignment between the shafts, sliding occurs between the external gears on the hub and its corresponding internal gears on the shroud.

Gear couplings are usually deployed on pumps above a rating of 75 kW.

5.4.8.2 Grid coupling

The grid coupling in some ways is similar to the gear coupling. It also consists of two hubs mounted on the driver and driven shafts. The hubs are slotted and house a flexible grid member. Grease is applied to lubricate any sliding that may occur between the grid member and the slots of the hub.

A cover contains the lubrication. Grid couplings are usually not used in pumps with a power rating of 750 kW.

A grid coupling is shown in Figure 5.4.34.

5.4.8.3 Disk coupling

A disk coupling shown in Figure 5.4.35 comes under the category of metallic element coupling. Diaphragm

Fig. 5.4.34 Grid coupling.

Fig. 5.4.35 Metallic disk coupling.

couplings used on turbo machines also belong to this category.

Metallic disk coupling comprises of two hubs mounted on the driver and driven shafts. A set of flexible shims or metallic element is placed between the spacer and the hub.

The torque is transmitted by simple tensile force between alternate driving and driven bolts on a common bolt circle diameter.

Such couplings are used on pumps with a power rating of over 75 kW.

5.4.8.4 Elastomeric shear type coupling

All elastomeric couplings are classified according to how their elastomeric elements transmit torque between driving and driven hubs. The elements could be either in compression or in shear.

In shear type couplings, the driving and driven hubs operate in separate planes, while the driving hub pulls the driven hub through an elastomeric element suspended between them. Here, the element transmits and cushions

Fig. 5.4.36 Elastomeric shear type coupling.

the force between the hubs by being stretched between them (Figure 5.4.36).

Among all couplings, this type can probably take the maximum amount of parallel misalignment.

These coupling are used in pumps below a rating of 75 kW.

5.4.8.5 Elastomeric compression type coupling

In jaw couplings, the element (called as a spider) is loaded in compression between the jaws of mating hubs.

These jaws operate in the same plane, with the driving hub jaws pushing toward the driven hub jaws. Legs of the elastomeric spider transmit and cushion the force between the driving and driven jaws by being compressed between them.

Compression type couplings offer some advantages over the shear type of coupling (Figure 5.4.37). These include:

- higher load capacity,
- greater torsional stiffness,
- more safety,
- easier installation.

Fig. 5.4.37 Elastomeric compression type coupling.

The design of the coupling allows it transmit the torque even if the spider breaks. The driving jaws simply rotate until they contact the driven jaws directly, and the coupling continues to function, though it is accompanied by considerable noise and accelerated wear. In some cases, it can prevent an expensive downtime.

They typically accommodate angular shaft misalignment up to 1° and parallel misalignment up to 0.015 in.

The elastomeric material in the above two types of couplings is mostly NBR (Nitrile Butadiene Rubber), sometimes called Buna N. It is the most economical and widely used standard coupling element material. It resembles natural rubber in resilience and elasticity, and is resistant to oil, hydraulic fluid, and most chemicals.

The operating temperature ranges from −40 to +100 °C. With hardness of 80 Shore A, NBR provides the best damping capability among elastomeric elements.

Another material used is Urethane. It has 1.5 times the torque capacity of NBR with very good chemical and oil resistance. But has less damping capability (90 Shore A hardness) and narrower operating temperature range from −39 to +71 °C. Urethane spiders are good choices when the application calls for greater torque in a confined space, or for resistance to atmospheric effects such as ozone, sunlight, and hydrolysis in tropical conditions.

Hytrel *(registered trademark of E.I. DuPont de Nemours & Co)* is designed for high operating temperature range from -51 to +121 °C. It offers excellent resistance to oils and chemicals and can transmit 3 times the torque of standard NBR.

It also provides resistance to ozone, sunlight, and hydrolysis in tropical conditions. With hardness of 55 Shore D, however, Hytrel cuts angular misalignment ratings in half, and damping capacity is low.

The spiders come in the following four types:

1. Standard solid center spider.
2. Open center type.
3. *Snap wrap (with or without retainer ring)*: This flat-strip, open-end design connects the spider legs around the perimeter of the coupling rather than at the center. This allows for easy removal or installation without disturbing the alignment of either coupling hub. With no center connections, this design does not overlap into the bore, and therefore it allows shaft ends to extend at maximum bore diameter to a minimal distance between the shaft ends. This element is radially 'wrapped' around the jaws and needs to be held in place by either a ring or a collar. When retained by a ring, it has a maximum RPM limit of 1750. The collar configuration, on the other hand, achieves the same RPM rating as the standard coupling because the collar is attached to one hub.
4. *Load cushions (separate blocks)*: As small, separate blocks, these cushions can be installed easily and removed radially, which can be very helpful for maintenance in heavy-duty applications. In certain models of coupling, load cushions must be held in place by a collar.

Chapter 5.5

5.5

Pump hydraulics

Girdhar, Moniz and Mackay

We have seen earlier the evolution of pumps, their construction, and their wide applicability. The many requirements of pressurized liquid lead to a large variety of pumps, each designed and suited to a required application.

In spite of many different types of components, the basic mechanics and the principle of operation of the centrifugal pumps are similar.

Centrifugal pumps are hydraulic machines that are used to energize and transfer a fluid within a system, at a flow that is dependent upon system needs. In order to understand how the pumps perform this function, it is essential to get familiar with some of the hydraulic terms associated with centrifugal pumps and of the liquids that they handle.

5.5.1 Specific gravity

The term 'specific gravity' refers to the ratio of the density of liquid to the density of water at 4 °C (the density of water at this temperature is 1.000 kg/l). Specific gravity is a ratio and is hence a dimensionless quantity and is not expressed in any units.

$$\text{Specific gravity} = \frac{\rho_{\text{liquid}}}{\rho_{\text{water}} \text{ at } 4\ ^\circ\text{C}}$$

To find the specific gravity of a liquid, we must know its density in kilograms per meter cubed (kg/m^3) or in grams per millimeter cubed (g/mm^3). Then, divide this density by the density of pure water in the same units. If you use kg/m^3, divide by 1000. If we use g/mm^3, divide by 1 (that is, leave the number alone). It is important to use the same number of units in the numerator and denominator.

Materials with a specific gravity of less than 1 are less dense than water and therefore will float on it. Substances with a specific gravity of more than 1 are denser than water and will sink.

An object with a density of 100 kg/m^3 has a specific gravity of 0.1, and can float on the surface of the body of water. An object with a density of 10 g/mm^3 has a specific gravity of 10 and can sink rapidly.

5.5.2 Viscosity

Viscosity is best understood by imagining a styrofoam cup with a hole in the bottom. If honey is poured in this glass, it is noticed that the cup drains very slowly because the viscosity of honey is large compared to other liquids. If the same cup is filled with water, the cup will drain much more quickly. Viscosity is a measure of a fluid's resistance to flow.

It describes the internal friction of a moving fluid. A fluid with large viscosity resists motion because its molecular makeup gives it a lot of internal friction. A fluid with low viscosity flows easily because its molecular makeup results in very little friction when it is in motion.

In certain fluids called newtonian fluids, the shear stress that causes the flow is directly proportional to the shear strain (rate of deformation).

The ratio of this shear stress to the shear strain is constant for a given fluid at a fixed temperature.

This constant is called the dynamic or absolute viscosity (μ) and often simply the viscosity. The viscosity of

Practical Centrifugal Pumps: Design, Operation and Maintenance; ISBN: 9780750662734

liquids decreases rapidly with an increase in temperature. Thus, upon heating, liquids flow more easily.

The dimensions of dynamic viscosity are force times time divided by area. The unit of viscosity, accordingly, is newton-second per meter square ($N\text{-}s/m^2$).

For some applications, the kinematic viscosity is more useful than the absolute or dynamic viscosity.

Kinematic viscosity is obtained by dividing the absolute viscosity of a fluid by its mass density. (Mass density is the mass of a substance divided by its volume.)

$$\text{Kinematic viscosity } \nu = \frac{\mu}{\rho}$$

The dimensions of kinematic viscosity are area divided by time. Its units are meter squared per second (m^2/s).

Kinematic viscosity (ν) is often expressed in stokes, St, where 10^4 St = 1 m^2/s. However, a more common unit of measure is centistokes (cSt).

Some other common viscosity units and conversion factor are listed below:

Kinematic viscosity	× Specific gravity			Absolute viscosity
Centistokes	×	S.G.		Centipoise
SSU × 0.2198[a]	×	S.G.		Centipoise
SSU[a]	×	0.2198	=	Centistokes
Degree Engler[a]	×	7.45	=	Centistokes
Seconds Redwood[a]	×	0.2469	=	Centistokes

[a] For centistokes greater than 50.

5.5.3 Vapor pressure

The vapor pressure of a liquid, pure or mixed, is defined as the pressure exerted by those molecules that escape from the liquid to form a separate vapor phase above the liquid.

If a quantity of liquid is placed in an evacuated, closed container, the volume of which is slightly larger than that of the liquid, most of the container is filled with the liquid. After a period, a vapor phase forms in the space above the liquid surface. This space consists of molecules that have passed through the liquid surface from liquid to gas. The pressure exerted by that vapor phase is called the *vapor* (or *saturation*) *pressure*. For a pure liquid, this pressure depends only on the temperature.

Following are some examples of vapor pressures for a few common liquids. The vapor pressure is 1 atm at 100 °C for water, at 78.5 °C for ethyl alcohol, and at 125.7 °C for octane.

Similarly, at 20 °C, water has a vapor pressure of 0.023 atm (17.5 mm Hg). Isopropyl alcohol (rubbing alcohol) has a vapor pressure of 0.043 atm (33 mm Hg) at 20 °C.

In a liquid solution, the component with the higher vapor pressure is called the light component (tendency to vaporize quicker), and that with the lower vapor pressure is called the heavy component.

5.5.4 Flow

The first and most important point to consider is that centrifugal pumps are volumetric machines. The liquid pumped is measured in terms of the volume flow rate. The units used are m^3/h or gpm (US or Imperial).

It is worthy to note that any pump, for a single point of operation, would always give the same volumetric flow rate for any liquid, be it hydrocarbon, water, or any other. Depending on the density of the liquid, the mass flow rate changes.

If we have a pump whose capacity is 20 m^3/h, then the mass flow rate would pump 20 tph of water. However, the same pump when handling a hydrocarbon with a specific gravity of 0.8 would pump only 16 tph.

5.5.5 Head

The pressure of the liquid can be stated in terms of meters (feet) of head of the liquid column (mlc). As in case of volumetric flow rate, the head generated by the pump in mlc for a single point of operation is the same for any liquid. Depending on the density of the liquid, what changes is the reading on the pressure gage.

Any pump raises the liquid from one gradient (head) to another. Thus, the difference between the discharge head and the suction head is termed as 'differential head'.

The differential head developed by a pump is expressed in 'm' of liquid:

$$H_m = \frac{(P_d - P_s) \times 10}{\rho}$$

where P_d is the discharge pressure (kg/cm^2), P_s the suction pressure (kg/cm^2), ρ the specific gravity of the liquid.

5.5.6 System resistance

The flow rate delivered by the centrifugal pump is dependent on the total *frictional and static* head that it has to overcome.

The required head comprises of two components. These are:

1. *A static component*: h_s in meters 'm', which is independent of the flow through the pump. For example, if the liquid has to be raised from one height to another, then it is the difference of the two heights.

2. *A friction head loss component*: H_f in meters 'm'. This head is proportional to the square of the flow rate – 'Q' in l/s.

The friction component is the summation of losses that occurs as the liquid flows through the pipes and various equipment like heat exchangers.

To account for the losses, the entire flow path from the suction vessel to the discharge vessel is considered.

If this path has a large number of fittings, such as elbows (more bends), reducers, valves, and orifices, the losses are higher. To ease the calculations, nomogram of equivalent length of valves and fittings is used. An equivalent length is the length of pipe that would offer the same losses for a flow rate as offered by the fitting.

Thus, the suction and discharge paths, which may have a few fittings, are converted to an equivalent length. Using another nomogram, the friction loss due to the flow rate can be estimated.

Usually the total losses (static + frictional) on the suction side of the pump are calculated separately to establish the fact that there is adequate NPSH-a available compared to NPSH-r required, for the pump to operate satisfactorily. If NPSH-a is less than NPSH-r, the pump will operate under conditions of cavitation which is undesirable. The next step is to evaluate the total system resistance H_t, which is the summation of losses on both the suction and discharge sides and includes the static lift H_{st}.

To compute the system resistance, consider the system as shown in Figure 5.5.1 where the pump flow rate is 100 m^3/h (27.8 l/s) of water, through steel pipes.

5.5.6.1 Evaluate the suction side

Step 1: Calculate the velocity in the suction pipe – 6 in. (152.4 mm)

$$\text{Velocity} = \frac{\text{Flow}}{\text{Area}} = \frac{(100\ /3600)}{[(\pi\ /\ 4)\ \times\ (0.1524)^2]} = 1.52\ \text{m/s}$$

Step 2: Compute the suction head

The suction is from atmospheric vessel – $h_a = 10.34$ m

Suction height – $H_s = 2$ m

Step 3: Compute the equivalent pipe length

Pipe length = 12 m (assume almost all length is 6″ or 150 mm nom. bore)

6″ Gate valve (fully open) *V1* – equivalent length = 1 m

6″ × 4″ eccentric reducer *V4* – equivalent length = 1.4 m

Entry losses – equivalent length = 6 m

Total equivalent pipe length = 14.4 m.

Step 4: Compute friction loss from pipe friction tables

Friction loss for pipe length of 100 m = 1.52 m/ 100 m

Friction loss for pipe length of 20.4 m = 0.31 m/ 100 m

Step 5: Compute total suction head

$$h_a + H_s - P_l = 10.34 + 2 - 0.31 = 12.03\ \text{m}$$

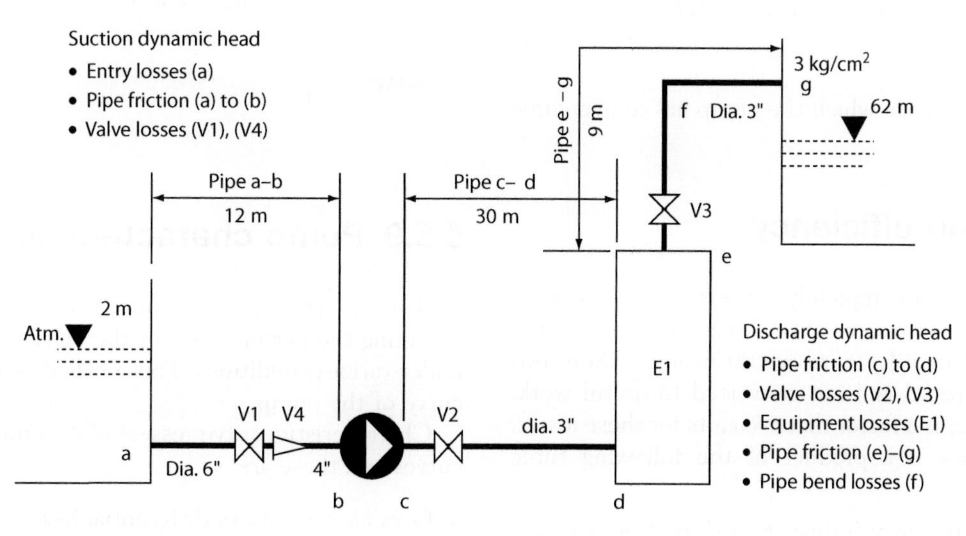

Fig. 5.5.1 A typical pumping system.

5.5.6.2 Evaluate the discharge side

Step 1: Calculate the velocity in the discharge pipe – 3.5 in. (90.1 mm)

$$\text{Velocity} = \frac{\text{Flow}}{\text{Area}} = \frac{(100/3600)}{[(\pi/4)\times(0.0901)^2]} = 4.35\ \text{m/s}$$

Step 2: Compute the discharge head

The discharge head in the vessel – $h_d = 30$ m

Discharge height – $H_d = 62$ m

Step 3: Compute the equivalent pipe length

Pipe length = 39 m (assume almost all length is 3.5″ or 90 mm nom. bore)

2 nos of 3.5″ Gate valve (fully open) $V3$ and $V4$ – equivalent length = 0.5 × 2 = 1 m

One 90° bend – equivalent length = 1 m

Total Equivalent pipe length = 40 m.

Step 4: Compute friction loss from pipe friction tables

Total Friction loss:

20.4 × (1.33/100) for 150 mm nom.

Bore pipe + 41 × (20.06/100) for 90 mm nom.

Bore pipe = 0.271 + 8.23 = 8.5 m

Step 5: Compute total discharge head

$$h_d + H_d + P_l = 30 + 62 + 8.5 = 100.5\ \text{m}$$

Thus, the pump sees a differential head of:

Discharge head – suction head = 100.5 – 12.03 = 88.47 m.

The system resistance which the pump has to overcome is 88.47 m.

5.5.7 Pump efficiency

The pump does not completely convert kinetic energy to pressure energy since some of the kinetic energy is lost in this process. Primarily, there are three areas where this energy is dissipated and not converted to useful work. Pump efficiency is a factor that accounts for these losses. Pump efficiency is a product of the following three efficiencies:

1. Hydraulic efficiency (primarily, disk friction, which is the friction of the liquid with the impeller shrouds. This is a function of speed and impeller geometry. Other losses are shock losses during rapid changes in direction along the impeller and volute).
2. Volumetric efficiency (recirculation losses at wear rings, interstage bushes, and other).
3. Mechanical efficiency (friction at seals or gland packing and bearings).

Some texts call the product of the first two efficiencies as internal efficiency of the pump.

Every pump is designed for a specific flow and a corresponding differential head, though it is possible to operate at certain percentage points away from the designed values.

However, the efficiency of the pump at the designed point is maximum and is called as the BEP. Efficiency at flows lower or higher than this design point is lower.

The efficiency of the pump has a close relationship to an important pump number called as the specific speed. This we shall cover in Section 5.5.11.

5.5.8 Hydraulic power

If a pump were an ideal machine, the required input power to drive the pump would entirely lift the mass flow rate from one elevation to another. This power is called as the hydraulic power.

$$P_{H(kW)} = \frac{Q \times \rho \times g \times H}{3.6 \times 10^6}$$

where Q, capacity in m^3/h, ρ, liquid density in kg/m^3 at pumping temperature; H, differential head in m (meters of liquid column); g, gravitational acceleration in m/s^2.

When this hydraulic power is divided by pump efficiency, we get the shaft power.

$$P_{S(kW)} = \frac{P_H}{\eta_p}$$

5.5.9 Pump characteristic curve

With every pump, the manufacturer provides a curve depicting the performance or the behavior of the pump under various conditions. This is called as a characteristic curve of the pump.

Characteristic curve essentially comprises of four curves and these are:

1. Q vs H: capacity vs differential head.
2. Q vs efficiency: capacity vs pump efficiency.

3. Q vs power: capacity vs shaft power.

4. Q vs NPSH-r: capacity vs Net Positive Suction Head – required.

5.5.9.1 Flow rate (*Q*) vs differential head (*H*) curve

The Q vs *H* curve is a continuously drooping curve from shut-off (no flow) condition to BEP. API recommends that the curve from BEP to shut-off should rise by at least 10% for single-stage, single pump operation.

Any pump model can be assembled with trimmed impellers (usually not smaller than 20% of the maximum possible diameter).

The Q vs *H* curve of trimmed impellers in the operating range are parallel and below the Q vs *H* curve of the maximum diameter impeller.

The characteristic curves encompassing performance of all the possible impeller diameters for that model have efficiency depicted as iso-efficiency curves on the Q vs *H* curve.

On every Q–*H* curve, a small triangle is plotted to indicate the *rated* point of operation. The pump manufacturer guarantees this flow and the corresponding differential head.

Usually, the flow at rated point is in excess by 5% or 10% of the flow at which the pump will operate at most of the times or as specified by process demands. This operating point is called as the *normal* operating point.

Centrifugal pumps with radial impellers are started with discharge valves closed. At this point, there is no flow supplied by the pump and the entire liquid keeps churning in the casing. This point of operation is termed shut-off head and run time in this mode must be minimized.

5.5.9.2 Flow rate (*Q*) vs pump efficiency (η_p)

The Q vs pump efficiency of the pump is an inverted 'U' shaped curve. At no flow, the efficiency is zero and then rises to a maximum value at a flow rate, which is termed as the BEP. Beyond this, the curve again drops.

The pumps operate in a range of flows but it has to be kept in mind that they are designed only for one flow rate point. Flow rates above and below this value result in higher hydraulic losses and hence lesser efficiency. The design point is the BEP.

5.5.9.3 Flow rate (*Q*) vs power (P_s)

The pump trial is carried using cold water as the liquid. As volumetric flow in m^3/h, differential head in m, and pump efficiency are independent of the liquid pumped, the results obtained are valid for all service liquids.

Power obtained is for water and can be easily extrapolated for the liquid by multiplying it with the specific gravity of the service liquid.

5.5.9.4 Flow rate (*Q*) vs NPSH-r

NPSH-r is covered in detail in Section 5.5.12. However to initiate, it is the Net Positive Suction Head required by a pump to avoid a phenomenon called as cavitation.

NPSH-r on the characteristic curves is the measured suction head obtained while throttling the suction flow until a 3% drop in the differential head is observed at any particular flow rate (see Figures 5.5.2 and 5.5.3).

NPSH-r is dependant on the service liquid but it is known that cavitation resulting from cold water is most damaging as compared with most commonly pumped liquids (hydrocarbons, hot water) and so NPSH-r results obtained with cold water can be safely applied to other service liquids as well.

5.5.10 Curve corrections

The pump curves are generated while testing the pump using cold water as the liquid. The curve is fixed for a particular speed, impeller diameter, and water.

It is not necessary that the pumps' actual operation throughout its life will be for the same speed or impeller diameter and service. When any of these change, the pump flow and head generated will differ.

In certain cases, it is possible to predict the flow and head for alternate conditions using factors.

Thus, with the help of these factors, the curves can be corrected to obtain a performance map without retesting pump with modified conditions.

5.5.10.1 Affinity laws

The 'Affinity laws' are mathematical expressions that best define changes in pump capacity, head, and power absorbed by the pump when a change is made to pump speed, with all else remaining constant.

According to affinity laws

Capacity Q changes in direct proportion to the change in pump speed *N* ratio:

$$Q_2 = Q_1 \times \left(\frac{N_2}{N_1}\right)$$

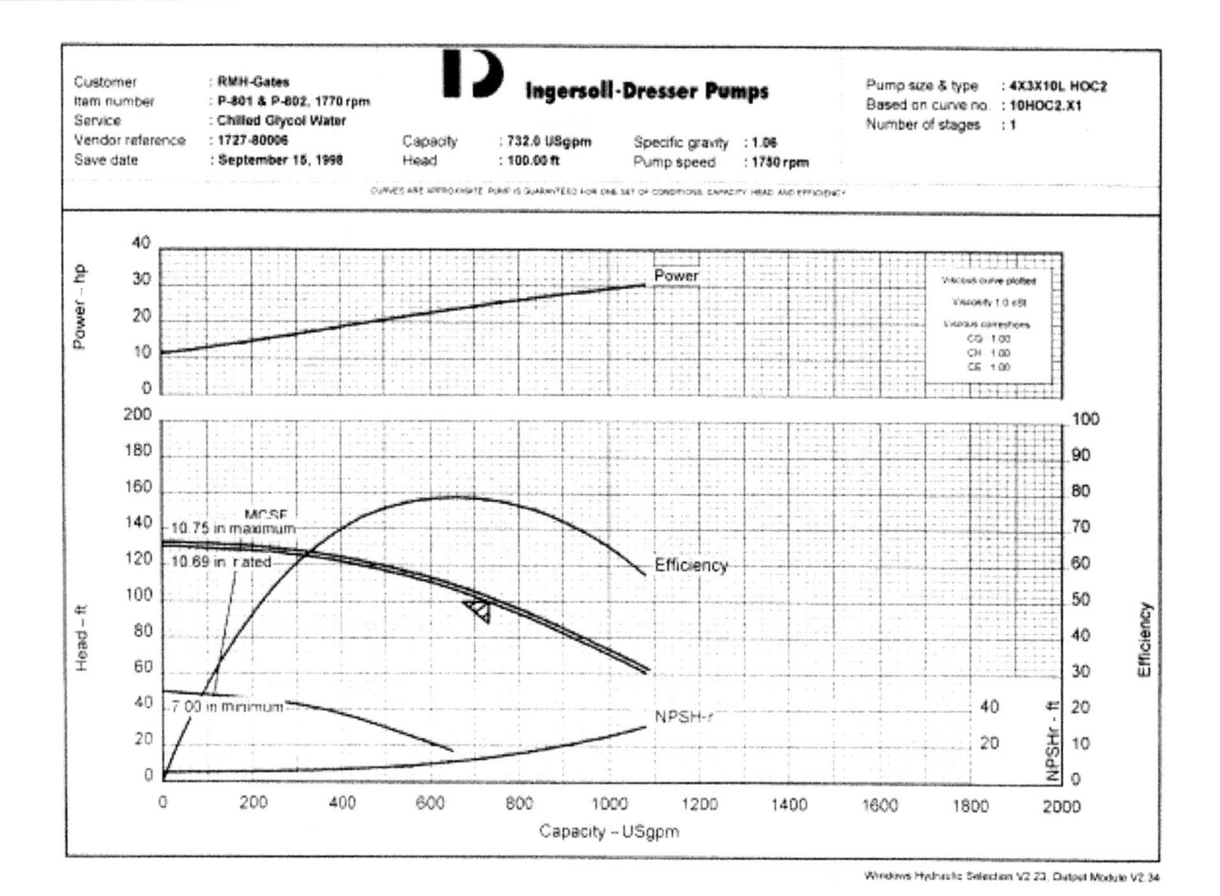

Fig. 5.5.2 A typical pump characteristic curve.

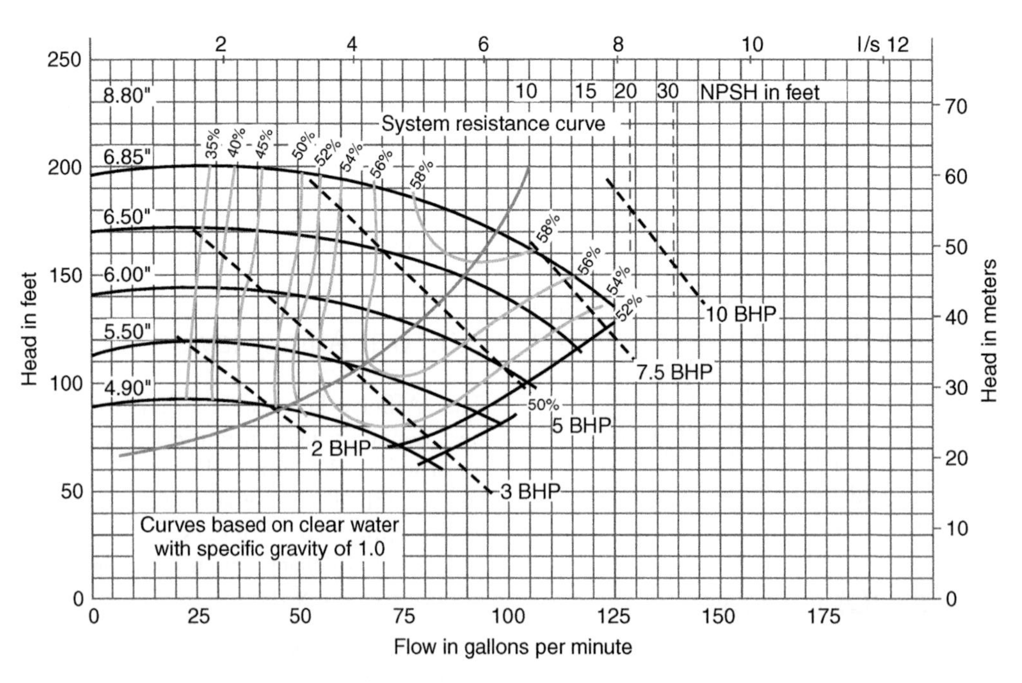

Fig. 5.5.3 Pump curve at various diameters with system resistance and iso-efficiency curves.

Head H changes in direct proportion to the square of the speed N ratio:

$$H_2 = H_1 \times \left(\frac{N_2}{N_1}\right)^2$$

Power P changes in direct proportion to the cube of the speed N ratio:

$$P_2 = P_1 \times \left(\frac{N_2}{N_1}\right)^3$$

where the subscript 1 refers to initial condition and 2 refers to new condition.

Important: the Affinity laws are valid only under conditions of constant efficiency.

The pump affinity laws mentioned above maybe utilized to determine the relationship between flow '*Q*' and impeller diameter as well as to predict head '*H*' and power '*P*' values with change in impeller diameter, whilst speed is kept constant.

The results obtained however are approximate as these formulae are analogous to the centrifugal pump affinity laws. Hence we have;

$$Q_2 = Q_1 \times \left(\frac{D_2}{D_1}\right), \quad H_2 = H_1 \times \left(\frac{D_2}{D_1}\right)^2,$$

$$P_2 = P_1 \times \left(\frac{N_2}{N_1}\right)^3$$

The affinity laws described above require correction when performance is to be predicted following a change in impeller diameter. Due to the above-mentioned constant efficiency factor, there is a discrepancy between the calculated impeller diameter and the achieved performance. This error becomes larger with the increase in cut of the impeller.

If C is the calculated required percentage of impeller diameter and A is the actual required diameter percentage of the original diameter then:

$A = 16.2 + 0.838 \times C$

Thus, by calculation using affinity laws, it is computed that the impeller has to be trimmed to 84% of the original diameter then the actual trimming should be limited to 86.6% of the original diameter.

Note: there are a number of recommended empirical formulae to calculate impeller diameters to match reduced or increased pump flow rates. KSB – centrifugal pump design recommends the following approximate formula for KSB pump impeller trim.

$$\left(\frac{D_2}{D_1}\right)^2 \text{approx.} = \frac{Q_2}{Q_1} \text{approx.} = \frac{H_2}{H_1}$$

In all cases, however, proceed with caution.

5.5.10.2 Viscosity corrections

Under Section 5.5.2, we have discussed viscosity as a property of any fluid that is measure of its resistance to flow.

As the liquid flows through the pump, hydrodynamic losses are increased due to higher viscosity, as a result it is observed that when a viscous fluid is handled by a centrifugal pump:

- The brake horsepower requirement increases.
- There is a reduction in the head generated by the pump.
- Capacity reduction occurs with moderate and high viscosities.
- There is a decrease in the pump efficiency.

This is more evident in smaller pumps. For higher viscosities, larger pumps are used.

A viscosity correction chart from the Hydraulic Institute (as shown in Figure 5.5.4) provides coefficients for flow C_q, head C_h, and efficiency C_η).

These coefficients are used to modify the values of flow, head, and efficiency from the original curve. The new flow, head, and efficiency are obtained using the equations mentioned below.

$$\begin{aligned} Q_{vis} &= C_q \times Q_w \\ H_{vis} &= C_h \times H_w \\ \eta_{vis} &= C_\eta \times \eta_w \end{aligned}$$

Usually fluids more than 2 centipoise should be considered for viscosity correction. e.g., – Q = 500 US-gpm, $H = 80$ ft, viscosity = 1000 SSU.

5.5.11 Specific speed

Specific speed is a number characterizing the type of impeller in a unique and coherent manner.

Specific speed is defined by the equation:

$$N_s = \frac{N\sqrt{Q}}{(H)^{3/4}}$$

where N, pump speed; Q, flow at BEP at maximum impeller diameter (no corrections even if it is a double suction impeller); and H, head per stage at BEP at maximum impeller diameter.

It states that N_s is the speed in rpm at which a pump, if sufficiently reduced in size, would deliver (in US units) 1 gpm at a head of 1 ft. This definition is of little practical utility.

Specific speed is for an impeller, hence for multistage pumps only the first impeller is considered (in the equation, $H = H_{(total)}$/number of stages).

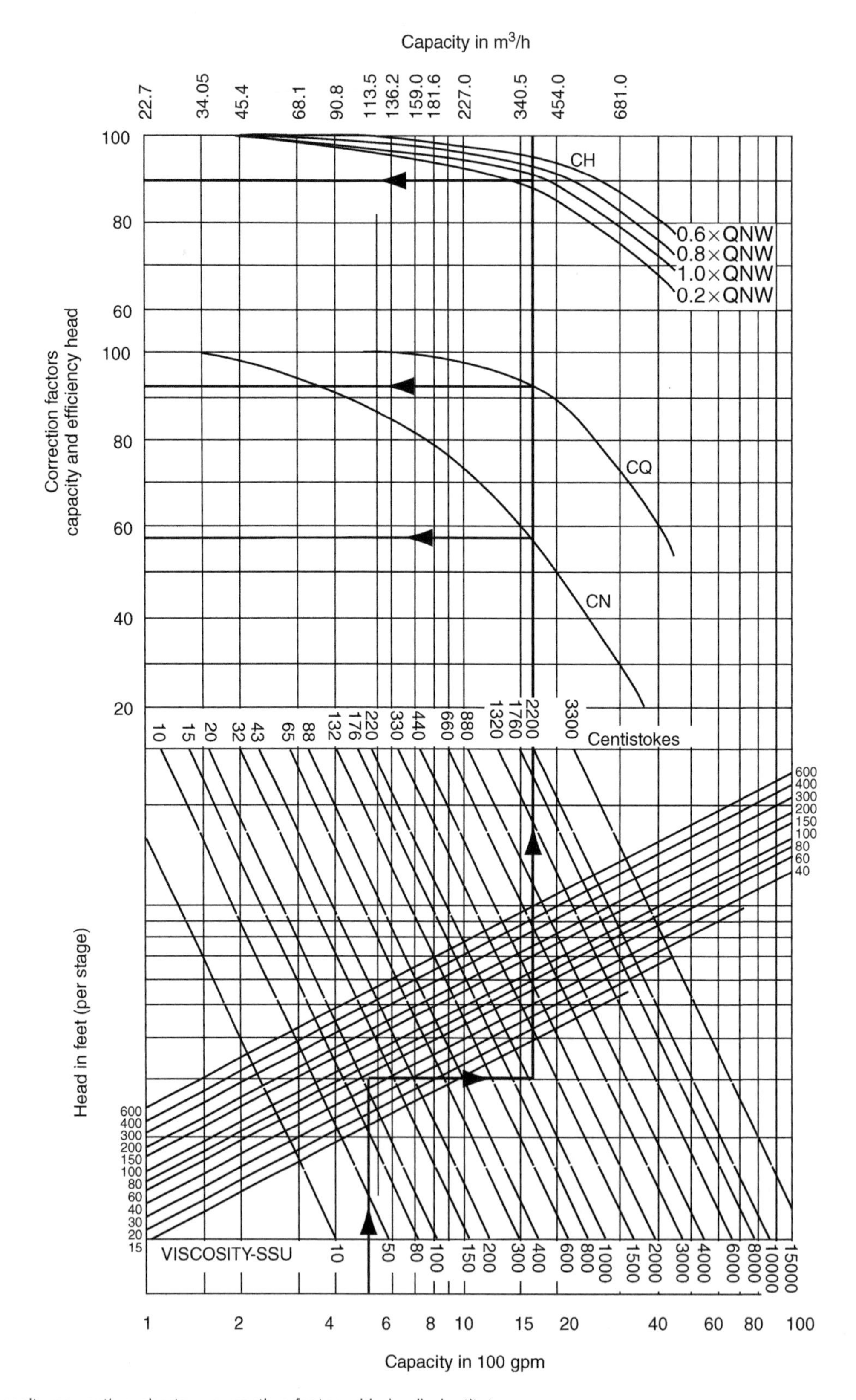

Fig. 5.5.4 Viscosity correction charts – correction factors, Hydraulic Institute.

An index identifies the geometric similarity of pumps. Pumps of the same N_s but have different sizes are considered geometrically similar, one pump being a size-factor of the other.

However, many critical parameters used for impeller design and geometry, volute design, pumps efficiency, layout of pump model performance charts, as required by pump manufacturers are based on the specific speed.

Though, in principle, this text uses SI units as a base, an exception has been made in the case of specific speed. There is so much empirical work done with specific speed in US-gpm, ft, and rpm units that it is considered prudent to get acquainted with FPS units than to persist with SI units and create confusion and errors.

To assist in getting familiar, the conversion from FPS to metric units is given below:

1 US-gpm = 0.2271 m^3/h → 1 British gpm = 0.2728 m^3/h
1 ft = 0.3048 m
US N_s = 1.63 N_s (metric N_s)
Metric N_s = 0.614 US N_s → British N_s = 1.5 metric N_s

Since the 1900s, it was known that there existed a correlation between efficiency and the pump flow, head, and speed. In 1947, George Wislicenus generated curves (shown in Figure 5.5.5) of specific speed vs the pump efficiency. This was a statistical average of the data made available from the commercial pumps in those times.

A more detailed study was generated taking into account various factors such as dimensional tolerance, surface roughness of wet parts, specified wearing ring clearances, and designs not exceeding specified suction-specific speed limits and similar curves.

The curves shown in Figure 5.5.6 is based on the work conducted by Eugene P. Sabini and Warren H. Fraser in their paper, 'The Effect of Specific Speed on the Efficiency of Single-Stage Pumps' presented at the 1986 Pumps Users Symposium.

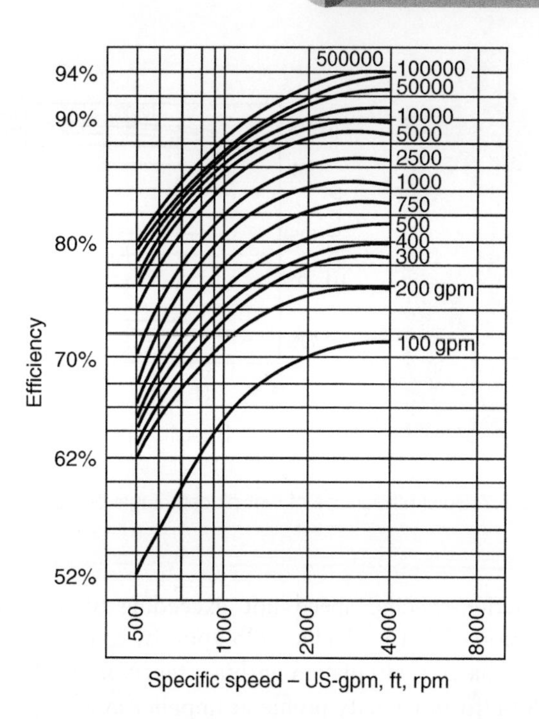

Fig. 5.5.6 Efficiencies of single-stage end-suction and double suction impeller pumps.

They considered the following in the preparation of the curves:

- Single-stage pumps only.
- Finish and dimensional tolerance to within ±1% for vanes and hydraulic passages.
- Surface finish of wet surfaces to be 2×10^{-6} per inch of impeller diameter or better.
- Wearing ring clearances to be 0.0015 in. of ring diameter.

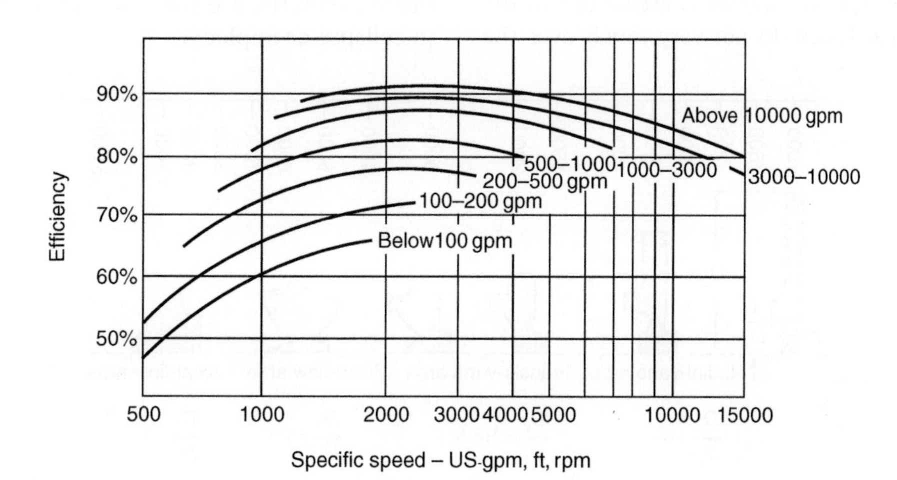

Fig. 5.5.5 An early chart – relating specific speed with single-stage pump efficiency (Original by George Wislicenus – 1947).

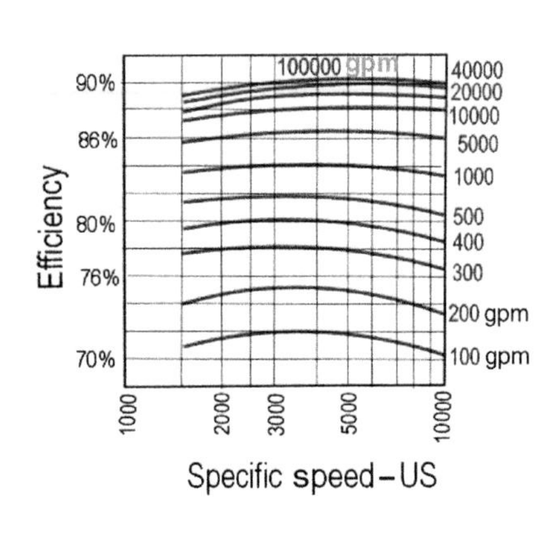

Fig. 5.5.7 Bowl efficiencies of wet pit centrifugal pumps.

- Suction-specific speed not exceeding 8500 (refer Section 5.5.13 – Units – US-gpm, ft, rpm).
- Discharge recirculation within a range of 80–90%.
- A uniform velocity profile at impeller inlet.
- Fluid used was clean water at a temperature of 150 °F or less.
- Efficiencies were based on maximum impeller diameter.
- Wet pit pump efficiencies were based on impellers with no back rings or balancing holes (Figure 5.5.7).

A similar study was conducted by Lobanoff and Ross on pumps having six stages or less and operating at 3560 rpm. The study indicated that the efficiency for multi-stage pumps increases very rapidly to a specific speed of 2000 (US-gpm, ft, rpm) and stays constant until 3500 rpm. Then it begins to taper off a bit.

This is explained on the basis that hydraulic friction and shock losses for high specific speed pumps contribute greater percentage of total head than for low specific speed pumps.

The drop at low specific speeds is attributed to the fact that mechanical losses do not vary much over the range of specific speeds and are therefore a greater percentage of the total power consumption at the lower specific speeds.

Specific speed is a reference number that describes the hydraulic features of a pump, whether radial, semi-axial, or propeller type.

Another index related to the specific speed of the pump is the modeling law. It is usually applied to very large pumps in hydroelectric applications. It states that two geometrically similar pumps working against the same head will have similar flow conditions (same velocities at all sections) if they run at speeds inversely proportional to their size. In this case, the capacity will vary the square of the size.

The optimum laying out of the performance chart or the family curves of a pump model is based on the specific speed (Figure 5.5.8). The BEPs of the family of pumps are usually lined up along the same specific speed. Their size is factored upward for higher flows and heads.

By far, it is the most important number of any pump model.

5.5.12 Cavitation, recirculation, and net positive suction head (NPSH)

5.5.12.1 Cavitation

Gases under pressure can dissolve in a liquid. When the pressure is reduced, they bubble out. Opening of a soda-water bottle is a good example.

In a somewhat similar way, when the liquid is sucked in the pump inlet, the pressure acting on the liquid surface drops. Under conditions, when the reduced pressure approaches the vapor pressure of the liquid (at that temperature), it causes the liquid to vaporize (see Figure 5.5.9). As these vapor bubbles travel further into the impeller, the pressure rises again causing the bubbles to collapse or implode.

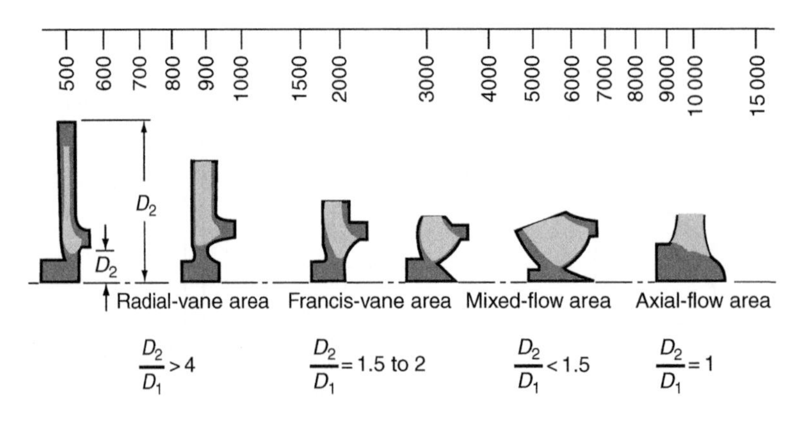

Fig. 5.5.8 Flow area chart.

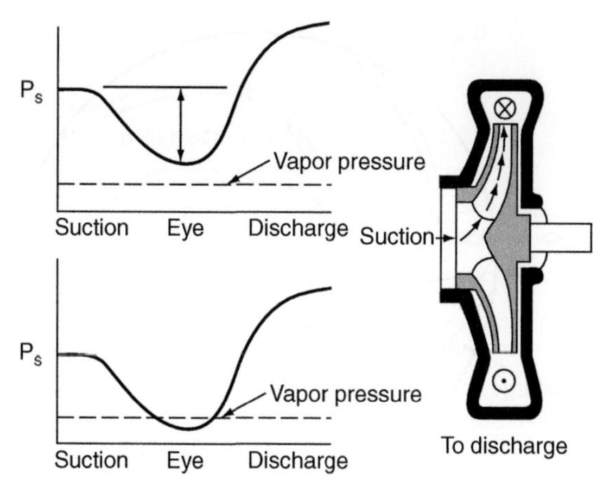

Fig. 5.5.9 Suction pressure falling below vapor pressure causes bubble formation.

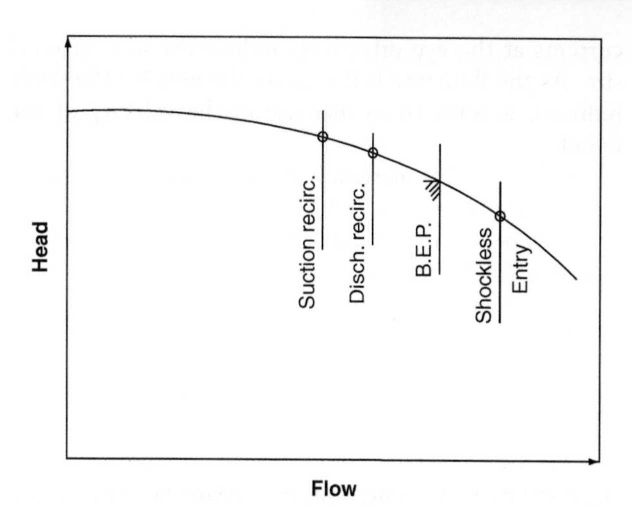

Fig. 5.5.10 Points on curve where recirculation can be expected.

This implosion adversely affects pump performance and could cause severe damage to pump internals. This phenomenon is called as 'cavitation'.

Cavitation damage to a centrifugal pump may range from minor pitting to catastrophic failure and depends on the pumped fluid characteristics, energy levels, and duration of cavitation.

Most of the damage usually occurs within the impeller; specifically, to the leading face of the non-pressure side of the vanes. This is the area where the bubbles normally begin to collapse and release energy on the vane. The net effect observed on the impeller vane will be a pockmarked, rough surface, and severe thinning of the vanes from metal erosion.

5.5.12.2 Recirculation

Another type of cavitation seen in pumps is due to a phenomenon called as recirculation.

One of the complex problems associated with operation of pumps is that of recirculation. Recirculation is defined as flow reversal either at the inlet or at the outlet tips of the impeller vanes.

It is well established that cavitation type of damage seen on the inlet vanes and not associated with inadequate NPSH can be directly linked to the pump operation in the suction recirculation zone. Similar damage seen on the discharge vane tips too can be associated with pump operation in the discharge recirculation zone.

The suction and discharge recirculation may occur at different points as shown in Figure 5.5.10.

The capacities at which the suction and discharge recirculation occurs are dependent on the design of the impeller at the inlet and outlet, respectively. The casing has an influence on the intensity of the discharge recirculation but not on its inception.

Another observation made during extensive research indicated that when the inlet-to-outlet diameter ratio of the impeller equals or exceeds 0.5, the suction recirculation is in effect the capacity at which discharge recirculation occurs.

There are many explanations put forth to explain the phenomena of recirculation.

Recirculation can occur at the suction as well as in the discharge as shown in Figure 5.5.11.

One theory suggests that *recirculation cavitation* (rotating stall or separation) is the formation of vapor-filled pockets. This type of cavitation is different from the classical cavitation described earlier.

In suction recirculation, as the pump is operated to the left of the BEP, eddy currents begin to form at the eye of the impeller.

At this point of operation on the curve, there is no reduction in the flow rate through the pump. The eddy

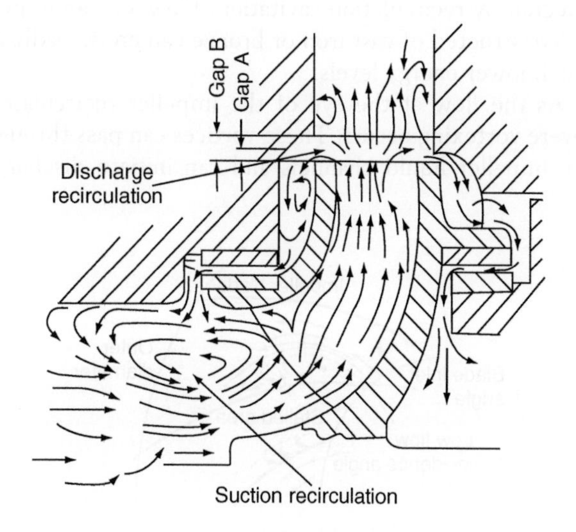

Fig. 5.5.11 Suction and discharge recirculation.

currents at the eye effectively reduce the flow channel size. As the flow rate is the same, the area is effectively reduced; it leads to an increase in the velocity of the liquid.

As the velocity increases, the pressure drops due to friction also increases. When there is a large drop in pressure below the liquid's vapor pressure, the pump experiences classical cavitation because of the initiating action of recirculation cavitation.

Another explanation provided for recirculation is that as the fluid flows over an impeller vane, the pressure near the surface is lowered and the flow tends to separate.

This separated region occurs when the incidence angle (see Figure 5.5.12), which is the difference between flow angle and pump impeller vane inlet angle, increases above a specific critical value.

The stalled area eventually washes but as the rotation continues, it is reformed. The area contains a vapor surrounded by a turbulent flowing liquid at a higher pressure than the vapor pressure. This separated region will then fill with liquid from the downstream end.

The vapor pocket collapses, which causes damage to the surface of the impeller vane. This may occur up to 200–300 times per second.

The damage due to recirculation occurs on the opposite side of the vane where classical cavitation occurs.

This continuous recycling results in noise, vibration, and pressure pulsations. These results imitate classical cavitation, and thus recirculation is often incorrectly diagnosed as cavitation. Figure 5.5.13 shows regions within an impeller that are affected by cavitation and recirculation.

It is often believed that only high-energy pumps (as per API 610 – sixth edition states: High-energy pumps are defined as pumping to a head greater than 650 ft (198 m) and more than 300 HP (224 kW) per stage) are affected by recirculation cavitation. However, an impeller constructed of cast iron or bronze can erode badly at much lower energy levels.

As the flow at the eye of the impeller recirculates, severe vortexing occurs. These vortices can pass through the impeller liquid channels and can initiate discharge

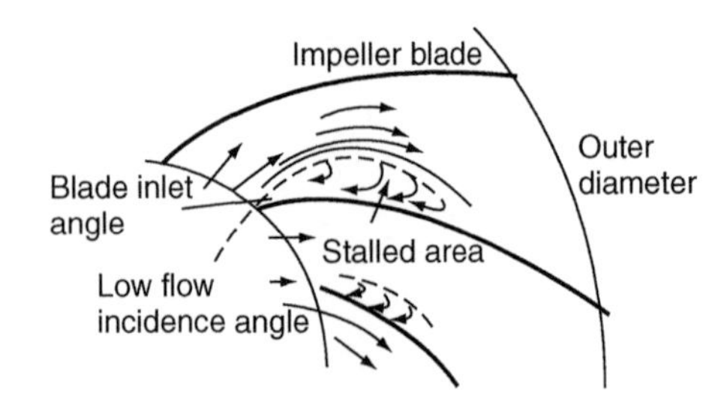

Fig. 5.5.12 Region of stall in the impeller.

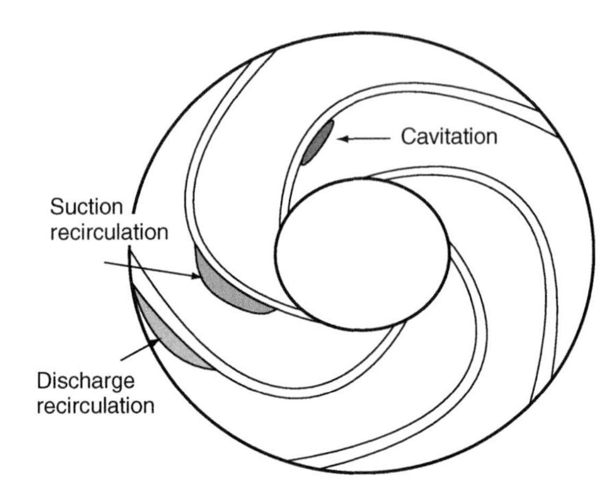

Fig. 5.5.13 Regions within an impeller that are affected by cavitation and recirculation.

recirculation, as shown in the figure on recirculation. The further the pump operates from its BEP, the greater the amount of vortexing.

Most pumps operate in either continuous or intermittent suction or discharge recirculation, especially those designed with high N_{ss} – suction-specific speeds (above about 11,000). (Suction-specific speed is covered in Section 5.5.13.)

Impeller internal circulation usually shows up as cavitation noise and erosion damage, rotor oscillation, shaft breakage, or surging in varying degrees depending on the pump design and application. Many of these problems can be avoided by designing the pump for lower suction-specific speed values and limiting the range of operation to capacities above the point of recirculation.

5.5.12.3 Net positive suction head (NPSH)

The concept of NPSH involves two terms:

1. NPSH-r, called as the Net Positive Suction Head as required by the pump in order to prevent the inception of cavitation and for safe and reliable operation of pump.
2. NPSH-a, called as the Net Positive Suction Head as made available by the suction system of the pump.

NPSH and its correlation to inception of cavitation has been a matter of great research and many theories.

This field is still misunderstood, misapplied, and misused that results in costly over design of new systems or unreliable operation of existing installations of pumps.

5.5.12.3.1 Net Positive Suction Head (NPSH) – required

The Hydraulic Institute defines NPSH-r of a pump as the NPSH that causes the total head (first stage head of multistage pumps) to be reduced by 3% due to flow blockage from cavitation vapor in the impeller vanes.

The above term is a practical method of exactly determining the point of minimum suction head for a pump.

The rated pump head is not achieved when the NPSH-a equals the NPSH-r of the pump. The head will be 3% less than the fully developed head value as shown in Figure 5.5.14.

Strange it may sound but NPSH-r by the above definition does not necessarily imply that this is the point at which cavitation starts; that level is referred to as incipient cavitation.

The NPSH at incipient cavitation can be from 2 to 20 times the 3% NPSH-r value, depending on pump design. The higher ratios are normally associated with high-suction energy pumps or pumps with large impeller inlet areas.

The suction energy level of a pump increases with:

- The casing suction nozzle size.
- The pump speed.
- The suction-specific speed.
- Specific gravity of the pumped liquid.

Anything that increases the velocity in the pump impeller eye, the rate of flow of the pump, or the specific gravity, increases the suction energy of the pump (Figure 5.5.15).

Most standard low-suction energy pumps can operate with little or no margin above the NPSH-r value, without seriously affecting the service life of the pump.

Thus, we see that NPSH-r as per Hydraulic Institute's definition of 3% head drop is not an indicator of inception of cavitation and consequent pump damage.

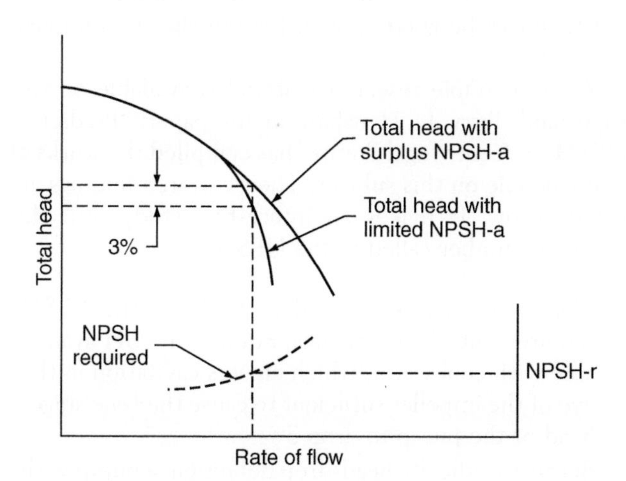

Fig. 5.5.14 When the margin of NPSH-a and NPSH-r lowers, the differential head drops.

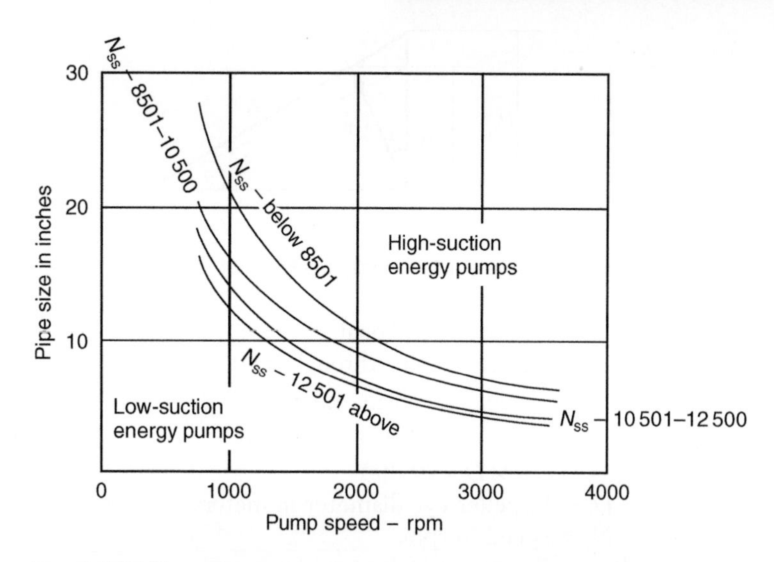

Fig. 5.5.15 Flow rate chart.

It then becomes necessary to come up with a theoretical NPSH-r that can indicate the inception of cavitation that can then ensure a cavitation-free operation.

The theoretical derivation of NPSH-r or 'Cavitation-Free NPSH' is based on factors such as:

- Head loss due to friction.
- Head drop due to fluid acceleration.
- Head loss due to improper fluid entry into the impeller blade.

As the liquid in the suction pipe approaches the impeller eye, it has velocity and acceleration. In addition, it has to change its direction to enter the impeller. Losses in terms of liquid head occur due to each of the above and because of friction.

The pump inlet nozzle and impeller inlet vane geometry are designed to minimize the losses largely but cannot be eliminated.

Other factors like higher flow rates and recirculation due to higher clearance at wear rings and use of smaller diameter impellers in volutes can increase the losses.

The summation of the above losses is termed as Net Positive Suction Head as required by the pump or NPSH-r. In other words, NPSH-r is the summation of losses in the critical area between the suction nozzle and the leading edge of the first stage impeller blades.

Mathematically, the NPSH-r is expressed in the following equation:

$$\text{NPSH-r} = \frac{K_1 \times C_{M_1}}{2g} + \frac{K_2 \times W^2}{2g}$$

The first term represents the friction and acceleration losses and the second term represents the blade entry losses.

To understand this equation, we need to learn the inlet flow velocity triangle of the pump impeller (Figure 5.5.16).

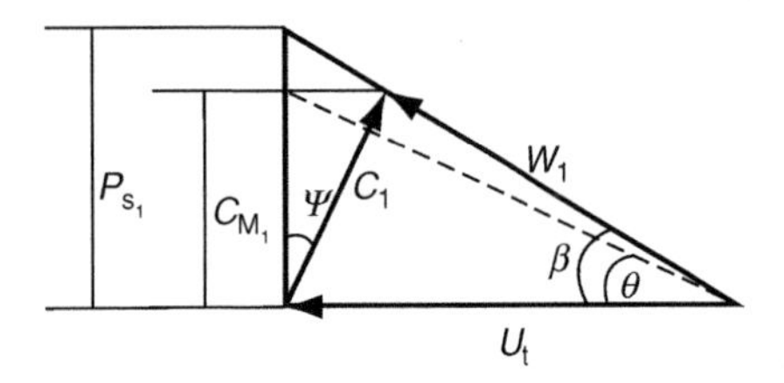

Fig. 5.5.16 Impeller inlet velocity triangle.

C_{M_1} = Average meridional (plane passing through shaft axis) velocity at blade inlet $= \frac{Q\ m^3/gec}{A_{m2}}$ (m/s)
U_t = Peripheral blade velocity (m/s) = $(\pi \times D \times N)/60$
D = Impeller eye diameter in meters
N = speed in rpm
W_1 = Relative velocity (m/s)
C_1 = Absolute velocity of flow (m/s)
P_{S_1} = C_{M_1}/R_1 (R_1 is a factor used to determine the vane outlet angle)
θ = Angle of flow approaching the impeller = Blade angle at outer radius of impeller eye
Ψ = Pre-rotation angle (usually not more than 30°)
α = Incidence angle $= (\beta - \theta)$.

The angle of the inlet edge of the impeller vane, at the point where the vane joins the front shroud, measured in a plane tangent to the shroud surface, is β.

U_t is the peripheral velocity of that same point.

W_1 is the relative velocity (relative to the impeller) of the liquid just before entering the vanes.

C_{M_1} is the meridional velocity of the liquid just before entering the vanes. The meridional velocity C_{M_1} is the velocity relative to the casing. It lies in the meridional plane (the plane that passes through the shaft centerline).

(If you were standing in the suction nozzle of the casing, facing the impeller, C_{M_1} velocity would hit you squarely in the back.) The capacity is such that these three vectors create a right triangle.

If there is 'no pre-rotation' or 'no pre-swirl' or 'shock less-entry', then the included angle $\theta = \beta$.

Going back to the NPSH-required equation, the first term of $K_1 C_{M_1}^2/2g$ represents the friction and acceleration losses and the second term $K_2 W_1^2/2g$ represents blade entry losses.

In low-suction energy pumps, the first term is the prime factor and in the high-suction energy pumps the latter term gains prominence.

The constant K_1 is influenced largely by the pump suction nozzle approaching the impeller eye.

The angle of incidence α that influences K_2 is the difference between the inlet angle β and the flow angle θ. The angle β is determined from C_{M_1} multiplied by the factor R_1 that allows for the effects of recirculated flow Q_L and non-uniform velocity distribution.

The flow Q_L may vary depending on the wear that has occurred causing additional leakages from wearing ring clearances and balance lines of a multistage pump. In lower specific speed pumps, the percentage of leakage flow is a higher percentage of the total flow and hence there is a greater impact on the NPSH-r of such pumps.

All the discussions on cavitation and recirculation were based on phenomena that lead to vapor formation and implosion.

The boiling of the liquid or vapor formation during cavitation or recirculation is a thermal process and is dependent on the properties of the liquid. These properties include pressure, temperature, specific and latent heats of vaporization. If liquid has to boil, the latent heat of vaporization has to be derived from the liquid flow.

The extent of cavitation depends on the proportion of vapor released, the rapidity of liberation, and the vapor specific volume.

This is accounted by a gas to liquid ratio factor C_b. In a paper by D.J. Vlaming, 'A Method of Estimating the Net Positive Suction Head Required by Centrifugal Pumps', ASME 81-WA/FE-32 has provided values of C_b that indicate that cold water has the potential for causing most damage by way of cavitation.

Taking all these into consideration, it is found that cold water is the most damaging of the commonly pumped liquids. Similarly, this difference applies to water at different temperatures.

A review of the properties of water and its vapor at several temperatures shows that the specific volume of vapor decreases rapidly as pressure and temperature increase. Hence, due to this, cold water is again more damaging than hot water.

Even field experience tends to corroborate the above study. Pumps handling certain hydrocarbons or hot water operate satisfactorily with lower NPSH-a than does cold water.

Thus, pumps tested with cold water to detect NPSH-r are found to be good enough for the above-mentioned services.

A considerable research material is available on this topic and Terry L. Henshaw in his paper; 'Predicting NPSH for Centrifugal Pumps' has compiled the works of many people on this subject. Their research findings are listed below. It makes an interesting study into this complex number called as the NPSH-r.

- The current industry standard to define the NPSH requirement of a centrifugal pump as NPSH available with cool water, which creates cavitation in the eye of the impeller sufficient to cause the (one stage) head of the pump to drop 3%.
- Because of the 3% head-drop definition, a pump with a larger-diameter impeller will require less NPSH than the same pump with a smaller diameter impeller.

- The amount of NPSH required to achieve 100% head is typically 1.05–2.5 times the NPSH-r for the 3% head drop.
- The amount of NPSH required to suppress all cavitation is typically 4–5 times the NPSH-r for the 3% head drop, although this ratio can vary from 2 to 20.
- 'Incipient' cavitation causes minimal damage to the impeller.
- The peak cavitation erosion rate occurs at an NPSH-a value above that of the 3% NPSH-r and below that coincident with incipient cavitation.
- Cool water among the many liquids is the most damaging to a cavitating pump.
- For cool water services, the 3% head-drop NPSH is not sufficient to prevent cavitation erosion to the impeller.
- For most pumps, at best efficiency flow rate (BEP), the NPSH-r, based on the 3% head drop, does not vary with speed to the exponent of 2. The exponent is more typically 1.5. Therefore, suction-specific speed at BEP, $N_{SS\text{-}BEP}$ increases as speed increases.
- The shape of the NPSH-r curve varies with the percentage head drop. The NPSH-r3 decreases as flow rate decreases, reaching a minimum value, normally at or below 40% of the BEP. The NPSH-r curves for 1 and 0% head drop increase as the flow rate decreases below BEP.
- Field experience has revealed that an increase in pump failure rates occur when suction-specific speeds (calculated at BEP and in US units) exceed around 10,000, with a pronounced increase at 11,000.
- The Hydraulic Institute Standards uses N_{ss} value of 8500 (US) as the basis for their maximum speed recommendations.
- At no-prerotation flow rate, the NPSH-a for incipient cavitation is equal to the NPSH-a for the 3% head drop plus 'peripheral velocity head' ($U_1^2/2g$).
- To reduce the NPSH-r3 at BEP, impellers are typically designed, at the BEP flow rate, such that $P_{s_1} > C_{M_1}$ (refer Figure 5.5.16). The $P_{s_1} > C_{M_1}$ ratio is typically about 1.25. Therefore, the flow rate coincident with no prerotation is typically about 25% larger than the BEP flow rate.

5.5.12.3.2 Net positive suction head (NPSH) – available

Every pump has an associated inlet system comprising of vessel, pipes, valves, strainers, and other fittings. The liquid, which has a certain suction pressure, experiences losses as it travels through the inlet system.

Thus, the net inlet pressure (in absolute terms) of the pipe and fitting losses is what is available at pump inlet and this is called as the Net Positive Suction Head – available or NPSH-a.

Now knowing what is NPSH-a and NPSH-r, it becomes clear that their difference has to be greater than the vapor pressure of liquid at that temperature to avoid vaporization of liquid.

As a convention, a mathematical simplification is done. The vapor pressure of the liquid is subtracted from the NPSH-a. In pump terminology, NPSH-a includes vapor pressure correction.

Thus, all we need to insure is that the NPSH-a is greater than NPSH-r. Their difference in such a case would then be in the true sense of the word *Net Positive Suction Head*. Some common examples of NPSH calculations are provided below.

5.5.12.3.3 Calculating NPSH-available: pressurized flooded suction

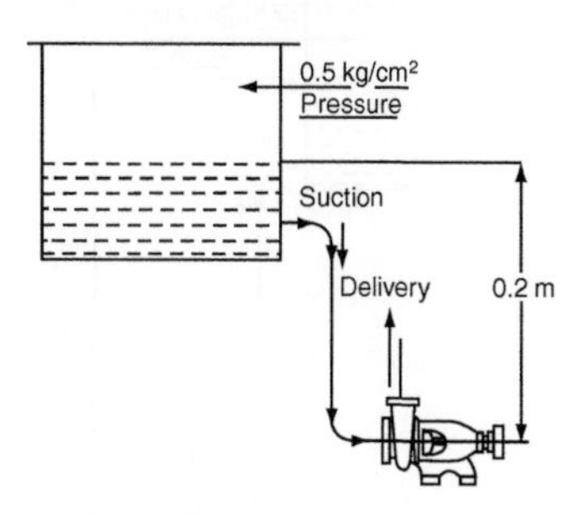

Vapor pressure – $P_{vap} = 0.45$ kg/cm^2

Pipe losses – $H_f = 1.5$ m

Specific gravity = 0.8

P_g – gage pr. = 0.5 kg/cm^2

H_g in meters 'm' = $0.5 \times 10.2/0.8 = 6.4$ m

H_{st} in meters 'm' = + 0.2 m

$H_f = 1.5$ m

H_a – atmospheric pressure = $10.325/0.8 = 12.9$ m

H_{vap} in meters 'm' = $0.45 \times 10/0.8 = 5.7$ m

$H_a + H_s + H_{st} - H_f - H_{vap}$

NPSH-a in meters 'm' = $12.9 + 6.4 + 0.2 - 1.5 - 5.7 = 12.3$ m

5.5.12.3.4 Calculating NPSH-available: atmospheric flooded suction

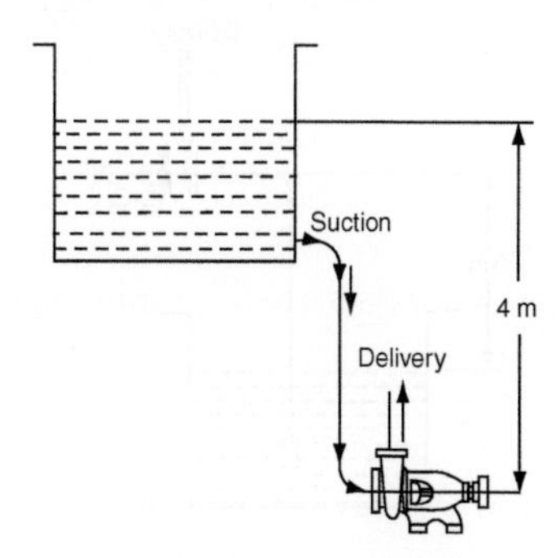

Vapor pressure – P_{vap} = 0.45 kg/cm^2
Pipe losses – H_f = 1.5 m
Specific gravity = 0.8
P_g – gage pressure = 0 kg/cm^2
H_g in meters 'm' = 0 × 10.2/0.8 = 0 m
t H_{st} in meters 'm' = +4 m
H_a – atmospheric pressure = 10.325/0.8 = 12.9 m
H_f in meters 'm' = 1.5 m
H_{vap} in meters 'm' = 0.45 × 10/0.8 = 5.7 m
$= H_a + H_g + H_{st} - H_f - H_{vap}$
NPSH-a in meters 'm' = 12.9 + 0 + 4 − 1.5 − 5.7
= 9.7 m

5.5.12.3.5 Calculating NPSH-available: vacuum flooded suction

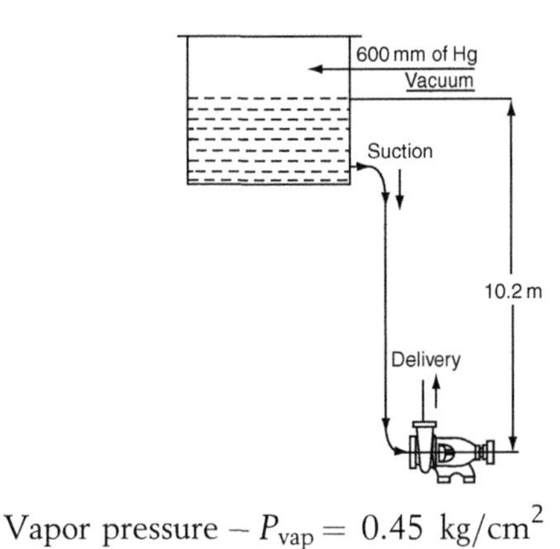

Vapor pressure – P_{vap} = 0.45 kg/cm^2
Pipe losses – H_f = 1.5 m;
Specific gravity = 0.9
P_g – gage pressure = −600 mm-Hg
H_g in meters 'm' = −(600/1000) × 13.6/0.9
= −9.1 m
H_{st} in meters 'm' = +10.2 m
H_a – atmospheric pressure = 1.033 × 10/0.9
= 11.5 m
H_f in meters 'm' = 1.5 m
H_{vap} = 0.45 × 10/0.9 = 5.1 m
$= H_a + H_g + H_{st} - H_f - H_{vap}$
NPSH-a = 11.5 − 9.1 + 10.2 − 1.5 − 5.1 = 6.0 m

5.5.12.3.6 Calculating NPSH-available: negative suction

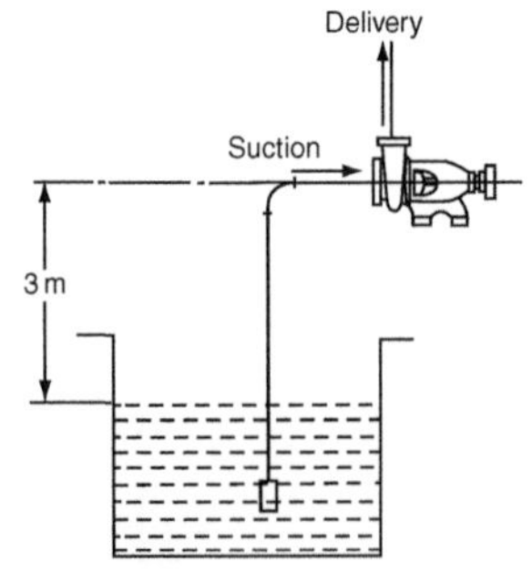

Vapor pressure–P_{vap} = 0.45 kg/cm^2
Pipe losses–H_f = 1.5 m;
Specific gravity = 0.8
P_g–gage pressure = 0 kg/cm^2
H_g in meters 'm' = 0 × 10.2/0.8 = 0 m
H_a – atmospheric pressure = 10.325/0.8 = 12.9 m
H_{st} in meters 'm' = −3 m
H_f in meters 'm' = 1.5 m
H_{vap} in meters 'm' = 0.45 × 10.2/0.8 = 5.7 m
$= H_a + H_g + H_{st} - H_f - H_{vap}$
NPSH-a in meters 'm' = 12.9 + 0 + (−3) − 1.5 − 5.7
= 2.7 m (Satisfactory)

12.3.7 Net Positive Suction Head (NPSH) – margin

The simple approach considered with regard to NPSH margin is the net between the available and required NPSH.

It is a requirement that the NPSH-a available must be equal to or greater than the NPSH-r stipulated by the pump manufacturer. Most pump specifications quote a margin of not less than 1 to 1.5 m over the entire range of pump operation.

When the difference between NPSH-a and NPSH-r is less than the stated margin, it calls for an exact determination of the NPSH-r by carrying out the NPSH-r test.

Another approach adopted to define the margin is by taking the ratio of NPSH-a and NPSH-r.

The table given below offers suggested minimum NPSH margin ratio guidelines (NPSH-a/NPSH-r), within the allowable operating region of the pump (with standard materials of construction). It is based on the experience of the many pump manufacturers with many different pump applications.

	Minimum NPSH margin ratio guidelines (NPSH-a/NPSH-r) Suction energy levels		
Application	**Low**	**Medium**	**High**
Petroleum	1.1-a	1.3-c	
Chemical	1.1-a	1.3-c	
Electrical power	1.1-a	1.5-c	2.0-c
Nuclear power	1.5-b	2.0-c	2.5-c
Cooling towers	1.3-b	1.5-c	2.0-c
Water/waste water	1.1-a	1.3-c	2.0-c
General industry	1.1-a	1.2-b	
Pulp and paper	1.1-a	1.3-c	
Building services	1.1-a	1.3-c	
Slurry	1.1-a		
Pipeline	1.3-b	1.7-c	2.0-c
Water flood	1.2-b	1.5-c	2.0-c

Source – www.pumps.org: 'a' – Or 0.6 m (2 ft) whichever is greater; 'b' – Or 0.9 m (3 ft) whichever is greater; 'c' – Or 1.5 m (5 ft) whichever is greater.

Vertical turbine (misnomer) pumps often operate without NPSH margin without damage, but with slightly reduced discharge head. Such pumps generally have low-suction energy, and cavitation noise is normally not an issue. NPSH-a has to be equal to or larger than the NPSH-r over the allowable operating region of the pump, including a low water level.

High and very high suction energy pumps operating with the margins specified in the table will have acceptable bearing and seal life. However, they may still be susceptible to impeller erosion and higher noise levels.

In addition to the margins specified in the table additional requirements of suction head arise due to:

- Increase in wearing ring clearances due to wear. This increases the leakage flow to the impeller eye and disturbs the inlet flow pattern.
- Gas content in the liquid.
- Improperly designed inlet piping and pump casing that cause non-uniform suction flow or turbulence.
- Operation of the pump on the farther right-hand side of BEP. In this region, the NPSH-a reduces and NPSH-r increases.

5.5.13 Suction-specific speed

Suction-specific speed is defined by the equation;

$$N_{ss} = \frac{N\sqrt{Q}}{(\text{NPSH-r})^{3/4}}$$

where N, pump speed; Q, capacity at BEP at maximum impeller diameter (it gets halved for a double suction impeller); NPSH-r, Net Positive Suction Head (required) at BEP at maximum impeller diameter.

Studies carried out have empirically established that pump models with N_{ss} less than 11,000 (US units: Q – US-gpm, N – rpm, NPSH-r – feet) have a more stable operation and are more reliable.

Therefore, it is commonly used as a basis for estimating the safe operating range of capacity for a pump. The higher the N_{ss} is, the narrower is its safe operating range from its BEP. Most users prefer that their pumps have N_{ss} in the range of 8000–11 000 for optimum and trouble-free operation.

It is usually recommended that such pumps ($N_{ss} >$ 11,000) should not be operated at flow rates below 60–70% of the BEP. When the pump is operated below this range, it may experience:

- Impeller and casing erosion.
- Shaft deflection and stress.
- Radial and thrust bearing failures.
- Seal problems.

The above are attributed to the recirculation of liquid at the impeller inlet, which has been covered in the earlier section.

For a smooth operation, the liquid enters the impeller at a particular designed angle. This inlet angle is meant for flows at the BEP, however, at lower flows, the liquid enters the impeller at a much different angle and is unable to make an entry into the impeller.

As a result, it is forced back into the pump suction pipe. The liquid keeps recirculating in front of the impeller.

Evidence of recirculation at impeller inlet is:

- Suction pressure gage fluctuations.
- Noisy operation.
- High vibrations at low flow rates.

If a higher N_{ss} pump model is thus encountered, users prefer to buy a lower speed pump even though it may cost more.

5.5.14 Performance calculation procedure

For a centrifugal pump, the performance calculation's aim is to determine the pump efficiency. This value can be read on the characteristic curves provided by the manufacturer. The deviation of the calculated efficiency from the rated efficiency indicates the performance degradation of the pump.

5.5.14.1 Flow measurement

Flow measurement can be taken from a flow measuring device, if fitted. In cases where flow measurement devices are not installed, non-invasive ultrasonic flow meters can be used to measure the flow from the pump.

The flows are usually indicated as mass flow in kg/h. It is recommended to convert this to volume flow in m^3/h.

M = mass flow (kg/h)
Q = volumetric flow (m^3/h)
δ = density at pumping temperature (kg/m^3)

$$Q = \frac{M}{\delta}$$

5.5.14.2 Differential head

h_s = suction head (m)
P_s = suction pressure (kg/cm^2)
ρ = specific gravity at pumping temperature

$$h_s = \frac{10 \times P_s}{\rho}$$

Discharge head h_d in m is calculated in a similar manner.

Differential head – h in m is further calculated as

$$H = h_d - h_s$$

5.5.14.3 Hydraulic power

The next step in this process is to calculate the hydraulic power. This is calculated in the following manner.

g = gravitational acceleration – 9.81 m/s^2.

$$P_{H(kW)} = \frac{Q \times \delta \times g \times H}{3.6 \times 10^6}$$

The hydraulic power is minimum power required to pump the fluid. This will be the power if the pump had an efficiency of 100%. However, this is not possible in practice. To obtain the actual pump efficiency we need to go to the next step of calculating the energy being provided to the pump by the prime mover. Let us assume that in this case the prime mover is an electrical motor.

5.5.14.4 Motor power

The electrical power is fed to the motor to its terminals. However, we are interested in the power that is delivered by the motor at its coupling with the pump. Thus, we need to consider the efficiency of the motor too.

Efficiency of the motor is not a fixed number but changes with the load on the motor. A part load efficiency of the motor is much lower than its efficiency at full load. This table of motor efficiency with respect to its load is provided by the motor manufacturer.

V = Measured voltage in volts
I = Measured current in ampere
$\cos\phi$ = Measured power factor
η_e = Motor efficiency

Motor power, P_M in kW at its coupling is:

$$P_M = \frac{\sqrt{3} \times V \times I \times \cos\phi \times \eta_e}{1000}$$

5.5.14.5 Pump efficiency

Having performed the above calculations, we are now in a position to derive the pump efficiency. The ratio of pump hydraulic power to the motor power gives the pump efficiency.

η_p = pumping efficiency (hydraulic efficiency)

$$\eta_p = \frac{P_H}{P_M}$$

Chapter **5.6**

5.6

Hydraulic pumps and pressure regulation

Parr

A hydraulic pump (Figure 5.6.1) takes oil from a tank and delivers it to the rest of the hydraulic circuit. In doing so, it raises oil pressure to the required level. The operation of such a pump is illustrated in Figure 5.6.1a. On hydraulic circuit diagrams, a pump is represented by the symbol of Figure 5.6.1b, with the arrowhead showing the direction of flow.

Hydraulic pumps are generally driven at constant speed by a three-phase AC induction motor rotating at 1500 rpm in the UK (with a 50 Hz supply) and at 1200 or 1800 rpm in the USA (with a 60 Hz supply). Often, pump and motor are supplied as one combined unit. As an AC motor requires some form of starter, the complete arrangement illustrated in Figure 5.6.1c is needed.

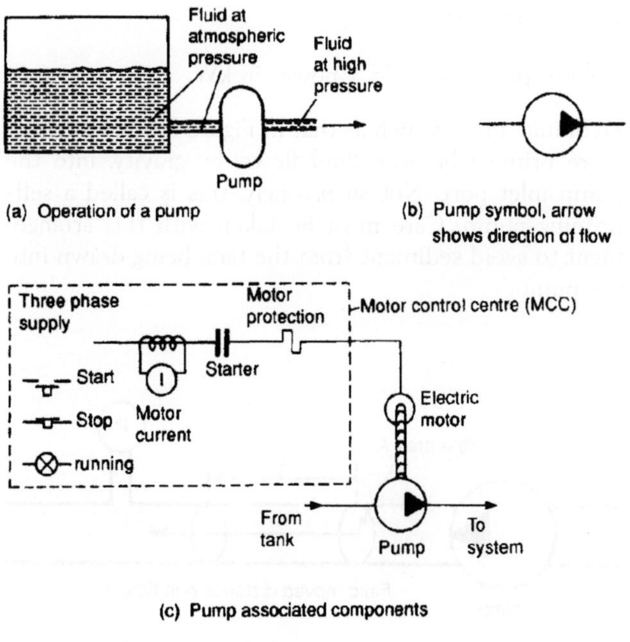

There are two types of pump (for fluids) or compressor (for gases) illustrated in Figure 5.6.2. Typical of the first type is the centrifugal pump of Figure 5.6.2a. Fluid is drawn into the axis of the pump, and flung out to the periphery by centrifugal force. Flow of fluid into the load maintains pressure at the pump exit. Should the pump stop, however, there is a direct route from outlet back to inlet and the pressure rapidly decays away. Fluid leakage will also occur past the vanes, so pump delivery will vary according to outlet pressure. Devices such as that shown in Figure 5.6.2a are known as hydrodynamic pumps, and are primarily used to shift fluid from one location to another at relatively low pressures. Water pumps are a typical application.

Fluid at atmospheric pressure
Fluid at high pressure
Pump
(a) Operation of a pump
(b) Pump symbol, arrow shows direction of flow

Three phase supply
Motor protection
Motor control centre (MCC)
Starter
Start
Stop
running
Motor current
Electric motor
From tank
Pump
To system
(c) Pump associated components

Fig. 5.6.1 The hydraulic pump.

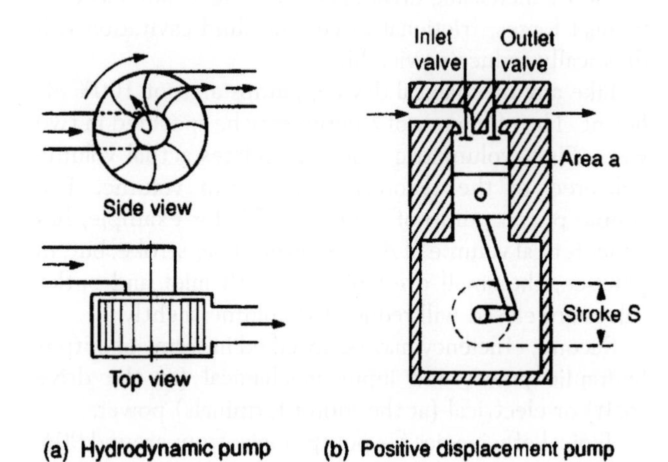

Fig. 5.6.2 Types of hydraulic pump.

Hydraulics and Pneumatics; ISBN: 9780750644198

Figure 5.6.2b shows a simple piston pump called a positive displacement or hydrostatic pump. As the piston is driven down, the inlet valve opens and a volume of fluid (determined by the cross-sectional area of the piston and the length of stroke) is drawn into the cylinder. Next, the piston is driven up with the inlet valve closed and the outlet valve open, driving the same volume of fluid to the pump outlet.

Should the pump stop, one of the two valves will always be closed, so there is no route for fluid to leak back. Exit pressure is therefore maintained (assuming there are no downstream return routes).

More important, though, is the fact that the pump delivers a fixed volume of fluid from inlet to outlet each cycle regardless of pressure at the outlet port. Unlike the hydrodynamic pump described earlier, a piston pump has no inherent maximum pressure determined by pump leakage: if it drives into a dead end load with no return route (as can easily occur in an inactive hydraulic system with all valves closed) the pressure rises continuously with each pump stroke until either piping or the pump itself fails.

Hydraulic pumps are invariably hydrostatic and, consequently, require some method of controlling system pressure to avoid catastrophic pipe or pump failure. This topic is discussed further in a later section.

A hydraulic pump is specified by the flow rate it delivers (usually given in litres min^{-1} or gallons min^{-1}) and the maximum pressure the pump can withstand. These are normally called the pump capacity (or delivery rate) and the pressure rating.

Pump data sheets specify required drive speed (usually 1200, 1500 or 1800 rpm corresponding to the speed of a three-phase induction motor). Pump capacity is directly related to drive speed; at a lower than specified speed, pump capacity is reduced and pump efficiency falls as fluid leakage (called slippage) increases. Pump capacity cannot, on the other hand, be expected to increase by increasing drive speed, as effects such as centrifugal forces, frictional forces and fluid cavitation will drastically reduce service life.

Like any mechanical device, pumps are not 100% efficient. The efficiency of a pump may be specified in two ways. First, volumetric efficiency relates actual volume delivered to the theoretical maximum volume. The simple piston pump of Figure 5.6.2b, for example, has a theoretical volume of $A \times s$ delivered per stroke, but, in practice, the small overlap when both inlet and outlet valves are closed will reduce the volume slightly.

Second, efficiency may be specified in terms of output hydraulic power and input mechanical (at the drive shaft) or electrical (at the motor terminals) power.

Typical efficiencies for pumps range from around 90% (for cheap gear pumps) to about 98% for high-quality piston pumps. An allowance for pump efficiency needs to be made when specifying pump capacity or choosing a suitable drive motor.

The motor power required to drive a pump is determined by the pump capacity and working pressure:

$$\text{Power} = \frac{\text{work}}{\text{time}} = \frac{\text{force} \times \text{distance}}{\text{time}} \tag{5.6.1}$$

In Figure 5.6.3, a pump forces fluid along a pipe of area A against a pressure P, moving fluid a distance d in time T. The force is PA, which, when substituted into expression (5.6.1), gives

$$\text{Power} = \frac{P \times A \times d}{T}$$

but $A \times d/T$ is flow rate, hence:

$$\text{Power} = \text{pressure} \times \text{flow rate}. \tag{5.6.2}$$

Unfortunately, expression (5.6.2) is specified in impractical SI units (pressure in pascal, time in seconds, flow in cubic metres). We may adapt the expression to use more practical units (pressure in bar, flow rate in litres min^{-1}) with the expression:

$$\text{Power} = \frac{\text{pressure} \times \text{flow rate}}{600} \text{ kW}. \tag{5.6.3}$$

For Imperial systems (pressure in psig, flow rate in gallons min^{-1}), the expression becomes:

$$\text{Power} = \frac{\text{pressure} \times \text{flow rate}}{1915} \text{ kW}. \tag{5.6.4}$$

For fully imperial systems, motor power in horsepower can be found from

$$\text{Horsepower} = 0.75 \times \text{power in kW}. \tag{5.6.5}$$

Hydraulic pumps such as that in Figure 5.6.1 do not require priming because fluid flows, by gravity, into the pump inlet port. Not surprisingly, this is called a self-priming pump. Care must be taken with this arrangement to avoid sediment from the tank being drawn into the pump.

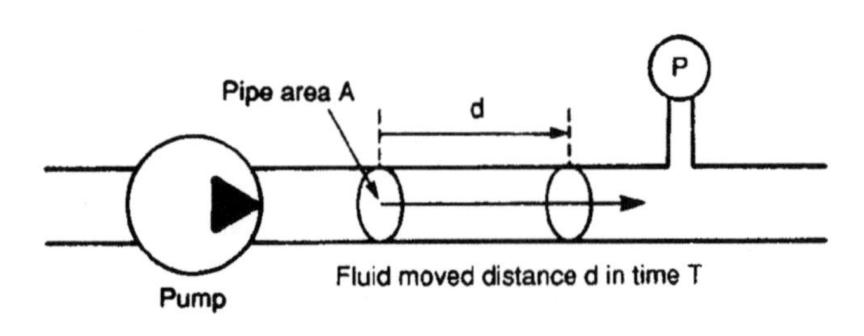

Fig. 5.6.3 Derivation of pump power.

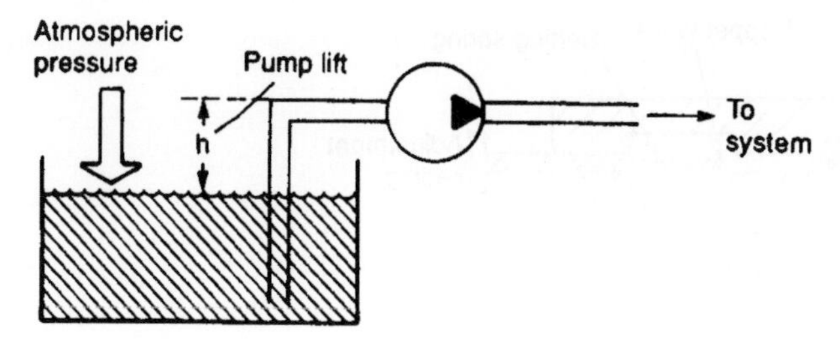

Fig. 5.6.4 Pump lift.

The pump in Figure 5.6.4 is above the fluid in the tank. The pump creates a negative (less than atmospheric) pressure at its inlet port causing fluid to be pushed up the inlet pipe by atmospheric pressure. This action creates a fluid lift which is, generally, incorrectly described as arising from pump suction. In reality fluid is *pushed* into the pump.

Maximum pump lift is determined by atmospheric pressure. In theory, a lift of about 8 m is feasible but, in practice, would be accompanied by undesirable side effects such as cavitation (formation and destructive collapse of bubbles from partial vaporisation of fluid). The lift should be as small as possible and around 1 m is a normal practical limit.

Fluid flow in the inlet line always takes place at negative pressure, and a relatively low flow velocity is needed to reduce these side effects. The design should aim for a flow velocity of around $1\ \text{m s}^{-1}$. Examination of any hydraulic system will always reveal pump inlet pipes of much larger diameters than outlet pipes.

5.6.1 Pressure regulation

Figure 5.6.5a shows the by-now familiar system where a load is raised or lowered by a hydraulic cylinder. With valve V_1 open, fluid flows from the pump to the cylinder, with both pressure gauges P_1 and P_2 indicating a pressure of F/A. With valves V_1 closed and V_2 open, the load falls with fluid being returned to the tank. With the load falling, gauge P_2 will still show a pressure of F/A, but at P_1 the pump is dead-ended leading to a continual increase in pressure as the pump delivers fluid into the pipe.

Obviously some method is needed to keep P_1 at a safe level. To achieve this, pressure regulating valve V_3 has been included. This is normally closed (no connection between P and T) while the pressure is below some preset level (called the cracking pressure). Once the cracking pressure is reached valve V_3 starts to open, bleeding fluid back to the tank. As the pressure increases, valve V_3 opens more until, at a pressure called the full flow pressure, the valve is fully open. With valve V_1 closed, all fluid from the pump returns to the tank via the pressure regulating valve, and P_1 settles somewhere between the cracking and full flow pressures.

Cracking pressure of a relief valve *must* be higher than a system's working pressure, leading to a fall in system

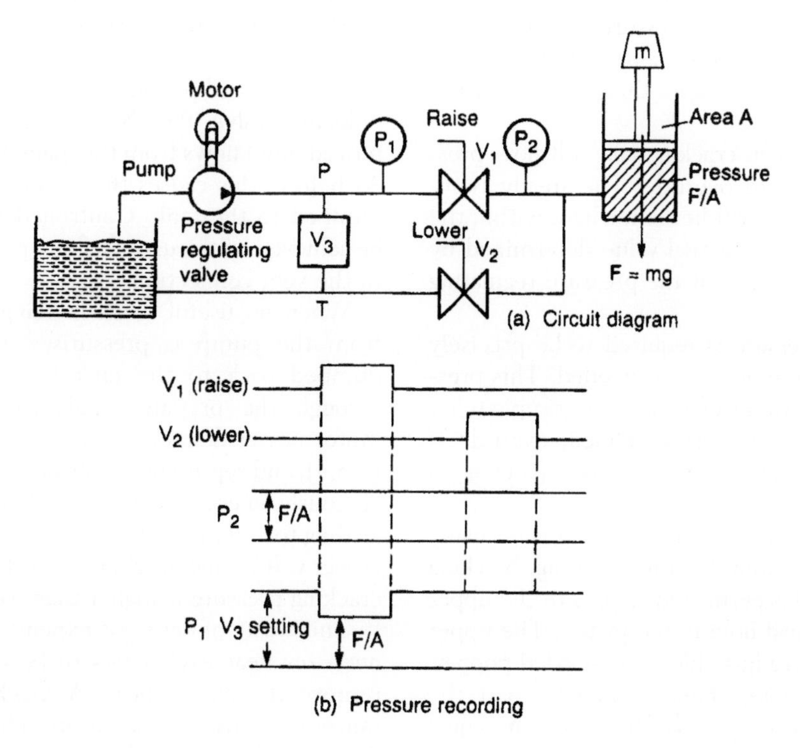

Fig. 5.6.5 Action of pressure regulation.

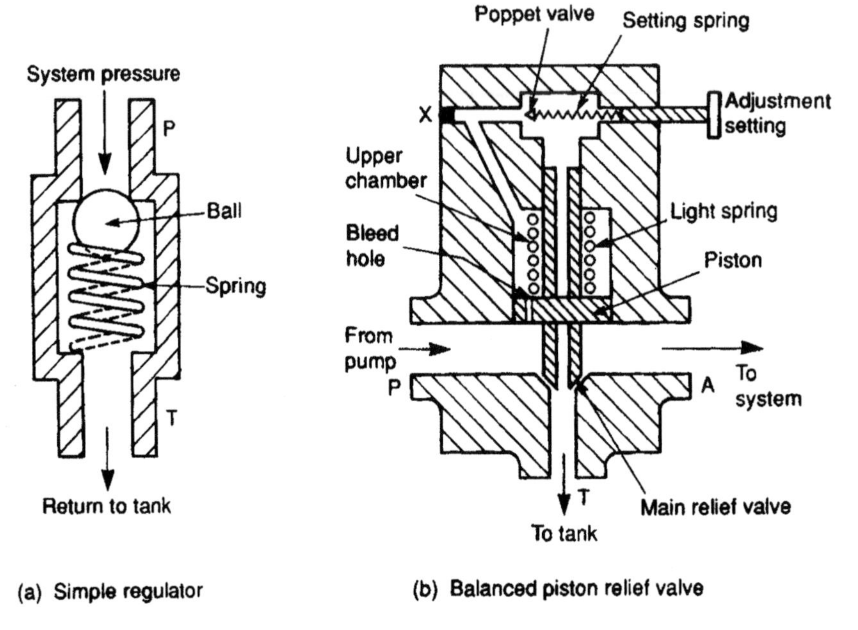

Fig. 5.6.6 Pressure regulation.

pressure as valve V_1 opens and external work is performed. Valve positions and consequent pressure readings are shown in Figure 5.6.5b.

The simplest form of pressure regulation valve is the ball-and-spring arrangement of Figure 5.6.6a. System pressure in the pipe exerts a force of $P \times A$ on the ball. When the force is larger than the spring compressive force the valve will crack open, bypassing fluid back to the tank. The higher the pipe pressure, the more the valve opens. Cracking pressure is set by the spring compression and in practical valves this can be adjusted to suit the application.

The difference between cracking and full flow pressure is called the pressure override. The steady (non-working) system pressure will lie somewhere within the pressure override, with the actual value determined by pipe sizes and characteristics of the pressure regulating valve itself.

If the quiescent pressure is required to be precisely defined, a small pressure override is needed. This pressure override is related to spring tension in a simple relief valve. When a small, or precisely defined, override is required, a balanced piston relief valve (shown in Figure 5.6.6b) is used.

The piston in this valve is free moving, but is normally held in the lowered position by a light spring, blocking flow to the tank. Fluid is permitted to pass to the upper chamber through a small hole in the piston. The upper chamber is sealed by an adjustable spring-loaded poppet. In the low-pressure state, there is no flow past the poppet, so pressure on both sides of the piston are equal and spring pressure keeps the valve closed.

When fluid pressure rises, the poppet cracks and a small flow of fluid passes from the upper chamber to the tank via the hole in the piston centre. This fluid is replenished by fluid flowing through the hole in the piston. With fluid flow there is now a pressure differential across the piston, which is acting only against a light spring. The whole piston lifts, releasing fluid around the valve stem until a balance condition is reached. Because of the light restoring spring, a very small override is achieved.

The balanced piston relief valve can also be used as an unloading valve. Plug X is a vent connection and, if removed, fluid flows from the main line through the piston. As before, this causes the piston to rise and flow to be dumped to the tank. Controlled loading/unloading can be achieved by the use of a finite position valve connected to the vent connection.

When no useful work is being performed, *all* fluid from the pump is pressurised to a high value then dumped back to the tank (at atmospheric pressure) through the pressure-regulating valve. This requires motor power defined earlier by expressions (5.6.3) and (5.6.4), and represents a substantial waste of power. Less obviously, energy put into the fluid is converted to heat leading to a rise in fluid temperature. Surprisingly, motor power will be higher when no work is being done because cracking pressure is higher than working pressure.

This waste of energy is expensive, and can lead to the need for heat exchangers to be built into the tank to remove the excess heat. A much more economic arrangement uses loading/unloading valves, a topic discussed further in a later section.

5.6.2 Pump types

There are essentially three different types of positive displacement pump used in hydraulic systems.

5.6.2.1 Gear pumps

The simplest and most robust positive displacement pump, having just two moving parts, is the gear pump. Its parts are non-reciprocating, move at constant speed and experience a uniform force. Internal construction, shown in Figure 5.6.7, consists of just two close-meshing gear wheels which rotate as shown. The direction of rotation of the gears should be carefully noted; it is the *opposite* of that intuitively expected by most people.

As the teeth come out of mesh at the centre, a partial vacuum is formed which draws fluid into the inlet chamber. Fluid is trapped between the outer teeth and the pump housing, causing a continual transfer of fluid from inlet chamber to outlet chamber where it is discharged to the system.

Pump displacement is determined by volume of fluid between each pair of teeth, number of teeth and speed of rotation. Note the pump merely delivers a fixed volume of fluid from inlet port to outlet port for each rotation; outlet port pressure is determined solely by the design of the rest of the system.

Performance of any pump is limited by leakage and the ability of the pump to withstand the pressure differential between inlet and outlet ports. The gear pump obviously requires closely meshing gears, minimum clearance between teeth and housing, and also between the gear face and side plates. Often the side plates of a pump are designed as deliberately replaceable wear plates. Wear in a gear pump is primarily caused by dirt particles in the hydraulic fluid, so cleanliness and filtration are particularly important.

The pressure differential causes large side loads to be applied to the gear shafts at 45° to the centre line as shown. Typically, gear pumps are used at pressures up to about 150 bar and capacities of around 150 gpm (6751 min^{-1}). Volumetric efficiency of gear pumps at 90% is lowest of the three pump types.

There are some variations of the basic gear pump. In Figure 5.6.8, gears have been replaced by lobes giving a pump called, not surprisingly, a lobe pump.

Figure 5.6.9a is another variation called the internal gear pump, where an external driven gear wheel is connected to a smaller internal gear, with fluid separation as gears disengage being performed by a crescent-shaped moulding. Yet another variation on the theme is the gerotor pump of Figure 5.6.9b, where the crescent moulding is dispensed with by using an internal gear with one less tooth than the outer gear wheel. Internal gear pumps operate at lower capacities and pressures (typically 70 bar) than other pump types.

5.6.2.2 Vane pumps

The major source of leakage in a gear pump arises from the small gaps between teeth, and also between teeth and pump housing. The vane pump reduces this leakage by using spring (or hydraulic) loaded vanes slotted into a driven rotor, as illustrated in the two examples of Figure 5.6.10.

In the pump shown in Figure 5.6.10a, the rotor is offset within the housing, and the vanes constrained by a cam ring as they cross inlet and outlet ports. Because

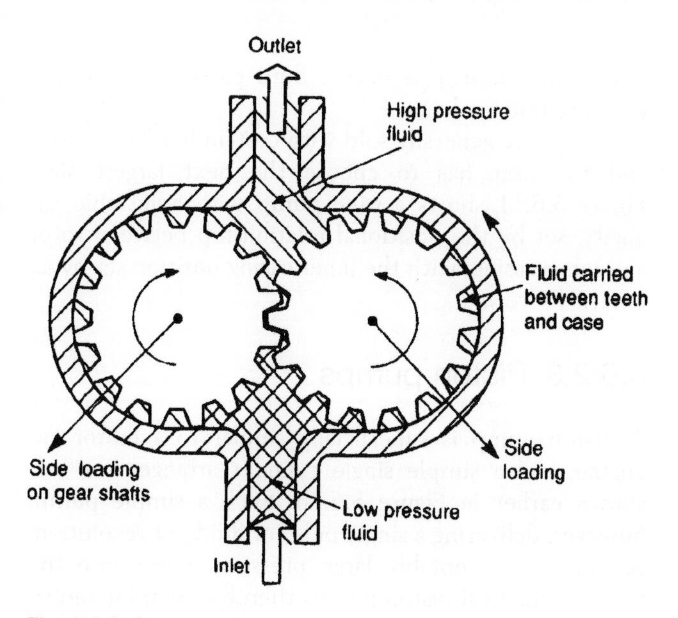

Fig. 5.6.7 Gear pump.

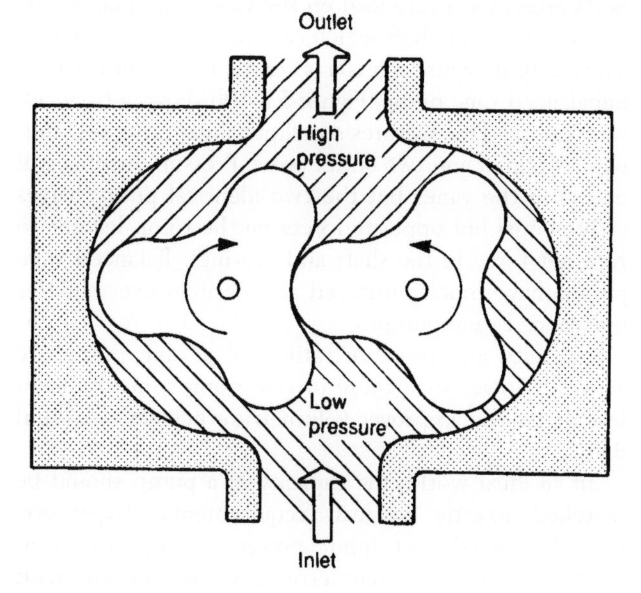

Fig. 5.6.8 The lobe pump.

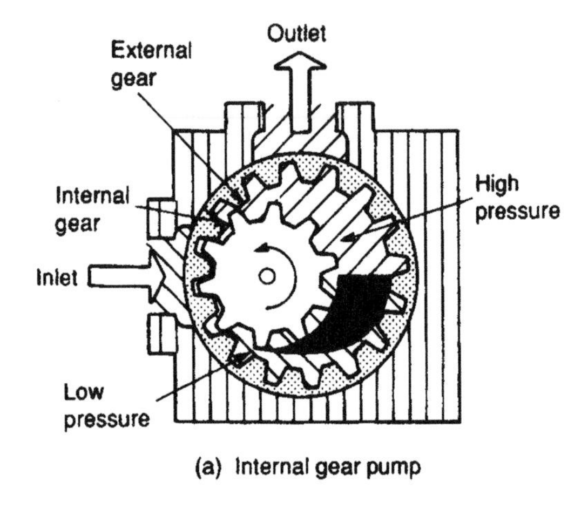

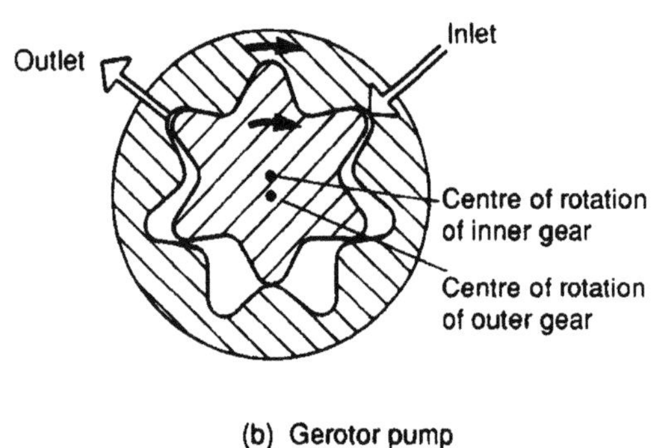

Fig. 5.6.9 Further forms of gear pump.

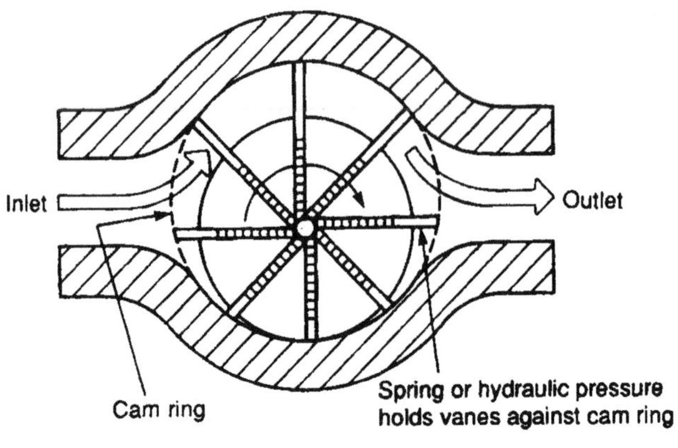

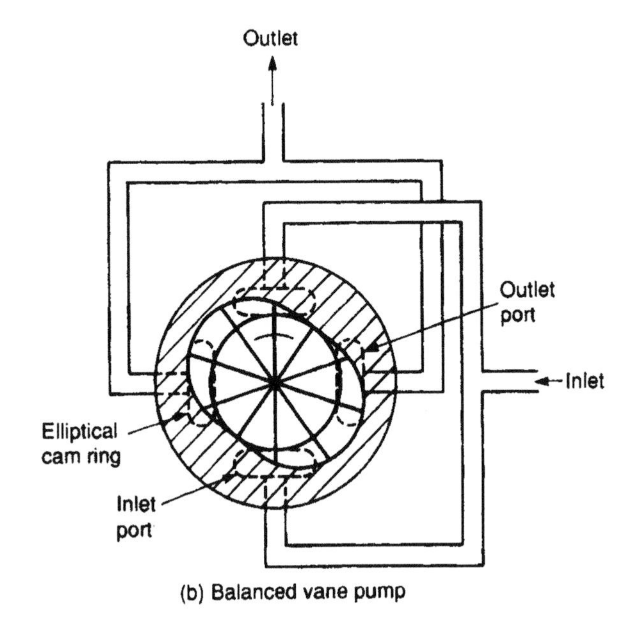

Fig. 5.6.10 Vane pumps.

the vane tips are held against the housing, there is little leakage and the vanes compensate to a large degree for wear at vane tips or in the housing itself. There is still, however, leakage between rotor faces and body sides. Pump capacity is determined by vane throw, vane cross-sectional area and speed of rotation.

The difference in pressure between outlet and inlet ports creates a severe load on the vanes and a large side load on the rotor shaft which can lead to bearing failure. The pump in Figure 5.6.10a is consequently known as an unbalanced vane pump. Figure 5.6.10b shows a balanced vane pump. This features an elliptical cam ring together with two inlet and two outlet ports. Pressure loading still occurs in the vanes but the two identical pump halves create equal but opposite forces on the rotor, leading to zero net force in the shaft and bearings. Balanced vane pumps have much improved service lives over simpler unbalanced vane pumps.

Capacity and pressure ratings of a vane pump are generally lower than those of gear pumps, but reduced leakage gives an improved volumetric efficiency of around 95%.

In an ideal world, the capacity of a pump should be matched exactly to load requirements. Expression (5.6.2) showed that input power is proportional to system pressure and volumetric flow rate. A pump with too large a capacity wastes energy (leading to a rise in fluid temperature) as excess fluid passes through the pressure relief valve.

Pumps are generally sold with certain fixed capacities and the user has to choose the next largest size. Figure 5.6.11 shows a vane pump with adjustable capacity, set by the positional relationship between rotor and inner casing, with the inner casing position set by an external screw.

5.6.2.3 Piston pumps

A piston pump is superficially similar to a motor car engine, and a simple single cylinder arrangement was shown earlier in Figure 5.6.2b. Such a simple pump, however, delivering a single pulse of fluid per revolution, generates unacceptably large pressure pulses into the system. Practical piston pumps therefore employ multiple cylinders and pistons to smooth out fluid delivery, and

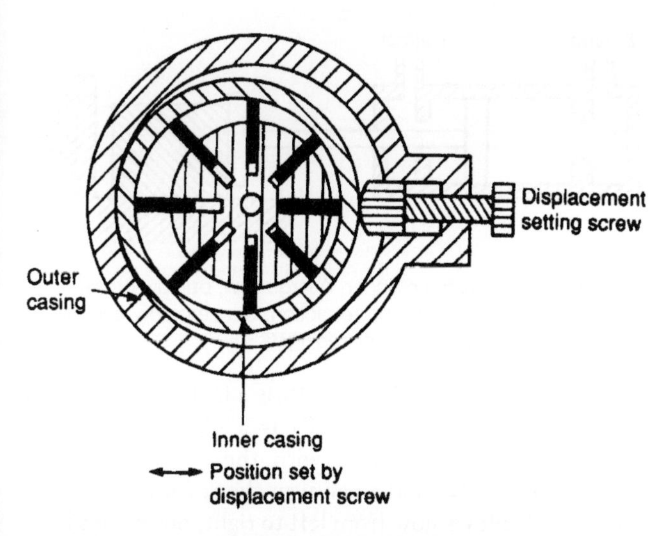

Fig. 5.6.11 Variable displacement vane pump.

much ingenuity goes into designing multicylinder pumps which are surprisingly compact.

Figure 5.6.12 shows one form of radial piston pump. The pump consists of several hollow pistons inside a stationary cylinder block. Each piston has spring-loaded inlet and outlet valves. As the inner cam rotates, fluid is transferred relatively smoothly from inlet port to the outlet port.

The pump of Figure 5.6.13 uses the same principle, but employs a stationary cam and a rotating cylinder block. This arrangement does not require multiple inlet and outlet valves and is consequently simpler, more reliable, and cheaper. Not surprisingly, most radial piston pumps have this construction.

An alternative form of piston pump is the axial design of Figure 5.6.14, where multiple pistons are arranged in

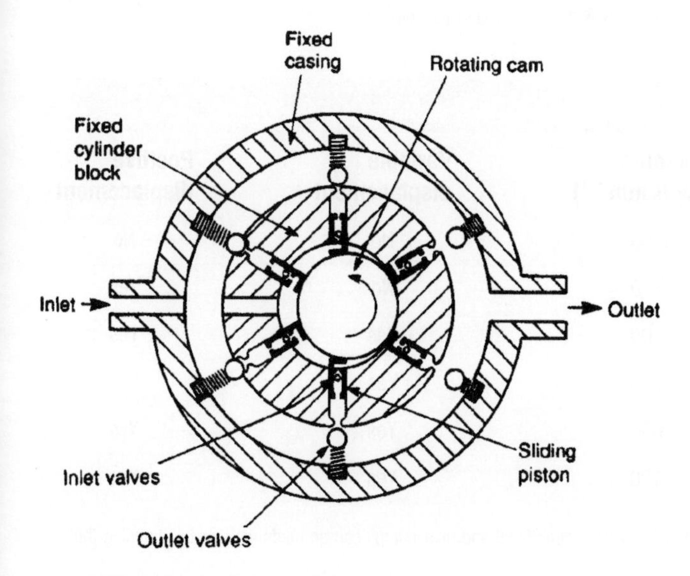

Fig. 5.6.12 Radial piston pump.

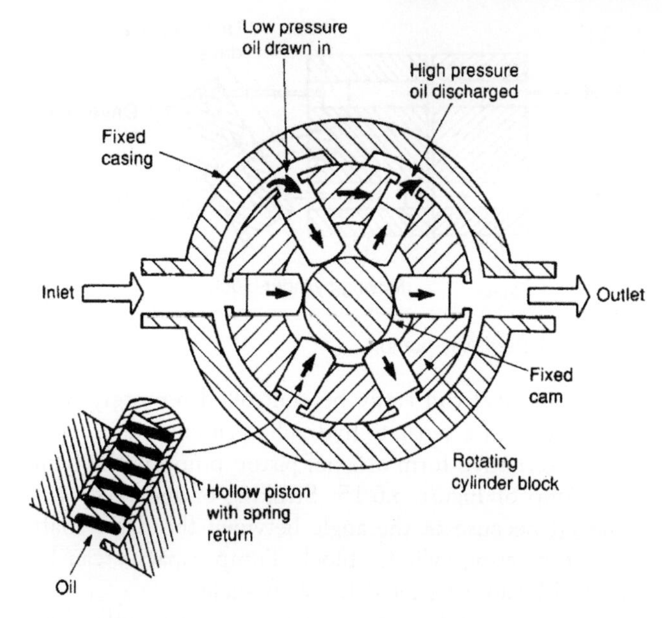

Fig. 5.6.13 Piston pump with stationary cam and rotating block.

a rotating cylinder. The pistons are stroked by a fixed angled plate called the swash plate. Each piston can be kept in contact with the swash plate by springs or by a rotating shoe plate linked to the swash plate.

Pump capacity is controlled by altering the angle of the swash plate; the larger the angle, the greater the capacity. With the swash plate vertical capacity is zero, and flow can even be reversed. Swash plate angle (and hence

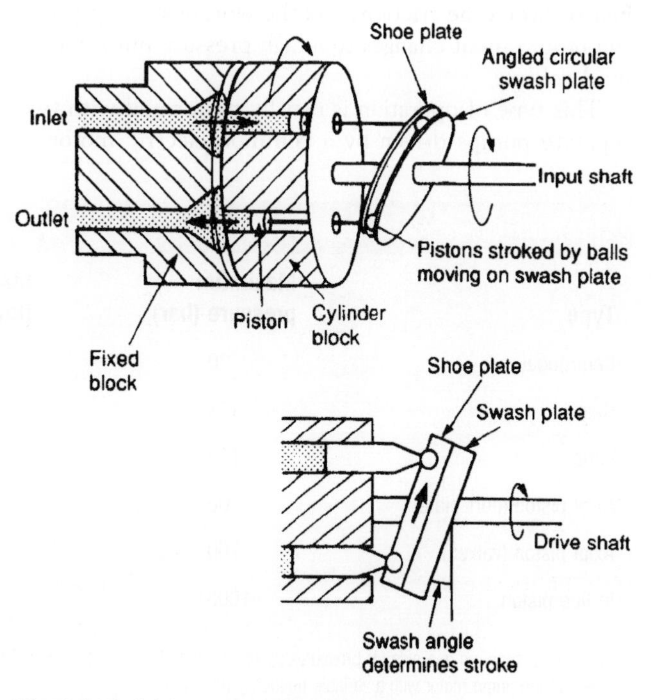

Fig. 5.6.14 Axial pump with swash plate.

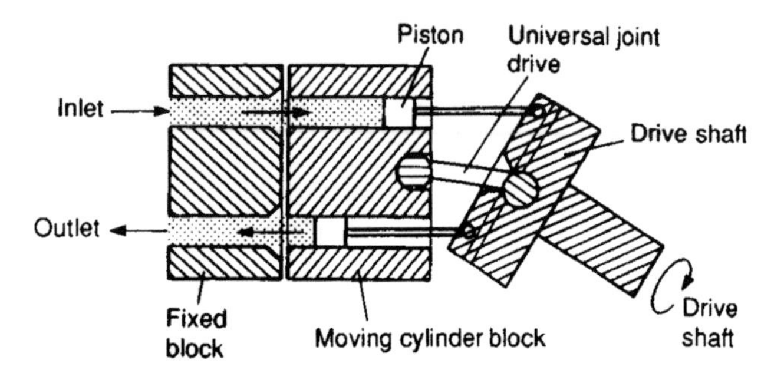

Fig. 5.6.15 Bent axis pump.

pump capacity) can easily be controlled remotely with the addition of a separate hydraulic cylinder.

An alternative form of axial piston pump is the bent axis pump of Figure 5.6.15. Stroking of the pistons is achieved because of the angle between the drive shaft and the rotating cylinder block. Pump capacity can be adjusted by altering the drive shaft angle.

Piston pumps have very high volumetric efficiency (over 98%) and can be used at the highest hydraulic pressures. Being more complex than vane and gear pumps, they are correspondingly more expensive. Table 5.6.1 gives a comparison of the various types of pumps.

5.6.2.4 Combination pumps

Many hydraulic applications are similar to Figure 5.6.16, where a workpiece is held in place by a hydraulic ram. There are essentially two distinct requirements for this operation. As the cylinder extends or retracts, a large volume of fluid is required at a low pressure (sufficient just to overcome friction). As the workpiece is gripped, the requirement changes to a high pressure but minimal fluid volume.

This type of operation is usually performed with two separate pumps driven by a common electric motor as

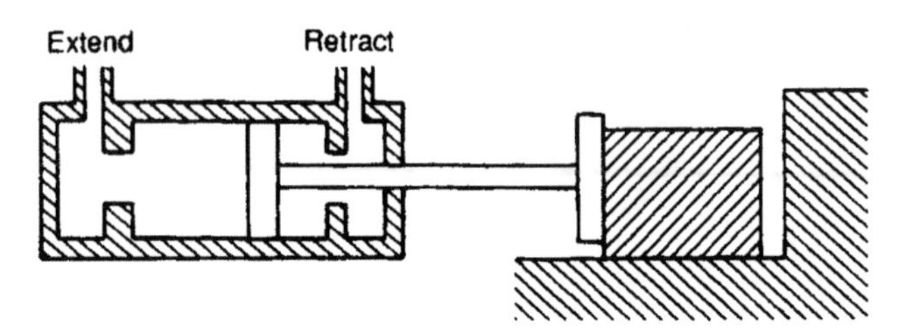

Fig. 5.6.16 A clamping cylinder. A large flow, but low pressure, is needed during extension and retraction, but zero flow and high pressure are needed during clamping.

shown in Figure 5.6.17. Pump P_1 is a high-pressure low-volume pump, while pump P_2 is a high-volume low-pressure pump. Associated with these are two relief valves RV_1 and RV_2 and a one-way check (or non-return) valve which allows flow from left to right, but blocks flow in the reverse direction.

A normal (high pressure) relief valve is used at position RV_1 but relief valve RV_2 is operated not by the pressure at point X, but remotely by the pressure at point Y. This could be achieved with the balanced piston valve

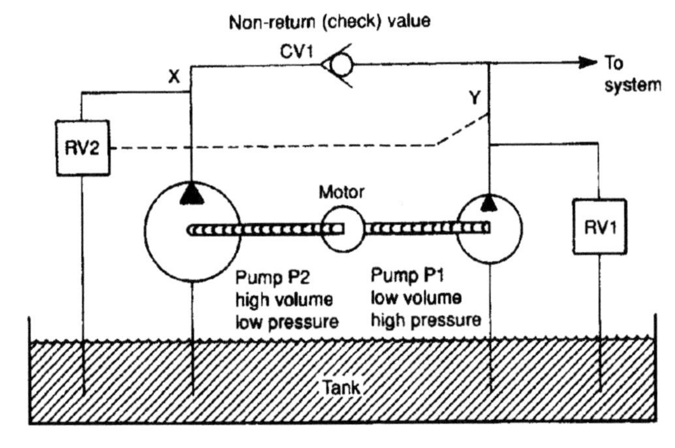

Fig. 5.6.17 Combination pump.

Table 5.6.1 Comparison of hydraulic pump types

Type	Maximum pressure (bar)	Maximum flow (l min^{-1})	Variable displacement	Positive displacement
Centrifugal	20	3000	No	No
Gear	175	300	No	Yes
Vane	175	500	Yes	Yes
Axial piston (port-plate)	300	500	Yes	Yes
Axial piston (valved)	700	650	Yes	Yes
In-line piston	1000	100	Yes	Yes

Specialist pumps are available for pressures up to about 7000 bar at low flows. The delivery from centrifugal and gear pumps can be made variable by changing the speed of the pump motor with a variable frequency (VF) drive.

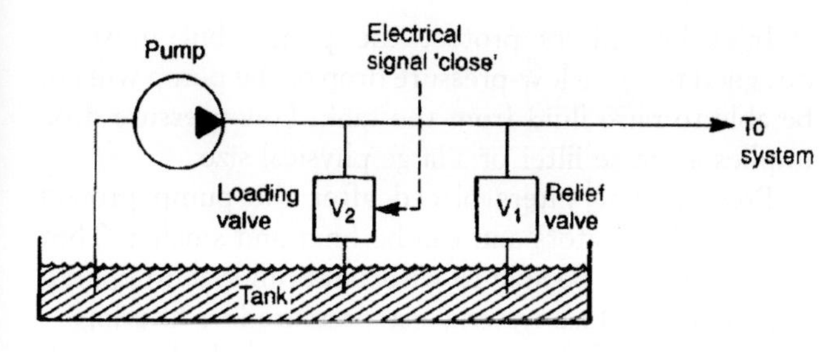

Fig. 5.6.18 Loading valve.

of Figure 5.6.6. In low-pressure mode both relief valves are closed and both pumps P_1 and P_2 deliver fluid to the load, the majority coming from pump P_2 because of its higher capacity.

When the workpiece is gripped, the pressure at Y rises, and relief valve RV_2 opens causing all the fluid from pump P_2 to return straight to the tank and the pressure at X to fall to a low value. Check valve CV_1 stops fluid from pump P_1 passing back to the tank via relief valve RV_2; consequently, pressure at Y rises to the level set by relief valve RV_1.

This arrangement saves energy as the large volume of fluid from pump P_2 is returned to the tank at a very low pressure, and only a small volume of fluid from pump P_1 is returned at a high pressure. Pump assemblies similar to that shown in Figure 5.6.17 are called combination pumps and are manufactured as complete units with motor, pumps, relief and check valves prefitted.

5.6.3 Loading valves

Expression (5.6.2) shows that allowing excess fluid from a pump to return to the tank by a pressure relief valve is wasteful of energy and can lead to a rapid rise in temperature of the fluid as the wasted energy is converted to heat. It is normally undesirable to start and stop the pump to match load requirements, as this causes shock loads to pump, motor and couplings.

In Figure 5.6.18, valve V_1 is a normal pressure relief valve regulating pressure and returning excess fluid to the tank as described in earlier sections. The additional valve V_2 is opened or closed by an external electrical or hydraulic signal. With valve V_2 open, all the pump output flow is returned to the tank at low pressure with minimal energy cost.

When fluid is required in the system the control signal closes valve V_2, pressure rises to the setting of valve V_1, and the system performs as normal. Valve V_2 is called a pump loading or a pump unloading valve according to the interpretation of the control signal sense.

5.6.4 Filters

Dirt in a hydraulic system causes sticking valves, failure of seals and premature wear. Even particles of dirt as small as 20 μm can cause damage (1 μm is one millionth of a metre; the naked eye is just able to resolve 40 μm). Filters are used to prevent dirt entering the vulnerable parts of the system, and are generally specified in microns or meshes per linear inch (sieve number).

Inlet lines are usually fitted with strainers inside the tank, but these are coarse wire mesh elements only suitable for removing relatively large metal particles and similar contaminants. Separate filters are needed to remove finer particles and can be installed in three places as shown in Figures 5.6.19a–c.

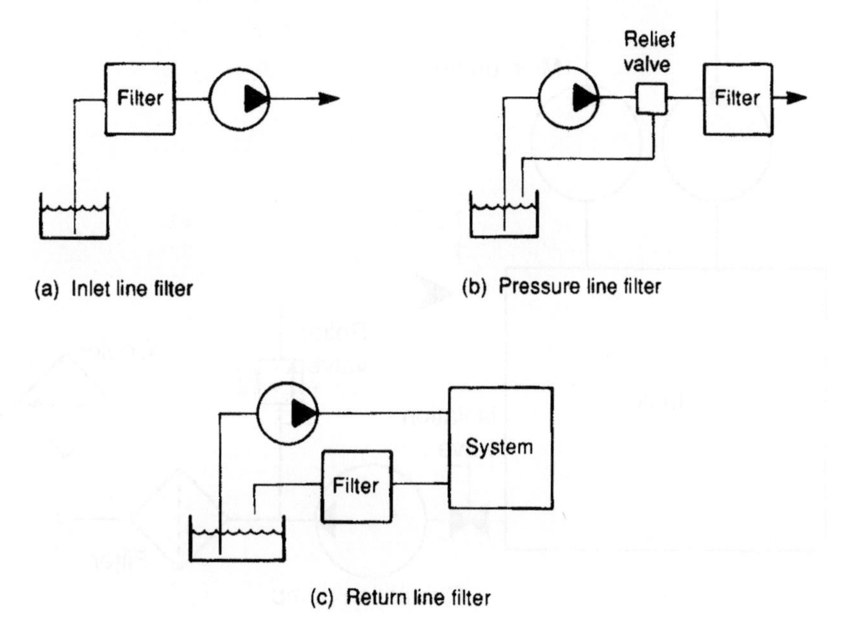

Fig. 5.6.19 Filter positions.

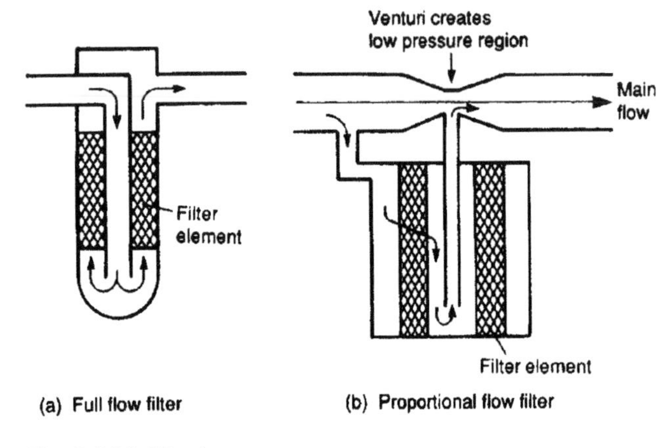

Fig. 5.6.20 Filter types.

Inlet line filters protect the pump, but must be designed to give a low-pressure drop or the pump will not be able to raise fluid from the tank. Low-pressure drop implies a coarse filter or a large physical size.

Pressure line filters placed after the pump protect valves and actuators and can be finer and smaller. They must, however, be able to withstand full system operating pressure. Most systems use pressure line filtering.

Return line filters may have a relatively high pressure drop and can, consequently, be very fine. They serve to protect pumps by limiting size of particles returned to the tank. These filters only have to withstand a low pressure. Filters can also be classified as full or proportional flow. In Figure 5.6.20a, all flow passes through the filter. This is obviously efficient in terms of filtration,

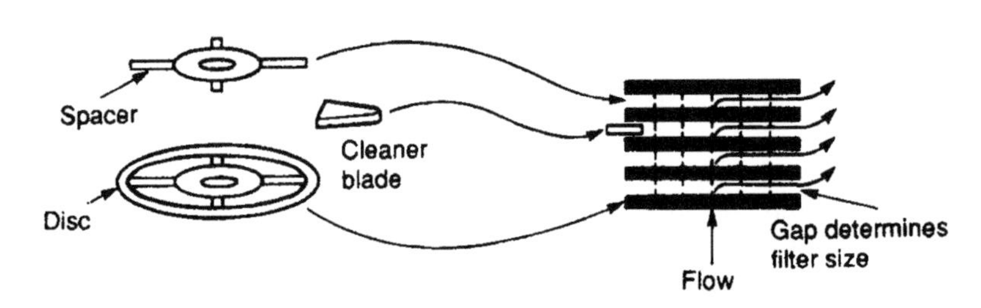

Fig. 5.6.21 Edge type filter.

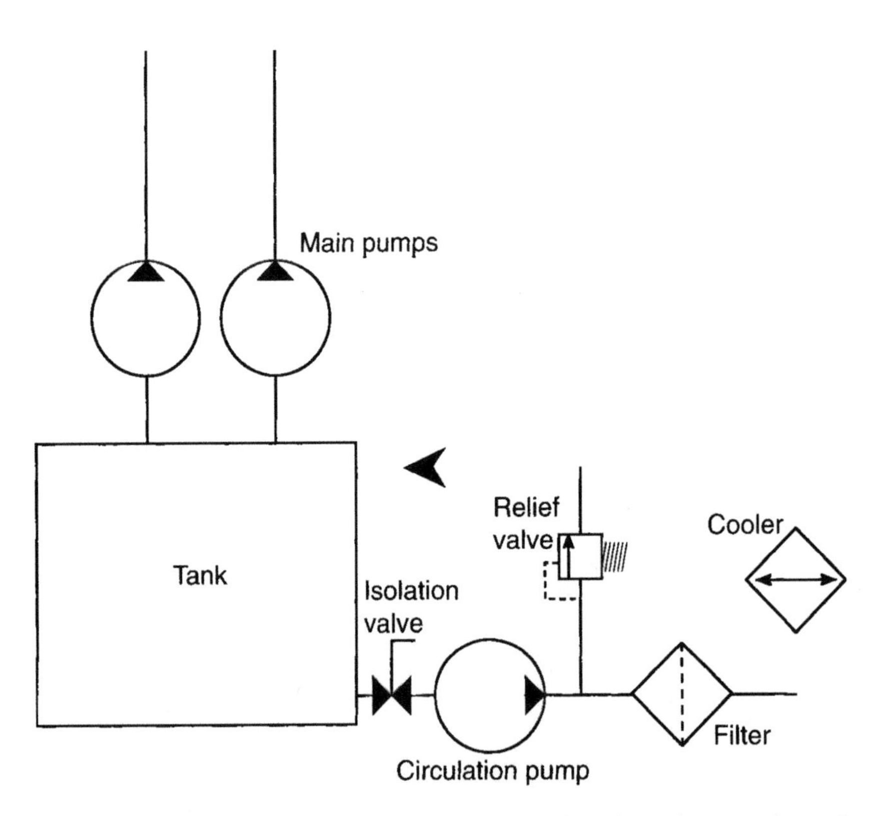

Fig. 5.6.22 A circulation pump used to filter and clean the fluid when the draw from the main pumps is small.

but incurs a large pressure drop. This pressure drop increases as the filter becomes polluted, so a full flow filter usually incorporates a relief valve which cracks when the filter becomes unacceptably blocked. This is purely a safety feature, though, and the filter should, of course, have been changed before this state was reached as dirty unfiltered fluid would be passing round the system.

In Figure 5.6.20b, the main flow passes through a venturi, creating a localised low pressure area. The pressure differential across the filter element draws a proportion of the fluid through the filter. This design is accordingly known as a proportional flow filter, as only a proportion of the main flow is filtered. It is characterized by a low pressure drop, and does not need the protection of a pressure relief valve.

Pressure drop across the filter element is an accurate indication of its cleanliness, and many filters incorporate a differential pressure meter calibrated with a green (clear), amber (warning), red (change overdue) indicator. Such types are called indicating filters.

Filtration material used in a filter may be mechanical or absorbent. Mechanical filters are relatively coarse, and utilise fine wire mesh or a disc/screen arrangement as shown in the edge-type filter of Figure 5.6.21. Absorbent filters are based on porous materials such as paper, cotton or cellulose. Filtration size in an absorbent filter can be very small as filtration is done by pores in the material. Mechanical filters can usually be removed, cleaned and re-fitted, whereas absorbent filters are usually replaceable items.

In many systems where the main use is the application of pressure the actual draw from the tank is very small reducing the effectiveness of pressure and return line filters. Here a separate circulating pump may be used as shown on Figure 5.6.22 to filter and cool the oil. The running of this pump is normally a pre-condition for starting the main pumps. The circulation pump should be sized to handle the complete tank volume every 10–15 min.

Note the pressure relief valve – this is included to provide a route back to tank if the filter or cooler is totally blocked. In a real-life system, additional hand isolation and non-return valves would be fitted to permit changing the filter or cooler with the system running. Limit switches and pressure switches would also be included to signal to the control system that the hand isolation valves are open and the filter is clean.

5.6a Air compressors, air treatment and pressure regulation

The vast majority of pneumatic systems use compressed atmospheric air as the operating medium (a small number of systems use nitrogen obtained commercially from liquid gas suppliers). Unlike hydraulic systems, a pneumatic system is 'open'; the fluid is obtained free, used and then vented back to atmosphere.

Pneumatic systems use a compressible gas; hydraulic systems an incompressible liquid, and this leads to some significant differences. The pressure of a liquid may be raised to a high level almost instantaneously, whereas pressure rise in a gas can be distinctly leisurely. In Figure 5.6a.1a, a reservoir of volume 2 m^3 is connected to a compressor which delivers 3 m^3 of air (measured at atmospheric pressure) per minute. Using Boyle's law, the pressure rise shown in Figure 5.6a.1b can be found.

Pressure in a hydraulic system can be quickly and easily controlled by devices such as unloading and pressure-regulating valves. Fluid is thus stored at atmospheric pressure and compressed to the required pressure as needed. The slow response of an air compressor, however, precludes such an approach in a pneumatic system and necessitates storage of compressed air at the required pressure in a receiver vessel. The volume of this vessel is chosen so there are minimal deviations in pressure arising from flow changes in loads and the compressor is then employed to replace the air used, averaged over an extended period of time (e.g. a few minutes).

Deviations in air pressure are smaller, and compressor control is easier if a large receiver feeds many loads. A large number of loads statistically results in a more even flow of air from the receiver, also helping to maintain a steady pressure. On many sites, therefore,

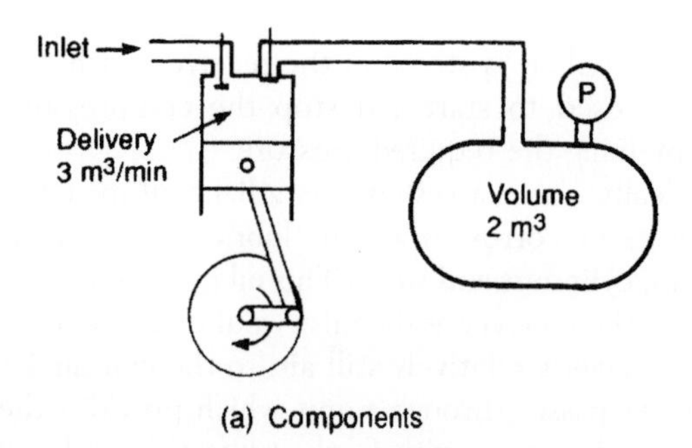

(a) Components

t (min)	Volume (at NTP)	P Abs	P gauge
0	2	1	0
1	5	2.5	1.5
2	8	4	3
3	11	5.5	4.5

(b) Response

Fig. 5.6a.1 Compressibility of a gas.

compressed air is produced as a central service which is distributed around the site in a similar manner to electricity, gas and water.

Behaviour of a gas subjected to changes in pressure, volume and temperature is governed by the general gas equation and reproduced here:

$$\frac{P_1V_1}{T_1} = \frac{P_2V_2}{T_2} \qquad (5.6a.1)$$

where pressures are given in absolute terms and temperatures are measured in degrees kelvin.

A compressor increases air pressure by reducing its volume, and expression (5.6a.1) predicts a resultant rise in temperature. A pneumatic system must therefore incorporate some method of removing this excess heat. For small systems, simple fins on the compressor (similar in construction to an air-cooled internal combustion engine) will suffice. For larger systems, a separate cooler (usually employing water as the heat-removing medium) is needed.

Atmospheric air contains water vapour, the actual amount varying from day to day according to humidity. The maximum amount of water vapour held in a given volume of air is determined by temperature, and any excess condenses out as liquid droplets (commonly experienced as condensation on cold windows). A similar effect occurs as compressed air is cooled, and if left the resultant water droplets would cause valves to jam and corrosion to form in pipes. An aftercooler must therefore be followed by a water separator. Often aftercoolers and separators are called, collectively, primary air treatment units.

Dry cool air is stored in the receiver, with a pressure switch used to start and stop the compressor motor, maintaining the required pressure.

Ideally, air in a system has a light oil mist to reduce chances of corrosion and to lubricate moving parts in valves, cylinders and so on. This oil mist cannot be added before the receiver as the mist would form oil droplets in the receiver's relatively still air, so the exit air from the receiver passes through a unit which provides the lubricating mist along with further filtration and water removal. This process is commonly called secondary air treatment.

Often, air in the receiver is held at a slightly higher pressure than needed to allow for pressure drops in the pipe lines. A local pressure regulation unit is then employed with the secondary air treatment close to the device using air. Composite devices called service units comprising water separation, lubricator and pressure regulation are available for direct line monitoring close to the valves and actuators of a pneumatic system.

Figure 5.6a.2 thus represents the components used in the production of a reliable source of compressed air.

5.6a.1 Compressor types

Like hydraulic pumps, air compressors can be split into positive displacement devices (where a fixed volume of air is delivered on each rotation of the compressor shaft) and dynamic devices such as centrifugal or axial blowers. The vast majority of air compressors are of the positive displacement type.

A compressor is selected by the pressure it is required to work at and the volume of gas it is required to deliver. As explained in the previous section, pressure in the receiver is generally higher than that required at the operating position, with local pressure regulation being used. Pressure at the compressor outlet (which for practical purposes will be the same as that in the receiver) is called the working pressure and is used to specify the compressor. Pressure at the operating point is called, not surprisingly, the operating pressure and is used to specify valves, actuators and other operating devices.

Care should be taken in specifying the volume of gas a compressor is required to deliver. Expression (5.6a.1) shows the volume of a given mass of gas to be highly dependent on pressure and temperature. Delivery volume of a compressor is defined in terms of gas at normal atmospheric conditions. Two standards known as *standard temperature and pressures* (STP) are commonly used, although differences between them are small for industrial users.

The *technical normal condition* is:

$$P = 0.98 \text{ bar absolute}, \quad T = 20\ ^\circ\text{C}$$

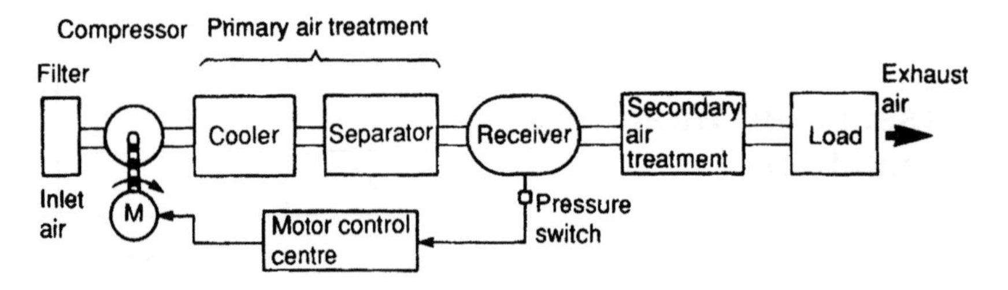

Fig. 5.6a.2 Component parts of a pneumatic system.

and the *physical normal condition* is:

$$P = 1.01 \text{ bar absolute}, \ T = 0\ ^\circ\text{C}$$

The term *normal temperature and pressure* (NTP) is also used.

Required delivery volume of a compressor (in m^3 min^{-1} or ft min^{-1}, according to the units used) may be calculated for the actuators at the various operating positions (with healthy safety margins to allow for leakage) but care must be taken to ensure this total volume is converted to STP condition before specifying the required compressor delivery volume.

A compressor delivery volume can be specified in terms of its theoretical volume (swept volume multiplied by rotational speed) or effective volume which includes losses. The ratio of these two volumes is the efficiency. Obviously the effective volume should be used in choosing a compressor (with, again, a safety margin for leakage). Required power of the motor driving the compressor is dependent on working pressure and delivery volume, and may be determined from expressions (5.6.2) and (5.6.5). Allowance must be made for the cyclic on/off operation of the compressor with the motor being sized for on-load operation and not averaged over a period of time.

5.6a.1.1 Piston compressors

Piston compressors are by far the most common type of compressor, and a basic single cylinder form is shown in Figure 5.6a.3. As the piston descends during the inlet stroke (Figure 5.6a.3a), the inlet valve opens and air is drawn into the cylinder. As the piston passes the bottom of the stroke, the inlet valve closes and the exhaust valve opens allowing air to be expelled as the piston rises (Figure 5.6a.3b).

Figure 5.6a.3 implies that the valves are similar to valves in an internal combustion engine. In practice, spring-loaded valves are used, which open and close under the action of air pressure across them. One common type uses a 'feather' of spring steel which moves above the inlet or output port, as shown in Figure 5.6a.3c.

A single cylinder compressor gives significant pressure pulses at the outlet port. This can be overcome to some extent by the use of a large receiver, but more often a multicylinder compressor is used. These are usually classified as vertical or horizontal in-line arrangements and the more compact V, Y or W constructions.

A compressor which produces one pulse of air per piston stroke (of which the example of Figure 5.6a.3 is typical) is called a single-acting compressor. A more even air supply can be obtained by the double-acting action of the compressor in Figure 5.6a.4, which uses two sets of valves and a crosshead to keep the piston rod square at all times. Double-acting compressors can be found in all configurations described earlier.

Piston compressors described so far go direct from atmospheric to required pressure in a single operation. This is known as a single-stage compressor. The general gas law showed compression of a gas to be accompanied by a significant rise in gas temperature. If the exit pressure is above about 5 bar in a single-acting compressor, the compressed air temperature can rise to over 200 °C and the motor power needed to drive the compressor rises accordingly.

For pressures over a few bar, it is far more economical to use a multistage compressor with cooling between stages. Figure 5.6a.5 shows an example. As cooling (undertaken by a device called an intercooler) reduces the volume of the gas to be compressed at the second stage, there is a large energy saving. Normally, two stages are used for pneumatic pressures of 10–15 bar, but multistage compressors are available for pressures up to around 50 bar.

Multistage compressors can be manufactured with multicylinders as shown in Figure 5.6a.5 or, more compactly, with a single cylinder and a double diameter piston as shown in Figure 5.6a.6.

There is contact between pistons and air, in standard piston compressors, which may introduce small amounts of lubrication oil from the piston walls into the air. This

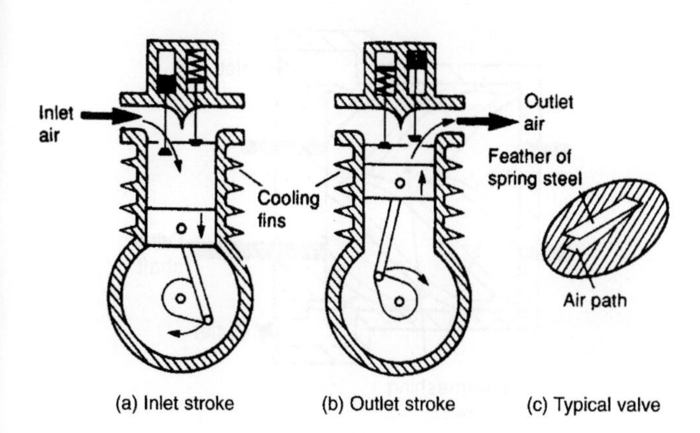

Fig. 5.6a.3 Single cylinder compressor.

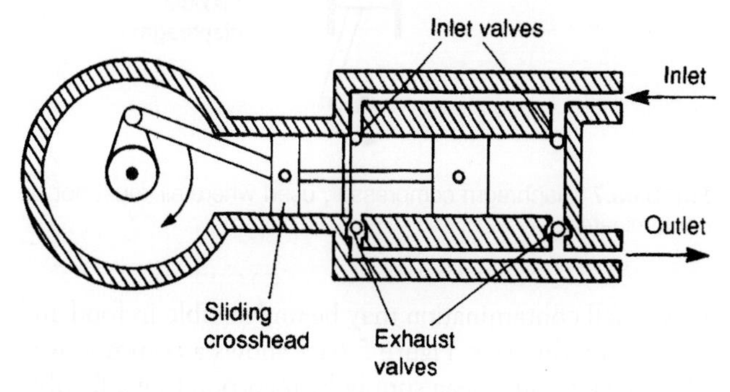

Fig. 5.6a.4 Double-acting compressor.

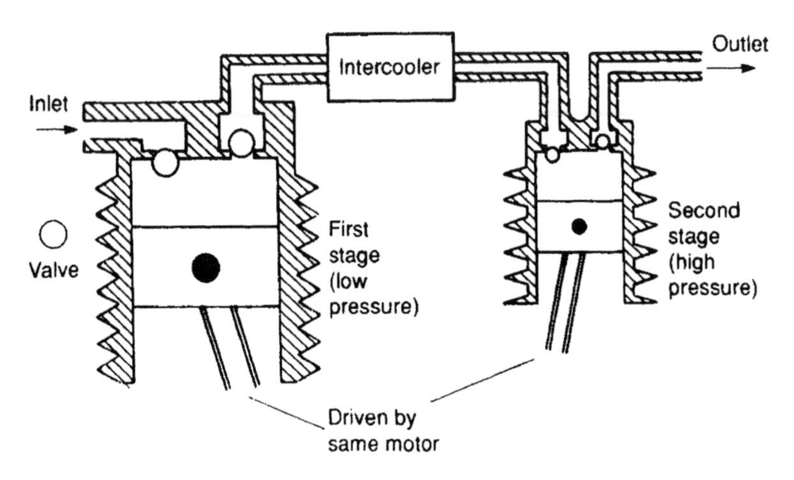

Fig. 5.6a.5 Two-stage compressor.

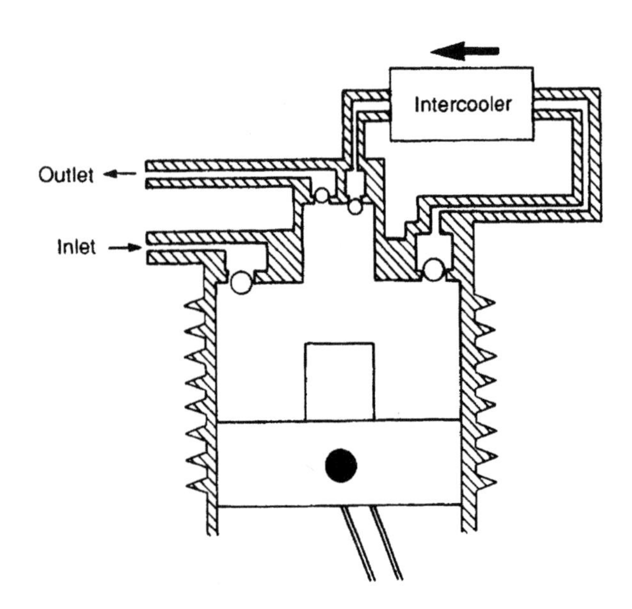

Fig. 5.6a.6 Combined two-stage compressor.

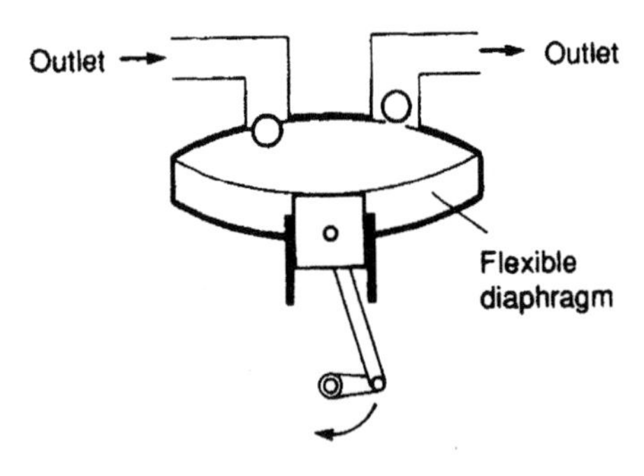

Fig. 5.6a.7 Diaphragm compressor, used where air must not be contaminated.

very small contamination may be undesirable in food and chemical industries. Figure 5.6a.7 shows a common way of giving a totally clean supply by incorporating a flexible diaphragm between piston and air.

5.6a.1.2 Screw compressors

Piston compressors are used where high pressures (>20 bar) and relatively low volumes (<10,000 $m^3\ h^{-1}$) are needed, but are mechanically relatively complex with many moving parts. Many applications require only medium pressure (<10 bar) and medium flows (around 10,000 $m^3 h^{-1}$). For these applications, rotary compressors have the advantage of simplicity, with fewer moving parts rotating at a constant speed, and a steady delivery of air without pressure pulses.

One rotary compressor, known as the dry rotary screw compressor, is shown in Figure 5.6a.8 and consists of two intermeshing rotating screws with minimal (around 0.05 mm) clearance. As the screws rotate, air is drawn into the housing, trapped between the screws and carried along to the discharge port, where it is delivered in a constant pulse-free stream.

Screws in this compressor can be synchronised by external timing gears. Alternatively, one screw can be driven, and the second screw rotated by contact with the drive screw. This approach requires oil lubrication to be sprayed into the inlet air to reduce friction between

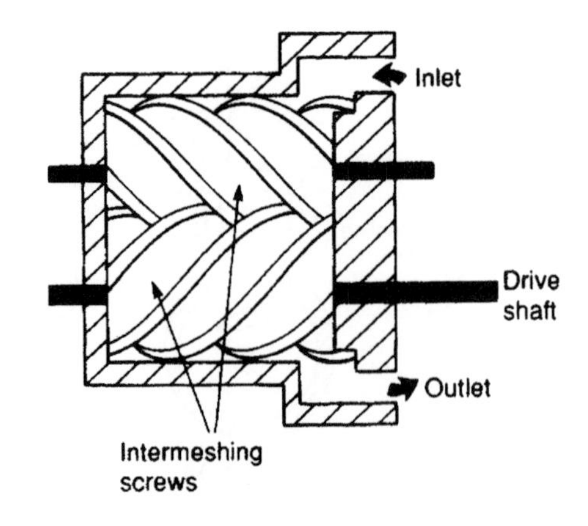

Fig. 5.6a.8 Dry screw rotary compressor.

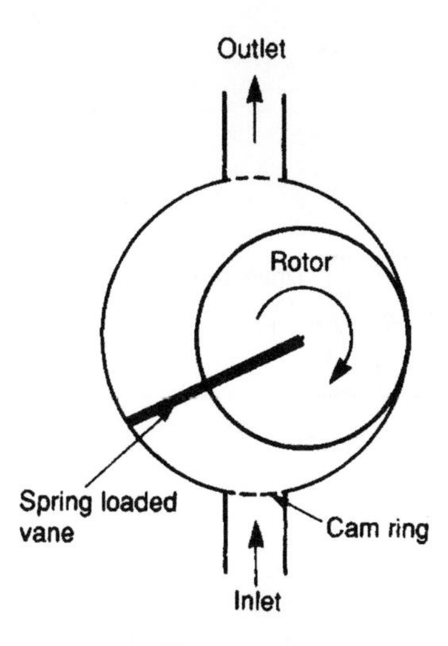

Fig. 5.6a.9 Vane compressor.

screws, and is consequently known as a wet rotary screw compressor. Wet screw construction though, obviously introduces oil contamination into the air which has to be removed by later oil separation units.

5.6a.1.3 Rotary compressors

The vane compressor, shown in Figure 5.6a.9, operates on similar principles to the hydraulic vane pump described in Chapter 5.6, although air compressors tend to be physically larger than hydraulic pumps. An unbalanced design is shown, balanced versions can also be constructed. Vanes can be forced out by springs or, more commonly, by centrifugal force.

A single-stage vane compressor can deliver air at up to 3 bar, a much lower pressure than that available with a screw or piston compressor. A two-stage vane compressor with large low-pressure and smaller high-pressure sections linked by an intercooler allows pressures up to 10 bar to be obtained.

Figure 5.6a.10 shows a variation on the vane compressor called a liquid ring compressor. The device uses many vanes rotating inside an eccentric housing and contains a liquid (usually water) which is flung out by

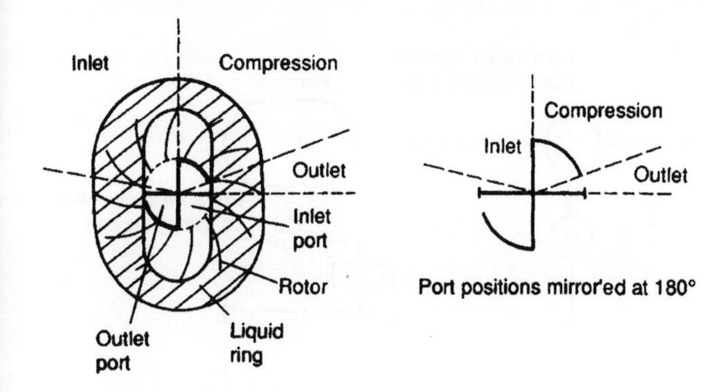

Fig. 5.6a.10 Liquid ring compressor.

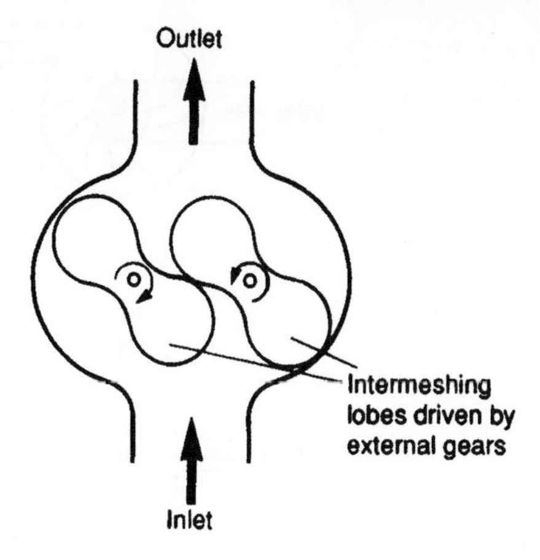

Fig. 5.6a.11 Lobe compressor.

centrifugal force to form a liquid ring which follows the contour of the housing to give a seal with no leakage and minimal friction. Rotational speed must be high (typically 3000 rpm) to create the ring. Delivery pressures are relatively low, at around 5 bar.

The lobe compressor of Figure 5.6a.11 (often called a Roots blower) is often used when a positive displacement compressor is needed with high delivery volume but low pressure (typically 1–2 bar). Operating pressure is mainly limited by leakage between rotors and housing. To operate efficiently, clearances must be very small, and wear leads to a rapid fall in efficiency.

5.6a.1.4 Dynamic compressors

A large volume of air (up to 5000 $m^3\ min^{-1}$) is often required for applications such as pneumatic conveying (where powder is carried in an air stream), ventilation or where air itself is one component of a process (e.g. combustion air for gas/oil burners). Pressure in these applications is low (at most a few bar) and there is no need for a positive displacement compressor.

Large volume low pressure air is generally provided by dynamic compressors known as blowers. They can be subdivided into centrifugal or axial types, shown in Figure 5.6a.12. Centrifugal blowers (Figure 5.6a.12a) draw air in, then fling it out by centrifugal force. A high shaft rotational speed is needed and the volume to input power ratio is lower than any other type of compressor.

An axial compressor comprises a set of rotating fan blades as shown in Figure 5.6a.12b. These produce very large volumes of air, but at low pressure (less than 1 bar). They are primarily used for ventilation, combustion and process air.

Output pressures of both types of dynamic compressor can be lifted by multistage compressors with intercoolers between stages. Diffuser sections reduce air entry velocity to subsequent stages, thereby converting air kinetic energy to pressure energy.

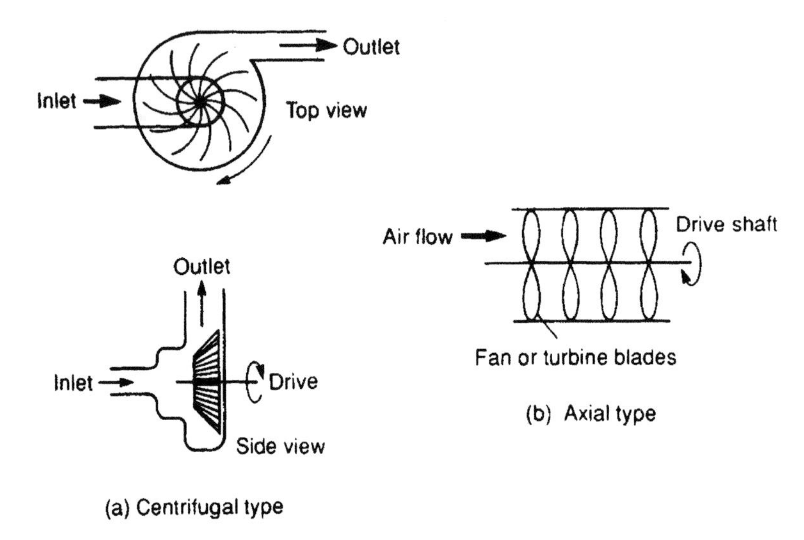

Fig. 5.6a.12 Non-positive displacement compressors (blowers).

Positive displacement compressors use oil to lubricate the close machined parts and to maintain the air seal. Dynamic compressors have no such need and, consequently, deliver very clean air.

5.6a.2 Air receivers and compressor control

An air receiver is used to store high-pressure air from the compressor. Its volume reduces pressure fluctuations arising from changes in load and from compressor switching.

Air coming from the compressor will be warm (if not actually hot!) and the large surface area of the receiver dissipates this heat to the surrounding atmosphere. Any moisture left in the air from the compressor will condense out in the receiver, so outgoing air should be taken from the receiver top.

Figure 5.6a.13 shows essential features of a receiver. They are usually of cylindrical construction for strength, and have a safety relief valve to guard against high pressures arising from failure of the pressure control scheme. Pressure indication and, usually, temperature indication are provided, with pressure switches for control of pressure and high-temperature switches for remote alarms.

A drain cock allows removal of condensed water, and access via a manhole allows cleaning. Obviously, removal of a manhole cover is hazardous with a pressurised receiver, and safety routines must be defined and followed to prevent accidents.

Control of the compressor is necessary to maintain pressure in the receiver. The simplest method of achieving this is to start the compressor when receiver pressure falls to some minimum pressure, and stop the compressor when pressure rises to a satisfactory level again, as illustrated in Figure 5.6a.14. In theory, two pressure switches are required (with the motor start pressure lower than the motor stop pressure) but, in practice, internal hysteresis in a typical switch allows one pressure switch to be used. The pressure in the receiver cycles between the start and stop pressure settings.

In Figure 5.6a.15, another method of pressure control is shown, where the compressor runs continuously and an exhaust valve is fitted to the compressor outlet. This

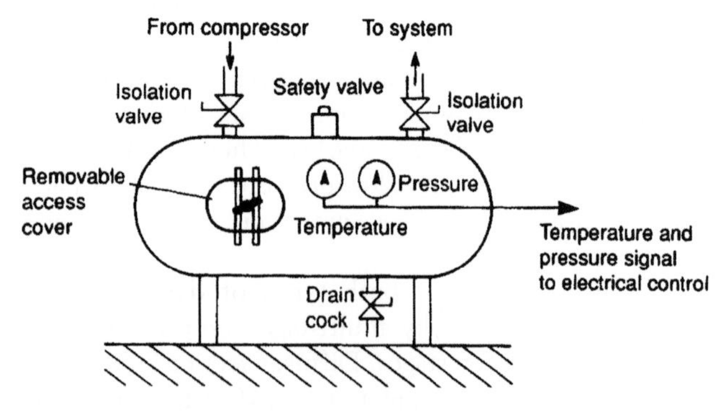

Fig. 5.6a.13 Compressed air receiver.

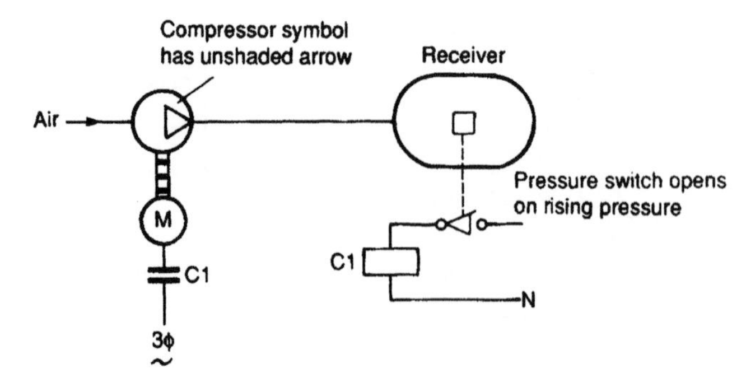

Fig. 5.6a.14 Receiver pressure control via motor start/stop.

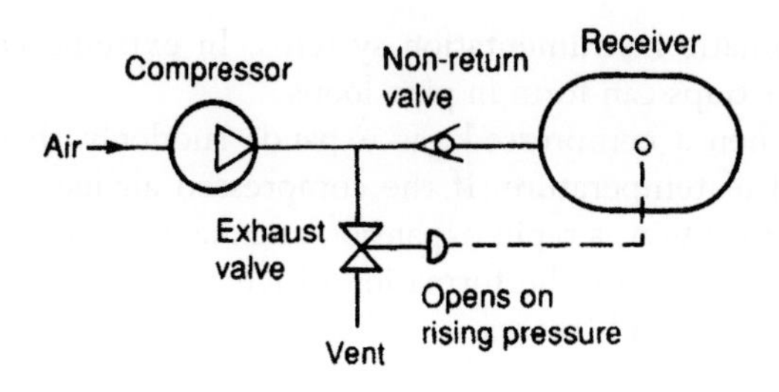

Fig. 5.6a.15 Receiver pressure control using compressor outlet valve.

valve opens when the required pressure is reached. A non-return valve prevents air returning from the receiver. This technique is known as exhaust regulation.

Compressors can also be controlled on the inlet side. In the example of Figure 5.6a.16, an inlet valve is held open to allow the compressor to operate, and is closed when the air receiver has reached the desired pressure, (the compressor then forms a near vacuum on its inlet side).

The valves in Figures 5.6a.15 and 5.6a.16 can be electrically-operated solenoid valves controlled by pressure switches, or can be pneumatic valves controlled directly by receiver pressure.

The control method is largely determined by flow rates from receiver to the load(s) and the capacity of the compressor. If the compressor has significant spare capacity, for example, start/stop control is commonly used.

If compressor capacity and load requirements are closely matched, on the other hand, short start/stop cycling may cause premature wear in the electrical starter for the compressor motor. In this situation, exhaust or inlet regulation is preferred.

Air receiver size is determined by load requirements, compressor capacity, and allowable pressure deviations in the receiver. With the compressor stopped, Boyle's law gives the pressure decay for a given volume of air delivered from a given receiver at a known pressure. For example, if a receiver of 10 m^3 volume and a working pressure of 8 bar delivers 25 m^3 of air (at STP) to a load, pressure in the receiver falls to approximately 5.5 bar.

With the compressor started, air pressure rises at a rate again (with the air mass in the receiver being *increased* by the difference between the air delivered by the compressor and that removed by the load).

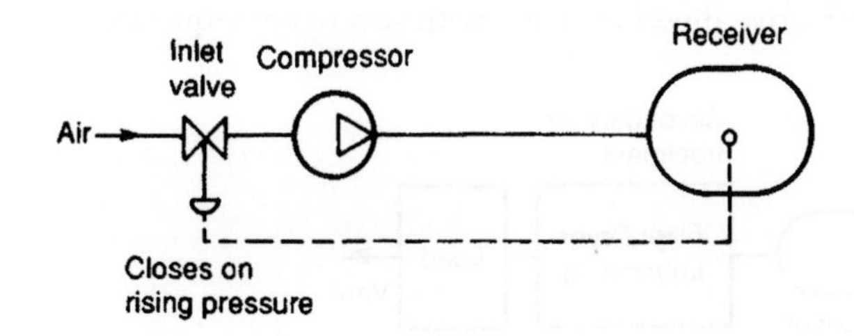

Fig. 5.6a.16 Receiver pressure control using compressor inlet valve.

These two calculations give the cycle time of the compressor when combined with settings of the cut-in and drop-out pressure switches. If this is unacceptably rapid, say less than a few minutes, then a larger receiver is required. Manufacturers of pneumatic equipment provide nomographs which simplify these calculations.

An air receiver is a pressure vessel and as such requires regular visual and volumetric pressure tests. Records should be kept of the tests.

5.6a.3 Air treatment

Atmospheric air contains moisture in the form of water vapour. We perceive the amount of moisture in a given volume of air as the humidity and refer to days with a high amount of water vapour as 'humid' or 'sticky', and days with low amounts of water vapour as 'good drying days'. The amount of water vapour which can be held in a given volume depends on temperature but does *not* depend on pressure of air in that volume. One cubic metre at 20 °C, for example, can hold 17 g of water vapour. The amount of water vapour which can be held in a given volume of air rises with temperature as shown in Figure 5.6a.17.

If a given volume of air contains the maximum quantity of water vapour possible at the air temperature, the air is said to be *saturated* (and we would perceive it as sticky because sweat could not evaporate from the surface of the skin). From Figure 5.6a.17, air containing 50 g of water vapour per cubic metre at 40 °C is saturated.

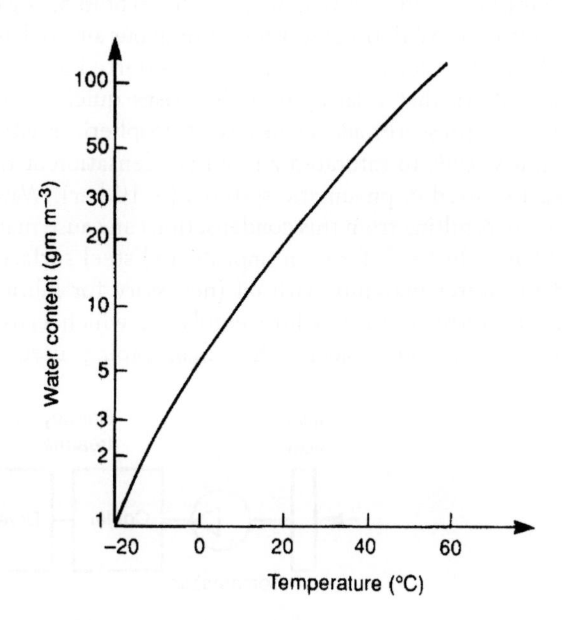

Fig. 5.6a.17 Moisture content curve.

Moisture content of unsaturated air is referred to by relative humidity, which is defined as

$$\text{Relative humidity} = \frac{\text{water content per cubic metre}}{\text{maximum water content per cubic metre}} \times 100\% \qquad (5.6a.3.1)$$

Air containing 5 g of water vapour per cubic metre of air at 20 °C has, from Figure 5.6a.17, a relative humidity of 30%.

Relative humidity is dependent on both temperature and pressure of the air. Suppose air at 30 °C contains 20 g of water vapour. From Figure 5.6a.17, this corresponds to 67% humidity. If the air is allowed to cool to 20 °C, it can only hold 17 g of water vapour and is now saturated (100% relative humidity). The excess 3 g condenses out as liquid water. If the air is cooled further to 10 °C, a further 8 g condenses out.

The temperature at which air becomes saturated is referred to as the 'dew point'. Air with 17.3 g of water vapour per cubic metre has, for example, a dew point of 20 °C.

To see the effect of pressure on relative humidity, we must remember the amount of water vapour which can be held in a given volume is fixed (assuming a constant temperature). Suppose a cubic metre of air at atmospheric pressure (0 bar gauge or 1 bar absolute) at 20 °C contains 6 g of water vapour (corresponding to 34% relative humidity). If we wish to increase air pressure while maintaining its temperature at 20 °C, we must compress it. When the pressure is 1 bar gauge (or 2 bar absolute) its volume is 0.5 m^3, which can hold 8.6 g of water vapour, giving us 68% relative humidity. At 2 bar gauge (3 bar absolute) the volume is 0.33 m^3, which can hold 5.77 g of water vapour. With 6 g of water vapour in our air, we have reached saturation and condensation has started to occur.

It follows that relative humidity rises quickly with increasing pressure, and even low atmospheric relative humidity leads to saturated air and condensation at the pressures used in pneumatic systems (8–10 bar). Water droplets resulting from this condensation can cause many problems. Rust will form on unprotected steel surfaces, and the water may mix with oil (necessary for lubrication) to form a sticky white emulsion, which causes valves to jam and blocks the small piping used in pneumatic instrumentation systems. In extreme cases, water traps can form in pipe loops.

When a compressed gas expands suddenly there is a fall of temperature. If the compressed air has a high water content, a rapid expansion at exhaust ports can be accompanied by the formation of ice as the water condenses out and freezes.

5.6a.4 Stages of air treatment

Air in a pneumatic system must be clean and dry to reduce wear and extend maintenance periods. Atmospheric air contains many harmful impurities (smoke, dust, water vapour) and needs treatment before it can be used.

In general, this treatment falls into three distinct stages, shown in Figure 5.6a.18. First, inlet filtering removes particles which can damage the air compressor. Next, there is the need to dry the air to reduce humidity and lower the dew point. This is normally performed between the compressor and the receiver and is termed primary air treatment.

The final treatment is done local to the duties to be performed, and consists of further steps to remove moisture and dirt and the introduction of a fine oil mist to aid lubrication. Not surprisingly, this is generally termed secondary air treatment.

5.6a.5 Filters

Inlet filters are used to remove dirt and smoke particles before they can cause damage to the air compressor, and are classified as dry filters with replaceable cartridges (similar to those found in motor car air filters) or wet filters where the incoming air is bubbled through an oil bath, then passed through a wire mesh filter. Dirt particles became attached to oil droplets during the bubbling process and are consequently removed by the wire mesh.

Both types of filters require regular servicing: replacement of the cartridge element for the dry type; cleaning for the wet type. If a filter is to be cleaned, it is essential the correct detergent is used. Use of petrol or similar petrochemicals can turn an air compressor into an effective diesel engine – with severe consequences.

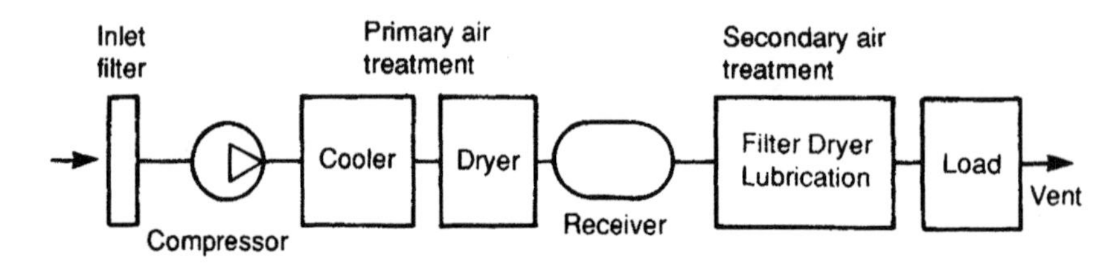

Fig. 5.6a.18 Three stages of air treatment.

Filters are classified according to the size of particles they will stop. Particle size is measured in SI units of micrometres (the older metric term *microns is* still common), one micrometre (1 μm) being 10^{-6} m or 0.001 mm. Dust particles are generally larger than 10 μm, whereas smoke and oil particles are around 1 μm. A filter can have a nominal rating (where it will block 98% of particles of the specified size) or an absolute rating (where it blocks 100% of particles of the specified size).

Microfilters with removable cartridges passing air from the centre to the outside of the cartridge case will remove 99.9% of particles down to 0.01 μm, the limit of normal filtration. Coarse filters, constructed out of wire mesh and called strainers, are often used as inlet filters. These are usually specified in terms of the mesh size which approximates to particle size in micrometres as follows:

Mesh size	μm
325	30
550	10
750	6

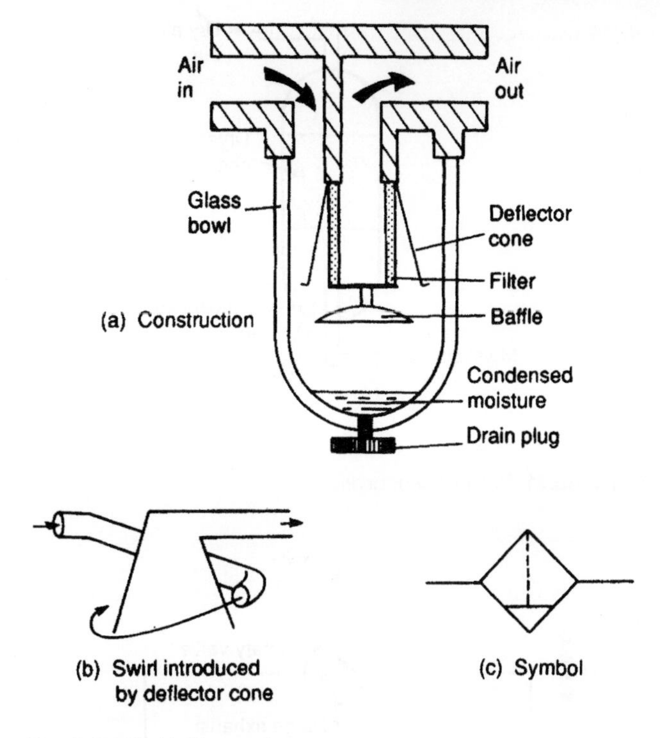

Fig. 5.6a.19 Air filter and water trap.

5.6a.6 Air dryers

An earlier section described how air humidity and dew point are raised by compression. Before air can be used, this excess moisture has to be removed to bring air humidity and dew point to reasonable levels.

In bulk air systems, all that may be required is a simple aftercooler similar to the intercoolers described earlier, followed by a separator unit where the condensed water collects and can be drained off.

Figure 5.6a.19a shows a typical water trap and separator. Air flow through the unit undergoes a sudden reversal of direction and a deflector cone swirls the air (Figure 5.6a.19b). Both of these cause heavier water particles to be flung out to the walls of the separator and to collect in the trap bottom from where they can be drained. Water traps are usually represented on circuit diagrams by the symbol of Figure 5.6a.19c.

Dew point can be lowered further with a refrigerated dryer, the layout of which is illustrated in Figure 5.6a.20. This chills the air to just above 0 °C, condensing almost all the water out and collecting the condensate in the separator. Efficiency of the unit is improved with a second heat exchanger in which cold dry air leaving the dryer pre-chills incoming air. Air leaving the dryer has a dew point similar to the temperature in the main heat exchanger.

Refrigerated dryers provide air with a dew point sufficiently low for most processes. Where absolutely dry air is needed, chemical dryers must be employed. Moisture can be removed chemically from air by two processes.

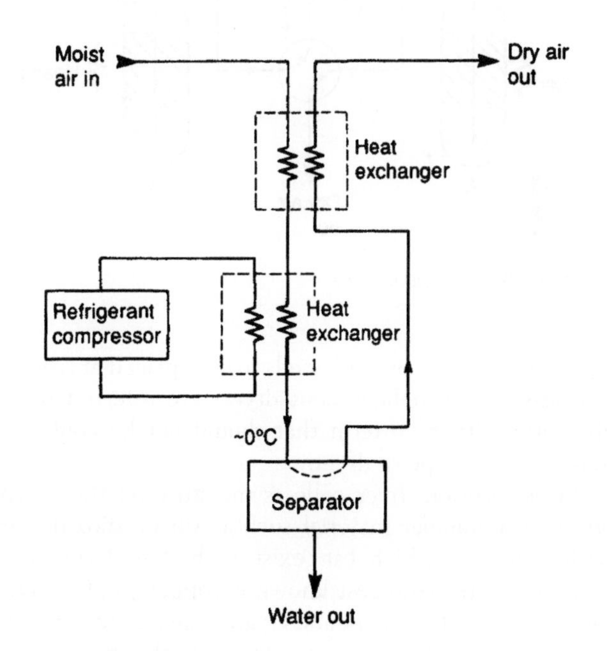

Fig. 5.6a.20 Refrigerated dryer.

In a deliquescent dryer, the layout of which is shown in Figure 5.6a.21, a chemical agent called a desiccant is used. This absorbs water vapour and slowly dissolves to form a liquid which collects at the bottom of the unit where it can be drained. The dessicant material is used

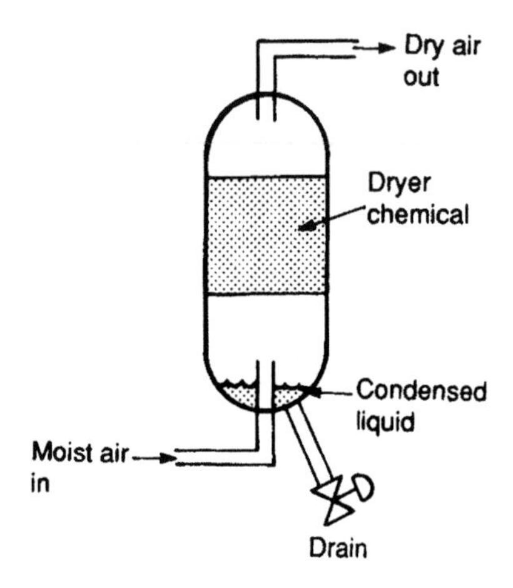

Fig. 5.6a.21 Deliquescent dryer.

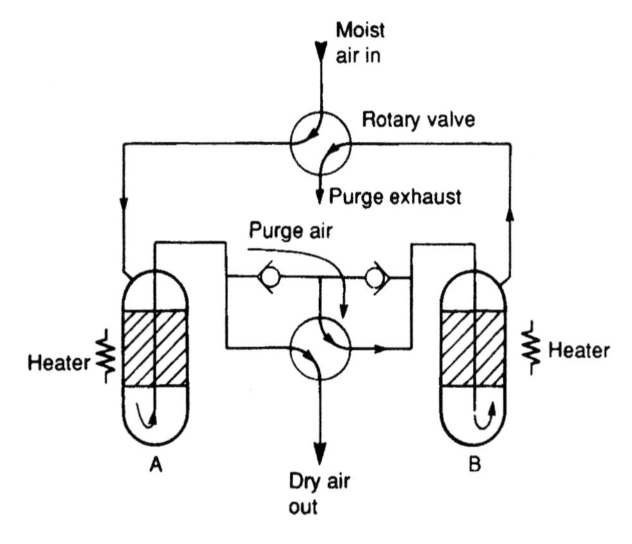

Fig. 5.6a.22 Adsorption dryer.

up during this process and needs to be replaced at regular intervals. Often deliquescent dryers are referred to as absorbtion dryers, a term that should not be confused with the next type of dryer.

An adsorption dryer collects moisture on the sharp edges of a granular material such as silicon dioxide, or with materials which can exist in hydrated *and* dehydrated states (the best known is copper sulphate but more efficient compounds are generally used). Figure 5.6a.22 shows the construction of a typical adsorption dryer. Moisture in the adsorption material can be released by heating, so two columns are used. At any time, one column is drying the air while the other is being regenerated by heating and the passage of a low purge air stream. As shown, column A dries the air and column B is being regenerated. The rotary valves are operated automatically at regular intervals by a time clock. For obvious reasons, adsorption dryers are often referred to as regenerative dryers.

5.6a.7 Lubricators

A carefully controlled amount of oil is often added to air immediately prior to use to lubricate moving parts (process-control pneumatics are the exception as they usually require dry unlubricated air). This oil is introduced as a fine mist, but can only be added to thoroughly clean and dry air or a troublesome sticky emulsion forms. It is also difficult to keep the oil mist-laden air in a predictable state in an air receiver, so oil addition is generally performed as part of the secondary air treatment.

The construction of a typical lubricator is shown with its symbol in Figure 5.6a.23. The operation is similar to the principle of the petrol air mixing in a motor car carburettor. As air enters the lubricator, its velocity is increased by a venturi ring causing a local reduction in pressure in the upper chamber. The pressure differential

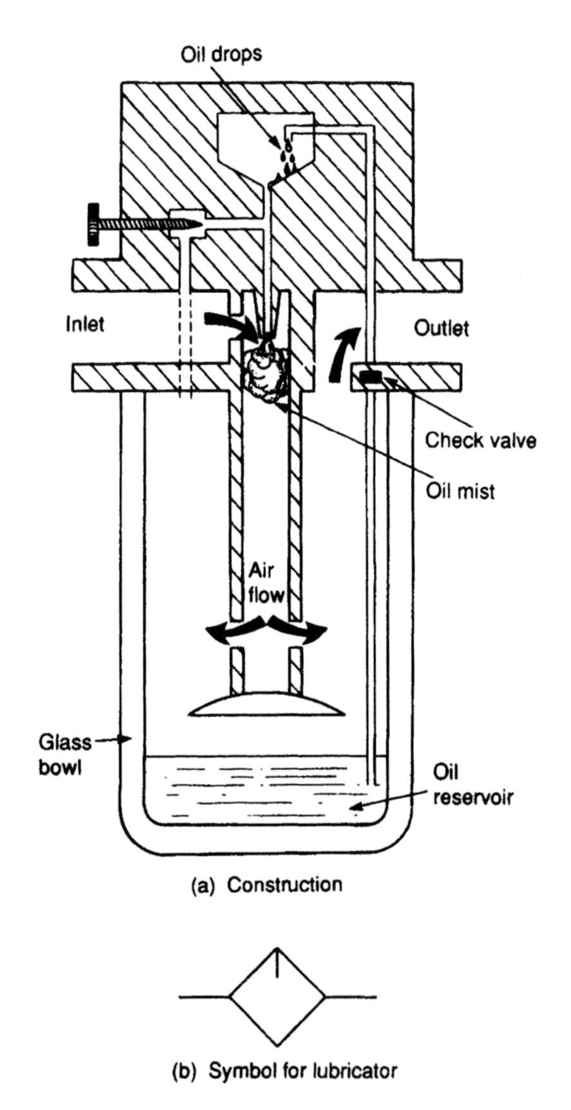

Fig. 5.6a.23 Lubricator.

between lower and upper chambers causes oil to be drawn up a riser tube, emerging as a spray to mix with the air. The needle valve adjusts the pressure differential across the oil jet and hence the oil flow rate.

The air–oil mixture is forced to swirl as it leaves the central cylinder, causing excessively large oil particles to be flung out of the air stream.

5.6a.8 Pressure regulation

Flow velocities in pneumatic systems can be quite high, which can lead to significant flow-dependent pressure drops between the air receiver and the load.

Generally, therefore, air pressure in the receiver is set higher than the required load pressure and pressure regulation is performed local to loads to keep pressure constant regardless of flow. Control of air pressure in the receiver was described in an earlier section. This section describes various ways in which pressure is locally controlled.

There are essentially three methods of local pressure control, illustrated in Figure 5.6a.24. Load A vents continuously to atmosphere. Air pressure is controlled by a pressure regulator which simply restricts air flow to the load. This type of regulator requires some minimum flow to operate. If used with a dead-end load which draws no air, the air pressure will rise to the main manifold pressure. Such regulators, in which air must pass through the load, are called non-relieving regulators.

Load B is a dead-end load, and uses a pressure regulator which vents air to atmosphere to reduce pressure. This type of regulator is called a three-port (for the three connections) or relieving regulator. Finally, load C is a large capacity load whose air–volume requirements are beyond the capacity of a simple in-line regulator. Here, a pressure control loop has been constructed comprising pressure transducer, electronic controller and separate vent valve. This technique can also be used if the pressure regulating valve cannot be mounted locally to the point at which the pressure is to be controlled.

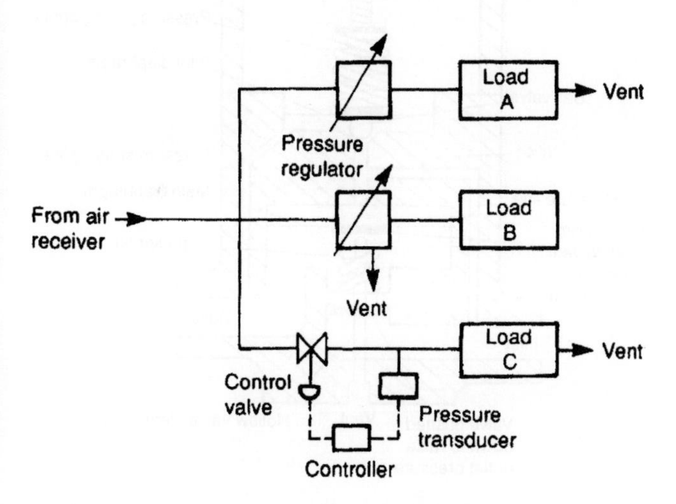

Fig. 5.6a.24 Three types of pressure regulator.

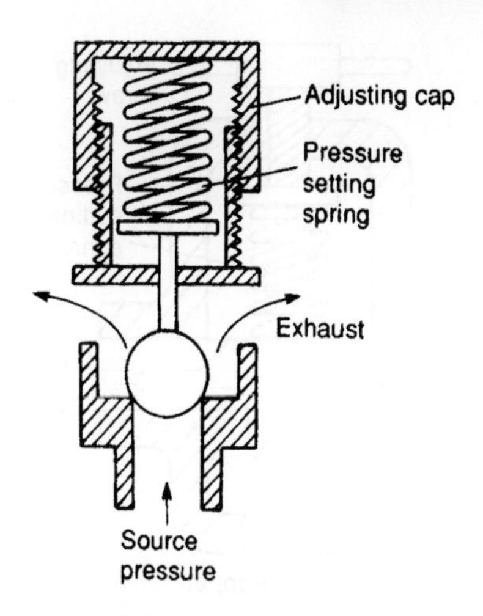

Fig. 5.6a.25 Relief valve.

5.6a.8.1 Relief valves

The simplest pressure regulating device is the relief valve shown in Figure 5.6a.25. This is not, in fact, normally used to control pressure but is employed as a backup device should the main pressure-control device fail. They are commonly fitted, for example, to air receivers.

A ball valve is held closed by spring tension, adjustable to set the relief pressure. When the force due to air pressure exceeds the spring tension, the valve cracks open releasing air and reducing the pressure. Once cracked, flow rate is a function of excess pressure; an increase in pressure leading to an increase in flow. A relief valve is specified by operating pressure range, span of pressure between cracking and full flow, and full flow rate. Care is needed in specifying a relief valve because, in a fault condition, the valve may need to pass the entire compressor output.

A relief valve has a flow/pressure relationship and self-seals itself once the pressure falls below the cracking pressure. A pure safety valve operates differently. Once a safety valve cracks, it opens fully to discharge all the pressure in the line or receiver, and it does not automatically reclose, needing manual resetting before the system can be used again.

5.6a.8.2 Non-relieving pressure regulators

Figure 5.6a.26 shows construction of a typical non-relieving pressure regulator. Outlet pressure is sensed by a diaphragm which is preloaded by a pressure setting spring. If outlet pressure is too low, the spring forces the

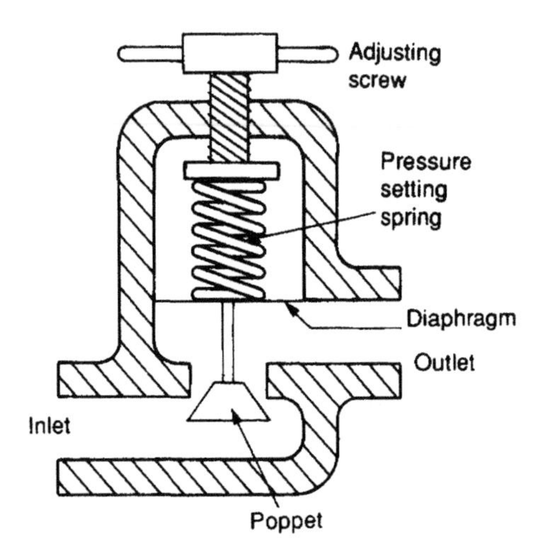

Fig. 5.6a.26 Non-relieving pressure regulator.

diaphragm and poppet down, opening the valve to admit more air and raise outlet pressure.

If the outlet pressure is too high, air pressure forces the diaphragm up, reducing air flow and causing a reduction in air pressure as air vents away through the load. In a steady state, the valve will balance with the force on the diaphragm from the outlet pressure just balancing the preset force on the spring.

5.6a.8.3 Relieving pressure regulators

A relieving regulator is shown in Figure 5.6a.27. Outlet pressure is sensed by a diaphragm preloaded with an adjustable pressure setting spring. The diaphragm rises if the outlet pressure is too high, and falls if the pressure is too low.

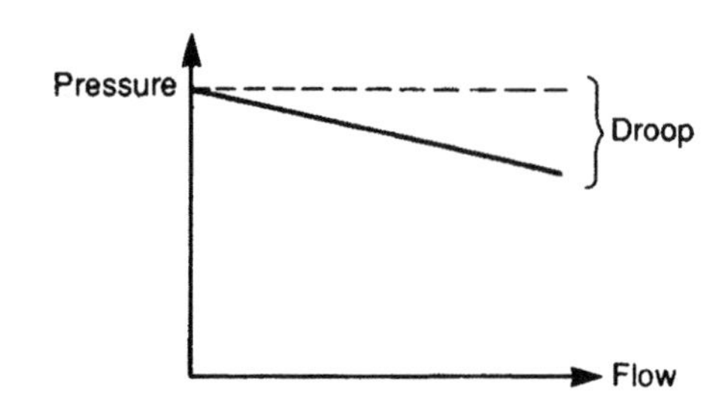

Fig. 5.6a.28 Response of simple pressure regulators.

If outlet pressure falls, the inlet poppet valve is pushed open admitting more air to raise the pressure. If the outlet pressure rises, the diaphragm moves down closing the inlet valve and opening the central vent valve to allow excess air to escape from the load, thereby reducing pressure.

In a steady state the valve will balance, *dithering* between admitting and venting small amounts of air to keep load pressure at the set value.

Both the regulators in Figures 5.6a.26 and 5.6a.27 are simple pressure regulators and have responses similar to that shown in Figure 5.6a.28, with outlet pressure decreasing slightly with flow. This droop in pressure can be overcome by using a pilot-operated regulator, shown in Figure 5.6a.29.

Outlet pressure is sensed by the pilot diaphragm, which compares outlet pressure with the value set by the pressure setting spring. If outlet pressure is low the

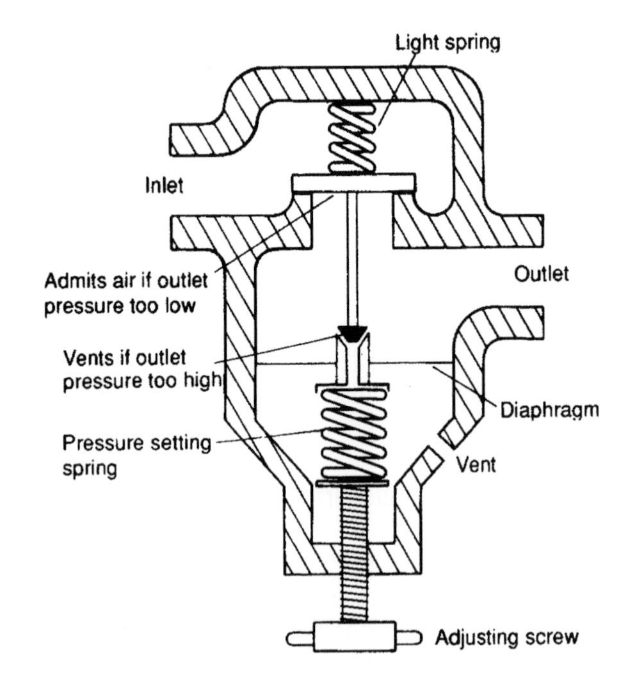

Fig. 5.6a.27 Relieving pressure regulator.

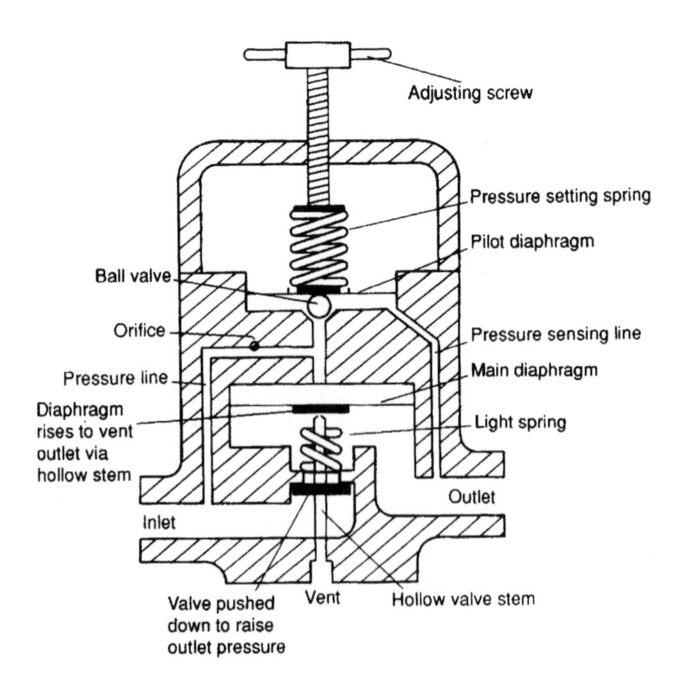

Fig. 5.6a.29 Pilot-operated regulator.

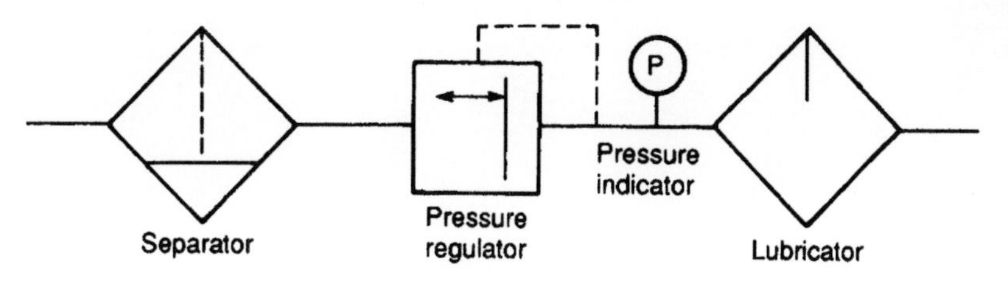

(a) Symbols for individual components

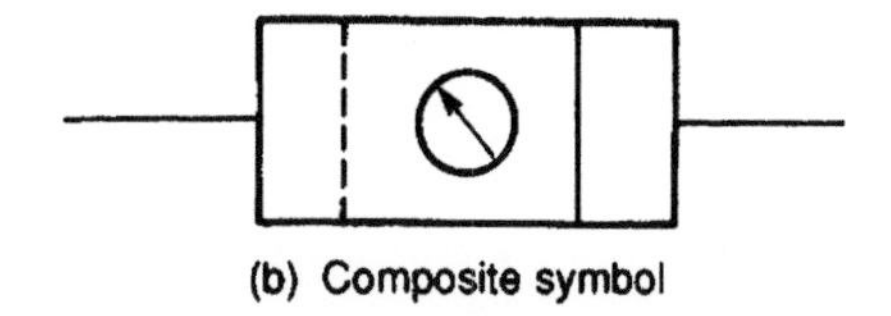

(b) Composite symbol

Fig. 5.6a.30 The service unit.

diaphragm descends, while if outlet pressure is high the diaphragm rises.

Inlet air is bled through a restriction and applied to the top of the main diaphragm. This space can, however, be vented to the exit side of the valve by the small ball valve.

If outlet pressure is low, the pilot diaphragm closes the ball valve causing the main diaphragm to be pushed down and more air to be admitted to the load.

If outlet pressure is high, the pilot diaphragm opens the ball valve and the space above the main diaphragm de-pressurises. This causes the main diaphragm to rise, opening the central vent allowing air to escape from the load and pressure to be reduced.

Action of pilot diaphragm and inlet air bleed approximates to integral action, giving a form of P + I (proportional plus integral) control. In the steady state, the outlet pressure equals the set pressure and there is no pressure droop with increasing flow.

5.6a.9 Service units

In pneumatic systems a moisture separator, a pressure regulator, a pressure indicator, a lubricator and a filter are all frequently required, local to a load or system. This need is so common that combined devices called *service units* are available. Individual components comprising a service unit are shown in Figure 5.6a.30a, while the composite symbol of a service unit is shown in Figure 5.6a.30b.

Chapter 5.7

5.7

Actuators

Parr

A hydraulic or pneumatic system is generally concerned with moving, gripping or applying force to an object. Devices which actually achieve this objective are called actuators, and can be split into three basic types.

Linear actuators, as the name implies, are used to move an object or apply a force in a straight line. Rotary actuators are the hydraulic and pneumatic equivalent of an electric motor. This chapter discusses linear and rotary actuators.

The third type of actuator is used to operate flow control valves for process control of gases, liquids or steam. These actuators are generally pneumatically operated.

5.7.1 Linear actuators

The basic linear actuator is the cylinder, or ram, shown in schematic form in Figure 5.7.1. Practical constructional details are discussed later. The cylinder in Figure 5.7.1 consists of a piston, radius R, moving in a bore. The piston is connected to a rod of radius r which drives the load. Obviously, if pressure is applied to port X (with port Y venting) the piston extends. Similarly, if pressure is applied to port Y (with port Z venting), the piston retracts.

The force applied by a piston depends on both the area and the applied pressure. For the extend stroke, area A is given by πR^2. For a pressure P applied to port X, the extend force available is

$$F_c = P\pi R^2 \tag{5.7.1}$$

The units of expression (5.7.1) depend on the system being used. If SI units are used, the force is in newtons.

Expression (5.7.1) gives the maximum achievable force obtained with the cylinder in a stalled condition. One example of this occurs where an object is to be gripped or shaped.

In Figure 5.7.2, an object of mass M is lifted at constant speed. Because the object is not accelerating, the upward force is equal to Mg newtons (in SI units) which from expression (5.7.1) gives the pressure, in the cylinder. This is lower than the maximum system pressure, the pressure drop occurring across flow control valves and system piping. Dynamics of systems similar to this are discussed later.

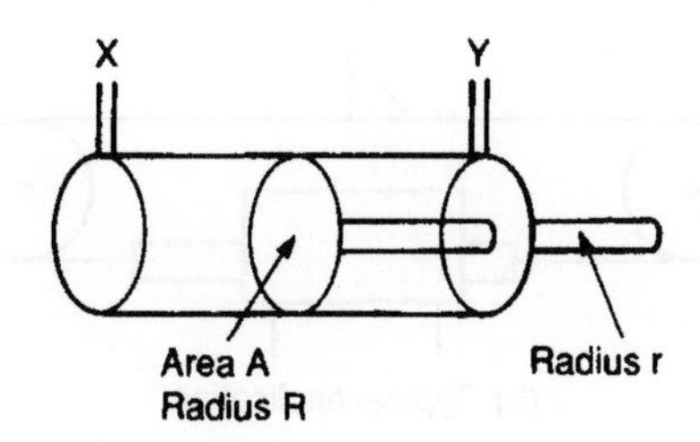

Fig. 5.7.1 A simple cylinder.

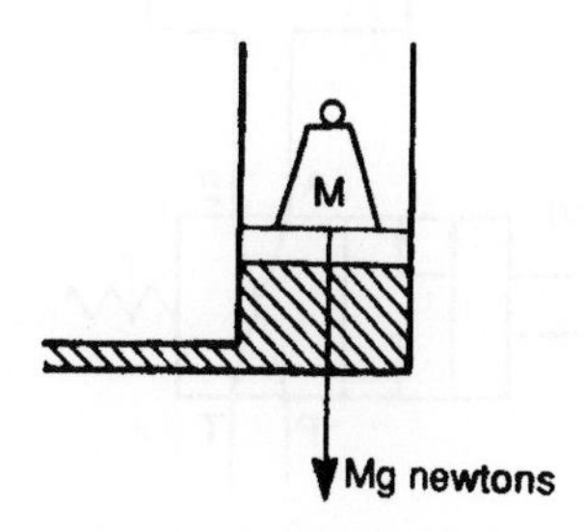

Fig. 5.7.2 A mass supported by a cylinder.

Hydraulics and Pneumatics; ISBN: 9780750644198

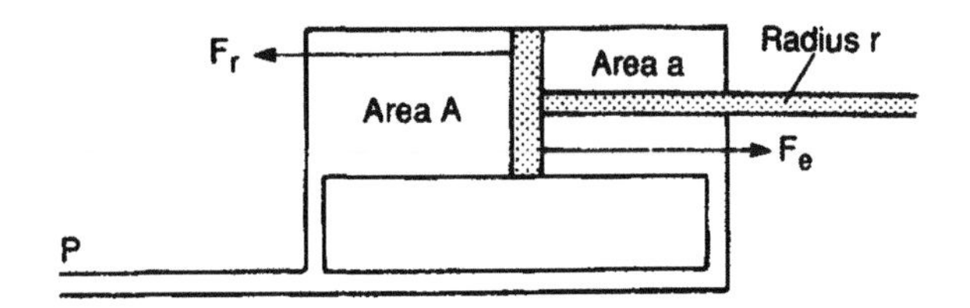

Fig. 5.7.3 Pressure applied to both sides of piston.

When pressure is applied to port Y, the piston retracts. Total piston area here is reduced because of the rod, giving an annulus of area A_a where

$$A_a = A - \pi r^2$$

and r is the radius of the rod. The maximum retract force is thus

$$F_r = PA_a = P(A - \pi r^2) \qquad (5.7.2)$$

This is lower than the maximum extend force. In Figure 5.7.3 identical pressure is applied to both sides of a piston. This produces an extend force F_c given by expression (5.7.1), and a retract force F_r given by expression (5.7.2). Because F_c is greater than F_r, the cylinder extends.

Normally the ratio A/A_a is about 6/5. In the cylinder shown in Figure 5.7.4, the ratio A/A_a of 2:1 is given by a large diameter rod. This can be used to give an equal extend and retract force when connected as shown.

Cylinders shown so far are known as double-acting, because fluid pressure is used to extend and retract the piston. In some applications, a high extend force is required (to clamp or form an object) but the retract force is minimal. In these cases, a single-acting cylinder (Figure 5.7.5) can be used, which is extended by fluid but retracted by a spring. If a cylinder is used to lift a load, the load itself can retract the piston.

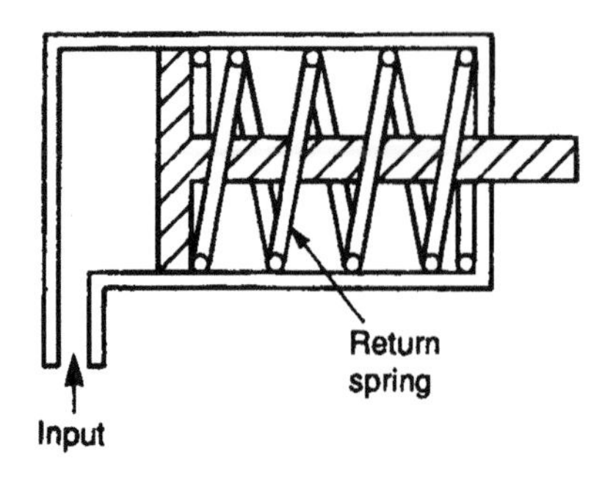

Fig. 5.7.5 Single-acting cylinder.

Single-acting cylinders are simple to drive (particularly for pneumatic cylinders with quick exhaust valves) but the extend force is reduced and, for spring-return cylinders, the figure length of the cylinder is increased for a given stroke to accommodate the spring.

A double-rod cylinder is shown in Figure 5.7.6a. This has equal fluid areas on both sides of the piston, and hence can give equal forces in both directions. If connected as shown in Figure 5.7.3, the piston does not move (but it can be shifted by an outside force). Double-rod cylinders are commonly used in applications similar to Figure 5.7.6b where a dog is moved by a double-rod cylinder acting via a chain.

The speed of a cylinder is determined by volume of fluid delivered to it. In the cylinder in Figure 5.7.7, the piston, of area A, has moved a distance d. This has required a volume V of fluid where

$$V = Ad \qquad (5.7.3)$$

If the piston moves at speed v, it moves distance d in time t where

$$t = d/v$$

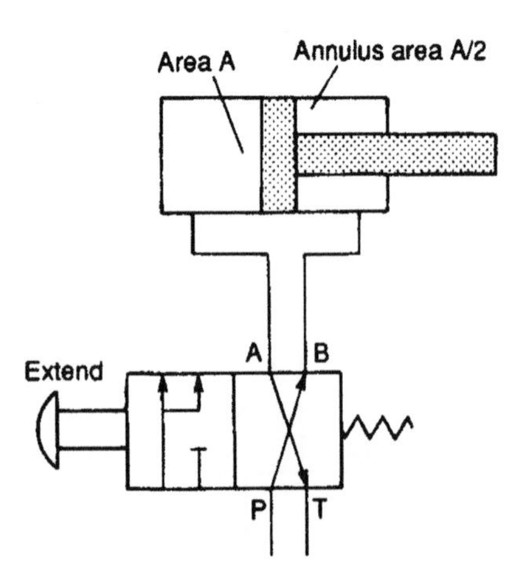

Fig. 5.7.4 Cylinder with equal extend/retract force.

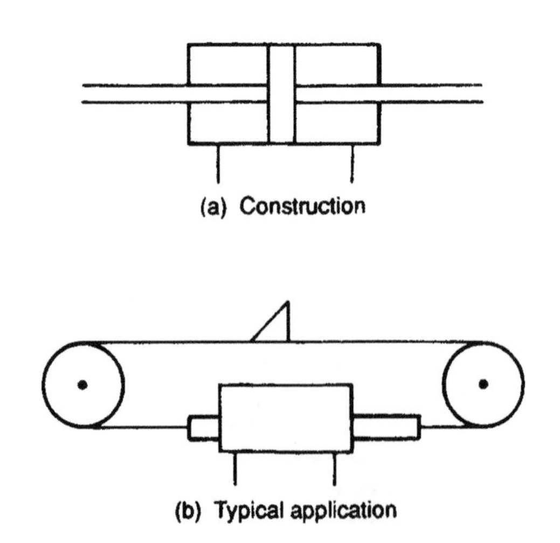

Fig. 5.7.6 Double-rod cylinder (with equal extend/retract force).

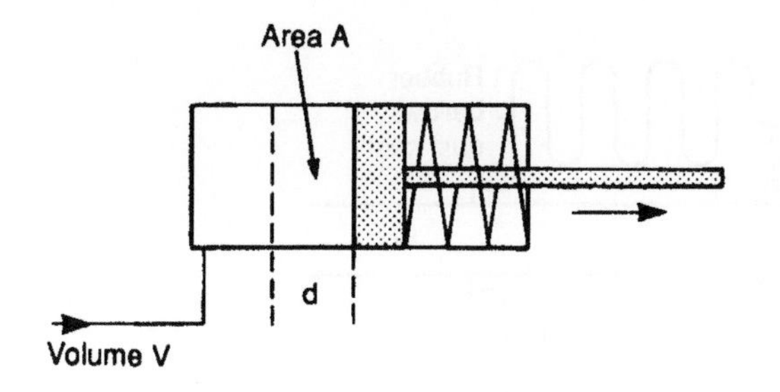

Fig. 5.7.7 Derivation of cylinder speed.

Flow rate, V_f, to achieve speed v is thus

$$V_f = \frac{Ad}{t} = Av \quad (5.7.4)$$

The flow rate units of expression (5.7.4) depend on the units being used. If d is in metres, v in metres min^{-1} and A in $metres^2$, flow rate is in $metres^3$ min^{-1}.

In pneumatic systems, it should be remembered, it is normal to express flow rates in STP (see Chapter 5.6a). Expression (5.7.4) gives the fluid volumetric flow rate to achieve a required speed at working pressure. This must be normalised to atmospheric pressure by using Boyle's law.

The air consumption for a pneumatic cylinder must also be normalised to STP. For a cylinder of stroke S and piston area A, normalised air consumption is

$$\text{volume/stroke} = SA\frac{(P_a + P_w)}{P_a} \quad (5.7.5)$$

where P_a is atmospheric pressure and P_w the working pressure. The repetition rate (e.g. 5 strokes min^{-1}) must be specified to allow mean air consumption rate to be calculated.

It should be noted that fluid pressure has no effect on piston speed (although it does influence acceleration). Speed is determined by piston area and flow rate. Maximum force available is unrelated to flow rate, instead being determined by line pressure and piston area. Doubling the piston area while keeping flow rate and line pressure constant, for example, gives half speed but doubles the maximum force. The ways in which flow rate can be controlled are discussed later.

5.7.1.1 Construction

Pneumatic and hydraulic linear actuators are constructed in a similar manner, the major differences arising out of differences in operating pressure (typically 100 bar for hydraulics and 10 bar for pneumatics, but there are considerable deviations from these values).

Figure 5.7.8 shows the construction of a double-acting cylinder. Five locations can be seen where seals are required to prevent leakage. To some extent, the art of cylinder design is in the choice of seals, a topic discussed further in a later section.

There are five basic parts in a cylinder: two end caps (a base cap and a bearing cap) with port connections, a cylinder barrel, a piston and the rod itself. This basic construction allows fairly simple manufacture as end caps and pistons are common to cylinders of the same diameter, and only (relatively) cheap barrels and rods need to be changed to give different length cylinders. End caps can be secured to the barrel by welding, tie rods or by threaded connection. Basic constructional details are shown in Figure 5.7.9.

The inner surface of the barrel needs to be very smooth to prevent wear and leakage. Generally, a seamless drawn steel tube is used which is machined (honed) to an accurate finish. In applications where the cylinder is used infrequently or may come into contact with corrosive materials, stainless steel, aluminium or brass tube may be used.

Pistons are usually made of cast iron or steel. The piston not only transmits force to the rod, but must also act as a sliding beating in the barrel (possibly with side

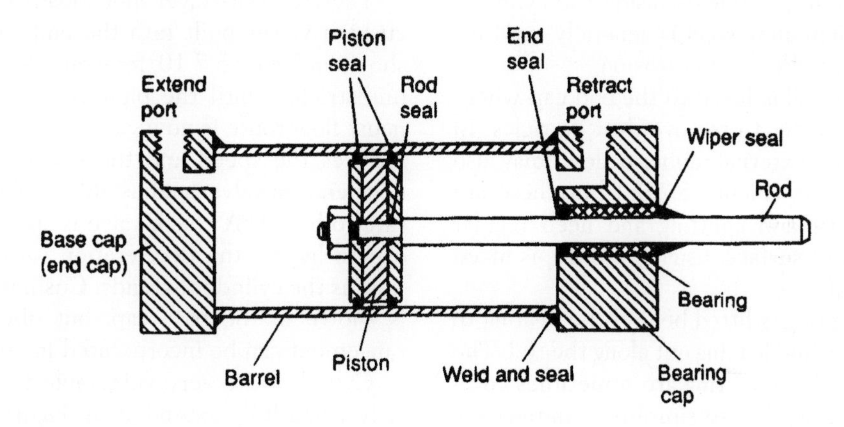

Fig. 5.7.8 Construction of a typical cylinder.

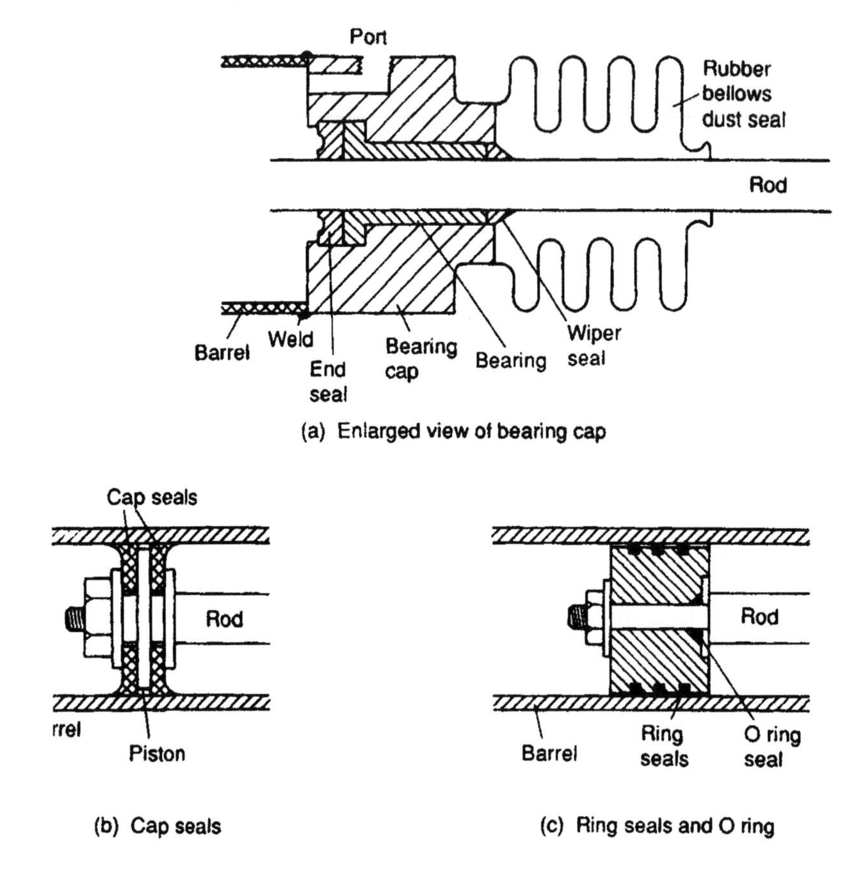

Fig. 5.7.9 Cylinder constructional details.

forces if the rod is subject to a lateral force) and provide a seal between high- and low-pressure sides. Piston seals are generally used between piston and barrel. Occasionally small leakage can be tolerated and seals are not used. A beating surface (such as bronze) is deposited on to the piston surface then honed to a finish similar to that of the barrel.

The surface of the cylinder rod is exposed to the atmosphere when extended, and hence is liable to suffer from the effects of dirt, moisture and corrosion. When retracted, these corrosive materials may be drawn back inside the barrel to cause problems inside the cylinder. Heat-treated chromium alloy steel is generally used for strength and to reduce effects of corrosion.

A wiper or scraper seal is fitted to the end cap where the rod enters the cylinder to remove dust particles. In very dusty atmospheres external rubber bellows may also be used to exclude dust (Figure 5.7.9a) but these are vulnerable to puncture and splitting and need regular inspection. The beating surface, usually bronze, is fitted behind the wiper seal.

An internal sealing ring is fitted behind the beating to prevent high-pressure fluid leaking out along the rod. The wiper seal, bearing and sealing ring are sometimes combined as a cartridge assembly to simplify maintenance. The rod is generally attached to the piston via a threaded end as shown in Figures 5.7.9b and c. Leakage can occur around the rod, so seals are again needed. These can be cap seals (as in Figure 5.7.9b) which combine the roles of piston and rod seal, or a static O ring around the rod (as in Figure 5.7.9c).

End caps are generally cast (from iron or aluminium) and incorporate threaded entries for ports. End caps have to withstand shock loads at extremes of piston travel. These loads arise not only from fluid pressure, but also from kinetic energy of the moving parts of the cylinder and load.

These end-of-travel shock loads can be reduced with cushion valves built into the end caps. In the cylinder shown in Figure 5.7.10, for example, exhaust fluid flow is unrestricted until the plunger enters the cap. The exhaust flow route is now via the deceleration valve which reduces the speed and the end of travel impact. The deceleration valve is adjustable to allow the deceleration rate to be set. A check valve is also included in the end cap to bypass the deceleration valve and give near full flow as the cylinder extends. Cushioning in Figure 5.7.10 is shown in the base cap, but obviously a similar arrangement can be incorporated in bearing cap as well.

Cylinders are very vulnerable to side loads, particularly when fully extended. In Figure 5.7.11a, a cylinder with a 30 cm stroke is fully extended and subject to a 5 kg

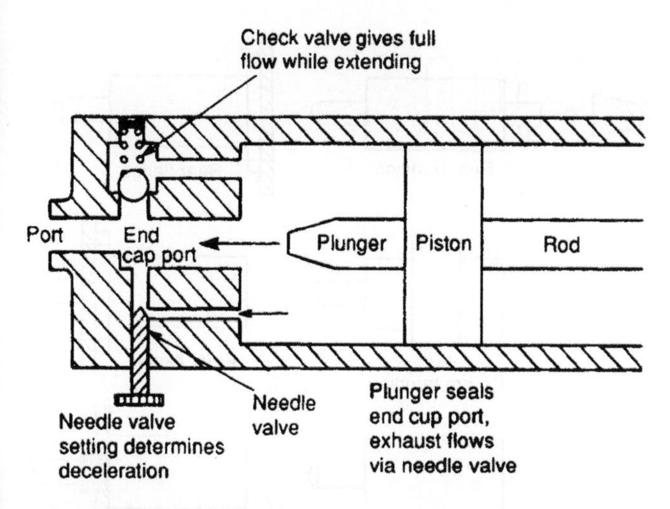

Fig. 5.7.10 Cylinder cushioning.

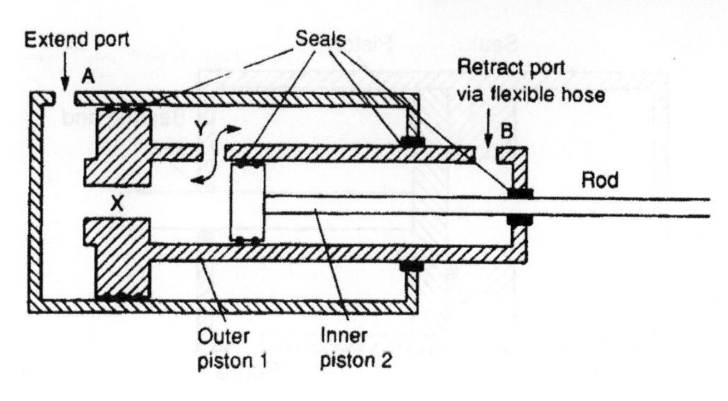

Fig. 5.7.12 Two-stage telescopic piston.

side load. When extended there is typically 1 cm between piston and end beating. Simple leverage will give side loads of 155 kg on the bearing and 150 kg on the piston seals. This magnification of side loading increases cylinder wear. The effect can be reduced by using a cylinder with a longer stroke, which is then restricted by an internal stop tube as shown in Figure 5.7.11b.

The stroke of a simple cylinder must be less than barrel length, giving at best an extended/retracted ratio of 2:1. Where space is restricted, a telescopic cylinder can be used. Figure 5.7.12 shows the construction of a typical double-acting unit with two pistons. To extend, fluid is applied to port A. Fluid is applied to both sides of piston 1 via ports X and Y, but the difference in areas between sides of piston 1 causes the piston to move to the right.

To retract, fluid is applied to port B. A flexible connection is required for this port. When piston 2 is driven fully to the left, port Y is now connected to port B, applying pressure to the right-hand side of piston 1 which then retracts.

The construction of telescopic cylinders requires many seals which makes maintenance complex. They also have smaller force for a given diameter and pressure, and can only tolerate small side loads.

Pneumatic cylinders are used for metal forming, an operation requiring large forces. Pressures in pneumatic systems are lower than in hydraulic systems, but large impact loads can be obtained by accelerating a hammer to a high velocity then allowing it to strike the target.

Such devices are called impact cylinders and operate on the principle illustrated in Figure 5.7.13. Pressure is initially applied to port B to retract the cylinder. Pressure is then applied to both ports A and B, but the cylinder remains in a retracted state because area X is less than area Y. Port B is then vented rapidly. Immediately, the full

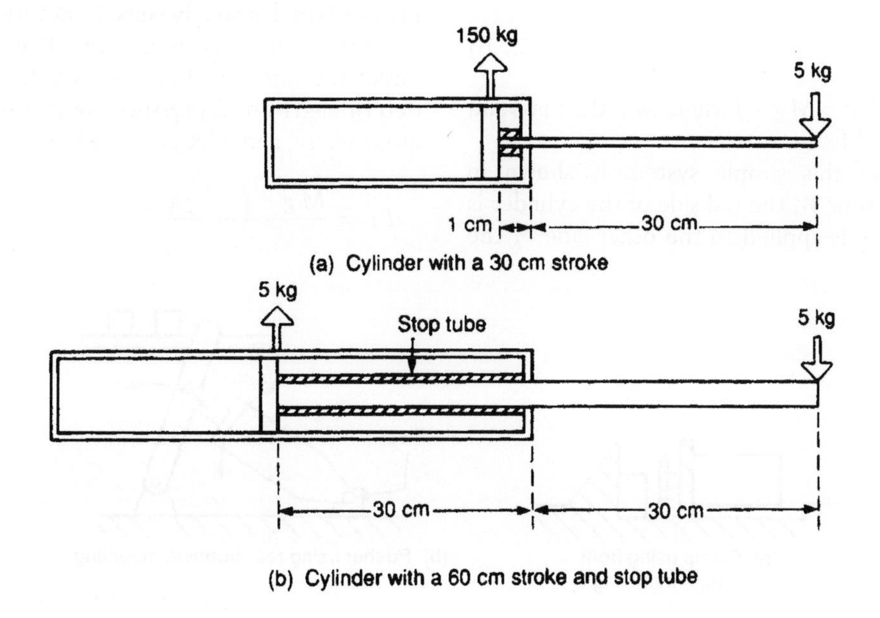

Fig. 5.7.11 Side loads and the stop tube.

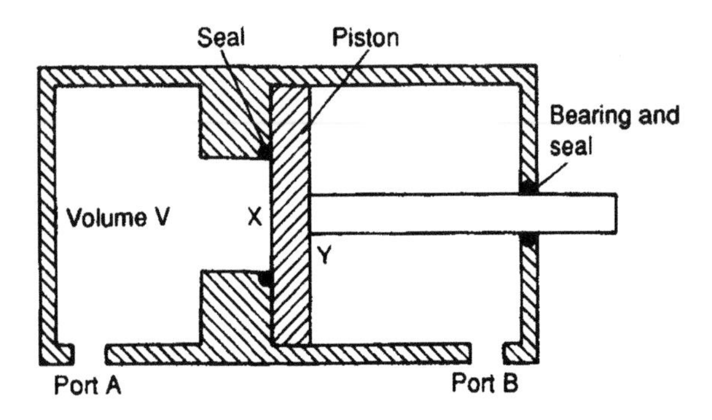

Fig. 5.7.13 An impact cylinder.

piston area experiences port A pressure. With a large volume of gas stored behind the piston, it accelerates rapidly to a high velocity (typically 10 m s^{-1}).

5.7.1.2 Mounting arrangements

Cylinder mounting is determined by the application. Two basic types are shown in Figure 5.7.14. The clamp of Figure 5.7.14a requires a simple fixed mounting. The pusher of Figure 5.7.15b requires a cylinder mount which can pivot.

Figure 5.7.15 shows various mounting methods using these two basic types. The effects of side loads should be considered on non-centreline mountings such as the foot mount. Swivel mounting obviously requires flexible pipes.

5.7.1.3 Cylinder dynamics

The cylinder in Figure 5.7.16a is used to lift a load of mass M. Assume it is retracted, and the top portion of the cylinder is pressurised. The extending force is given by the expression:

$$F = P_1A - P_2a \tag{5.7.6}$$

To lift the load at all, $F > Mg + f$ where M is the mass and f the static frictional force.

The response of this simple system is shown in Figure 5.7.16b. At time W, the rod side of the cylinder is vented and pressure is applied to the other side of the

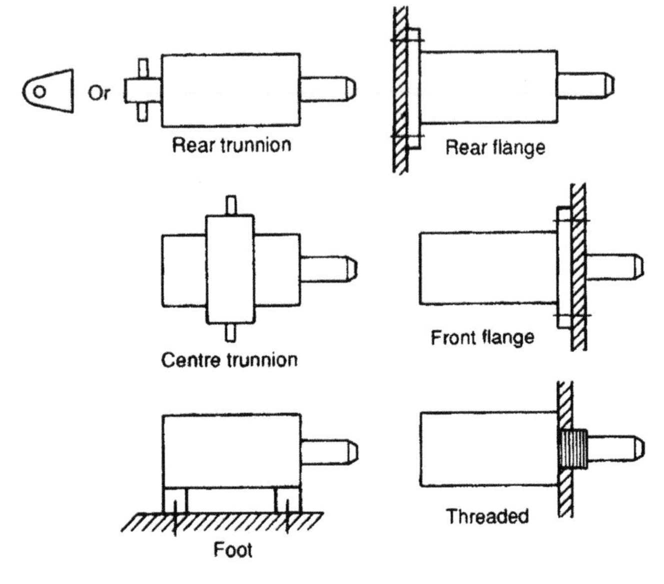

Fig. 5.7.15 Methods of cylinder mounting.

piston. The pressure on both sides of the piston changes exponentially, with falling pressure P_2 changing slower than inlet pressure P_1, because of the larger volume. At time X, extension force P_1A is larger than P_2a, but movement does not start until time Y when force, given by expression (5.7.6), exceeds mass and frictional force.

The load now accelerates with acceleration given by Newton's law:

$$\text{acceleration} = \frac{F_a}{M} \tag{5.7.7}$$

where $F_a = P_1A - P_2a - Mg - f$.

It should be remembered that F_a is not constant, because both P_1 and P_2 will be changing. Eventually, the load will reach a steady velocity, at time Z. This velocity is determined by maximum input flow rate or maximum outlet flow rate (whichever is lowest). Outlet pressure P_2 is determined by back pressure from the outlet line to tank or atmosphere, and inlet pressure is given by the expression:

$$P_1 = \frac{Mg + f + P_2a}{A}$$

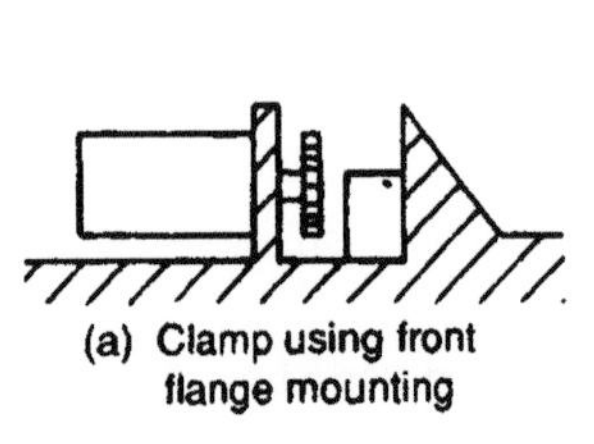

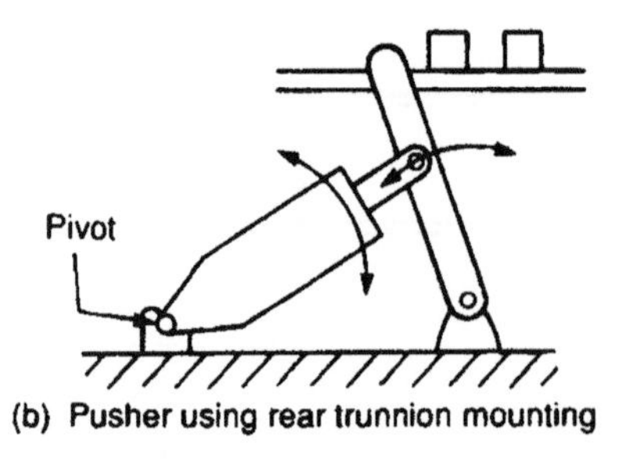

Fig. 5.7.14 Basic mounting types.

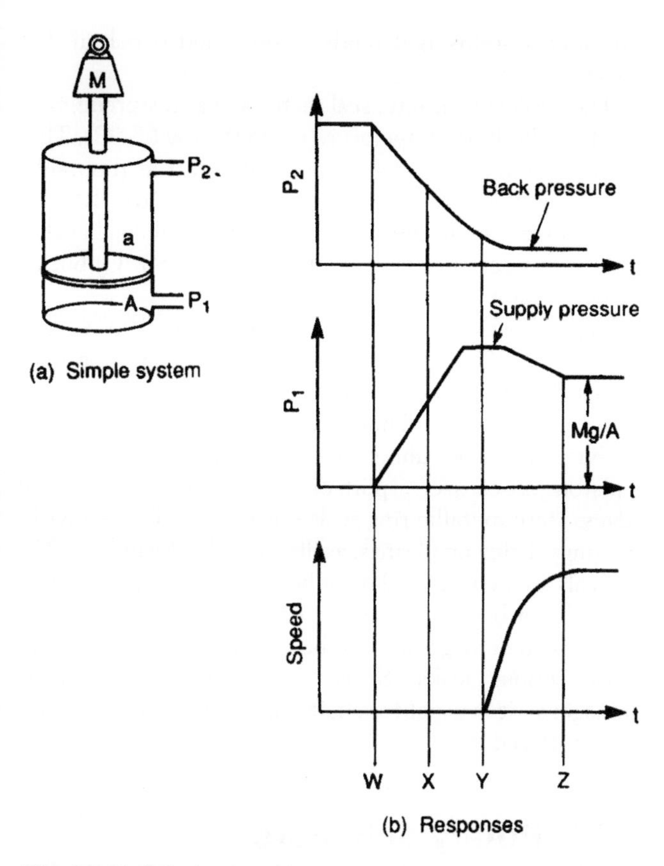

Fig. 5.7.16 Cylinder dynamics.

The time from W to Y, before the cylinder starts to move, is called the 'dead time' or 'response time'. It is determined primarily by the decay of pressure on the outlet side, and can be reduced by de-pressurising the outlet side in advance or (for pneumatic systems) by the use of quick exhaust valves.

The acceleration is determined primarily by the inlet pressure and the area of the inlet side of the piston (term P_1A in expression (5.7.6)). The area, however, interacts with the dead time – a larger area, say, gives increased acceleration but also increases cylinder volume and hence extends the time taken to vent fluid on the outlet side.

5.7.2 Seals

Leakage from a hydraulic or pneumatic system can be a major problem, leading to loss of efficiency, increased power usage, temperature rise, environmental damage and safety hazards.

Minor internal leakage (round the piston in a double-acting cylinder, for example) can be of little consequence and may even be deliberately introduced to provide lubrication of the moving parts.

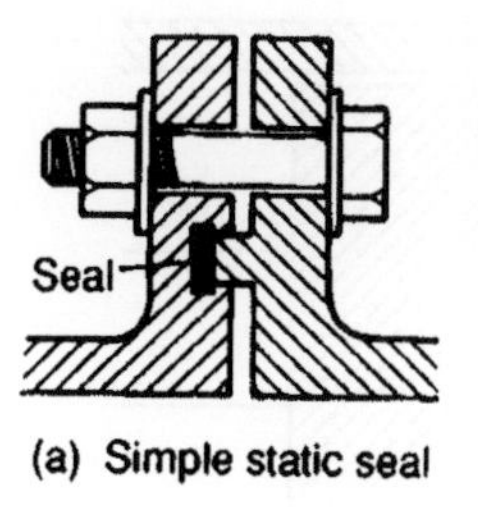

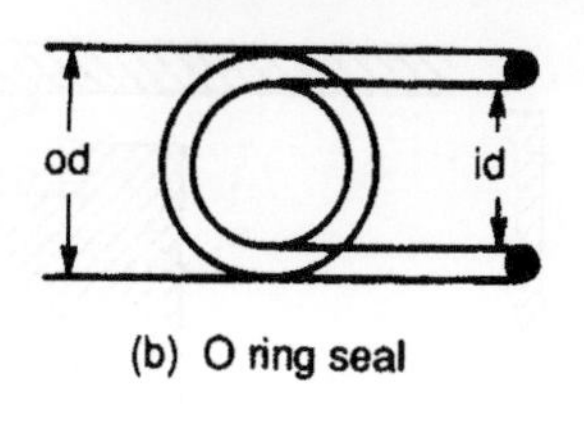

Fig. 5.7.17 Static seals.

External leakage, on the other hand, is always serious. In pneumatic systems, external leakage is noisy; with hydraulic systems, external loss of oil is expensive as lost oil has to be replaced, and the resulting pools of oil are dangerous and unsightly.

Mechanical components (such as pistons and cylinders) cannot be manufactured to sufficiently tight tolerances to prevent leakage (and even if they could, the resultant friction would be unacceptably high). Seals are therefore used to prevent leakage (or allow a controlled leakage). To a large extent, the art of designing an actuator is really the art of choosing the right seals.

The simplest seals are 'static seals' (Figure 5.7.17) used to seal between stationary parts. These are generally installed once and forgotten. A common example is the gasket shown in a typical application in Figure 5.7.17a. The O ring of Figure 5.7.17b is probably the most-used static seal, and comprises a moulded synthetic ring with a round cross section when unloaded. O rings can be specified in terms of inside diameter (id) for fitting onto shafts, or outside diameter (od) for fitting into bores.

When installed, an O ring is compressed in one direction. Application of pressure causes the ring to be compressed at right angles, to give a positive seal against two annular surfaces and one flat surface. O rings give effective sealing at very high pressures.

O rings are primarily used as static seals because any movement will cause the seal to rotate, allowing leakage to occur.

Where a seal has to be provided between moving surfaces, a dynamic seal is required. A typical example is the end or cup seal shown, earlier, in Figure 5.7.9a. Pressure in the cylinder holds the lip of the seal against the barrel to give zero leakage (called a 'positive seal'). Effectiveness of the seal increases with pressure, and leakage tends to be more of a problem at low pressures.

The U ring seal of Figure 5.7.18 works on the same principle as the cap seal. Fluid pressure forces the two lips apart to give a positive seal. Again, effectiveness of the seal is better at high pressure. Another variation on

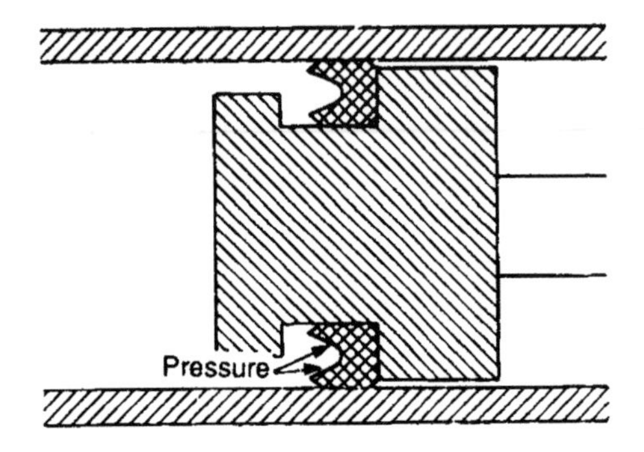

Fig. 5.7.18 The U ring seal.

the technique is the composite seal of Figure 5.7.19. This is similar in construction to the U ring seal, but the space between the lips is filled by a separate ring. Application of pressure again forces the lips apart to give a positive seal.

At high pressures, there is a tendency for a dynamic seal to creep into the radial gap, as shown in Figure 5.7.20a leading to trapping of the seal and rapid wear. This can be avoided by the inclusion of an anti-extrusion ring behind the seal, as in Figure 5.7.20b.

Seals are manufactured from a variety of materials, the choice being determined by the fluid, its operating pressure and the likely temperature range. The earliest material was leather and, to a lesser extent, cork but these have been largely superseded by plastic and synthetic rubber materials. Natural rubber cannot be used in hydraulic systems as it tends to swell and perish in the presence of oil.

The earliest synthetic seal material was neoprene, but this has a limited temperature range (below 65 °C). The most common present-day material is nitrile (buna-N) which has a wider temperature range (−50 °C to −100 °C and is currently the cheapest seal material. Silicon has the highest temperature range (−100 °C to +250 °C) but is expensive and tends to tear.

In pneumatic systems, viton (−20 °C to −190 °C and teflon (−80 °C to +200 °C) are the most common materials. These are more rigid and are often used as wiper or scraper seals on cylinders.

Synthetic seals cannot be used in applications where a piston passes over a port orifice which nicks the seal edges. Here metallic ring seals must be used, often with the tings sitting on O rings, as illustrated in Figure 5.7.21.

Seals are delicate and must be installed with care. Dirt on shafts or barrels can easily nick a seal as it is slid into place. Such damage may not be visible to the eye but can cause serious leaks. Sharp edges can cause similar damage; so it is usual for shaft ends and groove edges to be chamfered.

5.7.3 Rotary actuators

Rotary actuators are the hydraulic or pneumatic equivalents of electric motors. For a given torque, or power, a rotary actuator is more compact than an equivalent

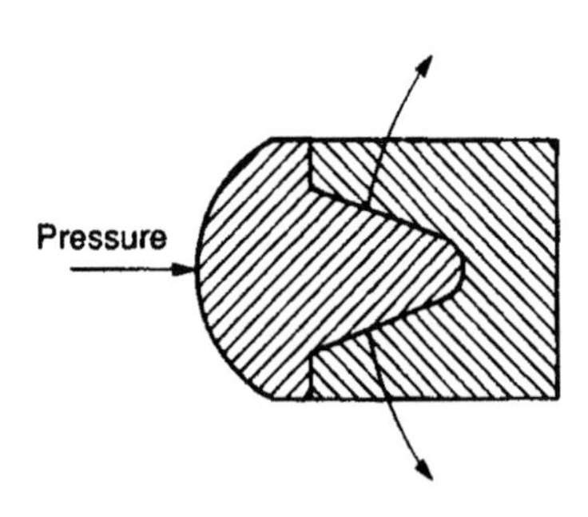

Fig. 5.7.19 The composite seal.

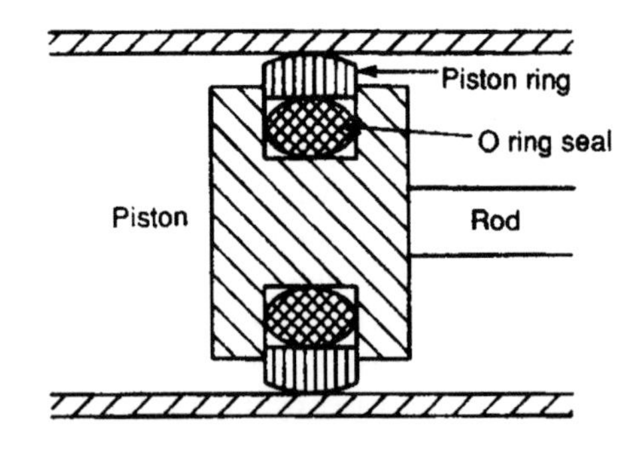

Fig. 5.7.21 Combined piston ring and O ring seal (not to scale).

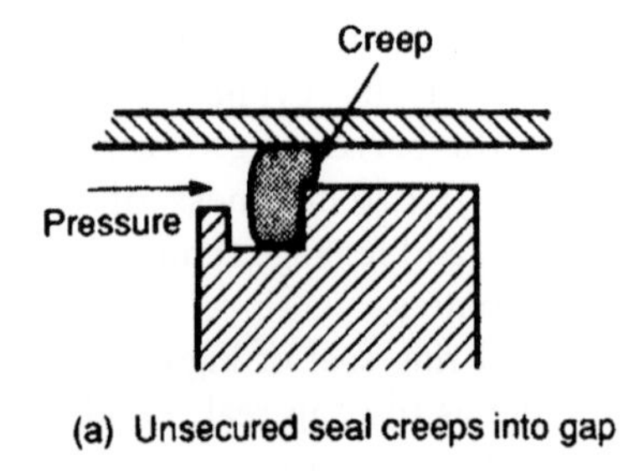

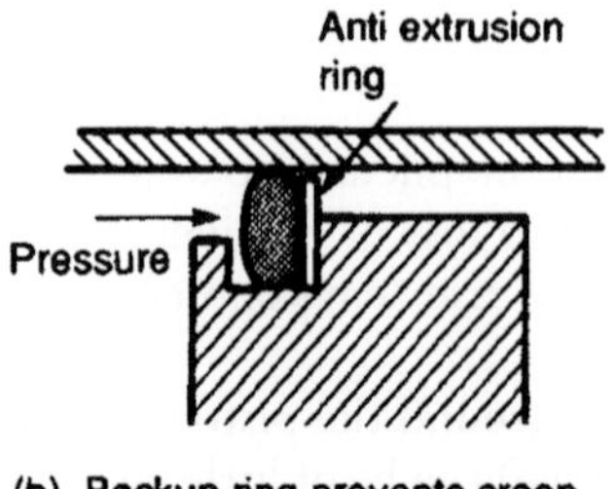

Fig. 5.7.20 Anti-extrusion ring.

motor, cannot be damaged by an indefinite stall and can safely be used in an explosive atmosphere. For variable speed applications, the complexity and maintenance requirements of a rotary actuator are similar to a thyristor-controlled DC drive, but for fixed speed applications, the AC induction motor (which can, for practical purposes, be fitted and forgotten) is simpler to install and maintain.

A rotary actuator (or, for that matter, an electric motor) can be defined in terms of the torque it produces and its running speed, usually given in revolutions per minute (rpm). Definition of torque is illustrated in Figure 5.7.22, where a rotary motion is produced against a force of F newtons acting at a radial distance d metres from a shaft centre. The device is then producing a torque T given by the expression:

$$T = Fd \text{ Nm} \tag{5.7.8}$$

In imperial units, F is given in pounds force, and d in inches or feet to give T in lbf ins or lbf ft. It follows that 1 Nm = 8.85 lbf ins.

The torque of a rotary actuator can be specified in three ways. Starting torque is the torque available to move a load from rest. Stall torque must be applied by the load to bring a running actuator to rest, and running torque is the torque available at any given speed. Running torque falls with increasing speed, typical examples being shown in Figure 5.7.23. Obviously, torque is dependent on the applied pressure; increasing the pressure results in increased torque, as shown.

The output power of an actuator is related to torque and rotational speed, and is given by the expression

$$P = \frac{TR}{9550} \text{ kW} \tag{5.7.9}$$

where T is the torque in newton metre and R is the speed in rpm. In imperial units, the expression is

$$P = \frac{TR}{5252} \text{ hp} \tag{5.7.10}$$

where T is in lbsf ft (and R is in rpm) or

$$P = \frac{TR}{63,024} \text{ hp} \tag{5.7.11}$$

where T is in lbsf ins.

Figure 5.7.23 illustrates how running torque falls with increasing speed, so the relationship between power and speed has the form of Figure 5.7.24, with maximum power at some (defined) speed. Power, like torque, is dependent on applied pressure.

The torque produced by a rotary actuator is directly related to fluid pressure; increasing pressure increases maximum available torque. Actuators are often specified by their torque rating, which is defined as

$$\text{torque rating} = \frac{\text{torque}}{\text{pressure}}$$

In imperial units, a pressure of 100 psi is used, and torque is generally given in lbf ins.

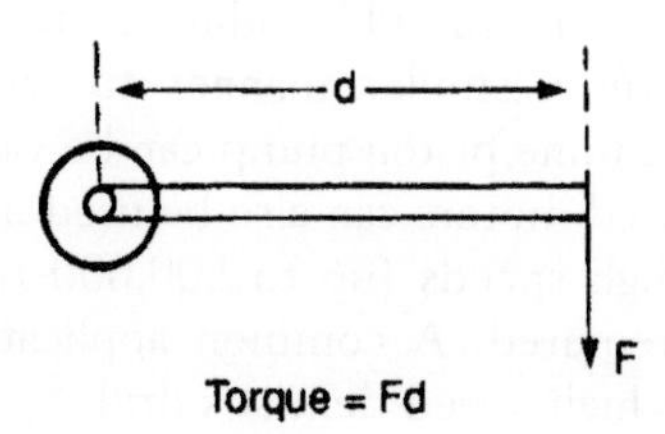

Fig. 5.7.22 Definition of torque.

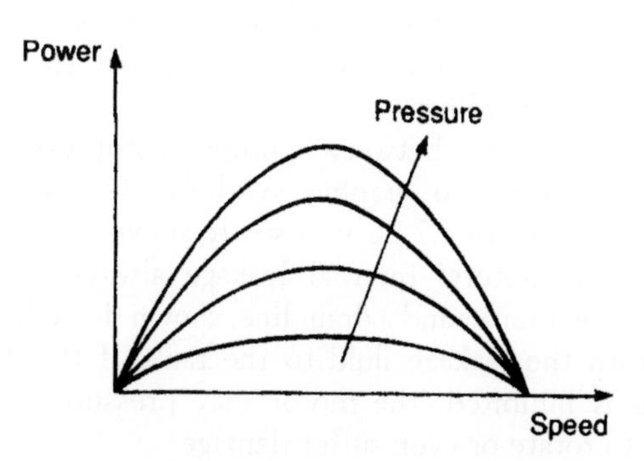

Fig. 5.7.24 Power/speed curve for pneumatic rotary actuator.

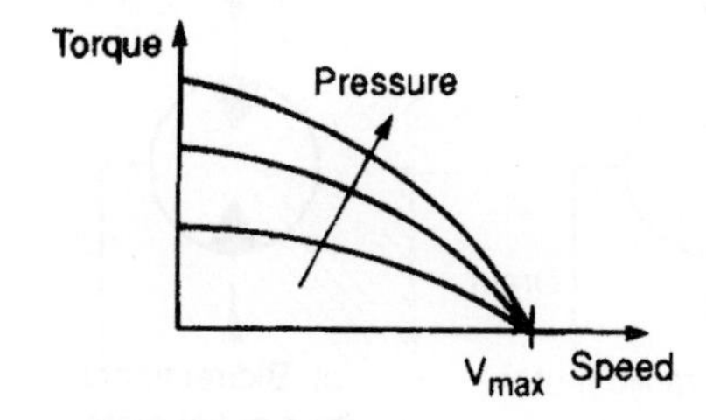

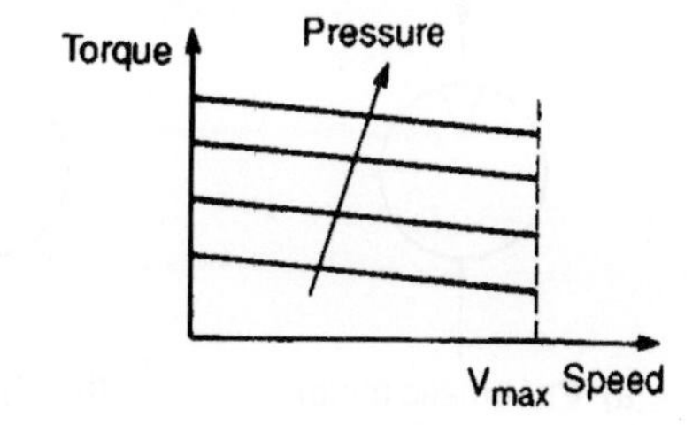

Fig. 5.7.23 Torque/speed curves for rotary actuators.

The allowable pressure for an actuator is defined in terms of pressure rating (maximum applicable pressure without risk of permanent damage), and pressure range (the maximum and minimum pressures between which actuator performance is defined).

Fluid passes through an actuator as it rotates. For hydraulic actuators, displacement is defined as the volume of fluid used for one motor rotation. For a given design of motor, available torque is directly proportional to displacement. For pneumatic actuators, the air usage per revolution at a specified pressure is generally given in terms of STP (see Chapter 5.6a).

Rotational speed is given by the expression:

$$\text{rotational speed} = \frac{\text{fluid flow rate}}{\text{displacement}}$$

With the torque rate and displacement fixed for a chosen motor, the user can control maximum available torque and speed by adjusting, respectively, pressure setting and flow rate of fluid to the actuator.

5.7.3.1 Constructional details

In electrical systems, there are many similarities between electrical generators and electric motors. A DC generator, for example, can be run as a motor. Similarly, a DC motor can be used as a generator. Similar relationships exist between hydraulic pumps and motors and between pneumatic compressors and motors. This similarity is extended as manufacturers use common parts in pumps, compressors and motors to simplify users' spares holdings.

The similarity between pumps, compressors and motors extends to graphic symbols. The schematic symbols of Figure 5.7.25 are used to show hydraulic and pneumatic motors. Internal leakage always occurs in a hydraulic motor, and a drain line, shown dotted, is used to return the leakage fluid to the tank. If this leakage return is inhibited, the motor may pressure lock and cease to rotate or even suffer damage.

There are three basic designs of rotary or pump compressor; the gear pump, the vane pump and various designs of piston pump or compressor described earlier in Chapter 5.6. These can also be used as the basis of rotary actuators. The principles of hydraulic and pneumatic devices are very similar, but the much higher hydraulic pressures give larger available torques and powers despite lower rotational speeds.

Figure 5.7.26 shows the construction of a gear motor. Fluid enters at the top and pressurises the top chamber. Pressure is applied to two gear faces at X, and a single gear face at Y. There is, thus, an imbalance of forces on the gears, resulting in rotation as shown. Gear motors suffer from leakage which is more pronounced at low speed. They thus tend to be used in medium-speed, low-torque applications.

A typical vane motor construction is illustrated in Figure 5.7.27. It is very similar to the construction of a vane pump. Suffering from less leakage than the gear motor, it is typically used at lower speeds. Like the vane pump, side loading occurs on the shaft of a single vane motor. These forces can be balanced by using a dual design similar to the pump shown in Figure 5.6.10b. In a vane pump, vanes are held out by the rotational speed. In a vane motor, however, rotational speed is probably quite low and the vanes are held out, instead, by fluid pressure. An in-line check valve can be used, as in Figure 5.7.28, to generate a pressure which is always slightly higher than motor pressure.

Piston motors are generally most efficient and give highest torques, speeds and powers. They can be of radial design similar to the pump of Figures 5.6.2 and 5.6.3, or in-line (axial) design similar to those of Figures 5.6.14 and 5.6.15. Radial piston motors tend to be most common in pneumatic applications, with in-line piston motors most common in hydraulics. The speed of the piston motor can be varied by adjusting the angle of the swash plate (in a similar manner to which delivery volume of an in-line piston pump can be varied).

Turbine-based motors can also be used in pneumatics where very high speeds (up to 500,000 rpm) but low torques are required. A common application of these devices is the high-speed dentist's drill.

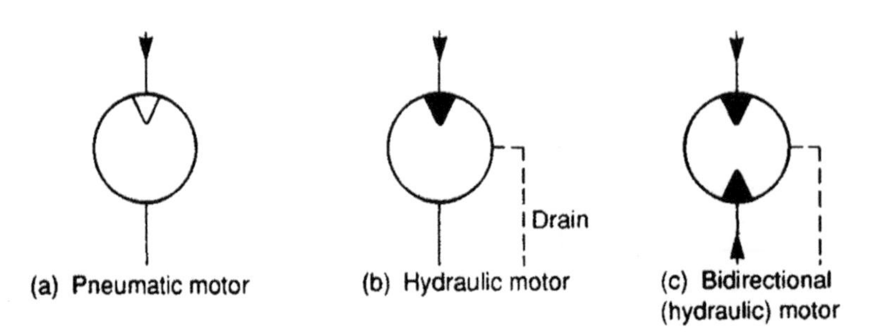

Fig. 5.7.25 Rotary actuator symbols.

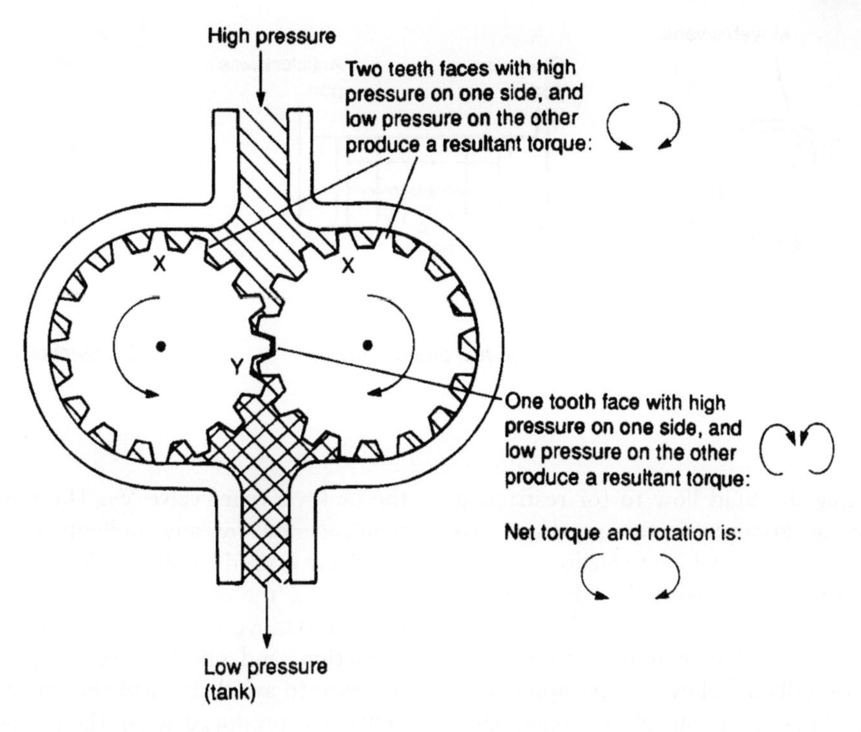
High pressure
Two teeth faces with high pressure on one side, and low pressure on the other produce a resultant torque:
X
X
Y
One tooth face with high pressure on one side, and low pressure on the other produce a resultant torque:
Net torque and rotation is:
Low pressure (tank)

Fig. 5.7.26 A gear motor.

All the rotary actuators described so far have been pneumatic or hydraulic equivalents of electric motors. However, rotary actuators with a limited travel (say 270°) are often needed to actuate dampers or control large valves. Some examples are illustrated in Figure 5.7.29. The actuator in Figure 5.7.29a is driven by a single vane coupled to the output shaft. In that of Figure 5.7.29b, a double-acting piston is coupled to the output shaft by a rack and pinion. In both cases, the shaft angle can be finely controlled by fluid applied to the ports. These have the graphic symbols shown in Figure 5.7.29c.

5.7.4 Application notes

5.7.4.1 Speed control

The operational speed of an actuator is determined by the fluid flow rate and the actuator area (for a cylinder) or the displacement (for a motor). The physical dimensions are generally fixed for an actuator, so speed is

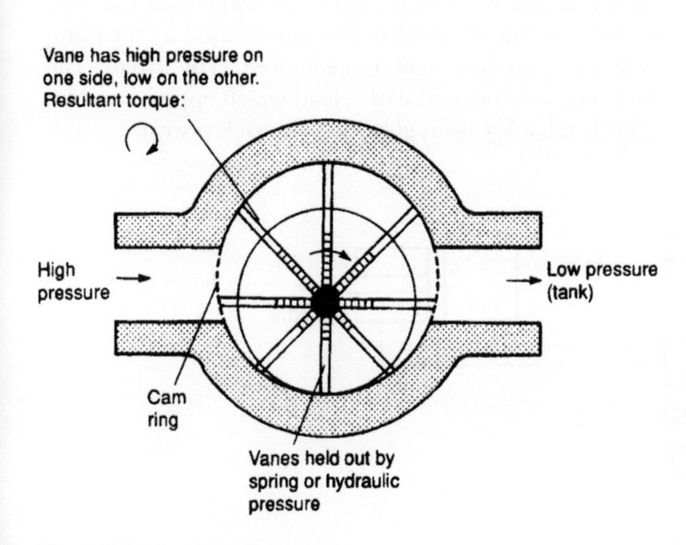
Vane has high pressure on one side, low on the other. Resultant torque:
High pressure
Low pressure (tank)
Cam ring
Vanes held out by spring or hydraulic pressure

Fig. 5.7.27 A vane motor.

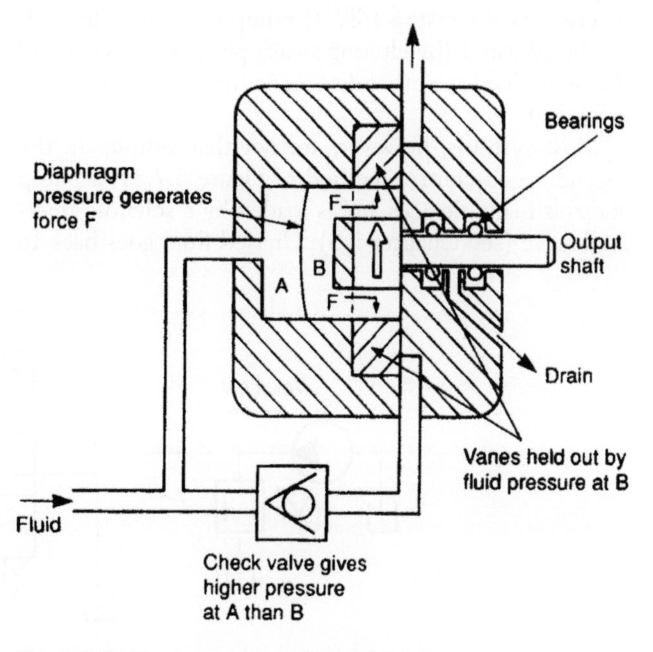
Bearings
Diaphragm pressure generates force F
F
F
A
B
Output shaft
Drain
Vanes held out by fluid pressure at B
Fluid
Check valve gives higher pressure at A than B

Fig. 5.7.28 Vane operation in hydraulic motor.

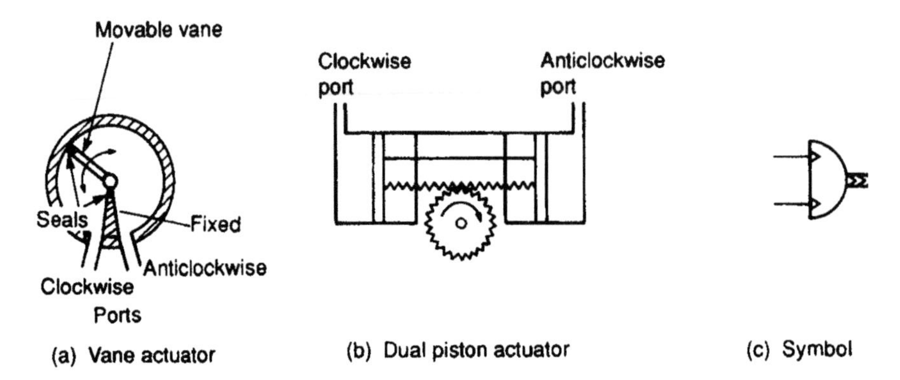

Fig. 5.7.29 Limited motion rotary actuators.

controlled by adjusting the fluid flow to (or restricting flow from) the actuator. Rotary actuator speed can also be controlled by altering the swash plate angle.

The compressibility of air, normally advantageous where smooth operation is concerned, makes flow control more difficult for pneumatic than hydraulic systems. Although techniques described below can be applied in pneumatics, precise slow-speed control of a pneumatic actuator is achieved with external devices described later.

There are essentially four ways in which fluid flow can be controlled. The first is shown in Figure 5.7.30, where a pump delivers a fluid volume V per minute. Because the pump is a fixed displacement device, this volume of fluid *must* go either back to the tank or to the actuator. When the control valve moves from its centre position, the actuator moves with a velocity

$$v = \frac{V}{A}$$

where A is the piston area. If pump delivery volume V can be adjusted (by altering swash plate angle, say,) *and* the pump feeds no other device, no further speed control is needed.

Most systems, however, are not that simple. In the second speed-control method of Figure 5.7.31, a pump controls many devices and is loaded by a solenoid-operated valve (see Chapter 5.6). Unused fluid goes back to the tank via relief valve V_3. The pump output is higher than needed by any individual actuator, so a flow restrictor is used to set the flow to each actuator. This is known as a 'meter in' circuit, and is used where a force is needed to move a load. Check valve V_1 gives a full-speed retraction, and check valve V_2 provides a small back pressure to avoid the load running away. The full pump delivery is produced when the pressure reaches the setting of relief valve V_3, so there is a waste of energy and unnecessary production of heat in the fluid.

If the load can run away from the actuator, the third speed control method, the 'meter out' circuit of Figure 5.7.32 must be used. As drawn, this again gives a controlled extension speed, and full retraction speed (allowed by check valve V_1). As before, the pump delivers fluid at a pressure set by the relief valve, leading to heat generation.

Finally, in the fourth speed control method of Figure 5.7.33, a bleed-off valve V_1 is incorporated. This returns a volume v back to the tank, leaving a volume $V - v$ to go to the actuator (where V is the pump delivery volume). Pump pressure is now determined by the required actuator pressure, which is lower than the pressure set on the relief valve. The energy used by the pump is lower, and less heat is generated. The circuit can, however, only be used with a load which opposes motion. Check valve V_2 again gives a small back pressure.

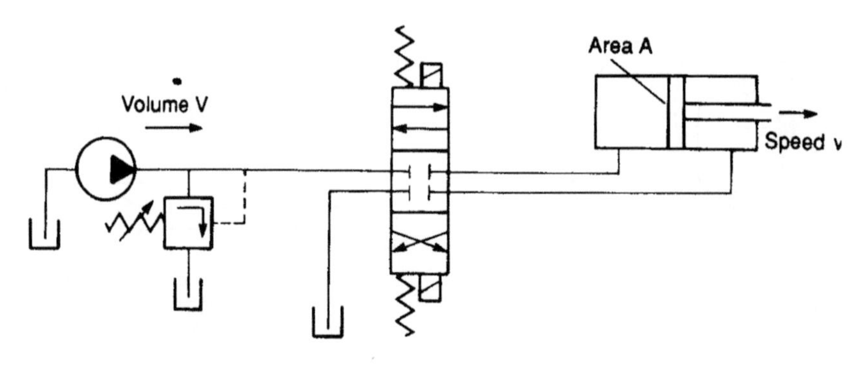

Fig. 5.7.30 Speed control by pump volume.

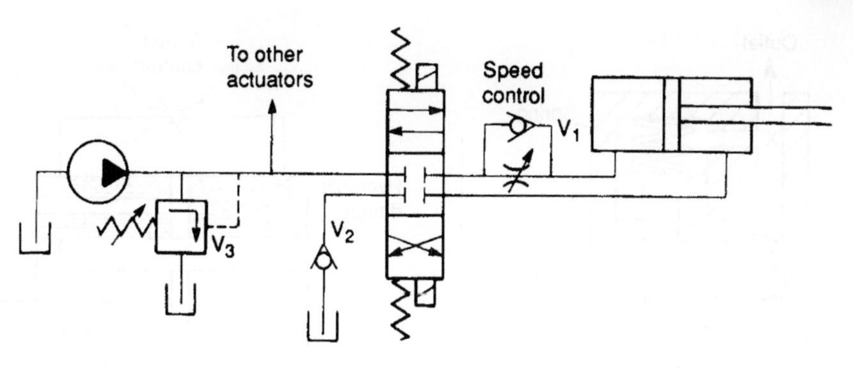

Fig. 5.7.31 Meter in speed control.

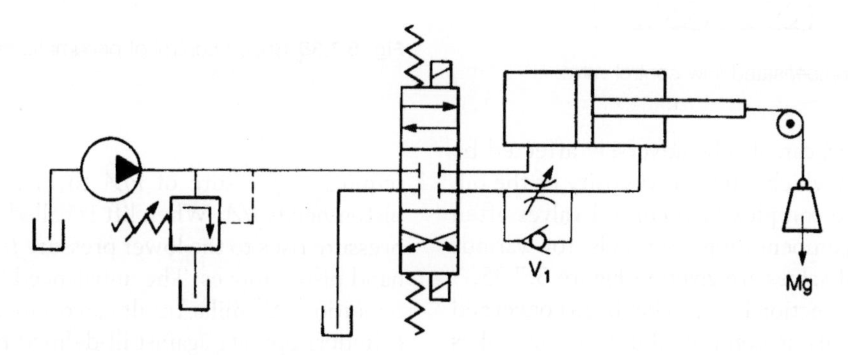

Fig. 5.7.32 Meter out speed control for overhauling load.

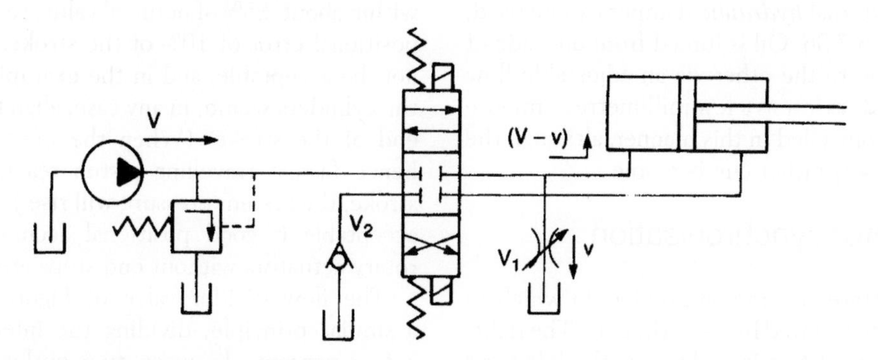

Fig. 5.7.33 Bleed-off speed control.

Any unused fluid from the pump is returned to the tank at high pressure leading to wasted energy, even with the more efficient 'bleed'-off circuit. One moral, therefore, is to have a pump delivery volume no larger than necessary.

Figures 5.7.31–5.7.33 imply flow, and hence speed, is set by a simple restriction in piping to the actuator. While a simple restriction reduces flow and allows speed to be reduced, in practice, a true flow control valve is needed which delivers a fixed flow regardless of line pressure or fluid temperature.

An ideal flow controller operates by maintaining a constant pressure drop across an orifice restriction in the line, the rate being adjusted by altering orifice size. The construction of such a device is shown in Figure 5.7.34. The orifice is formed by a notch in a shaft which can be rotated to set the flow. The pressure drop across the orifice is the difference in pressure between points X and Y, and is applied to the moveable land. The pressure at X, in conjunction with the spring pressure, causes a downward force, while pressure at Y causes an upward force. If the land moves up the flow reduces; if the land moves down the flow increases. The piston thus moves up and down until the pressure differential between X and Y matches the spring compressive force. The device thus maintains a constant pressure drop across the orifice, which implies constant flow through the valve, and is known as a pressure-compensated flow-control valve.

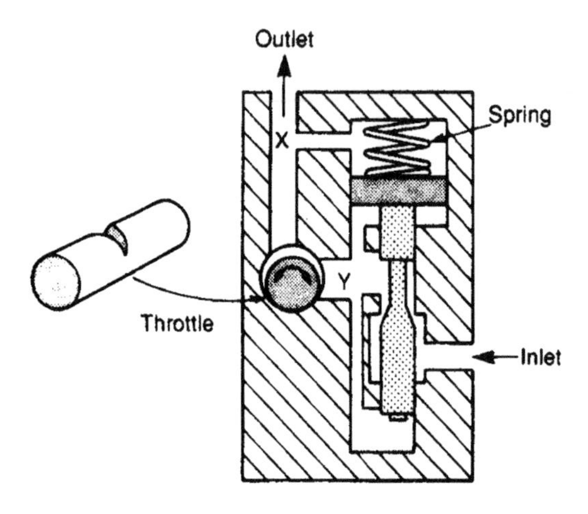

Fig. 5.7.34 Pressure compensated flow control valve.

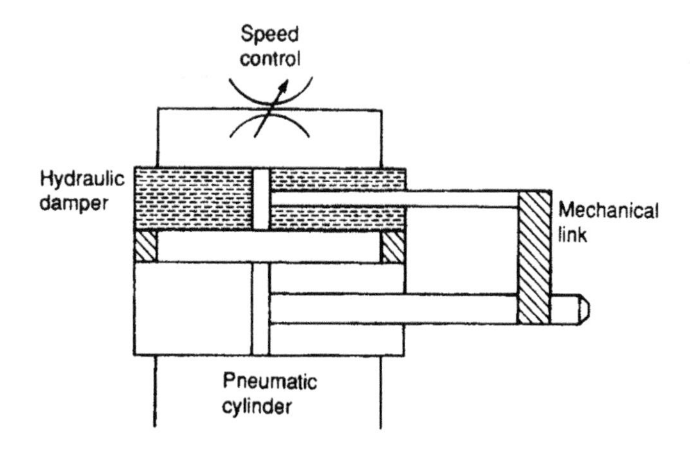

Fig. 5.7.36 Speed control of pneumatic cylinder.

Flow control valves can also be adversely affected by temperature changes which alter the viscosity of the oil. For this reason, more complex flow control valves often have temperature compensation. Symbols for various types of flow control valves are given in Figure 5.7.35.

Discussions in this section have, so far, been concerned with hydraulic systems as compressibility of air makes speed control of pneumatic actuators somewhat crude. If a pneumatic actuator is required to act at a slow, controlled speed an external *hydraulic* damper can be used, as shown in Figure 5.7.36. Oil is forced from one side of the hydraulic piston to the other via an adjustable flow control valve. Speeds as low as a few millimetres a minute can be accurately controlled in this manner, although the technique is physically rather cumbersome.

5.7.4.2 Actuator synchronisation

Figure 5.7.37 illustrates a common problem in which an unbalanced load is to be lifted by two cylinders. The right-hand cylinder is subject to a large force *F*, the left-hand cylinder to a smaller force *f*. The right-hand piston requires a pressure of F/A to lift, while the left-hand piston needs f/A. When lift is called for on valve V_1, the pressure rises to the lower pressure f/A, and only the left-hand piston moves. The unbalanced load results in faulty operation. A similar result can occur where two, or more, cylinders operate against ill-defined frictional forces.

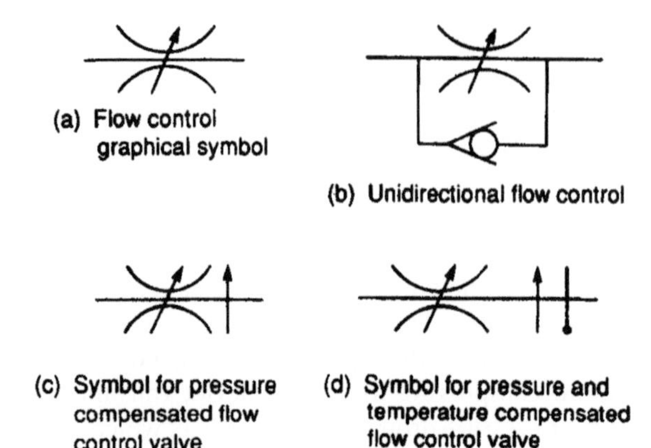

Fig. 5.7.35 Flow control valves.

One simple solution is the inclusion of flow regulating valves. A flow control valve can set, and hold, fluid flow to within about $\pm5\%$ of nominal value, resulting in a possible positional error of 10% of the stroke. This may, or may not, be acceptable, and in the example of Figure 5.7.37 the cylinders would, in any case, align themselves at each end of the stroke. (When the most lightly laden and hence fastest travelling piston reaches the end of its stroke, the system pressure will rise.) This solution is not acceptable if good positional accuracy is required or rotary actuators without end stops are being driven.

The flow divider valve of Figure 5.7.38 works on a similar principle, dividing the inlet flow equally (to a few percent) between two outlet ports. The spool moves to maintain equal pressure drops across orifices X and Y, and hence equal flow through them.

The displacement of a hydraulic or pneumatic motor can be accurately specified, and this forms the basis of an alternative flow divider circuit of Figure 5.7.39. Here, fluid for two cylinders passes through two mechanically coupled motors. The mechanical coupling ensures the two motors rotate at the same speed, and hence equal flow is passed into each cylinder.

The two cylinders in Figure 5.7.40 are effectively in series with fluid from the annulus side of cylinder 1 going to the full bore side of cylinder 2. The cylinders are chosen, however, so that full bore area of cylinder 2 equals the annulus area of cylinder 1. Upon cylinder extension, fluid exits from cylinder 1 and causes cylinder 2 to extend. The two cylinders move at equal speed because of the equal areas.

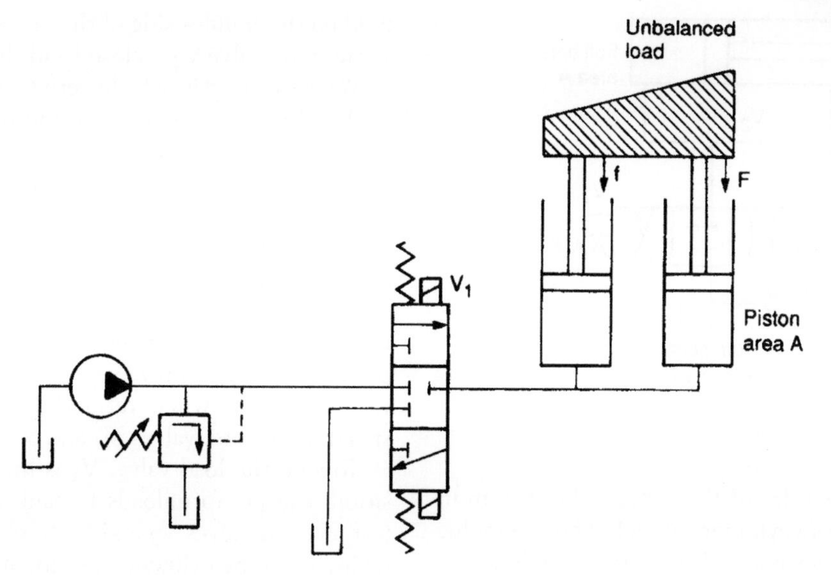

Fig. 5.7.37 Linked cylinders with unbalanced load.

There is, though, an unfortunate side effect. Pressure P_2 in cylinder 2 is F/a. Fluid on the full bore side of cylinder 1 has to lift the piston against force f plus the force from P_2 acting on the annulus side of the piston. Pressure P_1 is $(F + f)/A$, higher than would be required by two independent cylinders acting in parallel. The rotational speed of motors with equal displacement can similarly be synchronised by connecting them in series. Inlet pressure of the first motor is again, however, higher than needed to drive the two motors separately or in parallel.

None of these methods gives absolute synchronisation, and if actuators do not self-align at the ends of travel, some method of driving actuators individually should be included to allow intermittent manual alignment. The best solution, however, is usually to include some form of mechanical tie to ensure actuators experience equal loads and cannot get out of alignment.

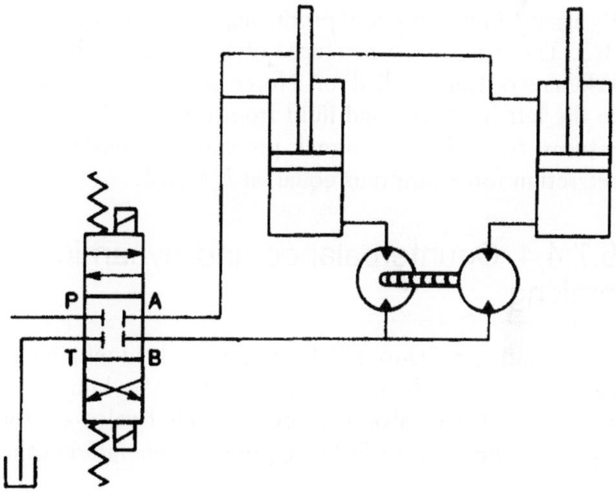

Fig. 5.7.39 Cylinder synchronisation with linked hydraulic motors.

5.7.4.3 Regeneration

A conventional cylinder can exert a larger force extending than retracting because of the area difference between

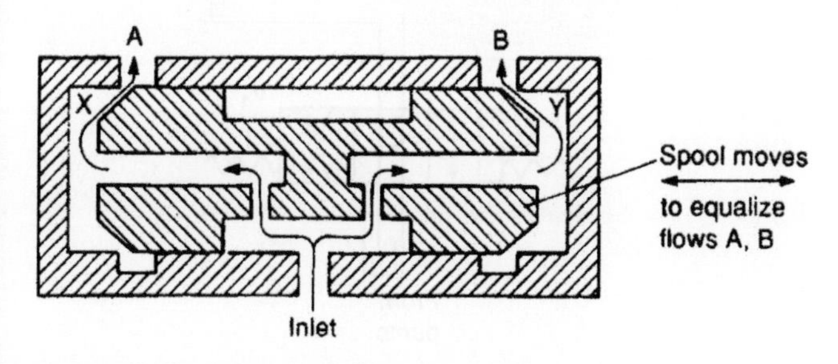

Fig. 5.7.38 Flow divider valve.

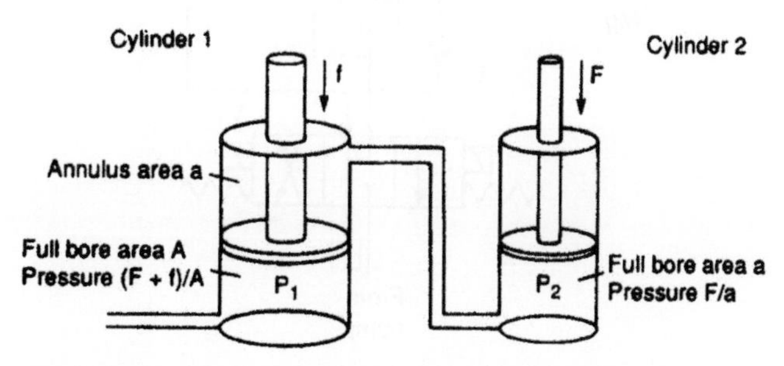

Fig. 5.7.40 Cylinder synchronisation with series connection.

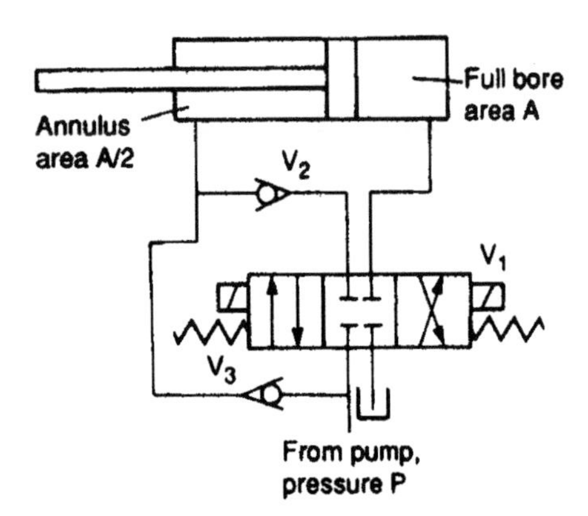

Fig. 5.7.41 Regeneration circuit.

full bore and annulus sides of the piston. The system in Figure 5.7.41 employs a cylinder with a full bore/annulus ratio of 2:1, and is known as a differential cylinder.

Upon cylinder extension, line pressure P is applied to the right-hand side of the piston giving a force of $P \times A$, while the left-hand side of the piston returns oil via valve V_3 against line pressure P producing a counter force $P \times A/2$. There is thus a net force of $P \times A/2$ to the left. When retraction is called for, a force of $P \times A/2$ is applied to the left-hand side and fluid from the right-hand side returns to tank at minimal pressure. Extension and retraction forces are thus equal, at $P \times A/2$.

5.7.4.4 Counterbalance and dynamic braking

The cylinder in Figure 5.7.42 supports a load which can run away when being lowered. Valve V_2, known as a counterbalance valve, is a pressure-relief valve set for a pressure higher than $F/2$ (the pressure generated in the fluid on the annulus side of the piston by the load). In the static state, valve V_2 is closed and the load holds in place.

When the load is to be lowered, line pressure is applied to the full bore side of the piston through valve V_1. The increased pressure causes valve V_2 to open and the load to lower. Check valve V_{2a} passes fluid to raise the load.

Counterbalance valves can also be used to brake a load with high inertia. Figure 5.7.43 shows a system where a cylinder moves a load with high inertia. Counterbalance valves V_2 and V_3 are included in the lines to both ends of the cylinder. Cross-linked pilot lines (shown dotted as per convention) keep valve V_2 open when extending and valve V_3 open when retracting. At constant cylinder speed, therefore, valves V_2 and V_3 have little effect.

To stop the load valve, V_1 is moved to its centre position, the pump unloads to tank and pilot pressure is lost, causing valves V_2 and V_3 to close. Inertia, however, maintains some cylinder movement. If, for example, the cylinder had been extending, inertia keeps it moving to the left – raising pressure on the piston's annulus side until valve V_2 reaches its pressure setting and opens. A constant deceleration force Pa (where P is the setting of valve V_1 and a is the annulus area) is applied to the load. On deceleration, fluid passes to the full bore side of the cylinder through check valve V_{3a}.

5.7.4.5 Pilot-operated check valves

Directional control valves and deceleration valves have a small, but definite, leakage and can only be used to hold

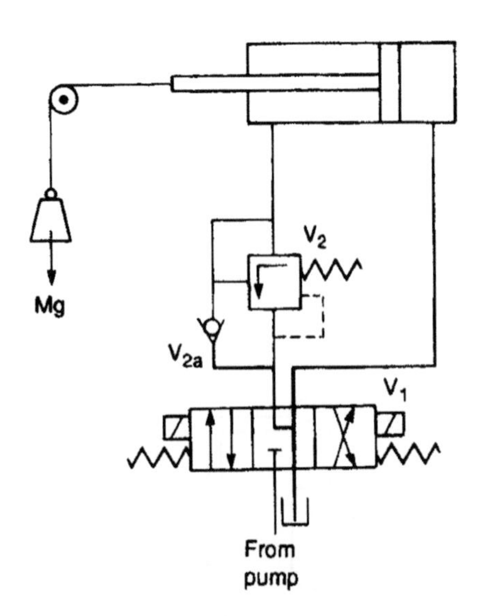

Fig. 5.7.42 Counterbalance circuit

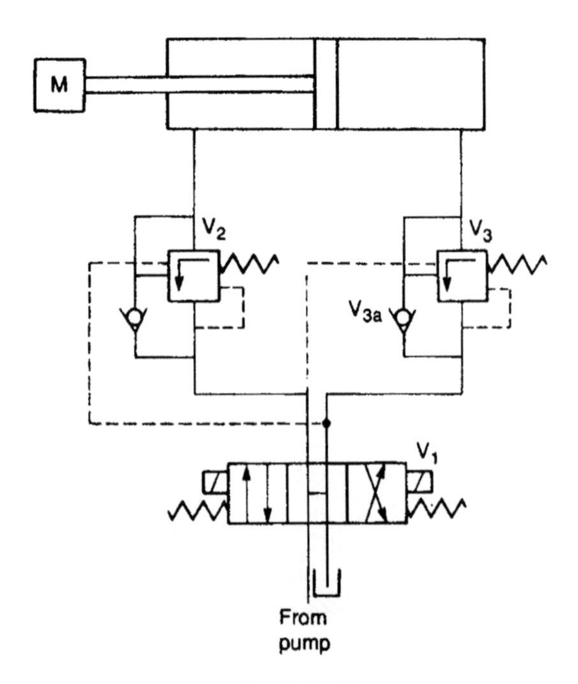

Fig. 5.7.43 Braking a high inertia load.

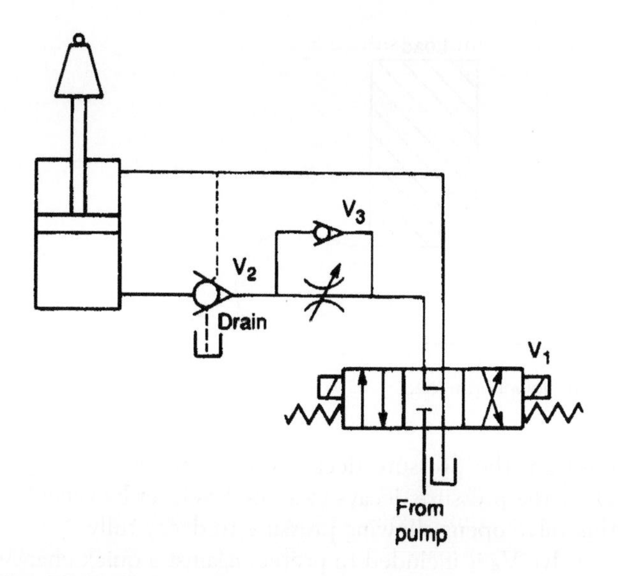

Fig. 5.7.44 Pilot-operated check valve used to hold an overhauling load.

an opposing load in position for short periods (of the order of minutes rather than hours) without the energy-wasting procedure of permanently applying pressure to the cylinder.

A check valve can be constructed with zero leakage. The pilot-operated check valve can thus be used to 'lock' an actuator in position. Figure 5.7.44 shows a typical example. Valve V_2 passes fluid normally when extending, but closes when valve V_1 is in its centre position. In this state, energy is saved by unloading the pump to tank. Pilot line pressure opens valve V_2 when the load is to be lowered. Counterbalance valve V_3 gives a controlled lowering but also ensures sufficient line pressure exists on the annulus side of the cylinder to give the pilot pressure needed to open valve V_2.

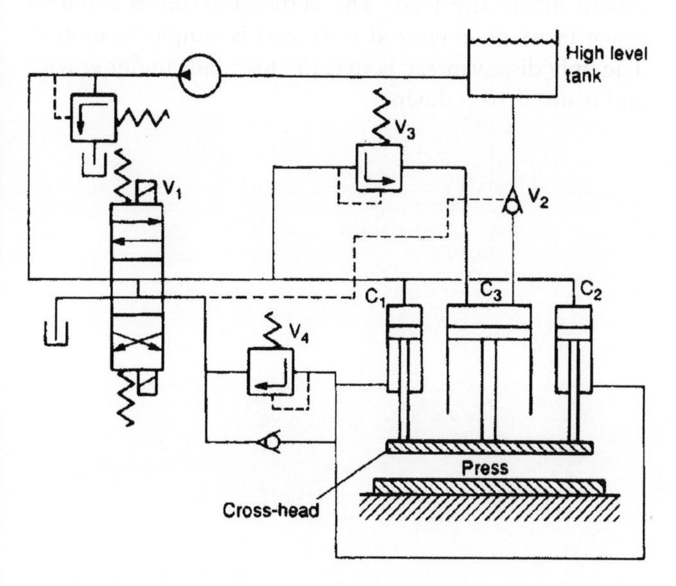

Fig. 5.7.45 Pre-fill circuit

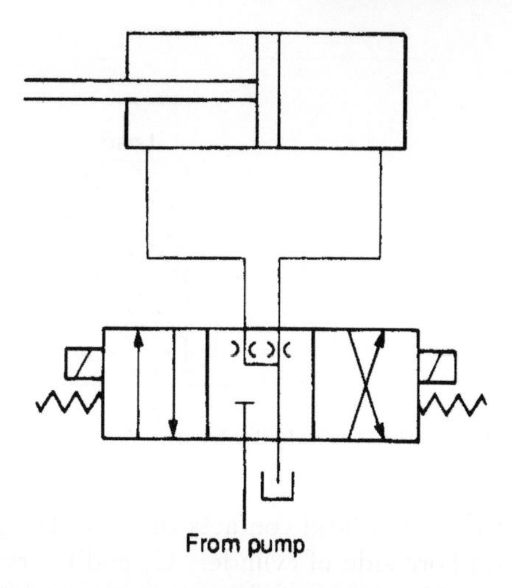

Fig. 5.7.46 Valve with central damping block.

5.7.4.6 Pre-fill and compression relief

Figure 5.7.45 shows the hydraulic circuit for a large press. To give the required force, a large diameter cylinder is needed and, if this is driven directly, a large capacity pump is required. The circuit shown (known as a pre-fill circuit) uses a high-level tank and pilot-operated check valve to reduce the required pump size.

The cross-head of the press is raised and lowered by small cylinders C_1 and C_2. When valve V_1 is switched to lower, the pressure on the full bore sides of cylinders C_1 and C_2 is low and valve V_3 is closed. Valve V_4 is a counterbalance valve, giving a controlled lower. As cylinders C_1 and C_2 extend, cylinder C_3 also extends because it is mechanically coupled, drawing its fluid direct from the high-level tank via pilot valve V_2.

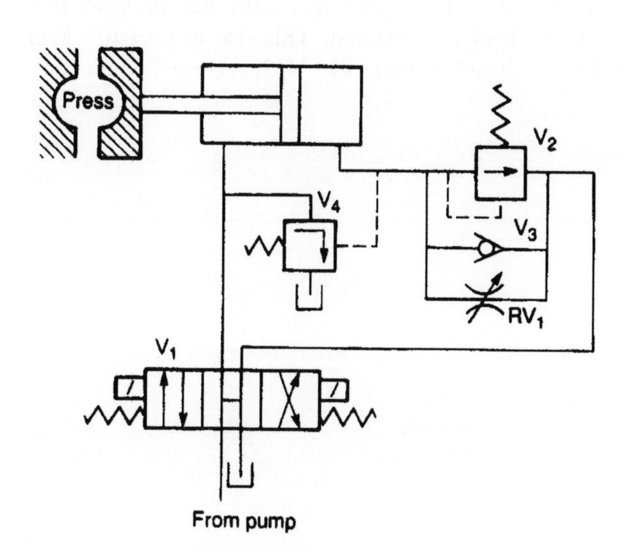

Fig. 5.7.47 Controlled decompression circuit.

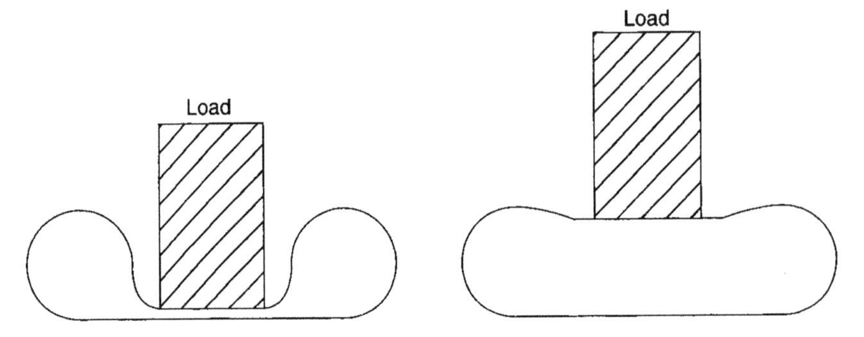

Fig. 5.7.48 The use of pneumatic bellows gives a simple way of raising and lowering a load.

When the cross-head contacts the load, the pressure on the full bore side of cylinders C_1 and C_2 rises. This causes valve V_3 to open, full-line pressure to be applied to cylinder C_3 and check valve V_2 to close. Full operating force is now applied to the load via cylinder C_3.

When the cross-head is raised, pressure is applied to the annulus side of cylinders C_1 and C_2. This opens check valve V_2, allowing fluid in cylinder C_3 to be returned directly to tank.

High-pressure hydraulic circuits like this require care both in design and in maintenance. For most practical purposes, hydraulic fluid can be considered incompressible. In reality, it compresses by about 0.8% per 100 bar applied pressure. When high pressure and large volumes of oil are present together, sudden release of pressure can result in an explosive release of fluid. The design must, therefore, allow for the gradual release of high-pressure, high-volume fluid.

Large-volume, high-pressure valves are thus fitted with a central damping block as illustrated in Figure 5.7.46 to return fluid to tank slowly.

Figure 5.7.47 shows a common decompression circuit. When the cylinder extends, fluid passes to the full bore side via check valve V_3 as usual, with fluid pressure rising once the load is contacted. This rise in pressure keeps valve V_2 closed. When valve V_1 is returned to its centre position, the pressure decays via restriction valve RV_1. Once the pressure decays to a safe level, set by valve V_2, this valve opens allowing pressure to decay fully.

Valve V_4 is included to protect against a quick change from high pressure extend to retract, without a pause to allow the pressure to decay. When the full bore side of the cylinder is pressurised, valve V_4 is held open causing the pump to unload to tank if retract is requested before decompression is complete. Once pressure on the full bore side decays, valve V3 closes and the cylinder can retract as normal.

5.7.5 Bellows actuator

Many applications require a simple lift function, for example to raise a disappearing stop on a set of rollers. This function is usually provided by a pneumatic cylinder which requires space and mounting lugs. A simple alternative is the bellows of Figure 5.7.48. In the de-energized state, the bellows are deflated and the load falls under gravity. When air is passed to the bellows, they inflate lifting the load. The actuator requires minimal space in its de-energized state and is simple to mount. The only disadvantage is that the load falls under gravity and is not driven down.

Index

I

T

U

V

PHYSICAL CONSTANTS IN SI UNITS

Absolute zero temperature	$-273.3°C$
Acceleration due to gravity, g	$9.807\ m\ s^{-2}$
Avogadro's number, N_A	6.022×10^{23}
Base of natural logarithms, e	2.718
Boltsmann's constant, k	$1.381 \times 10^{-23}\ JK^{-1}$
Faraday's constant, k	$9.648 \times 10^{4}\ C\ mol^{-1}$
Gas constant, R	$8.314\ J\ mol^{-1}\ K^{-1}$
Permeabillity of vacuum, μ_0	$1.257 \times 10^{-6}\ H\ m^{-1}$
Permittivity of vacuum, ε_0	$8.854 \times 10^{-12}\ F\ m^{-1}$
Planck's constant, h	$6.626 \times 10^{-34}\ J\ s^{-1}$
Velocity of light in vacuum, c	$2.998 \times 10^{8}\ m\ s^{-1}$
Volume of perfect gas at STP	$22.41 \times 10^{-3}\ m^3\ mol^{-1}$

CONVERSION OF UNITS

Angle, θ	1 rad	57.30°
Density, ρ	$1\ lb\ ft^{-3}$	$16.03\ kg\ m^{-3}$
Diffusion Coefficient, D	$1\ cm^3\ s^{-1}$	$1.0 \times 10^{-4}\ m^2\ s^{-1}$
Force, F	1 kgf	9.807 N
	1 lbf	4.448 N
	1 dyne	$1.0 \times 10^{-5}\ N$
Length, L	1 ft	304.8 mm
	1 inch	25.40 mm
	1 Å	0.1 nm
Mass, M	1 tonne	1000 kg
	1 short ton	908 kg
	1 long ton	1107 kg
	1 lb mass	0.454 kg
Specific Heat, Cp	$1\ cal\ gal^{-1}\ °C$	$4.188\ kJ\ kg^{-1}\ °C$
	$Btu\ lb^{-1}\ °F$	$4.187\ kJ\ kg^{-1}\ °C$
Stress Intensity, K_{IC}	$1\ ksi\sqrt{in}$	$1.10\ MN\ m^{-3/2}$
Surface Energy, γ	$1\ erg\ cm^{-2}$	$1\ mJ\ m^{-2}$
Temperature, T	1 °F	0.556 K
Thermal Conducitivity, λ	$1\ cal\ s^{-1}\ cm\ °C$	$418.8\ W\ m^{-1}\ °C$
	$1\ Btu\ h^{-1}\ ft\ °F$	$1.731\ W\ m^{-1}\ °C$
Volume, V	1 Imperial gall	$4.546 \times 10^{-3}\ m^3$
	1 US gall	$3.785 \times 10^{-3}\ m^3$
Viscosity, η	1 poise	$0.1\ N\ s\ m^{-2}$
	1 lb ft s	$0.1517\ N\ s\ m^{-2}$

CONVERSION OF UNITS—STRESS AND PRESSURE*

	MN m^{-2}	**dyn cm^{-2}**	**lb in^{-2}**	**kgf mm^{-2}**	**bar**	**long ton in^{-2}**
MN m^{-2}	1	10^7	1.45×10^2	0.102	10	6.48×10^{-2}
dyn cm^{-2}	10^{-7}	1	1.45×10^{-5}	1.02×10^{-8}	10^{-8}	6.48×10^{-9}
lb in^{-2}	6.89×10^{-3}	6.89×10^4	1	703×10^{-4}	6.89×10^{-2}	4.46×10^{-4}
kgf mm^{-2}	9.81	9.81×10^7	1.42×10^3	1	98.1	63.5×10^{-2}
bar	0.10	10^6	14.48	1.02×10^{-2}	1	6.48×10^{-3}
long ton in^{-2}	15.44	1.54×10^8	2.24×10^3	1.54	1.54×10^2	1

CONVERSION OF UNITS—ENERGY*

	J	**erg**	**cal**	**eV**	**Btu**	**ft lbf**
J	1	10^7	0.239	6.24×10^{18}	9.48×10^{-4}	0.738
erg	10^{-7}	1	2.39×10^{-8}	6.24×10^{11}	9.48×10^{-11}	7.38×10^{-8}
cal	4.19	4.19×10^7	1	2.61×10^{19}	3.97×10^{-3}	3.09
eV	1.60×10^{-19}	1.60×10^{-12}	3.38×10^{-20}	1	1.52×10^{-22}	1.18×10^{-19}
Btu	1.06×10^3	1.06×10^{10}	2.52×10^2	6.59×10^{21}	1	7.78×10^{-2}
ft lbf	1.36	1.36×10^7	0.324	8.46×10^{18}	1.29×10^{-3}	1

CONVERSION OF UNITS—POWER*

	kW (kj s^{-1})	**erg s^{-1}**	**hp**	**ft lbf $^{s-1}$**
kW (kJ s^{-1})	1	10^{-10}	1.34	7.38×10^2
erg s^{-1}	10^{-10}	1	1.34×10^{-10}	7.38×10^{-8}
hp	7.46×10^{-1}	7.46×10^9	1	15.50×10^2
ft lbf s^{-1}	1.36×10^{-3}	1.36×10^7	1.82×10^{-3}	1

*To convert row unit to column unit, multiply by the number at the column-row intersection, thus 1 MN m^{-2} =10 bar

Printed and bound by CPI Group (UK) Ltd, Croydon, CR0 4YY
08/05/2025
01864924-0001